# 9th International Symposium on Cavitation (CAV2015)



Journal of Physics: Conference Series Volume 656

Lausanne, Switzerland
6 - 10 December 2015

ISBN: 978-1-5108-1730-2
ISSN: 1742-6588

**Printed from e-media with permission by:**

Curran Associates, Inc.
57 Morehouse Lane
Red Hook, NY 12571

proceedings.com

**Some format issues inherent in the e-media version may also appear in this print version.**



Printed by Curran Associates, Inc. (2016)

For permission requests, please contact the Institute of Physics
at the address below.

Institute of Physics
Dirac House, Temple Back
Bristol BS1 6BE UK

Phone: 44 1 17 929 7481
Fax: 44 1 17 920 0979

techtracking@iop.org

**Additional copies of this publication are available from:**

Curran Associates, Inc.
57 Morehouse Lane
Red Hook, NY 12571 USA
Phone: 845-758-0400
Fax: 845-758-2634
Email: curran@proceedings.com
Web: www.proceedings.com

# Table of contents

**Volume 656**

**9th International Symposium on Cavitation (CAV2015)**
**6–10 December 2015, Lausanne, Switzerland**

**Accepted papers received: 21 October 2015**
**Published online: 3 December 2015**

## Preface



## Papers

### Biomedical applications

**Cavitation erosion**

**Cavitation in fluid machinery**

**Cavitation in fuel injection systems, fuels and lubricants**

**Sheet, cloud and supercavitation**

## Sonoluminescence and sonochemistry



## Thermodynamic effects



## Tip vortex cavitation

Journal of Physics: Conference Series **656** (2015) 011001 doi:10.1088/1742-6596/656/1/011001

# Welcome Message

It is our pleasure and privilege to welcome all the participants of the 9th International Symposium on Cavitation (CAV2015) to Lausanne.

Since its initiation in 1986 in Sendai, Japan, the CAV symposium has grown to become the world's foremost event dedicated to cavitation. Hosted by EPFL (Ecole Polytechnique Fédérale de Lausanne) and staged at the SwissTech Convention Center, CAV2015 is a unique opportunity to exchange with leading scientists and industry experts about the latest advances in theoretical modelling, numerical simulation and experimentation related to cavitation phenomena with a special emphasis on practical applications.

The topics covered by CAV2015 include cavitation in fluid machinery and fuel systems, bubble dynamics, cavitation erosion, advanced numerical simulation, sonochemistery, biomedicine and experimental techniques.

CAV2015 will also host an exhibition of leading providers of state of the art measurement equipment, including high-speed imaging systems, non-intrusive velocimetry, pressure sensors, as well as numerical solvers.

We have accepted over 190 papers, which will be presented in four parallel sessions. The proceedings will appear in the open access *Journal of Physics: Conference Series* (JPCS), which is part of the IOP Conference Series. All published papers are fully citable and upon publication will be free to download in perpetuity. We would like to thank all the reviewers for their great help during the selection process.

We will also propose six plenary speakers to highlight cavitation issues in different fields.

Finally, we would like to warmly thank our sponsors for their valuable support and the local Organizing Committee for the efforts in setting up this important event.

We look forward to seeing you in Lausanne!

Dr Mohamed Farhat, EPFL
Chairman of CAV2015

Dr Andres Müller, EPFL
Head of CAV2015 Local Organizing Committee

# Peer review statement

All papers published in this volume of *Journal of Physics: Conference Series* have been peer reviewed through processes administered by the proceedings Editors. Reviews were conducted by expert referees to the professional and scientific standards expected of a proceedings journal published by IOP Publishing.

# Estimation of mechanical properties of gelatin using a microbubble under acoustic radiation force

**Eriko Shirota and Keita Ando**

Department of Mechanical Engineering, Keio University 3-14-1 Hiyoshi, Kohoku-ku, Yokohama, 223-8522, Japan

E-mail: eco.shirota@keio.jp, kando@mech.keio.ac.jp

**Abstract.** This paper is concerned with observations of the translation of a microbubble (80 $\mu$m or 137 $\mu$m in radius) in a viscoelastic medium (3 w% gelatin), which is induced by acoustic radiation force originating from 1 MHz focused ultrasound. An optical system using a high-speed camera was designed to visualize the bubble translation and deformation. If the bubble remains its spherical shape under the sonication, the bubble translation we observed can be described by theory based on the Voigt model for linear viscoelastic solids; mechanical properties of the gelatin are calculated from measurements of the terminal displacement under the sonication.

## 1. Introduction

It is desirable to obtain mechanical properties of ocular tissues in ophthalmic diagnostics and surgeries. A bubble-based acoustic radiation force technique has been developed to determine mechanical properties of viscoelastic tissues [1]. In principle, a bubble in viscoelastic media can translate under the action of the radiation force that arises from interaction, via acoustic impedance mismatching, between ultrasound and the bubble. We may model the translational motion based on the Voigt model (for linear viscoelastic solids); existing theories assume the case with a spherical bubble to relate the radiation force and evolution of the resulting bubble translation [2]. In the experiment of Yoon *et al.* [1], the elasticity of gelatin was inferred by fitting their acoustic measurement to the theory. However, optical visualization of ultrasound-induced bubble translation (and deformation) has not been provided in previous studies.

In this study, we design an optical system to visualize translation and deformation of a microbubble in gelatin under megasonic wave irradiation and estimate the mechanical properties of the gelatin.

## 2. Theory

Supposing that a spherical bubble in a linear viscoelastic solid is subjected to constant acoustic radiation force applied in step-wise manner from time $t = 0$, the transient motion of the spherical bubble can be derived based on the Voigt model [2]:

$$x(t) = x_{max}\left[1 - \exp\left(\frac{t}{\tau}\right)\right] \quad (1)$$

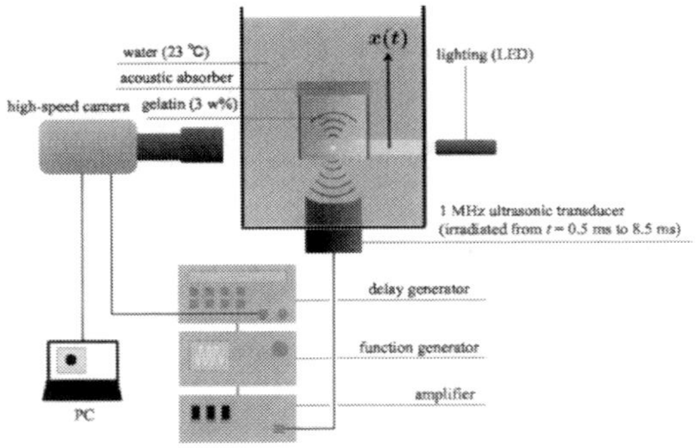


**Figure 1.** Experimental setup for observing the translation and deformation of a microbubble in gelatin under megasonic irradiation.

where $x$ is the displacement from the original position before applying the radiation force, $x_{max}$ is the terminal displacement, and $\tau$ is the time constant that can be related to the ratio of viscosity to the elasticity of the surrounding medium. Note that the application of the Voigt model implies sufficiently small $x_{max}$ (or equivalently small strain in the medium) for its linear assumption to hold in experiments.

The bubble under constant radiation force eventually reaches a (static) mechanical equilibrium at which the radiation force is balanced with restoring force from elasticity of the surrounding medium. At the static equilibrium, the terminal displacement in Eq. (1) is given by [3]

$$x_{max} = \frac{F}{4\pi RG} \tag{2}$$

where $F$ is the (constant) acoustic radiation force, $R$ is the variable radius of the deformed bubble, and $G$ is the shear modulus of the viscoelastic solid surrounding the bubble. Now that acoustic waves of intensity $I$ are exerted on the projected area $\pi R^2$, the radiation force $F$ is calculated as

$$F = \frac{2I\pi R^2}{c} \tag{3}$$

where $c$ is the speed of sound in the viscoelastic solid and the prefactor 2 on the right-hand side means perfect wave reflection at the bubble wall (with no waves transmitted into the bubble). Assuming that the bubble deformation is negligible, we may normalize the terminal displacement by the initial bubble radius $R_0$. That is,

$$\frac{x_{max}}{R_0} = \frac{I}{2cG} \tag{4}$$

It follows from Eq. (4) that the shear modulus may be estimated only from measurements of the maximum bubble displacement, provided the bubble keeps its spherical shape.

## 3. Experimental methods

A gelatin phantom was selected as a (tissue-like) viscoelastic solid; the phantom was prepared with 3 w% gelatin (G2500, Type-A, Sigma-Aldrich) mixed with water. A spherical microbubble in the gealtin phantom can be produced by focusing an infrared laser pulse (ULTRA 50 GRM, Quantel) accoring to [4]; after vapor condenses back into the surrounding medium, noncodensible gas contents (air) are left as a spherical cavity whose radius is varied around 100 $\mu$m for observations.

The experimental setup is depicted in Fig. 1. The gelatin phantom containing a spherical microbubble was placed inside an acrylic case filled with tap water (23 °C). Acoustic radiation force acting on the microbubble was generated by a 1.0 MHz, focus-type transducer (IZ30I R30, JAPAN PROBE) located at the case bottom, which is connected to a function generator (WF1973, NF Corporation) driven at 10 V, a delay generator (DG535, Stanford Research Systems Inc.), and an amplifier (GAIN ×10, HSA4011, NF Corporation). The microbubble was located in the focal zone in which the constant acoustic intensity was achieved soon after the irradiation starts and the pressure amplitude was recorded as 1.82 ± 0.03 MPa (HPM02/1 , Precious Acoustics Ltd.). The focal zone is confirmed to be larger than the bubble size, so that the pressure may be assumed to act uniformly on the bubble. Since the forced frequency is much higher than the natural frequency of bubbles of our concern, the applied acoustic force is expected to drive the translation of the bubble, not oscillation in the bubble volume. To minimize wave reflections from boundaries, an acoustic absorber (EUA201, EASTEK) was placed at the top of the gelatin phantom. The ultrasound was irradiated from time $t = 0.5$ ms to 8.5 ms in Figs. 2 to 3. The motion of the microbubble was illuminated by LED light (SLG-150V, REVOX) and recorded through 8× magnification by a high-speed camera (FASTCAM SA-X2, Photron) whose frame rate and exposure time are 40,000 frames/s and 1/42105 s, respectively. The MATLAB image processing was used to extract the bubble information such as displacement of the centroid and trasverse radius of the deformed bubble [4].

## 4. Results and Discussion

Figure 2 shows an example of sequential images of the translation of a bubble with $R_0 = 137$ $\mu$m in the gelatin under the megasonic irradiation. As expected, the translation of the bubble is induced in the direction of the megasonic wave propagation. After the sonication, the bubble returns to its original position due to the gelatin elasticity, as seen in the last frame of Fig. 2. In this example, the bubble seems sufficiently small, meaning that surface tension and elasticity of the surrounding medium are large enough to keep its spherical shape against the acoustic radiation force.

A typical example of the temporal evolution of the translation for the case of $R_0 = 87$ $\mu$m (where the surface tension and elasticity are expected to be even stronger than in the previous example of $R_0 = 137$ $\mu$m) is presented in Fig. 3 (see the left vertical axis). To examine deviation from the spherical shape, that of the traverse radius $R_{trav}$ (defined in Fig. 2) normalized by the initial radius $R_0$ is also plotted in Fig. 3 (see the right vertical axis). The initially spherical bubble is found to be flattened under sonication. In particular, the non-sphericity is emphasized by encountering unsteady motion when the sonication is turned on or off and inertia of the surrounding medium plays a more important role. However, the increase in the transverse radius is bounded within a few percents from the unperturbed value, meaning that surface tension defeats pressure and inertial forces, which make the bubble non-spherical, in the overall process.

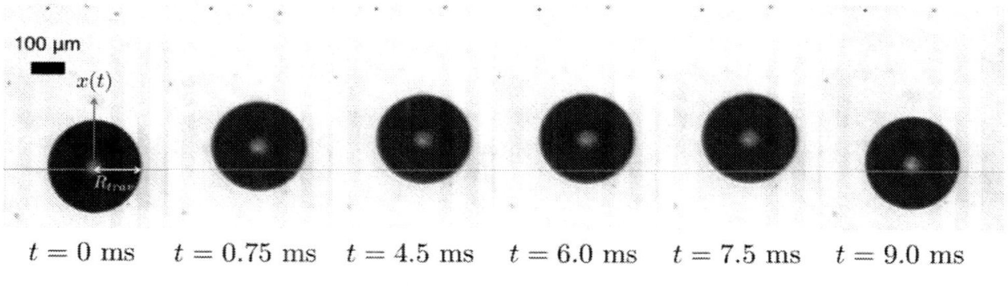


**Figure 2.** Translation of the microbubble ($R_0 = 137$ $\mu$m) in the gelatin. The megasonic waves propagating upward are applied from $t = 0.5$ ms to 8.5 ms.

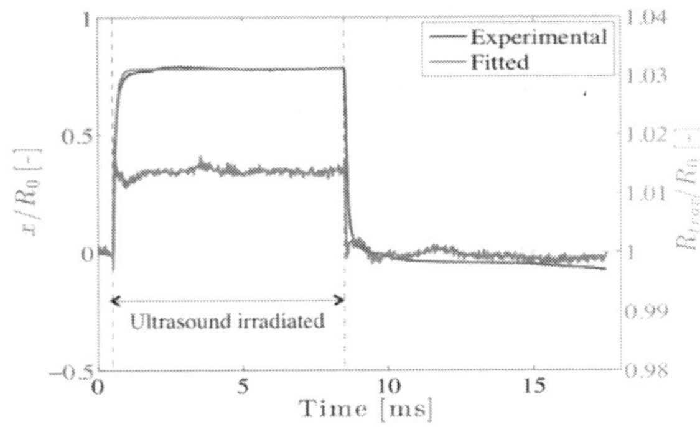


**Figure 3.** Displacement and traverse radius of the microbubble ($R_0 = 87\ \mu$m).

Now that the bubble remains spherical in Fig. 3, the measured evolution of the bubble translation is fitted to the theory based on the Voigt model represented by Eq. (1); the fitting is denoted by the red line in Fig. 3. The good agreement between the measurement and the theory means that the deformation of the gelatin surrounding the translating bubble is sufficiently small for the linear Voigt assumption to be valid. It follows from the fitting that the shear modulus is predicted to be $650 \pm 20$ Pa, which agrees with values reported in the previous study [5].

## 5. Conclusions

In this study, the experimental technique has been developed to visualize the bubble translation and deformation under acoustic radiation force in a viscoelastic solid (gelatin). The observed evolution of bubble translation where the bubble remains its spherical shape was well explained by the theoretical model for spherical bubbles in the (linear) Voigt solid. The comparison between the measurement and theory leads to an estimate of the gelatin shear modulus, which agrees well with the previous study.

In future, a series of the experiment with varying $R_0$ will be conducted to see how the bubble size affects the assumption of spherical shape based on dimensionless numbers including the Weber number.

## Aknowledgements

We thank Fumiya Hamaguchi for his support in the experiment and analyses. This study is supported by JSPS KAKENHI Grant Number 25709008.

## References

[1] S. Yoon *et al.*, "Estimation of mechanical properties of a viscoelastic medium using a laser-induced microbubble interrogated by an acoustic radiation force," *J. Acoust. Soc. Am.*, vol. 130, no. 4, pp. 2241–2248, 2011.

[2] T. N. Erpelding and M. O'Donnell, "Bubble-based acoustic radiation force elasticity imaging," *IEEE Trans. Ultrason. Ferroelectr. Freq. Contr.*, vol. 52, no. 6, pp. 971–979, 2005.

[3] Y. A. Ilinskii *et al.*, "Gas bubble and solid sphere motion in elastic media in response to acoustic radiation force," *J. Acoust. Soc. Am.*, vol. 117, no. 4, pp. 2338–2346, 2005.

[4] F. Hamaguchi and K. Ando, "Linear oscillation of gas bubbles in viscoelastic material under ultrasound irradiation," *Phys. Fluids*, Submitted.

[5] A. B. Karpiouk *et al.*, "Assesment of shear modulus of tissue using ultrasound radiation force acting on a spherical acoustic inhomogeneity," *IEEE Trans. Ultrason. Ferroelectr. Freq. Contr.*, vol. 56, no. 11, pp. 2380–2387, 2009.

# The role of encapsulated microbubbles in the diagnosis of stenosis in arteries

**Karuna Agarwal and Dhiman Chatterjee**[1]
Department of Mechanical Engineering, Indian Institute of Technology Madras, Chennai, India 600036

E-mail: dhiman@iitm.ac.in

**Abstract.** Numerical modelling of sound scattered by ultrasound contrast agent bubbles in arteries is presented. Nonlinear dynamics of an encapsulated microbubble coupled with transmission of sound through liquid-filled flexible tubing is developed. Based on the frequency response, a normal artery can be distinguished from that of a stenosed artery. Effects of parameters like incident pressure amplitude, driving frequency and degree of stenosis are studied. It is further found that the effect of variable pressure produced by blood flow does not have a significant effect on the observed sound pressure levels.

## 1. Introduction

Stenosis is an abnormal narrowing of the artery, which could lead to stroke. At present angiogram is the common diagnostic technique for detecting stenosis. Several papers have pointed out the potential use of ultrasound contrast microbubbles to detect stenosis replacing more expensive existing method. However, to the best of our knowledge, an extensive study has not yet been done [1]. This forms the focus of this work.

Response of bubbles, both free and encapsulated, to acoustic waves in an infinite liquid have been modelled in the past using different variants of Rayleigh-Plesset like equation [2-6]. However, only few papers have discussed dynamics of bubble in constrained media [7]. This influence of neighbouring walls becomes more important in the case of stenosed arteries. In this paper, we use a Newtonian rheological model to estimate the dynamics of encapsulated microbubbles. Modelling the dynamics of bubbles in presence of flexible blood vessel wall poses formidable challenge; this problem is solved in two stages.

## 2. Theoretical framework of the problem

The problem of the dynamics of microbubbles inside a blood vessel under insonication from outside is solved in two steps using combined use of Matlab© and COMSOL© software packages. When an acoustic wave is incident on an elastic blood vessel, only a part of the incident energy travels inward and reaches the microbubble. The sound radiated by oscillating microbubbles under the influence of this incident wave is modelled and this radiated sound travels back to the source. The passage (transmission and reflection at the wall) of sound is modelled using acoustic module in COMSOL©. Dynamics of microbubbles has been modelled using both Matlab© and COMSOL©.

### *2.1 Modelling of encapsulated microbubble*

Rayleigh-Plesset equation is modified to include the effects of the surface rheology of an encapsulated bubble [4-5]:

1. To whom any correspondence should be addressed.

$$\rho\left(R\ddot{R}+\frac{3}{2}\dot{R}^2\right)=p_{B0}\left(\frac{R_0}{R}\right)^{3k}-4\mu\frac{\dot{R}}{R}-\frac{4\kappa^S\dot{R}}{R^2}-\frac{2\gamma}{R}-P_0+P_A\sin(\omega t)\,, \quad (1)$$

where $\rho$ and $\mu$ are the density and the viscosity of the surrounding medium, $\gamma$ is the surface tension, $\kappa^s$ is the surface dilatational viscosity, $R_0$ is the initial radius of bubble, $k$ is the polytropic constant for gas, $p_{B0}$ is the initial pressure in the bubble, $P_0$ is the ambient pressure and $P_A$ is the insonication pressure amplitude.

The bubble size distribution of Definity (figure 1) is taken from Streeter et. al. [8] and discretized at various sizes, that are supposed to be representative of the near-by sizes. Individual scattered pressure intensity is obtained for each size and multiplied with number of bubbles of those sizes to get the total intensity. This assumes no bubble-bubble interaction. The largest diameter of bubble is 4.178 microns, so the maximum allowable concentration of bubbles for our artery size as 1.028 x $10^{10}$/ml, which is significantly larger than the actual concentration of 1.64 x $10^9$/ml.

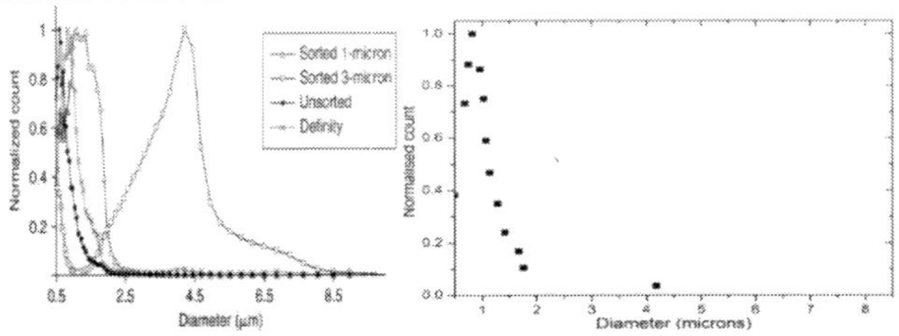


**Figure 1.** Definity size distribution (a) from Streeter et. al. [8] (b) used for present simulations

*2.2 Transmission of acoustic wave*

We know that the dynamics of a bubble is non-linear in nature and the frequency response of its dynamics is spread over a range of frequencies. The acoustic model in COMSOL© is unable to capture this. So to include this effect, a line transmission model of acoustic waves is set up. In this model, an acoustic wave is incident on a solid and the boundary condition is continuity of pressure and displacement. Initially rigid cylinder was simulated to validate the presently adopted finite element based package (COMSOL©) and the results obtained with COMSOL© simulation [9] compares well with the theoretical analysis given in Morse and Ingard [10]. Effect of grid size was also discussed by Agarwal [9] and is not shown here for brevity. Based on grid independence study, a maximum mesh size of 0.01 mm was chosen for all the simulations. Complete geometry with different components is clearly brought out in figure 2. Floquet periodic boundary conditions at the edges parallel to the direction of incidence.

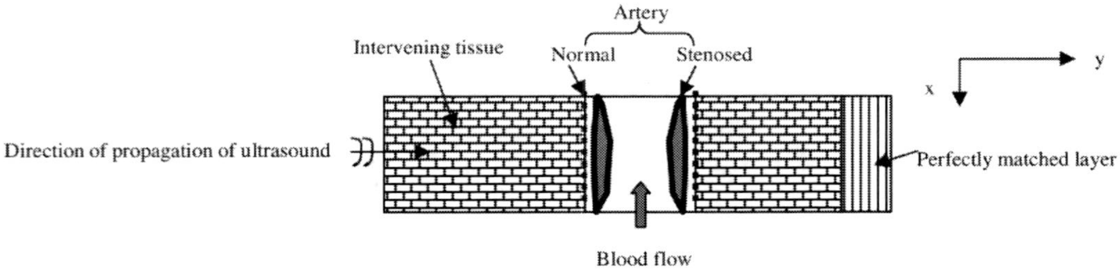


**Figure 2.** Schematic of the computational domain used to carry out simulation of acoustic propagation

Blood is modelled as a fluid with density of 1040 kg/m$^3$, viscosity of 0.004 Pa-s and velocity of sound in the medium is 1500 m/s. Normal arterial tissue is assumed to be linearly elastic material with density 1160 kg/m$^3$, Young's modulus 0.4 MPa and Poisson's ratio 0.499. The blocking tissue had Young's modulus 4 times that of normal tissue. An attenuation of energy by 0.8 dB/cm/MHz in the tissues is also considered. For the case of the bubbles injected in the stream, the density and speed of sound in the blood is modified according to formulae for bubbly liquids [2]. The pressure in the blood stream is noted.

For 1000 Pa driving pressure, pressure in the stream is about 400 Pa for a healthy artery and ~250 Pa for a stenosed artery (figure 3). Results showing dynamics of encapsulated bubble are obtained for this pressure inside the blood vessel experienced by the bubble and the resulting scattered pressure by bubbles is added to the line transmission model as a point source intensity of varying frequencies. This is now taken up for discussion.

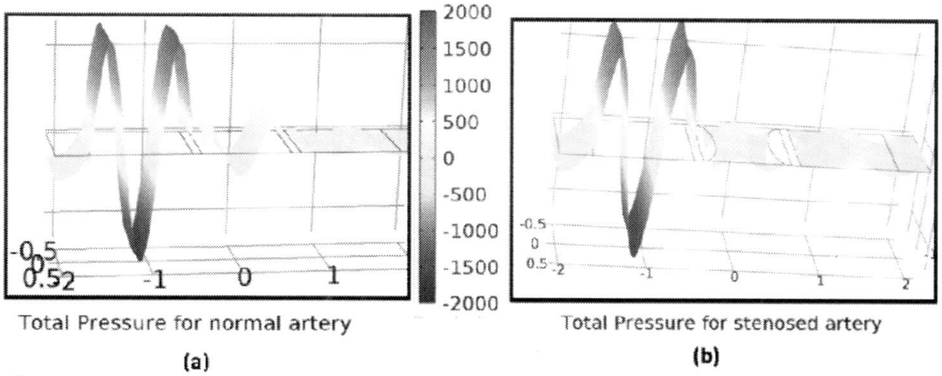


**Figure 3**. Total pressure (in Pa) inside: (a) normal artery without bubble (b) stenosed artery without bubble

## 3. Results and discussion

From figure 4, we can see that at 2 MHz, there is not much difference between stenosed and normal artery with or without bubbles. In these simulations, it is assumed that artery (normal/stenosed) is a linear scatterer and hence the only frequency produced is that of driving frequency. This appears to be a reasonable assumption which can be corroborated with further experiments. However, addition of bubbles results in the observation of sound pressure levels at frequencies other than the insonication frequency and for these, the difference in sound for normal and stenosed artery is significant. The difference stems from greater transmission losses and resultant reduced sound scattered by the bubbles and the increased elastic modulus of the stenosed artery. For two pressure levels, namely, 1000 Pa and 0.5 MPa as expected, the difference is more significant (~10 dB) at higher pressure amplitude when non-linear oscillation of the bubble is expected to be more significant. Since there exists a difference in results of solvers in Matlab© and COMSOL©, to verify the theory of harmonic imaging being useful, the observed sound pressure at the end of the line transmission set-up in both the cases is also plotted together. The results from both follow the same trend.

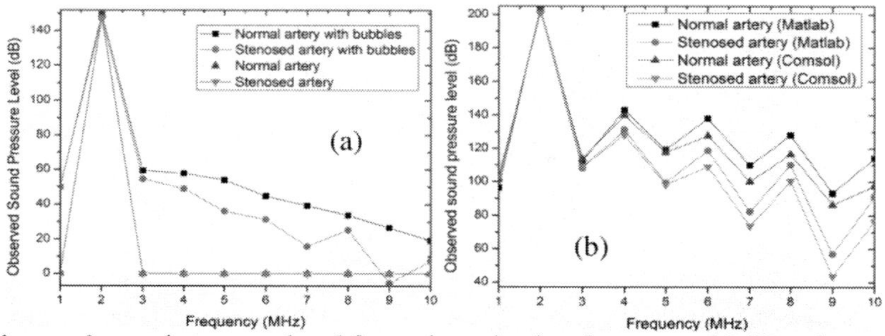


**Figure 4.** The observed sound pressure level for an insonication frequency 2 MHz when pressure amplitude is (a) 1000 Pa (simulated in Matlab©) and (b) 0.5 MPa (simulated using both Matlab© and COMSOL©)

The flow of blood affects the pressure distribution inside the blood vessel and as a result might affect the observed sound pressure levels. On addition of a velocity of 0.1 m/s to the acoustic model the sound observed was not much affected in case of the normal artery. In stenosis, because of the narrowing and resulting pressure rise, there was a slight difference, but not enough to be significant.

We also noticed the effect of stenosis (figure 5). Even a small degree of stenosis such as 36% resulted in significantly different sound levels. The difference grew with degree of stenosis, but not linearly. The frequency- averaged sound level was 126 dB for a normal artery, 110 dB for 36% blocked artery and 106 dB for 62% block. However at 4MHz, the variation is almost linear, and hence distinguishing higher degrees of stenosis is easier.

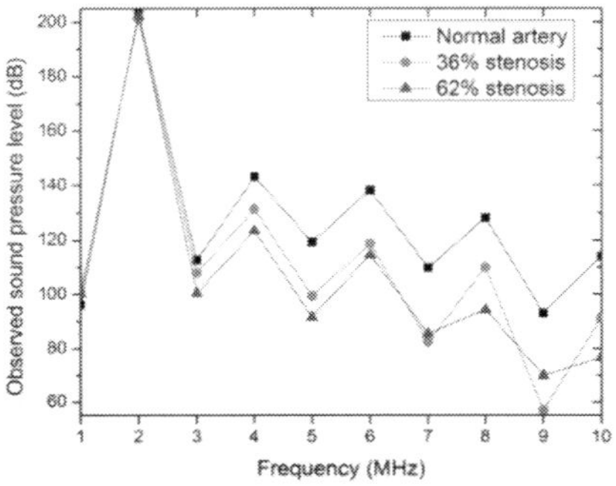


**Figure 5.** The observed sound pressure level for different degrees of stenosis in presence of microbubble when driving pressure is 0.5 MPa, frequency is 2 MHz

Similar study was carried out for a frequency of 5 MHz (not presented here). Though the trend is similar, the sound levels were lower due to greater attenuation of higher frequency sound. Care also should be taken to consider the size distribution of bubbles while selecting driving frequency because size of bubbles determine resonant frequency of the bubbles.

## 4. Conclusion

Harmonic imaging may be useful for ultrasound detection of stenosis, for even mild cases, in a minimally non-invasive and cheap manner. Further experiments are being set-up to verify this.

## References

[1] de Korte C L, Hansen H H and van der Steen A F 2011 Vascular ultrasound for atherosclerosis imaging *Interface Focus* 1(4) 565-575

[2] Brennen C E 1995 Cavitation and Bubble Dynamics (Oxford University Press)

[3] Church C C 1995 The effects of an elastic solid surface layer on the radial pulsations of gas bubbles *J. Acoust. Soc. Am.* 97 1510–1521

[4] Chatterjee D and Sarkar K 2003 A Newtoninan rheological model for the interface of microbubble contrast agents *Ultrasound Med. Biol.* 29 1749

[5] Sarkar K, Shi W T, Chatterjee D and Forsberg F 2005 Characterization of ultrasound contrast microbubbles using in vitro experiments and viscous and viscoelastic interface models for encapsulation *J. Acoust. Soc. Am.* 118 539

[6] Wang Y C and Brennen C E 1995 The noise generated by the collapse of a cloud of cavitation bubbles *American Society of Mechanical Engineers* 226 17-29

[7] Zhong P, Zhou Y and Zhu S 2001 Dynamics of bubble oscillation in constrained media and mechanisms of vessel rupture in SWL *Ultrasound in medicine & biology* 27(1) 119-134

[8] Streeter E J, Gessner R, Miles I and Dayton P A 2010 Improving sensitivity in ultrasound molecular imaging by tailoring contrast agent size distribution: In vivo studies *Mol Imaging* 9(2) 87–95

[9] Agarwal, K, "Investigating the role of encapsulated microbubbles in diagnosis of stenosis in arteries", Dual Degree Project Report, Dept of Mech Engg, IIT Madras

[10] Morse, P M and Ingard K U 1987 Theoretical Acoustics, Princeton University Press

# Development of a confocal ultrasound device using an inertial cavitation control for transfection *in-vitro*.

**J L Mestas[1], K Chettab[2], S Roux[2], F Prieur[1], M Lafond[1], C Dumontet[2], C Lafon[1]**
[1] INSERM, U1032, *LabTau, Lyon, F-69003, France ; Université de Lyon, Lyon, F-69003, France ; Université Lyon 1, Lyon, F-69003, France*
[2] Université de Lyon, *69000, Lyon, France; Université de Lyon 1, 69000, Lyon, France;* INSERM U1052, *69008, Lyon, France; CNRS UMR 5286, 69008, Lyon, France*

E-Mail: jean-louis.mestas@inserm.fr

**Abstract.** Sonoporation using low-frequency high-pressure ultrasound (US) is a non-viral approach for *in vitro* and *in vivo* gene delivery. We developed a new sonoporation device designed for spatial and temporal control of ultrasound cavitation. This device was evaluated for the *in vitro* transfection efficiency of a plasmid coding for Green Fluorescent Protein (peGFP-C1) in adherent and non-adherent cell lines. The frequency spectrum of the signal receive by a hydrophone is used to compute a cavitation index (CI) representative of the inertial cavitation activity. The influence of the CI on transfection efficiency, as well as reproducibility were determined. A real-time feedback loop control on CI was integrated in the process to regulate the cavitation level during sonoporation. In both adherent and non-adherent cell lines, the sonoporation device produced a highly efficient transfection of peGFP-C1 (40-80%), as determined by flow cytometry analysis of GFP expression, along with a low rate of mortality assessed by propidium iodide staining. Moreover, the sonoporation of non-adherent cell lines Jurkat and K562 was found to be equivalent to nucleofection in terms of efficiency and toxicity while these two cell lines were resistant to transfection with lipofection.

## Introduction

Transfection is a widely used tool for the analysis of the role of a target gene, determination of its function, identification of regulatory sequences controlling its expression as well as for the demonstration of biological processes involved in molecular interactions causing diseases. Depending on the cell type, different techniques of nucleic acid transfection into eukaryotic cells can be used. Viral vectors provide the most efficient means for *in vivo* and *in vitro* gene delivery. However, the use of viral vectors carries significant biological risks and difficulty of vector production [1]. Chemical methods using natural or synthetic carriers to deliver genes into cells are safe, effective for transfection of adherent cell lines but suffer from their low efficiency in the case of non-adherent cell lines as well as *in vivo* [2]. Physical methods for gene delivery such as micro-injection, electroporation, gene gun and ultrasound-mediated methods have been the object of considerable research during the last decade. Physical methods involve the use of physical force to increase the permeability of the cell membrane, allowing genetic materials and therapeutic molecules to enter into the cell [3]. Among them ultrasound-

[1] Jean-Louis Mestas - INSERM U1032-LabTAU – 151 cours Albert Thomas – 69424 Lyon Cedex.

induced biological effects are commonly assumed to be caused by acoustic cavitation which results in the creation of transient holes in the cell membrane (sonoporation).
The objective of this work is to develop a device to transfect with increased efficiency and acceptable viability in a reproducible manner. The device should work without adjuvant or contrast agent and should be adapted to routine use in research laboratories.

## Material and methods

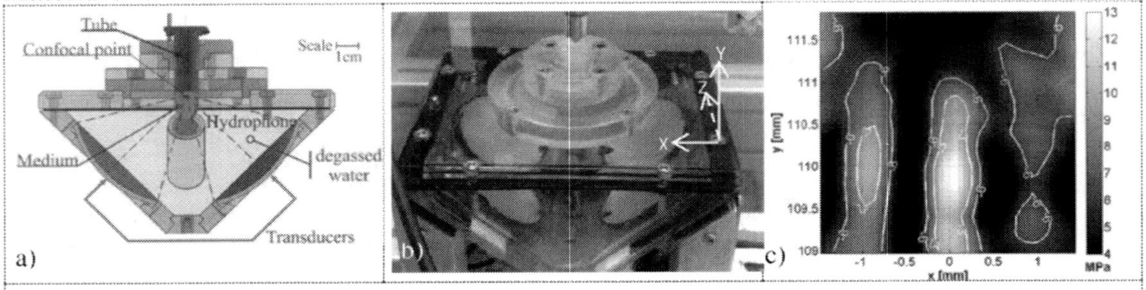


**Figure 1**: a) Confocal device; b) Fiber Optic Probe Hydrophone FOPH 2000 (RP Acoustic, Leutenbach, Germany) positioned for pressure measurements. c) Interference pressure pattern in degassed water (-3 and -6 dB curves)

The ultrasonic set-up is based on two focused transducers PZ28 (Ferroperm, Kvistgaard, Denmark), operated at the frequency of 1.1 MHz. The diameter and the radius of curvature of these transducers were 50 mm. The transducers were positioned in a confocal manner with a 90° angle between the acoustic axes (figure 1(a)). Due to the interference between the two ultrasound beams, the pressure field presents a pattern around focus that traps the bubbles of various size on the pressure nodes and antinodes (figure 1(b-c)). The bubble cloud in a 2 ml Eppendorf tube filled of water is therefore precisely localized as presented by ultrasound imaging (figure 3(a)); Pulse Repetition Frequency PRF = 250Hz, Duty cycle DC = 1%, Peak negative pressure P- = 8MPa).
For transfection, longer pulse are applied to the culture medium (650 µl; PRF = 100Hz; DC = 25%). The placement of the 2-mL tube and the volume of fluid are fixed so that the focal point is situated on the symmetrical axis of the tube 2 mm below the air medium interface (figures 1(a), 2). The coupling media between the transducers and the tube is degassed water ($O_2$ concentration <3 mg/L). This setup is appropriate for easily initiating cavitation with no more than 4 MPa peak negative pressure inside the tube.

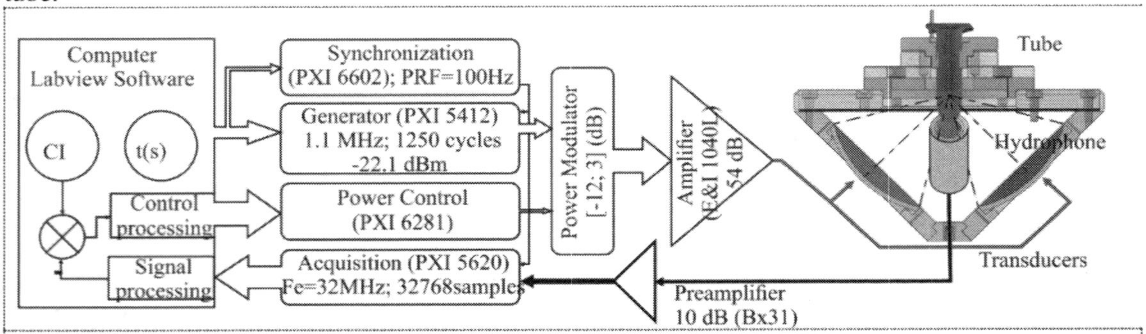


**Figure 2**. Experimental setup; Software and controlled hardware are from National Instruments (Austin, Texas).

To characterize the inertial cavitation phenomenon, we have chosen to study the evolution of the broadband noise induced by the bubble collapse. A hydrophone is used to record the signal back-scattered by the cavitating bubbles. A cavitation index (CI) is determined by computing the mean of this signal frequency spectrum in dB in the frequency range 0.1 to 7.1 MHz [4]. To this value a reference

Journal of Physics: Conference Series **656** (2015) 012003 doi:10.1088/1742-6596/656/1/012003

value representative of the electronic noise (~-93 dB measured when the ultrasound power is switched off) is removed.

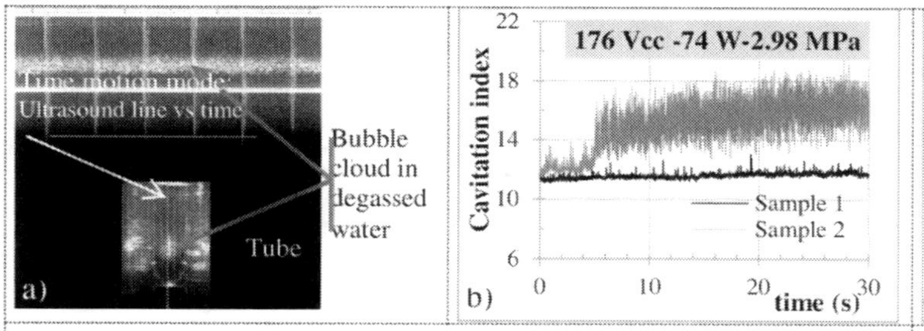


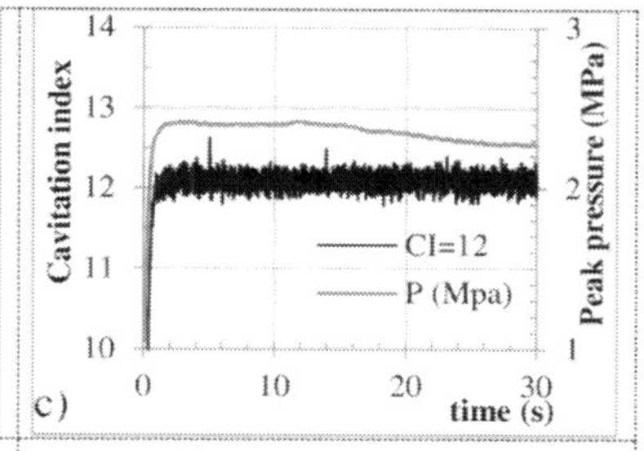


**Figure 3**: a) Ultrasound images showing a cavitation-induced bubble cloud in a 2 ml Eppendorf tube (PRF=250Hz; DC=1%;P- =8MPa in tube) b) Two examples of cavitation index measurement for 30s-exposures at constant power (PRF=100Hz; DC=25%; record length: last 1.024 ms of signal)

**Figure 4:** CI measurement and applied pressure using the closed-loop control process with a target CI of 12 (100Hz; 25%; record length: id figure 3b).

The signal applied to the transducers is first generated numerically before being amplified by a voltage-gain controlled pre-amplifier. This pre-amplifier has a short response time and allows a feedback on its amplification gain at each burst (up to 250 Hz). Since the inertial cavitation activity can evolved differently for the same excitation parameters (figure 3(b)) a LabVIEW-based program computes after each burst a new gain using the past CIs and the target CI and the pre-amplifier gain is updated. The pre-amplified signal is then fed to a RF power amplifier (E&I 1040L Electronics & innovation Ltd, 400W, 54dB) and applied to the transducers. This closed control loop provides a tighter control of the inertial cavitation activity (figure 4). The electrical peak power ranges from 3 to 100 W corresponding to a peak pressure up to 8 MPa in water and from 0.4 to 3.5 MPa inside the tube.

To demonstrate that the set-up is able to maintain a constant inertial cavitation activity experiments were performed with a chemical dosimeter of inertial cavitation [5]. Acoustic cavitation generates free radicals from the breakdown of water molecules. The initial step of water decomposition is the production of hydroxyl and hydrogen radicals. It has been shown that terephthalic acid (TA) [benzene-1, 4-dicarboxylic acid] is suitable for detecting and quantifying free hydroxyl radicals (OH•) generated by ultrasound-induced inertial cavitation. During this process, the TA solution (20 mM of TA in PBS) reacts with OH• generated through water sonolysis to produce 2-hydroxyterephthalic acid (HTA that can be detected using spectroscopy with excitation and emission wavelengths at 318 and 426 nm, respectively.

As the non-adherent cell lines are considered hard to transfect we proposed to restrict the study presentation to only 3 non-adherent cell lines: K562 (human chronic myelogenous leukemia), Jurkat (T-lymphoblastoid), RL (human follicular lymphoma) and 1 adherent cell line BT474 (human breast tumour). Prior to treatment, non-adherent or trypsinized adherent cell lines were harvested and prepared for sonoporation (650 µl in 2 mL tube containing $10.10^6$cells/ml in Opti-MEM® medium with 10µg of peGFP-C1 plasmid) or for nucleofection using a procedure according to the manufacturer's recommendations (Nucleofector® Lonza, Basel, Switzerland; $10.10^6$ cells/ml in 100µl solution with 2 µg of peGFP-C1 plasmid). 24 hours post-treatment the transfection efficiency and cell viability were evaluated by flow cytometry after propidium iodide (Invitrogen Carlsbad, CA, USA) staining. For sonoporation 2 or 3 experiments in triplicate were analysed.

**Results and discussion**

Figure 5 shows that the cavitation index is positively correlated with the chemical dosimeter and is representative of the inertial cavitation activity. The saturation observed for CI upper than 14 is due to a strong mechanical agitation of the liquid medium. In this case the liquid is projected by the ultrasound radiation forces and the OH• production by inertial cavitation seems to be altered. This mechanical agitation was also observed for the open loop in figure 3b. To complete these observations cells did not supported this regimen and were all destroyed after 30s treatment. Figures 6 and 7 show that the device is able to transfect adherent and non-adherent cells with an acceptable level of efficiency. They also show a correlation between the CI, the transfection efficiency, and the viability. Finally, Figure 8 shows that the device efficiency is comparable to what is attainable using nucleofection on two cell lines. Notice that nucleofection is known as one of the most effective non-viral methods for in vitro gene delivery [6, 7].

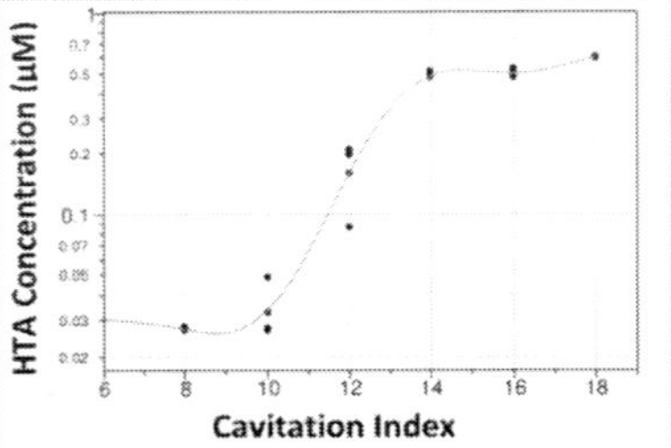


**Figure 5**: HTA concentration vs CI on 650μl of 20mM TA solution sonicated during 60s

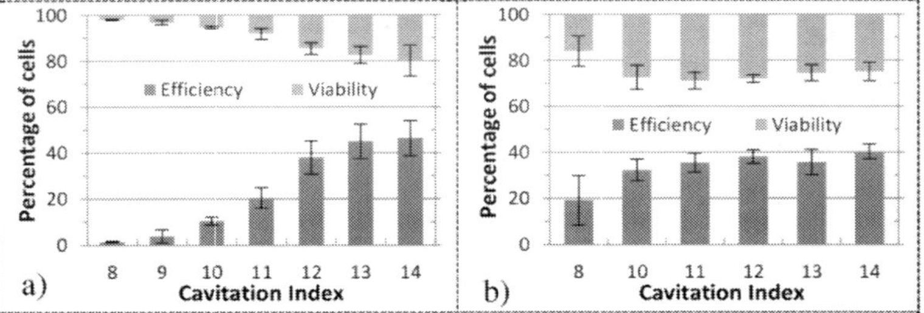


**Figure 6:** Transfection efficiency and viability vs CI of (a) Jurkat and (b) K562 cell lines 24h after ultrasound treatment (30s). The error bars correspond to standard deviations

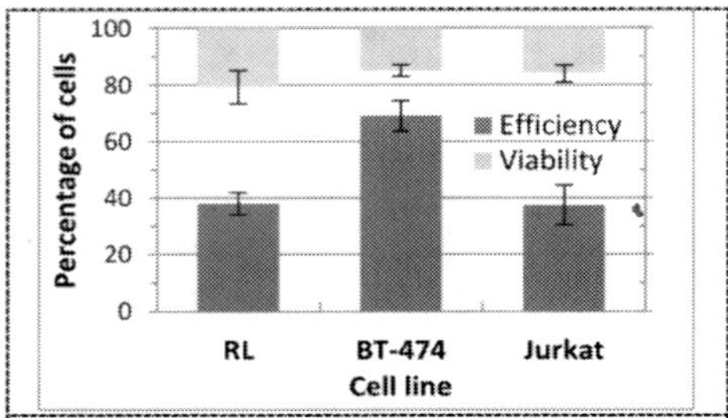


**Figure 7**: Transfection efficiency and viability for CI=12 during 30s for 3 cell lines

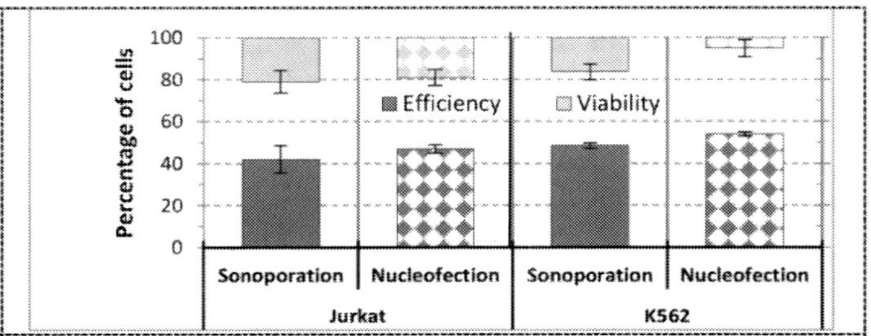


**Figure 8**: Comparison of sonoporation method with nucleofection on 2 non-adherent cell lines hard to transfect

## Conclusion

We developed a user-friendly and cost-effective ultrasound device, well adapted for routine *in vitro* high-yield transfection experiments and which does not require the use of any transfection reagent or gas micro-bubbles. It provides a well-adapted method for low cost routine pDNA in vitro delivery for both adherent and non-adherent cell lines. Our results confirm ultrasound as an alternative non-viral technology for the efficient transient transfection of a wide range of different cells including non-adherent cells or fresh human cells, and the preparation of stably transfected cells.

## References

[1] Marshall E. (1999) Science 286: 2244-2245
[2] Pack DW & all (2005) Nat Rev Drug Discov 5: 581-593
[3] Hahn P & all (2010) Top Curr Chem 296: 1-13
[4] Sabraoui A & all (2011) Ultrason Sonochem. 18(2):589-94
[5] Villeneuve L & all (2009) Ultrason Sonochem. 16: 339–344
[6] Robenek & all (2009) J Cell Mol Med 7: 1381–1390
[7] Grzesik BA & all (2013) Am J Physiol Lung Cell Mol Physiol 305: L786– L794

## Acknowledgments

This work was supported by Caviskills SA and a European Project Eurostars E! 6173 named "Oncoson". It was performed within the framework of the LabEx DevWeCan (ANR-10-LABX-0061) and CeLyA (ANR-10-LABX-0060) of Université de Lyon.

# Bubble Dynamics in Laser Lithotripsy

**Milad Mohammadzadeh, Julian Martinez Mercado, and Claus-Dieter Ohl**

Division of Physics and Applied Physics, School of Physical and Mathematical Sciences, Nanyang Technological University, Singapore

E-mail: cdohl@ntu.edu.sg

**Abstract.** Laser lithotripsy is a medical procedure for fragmentation of urinary stones with a fiber guided laser pulse of several hundred microseconds long. Using high-speed photography, we present an *in-vitro* study of bubble dynamics and stone motion induced by Ho:YAG laser lithotripsy. The experiments reveal that detectable stone motion starts only after the bubble collapse, which we relate with the collapse-induced liquid flow. Additionally, we model the bubble formation and dynamics using a set of 2D Rayleigh-Plesset equations with the measured laser pulse profile as an input. The aim is to reduce stone motion through modification of the temporal laser pulse profile, which affects the collapse scenario and consequently the remnant liquid motion.

## 1. Introduction

As a minimally invasive method of fragmenting urinary stones, intracorporeal laser lithotripsy has been in use for over two decades [3]. In this procedure, an endoscope is inserted into the urethra to locate calculi and a fiber optic is inserted through the working channel of the scope. A pulsed laser is coupled into the fiber to irradiate the stone, causing the stone to disintegrate into smaller fragments which are later on washed out of the urinary tract. The most commonly used laser type is Holmium:YAG ($\lambda = 2.01\,\mu m$) with a long pulse duration (e.g. $\tau = 300\,\mu s$). In Ho:YAG laser lithotripsy, the fundamental mechanism of transfer of laser energy to the stone is through formation of a vapor bubble that bridges the fiber optic with the urinary stone. Then the calculus absorbs the laser energy and is fragmented due to a photothermal mechanism [6].

The reported clinical challenge involved with laser lithotripsy is the retropulsive movement of the stone during laser ablation. Stone motion can prolong the operation time and cause clinical complications [5]. In order to reduce stone retropulsion, the effect of several parameters has been experimentally investigated, including optical pulse energy, fiber diameter, and the pulse length. Recent studies have shown that longer pulses reduce stone retropulsion [4].

Using high-speed photography, we have conducted *in-vitro* experiments for different settings of a clinical Ho:YAG laser while recording the bubble dynamics, stone motion, and the temporal pulse profile. Also we present a simple model to relate the laser pulse with the observed bubble dynamics.

## 2. Experimental setup

The experimental set-up is shown from the side and top view in figure 1. A transparent acrylic container (6 mm) is filled with water. An acrylic cube (100 x 50 x 57 mm) serves as stone phantom. The 270 µm fiber from the clinical Ho:YAG (Dornier, Ho:YAG $\lambda$ =2.01 µm) is inserted from the right into the container. The dynamics of the bubble and stone movement is observed with a high-speed camera (Photron FASTCAM SA-X2). Illumination (200 W metal halide lamp) is provided from the back through a glass diffuser. The temporal profile of the laser pulse

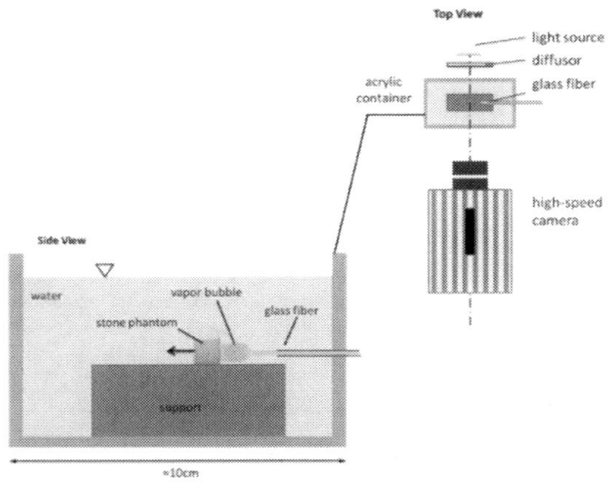


**Figure 1.** Side and top view of the experimental set-up. The high-speed camera records the bubble dynamics and stone movement, and a photodiode measures the laser pulse profile.

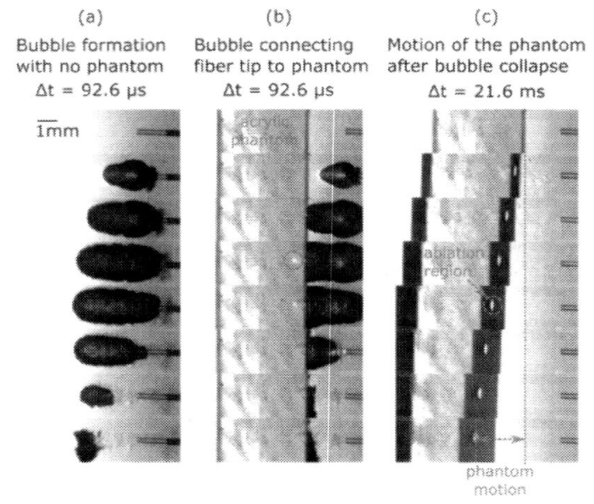


**Figure 2.** High-speed photography of bubble dynamics (a), transmission of energy to the stone (b), and stone motion in Ho:YAG laser lithotripsy (c). Ablation region in the phantom is marked with a red circle.

is measured with a reverse-biased InGaAs photodiode (Thorlabs, FGA20). The experiments were conducted at Tan Tock Seng Hospital, Singapore.

## 3. Experimental results: bubble dynamics and stone movement

Figure 2 depicts three series of images of bubble dynamics and stone movement. The first set, figure 2a, displays the growth and collapse of the elongated bubble during the Ho:YAG laser pulse. There the bubble grows from a pear shape to an elongated shape and collapses within 700 μs. The photodiode recordings of the laser pulse profile (see figure 3) reveal that the bubble grows as long as the laser is on, thus reaching its maximum size just before the laser pulse is switched off. After the laser energy is no longer deposited in the medium, the bubble starts shrinking and collapses within a few hundreds of microseconds.

The physical process that leads to formation of such elongated bubbles during mid-infrared laser irradiation is the metastable vaporization of superheated liquid [7]. During laser lithotripsy, the pulse width, $\tau$, is significantly larger than the acoustic relaxation, but smaller than thermal relaxation time:

$$t_{st} = \frac{\delta}{c} < \tau < t_{th} = \frac{\delta^2}{4k}. \tag{1}$$

Here $c$ is the speed of sound, $k$ is the thermal diffusivity of the liquid medium, and $\delta$ is the length scale of the irradiation volume, in this case the fiber diameter. For a Ho:YAG laser pulse of $\tau$=300 μs delivered into water with a fiber diameter $d$ =300 μm, since $t_{st}$ =100 ns $< \tau$, stress confinement does not occur and therefore no significant shock waves are produced during bubble formation. However, since $\tau < t_{th}$ =1 ms, the liquid is heated under thermal confinement, and maximum possible temperature is obtained inside the irradiated volume. In this condition, the liquid could be heated up to the spinodal limit before phase explosion leads to the formation of a vapor bubble. Optical temperature probing techniques show that water can be heated up to 250 °C before phase transition occurs and the bubble is formed [2].

In the experiments with a phantom stone in front of the fiber optic, we observe that the formation of a vapor bubble is necessary for transmission of laser energy to the stone. Once the bubble touches the stone, ablation of acrylic material occurs. Figure 2b reveals that the water separating the phantom from the fiber tip is vaporized and the ablation process begins once the vapor bubble reaches the phantom. Having a transparent phantom, transmission of the laser light into the phantom and resulting ablation of the material becomes visible as a bright region within the phantom, see circle in figure 2b. During this process the stone does not move.

As it is shown in figure 2c, the stone motion happens after the bubble collapse, and for a much longer time scale (80 ms) compared to the life time of the bubble (0.7 ms). We speculate that a ring vortex is created during the bubble collapse, the origin of which could be the liquid jet moving towards the stone. The heat induced Schlieren in the water allows us to reveal the long lasting circulatory liquid motion.

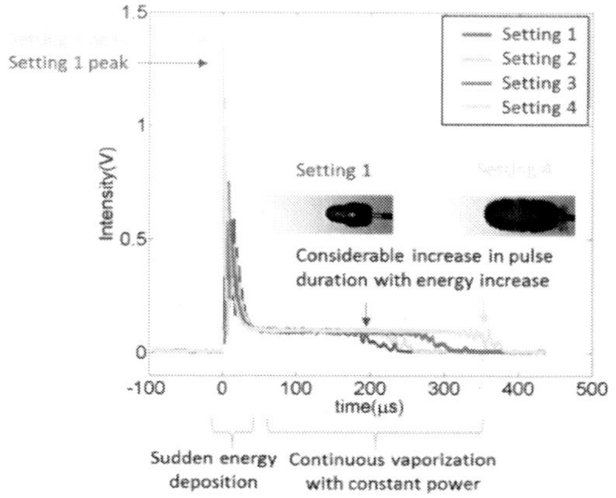


**Figure 3.** Photodiode recordings of the Ho:YAG laser pulse. Energy of the laser increases from setting 1 to setting 4. Although the peak power does not differ much between different settings, the pulse duration varies considerably.

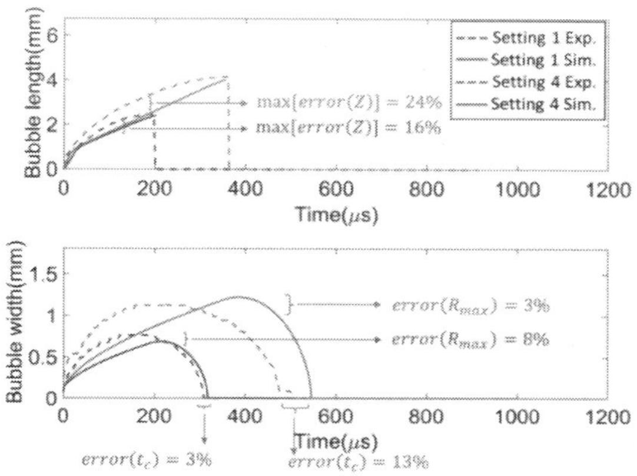


**Figure 4.** Simulation results for two different laser settings compared with experiments. Bubble length ($z$) is along the axis of the fiber optic, while bubble width ($R(z,t)$) is normal to the fiber axis.

## 4. Mathematical modeling of the bubble shape

Similar to the modeling of elongated cavities for the classical water entry problem [1], we model the shape of the elongated bubble in laser lithotripsy by discrete layers, where each layer is governed by a two dimensional Rayleigh-Plesset equation. This model is based on the assumption of pure radial and potential flow. The full form of 2D Rayleigh-Plesset equation for a non-confined bubble is

$$(R\ddot{R} + \dot{R}^2)\log(\frac{R_\infty}{R}) - \frac{1}{2}\dot{R}^2 = \frac{-p_\infty(t) + p_B(t)}{\rho_l} - \frac{2\nu\dot{R}}{R} - \frac{\sigma}{\rho_l R}, \quad (2)$$

where all the variables are defined similar to [1]. We model the radius of the elongated cavity at any distance $z$ away from the fiber tip ($z = 0$) and time $t$, i.e. $R(z,t)$.

Based on the above model, two initial conditions are necessary to find the bubble shape at any time: initial radial velocity of each layer, and the time each layer starts expanding. These conditions are found through an energy balance argument based on hemispherical vaporization of the liquid at the fiber tip. At each layer, the bubble starts expanding with an initial radial velocity that is linearly correlated with the laser power $\dot{R}(z, t = t_z) = K_{exp}p_l(t = t_z)$. Here $K_{exp}$ is the correlation constant which could be found experimentally or estimated based on thermodynamics of superheated vaporization, and $p_l(t)$ is the laser power as a function of time. The time $t_z$ is the time when the laser has down to depth $z$, which is found from $t_z = E^{-1}(\frac{z}{K_{exp}})$, where $E(t) = \int_0^t p_l(t)dt$ is the laser energy deposited in the liquid up to time $t$, and $E^{-1}$ denotes the inverse of this function.

The photodiode measurements, see figure 3, are used as input for the bubble dynamics simulations for two different settings of a Ho:YAG laser. The results are compiled in figure 4. Although the peak power between both settings differs by less than 10% (see figure 3) their total pulse duration varies by 80%. In order to find how these different pulse profiles affect the bubble dynamics, we have plotted the bubble length as a function of time for the duration of laser irradiation, and maximum bubble radius for each setting in figure 4. The bubble length shows fair agreement with an average error of 20% with the simulation, while the bubble width ($R_{max}$) and collapse time ($t_c$) have a discrepancy of 10%. We understand that the model has shortcomings, in particular in the beginning of the pulse where a spherical bubble is generated. We are currently accounting for this. Nevertheless, the shape, temporal and spatial dynamics provide overall agreement.

## 5. Conclusion

While laser lithotripsy has been used as a standard method for fragmentation of urinary stones for over two decades, there are still unresolved problems regarding the physical process that takes place during this procedure. Regarding one of the main challenges in laser lithotripsy, stone movement, it should be noted that the motion happens after the bubble collapse and lasts for a long time, for about 100 ms, while the bubble life time is about 500 µs. Therefore, by controlling the bubble dynamics, we aim to reduce the liquid motion and stone movement after the collapse. Using a mathematical model for an elongated bubble, we show that it is possible to correlate the bubble dynamics with the laser pulse. We are currently developing a suitable laser system which may allows us to mitigate stone motion.

## References

[1] Jeffrey M Aristoff and John WM Bush. Water entry of small hydrophobic spheres. *Journal of Fluid Mechanics*, 619:45–78, 2009.

[2] Tobias Brendel and Ralf Brinkmann. Mid-infrared laser-induced superheating of water and its quantification by an optical temperature probe. *Applied optics*, 43(9):1856–1862, 2004.

[3] Stephen P Dretler. Laser lithotripsy: a review of 20 years of research and clinical applications. *Lasers in surgery and medicine*, 8(4):341–356, 1988.

[4] Hyun Wook Kang, Ho Lee, Joel MH Teichman, Junghwan Oh, Jihoon Kim, and Ashley J Welch. Dependence of calculus retropulsion on pulse duration during ho: Yag laser lithotripsy. *Lasers in surgery and medicine*, 38(8):762–772, 2006.

[5] LI Kuznetsov. Recoil momentum at a solid surface during developed laser ablation. *Quantum Electronics*, 23(12):1035, 1993.

[6] George J Vassar, Kin F Chan, Joel MH Teichman, Randolph D Glickman, Susan T Weintraub, T Joshua Pfefer, and Ashley J Welch. Holmium: Yag lithotripsy: photothermal mechanism. *Journal of endourology*, 13(3):181–190, 1999.

[7] Alfred Vogel and Vasan Venugopalan. Mechanisms of pulsed laser ablation of biological tissues. *Chemical reviews*, 103(2):577–644, 2003.

# Intensely oscillating cavitation bubble in microfluidics

**Siew-Wan Ohl[1,*], Tandiono[1], Evert Klaseboer[1], Dave Ow[2], Andre Choo[2], Claus-Dieter Ohl[3]**

[1]Institute of High Performance Computing, 1 Fusionopolis Way, #16-16 Connexis North, Singapore 138632.
[2]Bioprocessing Technology Institute, 20 Biopolis Way, #06-01, Singapore 138668.
[3]School of Physical and Mathematical Sciences (SPMS), Nanyang Technological University, SPMS-05-07, 21 Nanyang Link, Singapore 637371.

Corresponding author's e-mail address: ohlsw@ihpc.a-star.edu.sg

**Abstract**. This study reports the technical breakthrough in generating intense ultrasonic cavitation in the confinement of a microfluidics channel [1], and applications that has been developed on this platform for the past few years [2,3,4,5]. Our system consists of circular disc transducers (10-20 mm in diameter), the microfluidics channels on PDMS (polydimethylsiloxane), and a driving circuitry. The cavitation bubbles are created at the gas-water interface due to strong capillary waves which are generated when the system is driven at its natural frequency (around 100 kHz) [1]. These bubbles oscillate and collapse within the channel. The bubbles are useful for sonochemistry and the generation of sonoluminescence [2]. When we add bacteria (*Escherichia coli*), and yeast cells (*Pichia pastoris*) into the microfluidics channels, the oscillating and collapsing bubbles stretch and lyse these cells [3]. Furthermore, the system is effective (DNA of the harvested intracellular content remains largely intact), and efficient (yield reaches saturation in less than 1 second). In another application, human red blood cells are added to a microchamber. Cell stretching and rapture are observed when a laser generated cavitation bubble expands and collapses next to the cell [4]. A numerical model of a liquid pocket surrounded by a membrane with surface tension which was placed next to an oscillating bubble was developed using the Boundary Element Method. The simulation results showed that the stretching of the liquid pocket occurs only when the surface tension is within a certain range.

## 1. Introduction: bubbles in microfluidics channel

### *1.1. Bubble dynamics in confinement*

Ultrasonic cavitation is created by ultrasound in water. These bubbles oscillate and collapse creating high energy concentration (up to 1000 bar and 5000 K). They are used in industrial and bioprocessing applications such as mixing, emulsification, rupturing of cell membrane, and catalyzing chemical reactions. However, in microfluidics, the use of ultrasonic cavitation has been mainly restricted to mixing and pumping. The ultrasound is used to cause pre-existing air pockets in a microfluidics channel to oscillate so as to facilitate liquid flow. In our study, strongly oscillating cavitation bubbles are created instead. They do not oscillate spherically but collapse violently with high speed jets.

## 2. Cavitation bubbles in microfluidics

Figure 1 shows the experimental setup to generate intense ultrasonic cavitation in our microfluidics channels. The microfluidics channel is made with polydimethylsiloxane (PDMS). It is then attached to a microscope glass slide. A Lead Zirconium Titanate (PZT) transducer is placed next to the microfluidics channel. It is driven by a linear amplifier and a function generator. The experimental observation is recorded by a high speed camera via microscope lens. Details about the setup are found in [1].

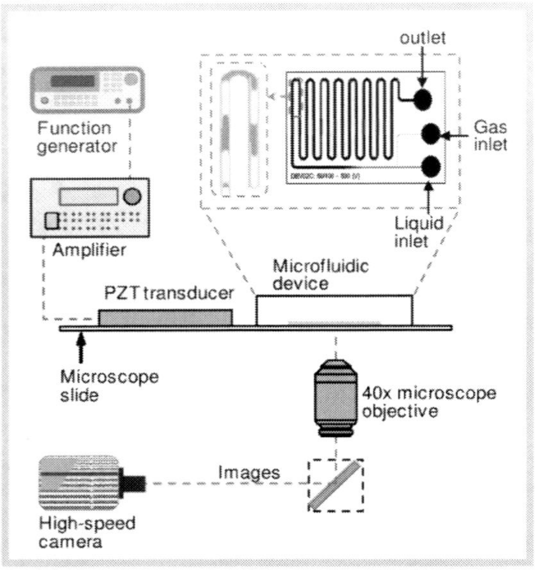


Fig. 1 The experimental setup to generate intensely oscillating ultrasonic bubbles in microfluidics channels. The microfluidics channel has two inlets (one for liquid and one for gas), and one outlet. This creates a liquid with gas pockets in the microfluidics channel. The ultrasound is transmitted from the oscillating transducer through the glass slide into the microfluidics channel.

### 2.1. *Generation of intensely oscillating bubble by surface waves*

The volume of liquid in a microfluidics channel is small. Thus to create cavitation bubbles in the microfluidics liquid is difficult even with high acoustic pressure. Thus air pockets which act as bubble nuclei are introduced into the microfluidics channel. The acoustic energy is transmitted to the PDMS via the glass slide. It causes surface waves to form at the air-water interface inside the channel. Figure 2 shows the process at which the cavitation bubble is formed [2]. The first frame (time = 0) shows two crests coalescing and then forming a gas pocket (as indicated by white arrows). Subsequently the gas pocket oscillates and moves away from the interface. More oscillating bubbles are formed this way and eventually fill up the whole channel.

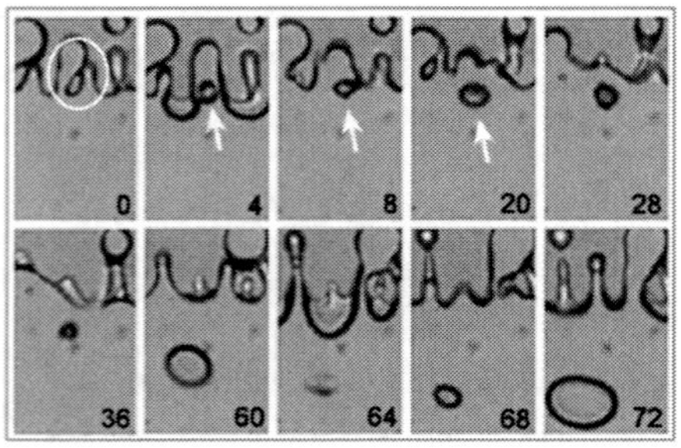


Fig. 2 The entrapment of a gas bubble at the air-water interface (air on top, water below). The surface wave generated by the ultrasound causes two crests to coalesce (time = 0) and a gas bubble is formed as indicated by the white arrows. Driving voltage and frequency is 50 V and 100 kHz, respectively. The video was recorded at 250,000 frames per second with an exposure time 1μs. Width of each frame is 100 μm.

### 2.2. *Sonochemistry and sonoluminescence*

The ultrasonic bubble in our microfluidics channel collapses and heats up the liquid locally. This process generates an intense concentration of energy which is able to trigger chemical reactions (sonochemistry) and to emit light (sonoluminescence) [3]. Figure 3 shows the oxidation of luminal in a sodium carbonate base solution. Radicals H and OH are produced and they trigger the formation of an amino phthalate derivative with electrons in an excited state. When these electrons relax to lower energy states, excess energy is emitted as visible bluish light (Fig. 3). It is noted that the luminal emission only occurs in the liquid, and close to the gas-liquid interface where the cavitation occurs.

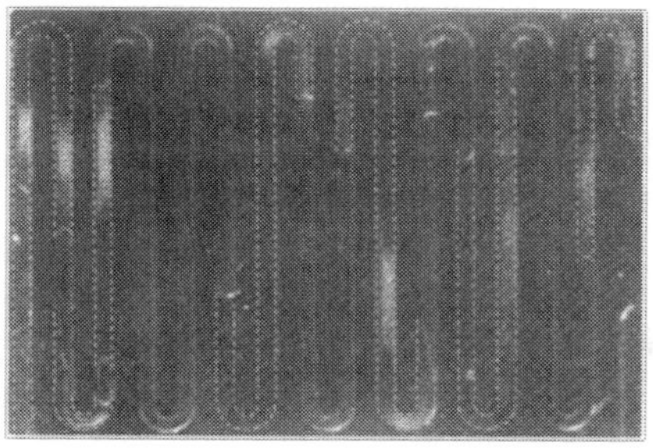

Fig. 3 Luminol chemiluminescence from cavitation bubbles in microfluidics channels as captured by an intensified Electron Multiplying Charge Coupled Device (EMCCD) camera. The overlaid green lines indicate the microfluidics channels. The blue light is emitted from the chemical reaction of luminal oxidation.

### 2.3. *Applications in biotechnology*

Cavitation bubbles in microfluidics channels are useful for the harvesting of inter-cellular contents, and for the study of rheology of the red blood cell [5]. Often in biotechnology, bacteria or yeast cells with modified DNA are utilized. Their intracellular contents need to be extracted for analysis. We report the use of ultrasonic cavitation in a microfluidics channel for this purpose [3].

*2.3.1. Lysis of cells: Yeast cells and E. Coli.* The intensely oscillating bubbles are capable of lysing *Escherichia coli* (*E. Coli*), and *Pichia pastoris* (yeast cell) when they are placed in the microfluidics channel [3]. As seen in Fig. 4, the rod-shaped bacteria (*E. Coli*) is broken down into small fragments after being sonicated at 128.7 kHz, 200 V for 389 ms. The more robust yeast cells are completely lysed within 1 s. Flourescence intensity measurements of the green fluorescence protein (GFP) expressing bacteria and cells show that the functionality of GFP is maintained after the treatment. Real-time polymerase chain reaction (qRT-PCR) analysis confirms that the genomic DNA of the bacteria and cells remain intact after sonication. This technique provides a gentle yet efficient way for the lysis of *E. Coli* and yeast cells in microfludic channels.

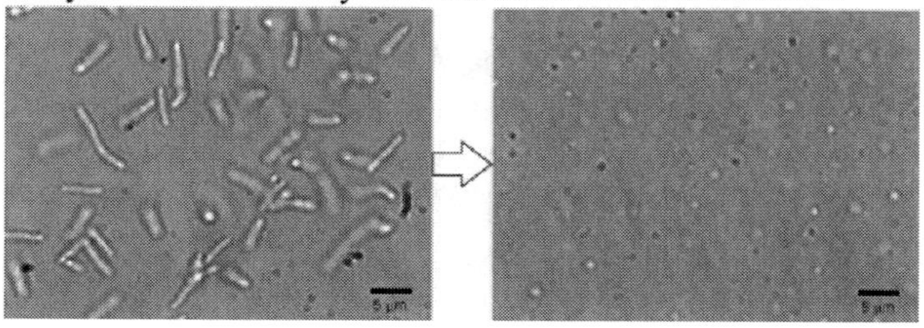

Fig. 4 The lysis of *E. Coli* in a microfluidics channel as viewed under a microscope. The rod-shaped bacteria are broken into pieces after sonication (389 ms at 128.7 kHz).

*2.3.2. Stretching of red blood cell.* The phenomenon of stretched cells in the vicinity of a single cavitation bubble in a micro-chamber is studied [4]. The bubble is generated by heating up the liquid using a pulsed laser. Figure 5 shows the bubble-cell interaction. In the experiment, the bubble is generated at a distance of 84 μm from the cell (Fig. 4A). It expands to its maximum size of 100 μm in radius at frame 4. The cell is pushed up and is slightly flattened. The bubble then collapses. The liquid flow around the cell generated by the collapsing bubble causes the cell to be strongly elongated.

Figure 4B shows a numerical simulation of an oscillating bubble next to an elastic liquid vesicle (mimicking a cell) using the Boundary Element Method. In this method only the boundaries of the bubble and cell are meshed. The cell surface has elasticity as it is modeled with membrane tension. It is found from parametric study of the membrane tension and stand-off distance (the separation between the bubble and cell centers) that cell stretching can only if the cell elasticity is within certain threshold. Also the maximum elongation occurs when the bubble oscillates at half of the oscillation time of the cell.

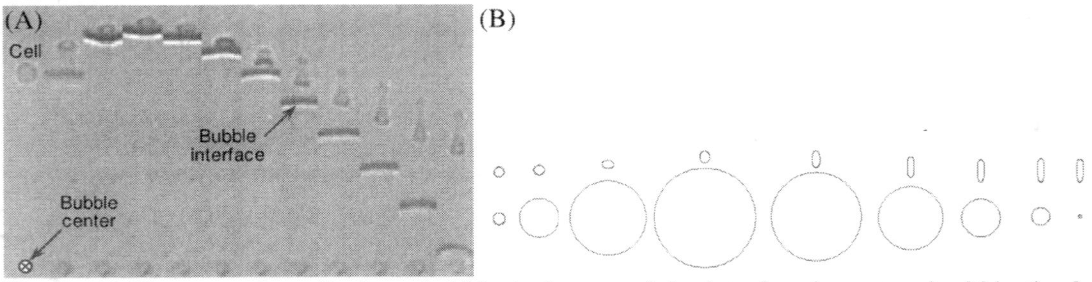


Fig. 5 (A) The expansion of a large bubble (only part of the interface is captured within the frame) near a human red blood cell. The bubble expands to its maximum size (frame 4) and then collapses. The cell is stretched as the bubble collapses. The initial cell radius is 4 μm, and the maximum bubble radius is 100 μm. The width of each frame is 16 μm. The framing rate is 360,000 frames per second. (B) Numerical simulation of an oscillating bubble near an elastic liquid pocket (cell-like object). The 'cell' stretches a lot as the bubble collapses.

## 3. Conclusion

We present a series of studies involving the use of cavitation bubbles in microfluidics. We have developed a technique to generate strongly oscillating bubble in a microfluidics system. These bubbles are capable of inducing sonochemistry and sonoluminescence. The strong shear flow generated by the collapsing bubbles could be utilized to lyse *E.Coli* and yeast cells effectively and efficiently. The intracellular contents remain intact after the brief sonication. The same flow created by a collapsing laser-generated cavitation bubble is seen to stretch a human red blood cell. The phenomenon is modeled successfully using the Boundary Element Method. The numerical study indicates that elastic object with certain elastic properties is much elongated next to a collapsing bubble. In conclusion, our ultrasonic microfluidics system is versatile and useful for many biotechnology applications. We are currently investigating the use of the system for emulsification, gene transfection, and lysis of endospores.

## References

[1] Tandiono, S. W. Ohl, D. S. W. Ow, E. Klaseboer, V. V. T. Wong, A. Camattari, C. D. Ohl, "Creation of cavitation activity in a microfluidic device through acoustically driven capillary waves", Lab Chip 10(14), 1848-1855 (2010).

[2] Tandiono, S. W. Ohl, D. S. W. Ow, E. Klaseboer, V. V. Wong, R. Dumke, C. D. Ohl, "Sonochemistry and sonoluminescence in microfluidics", Proc Natl Acad Sci USA 108(15), 5996-5998 (2011).

[3] T. Tandiono, D. S. W. Ow, L. Driessen, C. S. H. Chin, E. Klaseboer, A. B. H. Choo, S. W. Ohl, C. D. Ohl, "Sonolysis of Escherichia coli and Pichia pastoris in microfluidics", Lab Chip 12 (4), 780-786 (2012).

[4] T. Tandiono, E. Klaseboer, S. W. Ohl, D. S. W. Ow, A. B. H. Choo, F. Li, C. D. Ohl, "Resonant stretching of cells and other elastic objects from transient cavitation", Soft Matter 9 (36), 8687-8696 (2013).

[5] P. A. Quinto-Su, K. Claudia, P. R. Preiser, C. D. Ohl. "Red blood cell rheology using single controlled laser-induced cavitation bubbles." Lab Chip 11, no. 4 (2011): 672-678.

Journal of Physics: Conference Series **656** (2015) 012006 doi:10.1088/1742-6596/656/1/012006

# Tracking the deformation of a tissue phantom induced by ultrasound-driven bubble oscillations

**M Tinguely, O K Matar, V Garbin**

Department of Chemical Engineering, Imperial College London, London SW7 2AZ, United Kingdom

E-mail: m.tinguely@imperial.ac.uk

**Abstract.** Microbubbles are used as contrast agents in ultrasound medical imaging. Once the microbubbles are injected into the body, they flow through the vascular system, confined by viscoelastic boundaries. The proximity of the boundaries affects the dynamics of the bubbles in ultrasound, in a manner that depends on the boundary's viscoelastic properties. Experiments on violently collapsing bubbles have revealed the dynamics of deformation of blood vessel walls. However, the deformation field induced by a bubble undergoing small-amplitude oscillations, relevant for ultrasound imaging, is difficult to access in experiment, and has not been reported yet. We present an experimental method to measure the deformation field induced by a bubble oscillating inside a microchannel within a tissue phantom. We use high-speed video microscopy to track the displacement of tracer particles embedded in the phantom, along with the dynamics of the bubble.

## 1. Introduction

Microbubbles have gained a lot of interest in the biomedical field for their ability to enhance contrast in ultrasound imaging, and for their potential to promote the uptake of drugs at ultrasound-targeted locations in the body [1]. Once the microbubbles are injected into a patient's body, they flow through the vascular system, confined by viscoelastic walls. This configuration is quite different from the classical problem of a bubble near a rigid boundary or a free surface [2] because both the geometric confinement and the viscoelastic properties of the walls have to be taken into account to accurately predict the dynamics of the bubble, and the effect on the deformation of walls. Experiments on the growth and collapse of a microbubble inside a blood vessel show vessel distension and invagination [3]. Numerical simulations have investigated similar configurations to predict the bubble dynamics, and the displacement and the stress induced on the surface of the vessel wall [4, 5, 6]. These phenomena are strongly dependent on the properties of the confining walls. To accurately model the interaction of ultrasound-driven bubbles inside vessels with confining walls, it is crucial to characterize the deformation of the surrounding tissue induced by the bubble dynamics. Experimental investigations, however, usually only report the deformation of the solid surface [7, 8].

We present a method to measure the deformation of both the surface and the bulk of a tissue phantom induced by an oscillating bubble. We use side-view and bottom-view high-speed video microscopy to track the displacement of particles embedded in the bulk of the tissue phantom. The ultrasound-driven bubble oscillates in a microchannel inside the phantom, in order to mimic a vessel within a tissue. Our setup allows extracting the deformation of the phantom as a function of time in horizontal and vertical cross-section of the phantom, providing information to reconstruct the local deformation in four dimensions, $(x, y, z, t)$.

## 2. Methods

*Preparation of tissue phantoms* - We used agarose gel as tissue phantom. The viscoelastic properties are controlled by tuning the concentration of the gels. We can thus mimic different tissues by producing gels with the corresponding properties [9]. To produce a gel, a solution of agarose powder (A9539, Sigma Aldrich) was dissolved in water, heated to 95°C, and stirred for 30 min. Tracer particles for particle image velocimetry (9-13 $\mu$m hollow glass spheres, 110P8, LaVision GmbH) were then added to the solution. The solution was poured into a mold where we previously positioned a square capillary (8290, CM Scientific, 1.2×1.2×5 mm) on two spacers to create the channel inside the gel. The solution was left to cool and set for 24 hours at room temperature. The gel was removed from the mold, and the capillary was gently removed from the gel, leaving an empty channel inside the gel. The gel sample was 40 mm long, 20 mm wide, and 7 mm thick, and the bottom of the 1.2×1.2×40 mm long channel ran 2 mm above the bottom of the gel sample (see fig. 1). We present results for a gel with an agarose concentration of 2% w/v. The shear modulus $G = 43 \pm 2$ kPa was measured from a creep test using a stress-controlled rotational rheometer (Discovery HR-1, TA Instruments).

*Experimental setup* - Figure 1 shows the experimental setup. A piezoelectric transducer (Physik Instrumente, P-121.05) was glued on the bottom surface of a glass dish placed on the stage of an inverted microscope (Olympus, IX71). The gel sample was positionned on the bottom of the dish with the direction of the channel along the $x$ axis, and submerged in ultrapure water. A 45° mirror was placed in water for side-view imaging. The objective could be placed either below the channel, or below the mirror for respectively bottom-view and side-view visualizations in transmission mode. The deformation of the channel by the oscillating bubble was imaged at 4× magnification and recorded using a high-speed camera (Photron, FASTCAM SA5) at 60,000 frames per second. A waveform generator (Agilent, 33220A), and a power amplifier (T&C Power Conversion, AG 1021) actuated the piezoelectric transducer. The amplitude of the pressure oscillations inside the channel was recorded with a hydrophone (RP acoustics, PVDF RP 33 s). We measured the pressure amplitude by positioning the hydrophone tip inside the channel, through the gel, after the experiments. Nitrogen microbubbles of controlled size were generated with a co-flowing microfluidic device [10]. By moving the outlet of the device, we were able to inject only one bubble into the channel. We present results for a bubble having a radius $a = 350$ $\mu m$. We applied a single ultrasound burst of 10 cycles at 10 kHz. For a peak-to-

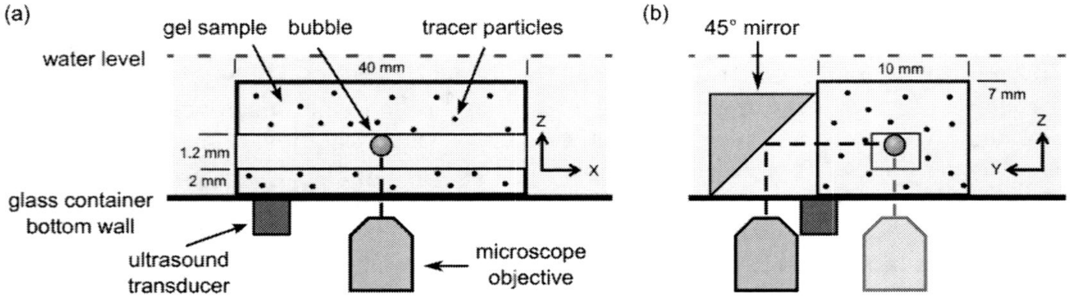


Figure 1: Schematic of the experimental setup and definition of the coordinate system. The black dash lines represent the optical path. (a) $(x, z)$ plane, with microscope objective below the bubble for bottom-view visualizations. (b) $(y, z)$ plane, with microscope objective below the mirror for side-view visualizations.

peak pressure amplitude of 60 kPa, the maximum amplitude of oscillations of the bubble was $\Delta a = (a_{max} - a_{min})/2a_0 = 5\%$ of the initial radius. Due to the confinement, the bubble slightly deviated from a spherical shape towards an ellipsoid during the oscillations. The deviation from spherical shape remained small: the maximum deformation parameter, $D = |d_1 - d_2|/(d_1 + d_2)$, where $d_1$ and $d_2$ are the major and minor axis of the ellipsoid in the vertical plane through the center of the bubble, did not exceed 2.3 %.

*Particle tracking* - The gel deformation was measured by tracking the displacement of the tracer particles with sub-pixel accuracy. The positions of the particles were determined in each frame to obtain trajectories. First, the positions of the particles were obtained with accuracy of

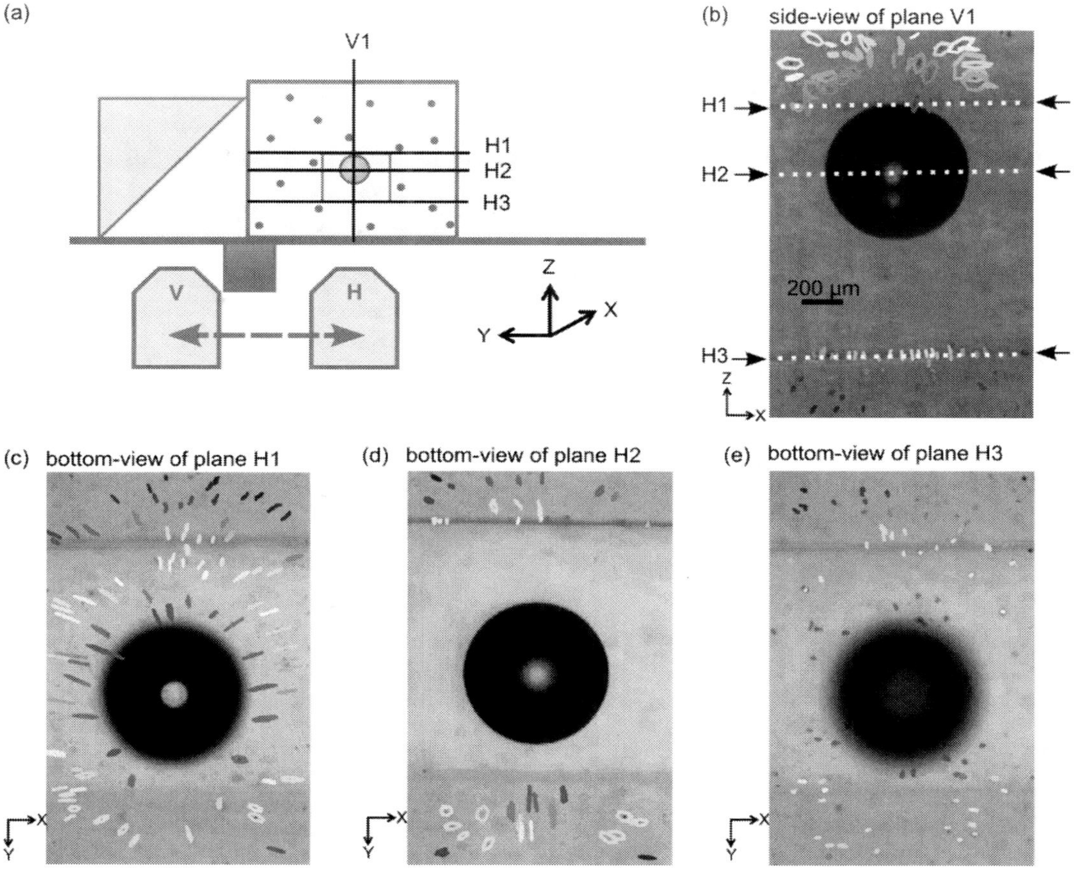


Figure 2: (a) Schematic of the experiment showing the planes where the visualizations were recorded. The microscope objective is placed either below the bubble (H) or below the mirror (V) for bottom-view or side-view visualization. (b) Trajectories of tracer particles from a side-view high-speed movie of the vertical plane V1 through the center of the bubble. The white dashed lines represent the horizontal planes H1, H2, and H3. (c) Trajectories of tracer particles from a bottom-view high-speed movie of the horizontal plane H1 containing the top surface of the channel. (d) Trajectories of tracer particles from a bottom-view high-speed movie of the horizontal plane H2 through the center of the bubble. (e) Trajectories of tracer particles from a bottom-view high-speed movie of the horizontal plane H3 containing the bottom surface of the channel. In (b-e), the particle displacement is magnified by a factor of 50 for clarity, and the color changes with the distance from the bubble.

one pixel by finding the local maxima from binarized images [11]. The grayscale values of the pixels in the original image were then used to improve the accuracy. For each particle position obtained from the first step, we calculated the center of mass of the particle using as weight the brightness of the pixels within a radius of 4 pixels.

## 3. Results

We present four visualizations, shown in fig. 2 (b)-(e), which reveal the deformation field in the horizontal planes containing the top and the bottom of the channel: H1 and H3, respectively, and in the vertical and the horizontal planes through the center of the bubble: V1 and H2, respectively. Because of buoyancy, the bubble is in contact with the top wall of the channel, but does not touch the other walls. The trajectories within the top planes H1 (c) and the bottom plane H3 (e) are mostly radial, centered at the (x,y) position of the center of the bubble. In contrast, within the planes passing through the center of the bubble, V1 (b) and H2 (d), the particles describe elliptic trajectories. The deformation occurs on all four walls, with a larger amplitude of deformation for the wall that is in contact with the bubble. The high spatial and temporal resolution of these measurements enable us to track the deformation induced by the localized oscillating pressure applied by the bubble on the top wall, but also the complex deformation field induced by the fluid flow generated by the volumetric oscillations of the bubble. A quantitative analysis of the deformation field in the phantom will be the subject of future work. Using this setup we have recently investigated the deformation induced by a bubble oscillating near a planar viscoelastic surface [12].

## 4. Conclusion

We presented an experimental method to measure the deformation induced by an oscillating bubble in a channel inside a tissue phantom. Using high-speed video microscopy, we tracked the displacement of tracer particles embedded in the phantom, and extracted the deformation field on the surface of the channel and in the bulk. Accurate measurements of the deformation field will help develop models to predict the effect of viscoelastic confinement on the dynamics of bubbles for different geometries.

## Acknowledgments

M. T. acknowledges support from the Swiss National Science Foundation under grant P2ELP2_151953. O. K. M. acknowledges support from the Engineering and Physical Sciences Research Council (UK) through the MEMPHIS Programme Grant (EP/K003976/1).

## References

[1] Ferrara K, Pollard R and Borden M 2007 *Annual Review of Biomedical Engineering* **9** 415–447
[2] Blake J R and Gibson D C 1987 *Annual Review of Fluid Mechanics* **19** 99–123
[3] Chen H, Kreider W, Brayman A A, Bailey M R and Matula T J 2011 *Physical review letters* **106** 034301
[4] Hay T A, Ilinskii Y A, Zabolotskaya E A and Hamilton M F 2012 *The Journal of the Acoustical Society of America* **132** 124–137
[5] Coralic V and Colonius T 2013 *European Journal of Mechanics-B/Fluids* **40** 64–74
[6] Hosseinkhah N, Chen H, Matula T, Burns P and Hynynen K 2013 *The Journal of the Acoustical Society of America* **134** 1875–1885
[7] Van Wamel A, Kooiman K, Harteveld M, Emmer M, Ten Cate F J, Versluis M and De Jong N 2006 *Journal of controlled release* **112** 149–155
[8] Yoshida K, Nakatani S, Tsukamoto A, Ushida T and Watanabe Y 2008 *Japanese Journal of Applied Physics* **47** 4200
[9] Duck F A 1990 *Physical properties of tissues: a comprehensive reference book* (Academic Press)
[10] Benson B R, Stone H A and Prud'homme R K 2013 *Lab on a chip* **13** 4507–4511
[11] Crocker J C and Grier D G 1996 *Journal of colloid and interface science* **179** 298–310
[12] Tinguely M, Hennessy M G, Pommella A, Matar O K and Garbin V 2015 *submitted*

# Acoustic methods for cavitation mapping in biomedical applications

**M Wan[1], S Xu, T Ding, H Hu, R Liu, C Bai and S Lu**

The Key Laboratory of Biomedical Information Engineering of Ministry of Education
Department of Biomedical Engineering, School of Life Science and Technology
Xi'an Jiaotong University, Xi'an 710049, People's Republic of China

E-mail:mxwan@mail.xjtu.edu.cn

**Abstract.** In recent years, cavitation is increasingly utilized in a wide range of applications in biomedical field. Monitoring the spatial-temporal evolution of cavitation bubbles is of great significance for efficiency and safety in biomedical applications. In this paper, several acoustic methods for cavitation mapping proposed or modified on the basis of existing work will be presented. The proposed novel ultrasound line-by-line/plane-by-plane method can depict cavitation bubbles distribution with high spatial and temporal resolution and may be developed as a potential standard 2D/3D cavitation field mapping method. The modified ultrafast active cavitation mapping based upon plane wave transmission and reception as well as bubble wavelet and pulse inversion technique can apparently enhance the cavitation to tissue ratio in tissue and further assist in monitoring the cavitation mediated therapy with good spatial and temporal resolution. The methods presented in this paper will be a foundation to promote the research and development of cavitation imaging in non-transparent medium.

## 1. Introduction

Acoustic cavitation is the phenomenon observed when ultrasound of sufficient intensity is transmitted through a liquid causing micron-sized gas bubbles, termed as active cavitation bubbles, to oscillate, grow, and violently implode giving rise to extreme, but localized, conditions within the collapsed cavities. When ultrasound exposure is turned off, these residual cavitation bubbles tend to dissolve. With pulsed ultrasound, residual cavitation bubbles between successive ultrasound exposures may be acted as cavitation nucleus to enhance cavitation effects. An understanding of spatial-temporal distribution of both active and residual bubbles is essential to optimize cavitation induced various applications. In this paper, several acoustic methods for cavitation mapping developed or modified on the basis of previous study will be summarized to help to depict spatial-temporal distribution of both active and residual cavitation bubbles in non-transparent medium.

## 2. Methods and Results

### *2.1. Microsecond Resolution Ultrasound Line-by-line 2D Mapping*

As cavitation is highly complex with transient characteristics, it is essential to develop a standard cavitation mapping similar to the exact acoustic field characterization with both high spatial and

[1] To whom any correspondence should be addressed.

Journal of Physics: Conference Series **656** (2015) 012007 doi:10.1088/1742-6596/656/1/012007

temporal resolutions. A novel precise active cavitation mapping (ACM) by a modified ultrasound line-by-line scanning method is proposed by Ding et al [1]. Scattered signals from cavitation bubbles are firstly obtained by a scan line immediately after one high intensity focused ultrasound (HIFU) exposure, and then there is a waiting time (2s) to make the liquid back to the original state. As this pattern extended, an image can be built up by sequentially acquiring a series of such lines. The synchronization of HIFU exposure and radio frequency (RF) data acquisition by a programmable ultrasound scanner (Sonix-RP, Ultrasonix, Burnaby, BC, Canada) is illustrated in figure 1 (a). T and D which can vary from microseconds to seconds are HIFU exposure time and time delay between RF data acquisition and the end of HIFU exposure, respectively.

The temporal evolution of cavitation bubble cloud in tap-water is shown figure 1 (b). The principle of this technique is very simple while it may be a useful tool in cavitation mapping for different cavitation sources which can be repeated.

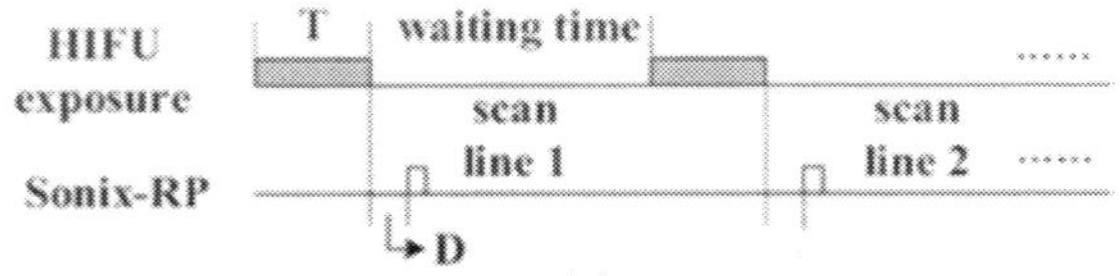


(a)

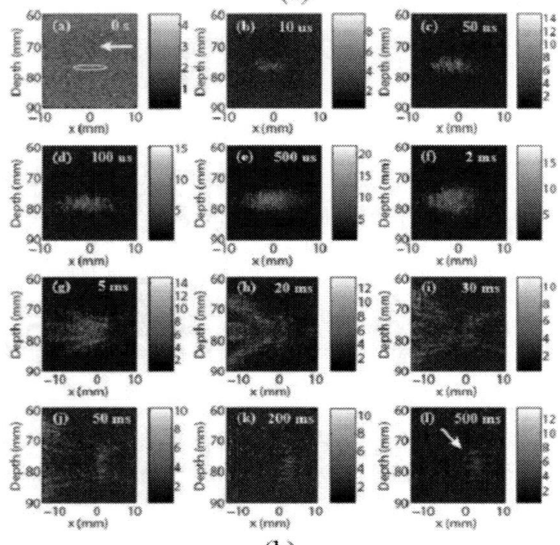


(b)

Figure 1. (a)Timing-sequences for modified ultrasound line-by-line scanning method. (b) Temporal evolution of the cavitation bubble cloud in tap-water. The pointer and the solid circle in top left subplot in (b) show the transmission direction of HIFU and the position of the focal region respectively. The various T are indicated in the right corner of each image. D is set to 1.0 ms. Zero of the x-axis is set to the centre element of the linear array [1].

## 2.2. *Ultrafast Active Cavitation Mapping with Plane Wave*

The method introduced in section 2.1 is time-consuming and restricted within the medium with repeatable cavitation distribution. Nevertheless, the due to the transient characteristics, mapping of cavitation should be achieved with high frame rate, preferably at one time over the entire region. Hu et al. employed an ultrafast ACM (UACM) with high temporal resolution and relatively good signal-to-noise ratio (SNR) by combining plane wave transmission, minimum variance (MV) beamforming and coherence factor (CF) weighting to monitor the evolution of residual cavitation bubbles during histotripsy process [2]. The schematic representation of the process is shown in figure 2 (a).

Figure 2 (b) shows the representative image sequence of residual bubbles around the water-tissue interface in histotripsy (Pulse Duration = 10 μs, PRF = 2 kHz, total HIFU pulses = 840000), from which the residual bubbles are distributed at the interface of the tissue initially, then gradually expanded to the internal region of the tissue with an increasing HIFU pulse number, and finally the entire bubble cloud gets distributed within the tissue. However, there are no significant differences between the B-mode images before and after histotripsy. A direct relationship with a coefficient of 1.0365 between histotripsy lesion area and inner residual bubble area was found, as shown in figure 2 (c). These results can assist in monitoring and optimization of the other cavitation enhanced therapy further.

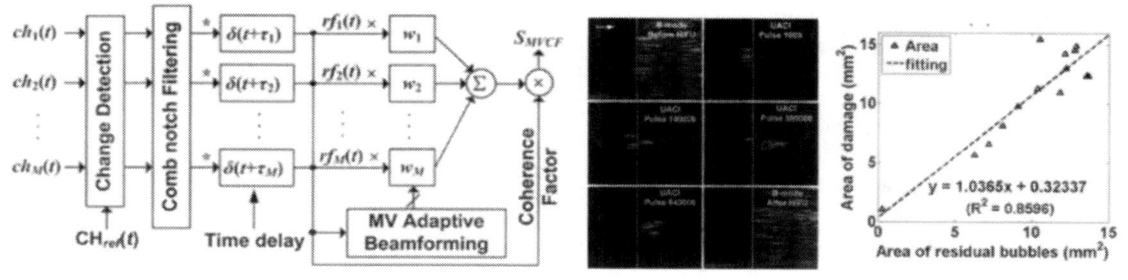


(a) (b) (c)

Figure 2. (a) Schematic representation of UACM. (b) Representative sequence of residual bubble images (20 mm×15 mm). (c) Correlation curve between damage areas and residual bubble cloud areas [2].

*2.3. Microsecond Resolution Ultrasound Plane-by-plane 3D Mapping*

Combining the methods presented in section 2.1 and 2.2, an ultrasound plane-by-plane method for 3D cavitation mapping is proposed by our group [3]. Scattered signal from cavitation bubble cloud immediately after one HIFU exposure was obtained by plane wave. Then a series of spatial channel-domain raw RF data were obtained in plane-by-plane mode. The cavitation bubble distribution in 3D was reconstructed by reconstruction algorithm. MV adaptive beamforming, CF weighting and compressive sensing (CS) theory were used to process the raw RF data to improve the spatial resolution, SNR and calculation speed.

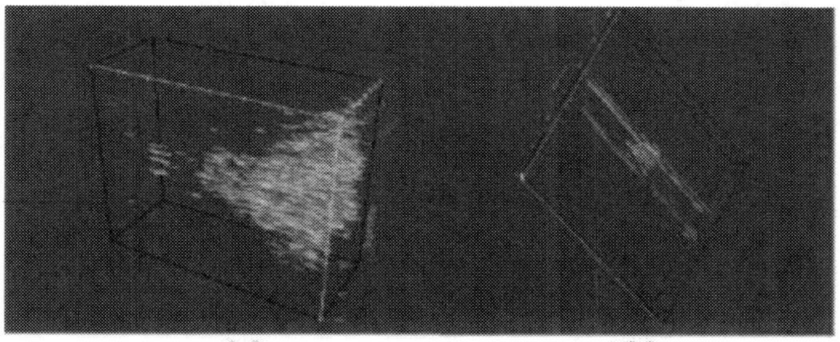

(a) (b)

Figure 3. (a)3D cavitation mapping in tap-water when PD was 20 ms. (b)3D cavitation mapping in a phantom vessel with the pulsatile flow when PD was 20 μs [3].

*2.4. UACM with Enhanced Cavitation to Tissue Ratio Based on Bubble Wavelet Transform and Pulse Inversion Technique*

Due to the low transmission pressure of plane wave, a reduction in the backscattering signals occurs, leading to poor image quality aggravating cavitation detection. To solve this problem, a UACM method that combined wavelet transform with pulse inversion (PI) technique was modified on the basis of our previous work on ultrasound contrast imaging [4]. A mother wavelet named "cavitation bubble wavelet" was constructed according to Rayleigh-Plesset-Noltingk-Neppiras-Poritsky model. A high correlation between cavitation bubble wavelet and backscattering signals was expected to obtain [5].

Results showed that the image quality was associated with the initial radius of bubble and the scale. The SNR of the best cavitation bubble wavelet transform (CBWT) mode image (right panel of figure 4(a)) was improved by 3.2 dB compared with that of the B-mode image (left panel of figure 4(a)) in free-field experiments. Figures 4(c) and 4(d) showed that cavitation to tissue ratio (CTR) of the optimum PI-based CBWT mode image was improved by 2.2 dB compared with that of the PI-based B mode image in tissue experiments. Based on this method, the spatial distribution of the cavitation bubbles was clearly observed.

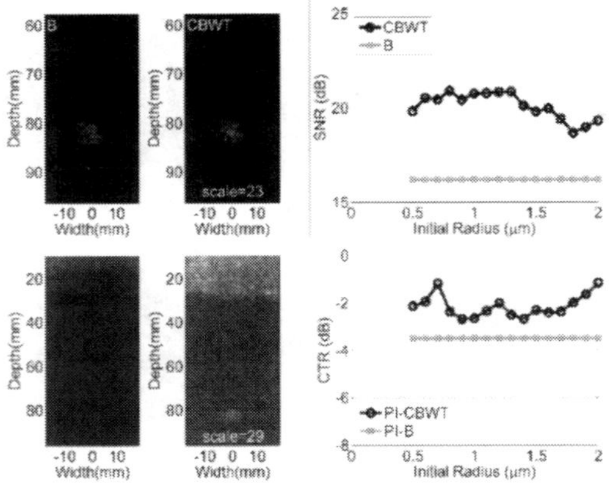


Figure. 4 (a)Left panel was the B mode image using plane wave and right panel was the best CBWT mode image using plane wave with scale = 23 when radius = 1.3 μm. (b) SNR versus initial radius curve. (c) Left panel was the B mode image by using plane wave and the right panel was the most optimal CBWT image by using plane wave with scale = 29 when radius = 0.7 μm. (d) CTR versus initial radius curve [5].

### *2.5. Passive Cavitation Mapping*

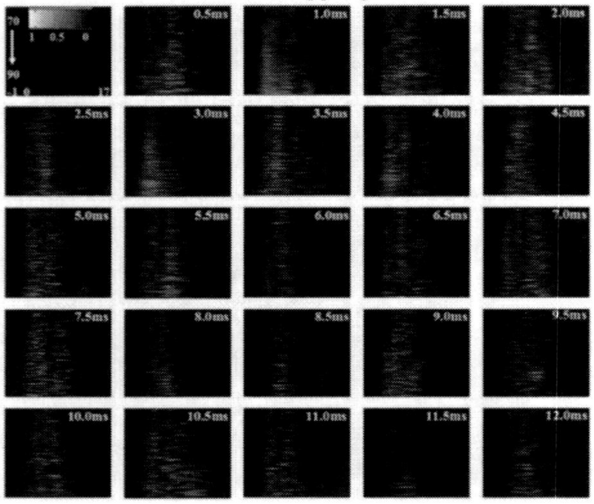


Conventional algorithm of PCM is to calculate short-term energy accumulation of channel data in time domain or frequency domain. Figure 5 is spatial distribution of active bubbles in free field during HIFU exposure with a temporal resolution of 0.5ms. Through calculating energy accumulation of acoustic emission received by a linear array, process of bubble cluster evolution within 12ms can be tracked. Besides, sonoluminescence, the most direct indicator of bubble inertial collapse, has been employed to map the active cavitation bubbles. However, it is limited in the transparent medium and might not be applied to investigate the active bubble distribution in tissue.

Figure.5 Passive cavitation mapping within 12ms.

## 3. Conclusions

The proposed novel ultrasound line-by-line/plane-by-plane method can map residual cavitation bubbles with high spatial and temporal resolution and may be developed as a potential standard 2D/3D cavitation field mapping method. The modified UACM based upon plane wave transmission and reception together with bubble wavelet and PI technique can apparently enhance the CTR in tissue and further assist in monitoring the cavitation mediated therapy with good spatial and temporal resolution. All the developed or modified acoustic methods for cavitation mapping established solid foundation for cavitation imaging in non-transparent medium.

## References

[1] Ding T, Zhang S, Fu Q, Xu Z, and Wan M 2014 *Ultrasonics* **54** 147
[2] Hu H, Xu S, Yuan Y, Liu R, Wang S, and Wan M 2015 *J. Acoust. Soc. Am.* **137** 2563
[3] Ding T, Hu H, Bai C, Guo S, Yang M, Wang S and Wan M 2015 *Int. Symp. on Therapeutic Ultrasound (Utrecht)*
[4] Zhao X, Zhong H, Wan M and Shen L 2011 *Ultrasound Med. Biol.* **37** 1292
[5] Liu R, Hu H, Xu S, Huo R, Wang S, and Wan M 2015 *J. Acoust. Soc. Am.* **137** 3099

Journal of Physics: Conference Series **656** (2015) 012008 doi:10.1088/1742-6596/656/1/012008

# Photoacoustic cavitation for theranostics: mechanism, current progress and applications

**Y Feng, D Qin and M Wan**[1]

The Key Laboratory of Biomedical Information Engineering of Ministry of Education
Department of Biomedical Engineering, School of Life Science and Technology
Xi'an Jiaotong University, Xi'an 710049, People's Republic of China

E-mail:mxwan@mail.xjtu.edu.cn

**Abstract**. As an emerging cavitation technology, photoacoustic cavitation (PAC) means the formation of bubbles in liquids using focused laser and pre-established ultrasound synchronously. Its significant advantages include the decreased threshold of each modality and the precise location of cavitation determined by the focused laser. In this paper, a brief review of PAC is presented, including the physical mechanism description, the classic experimental technology, the representative results in variety of media, and its applications in biomedical imaging and therapy. Moreover, some preliminary results of PAC in perfluoropentane (PFP) liquid and PFP droplets investigated by passive cavitation detection (PCD) in our group are also presented.

## 1. Introduction

Cavitation is the formation of cavities or bubbles in liquids, which can be induced either by lowering the liquid pressure or by elevating the liquid temperature. Acoustic cavitation works through lowering the liquid pressure, while optical cavitation in which laser-induced cavitation (LIC) is a well-established and highly precise approach, forms plasma to elevate the liquid temperature. Both acoustic and optical cavitation occurs above certain characteristic threshold for the respective forms of radiation and the cavitation threshold is always high *in vivo*. By synchronously using a laser and a pre-established ultrasound, photoacoustic cavitation (PAC) allows for bubble formation at a lower laser fluence and peak negative acoustic pressure than when using either modality alone. Thus, it is an effective method to avoid the unsafe high-energy radiation and to control the occurrence of cavitation at intended locations. In this paper, studies on PAC have been reviewed briefly, including its physical mechanisms, representative results in variety of media, and its applications in biomedical imaging and therapy.

## 2. Description of the Physical Process

The principle of PAC is illustrated in Fig. 1, including the formation and dynamics of a photoacoustic cavitation bubble in water. A focused laser pulse creates a strong optical field to ionize water molecules and generates plasma in the confocal volume of the laser and ultrasound. The water in the confocal volume is moved to a metastable region because both the liquid temperature $\boldsymbol{T_l}$ is elevated

[1] To whom any correspondence should be addressed.

due to energy absorption by the plasma and the liquid pressure $\boldsymbol{p_l}$ is reduced due to the tensile effects of ultrasound and thermoelastic waves. Then a nucleus is generated in the metastable water due to spontaneous fluctuations, accompanied by shock wave generation. The nucleus will grow explosively to a macroscopic cavitation bubble under both the vapor pressure $\boldsymbol{p_v}$ inside the bubble and the negative liquid pressure $\boldsymbol{p_l}$ outside the bubble. Moreover, the introduction of nanoparticles (NPs) can simplify this process because NPs have the light absorption characteristics and can be used as pre-existing nucleation nuclei[1].

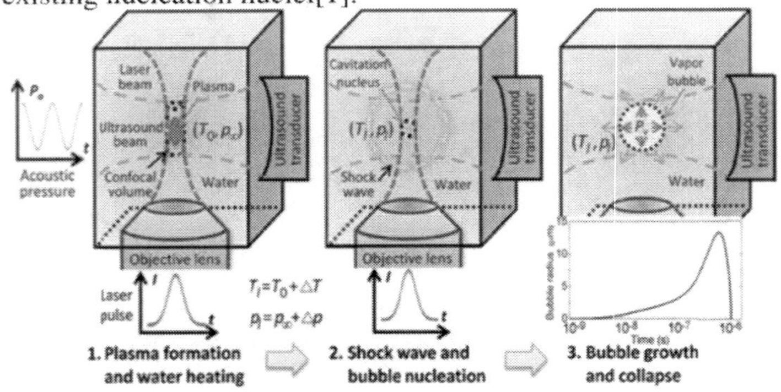


**Figure 1.** Schematic diagram of PAC.

## 3. Methods, Results and Applications

*3.1. Experimental System*

A schematic of a classic experimental PAC system is shown in Fig. 2. The ultrasound is generated by a focused ultrasound transducer. The laser is focused by the objective of the reconstructed inverted microscope system, and the high-speed photography is confocal to the laser. A hydrophone is used to passively detect the acoustic signals emitted during this process. Moreover, to ensure synchronization of the laser pulse, flash illumination, and operation of the high-speed camera, each component is electronically triggered with appropriately delayed TTL pulses from a timing system.

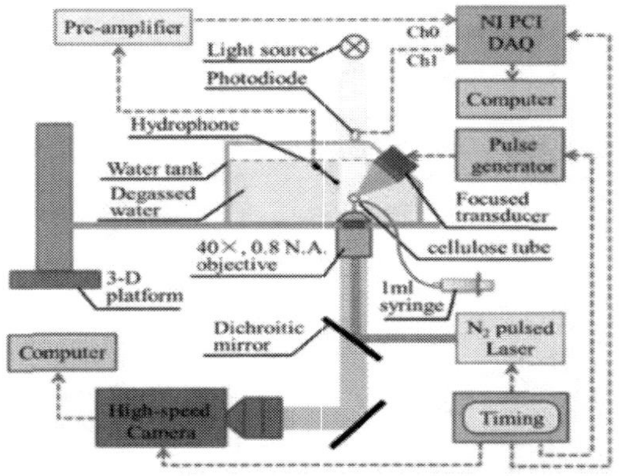


**Figure 2.** A classic experimental system of PAC.

*3.2. PAC in Water.*

Gerold *et al.* introduced an instrument in which acoustic cavitation can be forced to occur at a specific location in water using a laser to nucleate a microcavity in a pre-established ultrasound field [2]. The high-speed photographic observations of their LIC and PAC studies proved that cavitation can be nucleated below the optical cavitation threshold under ambient pressure conditions in a pre-established ultrasound field. Meanwhile, the emitted acoustic signals detected by a hydrophone suggested that the PAC is inertial due to a strong increase from 100 kHz to 1 MHz in frequency-domain. Kurz *at al* investigated the optic cavitation bubble dynamics and luminescence in an ultrasonic field. The results demonstrated that the initially small laser-generated bubbles can be expanded significantly by the sound field and that weak cavitation luminescence can be observed in two small intervals of the seeding phase [3].

Nanoparticles were usually introduced to enhance the optical absorption and to nucleate a cavitation event. Farny *et al.* reported that laser illumination on the gold nanoparticles proved effective in significantly lowering the HIFU cavitation threshold pressure, decreasing from 4.50 MPa to 0.92 MPa with the laser energy 0.10 mJ/pulse [4]. Ju *et al.* examined the photoacoustic cavitation events for a broad range of ultrasound pressures and nanoparticle concentrations for gold nanorods and nanospheres [5]. The cavitation thresholds for both nanoparticle types are drastically reduced in the presence of an ultrasound field. McLaughlan *et al.* enhanced the photoacoustic emissions from laser-heated nanoparticles subjected to a ultrasound field [6]. Moreover, the strong acoustic emission and low-threshold fluence associated with ultrasound-assisted cavitation make PAC well suited for biological imaging and tissue therapy.

### *3.3. PAC in a Phase-shift Liquid.*
The threshold of PAC in liquid is closely related to the liquid boiling point. Thus perfluorocarbons (PFCs) with a relatively low-boiling point, such as perfluoropentane (PFP, 29 °C) and perfluorohexane (PFH, 56 °C) were introduced in PAC considering that less energy is needed to perform the cavitation [7].

Figure 3 shows the differential PCD signal of PAC in PFP, with a 532-nm laser pulse and a 5 MHz ultrasound pulse, using the experimental system introduced above. There are four peaks in the time domain as shown in Fig. 3(a), which corresponds to the observation of bubble formation in high-speed photographs. The two larger peaks represent the primary signals emitted during breakdown and the first bubble collapse in PAC, respectively. The two smaller peaks are the corresponding reflections of the primary signals from the objective lens deduced from the distance between the sample and the objective lens. Broadband noises in the low frequency region and small peaks at 5, 10 and 15 MHz are presented in figures 3(b) and (c). These small peaks at the driving frequency and its harmonics indicate the presence of stable cavitation that is induced by the synergistic effects of ultrasound and laser. It also reveals that during the PAC nucleation, the ultrasonic field used to form the PAC bubble has a significant effect on the characteristics of bubble dynamics.

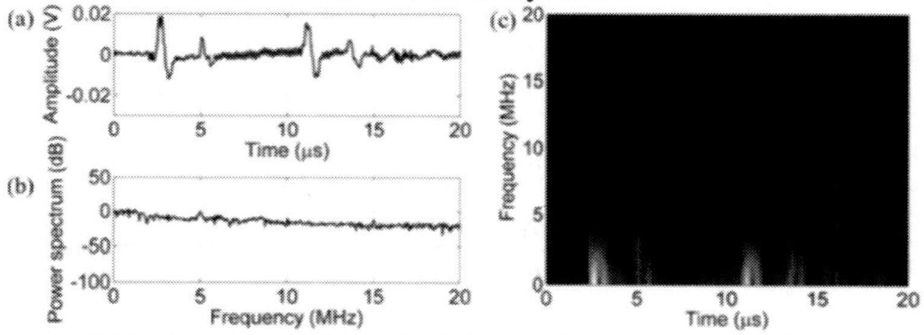


**Figure 3.** PCD signals of PAC in PFP liquid, including (a) differential signal in time-domain, (b) frequency spectrum and (c) time-frequency diagram.

### *3.4. PAC of PFC Nanodroplets.*
Liquid PFCs nanodroplets can undergo a transition into a gaseous state if they are triggered by synchronous laser and ultrasound above a certain threshold. [7]. The differential PCD signal of PAC for the PFP droplets is shown in Fig. 4. Two peaks in the time domain represent the primary signal emitted during breakdown in PAC and its reflection. Broadband noise at a low frequency is shown in the spectrum analysis. Similar to the PAC in PFP liquid, the small peaks appear at 5, 10 and 15 MHz, revealing that the PAC of PFC droplets occurred and the close relationship between PAC bubble dynamics and ultrasound pulse.

Recently, nanoemulsion beads consisted of a perfluorohexane core and coated with gold nanopsheres (NEB-GNS) were also developed and excited with simultaneous light and ultrasound, resulting in a phase transition achievable under different conditions, enhancing laser-induced acoustic signals and enabling specific detection of nanoprobes at lower concentration [8]. It was demonstrated that nanodroplets vaporization could be achieved at lower thresholds than when using either modality alone and at a reduced exposure compared to water surrounding single gold nanosphere.

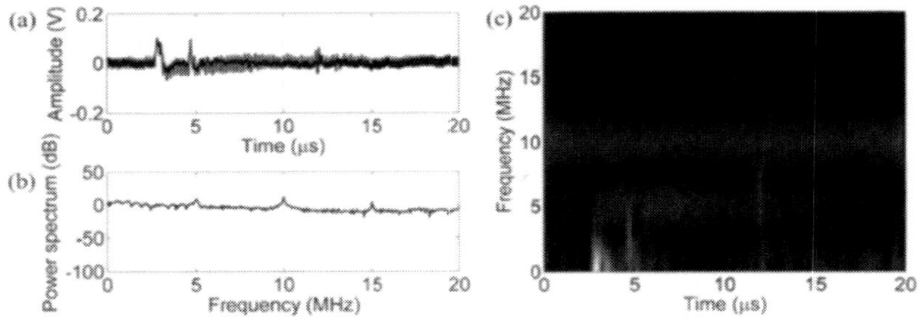


**Figure 4.** PCD signals of PAC in PFP droplets, including (a) differential signal in time-domain, (b) frequency spectrum and (c) time-frequency diagram.

*3.5. Applications of PAC*

In PAC, the laser was employed to illuminate the sample concurrently with ultrasound radiation, so it has the potential to reduce cavitation threshold *in vivo* and enhance HIFU treatments [9-11]. When laser was focused, cavitation can be controllably introduced by using the laser to nucleate a microcavity in the ultrasound focus [2]. Furthermore, the combined laser nucleation and ultrasound produced acoustic emissions and resulted in large hemolytic effects[12]. In addition, PAC with nanoparticles can enhance the photoacoustic emissions associated with bubble collapse, which is well suited for biological imaging [6]. A new imaging mode was presented with the nanodroplets coated with gold nanospheres at laser fluences and negative ultrasound pressures far below maximum permissible exposure limits [13]. This technology showed great potential for molecular imaging with ultrasensitive detection of targeted gold coated nanodroplets and cavitation-assisted theranostic approaches.

## 4. Conclusions

To avoid unsafe high-energy exposures in either acoustic cavitation or LIC, PAC shows great advantage because the combination of laser and pre-established ultrasound reduced threshold of each energy modality significantly. In the study, the physical process of PAC was first described, and then the characteristics of PAC in a variety of media, such as water, PFP, PFP droplets, and NEB-GNS were presented. Finally, application of PAC in biomedical imaging and therapy was reviewed briefly. By developing multifunctional and targeted nanoprobes, PAC would have great potential for targeted theranostic applications at cellular and molecular level.

## References

[1] Wu T, Farny C H, Roy R A and Holt R G 2011 *J. Acoust. Soc. Am.* **130** 3252-63
[2] Gerold B, Kotopoulis S, McDougall C, McGloin D, Postema M and Prentice P 2011 *Rev. Sci. Instrum.* **82** 044902
[3] Kurz T, Kröninger D, Geisler R and Lauterborn W 2006 *Phys. Rev. E* **74** 066307
[4] Farny C H, Wu T, Holt R G, Murray T W and Roy R A 2005 *Acoust. Res. Lett. Onl.* **6** 138-43
[5] Ju H, Roy R A and Murray T W 2013 *Biomed. Opt. Express.* **4** 66–76
[6] McLaughlan J R, Roy R A, Ju H and Murray T W 2010 *Opt Lett.* **35** 2127-9
[7] Krasovitski B, Kislev H and Kimmel E 2007 *Ultrasonics.* **47** 90–101
[8] Arnal B, Perez C, Wei C-W, Xia J, Lombardo M, Pelivanov I, Matula T J, Pozzo L D and O'Donnell M 2015 *Photoacoustics.* **97**
[9] Cui H and Yang X 2013 *J. Acoust. Soc. Am.* **133** EL123
[10] Cui H, Zhang T and Yang X 2013 *Appl. Phys. Lett.* **102** 133702
[11] Cui H and Yang X 2011 *Appl. Phys. Lett.* **99** 231113
[12] Miller D L, Spooner G and Williams A R 2001 *J. Biomed. Opt.* **6** 351-8
[13] Arnal B, Wei C-W, Perez C, Nguyen T-M, Lombardo M, Pelivanov I, Pozzo L D and O'Donnell M 2015 *Photoacoustics.* **3** 11-9

# Relationship between thrombolysis efficiency induced by pulsed focused ultrasound and cavitation bubble size

**S Xu, X Liu, S Wang and M Wan**[1]

The Key Laboratory of Biomedical Information Engineering of Ministry of Education
Department of Biomedical Engineering, School of Life Science and Technology
Xi'an Jiaotong University, Xi'an 710049, People's Republic of China

E-mail:mxwan@mail.xjtu.edu.cn

**Abstract**. In this study, the relationship between the efficiency of pulsed focused ultrasound (FUS)-induced thrombolysis and the size distribution of cavitation bubbles has been studied. Firstly, the thrombolysis efficiency, evaluated by degree of mechanical fragmentation was investigated with varying duty cycle. Secondly, the size distribution of cavitation bubbles after the 1st, $10^3$th and $10^5$th pulse during experiments for various duty cycles was studied. It was revealed that the thrombolysis efficiency was highest when the cavitation bubble size distribution was centred around linear resonance radius of the emission frequency of the FUS transducer. Therefore, in cavitation enhanced therapeutic applications, the essential of using a pulsed FUS may be controlling the size distribution of cavitation nuclei within an active size range so as to increase the treatment efficiency.

## 1. Introduction

As a completely novel, non-pharmacological approach, focused ultrasound (FUS) thrombolysis has been extensively studied for several decades, which offers the potential to shorten therapy time, increase reperfusion and limit bleeding complications. However, there is still the clinical limit that any clot debris produced by the fragmentation of the original clot may block micro vessels in distal vascular beds. Therefore, it is essential to measure the size distribution of clot debris following FUS-mediated lysis and to evaluate how the size distribution of the clot debris may be related to the acoustic parameters of FUS so that the FUS-induced thrombolytic effect may be optimized. Current research on FUS-induced thrombolysis has been focused on the feasibility of using this technique to break clots [1], while there have only been very limited studies on the size distribution of the resulting clot debris.

Previous studies have shown that thrombolysis is more efficient using pulsed FUS rather than continuous FUS [2], as the degree of acoustic shielding is minimized by the relatively long pauses between pulses, allowing intense cavitation to take place in the focus, which could disintegrate clots mechanically. More specifically, each FUS pulse creates a cluster of microbubbles localized at the transducer focus, whose collapse will cause local stress to remove a portion of the targeted clot. Residual microbubbles may also act as nuclei that can be excited by subsequent pulses, predisposing clot in the focal region to further damage. Therefore, it may be possible to achieve a synergistic effect between residual cavitation nuclei by adjusting the timing between two consecutive pulses. It can be

[1] To whom any correspondence should be addressed.

hypothesized then that in pulsed FUS-induced thrombolysis, if the size distribution of microbubbles is within an active size range in which the bubbles oscillate strongly in response to the applied acoustic pressure, the cavitation activity will be enhanced and result in optimal thrombolysis. To test this hypothesis, we must know the size of microbubbles generated by the treatment process and how they relate to changes in FUS acoustic parameters.

In this study, the relationship between the efficiency of FUS-induced thrombolysis, the duty cycle and the size distribution of cavitation bubbles will be investigated. The efficiency of thrombolysis was evaluated through the degree of mechanical fragmentation and the size distribution of cavitation bubbles was calculated through the method previously proposed by our group [3].

## 2. Materials and methods

A schematic representation of the experimental setup is shown in figure 1(a), including US generation, clot sample, passive cavitation detection (PCD), and cavitation bubble size estimation. Exposures were performed in degassed water maintained at 37℃ within a tank. Experimental sample was fresh non-heparinized blood obtained from domestic swine, which will be coagulated at room temperature for 3 h and maintained at 4℃for up to 3 days to allow for maximal clot retraction, lytic resistance, and stability before experiment.

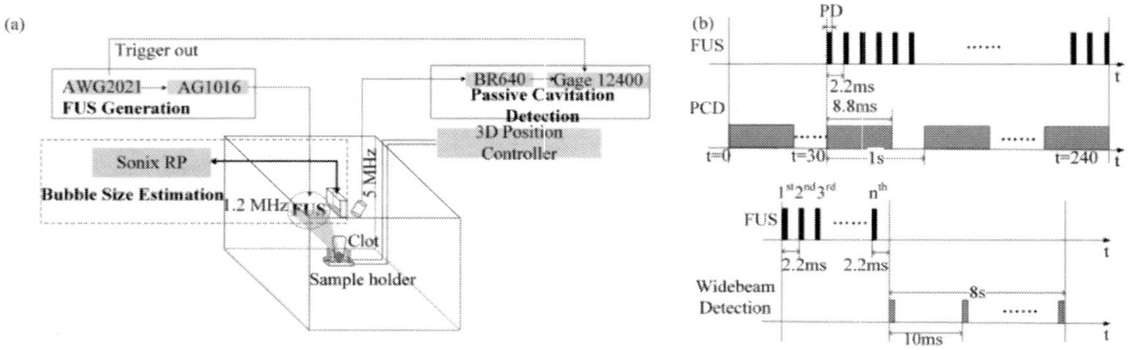


**Figure 1.** (a) The experimental setup includes FUS generation, sample clot, and PCD. (b) Time sequence between FUS generation and PCD.

Destructive US were achieved using a 1.2-MHz single element, high-intensity focused US (HIFU) transducer(15 cm aperture, 12 cm focal length), which was pulsed at an acoustic power of 60 W at a PRF of 454 Hz throughout this study(figure 1 (b)). A 5 MHz single element transducer, adjusted to confocal with HIFU transducer through a needle hydrophone, was used to receive acoustic backscatter signals. Each signal will then be used to calculate the inertial cavitation dose (ICD) through the conventional PCD method [4].

In order to investigate the relationship between the efficiency of thrombolysis and the duty cycle, three pulse durations (PDs) (50μs, 200μs, and 400μs) per pulse period (1/454 Hz≈2.2 ms) were evaluated, corresponding to three duty cycles (2.3%, 9%, and 18%), which were chosen based on preliminary experiments that showed obvious thrombolytic effects at these settings. The total treatment time was 210 s and the PCD was started 30 s prior to treatment time to determine the baseline noise level (upper in figure 1 (b)). Clot debris resulting from the treatment group was also compared to the control group, in which the clot was submerged in saline without FUS exposure for the same period of time.

For each duty cycle, the size distribution of residual cavitation bubbles after the $n$th pulse ($n$=1, $10^3$, $10^5$), which can be regarded as the size distribution of cavitation nuclei for the ($n$+1)th pulse, was estimated using the wide beam ultrasonic method described in our previous paper [3]. The time sequence between the FUS transducer and the wide beam ultrasonic detection using the linear array is illustrated in the bottom in figure 1 (b).

After each treatment, saline samples containing suspended clot debris were removed from the condom and the size distribution of debris particles was measured using the Coulter Counter (Multisizer™ III, Beckman Coulter, USA) particle sizing system.

## 3. Results and discussion

Figure 2 shows the number and number percentage of clot debris particles induced by FUS at three duty cycles, as well as the control group. These results demonstrate that the number of debris using a 9% duty cycle was higher than that when the lower (2.3%) or higher (18%) duty cycle was used. Furthermore, the mean size of clot debris particles generated using a 9% duty cycle was smaller than that generated at the two other duty cycles.

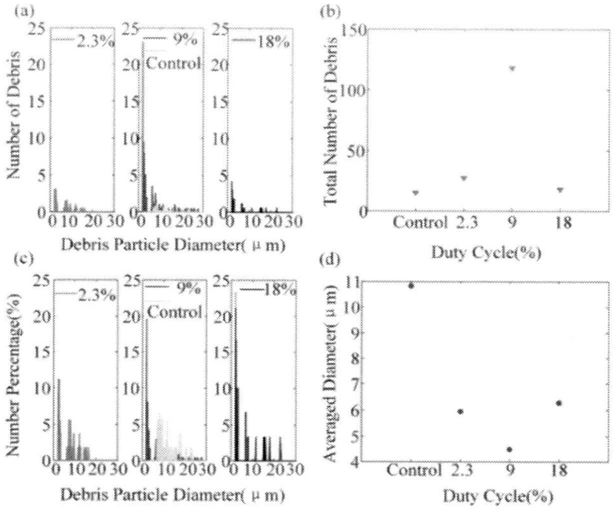


**Figure 2.** Histograms of the (a) number and (c) number percentage of clot debris particles with sizes ranging from 2 to 30μm. (b)Total number and (d) average diameter of clot debris particles as a function of duty cycle.

Figure 3(a) shows the inertial cavitation activity for the three tested duty cycles over time. For the shorter duty cycle (2.3%), the amplitude of the broadband noise is low but increases steadily with time. The cavitation activity slowly accumulates over the course of the treatment. For the longer duty cycle, the cavitation activity initially rapidly reaches steady state, but then gradually decays to a steady state that is higher than the baseline noise (18%). The ICD from the 9% duty cycle treatment was greater than that measured using the other two duty cycles (figure 3(b)).

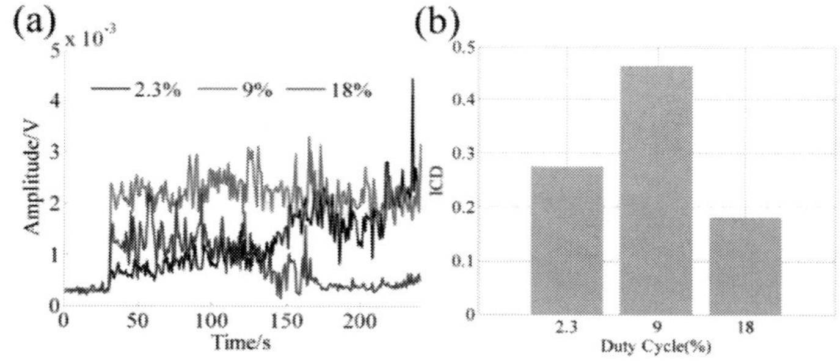


**Figure 3.** (a) Passively detected broadband noise emission over time and (b) ICD generated during the total treatment time as a function of duty cycle.

Figure 4 shows the evolution of the size distribution of cavitation bubbles for all three tested duty cycles. In figure 4 (a), when the duty cycle was 2.3%, the size distribution of the cavitation bubble from the 1st, $10^3$th, $10^5$th pulse increased slightly, but appears fairly smaller than the resonance radius of the emission frequency of the FUS transducer ($R_{res}\approx 3/f_0$, $f_0$= 1.2 MHz, vertical solid line in figure 4). When the duty cycle was 9%, the size distribution of the cavitation bubble remained nearly unchanged, right around the resonance radius (figure 4(b)). When the duty cycle was increased to 18%, the mean size of the cavitation bubble increased to a size that is substantially larger than the resonance radius (figure 4(c)).

It is known that periodic switching on and off of the sound field is a way to increase the efficiency of these cavitation-enhanced applications, the rationale behind which is to stabilize the bubble size distribution within the "active" radius range, which spans from the Blake-threshold to a radius within an order of magnitude of the linear resonance radius. When the duty cycle was too small, such as 2.3%, cavitation bubbles either dissolve completely or to a size below the threshold required for nucleation. Conversely, when duty cycle is too large, such as 18%, cavitation bubbles may grow too large during US sonication and reach a size beyond the range of "active" radii, resulting in a decrease of subsequent cavitation activity [5]. Therefore, determining whether bubbles induced by the previous round of US sonication are within the ''active'' size interval is helpful in optimizing the sonication scheme.

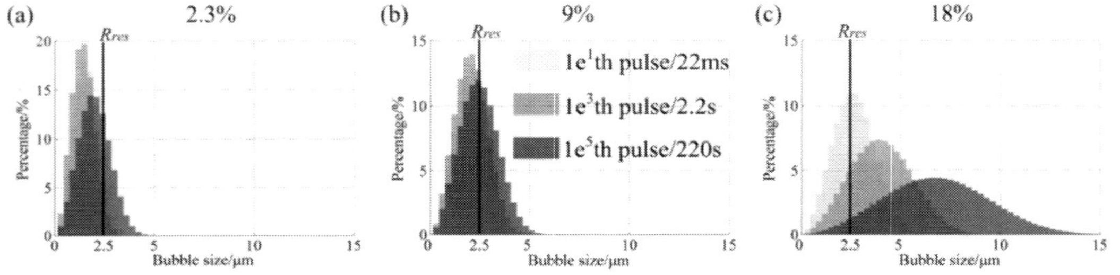


**Figure 4.** Size distributions of the induced cavitation bubbles after the 1st, $10^3$th, $10^5$th pulse at a duty cycle of (a) 2.3%, (b) 9% and (c) 18%.

## 4. Conclusions

It has been known that the number of clot debris particles was largest with a 9% duty cycle, while the mean diameter was smallest. Moreover, the size distribution of cavitation bubbles was mainly centred around the linear resonance radius of the emission frequency of the FUS transducer at 9% duty cycle, while the size distribution was centred around a smaller value for the 2.3% duty cycle, and a larger value for the 18% duty cycle. In addition, the ICD from the 9% duty cycle treatment was much higher than the ICD from the other two duty cycles. The results indicated that pulsing ultrasound enhances cavitation via controlling the size distribution of cavitation bubbles within the active size range, which was found to be near the linear resonance radius of the emission frequency of the FUS transducer.

## References

[1] Maxwell AD, Cain CA, Duryea AP, Yuan L, Gurm HS and Xu Z 2009 Ultrasound Med. Biol. **12** 1982
[2] Westermark S, Wiksell H, Elmqvist H, Hultenby K and Berglund H 1999 *Clin. Sci.* **97** 67
[3] Xu S, Zong Y, Li W, Zhang S and Wan M 2014 *Ultrason. Sonochem.* **21** 975
[4] Frohly J, Labouret S, Bruneel C, Looten-Baquet I, Torguet R 2000 *J.Acoust. Soc. Am.* **108** 2012
[5] Hauptmann M, Struyf H, Gendt SD, Glorieux C and Brems S 2013 *J. Appl. Phys.* **113** 184902

# Growth and Collapse of a Single Bubble near a Plate by Spark Discharge in Water

**I Akcam**[1, 2] **K Inaba**[1] **K Takahashi**[1] **K Kishimoto**[1]
[1] Tokyo Institute of Technology, 2-12-1, Ookayama, Meguro-ku, Tokyo 152-8550, JAPAN

E-mail: akcam.k.aa@m.titech.ac.jp

**Abstract.** Single bubble dynamics in the vicinity of a solid boundary submerged in water were studied experimentally. Single bubble inside a water tank was generated by a spark discharge of capacitor into a couple of copper wires closing a simple circuit. A circular polycarbonate plate was placed horizontally above the bubble creation site. Polycarbonate plates with two different thicknesses were tested by changing the distance between the plate and the creation site. The effects of distance to the wall and wall thickness on the bubble motion is observed by considering the fluid-structure interaction. It is shown that motion of the two boundaries during the bubble generation differ from each other. Jetting behavior of two cases is also different.

## 1. Introduction

Establishing the interaction between the bubbly liquid and solid boundary is important in order to estimate damage mechanisms and potential and there have been many investigations on the bubble behavior or bubble dynamics near solid boundaries [1-7]. Emission of shock waves from a collapsing bubble, which are high-pressure pulses originate from the compression of bubble contents due to pressure difference with the outer surface of the bubble, was firstly demonstrated by the famous work of Rayleigh [1]. It is recently predicted that profile of a shockwave passing through a bubbly liquid is not uniform but gradually changing [8]. In order to estimate such profile, bubble motion should be known. It is known that if the boundary is flexible, the pressure variation of the fluid near the boundary will be different than that of a rigid boundary case. In order to reveal the effect of a flexible boundary, firstly single bubble motion by considering the fluid-structure interaction should be addressed.

In the present study, effects of a thin solid wall on cavitation bubble dynamics were investigated by observation of a single bubble generated by spark-induction technique near the wall. The purpose of this research was to investigate the effect of the motion of a solid wall, placed near bubble-generation site, on bubble dynamics by the pressure reduction around the bubble due to the Joukowsky pressure that was created by the motion of the plate.

## 2. Experiments

Polycarbonate plates with different thicknesses were used in the experiment. High speed video camera, KEYENCE VW-9000 is used to record bubble generation and collapse. Matlab® Image Processing Toolbox is used in order to find the maximum radius and oscillation period. Polycarbonate circular plates of thickness 1 mm and 2 mm are clamped by a single steel annular disc of 4 mm thickness which is bolted to the main frame. Downward face of the polycarbonate plates are in contact with the water

[2] Author to whom any correspondence should be addressed

and four strain gages, shown as g1, g2, g3 and g4 in Figure 1, of 10 mm in distance to each other are applied to the upward face in order to observe the plate motion. Theoretically the radial and tangential strains must be equal due to the equi-biaxial membrane stretching. It is assumed that the plates are thin and deform like a membrane by Rajendran and Narasimhan [9]. All the effects of bending are neglected. Boundary condition of the plate is assumed as fixed boundary. Bubble is created directly under the center of the plate, which corresponds to the location of gage number 1, indicated as g1.

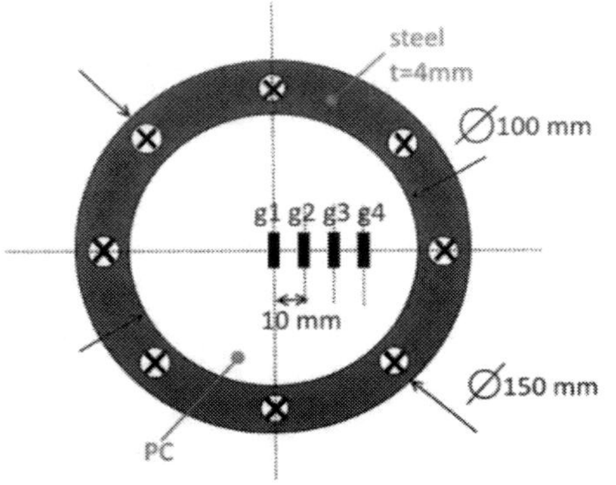


**Figure 1.** Schematic representation of strain-gage arrangement on clamped circular plate.

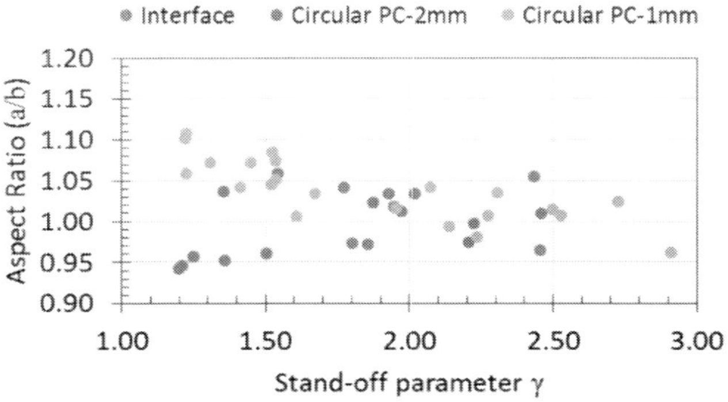


**Figure 2.** Aspect ratio vs. stand-off parameter of the bubbles near different boundary conditions.

Electrical discharge technique was employed in order to create a single bubble. Spark was generated by two electrode wires slightly in touch under water. Spark-generated bubbles are generally produced by high voltage, however, low-voltage can also be used provided that the electrodes in direct contact. This technique produces spherical bubbles of different sizes. The maximum bubble diameter depends on energy discharge from capacitor and diameter of generating wires.

## 3. Results & Discussion

Dynamical behavior of collapsing bubble near a boundary strongly depends on the non-dimensional stand-off parameter $\gamma$ [2-3]. It is observed that as stand-off decreases, bubble takes a nearly ellipsoid shape due to interaction with the wall. In this study, bubble shape is assumed as an oblate ellipsoid and an equivalent radius of same volume of perfectly spherical case is approximated for the sake of simplicity. Non-dimensionless stand-off parameter was defined as initial distance of boundary wall to bubble generation site divided by the maximum radius of the equivalent spherical bubble ($\gamma = L_i/R_{max}$). In order to indicate amount of deformation in the bubble symmetry, which is related to proximity of the bubble generation site to the solid wall, the aspect ratio ($a/b$) of the elliptically approximated cavitation bubble versus stand-off parameter is plotted for each boundary condition. Aspect ratio is defined as the ratio of major radius to minor radius, major radius being the smaller angle to horizontal, when bubble reaches its maximum equivalent radius $R_{max}$. It is seen that aspect ratio is larger and smaller than 1.0 up to $\gamma = 1.5$ for PC boundary plates and water-air interface respectively. After $\gamma = 1.5$, bubble converges to a spherical shape. Since 1 mm and 2 mm plates indicate similar behavior, plate thickness may not be dominant for the first expansion of the bubble.

### *3.1 Effect of Boundary Motion*

Experimental results display the strain on the four points on the polycarbonate plates. Strain histories of the two plates are given in Figure 3. Figure 4 shows the strain signals with respect to specific times when

Journal of Physics: Conference Series **656** (2015) 012010    doi:10.1088/1742-6596/656/1/012010

each gage reach its first peak value. According to Figure 3 and Figure 4, thinner 1 mm plate deforms slower than 2 mm plate and larger strains were observed with 1 mm plate.

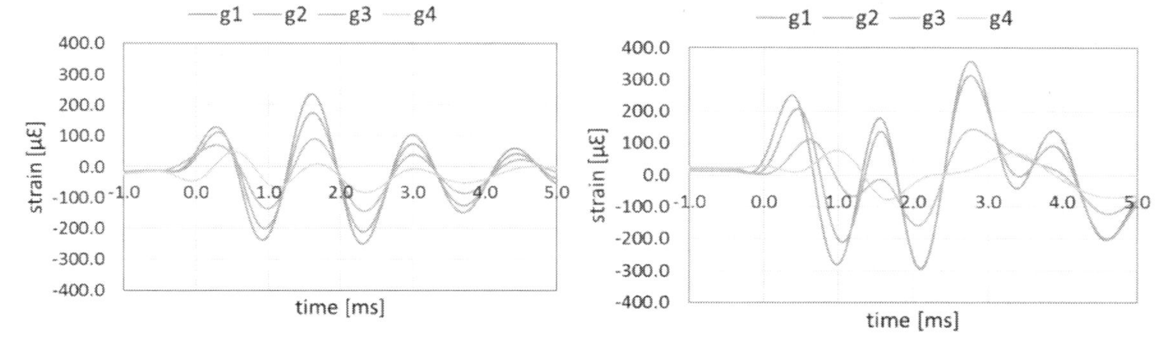


**Figure 3**. Strain signals from two plates (left) 2 mm (right) 1mm, Li=6.0 mm, $\gamma$ =1.5

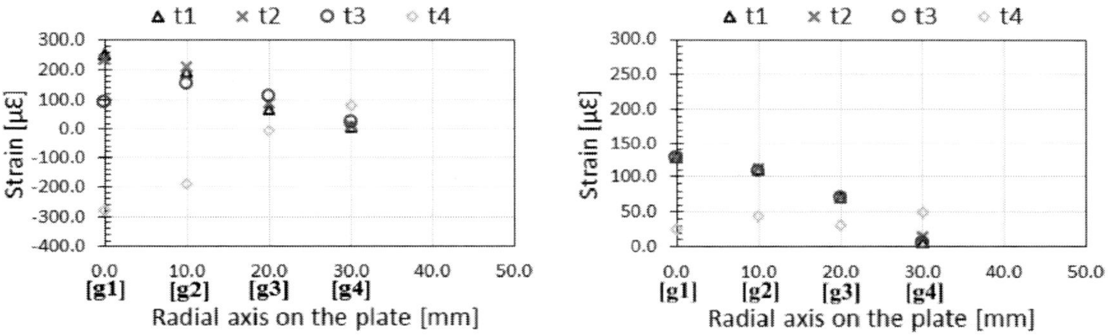


**Figure 4.** Strain signals from the gages along the radial axis for two plates (left) 2mm (right) 1 mm. t1,t2,t3 and t4 indicates the time gages g1, g2, g3 and g4 reaches maximum value,respectively.

*3.2 Overview of Bubble-Boundary Interaction*

Figure 5 illustrates the consecutive evaluation of bubble at a frame rate of 23,000 fps near two circular polycarbonate plates. It is observed that directions of the jets created during the collapse phase directed to opposite directions.

Jetting behavior of the bubbles are observed using different stand-off parameters for polycarbonate plates and no boundary case, i.e. water-air. During the experiments, it is seen that direction of the jet is always away from the boundary for the water-air interface. For PC plate boundaries, jet changes its direction towards to the boundary as the stand-off parameter decreases. The value of stand-off parameter to change jet direction is around $\gamma$=1.80 and $\gamma$=1.20 for 2 mm and 1 mm circular plate, respectively. Considering this behavior together with the different motion of the plates during the bubble generation, it can be said that fluid-structure interaction and plate flexibility plays an important role on collapse motion of the bubble.

Jetting behavior of the bubbles are observed using different stand-off parameters for polycarbonate plates and no boundary case, i.e. water-air. During the experiments, it is seen that direction of the jet is always away from the boundary for the water-air interface. For PC plate boundaries, jet changes its direction towards to the boundary as the stand-off parameter decreases. . The value of stand-off parameter to change jet direction is around $\gamma$=1.80 and $\gamma$=1.20 for 2 mm and 1 mm thick circular plate, respectively. Considering this behavior together with the different motion of the plates during the bubble generation, it can be said that fluid-structure interaction and plate flexibility plays an important role on collapse motion of the bubble.

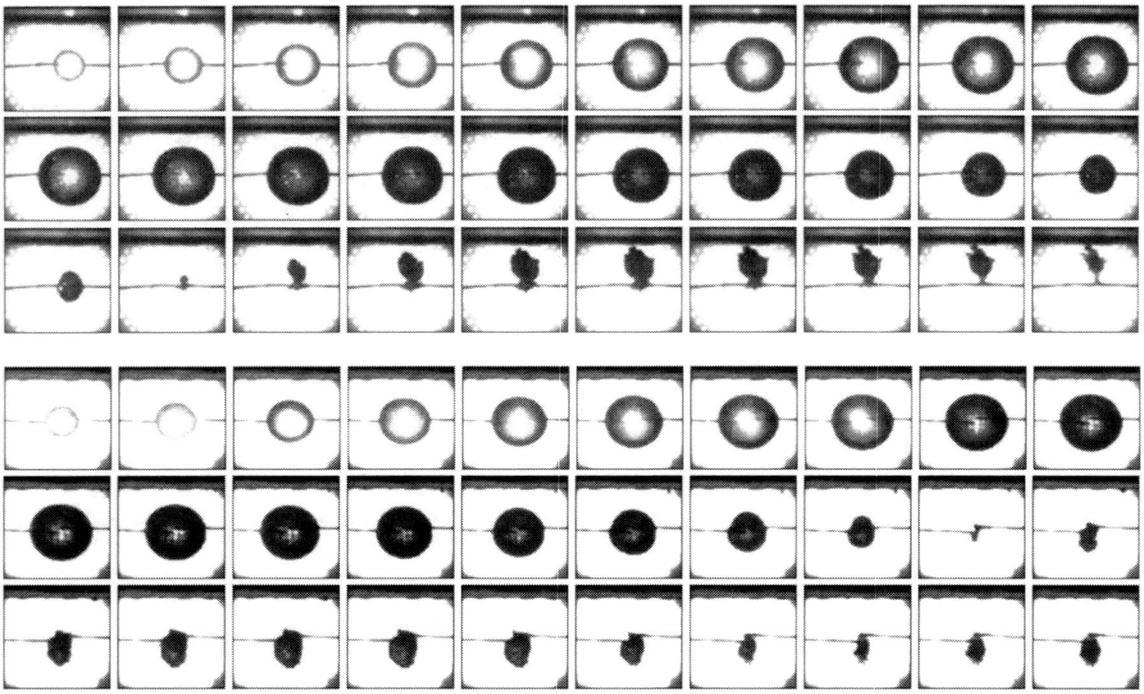

**Figure 5.** Bubble growth and collapse near polycarbonate plate of thickness (up) 2mm (down) 1 mm, $R_{max}$=4.0 mm, $\gamma$=1.5.

## 4. Summary

In this work, a single bubble created by spark-discharge near a polycarbonate plate. Bubble behavior for two different plate thickness and several initial stand-offs are observed. Plate motion is evaluated using strain gages. It is seen that for two different circular plates, plate motion is considerably different from each other. Direction of the jet is away from the boundary for both plates when the stand-off parameter is higher. Jet of the bubble changes its direction as the stand-off parameter gets smaller. Future work includes evaluation pressure reduction near the plate due to the different motions and its direct relation with the bubble behavior.

## References

[1] Rayleigh L VIII 1917 *The London, Edinburgh, and Dublin Philosophical Magazine and Journal of Science* **34**(200) 94-98
[2] Benjamin T B, Ellis A T 1966 *Philosophical Transactions for the Royal Society of London. Series A, Mathematical and Physical Sciences* **260** 221-240
[3] Tomita Y, Shima A **1986** *Journal of Fluid Mechanics* **169** 535-564
[4] Brujan E A, Keen G S, Vogel A, Blake J R **2002** *Physics of Fluids* **14(1)** 85-92
[5] Philipp A, Lauterborn W 1998 *Journal of Fluid Mechanics* **361** 75-116
[6] Brujan E A, Nahen K, Schmidt, Vogel A 2001 *Journal of Fluid Mechanics* **433** 251-281.
[7] Müller M, Hujer J, Kotek M, Zima P 2013 *EPJ Web of Conferences* **45** 01120.
[8] Ando, Keita, et al. *Journal of Fluid Mechanics* **671** (2011) 339-363.
[9] Rajendran R, Narasimhan K 2001 *International Journal of Impact Engineering* **25.5** 493-506.

# Numerical and experimental study of shock-driven cavity collapse

**MR Betney[1,4], PA Anderson[1], H Doyle[2], B Tully[2], NA Hawker[2], Y Ventikos[3]**

[1]Department of Engineering Science, University of Oxford, Parks Road, Oxford, OX1 3PJ, UK, [2]First Light Fusion Ltd., Begbroke Science Park, Begbroke, Oxfordshire, OX5 1PF, UK, [3]Department of Mechanical Engineering, University College London, London, WC1E 7JE, UK

E-mail: [4]matthew.betney@eng.ox.ac.uk

**Abstract.** A study is presented of the interaction of a shock wave with gas cavities cast in a hydrogel. Simulations are conducted using a front-tracking approach, whereby Lagrangian hypersurfaces are used to model the interface between materials. These 'fronts' are overlaid on an Eulerian grid which is used to model the bulk flow. Results are compared to an experimental investigation, in which a light gas gun is used to create ~600 MPa shock waves in hydrogel blocks, within which air cavities have been cast. Experimental results are presented, including temporally resolved measurements of light emission. Comparison between experimental and simulated results shows good agreement.

## 1. Introduction

The collapse of cavities or bubbles by shock waves in fluids is a common occurrence in nature and technology, with applications ranging from shock wave lithotripsy through to cavitation damage and the sensitisation of explosives. Numerous studies of this process[1, 2] have shown that the impact of a shock wave on a cavity leads to the development of a high-speed jet. This jet propagates across the cavity and strikes the far wall, compressing a section of gas to high temperatures and pressures, and leading to the formation of a strong water hammer shock in the liquid. The action of the jet causes the remains of the cavity to form a torus, in which further regions of high temperature and density are formed. Of particular interest, it has been shown experimentally that this process leads to two distinct bursts of light emission from the cavity[3]. These bursts of light have been correlated with the impact of the jet on the leeward cavity wall, and the collapse of the toroidal cavity remains.

## 2. Methods

### *2.1. Numerical Methods*

Simulations in this work are conducted using a front-tracking methodology[4]. This method uses Lagrangian hyper-surfaces or 'fronts' to model contact discontinuities between different materials. These fronts are overlaid on an underlying Eulerian mesh, which is used to model the bulk flow. The fronts break the simulation up into a number of separate regions and act as a set of boundary conditions, across which ghost cell extrapolation is performed. The propagation of waves between components is completely handled by the update to the position of the fronts. This method effectively eliminates numerical diffusion of the interface; this is a key challenge in modelling problems such as shock-driven cavity collapse. A stiffened polytropic equation of state is use to model the hydrogel, and the air is modelled as an ideal gas. Simulations are conducted using rotationally-symmetric coordinates. For a more comprehensive description of the methods used in this work, the reader is referred to[2, 5].

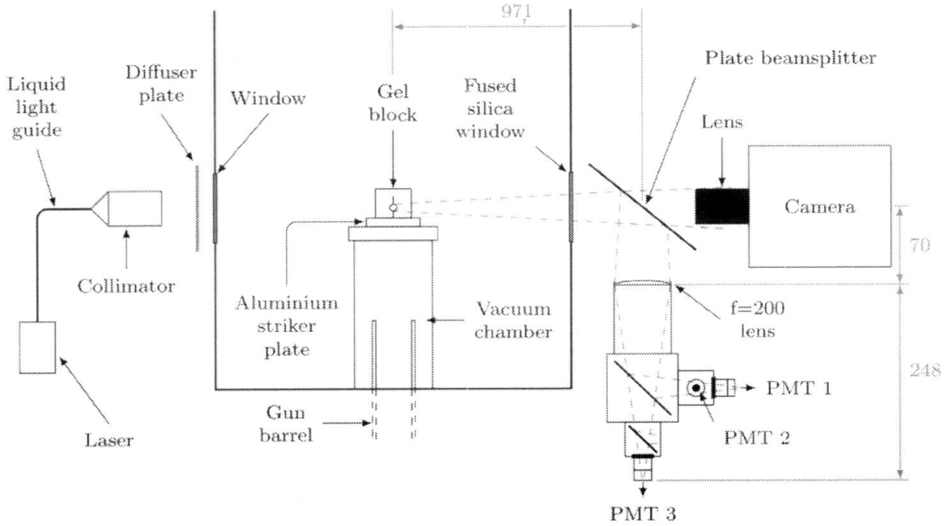


**Figure 1.** Experimental setup. A 70 mm bore light gas gun is used to create shocks waves in a hydrogel block. A SimX16 high speed camera is used to image the collapse process, and a calibrated array of PMTs are used to capture light emission.

### *2.2. Experimental Methods*

The experimental methods used in this work build upon previous research on the subject, in which a single-stage gas gun or explosives were used to create shock waves in gelatin blocks into which cylindrical cavities were cast[1]. Figure 1 shows the experimental setup used in this study. A 70 mm bore single-stage light-gas gun is used to accelerate projectiles into an aluminium striker plate, behind which a hydrogel block is placed. Shock waves (pressure ~600 MPa) are created in the gel block, which go on to collapse spherical cavities cast within the block. The projectile used consists of a 50 mm diameter, 15 mm deep mild steel disk, which is placed within a nylon sabot. The gas gun is capable of accelerating this projectile, which has a mass of approximately 450 g, to velocities of up to 700 $ms^{-1}$. The aluminium striker plate seals a vacuum chamber placed over the gun barrel, allowing the barrel to be evacuated prior to the shot to a pressure of approximately 0.8 Torr. The hydrogel used for the target is made by dissolving 0.5 % by mass gellan gum (Sigma Aldrich) in deionised water heated to 90 °C, with 0.05 % CaCl used as a setting agent. This mixture is allowed to cool before being poured into moulds to set. Half-spherical depressions are cast into the walls of two gel blocks, which are then joined together to form spherical cavities within the combined block. By this method, complex multi-cavity arrangements are easily implemented. Completed gel blocks are of depth 42 mm, width 50 mm and height 50 mm. Cast cavities are of diameter 5 mm.

Imaging of cavity collapse is conducted using a SimX16 high speed camera (Specialised Imaging Ltd., UK). This camera is capable of capturing 16 frames at a maximum frame rate of 200 Mfps. Images may be backlit using a 200 W pulsed diode laser (Cavilux SMART, Cavitar Ltd., Finland). Any light emission from cavity collapse is focussed through a series of beamsplitters on to 3 photo-multiplier tube modules (PMTs, Hamamatsu H11526-20-NF), each of which is filtered to detect a particular wavelength of light. The wavelength of these filters (405 nm, 460 nm and 760 nm) are chosen such as to avoid the 640 nm central wavelength of the laser. This setup is calibrated using a Bentham CL6 II calibrated light source.

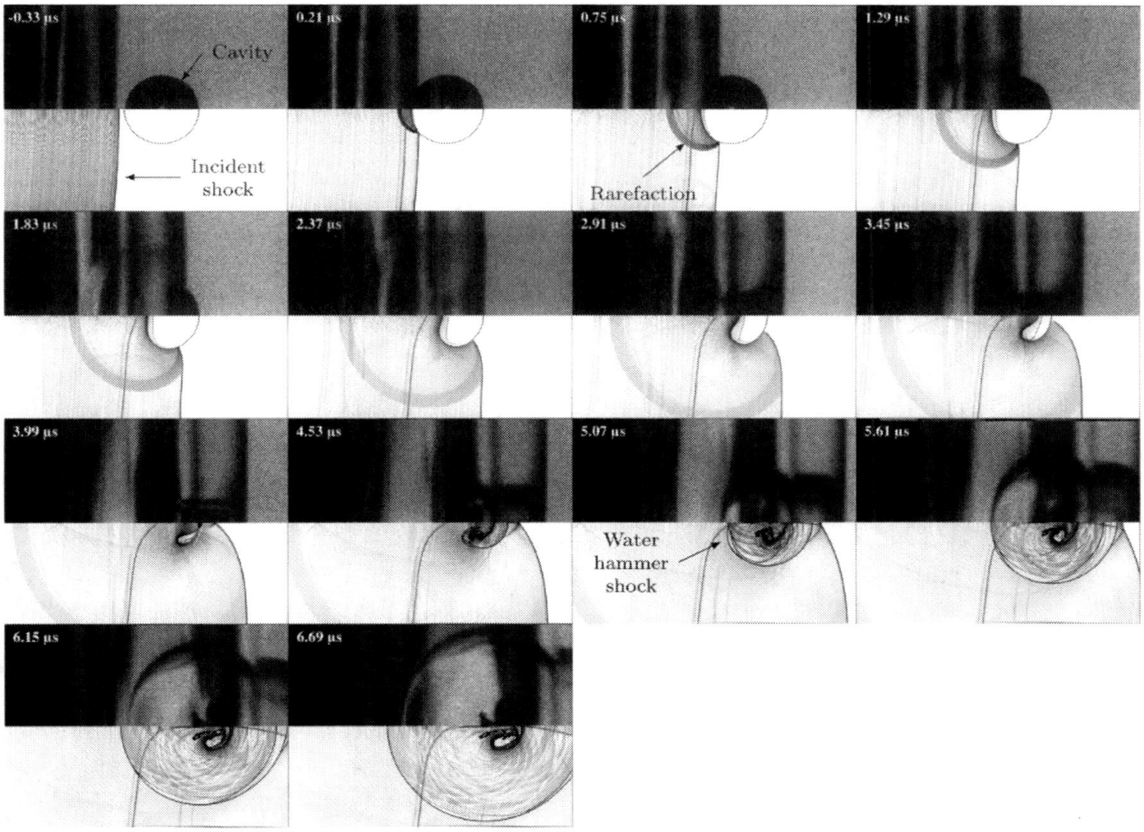


**Figure 2.** Image array showing the interaction of a shock wave with a single 5 mm diameter spherical cavity. Approximate shock pressure 600 MPa. Top half of images show experimental results, bottom half the results of a matching simulation.

## 3. Results

The impact of a 600 MPa incident shock wave on a single spherical cavity is shown in Figure 2. The top half of the images show the experimental results found using the setup of Figure 1, while the bottom half shows the results of a matching simulation. The simulated results are shown using a plot of the magnitude of the gradient of the density, giving a numerical representation of the experimental result, in which large gradients in density appear as dark lines. The images show a central section of the gel block, with the gas cavity positioned 10 mm from the aluminium striker plate. The shock wave propagates across the images from left to right, and the times shown are relative to the time of shock impact on the cavity. Although the high-speed jet is not visible in the experimental images, its formation may be inferred from other phenomena, such as the water hammer shock formed by the impact of the jet on the leeward cavity wall. Also clearly seen in the experimental images is the expansion of the rarefaction wave, formed in the liquid by the impact of the shock wave on the cavity. Comparing the simulated and experimental results, there appears to be reasonable agreement in the formation and propagation of these features between the two result sets.

Light emission from the collapse of the cavity as recorded by the PMTs is shown in Figure 3. Times shown are as in Figure 2. As can be seen, the light emission is divided into three distinct bursts. Analysis of shock propagation in the experimental images suggests that the first peak

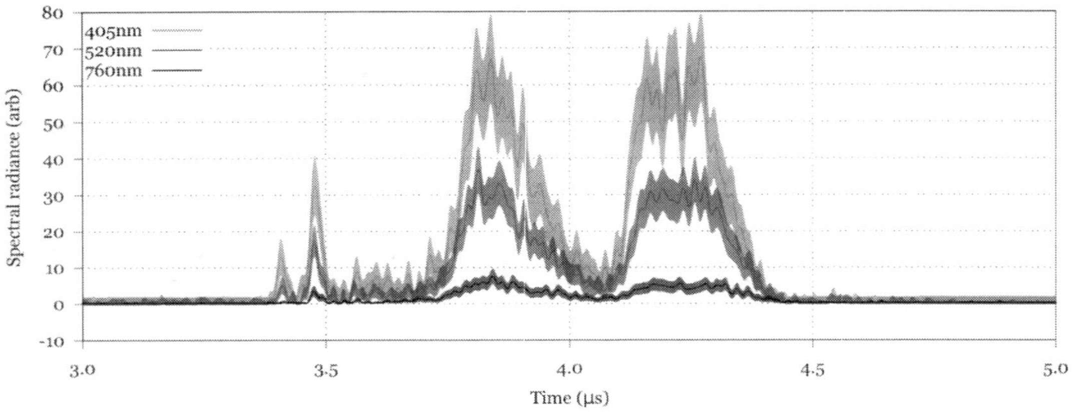


**Figure 3.** Light emission from cavity collapse as measured by the PMTs. Times as in Figure 2.

corresponds to the impact of the high-speed jet on leeward cavity wall, while the second and third peaks correspond to the collapse of the toroidal cavity remains. This gives a difference in time between the impact of the incident shock on the cavity and jet impact of 3.4 $\mu$s. This compares to an equivalent simulated time difference of 4.1 $\mu$s. Currently our best estimate for this 20% discrepancy relates to the difficulty in obtaining an accurate equation of state for the hydrogel, at the conditions relevant to this experiment. Comparing the relative amplitudes of the PMT traces, the light emission can be seen to be significantly more intense at shorter wavelengths. These relative amplitudes are consistent with the cavity emitting light as a black body, with peak temperatures of approximately 20,000 K. More detailed analysis will be required to improve the accuracy of this result.

## 4. Conclusion

This paper has presented initial results from an investigation into the interaction of an incident shock wave with air cavities cast in a hydrogel. Comparison has been made between experimental and simulated results, with reasonable agreement shown. By using an array of photomultiplier tubes, it has been shown that under the influence of the incident shock wave, cavities collapse and emit bursts of light. This light is concentrated into three distinct peaks, with the intensity of the light greater in the ultraviolet than the visible.

## References

[1] Dear J P and Field J E 1988 *Journal of Fluid Mechanics* **190** 409–425 URL `http://journals.cambridge.org/abstract_S0022112088001387`

[2] Hawker N A and Ventikos Y 2012 *Journal of Fluid Mechanics* **701** 59–97 ISSN 0022-1120

[3] Bourne N K and Field J E 1999 *Philosophical Transactions of the Royal Society A: Mathematical, Physical and Engineering Sciences* **357** 295–311 ISSN 1364-503X URL `http://rsta.royalsocietypublishing.org/cgi/doi/10.1098/rsta.1999.0328`

[4] Glimm J, Grove J W and Zhang Y 2002 *SIAM Journal on Scientific Computing* **24** 208–236 URL `http://www.ams.sunysb.edu/ yzhang/papers/siam2001/36669.pdf`

[5] Betney M R, Hawker N A, Tully B and Ventikos Y 2015 *Physics of Fluids* **27** 036101 ISSN 1070-6631

# Dynamics of a vapor nanobubble collapsing near a solid boundary

**Francesco Magaletti, Mirko Gallo, Luca Marino, Carlo Massimo Casciola**

Dipartimento di Ingegneria Meccanica e Aerospaziale, Università di Roma "La Sapienza", via Eudossiana 18, 00184 Roma Italy

E-mail: `carlomassimo.casciola@uniroma1.it`

**Abstract.** In the present paper a diffuse interface approach [1] is used to address the collapse of a sub-micron vapor bubble near solid boundaries. This formulation enables an unprecedented description of interfacial flows that naturally takes into account topology modification and phase changes (both vapor/liquid and vapor/supercritical fluid transformations). Results from numerical simulations are exploited to discuss the complex sequence of events associated with the bubble collapse near a wall, encompassing shock-wave emissions in the liquid and reflections from the wall, their successive interaction with the expanding bubble, the ensuing asymmetry of the bubble and the eventual jetting phase.

## 1. Introduction

The collapse of vapor bubbles near solid boundaries started being investigated in the last century, motivated by the destructive effects of cavitation on ship propellers and hydraulic turbines and pumps and is still today receiving an increasingly significant attention [2, 3, 4]. Only recently the increasing impact of the micro and nano technologies shifted the attention from millimeter-sized to micrometer or sub-micrometer bubbles. In a *lab on a chip*, cavitation can be employed for microfluidic pumping [5], to enhance mixing and for surface cleaning purposes [6]. Cavitation is also exploited in medicine, e.g. in high intensity focused ultrasound (HIFU) and extracorporeal shock wave lithotripsy (ESWL) [7] to enhance drug delivery or localized heat deposition deep within the body, and to induce cell membrane poration [8] or to comminute kidney stones [9].

The cornerstone work of Rayleigh [10], refined in [11, 12, 13, 14, 15], is still today impressive for the correspondence with experimental results despite substantial simplifying assumptions. Successively, nearby boundaries were addressed by numerical simulations [16, 17, 18].

In the present paper, by extending the analysis of symmetric bubble collapse discussed in [1], we use the diffuse interface approach to address the collapse of a sub-micron vapor bubble near solid boundaries. This formulation enables a natural description of interfacial flows, changes of topology, and vapor/liquid and vapor/supercritical fluid phase changes which have been shown to be crucial for the correct description of the final stages of the bubble collapse.

## 2. Mathematical model

We exploit an unsteady diffuse interface description [19, 1] of the two-phase (liquid/vapor) flow in a domain $\mathcal{D}$ based on the van der Waals gradient approximation of the free energy functional

$F[\rho,\theta]$ [20]:

$$F[\rho,\theta] = \int_{\mathcal{D}} \hat{f} dV = \int_{\mathcal{D}} \left( \hat{f}_0(\rho,\theta) + \frac{\lambda}{2}|\nabla\rho|^2 \right) dV ,$$

where $\hat{f} = \hat{f}_0 + \lambda/2|\nabla\rho|^2$ with $\hat{f}_0(\rho,\theta)$ the classical Helmholtz free energy density per unit volume of the homogeneous fluid at temperature $\theta$ and mass density $\rho$. Together with $\hat{f}_0$, the coefficient $\lambda(\rho,\theta)$ embodies all the information on the interfacial properties of the liquid-vapor system, namely surface tension $\sigma$ and interface thickness $\epsilon$. For a van der Waals fluid, the free energy reads

$$\hat{f}_0(\rho,\theta) = \bar{R}\rho\theta \left[ -1 + \log\left( \frac{\rho K \theta^{1/\delta}}{1-b\rho} \right) \right] - a\rho^2 ,$$

with $\delta = \bar{R}/c_v$, $\bar{R}$ the gas constant, $c_v$ the constant volume specific heat, $a$ and $b$ the van der Waals coefficients and $K$ a constant related to the de Broglie length [21].

The dynamics of the inhomogeneous system is described by the equations for mass $\rho$, momentum $\rho\mathbf{u}$, and total energy density $E = 1/2\rho|\mathbf{u}|^2 + \mathcal{U}$, with $\mathcal{U} = \bar{R}\theta/\delta - a\rho + \lambda|\nabla\rho|^2/(2\rho)$ the internal energy density,

$$\frac{\partial\rho}{\partial t} + \nabla\cdot(\rho\mathbf{u}) = 0 ,$$

$$\frac{\partial\rho\mathbf{u}}{\partial t} + \nabla\cdot(\rho\mathbf{u}\otimes\mathbf{u}) = \nabla\cdot\boldsymbol{\tau} ,$$

$$\frac{\partial E}{\partial t} + \nabla\cdot(\mathbf{u}E) = \nabla\cdot[\boldsymbol{\tau}\cdot\mathbf{u} - \mathbf{q_e}] .$$

This system of conservation equations is closed with thermodynamically consistent constitutive relations for the stress tensor $\boldsymbol{\tau}$ and the energy flux $\mathbf{q_e}$. Under the assumptions of local equilibrium and linear dependence of thermodynamic fluxes on thermodynamic forces – $\nabla\theta$ and $\nabla\mathbf{u}$ – the general approach outlined in [22] leads to identify the stress tensor as

$$\boldsymbol{\tau} = \left( -p_0 + \frac{\lambda}{2}|\nabla\rho|^2 + \rho\nabla\cdot(\lambda\nabla\rho) \right)\boldsymbol{I} - \lambda\nabla\rho\otimes\nabla\rho + \mu\left[ (\nabla\mathbf{u} + \nabla\mathbf{u}^T) - \frac{2}{3}\nabla\cdot u\boldsymbol{I} \right] ,$$

where the usual viscous terms with $\mu > 0$ in the last term are the source of irreversibility (the second viscosity coefficient is assumed to be $-2\mu/3$). From the van der Waals free energy density $\hat{f}_0$, Eq. (1), the pressure is $p_0 = \bar{R}\rho\theta/(1-b\rho) - a\rho^2$. Concerning the energy flux, one finds

$$\mathbf{q_e} = \lambda\rho\nabla\rho\nabla\cdot\mathbf{u} - k\nabla\theta ,$$

where $k > 0$ is the thermal conductivity. We like to stress that the adoption of a consistent thermodynamic model, like the van der Waals fluid, embodies the model with all the features related to phase change, i.e. latent heat release/adsorption during condensation/vaporization and transition to supercritical conditions at increasing temperature and pressure.

## 3. Results and discussion

The simulations are performed using an axisymmetric code that is the extension of the one developed to deal with the spherically symmetric case, see [1] for details on the numerical technique. The system, consisting of a pure van der Waals fluid, is initialized with a vapor bubble of radius $R_{eq}$ centered at $z_0$, the distance between the planar wall and the bubble center. The vapor bubble is in equilibrium with the confining liquid at temperature $\theta/\theta_c = 0.6$, with $\theta_c$ the critical temperature of the fluid. A shock wave with intensity $I = (p_2 - p_1)/p_1$, with $p_2$ and

$p_1$ the pressure in the perturbed and the unperturbed state, respectively, is initialized in the liquid phase to propagate toward the wall and hit the bubble, thus triggering its collapse. The fluid domain with dimension $4R_{eq} \times 4R_{eq}$ is discretized with a uniform grid $2048 \times 2048$. The dimensionless parameters of the simulations are $Re = R_{eq}\sqrt{p_c\rho_c}/\mu = 50$, $Pr = 3\mu\bar{R}/(8k) = 0.2$, and $\mathcal{C} = \lambda\rho_c^2/(p_c R_{eq}^2) = 1.6 \times 10^{-4}$, which correspond, e.g., to a bubble radius order 100 nm with typical viscosity, thermal conductivity, surface tension and critical values of water.

In our simulations we invariably observe bubble rebounds induced by the vapor temperature raise inside the cavity which, during collapse, exceeds the critical temperature to form an incondensable phase. This feature was most unexpected when first observed during numerical experiments of spherical collapse of a pure vapor nanobubble [1]. Two examples of the evolution are represented in Fig. 1 for two different initial wall distances, when the collapse is triggered by a shock wave with intensity $I = 75$. The relatively weak impinging shockwave is not sufficient to immediately break the spherical symmetry and to produce the classical liquid jet that porates the bubble, as observed in millimeter-bubble experiments [23, 24, 6, 4]. At sub-micron scale the surface tension is, in fact, predominant and preserves the nearly spherical shape during the first part of the evolution. Symmetry breaking eventually occurs when the bubble shrinks to its minimum volume and a non-spherical shockwave is emitted. This effect is more pronounced for bubbles closer to the wall. In these conditions the shockwave produced at collapse consists

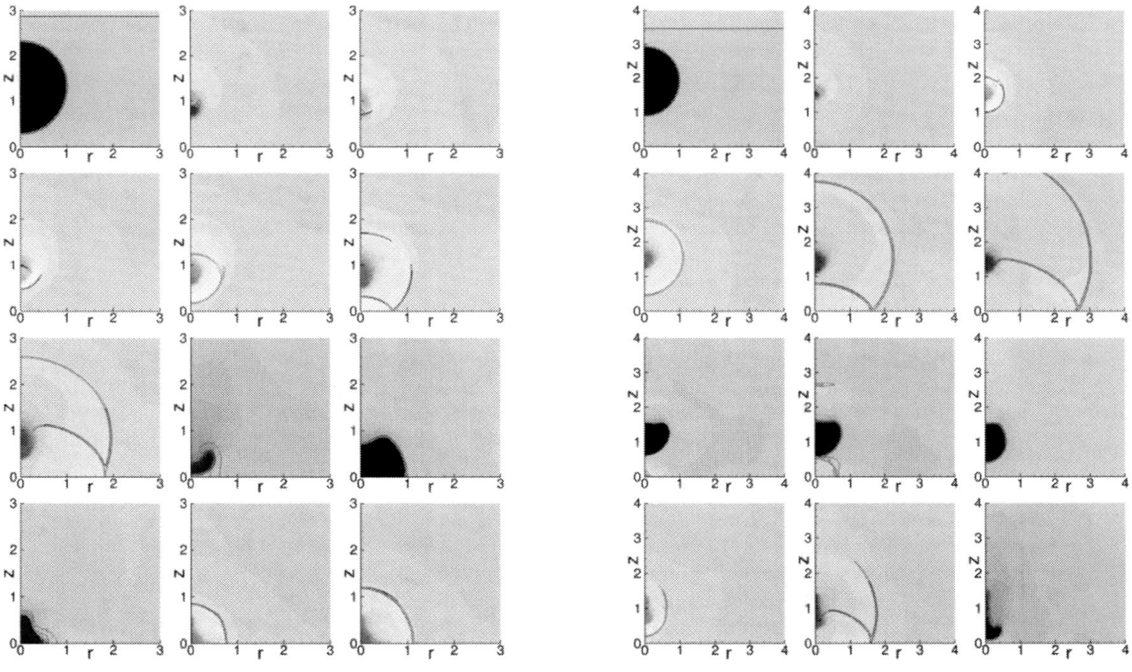

**Figure 1.** Snapshots during the evolution of a collapsing bubble with $z_0 = 1.3$ (left) and $z_0 = 1.9$ (right). The sequence runs from left to right and from top to bottom and is not uniformly spaced in time (left: $t = 0, 1.887, 1.907, 1.923, 1.953, 2.004, 2.131, 3.502, 5.403, 6.303, 6.480, 6.524$; right: $t = 0, 1.911, 1.956, 2.023, 2.202, 2.337, 3.833, 3.902, 5.937, 6.353, 6.524, 7.997$). The grey tones from darker to lighter represent the density field from smaller (vapor phase) to higher (liquid phase). The black lines are Schlieren-like iso-lines obtained as $S = \exp(-|\nabla p_0|/|\nabla p_0|_{max})$. The drawn iso-levels are $S = 0.9$ and $S = 1$ in order to highlight the regions with the highest pressure gradients, i.e. the vapor-liquid interface and the shockwaves.

of two curved shock fronts that propagate toward and away from the wall, respectively. The former is eventually reflected by the solid wall and interacts back with the re-expanding bubble. During this stage the bubble flattens out and moves toward the wall. The expansion stage is strongly affected by the bubble–wall distance, with the closest one (left panel in Fig. 1) touching the wall and the farthest one (right panel in Fig. 1) remaining detached. During the bubble expansion, the liquid in the thin layer between vapor and wall is compressed and a new shockwave is observed, plots in the third row on the right of Fig. 1. This sequence of bubble–wall interactions completely breaks the spherical symmetry, thereby reducing the strength of the successive collapse. Eventually, the field becomes more and more complex, until dissipation prevails.

For stronger shocks the nanobubble becomes asymmetric much earlier, originating a strong jet focused towards the walls. A complete simulation campaign is presently under way to investigate the role of the different parameters entering the problem, focusing in particular on the pressure pulse induced at the wall during the collapse, a major concern in most applications of collapsing micro/nano bubbles. A further crucial parameter is the maximum achieved temperature which, according to our preliminary results, typically ranges between 5 and 15 times the critical temperature and is strongly depend on the specific initial conditions.

The present results preliminarily confirm that the thermodynamically consistent description of cavitation bubbles provided through the diffuse interface model we have briefly outlined above allows to follow in full detail the complex dynamics and thermodynamics of the non-spherical collapse, accounting for changes of topology in the bubble. More work is under way to include additional features to the model, like chemical effects related to molecular dissociation and ionization and the presence of dissolved gas in the otherwise pure liquid.

## References

[1] Magaletti F, Marino L and Casciola C 2015 *Physical Review Letters* **114** 064501
[2] Silberrad D 1912 *Engineering* 33–35
[3] Leighton T 2012 *The acoustic bubble* (Academic press)
[4] Obreschkow D, Tinguely M, Dorsaz N, Kobel P, de Bosset A and Farhat M 2011 *Physical Review Letters* **107**(20) 204501
[5] Dijkink R and Ohl C D 2008 *Lab on a Chip* **8** 1676–1681
[6] Ohl C D, Arora M, Dijkink R, Janve V and Lohse D 2006 *Applied Physics Letters* **89** 074102
[7] Coussios C C and Roy R A 2008 *Annu. Rev. Fluid Mech.* **40** 395–420
[8] Sankin G, Yuan F and Zhong P 2010 *Physical Review Letters* **105** 078101
[9] Zhu S, Cocks F H, Preminger G M and Zhong P 2002 *Ultrasound in Medicine & Biology* **28** 661–671
[10] Rayleigh L 1917 *The London, Edinburgh, and Dublin Philosophical Magazine and Journal of Science* **34** 94–98
[11] Keller J B and Kolodner I I 1956 *Journal of Applied Physics* **27** 1152–1161
[12] Hickling R and Plesset M S 1964 *Physics of Fluids (1958-1988)* **7** 7–14
[13] Plesset M S and Chapman R B 1971 *Journal of Fluid Mechanics* **47** 283–290
[14] Plesset M S and Prosperetti A 1977 *Annual Review of Fluid Mechanics* **9** 145–185
[15] Shima A and Sato Y 1981 *Journal de Mecanique* **20** 253–271
[16] Blake J R and Gibson D 1981 *Journal of Fluid Mechanics* **111** 123–140
[17] Tipton R E, Steinberg D J and Tomita Y 1992 *Jsme International Journal. Ser. 2, Fluids Engineering, Heat Transfer, Power, Combustion, Thermophysical Properties* **35** 67–75
[18] Ding Z and Gracewski S 1996 *Journal of Fluid Mechanics* **309** 183–209
[19] Anderson D, McFadden G and Wheeler A 1998 *Annual Review of Fluid Mechanics* **30** 139–165 ISSN 0066-4189
[20] Jamet D, Lebaigue O, Coutris N and Delhaye J 2001 *Journal of Computational Physics* **169** 624–651
[21] Zhao N, Mentrelli A, Ruggeri T and Sugiyama M 2011 *Physics of Fluids* **23** 086101
[22] De Groot S R and Mazur P 2013 *Non-equilibrium thermodynamics* (Courier Dover Publications)
[23] Benjamin T B and Ellis A T 1966 *Philosophical transactions of the Royal Society of London. Series A, Mathematical and physical sciences* **260** 221–240
[24] Lauterborn W and Bolle H 1975 *Journal of Fluid Mechanics* **72** 391–399

# Acting Force Mechanism of a Body Crossing the Water Layer of a Cavity

**Weiqi Chen, Baoshou Wang, Kai Yan and Shiming Sun**

National Key Lab of Science and Technology on Hydrodynamics, China Ship Scientific Research Centre, Wuxi 214082, P.R. China

E-mail: tiger_cwq @aliyun.com

**Abstract**. The acting force mechanism of the water layer of the cavity on a body when it crosses the water surface is studied. A simplified mechanical model is proposed to explain the mechanism of the impact of the water layer onto the body. The estimating formula for the maximum pressure in the water layer and on the body surface impacted by the water layer is derived. It is proved that the maximum pressure is always in proportion to the square of the moving velocity of the maximum pressure position.

## 1. Introduction

The collapse of an underwater cavity or bubble can cause damages to underwater structures such as propellers and ships. Many works have been done to understand the damage mechanism of the cavity collapse. Besant [1] was the first to consider the problem of bubble dynamics in history and Rayleigh [2] established the dynamic equation for a spherical bubble in an unbounded flow field, as

$$R\ddot{R}+\frac{3}{2}\dot{R}^2=\frac{p_c-p_\infty}{\rho} \tag{1}$$

Subsequently, Plesset [3] built the well-known Rayleigh-Plesset (RP) equation by means of adding the viscosity coefficient $\mu$ and the surface tension coefficient $\sigma$ to equation (1), as

$$R\ddot{R}+\frac{3}{2}\dot{R}^2=\frac{p_c-p_\infty}{\rho}-\frac{2\sigma R}{\rho}-\frac{4\mu\dot{R}R}{\rho} \tag{2}$$

RP equation is only applicable for describing the expansion or collapse of a single spherical bubble in an unbounded flow field. It is found that when the bubble moves near to the body surface, the collapse will lead to the bubble deformation and develop a jet flow. The impact force by the jet flow is considered as the main reason for the body damage. Plesset [4] simulated the collapsing process of a bubble near the solid wall surface with a special numerical method, and the results showed that the bubble will develop a jet directed towards the plane solid wall before the bubble radius decreases to its minimum. Nakajama and Shima [5] analyzed the behavior of a non-spherical bubble in a viscous incompressible liquid under axial conditions with finite element method, and made it clear that a jet formed on the bubble is decelerated by the effect of liquid viscosity.

The researches mentioned above have just focused on the impact of the bubble collapse onto the the body outside the bubble, while few considered the impact of the bubble collapse onto the body inside the bubble which is crossing the bubble surface. This kind of cavity can be seen in the process of water exit and entry. When a body enters water, the splash sheath is gathering on the water surface to form a closed cavity, which is called the surface closure. However, if the afterbody is long enough, the splash sheath may close on the body surface (figure 1a). Similarly, when the body exits from water with attached cavity, the part of cavity bulging water surface will also close on the body surface (figure 1b).

These two kinds of cavities will also collapse under the atmospheric pressure and produce very high pressure on the closure position. Based on this point, we considered these two cases as the acting force problem of a body crossing the cavity near the water surface. As the liquid layer of the cavity near the water surface is thin and the collapsing time is very short, it is impossible to observe the movement of the water layer of the collapsing cavity on the body by experiments. To understand the mechanism of the impact of the liquid layer of the cavity onto the body surface, this paper proposed a simplified theoretical model to explain the formation of the collapsing pressure of the cavity near the water surface and to estimate its value.

The characteristics of cavity collapse without body crossing the cavity are analyzed in section 2. Then, in section 3 the acting force mechanism of a body crossing the cavity near the water surface is studied. At last, the conclusion is given in section 4.

## 2. Collapse of a single bubble with finite thickness water layer

As the upper part of the cavity near the water surface is formed by the closure of the splash sheath, the thickness of the water layer is finite. For convenience of the mathematical treatment, the upper part of the cavity near the water surface can be simplified as a spherical cavity and the thickness of the water layer outside the cavity is finite, as seen in figure 2.

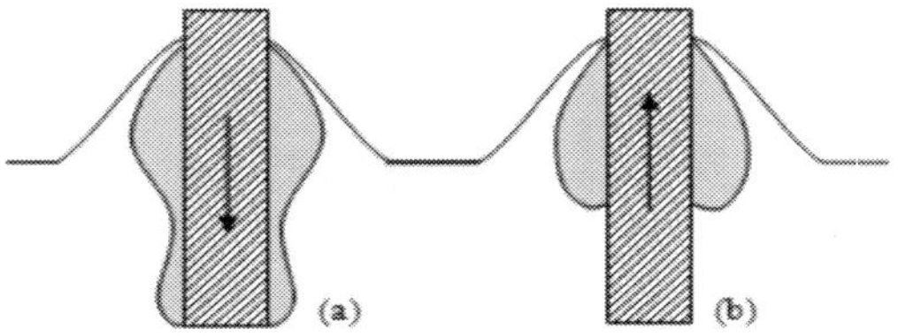


**Figure 1.** The cavity formed on the water surface in the process of water exit and entry.

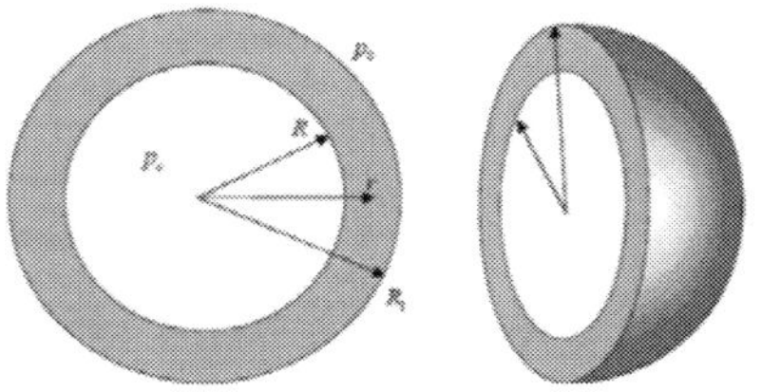

**Figure 2.** Schematic view of a spherical bubble closed by water sheath with finite thickness.

Let $R(t)$ and $R_1(t)$ be the radius of the inner and outer sphere respectively, $\rho$ be the fluid density, $p_c$ and $p_0$ be the pressure inside and outside the bubble respectively, $p(r,t),(R \le r \le R_1)$ be the pressure in the water layer. Obviously, $p(R,t) = p_c$ and $p(R_1,t) = p_0$. The fluid is assumed to be ideal, incompressible, and irrotational.

Due to a spherical symmetrical flow, the momentum equation can be written as [6]

$$\ddot{R}\frac{R^2}{r^2} + 2\dot{R}^2\left[\frac{R}{r^2} - \frac{R^4}{r^5}\right] = -\frac{1}{\rho}\frac{\partial p}{\partial r} \tag{3}$$

For a single bubble with finite thickness water layer, integrating (3) from $R$ to $R_1$ and taking the boundary conditions that $p(R,t) = p_c$ and $p(R_1,t) = p_0$ into account, we get

$$\left(R_1\ddot{R}_1 + \frac{3}{2}\dot{R}_1^2\right) - \left(R\ddot{R} + \frac{3}{2}\dot{R}^2\right) = \frac{p_0 - p_c}{\rho} \tag{4}$$

The pressure distribution in water layer can be described as

$$\frac{p(r) - p_c}{\rho} = \left(\frac{R}{r} - 1\right)\left(R\ddot{R} + 2\dot{R}^2\right) - \frac{1}{2}\dot{R}^2\left(\frac{R^4}{r^4} - 1\right) \tag{5}$$

Let $r_{max}$ represent the position with maximum pressure $p_{max}$ .If the cavity is an vapor bubble, namely $p_c = p_v = const$ , it can be proved easily by mathematical derivation that the following equality holds when the cavity radius tends to be zero ( $R \to 0$ ), as

$$r_{max}/R \to 4^{1/3}, \quad r_{max}/R_1 \to 0, \quad \dot{r}_{max}/\dot{R} = R^2/r_{max}^2 \to 1/2^{4/3} \tag{6}$$

According to (4) and (5), the maximum pressure can be expressed as

$$\frac{p_{max} - p_0}{\rho} = \frac{\dot{r}_{max}^2}{2}\left(3 - \frac{4r_{max}}{R_1} + \frac{r_{max}^4}{R_1^4}\right) \tag{7}$$

So, if $R$ is small enough, the maximum pressure can be estimated by substituting (6) into (7), as

$$\frac{p_{max} - p_0}{\rho} \approx \frac{3}{2}\dot{r}_{max}^2 \tag{8}$$

In reference [7], the formula to calculate the maximum pressure in the unsteady cavity closure region in the infinite flow is proposed, as

$$\frac{p_{max} - p_0}{\rho} = \frac{1}{2}\dot{r}_{max}^2 \tag{9}$$

Equation (8) and (9) have the same form but with different coefficients for the $\dot{r}_{max}$ velocity item.

## 3. A body crossing the cavity near the water surface

When a body crosses the cavity near the water surface, it will certainly influence the liquid layer of cavity. Therefore, it can't be described directly by the spherical cavity model. Figure 3a shows a simplified flow scheme of the liquid layer of cavity on a cylindrical body.

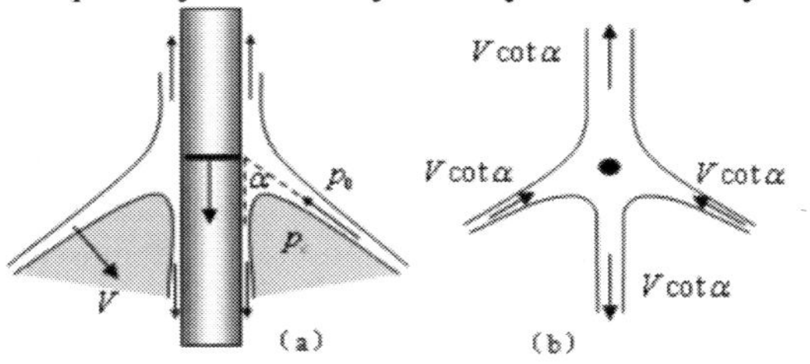


**Figure 3.** Scheme of the cavity flow on the body surface.

Obviously, the impact of the cavity onto the body is only related with the local flow of the liquid layer on the body surface, rather than with the shape of other cavity parts. The cavity shape near the body surface can be assumed to be a taper with half cone angle of $\alpha$ . The thick black line in figure 3 shows the position with the maximum pressure. On the condition that the cavity contacts under the pressure difference $\Delta p$ , the maximum pressure position of $r_{max}$ will move down along the axis of the cylindrical body with the velocity of $\dot{r}_{max}$ .For convenience, we proposed three hypotheses: (A) the moving velocity of the maximum pressure position is constant ( $\dot{r}_{max} = const$ ), (B) the thickness of the water layer is constant, (C) the cone angle is constant ( $\alpha = const$ ). So in the inertial coordinate system moved with $\dot{r}_{max}$ , we can get the steady flow state as shown in figure 3b. This flow state has some similarities with the symmetrical jet flow in the shaped charge theory [8].

The value of the maximum pressure can be estimated according to the shaped charge theory. Obviously, the maximum pressure position of $r_{max}$ is the stagnation point that the upper and lower jets intersect at. Let the cavity contracting velocity be $V$ , then we get the jet velocity as $V\cot\alpha$ and the moving velocity of the coordinate origin as $\dot{r}_{max} = V/\cos\alpha$ . Based on the Bernoulli equation, we get the maximum pressure expression, as

$$\frac{p_{max} - p_0}{\rho} = \frac{1}{2}(V\cot\alpha)^2 = \frac{\sin^2\alpha}{2}\cdot\dot{r}_{max}^2 \tag{10}$$

Compared to (8) and (9), it can be found that (10) has the similar form. But the coefficient in (9) is decided by the half cone angle of $\alpha$ , which means that the smaller the angle between the water surface cavity and the body surface, the smaller the maximum pressure.

The model (8) and (10) are very simplified and mainly used to explain the mechanical mechanism of the cavity collapse pressure near the water surface, so the model results can be only used to estimate the order of magnitude. In experiment, the maximum pressure $p_{max}$ and its position $r_{max}$ were measured by the pressure sensors. Figure 4 shows the measured $t \sim r_{max}$ curve within $T = 0.05$ s, which is approximated as a straight line. As the cavity collapses so fast that it is very difficult to obtain the speed $\dot{r}_{max}$ for every moment, only the average speed is got (the slope of line in figure 4). Similarly, the average cone angle $\alpha$ can be obtained by high-speed photography. The scope of maximum pressure can be estimated by taking the average speed and angle into model (8) and model (10).The comparisons between calculation results and experimental results are shown in figure 5.

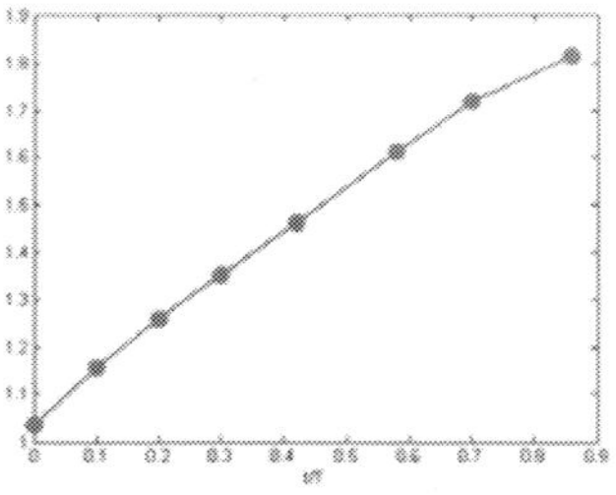

**Figure 4.** The position of the maximum pressure from experiments.

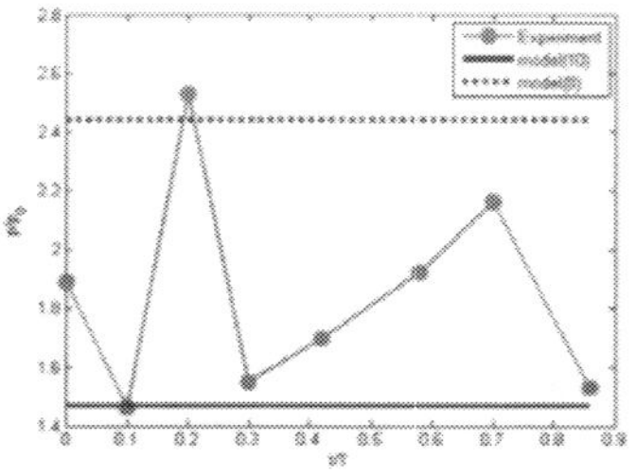

**Figure 5.** Comparison between model results and experimental results.

According to the experiment results, the hypothesis (A) and (B) are approximately acceptable while the hypothesis (C) is defective. It needs to be pointed out that as the pressure in the cavity is different form that outside the cavity, the flow velocity on the internal boundary and the external boundary are different too. So the jet velocity mentioned above should be understood as the mean velocity.

## 4. Conclusion

If the body crossing the cavity is relatively small compared to the cavity itself, which means that the interference of the body to the cavity can be ignored, the acting force on the body can be calculated by maximum pressure formula (8). However, if the body crossing the cavity changed the flow state in the water layer of the cavity, the acting force needs to be calculated by formula (10). In both cases, the maximum pressure is always in proportion to the square of the moving velocity of the maximum pressure position, and the difference of coefficients just reflects the different action way between the body and the cavity.

## References

[1] W. B. 1859 *Hydrostatics and Hydrodynamics* ( Cambridge University Press)

[2] Rayleigh L 1917 *The London, Edinburgh, and Dublin Philosophical Magazine and Journal of Science.*, **34**(200) 94-98

[3] Plesset M S 1949 *Journal of applied mechanics.* **16** 277-282

[4] Plesset M S, Chapman R B 1971 *Journal of Fluid Mechanics* **47**(02) 283-290

[5] Nakajima K, Shima A 1977 *Ingenieur-Archiv.* **46**(1) 21-34

[6] Franc J P, Michel J M 2004 *Fundamentals of Cavitation* (Kluwer Academic Publishers)

[7] Chen W.Q, Wang B.S, Yi S.Q, et al. 2012 *Chinese Journal of theoretical and applied mechanics* **44**(4) 701-708 (in Chinese).

[8] Batchelor G K 1994 *An introduction to fluid dynamics* ( London: Cambridge University Press)

# Cavitation due to an impacting sphere

**K L de Graaf, P A Brandner, B W Pearce and J Y Lee**

Australian Maritime College, University of Tasmania, Launceston, Tasmania, Australia 7250

E-mail: kdegraaf@amc.edu.au

**Abstract.** Cavitation associated with the impact of a sphere on a flat surface is investigated using high-speed photography. The sphere, of diameter 15 or 45 mm and made from Ertacetal® or stainless steel, was fully submerged and accelerated using a spring-loaded mechanism to achieve Reynolds numbers based on impact velocity and sphere radius of up to $7.2\times10^4$. The static pressure and impact velocity were varied to achieve cavitation numbers ranging from 8.9 to 120.9. High-speed photography of the impacting sphere and induced cavitation bubble was filmed at 105-140 kHz. A log law relationship was found between the non-dimensional maximum bubble radius and the cavitation number. The relationship was modulated by the material properties. Interaction between the sphere and the bubble was also noted.

## 1. Introduction

Cavitation induced by impact forces is of interest for basic research into inception and has been the subject of many studies. Kleine *et al.* [1] dropped glass and stainless steel spheres onto a submersed target and used Schlieren photography to identify pressure waves in the water which were assumed to be the cause of cavitation. Marston *et al.* [2] dropped stainless steel spheres onto a glass plate covered in a film of liquid. Estimates of the tensile stress for onset of cavitation compared well to the maximum tension criterion proposed by Joseph [3]. Seddon *et al.* [4] observed vapour cavities using high-speed imaging during approach of a steel ball onto a liquid covered glass plate. Uddin *et al.* [5] also used high-speed imaging, and particle tracking, to investigate squeeze flow around a sphere impacting a non-Newtonian film covered plate. High shear regions led to significant viscosity reduction which may have increased sphere penetration into the film and induced cavitation on sphere rebound. Mansoor *et al.* [6] investigated dropping a tungsten carbide sphere onto a liquid covered glass surface, but were not able to reproduce results of Seddon *et al.* [4] of cavitation forming before impact. In nature, cavitation occurs after impact of the raptorial appendage of a peacock mantis shrimp (*Odontodactylus scyllarus*), which can generate forces up to 1501 N at peak speeds of 14-23 m/s, aiding the shrimp's attack on its prey.

The majority of investigations published on cavitation due to impacting spheres involve an initial approach of the sphere through air before entering the liquid. The present study is undertaken in a fully immersed environment, with the sphere accelerated towards the plate with a velocity greater than free-fall.

## 2. Experimental Setup

The experiments were performed in the variable-pressure Bubble Dynamics Chamber (BDC) at the Cavitation Research Laboratory (CRL), University of Tasmania. The impact spheres were

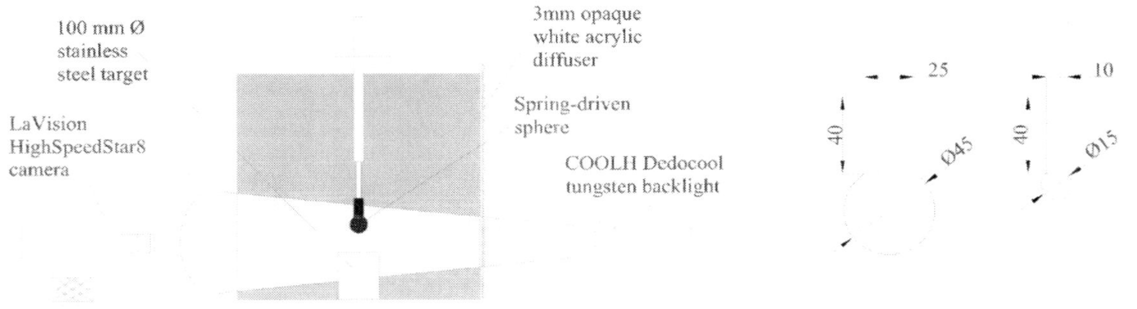


Figure 1: Schematic of the experimental setup showing location of sphere, target, high-speed camera, backlight and diffuser. The compression spring to accelerate sphere is housed within the sphere-tube assembly and not shown. The dimensions of the spheres are shown at right.

15 and 45 mm in diameter, constructed of Ertacetal® plastic and stainless steel. Each in turn were attached to a lightweight rod, fitted to a spring-loaded mechanism mounted in the top of the BDC. The spheres were accelerated into a 100 mm diameter stainless steel target located on the BDC floor at Reynolds numbers, $Re = UR_S/\nu$, ranging from 1.3 to $7.2\times10^4$ (or up to $U =$ 3.2 m/s for 45 mm spheres and 5 m/s for 15 mm spheres), where $U$ is the impact velocity, $R_S$ the sphere radius and $\nu$ the kinematic viscosity of water. Tests were conducted at cavitation numbers, $\sigma = (P - P_v)/(0.5\rho U^2)$, ranging from 8.9 to 120.9, where $P$ is the tank pressure, $P_v$ is the vapour pressure and $\rho$ is the water density. The release of the sphere triggered the acquisition of the high-speed photography taken at 105 kHz (45 mm spheres) and 140 kHz (15 mm spheres). A schematic of the setup and dimensions of the spheres are shown in figure 1 and further detail can be found in [7].

## 3. Results

Images from the high-speed photography for the 45 mm steel sphere are shown in figure 2. Upon impact, cavitation inception occurs due to the liquid stress/pressure singularity created [3]. The annular bubble grows rapidly in the confined space between the sphere and the target. As the bubble approaches its maximum, small annular protrusions form towards the top and bottom of the bubble. At the bubble maximum radius, surface instabilities are evident. These instabilities are also evident in the forward-lit photo shown in figure 3. As the bubble collapses, an annular re-entrant jet forms. The bubble rebounds a second time and becomes more glassy in nature, as evidenced by the backlighting that can be seen through the bubble. The bubble grows a third time, this time in an elongated vertical shape as the top and bottom of the bubble remains attached to the sphere and target respectively.

Figure 4 shows the displacement of the impacting sphere and growth of the cavitation bubble. Both the sphere displacement, $d_S$, and bubble radius, $R$, are non-dimensionalised based on the sphere radius, $R_S$. Time, $t$, is non-dimensionalised on the first bubble period, $T_B$, and sphere displacements have been zeroed at the point of impact. For these cases, the deformation of the spheres was calculated from comparison of high-speed movies with images taken of the sphere at rest. The plastic spheres exhibited a maximum deformation (in the vertical direction) of about $0.03R_S$ for a total period of $0.72T_B$ (15 mm sphere) and $0.01R_S$ for $0.30T_B$ (45 mm sphere). The steel spheres deformed about $0.01R_S$ for $0.61T_B$ (15 mm sphere) and $0.004R_S$ for $0.31T_B$ (45 mm sphere). The magnitude of deformation of the different sized spheres is similar for the same material type; however, the deformation time period is similar for the same sized spheres despite the material. It can also be seen that the rebound of the smaller spheres is affected by the bubble dynamics (for the Ertacetal® sphere this is more evident in other instances, not

Figure 2: Images of 45 mm stainless steel sphere taken at 105 kHz, every $20^{th}$ frame shown. $Re = 4.0\times10^4$ ($U$ = 1.8 m/s), $\sigma$ = 63.74.

Figure 3: Forward-lit image of the 45 mm Ertacetal® sphere impacting the stainless steel target (left) and an enlarged view of the cavitation (above). Surface instabilities are evident at the vapour-water interface. From comparison with high-speed shadowograph movies, the bubble is in its first growth phase.

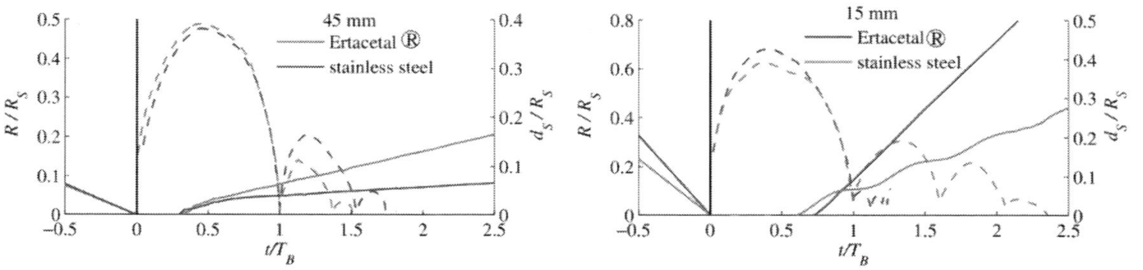


Figure 4: Displacement of sphere and growth of cavitation bubble. 45 mm stainless steel sphere: $Re = 5.1\times10^4$ ($U$ = 2.3 m/s), $\sigma$ = 37.9. 45 mm Ertacetal® sphere: $Re = 5.4\times10^4$ ($U$ = 2.4 m/s), $\sigma$ = 35.37. 15 mm stainless steel sphere: $Re = 2.6\times10^4$ ($U$ = 3.4 m/s), $\sigma$ = 20.3. 15 mm Ertacetal® sphere: $Re = 3.1\times10^4$ ($U$ = 4.4 m/s), $\sigma$ = 11.7.

plotted here). The 45 mm spheres are largely unaffected by the bubble dynamics, presumably due to the larger inertia.

Figure 5 shows $R_{max}/R_S$ against $\sigma$. A log law relationship is evident. A line of best fit for each data set gives a difference in gradient (on the semi-log plot) of approximately 3% and a vertical offset of 0.08. The difference does not change substantially when $\sigma$ is calculated using the rebound velocity, suggesting that the sphere deformation accounts for the difference in the

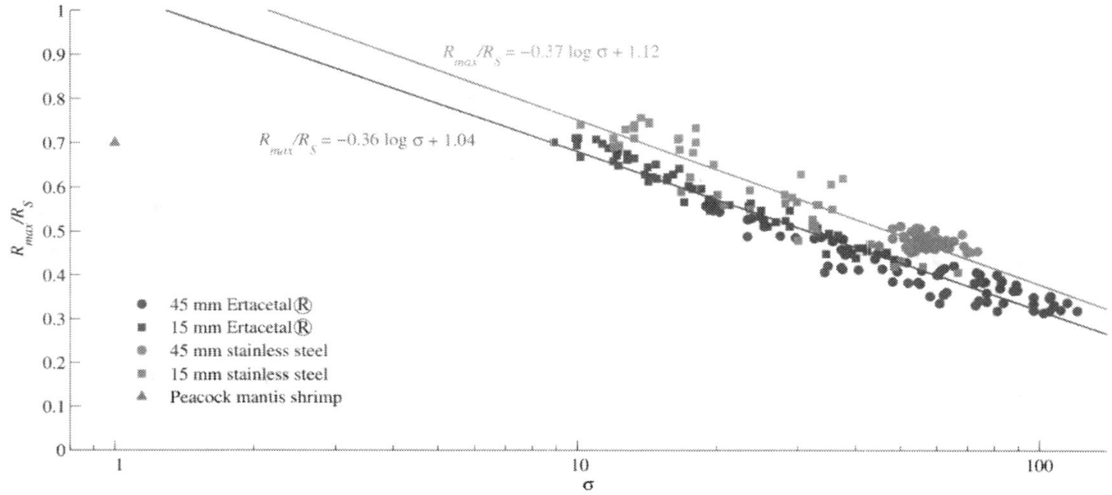


Figure 5: Non-dimensional maximum bubble radius, $R_{max}/R_S$ vs. cavitation number, $\sigma$. Peacock mantis shrimp data calculated from Fig. 1 of [8].

maximum bubble radius produced by the Ertacetal® and stainless steel spheres for the same $\sigma$. The bubble formed by a peacock mantis shrimp, calculated from Fig. 1 of [8] (assuming the shrimp is not at great depth) is smaller than those produced here. This may be attributable to the difference in material properties of the dactyl heel. Changes in bubble dynamics may also be affected by the differences in seawater due to chemical properties and contaminants affecting surface tension [9].

## 4. Conclusions

Cavitation produced after impact, in analogy with the mantis shrimp strike, has been investigated using two materials at various cavitation and Reynolds numbers in fully submerged conditions. The maximum dimensionless radius of the cavitation bubble produced scales with a log law relationship with the cavitation number. The offset in radius vs. cavitiation number relationship between the two sphere materials suggests there are other factors (such as material compressibility) that affect the cavitation bubble dynamics, with the more compliant material producing a smaller bubble. Shockwaves in both the liquid and solid media may have an influence and experiments incorporating the schlieren technique are planned to investigate this aspect. There is evidence for the bubble dynamics influencing the sphere rebound, which is more prevalent for the smaller spheres.

## References

[1] Kleine H, Tepper S, Takehara K, Etoh T G and Hiraki K 2009 Cavitation induced by low-speed underwater impact *26th International Symposium on Shock Waves* vol 1 (Göttingen, Germany)
[2] Marston J O, Yong W, Ng W K, Tan R B H and Thoroddsen S T 2011 *Exp. Fluids* **50** 729–746
[3] Joseph D D 1998 *J. Fluid Mech.* **366** 367–378
[4] Seddon J R T, Kok M P, Linnartz E C and Lohse D 2012 *Europhys. Lett.* **97** 24004
[5] Uddin J, Marston J O and Thoroddsen S T 2012 *Phys. Fluids* **24** 073104
[6] Mansoor M M, Uddin J, Marston J O, Vakarelski I U and Thoroddsen S T 2014 *Exp. Fluids* **55** 1648
[7] de Graaf K L, Brandner P A, Lee J Y and Pearce B W 2014 Cavitation about a sphere impacting a flat surface *19th AFMC* (Melbourne, Australia)
[8] Patek S N and Caldwell R L 2005 *J. Exp. Bio.* **208** 3655–3664
[9] Chanson H, Aoki S and Hoque A 2006 *J. Coast. Res.* **22** 664–677

# Bubble breakup in a turbulent shear layer

**P A Brandner[1] , A D Henderson[2], K L de Graaf[1] and B W Pearce[1]**

[1] Australian Maritime College, University of Tasmania Launceston Tasmania Australia 7250

[2] School of Engineering and ICT, University of Tasmania Hobart Tasmania Australia 7250

E-mail: kdegraaf@amc.edu.au

**Abstract.** The breakup of a millimetre size buoyantly rising bubble encountering a horizontal plane turbulent jet is experimentally investigated using high-speed shadowgraphy and acoustic techniques. The bubble diameter to jet height ratio is 0.75 and the jet height based Reynolds number is 4000. The high-speed imaging was recorded at 7 kHz simultaneous with hydrophone output at 100 kHz. Bubble breakup events were seen to produce simple binary divisions into products of similar size as well as three products where at least one was much smaller than the other products. Coalescence of products was also observed. In almost all cases time-frequency analysis of the acoustic emissions enabled the products to be identified and sized.

## 1. Introduction

Bubble breakup in turbulent and other high-shear flows is of interest in a range of physical processes. Typical flows include bubble columns in chemical processing [1] bubble formation and breakup at free surfaces relating to climate science (e.g. dissolution of atmospheric gases) [2], ship flows [3] and breakup of vortex cavitation bubbles [4, 5, 6]. There is an extensive literature of basic work in the field recent examples include single bubble breakup in a simple shear flow [7], breakup on the axis of a round turbulent jet [8, 9], breakup in a turbulent field [10, 11] and turbulent field in microgravity [12]. In the present work preliminary results are presented on a basic experiment of a rising bubble encountering a horizontal plane jet. This setup permits the study of the physics of a single bubble breakup whereby the bubble size and turbulent parameters including intensity, length-scale and dissipation can be easily controlled and varied (i.e: ranges of Reynolds and Weber numbers and Hinze scale [2]). Of particular interest in this work also is the potential use of acoustic emissions from breakup events to characterise the products. The jet has been investigated using particle imaging velocimetry and comparisons made with other plane jet results [13].

## 2. Experimental Overview

Figure 1 shows a schematic of the experimental setup, jet nozzle geometry, pre-flow conditioning and instrumentation. The plenum side walls extend beyond the nozzle exit $50h$ to laterally constrain the jet. A train of bubbles, with nominal diameter $d = 3$ mm ($d/h = 0.75$) is released from a 0.7 mm outside and 0.4 mm inside diameter hypodermic located mid-span. The entire arrangement is held within a tank of 1.2 m width, depth and height about mid-depth, mid-width and 200 mm from the wall through which optical access is made. Optical access is via a 70 mm thick 160 mm diameter glass window. The hydrophone is conditioned using a Bruel & Kjaer model 2626 charge amplifier. The jet supply is taken from the surrounding water, remote from

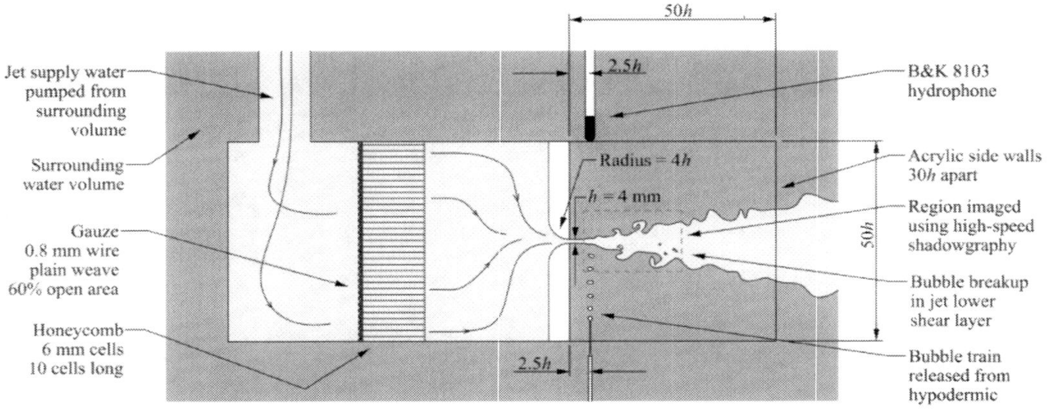


Figure 1: Schematic side view of plane jet immersed in a surrounding water volume showing the extent of confining side walls, flow conditioning and locations of the bubble train release, region imaged and hydrophone for measurement of acoustic emissions.

the jet location. The pump is controlled using closed loop feedback, based on jet Reynolds number, calculated real-time from the plenum pressure differential and nozzle calibration curve.

The high-speed shadowgraphy was carried out using a LaVision HighSpeedStar 5 camera with constant backlighting from a Dedolight DLH650 lamp diffused through a 3 mm white acrylic sheet. The camera was operated using DaVis 8.2 software at 7 kHz frame rate with 14 bit grey scale dynamic range and resolution of 1024 $\times$ 1024 pixels providing a spatial resolution of 21.5 pixel/mm. The hydrophone output was simultaneously recorded at 100 kHz (100 Hz high-pass 4th order Butterworth filter). For all tests reported here the jet was operated at a Reynolds number based on $h$ of 4000.

## 3. Results

Figures 2 and 3 show typical image sequences, at 2 ms intervals, from the high-speed shadowgraphy and corresponding time-series of the hydrophone outputs with image timings marked. Time-frequency analysis of the hydrophone outputs has been made using the continuous wavelet transform with a Morlet wavelet [14] [1]. In both cases the image sequences start with the rising spheroidal shaped bubble just encountering the lower shear layer of the jet. Figure 2 shows the typical bubble distortion and stretching due to the jet turbulence causing the bubble to break twice, each a binary division, between images e and f, and k and l, respectively. The first division corresponds to the first relatively large acoustic impulse whereas the second produces no discernable output. The wavelet analysis shows the first impulse to be composed of two frequencies of about 3025 and 5087 Hz which give corresponding bubble diameters of 2.2 and 1.3 mm respectively from Minneart's equation. Equivalent bubble sizes measured from the high-speed imaging give comparable diameters of about 2.3 and 1.5 mm. The impulse generated by this first division is attributable to the excitation and resonances of the products. In contrast the second division would appear to occur without excitation of the products. The impulse that occurs between frames v and w corresponds to a stretching event that did not produce breakup but created sufficient excitation to cause some resonance. Figure 3 shows a more complicated breakup process for the same conditions as above. The first breakup between frames d and e is an apparent binary division similar to the first described above although a third small product is evident in later frames. The use of an additional camera would provide more information on the origin of the third bubble which is beyond the resolution of the imaging and range of acoustic

[1] Software available: http://paos.colorado.edu/research/wavelets/

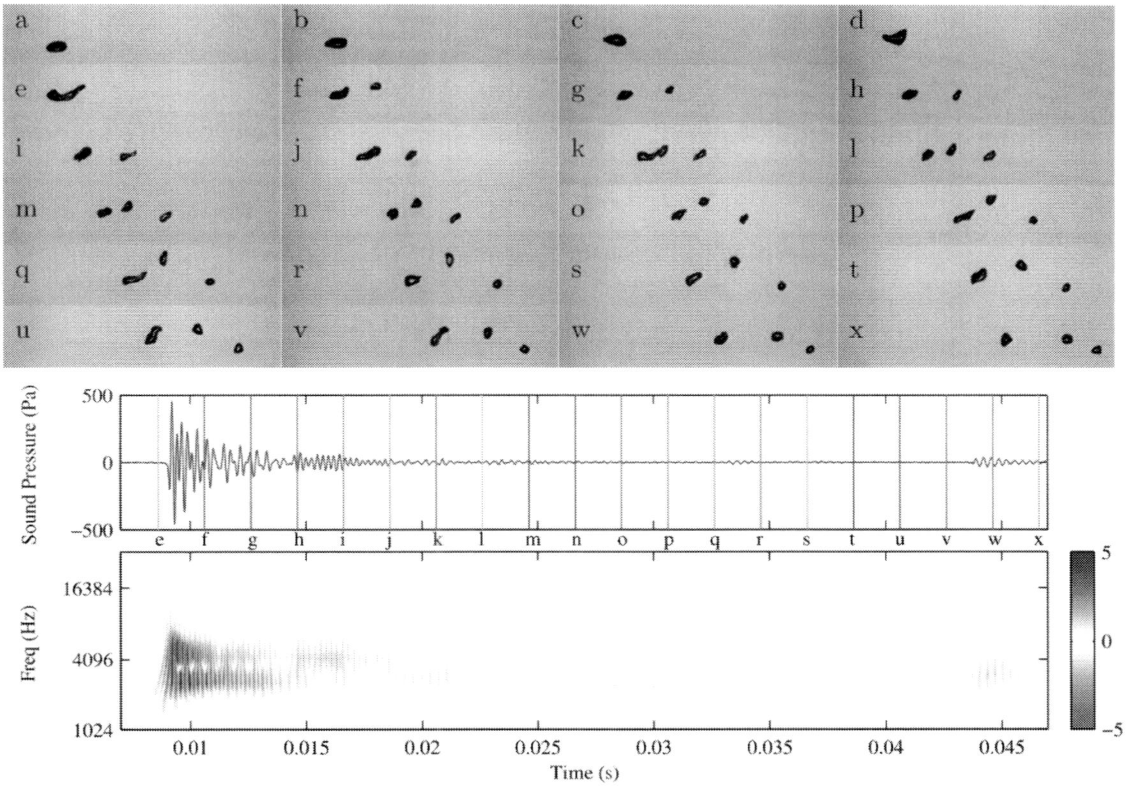


Figure 2: Typical bubble breakup sequence for initial bubble size of 3 mm and jet Reynolds number of 4000 in which the bubble undergoes an initial binary division followed by one of the products undergoing a second binary division.

measurement. A second event occurs between frames e and f that results in two additional small bubbles (each from a binary division) with corresponding distinct impulses. These measure about 0.8 and 0.4 mm from the imaging and give frequencies of 8556 and 17112 Hz from wavelet analysis, corresponding to diameters of 0.8 and 0.4 mm. The third much weaker event between frames k and l corresponds to coalescence between one of the smaller bubbles and one of the larger after which it breaks into two bubbles of similar size. The fourth event between frames n and o corresponds to non-binary breakup producing two new products to make a total of seven. In this case a ligament developed sufficient to form three bubbles. One of these measures 0.3 mm from the imaging and gives a frequency of 20349 Hz, which also gives 0.3 mm diameter. The last event is a breakup of one of the larger bubbles (total 8 products) again providing two distinct frequencies from the wavelet analysis.

## 4. Conclusions

The two examples presented of bubble breakup show that breakup may be into few products of similar size or a larger number of significantly different sizes. The breakups observed may be either binary or to produce three products which in all cases produced at least one much smaller product. Products were also seen to coalesce. Apart from one event all breakup events produced discernible acoustic emissions using a single hydrophone from which the product sizes could be derived. An additional camera and additional hydrophones may provide more information on the nature of the breakup events. A future more detailed characterisation of the jet flow and a statistical analysis of breakup events will give product diameter as a function of Reynolds and

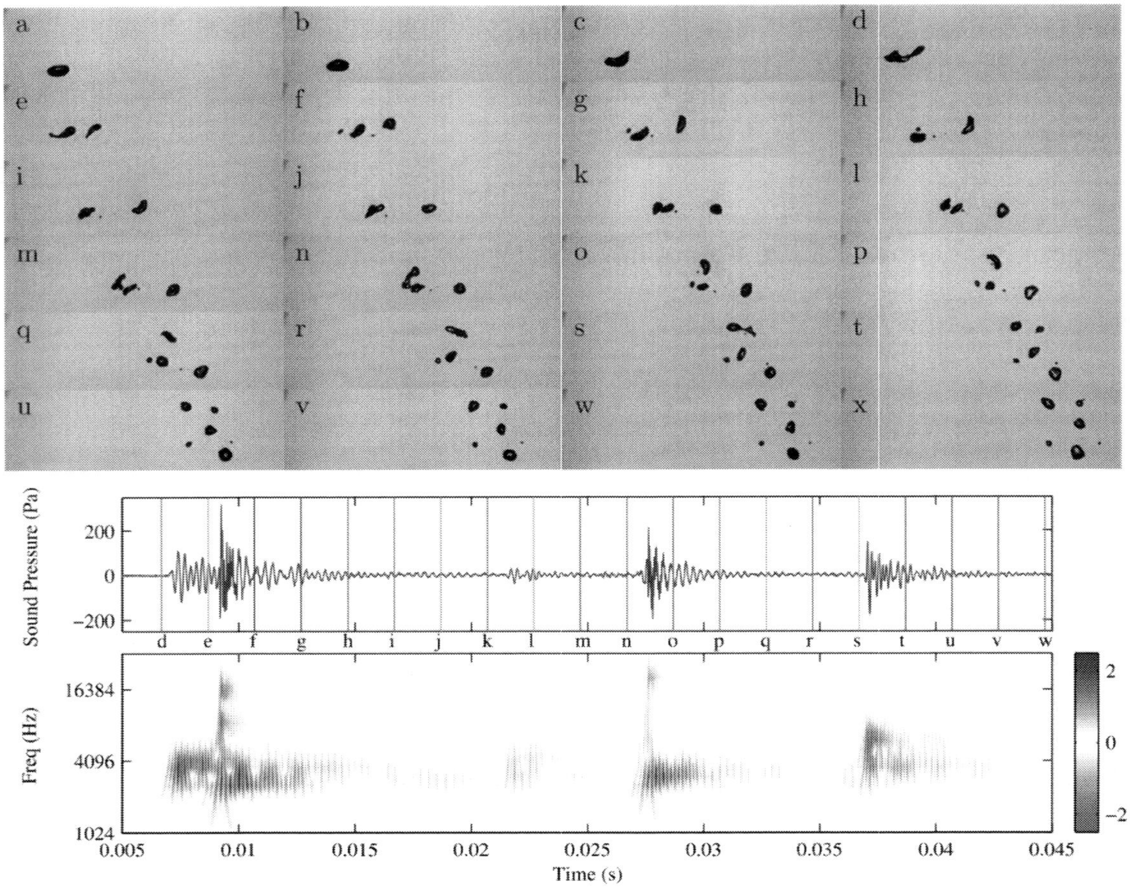


Figure 3: Alternate breakup sequence for the same conditions as for figure 2 but by a more complicated process. The bubble undergoes a succession of breakup (and one coalescence) events, not all of which are binary divisions, to produce 8 products of significantly varying size.

Weber numbers and a measurement of the Hinze scale.

## References

[1] Lehr F, Millies M and Mewes D 2002 *AIChE J.* **48** 2426–2443
[2] Deane G B and Stokes M D 2002 *Nat.* **418** 839–844
[3] Castro A M 2011 *Polydispersed bubble flow model for ship hydrodynamics with application to Athena R/V* PhD thesis University of Iowa
[4] Oweis G F, Jaehyug C and Ceccio S L 2004 *J. Acoust. Soc. Am.* **115** 1049–1058
[5] Chang N A and Ceccio S L 2011 *J. Acoust. Soc. Am.* **130** 3209–3219
[6] Choi J and Ceccio S L 2007 *J. Fluid Mech.* **575** 1–26
[7] Muller-Fischerü N, Tobler P, Dressler M, Fischer P and Windhab E J 2008 *Exp. Fluids* **45** 917–926
[8] Martínez-Bazán C, Montañés J L and Lasheras J C 1999a *J Fluid Mech.* **401** 157–182
[9] Martínez-Bazán C, Montañés J L and Lasheras J C 1999b *J. Fluid Mech* **401** 183–207
[10] Ravelet F, Colin C and Risso F 2011 *Phys. Fluids* **23** 103301 (12 pages)
[11] Qian D, McLaughlin J B, Sankaranarayanan K, Sundaresan S and Kontomaris K 2006 *Chemical Engineering Communications* **193** 1038–1063
[12] Risso F and Fabre J 1998 *J. Fluid Mech.* **372** 323–355
[13] Zarruk G A, Henderson A D, de Graaf K L, Brandner P A and Pearce B W 2014 Validation of a microbubble seeding generator for particle image velocimetry applications *19th AFMC* (Melbourne, Australia)
[14] Torrence C and Compo G P 1998 *Bulletin of the American Meteorological Society* **79** 61–78

# First Iterative Solution of the Thermal Behaviour of Acoustic Cavitation Bubbles in the Uniform Pressure Approximation

**Can F Delale**[1,2] and **Şenay Pasinlioğlu**[3]
[2]Department of Mechanical Engineering, MEF University, Ayazağa Caddesi No 4, 34980 Maslak, Sarıyer, Istanbul, Turkey
[3]Department of Mathematics, Istanbul Technical University, Büyükdere Caddesi, 34465 Maslak, Sarıyer, Istanbul,Turkey
E-mail: canfuat.delale@mef.edu.tr

**Abstract.** The thermal behaviour of a spherical gas bubble in a liquid driven by an acoustic pressure is investigated in the uniform pressure approximation by employing an iterative method to solve the energy balance equations between the gas bubble and the surrounding liquid for the temperature distribution and the gas pressure inside the bubble. It is shown that the first iterative solution leads to the first order law of the gas pressure as a polytropic power law of the bubble wall temperature and of the bubble radius, with the polytropic index given as an explicit function of the isentropic exponent of the gas. The resulting first order law of the gas pressure reduces to the classical isothermal and adiabatic laws in the appropriate limits. The first order gas pressure law is then applied to an acoustically driven cavitation bubble by solving the Rayleigh-Plesset equation. Results obtained show that the bubble wall temperature pulsations during collapse and rebound can become a few orders of magnitude higher than the bulk liquid temperature.

## 1. Introduction

Thermal effects play an important role in the final stage of collapse of inertially controlled bubbles. The pressures and temperatures can reach very high values in applications such as cavitation [1-3], sonochemistry [4] and single-bubble sonoluminescence [5-6]. Various sophisticated models that address the effect of the thermal behaviour of the bubble on the gas pressure inside the bubble in the uniform pressure approximation have been constructed ([7-10] and references therein). Although full numerical computations of these models are available for single bubbles, the complexity of the numerical solutions demonstrates the need of simplified expressions for the gas pressure and temperature. The attempt of modifying the adiabatic approximation for the gas pressure in this direction either by artificially increasing the liquid viscosity or by employing a variable isentropic index was not found to be satisfactory [11]. The aim of this investigation is to derive a simple relation between the gas pressure, the bubble wall temperature, the bubble radius and the isentropic exponent of the gas for use in various applications. For this reason we consider the energy balance between a spherical gas bubble and the surrounding liquid. We investigate the thermal behaviour inside the bubble in the uniform pressure approximation by studying the well-known coupled equations for the gas pressure and temperature, which we attempt to solve iteratively. In particular, in the first iteration we show that the temporal evolution of the gas pressure can be decoupled from that of the gas temperature, resulting in the first order law for the gas pressure as a polytropic power law of the bubble wall temperature and of the bubble

[1] To whom any correspondence should be addressed.

radius, with the polytropic index given explicitly in terms of the isentropic exponent of the gas. The bubble wall temperature is obtained from the Plesset-Zwick solution [12] of the temperature distribution of the liquid side in the thin boundary layer approximation. Results obtained for acoustically driven cavitation bubbles by use of the Rayleigh-Plesset equation of spherical bubble dynamics show that bubble wall temperature pulsations can reach values a few orders of magnitude higher than the bulk liquid temperature during the final stage of bubble collapses.

## 2. Energy Balance for a Spherical Gas Bubble

In this section we discuss the energy balance between a spherical gas bubble and its surrounding liquid. We focus on the thermal behaviour inside the bubble in the uniform pressure approximation as well as inside the surrounding liquid.

### *2.1 Thermal Diffusion Through the Gas Bubble in the Uniform Pressure Approximation*

The partial differential equation for the temperature field inside a gas bubble in the uniform pressure approximation is given by Prosperetti [13], and in normalized form takes the form [14]

$$\frac{p}{T}\left\{\frac{\partial T}{\partial t}+\frac{D}{p\,R^2}\left[\lambda(T)\frac{\partial T}{\partial y}-y\left(\frac{\partial T}{\partial y}\right)_{y=1^-}\right]\frac{\partial T}{\partial y}\right\}=\frac{(\gamma-1)}{\gamma}\frac{dp}{dt}+\frac{D}{R^2\,y^2}\frac{\partial}{\partial y}\left[y^2\lambda(T)\frac{\partial T}{\partial y}\right] \tag{1}$$

In equation (1) $T$ is the temperature field inside the bubble, normalized with respect to the bulk liquid temperature $T_0'$ ; $p$ is the bubble gas pressure in the uniform pressure approximation, normalized with respect to a reference pressure $p_0'$; $R$ is the bubble radius, normalized with respect to the initial bubble radius $R_0'$ ; $\lambda(T)$ is the thermal conductivity of the gas, normalized with respect to its value $\lambda_R'$ at the bubble wall, $\gamma$ is the isentropic exponent of the gas; $y$ is the radial coordinate, measured from the bubble center and normalized with respect to the instantaneous value of the bubble radius; $t$ is the time, normalized with respect to a characteristic time $\Theta'$ and $D$ is the square of the ratio of the penetration length to the initial bubble radius and is given by

$$D=\frac{(\gamma-1)\,\lambda_R'\,T_0'\,\Theta'}{\gamma\,p_0'\,(R_0')^2} \tag{2}$$

The gas pressure in the uniform pressure approximation can then be found by solving the equation

$$\frac{dp}{dt}=\frac{3\,\gamma}{R}\left[\frac{D}{R}\left(\frac{\partial T}{\partial y}\right)_{y=1^-}-p\,\frac{dR}{dt}\right] \tag{3}$$

which requires the solution of equation (1) together with the initial and boundary conditions

$$T(t=0,y)=1\ ;\quad T(t,y=1^-)=T_R(t)\ ;\quad \left(\frac{\partial T}{\partial y}\right)_{y=0}=0 \tag{4}$$

for the evaluation of the bubble wall temperature gradient $c(t)=(\partial T/\partial y)_{y=1^-}$. For this reason we adopt an iterative method of solution by first carrying out an expansion for the temperature field of the gas inside the bubble near the bubble wall as

$$T(t,y)=T_R(t)+\left(\frac{\partial T}{\partial y}\right)_{y=1^-}(y-1^-)+\frac{1}{2}\left(\frac{\partial^2 T}{\partial y^2}\right)_{y=1^-}(y-1^-)^2+O[(y-1^-)^3] \tag{5}$$

together with a similar expansion for the thermal conductivity of the gas. Substitution of the expansions into equation (1) and taking the limit $y$=1$^-$ leads to the below exact expression for the bubble wall gas pressure gradient

$$c(t)=\left(\frac{\partial T}{\partial y}\right)_{y=1^-}=\frac{R^2}{2D}\left[\frac{p}{T_R}\frac{dT_R}{dt}-\frac{(\gamma-1)}{\gamma}\frac{dp}{dt}\right]-\frac{1}{2}\left(\frac{\partial^2 T}{\partial y^2}\right)_{y=1^-} \tag{6}$$

where the temperature variation of the thermal conductivity $\lambda(T)$ at the bubble wall is neglected (accounting for the derivative of the thermal conductivity with respect to temperature leads to a quadratic equation for c(t) ). Equations (1) and (3) together with equation (6) can now be solved iteratively for the gas temperature and gas pressure, utilizing the initial and boundary conditions for the gas temperature and an initial condition for the gas pressure provided that the bubble wall temperature $T_R(t)$ is known (the bubble wall temperature will be obtained from the solution of the thermal diffusion equation of the liquid side taking into account the interface conditions).

*2.1.1. First Iterative Solution: The First Order Law of the Gas Pressure during Growth and Collapse*
In the first iteration we neglect the effect of the second radial derivative of the gas temperature at the bubble wall in equation (6), i.e. we assume that $(\partial^2 T/\partial y^2)_{y=1^-} = 0$. This results in decoupling of equation (3) for the gas pressure from the solution of equation (1) for the temperature field inside the bubble. Substituting the first iterative approximation of $c(t)$ into equation (3), upon exact integration, leads to the first approximation for the law of the gas pressure during the growth and collapse of the bubble in the form

$$p = p_i \left(\frac{\sqrt{T_R}}{R}\right)^{3\kappa} \quad ; \quad \kappa = \frac{2\gamma}{(3\gamma - 1)} \tag{7}$$

where $p_i$ is the initial gas pressure to be obtained from the initial equilibrium of the bubble and κ is a polytropic index. It can further be demonstrated that the gas pressure law given by equation (7) reduces to the isothermal law whenever $\gamma=1$ and $T_R=1$ and to the adiabatic law when the bubble wall temperature satisfies the isentropic relation $T_R = (p/p_i)^{(\gamma-1)/\gamma}$. For the temperature distribution inside the liquid, we use the Plesset-Zwick solution [12] in the thin boundary layer approximation which provides the normalized bubble wall temperature $T_R$.

## 3. Results for Acoustic Cavitation Bubbles and Conclusion

In this section we consider acoustic cavitation (gas-vapor) bubbles where the pressure inside the bubble is taken as the sum of the partial vapor pressure and the partial gas pressure, which in the first iterative approximation in normalized form can be written as

$$p = p_v + p_i \left(\frac{\sqrt{T_R}}{R}\right)^{3\kappa} \quad . \tag{8}$$

Using the interfacial condition for the heat flux at the bubble wall of a vapor-gas bubble and assuming that the thermal conductivity of the liquid is much greater than that of the gas-vapor mixture, the phase change at the bubble wall can be thought to be dominated by the liquid side. In this case the Plesset-Zwick solution for the bubble wall temperature from the liquid side becomes [1, 14]

$$T_R(t) = 1 - \mathcal{B} \int_0^t \frac{R^2(\xi)\,(dR/d\xi)}{\left[\int_\xi^t R^4(\tau)d\tau\right]^{1/2}}\, d\xi \tag{9}$$

where the constant $\mathcal{B}$ is given by

$$\mathcal{B} = \left(\frac{L'}{T_0'\, c_{p\ell}'}\right)\left(\frac{\rho_v'}{\rho_\ell'}\right)\frac{R_0'}{(\pi\,\Theta'\alpha_\ell')^{1/2}} \tag{10}$$

with $L'$ denoting the latent heat of condensation, $\rho_v'$ denoting the saturated vapor density, $\rho_\ell'$ denoting the liquid density, $\alpha_\ell'$ denoting the thermal diffusivity of the liquid and $c_{p\ell}'$ denoting the specific heat of the liquid. To close the system of equations, we employ the classical Rayleigh-Plesset equation to describe the radial dynamics of the bubble by implementing the first order gas pressure law for the partial gas pressure inside the bubble. For the driving acoustic pressure we use the equation

$$p_\infty(t) = -0.25\left[1 - \cos\left(\frac{2\pi t}{500}\right)\right] \quad ; \quad 0 < t < 500 \tag{11}$$

where $p_\infty(t)$ is the normalized driving acoustic pressure. For applications we consider the acoustic cavitation of water-vapor/air bubbles in water at $T_0'$=20°C with saturated vapor pressure $p_v'$= 0.0234 bar, surface tension coefficient $S'$ = 0.071 N/m, viscosity of water $\mu_\ell'$ = $10^{-3}$ kg/m-s, cavitation number σ = 0.492, initial equilibrium bubble radius $R_0'$=100 μm and a characteristic time $\Theta'$ = $10^{-5}$ s under the normalized driving acoustic pressure given by equation (11). The Rayleigh-Plesset equation incorporating the first order gas pressure law is then integrated using the Runge-Kutta-Fehlberg method together with the solution of equation (9) obtained by Simpson's 3/8-rule of numerical integration. Figure 1(a) shows a comparison of the temporal behaviour of the bubble radius using the present first order gas pressure law with thermal damping against the classical isothermal and adiabatic laws for the gas pressure. It is seen that the bubble radius grows to values greater than those of the isothermal case with a noticeable thermal damping effect. Figure 1(b) shows the detailed temporal evolution of the bubble wall temperature during the first collapse period, with bubble wall absolute gas temperature growing twenty three times the absolute bulk liquid temperature.

The present paper presents results for acoustic cavitation bubbles with the inclusion of thermal effects by a reduced order gas pressure law in spherical bubble dynamics. The inclusion of higher iterative approximations for the reduced gas pressure law together with comparison of the results with those of full numerical simulation of the radial PDE system for possible error analysis are left out for future work.

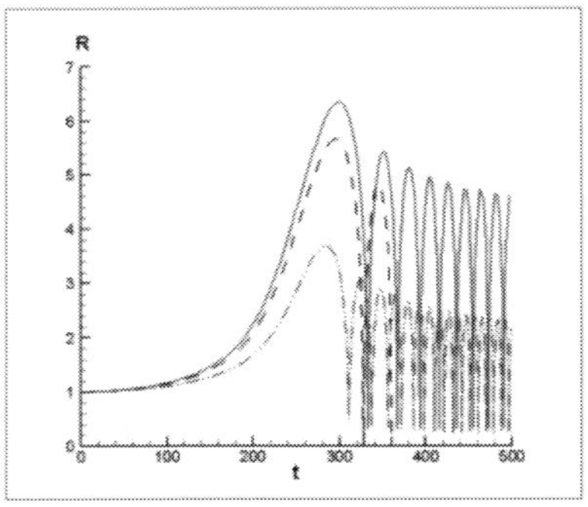


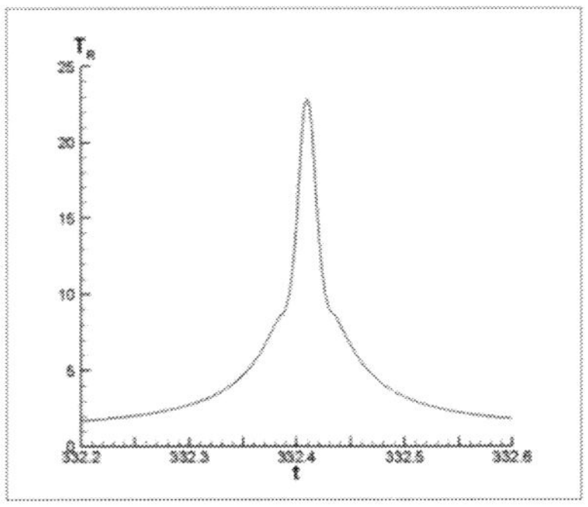


**(a)** **(b)**

**Figure 1. (a)** The temporal evolution of the normalized bubble radius driven by the acoustic pressure given by equation (11) for water-vapor/air bubbles in water with initial equilibrium radius $R_0'$=100 μm, cavitation number $\sigma = 0.492$ using the first order gas pressure law given by equation (8) (red solid line), the isothermal law (black dashed line) and the adiabatic law (blue dotted dashed line), **(b)** detailed temporal evolution of the normalized bubble wall temperature during the first collapse and rebound.

## 4. References

[1] Brennen C E 1995 *Cavitation and Bubble Dynamics* (Oxford University Press, Oxford) 47-67
[2] Plesset M S and Prosperetti A 1977 *Annu.Rev.Fluid Mech.* **9** 145
[3] Leighton T G 1994 *The Acoustic Bubble* (Academic Press, London) 84-86
[4] Suslick K S 1990 *Science* **247** 1439
[5] Gaitan D F, Crum L A, Church C C and Roy R A 1992 *J. Acoust. Soc. Am.* **91** 3166
[6] Crum L A 1994 *Phys.Today* **47** 22
[7] Nigmatulin R I, Khabeev N S and Nagiev F B 1981 *Int. J. Heat Mass Transfer* **24** 1033
[8] Matsumoto Y and Takemura F 1994 *JSME Int. J. Ser. B* **37** 288
[9] Preston A T, Colonius T and Brennen CE 2007 *Phys. Fluids* **19** Art. No. 123302
[10] Stricker L, Prosperetti A and Lohse D 2011 *J. Acoust. Soc. Am.* **130** 3243
[11] Prosperetti A, Crum L A and Commander K W 1988 *J. Acoust. Soc. Am.* **83** 502
[12] Plesset M S and Zwick S A 1952 *J. Appl. Phys.* **23** 95
[13] Prosperetti A 1991 *J. Fluid Mech..* **222** 587
[14] Delale C F 2002 *J. Fluids Eng.* **124** 969

# Cavitation-induced fragmentation of an acoustically-levitated droplet

**Silvestre Roberto Gonzalez Avila and Claus-Dieter Ohl**

Nanyang Technological University, School of Physical and Mathematical Sciences
Division of Physics and Applied Physics, 21 Nanyang Link, Singapore, 637371

sroberto@ntu.edu.sg

**Abstract**. In this paper we investigate the initial sequence of events that lead to the fragmentation of a millimetre sized water droplets when interacting with a focused ns-laser pulse. The experimental results show complex processes that result from the reflection of an initial shock wave from plasma generation with the soft boundary of the levitating droplet; furthermore, when the reflected waves from the walls of the droplet refocus they leave behind a trail of microbubbles that later act as cavitation inception regions. Numerical simulations of a shock wave impacting and reflecting from a soft boundary are also reported; the simulated results show that the lowest pressure inside the droplet occurs at the equatorial plane. The results of the numerical model display good agreement with the experimental results both in time and in space.

## 1. Introduction

The fragmentation, or breakdown of a liquid volume into smaller droplets, is commonly found both in nature (mist formation due to the shear of the wind with the ocean waves and in the fragmentation of lava when a volcano erupts) as well as in engineering applications (spray paint, irrigation processes, internal combustion engines and ink-jet printers to name a few); the interested reader can refer to the reviews by Villermaux [1] and Eggers and Villermaux [2]. Recently, interest has arisen due to a deeper understanding of the sheet formation required in the extreme ultraviolet light sources [3,4]. The droplet fragmentation from focused laser pulses is the result of a complex sequence of events and little studied so far in the literature. To shed more light we report in this work on the initial stage of the fragmentation process by performing experiments just a few microseconds after the starting of the events that lead to the fragmentation process of sub-millimeter liquid droplets. To study the fragmentation process we levitate the droplet with an acoustic field.

## 2. Experimental setup

The experimental setup, depicted in figure 1a, consists of a Langevin-type transducer that is driven at its resonant frequency, $f_r$ = 27.4 kHz, and an aluminum block that acts as a reflector of the acoustic waves from the transducer. In our setup we used the acoustic radiation pressure to maintain a liquid droplet floating in air. Once the droplets are levitating between the transducer and the reflector a focused laser pulse from a Nd:YAG laser (Orion, New wave research, 532 nm, 6 ns pulse duration) is aimed at the droplet to fragment it. The initial stage of the fragmentation process is analyzed by capturing single frames after specific delays, $\Delta t$, from the time of the laser pulse; the images are recorded with a digital 12bit CCD camera (Sensicam QE, 500 ns minimum exposure). The scene is illuminated with red fluorescence emission from a second laser of the same model. The laser beams and the CMOS camera are all synchronized with a pulse delay generator (BNC, Berkeley Scientific). For each droplet two images are taken, one just before the fragmentation and one that shows the fragments after a specific time delay. The experimental parameters relevant to this study are depicted in figure 1b. Each droplet is positioned in the acoustic field such that the laser impacts it approximately at the equatorial plane, at $-1 \le \varepsilon_y \le 0.1$, where $\varepsilon_y = \delta_y / R_y$, is the vertical excentricity paramerter, as can be seen in figure 1b.

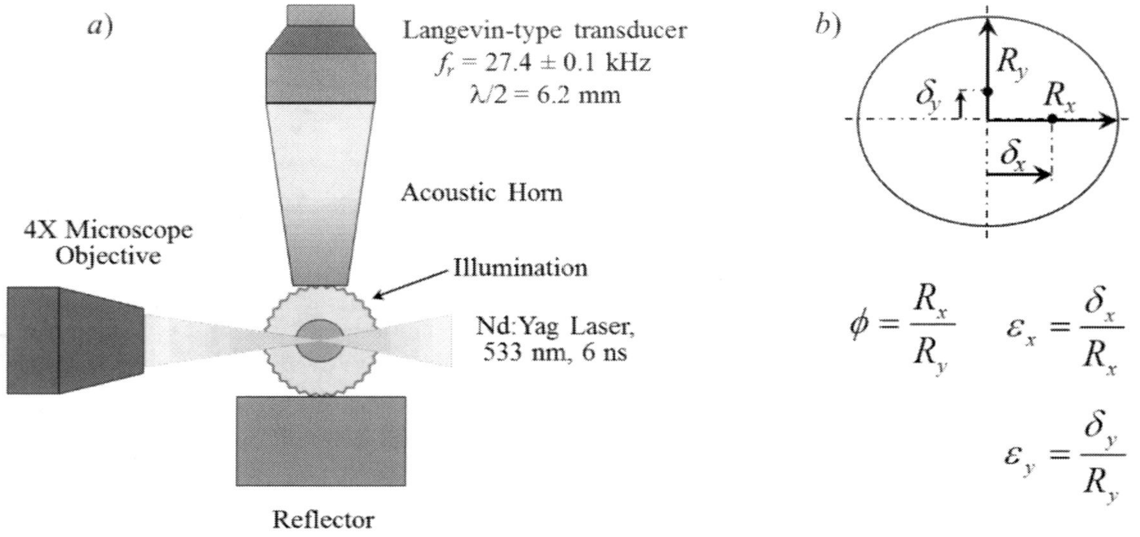


Figure 1. Experimental setup used for acoustic levitation and laser induced fragmentation of single millimetre-sized droplets.

## 3. Results

Figures 2a-c show images recorded 1, 3 and 10 μs after the laser is fired; here, the laser beam is generating a bubble at the right side of the droplet. During bubble expansion the thin film between the bubble and the atmosphere ruptures. Likely, the high pressure difference leads to shock wave emission and shear layers on the surface of the droplet. There small droplets are ejected, as can be seen from figure 2a; some of these small droplets travel even faster than the shock wave, visible as the circular object surrounding the droplet. the estimated velocity of the fastest droplets is 1360 ± 65 m/s, 4 time faster than the speed of sound in air at 20°C and 1 atm. Due to their smallness we cannot resolve their size; expecting that they need to be sufficiently small to remain stable at these fast accelerations we can make an estimate; the capillary length, $\lambda_c$, at an acceleration, $a_x$, imparted at $\Delta t$ = 1 μs is, $\lambda_c = \sqrt{\sigma / \rho a_x}$, where $\sigma$ and $\rho$ are the surface tension and the density of the liquid respectively; with $\sigma$ = 72 x$10^{-3}$ N/m, $\rho$ = 1000 kg/m$^3$ and the horizontal component of the droplet's velocity is ~ 1000 m/s, this velocity is attained in 1μs, thus the acceleration, $a_x$ ~ $10^3/10^{-6}$ ~ $10^9$ m/s$^2$; the capillary length of these fast traveling droplet fragments is ~ 200-300 nm. This is much smaller than the minimum features that can be resolve with the camera, i.e., 6.5 μm. Also, while the left side of the main droplet appears to be smooth the stretched droplet wall displays a rough surface with a complicated pattern. At

$\Delta t$ = 3 μs the fast ejected droplets are still flying faster than the shock wave, however, they are rapidly decelerating as in this case their speed has decreased to 820 m/s; finally, figure 2c depicts the state of the droplet 10 μs after the laser is fired; at this $\Delta t$ the droplet is significantly fragmented, the leading edge is thinner than in both figure 2a and 2b and we the shock wave has rebounded from the upper and lower walls.

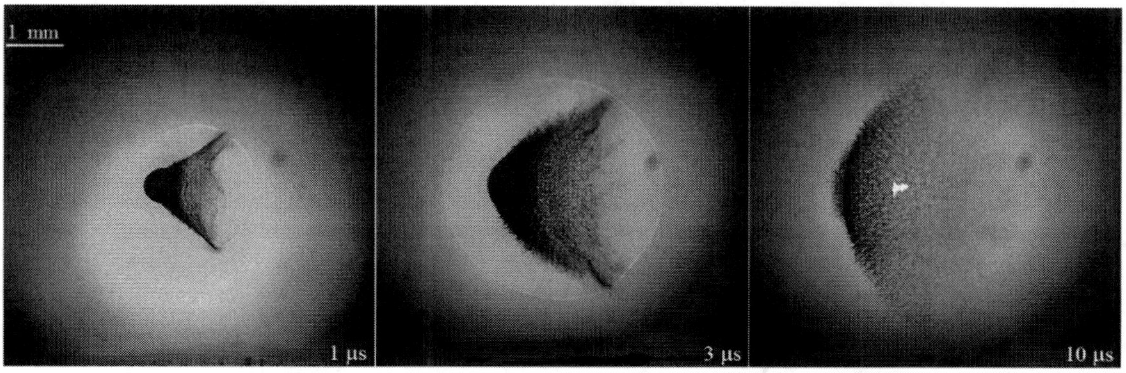


Figure 2. Fragmentation of a water droplet; single images captured at different time delays $\Delta t$; a) $\Delta t$ = 1 μs, $R_x$ = 198 μm, $R_y$ = 192 μm; b) $\Delta t$ = 3 μs, $R_x$ = 198 μm, $R_y$ = 192 μm; b) $\Delta t$ = 10 μs, $R_x$ = 163 μm, $R_y$ = 156 μm.

The small size of the droplet and fast times-scales make it difficult to understand what dynamic processes lead to fragmentation. In particular, small droplets show a variety of fragmentation patterns which we relate to the position and size of the initial vapour bubble generated. Next, we show experimental and numerical results which detail this initial stage. Therefore, we use larger droplets as they allow a better visualization of the bubble and shockwave dynamics within the droplet.

Figure 3a shows a sequence of images with the time of recording/simulation stated in in microseconds. The top image shows the experiment, while the lower is generated from a simulation of the acoustic wave propagation using a 3-dimensional finite element simulation of the linear wave equation in COMSOL. The wave is emitted from the surface of the bubble and modelled as a time dependent positive cycle from squared sine-function oscillating at 10MHz. The surfaces of the inner bubble and the droplet are perfect soft reflectors. The speed of sound is 1500 m/s and the CFL number is 0.2. In the first two experimental images, $\Delta t$ = 0 and 0.1 μs the shock wave is expanding spherically from the laser generated bubble, this is also reproduced numerically. At $\Delta t$ = 0.4 μs the shockwave is hidden in the curved parts of the droplet yet visible on the right (see arrow), the simulated image, $\Delta t$ = 0.5 μs, shows that the wave is reflected from the free surface of the droplet leading to an inversion of the acoustic, due to the very large difference in the acoustic impedance of water and air the wave is almost entirely reflected back inside the droplet [5];, upon reflection the trailing part of the wave can put the surface of the droplet under tension and induce secondary cavitation at the droplet's surface as has been observed before [6] . The reflected wave from the upper and lower portions of the droplet superimpose at the equatorial plane, as seen at $\Delta t$ = 1.10 and 1.25 and 1.30 μs; at $\Delta t$ = 1.25 and 1.30 μs a thin line of bubbles can be observed behind the point where the reflected waves merge; these cavitation bubbles may appear due to the encounter of shock waves propagating in opposite directions as has been reported before [7]; the numerical simulation also show good agreement with the location of the lowest pressure both in time and in space; in the last three experimental images at $\Delta t$ = 1.35 and 1.40 and 1.45 μs the point at which the counter propagating waves meet moves to the right in each frame, with the trail of bubbles increasing from left to right; similar trend is observed in the simulated images where the point where the waves meet propagates to the right; the darkest region in these images corresponds to the lowest pressure. Figure 3b displays simulated images that correspond to the images shown in figure 3a, however, the images portray a top view; surprisingly, the location where the lowest pressure values (most probable sites for cavitation inception) is not through the centre of the droplet but instead closed to its periphery. The image at the bottom, figure 3c, is a 3d representation of the simulated droplet with the location of the lowest pressure for all the time steps in the simulation.

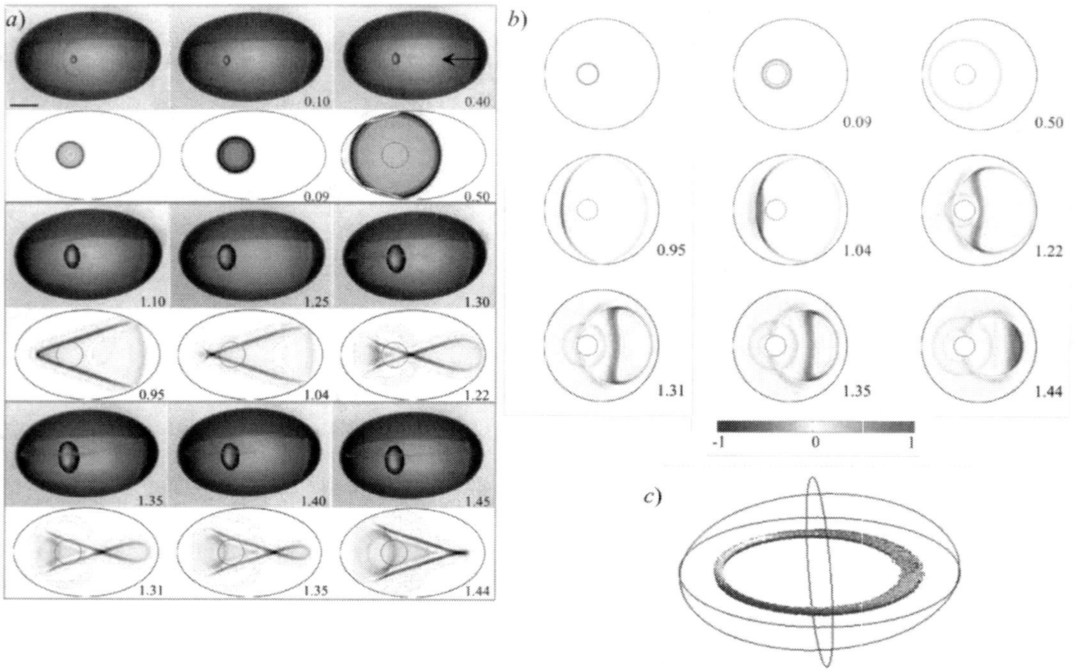

Figure 3. The position of the shock wave inside a levitated droplet and comparison with numerical simulations; a) side view; b) top view; c) position of lowest pressure during the wave propagation. The length of the bar in the lower left corner of the first image from the top is 500 μm.

## 4. Conclusions

Due to the reflection of the shock wave from the soft droplet boundary secondary cavitation is induced in a region confined close to the periphery of the droplet; the experimental results show good agreement with numerical simulations.

**Acknowledgements**
The authors thank Professor David Quéré for useful suggestions

## References

[1] Villermaux, E 2007 *Annu. Rev. Fluid Mech.* 39:419–46
[2] A Eggers, J & Villermaux, E 2008 *Rep. Prog. Phys.* 71, 036601.
[3] Klein A, L, Bouwhuis, W, Visser, C, W, Lhuissier, H et al. 2015, Phys. Rev. Applied 3, 044018
[4] Banine, Y, Koshelev, K, N & Swinkels, G, H, P, M 2011 J. Phys. D: Appl. Phys. **44** 253001
[5] Ohl, C-D, 2002 *Phys Fluids* 14, 10 3512-3521
[6] Robert, E, Lettry, J, Farhat, M, Monkewitz, P A & Avellan, F 2007 *Phys. Fl.* **19**, 067106
[7] Paltauf, G, Schmidt-Kloiber, H & Frenz, M 1998 J. Acoust. Soc. Am. **104**: 890-897

Journal of Physics: Conference Series **656** (2015) 012018 doi:10.1088/1742-6596/656/1/012018

# Transport of micro-bubbles in turbulent shear flows

**P. Gualtieri, F. Battista and C.M. Casciola**

Diaprtimento di Ingegneria Meccanica e Aerospaziale, Sapienza Università di Roma. Via Eudossiana 18, 00184 Roma, Italy

E-mail: paolo.gualtieri@uniroma1.it

**Abstract.** The dynamics of micro-bubbles, which are typical in many industrial applications, is addressed by means the Direct Numerical Simulations (DNS) of two prototypal flows, namely a homogeneous shear flow and a fully developed pipe flows. This preliminary study has a two-fold purpose. The homogenous turbulent shear flow is useful to characterize the bubble dynamics in terms of their eventual clustering properties which is expected to be controlled by the Stokes number. The time history of the fluid pressure experienced by the bubbles during their evolution is recorded and successively employed to force the Rayleigh-Plesset equation [1]. The ensuing data are used to address *a posteriori* the bubble diameter statistics in view of bubble collapse induced by strong and intermittent turbulent pressure fluctuations. The turbulent pipe flow simulations serve to address the bubble dynamics in wall bounded flows. Here the bubbles are observed to accumulate in the near-wall region with different intensity depending on the bubble dimensions.

## 1. Introduction

Many natural and technological applications present regimes where micro-bubbles are transported by a carried eventually turbulent flow. For instance, the cavitation events and the collapse of the bubbles are of primary importance in the propeller design due to the structural damages caused by cavitation.

This paper deals with the dynamics of a point-like disperse phase in prototypal turbulent shear flows, a homogeneous shear flow and a pipe flow, laden with *particles* of different density ratios ranging from neutrally buoyant particles with $\rho_p/\rho_f = 1$ to the typical density ratios of the bubbles, i.e. $\rho_p/\rho_f = 10^{-2} - 10^{-3}$. The turbulent carrier flow is resolved by means of the Direct Numerical Simulation of the incompressible Navier-Stokes equations at a moderate value of the Reynolds number. The carrier phase transports inertial particles which are assumed to small in comparison with the smallest hydrodynamical scale, i.e. the Kolmogorov dissipative scale $\eta$. In such conditions the particles dynamics is well captured by the equations reported in [2] and [3], namely

$$\frac{d\mathbf{x}_p}{dt} = \mathbf{v}_p(t); \qquad \tau_p \frac{d\mathbf{v}_p}{dt} = [\mathbf{u}(\mathbf{x}_p, t) - \mathbf{v}_p(t)] + \frac{3}{2}\frac{\tau_p}{\left(\frac{1}{2} + \frac{\rho_p}{\rho_f}\right)} \frac{D\mathbf{u}}{Dt}(\mathbf{x}_p, t)\,, \tag{1}$$

where $\mathbf{x}_p$ and $\mathbf{v}_p$ are the particle instantaneous position and velocity, respectively, $\mathbf{u}(\mathbf{x}_p, t)$ is the fluid velocity at the particle position, $\rho_p/\rho_f$ is the particle to fluid density ratio,

$\tau_p = (\rho_p/\rho_f)d_p^2/(18\nu_f)$ is the particle relaxation time (with $d_p$ the particle diameter and $\nu_f$ the fluid kinematic viscosity). Equations (1) hold for particles whose diameter is smaller than the Kolmogorov scale, $d_p/\eta \ll 1$ and particle Reynolds number smaller than one, $Re_p = (|\mathbf{u} - \mathbf{v}_p|)d_p/\nu_f \ll 1$. The first and second terms on the right hand side of eq. (1) are the Stokes drag and the added mass respectively. In the latter $D\mathbf{u}/Dt$ is the fluid acceleration evaluated at the particle position. Equations (1) are time-integrated following the Lagrangian trajectory of the disperse phase and, when needed, the fluid velocity and the fluid acceleration ensuing from the time-integration of the background turbulent field are interpolated at the particle position.

Panels (a) and (c) of figure 1 address the two geometries considered here, namely the homogeneous shear flow in panel (a) and the turbulent pipe flow in panel (c). Briefly the homogeneous shear flow consists in a 3D periodic box where turbulent fluctuations are forced by an imposed mean velocity gradient see [4]. The computational domain of physical size $4\pi \times 2\pi \times 2\pi$ is discretized by means of $288 \times 144 \times 144$ collocation points where equations are solved by means of a pseudo-spectral method. This allows to resolve the Kolmogorov scale at the selected Reynolds number of $Re_\lambda = 80$. The pipe flow consists in a cylindrical duct with periodic conditions in the axial direction. The motion of the fluid is obtained by enforcing a constant mean pressure gradient along the pipe, see [5]. Here the pipe length is $2\pi R$ where $R$ is the pipe radius. In this case the code exploits staggered finite difference method to solve the Navier-Stokes equations in cylindrical coordinates on a discrete grid of $256 \times 192 \times 256$ points in the azimuthal, radial and axial direction respectively. Both codes take advantage of a low-storage Runge-Kutta scheme for the temporal integration of the carrier and disperse phase.

Several simulations have been performed to span the particle space parameter. In fact, in the limit of small particle Reynolds number $Re_p \ll 1$, the only two parameters describing the different motion regimes of the disperse phase are the particle Stokes number, namely the fluid to particle characteristic time ratio, $St = \tau_p/\tau_f$ and the particle to fluid density ratio $\rho_p/\rho_f$. Concerning the homogeneous shear flow we have considered three different Stokes numbers, based on the Kolmogorov time scale, and four values of the density ratio, namely $St_\eta = 0.5, 0.1, 0.01$ and $\rho_p/\rho_f = 1, 0.1, 0.01, 0.001$ which amounts to twelve different cases. In the pipe flow two Stokes number, based on the wall-unit time-scale, and five values of the density ratio, namely $St^+ = 0.2, 2$ and $\rho_p\rho_f = 5, 1, 0.1, 0.01, 0.001$ are available amounting to ten cases. The turbulent Reynolds number based on the Taylor scale is $Re_\lambda = 80$ for the homogeneous shear flow and the friction Reynolds number is $Re_\tau = 310$ for the pipe flow configuration.

## 2. Results and Discussion

Figure 1 reports the instantaneous configuration of the disperse phase for the two geometries addressed in this paper, namely panel (b) for the homogeneous shear flow and panel (d) for the pipe flow. Some few slices are reported to give a visual representation of the underlying turbulent fluctuations in panels (a) and (c). In the homogeneous shear flow it is shown the instantaneous configuration of the disperse phase pertaining to $St_\eta = 0.5$ and $\rho_p/\rho_f = 0.001$. These are the typical values for bubbles which are well known to accumulate in the low pressure region of the turbulent field, i.e. in the coherent vortical structures. This feature clearly emerge from the scatter plot in panel (b) where the bubbles form filament-like structures which mark to the cores of the turbulent vortices. Concerning the pipe flow, at first sight, no significant spatial localization of particles appears in the case of $St^{=}0.2$ and $\rho_p/\rho_f = 0.001$ probably due to a relatively small value of the Stokes number.

More quantitative statistical results are shown in figure 2. Panels (a) and (b) show the the radial distribution function $g_{00}$ versus the spatial separation $r/\eta$. The radial distribution

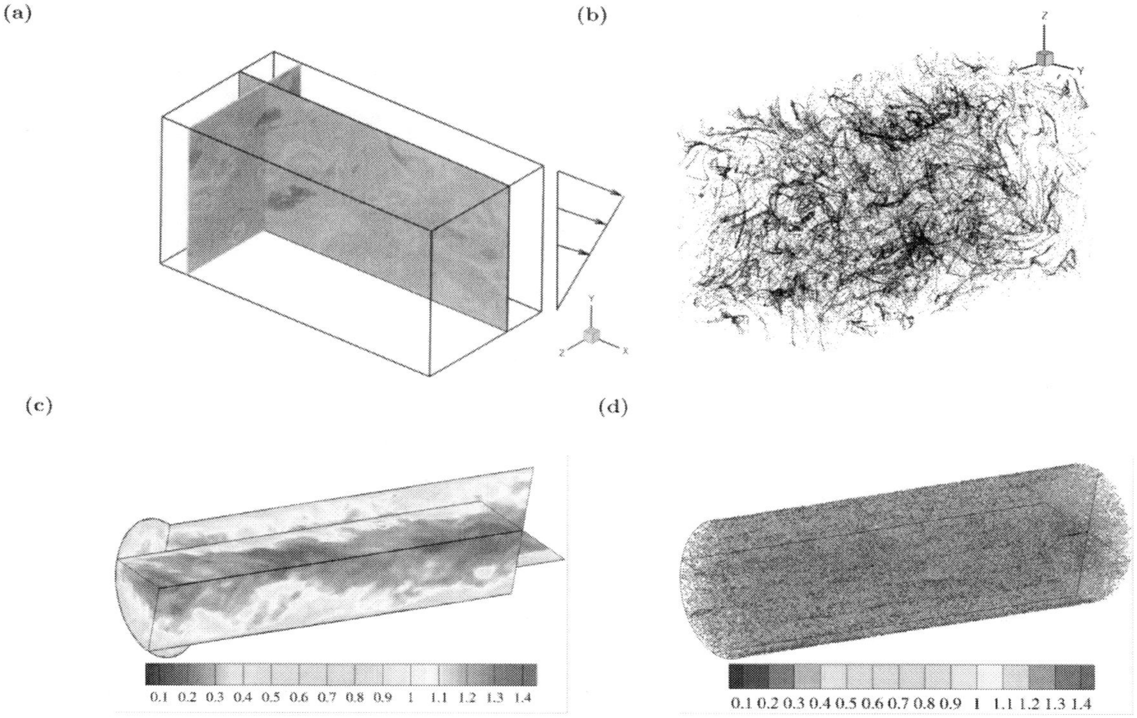


**Figure 1.** Instantaneous configuration of the fluid domain and for the homogeneous shear flow, panel (a), and for the pipe flow, panel (c). Colors denotes the stream-wise fluid velocity intensity. Panel (b): instantaneous configuration of the disperse phase in the homogeneous shear flow corresponding to $St_\eta = 0.5$ and $\rho_p/\rho_f = 0.001$. Panel (d): instantaneous configuration of the disperse phase in the pipe flow at $St$=0.2 and $\rho_p/\rho_f = 0.001$. Bubbles are colored with their stream-wise velocity intensity.

function actually measures the probability to find a particle pair at a certain distance $r$. The different curves in the panel address the effect of the Stokes number $St_\eta$ at a fixed density ratio $\rho_p/\rho_f = 0.001$ (panel (a)) and the effect of the density ratio at fixed Stokes number $St_\eta = 0.1$ (panel (b)). As expected, for vanishing Stokes number (Lagrangian tracers), the particle distribution behaves as a spatially homogeneous distribution and the radial distribution function assumes the unitary value at all separations. However, when the Stokes number is increased, small scale clustering occurs, i.e. $g_{00}(r)$ diverges at small separations, see figure 2 panel (a). The behavior of particles at fixed Stokes number, $St^+ = 0.1$, depending on the particle/fluid density ratio, shows that for bubbles $\rho_p/\rho_f = 0.001$ the clustering formation is more intense than the other cases $\rho_p/\rho_f \to 1$. Dealing with the turbulent pipe flow, panels (c) and (d) report the particle mean concentration profile normalized with the homogeneous concentration for two different values of the Stokes number, $St^+ = 2$ panel (c) and $St^+ = 0.5$ panel (d). The disperse phase appears to segregate in the near wall region. In particular the segregation intensity increases with the increase of the Stokes number and the maximum concentration peak moves away from wall by decreasing the particle density, staying in any cases in the viscous sublayer or the buffer layer.

A deeper analysis of the transport of bubbles in a turbulent flow calls into play the particle radius dynamics obtained by the Rayleight-Plesset [1] forced by the fluid pressure history experienced along their trajectory. In particular, due to the preferential sampling of turbulent

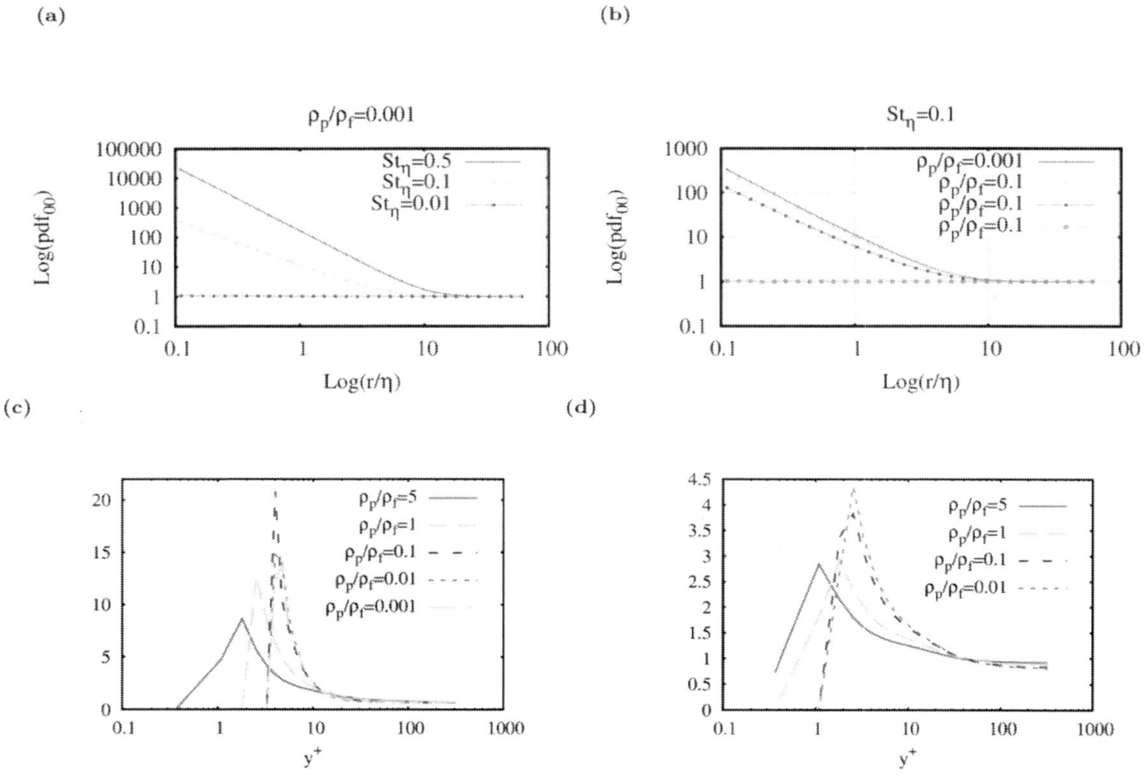


**Figure 2.** Top panels: radial distribution function $g_{00}$ versus normalized separation $r/\eta$ at fixed density ratio $\rho_p/\rho_f = 0.001$ and different Stokes number $St_\eta$ (panel (a)), and at fixed Stokes number $St_\eta = 0.1$ and different density ratios (panel (b)). Bottom panels: wall-normal mean particle concentration as a function of the normalized wall-normal distance $y^+ = (1-r)/y^*$ where $y^*$ is the wall unit. Two Stokes number are showed, $St^+ = 2$, panel (c), and $St^+ = 0.5$, panel (d).

fluctuations, the light bubbles accumulate in relatively low pressure region of the flow. It follow that the fluctuations of the radius are strongly intermittent and the tails of the ensuing p.d.f is characterized by intense and relatively probable events.

## 3. Acknowledgments
The present research has received funding from the European Research Council under the European Union's Seventh Framework Program (FP7/ 2007-2013)/ERC Grant Agreement No. [339446]. The authors are grateful to PRACE for awarding access to resource FERMI based in Italy at Casalecchio di Reno

## 4. References
[1] Brennen C E 2014 *Cavitation and Bubble Dynamics* (Cambridge University Press)
[2] Maxey M R and Riley J J 1983 *Physics of Fluids (1958-1988)* **26** 883–889
[3] Gatignol R 1983 *Journal de Mécanique théorique et appliquée* **2** 143–160
[4] Gualtieri P, Picano F and Casciola C 2009 *Journal of Fluid Mechanics* **629** 25
[5] Picano F, Sardina G and Casciola C 2009 *Physics of Fluids* **21** 093305

# Numerical investigations of single bubble oscillations generated by a dual frequency excitation

**Matthieu Guédra[1], Claude Inserra[1], Bruno Gilles[1] and Jean-Christophe Béra[1]**

[1] Inserm, U1032, LabTau, Lyon, F-69003, France ; Université Lyon 1, Lyon, F-69003, France

E-mail: matthieu.guedra@inserm.fr

**Abstract.** The oscillations of a single bubble excited with a dual frequency acoustic field are numerically investigated. Computations are made for an air bubble in water exposed to an acoustic field with a linearly varying amplitude. The bubble response to an excitation containing two frequencies $f_1 = 500$kHz and $f_2 = 400$kHz at the same amplitude is compared to the monofrequency case where only $f_1$ is present. Time-frequency representations show a sharp transition in the bifrequency case, for which the low frequency component $f_2$ becomes resonant while the high frequency component $f_1$ is strongly attenuated. The temporal evolution of the power spectra reveals that the resonance of the low frequency component is correlated with the time varying mean radius of the bubble. It is also observed that the total power of the bubble response in the bifrequency case can reach almost twice the power obtained in the monofrequency case, which indicates a strong enhancement of the cavitating behavior of the bubble for this specific frequency combination.

## 1. Introduction

Controlling the cavitation activity is of primary interest for specific clinical uses of ultrasound whose efficiency can be enhanced with the generation of implosive bubbles. For example, when used for thrombolysis, focused ultrasound-induced cavitation plays a major role in the destruction of blood clots. For this type of application, reducing the cavitation threshold is challenging as it promotes the mechanical effects of cavitation while limiting the heat deposition on tissues. A number of studies have already shown that this threshold can be significantly lowered by using a multifrequency excitation. Several frequency combinations have been tested such as fundamental and second harmonic superimposition [1, 2], low frequency component addition [3–6] or dual neighboring frequencies emission [7, 8]. While all these experiments demonstrated the beneficial effect of dual frequency emission on cavitation activity, quite few works have been devoted to the theoretical investigation of this phenomenon. In the present study, this effect is quantified through the analysis of the response of a single bubble to a dual frequency excitation based on the numerical solving of the Keller-Miksis equation [9]. Computations are made for an air bubble in water exposed to a mono- or bi-frequency acoustic field with a linearly time-varying amplitude, and the temporal evolutions of the power spectra are determined for both mono- and bi-frequency cases.

Journal of Physics: Conference Series **656** (2015) 012019 doi:10.1088/1742-6596/656/1/012019

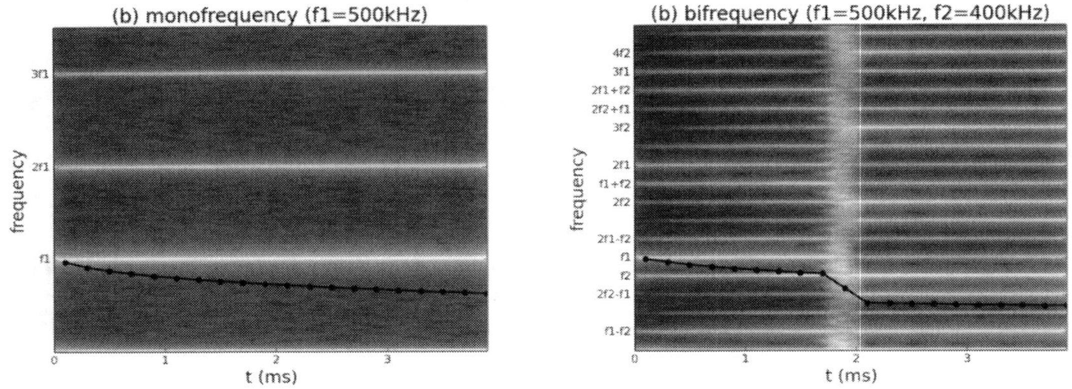


**Figure 1.** Time-frequency representations of the bubble response in the monofrequency (a) and bifrequency (b) cases when the acoustic pressure amplitude $p_a$ is linearly varying from 0 to 100kPa. On both graphs, black circles (● ) correspond to the evolution of the linear resonant frequency calculated from the static component (mean radius) of the signal (see ■ on Fig. 2).

## 2. Single bubble response to a dual frequency excitation

The present study deals with the nonlinear oscillations of a gas bubble in a slightly compressible liquid. Assuming that the size of the bubble is much smaller than the acoustic wavelength allows to keep spherical symetry of the problem, so that the bubble radius $R$ is governed by the Keller-Miksis equation [9]:

$$R\ddot{R}\left(1-\frac{\dot{R}}{c}\right)+\frac{3}{2}\dot{R}^2\left(1-\frac{\dot{R}}{3c}\right)=\left(1+\frac{\dot{R}}{c}\right)\frac{p_i-p(t)}{\rho}+\frac{R}{c}\frac{\dot{p}_i-\dot{p}(t)}{\rho}-\frac{2\sigma}{\rho R}-\frac{4\mu}{\rho}\frac{\dot{R}}{R}, \qquad (1)$$

where $c$ is speed of sound in the liquid, $\rho$ and $\mu$ are the liquid density and viscosity, and $\sigma$ denotes the gas-liquid surface tension. The motion of the bubble is driven by the pressure difference existing between its internal pressure $p_i$, and the liquid pressure $p(t)$. Assuming that the bubble is only filled with non-condensable gas, thus neglecting the partial vapor pressure, $p_i$ is given by the following equation:

$$p_i=\left(p_\infty+\frac{2\sigma}{R_0}\right)\left(\frac{R_0}{R}\right)^{3\gamma}, \qquad (2)$$

where $\gamma$ is gas polytropic exponent, $R_0$ is the static radius of the bubble and $p_\infty$ is the ambient pressure. In the specific case of a dual frequency acoustic excitation, the liquid pressure $p(t)$ is written as:

$$p(t)=p_\infty-\frac{p_a}{2}\big(\sin(2\pi f_1 t)+\sin(2\pi f_2 t)\big). \qquad (3)$$

The minus sign ensures that the motion begins at $t=0$ with a rarefaction phase and thus an expansion of the bubble. Note that in order to simplify further analysis, we state here that both frequency components are generated with the same amplitude $p_a/2$, corresponding to half the amplitude of the monofrequency case $p(t)=p_\infty-p_a\sin(2\pi f_1 t)$ (same peak amplitude for both mono- and bi-frequency cases).

## 3. Enhancement of cavitation activity: spectral analysis

The motion equation Eq. (1) is solved using a standard numerical method to obtain the temporal evolution of the bubble radius. In the following, the computations are made for an air bubble

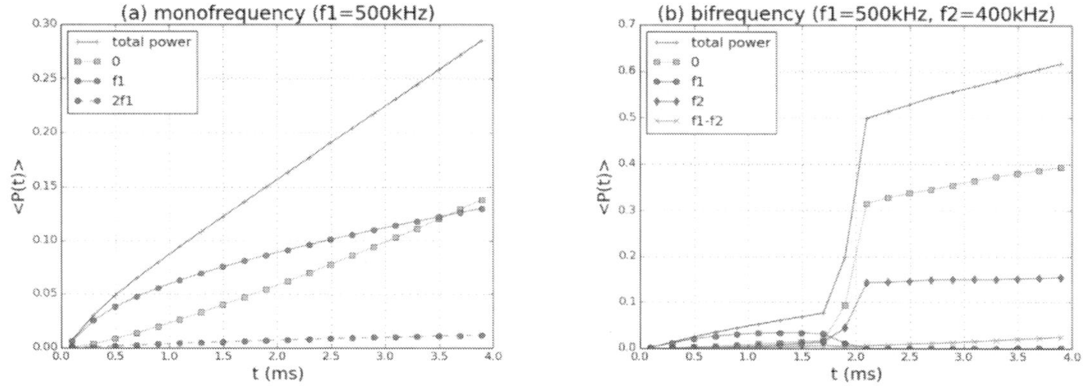


**Figure 2.** Time-averaged power of the spectral components of the bubble response signal in the monofrequency (a) and bifrequency (b) cases. On graph (a) are given the power amplitudes of the fundamental frequency $f_1$ and its first harmonic $2f_1$, as well as the power amplitude of the static component (mean radius). On graph (b) are given the power amplitudes of the primary frequencies $f_1$ and $f_2$ and low frequency modulation $f_1 - f_2$, as well as the power amplitude of the static component.

in water at ambient temperature $T = 300$K and atmospheric pressure $p_\infty = 1.013$bar. The initial bubble radius is fixed to $R_0 = 7.058\mu$m, corresponding to the linear damped resonant radius at frequency $f_1 = 500$kHz. The acoustic pressure amplitude $p_a$ is linearly varying from 0 to 100kPa (thus from 0 to 50kPa in bifrequency case) over 4ms, corresponding to 2000 acoustic cycles at frequency $f_1$. Figure 1 gives the time-frequency representations of the bubble response in the monofrequency case for $f_1 = 500$kHz (a) and in the bifrequency case when a frequency component $f_2 = 400$kHz is added in the excitation term (b). The typical nonlinear harmonics generation is recovered in the monofrequency case with the rising of $2f_1$, $3f_1$ and $4f_1$ components (Note that because of the proper choice of the bubble radius and pressure amplitude, no subharmonics are generated in the bubble response). When a second frequency $f_2$ is added to the excitation, a spectral broadening occurs with the generation of harmonics and combinations of both frequencies. The most important observation concerns the transition towards strongly nonlinear oscillations at $t \simeq 1.8$ms associated with the generation of a broadband spectrum, which is clearly absent in the monofrequency case. A second transition is observed at $t \simeq 2$ms when the bubble recovers a periodic behavior. After this transition, one can observe a prevailing of the $f_2$-component over the $f_1$-component which is considerably attenuated.

To better understand the role of the additional $f_2$ component on the bubble response, the time evolution of the power spectra is given in Fig. 2 in both monofrequency (a) and bifrequency (b) cases. First analyzing the monofrequency case, what is clearly not seen on the time-frequency representations in Fig. 1 is the emergence of the static component of the bubble motion (see ■ on Fig. 2) which carries an important part of the total energy, with the same order of magnitude than the fundamental frequency and much larger than the first harmonic. This static component arises due to the asymmetrical oscillations of $R(t)$ around its mean value $R_m$ and corresponds to a change of the bubble mean radius with time, to which one can associate a variation of the linear resonant frequency of the bubble [10]:

$$f_r = \frac{1}{2\pi R_m \sqrt{\rho}} \sqrt{3\gamma \left( \frac{p_\infty R_0}{R_m} + \frac{2\sigma}{R_m} \right) - \frac{2\sigma}{R_m} - \frac{4\mu^2}{\rho R_m^2}}. \qquad (4)$$

As the mean radius increases with time, the linear resonant frequency decreases, as shown by circles (●) in Fig. 1, being also responsible for the saturation of the fundamental and harmonics power amplitudes (Fig. 2) as $f_1$ deviates from the resonance. In the bifrequency case, the transition happens when the lower frequency $f_2$ becomes resonant. An important part of energy is then transferred to this component, which gives rise to the $\diamond$-curve in Fig. 2(b). One can also observe that the static component is prevailing and that the total power is clearly increased, being almost twice the total power obtained in the monofrequency case at $t = 4$ms. Finally, Fig. 2(b) also shows that the $f_1$-component completely vanishes at the transition and all the energy supplied at this frequency is transferred into others such as the low-frequency modulation $f_1 - f_2$ ($\times$) and the higher order combination $2f_2 - f_1$ (not shown for readability). The latter becomes resonant as soon as it matches with the shifted linear frequency given by Eq. (4) (see Fig. 1).

## 4. Conclusion

Numerical computations of the single bubble response to an acoustic field have been conducted for a monofrequency excitation ($f_1 = 500$kHz) and for a bifrequency excitation ($f_1 = 500$kHz and $f_2 = 400$kHz). In the bifrequency case, the time-frequency representation showed the emergence of multiple combinations of both frequencies, followed by a sharp transition towards strongly nonlinear oscillations and the generation of a broadband spectrum. After this transition, the recovered periodic behavior of the bubble was characterized by the rising of the lower frequency component ($f_2$) and the strong attenuation of the higher one ($f_1$). The emergence of the $f_2$-component was correlated with the increase of the mean bubble radius with time, associated with a decrease of its linear resonant frequency. Due to the resonant character of the $f_2$-component or of higher-order combinations, the total power reached twice the power obtained in the monofrequency case after the transition, meaning that the cavitating behavior of the bubble was largely enhanced by the application of this dual frequency combination. In conclusion, this study highlights some fundamental aspects of the bifrequency driven single bubble oscillations and thus constitutes a step towards a better understanding of the role of dual frequency excitations in the enhancement and control of cavitation activity.

## Acknowledgments

This work was supported by the French National Research Agency ANR project "ULysSE" (ANR-11-JSV5-0008). It was performed within the framework of the LabEx CeLyA (ANR-10-LABX-0060) of Université de Lyon, within the program "Investissements d'Avenir" (ANR-11-IDEX-0007) operated by the French National Research Agency (ANR).

## References

[1] Kawabata K and Umemura S 1996 *J. Phys. Chem.* **100** 18784-18789
[2] Holzfuss J, Rúggeberg M and Mettin R 1998 *Phys. Rev. Lett.* **81** 1961-1964
[3] Iernetti G, Ciuti P, Dezhkunov N V, Reali M, Francescutto A and Johri G K 1997 *Ultrason. Sonochem.* **4** 263-268
[4] Ciuti P, Dezhkunov N V, Francescutto A, Kulak A I and Iernetti G 2000 *Ultrason. Sonochem.* **7** 213-216
[5] Zhao Y, Zhu C, Feng R, Xu J and Wang Y 2002 *Ultrason. Sonochem.* **9** 241-243
[6] Feng R, Zhao Y, Zhu C and Mason T J 2002 *Ultrason. Sonochem.* **9** 231-236
[7] Gilles B, Béra J-C, Mestas J-L and Cathignol D 2006 *Appl. Phys. Lett.* **89** 094106
[8] Saletes I, Gilles B and Béra J-C 2011 *Ultrasonics* **51** 94-101
[9] Keller J B and Miksis M 1980 *J. Acoust. Soc. Am.* **68** 628-633
[10] Lauterborn W 1976 *J. Acoust. Soc. Am.* **59** 283-293

# 3D crown spike of free surface induced by two bubbles

**R Han**[1]**, X L Yao, A M Zhang**
College of Shipbuilding, Harbin Engineering University
#515 Chuanhai Building, Nantong Street No.145, Nangang District, Harbin, Heilongjiang Province, China

E-mail: hanruihope2008@126.com

**Abstract.** A specific physics called 'crown phenomenon' is discovered in the interaction between weak buoyancy bubbles and free surface. The 'crown phenomenon' is that a circle of the outer fluid appears to surround the middle spike of water after the jet impact of bubbles, and this kind of spike is defined as 'crown spike'. In this study, the crown spike due to the coupling effect between two bubbles and free surface is studied both experimentally and numerically. In the experiment, copper wires in series connection are used to generate two in-phase bubbles and the bubble and free surface shapes are recorded by high-speed photography. In the numerical study, a three-dimensional model is established to simulate the bubble-free-surface interaction with a boundary integral method and then the motion of free surface is further simulated without regard to the effect of bubbles after the jet impact. The computation also traces the 'crown phenomenon', which is considered as a second spike related to a large high-pressure region formed after the impact. The large high-pressure region leads to a thick column of water on the free surface and then the column of water gradually increases to surround the first spike. Both oblique jets and crown spike are observed in the experimental and numerical results, and the favorable agreements of bubbles and free surface shapes validate the present model. The effect of the inter-bubble distance on crown spike is also investigated.

## 1. Introduction

The interaction between bubble and free surface has applications in printing [1] and water barrier [2]. Experimental observations show that a downward jet is formed by weak buoyancy bubbles near the end of collapse phase, while the free surface produces a spike moving upward [3]-[5]. It's clear that the motion of free surface is significantly influenced by the bubble-free-surface distance. Zhang *et al.*[6] observed a 'crown spike' after the jet impact with a high-speed camera, that is a circle of the outer fluid rises to surround the middle column of water, appearing a crown-like shape. There is little progress in the study about this specific physics, the main reason is that the toroidal bubble would collapse for a short period [7]. The method of ignoring the effect of toroidal bubble [8]-[10] has been used in the previous studies to avoid the difficulties in computation. The previous numerical method is taken as a reference for the simulation of crown spike and the main reasons are described as follows. An energy loss of about 70% occurs primarily at the minimum volume of bubble and the toroidal bubble after the downward jet would collapse soon. It is reckoned that the bubble loses energy continuously and that its influence on the free surface gradually becomes weaker. Li *et al.*[11] established an axisymmetric model to investigate the effects of parameter variation on spike pattern.

[1]E-mail address: hanruihope2008@126.com

After the jet impact, they further simulated the motion of free surface ignoring the effect of toroidal bubble. Moreover, Han *et al.* [12] established a three-dimensional model to simulate the crown spike in both single- and two-bubble cases.

In this paper, crown spike of free surface is studied through both experimental and numerical method. The motion of two bubbles and free surface is observed experimentally using a high-speed photography. Meanwhile, a three-dimensional model is established using a boundary integral method to simulate the interaction between two bubbles and free surface. The computation also traces the motion of free surface after the jet impact. The numerical solutions agree well with the experimental results with regard to bubble and free surface shapes, which confirms the validity of the present model. Detailed analyses are made for crown phenomenon in the present investigation, aiming to provide information for studying this specific physics.

## 2. Method

Based on potential flow theory, a 3D numerical model is used to simulate the crown spike of free surface. The flow is assumed to be irrotational, inviscid and incompressible, thus the velocity potential satisfies the boundary integral equation [13]:

$$\vartheta\Phi(\boldsymbol{p}) = \iint_S \left( \frac{\partial\Phi(\boldsymbol{p}))}{\partial n} \frac{1}{r_{pq}} - \Phi(\boldsymbol{q}) \frac{\partial}{\partial n}\left(\frac{1}{r_{pq}}\right) \right) \mathrm{d}S \tag{1}$$

where the surfaces of two bubbles and free surface are denoted by $S$, $\boldsymbol{p}$ and $\boldsymbol{q}$ are the fixed point and the source point located on the liquid-gas interfaces, $\vartheta$ is the solid angle at $\boldsymbol{p}$ which equals $2\pi$, the normal derivative is defined as $\partial/\partial n = \boldsymbol{n}\cdot\nabla$ and $\boldsymbol{n}$ is the outward normal.

All the boundaries are discretized and finer grids are used in the central two thirds of the free surface for the simulation of crown spike, as shown in figure 1. The surface of each bubble is discretized into 1280 triangular elements and the free surface into 7792 triangular elements. The radius of free surface is 4.5 times of the maximum radius attained by the bubble [14].

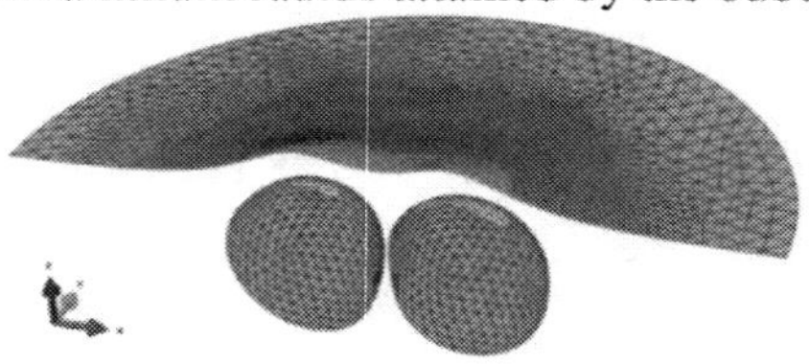

**Figure 1.** Discretization

Incompressible ideal fluid in unsteady irrotational flow domain satisfies Bernoulli equation, and dynamic boundary condition governing the motion of the bubbles and free surface is [7]:

$$\frac{\mathrm{d}\Phi}{\mathrm{d}t} = \frac{|\nabla\Phi|^2}{2} + \frac{P_\infty - P_c}{\rho} - \frac{P_{0,b}}{\rho}\left(\frac{V_{0,b}}{V_b}\right)^k - gz \text{ (for bubble surface)} \tag{2a}$$

$$\frac{\mathrm{d}\Phi}{\mathrm{d}t} = \frac{|\nabla\Phi|^2}{2} - g(z-d) \text{ (for free surface)} \tag{2b}$$

where $P_\infty$ is the ambient pressure on the plane of the bubble centre at inception, $P_c$ is the constant vapour pressure with $P_c = 0.5$bar [14], $P_{0,b}$ and $V_{0,b}$ are the initial pressure and volume of bubble, $k$ is the ratio of the specific heats of the gas which is taken as 1.25 [8], $\rho$ is the fluid density, $g$ is the gravitational acceleration, and $d$ is the initial distance of the bubble centroid from free surface.

The kinematic boundary condition governing motion of bubbles and free surface is [7]:

$$\frac{\mathrm{d}\boldsymbol{r}}{\mathrm{d}t} = \nabla\Phi \tag{3}$$

where $\boldsymbol{r}$ denotes the spatial position of the fixed point $\boldsymbol{p}$ on the bubble surface or the free surface. In the computation, all variables are normalized: length is scaled with respect to the maximum radius $R_m$ the bubble would attain in an infinite fluid; pressure with respect to $P_\infty$.

## 3. Results and discussion

Numerical results are compared with the experimental observations. Bubbles are generated via spark discharge using a 210V dc supply and copper wires with a diameter of 0.013mm are used as electrodes. Series connection must be assured to generate two in-phase bubbles. The details of the experimental arrangement and method can be found in Ref. [6]. The maximum radius of bubble is related to discharge voltage and the maximum radius of the bubbles is about 9.35 millimetres when the voltage is 210V (measured in an experiment where two bubbles don't influence each other). The inter-bubble distance is about 18.7 millimetres and the initial depth of bubble inception is $d \approx 9.35\text{mm}$. The numerical results are compared with the experimental observations and bubbles and free surface shapes are illustrated in figure 2.

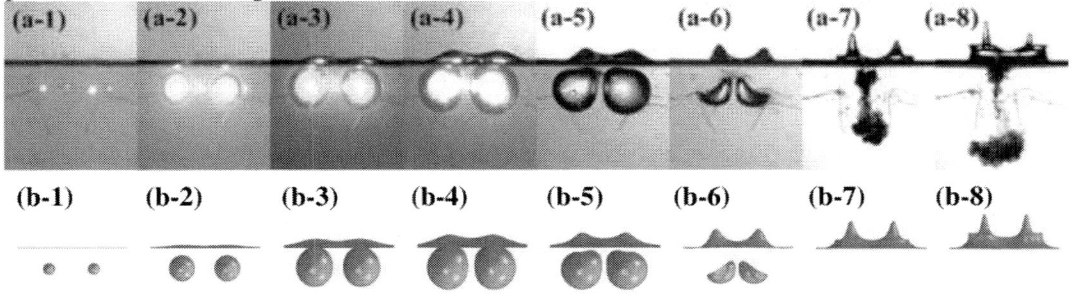


**Figure 2.** Comparison between (a) the experiment and (b) the BIM computation; the corresponding times in experiment are (a) 0.05ms, 0.45ms, 0.85ms, 1.60ms, 2.10ms, 3.00ms, 5.30ms, 8.60ms, and in computation are (b) 0.05ms, 0.31ms, 0.83ms, 1.46ms, 2.10ms, 3.06ms, 5.29ms, 7.11ms. The image width is 46mm and the height is 60mm.

It's shown that two bubbles are incepted simultaneously and then they begin to expand. After a while, free surface is slightly elevated by the expansion of bubbles which means the initial formation of the spike (shown in figure 2(a-3)). The bubbles attain the maximum radii at 1.600ms and the walls close to each other become flattened. Then the bubbles come into collapse phase and a downward jet is formed under the influence of the Bjerknes force. Meanwhile, the adjacent bubble restrains the bubble collapse and affects the direction of the water jet. Therefore, the water jet is directed away from the free surface with slight deviation to the neighbouring bubble, and oblique jets are obtained. The oblique jets are about to penetrate the bubbles at 2.10ms while the height of the spike keeps increasing affected by high-pressure region. It's noted that the jets hit the lower bubble wall (shown in figure 2(a-6) and figure 2(b-6)), because the presence of free surface dominates over the mutual interactions between two bubbles. In the experiment, each bubble collapses into two parts after the jets penetrate the bubbles. Then they continue to contract and attain a minimum volume at 3.20ms (results not shown here). Afterwards, most contents of bubbles move downward while the remnant contents slowly migrate upwards due to the weak buoyancy force. At 5.30ms a circle of the outer fluid rises to surround the middle spike, which is the 'crown phenomenon' mentioned above. And then the middle spike and the 'crown' continue to rise (see the profile of free surface at 8.60ms). The crown keeps rising and finally rolls up (results not shown here). In the BIM computation, the evolution of the free surface is further investigated after the jet impact and the 'crown phenomenon' in the experiment is obtained in the calculations. There is also a continuous increase in the height of the crown and the middle spike. It is reckoned that the 'crown phenomenon' is related to high-pressure region formed after the jet impact. Because the fluid is continually drawn in after the jets penetrate the bubbles, a large high-pressure region is formed but with a small pressure peak. A second spike is induced by the high-pressure region, which is the 'crown phenomenon' mentioned above. And the rim of the second spike rises fast, forming the crown around 8.60ms in the experiment.

The comparison is favourable with respect to the bubbles and free surface shapes and there are three reasons for the difference in the profiles and the time. Firstly, viscosity effect and surface tension are ignored. Next, after the jets penetrate the bubble wall, we neglect the influence of the bubbles on the evolution of the free surface. At last, the initial condition in the computation makes a difference in the results. Generally speaking, favourable comparisons of bubble and free surface shapes are observed and the validity of our model is confirmed.

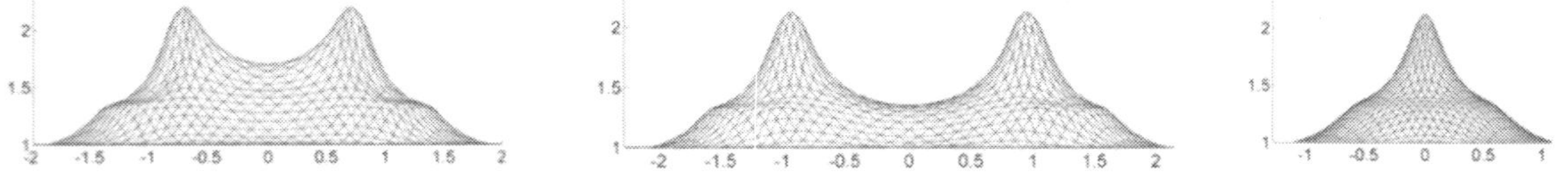

**Figure 3.** Crown spike induced by bubbles at the dimensionless time $T = 4.00$ with the inception depth $= R_m$. The dimensionless height of the central spike is (a) 1.18 when inter-bubble distance = 1.6 $R_m$ (b) 1.10 when inter-bubble distance = $2.0R_m$ and (c) 1.07 in single bubble case.

The crown spike shapes in single- and two-bubble cases at the same moment are illustrated in figure 3. In the computation, the crown phenomenon occurs earlier in two-bubble case than that in single-bubble case. If the inter-bubble distance decreases, a greater height of the central spike is noted. It's reckoned that the bubble-free-surface coupling effect is strengthened with the existence of another bubble and that more water volume is obtained in two-bubble case.

## 4. Conclusions

A 3D model is established to simulate crown spike of free surface in two-bubble case with a boundary integral method and the effect of bubbles after the jet impact is ignored. It's inferred that the crown is a second spike induced by the large high-pressure region formed after the jet impact. Experimental and numerical results are compared and the favorable agreements of bubbles and free surface shapes validate the present simplified model. Comparing the crown spike in single- and two-bubble cases, it's noted that the coupling effect between bubbles and free surface in two bubble case is stronger than that in single-bubble case. If the inter-bubble distance decreases, the coupling effect increases.

## 5. References

[1] Patrascioiu A, Florian C, Fernandez-Pradas J M, Morenza J L, Hennig G, Delaporte P and Serra P 2014 *Appl. Phys. Lett.* **105** 014010
[2] Kolsky H, Lewis J P, Sampson M T, Shearman A C and Snow C I 1949 *Proc. R. Soc. Lond. A* **196** 379
[3] Gibson D C 1968 *The 3rd Australasian Conf. on Hydraulic and Fluid Mechanics* November 25-29, 1968 (Sydney, Australia) p. 210
[4] Chahine G L 1977 *Trans. ASME Ser. I. J. Fluid Eng.* **99** 708
[5] Blake J R and Gibson D C 1981 *J. Fluid Mech.* **111** 123
[6] Zhang A M, Wang C, Wang S P and Cheng X D 2012 *Acta Phys. Sin.* **61** 084701 (in Chinese)
[7] Wang Q X, Yeo K S, Khoo B C and Lam K Y 1996 *Comput. Fluids* **25(7)** 607
[8] Dadvand A, Shervani-Taba M T and Khoo B C 2011 *Int. J. Adv. Manuf. Technol.* **56** 245
[9] Fong S W, Adhikari D, Klaseboer E and Khoo B C 2008 *Exp. Fluids* **46** 705
[10] Dadvand A, Khoo B C, Shervani-Tabar M T and Khalilpourazary S 2012 *Eng. Anal. Bound. Elem.* **36** 1595
[11] Li S, Zhang A M and Wang S P 2013 *Acta Phys. Sin.* **62(19)** 194703 (in Chinese)
[12] Han R, Zhang A M and Li S 2014 *Chin. Phys. B* **23(3)** 034703
[13] Newman J N 1977 Marine Hydrodynamics (1st Ed.) ( London: MIT Press) 131
[14] Zhang A M and Yao X L 2008 *Acta Phys. Sin.* **57** 339 (in Chinese)

**Acknowledgments**

This paper is funded by the International Exchange of Harbin Engineering University for Innovation-oriented Talents Cultivation. The authors would like to acknowledge the National Natural Science Foundation of China (Grant No. 51379039 and U1430236) for funding this work.

# Numerical simulation of bubble collapse and the transfer of vapor and noncondensable gas through the bubble interface using the ghost fluid method

**Y Jinbo[1*], K Kobayashi[1], M Watanabe[1] and H Takahira[2]**
[1] Division of Mechanical and Space Engineering, Hokkaido University, Sapporo 060-8628, Japan
[2] Department of Mechanical Engineering, Osaka Prefecture University, Sakai 599-8531, Japan
[*] E-mail: jinbo_y@eng.hokudai.ac.jp

**Abstract.** The ghost fluid method is improved to include heat and mass transfer across the gas–liquid interface during the bubble collapse in a compressible liquid. This transfer is due to both nonequilibrium phase transition at the interface and diffusion of the noncondensable gas across the interface. In the present method, the ghost fluids are defined with the intention of conserving the total mass, momentum, and energy, as well as the mass of each component while considering the heat and mass fluxes across the interface. The gas phase inside the bubble is a mixture of vapor and noncondensable gas, where binary diffusion between the mixture components is taken into account. The gas diffusion in the surrounding liquid is also considered. This method is applied to a simulation of a single spherical bubble collapse with heat and mass transfer across the interface in a compressible liquid. When noncondensable gas is present, it accumulates near the interface due to vapor condensation, thereby preventing further condensation. This results in a weaker bubble collapse than the case without noncondensable gas.

## 1. Introduction

Significant research efforts have recently been devoted to applying the extremely high pressure and temperature fields derived from bubble collapse to new environments for medical application and for the synthesis of new materials [1, 2]. Hatanaka [2] measured the amount of OH radicals dissolving into liquid from a single bubble oscillating in ultrasonic fields and showed that the amount increases under the "dancing bubble condition," where a single bubble is not stably located while oscillating, but rather moves around accompanied with splitting and coalescing. To clarify the mechanism behind the increase, it is necessary to construct a computational method that can simultaneously treat both nonspherical bubble collapse accompanied by complicated interfacial structure and heat and mass transfer through the interface. Although many numerical simulations have been conducted until now [3-5], these methods are considered to be insufficient for computing violent bubble collapse yielding chemical reactions whilst considering both the complicated structure and the heat and mass transfer.

In this study, a numerical method for compressible two-phase flows is presented that can treat nonspherical bubble collapse with complicated deformation of the interface accompanied by heat and mass transfer through the interface (i.e., diffusion of noncondensable gas and nonequiliblium phase transition of vapor). In our previous studies [6, 7], nonequilibrium phase transition following nonspherical bubble collapse was considered by developing the ghost fluid method (GFM) [8]. In this study, this method is extended to treating heat and mass transfer due to diffusion of noncondensable gas through an interface, while simultaneously treating the nonequilibrium phase transition.

A similar method has been recently constructed to solve the pressure and velocity differences at the interphase due to phase transition by Houim & Kuo [9]; their method employs the idea of the modified

GFM [10] which applies the solution of two-phase Riemann problem to the definition of the ghost fluid. Their method is considered to solve certain situations more appropriately than our method just as the modified GFM is known to perform better than the original GFM [10]. Here, however, we adopt the original GFM because of the simplicity of extending it to the multidimensional problem.

## 2. Numerical methods

Although the following presents the construction of the spherical 1-D computational method, it can be straightforwardly extended to the multidimensional simulation and treatment of the complicated interfacial structure, because the present method is constructed by extending the level set method (LSM) [11] and the GFM [8] which are respectively used for implicitly capturing the location of the interface and implicitly satisfying the boundary conditions at the interface.

### *2.1. Governing equations*

In the bubble interior, the Euler equations for the two species are solved with consideration for heat conduction and mutual diffusion of vapor and noncondensable gas. This is the same as in the computation of Matsumoto & Takemura [3] except that we neglect the fluid viscosity. For the bubble exterior, the Euler equations are solved; the distribution of the noncondensable gas dissolved in the liquid phase is obtained by solving the advection–diffusion equation with the velocity field of the liquid phase in a manner similar to the computation of Akhatov et al. [4]. The following stiffened gas equations of state [12] are employed for the noncondensable gas (air), the liquid (water), and its vapor:

$$p = (\gamma - 1)\rho(E - \epsilon) - \gamma\Pi, \qquad E = \frac{C_p}{\gamma}T + \frac{\Pi}{\rho} + \epsilon \tag{1}$$

where $p$ is the pressure, $\rho$ is the density, $E$ is the internal energy per unit mass, $T$ is the temperature and $\gamma$, $\Pi$, $\epsilon$, $C_p$ are the parameters for the characteristic of materials. The following values are used.

$$\text{water: } \gamma_l = 2.35, \Pi_l = 10^9\ \text{Pa}, \epsilon_l = -1167 \times 10^3\ \text{J/kg}, C_{pl} = 5.947 \times 10^3\ \text{J/(kgK)}$$
$$\text{vapor: } \gamma_v = 1.43, \Pi_v = 0\ \text{Pa}, \epsilon_v = 2030 \times 10^3\ \text{J/kg}, C_{pv} = 1.487 \times 10^3\ \text{J/(kgK)}$$
$$\text{air: } \gamma_g = 1.4, \Pi_g = 0\ \text{Pa}, \epsilon_g = 0\ \text{J/kg}, C_{pg} = 1.007 \times 10^3\ \text{J/(kgK)}$$

where subscript $l$, $v$ and $g$ represents the liquid, vapor and noncondensable gas, respectively.

Heat conduction obeys Fourier's law. The heat conductivity of each gas component linearly varies with temperature [13, 14]. The diffusion velocity is determined using Fick's law. The coefficient of binary diffusion of vapor and noncondensable gas is calculated using Fuller's equation [15]. The diffusion coefficient of noncondensable gas in liquid is $D_l = 1.76 \times 10^{-9}\ \text{m}^2/\text{s}$ [4]. The third-order TVD Runge–Kutta scheme and the third-order ENO-LLF scheme are used for time and space discretization of the Euler equations, respectively [16].

### *2.2. Numerical methods for gas–liquid interface*

The boundary conditions imposed on the gas–liquid interface are almost the same as the ones used in [9]; in the present simulation, heat and mass fluxes due to the nonequilibrium phase transition, $\dot{m}_v$ (positive for evaporation), and due to the diffusion of noncondensable gas through the interface, $\dot{m}_g$ (negative for dissolution), are considered while the fluid viscosity is neglected.

*2.2.1. Level set method [11].* The level set function $\varphi$ is used for capturing the interface location. When mass transfer through the interface is neglected, the level set function is usually advanced by the advection equation with the fluid velocity fields. When a phase transition is considered, the distribution of velocities normal to the interface is discontinuous at the interface due to the phase transition. Therefore, the level set function is advanced using the modified velocity fields $\boldsymbol{u}_R$ as follows:

$$\boldsymbol{u}_R = \boldsymbol{u} - \frac{\dot{m}_g + \dot{m}_v}{\rho_m^i}\boldsymbol{n}\ (\varphi > 0\text{: gas phase}), \qquad \boldsymbol{u}_R = \boldsymbol{u} - \frac{\dot{m}_g + \dot{m}_v}{\rho_l^i}\boldsymbol{n}\ (\varphi < 0\text{: liquid phase}) \tag{2}$$

where $\boldsymbol{u}$ is the velocity vector, $\boldsymbol{n}(=\nabla\varphi/|\nabla\varphi|)$ is the unit normal, subscripts $m$ represents the mixture, and superscript $i$ means the value at the interface.

*2.2.2. Ghost fluid method [8].* The GFM is used to prevent the fluid variables with discontinuous distributions across the interface from being numerically diffused. In the GFM, instead of explicitly applying the boundary conditions to the interface, the boundary conditions at the interface are implicitly satisfied by defining an artificial fluid (ghost fluid). The following definitions of the ghost fluid are obtained by extending the ghost fluid that is defined for modeling phase transitions to the treatment of the noncondensable gas diffusion through the interface.

$$\begin{aligned}
&\boldsymbol{u}_m^{\text{ghost}}\cdot\boldsymbol{n}=\boldsymbol{u}_l^{\text{real}}\cdot\boldsymbol{n}-\frac{\dot{m}_g+\dot{m}_v}{\rho_l^i}+\frac{\dot{m}_g+\dot{m}_v}{\rho_m^i}, p_m^{\text{ghost}}=p_m^{\text{ext}}, s_m^{\text{ghost}}=s_m^{\text{ext}},\\
&\boldsymbol{u}_l^{\text{ghost}}\cdot\boldsymbol{n}=\boldsymbol{u}_l^{\text{ext}}\cdot\boldsymbol{n}, p_l^{\text{ghost}}=p_m^{\text{real}}+\frac{(\dot{m}_g+\dot{m}_v)^2}{\rho_m^i}-\frac{(\dot{m}_g+\dot{m}_v)^2}{\rho_l^i}+\sigma\kappa, s_l^{\text{ghost}}=s_l^{\text{ext}},\\
&\boldsymbol{q}_m^{\text{ghost}}\cdot\boldsymbol{n}=\boldsymbol{q}_l^{\text{real}}\cdot\boldsymbol{n}-\dot{m}_vL_v-\dot{m}_gL_g+\frac{1}{2}\frac{(\dot{m}_g+\dot{m}_v)^3}{\rho_l^{i2}}-\frac{1}{2}\frac{(\dot{m}_g+\dot{m}_v)^3}{\rho_m^{i\,2}},\\
&\boldsymbol{q}_l^{\text{ghost}}\cdot\boldsymbol{n}=\boldsymbol{q}_m^{\text{real}}\cdot\boldsymbol{n}+\dot{m}_vL_v+\dot{m}_gL_g+\frac{1}{2}\frac{(\dot{m}_g+\dot{m}_v)^3}{\rho_m^{i\,2}}-\frac{1}{2}\frac{(\dot{m}_g+\dot{m}_v)^3}{\rho_l^{i2}},\\
&(\rho_g\boldsymbol{v}_g)_g^{\text{ghost}}\cdot\boldsymbol{n}=\frac{\rho_g}{\rho_m}\dot{m}_v-\frac{\rho_v}{\rho_m}\dot{m}_g, (\rho_g\boldsymbol{v}_g)_l^{\text{ghost}}\cdot\boldsymbol{n}=\dot{m}_g-c^i(\dot{m}_g+\dot{m}_v)
\end{aligned}\tag{3}$$

where $s$ is the entropy, $\sigma$ is the surface tension, $\kappa(=\nabla\cdot\boldsymbol{n})$ is the curvature of the interface, $\boldsymbol{q}$ is the heat flux vector, $L_v(=h_v^i-h_l^i$: $h$ is the specific enthalpy) is the latent heat accompanying the phase transition of the vapor, $L_g(=h_g^i-h_l^i)$ is the latent heat related to the dissolution and volatilization of the noncondensable gas, and $c$ is the concentration of the noncondensable gas dissolved in the liquid. The velocity of the mixture, $\boldsymbol{u}_m$, is the mass averaged velocity of each component and is related to the velocities of the individual components by $\boldsymbol{u}_g=\boldsymbol{u}_m+\boldsymbol{v}_g$ and $\boldsymbol{u}_v=\boldsymbol{u}_m+\boldsymbol{v}_v$, where $\boldsymbol{v}$ is the diffusion velocity. Superscripts "ghost," "real," and "ext" represent the ghost fluid value, the actual value at the point where the ghost fluid is defined, and the value extrapolated to the point from the other fluid across the interface, respectively. The ghost temperatures used for the calculation of $\boldsymbol{q}$ are defined considering the temperature jump, $\Delta T$, due to the phase transition as follows:

$$T_m^{\text{ghost}}=T_l^{\text{real}}-\Delta T,\qquad T_l^{\text{ghost}}=T_m^{\text{real}}+\Delta T\tag{4}$$

Although the mass flux can be determined from any phase transition model, the Hertz–Knudsen–Langmuir model [4] (with the accommodation coefficient of 0.055) is employed in this work. The temperature jump is determined by an equation derived from molecular gas dynamics [17]. The mass flux due to the noncondensable gas diffusion through the interface is determined as follows:

$$\dot{m}_g=D_l(c_l^i-c_{sat}),\qquad c_{sat}=H_np_g^i\tag{5}$$

where $H_n$ is Henry's law constant with the value of $1.481\times10^{-10}$ Pa$^{-1}$ [4].

## 3. Numerical results

The result obtained from the present method was compared with Fig. 7 in Matsumoto & Takemura [3]. It was confirmed that the present result was in good agreement with Matsumoto & Takemura.

Examples of the spherical symmetric computation are shown where a single vapor bubble containing a small amount of noncondensable gas collapses. Initial conditions are as follows: water pressure surrounding the bubble, $p_0$, is 0.1013 MPa; water density, $\rho_{l0}$, is 989.0 kg/m$^3$; initial temperature in the computational domain, $T_0$, is 296 K; and initial bubble radius, $R_0$, is 1 mm. Initially, the mass fraction of the mixture inside the bubble is spatially uniform and is equal to $\phi_0(=\rho_g/(\rho_g+\rho_v))$. The mass fraction is set to 0 (only vapor), 0.02, 0.05, and 0.1 in the following computations. Initial vapor pressure, $p_{v0}$, is the saturation pressure at temperature $T_0$. In the present simulation, adaptive zonal grids with 10 layers (where the finest grid spacing is $7.8125\times10^{-5}R_0$) is applied according to the bubble radius. The length of the computational domain is $200R_0$ in the radial direction. Symmetric boundary conditions are used at the origin of the spherical coordinates, and nonreflective boundary conditions are applied at the edge of the computational region.

Figure 1 presents the time histories of (a) bubble radii and (b) total mass of vapor inside the bubble, $\overline{M_v}$. Enlarged view of a specific time period is shown in the inset at the right-hand side of each figure. The four lines represent the data obtained for different initial mass fraction. After the bubbles oscillate for several periods, the energy of oscillation is dissipated and the bubbles stabilize at a certain radius when the bubbles contain a small amount of noncondensable gas. On the other hand, when the bubble contains only the vapor ($\phi_0 = 0$), the entire vapor is transformed into liquid due to condensation.

## 4. Conclusion

We demonstrated improved LSM and GFM to treat the gas–liquid interface followed by the diffusion of the noncondensable gas through the interface and the nonequilibrium phase transition. The spherical bubble collapse accompanied by the heat and mass fluxes through the bubble interface was reasonably simulated.

## References


[1] Umemura S, Kawabata K and Sasaki K 2007 *Elect. Commun. Japan Part III* **90** 63
[2] Hatanaka S 2011 *Proc. Symp. Ultrasonic Electronics (Kyoto, 8-10, Nov. 2011)* vol 32 p 41
[3] Matsumoto Y and Takemura F 1994 *JSME Int. J.* B **37** 288
[4] Akhatov I, Lindau O, Topolnikov A, Mettin R, Vakhitova N and Lauterborn W 2001 *Phys. Fluids* **13** 2805
[5] Szeri A J, Storey B D, Pearson A and Blake J R 2003 *Phys. Fluids* **15** 2576
[6] Jinbo Y and Takahira H 2012 *Trans. Jpn. Soc. Mech. Eng.* B **78** 1302
[7] Jinbo Y, Ogasawara T and Takahira H 2015 *Exp. Therm. Fluid Sci.* **60** 374
[8] Fedkiw R, Aslam T, Merriman B and Osher S 1999 *J. Comput. Phys.* **152** 457
[9] Houim R W and Kuo K K 2013 *J. Comput. Phys.* **235** 865
[10] Liu T G, Khoo B C and Yeo K S 2003 *J. Comput. Phys.* **190** 651
[11] Sussman M, Smereka P and Osher S 1994 *J. Comput. Phys.* **114** 146
[12] Saurel R, Petitpas F and Abgrall R 2008 *J. Fluid Mech.* **607** 313
[13] Prosperetti A, Crum L A and Commander K W 1988 *J. Acoust. Soc. Am.* **83** 502
[14] Spencer B B, Wang H and Anderson K K 2000 *Oak Ridge National Laboratory/TM-2000* **285**
[15] Fuller E N, Schettler P D and Giddings J C 1966 *Ind. Eng. Chem.* **58** 18
[16] Shu C -W and Osher S 1989 *J. Comput. Phys.* **83** 32
[17] Nigmatulin R I 1991 *Dynamics of Multiphase Media* vols 1 and 2 (New York: Hemisphere)


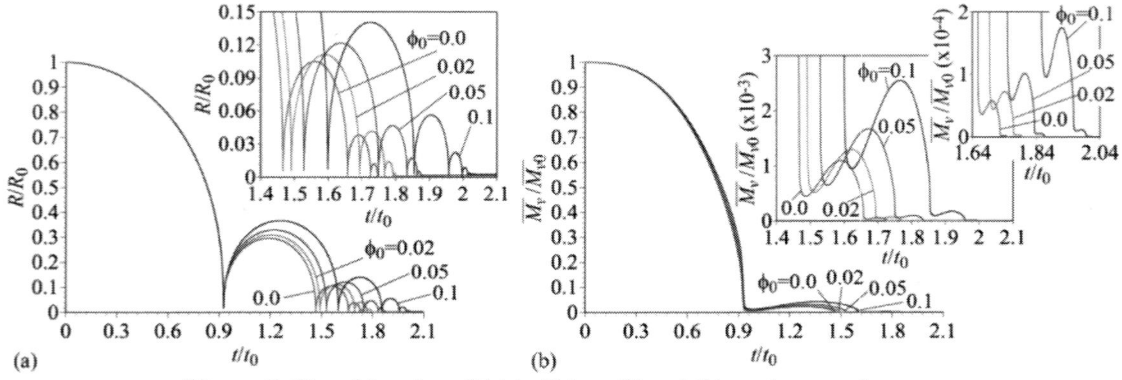


**Figure 1.** Time histories of (a) bubble radii and (b) total mass of vapor.

# Bubble dynamics in soft materials: Viscoelastic and thermal effects

**Eric Johnsen and Lauren Mancia**

Mechanical Engineering Department, University of Michigan, Ann Arbor, MI 48019, USA

E-mail: `ejohnsen@umich.edu`

**Abstract.**

The oscillations of a single spherical bubble in a soft surrounding medium are investigated numerically. In particular, the combined effects of the viscoelasticity and compressibility of the surroundings, as well as subsequent heating, on the bubble dynamics are quantified for forced and free collapse. In a Keller-Miksis framework, a Kelvin-Voigt viscoelastic model with full thermal effects is considered, in which the elastic term is represented by a Neo-Hookean model to account for the finite strains. The history and spatial distribution of the stresses and temperatures produced in the surroundings are examined and related to potential cavitation damage mechanisms.

## 1. Introduction

Over the past century, significant efforts were dedicated to understanding bubble dynamics in water [3], primarily with application to naval engineering. In particular, damage thereby inflicted to neighboring solids was one of the main motivating factors. More recently, applications have emerged in which cavitation occurs in soft materials, such as in chemical processing, botany and medicine. The knowledge acquired from our experience in cavitation in water does not directly translate to bubble dynamics in soft matter, due to additional physics (viscoelastic properties such as relaxation and elasticity that are not present in water) and corresponding time scales. Furthermore, experiemental characterization of cavitation in soft materials can be challenging [7]. In this work, our goal is to understand the effect of viscoelasticity on the bubble dynamics when taking into account thermal and compressibility effects, and assess potential damage mechanisms by examining the stress and temperature fields in the bubble surroundings.

## 2. Model

Our numerical approach to modeling the dynamics of a single spherical bubble in soft materials [5] is based on the compressible Keller-Miksis equation, alongside which the energy equation is solved inside and outside the bubble as in [4]. Furthermore, a generalized constitutive relation that includes linear and nonlinear elasticity and relaxation is implemented; in the most general form, a spectral collocation method is used to solve the partial differential equations for the stresses in the surrounding medium (and the energy equation). For this particular study, we assume that the surroundings are viscoelastic following a Kelvin-Voigt formulation – the total deviatoric stress consists of the sum of viscous and elastic stresses. Moreover, to account for

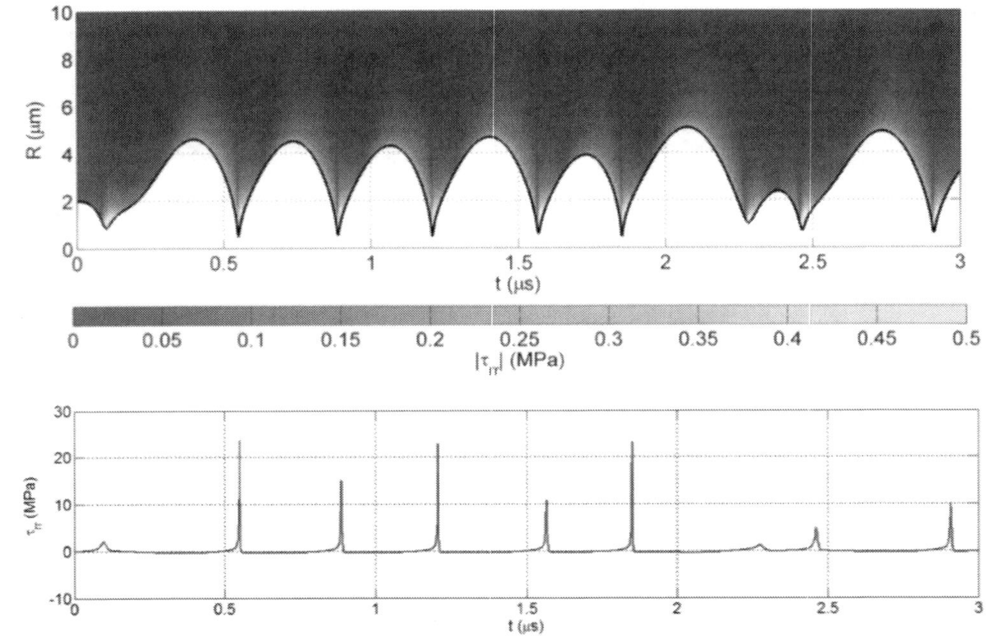


**Figure 1.** Time history of the total deviatoric stresses as a function of radial distance from the origin. The white outline denotes the radius vs. time. In the top, the absolute value of the stresses is reported on a logarithmic scale. In the bottom, the stresses in the medium just at the bubble wall are plotted.

the large deformations/finite strains experienced by the surrounding medium, a neo-Hookean (nonlinear elastic) model [1] is considered.

## 3. Results

An air bubble of initial radius 2 $\mu$m is driven by a sinusoidally varying far-field pressure waveform with amplitude 1 MPa and frequency 3 MHz. The viscosity of the surroundings is 5 mPa·s and the shear modulus 100 kPa, which fall within the range of values used to represent soft tissue [2, 6]. Fig. 1 shows the bubble response as a function of time (outlined in white), as well as the absolute value of the radial deviatoric stress in the surrounding medium (color contours), given by

$$\tau_{rr} = \tau_{rr}^{(visc)} + \tau_{rr}^{(elastic)} = -\frac{2}{3}G\left[\left(\frac{r_o}{r}\right)^4 - \left(\frac{r}{r_o}\right)^2\right] - 4\mu\frac{R^2\dot{R}}{r^3}, \tag{1}$$

where $G$ is the linear shear modulus, $\mu$ the viscosity and

$$r_o(r,t) = (r^3 - R^3 + R_o^3)^{1/3}, \tag{2}$$

with $r$ being the radial coordinate, $R$ the bubble radius and $R_o$ the initial radius. While large at collapse and near the bubble, the stresses die out rapidly into the surrounding medium.

The bottom plot in Fig 1 shows the time history of the stress in the surroundings right at the bubble wall, by placing a Lagrangian particle there initially and following it. Due to the incompressibility of the near field (and thus conservation of volume), these particles get "closer" to the bubble wall when the bubble radius is larger than initial, and "farther" when the bubble radius is smaller than initial. For these oscillation amplitudes, substantial stresses are observed,

Journal of Physics: Conference Series **656** (2015) 012022 doi:10.1088/1742-6596/656/1/012022

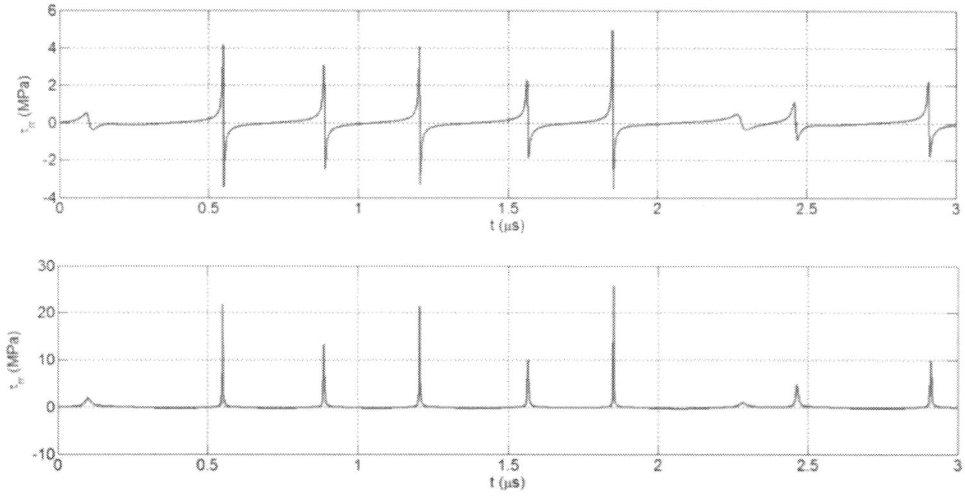


**Figure 2.** Time history of the viscous (top) and elastic (bottom) stresses.

mostly at collapse. It is worthwhile to note that for a given cycle the stresses are larger during the collapse phase than they are during growth.

Fig. 2 decomposes the radial deviatoric stresses at the bubble wall into viscous (top) and elastic (bottom) components. Overall, the elastic term dominates, particularly at collapse, and is primarily positive, thus reducing the tendency of the bubble to deviate from equilibrium. Due to its dependence on $\dot{R}$, the viscous term can exhibit relatively large positive and negative values. The relative importance of the viscous and elastic stresses thus strongly depend on the oscillations characteristics ($R$ and $\dot{R}$), which themselves depend on the viscoelastic properties, as well as the driving pressure (amplitude and frequency). Larger stresses further give rise to larger amounts of energy dissipation and localized heating.

Fig. 3 shows the time history of the temperature as a function of radial distance from the origin, as well as line-outs for Lagrangian particles starting at different distances away from the bubble. For relatively violent collapses, significant temperatures are achieved (nearly 80$^o$C) right at the bubble wall. However, these temperatures decrease rapidly away from the bubble wall. Nevertheless, appreciable heating is observed in the surroundings over time: for a particle initially 1 $\mu$m away from the bubble, the temperature increases by nearly 1$^o$C after 4 $\mu$s. Under these circumstances, the heating produced by the successive collapses accumulates.

## 4. Conclusions

The oscillations of a single spherical bubble in a soft surrounding medium are investigated numerically. In a Keller-Miksis framework, a Kelvin-Voigt viscoelastic model with full thermal effects is considered, in which the elastic term is represented by a Neo-Hookean model to account for the finite strains. For the problem under consideration, viscous and elastic contributions are substantial; their relative importance depends on the oscillations characteristics (bubble radius and velocity), which themselves depend on the forcing and material properties. A more detailed study of the dependence of the bubble dynamics and stress/temperature fields on these quantities will be presented at the meeting.

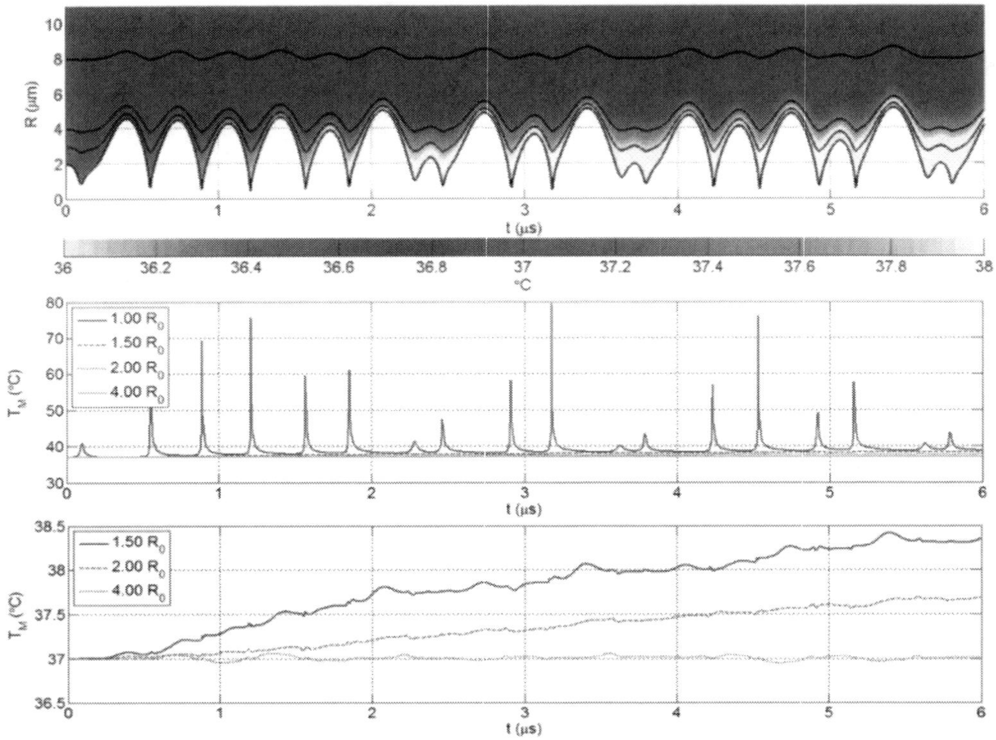


**Figure 3.** Time history of the temperature as a function of radial distance from the origin. The white outline denotes the radius vs. time in the top plot, and the radial distance is vertically upward; the color scale represents the temperature. The middle and bottom plots are line-outs of the temperature for Lagrangian particles starting at different initial distances away from the bubble, on different scales.

## 5. Acknowledgements

This work was supported in part by NSF grant number CBET 1253157, NIH grant number 1R01HL110990-01A1 and ONR grant N00014-12-1-0751 (under Dr. Ki-Han Kim).

## References

[1] R. Gaudron, M. T. Warnez and E. Johnsen, Bubble dynamics in a viscoelastic medium with nonlinear elasticity, J. Fluid Mech. 766 (2015) 54–75.
[2] C. Hua and E. Johnsen, Nonlinear oscillations following the Rayleigh collapse of a gas bubble in a linear viscoelastic (tissue-like) medium, Phys. Fluids 25 (2013) 083101.
[3] M. S. Plesset and A. Prosperetti, Bubble dynamics and cavitation, Ann. Rev. Fluid Mech. 9 (1977) 145–185.
[4] L. Stricker, A. Prosperetti and D. Lohse, Validation of an approximate model for the thermal behavior in acoustically driven bubbles, J. Acoust. Soc. Am. 130 (2011) 3243–3251.
[5] M. T. Warnez and E. Johnsen, Numerical modeling of bubble dynamics in viscoelastic media with relaxation, submitted (2015).
[6] X. Yang and C. C. Church, A model for the dynamics of gas bubbles in soft tissue. J. Acoust. Soc. Am. 118 (2005) 3595–3606.
[7] J. A. Zimberlin, N. Sanabria-DeLong, G. N. Tew and A. J. Crosby, Cavitation rheology for soft materials, Soft Matter 3 (2007) 763–767.

# Study on Pressure Wave Propagation in a Liquid Containing Spherical Bubbles in a Rectangular Duct

**Junya Kawahara, Masao Watanabe and Kazumichi Kobayashi**

Division of Mechanical and Space Engineering, Hokkaido University, N13W8, Kita-ku, Sapporo 060-8628, Japan

E-mail: junyakawahara@frontier.hokudai.ac.jp

**Abstract.** Pressure wave propagation in a liquid containing several bubbles is numerically investigated. We simulate liner plane wave propagation in a liquid containing 10 spherical bubbles in a rectangular duct with the equation of motion for $N$ spherical bubbles. The sound pressures of the reflected waves from the rigid walls are calculated by using the method of images. The result shows that the phase velocity of the pressure wave propagating in the liquid containing 10 spherical bubbles in the duct agrees well with the low-frequency speed of sound in a homogeneous bubbly liquid.

## 1. Introduction

The presence of gas bubbles dispersed in a liquid affects profoundly the acoustic behaviour of the liquid. For example, the low-frequency speed of sound in a bubbly liquid is lower than that in a pure liquid. In the previous studies [1, 2, 3], the physical mechanism of the reduction of the low-frequency speed of sound has been explained in a liquid containing bubbles where the assumption of homogeneous medium holds. In general, the speed of sound in a continuum medium is defined as $a = \sqrt{1/\kappa\rho}$, where $\kappa$ is the compressibility and $\rho$ is the density of the medium. The compressibility of a homogeneous bubbly liquid is primarily determined by the amount of bubbles and the density by the amount of liquid. Bubbles greatly increase the compressibility of a homogeneous bubbly liquid, but hardly affect the density. High compressibility and density yield a low speed of sound.

The purpose of the present paper is to answer the question: *will the low-frequency speed of sound be reduced in a liquid containing several bubbles where the assumption of homogeneous medium no longer holds?* In the present study, we simulate pressure wave propagation in a liquid containing 10 spherical bubbles in a rectangular duct with the equation of motion for $N$ spherical bubbles, in which the bubble-bubble interactions are taken into account. We assume that bubble size is much smaller than the wavelength of the pressure wave; hence, we restrict ourselves to low-frequency wave propagation in a bubbly liquid [3]. We compare the low-frequency speed of sound in a homogeneous bubbly liquid with the phase velocity of the pressure wave propagating in the duct.

## 2. Problem Statement

Let us simulate pressure wave propagation in a liquid containing 10 spherical bubbles in a rectangular duct [Fig. 1(a)]. The bubbles have the same initial radius $R_0$, and are placed at

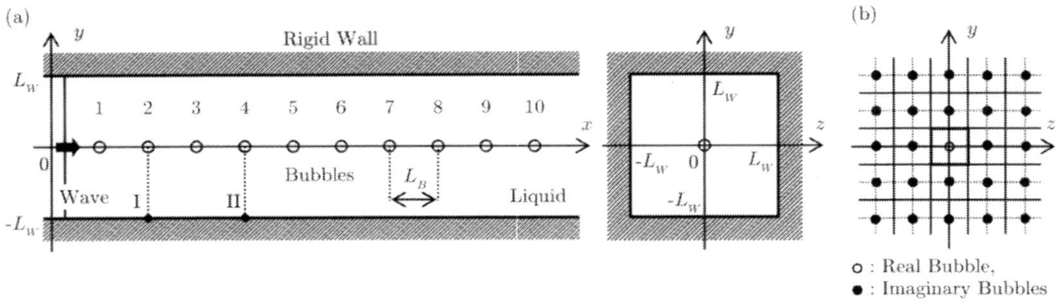


**Figure 1.** (a) 10 spherical bubbles are placed at equal intervals in a rectangular duct. (b) Schematic of distribution of imaginary bubbles in the $y - z$ plane.

equal intervals on the center axis of the duct. The bubbles are excited by a linear plane wave started to propagate from $x = 0$. The input pressure wave having $p_{\ell 0} + P_{\rm in}(t,x)$ propagates at the speed of sound in a liquid-phase, where $t$ is the time and $p_{\ell 0}$ is the pressure of the liquid in the initial undisturbed state. The sound pressure of the input pressure wave $P_{\rm in}(t,x)$ is $P_A[1-\cos\omega(t-x/a_\ell)]$, where $P_A$ is the amplitude of the input pressure wave ($P_A/p_{\ell 0} \ll 1$), $\omega$ is the angular frequency of the input pressure wave, and $a_\ell$ is the speed of sound in a liquid-phase. The delay time $x/a_\ell$ denotes the time it takes for the pressure wave to propagate the distance $x$ at the speed $a_\ell$. The angular frequency $\omega$ is much smaller than the natural angular frequency of the single bubble.

In the present study, the following assumptions are made: (i) the liquid flow is a potential flow. (ii) The translation and deformation of the bubbles are ignored. (iii) The bubble-bubble interactions are taken into account. (iv) The bubbles are filled with a non-condensable gas, which follows the isothermal process, without phase changes. (v) The bubble radius $R(t)$ is large enough so that the surface tension is negligible. (vi) All attenuations including the acoustic radiation are ignored.

We shall use the equation of motion for $N$ spherical bubbles in a sound field [4]. Under the present assumptions, we linearize the equation of motion,

$$\ddot{R}'_I(t) + \sum_{J=1,\ J\neq I}^{N} R_0 \frac{\ddot{R}'_J(\zeta_{JI})}{r_{IJ}} + \omega_B^2 R'_I(t) = -\frac{P_{\rm in}(t,x_I)}{\rho_\ell R_0}, \tag{1}$$

where $R'(t) = R(t) - R_0$ is the small fluctuation of the bubble radius ($|R'|/R_0 \ll 1$), the subscripts $I$ and $J$ are the bubble indices ($I, J = 1, 2, \cdots, N, I \neq J$), $x_I$ is the $x$ coordinate of the center of bubble $I$, $r_{IJ}$ is the distance between the centers of bubbles $I$ and $J$, $\rho_\ell$ is the density of a liquid-phase, and $\zeta_{JI} = t - [r_{IJ} - R_J(\zeta_{JI})]/a_\ell$. The delay time $[r_{IJ} - R_J(\zeta_{JI})]/a_\ell$ denotes the time it takes for the radiative wave generated by the radial oscillation of bubble $J$ to propagate the distance $r_{IJ}$ at the speed $a_\ell$. The natural angular frequency of the single bubble $\omega_B$ is $\sqrt{3p_{\ell 0}/(\rho_\ell R_0^2)}$ ($\omega/\omega_B \ll 1$). The second term in the left-hand side of Eq. (1) represents the bubble-bubble interactions. In the limit as $r_{IJ} \to \infty$, or $N \to 1$ (when the bubble-bubble interactions are ignored), Eq. (1) is reduced to the equation of linear oscillation for a single spherical bubble [4].

The radiative waves generated by the radial oscillations of the bubbles in the rectangular duct are reflected from the rigid walls. We calculate the sound pressures of the reflected waves by using the method of images. We place imaginary bubbles in a two-dimensional array in the $y - z$ plane as shown in Fig. 1(b).

Now, we calculate the pressure field at the point $Q(x, y, z)$. The total sound pressure $P_Q$ is given by the summation of $P_{\rm in}(t, x)$ and the sound pressures generated by the radial oscillations of the real and imaginary bubbles. It should be emphasized here that during a given period of time $T$, the radiative waves generated by the radial oscillations of the bubbles propagate only for a distance $a_\ell T$; therefore, we calculate only the sound pressures generated by the radial oscillations of the real and imaginary bubbles in the distance $a_\ell T$ from the point $Q$. The sound pressure generated by the radial oscillation of a real, or imaginary bubble $k$ is written by

$$P_{Qk}(t) = -\rho_\ell \frac{\partial \phi_k}{\partial t} = \rho_\ell R_0^2 \frac{\ddot{R}'_k(t - r_k/a_\ell)}{r_k}, \tag{2}$$

where $\phi_k$ is the velocity potential for the flow field generated by the radial oscillation of bubble $k$ given by ref. [5] and $r_k$ is the distance between the center of bubble $k$ and the point $Q$. Hence, the total sound pressure $P_Q$ can be calculated by using the following equation:

$$P_Q(t) = P_{\rm in}(t, x) + \sum_k \rho_\ell R_0^2 \frac{\ddot{R}'_k(t - r_k/a_\ell)}{r_k}. \tag{3}$$

The total sound pressure in a water at 20°C is calculated by numerically solving Eqs. (1) and (3) under the conditions with $p_{\ell 0}$, $a_\ell$, and $\rho_\ell$ of 101.3 kPa, 1483 m/s, and 998.2 kg/m$^3$, respectively. The bubble 1 is placed at $x = L_B$. The distance between adjacent bubbles $L_B$ is 50.0 mm. The nondimensional amplitude $P_A/p_{\ell 0}$ and the nondimensional angular frequency $\omega/\omega_B$ of the input pressure wave are $1.82 \times 10^{-3}$. The ordinary differential equations (1) are solved numerically by using the 4th order Runge-Kutta method.

## 3. Numerical Results and Discussion

We compare the low-frequency speed of sound in a homogeneous bubbly liquid with the phase velocity of the pressure wave propagating in the duct. The phase velocity is defined as follows. Let $P_{\rm I}(t)$ and $P_{\rm II}(t)$ be the total sound pressures at points I and II (points $x = 2L_B$ and $4L_B$ on the rigid wall), respectively. From the simulation results, we find $t'$ that satisfies the equation $P_{\rm I}(t) = P_{\rm II}(t')$ as a function of $t$, and define the phase velocity as $a_{\rm p}(t) = 2L_B/(t' - t)$ in a time $t = 10L_B/a_\ell$ when the input pressure wave arrived at the center of the bubble 10. The low-frequency speed of sound in a homogeneous bubbly liquid depends on the void fraction. In the present study, we define the void fraction $\beta$ as the ratio of the volume of the bubble and the domain around the middle point between adjacent bubbles:

$$\beta = \frac{4\pi R_0^3/3}{L_B \times 2L_W \times 2L_W}. \tag{4}$$

We conducted the simulations with various void fractions by changing either $L_W$ ($R_0 = 1.00$ mm) or $R_0$ ($L_W = 10.0$ mm). Figure 2(a) shows the time evolution of the total sound pressures at the points I and II in the case of $R_0 = 1.00$ mm and $L_W = 10.0$ mm. The present result is normalized by the natural period of the single bubble: $2\pi/\omega_B$. It should be emphasized here that we investigate the pressure wave development in the period of the order of $2\pi/\omega_B$, which corresponds to the only 0.1% of the period of the input pressure wave. In the case of $\beta = 2.09 \times 10^{-4}$, the phase velocity of the pressure wave propagating in the duct is 639 m/s.

The low-frequency speed of sound in a homogeneous bubbly liquid is often modeled by an equation attributed to Wood's equation [2, 3, 6],

$$a_{\rm mlf} = \left[ \frac{(1-\beta)^2}{a_\ell^2} + \frac{\beta^2}{a_g^2} + \beta(1-\beta)\frac{\rho_g^2 a_g^2 + \rho_\ell^2 a_\ell^2}{\rho_\ell \rho_g a_\ell^2 a_g^2} \right]^{-1/2}, \tag{5}$$

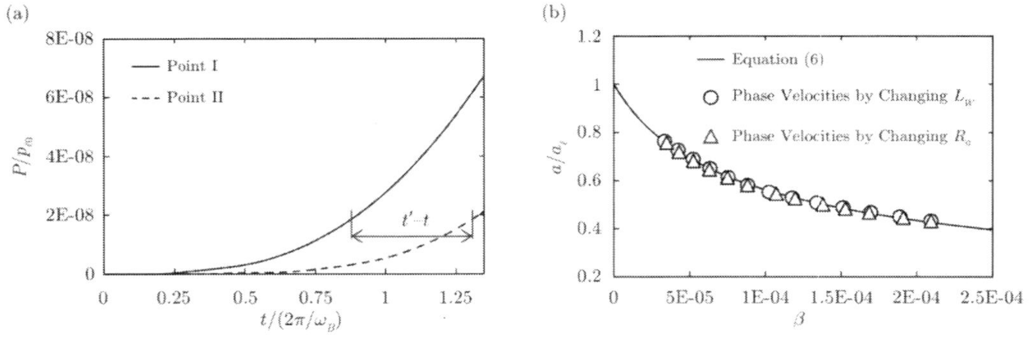


**Figure 2.** (a) Time evolution of the total sound pressures at the points I and II. (b) Relation between Eq. (6) and the phase velocities of the pressure wave.

where $a_g$ is the speed of sound in a gas-phase and $\rho_g$ is the density of a gas-phase. When the relation: $a_g^2 = p_{\ell 0}/\rho_g$ is used and the conditions: $\beta^2 \ll (\rho_g a_g^2)/(\rho_\ell a_\ell^2) \ll 1$ are imposed in Eq. (5), we derive the following equation,

$$a_{\mathrm{mlf}} = a_\ell \left(1 + \frac{\rho_\ell a_\ell^2}{p_{\ell 0}} \beta\right)^{-1/2}. \tag{6}$$

Figure 2(b) shows both the numerically obtained phase velocities and the low-frequency speed of sound by Eq. (6). The phase velocities of the pressure wave propagating in the duct agree well with Eq. (6). From these results, we conclude that in a liquid containing 10 spherical bubbles in a rectangular duct, the phase velocity of the pressure wave propagating in the media is reduced from the speed of sound in a liquid-phase with the increase in the void fraction defined by Eq. (4).

## 4. Conclusions

Pressure wave propagation in the liquid containing 10 spherical bubbles in the rectangular duct was simulated with the equation of motion for $N$ spherical bubbles. The sound pressures of the reflected waves from the rigid walls were calculated by using the method of images. The phase velocity of the pressure wave propagating in the liquid containing 10 spherical bubbles in the duct agrees well with the low-frequency speed of sound in a homogeneous bubbly liquid.

## Acknowledgments

This work was supported by JSPS KAKENHI Grant Numbers 26·1417, 26630043.

## References

[1] Mallock A 1910 *Proc. Roy. Soc.* A vol 84 (London: Royal Society) p 391
[2] Wood A B 1930 *A Textbook of Sound* (New York: Macmillan) p 326
[3] Wilson P S and Roy R A 2008 *Am. J. Phys.* **76** 975
[4] Takahira H, Yamane S and Akamatsu T 1995 *JSME Int. J.* B **38** 432
[5] Keller J B and Kolodner I I 1956 *J. Appl. Phys.* **27** 1152
[6] van Wijngaarden L 1972 *Annu. Rev. Fluid Mech.* **4** 369

Journal of Physics: Conference Series **656** (2015) 012024 doi:10.1088/1742-6596/656/1/012024

# Structure dynamics of cavitative magma flow in a volcano channel at explosive eruption

**V Kedrinskiy**

Lavrentyev Institute of Hydrodynamics, SB RAS, Novosibirsk, RF

E-mail: kedr@hydro.nsc.ru

**Abstract.** The effect of the density of cavitation nuclei on the state and flow structure of the magma melt in the case of its explosive decompression is numerically studied within the framework of the gas-dynamic model of mechanics of multiphase media. It is demonstrated that a zone with anomalously high values of the basic characteristics of the magma state is formed in the vicinity of the free surface of the magma column as the density of cavitation nuclei increases. A drastic increase in the particle velocity up to 150 m/s on the lower boundary of this zone testifies to a high probability of its separation from the main flow. Within the framework of the numerical scheme, this separation is "realized," and the dynamic behavior of the magma column remaining in the conduit is analyzed. It is shown that the detected effects of anomalous zone formation and possibility of separation turn out to be fairly stable, which allows one to define them as a "self-sustained regime of cyclic ejections."

## 1. Introduction

A comparison of results of experimental studies and numerical models of the dynamics of the states of conventional fluids under shock wave (SW) loading [1] with numerous investigations of natural processes (detailed reviews [2] should be noted) accompanying explosive volcanic eruptions allows an important conclusion to be drawn. The fundamental problems of the dynamics of the magma state in decompression waves [3], its transition from the state of a single-phase melt to the gas-particle state (formation of ash clouds), and the cyclic character of ejections are directly related to high-velocity hydrodynamics of multiphase media. Previous investigations show that the dynamics of the state of the magma melt in the gravity field is characterized by intense development of the cavitation process [3]. However, the problem of modeling the final stage of the process (destruction of the flow), which was described in [4] for conventional fluids under SW loading and in [5,6] for open volcanic systems, is still far from being solved despite numerous experimental studies based on various models of the viscous magma. Another aspect (cyclic character of ejections) is also rather problematic; some of the solutions are obtained on the basis of studying physically obvious models of the slug mode, effects of slipping/adhesion of the cavitating flow on the conduit walls, and elastic properties of the system consisting of the volcanic chamber and the conduit [2, 7-9]. In the proposed paper, the challenge is to solve a more general problem, i.e., to find mechanisms responsible for the formation of discontinuities in the magma flow and, as a consequence, for the discrete character of ejection of ash clouds.

Journal of Physics: Conference Series **656** (2015) 012024 doi:10.1088/1742-6596/656/1/012024

## 2. Two-phase mathematical model of state dynamics of cavitating magma

It is known that eruptions of closed volcanic systems can be initiated by a sudden decrease in pressure on the crater-conduit interface; as a result, a decompression wave (DW) starts to form in the high-pressure magma melt and propagates inward the conduit. As a consequence, the gas in the form of cavitation bubbles is released in the process of diffusion behind the DW front in the melt saturated by the gas: a cavitating magma flow is formed in the volcanic conduit. An interesting example for analyzing such systems is the St Helens volcano (eruption of May 1980). The point is that the scheme of its pre-explosion state reconstructed in [10] turned out to be absolutely adequate to the pulsed rarefaction tube (PRT) scheme. This fact is important, first of all, from the viewpoint of verification of results of experimental modeling of the magma state dynamics. It were the PRT-method that were used in the 1980s to study essentially nonlinear processes of cavitation destruction of distilled water under SW loading as a reference sample of heterogeneous media with free gas microbubbles. Experimental investigations by the diffraction-optical method of laser radiation scattering on microinhomogeneities combined with SW loading showed that their number contained in distilled water was of the order of $10^6$ $cm^{-3}$ [11]. A similar order ($10^{12-14}$ $m^{-3}$) of the density of micropores was observed in solidified lava samples.

Thus, unsteady high-velocity processes that occur under pulsed loading of distilled water [1, 11] can be considered under certain conditions as analogs of natural volcanic processes from the viewpoint of both possible mechanisms of their initiation and flow dynamics. In constructing the mathematical model of the magma state dynamics in the volcanic conduit, it is reasonable to take the basic model in the form of the Iordanskii-Kogarko-van Wijngaarden model [1,3] with the Navier-Stokes equation for the averaged velocity $v$, density $\rho$, and pressure $p$ at a variable viscosity of the medium $\mu$ (more detailed see in [3]):

$$\frac{\partial \rho}{\partial t} + \frac{\partial(\rho v)}{\partial z} = 0; \qquad \frac{\partial v}{\partial t} + v\frac{\partial v}{\partial z} = -\frac{1}{\rho}\frac{\partial p}{\partial z} - g + \frac{1}{\rho}\frac{\partial}{\partial z}\Big(\mu\frac{\partial v}{\partial z}\Big),$$

supplemented by the kinetics equations of dynamic processes inherent in volcanic eruptions:

by the Tait equation of state and by the definition of the magma melt viscosity

$$p = p_0 + \frac{\rho c^2}{n}\Big\{\Big(\frac{\rho}{\rho_0(1-k)}\Big)^n - 1\Big\}, \quad \mu = \mu^* \exp\{E_\mu(C)/(k_{\rm B}T)\}$$

by the definition of the frequency of homogeneous spontaneous nucleation and by the Rayleigh equation describing the dynamics of a single bubble in a viscous liquid

$$J = J^* \exp(-W_*/(k_{\rm B}T)), \quad R\ddot{R} + (3/2)\dot{R}^2 = \rho^{-1}(p_{\rm g} - p) + 4\nu R^{-1}\dot{R},$$

by the equations of state of the gas in the bubble (the gas is assumed to be ideal) and of the gas diffusion in the melt

$$(4/3)p_{\rm g}R^3 = (m_{\rm g}/M)k_{\rm B}T, \quad \frac{dm_{\rm g}}{dt} = 4\pi R^2\rho D\Big(\frac{dC}{dr}\Big)_R = 4\pi R\rho D(C_i - C^{\rm eq}(p_{\rm g})),$$

The effect of the density of cavitation nuclei on the state and flow structure of the magma melt in the case of its explosive decompression behind the DW front is numerically studied in a 1-km column of the compressed magma melt saturated by the gas. At the lower boundary of the magma column the pressure is set and maintained at a constant value of 170 MPa; the dissolved gas concentration is assumed to be 5.2 % wt. The distributions of these variables along the column (z axis) are determined by hydrostatics. At the time t = 0, the pressure on the upper boundary of the magma column is instantaneously reduced to 0.1 MPa.

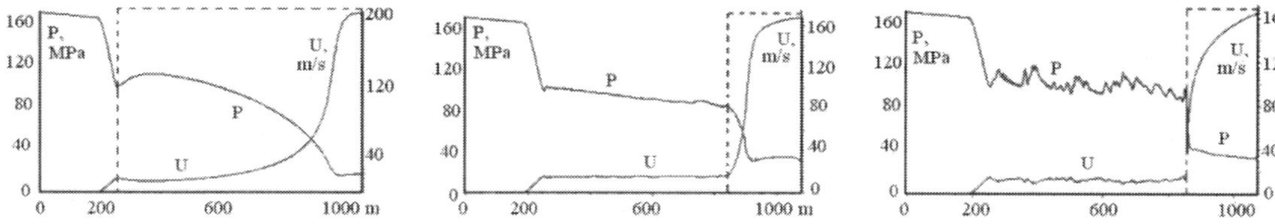


**Figure 1.** Profiles of decompression wave P(z) and structures of mass velocity field U(z) at t=0.5 s. Effect of the density of cavitation nuclei saturation $N_b$, $m^{-3}$ = $10^9$,$10^{11}$ and $10^{13}$ on the anomalous zone formation (from left to right, correspondingly).

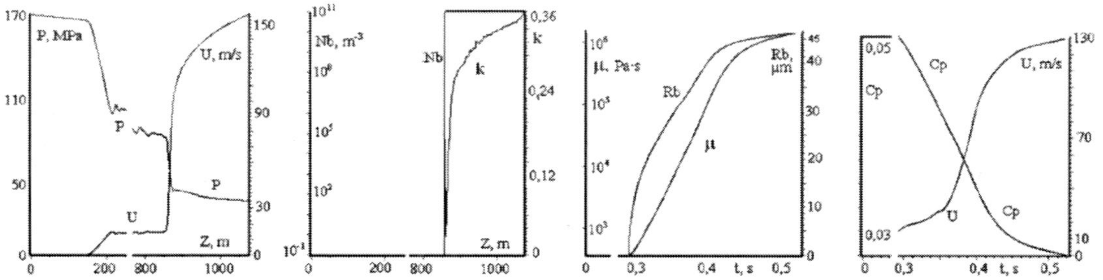


**Figure 2.** Decompression wave P(z), mass velocity U(z), density of cavitation nuclei $N_b$(z) and gas phase concentration k(z) distributions at t=0.54 s. Dynamics of bubble radii R(t), viscosity $\mu$(t), Pas, mass velocity U(t) and loses of dissolved gas $C_p$(t) by melt in a Lagrange-cross section.

The structures of the wave fields P(z), MPa, and particle velocity fields U(z), m/s, formed by the time t=0.5 s are shown in Fig. 1 for three limiting values of the cavitation nuclei density $N_b$,$m^{-3}$ (from left to right) : $10^9$ (spontaneous nucleation), $10^{11}$, and $10^{13}$ (homogeneous-heterogeneous nucleation). Here z, m is the abscissa axis; the values of the pressure (z) and particle velocity U(z) are plotted on the ordinate axes. The value z=0 corresponds to the lower boundary of the magma column. The dashed curves in each of the plots P(z)-U(z) show the saturation zones where the density of cavitation bubbles reaches the above-given limiting values.

It can be easily seen that the saturation zone size becomes substantially smaller (owing to redistribution of diffusion flows). The particle velocity distribution in the saturation zones becomes essentially different: from its gradient growth in the case of spontaneous nucleation to a jump in the particle velocity gradient ($N_b$= $10^{13}$ $m^{-3}$).

The data in Fig. 2 are presented in the distributions form of P(z), U(z), N(z), and k(z) at the time t=0.54 s and in the dynamic form as the cavitation bubble radii $R_b$(t), melt viscosity $\mu(t)$, loss of the dissolved gas from the melt $C_p$(t), and particle velocity U(t) in the Lagrangian cross section of the flow with the initial coordinate z = 900 m in the time interval up to 0.54 s. For a more compact presentation, the plots have breaks approximately 500 m long over the z axis and 0.1 s over the t axis.

The results obtained show that an increase in the cavitation nuclei density by 3-4 orders of magnitude with respect to spontaneous nucleation data leads to the formation (in the vicinity of the free surface of the magma column) of a saturation zone with anomalously high parameters (SZ) of the basic characteristics determining the state of the cavitating magma flow. In the SZ the gas mass in cavitative bubbles becomes stable, the diffusion processes come to a stop (see function $C_p$(t), Fig.2), the pressure reaches 30-40 MPa magnitude.

The presence of a drastic jump of velocity on the lower boundary gives grounds to say that

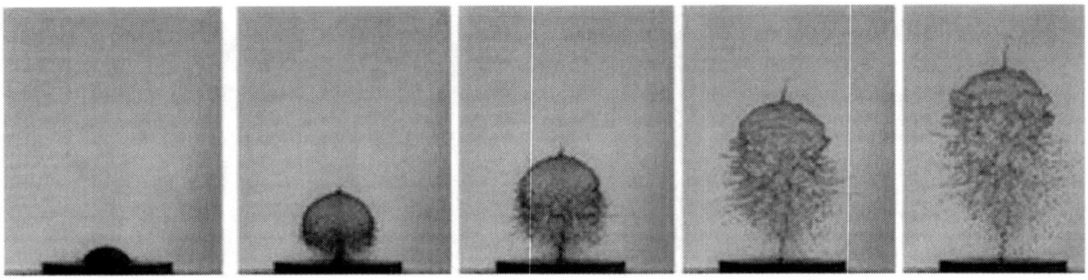

**Figure 3.** Cavitative explosion of 1 cm size drop at SW-loading (amplitude about 10 MPa, duration 5 $\mu s$): droplets cloud formation, t=0.15, 1.2, 2.0, 3.3 and 4 ms.

the formation of a discontinuity on this boundary and a new free surface is fairly realistic. A numerical analysis (with software implementation of "separation") of the dynamics of the state of the magma column remaining in the conduit shows that the formation of the anomalous zone with jumps of the values of the basic characteristics is a repeated process. The latter allows one to define the regime found as a "self-sustained regime of cyclic ejections."

## 3. Experimental simulation

The processes behind the DW-fronts are experimentally modeled in a dynamic statements at SW-loading of liquid samples with a free surface. SW-loading can be considered as a dynamic analogy of explosive volcanic eruption initiation: SW-compression of a sample simulates its pre-explosion state, and the rarefaction wave reflected from its free surface simulates the decompression wave inducing an intense cavitation process in a sample [1].

One can assume that the SZ mentioned above after the separation from main flow will have to be sprayed on the droplets cloud by the internal cavitative explosion. This effect is simulated on the electro-magnetic ST by SW-loading of 1 cm size drop of the distilled water (remind, that its 1 $cm^3$ contains about $10^6$ microinhomogeneities). Ultra-short SW (microseconds duration) in a drop is generated by duralumin membrane under pulse magnetic field (description can be found in [3]).

### *3.1. Acknowledgments*

The author is grateful to Dr. Davydov for modification of the numerical algorithm, and to the Russian Foundation for Basic Researches (Grant No. 15-05-03336) for financial support.

## 4. References

[1] Kedrinskii V K 2005 Hydrodynamics of Explosion: Experiments and Models (Germany: Springer) chapter 6 p 223-296
[2] Gonnermann H M and Manga M 2007 *Annu. Rev. Fluid Mech.* **39** 356
[3] Kedrinskiy V 2009 *Shock Waves* **18** (6) 464
[4] Kedrinskii V K 1993 *Appl. Mech. Tech. Physics* **34** (3) 91
[5] Woods W 1995 *Rev.Geophys.* **33** 4 530
[6] Dobran F 1992 *J. Volcanol. Geotherm. Res..* **49** 311
[7] Denlinger R P and Hobbitt R P 1999 *Geology* **27** 462
[8] Barmin A, Melnic O and Sparks S 2002 *Earth. Planet. Sci. Lett* **199** 184
[9] Costa A, Melnik O, Sparks R S J and Voight B 2007 *Geophys. Res. Lett.* **34** L02303 5
[10] Eichelberger J, Gordeev E, and Koyaguchi T 2006 Jun. 22 *http://www.uaf.edu/geology/ PIRE.pdf* 4
[11] Kedrinskii V K 2007 Shock Wave Science and Technology Reference Library vol 1, Multiphase Flows 1 ed M E H van Dongen (Springer Berlin Heidelberg) chapter 3 pp 67-97
[12] Gerst A, Horst M, Kyle P R, and Voege M 2006 *EOS Trans. Amer. Geophys. Union* **87** (52) V31G-04

# Experimental and numerical study on the dynamic pressure caused by the bubble jet

**S Li**[1]**, A M Zhang, X L Yao**
College of Shipbuilding Engineering, Harbin Engineering University, 145, Nantong Street, Harbin, China

E-mail: hgcls@163.com

**Abstract.** The high speed liquid jet is an important mechanism of damage to hydraulic machinery by cavitation bubbles, as well as damage to vessels by underwater explosion bubble. In this study, the bubble motion near a wall and the pressure impulse are investigated through experimental and numerical methods. In the experiment, the bubble is generated by the electric discharge, and a high speed camera is used to capture the bubble motion. Numerical studies are conducted using the boundary element method, and the vortex ring model is adopted to deal with the discontinued potential of the toroidal bubble. Calculated results show excellent agreement with experimental observations. Meanwhile, the dynamic pressure caused by the bubble in the flow domain is calculated by an auxiliary function, which improves the accuracy of the results. A highly localized pressure region will be generated on the wall by the bubble jet. The optimal stand-off parameter (the ratio of the distance the bubble center at inception from the wall to the maximum bubble radius) for a most damaging jet formation is around 0.9.

## 1. Introduction

Studying on the dynamic pressure caused by the bubble jet is of great importance in a wide range of engineering applications, such as cavitation on ship propellers and hydroturbines[1], underwater explosion bubbles[2], seismic airgun-generated bubbles for seabed geophysical exploration[3], bubble jets for increasing drilling rate[4], and micro-bubbles in medical applications[5]. A high speed liquid jet is induced during the collapse phase of a bubble because of the secondary Bjerknes force from a rigid wall. The jet impact is an important mechanism of damage to hydraulic machinery by cavitation bubbles, as well as damage to vessels by underwater explosion bubble [6].

In the recent four decades, the boundary element method (BEM) is widely and successfully used to study the dynamics of a violently-oscillating bubble [7]-[8]. The main characteristics of a bubble can be captured by BEM, such as expansion, collapse and jetting. Besides, the pressure and velocity in the flow domain can be calculated, providing more information to illustrate the mechanism of bubble motion and damage patterns. After the jet threading the bubble, Best [9] introduced a domain cut into the flow in order to compute the motion of the toroidal bubble, and Wang et al. [10] put forward a vortex ring model to deal with the discontinuous potential on the toroidal bubble surface.

In the present study, the dynamic behaviour of a bubble near a rigid wall is investigated through experimental and numerical methods. Much attention is paid on the dynamic pressure caused by the bubble jet. We adopt BEM to capture the bubble motion and use the auxiliary function [11] to

[1] Corresponding author: hgcls@163.com (S Li)

calculate the pressure in the flow domain. Pressure contours are shown to demonstrate the damage threat by the jet impact. In addition, an experiment is conducted to verify the validation of the numerical model. The effects of two dimensionless parameters on jet impact pressure are also discussed.

## 2. Theory and numerical method

The liquid surrounding the bubble is assumed inviscid and incompressible, and the motion irrotational. Thus, the velocity potential at any point in the domain could be expressed as an integral equation:

$$\iint_S \left( G(\boldsymbol{p},\boldsymbol{q}) \cdot \frac{\partial \Phi(\boldsymbol{q},t)}{\partial n} - \Phi(\boldsymbol{q},t) \frac{\partial G(\boldsymbol{p},\boldsymbol{q})}{\partial n} \right) dS(\boldsymbol{q}) = \lambda(\boldsymbol{p},t) \cdot \Phi(\boldsymbol{p},t) \tag{1}$$

where $\lambda(\boldsymbol{p},t)$ is the solid angle of a fixed point $\boldsymbol{p}$ on flow boundaries with the integration variable $\boldsymbol{q}$ also situated on boundaries, $\partial/\partial n$ is the normal outward derivative from the boundary $S$, $G(\boldsymbol{p},\boldsymbol{q})$ is the Green function.

The kinematic boundary condition and dynamic boundary condition on the bubble surface can be written as:

$$\frac{D\boldsymbol{r}}{\Delta t} = \left(\frac{\partial \Phi}{\partial n}\boldsymbol{n} + \frac{\partial \Phi}{\partial \tau}\boldsymbol{\tau}\right) \tag{2}$$

$$\frac{D\Phi}{Dt} = \frac{P_\infty - P_g}{\rho} + \frac{|\nabla \Phi|^2}{2} - gz \tag{3}$$

where $\rho$ is the density of the liquid , $P_\infty$ is the ambient pressure of the liquid around the bubble, $P_g$ is the pressure on the bubble surface, $g$ is the gravity acceleration.

The pressure inside the bubble is assumed to be uniform and consists of a constant vapor pressure and a volume-dependent noncondensable gas pressure [9]. Hence, the bubble pressure $P_g$ as a function of the volume is described as:

$$P_g = P_c + P_{ini}(V_{ini}/V)^{\vartheta} \tag{4}$$

where the subscript $ini$ denotes initial quantities, $\vartheta$ is the ratio of the specific heats for the gas, $P_c$ is the vapor pressure.

The vortex ring model is adopted to simulate the toroidal bubble [10]. The auxiliary function method [11] is employed to calculate the pressure on the rigid wall, which avoid making finite difference of the velocity potential. More details about these numerical methods can refer to Wang [10] and Wu [11], respectively.

Here we have scaled length with respect to the maximum radius which the largest bubble under consideration might achieve in an infinite body of fluid $R_m$ ; pressure by $\Delta P = P_\infty - P_c$ ; time by $R_m\sqrt{\rho/\Delta P}$ . Besides, three dimensionless parameters are given to describe a bubble: strength parameter $\varepsilon = P_{ini}/\Delta P$, stand-off parameter $\gamma = d/R_m$ and buoyancy parameter $\delta = (\rho g R_m/\Delta P)^{0.5}$.

## 3. Results and discussion

Experiments are conducted in a water tank with dimension 500mm×500mm×500mm, in which the water is filled up to 400mm in depth. A steel plate, 300mm×300mm×8mm, is placed at the bottom of the tank. The low-voltage spark bubble generation method can be found in Turangan's work [12]. The circuit employed in current bubble generation is based on Zhang [13], including a 6,600 $\mu F$ capacitor and a 220V DC power supply. A bubble is generated by burning the copper wire with its diameter about 0.25mm, and captured by the Phantom V12.1 high-speed camera. The camera works at 30010 frames per second with exposure time 10 $\mu s$. The whole experiment section is illuminated from the back with a 2kW light. In the experiment, the maximum radius of the bubble is about 15.6mm, and the distance between the initial bubble and the rigid wall is 15.7mm. The initial parameters of the bubble are set as: $\varepsilon = 30, \gamma = 1.01, \delta = 0.012$. In fact, the buoyancy has little effect on bubble dynamics in such a millimetre-sized bubble case. However, the buoyancy parameter is set according to the experiment for accuracy in calculation. Figure 1 shows the numerical and experimental results of the bubble motion near a rigid wall.

As shown in figure 1, each sequence shows the bubble just after it is created (frame a), during expansion (frame b), at maximum expansion (frame c), during collapse (frame d) and jet formation (frame e). In frame (e), it is clearly seen that a high pressure region is located above the bubble, which drives a high speed liquid jet into the bubble. These phenomena before the formation of a toroidal bubble have been discussed in many literatures. A toroidal bubble is formed after the jet threading the opposite side of the bubble (frame f). Due to inertia, a protrusion appears at the position of jet impact. At the same time, a high pressure region is caused at the wall center by the jet impact. The maximum dimensionless pressure reaches about 33.6. Then, a good portion of the bubble is 'stretched out' in the vertical direction (frame g) and contact with the wall directly. The numerical results show that the width of the jet increases as the bubble continues to shrink. The rigid wall will redirect jet to form a radial flow outward along the boundary from the jet axis, which results in a broader jet and a larger high pressure region. As the 'sideways jet' propagates along the bubble surface, the splitting may happen in the late stage of the toroidal bubble collapse, which can refer to our latest work [14]. It should be stated that the width of the bubble jet is limited to several millimetres, which is much smaller than regular pressure gauge. So it is very difficult to measure the pressure caused by the bubble jet [15] and there's no measured data of the jet impact pressure in this work. More effective and accurate measurement techniques need further developing.

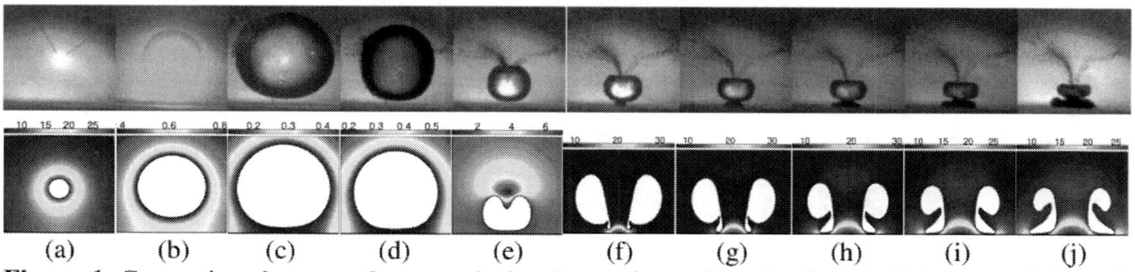


**Figure 1.** Comparison between the numerical and experimental results for a bubble near a rigid wall. The location of the rigid boundary is the lower limit of the frames. The frame width of the experiment is 18mm. The times (unit: ms) of the experiment are: 0, 0.529, 1.882, 2.852, 3.352, 3.441, 3.470, 3.499, 3.529, 3.588, respectively. The dimensionless times of the numerical results are: 0, 0.369, 1.307, 1.699, 2.308, 2.397, 2.415, 2.441, 2.459, and 2.478, respectively. Horizontal and vertical axes are: $r \leq 1.2$, $-1.01 \leq z \leq 1.2$ in (a)~(e), and $r \leq 0.6$, $-1.01 \leq z \leq -0.1$ in (f)~(j). The pressure legends represent the magnitudes of the dimensionless pressures.

Obviously, the stand-off parameter and strength parameter have much effect on the dynamic pressure induced by the bubble jet. In the following, the effect of $\gamma$ and $\varepsilon$ will be discussed in the following. Firstly, we fix $\varepsilon$ as 30, and $\gamma$ ranges from 0.6 to 1.2. The dynamic pressures at the wall center during the collapse phase of the bubble are shown in figure 2. The pressure increases rapidly soon after the jet impact and reaches a peak value afterwards. Then, the pressure decreases gradually due to the decrease of the jet velocity [8]. It is noted that the peak pressures of different stand-off parameter cases vary a lot from each other. There exists an optimal $\gamma$ for which the liquid jet thus formed is most damaging, denoted by $\gamma_{opt}$. The optimum stand-off is found to be around 0.9 when $\varepsilon$ is fixed as 30, which can be explained as follows. Generally, when $\gamma$ is greater than 1, there exists a water layer between the bubble and the rigid boundary, which would decrease the jet impact pressure. When $\gamma$ is less than 1, the jet hits the boundary directly without being decelerated by a water layer. Nevertheless, the maximum jet velocity becomes slightly smaller when $\gamma$ keeps decreasing. Then, we fix $\varepsilon$ as 100, the dynamic pressures at the wall center during the collapse phase of the bubble are shown in figure 3. The magnitudes of the pressures become larger, and the profile of the curves are similar to that for $\varepsilon = 30$. Meanwhile, $\gamma_{opt}$ is also around 0.9. After the jet impact, several pressure peaks may appear on pressure curves, which are associated with migration of the bubble, splashing effect and rebounding of the toroidal bubble [8].

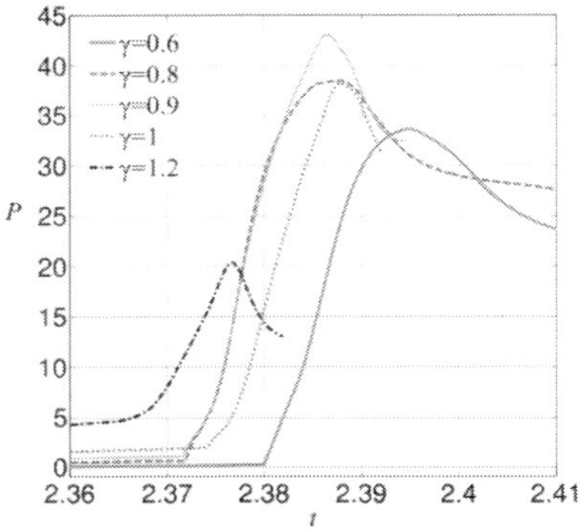


**Figure 2.** Evolutions of the dynamic pressure caused by the bubble jet versus time at different stand-off parameters with ε fixed as 30.

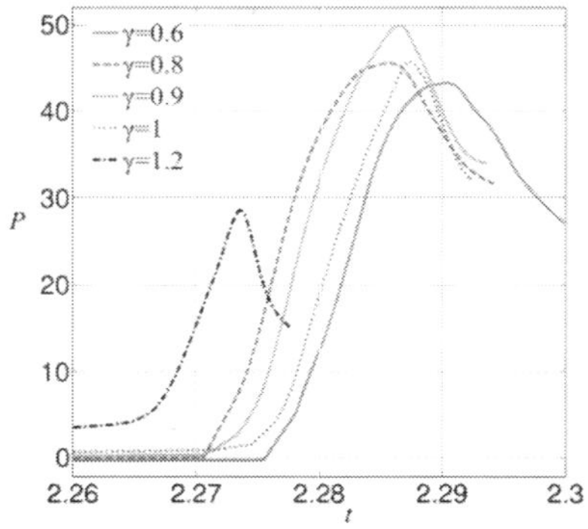


**Figure 3.** Evolutions of the dynamic pressure caused by the bubble jet versus time at different stand-off parameters with ε fixed as 100.

## 4. Conclusions

The dynamic pressure caused by the bubble jet is studied in this paper. The BEM is used to simulate the bubble motion and the results show excellent agreement with experimental observations. Meanwhile, the auxiliary function is adopted to calculate the pressure in the flow domain. A highly localized pressure region is generated on the wall by the bubble jet. The optimal stand-off parameter for a most damaging liquid jet formation is around 0.9. As the strength parameter of the bubble increases, the maximum pressure caused by the jet will increase but the pressure curve profile has little change.

## 5. References

[1] Hsiao C T and Chahine G L 2012 *J. Hydr. Ser. B* **24** 809
[2] Klaseboer E, Hung K C, Wang C, Wang C W, Khoo B C, Boyce P, Debono S and Charlier H 2005 *J. Fluid Mech.* **537** 387
[3] Graaf K L, Penesis I and Brandner P A 2014 *Ocean Eng.* **76** 32
[4] Johnson V E, Chahine G L, Conn A F, Frederick G S, Lindenmuth W T and Giacchino G J Jr. J. 1984 *Energ. Resour. ASME.* **106** 282
[5] Phillips L C, Klibanov A L, Wamhoff B R and Hossack J A 2010 *Ultrasound Med. Biol.* **36** 1470
[6] Plesset M S and Prosperetti A 1977 *Annu. Rev. Fluid Mech.* **9** 145
[7] Plesset M S and Chapman R B 1971 *J. Fluid Mech.* **47** 283
[8] Li S, Li Y B and Zhang A M. 2015 *Appl. Ocean. Res.* **50** 227
[9] Best J P 1993 *J. Fluid Mech.* **251** 79
[10] Wang Q X, Yeo K S, Khoo B C and Lam K Y 1996 *Comput. Fluids* **25** 607
[11] Wu G X and Hu Z Z 2004 *Proc. R. Soc.Lond.* **460** 2797
[12] Turangan C K, Ong G P, Klaseboer E and Khoo B C 2006 *J. Appl. Phys.* **100** 054910
[13] Zhang A M, Cui P and Wang Y 2013 *Exp. Fluid.* **54** 1062
[14] Zhang A M, Li S and Cui J 2015 *Phys. Fluids* **27** 062102
[15] Tong R P, Schiffers W P, Shaw S J, Blake J R and Emmony D C 1999 *J. Fluid Mech.* **380** 339

**Acknowledgments**

This paper is funded by the International Exchange Program of Harbin Engineering University for Innovation-oriented Talents Cultivation. The authors are also grateful to the support by the National Science Foundation of China (51379039, U1430236).

# Effects of the Encapsulating Membranes on the Translations of Pairs of Non-spherical Bubbles

**Yunqiao Liu**
MOE Key Laboratory of Hydrodynamics, School of Naval Architecture, Ocean and Civil Engineering, Shanghai Jiao Tong University, Shanghai 200240, China
E-mail: yunqiaoliu@sjtu.edu.cn

**Kazuyasu Sugiyama**
Department of Mechanical Science and Bioengineering, Graduate School of Engineering Science, Osaka University, 1-3 Machikaneyama, Toyonaka, Osaka 560-0043, Japan

**Shu Takagi**
Department of Mechanical Engineering, School of Engineering, The University of Tokyo, 7-3-1 Hongo, Bunkyo-ku, Tokyo 113-8656, Japan

**Abstract.** In this paper, we perform a theoretical analysis on the translational motions of gas and encapsulated bubbles undergoing deformation. We verify our theoretical model for the case of a gas bubble collapsing near a wall. By comparison, the encapsulated bubble near the wall is less unstable in shape. In the second case, the translations of two gas bubbles subjected to an oscillating driving pressure agree with Bjerknes' theory, that is, the two bubbles repel each other when the driving frequency is between the two natural frequencies $\omega_{01} < \omega_d < \omega_{02}$; while the two bubbles move towards each other at $\omega_d < \omega_{01}$ or $\omega_d > \omega_{02}$. For encapsulated bubbles, the translational displacements are smaller than those of gas bubbles. The encapsulating membrane and the deformation affect the translational direction. The encapsulated bubbles are prone to experience repulsive motion at resonance.

## 1. Introduction

In the applications of contrast-enhanced agents and drug-carrier capsules, encapsulated bubbles do not exist independently. Surrounding bubbles affect the behavior of each bubble. The interaction of two gas bubbles without encapsulation has been well studied. The pioneer work traces back to that of Bjerknes in 1906 [1]. The famous secondary Bjerknes force is the reason of motion of a bubble due to fluid acceleration caused by the volume oscillations of another bubble. The coupled pulsation and translation were studied in recent decades [2,3]. The effects of the shape oscillation are included in the theoretical work of Takahira et al. [4] and in numerical work by Pelekasis et al. [5,6] and Chatzidai et al. [7]. Most of the previous work on encapsulated bubbles, which are widely used in medical ultrasound, focused on the single encapsulated bubble [8-10]. The interaction of two encapsulated bubbles is seldom studied. In this work, we study the translation of two encapsulated bubbles. We consider the deformation of the bubbles to investigate the effects of the shape oscillation. To this aim, we perform a linear analysis of the dynamics of the radial oscillation, the translational motion and the shape oscillations of two encapsulated bubbles.

## 2. Basic equations

The dynamics of two bubbles suspended in an unbounded fluid are studied. We establish an axisymmetric coordinate system with the symmety axis connecting the centres of the two bubbles (figure 1). In the following equations, the subscript $i$ denotes the one bubble, and $j = 3 - i$ denotes the other one. Only the equations for the bubble $i$ are written, while the equations for the bubble $j$ are equivalent with the $i$ and $j$ exchanged.

The bubbles are coated by a viscoelastic membrane. We adopt the neo-Hookean law characterized by the surface modulus $G_{si}$, the bending modulus $G_{bi}$, the membrane viscosity $\mu_{si}$, and the Poisson ratio $v_{si}$.

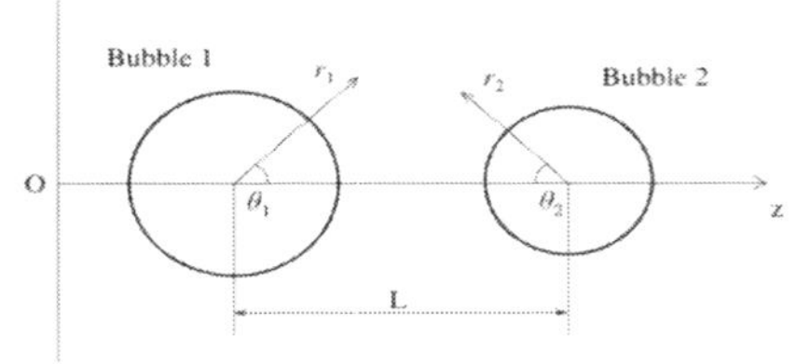


**Figure 1.** The coordinates of the two bubbles

We consider nonlinearity in radial motions $R_i$, which yields the Rayleigh-Plesset like equation:

$$R_i\ddot{R}_i + \frac{3}{2}\dot{R}_i^2 + \frac{1}{\rho}\Big[p_\infty - p_{g0}\big(\frac{R_{0i}}{R_i}\big)^{3\Gamma} + \frac{2\gamma}{R_i} + 4\mu_l\frac{\dot{R}_i}{R_i}\Big] - \frac{\mathrm{d}}{\mathrm{d}t}\Big(\frac{q_{0j}}{L}\Big) + \frac{R_j^3\dot{q}_{0i}}{L^4} - \frac{3v_{ri}^2}{4} + \frac{v_i^2}{2} \tag{1}$$
$$= -\frac{1}{\rho}\Big[\frac{2G_{si}(R_i^6 - R_{0i}^6)}{R_i^7} + 4\mu_{si}\frac{\dot{R}_i}{R_i^2}\Big],$$

where $\rho$ is the liquid density, $\gamma$ the surface tension, $\mu_l$ the liquid viscosity, $\Gamma$ the polytropic index. The last four terms on the left hand side of the equation represent the effect of the other bubble and the translations. $q_{0i} = -\dot{R}_iR_i^2$ is a coefficient from velocity potential. $L$ is the distance between two bubbles. $v_i$ is the velocity of the bubble center. $v_{ri} = v_i - q_{0j}/L^2$ is a reference velocity. The right hand side of the equation is the membrane force.

The equation of translational motion is:

$$\frac{\mathrm{d}}{\mathrm{d}t}\big(\frac{1}{2}\rho V_b v_{ri}\big) - \rho V_b\frac{\mathrm{d}}{\mathrm{d}t}\Big(\frac{q_{0j}}{L^2}\Big) + \frac{C_D}{2}\pi\rho R_i^2 v_{ri}|v_{ri}| + \rho V_b\frac{\dot{R}_i}{R_i^2}\int_{R_i}^{\infty}\Big[\Big(\frac{R_i}{s}\Big)^3 - 1\Big]\Big(\frac{R_i}{s}\Big)T_i(s,t)ds$$
$$-8\pi\mu_l R_i\int_{R_i}^{\infty}\frac{T_i(s,t)}{s}ds - \frac{1}{2}\rho V_b\Big[v_{ri}\Big(\frac{9\dot{a}_{2i}}{5R_i} - \frac{9\dot{R}_i a_{2i}}{5R_i^2} - \frac{6q_{0j}}{L^3}\Big) - \frac{6R_j^3\dot{q}_{0i}}{L^5} + \frac{18a_{2i}\dot{q}_{0j}}{5R_iL^2} - \frac{3R_j^3\dot{R}_i v_{jr}}{R_iL^3}\Big]$$
$$= V_b\Big[18G_{si}\frac{R_{0i}^6}{R_i^9}b_{1i} + 3(1+v_i)G_{bi}\frac{b_{1i}}{R_i^5} - \frac{6\mu_{si}}{R_i^4}(\dot{R}_i b_{1i} - R_i\dot{b}_{1i})\Big], \tag{2}$$

where

$$V_b = \frac{4\pi R_i^3}{3},\quad C_D = \frac{48}{Re},\quad Re = \frac{2R_i\rho|v_{ri}|}{\mu_l}. \tag{3}$$

In the above translation equation, the first term represents the inertial term of the added mass. The second term is the Bjerknes force. The third term is the drag force of a sphere with radius $R_i$ at the velocity $v_{ri}$. The coefficient of drag force $C_D$ is consistent with that of a rigid sphere. The fourth and fifth terms are the effect of viscous toroidal field. The sixth term is the perturbed term related to the shape oscillation and the translational motion of the other bubble. The right hand side is the membrane force due to the movement of the material points along the surface.

The shape deformation is described by the radial and tangential disturbances $a_k$ and $b_k$, respectively, with respect to the $k$th-order (associated) Legendre polynomial. The dynamic balances in the normal and tangential directions yield the equations for $a_k$ and $b_k$. The normal dynamic balance equation is

$$\frac{R_i}{k+1}\ddot{a}_{ki} + \dot{a}_{ki}\Big[\frac{3\dot{R}_i}{k+1} + \frac{2(k+2)\mu_l}{\rho R_i}\Big] + a_{ki}\Big[-\frac{k-1}{k+1}\ddot{R}_i + \frac{(k+2)(k-1)\gamma}{\rho R_i^2} + \frac{4(k-1)\mu_l\dot{R}_i}{\rho R_i^2}\Big]$$
$$-\frac{2k+1}{(k+1)L^{k+1}}\frac{\mathrm{d}}{\mathrm{d}t}(R_i^k q_{0j}) - 2k(2k+1)\frac{\mu_l R_i^{k-2}q_{0j}}{\rho L^{k+1}} + G_k$$
$$+\frac{1}{\rho}\Big\{k\mu_l\frac{T_i(R_i,t)}{R_i} + k\rho\frac{\dot{R}_i}{R_i}\int_{R_i}^{\infty}\Big[\Big(\frac{R_i}{s}\Big)^3 - 1\Big]\Big(\frac{R_i}{s}\Big)^k T_i(s,t)ds - 2k(k+1)\mu_l R_i^{k-2}\int_{R_i}^{\infty}s^{-k}T_i(s,t)ds\Big\}$$
$$= -\frac{1}{\rho}\Big\{-\frac{G_{si}}{R_i^8}[2(R_i^6 - 7R_{0i}^6)a_{ki} + 6k(k+1)R_{0i}^6 b_{ki}] + \frac{G_{bi}}{R_i^4}[k(k+1)(k^2+k-1+v_i)(a_{ki} - b_{ki})]$$
$$+\frac{2\mu_{si}}{R_i^3}[-4a_{ki}\dot{R}_i + 2R_i\dot{a}_{ki} + k(k+1)(\dot{R}_i b_{ki} - R_i\dot{b}_{ki})]\Big\}, \tag{4}$$

where

$$
\begin{aligned}
&k = 2, \quad G_k = \frac{3}{4} v_{ri}^2; \\
&k \geq 3, \quad G_k = \frac{3}{2} v_{ri} \left[ \dot{a}_{k-1,i} + \frac{2k-2}{2k-1} \frac{\dot{R}_i a_{k-1,i}}{R_i}, - \frac{(k-1) R_i^{k-2} q_{0j}}{L^k} \right]
\end{aligned} \tag{5}
$$

And the tangential dynamic balance equation is

$$
\begin{aligned}
&2\mu_l \Big[ \frac{k+2}{k+1} \frac{\dot{a}_{ki}}{R_i} - \frac{k-1}{k+1} \frac{\dot{R}_i a_{ki}}{R_i^2} - \frac{2k+1}{k+1} \frac{R_i^{k-2} q_{0j}}{L^{k+1}} + \frac{3k(k+2)}{2(2k-1)} \frac{a_{k-1,i} v_{ri}}{R_i^2} - R_i^{k-2} \int_{R_i}^{\infty} s^{-k} T_i(s,t) ds \\
&- \frac{T_i(R_i,t)}{2R_i} \Big] = \frac{G_{si}}{R_i^8} [(R_i^6 - 7R_{0i}^6) a_{ki} + (3k(k+1) R_{0i}^6 + (k-1)(k+2) R_i^6) b_{ki}] \\
&- \frac{G_{bi}}{R_i^4} (k^2 + k - 1 + \upsilon_i)(a_{ki} - b_{ki}) + \frac{2\mu_{si}}{R_i^3} [2\dot{R}_i a_{ki} - R_i \dot{a}_{ki} + (k^2 + k - 1)(R_i \dot{b}_{ki} - \dot{R}_i b_{ki})].
\end{aligned} \tag{6}
$$

In addition, the continuity in tangential velocity is needed:

$$
-\frac{1}{k+1} \left[ \dot{a}_{ki} + \frac{2\dot{R}_i a_{ki}}{R_i} - (2k+1) \frac{q_{0j} R_i^{k-1}}{L^{k+1}} \right] - \frac{3k v_{ri} a_{k-1,i}}{2(2k-1) R_i} + R_i^{k-1} \int_{R_i}^{\infty} s^{-k} T_i(s,t) ds = \dot{b}_{ki} - \frac{\dot{R}_i b_{ki}}{R_i} \tag{7}
$$

In the above equations, the toroidal field $T_i$ is solved by

$$
\rho \frac{\partial T_i}{\partial t} + \rho \frac{\partial}{\partial r_i} [\dot{R}_i (R_i / r_i)^2 T_i] - \mu_l \frac{\partial^2 T_i}{\partial r_i^2} + \mu_l k(k+1) r_i^{-2} T_i = 0, \tag{8}
$$

with the boundary condition $T_i \to 0$ as $r_i \to \infty$. $T_i$ on the bubble surface is determined by satisfying the tangential force balance.

For the above equations, the readers are referred to [11] for a detailed derivation.

## 3. Results

In this part, we will show some results of both gas bubbles and encapsulated bubbles. For gas bubbles, the coefficient of surface tension is chosen as $\gamma = 0.0729$ N/m. For encapsulated bubbles, the elastic modulus is $G_s = 0.5$N/m, the bending modulus is $G_b = 2 \times 10^{-13}$ N·m, and the membrane viscosity is $\mu_m = 10^{-8}$ kg/s.

The case of two bubbles with the same size is equivalent to that of a single bubble near a wall which is located at the midpoint between the centres of the two bubbles. In order to compare with the experimental result of Lauterborn & Bolle[12], we choose the same parameters. The initial radius is $R_{01} = 2.6 \times 10^{-3}$ m. The initial distance to the wall is $1.5R_{01}$. The pressure outside the bubbles is set to be $p_\infty = 1 \times 10^5$ Pa, and the pressure inside the bubble is $p_{g01} = 2337$ Pa initially. The solid line in Fig.2 (a) shows the bubble shapes when $\bar{t} = t/[R_{01}(\rho/\Delta p)^{1/2}]$ =0.725, 0.825, 0.961, 0.991, 1.016, progressing in inwards direction. The dots denote the experimental results. The bubble becomes elongated in the direction perpendicular to the wall and moves toward the wall. The third-order shape mode appears in the final time step, predicting a jet pointing towards the wall.

Figure 2 (b) shows the process of shrinking for an encapsulated bubble near a wall. We choose the same parameters as those in Fig. 2 (a), except for the parameters of the membrane which are added. Coincidentally, the natural frequency of this encapsulated bubble is 1258 Hz, close to that of the gas bubble in Fig. 2 (a) 1255 Hz. The deformation of the encapsulated bubble at $\bar{t}$ =1.016 is smaller than that of the gas bubble, implying that the encapsulated bubble is more stable.

According to Bjerknes' theory, the two gas bubbles will move away from each other when the driving frequency is between the two natural frequencies $\omega_{01} < \omega_d < \omega_{02}$; otherwise, the two bubble will attract to each other at $\omega_d < \omega_{01}$ or $\omega_d > \omega_{02}$. To validate this theory, we choose two bubbles with $R_{01} = 10$ $\mu$m and $R_{02} = 5$ $\mu$m. We plot the final distances between the two bubbles after 20 driving cycles versus the driving frequencies in Fig. 3. For gas bubbles, the zeroth-order natural frequencies are $\omega_{01} = 2\pi \times 0.34$ MHz and $\omega_{02} = 2\pi \times 0.72$ MHz, respectively. We can see that the two bubbles attract each other when the driving frequency is smaller than both of the natural frequencies. When the driving frequency approximates to the smaller natural frequency, i.e. $\omega_d = \omega_{01}$, the distance has a minimum. After this minimum, the attraction becomes repulsive. When the driving frequency approaches the other natural frequency, i.e. $\omega_d = \omega_{02}$, the

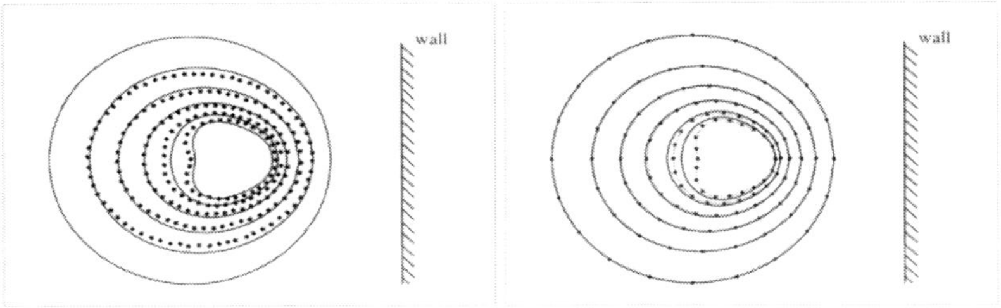


(a) Gas bubbles (b) Encapsulated bubbles

**Figure 2.** A bubble near a wall. The lines denote the bubble shapes at $\bar{t}$ =0, 0.725, 0.825, 0.961, 0.991 and 1.016, from outside and inwards. The dots in (a) are the experimental results by Lauterborn and Bolle. The dotted lines with symbol in (b) are reproduced from the gas bubble calculation in (a) for comparison.

two bubbles repel until a maximal distance. After this maximum, the translational motion of the bubbles shift to a weak attraction. For encapsulated bubbles, the zeroth-order natural frequencies are $\omega_{01} = 2\pi \times 0.51$ MHz and $\omega_{02} = 2\pi \times 1.28$ MHz, respectively. We can see that the repulsive and attractive motions are different from those of the Bjerknes theory (Fig. 3(b)). We predict the effects of the membrane force, and shape oscillations play an important role in these translations. A further investigation is in progress.

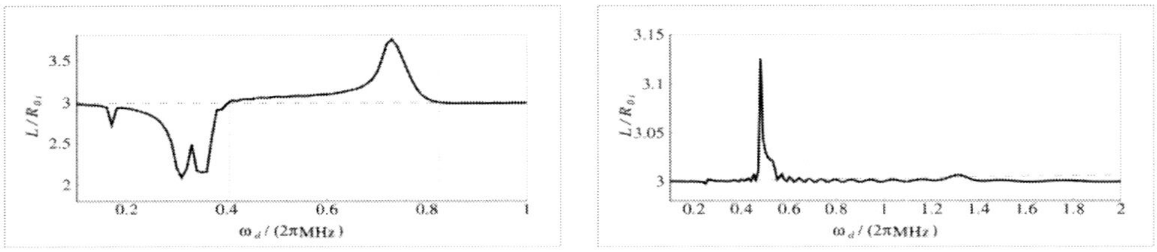

(a) Gas bubbles (b) Encapsulated bubbles

**Figure 3.** The final distance after 20 driving cycles at different driving frequencies with driving amplitude $\varepsilon = 0.1$. $R_{01} = 10\ \mu$m, $R_{02} = 5\ \mu$m.

## 4. Summary

We investigated a bubble with and without an encapsulating membrane when collapsing near a wall by means of theoretical analysis. The encapsulated bubble behaves more stably in shape than the gas bubble when going towards the wall. The translations of the two bubbles subjected to a pressure wave at different frequencies were also studied. The behavior of gas bubbles agrees with Bjerknes' theory, that is, the two bubbles repel from each other when the driving frequency is between the two natural frequencies $\omega_{01} < \omega_d < \omega_{02}$, otherwise, at $\omega_d < \omega_{01}$ or $\omega_d > \omega_{02}$, the two bubbles will move towards each other. However, the encapsulated bubbles show a different translational direction due to the effects of the membranes and the shape oscillations.

## 5. Acknowledgments

This work was supported by the National Natural Science Foundation of China (11302128).

## References

[1] Bjerknes V F K 1906 *Fields of Force*, Columbia University Press
[2] Barbat T, Ashgriz N and Liu C 1999 *J. Fluid Mech.*. **389** 137
[3] Harkin A, Kaper T J and Nadim A 2001 *J. Fluid Mech.* **445** 377
[4] Takahira H, Akamatsu T and Fujikawa S 1991 *Trans. JSME B* **57** 447
[5] Pelekasis N A and Tsamopoulos 1993 *J. Fluid Mech.* **254** 467
[6] Pelekasis N A and Tsamopoulos 1993 *J. Fluid Mech.* **254** 501
[7] Chatzidai N, Dimakopoulos Y and Tsamopoulos 2011 *J. Fluid Mech.* **673** 513
[8] Tsiglifis K and Pelekasis N A 2011 *Phys. Fluids* **23** 012102
[9] Liu Y, Sugiyama K, Takagi S and Matsumoto Y 2011 *Phys. Fluids* **23** 041904
[10] Liu Y, Sugiyama K, Takagi S and Matsumoto Y 2012 *J. Fluid Mech.* **691** 315
[11] Liu Y, Sugiyama K and Takagi S *in preparation*
[12] Lauterborn W and Bolle H 1975 *J. Fluid Mech.* **72** 391

# Modeling and experimental analysis of acoustic cavitation bubbles for Burst Wave Lithotripsy

**Kazuki Maeda[1], Wayne Kreider[2], Adam Maxwell[2,3], Bryan Cunitz[2], Tim Colonius[1], and Michael Bailey[2,3]**

[1]Devision of Engineering and Applied Science, 1200 E California Blvd, Pasadena, California 91125

[2]Center for Industrial and Medical Ultrasound, Applied Physics Laboratory, University of Washington, 1013 NE 40th St, Seattle, WA 98105.

[3]Department of Urology, University of Washington School of Medicine, 1959 NE Pacific St, Seattle, WA 98195

E-mail: maeda@caltech.edu

**Abstract.** Cavitation bubbles initiated by focused ultrasound waves are investigated through experiments and modeling. Pulses of focused ultrasound with a frequency of 335 kHz and a peak negative pressure of 8 MPa is generated in a water tank by a piezoelectric transducer to initiate cavitation. The pressure field is modeled by solving the Euler equations and used to simulate single bubble oscillation. The characteristics of cavitation bubbles observed by high-speed photography qualitatively agree with the simulation results. Finally, bubble clouds are captured using acoustic B-mode imaging that works synchronized with high-speed photography.

Burst Wave Lithotripsy (BWL) is a newly proposed non-invasive medical procedure to fragment kidney stones by focused ultrasound pulses of amplitude $O(1)$ MPa[1]. BWL is a potential alternative to Shock Wave Lithotripsy (SWL), a widely used procedure in clinical treatments that typically utilizes shock waves of $O(10)$ MPa for stone fragmentation; BWL enables comminution of stones into smaller fragmentations with lower peak pressure[1]. In both procedures, acoustic cavitation can be initiated during the passage of the tensile parts of the pressure wave of amplitude $O(1)$ MPa inside human body. Cavitation activity is important in such treatments because it can both erode kidney stones and cause tissue injury[2]. Because bubbles strongly interact with the incident sound field, cavitation can also be counterproductive by shielding the stone from the incident waves[3]. However, the fundamental physics of cavitation bubbles in strong interaction with the focused ultrasound field has not been fully studied in detail.

We report on combined modeling and experimental study of cavitation bubbles that are initiated by BWL-type ultrasound pulses in an open water tank. We generate waves using a piezoelectric transducer that has been used to break stones in both in vitro and in vivo environments[1]. In the modeling, we analytically emulate the transducer and simulate the sound field generated by the transducer in detail by solving the Euler equations. In the experiment, we observe the evolution of the cavitation bubble cloud using high-speed photography combined with B-mode acoustic imaging. The captured bubble clouds are in a qualitative agreement with the radial evolution of single spherical bubble calculated using Keller-Miksis equation combined with the pressure field from the simulations. Finally, through synchronization with the high-speed photography, we confirmed the potential of B-mode ultrasound imaging to effectively characterize bubble activity. The agreement among the modeling and the experimental measurements would lead to further understanding of the cavitation bubbles for BWL.

(a) Experimental setup　(b) Transducer　(c) Schematic

**Figure 1.** The images and the schematic of the experimental setup.

Figure 1 shows the experimental setup used for the generation and observation of cavitation bubbles. The piezoelectric transducer shown in figure 1b has 120 mm focal length and generates waves of frequency 335 kHz. An imaging probe (ATL HDI P4-2) is mounted in the center of the transducer. The high-speed camera (Model APX-RS camera, Photron USA, Inc., San Diego, CA) is triggered to captures images around the focal region of the transducer from the side of the tank (figure 1c) at 20,000 fps, when the waves arrive at the region. Tap water which was filtered and degassed to 65% $O_2$ level was used. The BWL pulses are characterized by the pulse repetition frequency (PRF), wave cycles per pulse ($N_c$), the total number of pulses ($N_p$) and maximum focal pressure $P_f$. In the present study, we use sinusoidal form of waves characterized by PRF= 200 Hz, $N_c = 10$, $N_p = 100$ and $P_f = 8$ MPa.

Figure 2 shows the pressure field at instances during the passage of the 10 cycles of left-going wave sent from the model transducer. The transducer is modeled as a spherical cup of 100 mm with its center located at the origin. The water is modeled as an inviscid fluid that follows stiffened gas equation of state. Acoustic monopoles and dipoles are distributed on the transducer surface such left-going waves of designated amplitude are generated. The simulation algorithm[4] uses a high-order accurate finite-volume weighted essentially non-oscillatory (WENO) reconstruction, to obtain the primitive flow variables at the cell-boundaries, and a Harten-Lax-van Leer-Contact (HLLC) approximate Riemann solver, to compute the

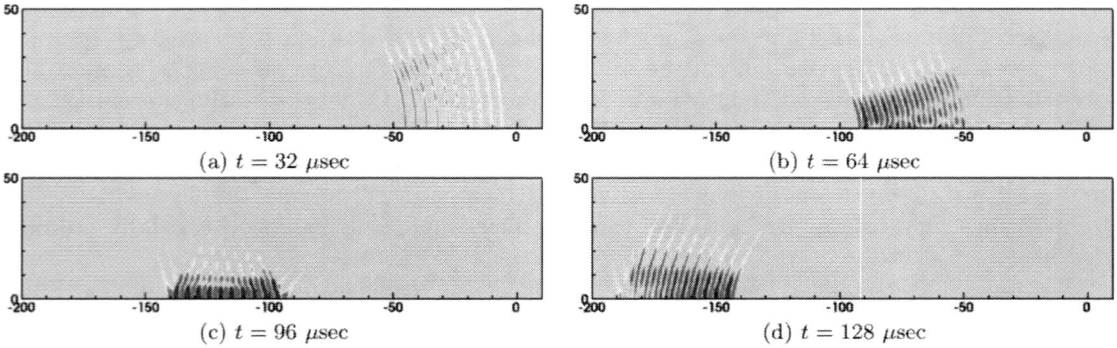


(a) $t = 32$ $\mu$sec　(b) $t = 64$ $\mu$sec

(c) $t = 96$ $\mu$sec　(d) $t = 128$ $\mu$sec

**Figure 2.** The flooded pressure contour of the wave sent from the model transducer along the upper half of a plane of symmetry of the 2D cylindrical domain. The red and blue color denote high and low pressure regions, respectively. The number of grid points used in the simulation is $[N_x,\ N_y] = 1250 \times 300$. The unit of the axis labels is mm.

resulting fluxes. Time marching is handled by a total-variation-diminishing Runge-Kutta (TVD-RK) time-stepper.

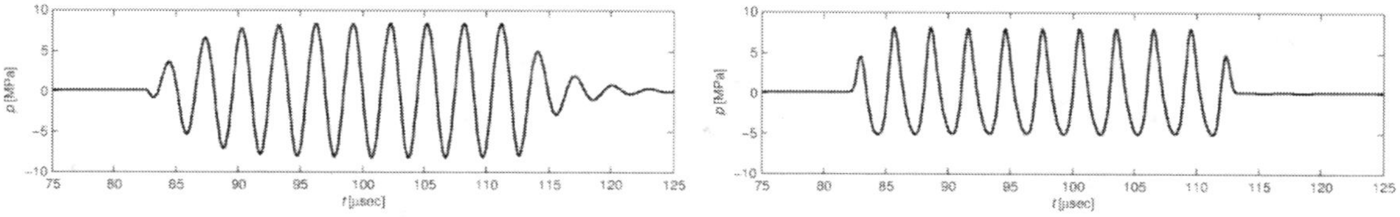


(a) Linear model (b) DNS

**Figure 3.** The focal pressure evolution of the waves sent from the transducer.

Figure 3 shows the focal pressure obtained from a linear transducer model that includes ring-up/ring-down effects, and that of the simulation. Though the simulated wave-form presents slight non-linear distortion and does not include the ring-up/ring-down effects, the values of peak maximum pressure as well as the qualitative wave forms agree.

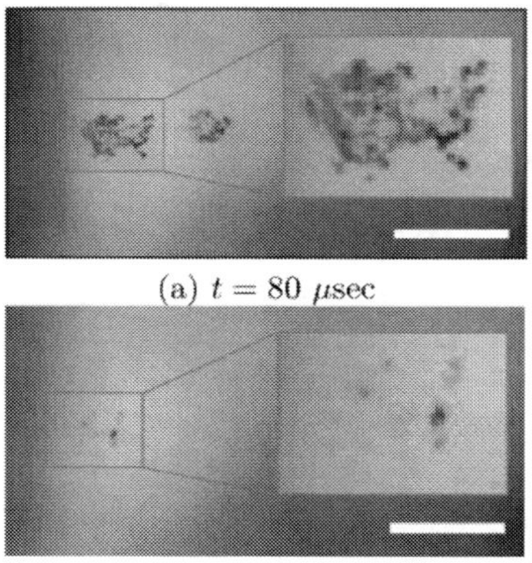

(a) $t = 80$ $\mu$sec

(b) $t = 130$ $\mu$sec

**Figure 4.** High-speed images taken at $t = 80$ $\mu$sec and 130 $\mu$sec during the 3rd pulse. The scale bars denote 5 mm.

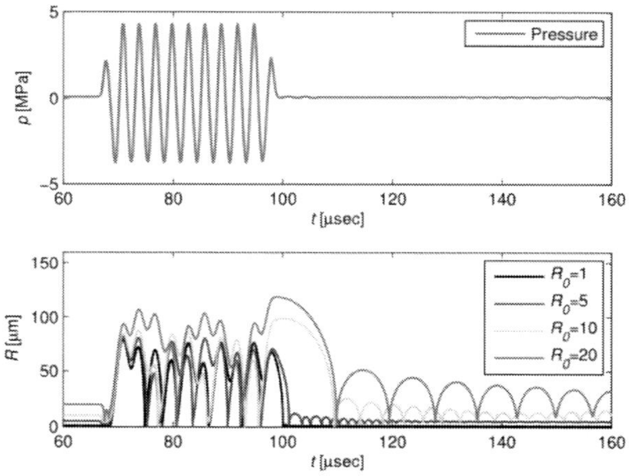


**Figure 5.** Simulated pressure evolution at $(x, y) = (-98, 0)$ and the corresponding radial evolution of isolated bubbles responding to the focal pressure with various initial radii.

Figure 4 shows the representative images of cavitation bubbles captured by the high-speed camera at $t = 80$ $\mu$sec and 130 $\mu$sec during the 3rd pulse, where $t = 0$ is the moment when the transducer was triggered to generate the pulse. The location of the center of the images correspond to (-98, 0) in the images of figure 2, thus 98 mm distant from the center of the transducer. At $t = 80$ $\mu$sec, we see two bubble clouds as large as 3 mm. In the magnified inset, the presence of individual bubble can be confirmed in the leftmost cloud. At $t = 130$ $\mu$sec, the clouds disappear and a few bubbles are discretely observed. We note that this tendency of the shrinkage of the cloud was confirmed in the images from different pulses, though the locations and the sizes of the bubble clouds were not consistent.

Figure 5 shows the pressure evolution at (-98, 0) obtained using the simulated pressure field and the corresponding radial evolution of pre-existing bubble nuclei with typical size, that are computed using the Keller-Miksis equation[5]. Around 80 $\mu$sec, bubbles oscillate to reach the maximum radius around 100 $\mu$m, while at $t = 130$ $\mu$sec, only bubbles with large initial radius (10 $\mu$m and 20 $\mu$m) remain and collapse and rebound to radii less than 50 $\mu$m. Bubbles with initial radius 1 and 5 $\mu$m are immediately damped after the passage of the wave, due to strong acoustic radiations. Thus at $t = 130$ $\mu$m we would assume to see fewer number of bubbles of smaller size. This assumption corresponds to the observation in figure 4.

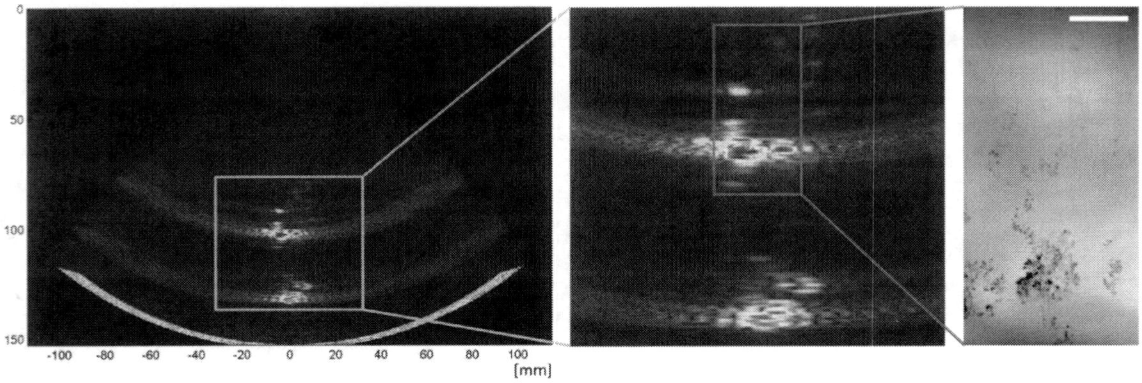


**Figure 6.** The images of cavitation bubbles taken by synchronized B-mode acoustic imaging (the left and middle images) and high-speed photography (the right image) at $t = 100$ $\mu$sec. The waves propagate from the top to the bottom of the images. The scale bar in the photographic image denotes 5 mm.

Finally, figure 6 shows representative images of cavitation bubbles obtained using synchronized B-mode imaging and high-speed photography. B-mode images were captured at 400 frames per BWL-pulse, using the imaging probe with a Verasonics ultrasound engine (Kirkland, WA, USA). Because the open water tank is devoid of scatterers other than bubbles, the intense regions in reconstructed images should correspond to bubbles. Though the fps of the imaging pulse is much lower than that of high-speed photography, the size and location of bubble clouds in the B-mode image agrees well with those of the high-speed photography, suggesting its potential for detecting bubble activity during treatment.

In conclusion, we have confirmed the qualitative agreement between modeling and experiments in terms of the radial evolution of isolated bubbles responding to focal pressure field simulated by solving the Euler equations. B-mode imaging and high-speed photography were successfully synchronized during the pulses. For further quantitative analysis of the acoustic cavitation, we plan to extend the present study by modeling bubble clouds taking inter-bubble interactions[6] into account and by measuring the pressure field simultaneously with higher-speed photography to capture the dynamics of the cloud in detail.

## Acknowledgment

The authors thank Dr. Julianna Simon and Barbrina Dunmire for support in the experimental measurements, Prof. Jonathan Freund, Dr. Pooya Movahed, Dr. Vedran Coralic, and Jomela Meng for helpful discussions. The work was supported by the National Institutes of Health under grant 2PO1-DK043881.

## References

[1] Maxwell A D, Cunitz B W, Kreider W, Sapozhnikov O A, Hsi R S, Harper J D, Bailey M R and Sorensen M D 2015 *J Urol* **193** 338–344
[2] Maxwell A D, Cain C A, Duryea A P, Yuan L and Gurm H S 2009 *Ultrasound Med. Biol.* **12** 1982–1994
[3] Tanguay M 2004 Ph.D. thesis California Institute of Technology
[4] Coralic V and Colonius T 2014 *J. Comput. Phys.* **274** 95–121
[5] Keller J B and Miksis M 1980 *J. Acoust. Soc. Am.* **68** 628–633
[6] Fuster D and Colonius T 2011 *J. Fluid Mech.* **688** 352–389 ISSN 0022-1120

# Diffuse interface modeling of a radial vapor bubble collapse

**Francesco Magaletti, Luca Marino, Carlo Massimo Casciola**

Dipartimento di Ingegneria Meccanica e Aerospaziale, Università di Roma "La Sapienza", via Eudossiana 18, 00184 Roma Italy

E-mail: `francesco.magaletti@uniroma1.it`

**Abstract.** A diffuse interface model is exploited to study in details the dynamics of a cavitation vapor bubble, by including phase change, transition to supercritical conditions, shock wave propagation and thermal conduction. The numerical experiments show that the actual dynamic is a sequence of collapses and rebounds demonstrating the importance of non-equilibrium phase changes. In particular the transition to supercritical conditions avoids the full condensation and leads to shockwave emission after the collapse and to successive bubble rebound.

## 1. Introduction

The collapse of a vapor bubble is a classical but still crucial phenomenon in the realm of the physics of fluids [1]. The interest is twofold, being aimed at the analysis and a better comprehension of the complex physical mechanisms involved in the processes and to the important possible applications which range from the cavitation over ship propellers to the drug delivery strategy in biological systems.

Collapse bubble dynamics is difficult to model since several spatial and time scales can be involved in the same process. Moreover non trivial effects like phase transition to and from supercritical conditions or shock formation can easily be encountered. As the bubble radius decreases the surface tension at the interface becomes more and more relevant and the condensation and compression phenomena can lead to extreme pressure and temperatures.

Experimental results are usually based on visual analysis carried out with high speed cameras which can give a quantitative description or indirect information [2]. On the other hand, numerical results can be particularly demanding and very sensitive to the mathematical model adopted. Recently more detailed physical models have been proposed to study the dynamics of the bubble interface [3, 4], possibly taking into account the presence of a dissolved gas [5].

We present the results obtained by a diffuse interface model based on the van der Waals free energy functional. Following this approach the effects of the capillarity are naturally embedded in the physical model and it is possible to deal with the interface dynamic as well as with complex phase change phenomena. The collapse of the vapor bubble has been thoroughly studied, in particular in the case where the interface speed may exceed the speed of sound. In this peculiar condition, the complex dynamic of shock waves, focused towards the bubble and successively reflected back in the liquid, has been analyzed and discussed. Furthermore we considered in details the effects of condensation and of rapid compression which locally bring the vapor in supercritical conditions.

## 2. Mathematical model and numerical solution

Vapor bubble dynamics is a particular case of the most general two-phase interfacial flows phenomena. The fluid dynamic modeling of interfacial problems has been a challenge for theoreticians due to their intrinsic multi-scale nature. Indeed the interfacial (i.e. capillary) properties of a liquid-vapor system, acting on the molecular length scale, strongly interact with the characteristic macroscopic length scale of the fluid dynamic problem. In the last decades several different models have been proposed to describe liquid-vapor systems endowing capillary stresses and they can be grouped into the two main categories of sharp and diffuse interface approaches (see the review [6]).

In this work we adopt an unsteady diffuse interface model coupled with the choice of van der Waals state equation (see [7, 8] for details). This model can be deduced introducing a suitable non-local free energy functional $F[\rho, \theta]$ with a gradient type excess energy:

$$F[\rho, \theta] = \int_{\Omega} \left( \hat{f}_0 (\rho, \theta) + \frac{\lambda}{2} |\boldsymbol{\nabla} \rho|^2 \right) dV \,, \tag{1}$$

where $\lambda$ is a coefficient strongly related to the interfacial properties of the liquid-vapor system (i.e. surface tension and interface thickness) and $\hat{f}_0 (\rho, \theta)$ the bulk free energy density per unit volume of a uniform fluid at temperature $\theta$ and density $\rho$. In particular, we choose the van der Waals free energy for a polytropic fluid (constant volume specific heat $c_v$) whose expression follows:

$$\hat{f}_0 (\rho, \theta) = \bar{R} \rho \theta \left[ -1 + \log \left( \frac{\rho \, K \, \theta^{1/\delta}}{1 - b\rho} \right) \right] - a \rho^2 \,, \tag{2}$$

with $\delta = \bar{R}/c_v$, $\bar{R}$ the gas constant, $a$ and $b$ the van der Waals gas constants and $K$ a constant related to the de Broglie length [9].

Given the thermodynamic behavior of the inhomogeneous system, its fluid dynamics, under the assumption of spherical symmetry, is described by the following conservation laws for mass, momentum and total energy densities

$$\frac{\partial \rho}{\partial t} + \frac{1}{r^2} \frac{\partial}{\partial r} \left( r^2 \rho u_r \right) = 0 \,, \tag{3}$$

$$\frac{\partial \rho u_r}{\partial t} + \frac{1}{r^2} \frac{\partial}{\partial r} \left( r^2 \rho u_r^2 \right) = - \frac{\partial p_0}{\partial r} + \frac{1}{r^2} \frac{\partial}{\partial r} \left( r^2 T_{rr} \right) - \frac{2 T_{\phi\phi}}{r} \,, \tag{4}$$

$$\frac{\partial E}{\partial t} + \frac{1}{r^2} \frac{\partial}{\partial r} \left[ r^2 u_r \left( E + p_0 \right) \right] = \frac{1}{r^2} \frac{\partial}{\partial r} \left[ r^2 \left( T_{rr} u_r - \lambda \rho \frac{\partial \rho}{\partial r} \frac{1}{r^2} \frac{\partial}{\partial r} \left( r^2 u_r \right) + k \frac{\partial \theta}{\partial r} \right) \right] \,, \tag{5}$$

with $u_r$ the radial component of the fluid velocity and $k$ the thermal conductivity. The stress tensor $\boldsymbol{T}$ in the equations takes into account both the viscous and diffused capillary stresses and its relevant components read:

$$T_{rr} = \lambda \left[ -\frac{1}{2} \left( \frac{\partial \rho}{\partial r} \right)^2 + \frac{\rho}{r^2} \frac{\partial}{\partial r} \left( r^2 \frac{\partial \rho}{\partial r} \right) \right] + 2\mu \left[ \frac{\partial u_r}{\partial r} - \frac{1}{3 r^2} \frac{\partial}{\partial r} \left( r^2 u_r \right) \right] \,, \tag{6}$$

$$T_{\phi\phi} = \lambda \left[ \frac{1}{2} \left( \frac{\partial \rho}{\partial r} \right)^2 + \frac{\rho}{r^2} \frac{\partial}{\partial r} \left( r^2 \frac{\partial \rho}{\partial r} \right) \right] + 2\mu \left[ \frac{u_r}{r} - \frac{1}{3 r^2} \frac{\partial}{\partial r} \left( r^2 u_r \right) \right] \,, \tag{7}$$

Journal of Physics: Conference Series **656** (2015) 012028 doi:10.1088/1742-6596/656/1/012028

The classical thermodynamic pressure $p_0$ for a van der Waals fluid is readily obtained from Eq. (2):

$$p_0 = -\frac{\partial f_0}{\partial v} = -\frac{\partial \hat{f}_0/\rho}{\partial v} = \bar{R}\frac{\rho\theta}{1-b\rho} - a\rho^2 , \tag{8}$$

with $f_0 = \hat{f}_0/\rho$ the specific bulk free energy and $v = 1/\rho$ the specific volume.

The system of equations in (3-5) is numerically challenging because it contains several phenomena, asking for specific numerical techniques. The first aspect is the presence of the moving and extremely thin liquid-vapor interface that requires a great amount of grid points in order to correctly capture the capillary stress. Another issue is related to the formation of shock waves, hence the numerical scheme needs to be based on shock-capturing method, such as ENO or WENO schemes [10]. These methods work only with hyperbolic equations, but the presence of viscosity and capillarity implies a non-hyperbolic nature of the equations. A classical Strang operator-splitting addresses this issue, isolating the viscous and capillary operators discretized with classical central finite difference from the hyperbolic part where a WENO scheme is used. The boundary conditions of fixed pressure and temperature, far away from the bubble, are imposed with a characteristic-wave approach [11]. Further details can be found in [8].

## 3. Results and discussion

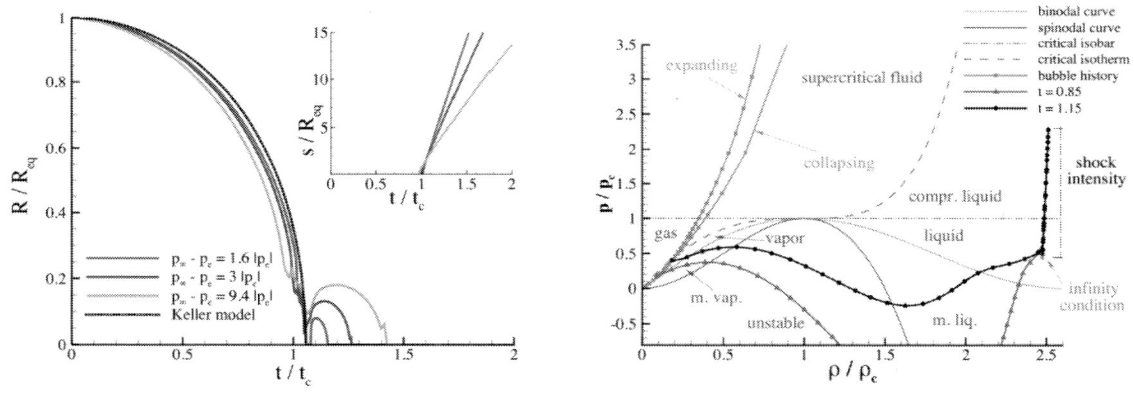


**Figure 1.** In the left panel, the radius evolution for different overpressure in the liquid. The inset shows the position of the shock radiated after the collapse with the same colors of the main plot. In the right panel, the evolution of the thermodynamic condition in the $p-\rho$ plane (with $p_c$ and $\rho_c$ the critical pressure and density, respectively). The green curve represents the conditions at the bubble center along the evolution. The red and black curves represent the conditions in the whole domain at two different time instant, before and after the collapse, respectively

The system is initially in equilibrium with a pure vapor bubble of radius $R_{eq}$ immersed in its liquid at pressure $p_e$. The radial collapse is initiated by imposing an overpressure in the liquid $\Delta p = p_\infty - p_e$. Along the evolution the bubble radius is defined as the distance of the liquid from the bubble center, see the left panel of Fig. 1 where the radius evolution is plotted for different overpressures and is compared with the prediction of the Keller model [3]. The actual dynamic consists in a sequence of collapses and rebounds (that resembles the evolution of a pure gas bubble), while the simplified Keller model predicts a full condensation after the first collapse if no dissolved gas is added in the description. The key features that explains the rebounds is the transition of the vapor into supercritical state. The phase change into supercritical conditions, indeed, avoids the full condensation and allows the bubble rebound. When the bubble reaches its minimum radius, a shockwave is radiated into the liquid phase, as shown in the inset of Fig. 1.

The thermodynamic conditions reached along the evolution are reported in the right panel of Fig.1. As background information (with blue lines), the figure provides the main thermodynamic features of the van der Waals system, e.g. critical isotherm and isobar, binodal and spinodal curve. The region between binodal and spinodal lines corresponds to metastable states (matestable liquid/vapor to the right/left of the critical density). In a finite size, equilibrium bubble, the vapor is in the corresponding metastable region, while the liquid state is inside the stable region to the right of the binodal line. The square symbols (green on line) show the thermodynamic state of the fluid occupying the bubble center at successive time instants during the first collapse/expansion phase. The evolution is towards increasing (decreasing) pressures during collapse (expansion). The two branches match at pressure higher than those reported in the diagram. During collapse, the fluid at the center of the bubble starts in the vapor region and follows the saturation line (binodal curve) only during the first stage of the collapse. Successively it enters the gas region to finally become supercritical reaching extremely high pressures and temperatures. The reverse sequence is observed during expansion. Triangles and circles (red and black lines, respectively) show the states of the fluid at changing radial position at two time instants, $t/t_c = 0.85$ during the collapse phase before the first rebound, and $t/t_c = 1.15$ during the successive expansion phase. In the latter curve, the steep pressure increase at high density (corresponding to liquid) is the pressure jump across the shockwave generated at rebound.

The transition to supercritical conditions is the key feature of the collapse of a vapor bubble. The rapid compression experienced by the vapor avoid the full condensation and leads to the transition to incondensable gaseous state. As a main consequence, the dynamic of a pure vapor bubble shows typical features of a pure gas bubble collapse such as shockwave emission and bubble rebounds, implying the importance of including phase change phenomena into the modeling of cavitation problems.

## References

[1] Rayleigh L 1917 *The London, Edinburgh, and Dublin Philosophical Magazine and Journal of Science* **34** 94–98

[2] Taleyarkhan R P, West C, Cho J, Lahey R, Nigmatulin R and Block R 2002 *Science* **295** 1868–1873

[3] Keller J B and Kolodner I I 1956 *Journal of Applied Physics* **27** 1152–1161

[4] Plesset M S and Chapman R B 1971 *Journal of Fluid Mechanics* **47** 283–290

[5] Akhatov I, Lindau O, Topolnikov A, Mettin R, Vakhitova N and Lauterborn W 2001 *Physics of Fluids* **13** 2805–2819

[6] Anderson D, McFadden G and Wheeler A 1998 *Annual Review of Fluid Mechanics* **30** 139–165 ISSN 0066-4189

[7] Jamet D, Lebaigue O, Coutris N and Delhaye J 2001 *Journal of Computational Physics* **169** 624–651

[8] Magaletti F, Marino L and Casciola C 2015 *Physical Review Letters* **114** 064501

[9] Zhao N, Mentrelli A, Ruggeri T and Sugiyama M 2011 *Physics of Fluids* **23** 086101

[10] Shu C W 1998 *Essentially non-oscillatory and weighted essentially non-oscillatory schemes for hyperbolic conservation laws* (Springer)

[11] Poinsot T J and Lele S 1992 *Journal of computational physics* **101** 104–129

# Beer tapping: dynamics of bubbles after impact

**V Mantič-Lugo, A Cayron, P-T Brun, F Gallaire**

Laboratory of Fluid Mechanics and Instabilities, École Polytechnique Fédérale de Lausanne. EPFL-STI-IGM-LFMI, CH-1015, Switzerland

E-mail: vladislav.manticlugo@epfl.ch

**Abstract.** Beer tapping is a well known prank where a bottle of beer is impacted from the top by a solid object, usually another bottle, leading to a sudden foam overflow. A description of the shock-driven bubble dynamics leading to foaming is presented based on an experimental and numerical study evoking the following physical picture. First, the solid impact produces a sudden downwards acceleration of the bottle creating a strong depression in the liquid bulk. The existing bubbles undergo a strong expansion and a sudden contraction ending in their collapse and fragmentation into a large amount of small bubbles. Second, the bubble clouds present a large surface area to volume ratio, enhancing the $CO_2$ diffusion from the supersaturated liquid, hence growing rapidly and depleting the $CO_2$. The clouds of bubbles migrate upwards in the form of plumes pulling the surrounding liquid with them and eventually resulting in the foam overflow. The sudden pressure drop that triggers the bubble dynamics with a collapse and oscillations is modelled by the Rayleigh-Plesset equation. The bubble dynamics from impact to collapse occurs over a time ($t_b \simeq 800$ $\mu s$) much larger than the acoustic time scale of the liquid bulk ($t_{ac} = 2H/c \simeq 80$ $\mu s$), for the experimental container of height $H = 6$ cm and a speed of sound around $c \simeq 1500$ m/s. This scale separation, together with the comparison of numerical and experimental results, suggests that the pressure drop is controlled by two parameters: the acceleration of the container and the distance from the bubble to the free surface .

## 1. Introduction

The beer tapping prank, which consists in impacting a bottle of beer at its mouth, involves several physical mechanisms. These multi-physics processes in supersaturated liquids are of interest in physics and chemistry, and are here combined in a simple model experiment, namely strong pressure variations, bubble dynamics, collapse, bubble clouds formation, bubble growth due to gas diffusion and motion of plumes. These processes are relevant to the understanding and description of certain geological phenomena as limnic gas driven eruptions [1] or volcanic eruptions [2], while also relevant, in industry and commercial purposes like in wine and champagne [3]. Simple experimental setups, with daily life products, have proven useful to explain complex phenomena [4, 5]. In this paper, we revisit the setup of [5] exacerbating the initial bubble growth.

Building on the experimental observation of a train of bubbles as illustrated in Fig. 1, the following physical picture appears. First, the solid impact produces a sudden downwards acceleration of the bottle creating a strong depression in the liquid bulk. The existing bubbles undergo a strong expansion and a sudden contraction ending in a collapse and fragmentation in a large amount of small bubbles. Second, because the bubble clouds present a large surface area to volume ratio, the $CO_2$ diffusion from the supersaturated liquid is enhanced and the clouds grow rapidly, depleting the $CO_2$. The clouds of bubbles then migrate upwards in the form of plumes pulling the surrounding liquid with them and eventually producing the foam overflow. In this study, we will focus on the first part of the whole process leading to the foaming and overflow.

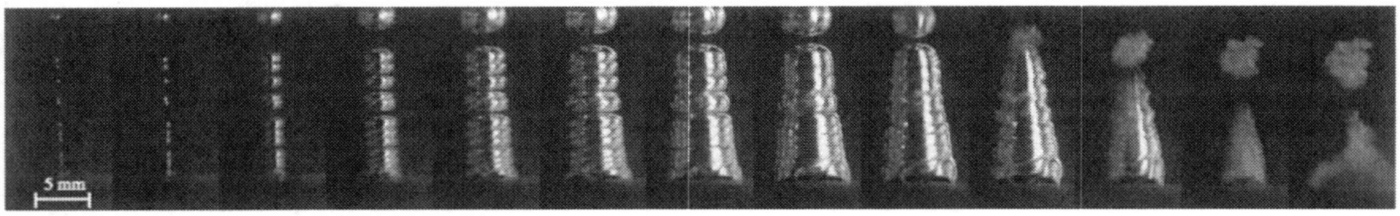


**Figure 1.** Growth and collapse of a train of ascending bubbles. Note the pyramidal shape of the maximum radii, and a cascade of collapses starting from the top and moving downwards. The collapse leads to bubble fragmentation into a cloud of bubbles that eventually grows and migrates upstream in the form of plumes. The time step between images is $100\mu s$.

## 2. Experimental setup and numerical model

The experimental setup is sketched in Fig. 2(a) and consists in a 10 cm glass beaker filled with 6 cm of lager beer (150 mL). The beaker is placed on top of a shock absorbing foam layer. The container is then impacted from the top using a piece of wood, released from a chosen height. The experiment is recorded at a frame rate ($\sim 10^4$ fps) with a high-speed camera (Phantom Micro M310).

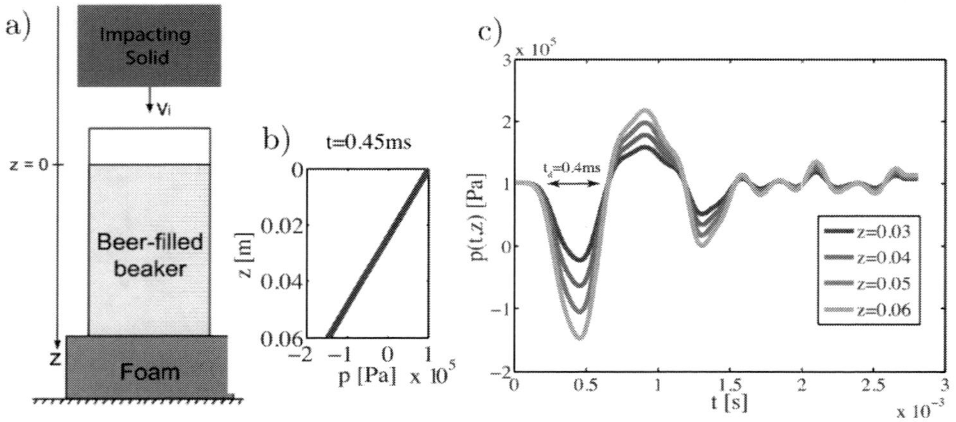


**Figure 2.** (a) Sketch of the experimental configuration. (b) Shown is the instantaneous pressure as function of depth $p(z)$ for the maximum pressure drop at $t = 0.45 \cdot 10^{-3}s$. (c) Pressure variation with time $t$ calculated from the measured beaker displacement during the vertical impact (see (1)) for several depth $z$.

The bubble dynamics is assumed to be driven by the surrounding pressure variations over time. The pressure in the liquid was derived from own measurements of the acceleration of the liquid bulk using Navier-Stokes equations for the liquid phase solely. Assuming that the fluid moves as a whole, the convective terms and viscosity effects can be neglected [6], yielding

$$\rho \frac{\partial \boldsymbol{u}}{\partial t} = -\nabla p + \rho \mathbf{g} \;\Rightarrow\; p(z,t) = p_{atm} - \rho \frac{\partial u_z}{\partial t} z + \rho g z, \tag{1}$$

where $\rho$ is the liquid density, $\mathbf{g}$ is the acceleration from gravity and $\boldsymbol{u}$ is the velocity field. For an unidirectional velocity $u_z$, the pressure $p(t, z)$ can simply be extracted as function of time $t$ and height $z$. The amplitude of the pressure drop is linearly related to the depth ($z$) (see Fig. 2(b,c)) as well as to the acceleration of the liquid (1). The acceleration of the liquid bulk is obtained from the beaker displacement, which is measured experimentally from the recorded images and then derivated twice numerically. A verification of the methodology was performed based on momentum conservation as well as its reproducibility. Subsequently, knowing the pressure in time at any point of the liquid $p(t, z)$, the Rayleigh-Plesset equation is used to model the bubble dynamics as standard in literature [7, 8].

## 3. Discussion

The numerical integration of Rayleigh-Plesset equation was performed using Runge-Kutta method with an initial radius $R_0 = 2.8e^{-4}$ m based on recorded images and the pressure evolution from Fig. 2(c). It provided the bubble time evolution at different depths $z$ as plotted in Fig. 3(a). Furthermore, the comparison with the experimental results shows an excellent agreement. An increase in the depth of the bubble entails a stronger pressure drop (Fig. 1(c)) yielding longer collapse times (Fig. 3(b)) and larger maximum radii (Fig. 3(c)). Notice that the bubble dynamics from growth to collapse occurs over a time $t_b \simeq 300 - 800$ $\mu s$ (Fig. 3), which depends on the depth through the pressure drop, despite a constant duration of the pressure oscillation $t_d = 400$ $\mu s$, as defined in Fig. 2(c).

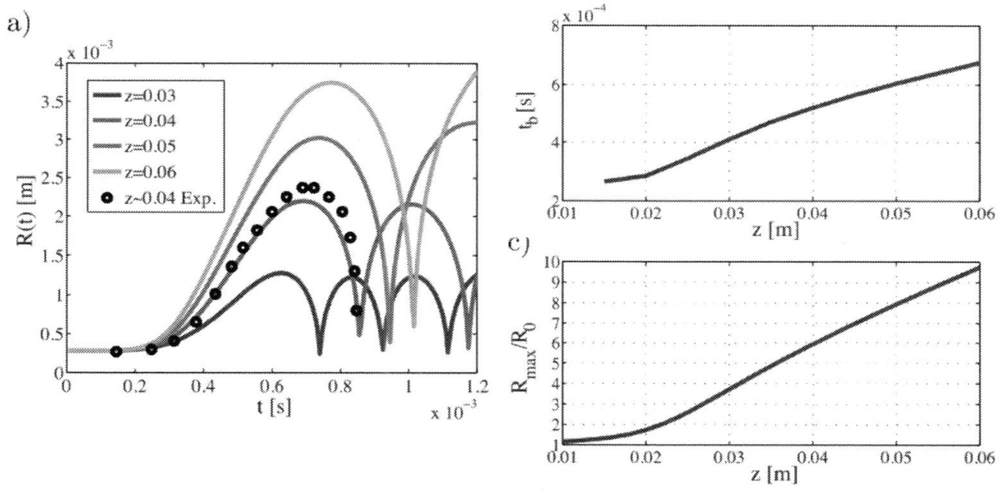


**Figure 3.** (a) Variation of the bubble radius with time for various depths compared to experiments denoted with black circles. (b) Variation of the impact to collapse time $t_b$ with the depth $z$ and (c) variation of the maximum radius $R_{\max}/R_0$ with the depth $z$. All results are integrated from the Rayleigh-Plesset equation using the pressure drop from Fig. 2 (c).

The experimental observations illustrated in Fig. 1 are in line with the numerical integration of the Rayleigh-Plesset equation (Fig. 3), showing the anticipated influence of the depth ($z$) in the bubble dynamics. The larger $z$ the larger is the maximum radius and longer collapse time. The distance from the surface is key and it influences the pressure drop and the maximum radius controlling the strength of the oscillation. If the bubble oscillation is not strong enough, collapse will not occur. This was observed for bubbles very close to the free surface. The linear vertical pressure distribution rationalises the cascade of bubble collapses (Fig. 1).

The bubble dynamics from impact to collapse occurs in a time scale $t_b \simeq 800\ \mu s$ much larger than the acoustic time of the liquid bulk, $t_{ac} = 2H/c \simeq 80\ \mu s$ (own container of height $H$ =6 cm and a speed of sound around $c \simeq 1500$ m/s), so that a train of bubbles such as the one in Fig. 1 grows in apparent synchrony. These results and observations are consistent with the idea that the acceleration of the container drives the bubble dynamics. This contrasts with the travelling acoustic waves interpretation mentioned in [5] (no time scale separation was reported).

Our development highlights the importance of using a rigid container to produce high accelerations. Soft and elastic materials damp the impact acceleration reducing the pressure drop. In contrast, the supporting material should allow the displacement of the container, to ensure a soft absorption of the downwards acceleration. Materials such as synthetic foam are well suited for this purpose. Holding the beer would lead to the same result. We have verified that a similar experiment performed on top of a rigid support does not lead to foam overflow. The impact is absorbed in the walls of the container. The container cannot move and there is no acceleration downwards entailing the required pressure drop for the growth and collapse of bubbles.

As a final remark, note that this simple experimental setup brings together a rich physical insight and accounts for complex phenomena such as jetting during bubble collapse. In the present case, the jetting is observed downwards towards the bottom of the container, which is in opposite direction to a gravity-induced jetting as illustrated in Fig. 4. This difference can be interpreted by the inverse pressure gradient created by the container acceleration, opposite to the pressure gradient induced by gravity (Fig. 2(a)).

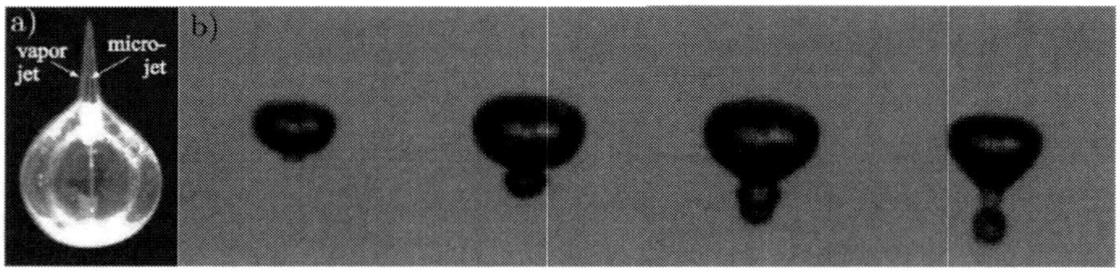


**Figure 4.** (a) Gravity-induced jetting, directed upwards, from [9]. (b) Downwards jetting observed during the collapse of a beer bubble under the reverse acceleration due to the vertical impact. The time step between images is 24 $\mu s$.

## 4. Conclusions

A thorough description of the physics involved in the beer tapping process is presented based on numerical and experimental investigations. Two key parameters have been identified to maximise the beer overflow: the acceleration of the container due to the impact, and the distance of the bubble from the free surface.

## References

[1] Zhang Y and Kling G W 2006 *Annu. Rev. Earth Planet. Sci.* **34(1)**, 293–324.
[2] Cashman K V, Stephen R, and Sparks, J 2013 *Bull. Geol. Soc. Am.* **125(5-6)**, 664–690.
[3] Liger-Belair G 2012 *Eur. Phys. J. Spec. Top.* **201(1)**, 1–88.
[4] Mott R W and Woods A W 2010 *J. Volcanol. Geotherm. Res.* **192(3-4)**, 151–158.
[5] Rodriguez-Rodriguez J, Casado-Chacon A, and Fuster D, 2014 *Phys. Rev. Lett.* **113( 21)**.
[6] Batchelor G K 1967 An Introduction to Fluid Dynamics, Cambridge University Press.
[7] Brennen C E 1995 Cavitation and bubble dynamics, Oxford University Press.
[8] Franc J-P and Michel J-M 2005 Fundamentals of Cavitation, Kluwer Academic Publishers.
[9] Obreschkow D, Tinguely M, Dorsaz N, Kobel P, de Bosset A and Farhat M, 2011 *Phys. Rev. Lett.* **107(20)**.

# Theoretical evidence for inconsistencies in experimental bubble nucleation rates of propane/$CO_2$ and R22/$CO_2$ mixtures

**Tomáš Němec**
Institute of Thermomechanics of the CAS, v.v.i., Dolejškova 5, 18200 Praha 8, Czech Republic
E-mail: nemec@it.cas.cz

**Abstract.** A binary formulation of the classical nucleation theory (CNT) is developed for homogeneous bubble nucleation in systems composed of a liquid solvent and a dissolved gas. The CNT predictions coincide with experimental nucleation data from the literature for diethylether/$N_2$ and isobutane/$CO_2$ mixtures, while several inconsistencies are identified for propane/$CO_2$ and R22/$CO_2$ experimental datasets.

## 1. Introduction

The homogeneous nucleation of bubbles in liquids [1] can be influenced considerably by the presence of a dissolved gas. The effects of the dissolved gas come into play in many natural and technological processes where the liquid mixture is put in a metastable state, e.g. during geyser eruptions, cavitation in fluid machinery, or production of carbonated beverages.

In 1970, Ward *et al* [2] formulated the nucleation theory for the case of a liquid solvent and a dissolved gas, which was based on the assumption of a weak solution. In more detail, the differences of chemical potentials between the gas phase and the liquid phase required for the evaluation of the nucleation work were written in an approximate form appropriate for weak solutions. And the surface tension of the pure solvent was used instead of the unknown surface tension of the liquid mixture. These assumptions enabled to simplify the resulting expressions for the binary nucleation work and the nucleation rate, resembling the simpler unary theoretical formulation. The theory was validated by the authors for the experimental bubble nucleation rates of the diethylether/$N_2$ mixture. Since the mole fraction of $N_2$ in diethylether was rather low, reaching roughly 5 %, the weak-solution assumption of the Ward theory was satisfied. The Ward theory, due to its simplicity, was used by many authors to describe the effect of a dissolved gas on bubble nucleation [3, 4, 5].

The Ward theory was later applied to the experimental nucleation rates measured in the propane/$CO_2$ system by Mori *et al* [6]. Their calculations show a very good agreement between the theoretical predictions and the measured nucleation rates, roughly within the experimental uncertainty of the bubble column experiment of $\pm 1.5$ K. However, due to mole fractions of $CO_2$ reaching up to 30 %, it is surprising that the weak-solution assumption within the Ward theory lead to such precise predictions of the nucleation rate. It is the value of the surface tension that is crucial for the calculation of the theoretical nucleation rate since it is located in the exponent and raised to the power of 3 in the theoretical expression for the nucleation rate. As the surface

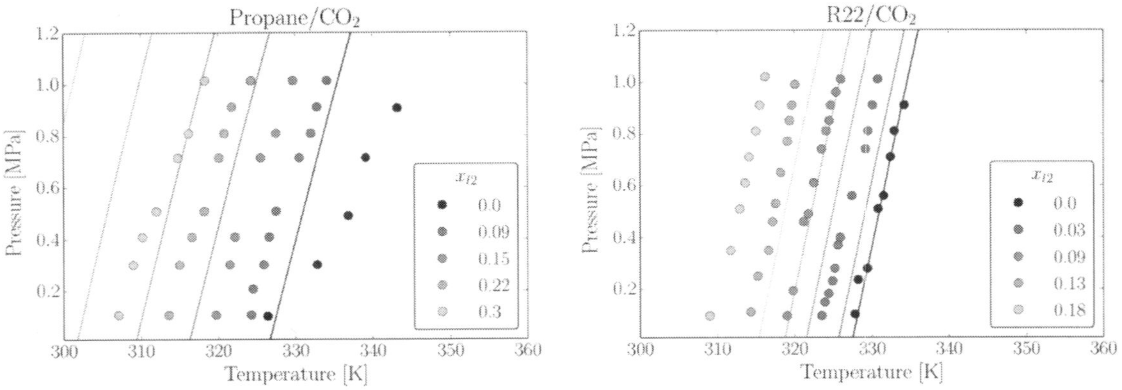


**Figure 1.** Experimental nucleation rates of $10^{10}$ m$^{-3}$s$^{-1}$ [6] in propane/$CO_2$ (left) and R22/$CO_2$ (right) mixtures (dots) compared to the predictions of the binary CNT (full lines).

tension of propane/$CO_2$ at the high concentrations of $CO_2$ studied in Mori's experiment can drop to almost one half of the surface tension of pure propane [7], the nucleation rates predicted with the Ward theory using the real surface tension of the mixture no longer coincide with the experimental data.

In the attempt to approach the "accuracy paradox" described in the previous paragraph from an independent theoretical viewpoint, a binary CNT formulation was developed focusing on the case of a liquid solvent with a dissolved gas [8]. The binary CNT approach presents an evolution of the Ward theory since the chemical potentials of both components are evaluated properly for the concentrated solution, and the real surface tension of the concentrated liquid mixture is used. The predictions of the binary CNT are shown in Fig. 1 and compared to the experimental data. For propane/$CO_2$, the CNT nucleation rates overestimate the reported experimental data considerably. The overestimation is manifested in the fact that the isolines of the constant theoretical nucleation rate of $10^{10}$ m$^{-3}$s$^{-1}$are shifted to lower temperatures compared to the measured data. This discrepancy between theory and experiment increases as the mole fraction $x_{l2}$ of $CO_2$ in propane grows, reaching almost 20 K difference for $x_{l2} = 0.3$. Moreover, the unary nucleation rates in the pure propane are not predicted correctly as well; the slope of the experimental datapoints in Fig. 1 for $x_{l2} = 0.0$ is significantly lower than the slope predicted by the theory. For R22/$CO_2$, opposite to the propane/$CO_2$ case, the predictions of the binary CNT shown in Fig. 1 underestimate the experimental data. An agreement between theory and experiment is observed for pure R22 data.

There are only two more bubble nucleation rate datasets in the literature that are large enough to allow for a comprehensive analysis in terms of pressure, temperature, and concentration dependence of the nucleation rate, namely the diethylether/$N_2$ [2] and isobutane/$CO_2$ [9]. Both of these datasets are predicted correctly by the binary CNT [8].

## 2. Scaled analysis of unary nucleation data

To investigate the difference in the slope of the theoretical $10^{10}$ m$^{-3}$s$^{-1}$ isoline and the slope of experimental nucleation rates for pure propane ($x_{l2} = 0.0$) as observed in Fig. 1, a comparison with nucleation data of similar substances in terms of the scaled theory [10, 11] can be performed. In the scaled approach, all substances behave similarly in the way that the nucleation data of a constant nucleation rate align on a single line in the scaled plot. The scaling behavior was verified for a set of 14 substances including normal alkanes, halogenated hydrocarbons, and

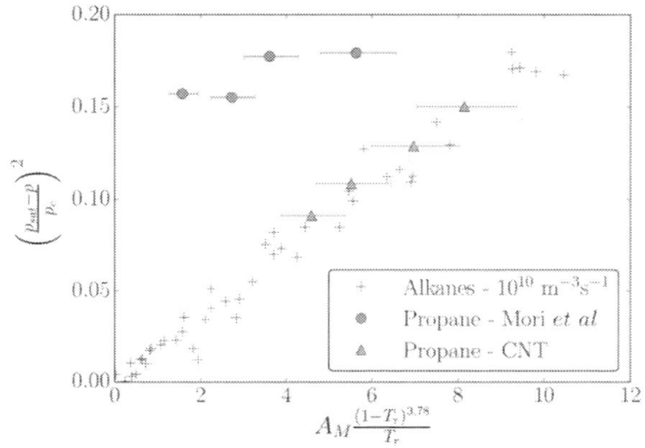


**Figure 2.** Scaled plot of experimental nucleation rates for alkanes shows the relation between the scaled temperature (horizontal axis) and the scaled pressure (vertical axis) for a fixed value of the measured nucleation rate; see [11] for details of the scaled variables. The $10^{10}$ $m^{-3}s^{-1}$ dataset includes nucleation rate data of $n$-alkanes for $n = (1, \ldots ,8)$. The errorbars show the uncertainty in the temperature of $\pm 1.5$ K.

aromatic hydrocarbons [11]. The propane data of Mori *et al* [6] behave anomalously in the scaled plot as shown in Fig. 2 (red dots); the data clearly deviate from the scaling behavior of other alkane experimental data for the nucleation rate of $10^{10}$ $m^{-3}s^{-1}$. However, the predictions of CNT for pure propane follow properly the scaling law in Fig. 2 (green triagles).

## 3. Inconsistency of binary experimental data

As shown in Fig. 1, the experimental nucleation rates of propane/$CO_2$ are overpredicted considerably by CNT, while the nucleation rates of R22/$CO_2$ are underpredicted by CNT. To explain such contradictory behavior, the thermophysical properties of both mixtures related to the nucleation rate can be analyzed. Actually, propane and R22 have almost identical thermophysical properties, their mixtures with $CO_2$ have also similar densities and surface tensions, and the only considerable discrepancy lies in the Henry constant of $CO_2$, which is roughly twice as large in propane than in R22. A higher Henry constant means that the dissolved gas for a given concentration of the liquid mixture creates a higher gas pressure above the liquid, and, therefore, results in a higher tendency of the liquid mixture to nucleate, see [8] for a detailed discussion. In other words, the nucleation rates in propane/$CO_2$ should be higher than in R22/$CO_2$ for a given $CO_2$ concentration. However, a closer look at the experimental nucleation rates in Fig. 1 reveals that the nucleation rates measured for the $CO_2$ mole fraction of $x_{l2} = 0.09$ are lower for propane/$CO_2$ than for R22/$CO_2$, i.e. the experimental datapoints for $x_{l2} = 0.09$ are located at higher temperatures for propane/$CO_2$. This relation between the measured nucleation data is in a direct opposition to the expectation based on the relation between the Henry constants.

The above contradiction between the experimental nucleation rates and the respective Henry constants supports a hypothesis of an accidental interchange of the experimental datasets reported in the original work of Mori *et al* [6]. Indeed, a very good agreement can be observed when the original R22/$CO_2$ experimental data are utilized as propane/$CO_2$ data in the comparison with CNT and vice-versa as shown in Fig. 3.

## 4. Conclusion

A thorough CNT analysis of bubble nucleation data reported by Mori *et al* [6] for two binary mixtures, propane/$CO_2$ and R22/$CO_2$, revealed essential inconsistencies in these experimental datasets. First, the precise predictions of the nucleation rate with the Ward theory for propane/$CO_2$ contrasts with the fact that the inaccurate surface tension of pure propane is used in the calculations. Second, the pressure and temperature dependence of the measured

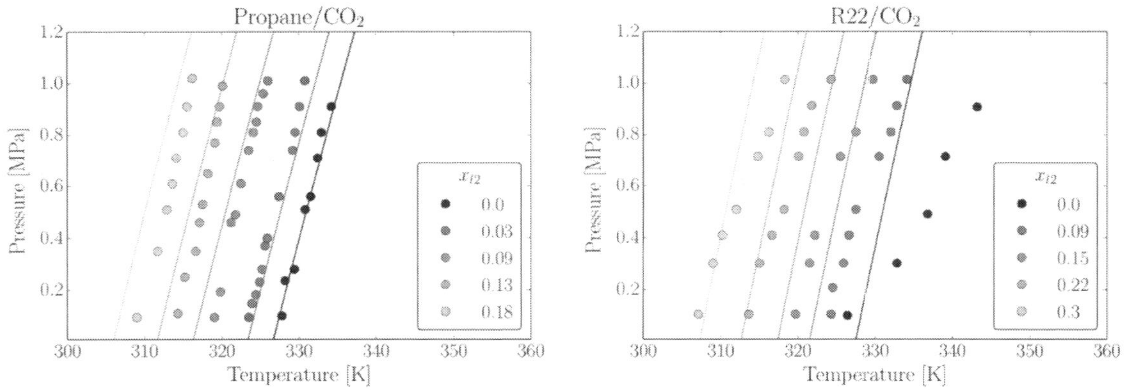


**Figure 3.** Nucleation rates of $10^{10}$ $m^{-3}s^{-1}$ predicted by the binary CNT (lines) for propane/$CO_2$ (left) and R22/$CO_2$ (right) compared to experimental data interchanged between the two respective mixtures (dots).

nucleation rate for pure propane violates a general scaling law, while the CNT nucleation rates obey the scaling relation correctly. Third, the experimental nucleation rates in propane/$CO_2$ are lower than in R22/$CO_2$, while a higher nucleation rate should be observed due to the higher Henry constant of $CO_2$ in propane. The higher nucleation rate in propane/$CO_2$ is indeed predicted by the CNT.

It is the goal of this work to point out that more bubble nucleation data should be acquired experimentally for the binary systems of a liquid solvent and a dissolved gas. Obtaining experimental data complementary to the existing measurements of the nucleation rate in the propane/$CO_2$ and R22/$CO_2$ systems would help to clarify the inconsistencies presented in this work, and measuring new nucleation data for other binary mixtures would stimulate further development of the binary nucleation theory.

## Acknowledgments

The author would like to acknowledge the institutional support RVO:61388998 of the Institute of Thermomechanics of the CAS, v. v. i., and the support by the Czech Science Foundation (projects GAP101/10/1819 and GA13–23550S), and by the Technology Agency of the Czech Republic (project TA03030978).

## References

[1] Baidakov V G 2013 *Low Temperature Physics* **39** 643–664
[2] Ward C A, Balakrishnan A and Hooper F C 1970 *J. Basic Eng.* **92** 71–80
[3] Hemmingsen E A 1975 *J. Appl. Phys.* **46** 213–218
[4] Wilt P M 1986 *J. Colloid Interface Sci.* **112** 530–538
[5] Lubetkin S D 2003 *Langmuir* **19** 2575–2587
[6] Mori Y, Hijikata K and Nagatani T 1976 *Int. J. Heat Mass Transfer* **19** 1153–1159
[7] Němec T 2014 *Experimental Fluid Mechanics 2014* ed Dančová P and Vít T (Český Krumlov (CZ), 18. 11. 2014 – 21. 11. 2014) pp 383–388
[8] Němec T 2015 *to be submitted*
[9] Avedisian C T 1985 *J. Phys. Chem. Ref. Data* **14** 695–729
[10] Hale B N 1986 *Physical Review A* **33** 4156–4163
[11] Němec T 2014 *Eur. Phys. J. E* **37** 111

# Experimental and numerical investigations of the bubble collapse at the center between rigid walls

**T Ogasawara[1], N Tsubota[1], H Seki[1], Y Shigaki[1] and H Takahira[1]**

Osaka Prefecture University, 1-1 Gakuen-cho, Naka-ku, Sakai-shi, Osaka 599-8531, Japan

oga@me.osakafu-u.ac.jp

**Abstract.** The collapsing behavior of a bubble generated at the center of the rigid walls and its effect on the wall are investigated experimentally and numerically. The bubble collapse drastically depends on the ratio of the gap width to the maximum bubble radius $w^*$. In case of $2.66 < w^* < 1.17$, by the decrease in $w^*$ to 2.5, the splitting collapse occurs; the bubble splits into two at the center of the gap during collapse and each bubble collapses near the wall with the translation toward nearer wall. Further decrease in $w^*$ to 1.4 causes neutral collapse where the bubble collapses at the center of the gap without splitting and translation. These collapsing behaviors are successfully simulated by considering the bubble shape at the maximum bubble volume. The peak pressure on the wall decreases by the transition of the collapsing behavior from the splitting collapse to the neutral collapse due to the decrease in $w^*$, which indicates the wall damage reduction due to the neutral bubble collapse in thinner gap.

## 1. Introduction

The collapse of cavitation bubbles near a solid wall cause the material damage and have been studied for many years [1]. Recently, in liquid mercury target systems, suppressing cavitation damages on the vessel walls is an important issue [2]. When the proton beams are injected, high pressure wave are generated by a sudden thermal expansion around the focus. One of the approaches to reduce the cavitation erosion is to adopt a double wall structure at a beam window, which is confirmed to have a positive effect [3]. On the other hand, it is indicated that the double wall structure can adversely cause larger pits on the wall of the narrow channel under a certain condition [4]. Bubble collapse within a narrow gap has been investigated by several researchers (ex. [5, 6]). However, the details about its dependence on the ratio of the gap width to the bubble size and its effect on the walls have not been clarified yet. In particular, the collapsing behavior under the condition that the bubble size is larger than the gad width becomes more important for the liquid mercury target system.

It is necessary to clarify the bubble collapse with the variation of the bubble size and its effect on the walls for mitigating the material damage in the narrow gap. To do this, the laser-bubble experiment and the visualization of corresponding shock waves have been conducted for the bubble between two parallel walls. Numerical simulations for the growth and collapse of bubbles have been also conducted and the pressure on the wall is evaluated.

## 2. Methods and conditions

### 2.1. Experimental setup

Figure 1 shows a schematic diagram of the experimental setup to generate a cavitation bubble within a narrow gap between two parallel rigid walls and to record a sequence of photographs of the bubble collapse and its corresponding shock waves by shadowgraph. The container is filled with distillated water and the U-shaped wall made of stainless steel is installed at the center of the container. The bubble is generated with expanded and refocused beam from a Q-switched Nd:YAG laser. The widely collimated beam and the refocusing by using an achromatic lens contribute to narrow focusing area. The bubble shape and the shock wave are recorded simultaneously by using a high-speed video camera with the frame rate of up to 1.25 Mfps. As a light source for the shadowgraphy, a collimated beam from CW laser is utilized. $w$ indicates a gap width (a distance between two parallel walls), and is 2.5 mm in this experiment. Another length scale related to the bubble size is the maximum value of the volume equivalent radius $R_{max}$. Therefore, a dimensionless parameter $w^*$ is defined as $w^* = w/R_{max}$. $R_{max}$ is controlled by changing the diameter of an iris diaphragm and the pressure inside the container. In the present experiment, the bubbles are precisely generated at the center between two parallel walls. The characteristic time of the bubble collapse in experiment is $\tau = R_{max}/(p_\infty/\rho)^{1/2}$ where $p_\infty$ and $\rho$ are the pressure and density of water inside the container, respectively. $t = 0$ represents the time when the bubble takes its maximum volume, and the dimensionless time $t^*$ is defined as $t^* = t/\tau$.

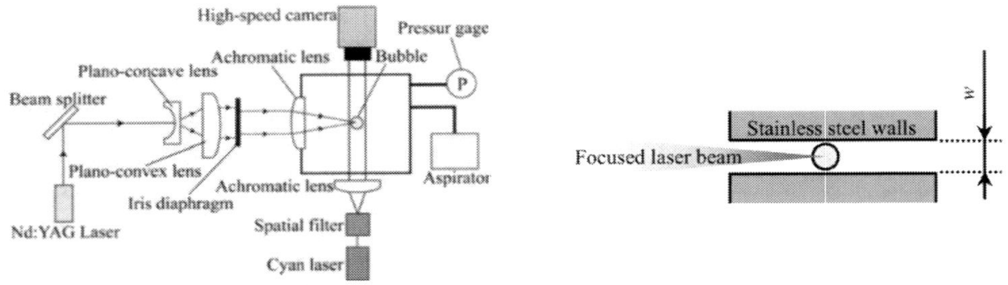


**Figure 1.** Experimental setup and the configuration of the rigid walls with a gap width of $w$.

### 2.2. *Numerical methods and conditions*

The initial shapes of bubbles in numerical analysis, shown in Figure 3 at $t^* = 0$, are set to be similar to those in the experimental results when the bubble volume is maximum. The bubble shape is assumed to be symmetric to the axis perpendicular to the walls. The initial volume $V_0$ is the same as the maximum volume in the experiments; $R_{max}$ in the simulations, which is identical to the initial equivalent radius $R_0$, is taken to be $R_{max}=R_0 = (3V_0/4)^{1/3}$. The bubble is initially at rest in a water-filled gap and thin liquid films exist between the bubble surface and the walls. The initial internal gas pressure $p_{g0}$ and liquid pressure around the bubble $p_\infty$ are taken to be $p_{g0} = 1.013 \times 10^5$ Pa and $p_\infty = 10^8$ Pa, respectively. Under the present numerical condition, the characteristic time scale in numerical analysis is defined as $\tau = R_0/(\Delta p/\rho_0)^{1/2}$, where $\Delta p = p_\infty - p_{g0}$, and $\rho_0$ is the initial density of ambient fluid. The governing equations are the two-dimensional axisymmetric Euler equations. The internal gas is assumed to be an ideal gas. Both gas and water are modeled with the equation of state for stiffened gas. Heat and mass transfer through the bubble interface and fluid viscosity are disregarded in the analysis. The ghost fluid method is used to treat a two-phase flow as two single-phase flows with satisfying the boundary conditions at the interface which is captured with the level set function [7].

## 3. Results and discussions

Figures 2 and 3 show some typical experimental results of the bubble shape for (a) $w^* = 1.61$ and (b) 1.25 and numerical schlieren images with bubble shapes correspond to the experimental conditions of (a) $w^* = 1.61$ and (b) $w^* = 1.25$, respectively. The typical shadowgraph images with shock wave emission are shown in figure 4.

As can been seen in figure 2 (a), the bubble becomes horizontally long ellipsoidal shape at its maximum volume in the case of $w^* = 1.61$. The shrinkage of the bubble surface around the middle of the gap towards the axis of symmetry occurs ahead. On the other hand, the bubble surface near the walls continues to expand outward ($t^* = 0.791$). The shrinking velocity parallel to the wall toward the axis of symmetry at the middle of the gap is much higher than that near the walls; the bubble gets constricted in the middle ($t^*$ = 1.172). The constriction develops and the bubble surface at the tip of the constriction collides at the axis of symmetry, and the bubble is split in two for the upper and lower sides ($t^*$ = 1.362). The split bubbles continue to shrink with the translation toward the nearer walls and collapse in the vicinity of the walls ($t^*$ = 1.407). This splitting collapse have been reported by the previous study [6] when $w^*$ takes a value of slightly larger than 2, although the phenomenon in case of smaller $w^*$ was not investigated. In the case of $w^*$ = 1.25 (figure 2 (b)), the bubble takes the flattened hourglass shape and reaches the maximum expansion ($t^*$ = 0). Once the bubble surface near the walls changes its motion from the expansion to the shrinkage, the surface velocity toward the axis of symmetry is accelerated and gets higher than that around the middle of the gap where the surface started to shrink earlier ($t^*$ = 0.917). After that, surface collisions occur near the walls in first ($t^*$ = 1.331), the bubble is left at the middle of the gap as a single bubble and neutrally collapses without translation ($t^*$ = 1.452).

The above mentioned bubble behaviors in the experiments are successfully simulated in figure 3. When $w^*$ = 1.61 with the initial bubble shape of the disc-like shape, the convex part around the middle of the gap shrinks faster and becomes cylindrical shape with a constriction in the middle at $t^*$ = 1.295. After that the surface collision occurs at the tip of the constriction ($t^*$ = 1.340) and the bubble is split in two. At this moment, a shock wave is emitted where the surface collision occurs. The split bubbles collapse with the translation toward the nearer walls and the small liquid jets are formed along the axis of symmetry toward the walls ($t^*$ = 1.364). These liquid jets penetrate the bubble near the walls and the bubbles become a toroidal shape ($t^*$ = 1.407). The liquid jet impingement causes the shock wave emission near the wall. This liquid-jet related shock wave is actually observed in the shadowgraph image in figure 4 (a). The toroidal bubbles continue to shrink and collapse just before $t^*$ = 1.451, the shock wave is emitted again due to the rebound of the bubble volume. On the other hands, in the case of the flattened hour glass ($w^*$ = 1.25) in figure 3 (b), the convex parts near the walls of the bubble surface prominently shrinks and the bubble shape becomes cylindrical ($t^*$ = 0.954) while the constriction initially existing around the middle of the gap does not develop so much. At this moment, the surface instability occurs near the wall, which leads to the further acceleration of the flow concentration toward the axis of symmetry. Subsequently, surface collisions occur and the main body of the bubble is

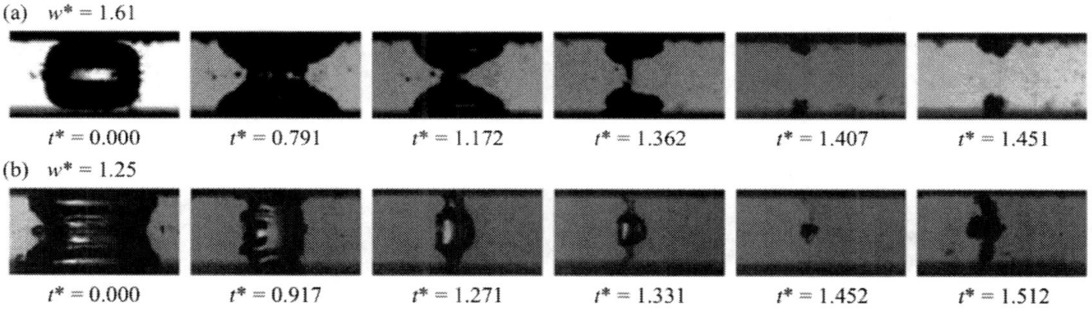


**Figure 2.** Typical snapshots of the collapsing bubble at (a) $w^*$ = 1.61 and (b) $w^*$ = 1.25.

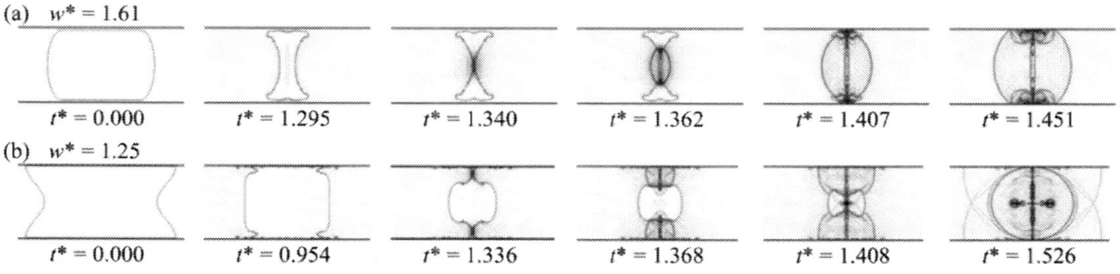


**Figure 3.** Numerical schlieren images with bubble shapes of the numerical simulation at (a) $w^*$ = 1.61 and (b) $w^*$ = 1.25. The bubble shape at $t^*$ = 0 represents the initial shape in computation.

(a) Splitting bubble collapse $w^* = 1.60$

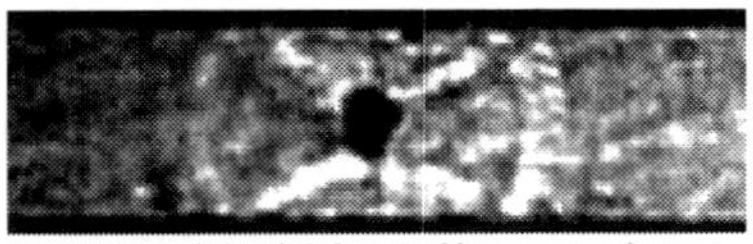

(b) Neutral bubble collapse $w^* = 1.17$

**Figure 4.** Visualization of the shock waves emissions; (a) splitting collapse, (b) neutral collapse.

remained at the center of the gap as a single bubble ($t^* = 1.336$). These surface collisions cause the shock wave emissions near the wall. The remained bubble collapses neutrally at the center of the gap without translation; the corresponding shock wave is generated and propagates spherically ($t^* = 1.526$). The rebound shock wave is observed in the experiments as evident in the shadowgraph image in figure 4 (b). It should be noted that the collapse times in figures 3 (a) and (b) are in good agreement with those in the corresponding experiments.

We also investigate the pressure on the solid walls in the simulations. It is found that in the case of the splitting bubble collapse ($w^* = 1.61$), the pressure increases suddenly at $t^* = 1.40$ and takes the maximum value. This pressure rise is found to be caused by the shock wave emission due to the jet impingement on the split bubble near the wall. On the other hand, in the case of the neutral collapse of the initially flattened hourglass shaped bubble ($w^* = 1.25$), two characteristic pressure peaks are observed; former pressure peaks are caused by the surface collisions near the wall and latter one by shock wave emission due to the rebounding motion of the collapsing bubble at the center of the gap. It is confirmed that the maximum value of the pressure peak in the case of the splitting collapse becomes much higher than that in the case of the neutral collapse.

## 4. Conclusion

The collapsing behavior of a bubble generated at the center of two rigid walls and its effect on the wall was investigated experimentally and numerically. We demonstrated two typical bubble behaviors depending on $w^*$ which was the ratio of the maximum bubble radius and the gap width: one was a spilling bubble collapse and the other was a neutral bubble collapse. In the splitting bubble collapse where the bubble becomes ellipsoidal shape at its maximum volume, the liquid-jets occurred toward both the upper and lower walls from the split bubbles. The shock wave associated with these jets were observed both in the experiments and simulations. In the neutral bubble collapse where the bubble shape became flattened hourglass shape at its maximum volume, the liquid-jets did not occur and the bubble collapsed at the center of the gap without translational motion; the rebound shock wave was emitted there. These collapsing behaviors were successfully simulated by using the ghost fluid method. It was shown that the magnitude of wall pressure by the shock waves was much higher in the splitting bubble collapse than that in the neutral bubble collapse.

**Acknowledgment**

This work was partly supported by JSPS KAKENHI Grant Number 25289034.

**References**

[1] Brennen C 1995 *Cavitation and bubble dynamics* (Oxford University Press)

[2] Futagawa M, Kogawa H, Hasegawa S, Naoe T, Ida M, Haga K, Wakui T, Tanaka N, Matsumoto Y and Ikeda Y 2008 *J. Nucl. Sci. Technol.* **45**(10) 1041

[3] McClintock D A, Riemer B W, Ferguson P D, Carroll A J and Dayton M J 2012 *J. Nucl. Mater.* **431** 147

[4] Riemer B, Haines J, Wendel M, Bauer G, Futakawa M, Hasegawa S and Kogawa H 2008 *J. Nucl. Mater.* **377** 162

[5] Gonzalez-Avila S R, Klaseboer E, Khoo B C and Ohl C-D 2011 *J. Fluid Mech.* **682** 241

[6] Ishida H, Nuntadusit C, Kimoto H, Nakagawa T and Yamamoto T 2001 *Proc. CAV 2001: Fourth Intl Symp. on Cavitation (Pasadena, CA)* CAV2001:sessionA5.003

[7] Kobayashi K, Kodama T and Takahira H 2011 *Physics in Medicine and Biology* **56** 6421

# Cloud cavitation induced by shock-bubble interaction in a viscoelastic solid

**Ryota Oguri[1] & Keita Ando[2]**

Department of Mechanical Engineering, Keio University, 3-14-1 Hiyoshi, Kohoku-ku, Yokohama, 223-8522, Japan

E-mail: [1]oguryo1-3@keio.jp,[2]kando@mech.keio.ac.jp

**Abstract.**

We experimentally study a shock-bubble interaction problem in a viscoelastic solid, which is relevant to shock wave lithotripsy. A gas bubble is produced by focusing an infrared laser pulse into gelatin. A spherical shock is then created, through rapid expansion of plasma that results from the laser focusing, in the vicinity of the gas bubble. The shock-bubble interaction is recorded by a CCD camera with flash illumination of a nanosecond green laser pulse. The observation captures cavitation inception in the gelatin under tension that results from acoustic impedance mismatching at the bubble wall. Namely, the shock reflects at the bubble interface as a rarefaction wave, which induces the nucleation of cavitation bubbles as a result of rupturing the gelatin.

## 1. Introduction

Cavitation is utilized in non-invasive medical treatment such as shock wave lithotripsy (SWL) and high-intensity focused ultrasound (HIFU) [1, 2]. When these acoustic waves are focused into the diseased part of patients (e.g., kidney stones or cancer cells), cloud cavitation will occur in the targeted region. The therapy is expected through mechanical and/or thermal actions of the cavitation bubble dynamics [3, 4]. However, controlling the extent of the acoustic cavitation is difficult and may give rise to side effects including unwanted tissue damage.

To minimize the side effects while maintaining the efficacy of cavitation treatment, controlling the nucleation of cavitation bubbles is desirable. One of difficulties in controlling the cavitating region is due to secondary cavitation resulting from interaction between incoming acoustic waves and the pre-existing cavitation bubbles. Such secondary cavitation is reported in the HIFU experiment where cavitation bubbles are nucleated by the backscattering of focused ultrasound from a pre-existing bubble in a gelatin-based tissue-mimicking phantom [5]; the compression part of the ultrasound is reflected as tension through acoustic impedance mismatching at the bubble wall [6], rupturing the material (i.e., the nucleation of cavitation bubbles). Similarly, secondary cavitation in water is obtained by interaction between laser-induced shocks and bubbles [7]. However, experimental observation of cavitation resulting from shock-bubble interaction in tissue-like materials is still lacking.

In this paper, we observe cloud cavitation induced by shock-bubble interaction in a tissue-mimicking material. The shock-bubble interaction and the subsequent cavitation in gelatin are captured by stroboscopic photography based on a nanosecond pulse laser. In what follows, we introduce the experimental procedure to visualize laser-induced phenomena and present analyses of the shock and cavitation dynamics.

## 2. Experiments

### Sample preparation

A gelatin phantom was used as a tissue-mimicking medium. The phantoms were made by mixing 10 wt% gelatin (Type-A, Sigma Aldrich, St. Louis, MO, USA) with water. The solution was poured into a cylindrical dish (polystyrene) of 56 mm diameter and 16 mm height. The solution was stored in refrigeration at 4°C overnight and placed in room temperature before experiments.

### Optical set-up

The experimental set-up (similar to that of Ando *et al.* [8]) was shown in Fig. 1. A Q-switched Nd:YAG laser (ULTRA 50 GRM, Quantel) emits single laser pulses simultaneously at wavelengths of 532 nm (green) and 1064 nm (infrared or IR) with a duration of 6 ns. An infrared pulse delivered into the gel was $1.58 \pm 0.08$ mJ (the $\pm$ sign denotes a standard deviation from 10 measurements). Optical breakdown was induced by focusing an IR laser pulse through a microscope objective (10×, NA = 0.25). After the optical breakdown, a spherical shock wave is radiated and a bubble arises from vaporization in the gelatin behind the shock [9]. Shadowgraphs (from top view) of the laser-induced phenomena (including shock-bubble interaction and the subsequent secondary cavitation) in the gelatin were captured using a CCD camera (Pixcelfly, PCO; one pixel = 0.6 μm) with stroboscopic illumination of a green laser pulse. The illumination timing was tuned by changing the length of the single-mode optical fiber through which the green laser pulse propagates.

### Shock evolution measurement

The evolution of the laser-induced shock was first visualized in the gelatin (without bubbles inside) so as to estimate the shock pressure. To avoid wave reflections within the gelatin case, the IR laser pulse was focused sufficiently away from any boundaries.

### Visualizing the shock-bubble interaction

A pre-existing, spherical gas bubble was created by focusing the IR laser pulse into the gelatin; even after the laser bubble collapses, non-condensible gas contents (air) are left. The size of the gas bubble slowly changes due to mass transfer driven by Laplace pressure effects [10]. When the bubble radius reaches 150 μm, the IR laser pulse was shot in the vicinity of the gas bubble in order to observe the interaction with the laser-induced shock. The distance from the laser focus to the proximal bubble wall is set at $d$ = 121 μm (Fig. 2).

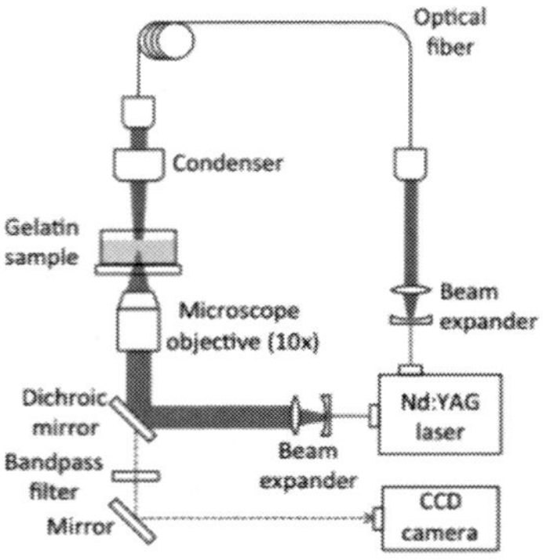


**Figure 1.** The optical system for visialization of laser-induced phenomena in gelatin.

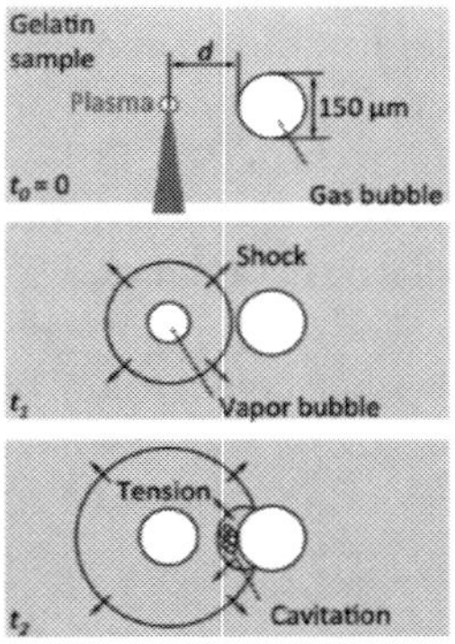


**Figure 2.** Sequence of the interaction of the laser-induced shock with a pre-existing gas bubble.

Journal of Physics: Conference Series **656** (2015) 012032 doi:10.1088/1742-6596/656/1/012032

## 3. Results and discussion

### Shock pressure estimation

Figure 3 shows sequential images of the laser-induced shock and bubble. The shock is identified as a dark annulus and its front is defined as the midpoint of the annulus. Figure 4(A) shows the the temporal evolution of the shock front whose position, $r_s$, is measured from the laser focus; each point (denoted by the circle) is the average out of 10 measurements and the error bar represents the standard deviations. Noting that the shock speed is eventually reduced to the speed of sound, the shock front positions in Figure 4(A) are found to be well fitted to

$$r_s(t) = a_1 + C_0 t - a_2 \exp\left(-\frac{t - a_3}{a_4}\right) \tag{1}$$

where $C_0$ is the speed of sound in the gelatin phantom at room temperature (1520 m/s [11]), and $a_1$, $a_2$, $a_3$ and $a_4$ are parameters to be fitted by the least-squares method. Substituting the shock speed calculated from the fitting function (1) into the Rankine–Hugoniot relation for stiffened-gas-type materials [12], the shock pressure that depends on the distance from the laser focus is estimated (see Fig. 4(B)).

### Shock-bubble interaction and cavitation

Figure 5 shows the sequential images of the laser-induced shock interacting with the pre-existing gas bubble in the gelatin. The shock wave reaches at the bubble wall in Fig. 5(A); the shock pressure is predicted at 81 MPa from Fig. 4(B). After the shock collision, it is reflected as a tension wave due to acoustic impedance mismatch between the gelatin and the bubble. This tension ruptures the gelatin, thus nucleating a cloud of cavitation bubbles as seen in Fig. 5(B). The secondary cloud cavitation is extending at a later time in Fig. 5(C). The pressure in the gelatin after the tension wave passage is determined by superimposing the incoming (shock) and reflected (tension) waves. Assuming that the incident shock decays significantly after its collision with the bubble, the gelatin under tension can in principle be subjected to very large negative

(A) 48.1 ns (B) 98.1 ns

**Figure 3.** Evolution of the laser-induced shock and bubble in the gelatin different times after the laser focusing. The time is measured from the laser focusing.

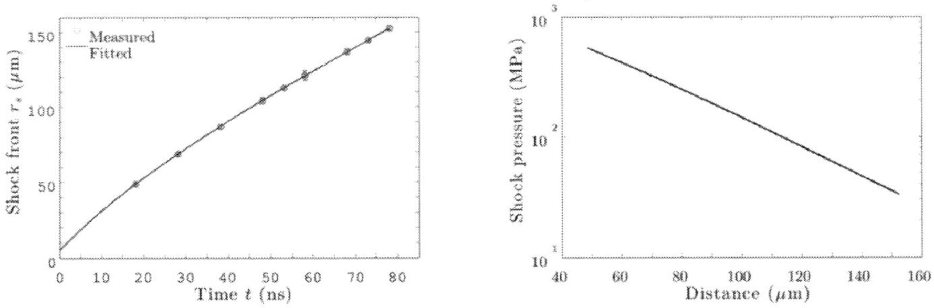


(A) (B)

**Figure 4.** (A) Temporal evolution of the shock front. (B) Shock pressure as a function of distance from the laser focus.

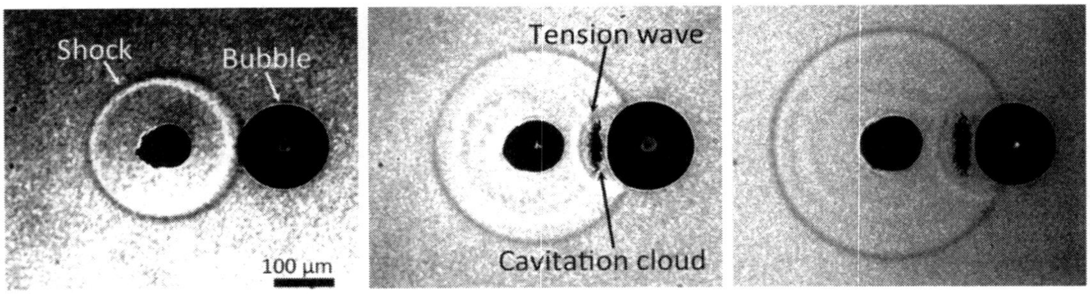


(A) 58.1 ns (B) 78.1 ns (C) 98.1 ns

**Figure 5.** Evolution of shock-bubble interaction and its accompanying cavitation in the gelatin. The time is measured from the laser focusing.

pressures (say, down to −81 MPa). However, before reaching such negative pressures, the gelatin would be ruptured with bubble nucleation at cavitation thresholds as seen in Figs. 5(B) and (C). This observation indicates that cavitating regions may possibly be extended due to the interaction between shocks and pre-existing cavitation bubbles in the application of SWL.

## 4. Conclusion

In this paper, we have demonstrate the possibility of having secondary cavitation in the viscoelastic solid induced by the interaction of a laser-induced shock wave and a pre-exsiting gas bubble, indicating a need to account for shock-bubble interaction in order to control cavitating regions in the application of SWL.

## Aknowledgements

We thank Takahiro Oda for his help in the experiment. This study is supported by JSPS KAKENHI Grant Number 25709008.

## References

[1] M. R. Bailey, J. A. McAteer, Y. A. Pishchalnikov, Y. A. Hamilton, and T. Colonius. Progress in lithotripsy research. *Acoustics Today*, 2:18–29, 2006.
[2] C. C. Coussios and A. R. Ronald. Applications of acoustics and cavitation to noninvasive therapy and drug delivery. *Annu. Rev. Fluid Mech.*, 40:395–420, 2008.
[3] T. Kodama and K. Takayama. Dynamic behavior of bubbles during extracorporeal shock-wave lithotripsy. *Ultrasound in Med. & Biol.*, 24:723–738, 1998.
[4] Y. Matsumoto, J.S. Allen, S. Yoshizawa, T. Ikeda, and Y. Kaneko. Medical ultrasound with microbubbles. *Experimental Thermal and Fluid Science*, 29:255–265, 2005.
[5] A. D. Maxwell, T. Y. Wang, C. A. Cain, J. B. Fowlkes, O. A. Sapozhnikov, M. R. Bailey, and Z. Xu. Cavitation clouds created by shock scattering from bubbles during histotripsy. *J. Acoust. Soc. Am.*, 130:1888–1898, 2011.
[6] T. G. Leighton. *The Acoustic Bubble.* Academic Press, 1994.
[7] P. A. Quinto-Su and K. Ando. Nucleating bubble clouds with a pair of laser-induced shocks and bubbles. *J. Fluid Mech.*, 733(R3), 2013.
[8] K. Ando, Q. Liu, and C. D. Ohl. Homogeneous nucleation in water in microfluidic channels. *Phys. Rev. Lett.*, 109(044501), 2012.
[9] A. Vogel, S. Busch, and U. Parlitz. Shock wave emission and cavitation bubble generation by picosecond and nanosecond optical breakdown in water. *J. Acoust. Soc. Am.*, 100:148–165, 1996.
[10] F. Hamaguchi. The Effect of Viscoelasticity on Bubble Dynamics under Ultrasound Irradiation. Master's thesis, Keio University, 2015.
[11] M. Nakahara, K. Nagayama, and Y. Mori. Shockwave dynamics of high pressure pulse in water and other biological materials based on Hugoniot data. *Japanese Journal of Applied Physics*, 47:3510–3517, 2008.
[12] E. Johnsen. Numerical Simulations of Non-spherical Bubble Collapse. PhD thesis, California Institute of Technology, 2007.

# Spark bubble interaction with a suspended particle

**Siew-Wan Ohl[1,*], Di Wei Wu[2], Evert Klaseboer[1], Boo Cheong Khoo[2]**

[1]Institute of High Performance Computing, 1 Fusionopolis Way, #16-16 Connexis North, Singapore 138632.
[2]Department of Mechanical Engineering, National University of Singapore, 10 Kent Ridge Crescent, Singapore 119260.

Corresponding author's e-mail address: ohlsw@ihpc.a-star.edu.sg

**Abstract**. Cavitation bubble collapse is influenced by nearby surfaces or objects. A bubble near a rigid surface will move towards the surface and collapse with a high speed jet. When a hard particle is suspended near a bubble generated by electric spark, the bubble expands and collapses moving the particle. We found that within a limit of stand-off distance, the particle is propelled away from the bubble as it collapses. At a slightly larger stand-off distance, the bubble collapse causes the particle to move towards the bubble initially before moving away. The bubble does not move the particle if it is placed far away. This conclusion is important for applications such as drug delivery in which the particle is to be propelled away from the collapsing bubble.

## 1. Introduction

The interaction of cavitation bubbles with particles in suspension is of importance in erosion studies. It is proposed that the existence of cavitation bubble enhanced erosion effect [1]. However an experimental work done by Soh and Willis [2] report that the spark-generated cavitation bubble does not move the particles much during its expansion and collapse. In another study using shock wave generated bubbles, the bubble formed on the surface of a particle propels the particle forward as it collapses [3]. In this study, a spark-generated bubble is used to move the suspended particle as it collapses. Our cavitation bubble is much smaller (relative to the particle) as compared to that used in [2]. Observations are recorded using high speed photography.

## 2. Experimental setup

Figure 1 shows the experimental setup used in this study. It consists of a circuit to generate the spark bubble, a suspended particle (on string), and a high speed camera system. The details of the circuit are found in [4]. The particle is made of Perspex, and has a radius of 1.50 mm. It has a density of 2.5 g/cm$^3$. It is suspended by a cotton thread of 0.12 mm in diameter. Different threads have been tested, and this thin thread is found to be affecting the movement of the particle to the smallest degree. We have also performed experiments where the particle is attached to a fixed thread by silicon oil, similar to [5]. In these cases, the string does not hinder the particle movement. Due to space limit, the results are not presented here. The spark bubble is generated at the crossing of two electrodes. The high speed photography system is made up of a high speed camera (Photron, SA1), and a LED spotlight. It is operated at 20,000 frames per second.

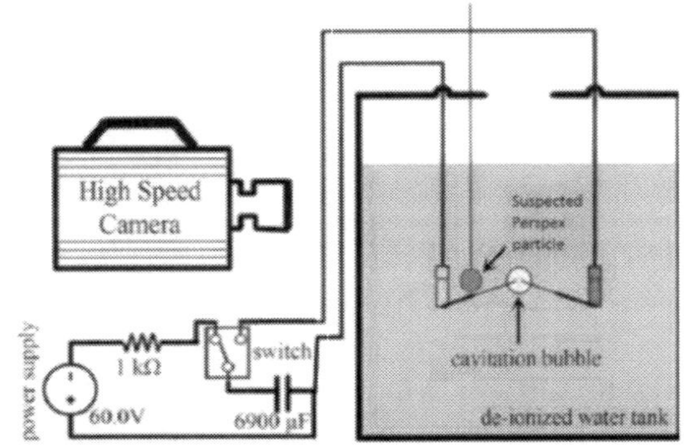


Fig. 1 The experimental setup in this study. The circuit to generate the spark bubble is given in details in [4]. Deionized water is used.

## 3. Results

### *3.1. Particle is pushed away*

A series of bubble particle interaction experiment has been performed. Figure 2 shows the case where the particle is propelled away (to the left) from the cavitation bubble. The bubble (on the right) was generated at a stand-off distance of 2.73 mm from the particle. The initial expansion of the bubble pushes the particle away from its initial position. When it collapses, the bubble forms a 'mushroom' shape (t = 1.20 ms), and then split into two (t = 1.25 ms). The part of the bubble on the particle pushes the particle to the left as it re-expands and collapses again. The other part of the bubble collapses and jets away from the particle.

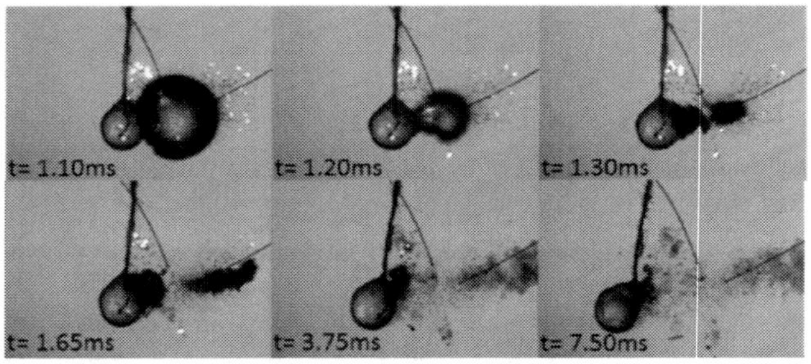


Fig. 2 Interaction of a cavitation bubble with a suspended particle during the bubble's collapse phase. The particle moves away from its initial position to the left. The maximum bubble radius is 3.99 mm (not shown). Framing rate is 20,000 frames per second

### *3.2. Particle moves towards the collapsing bubble*

However, if the spark bubble is generated slightly further from the suspended particle, the particle is initially pushed away as the bubble expands. When the bubble collapses, the particle is attracted towards the collapsed bubble. The movement of a suspended particle which is initially 5.16 mm away from a spark bubble is presented in Figure 3. The bubble obtains its maximum size of 4.05 mm at t = 0.95 ms. Then it collapses forming a 'neck' on the particle (t = 1.35 ms). The bubble collapses at t = 1.40 ms (not shown). The particle moves to the right after the bubble has collapsed (t = 12.5 ms).

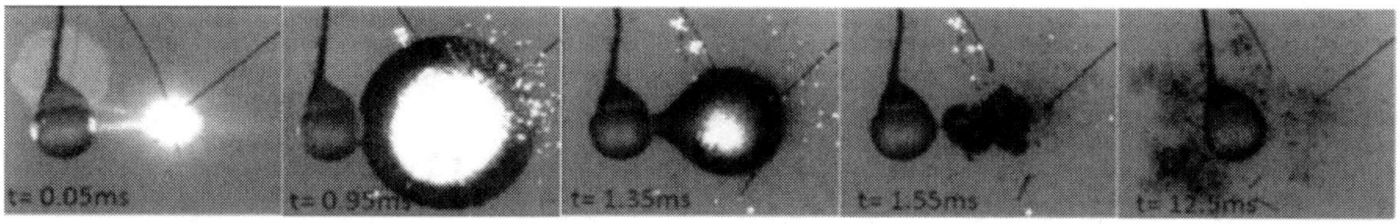


Fig. 3 Interaction of a cavitation bubble with a suspended particle where the particle moves towards the bubble after it has collapsed. The stand-off distance is 5.16 mm. Framing rate is 20,000 frames per second.

A larger Perspex particle of 2.0 mm radius is also tested. Its movement is dependent on the stand-off distance, $S$. Figure 4 shows the initial velocity of the particle after the bubble collapse. The stand-off distance is non-dimensionalized to $S'$ by dividing $S$ by the maximum bubble radius, $R_{max}$, which varies slightly around 4 mm. The cases in Figure 2 and 3 have $S'$ = 0.68 and 1.27, respectively. Within experimental error, the initial velocities for both particles are similar. This has been confirmed by a recent study [5] where 6 different particles are used.

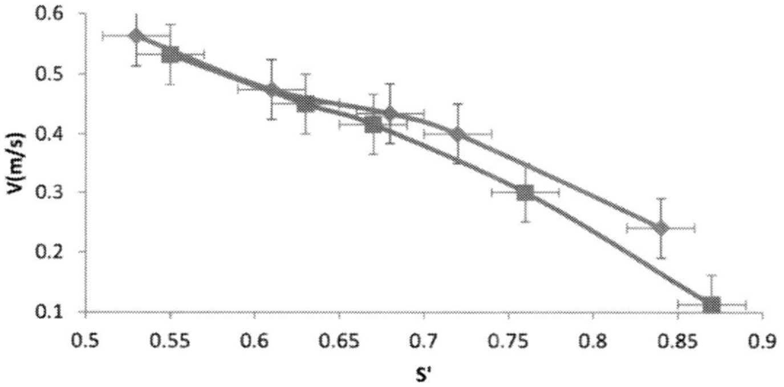


Fig. 4 The initial velocity $V$ of a suspended particle of radius 1.50 mm (blue), and 2.0 mm (red) after the oscillating bubble next to it has collapsed.

### *3.3. Effect of the stand-off distance*

The effect of the stand-off distance on the bubble particle interaction is summarized in Figure 5. The distance is non-dimensionalized with $R_{max}$. It is shown that for a cavitation bubble which is generated about one bubble radius away (1.14), the particle will be displaced away from the spark bubble as it collapses. However a non-dimensionalized stand-off distance between 1.14 to 1.85, the particle will move towards the collapsing bubble. If the particle is placed far from the spark bubble (non-dimensionalized stand-off distance larger than 1.85), it is not moved as the bubble expands and collapses spherically.

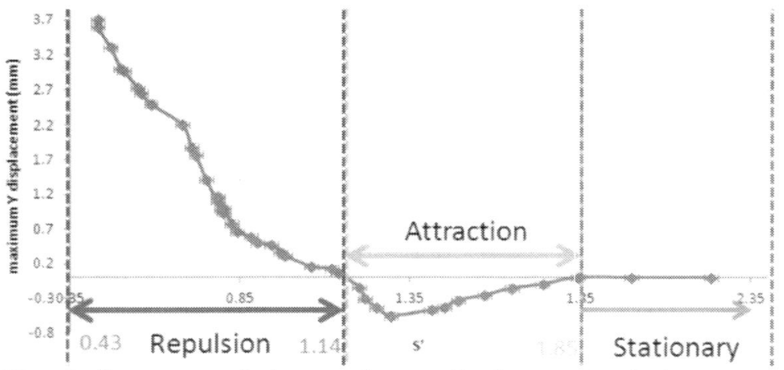


Fig. 4 Summary of the maximum displacement of the suspended particle with different non-dimensionalized stand-off distances from the spark bubble. For the particles that are close to the bubble (S' < 1.14), the particles are pushed away from the bubble after its collapse. For particles that are a bit further from the bubble, they move towards the bubble after its collapse. Similar observation is reported in [5]. For S' > 1.85, the bubble has no influence on the suspended particle, and it remains unmoved.

## 4. Conclusion

This study shows that the movement of a particle next to a cavitation bubble is dependent on the stand-off distance between the two. We have also performed experiments using different particles and a slightly different suspension method. In all cases, the same phenomenon of repulsion and attraction within a range of stand-off distances is observed. Due to the space constrains, only results from one particle is presented. However in a recent article [5], it is found that the initial velocity of the particle is independent of the particle material and to some extent its size. In conclusion, we deem that it is important to place the cavitation bubble very close to the particle if the objective is to 'shoot' the particle away, for example in drug delivery [6, 7] or erosion enhancement.

**References**

[1] Dunstan, P. J., and S. C. Li. "Cavitation enhancement of silt erosion: Numerical studies." Wear 268, no. 7 (2010): 946-954.

[2] Soh, W. K., and B. Willis. "A flow visualization study on the movements of solid particles propelled by a collapsing cavitation bubble." Experimental thermal and fluid science 27, no. 5 (2003): 537-544.

[3] Arora, Manish, Claus-Dieter Ohl, and Knud Aage Mørch. "Cavitation inception on microparticles: A self-propelled particle accelerator." Physical review letters 92, no. 17 (2004): 174501.

[4] Goh, B. H. T., Y. D. A. Oh, E. Klaseboer, S. W. Ohl, and B. C. Khoo. "A low-voltage spark-discharge method for generation of consistent oscillating bubbles." Review of Scientific Instruments 84, no. 1 (2013): 014705.

[5] Poulain, Stéphane, Gabriel Guenoun, Sean Gart, William Crowe, and Sunghwan Jung. "Particle Motion Induced by Bubble Cavitation." *Physical Review Letters* 114, no. 21 (2015): 214501.

[6] Pavard, Delphine, Evert Klaseboer, Siew-Wan Ohl, and Boo Cheong Khoo. "Removal of particles from holes in submerged plates with oscillating bubbles."*Physics of Fluids* 21, no. 8 (2009): 083304.

[7] Borkent, Bram M., Manish Arora, Claus-Dieter Ohl, Nico De Jong, Michel Versluis, Detlef Lohse, Knud Aage Mørch, Evert Klaseboer, and Boo Cheong Khoo. "The acceleration of solid particles subjected to cavitation nucleation."*Journal of Fluid Mechanics* 610 (2008): 157-182.

# Experimental evidence of nucleation from wall-bounded nuclei in a laminar flow

**T F Groß, G Ludwig, P F Pelz**
Chair of Fluid Systems, TU Darmstadt, 64289 Darmstadt, Germany
E-mail: peter.pelz@fst.tu-darmstadt.de

**Abstract.** The importance of nucleation from wall-bounded nuclei for cavitation and especially cavitation inception is undisputed. Although various theories and models found their way to standard literature, there is a lack of experiments allowing a closer look onto the process of nucleation. In the present paper we present a new experimental set-up that allows the investigation and analysis of nucleation from wall-bounded nuclei. The experimental findings support an extended understanding of nucleation as a self excited cyclic process. Impressive high-speed visualisations can be found in the supplementary material.

## 1. Introduction

Investigations about the role of nuclei and nucleation for the inception and formation of cavitation are part of cavitation research since Harvey postulated the existence of gas filled crevices on surfaces and particles in liquids [1]. In a supersaturated solution these wall-bounded nuclei can produce free nuclei due to mass transfer of gas or themselves work as weak-spots in the liquid that are necessary for the occurrence of a phase change under technically relevant pressures. The former process is referred to as nucleation and is examined in this work. In case of hydrophobic surfaces the wall-bounded nuclei are stabilised by the effect of surface tension that lowers the pressure in the gas pocket below the pressure of the surrounding liquid and thus impedes diffusion of gas out of the crevice, unless the liquid is undersaturated [2]. In the recent decades many important works on this subject have been published. Mørch gives an extensive overview of investigations about cavitation nuclei both theoretical and experimental ones [3][4]. The review article of Jones et. al [5] focuses on nucleation from wall-bounded nuclei.

The present paper shows new experimental results gained in a benchmark experiment that has been suggested by Peters et al. [6]. Furthermore we provide new insights into the mechanism of nucleation. A laminar radial gap flow is used to observe nucleation at artificial nucleation sites and allows to investigate the physics of nuclei production. Since the experimental set-up is reduced to the essentials we can easily control the flow parameters and also benefit from the optical accessibility to generate reproducible data. The paper can be outlined as follows. In section 2 we explain the experimental set-up shortly and describe the governing equations. In section 3 we show first experimental results. Section 4 closes the article with a conclusion.

## 2. Experimental set-up and governing equations

Experiments on cavitation are often carried out in complex test facilities since the results are to be applied on likewise complex machines such as pumps, propellers or valves. Due

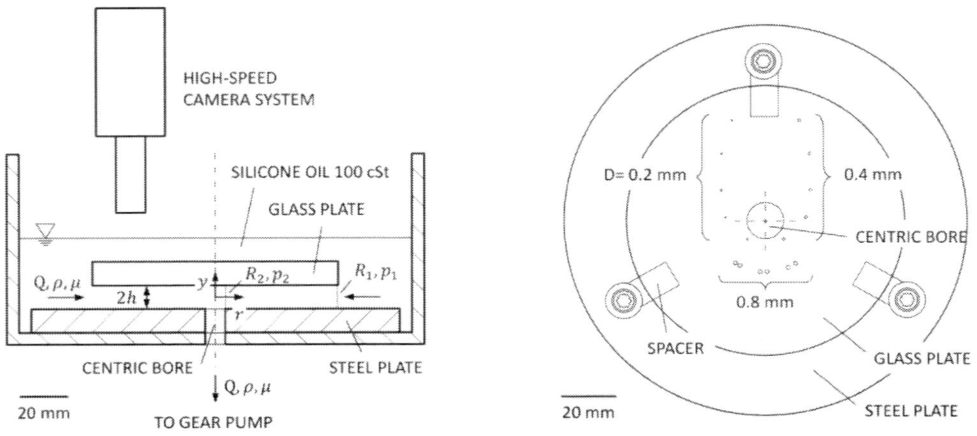


**Figure 1.** Test section of experimental set-up. Laminar radial flow between two closely spaced parallel plates with a sink at the centre. We consider the inflow case. Blind holes in the steel plate work as wall-bounded nuclei.

to the complexity and size of these test facilities (e.g. cavitation tunnels) there is often a limited accessibility to the working section resulting in extensive experimental procedures and time-consuming conversion work. Without a doubt studies of small-scale phenomena such as nucleation are difficult to study in such experimental set-ups. The experimental set-up we use for our nucleation experiments is kept simple so that a good accessibility (mechanical and optical) is ensured and we can easily control the experimental parameters.

Inside of a cylindrical vessel a laminar radial gap flow is realised, see figure 1. The gap is formed by spacers which are attached at three points between a steel plate and a glass plate with radius $R_1$. The half gap height $h$ can be adjusted between 0.05 and 1 mm. Blind holes with a depth of less than 1 mm and different diameters $D$ (0.2, 0.4, 0.6, and 0.8 mm) on different radii $r$ (10, 15, 20, 30, and 40 mm) drilled in replaceable steel plates serve as artificial nucleation spots which we observe in our experiments. The glass plate allows the use of high-speed imaging. The flow is driven by a speed-controlled gear pump that is connected to a bore (radius $R_2$) in the centre of the steel plate. A tank is used as a reservoir and decelerates the flow so that bubbles can rise to the surface. The liquid flows back to the vessel making a closed loop. We measure the pressure in the bore and the volume flux with an ultrasonic flow meter that is non-sensitive to small amounts of free gas. In addition, we measure the temperature and the oxygen content of the liquid in the vessel. Before starting the experiment the blind holes are filled with air by a syringe. It is ensured that the liquid is at least saturated at ambient pressure. Otherwise the gas in the crevices would dissolve in the surrounding liquid. The liquid we use is a silicone oil with a density of 960 $\mathrm{kg/m^3}$ and a kinematic viscosity of 100 $\mathrm{mm^2/s}$ at 25 °C. The Bunsen coefficient is 0.168 at 25 °C and 1 bar [6]. For the high-speed visualisation we use an IDT Motion Pro Y7 S3 camera with a resolution of 1920 x 1080 pixels and a maximum frame rate of 10600 frames per second. The frame rate can be increased by reducing the resolution. We measure the nucleation rate by means of image processing.

Figure 2 shows the principle of the nucleation process that includes the recurring grows and detachment of gas bubbles from wall-bounded nuclei we introduce here. It is a self excited cyclic process triggered by the bubble detachment. According to the convection-diffusion characteristics of the nucleation process the mass flux of gas that diffuses into a nucleus depends on both the supersaturation of the liquid and the fluid velocity. The critical size corresponding to

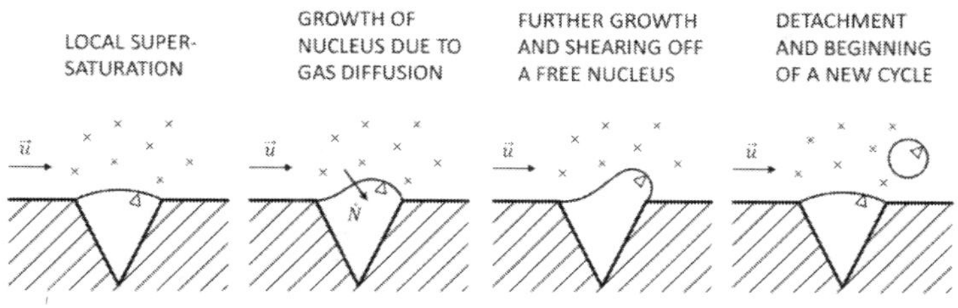


**Figure 2.** Principle sketch of growth and detachment of bubbles from wall-bounded nucleus.

the detachment depends on the velocity and the wall shear stress. In the present experiment the supersaturation occurs because of a pressure decrease along the flow path in inward direction. Following a flow perturbation theory, Savage [7] derives an approximate solution of the Navier-Stokes equation for a laminar-radial gap flow that allows the calculation of the local pressure

$$\frac{p(r) - p_1}{\frac{3}{4}\frac{\mu Q}{\pi h^3}} = -\ln\left(\frac{R_1}{r}\right) + \mathcal{O}\left(\frac{R_1^2}{r^2}\right) + \ldots$$

with an outer radius $R_1$ and a corresponding pressures $p_1$ and the dynamic viscosity of the liquid $\mu$. This approximation is only valid for small Reynolds numbers. The volume flux is set to be positive for the inflow case. The concentration $c$ of gas dissolved in a liquid in an equilibrium is given by Henry's law [8] $c(r) = p(r)k_H$ where $k_H$ is denoted as Henry constant. Assuming that the liquid in the vessel is saturated with gas in equilibrium at the inlet $c_1 = p_1 k_H$, the liquid becomes supersaturated due to the pressure decrease in inward direction. The local supersaturation $\chi(r) = c_1/c(r)$ can be specified with the equation above and Henry's law. Equation (2) shows that it is necessary to use a liquid with a relatively large viscosity to achieve relevant supersaturations. Otherwise, the gap would have to be very small. The area-averaged velocity $u(r) = Q/4\pi rh$ can be calculated with the continuity equation. Using the velocity profile derived by Savage [7] one eventually gets the wall shear stress as $\tau(r) = 3Q\mu/4\pi rh^2$.

## 3. Experimental results

As already mentioned the three independent parameters including supersaturation, fluid velocity, and wall shear stress influence the nucleation process and in particular the nucleation rate. In Figure 3 the nucleation rate is plotted versus the local flow velocity for different gap heights and different supersaturation. The nucleation rate increases with increasing velocity and supersaturation. The exact influence of the parameters remains hidden since the gap height also influences the nucleation rate. In future experiments the gap height has to be increased to suppress this influence. Figure 3 also demonstrates that nucleation rates over several orders of magnitude are achievable. Up to now the largest nucleation rate we measured is 1000 Hz which is of the order of nucleation rates observed in real cavitating flows [9].

Figure 4 shows the growth and detachment of a gas bubble from a wall-bounded nucleus. The process works as follows. A small part of the nucleus extends beyond the edge of the bore in the direction of the flow (Fig. 4 a-c). In a short amount of time a larger bubble is formed (Fig. 4 d) which is still connected to the wall-bounded nucleus. The bubble finally detaches due to the acting forces and is carried away with the liquid flow (Figs. 4 e-f). A high-speed video of the process can be found in the supplementary material.

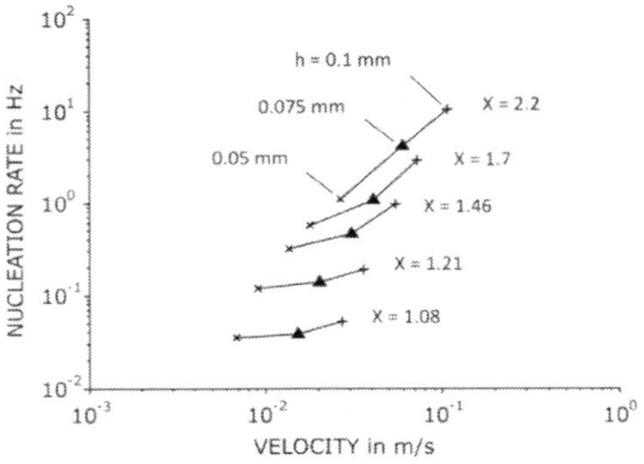


**Figure 3.** Nucleation rate depending on velocity, supersaturation and gap height ($D$ = 0.6 mm). Different symbols mark different gap heights. The solid lines indicate constant supersaturation.

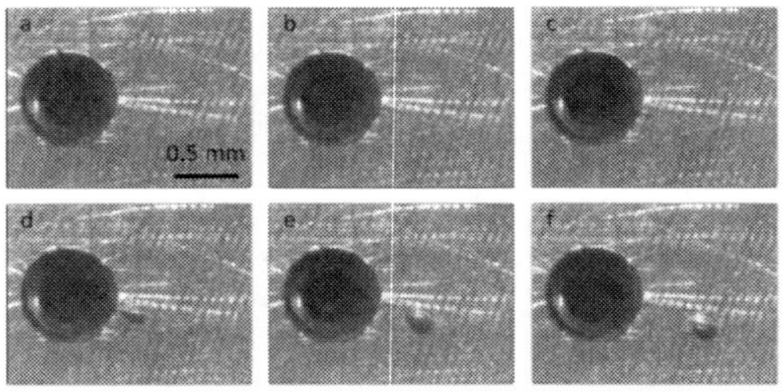


**Figure 4.** Growth and detachment of a gas bubble from an artificial wall-bounded nucleus ($D$ = 0.8 mm). The time interval between the images is 1 ms. The nucleation rate is 10.2 Hz. Flow direction is from left to right.

## 4. Conclusion

Although there is no doubt about the relevance of nuclei and nucleation for cavitation and cavitation inception, there are many unanswered questions. To gain new insights into the physics of nucleation we use a novel experimental set-up that allows the investigation of nucleation. The following list summarises the results and gives suggestions for future works.

- The nucleation rates depend on the level of supersaturation, flow velocity and wall shear stress.
- The nucleation rates are remarkably constant over time and show only small fluctuations.
- Observation of nucleation rates from 0.1 up to 1000 Hz are possible.
- In the present experiment the nucleation process depends on the length ratio $h/D$. In order to make a better correspondence to technical flows this ratio has to be increased.
- It is also necessary to investigate smaller bore diameters. Diameters in the size range of 1 $\mu$m to 10 $\mu$m are relevant for this issue.
- Due to the simplicity of the experiment, it can also be used as a benchmark for numerical simulations.

## Acknowledgments

We would like to thank Professor Dr.-Ing. F. Peters for his help and support regarding the experimental set-up.

## References

[1] Harvey E, Barnes D, McElroy W, Whiteley A, Pease D and Cooper K 1944 *J. Cell. Physiol.* **24** 1–22
[2] Atchley A and Prosperetti A 1989 *J. Acoust. Soc. Am.* **86** 1065–1084
[3] Mørch K 2007 *Phys. Fluids* **19**
[4] Mørch K 2008 *J. Hydrodyn.* **21** 176–189
[5] Jones S, Evans G and Galvin K 1999 *Adv. Colloid Interface Sci.* **80** 27–50
[6] Peters F and Honza R 2014 *Exp. Fluids* **55** 1786
[7] Savage S 1964 *J. Appl. Mech.* **31** 594–596
[8] Henry W 1803 *Philos. Trans. R. Soc. London* **93** 29–274 ISSN 0261-0523
[9] Guennoun F, Farhat M, Bouziad Y A and Avellan F 2003 *CAV 2003: Fifth International Symposium on Cavitation*

# Interaction of two pulsating spherical bubbles in external pressure field near the contact

**S V Sanduleanu**[1,2] **and A G Petrov**[1,2]

[1] Theoretical Mechanics Department, Moscow Institute of Physics and Technology, 9 Institutskiy Per., Dolgoprudny 141700, Russia
[2] Mechanics of Systems Laboratory, Institute for Problems in Mechanics of the Russian Academy of Sciences, 101-1 Pr. Vernadskogo, Moscow 119526, Russia

shtefan.sanduleanu@gmail.com

**Abstract**. The problem of coalescence of two pulsating spherical bubbles is studied using Lagrangian formalism in assumptions that the acoustic field is weak, the frequency of the external impact is appreciably less than natural oscillation frequency and influence of viscosity on phase of radii pulsation is small. The obtained necessary condition for coalescence of the bubbles is determined by the dimensionless parameter, whose boundary value is nonlinearly depending on the quotient of the bubbles' radii.

## 1. Introduction

The studies on the interaction of two oscillating spheres with periodically varying radii were first reported by V.F.K. Bjerknes [1].

When the distance $x$ (figure 1, $x = x_2 - x_1$, $u_1 = \dot{x}_1$, $u_2 = -\dot{x}_2$) between the centers of the spheres is large, the force averaged over the period (secondary Bjerknes force) is inversely proportional to the squared distance and equals

$$F_B = -\frac{4\pi\rho}{x^2} < a_1^2 a_2^2 \dot{a}_1 \dot{a}_2 >, \qquad (1)$$

where $a_1$, $a_2$ are the radii of bubbles, and we consider that $a_1 \le a_2$, $\rho$ is the fluid density and the dot signifies the time derivation.

In this paper we use the model of two interacting gas bubbles in the presence of a weak acoustic field

$$p = p_\infty + \Delta p \cos(\omega t), \ \Delta p << p_\infty. \qquad (2)$$

Assuming that the frequency of the external impact $\omega$ is considerably smaller than the natural oscillation frequency $\omega_0$ and satisfies the condition $\omega\gamma << \omega_0^2$, where $\gamma = 4\mu / \rho a^2$, we obtain that [2] $a_i = a_{0i}(1 + \alpha_0 \cos \omega t)$, $\alpha_0 = -\Delta P / 3\kappa p_\infty$, where $\kappa$ is the polytropic exponent and $\mu$ is the fluid's viscosity . In this case, as the bubbles are pulsating in phase, the secondary Bjerknes force between them is attractive.

When the distance $x$ is small, the attraction force is [3]

$$F_B \approx -\frac{\pi}{8}\rho a^2 < \dot{h}^2 > \ln\frac{a}{h}, \quad a = \frac{a_1 a_2}{a_1 + a_2}, \quad h = x - (a_1 + a_2). \tag{3}$$

The classical formula for the viscous interaction force for the bubbles with fixed radii which are nearly touching is [4]

$$F_\mu = -\frac{6\pi\mu}{\left(a_1^{-1} + a_2^{-1}\right)^2}\frac{\dot{h}}{h} = -\frac{6\pi\mu}{\left(a_1^{-1} + a_2^{-1}\right)^2}\frac{d\ln(h)}{dt}. \tag{4}$$

The force of the viscous interaction averaged over the period is equal to zero. So the sum

$$\left\langle F_B + F_\mu \right\rangle < 0. \tag{5}$$

It means that bubbles should always coalesce. But in some cases the bubbles coalesce, in others do not. The purpose of this work was finding the necessary conditions for the coalescence of the bubbles.

## 2. Theoretical part

The initial point of our research is the kinetic energy of the fluid $T$. We consider the surrounding medium to be incompressible and the bubble shape to be spherical. The exact solution of the problem of interaction of two solid spheres moving along the line of their centres was first obtained by Hicks [5]. The general expression for $T$ is [3,6]

$$\frac{T}{2\pi\rho} = A_1 u_1^2 + 2Bu_1 u_2 + A_2 u_2^2 + D_1\dot{a}_1^2 + 2E\dot{a}_1\dot{a}_2 + D_2\dot{a}_2^2 + [C_{12}\dot{a}_2 + C_{11}\dot{a}_1]u_1 + [C_{21}\dot{a}_1 + C_{22}\dot{a}_2]u_2$$

$$2A_i = \frac{a_i^3}{3} + \sum_{n=1}^{\infty}\left(\frac{a_i}{A_n^i}\right)^3, \quad 2B = \sum_{n=1}^{\infty}\left(\frac{a_k}{B_n^i}\right)^3, \quad D_i = a_i^3 + \sum_{n=1}^{\infty}\frac{a_i^3}{A_n^i}\left[1 + ((B_n^k)^2 - 1)\ln\left(1 - \frac{1}{(B_n^k)^2}\right)\right], \tag{6}$$

$$E = \frac{(a_1 a_2)^2}{a_k}\sum_{n=0}^{\infty}\left(\frac{1}{B_{n+1}^k} - B_n^i \ln\left(1 + \frac{1}{B_{n+1}^k B_n^i}\right)\right), C_{ii} = \sum_{n=1}^{\infty}\frac{a_i^3}{(A_n^i)^2 B_n^k}, C_{ki} = \sum_{n=1}^{\infty}\frac{a_k^3}{(B_n^i)^2 A_{n-1}^k}, i \neq k,\ i,k = 1,2$$

The coefficients $A_n^i$ and $B_n^i$ can be calculated from recurrence relations with the initial conditions

$$A_0^i = 1, B_0^i = 0, \ A_n^i = \frac{x}{a_i}B_n^k - \frac{a_k}{a_i}A_{n-1}^i, \quad B_n^i = \frac{x}{a_i}A_{n-1}^k - \frac{a_k}{a_i}B_{n-1}^i. \tag{7}$$

The Lagrange function for this system equals

$$L = L(x_i, a_i, \dot{x}_i, \dot{a}_i, t) = T - \Pi, \quad \Pi = \sum_{i=1}^{2}\left\{\frac{4\pi}{3}a_i^3\left[p(t) + \frac{p_{0i}}{\kappa - 1}\left(\frac{a_{0i}}{a_i}\right)^{3\kappa}\right] + 4\pi\sigma a_i^2\right\} \tag{8}$$

Using the thin-layer approximation, the more accurate formula for the force of the viscous interaction of the bubbles is [3]:

$$F_\mu = -\frac{6\pi\mu}{\left(a_1^{-1} + a_2^{-1}\right)^2}\left[\frac{\dot{h}}{h} + \left(\frac{\dot{a}_1}{a_1^2} + \frac{\dot{a}_2}{a_2^2}\right)\frac{\ln[h/(a_1 + a_2)]}{a_1^{-1} + a_2^{-1}}\right]. \tag{9}$$

The system of Lagrange equations is of 4th order. Assuming that $\omega^2 << \omega_0^2$, $\omega\gamma << \omega_0^2$ the system is reduced to 2nd order. In transition from the coordinates $x_1$ and $x_2$ to $h = x - a_1 - a_2$ and $y = x_1 + x_2$ it can be noticed that $y$ is a cyclic coordinate. In this case, using the Routh transformation, the problem is reduced to a single differential equation

$$\frac{d}{dt}\frac{\partial R}{\partial \dot{\delta}} = \frac{\partial R}{\partial \delta} + F_\mu \frac{\partial h}{\partial \delta}, \tag{10}$$

where $R = \pi\rho\left(\mu_{11}a_{01}^5(1+\alpha)^5\dot{\delta}^2 + \mu_{12}a_{01}^5(1+\alpha)^4\dot{\alpha}\dot{\delta} + \mu_{22}a_{01}^5(1+\alpha)^3\dot{\alpha}^2\right)$, $\alpha = \alpha_0\cos(\omega t)$, $\delta = h/a_1$, $\mu_{11}, \mu_{12}, \mu_{22}$ – are functions of variables $\delta$ and $a_1/a_2$.

Assuming that $\alpha << 1$, the differential equation takes the form

$$\frac{d}{dt'}[\mu_{11}\dot{\delta} + \mu_{12}\dot{\alpha}] = f_B - \dot{\alpha}(\mu_{11}\dot{\delta} + \mu_{12}\dot{\alpha}) + f_\mu, \quad (11)$$

$$f_B = \frac{1}{2}(d_{11}\dot{\delta}^2 + 2d_{12}\dot{\alpha}\dot{\delta} + d_{22}\dot{\alpha}^2),\ d_{ij} = \frac{\partial\mu_{ij}}{\partial\delta}, i, j = 1,2, \quad (12)$$

$$f_\mu = \frac{F_\mu}{2\pi\rho\omega^2 a_{01}^4} = -M\left(\frac{\dot{\delta}}{\delta} + \dot{\alpha}\ln\left(\delta\frac{a_{01}}{a_{01}+a_{02}}\right)\right),\ M = \frac{3\mu a_{02}^2}{\rho\omega a_{01}^2(a_{01}+a_{02})^2} \propto \gamma/\omega. \quad (13)$$

Here and hereafter, the dots signify the derivatives with respect to the dimensionless time $t' = \omega t$.

If the bubbles do not coalesce, they will pulsate near a certain distance $\delta_0$. In this case, the average force of the interaction is equal to zero. To find this force we have looked for the steady-state solution of (11) in the form $\delta = \delta_0 + \Delta(t')$. We take into account the linear and quadratic terms with respect to the oscillation amplitude $\alpha$. Transferring the linear terms to the left-hand side and the quadratic terms to the right-hand side, we obtain, in linear approximation, the amplitude and the phase for $\Delta$. The dimensionless average total force $f$ was obtained by averaging the quadratic terms in the right side of (11).

## 3. Results and discussion

We have obtained the dependence of the dimensionless average total force $f$ on the relative equilibrium distance $\delta_0$ for different quotients of bubbles' radii. These results are presented in figures 2-4.

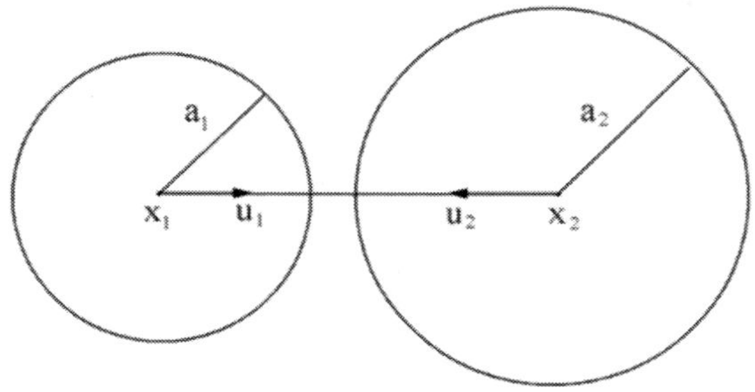


**Figure 1.** Problem statement

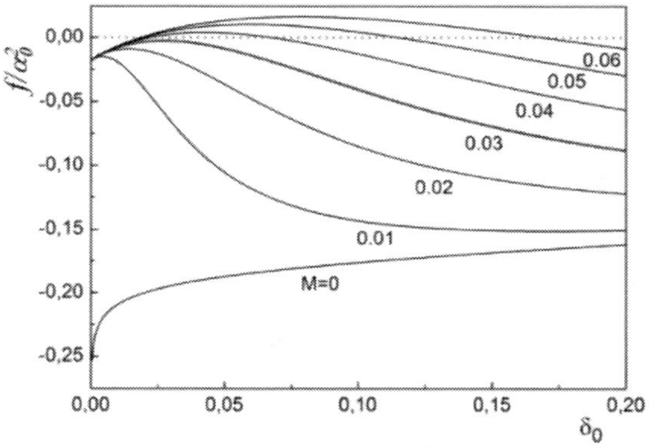


**Figure 2.** The dependences of the dimensionless average total force $f/\alpha_0^2$ on the relative distance $\delta_0$ for $a_2/a_1 = 1$.

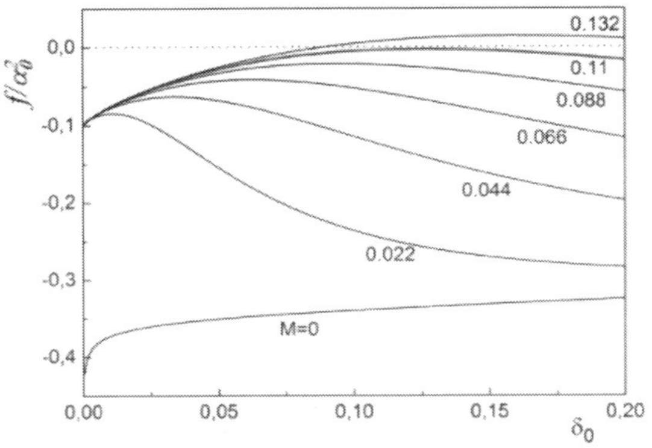


**Figure 3.** The dependences of the dimensionless average total force $f\ /\ \alpha_0^2$ on the relative distance $\delta_0$ for $a_2\ /\ a_1 = 1.5$

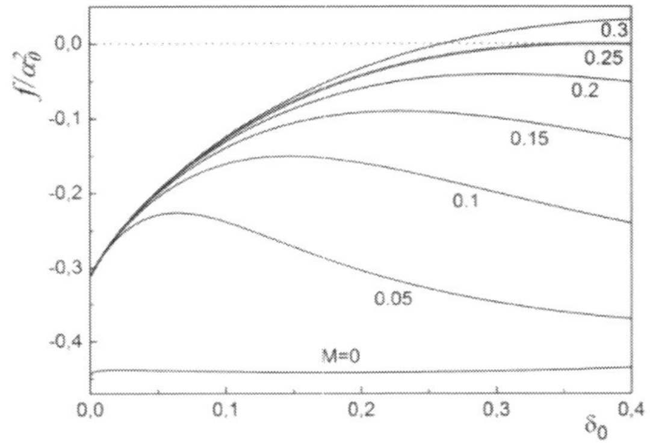


**Figure 4.** The dependences of the dimensionless average total force $f\ /\ \alpha_0^2$ on the relative distance $\delta_0$ for $a_2\ /\ a_1 = 2$

The range of numbers $M$, in which the averaged force of the interaction is negative ($\mu_{11}$ – is a positive function and may be considered as effective mass), determines the necessary condition for epy coalescence of the bubbles. In case of equal radii (figure 2), the coalescence takes place at $M \leq 0.03$. When the quotient of the radii is $a_2\ /\ a_1 = 1.5$ (figure 3), the area of the parameter $M$, for which the attraction is observed, expands to $M \leq 0.11$. When $a_2\ /\ a_1 = 2$ (figure 4), the attraction occurs at $M \leq 0.25$.

The obtained results are consistent with the data of recent experiments with two bubbles ($a_1 \approx a_2 \approx 25\mu m$) in an ultrasonic standing wave at 22.4 kHz [7] witch coalescence. For this values the parameter $M$ is about $0.015$ and it is less than $0.03$ and therefore it satisfies the necessary condition for coalescence. To be noted, in our model we have neglected the buoyancy force, as it can be assumed to be compensated by the first Bjerknes force. This fact was also investigated in [8].

## 4. Conclusion

In this work we studied the forces of bubbles' interaction that depend on their size, the parameters of the fluid and the periodic external pressure field. The necessary condition for coalescence of the bubbles is determined by the dimensionless parameter *M*. With the growth of the quotient of the bubbles' radii the boundary value of this parameter is rising nonlinearly.

The work was supported by the Russian Science Foundation, project No 14-19-01633.

## References

[1] Bjerknes V F K 1906 *Fields of Force* (Columbia University Press)
[2] Harkin A, Kaper T J and Nadim A 2001 *J. Fluid. Mech.* **445** 377 – 411
[3] Petrov A G 2011 *Fluid Dynamics* **45** (4) 579-595
[4] Batchelor G K 2000 *An Introduction to Fluid Dynamics* (Cambridge : Cambridge Univ. Press)
[5] Hicks W M 1880 *Phil. Trans. R. Soc. Lond.* **171,** 455-492
[6] Voinov O V and Petrov A G 1976, "Motion of Bubbles in a Fluid," in: *Itogi Nauki i Tekhniki. Gidromekhanika.* **10** 86-177
[7] Jiao J, He Y, Yasui K, Kentish S E, Ashokkumar M, Manasseh R and Lee J 2015 *Ultrasonics Sonochemistry* **22** 70-77
[8] Jiao J, He Y, Leong T, Kentish S E , Ashokkumar M, Manasseh R and Lee J 2013 *J. Phys. Chem.* **117** 12549–12555

# Study on Formation of Plasma Nanobubbles in Water

**Takehiko Sato[1], Tatsuyuki Nakatani[2], Takashi Miyahara[3], Shiroh Ochiai[4,a], Masanobu Oizumi[4,b], Hidemasa Fujita[4,c] and Takamichi Miyazaki[5]**

[1] Institute of Fluid Science, Tohoku University, 2-1-1 Katahira, Aoba-ku, Sendai 980-8577, Japan
[2] Research Institute of Technology, Okayama University of Science, 1-1 Ridaichou, Kita-ku, Okayama 700-0005, Japan
[3] Graduate School of Integrated Science and Technology, Shizuoka University, 3-5-1 Johoku, Naka-ku, Hamamatsu 432-8561, Japan
[4] Graduate School of Engineering, Tohoku University, 2-1-1 Katahira, Aoba-ku, Sendai 980-8577, Japan (Currently, [a]Isuzu Motors Limited, [b]Nippon Steel & Sumikin Eng. Co. Ltd., [c]Mitsubishi Motors Co.)
[5] School of Engineering, Tohoku University, 6-6-04 Aoba, Aramaki, Aoba-ku, Sendai 980-8579, Japan

sato@ifs.tohoku.ac.jp

**Abstract**. Nanobubbles of less than 400 nm in diameter were formed by plasma in pure water. Pre-breakdown plasma termed streamer discharges, generated gas channels shaped like fine dendritic coral leading to the formation of small bubbles. Nanobubbles were visualized by an optical microscope and measured by dynamic laser scattering. However, it is necessary to verify that these nanobubbles are gas bubbles, not solid, because contamination such as platinum particles and organic compounds from electrode and residue in ultrapure water were also observed.

## 3. Introduction

Micro- and nano-bubbles have potential applications in many fields such as medical treatment, medicine, food, cosmetic, health care and machines. For example, they have been applied as contrast media for ultrasound diagnosis, media of targeted drug delivery, bubble washing for beauty treatment, homogeneous cream for improvement of taste, micro pore material for creation of new materials, a power source for micro machines and reduction of drag force around submerged objects in water. For these applications, size control and stability of bubbles are required [1].

Generally, bubbles grow in supersaturated liquid and shrink in unsaturated liquid due to diffusion through the bubble interface. It is known that nanobubbles of 100 nm in radius disappear within micro second, the disappearance time becoming shorter with smaller bubbles. However, it has been reported that nanobubbles in water exist for several hours, the plausible mechanism of the reduction of the surface tension being due to the existence of nano particles, OH ions and surface-active agents. This reduction of surface tension causes reduction of internal pressure of nanobubbles, leading to long lifetime. In the field of a lot of researches on interfacial water, the existence and lifetime of nanobubbles are still under discussion [2].

Regarding interfacial phenomena of microbubbles, it has also been reported that the interface of the microbubbles have a positive in the case of a pH less than 3 though they have a negative in the case of a pH greater than 3 [3]. The nanobubbles become stable in a high pH solution but unstable in a saline solution with a high concentration. It is considered that the negative charge prevents aggregation and combination of nanobubbles and that nanobubbles become more stable which leads to a longer lifetime.

Methods using electrolysis, a supersonic wave [4], and pore materials [5] have been reported to produce nanobubbles in solution in large quantities. In the case of electrolysis, the size of nanobubbles is influenced by the concentration of the solution, pH, polarity and potential of the electrode. It decreases with an increase of the current density and flow velocity around the electrode. In the case of ultrasonic waves, the size increases with the increase of the power of the ultrasonic waves and with the concentration of the solution, and decreases with the addition of a surface-active agent and an increase in the length of its chain. In the case of porous materials, the size of nanobubbles is influenced by the pore sizes and hydrophilicity of the material surface. However, these methods require electrolyte solutions to achieve longer lifetime and the nanobubbles in pure water disappear in a short time.

We herein have proposed a new method of nanobubble formation using plasma to achieve greater stability in pure water without any electrolyte solution. The formation processes and visualization methods of "plasma nanobubbles" were investigated to develop a new method of nanobubble formation and to understand the physical features of nanobubbles.

## 4. Experimental methods

### *4.1. Experimental setup*

Figure 1 shows the experimental setup. Plasma was generated at a tip of a platinum wire electrode of 0.3 mm,, which, except for the end force of the electrode, was covered with silicone. A ring-shaped grounded platinum wire electrode was installed at the bottom of ultrapure water. The water was exposed to air for 40 min, leading to electrical conductivity of 2.5 MΩ·cm. Applied voltage was -5.5 kV at maximum and -3.5 kV at minimum with a frequency of 10 kHz.

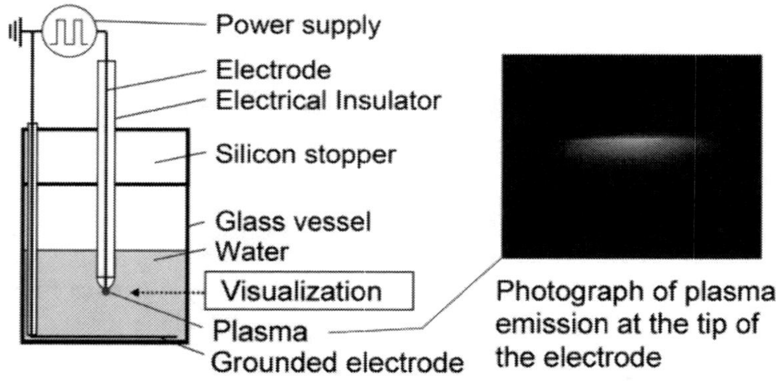


Fig. 1 Experimental setup and photograph of the plasma emission.

### *4.2. Measurement methods*

Underwater plasma generates streamer discharges known as pre-breakdown phenomena, and gas channels are formed along the streamer discharges due to Joule heating. The process of the formation and collapse of the gas channels was visualized with a high-speed camera (Nac Image Technology, MEMRECAM). Nanobubbles in a droplet of several mm in diameter on a clean glass plate were visualized by a digital microscope (Keyence, VX-9700). The digital microscope was capable of magnifying images up to 5000 times with a working distance of 4 mm. Visualization was performed in a clean bench to avoid contamination.

Size distributions of nanobubbles were analyzed by dynamic light scattering devices (Otsuka Electronic, DLS-6500 and Sympatec, NANOFOX). To analyze substances in plasma-generated water, a scanning electron microscope (SEM) and a transmission electron microscope (TEM) were used for visualization, and X-ray analysis and ICP-MS were used for analysis of the elements.

## 5. Results and discussion

### *5.1. Visualization of plasma-generated gas channels*

Figure 2 shows the formation and collapse process of gas channels shown as shadows and Fig. 3 shows a magnified snapshot of the gas channels. The gas channels, shaped like fine dendritic coral, were generated by streamer discharges due to Joule heating. These narrow gas channels collapsed while combining with each other at each position. After collapse, small bubbles were generated and remained. These results imply that plasma in water is capable of generating fine bubbles.

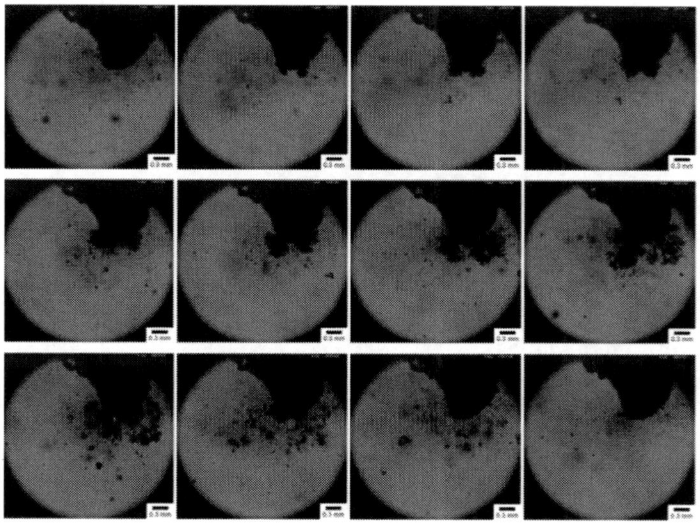

Fig. 2 Formation process of gas channels by plasma generation at the tip of the electrode. Residual fine bubbles were also observed after collapse of the gas channels. The process is not a series. The scale bar is 1 mm.

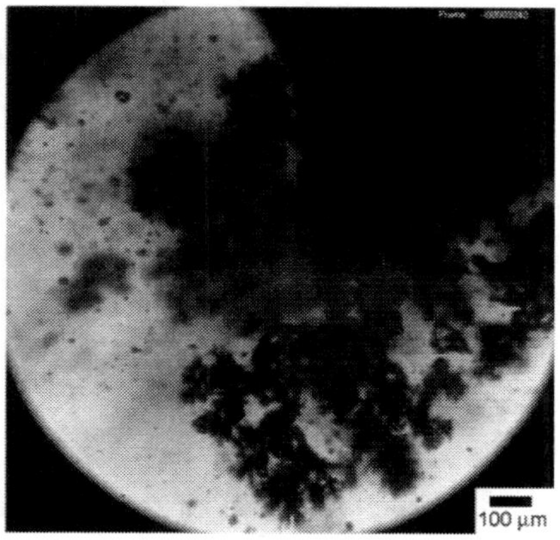


Fig. 3 Magnified snapshot of gas channels at the tip of the electrode. Many fine gas channels were generated by streamer discharges. The scale bar is 100 µm.

### *5.2. Visualization of plasma-generated nanobubbles*

Figure 4 shows the plasma-generated nanobubbles (a) and spherical polymer particles of 200 nm in diameter (b) visualized by the optical microscope. The nanobubbles were observed along the edge of the droplet. This could be caused by the fact that most nanobubbles were located in the focal point of the microscope because the depth of the water became shallow near the edge and the limitation of nanobubbles movement by Brownian motion occurred at the edge. The size of nanobubbles was less than several hundred nanometers. Although the outlines of nanobubbles of less than 200 nm were not clear, many such nanobubbles were observed. Here, spherical polymer particles of 200 nm in diameter were visualized using the optical microscope, as shown in Fig. 4 (b). Although their outlines were also not clear, they were observed. These nanobubbles were located along the entire outline of the droplet and were not observed in the ultrapure water without the generation of plasma. Furthermore, these nanobubbles were also observed even after 3 days.

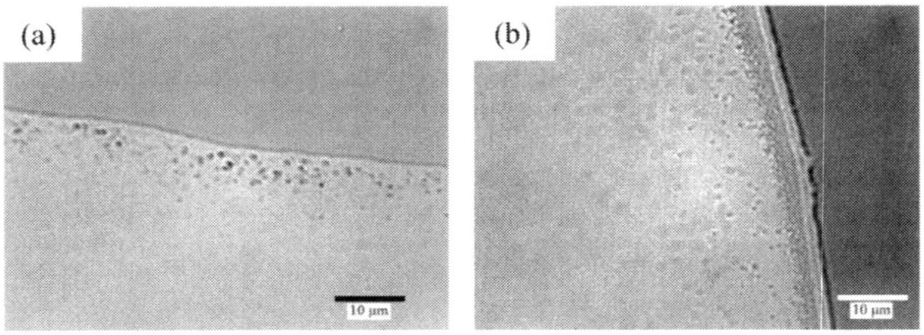


Fig. 4 (a) Plasma-generated nanobubbles near the edge of the droplet. (b) Spherical polymer particles (Latex) of 200 nm in diameter. Both images were observed by the optical microscope. The scale bar is 10 μm.

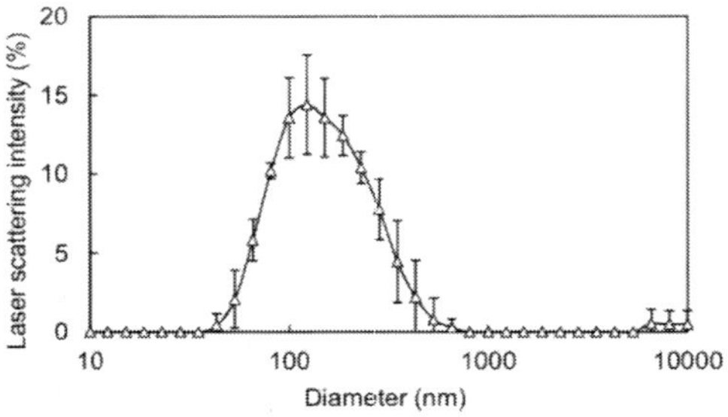


Fig. 5 Size distribution of the plasma-generated nanobubbles.

### *5.3. Size distribution of plasma-generated nanobubbles*

Figure 5 shows the size distribution of the plasma-generated nanobubbles. The size was distributed from 50 nm to 400 nm and the peak frequency was around 120 nm. These results mean that nanobubbles of less than 400 nm in diameter can probably be generated by plasma in water [3]. Here, the generated bubbles of greater than micro meter size moved up and were released into air.

## 6. Conclusion

In this study, a new method of nanobubble formation using plasma in water was developed. "Plasma nanobubbles" of less than 400 nm were visualized and measured experimentally and it was shown that those nanobubbles were quite stable in pure water. It is necessary to verify that these nanobubbles are gas bubbles, not solid, because contamination such as platinum particles and organic compounds from the electrode and residue in ultrapure water were also observed by SEM, TEM and ICP-MS.

This study was partly supported by a Grant-in-Aid for Scientific Research from JSPS (No. 25630045, 15K13864) and a Grant of the Collaborative Research Project of IFS. The authors wish to thank M. Farhat, M. Tinguely and A. Mueller, EPFL, T. Ohara, G. Kikugawa, T. Nakajima, Tohoku Univ., H. Matsuno, Japan Laser Co., S. Mine, Otsuka Electronics Co. Ltd. and Y. Kuramoto, Keyence Co. for their discussion and technical support.

## References

[4] A. Agarwal *et al.*, Chemosphere, **84** (2011), 1175.
[5] K. Kikuchi *et al.*, J. Electroanalytical Chem., **506** (2001), 22.
[6] M. Takahashi, J. Phys. Chem. B, **109** (2005), 21858.
[7] J.-Y. Kim *et al.*, J. Colloid Interf. Sci., **223** (2000), 285.
[8] M. Kukizaki and M. Goto, J. Membrane Sci., **281** (2006) 386.

# Heterogeneous nucleation of bubbles by molecular dynamics

**Donguk Suh**[1], **Mitsuki Nakamura**[1] **and Kenji Yasuoka**[1]

[1] Keio University, Department of Mechanical Engineering, Hiyoshi 3-14-1, Kohokuku, Yokohama 223-8522, Japan

E-mail: suh@mech.keio.ac.jp

**Abstract.** A homogeneous liquid system and a heterogeneous system with an impurity inserted inside it were used for investigation of bubble nucleation by molecular dynamics simulation. A constant particle number, volume, and temperature ensemble was used. The systems with the impurities showed an overall increase in bubble formation, which is consistent with previous studies. The shape of the impurities was changed to see if there was any direct influence on the bubble nucleation rate. With the limited number of systems investigated, the occurrence of a shape effect was inconclusive. As observed in previous heterogeneous nucleation studies with walls, the bubble initially forms remotely from the impurity and remains at some distance from the seed.

## 1. Introduction

Bubble nucleation is an important phenomenon in nature and industry, and studies have shown that the critical nucleus size can be as small as that of nano scale [1-3]. With the advent of lab-on-a-chips and nanomachinery, fluid flow and plugging of nanochannels due to bubbles is of great concern to the scientists that develop nano scale products [4]. Therefore, many studies have used molecular dynamics (MD) to understand the phenomenon for both homogeneous and heterogeneous systems [5-10]. The latter systems are known to produce higher nucleation rates (more bubbles) because these systems have a lower free energy barrier. However, most of the heterogeneous nucleation studies by MD on bubbles by MD are on two-dimensional surfaces [11,12]. The number of studies of bubble formation with a freely moving three-dimensional impurity in the system has been limited [13]. The scarcity in studies with regard to heterogeneous nucleation of impurities is unfortunate, since impurities with various shapes and sizes usually exist within a liquid. An understanding of the mechanism of how an impurity actually affects bubble formation is necessary.

In this study, an impurity or seed particle as in Fig. 1 is initially placed inside a subcooled liquid to examine by MD how the addition of the seed will influence bubble formation. The following section will explain the simulation setup and the results of the findings will ensue.

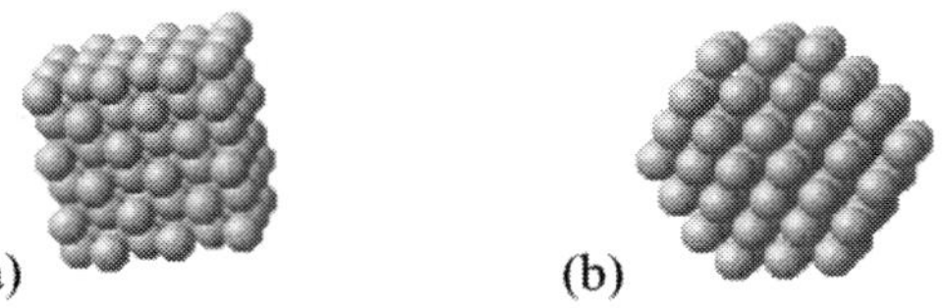

(a) (b)

**Figure 1.** Representation of impurity that will be inserted into a liquid system (a) cube and (b) sphere.

## 2. Simulation Setup

For the heterogeneous system, a spherical and a cubic seed particle are each constructed by cutting out such volumetric regions from an aluminum block of a (100) face-centered cubic lattice. The aluminum atoms have Lennard-Jones parameters ($\sigma_{Al}$=2.551Å, $\varepsilon_{Al}$=0.65 x$10^{-19}$ J) based on ref. [14] and are constrained by the RATTLE algorithm [15]. Further details of the process are in ref. [16,17]. Once the seed particle is made, it is immersed inside a cubic periodic system with liquid argon. Brownian motion of the impurity will take place inside the liquid. A homogeneous system is basically a heterogeneous system without the seed.

A constant number, volume, and temperature ensemble was used for the molecular dynamics simulation. The number of argon molecules in the system was 8788, and 108 or 110 aluminum atoms were added for creating a system with a cube or a sphere, respectively. The system volume was scaled with the addition of the seed molecules to keep the system isochoric. The initial system volume was set so the initial density of the system was at 1176 kg/m$^3$. The system was initially equilibrated to 144 K for 1.0 ns and the target temperature was decreased to 90.5 K, 95.0 K, and 99.9 K by velocity rescaling. Note that in an isochoric system, a temperature reduction is required for bubble nucleation to occur. The temperatures were set below the liquefaction temperature, where bubbles could be observed within the restricted time frame. The velocity rescaling method is a standard method that scales the molecular velocities to a target value depending on the temperature of the system based on statistical mechanics [15].

The Lennard-Jones 12-6 potential was applied for all interactions, and the Lorentz-Berthelot relations were used for argon and aluminum atom interactions. The time integration was performed by the velocity Verlet algorithm with a short $\Delta$t of 1.0 fs.

## 3. Results

Both homogeneous and heterogeneous systems were examined under the same thermodynamic conditions. The rate of nucleation is evaluated by the nucleation time, which is defined as the first instant, at which the bubble reaches its maximum volume within the system. The maximum volume is the largest volume to which the bubble grows within the observed time frame. This definition is necessary since the bubble size fluctuates continuously. A summary of the nucleation time for all target temperatures in the homogeneous and heterogeneous systems is presented in table 1. At 99.9K nucleation was not observed in any of the systems within the observed time frame of 1.5 ns.

**Table 1.** Summary of nucleation times for all systems

| | **HON** | **Cube** | **Sphere** |
|---|---|---|---|
| 90.5 K | 220 ps | 70 ps | 100 ps |
| 95.0 K | 590 ps | 700 ps | 450 ps |
| 99.9 K | - | - | - |

### *3.1. Homogeneous nucleation (HON)*

Figure 2 shows sliced snapshots of bubble nucleation for a homogeneous system. The slice thickness is 2.0 $\sigma_{Al}$. At the onset of quenching in Fig. 2(a) argon is randomly dispersed within the system, whereas once the bubble is formed, a clear void in the volume of argon molecules can be observed. Note that argon vapor molecules are seen within the void, as documented in previous studies.

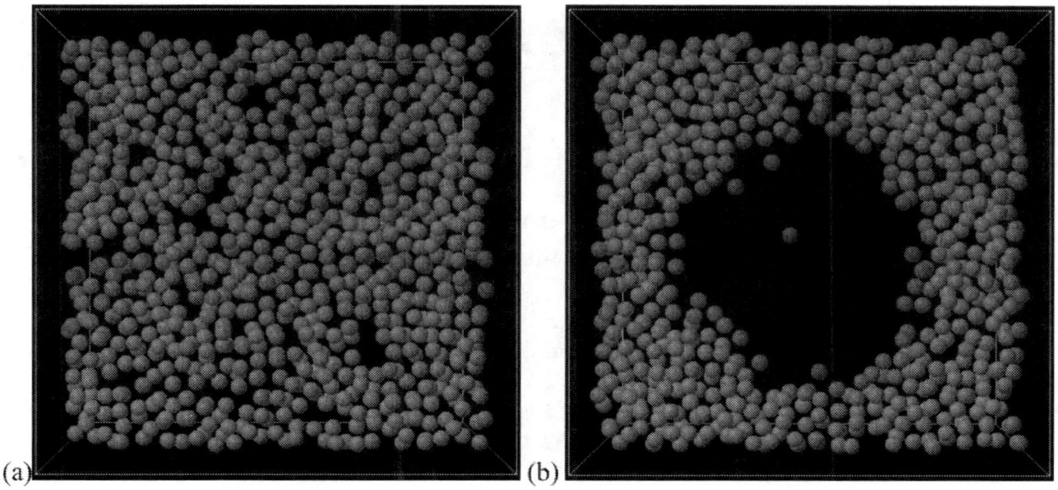


**Figure 2.** Homogeneous bubble nucleation of liquid argon (orange dots) (a) onset of quenching and (b) bubble formed at nucleation time.

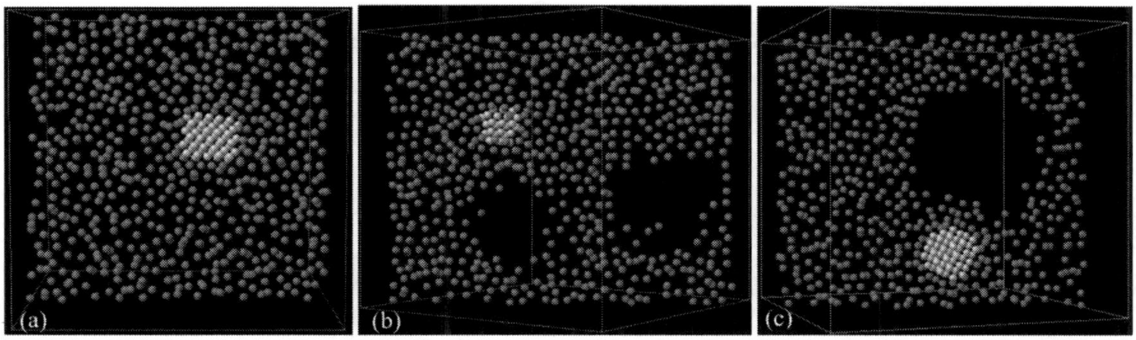


**Figure 3.** Sliced snapshots of heterogeneous nucleation, where yellow atoms are aluminium. (a) onset of quenching (b) before nucleation time, and (c) nucleation time.

### *3.2. Heterogeneous nucleation*

Impurities are added to the homogeneous system as in Fig. 3. Only a selection of plots are shown since the overall tendencies of the phenomenon did not drastically vary with the seed shape or thermodynamic condition. Figure 3 shows an example of nucleation with a cube. Fig. 3(a) is at the onset of quenching after equilibration of the system, which in Fig. 3(b) led to two voids that were created separately and subsequently in Fig. 3(c) merged into one large bubble. The voids form at a distance from the impurity, and once a bubble forms, there is no evidence that the impurity and the bubble directly interact.

The nucleation times are shown in table 1. Insertion of the impurity clearly produces faster nucleation at the strongest quenching (90.5 K). At moderate cooling (95.0 K) the results are mixed, the cube being slower than the homogeneous system. The irregularities in the results can be attributed to the lack of repetitions at this stage of the study. Nucleation is a highly stochastic phenomenon, so more samples need to be taken in order to obtain a clearer general tendency.

As previously stated, no clear shape effect has been observed at present due to the lack of statistics. Furthermore, a longer simulation time will most likely provide nucleation times for the 99.9 K systems.

**Acknowledgements**
This work was supported in part by grant-in-aid Mizuho Foundation for the Promotion of Sciences.

**References**

[1] Schmelzer J W P, Boltachev G S and Abyzov A S 2013 *Journal of Chemical Physics* **139**
[2] Xiao C, Heyes D M and Powles J G 2005 *Physica Status Solidi (B) Basic Research* **242** 749-61
[3] Lubetkin S D 2003 *Langmuir* **19** 2575-87
[4] Parthiban P and Khan S A 2012 *LChip* **12** 582-8
[5] Kinjo T, Ohguchi K, Yasuoka K and Matsumoto M 1999 *Computational Materials Science* **14** 138-41
[6] Novak B R, Maginn E J and McCready M J 2007 *Physical Review B* **75**
[7] Sekine M, Yasuoka K, Kinjo T and Matsumoto M 2008 *Fluid Dynamics Research* **40** 597-605
[8] Bazhirov T T, Norman G E and Stegailov V V 2008 *Journal of Physics-Condensed Matter* **20**
[9] Tsuda S, Takagi S and Matsumoto Y 2008 *Fluid Dynamics Research* **40** 606-15
[10] Wang Z J, Valeriani C and Frenkel D 2009 *Journal of Physical Chemistry B* **113** 3776-84
[11] Maruyama S and Kimura T 2000 *Heat and Technology* **18** 69-73
[12] Chen M, Yang J, Gao Y, Chen Y and Li D 2014 *Journal of Heat Transfer-Transactions of the Asme* **136**
[13] Hoyt J J and Potter A A 2012 *Metallurgical and Materials Transactions a-Physical Metallurgy and Materials Science* **43A** 3972-7
[14] Agrawal P M, Rice B M and Thompson D L 2002 *Surface Science* **515** 21-35
[15] Allen M P and Tildesley D J 1989 *Computer Simulation of Liquids* (New York: Oxford University Press, USA )
[16] Suh D and Yasuoka K 2011 *Journal of Physical Chemistry B* **115** 10631-45
[17] Suh D and Yasuoka K 2012 *Journal of Physical Chemistry B* **116** 14637-49

# Detailed Jet Dynamics in a Collapsing Bubble

**Outi Supponen[1], Danail Obreschkow[2], Philippe Kobel[1] and Mohamed Farhat[1]**

[1] Laboratory for Hydraulic Machines, Ecole Polytechnique Fédérale de Lausanne, Avenue de Cour 33bis, 1007 Lausanne, Switzerland

[2] International Centre for Radio Astronomy Research, University of Western Australia, M468 7 Fairway, Crawley, WA 6009, Australia

E-mail: `outi.supponen@epfl.ch`

**Abstract.** We present detailed visualizations of the micro-jet forming inside an aspherically collapsing cavitation bubble near a free surface. The high-quality visualizations of large and strongly deformed bubbles disclose so far unseen features of the dynamics inside the bubble, such as a mushroom-like flattened jet-tip, crown formation and micro-droplets. We also find that jetting near a free surface reduces the collapse time relative to the Rayleigh time.

## 1. Introduction

An important aspect of cavitation bubbles is their erosive power, which can be both harmful or beneficial, depending on the application. The erosion takes place during and shortly after the collapse of single bubbles, and is, in most cases, attributed to the formation of a strong shock or a fast, thin liquid jet, called the micro-jet. Such a micro-jet emerges when a cavitation bubble collapses aspherically due to anisotropies in the surrounding pressure field caused by, for example, nearby boundaries [1] or gravity [2]. The micro-jet usually becomes visible during the regrowth (rebound) of the bubble after the collapse as it pushes along a conical pocket of vapor sometimes called the vapor-jet [2]. The jet can also pierce the bubble well before the collapse when the bubble is particularly deformed [1,3]. Or, in cases of mild deformation, it might not even reach the opposite bubble wall and remains within the bubble throughout the collapse and rebound [2].

Here we study a single bubble collapsing near a free surface using high-speed imaging. This work extends on our recent visualizations of jet formation [1] and provides more quantitative insights. In the experiment (details in ref. [4]), a spherical cavitation bubble is created in water contained in a cubic ($18 \times 18 \times 18$ cm$^3$) test chamber. We use a green laser pulse (532 nm, 8 ns), first expanded to a diameter of 5 cm using a lens-system, and then refocussed onto a single point using a parabolic mirror with a high convergence angle (53°) to generate a point-like initial plasma [4]. In this way, we obtain a bubble of very high initial sphericity – impossible to achieve with a pure lens-system. The bubble centre is placed at a short distance of $h = 2.95$ mm below the free surface, causing the bubble to become vertically distorted during the collapse motion. This process is visualized with a high-speed camera (Photron Fastcam SA1.1) fitted with a 135 mm objective (Nikon Zoom-Micro 70–180mm f/4.5–5.6D) at a speed of 37,500 frames per second.

The experiment is conducted at room temperature and at low pressure (10.1 kPa=0.1 atm), implying large bubbles with long collapse times and thus increased spatio-temporal resolution.

## 2. Results and Discussion

A sequence of images showing a cavitation bubble collapsing in the vicinity of a free surface is shown in Figure 1. The bubble is initially spherical, yet upon reaching its maximum size ($R_{\mathrm{max}} \sim 5.3$ mm) it distorts to an egg-like shape while pushing liquid above it and deforming the free surface (frame 1 in Figure 1). The bubble then starts its collapse, during which a jet forms from the top of the bubble directed away from the free surface with a maximum speed of 8.2 m/s and an average width of 1 mm. The tip of the jet obtains an interesting "mushroom cap"-shape, which we believe is a result of interface instabilities. Figure 2 provides a more detailed visualization on the jet within the bubble, disclosing this peculiar jet shape in frames 1-4. The cap has a diameter of 2 mm. The jet then pierces the bubble well before the moment of collapse and entrains a pocket of the bubble's enclosed gases – this is the vapor-jet mentioned in Section 1. We discover that the boundary of this pocket has a complex shape, with a step leading from a diameter of 2.8 mm to 3.8 mm (frame 4 in Figure 1). These two diameters are associated with the jet and the "mushroom cap" diameters respectively. The inner part is the actual jet pushing the vapor whereas the outer part, which quickly slows down and smoothens out, is being pushed by the thin cap.

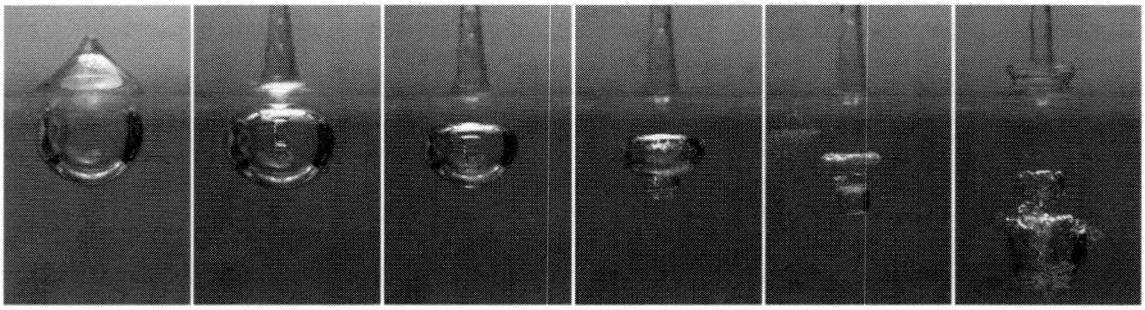

Figure 1: The collapse and rebound of a laser-induced cavitation bubble collapsing near a free surface. The frames are from times $t = 1.36, 1.96, 2.33, 2.60, 2.74$ and 3.40 ms after the bubble generation. See video in ref. [5].

The interface of the bubble quickly becomes opaque after the jet has pierced the opposite bubble wall. This is due to the shooting of micro-droplets onto the interface, which forms capillary waves on the bubble wall. These micro-droplets resulting from a crown generated within the bubble at the jet impact (frames 1-4 in Figure 2, bottom row). The jet impact occurs at spectacular Reynolds and Weber numbers ($Re \approx 10^4$ and $We \approx 10^3$ – note that the bubble interface velocity at the moment of impact is 1.5 m/s), providing conditions similar to those in investigations of a droplet impacting a flat free surface where the formation of a high-speed "ejecta"-sheet has been reported [6]. An ejecta-sheet is a thin jet that emerges rapidly after the impact horizontally between a drop and the free surface. For a sufficiently energetic impact ($We > 500$) this sheet disintegrates into tiny droplets [7].

The ejected micro-droplets mainly perturb the bubble boundary in the lower part of the bubble, near the impact point (last frame in Figure 3). However, as the collapse of the bubble continues, every point of its surface gets eventually sucked into the jet as shown theoretically in ref. [8]. By this process, the capillary surface perturbations that formed at the bottom of

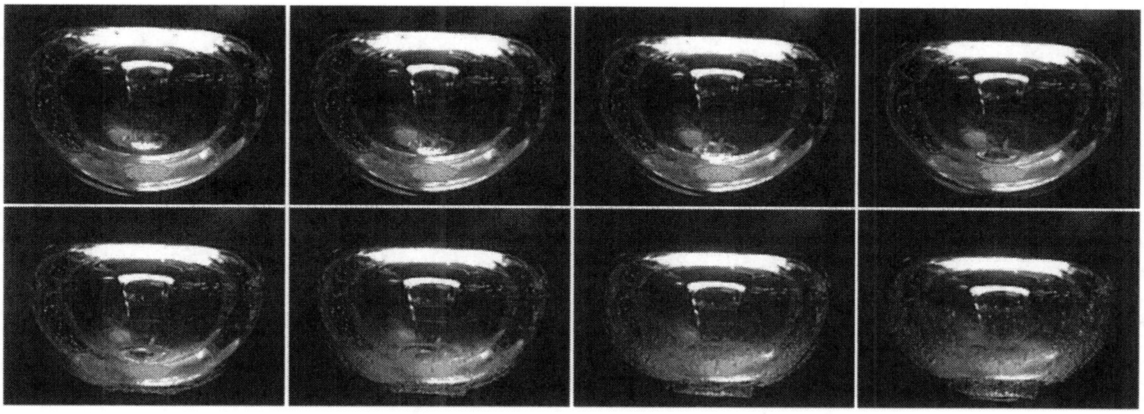

Figure 2: Visualization of jet impact within the collapsing cavitation bubble. First frame taken 2.25 ms after bubble generation. Inter-frame time 26 $\mu$s.

the bubble end up covering the entire bubble surface, including the jet. As a consequence, the bubble-jet system becomes less transparent to light and more difficult to observe.

It should be noted that mathematically the bubble maintains a spherical topology, as opposed to a toroidal one, until the vapor jet pocket detaches from the main bubble. At this moment, the bubble separates into two pockets, a torus associated with the main bubble, and a cylindrical pocket, still of spherical topology, associated with the jet (see frame 5 in Figure 1). These two bubble parts collapse individually with the cylindrical part breaking again into tori during its collapse. The initial micro-jet is still present within the rebound bubble, but it is barely visible due to the non-transparent interface of the rebound reminiscent of an upside-down mushroom cloud (frame 6 in Figure 1).

Figure 3 shows the evolution of the bubble radius with time from the bubble generation to

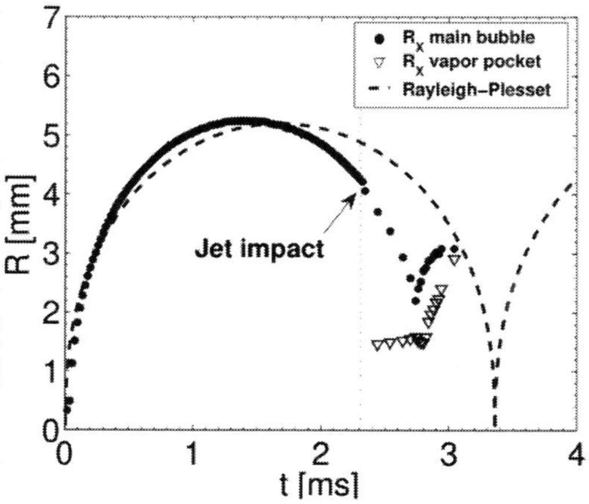


(a) $R_X$

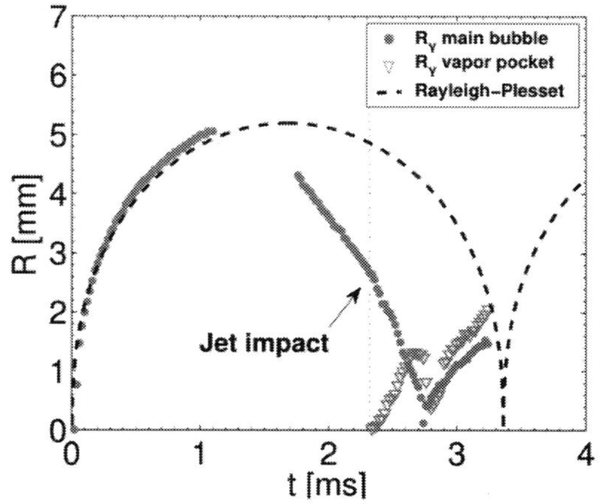


(b) $R_Y$

Figure 3: Time evolution of bubble radius in (a) horizontal and (b) vertical direction for the main bubble and the vapor pocket, compared with the Rayleigh-Plesset solution for a spherical bubble with $R_{\max} = R_X = 5.25$ mm and $p_\infty = 0.1$atm $= 10.15$ kPa, thus $\Delta p = p_\infty - p_{\text{vapor}} = 8.10$ kPa.

its collapse and first rebound. The bubble radius is measured separately in the horizontal (X) and vertical (Y) direction, since the bubble evolves into a shape that is not a sphere. These radii are calculated from the measured diameters, $R_X = D_X/2$ and $R_Y = D_Y/2$. Such a choice allows us to compare the measured bubble size with the Rayleigh-Plesset model describing the radial evolution of a corresponding spherical bubble [9]. In Figure 3b there is a region in time where the exact size of the bubble is difficult to extract due to the bubble being partly covered by the free surface. It is evident that the shrinking of the main bubble is accelerated after the jet impact at $t = 2.32$ ms (see the radius gradient change in Figure 3b), making it collapse faster than the Rayleigh model would predict. This is explained by part of the bubble contents being pushed away by the jet and participating in the growth of the external vapor pocket. The two bubbles collapse individually with a time difference of $\sim 60\mu$s, and the vapor pocket breaks up further during its collapse, releasing several individual tori.

A final feature worth discussing is the low velocity of the jet inside our bubble. We find typical values around 8.2 m/s, which are significantly smaller than usually reported values around 50-150 m/s [8]. This difference is, however, explainable by our low pressure settings of 0.1 atm and the close proximity of the bubble to the free surface, corresponding to a stand-off parameter $\gamma = h/R_{\max} = 0.56$, where $h$ is the distance of the initial bubble centre to the free surface. In fact, empirical models suggest jet velocities around $U_J = \xi(\Delta p/\rho)^{1/2}$ [8], where $\Delta p = p_\infty - p_{\text{vapor}}$ is the driving pressure, $\rho$ is the density of the liquid, and $\xi$ is a constant, numerically found to be around $\xi \sim 2.5$–$3$ for $\gamma = 0.56$ [10]. Using this constant, we then predict jet velocities of 7.1–8.6 m/s, consistent with our measurements.

## 3. Conclusion

The dynamics of the micro-jet impact within a laser-generated cavitation bubble collapsing in the vicinity of a free surface has been experimentally studied using high-speed imaging. Low pressure conditions and large bubble energies (obtained with a focussed laser) provide large and slowly moving bubbles and therefore excellent temporal and spatial resolutions to investigate detailed features of the collapse. The "mushroom cap"-shape of the jet tip, the crown-splash forming at the jet impact within the bubble and the micro-droplet shooting on the interface were discovered.

## Acknowledgements

We gratefully acknowledge the support of the Swiss National Science Foundation (Grant no. 513234), the European Space Agency, and the support of the University of Western Australia Research Collaboration Award obtained by co-authors DO and MF.

## References

[1] Supponen O, Kobel P, Obreschkow D and Farhat M 2015 *Phys. of Fluids* **27** 091101
[2] Obreschkow D, Tinguely M, Dorsaz N, Kobel P, De Bosset A and Farhat M 2011 *Phys. Rev. Letters* **107** 204501
[3] Zhang A M, Cui P and Wang Y 2013 *Exp. Fluids* **54** 1602
[4] Obreschkow D, Tinguely M, Dorsaz N, Kobel P, De Bosset A and Farhat M 2013 *Exp. Fluids* **54** 1503
[5] DOI: http://dx.doi.org/10.1103/APS.DFD.2014.GFM.V0084
[6] Yarin A L 2006 *Annu. Rev. Fluid Mech.* **38** 159-92
[7] Deegan R D, Brunet P and Eggers J 2007 *Nonlinearity* **21** C1
[8] Blake J R and Gibson D C 1987 *Ann. Rev. Fluid Mech.* **19** 99-123
[9] Rayleigh L 1917 *Phil. Magazine* **34** 94-98
[10] Robinson P B, Blake J R , Kodama T, Shima A, and Tomita Y, 2001 *J. Appl. Phys.* **89** 12

# Validation of Bubble Dynamics Equation for a Nano-scale Bubble via Molecular Dynamics Simulation

**S Tsuda**[1], **H Hyodo**[2], **S Watanabe**[1]
[1]Department of Mechanical Engineering, Kyushu University, 744 Motooka, Nishi-ku, Fukuoka 819-0395, Japan
[2]Department of Mechanical Engineering, Graduate School of Kyushu University, 744 Motooka, Nishi-ku, Fukuoka 819-0395, Japan

E-mail: tsudashin@mech.kyushu-u.ac.jp

**Abstract.** For a validation of the application of conventional bubble dynamics to a nano-scale bubble behaviour, we simulated a nano-scale bubble collapsing or vibration by Molecular Dynamics (MD) method and compared the result with the solution of Rayleigh-Plesset (RP) equation and that of Confined RP (CRP) equation, whose boundary condition was corrected to be consistent with that of MD simulation. As a result, a good coincidence was obtained between MD, RP, and CRP in the case of one-component fluid. In addition, also a good correspondence was obtained particularly in the comparison between MD and CRP in the case of two-component fluid containing non-condensable gas. The present results indicate that conventional bubble dynamics equation can be applied even to a nano-scale tiny bubble.

## 1. Introduction

We sometimes encounter the problem of cavitation in the development of fluid machinery such as pump and water turbine, and it causes deterioration of performance, unsteady vibration, noise, and erosion, all of which are not ignorable. Above all, preventing the fluid machinery from erosion, which originates from collapse of cavitation bubbles, is important for keeping the life time of the fluid machinery longer, and appropriate prediction of erosion in advance is required in the design phase. Such prediction method has progressed from continuum mechanics, such as fluid dynamics including bubble dynamics or CFD simulation, and also by solid mechanics. Many important and useful results have been obtained, but one of the essential questions is remained against the conventional method particularly for its application to the final process of collapsing bubble: To what extent can conventional bubble dynamics derived from fluid mechanics in the framework of continuum mechanics express the behaviour of a tiny bubble (just before the collapse) quantitatively? The clear answer against this question has not been obtained, and if conventional bubble dynamics were inappropriate against the description of collapsing bubble, some parts of the conventional prediction method for erosion might be revisited.

In this paper, we conducted a validation of bubble dynamics equations such as Rayleigh-Plesset (RP) equation throughout comparison with a result by Molecular Dynamics (MD) simulation, in which all molecular motion is numerically calculated based on a momentum equation for each molecule.

[1] Shin-ichi Tsuda: tsudashin@mech.kyushu-u.ac.jp

Bubble dynamics equations derived from fluid dynamics as continuum mechanics, which assume local equilibrium, might have some limitations for such small bubble dynamics because a tiny bubble just before collapse within very short time scale would not be necessarily in local equilibrium state, in which all molecular velocities must be Maxwell-Boltzmann distribution. On the other hand, at least theoretically, MD can rigorously describe both microscopic and macroscopic phenomena because it does not need any assumption such as local equilibrium. In this point of view, MD can principally offer more reliable results compared with continuum mechanics, and we can investigate whether macroscopic formulation such as RP can express the microscopic bubble dynamics or not using MD simulation. However, in the validation process of the applicability of conventional bubble dynamics equations, we must also deliberate items below carefully: (I) difference of boundary condition between bubble dynamics equation and MD, (II) influence of the feature of computational domain (e.g., spherical domain or cubic domain), (III) evaluation accuracy of thermophysical properties in MD, and so on. A few studies [1-3] already conducted comparison between bubble dynamics equations and MD results for a nano-scale bubble and the above points (II) and (III) were directly or indirectly investigated. As a result, the previous studies claim that bubble dynamics equations can fairly well describe the dynamic behaviour of a nano-scale bubble through the investigations. However, the influence of the point (I) has hardly been discussed. Point (I) is very important to obtain a clear answer against the question of "To what extent can bubble dynamics derived from the framework of fluid mechanics as continuum mechanics express the behaviour of a tiny bubble quantitatively?" because boundary condition of bubble dynamics equations such as RP and that of general MD are absolutely different as mentioned later in the subsection 2.2.

In this paper, we simulated a collapsing or vibrating bubble whose diameter is several nanometers using MD simulation, and the radius change was compared with not only RP equation but also a Confined RP (CRP) equation, which was derived in order to deliberate the influence of boundary condition. Throughout the comparison, we have reached closer answer against that question.

## 2. Method

### *2.1. Molecular Dynamics Simulation*

All molecular interactions were given by Lennard-Jones (LJ) potential [4] in a cubic computational domain on which a periodic boundary condition is imposed on all directions. Here, two kinds of fluid were simulated in this study; one is pure liquid oxygen, and the other is liquid oxygen including helium as non-condensable gas with the molar concentration of 0.05. Namely, the former is one-component fluid, and the latter is two-component. In both cases, 702,464 molecules were used to simulate, and the length on a side of the computational domain was around 30nm. The cut-off radius [4] was set to 3.5 in LJ unit as a standard value, and we have imposed an isothermal condition to keep the temperature constant based on so called *NVT*-constant MD method [4].

In the MD simulations, we first reproduced saturated liquid oxygen or the liquid oxygen containing helium gas at 90K in the computational domain, and slightly expanded the domain uniformly and isotropically in order to decrease the liquid density from the saturation liquid state. If we choose an appropriate density after the expansion, homogeneous nucleation occurs and finally one stable bubble is formed. This state is regarded as an initial condition, and the computational domain was next slightly compressed in order to cause the bubble collapsing. In this study, we focused on the bubble radius change with time. In the evaluation of the bubble radius, bubble region is recognized as "a cluster constituted by void points" and we extracted a 'cavity cluster' as a bubble [5]. Namely, the bubble radius was calculated from the volume of that cavity cluster.

Spherical shape of the bubble and incompressibility of liquid are the two important assumptions in the derivation process of RP. In advance, we confirmed that the bubble shape was sufficiently spherical at least in the present simulation condition. Also, we confirmed in advance that the bubble wall speeds observed in the MD simulations were clearly less than the Mach number of 0.3. Note that the present MD simulations were conducted so as to satisfy these two points.

*2.2. Rayleigh-Plesset (RP) and Confined Rayleigh-Plesset (CRP) Equations*

RP equation describes a spherical bubble motion which is placed in an incompressible liquid. In this equation, pressure as one of the boundary conditions is given at a place sufficiently far from the bubble. On the other hand, the computational region in MD is finite and a periodic condition is imposed. Namely, the liquid region surrounding the bubble in MD is very small and confined. To eliminate the influence of the difference in boundary condition between RP and MD as much as possible, we have derived a Confined RP (CRP) equation, whose boundary condition matches that in the present MD simulation. The derivation process is as follows.

In the derivation of RP, incompressible Navier-Stoke equation in a spherical polar coordinate is integrated from a positon infinitely far from centre of a bubble to the bubble wall corresponding to the bubble radius $R$. In the derivation of CRP, we have limited the integration range from a finite $R_{ref}$ whose radius equivalently corresponds to the volume of the computational domain in MD, to the bubble radius $R$. Here, fluid velocity is set to be zero at a place infinitely far from the bubble in RP while it was set to be zero at $R_{ref}$ in CRP. Rigorously speaking, the latter boundary condition for the velocity is inconsistent to that in MD, but the region corresponding to $R_{ref}$ is fairly close to the boundary of the computational domain in which periodic boundary condition is imposed. Taking into account that the periodic boundary condition keeps the number of molecules in the domain constant, the net velocity around the region is expected to be close to zero. In addition, pressure in RP is given at the positon infinitely far from the centre of the bubble while the pressure in CRP is given as $p_{ave}$, which denotes the spatially averaged pressure of liquid field at each time $t$. Note that pressure in MD is generally obtained in spatially averaged manner [4] while pressure in the original RP is given at sufficiently far from the bubble. This inconsistency could be also the possible reason of the difference between RP solution and MD result. Therefore, CRP was formulated so as to eliminate those differences as much as possible. The final form of CRP derived in this paper is given as follows.

$$\rho_l\left\{a(z)R\ddot{R}+b(z)\dot{R}^2\right\}=p(R)-p_{ave}(t)-\frac{2\gamma}{R}-4\mu_l\frac{\dot{R}}{R} \tag{1}$$

Here, $z$ is defined as $z = R_{ref}/R$ (note that $z$ as the ratio of equivalent domain radius $R_{ref}$ and bubble radius $R$ must be larger than 1 because $R_{ref}$ is of course larger than $R$) and

$$a(z)=\frac{(z-1)(5z^3+6z^2+3z+1)}{5(z^2+z+1)^2},\quad b(z)=\frac{1}{2}\left\{\frac{(5z^3+6z^2+3z+1)(4z^3+5)}{5(z^2+z+1)^3}-1\right\} \tag{2}$$

In Eq. (1), $R$ is the bubble radius, $\rho_l$ the liquid density, $p(R)$ the pressure at the bubble wall, $\gamma$ the surface tension, and $\mu_l$ the liquid viscosity coefficient. In L.H.S. of Eq. (1) as the inertial term, we can easily confirm the coefficients $a(z)$ and $b(z)$ become 1 and 3/2 respectively as same as those in original RP if $R_{ref}$ or $z$ becomes infinity (This situation almost corresponds to the boundary condition of original RP). Note that $a(z)$ is less than 1 and $b(z)$ is less than 3/2 for all $z > 1$. In a confined system such as the computational domain in MD, liquid field is narrow and the liquid inertia is clearly weaker than that in original RP. This effect is mainly reflected to Eq. (1). On the other hand, R.H.S. of Eq. (1) is almost the same as that in the original RP except $p_{ave}$ (In the original RP, $p_{ave}$ in Eq. (1) becomes pressure given at a place sufficiently far from the bubble).

In the numerical solution of RP or CRP, $p(R)$ was given as the saturation vapour pressure in the case of one-component fluid while it was estimated based on the assumption that the gas inside the bubble shows a polytropic change as an ideal gas in the case of two-component fluid. Equation (1) was numerically solved by 4th order Runge-Kutta method.

## 3. Results and Discussion

Figure 1(a) shows the bubble radius change of the collapsing bubble in the one-component case. As is shown, fairly good correspondences are obtained between MD, RP, and CRP. If we judge a bubble

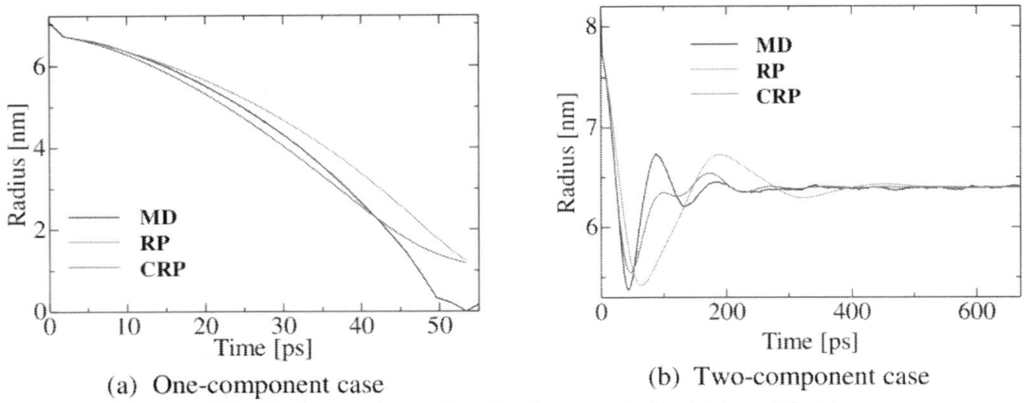


Fig. 1 Comparison of radius between MD, RP, and CRP

"collapsed" when its radius becomes around 1nm, difference of the collapsing time between MD, RP or CRP is estimated to be less than 15%. Here, we can figure that both RP and CRP quite well reproduce the bubble collapse in this case. Next, we compared the solutions of the two equations (RP and CRP) and MD result in liquid oxygen with helium gas. Figure 1(b) shows each radius change in the case of the two-component case, and as is shown, bubble radius in MD shows "rebound", and this phenomenon is also reproduced in RP and CRP qualitatively. And quantitatively, CRP reproduces the MD result better than RP. The first rebound time in CRP becomes shorter and closer to that in MD compared with RP, and it is originated from the appropriate evaluation of the inertial term in CRP (Weaker inertia reflected to CRP in Eq. (1) makes the rebound time shorter). This result indicates that bubble dynamics equation on the basis of fluid dynamics in the framework of continuum mechanics can be applied even to a nano-scale bubble if boundary condition is appropriately taken into account.

## 4. Conclusion

A validation of the application of bubble dynamics equation to a nano-scale bubble was conducted employing Molecular Dynamics (MD) simulation. In this study, not only oxygen as pure fluid but also oxygen including helium as non-condensable gas were simulated for the validation. As a result, the bubble radius changes observed in the MD simulation were reproduced fairly well by Rayleigh-Plesset (RP) equation, and by Confined Rayleigh-Plesset (CRP) equation, which was originally derived in this paper in order to compensate the essential difference of boundary condition between RP and MD simulation. The present results indicate that bubble dynamics equation on the basis of fluid dynamics as continuum mechanics can be applied even to a nano-scale bubble if we appropriately take the boundary condition into account.

## References

[1] Okumura, H. and Ito, N. 2003 Nonequilibrium molecular dynamics simulations of a bubble Nonequilibrium molecular dynamics simulations of a bubble, PHYSICAL REVIEW E 67, 045301(R)

[2] Hołyst, R., Litniewski, M., and Garstecki P. 2010 Large-scale molecular dynamics verification of the Rayleigh-Plesset approximation for collapse of nanobubbles PHYSICAL REVIEW E 82, 066309

[3] Hołyst, R., Litniewski, M., and Garstecki P. 2012 Collapse of a nanoscopic void triggered by a spherically symmetric traveling sound wave PHYSICAL REVIEW E 85, 056303

[4] Allen, M. P. and Tildesley D. J. 1987 Computer Simulation of Liquids, Clarendon Press

[5] Maruyama S. and Kimura T. 2000 A Molecular Dynamics Simulation of a Bubble Nucleation on Solid Surface Int. J. Heat & Technology 18-supplement1, 69-74

9th International Symposium on Cavitation (CAV2015) IOP Publishing
Journal of Physics: Conference Series **656** (2015) 012040 doi:10.1088/1742-6596/656/1/012040

# Transient bubble oscillations near an elastic membrane in water

**C K Turangan[1], B C Khoo[2]**

[1] Fluid Dynamics, IHPC, 1 Fusionopolis Way, #16-16, Connexis, Singapore 138632
[2] Department of Mechanical Engineering, National University of Singapore, Singapore 119260

E-mail: cary@ihpc.a-star.edu.sg

**Abstract.** We present a study of transient oscillating bubble-elastic membrane interaction by means of an experiment and a numerical simulation to study the dynamics of bubble's inertial collapse near an elastic interface. The bubble is generated very close to a thin elastic membrane using an electric spark, and their interaction is observed using high speed photography. The high pressure and temperature plasma from the dielectric breakdown precedes the bubble formation. The bubble then expands and creates a dimple on the membrane. After reaching its maximum size, the bubble begins to collapse. The membrane retracts back, transmitting a perturbation on the bubble surface. The coupling between bubble contraction and this perturbation strengthens the collapse and leads to the formation of a mushroom-shaped bubble, bubble pinching and splitting. Towards the end of the collapse, the water inertia surrounding the bubble pulls the membrane upwards forming a relatively sharp conical hump. The dynamics of this interaction is well predicted by the boundary element method (BEM) simulation.

## 1. Introduction

Studies of shock-wave lithotripsy (for kidney stone fragmentation) has shown that cavitation bubbles are generated in-vivo near the lithotripter shock focal point [1], and their collapse (either inertial or forced) are correlated to both stone fragmentation and tissue injury [2]. When bubbles oscillate near boundaries that exhibit elastic properties, they form mushroom-shapes that lead to bubble pinching and splitting [3, 4, 5]. Understanding this behaviour is important particularly in applications where bubbles may be induced near biological tissues e.g. in laser angioplasty, arthroscopic cartilage ablation and laser thrombolysis [6]. The experiment of a laser-generated bubble interaction with a thick elastic boundary made of polyacrylamide gel of 80% water concentration shows the formation of a mushroom-shape bubble, bubble pinching and splitting [7]. Bubble oscillation near an elastic membrane may also produce similar behaviour but only limited work have ever been reported in the literature. Turangan *et al.* [6] studied the behaviour of an electric-spark generated bubble near a thin elastic membrane and compared the observation with numerical simulations using the boundary element method (BEM) [8, 5]. The results show a perturbation on the bubble surface created by the membrane's repulsion, which strengthens the bubble collapse and forms the mushroom-shaped bubble that leads to bubble pinching and splitting. The impetus of the work presented here are to extend the methodologies presented by Turangan *et al.* [6] to study the dynamics of a single bubble-elastic membrane interaction where the bubble is generated very close to the membrane (small standoff distance) and to simulate the interaction using BEM to better understand the interaction behaviour.

## 2. An experimental technique of electric spark-generated bubble and the boundary element method for a transient bubble-elastic membrane interaction

### *2.1. An electric spark-generated bubble*

We have developed a low-voltage experimental technique to generate a single transient bubble in water using an electric spark via touching electrodes [6]. The set-up has 4 main components: an electric circuitry, a water tank, a high speed camera and a light source (figure 1). A thin elastic membrane (0.11 mm thickness) is clamped horizontally and stretched in one direction to have a tension of $\sigma = 43.6\,\mathrm{N/m}$. It is placed about 50 mm below the water surface to eliminate the free surface influence on the bubble dynamics. The copper-alloy electrodes of 0.11 mm in diameter are crossed manually. The experiment is started by charging the capacitors. The electricity is then discharged with the switch via the electrodes. This generates a single bubble with maximum radius $R_m$ of 2.8-4.0 mm range. To record the bubble-elastic membrane interaction, the electrical discharge is synchronized with a 12,500 frames/s ($\Delta t = 80\,\mu s$) high speed camera.

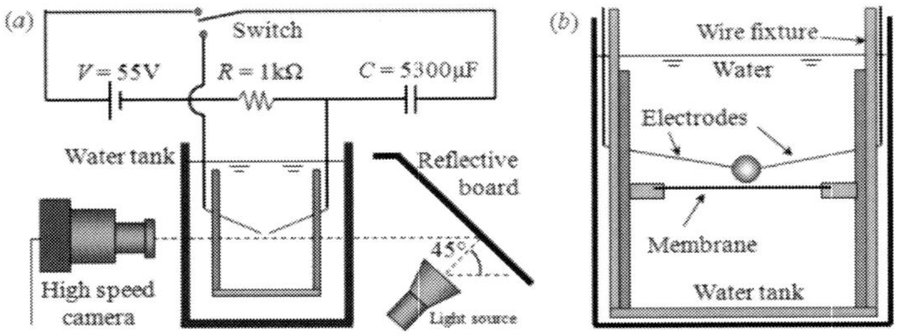


**Figure 1.** (a) The experimental setup. (b) The water tank ($17 \times 17 \times 17\,\mathrm{cm}^3$). A 55 V DC supply is connected to 5300 $\mu$F capacitors and 1 k$\Omega$ resistor. The water tank is equipped with an adjustable fixture and filled with preboiled water. Two 0.11 mm-dia copper-alloy electrode wires are placed above the membrane that is stretched in one direction and clamped horizontally.

### *2.2. The boundary element model for an oscillating bubble-elastic membrane interaction*

We consider a scenario where two fluids are separated by a thin elastic membrane (see ref. [8, 5, 6] for full detail). The model assumes that the fluids are inviscid and incompressible. The flow field satisfies the Laplace's equation ($\nabla^2\phi = 0$, where $\phi$ is the velocity potential), and the flow is driven by the fluid inertia. Assuming the fluid densities are the same (i.e. $\rho_1 = \rho_2 = \rho$, where 1 and 2 refer to fluid 1 and fluid 2, respectively), the fluid pressures on both sides of the membrane ($p_1$ and $p_2$) and on the bubble surface $p_b$ can be obtained using the Bernoulli equation, i.e.

$$p_1 = p_\infty - \rho\frac{D\phi_{i1}}{Dt} + \frac{1}{2}\rho|\bar{v}_1|^2,\ p_2 = p_\infty - \rho\frac{D\phi_{i2}}{Dt} + \rho\bar{v}_2\cdot\left(\bar{v}_1 - \frac{1}{2}\bar{v}_2\right),\ p_b = p_\infty - \rho\frac{D\phi_{b1}}{Dt} + \frac{1}{2}\rho|\bar{v}_b|^2 \quad (1)$$

where $t$ is time, $p_\infty$ is ambient pressure, $\bar{v}_1$ and $\bar{v}_2$ are velocity vectors of fluids 1 and 2 at the membrane surface and $\bar{v}_b$ is the velocity vector on the bubble surface. For a thin elastic membrane, the membrane does not support bending but acts like a surface tension. Here, it is only stretched in one direction. Therefore, the pressure difference between the fluids separated by the membrane is $p_1 - p_2 = \sigma K$, where $\sigma$ is the interfacial tension and $K$ is the membrane local curvature. We assume that the bubble internal pressure satisfies the adiabatic conditions. Therefore, bubble pressure $p_b = p_{g0}(V_0/V)^\gamma + p_\nu$, where $p_{g0}$ is the bubble pressure at the initial volume $V_0$, $V$ is bubble volume, $p_\nu$ is vapour pressure and $\gamma$ is the specific heat ratio.

Using the scaling factors ($R_m$ for lengths, $p_r = p_\infty - p_\nu$ for pressures, $t_0 = R_m\sqrt{\rho/p_r}$ for time, $\phi_0 = R_m\sqrt{p_r/\rho}$ for potentials, $1/R_m$ for curvature and $v_0 = \sqrt{p_r/\rho}$ for velocities), the pressure difference $p_1 - p_2 = \sigma K$ and the bubble surface pressure $p_b$ in dimensionless forms are

$$\frac{D}{Dt'}\left(\phi'_{i2} - \phi'_{i1}\right) = \bar{v}'_2 \cdot \left(\bar{v}'_1 - \frac{1}{2}\bar{v}'_2\right) - \frac{1}{2}|\bar{v}'_1|^2 + \kappa_m K', \quad \text{and} \quad \frac{D\phi'_b}{Dt'} = 1 - p'_b + \frac{1}{2}|\bar{v}'_b|^2 \tag{2}$$

where $\kappa_m$ is the elastic parameter of the membrane and is given by $\kappa_m = \sigma/(R_m\, p_r)$.

Boundary integral equations are used to calculate the normal component of velocity potential $\partial\phi/\partial n$. For fluid 1 that contains the bubble $b$ and elastic membrane interface $i$, and for fluid 2 that contains only the membrane interface $i$, they are respectively given by

$$c\,\phi'_1 = \int_{b+i} G\frac{\partial \phi'_1}{\partial n}dS - \int_{b+i} \phi'_1 \frac{\partial G}{\partial n}dS, \quad \text{and} \quad (4\pi - c)\,\phi'_2 = \int_i G\frac{\partial \phi'_2}{\partial n}dS + \int_i \phi'_2 \frac{\partial G}{\partial n}dS. \tag{3}$$

The variable $c$ denotes a solid angle that is a function of location of the computational nodes that represent the bubble and membrane surface and $G$ is the Green function (see ref. [8, 5]).

## 3. The growth and collapse of a single bubble near a thin elastic membrane

In the experiment, the electrodes are connected manually. Their contact point is placed about $d \approx 1.60$ mm above the membrane. The plots of the interaction is shown in figure 2 (frame-by-frame plots). Due to the limitation in temporal resolution, the exact time of bubble inception can only be estimated. After the inception, the bubble expands rapidly, and the flow is eventually dominated by the water inertia. The bubble's top hemisphere expands freely as it is not obstructed but the expansion of the bottom hemisphere is impeded by the membrane that forms a dimple as the bubble pushes it downwards. This exerts a localised tension in the membrane.

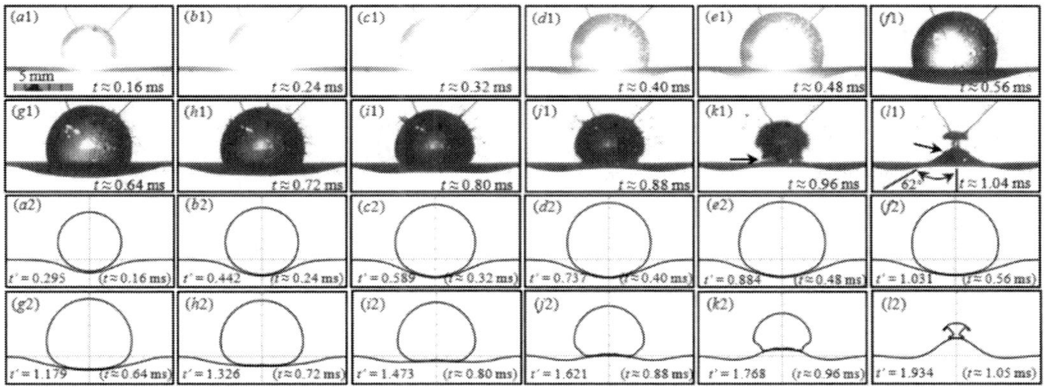


**Figure 2.** Frame-by-frame comparison of the interaction between experiment and simulation. $R_m \approx 3.84$ mm, $d \approx 1.60$ mm, $\xi = 0.42$ and $\kappa^*_m = 0.114$. The perturbation on bubble surface couples with bubble's contraction to form a mushroom-shape that leads to pinching and splitting. Dimensionless time $t'$ from the BEM simulation and the equivalent time $t = t' \cdot t_0$ are given.

The bubble maximum radius of $R_m \approx 3.84$ mm is estimated from figure 2($f1$), and at this moment, its pressure has fallen well below the pressure of the surrounding water. The expansion is halted and the bubble starts to contract. The membrane also begins to retract upwards, and the dimple will then flatten. The bubble's bottom surface is so close to the membrane that they move upwards together. As the membrane's retraction accelerates, a perturbation is imparted to the collapsing bubble's surface. This upwards propagating perturbation couples with the bubble's contraction to form the unique mushroom-shaped bubble ($t \approx 0.96$ ms). The thin film separating the bubble from the membrane seen in figure 2($k1$) is likely an artefact associated with the rapid movement of the membrane that the camera is unable to capture. At $t \approx 1.04\,\text{ms}$, the collapse intensifies. The perturbation eventually pinches the bubble, splitting it into top and bottom fractions with a slender 'bubble tubulus' connecting them. The top fraction forms a 'mushroom cap'. The bottom fraction is attached to the membrane's surface (figure 2($l1$)) that forms an upwards facing conical shape more profound than what was observed before [6].

In this experiment, the bubble collapse time ($t_c \approx 0.48\,\text{ms}$) is considerably higher than the Rayleigh collapse time ($t_{rc} \approx 0.35\,\text{ms}$) of a spherical bubble with $R_m = 3.84\,\text{mm}$. By comparing the radius-time histories of a spherical bubble generated using the same method and from computations based on the Rayleigh-Plesset equation, Turangan *et al.* [6] found that the one with $p_\nu = 50\,\text{kPa}$ matched well with the experiment. They argued that there might be a multiple dielectric breakdown (owing to the conventional way of connecting the electrodes) that leads to an excessive amount of hot metal vapour in the bubble and the prolonged heating of the bubble surface, and hence a higher vapour pressure and longer collapse time. The bright light that lasts up to $t \approx 0.56\,\text{ms}$ (see figure 2) may indicate so but further investigation is required to prove it.

A BEM simulation is carried out to complement the experimental observation. Figure 2 shows the comparison. The code solves the governing equations in an axisymmetric form. The bubble and the elastic membrane are represented using 51 computational nodes each. The bubble is assumed to initially contain a mixture of high pressured non-condensable gas and vapour, and thus $\gamma = 1.25$. The bubble initial radius $R_0 = 0.1458\,R_m$ at initial time $t' = 0$ is iterated using the Rayleigh-Plesset equation. This corresponds to the bubble's initial pressure of $p_{g0} = 10\,\text{MPa}$. For $d \approx 1.60\,\text{mm}$ and $R_m \approx 3.84\,\text{mm}$, the bubble's stand-off distance is $\xi = d/R_m = 0.42$. The membrane is only stretched in one direction, and therefore the elastic parameter is redefined to be $\kappa_m^* = \kappa/2 = 0.114$ (for $\sigma = 43.6\,\text{N/m}$). The plots for comparison are obtained by estimating the bubble's oscillation period and dividing it with the number of frames to be compared. The dimensionless time $t'$ of each plot is then multiplied with the scaling factor for time ($t_0$) to give the dimensional time $t$ (i.e. $t = t' \cdot t_0$, where $t_0 = R_m\sqrt{\rho/(p_\infty - p_\nu)}$, $\rho = 1000\,\text{kg/m}^3$, $p_\infty = 100\,\text{kPa}$ and $p_\nu = 50\,\text{kPa}$). The comparison shows that the simulation is able to model the interaction well including the membrane's response, the perturbation on the bubble surface, the mushroom shape and the pinching. However, there are several disparities that can be observed including the difference in bubble size in the early stage and the way the bubble splits.

## 4. Conclusions

The comparison between the experiment and simulation for a transient bubble generated very near to an elastic membrane matches well, showing a profound membrane's conical shape when the bubble splits. The experiment is easy to implement but manually connecting the electrodes is likely to lead to multiple electric breakdowns that cause high vapour pressures and longer collapse times. Nevertheless, the work demonstrates the efficacy of the method that can be easily applied to study the dynamics of bubbles near membrane materials with different elasticities, including those of biological materials. The numerical technique can be developed further to be a valuable tool to predict the dynamics of the interaction when parametric studies are required.

## References

[1] Coleman A J, Saunders J E, Crum L A and Dyson M 1987 *Ultrasound Med. Biol.* **13** 69–76
[2] Zhong P, Zhou Y and Zhu S 2001 *Ultrasound Med. Biol.* **27** 119–134
[3] Blake J R and Gibson D C 1987 *Annu. Rev. Fluid Mech.* **19** 99–123
[4] Shima A, Tomita Y, Gibson D C and Blake J R 1989 *J. Fluid Mech.* **203** 199–214
[5] Klaseboer E, Turangan C K and Khoo B C 2006 *Int. J. Multiphase Flow* **32** 1110–1122
[6] Turangan C K, Ong G P, Klaseboer E and Khoo B C 2006 *J. Appl. Phys.* **100** 054910
[7] Brujan E A, Nahen K, Schmidt P and Vogel A 2001 *J. Fluid. Mech.* **433** 251–281
[8] Klaseboer E and Khoo B C 2004 *J. Appl. Phys.* **96** 5808–5818

Journal of Physics: Conference Series **656** (2015) 012041 doi:10.1088/1742-6596/656/1/012041

# The condition of the resonant break-up of a gas bubble subjected to an acoustic wave in liquid

**V V Vanovskiy[1,2] and A G Petrov[1,2]**

[1] Theoretical Mechanics Department, Moscow Institute of Physics and Technology, 9 Institutskiy Per., Dolgoprudny 141700, Russia
[2] Mechanics of Systems Laboratory, Institute for Problems in Mechanics of the Russian Academy of Sciences, 101-1 Pr. Vernadskogo, Moscow 119526, Russia

E-mail: vovici@gmail.com

**Abstract**. The problem of a gas bubble break-up in liquid is considered in the conditions of the frequencies resonance of the radial and $n^{th}$ axially symmetric deformational mode 2:1. The nonlinear energy transfer between the modes is described using an efficient Krylov-Bogolyubov averaging technique. It is shown that the deformational mode magnitude can be some orders larger than the radial mode magnitude which is damped by the thermal, viscous and acoustic dissipation. The estimative criterion of bubble break-up is obtained in the cases of slow and fast acoustic wave start. The obtained pressure magnitudes in the wave for break-up are very small and the mechanism can have strong medical and technical applications.

## 1. Introduction

The mechanisms of break-up and the subharmonics' appearance are not yet fully understood and have gained much interest in present science. The bubble break-up problem has applications in medicine, oceanology and in all fields related to cavitation and sonoluminescence. For example one of the most promising applications of bubble break-up is the brain-blood-barrier opening by forcing break-up near barrier boundary with the focused ultrasound [1] and minimization of the ultrasound magnitude is very important for safeness.

There is much evidence that the bubbles in acoustic field emit subharmonic sound at frequency being the integer part of the driving frequency [2]. This fact can be attributed to the existence of nonspherical bubble harmonics resonant to driving frequency. There exist many experimental works that prove the possibility of bubble break-up because of its nonspherical shape oscillations [3]. Theory on this topic usually deals with the nonlinear interaction between oscillation modes and there are many works predicting the amplified energy transfer at frequency resonance 2:1 [4]. The possibility of break-up was mentioned in the latest work on this topic [5] that dealt with the problem of resonant energy transfer from radial oscillations mode to any deformation mode in case of free oscillations. In the latter the period of energy transfer and asymptotic trajectories were obtained and the enormous growth of deformation mode was attributed as a sign of break-up possibility. The main disadvantage of this work is that the damping was not considered at all. In this work a bubble in an acoustic wave is considered and for the purpose of this paper the case of axially symmetric deformation oscillations are considered (as shown in [5] the axially symmetric mode is the most dangerous for the break-up phenomenon).The problem is solved in a Hamiltonian approach by an effective Krylov-Bogolyubov averaging method [6]. In order to obtain radial bubble oscillations magnitude in linear case in an acoustic wave of known frequency and magnitude the different bubble damping mechanisms are considered in the linear approach similar to [7]. The deformation mode dissipation is considered to be negligibly small. The asymptotic for the bubble oscillations are obtained and the huge growth of the deformational mode magnitude is used to obtain the estimative criterion of bubble break-up in the cases of fast and slow driving wave start.

## 2. Main part

Let us consider a single gas bubble oscillating radially in an acoustic wave described by the pressure $p=p_\infty(1+Fe^{i\Omega t})$ where $p_\infty$ is the stationary pressure in liquid and $F$ the dimesionless wave magnitude. In the linear approximation the bubble radius is $a=a_0(1+Ae^{i\Omega t})$ where $a_0$ is the stationary bubble radius and $A$ the dimensionless magnitude of its oscillations. Making the key assumption of homogenous pressure inside the bubble the linear bubble response is described by:

$$Resp(\Omega)=\frac{A(\Omega)}{F}=\left(\frac{Dy}{1+\mathrm{i}\,Ma}+St-\mathrm{i}\,Vis-\frac{3}{(1+St)^{-1}-(1-1/\gamma)G}\right)^{-1},G=1-3\frac{\sqrt{\mathrm{i}Pe}\,\mathrm{cth}\sqrt{\mathrm{i}Pe}-1}{\mathrm{i}Pe}\quad(1)$$

Here we have introduced dimensionless parameters: $Dy=\rho a_0^2\Omega^2/p_\infty$ – the ratio of the dynamic pressure to the external pressure, $\rho$ – fluid density; $Ma=a_0\Omega/c$ – Mach number, $c$ – sound speed in fluid; $St=2\sigma/a_0p_\infty$ – the ratio of the Laplace pressure to the external pressure, $\sigma$ – surface tension; $Vis=4\mu\Omega/p_\infty$ – the ratio of the viscous pressure to the external pressure, $\mu$ –fluid viscosity; $Pe=a_0^2\Omega/k$ – Peclet number, $k$ – gas thermal diffusivity. In figure 1 the resonant curves for the bubbles of different radii and their enveloping curve (by dash line) are shown.

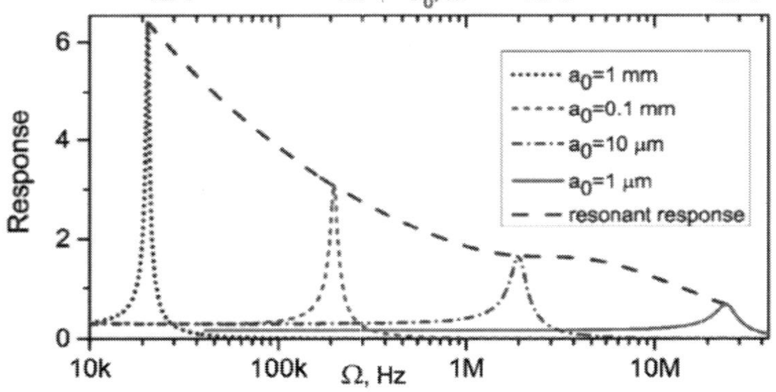


**Figure 1.** The resonant curves for the air bubbles of different radii in water

Following [5] the axially symmetric bubble with two degrees of freedom is considered and its surface is described by $r(\theta,\tau)=a_0(1+x(\tau)+\xi(\tau)P_n(\cos\theta))$. The bubble Hamiltonian obtained in [5] is used. The terms responsible for dissipation of the radial mode and driving force are added to the Hamilton equations:

$$\begin{aligned}&\dot{x}-u=-(n+3)yv,\ \dot{u}+4x=(3/2)v^2-4(n+1)y^2-\alpha\cos 2t-\beta u\dot{y}-v=-(n+3)yu-3xv,\\&\dot{v}+y=(n+3)uv-8(n+1)xy,\ \ y=\xi/\sqrt{(n+1)(2n+1)}\end{aligned}\quad(2)$$

Here n is the Legendre polynomial index, $x$, $u$ and $y$, $v$ the generalized coordinates and moments of radial and deformational oscillations, respectively. The two last terms in the second equation can be linked with the bubble physical properties by (1), and the driving force $\alpha=(4/3)F$ and the damping coefficient $\beta=2/(3Resp(\Omega))$ may be obtained. The driving force and the damping coefficient are considered to be small $\alpha=\varepsilon^2$, $\beta=\varepsilon\beta_1$. The stable asymptotic solution is obtained with the help of the Krylov-Bogolyubov averaging technique (the stability is proven by the second Bogolyubov theorem):

$$x(t) = -\alpha\left(\frac{8n(9n+16)+97}{(4n-1)^2}\cos 2t + \frac{8n+5}{4(4n-1)}\right) + \tilde{x}(t)e^{-\beta t/4}$$

$$y(t) = \pm\sqrt{\alpha}\left(\frac{2}{\sqrt{4n-1}}\sin t - \frac{10n(8n+15)+91}{(4n-1)^{5/2}}\beta\cos t\right) + \tilde{y}(t)e^{-\beta t/4}$$

Here $\tilde{x}(t), \tilde{y}(t)$ are quickly oscillating functions. In figure 2 the obtained analytical solution (bottom part) is compared with the numerical simulation of the initial system (2) (top part) with the same initial conditions. Functions $x(t)$ and $y(t)$ are plotted with red and blue colors. The left part is devoted to the $t<900$ and the right part to $4972<t<5003$. The initial conditions for simulation were: $\alpha=0.0001$, $\beta=0.01$, $n=19/8$, $x(0)=-0.004$, $u(0)=-0.01$, $y(0)=-0.002$, $v(0)=0.004$. The choice of initial conditions may be arbitrary in order to meet the conditions of smallness of driving force and dissipation. The obtained analytical formulas describe the qualitative behaviour of the system at first moment and are very precise in describing the steady oscillations. These oscillations magnitude practically doesn't change when we slightly modify the driving frequency and damping coefficient. Numerical simulations have shown that the possible frequency mismatch is proportional to $\sqrt{\alpha}$.

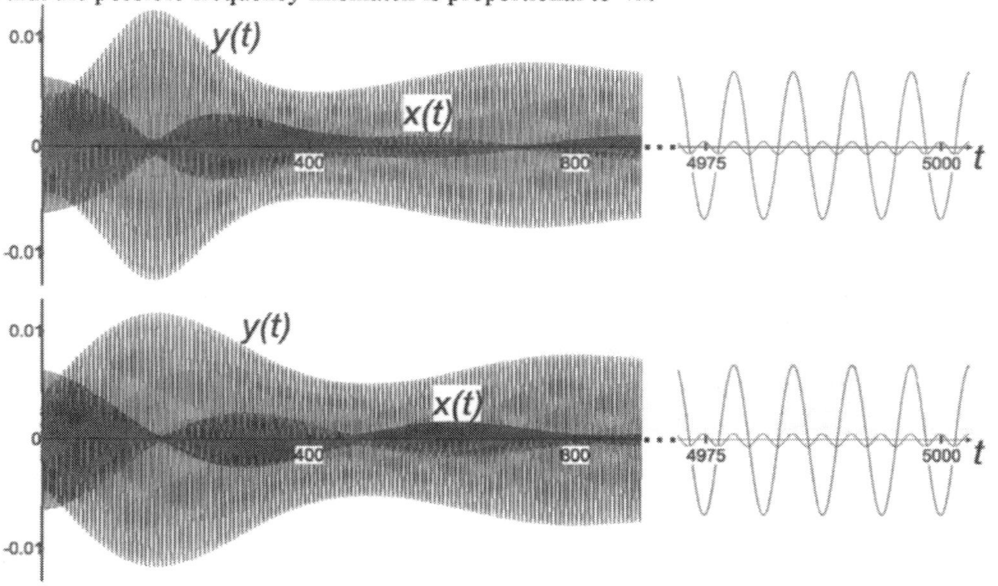


**Figure 2.** The comparison of the asymptotical solution (bottom part) and the numerical simulation (top part) of (2).

Some bubble oscillations modes associated with different Legendre polynomial indices $n$ are shown in figure 3. All the pictures except the bottom left one are drawn for $\xi\approx1$. Numerical simulations show that a neck separating parts of the bubble appears around $\xi= 0.5$ and this value may be considered critical for the break-up occurrence.

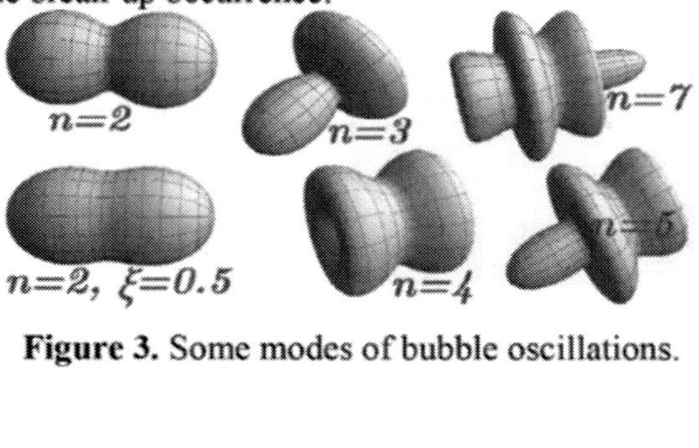


**Figure 3.** Some modes of bubble oscillations.

As may be observed in figure 2, the maximal value of the deformational mode is obtained during the transitional process and is approximately 1.7 times higher than the stationary magnitude. Thus, we obtain the estimative criterion for break-up in the cases of the fast and slow wave start. The latter is the critical acoustic wave magnitude $F_{cr} \approx 1/(5n)$. In figure 4 the break-up conditions for the air bubble in water at the atmospheric pressure is shown. The dashed lined stands for the maximal time of the fast acoustic wave start. Diamonds and circles stand for the critical acoustic wave magnitude in the cases of the fast and slow acoustic wave start for the bubbles satisfying the resonance 2:1 between radial and deformational modes condition:

$$a_0 = (2(n-1)(n+1)(n+2)+1)/(3\gamma)-1)\cdot(2\sigma)/p_\infty)$$

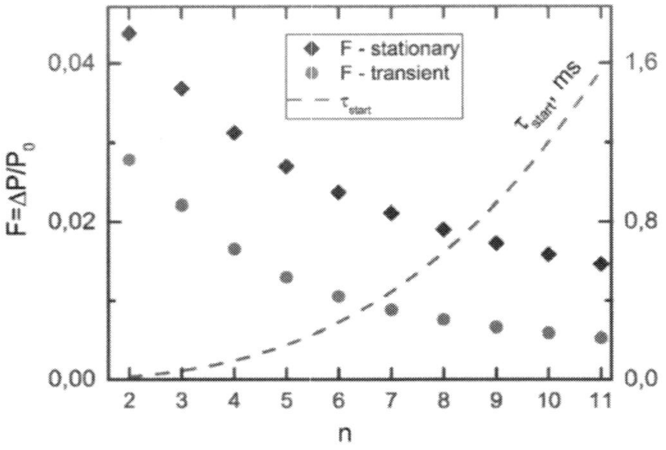


**Figure 4.** The break-up conditions and the time of the fast acoustic wave start.

One may observe that the break-up acoustic wave magnitude may be really small and this is possible due to two resonance conditions: resonance of frequencies 2:1 between the radial and the deformational modes and the frequency of the acoustic wave in resonance with the radial frequency. Surely, for proving this mechanism the strongly nonlinear deformations should be taken into account and the experiments should be carried out.

## 3. Conclusions

The estimative criterion of gas bubble break-up in liquid in case of the resonance of the radial and n-th deformational modes frequencies 2:1 and the resonant frequency of the driving force was obtained in cases of slow and fast acoustic wave start. The needed acoustic wave magnitude is very small and the proposed break-up mechanism may have numerous medical and technical applications.

The work was supported by the Russian Science Foundation, project No 14-19-01633.

## References

[1] Choi J J et al. 2007 Spatio-temporal analysis of molecular delivery through the blood–brain barrier using focused ultrasound *Phys. Med. Biol.* **52** 5509–5530

[2] Neppiras E M 1980 Acoustic cavitation *Phys. Rep.* **61** 159-251

[3] Leighton T G *Image intensifier studies of sonoluminiscence, with applications to the safe use of medical ultrasound* (*Ph.D. thesis*) (Cambridge Univ.)

[4] Longuet-Higgins M S 1991 Resonance in nonlinear bubble oscillations *J. Fluid Mech.* **225** 531

[5] Vanovskii V V and Petrov A G 2012 The Resonant Mechanism of Subdivision of a Gas Bubble in a Fluid *Doklady Physics* **57** **6** 238–242

[6] Bogolyubov N N and Mitropol'skii Yu A 1961 *Asymptotic methods in the theory of nonlinear oscillations* (Delhi: Gordon & Breach)

[7] Nigmatulin R I 1990 *Dynamics of Multiphase Media* vol 1 (New York: Hemisphere)

9th International Symposium on Cavitation (CAV2015) IOP Publishing
Journal of Physics: Conference Series **656** (2015) 012042 doi:10.1088/1742-6596/656/1/012042

# Three stages of bubble formation on submerged orifice under constant gas flow rate

**Xianxian Yu, Yiwei Wang*, Chenguang Huang, Tezhuan Du**

Key Laboratory for Mechanics in Fluid Solid Coupling Systems, Institute of Mechanics, Chinese Academy of Sciences
Hai dian District No.15 Beisihuanxi Road, Beijing, China
*wangyw@imech.ac.cn

**Abstract**. Bubble formation is involved in many engineering applications. It is important to understand the dynamics of bubble formation. This work reports experimental and numerical results of bubble formation on submerged orifice under constant gas flow rate. Compressible large eddy simulation combined volume of fluid (VOF) was adopted in simulation and results was validated by experiment. Bubble formation is divided into three stages in this paper, expansion stage, elongation stage and pinch-off stage. In expansion stage, The bubble grows radially due to the incoming gas flux, but the bubble base remains attached to the orifice. But as gas injected, the spherical bubble will go into the elongation stage when the downward resultant force is lager than upward resultant force. And when bubble neck's length is bigger than $\sqrt{2}R_o$, the bubble will go into pinch-off stage. Cylindrical Rayleigh-Plesset equation can be used to describe the pinch-off stage. Uncertain parameter $r$ in it is given reference value in this paper.

## 1. Introduction

The process of bubble formation at submerged orifice is very significant in many engineering applications, such as bubble column, gas-liquid separator, cavitation, etc.

Owning to its practical importance, the phenomenon of bubble formation and pitch-off has been widely studied theoretically and experimentally for decades.

Sufficient investigations of bubble formation at submerged orifice have been reported, as described in articles of Zhang et al(1), Kumar et al(2-4). Different models have been proposed to study the formation of bubble. In low gas flow rate, the formation of bubble is dominant by buoyancy and surface tension, which can be described by Young-Laplace equation (5). Gerlach(5; 6) et al. did a comprehensive study on bubble formation at a submerged orifice with low and medium constant gas flow rate. They brought up a theoretical model of the quasi-static bubble formation at very small flow-rate. It was observed that the strong vortex generated by a large leading bubble influences the growth of the succeeding bubble and leads the period-1 bubbling regime to the period-2 bubbling regime. In Zhang's work (1) the bubbling phenomena above a submerged orifice can be divided into four regimes, single bubbling, pairing, double coalescence and triple bubble formation. The experimental results and the theoretical prediction both demonstrate that bubble interaction is the primary reason for the aperiodic characteristics of bubble departure. Pinch-off in liquid is a complex and key point in bubble formation process which draws considerable interest(7, 8). Burton J C(8) used cylindrical version of the general Rayleigh-Plesset equation and a 100000 frame-per-second video to analyze the pinch-off of nitrogen gas bubbles in fluids with a wide range of viscosity. They found a power law solution in

The authors are grateful to the National Natural Science Foundation of China through grant numbers 11202215&11332011.

time, neglecting the pressure in the bubble. If the external fluid is highly viscous ($\eta_{ext} > 100\,cP$), the bubble neck radius is proportional to the time before break, $\tau$, and decreases smoothly to zero. If the external fluid has low viscosity ($\eta_{ext} < 10\,cP$), the radius scales as $\tau^{1/2}$.

There have been much evaluable theoretical and experimental results in this literature. In the present paper, we showed our numerical and experimental work on bubble formation with experiment and simulation, including analysis on three stages of formation process and discussion on Rayleigh-Plesset equation, etc. It is hoped make some new reference to this phenomenon.

## 2. Experiment and simulation

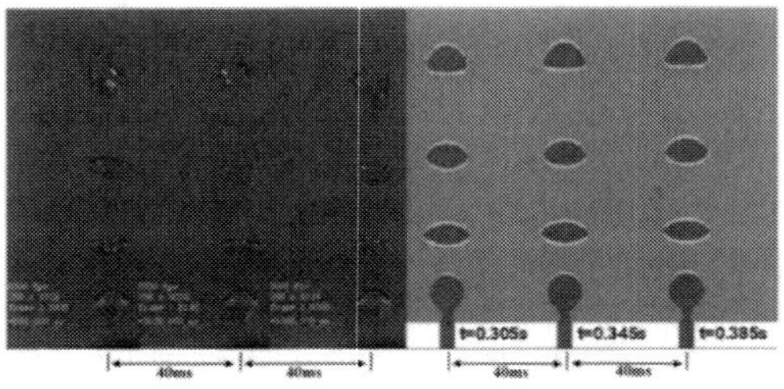


Fig. 1 Experiment and simulation

Bubble formation is a two phase phenomenon. For the numerical treatment of this problem, compressible large eddy simulation (LES) combined with a volume of fluid (VOF) method is used in this paper. Experiment is taken in a small water tank and recorded by a high-speed camera. Fig.1 shows the results of experiment and simulation. The orifice radius is 1mm and gsa flow rate is 100ml/min.

## 3. Results and discussion

In present paper, bubble formation ware divided into three stages, as shown in Figure 2, including stage I -bubble expansion, stage II -bubble elongation and stage III -bubble pinch-off. In stage I, sphere bubble takes shape and grows bigger with continuous injected gas through the channel, as Figure 2- I shows. When bubble becomes big enough, the bottom part begins to contract and a short cylindrical bubble appears , through which the whole bubble contacts with the orifice, as shown in Figure 2- II. When stretched to a certain length, the cylindrical bubble will lose stability, contracting inwards until detaches from the orifice. This stage is called pinch-off, as shown in Figure2-III. The detached bubble rises up in the water during the following time until reaches the water surface.

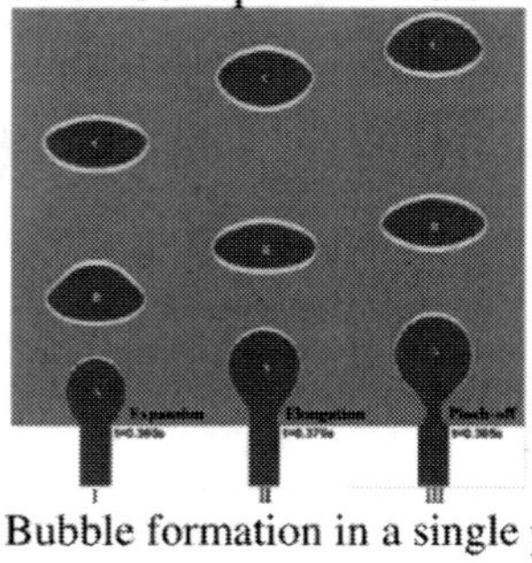


Fig. 2 Bubble formation in a single period

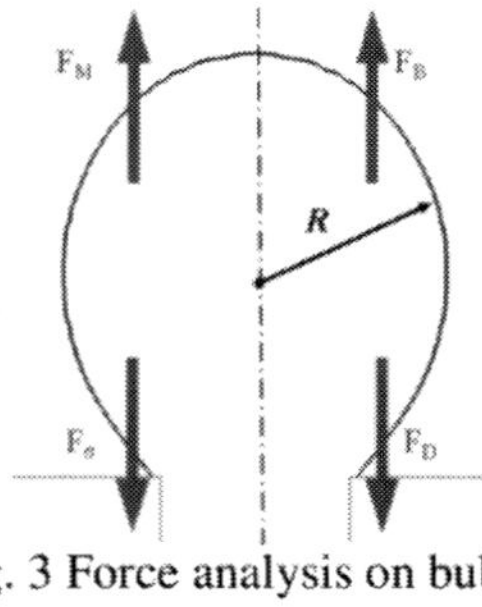


Fig. 3 Force analysis on bubble

**Expansion stage**. In expansion stage, as shown in Fig.3, the bubble grows radially due to the incoming gas flux, but the bubble base remains attached to the orifice. The growth rate of bubble volume nearly equals to the gas flow rate. As gas injected, the downward resultant force (surface tense $F_\sigma$ and drag force $F_D$) becomes lager than upward resultant force (buoyancy $F_B$ and gas momentum force $F_M$), ie $F_{B+}F_M \geq F_\sigma + F_D$, the spherical bubble goes into the elongation stage then.

**Elongation stage.** In elongation stage, the bubble rises away from the orifice, but still attached to it through a neck, which is thought to be a stable cylindrical bubble, represented by Figure 4. When the cylindrical bubble lose its stability, the bubble will go into next stage. The stability criterion of the neck is studied here. The cylindrical bubble is assumed to get a very small disturbance and its surface tense changes, as shown in Figure 5. Surface tense can be sorted into $F_{\sigma-out}$ and $F_{\sigma-in}$ . $F_{\sigma-out}$ could help the distortional neck to recover while $F_{\sigma-in}$ stops it. If $\Delta F_{\sigma-out} \ge \Delta F_{\sigma-in}$, the bubble can keep stable, or else it will lose stability. $\Delta F_{\sigma-out} = \frac{\sigma}{R_{out}}$, $\Delta F_{\sigma-in} = \frac{\sigma}{R_o - \Delta y} - \frac{\sigma}{R_o}$ where $R_o$ is radius of the orifice, $\sigma$ is surface tense coefficient and $R_{out}$ is curvature radius corresponding to $F_{\sigma-out}$ · $R_{out} = \frac{l^2 + \Delta y^2}{2\Delta y}$ where $\Delta y$ is relatively small compared with $l$ and $\Delta y^2$ can be neglected. Though analysis, we can get the stability criterion $l \le \sqrt{2}R_o$ where $R_o$ is radius of the orifice.

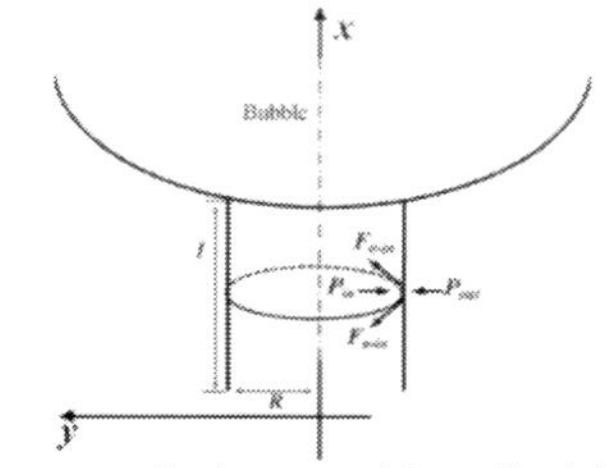

Fig.4 Force analysis on stable cylindrical bubbl

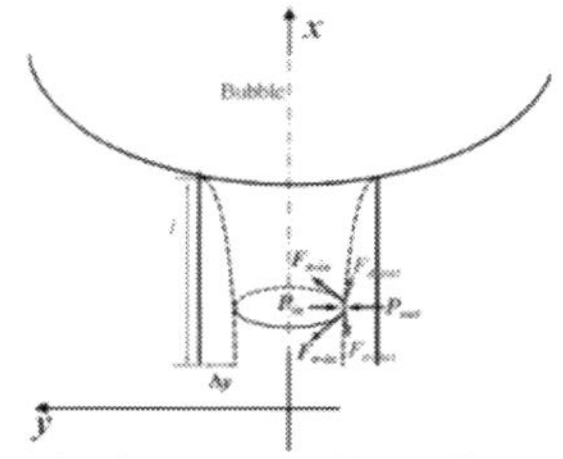

Fig. 5 Force analysis on unstable cylindrical bubble

**Pinch-off stage.** When bubble neck's length is bigger than $\sqrt{2}R_o$, the bubble will go into detachment stage, as shown in Fig.6. Cylindrical version of the general Rayleigh-Plesset equation can be used to describe bubble dynamics near the pinch region

$$\rho_{ext}[(\ddot{R}R + \dot{R}^2)\ln\frac{R}{r} + \frac{1}{2}\dot{R}^2] = [P_{out} - P_{in}] + (\frac{\sigma}{R} - \frac{\sigma}{R_{out}}) \quad (1)$$

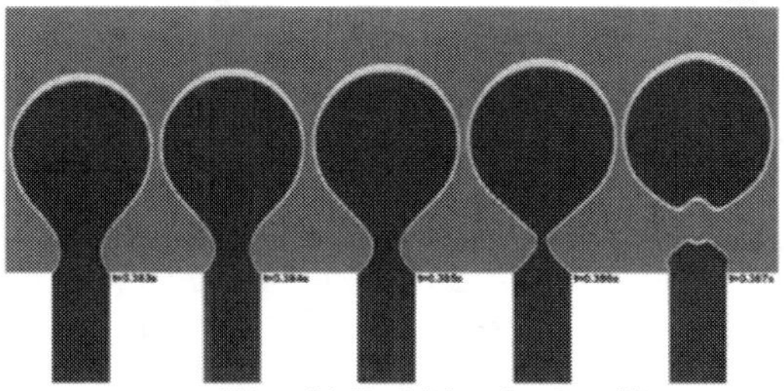

Fig. 61 Bubble pinch-off

But unknown $r$ in $\ln\frac{R}{r}$ leads to the equation can not be solved. We try to give appropriate value of r through simulation. We could obtain bubble profile and pressure field with time from simulation, as shown in Figure 7. From pressure field, $P_{out}$ and $P_{in}$ is known, which are both perpendicular to the bubble surface. From bubble profile, curvature of the neck can be got and thus surface tension $\frac{\sigma}{R}, \frac{\sigma}{R_{out}}$ is known where $\sigma$ is 0.07 for water. As a result, the right side of the equation in simulation can be determined. To left side of the equation, the minimum radius of the neck $R_{\min}$ is the minimum value of Y (bubble profile) in Figure 7. So $R_{\min}$ and its variation $\dot{R}_{\min}$, $\ddot{R}_{\min}$ can be obtained. If $r$ is given, the left side of the equation at $R_{\min}$ can be known . When both sides of the equation are close, the value of $r$ is what we want. This is the idea of getting appropriate $r$ in this paper.

According to $\ln\frac{R}{r}$ in the equation, $r$ is assumed as $r = R{\bullet}e^{n}$ , $n = 1, 1.5, 2, 3, 4$ . The comparison between the two sides of the equation at different $r$ is shown in Figure 8. We can see that value of both sides are close when n=1.5~2. So we think that $r$ in cylindrical version Rayleigh-Plesset equation for bubble pinch-off is about $R{\bullet}e^{1.5} \le r \le R{\bullet}e^{2}$ , ie $4.5R \le r \le 7.4R$ . Eq.(1) can be written as

$$\rho_{ext}[C{\bullet}(\ddot{R}R + \dot{R}^2) + \frac{1}{2}\dot{R}^2] = [P_{out} - P_{in}] + (\frac{\sigma}{R} - \frac{\sigma}{R_{out}} \qquad (2)$$

where $4.5 \le C \le 7.4$ . Eq.(2) could be used to predict dynamic behaviors of gas bubble formation in water.

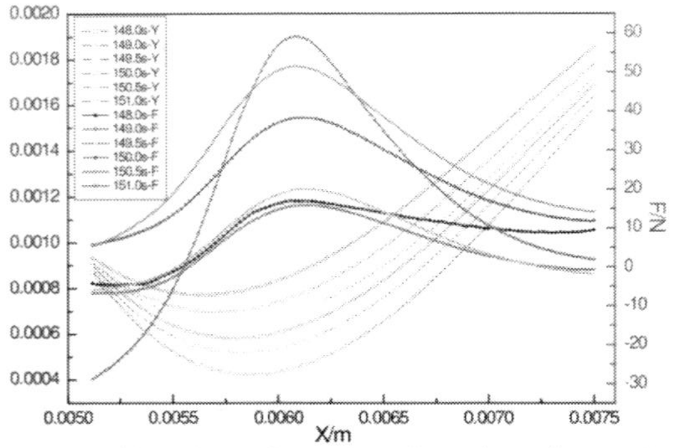

Fig.7 Profile of bubble neck and resultant force with time

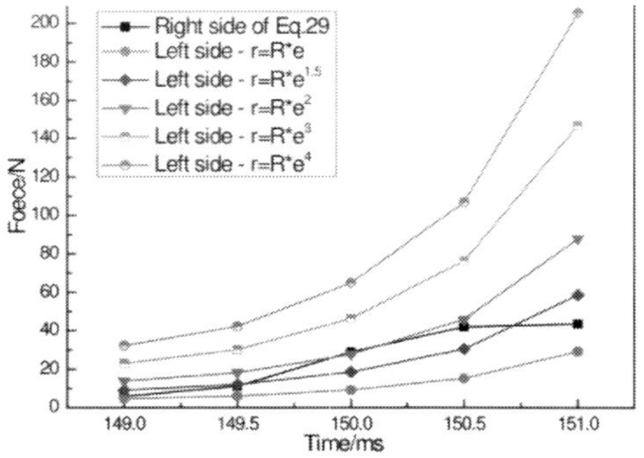

Fig. 8 Comparison between the two sides of the equation at different $r$

## 4. Conclusions

Experiment and simulation of bubble formation on submerged orifice under constant gas flow rate were taken in this paper. Compressible large eddy simulation combined volume of fluid (VOF) was adopted in simulation and results were validated by experiment. Bubble formation is divided into three stages in this paper, expansion stage, elongation stage and pinch-off stage.

In expansion stage, the bubble grows radially due to the incoming gas flux, but the bubble base remains attached to the orifice. The growth rate of bubble volume nearly equals to the gas flow rate. As gas injected, the spherical bubble will go into the elongation stage when the downward resultant force is lager than upward resultant force. In this stage, the bubble lifts away from the orifice, but still attached to it through a neck, which is thought to be a stable cylindrical bubble. When bubble neck's length is bigger than $\sqrt{2}$ times of the orifice radius, the bubble will go into pinch-off stage. Cylindrical Rayleigh-Plesset equation can be used to describe the pinch-off stage. Parameter $r$ in the equation is given as $R{\bullet}e^{1.5} \le r \le R{\bullet}e^{2}$ based on analysis of simulation results in this paper.

This research was sponsored by the Youth Innovation Promotion Association of CAS (2015015). The authors are grateful for the support.

## References

[1] Zhang L, Shoji M. 2001. Chem. Eng. Sci. **56**:5371-81
[2] Khurana A, Kumar R. 1969. Chem. Eng. Sci. **24**:1711-23
[3] Ramakrishnan S, Kumar R, Kuloor N. 1969. Chem. Eng. Sci. **24**:731-47
[4] Satyanarayan A, Kumar R, Kuloor N. 1969. Chem. Eng. Sci. **24**:749-61
[5] Gerlach D, Biswas G, Durst F, Kolobaric V. 2005. Int. J. Heat Mass Transf. **48**:425-38
[6] Gerlach D, Alleborn N, Buwa V, Durst F. 2007. Chem. Eng. Sci. **62**:2109-25
[7] Terasaka K, Tsuge H. 1993. Chem. Eng. Sci. **48**:3417-22
[8] Burton J, Waldrep R, Taborek P. 2005. Phys. Rev. Lett. **94**:184502

# Outcomes following single bubble collapse in a rigid tube

**C Ji[1], B Li[1], F Y Lin[1] and J Zou[1]**

[1]State Key Laboratory of Fluid Power Transmission and Control, Zhejiang University, Hangzhou 310027, China

E-mail: junzou@zju.edu.cn

**Abstract**. Following the collapse of a vapour bubble inside a rigid tube, various outcomes could be caused by the propagation and reflection of the shock wave. Using high-speed camera system and hydrophone, the secondary cavitation and acoustic pressure curve due to the collapse of single bubble in a rigid tube are studied experimentally. The secondary cavitation has a decisive effect on the pressure pulse sequence, which usually appears at a certain area symmetric to the first bubble. Two particular cases of first bubble generated near the midpoint and the endings of the tube are also investigated, where the strength, position and period of the secondary cavitation show different features. The mechanism behind these results are discussed briefly, which is related to the propagation and reflection of the shock wave.

## 1. Introduction

When cavitation occurs, the geometrical boundary constraints have a major effect on both the behaviour of cavitation bubbles and the outcomes after the collapse. A round-section rigid tube is a common boundary for cavitation bubbles, referring to lots of fields, such as hydraulic systems, MEMS and dental therapy. The dynamic behaviour of single bubble in a rigid tube has been investigated experimentally and numerically [1, 2], focusing on the deformation, the liquid jet, and the pressure distribution around it. Recently, the interaction phenomena between two vapour bubbles in a rigid tube were also reported [3]. Theoretically, a one-dimensional model was developed by Kornilovitch [4] to describe the inertial pump effect of the collapsing bubble, which produces a final momentum in the tube.

On the other hand, the mechanism of "shock wave-reflection-cavitation" has been widely observed in pipes following water hammer [5] or a manually excited shock wave [6], which could cause various outcomes. In a rigid tube, the collapse of vapor bubble could also emit shock wave, which will propagate along the tube, then be reflected and converge again. The outcomes could be greatly affected by the pressure waves, leading to secondary cavitation and different pressure pulses, similar to the bubble collapse near a free surface [7-9]. However, these outcomes were usually neglected or simply described in one sentence in past works about bubble collapse in rigid tube.

The present work aims to experimentally study the phenomena after the collapse of a single bubble in a rigid tube. Experimental results of the secondary cavitation and acoustic pressure sequence are presented, with a brief discussion on the mechanism behind the outcomes observed.

## 2. Experimental setup

As shown in figure 1, vapor bubbles are generated by low voltage electric spark method [10], which are named as "first bubble" in the present study. The circuit consists of a 50 V power source, three capacitors (3300 μF × 1, 1000 μF × 2), a resistance of 1 kΩ, and one pair or two pairs of electrodes (fine copper wires of 0.26 mm). An organic glass tank (400 × 200 × 200 $mm^3$) is used as the container to hold tap water, in which a glass tube is fixed. The liquid and laboratory temperature are maintained at 20°C ± 2°C, and the environment pressure is kept at 101.3 kPa. Before inducing the spark, the contact position of the electrodes is carefully adjusted to be inside the tube and at its radial centerline. The details of the first and secondary cavities are captured with a framing rate of 15,000 frames per second by a high-speed video camera (Photron FASTCAM-ultima APX), which is fitted with a Nikkor 60-mm microlens. Backlight is produced by a high-intensity LED lamp. A hydrophone (Reson TC4013) is employed to record the acoustic pressure curve during the experiments, which is placed at the radial center line, two centimeters away from the tube ending. The hydrophone has a frequency range up to 170 kHz, whose output is recorded by an acquisition board (KEYENCE NR-600).

The tubes employed in the experiments have unified interior diameter of $D_t$ = 6 mm, and length of $L_t$ = 110 mm. All the first bubbles occupy the entire section of the tube, therefore the problem could be considered one-dimensional. The axial position is scaled by $L_t$ into a non-dimensional coordinate $X$, which ranges from 0 to 1. The collapse position of the first bubble is noted as $X_F$. The equivalent maximum radius of the first bubble is noted as $R_e$, which is calculated by assuming a spherical bubble with a same volume.

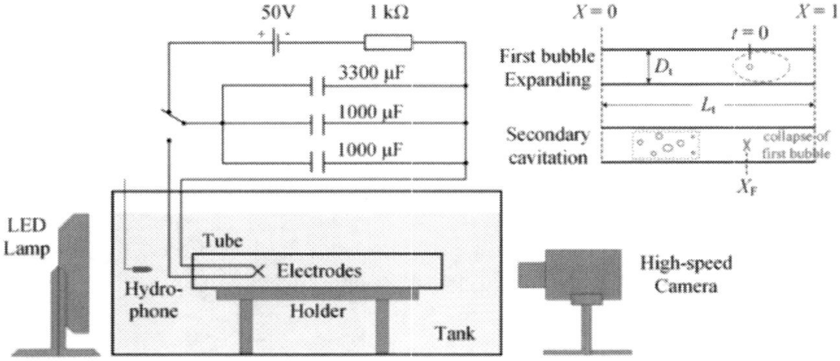


**Figure 1.** Sketch of the experimental setup

## 3. Results

A typical sequence of secondary cavitation in a rigid tube is shown in figure 2, where $X_F$ = 0.77, $R_e$ = 4.26 mm. The first bubble expands as an ellipsoid, before reaching its maximum volume at 2.2 ms. At 4.2 ms, the first bubble finishes its collapse. The collapse time of the first bubble is much longer than the Rayleigh time of 0.39 ms, which is calculated from $R_e$ with a same environment pressure. This difference results from the blocking effect of the tube wall, which could delay the collapse process dramatically, up to 7 times longer than Rayleigh time [1] and affected by $X_F$ [4]. After the collapse, the first bubble crushes into a cloud of tiny bubbles, whose rebound process is rather weak with a short period. A secondary cavitation appears in the frame next to the collapse of the first bubble, referring to the distributed small cavities in the picture of 4.33 ms. Once it appears, the small bubbles at both edges of the secondary cavitation area begin to collapse, leading to a contraction of the cavitation cluster till 7.13 ms. However, each of the cavities inside the cluster keeps expanding, till the edge of the cluster moves near it. Generally speaking, the behavior of the secondary cavitation cluster seems similar to a large bubble in a tube, while every single cavity inside it experiences an unusual sudden shift from expansion to contraction. Besides, the collapse position of the first bubble and secondary cavitation seems symmetric about the midpoint of the tube.

The acoustic pressure curve is also shown in figure 2. The first pressure pulse “A” corresponds to the low-voltage spark, while pulse “B” is caused by the collapse of the first bubble, 4.17 ms after pulse “A”. From the time interval of 2.88 ms, the third pulse “C” could be attributed to the collapse of the secondary cavitation. After that, the recorded pressure pulses are quickly weakened.

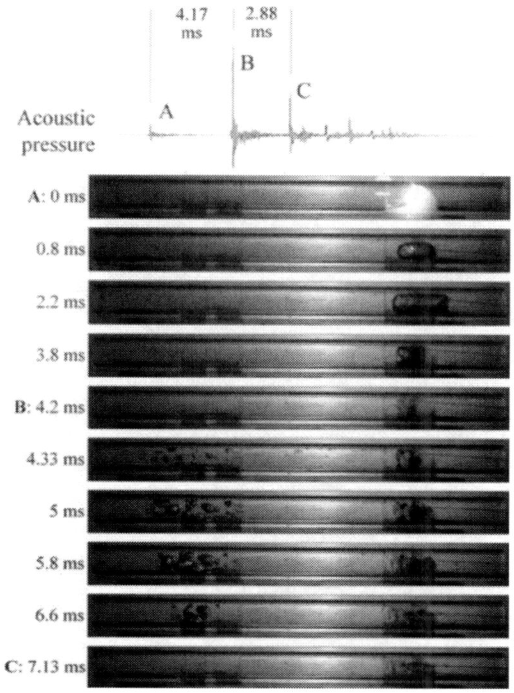


**Figure 2.** Symmetric secondary cavitation. ($X_F = 0.77$)

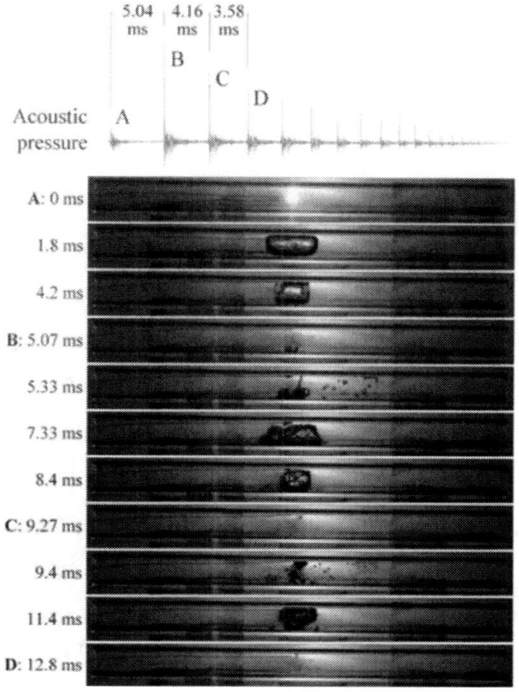


**Figure 3.** Large rebound bubble. ($X_F = 0.49$)

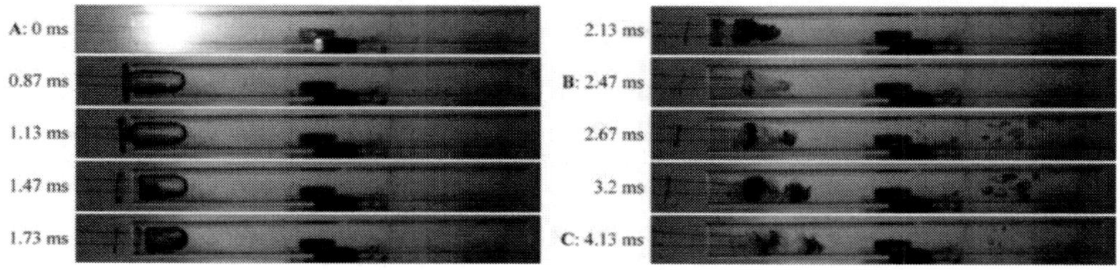


**Figure 4.** Asymmetric secondary cavitation. ($X_F = 0.10$)

In the experiments, the location of $X_F$ is found to dramatically affect the outcomes following the collapse of the first bubble, especially in the cases of first bubble generated near the axial midpoint ($X_F \rightarrow 0.5$) or near the tube endings ($X_F \rightarrow 0$ or $X_F \rightarrow 1$). Figure 3 shows a case of midpoint spark, where $X_F = 0.49$ and $R_e = 4.44$ mm. The first bubble experiences symmetric expansion and contraction, then it collapses to its minimum volume at 5.07 ms, which is also much longer than the period of 0.41 ms calculated by Rayleigh equation. The following outcomes have obvious difference with that in figure 2, including both secondary cavitation and the rebound cavity. From 5.07 ms to 9.27 ms in figure 3, a single large cavity could be obtained, which has an anomalistic shape. Besides this large rebound cavity, secondary cavitation also occurs. As shown in the picture of 5.33 ms, the secondary cavitation appears soon after the collapse of the first bubble. The secondary cavitation locates close to the rebound cavity, and disappears quickly. A similar oscillation period is obtained after the collapse of rebound cavity at 9.27 ms, where a smaller rebound cavity comes out, together with a weaker secondary cavitation around it. The midpoint spark produces more pressure peaks than those in figure 2. Besides, "C", "D" and other following pulses in figure 3 are all clearly related to the collapse of rebound cavities instead of secondary cavitation, which is similar to the vapor bubble near a wall [11].

Another particular case is a spark near the tube ending, as shown in figure 4, where $X_F = 0.10$ and

Journal of Physics: Conference Series **656** (2015) 012043 doi:10.1088/1742-6596/656/1/012043

$R_e$ = 4.15 mm. The first bubble expands and exceeds the left tube ending, forming a ring shaped bubble outside the tube. The secondary cavitation here is asymmetric with the collapse location of the first bubble, which is different from figure 2. The first two time intervals between acoustic pressure pulses are respectively 2.47 ms and 1.66 ms.

## 4. Discussion

As shown in figure 2 to figure 4, the preliminary results show that the collapse of single bubble in rigid tube could induce a secondary cavitation, which usually locates symmetric with the collapse position of the first bubble (about the midpoint $X$ = 0.5). Based on this position correlation, a mechanism of the secondary cavitation could be deduced as follows. The collapse of the first bubble is the energy source of the secondary cavitation, which emits shock wave towards both endings of the tube. Due to the violent section expansion at the tube endings, part of the energy in each shock wave is reflected as an expansion wave. If the wave speed is assumed unified in the entire tube, the expansion waves reflected from the two endings will meet at a certain position, symmetric to the collapse position of the first bubble. This interaction of two expansion waves causes an extremely low pressure at the symmetric position, where secondary cavitation happens. In the present experiments, the shock waves take most collapse energy of the first bubble, which is released to excite the secondary cavitation. Therefore, the rebounding process is obviously weakened.

When a vapour bubble collapses in free field, the strength of its rebounding cycles could be observed from the pressure pulses. However once the violent secondary cavitation appears, it is no longer appropriate to use the frequency of the pressure pulses to evaluate the size of the rebounding bubble. In this case, more energy dissipation during the collapse could lead to a larger interval between the pressure pulses, because that the pulses following the collapse are dominated by the secondary cavitation.The power of the rebound cavity or the secondary cavitation could be judged from the period ratio, i.e. comparison of the first two time intervals between the pressure pulses. In figure 2~4, the period ratios are respectively 0.69, 0.83 and 0.67. Comparing with the common case of figure 2, the tube ending case (figure 4) seems only to affect the position of secondary cavitation, while the strength of that changes slightly. However, the rebound in figure 3 is much stronger, which relates to the unique wave propagation in the case of $X_F \rightarrow 0.5$. The reflected expansion waves in midpoint cases converges just around $X_F$, so the major energy of the shock wave is transmitted back to the rebounding cavity, which could explain the increase of the rebound strength. The irregular shape of the rebound bubble results from the bubble fission during the last stage of the collapse.

## Acknowledgements

This work is supported by National Natural Science Foundation of China (Grant No. 51405429), and the National Natural Science Foundation for Excellent Youth Scholars of China (No. 51222501).

## References

[1] Ni B Y, Zhang A M, Wang Q X and Wang B 2012 *Acta Mech. Sin.* **28** 1248-60
[2] Hajizadeh Aghdam A, Khoo B C, Farhangmehr V and Shervani-Tabar M T 2015 *Exp. Therm. Fluid Sci.* **60** 299-307
[3] Zou J, Li B and Ji C 2015 *Exp. Therm. Fluid Sci.* **61** 105-12
[4] Kornilovitch P, Govyadinov A, Markel D and Torniainen E 2013 *Phys. Rev. E* **87** 023012
[5] Bergant A, Simpson A R and Tijsseling A S 2006 *J. Fluids Struct.* **22** 135-71
[6] Ranjan D, Oakley J and Bonazza R 2011 *Annu. Rev. Fluid Mech.* **43** 117-40
[7] Obreschkow D, Dorsaz N, Kobel P, de Bosset A, Tinguely M, Field J and Farhat M 2011 *Phys. Fluids* **23** 101702
[8] Tomita Y, Kodama T and Shima A 1991 *Appl. Phys. Lett.* **59** 274-6
[9] Ando K and Liu A Q 2012 *Phys. Rev. Lett.* **109** 044501
[10] Turangan C K, Ong G P, Klaseboer E and Khoo B C 2006 *J. Appl. Phys.* **100** 054910
[11] Yang Y X, Wang Q X and Keat T S 2013 *Ultrason. Sonochem.* **20** 1098-1103

# Temperature considerations in non-spherical bubble collapse near a rigid wall

**Shahaboddin Alahyari Beig, Eric Johnsen**
Mechanical Engineering Department, University of Michigan, Ann Arbor, MI 48109, USA
E-mail: alahyari@umich.edu, ejohnsen@umich.edu

**Abstract.** The inertial collapse of cavitation bubbles is known to be capable of damaging its surroundings. While significant attention has been dedicated to investigating the pressures produced by this process, less is known about heating of the surrounding medium, which may be important when collapse occurs near objects whose mechanical properties strongly depend on temperature (e.g., polymers). Using a newly developed computational approach that prevents pressure and temperature errors generated by naively implemented shock- and interface-capturing schemes, we investigate the dynamics of shock-induced collapse of gas bubbles near rigid surfaces. We characterize the temperature fields based on the relevant non-dimensional parameters entering the problem. In particular, we show that bubble collapse causes temperature rises in neighboring solid objects via two mechanisms: the shock produced at collapse and heat diffusion from the hot bubble close to the object.

## 1. Introduction

Cavitation bubble dynamics appear in a wide range of hydraulic applications, such as turbomachinery and naval structures. One of the important consequences of cavitation is the structural damage to neighboring surfaces due to bubble collapse [7, 8]. The modern description of cavitation erosion to metallic surfaces is based on a sequence of four steps [6]: production of small-scale vapor structures, impact loads due to bubble collapse, pitting, and failure. Based on this model, the origin of cavitation erosion lies in the mechanical loading produced on the material. However, most recent experiments show that high temperatures produced by bubble collapse near soft materials (e.g., polymer-based) may influence the damage mechanism(s).

Due to the complexity of the physics and the wide range of spatial and temporal scales in cavitation, numerical simulations have emerged as a reliable tool to complement analytical and experimental studies. However, direct simulations are challenging due to algorithmic difficulties. To overcome these difficulties, we employ a shock- and interface-capturing approach. However, a naive implementation of this approach has been known to give rise to spurious pressure and temperature oscillations across the interfaces [1, 2, 5]. Therefore, special care must be taken to prevent these errors [2, 5].

The objective of the present work is to better understand the potential damage caused by the shock-induced collapse of a single bubble near a rigid wall. We provide three-dimensional numerical simulations of this problem for different geometrical configurations to investigate the basic flow physics, with a particular emphasis on temperature effects. We further demonstrated that significant maximum wall temperatures can be achieved, depending on the heat diffusivity of the material.

Journal of Physics: Conference Series **656** (2015) 012044 doi:10.1088/1742-6596/656/1/012044

## 2. Numerical method

The compressible Navier-Stokes equations with no mass transfer or surface tension effects are solved using a shock- and interface-capturing approach that prevents spurious numerical errors at material interfaces. The time marching is handled by a third-order accurate explicit SSP Runge-Kutta scheme. For the spatial discretization, a three-dimensional solution-adaptive high-order accurate central difference/discontinuity-capturing method is used. Smooth regions are computed using central differences, a finite difference weighted essentially non-oscillatory (WENO) scheme with Lax-Friedrichs flux splitting handles shock waves, and the approach of Johnsen & Colonius [3] is used for material interfaces.

## 3. Results

The governing equations are non-dimensionalized by the density and sound speed of air at atmospheric pressure, initial bubble radius $R_o = 0.2$ mm, and $T = 300$ K. The schematic of the problem setup is shown in Fig. (1). An initially spherical bubble located a distance $H/R_o$ from a rigid wall is impacted by a Mach 1.018 shock (500:1 pressure ratio).

The simulations are done for different initial stand-off distance $H$ to study its effects on pressure and temperature along the wall. Fig. (2) shows the pressure and temperature contours during the collapse of a gas bubble for $H/R_o = 1.2$. The right-moving shock interacts with the bubble, thus producing a reflected rarefaction wave. As the bubble starts its collapse, the incoming shock hits the rigid wall and reflects back onto the bubble. This doubles the surrounding pressure and leads to a stronger collapse. During the collapse, a re-entrant jet directed toward the wall is produced, which, upon impact with the distal side, generates an outward propagating shock. Thereafter, the bubble takes the form of a vortex ring, which is hot due to the rapid compression and is convected toward the wall. This process produces regions of high pressure and temperature along the wall.

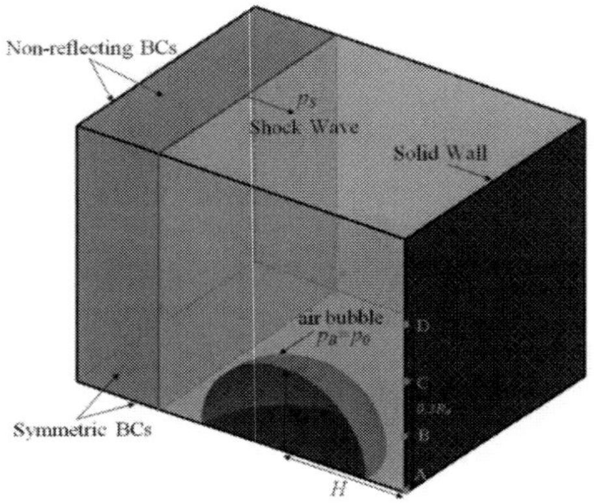


**Figure 1.** Problem setup.

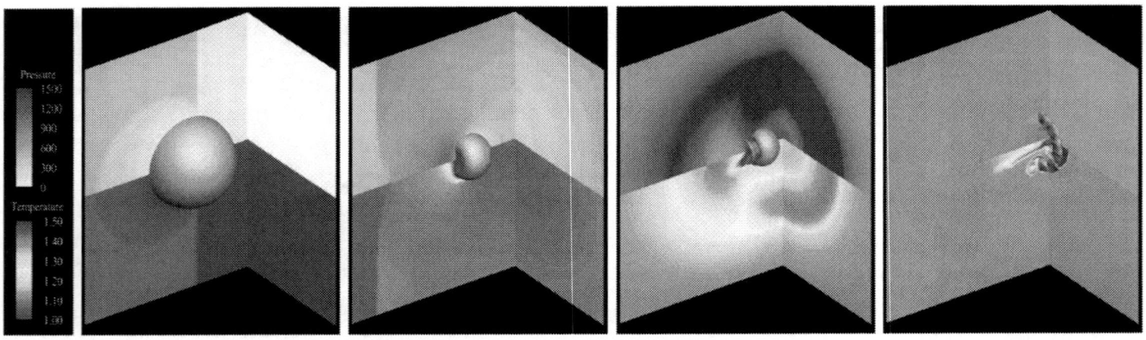

**Figure 2.** Shock-induced bubble collapse ($p_s/p_o = 500, H/R_o = 1.2$) at different times $t = 0.1, 0.6, 1.1, 1.9$. top: pressure; bottom: temperature.

Two mechanisms raise the temperature of the wall: the shock produced at collapse and heat diffusion from the hot bubble if sufficiently close to the solid wall. Fig. (3) shows the maximum temperature along the wall (from the simulations) and the maximum temperature caused by the shock wave (calculated based on the corresponding shock pressure and the stiffened equation of state for water) for different values of $H/R_o$. This figure shows that for the cases where the bubble is initially close to the wall ($H/R_o \lesssim 1.3$), the maximum temperature is higher than the shock temperature, indicating that its source is the hot vortex ring touching the wall. For larger $H/R_o$, the maximum temperature is caused by the shock produced at collapse.

This mechanism can be discussed further by considering the time evolution of the wall pressure and temperature at "point A" for three different $H/R_o$. Johnsen & Colonius [4] showed that proximity to the wall leads to a faster and stronger collapse. This phenomenon can be observed through the pressure plot in Fig. (4). For the largest $H/R_o$, the maximum temperature happens right at the time of maximum pressure. However, the story is different for bubbles initially close to the wall. The time lag between the maximum pressure and maximum temperature indicates that the source of high temperature is not the shock wave. A large rise in temperature is observed, which can be explained by noticing that the vortex ring is in contact with the wall. This can be seen better in the other temperature plots in Fig. (4) where increasing $H/R_o$ exhibits a larger time lag (e.g. for $H/R_o = 1.25$ the shock wave from the collapse results in the first peak, while the second peak is caused by the hot vortex ring).

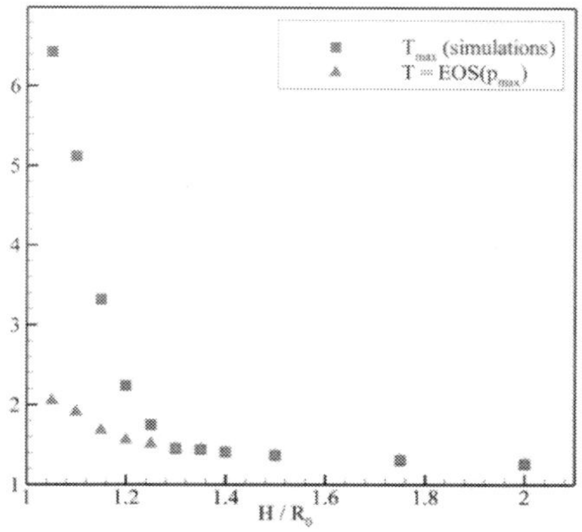


**Figure 3.** Maximum computed temperature and shock temperature of point A

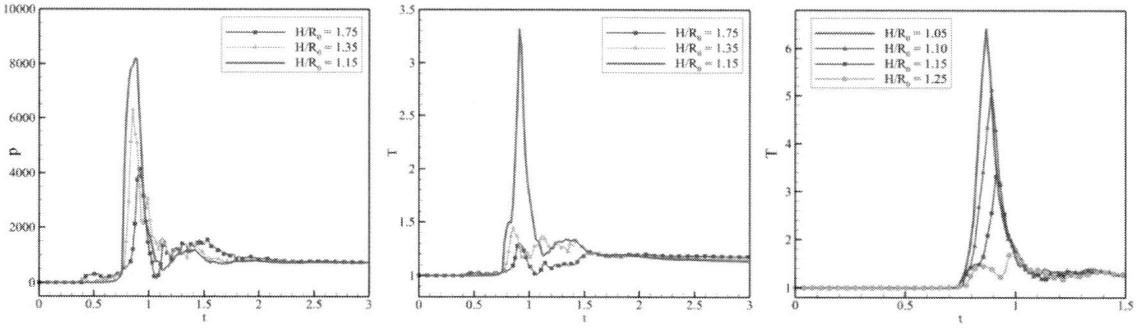


**Figure 4.** Time histories of the pressure (left) and temperature (center) at point A for different initial bubble locations and temperature at point A for small values of $H/R_o$ (right).

The maximum temperature in the whole domain happens inside the air bubble at the collapse time where we have the maximum pressure. After the collapse, the pressure decreases rapidly due to the bubble expansion, which corresponds to an immediate reduction in bubble temperature. This implies that if the bubble is initially close to the wall, the collapse is stronger, the air has a higher temperature, and the hot vortex ring touches the wall closer to the collapse time. The combination of all of these effects leads to a higher temperature measured in the liquid along the wall. However, this temperature is not the actual temperature of the wall. This is schematically illustrated in Fig. (5), which shows the thermal boundary layer between the fluid touching the wall and the wall surface. To estimate the wall temperature, we solve a simplified model problem based on the data from our simulations.

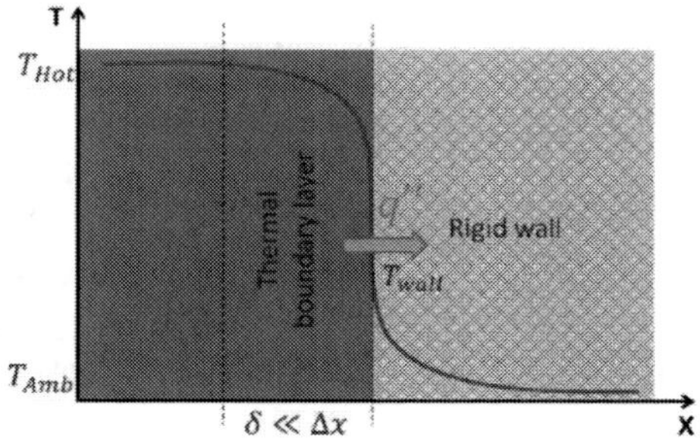


**Figure 5.** Schematic of the thermal boundary layer of the wall

We start with the heat equation $\partial T/\partial t = \alpha\, \partial^2 T/\partial {x_j}^2$ where $\alpha$ is the thermal diffusivity. We solve this equation in the wall with appropriate initial condition $T(x,0) = T_{Amb}$, and boundary conditions $T(\infty,t) = T_{Amb}$ and $q'' = k_f\,(T_{Hot} - T_{Amb})/\delta$, where $T_{Amb}$ is the ambient temperature, $T_{Hot}$ is the flow temperature, $k_f$ is the flow temperature, $q''$ is the heat flux, and $\delta$ is an estimate of the thermal boundary layer thickness. The wall temperature is highly dependent on the heat diffusivity of the solid. Fig. (6) shows the maximum temperature of the surface of the wall for different materials using this diffusion model. The results show that if the wall material has a low heat diffusivity (e.g., polymers) the surface temperature of the wall may increase significantly, while materials with high heat diffusivity dissipate the incoming heat flux through the wall and the surface temperature does not change significantly. However, this simple heat transfer model has some limitations: adiabatic and rigid wall assumptions are made in the simulations, as well as finite resolution. Heat transfer coupling between the wall and the flow, and also a compliant wall, may affect on the final results. In addition, because of the finite mesh size, one can never fully resolve the thermal boundary layer. Despite these limitations, this analysis provides us with a useful estimate of the wall temperature.

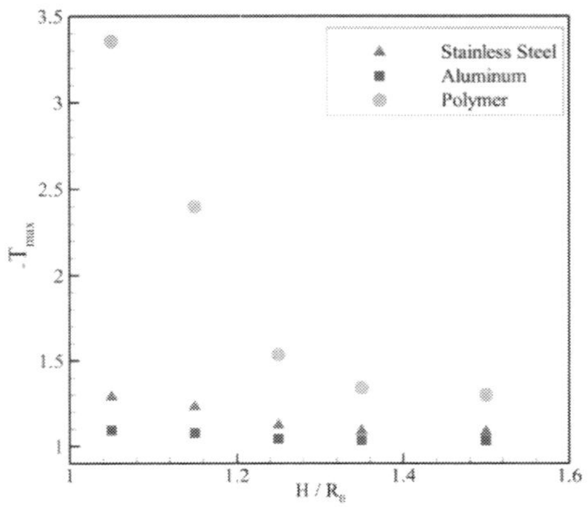


**Figure 6.** Maximum wall temperature for different materials

## 4. Conclusion

To provide a better understanding of cavitation erosion, we conducted numerical simulations of shock-induced collapse of a gas bubble near a rigid wall for different geometrical configurations. We introduced a simple heat transfer model to estimate the temperature of the wall, and showed that the initial location of the bubble plays a critical role in achieving a particular temperature.


## Acknowledgements

This research was supported by ONR grant N00014-12-1-0751 under Dr. Ki-Han Kim.


## References

[1] R. Abgrall, How to prevent pressure oscillations in multicomponent flow calculations: a quasi conservative approach, J. Comput. Phys. 125 (1996) 150–160.
[2] S. Alahyari Beig, E. Johnsen, Maintaining interface equilibrium conditions in compressible multiphase flows using interface capturing, (2015), submitted.
[3] E. Johnsen, T. Colonius, Implementation of WENO schemes in compressible multicomponent flow problems, J. Comput. Phys. 219 (2006) 715–732.
[4] E. Johnsen, T. Colonius, Numerical simulations of non-spherical bubble collapse, J. Fluid Mech. 629 (2009) 231–261.
[5] E. Johnsen, F. Ham, Preventing numerical errors generated by interface-capturing schemes in compressible multi-material flows, J. Comput. Phys. 231 (2012) 5705–5717.
[6] K. H. Kim, G. Chahine, J. P. Franc, A. Karimi, Advanced Experimental and Numerical Techniques for Cavitation Erosion Prediction, Vol. 106, Springer, Netherlands, 2014.
[7] A. Philipp, W. Lauterborn, Cavitation erosion by single laser-produced bubbles, J. Fluid Mech. 361 (1998) 75–116.
[8] Y. Tomita, A. Shima, Mechanisms of impulsive pressure generation and damage pit formation by bubble collapse, J. Fluid Mech. 169 (1986) 535–564.

# Quantitative evaluation of erosive cavitation pressure field from pits in material: fact or myth?

**J-K Choi[1] and G L Chahine**
DYNAFLOW, INC.
10621-J Iron Bridge Road, Jessup, Maryland, 20794, USA

E-mail: jkchoi@dynaflow-inc.com

**Abstract.** Material pitting in a cavitating flow has been used for a long time as an indicator of the vague 'cavitation intensity' concept. Periodically, some researchers suggest pitting tests as a "simple" means to provide quantitative measurements of the amplitude of the impulsive pressures in the cavitation field, especially when combined with Tabor's formula or with simple finite element computations with static loads. This paper examines the viability of such a method using fully coupled bubble dynamics and material response, and strongly concludes that the commonly accepted idea is a myth, as different loading scenarios with the same amplitude of the cavitation impulsive pressure result in different pit aspect ratios.

## 1. Introduction

In order to characterize cavitation erosive fields for different operating conditions (e.g. different flow speeds, cavitation sources, or small scale laboratory accelerated erosion tests versus full scale real operation), a reliable method to evaluate the cavitation field 'intensity' is needed. As expanded upon in [1] the cavitation field intensity is defined by a distribution function, $N(P_{peak})$, where $N$ is the number of impulsive pressure peaks per unit exposed area and unit time of given amplitude, $P_{peak}$ . To this, we should add the distribution of time widths of the pressure peaks [2]. Acoustic pressure measurements are probably the best that can be done presently to determine this intensity despite limitations in the high frequency response and in the size of the sensors [1,2]. However, with advances in instrumentation and miniaturization, these limitations will be reduced over time. Another method, which has been used for at least a whole century [3,4] is to "use the material as the pressure sensor". Short duration cavitation pitting tests within the incubation period, where non-overlapping pits are produced, are conducted and the peak pressures are deduced from the pit geometric characteristics using the Tabor relationship [1,5]. Recently, finite element method analysis was also conducted to obtain a substitute to Tabor equations [6] using however the same basis for the load, i.e. time independent idealized constant pressure loads. Since in such approaches the load is also characterized by over-simplified parameters, almost a one-to-one relationship between the loads and the pit geometry characteristics is also found.

In order to investigate more rigorously whether cavitation pit geometric characteristics can be used along with Tabor relationship to deduce the pressure amplitude responsible for the pit, we conducted fully coupled computations of bubble dynamics/material response and compared the Tabor-predicted

[1] To whom any correspondence should be addressed.

peak pressure from the resulting pit to the actually peak pressure generated by the bubble collapse and applied to the material. The numerical approach is documented with details in [7–9].

## 2. Analytic relation between strain and pit geometry

In Tabor's approach [5], a pit is assumed to be a spherical indentation characterized by its diameter $D$ and depth $h$ (figure 1) and the mean strain, $\varepsilon_m$, in the material around the pit can be estimated from:

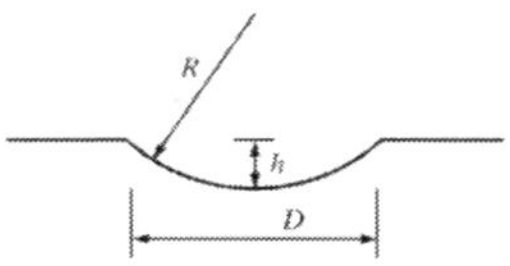


**Figure 1.** Sketch of idealized spherical pit.

$$\varepsilon_m = 0.1\frac{D}{R} = 0.8\frac{h/D}{1+4(h/D)^2}, \qquad (1)$$

where $R$ is the sphere radius that fits in the pit. The maximum strain is related to the mean strain after assuming an exponential strain distribution in the depth direction [10] by:

$$\varepsilon_{\max} = \varepsilon_m (1+\theta), \qquad (2)$$

where $\theta$ is the exponent of the exponential strain distribution, which can be obtained experimentally. The values reported in [7,10] are used in this study. From the maximum strain, the maximum stress can be deduced using the stress-strain curve of the material. The maximum stress occurs at the surface of the material, and in this approach is assumed to be equal to the peak of the impulsive pressure at the material surface.

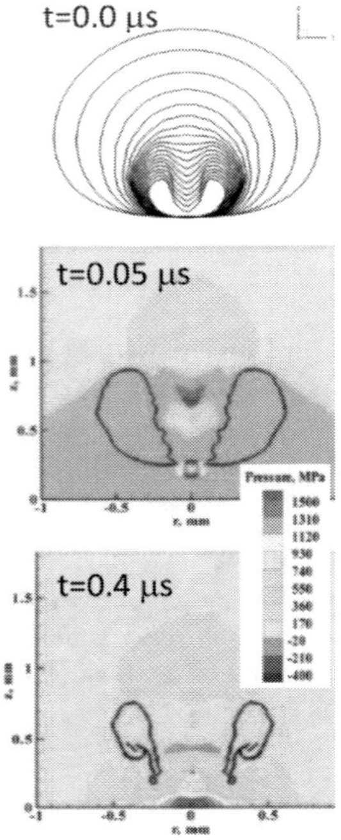


**Figure 2.** Example bubble shapes and pressure field during bubble collapse near a wall [8].

## 3. FSI simulations of pit formation and pit geometry

In order to examine the validity of inferring the magnitude of the impulsive pressure generated by the collapse of a bubble, which resulted in the analyzed pit, we conducted an extensive series of finite element simulations of pit formation in a material exposed to two types of loads: a) fully coupled fluid structure interaction (FSI) simulations of bubble dynamics and material response; b) material deformation due to synthetic loading configurations where we varied systematically the amplitude, space extent and duration of the load [11]. The FSI simulations were conducted using well-validated 3D bubble dynamics codes and procedures [7–9], which modelled bubble shape deformation, reentrant jet development, compressible impact of the jet on the material with generation of a water hammer impact and emission of a shock wave. Bubble size, standoff from wall, and collapse driving pressure functions were varied to cover a range of FSI conditions. Figure 2 illustrates this for a particular condition taken from [8]. Figure 2a shows bubble outlines at different instants during the bubble collapse. Figure 2b and figure 2c show bubble shapes and pressure contours with very high pressures generated respectively by liquid-liquid impact at reentrant jet touchdown and by jet-wall impact. The liquid-liquid impact event generates a localized high pressure region which expands quasi spherically and propagate to the wetted material surface (first pressure peak in figure 3). The bubble ring left after the jet touchdown shrinks, then collapses and generates another high pressure wave (second peak in figure 3).

The dynamics of the material is modeled by a non-linear finite element method. Figure 3 illustrates the FSI computations with the pressures generated at the material surface center by the collapsing bubble and the corresponding time dependent response of the material illustrated by the development of the indentation depth with time. Due to the high pressure generated

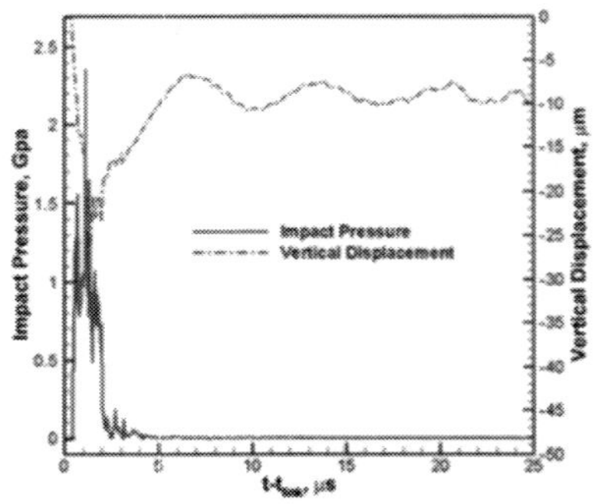


**Figure 3.** Time history of pressure and vertical displacement of Al 7075 following the collapse of a cavitation bubble [8].

Journal of Physics: Conference Series **656** (2015) 012045 doi:10.1088/1742-6596/656/1/012045

by the collapse, the material deforms elastically first, then exceeds the elastic limit to result in unrecoverable plastic deformation and a pit.

Four metals (Al 1100, Al 7075, NAB, and SS A2205) were modeled using elastic-plastic models. The stress-strain curves of these metals are given in [7]. An example of permanent deformation predicted from an FSI simulation is shown in figure 4. The profile of the permanent deformation generated on the surface resembles the experimentally observed pit shapes [1]. The figure also shows the circle, which corresponds to equations (1) and (2) spherical indentation assumption. In this case of a bubble-generated pit, the shape is closer to a cone rather than a sphere sector. Figure 5 also shows a pit shape resulting from a synthetic time and space Gaussian distribution pressure loading. The shape is rounder but it is still significantly different from a spherical shape.

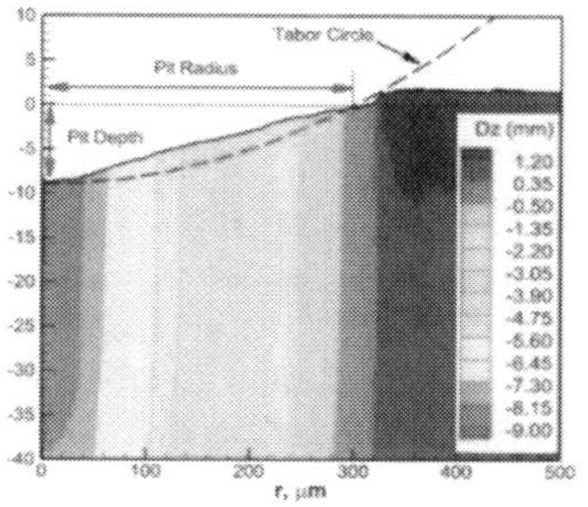


**Figure 4.** Pit on Al 7075 material following a bubble collapse [8].

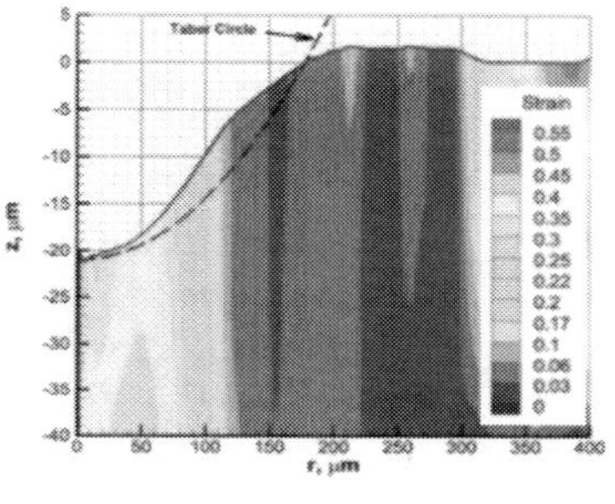


**Figure 5.** Pit shape on Al 7075 material due to a synthetic load [11]. The circular shape used for Tabor formula is also shown.

A synthesis of the results from the set of cases studied is shown in figures 6 through 8. Figure 6 collects pit depth to diameter ratios, $h / D$, which according to Tabors' formula determines the load input. The figure clearly shows that the same impulsive pressure obtained by different bubble collapse mechanisms can produce very different $h / D$ (see for example the data points along the dashed line in the figure). For the same impulsive pressure peak level, the shape of the pressure function can vary wildly, both in space and time, for different bubble sizes, standoffs, and collapse driving pressure functions. In addition, for large strains, strain and stress distributions in the material under the pit deviate from the simple relationship (2).

The $h / D$ ratios are converted to maximum strain using (2) then using the stress-strain curve of each material, the maximum stress is deduced from the maximum strain. For each $h / D$, this provides the supposed peak impulsive pressure applied to the material. Figure 7 summarizes the results and compares the actual input peak cavitation pressure to the one deduced from $h / D$ and (2). This brings the data closer to a common curve because the procedure has accounted for differences in the various material strengths. However, this highlight that for all data, the maximum stress predicted from the pit is very far from the actual load applied (good correspondence would be the purple line in the plot).

The ratio of the pressure predicted from the pit geometry ( $h / D$ ) and Tabor relation to the actually applied pressure peak is plotted in figure 8. The data by the Tabor relation forms roughly a curved band for all materials. Actual pressure over the pit-Tabor predicted pressure significantly increases with increasing loads exceeding 10 at the highest load values. The numerical computed maximum effective stress *inside* the material is also included for comparison. The stress from the Tabor relation agrees very well with this maximum effective stress (von Mises stress) recorded in the material (not at the surface). It is now evident that the stress from the Tabor relation is in agreement with the maximum effective stress in the material and *not with the load itself.* The maximum effective stress is only 10% - 60% of the peak load depending on the magnitude of the load.

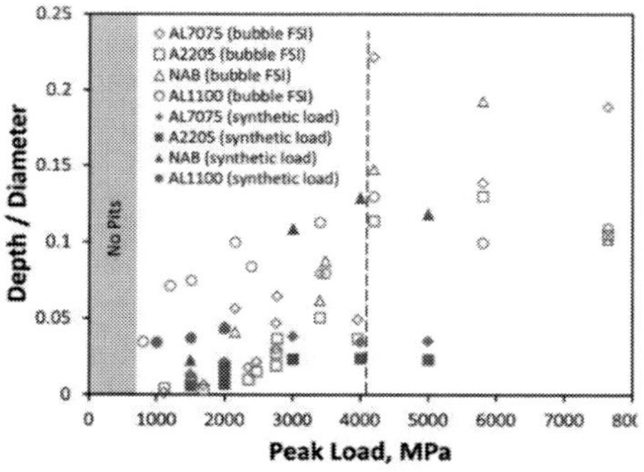


**Figure 6.** Pit depth to diameter ratio vs. peak load from FEM and FSI simulations.

## 4. Conclusions

Pits forming on a material from the impulsive pressure loading resulting from bubble collapse were simulated numerically. This enabled careful examination of the suitability of using material pits measurements and Tabor's analytic relation to infer pressure peaks from the pit geometry. The study clearly showed that this approach is very inaccurate and actually provides the maximum effective (von Mises) stress inside the material and not the amplitude of the cavitation pressure peak. Additional information such as the time and space dependent pit characteristics would be needed to attempt a better correlation.

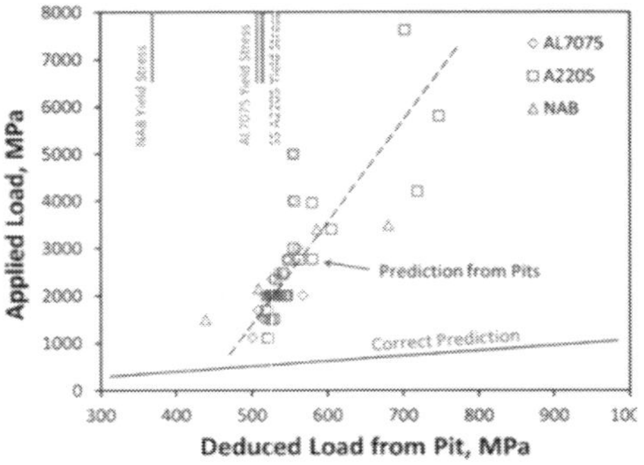


**Figure 7.** Comparison of the pressure deduced from Tabor pit analysis and the actual pressure applied to the material.

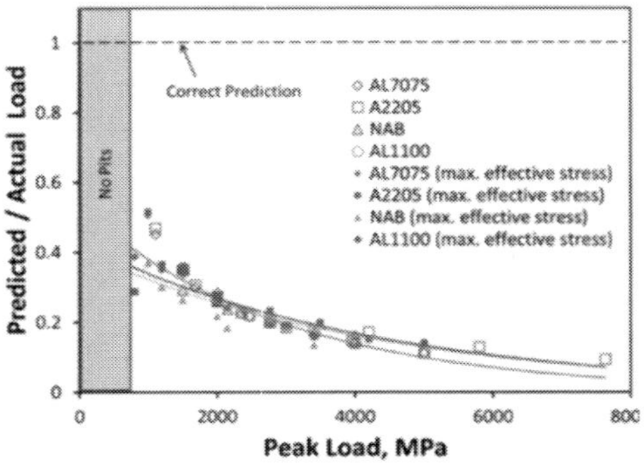


**Figure 8.** Ratio of the Tabor pit analysis pressure to the actual cavitation peak pressure.

### Acknowledgments

This work was supported by the Office of Naval Research, Contract N00014-11-C-0378 Dr. Ki-Han Kim monitor. We gratefully acknowledge this support.

### References

[1] Kim K-H, Chahine G L, Franc J-P and Karimi A 2014 *Advanced Experimental and Numerical Techniques for Cavitation Erosion Prediction, Series Fluid Mechanics and Its Applications* vol 106, ed K-H Kim, G L Chahine, J-P Franc and A Karimi (Springer, Berlin, Germany)

[2] Singh S, Choi J-K and Chahine G L 2013 Characterization of Cavitation Fields From Measured Pressure Signals of Cavitating Jets and Ultrasonic Horns *J. Fluids Eng.* **135** 9111–91302

[3] Knapp R T 1955 Recent investigations of the mechanics of cavitation and cavitation damage *Trans. Am. Soc. Mech. Eng.* **77** 1045–54

[4] Parsons C A and Cook S S 1919 Investigation into causes of corrosion or erosion of propellers *J. Am. Soc. Nav. Eng.* **31** 536–41

[5] Tabor D 1951 *The Hardness of Metals* (Oxford, United Kingdom: Clarendon Press)

[6] Pöhl F, Mottyll S, Skoda R and Huth S 2015 Evaluation of cavitation-induced pressure loads applied to material surfaces by finite-element-assisted pit analysis and numerical investigation of the elasto-plastic deformation of metallic materials *Wear*

[7] Chahine G L 2014 Modeling of Cavitation Dynamics and Interaction with Material *Advanced Experimental and Numerical Techniques for Cavitation Erosion Prediction* ed A Kim, K-H, Chahine, G L, Franc, J-P, Karimi (Berlin: Springer) pp 123–73

[8] Hsiao C-T, Jayaprakash A, Kapahi A and Choi J-K 2014 Modeling of Material Pitting from Cavitation Bubble Collapse *J. Fluid Mech.* **755** 142–75

[9] Chahine G L, Kapahi A, Choi J-K and Hsiao C-T 2015 Modeling of Surface Cleaning by Cavitation Bubble Dynamics and Collapse *Ultrason. Sonochem.* **10.1016/j.**

[10] Karimi A and Franc J-P 2014 Modeling of material response *Advanced Experimental and Numerical Techniques for Cavitation Erosion Predicition* ed K-H Kim, G . Chahine, J-P Franc and A Karimi (Springer) pp 163–80

[11] Choi J-K, Jayaprakash A, Kapahi A, Hsiao C-T and Chahine G L 2014 Relationship between space and time characteristics of cavitation impact pressures and resulting pits in materials *J. Mater. Sci.* **49** 3034–51

# New insights into the Mechanisms of Cavitation Erosion

**M Dular, M Petkovšek**

Laboratory for Water and Turbine Machines, Faculty of Mechanical Engineering, University of Ljubljana, Askerčeva 6, 1000 Ljubljana, Slovenia

E-mail: matevz.dular@fs.uni-lj.si

**Abstract**. Recently simultaneous observation of both cavitation structures and cavitation damage, has pointed to the fact that the small scale structures and the topology of the cavitation clouds play a significant role in cavitation erosive potential. Although this opened some new insights to the physics of cavitation damage, many new questions appeared. In the present study we attached a thin aluminium foil to the surface of a transparent Venturi section using two sided transparent adhesive tape. The surface was very soft – prone to be severely damaged by cavitation in a very short period of time. Using high speed cameras, which captured the images at 30000 frames per second, we simultaneously recorded cavitation structures (from several perspectives) and the surface of the foil. Analysis of the images revealed that five distinctive damage mechanisms exist – spherical cavitation cloud collapse, horseshoe cavitation cloud collapse, the "twister" cavitation cloud collapse and in addition it was found that pits also appear at the moment of cavitation cloud separation and near the stagnation point at the closure of the attached cavity.

## 1. Introduction

Cavitation and consequently cavitation erosion is one of the most ubiquitous problems at operation of turbines, pumps, ship propellers and valves.

In the past, cavitation shedding process and its relation to the erosive potential of cavitation was frequently discussed. Reisman et al. [1] suggested that the formation, focusing and propagation of shock waves, emitted at macroscopic structures collapse play a critical role in generating the pressure pulses which lead to cavitation noise and damage. Bark et al. [2] pointed to the spherical cavitation cloud collapse, where the converging flow enables efficient focusing of the energy which increases as the collapse proceeds and is finally converted into acoustic energy in the cavity collapse, associated with the radiation of shock waves. Foeth et al. [3] have shown that the cloud cavitation that is associated with the break up of sheet cavitation is essentially an organized mixture of cavitating vortices. Van Rijsbergen et al. [4] related the structures to the acoustic emission signals and suggested that the orientation of the shock wave should be considered.

Large scale primary spanwise vortices were studied by Pereira et al. [5] and Kawanami et al. [6], who have shown that the focusing of the collapse could play a role in the aggressiveness of erosion.

The present study is a logical continuation of our previous work [7]. Here the types of macro-cavities or the origins of damage occurrence are discussed. High speed cameras were used to simultaneously record the sequences of the cavitation dynamics and the damage sustained on the surface of the solid body. It is shown that the shedding process can evolve in different ways, leading to various hydrodynamic mechanisms that eventually converge to extreme conditions that can trigger events on a micro-scale that cause erosion. The results are valuable for future development of cavitation erosion prediction methods [8, 9].

## 2. Experimental set-up

Erosion was studied in a Venturi geometry, which was 10 mm wide with a converging angle of 18° and diverging angle of 8°. The throat dimensions were 10×10mm$^2$.

The idea of the experiment was to simultaneously record images of cavitation structures and cavitation erosion. The upper side of the foil is covered by vapour structures that obstruct the view, hence one needs to look at the foil from the bottom side to see the damage. Consequently the whole test section had to be made of transparent material and, equally important, the foil had to be thin enough so that the cavitation damage, which occurs on the side exposed to cavitation was also visible on the other side.

For the present experiment high speed cameras were synchronized and recorded at 30000fps at a reduced (608×86 pixels) resolution. One camera was used for observation of the aluminium foil while the other one captured images of cavitation structures.

Erosion was evaluated in image pairs: image at time t+Δt was subtracted from image at time t. This way the change between times t and t+Δt could be detected and recognized as damage [7].

## 3. Results

In the following we present the results of extensive image analysis. In the process, first the instant of the damage occurrence was determined by comparing pairs of successive images of the foil. Then the evolution of cavitation structures during this period was analyzed and correlations between the two were investigated. Figure 1 represents the time evolution and the variants of the shedding that lead to different mechanisms of cavitation induced damage, which were observed during our study (Fig. 2).

The instants (C, E, H, L and O) when the damage occurs are highlighted by the darker colour.

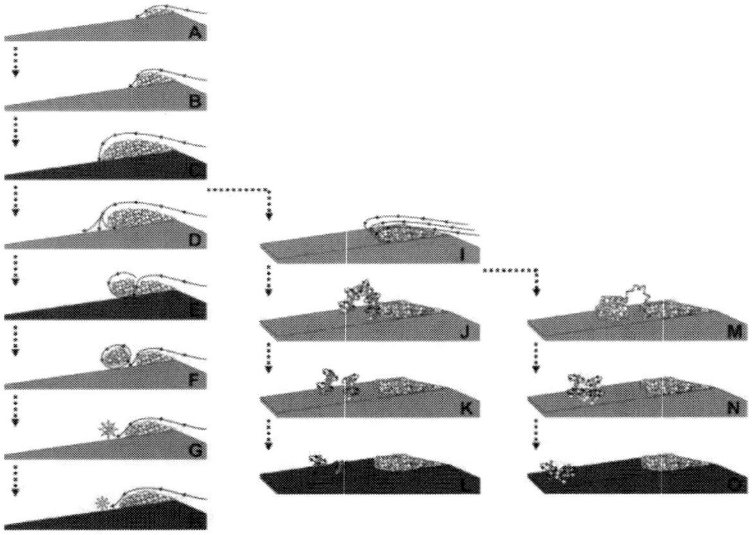


**Figure 1.** Representation of observed mechanisms that lead to occurrence of cavitation erosion.

Not less than 5 different mechanisms that lead to the formation of the pit were determined, although it is possible that the physical background of some is more or less the same.

The damage in Fig. 2 is enlarged (3 times) for better representation and appears smaller in reality (its position and shape are not altered). A deeper discussion of the mechanisms follows.

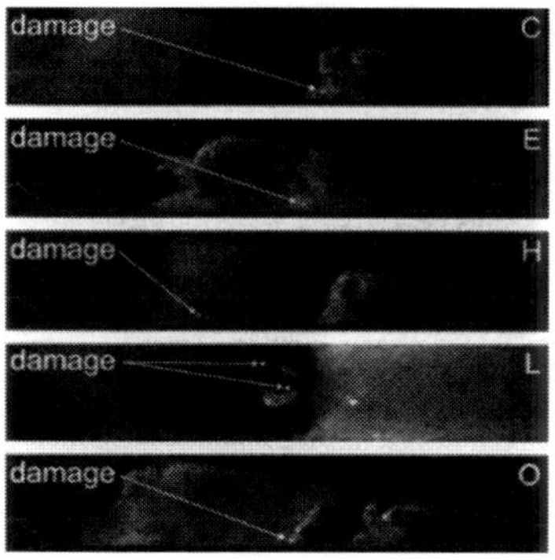


**Figure 2.** Cavitation structures at the moments at which the damage appeared on the foil. The position and the size of the damage were obtained from the images of the foil and were superimposed to the images of the structures.

### *3.1. Closure of the attached cavity (C in Figs. 1 and 2)*

During the shedding process the outer flow, due to the pressure difference between cavity exterior and interior, deviates towards the solid surface and creates a stagnation point with elevated pressure. For erosion to occur small scale cavitation structures (single bubbles, which sporadically break off from the attached cavity) must enter the region of higher pressure in the vicinity of the solid surface, and violently collapse. This, however, occurs very rarely due to the fact that the developing re-entrant jet blocks the shedding of small scale structures.

### *3.2. Cavitation cloud separation (E in Figs. 1 and 2)*

As the re-entrant jet looses momentum it turns upwards cutting the attached cavity into two parts. At this moment the gap between the cloud and the attached cavity is filled up with liquid causing an increase of the pressure. Again small scale structures that are present in the gap collapse and cause pit formations.

### *3.3. Spherical cavitation cloud collapse (H in Figs. 1 and 2)*

Most often erosion is associated with the collapse of primarily cloud cavitation, referring to the fact that collective collapses of interacting bubbles in the cloud are shown to be very effective in focusing collapse energy into a small spot, and typically generate significantly larger collapse pulse amplitudes than singly connected cavities. According to the energy cascade approach [2] the pressure waves emitted at the cloud collapse spread and cause implosion of micro-scale single bubbles in the vicinity of solid structures – causing pitting and eventually erosion.

### *3.4. Horseshoe cavitation structure collapse (L in Figs. 1 and 2)*

Cavitating vortices, which appear in a form of a horseshoe shaped structure, are known to be responsible for severe erosion in fluid machinery.

We could observe that the damage occurs as the vortex begins to break up. Pits appear in the region where the vortex leg touches the surface, thus implying that the breakup of the vortex "leg" and consequent focusing of acoustic energy in space and time (as postulated by van Terwisga et al. [10]) increases the aggressiveness and causes the formation of the pit.

### *3.5. "Twister" cavitation structure collapse (O in Figs. 1 and 2)*

We believe that the origin of the "twister" type of cavitation structure is the same as for the horseshoe structure. The vortex is produced at the breakup region, but it remains liquid. As it is carried downstream by the flow it interacts with the cavitation cloud, what causes it to cavitate. This, like in the case of a breakup of a horseshoe cavitation structure, focuses the acoustic energy towards the solid material.

We can again see that the damage is concentrated in the region where the twister touches the solid surface and that it generally occurs when the vortex breaks-up.

### *3.6. Some statistics*

The importance of individual mechanism can be defined by the product of its probability of occurrence and the damage (averaged damaged area sustained during individual events) it causes. A large ensemble of the data (over 3000 shedding periods) enables statistical evaluation of the probability of occurrence and the erosive potential of the mechanisms – Fig. 3 shows the results.

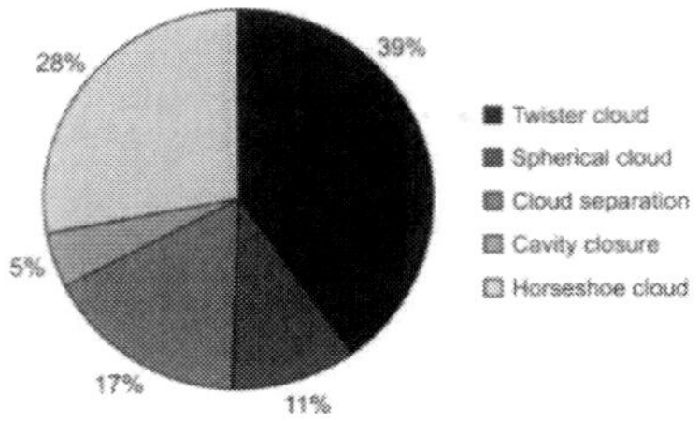


**Figure 3.** The relative importance of individual mechanism.

One can see that, the cloud implosion combined (the sum of the importance of the twister cloud, the horseshoe cloud and the spherical cloud) contributes to almost 80 % of the damage – it is clear why numerous researchers stressed so much importance to it. Between the mechanisms the twister cloud implosion is the one to be most considered. It appears frequently and is aggressive; hence it contributes to 39% of the sustained damage. A very important mechanism is also the horseshoe cloud implosion – mainly due to its aggressiveness. It is possible that its importance is even greater in channels with wider span, where the horseshoe vortex can more easily develop.

Surprisingly important is the mechanism of cavitation cloud separation from the attached part of the cavity, which was, prior to this study, probably mistaken for the mechanism of cavity closure, since they both contribute to damage in a similar region. The later is of an insignificant importance.

## 4. Conclusions

The study presents an original experimental technique where the cavitation structures and cavitation damage are observed simultaneously at a relatively high frame rate (30000 fps).

It was shown that the process of cavitation erosion is much more complex than previously thought. New phenomena were observed – for example the development and focused collapse of the twister cloud and damage occurrence in the region of cavitation cloud separation.

**Acknowledgments**

The study was performed under a grant by the Slovenian Research Agency (J2—6774). The test facility was developed with a continuous financial support by the European Space Agency (ESA).

## References

[1] Reisman G E, Wang Y C and Brennen C E 1998 J. Flu. Mech. **355** 255-83
[2] Bark G, Friesch J, Kuiper G and Ligtelijn J T 2004 Proc 9th Symposium on Practical Design of Ships and Other Floating Structures, Luebeck-Travemuende.
[3] Foeth E J, van Terwisga T and van Doorne C 2008 *ASME J. Fluids Eng.* **130(7)**,071303
[4] van Rijsbergen M, Foeth E J, Fitzsimmons P and Boorsma A 2012 Proc 7th International Symposium on Cavitation, Singapore, Singapore.
[5] Pereira F, Avellan F, and Dupont J M 1998 *ASME J. Fluids Eng* **120(4)**, 719-27
[6] Kawanami Y, Kato H, Yamaguchi H, Maeda M and Nakasumi S 2002 *JSME International Journal* **45(3)**, 655-61.
[7] Petkovšek M and Dular M 2013 *Wear* **300(1/2)**, 55-64
[8] Dular M and Coutier-Delgosha O 2009 *Int. J. Numer. Meth. Fluids* **61**,1388–410
[9] Wang J, Petkovšek M, Liu H, Sirok B and Dular M 2015 *ASME J. Fluids Eng* **137(5)**, 1-9.
[10] van Terwisga T, Fitzsimmons P A, Ziru and Foeth E J 2009 Proc 7th International Symposium on Cavitation, Ann Arbor, Michigan, USA

# Cavitation erosion - scale effect and model investigations

**F Geiger**[1] **and P Rutschmann**[2]

[1] Technical University of Munich, Hydraulic Laboratory Obernach, 82432 Walchensee, Germany
[2] Technical University of Munich, Chair of Hydraulic and Water Resources Engineering, Arcistrasse 21, 8033 Munich, Germany

E-mail: franz.geiger@tum.de

**Abstract:** The experimental works presented in here contribute to the clarification of erosive effects of hydrodynamic cavitation. Comprehensive cavitation erosion test series were conducted for transient cloud cavitation in the shear layer of prismatic bodies. The erosion pattern and erosion rates were determined with a mineral based volume loss technique and with a metal based pit count system competitively. The results clarified the underlying scale effects and revealed a strong non-linear material dependency, which indicated significantly different damage processes for both material types. Furthermore, the size and dynamics of the cavitation clouds have been assessed by optical detection. The fluctuations of the cloud sizes showed a maximum value for those cavitation numbers related to maximum erosive aggressiveness. The finding suggests the suitability of a model approach which relates the erosion process to cavitation cloud dynamics. An enhanced experimental setup is projected to further clarify these issues.

## 1. Introduction

The erosive effect of hydrodynamic cavitation is so far insufficiently understood for practical prediction. Corresponding research is based on experimental investigations of erosion pattern and erosion rates. The transferability of experimental results to different facilities or conditions has to account for scale effects, foremost the size scale effect and the velocity scale effect [1]. The common experimental techniques are based on the observation of damage on material probes and employ almost exclusively metal substances. The presented results of different methods for experimental erosion assessment with metal and mineral materials suggest a high dependency of erosion rates on the employed type of probe. Such material dependencies could be caused by different damage processes for these compounds [2].

Although single bubble experiments and respective theoretical considerations could obtain outstanding information about the related microscopic damage processes [3], these findings can barely be combined with the complex conditions of large scale hydrodynamic cavitation due to immense bubble numbers, intense dynamics and unspecified interactions. A model which relates the erosive aggressiveness of transient cloud cavitation to the macroscopic cloud dynamics appears more promising for future prediction applications and could already achieve good accordance with experimental observations [4]. In detail the model postulates coherent cloud collapses which emit strong pressure waves and consequently induce the formation of micro jets in bubbles near a solid boundary which cause the actual material damage. So far the model was only implemented using time averaged and pixel-wise greyscale fluctuations for the evaluation of the vapour fraction variations and only applied to one type of experiment, i.e. shedding cloud cavitation at hydrofoils. This paper presents a first application and assessment of the model for the conducted erosion test series.

## 2. Experimental Setup

The test series were conducted in the cavitation rig K26 at the hydraulic laboratory of the Technical University of Munich in Obernach. The closed circuit system contains about 12 m³ test fluid (filtered tab water) and features a 90 kW powered impeller for flow generation as well as vacuum pumps for the regulation of the system pressure. The test section had 30 cm x 30 cm cross section and 200 cm length. The approach flow velocity was controlled by LDA measurements (TSI LDA) in the center of the first third of the test section length. The system pressure was referenced at the equivalent position by piezo pressure transducers (Kistler Piezo). Furthermore the test fluid tensile strength was measured by vorto-tensimeter [5] and aeration and degasing procedures prior to the actual test assured equivalent tensile strength during the measurement campaigns. Additionally temperature and dissolved oxygen content were measured (WTW Oxi).
To create the intended cavitation phenomena a rectangular prism (height 97.5 mm) with an equilateral triangle base (side length 75 mm) was mounted in the middle of the test section with one corner pointing in opposition to the flow direction. The setup enabled the generation of minor tip vortex cavitation up to super cavitation with > 1 m length extension in the whole wake of the prism. For cavitation numbers of about 1.5 ± 0.5 transient cavitation clouds appeared in the triangle wake's shear layer vortices and resulted in cavitation erosion in the respective areas. Figure 1 illustrates the test setup and the resulting cavitation and cavitation erosion pattern. Test series were conducted for approach flow velocities between 9 m/s and 14 m/s.

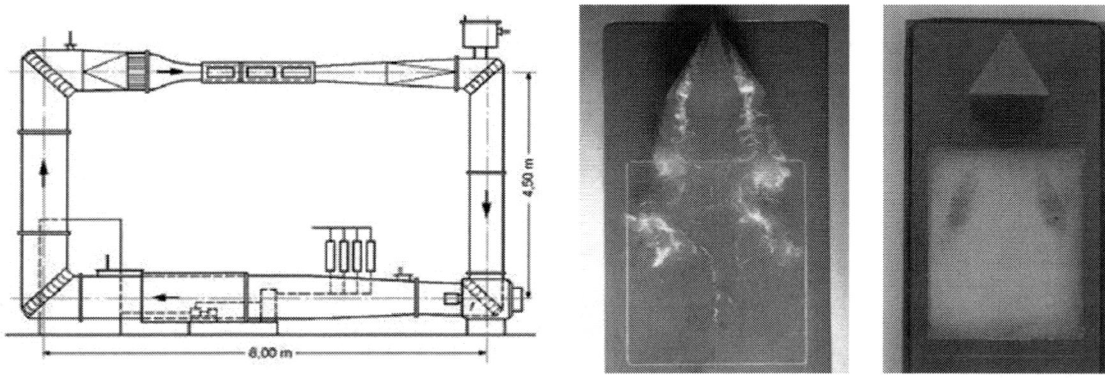


**Figure 1**: Schema of the test facility K26 (left), cloud cavitation behind the prismatic body in the test section (middle, flow direction top bottom) and the resulting erosion pattern on a cement probe (right)

The cavitation erosion was measured by several complementary methods. Using concrete material probes the position and size of the resulting erosion pattern was recorded by 3D laser profilometry (Micro-Epsilon measurement device and Berger-Lahr stepper motor positioning device and 0.1 mm overall accuracy). A competitive record of the eroded volumes was conducted using a volume substitution method [6], i.e. the filling of the erosion pattern by a compound of known density and the calculation of the respective volume via mass measurements (Sauter, 0.01 mg accuracy). Furthermore, a 2D pit count technique was employed [7]. The microscopic ductile deformations ("pits") of polished stainless steel (V4A, HV10: 181) were measured by microscopic image acquisition (1.8 µm resolution) for the whole eroded area (~ 15 cm²). Using an automated positioning (Berger-Lahr stepper motors, µm accuracy) and image processing (LabVIEW based) the pit areas and the pit numbers could be deduced for the whole probe surface in order to identify the location of cavitation erosion and to quantify the erosion rates in terms of the pitted area or equivalently the pit number per exposure time.

The quantification of the cavitation cloud volume and their dynamics was based on a simplified optical approach employing a CMOS digital camera (Edmund Optics, 1.3 MP, b/w) and stroboscopic illumination (DrellScop 2008). Optical access to the experiment was provided by acrylic glass in top view of the prism. The adjustment and synchronization of frame rate, exposure time and stroboscope frequency could achieve series of non-continuous but time resolved cavitation cloud images. In each tests series 1000 to 5000 images were taken. A subsequent image processing (LabVIEW based) was employed to obtain the cloud projection surface (i.e. the size of the cloud on the image) as gauge for the total cloud size.

Whereas the simplified cloud volume assessment on basis of only one perspective is in general not capable to deduce the total cloud volume because of the complex vapor structures and dynamics, the given type of cavitation cloud patterns were known from exemplary high-speed imaging [6] and the implemented approach was found suitable for an assessment of the cloud size characteristics.

## 3. Results

The cement based volume loss as well as metal based pit count method showed similar erosion patterns. The coordinates of the centers of erosion were constant throughout the exposure times and identical for both employed measurement techniques. In detail the exposure times were adapted to the probe material and to the respective hydraulic conditions in order to achieve erosion extents suitable for the employed measurement techniques and ranged from five minutes for high approach flow velocities up to two hours for low flow velocities. The erosion rate for volume loss of cement probes showed a magnitude from several hundred to several thousand $mm^3/h$. The pit count method showed average pit radius of 20 $\mu m$ and an exponentially decreasing relation between pit radius and pit numbers. The competitive analysis of erosion rates by pitted area and pit numbers yielded comparable results.

For tests with identical flow velocities both measurement techniques, cement based as well as metal based one, showed maximum erosion rates in the area of a cavitation number $\sigma = 1.5 \pm 0.1$ and monotonously decreasing values for higher or lower cavitation numbers. Considering the velocity scale effect both techniques confirmed a high increase of the erosion rate with increasing approach flow velocity. Figure 2 shows a 3D visualization (surface and contour lines) of the dependency of the erosion rate on flow velocity and cavitation number for the volume loss method. Considering the velocity scale effect the erosion rate of cement probes, i.e. the eroded volume per exposure time yielded best fits for the ninth power of the velocity. The equivalent relation for metal probes revealed best fits for the sixth power of the velocity.

The analysis of the cloud size was based on the average value of the cloud projection size. The standard deviation of the cloud projection for all clouds of one experimental setup was used to quantify the cloud size fluctuations. The average cloud projection size decreased with increasing cavitation number which is coherent with the cavitation number definition. The standard deviation of the cloud sizes i.e. the cloud fluctuations showed a local maximum in the vicinity of $\sigma = 1.4$ to $\sigma = 1.5$. Both relations are illustrated in figure 2 for an exemplary approach flow velocity of $v = 9.2$ m/s.

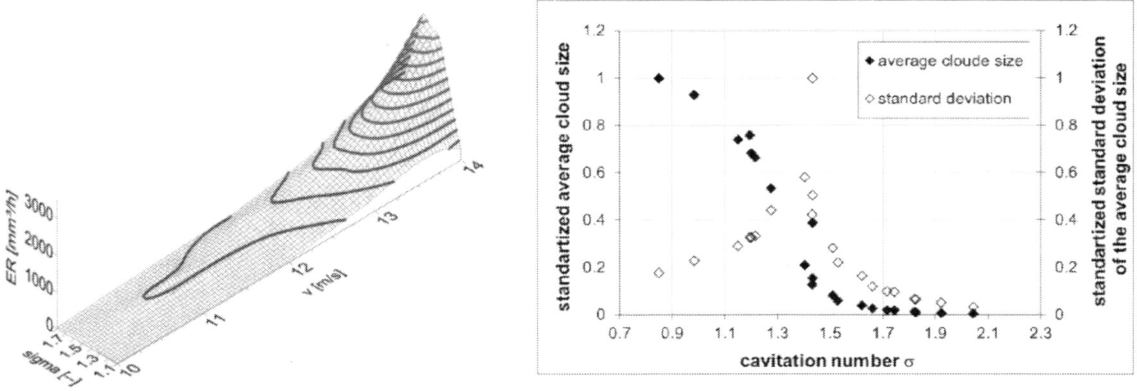


Figure 2: Dependency of erosion rate ER on flow velocity v and cavitation number sigma (left) and dependency of average cloud size and its fluctuations on the cavitation number for v = 9.2 m/s (right)

## 4. Conclusion

The obtained erosion data quantifies the velocity scale effect for the underlying cavitation type for both employed measurement techniques. The pit count results are coherent with literature references [2]. The diverging scale effect for the cement material indicates a significantly different damage process for this type of compounds and their higher vulnerability face to hydrodynamic cavitation load, similar to recent finding about damage process material dependencies for further materials [8]. The investigation of the cavitation cloud size and its fluctuations revealed maximum cavitation cloud fluctuations for those hydrodynamic conditions which were also related to the highest erosion rates. This finding corresponds to the cavitation cloud erosion model hypothesis [4] and thus suggests its suitability for the given cavitation phenomenon.

In following research projects the cavitation cloud size will be recorded time resolved and volumetric by use of several high speed cameras to verify the suitability of the erosion model in detail. The time resolved investigation will also gain insight to the underlying processes of the velocity scale effect. Furthermore a simultaneous microscopic observation of the cavitation load on the surface is intended to obtain information about the actual damage process.

## References

[1] Lecoffre Y 1999 *Cavitation - Bubbel Trackers* A.A. BalkemaRotterdem pp 244-290

[2] Franc J P, Avellan F, Belahadji B, Billard J Y, Briancon-Marjollet L, Frechou D, Fruman D H, Karimi A, Kueny J L and Michel J M 1995 *La Cavitation – Mechanismes Physiques et aspects industriels* Presses Universitaires de Grenoble pp 265-336

[3] Brennen C E 1995 *Cavitation and Bubble Dynamics* Oxford University Press

[4] Dular M, Stoffel B and Sirok B 2006 *Development of a cavitation erosion model* Wear 261(5) pp 642 - 655

[5] Keller A P 2001 *Cavitation Scale Effects – Empirically Found Relations and the Correlation of Cavitation and Hydrodynamic Coefficients* 4th International Symposium on Cavitation

[6] Huber R 2004 *Geschwindigkeitsmaßstabseffekte bei der Kavitationserosion in der Scherschicht nach prismatischen Kavitatoren* Berichte des Lehrstuhls und der Versuchsanstalt für Wasserbau und Wasserwirtschaft der Technischen Universität München

[7] Geiger F, Huber R and Rutschmann P 2009 *Dependency of Cavitation Erosion Scale Tests on Measurement Methods* 33rd IAHR Congress

[8] Escaler X, Farhat M, Avellan F and Egusquiza E 2003 *Cavitation erosion tests on a 2D hydrofoil using surface-mounted obstacles* Wear 254(5) pp 441-449

# Cavitation erosion of low-density polyethylene coatings for pipe liners

**S Hattori[1], E Benitani[1], W Ruan[1], Y Suda[2], R Takeuchi[2] and T Iwata[2]**

[1] Graduate School of Engineering, University of Fukui,
3-9-1 Bunkyo, Fukui-shi, Fukui 910-8507, Japan

[2] Mitsubishi Heavy Industries, LTD.
1-1, Wadasaki-cho 1-chome, Hyogo-ku, Kobe 652-8585, Japan

E-mail: hattori@u-fukui.ac.jp

**Abstract.** The relationship between mechanical properties and the erosion rate was examined for chloroprene rubber and a number of polyethylene materials produced by different methods. As electric power plants are in operation over long periods of time, the effect of aging was also examined by testing material intended for use in pipes in electric power plants. Cavitation erosion tests were carried out by using a flowing apparatus as specified in the American Society for Testing Materials G134-95 standard. A flow velocity of 150 m/s and a test time of 24hours, were the experimental conditions used for a cavitating liquid jet test on polyethylene. The maximum depth of erosion rate (MaxDER) of polyethylene was found to decrease with the increase in hardness. Among all the tested materials, the relatively high molecular weight polyethylene with low density (m-LLDPE-H), showed the best resistance to cavitation erosion in terms of MaxDER. The effect of aging on the erosion rate of polyethylene was limited.

## 1. Introduction

Resin lining is widely used as an anti-corrosion method of steel pipes for tap water and water conditioning in chemical plants and electric power plants. Since there are many valves for piping systems, and cavitation sometimes occurs near valves, it is necessary to consider not only the corrosion resistance but also the erosion resistance to cavitation. Barletta and Ball [1] carried out cavitation erosion tests for many kinds of polymeric materials and found their erosion resistances. Kobayashi et al. [2] carried out erosion tests of lined pipes with epoxy resin, tar epoxy, polyester, polyethylene and neoprene rubber using a butterfly valve, and found that chloroprene rubber has the best erosion resistance. The second best erosion resistant material was polyethylene, showing the loss of luster on bright surface without material removal. The authors [3] carried out the cavitation erosion tests for epoxy resin, polypropylene, high-density polyethylene and polyamide 66. It was found that the cavitation erosion resistance can be evaluated in terms of bubble collapse impact energy and the strain energy obtained from fatigue strength. Recently, Deplancke et al. [4] found that ultra-high molecular weight polyethylene showed a high cavitation erosion resistance. But this material is difficult to fabricate components. In general, the mechanical properties of plastics can vary with the catalyst, material density, and the mixture with elastomer. However, little cavitation erosion research has been done from this viewpoint.

In this study, nine kinds of polyethylene and one chloroprene rubber were fabricated to carry out the cavitation erosion tests using a cavitating liquid jet apparatus. The erosion rate was evaluated in terms of mechanical properties of the materials. Many pipes in electric power plants are expected to be used for 60 years. The specimens after aging treatment were used for the cavitation erosion test. The effect of aging on the erosion rate was examined.

**Table 1.** Physical and mechanical properties of polyethylene and chloroprene rubber.

| Material | Density ($g/cm^3$) | Melting point (℃) | Elonga-tion (%) | 100% modulus* (MPa) | Tensile strength (MPa) | Durometer hardness HDD |
|---|---|---|---|---|---|---|
| Z-LLDPE | 0.926 | 120 | 560 | 9.5 | 13 | 55 |
| LDPE | 0.921 | 109 | 510 | - | 14 | 53 |
| m-LLDPE | 0.933 | 122 | 560 | 10.3 | 14 | 59 |
| E-LLDPE-10% | 0.930 | 123 | 720 | - | 21 | 55 |
| E-LLDPE-15% | 0.929 | 123 | 560 | 9.6 | 14 | 54 |
| E-LLDPE-30% | 0.924 | 123 | 640 | 9.0 | 18 | 47 |
| CL-LLDPE-1 | - | - | 520 | 11.4 | 16 | 59 |
| CL-LLDPE-2 | - | - | 580 | 11.5 | 20 | 60 |
| m-LLDPE-H | 0.940 | 126 | 880 | - | 24 | 62 |
| Chloroprene rubber (CR) | - | - | 460 | 2.5 | 12 | 74 (HAD) |

*100% modulus indicates the stress at 100% strain.

## 2. Experimental procedure and materials

Cavitation erosion test were carried out by using the vibratory apparatus which is specified in the American Society for Testing Materials (ASTM) G32 standard. However, no erosion occurred for the polyethylene used currently. Therefore, the cavitation erosion tests were carried out using the cavitating liquid jet apparatus specified in ASTM G134 standard. Jet velocity $V$ and cavitation number $\sigma$ were calculated based on ASTM G134 standard.

$$V = \sqrt{\frac{2}{\rho}(p_u - p_d)} \quad (1) \qquad \sigma = \frac{p_d - p_v}{p_u - p_d} \quad (2)$$

where $\rho$ is a liquid density, $p_u$ is upstream pressure, $p_d$ is downstream pressure and $p_v$ is vapor pressure. Cavitation number $\sigma$ was set at 0.025, as defined by ASTM G134-95 for all tests, and the flow velocity was determined by adjusting the upstream and downstream pressures.

The physical and mechanical properties are listed in Table 1 for nine kinds of polyethylene and a chloroprene rubber. Z-LLDPE is a linear-chain low-density polyethylene, polymerized by Ziegler catalyst. LDPE is a low-density polyethylene polymerized by a high pressure method. m-LLDPE a linear-chain low-density polyethylene polymerized by metallocene catalyst m-LLDPE (reference resin). Polyethylene E-LLDPE-10%, 15% and 30%, with various additions of elastomer, were also among the test materials. Polyethylene CL-LLDPE-1 and -2, with the different molecular weights, was also used. Also included was m-LLDPE-H, which is the same as m-LLDPE but with increased density. Chloroprene rubber was also used. 100% modulus is the stress at 100% strain. Therefore, the high value of 100% modulus means that the material is hard to deform. Hattori et al. [3] reported that bubble collapse impact loads decrease on the low value of Young's modulus materials. The polyethylene with addition of elastomer was expected to improve the erosion resistance because of low 100% modulus. On the other hand, it is reported that high hardness material is erosion resistant. The high molecular weight materials were expected to improve the erosion resistance.

## 3. Experimental results and discussion

### *3.1. Cavitation erosion of polyethylene and chloroprene rubber*

A cavitation erosion test of E-LLDPE-30% was carried out at a velocity of 184 m/s (upstream absolute pressure $p_u$=17.4 MPa and downstream absolute pressure $p_d$=0.44 MPa) for two hours. Heat melting

Journal of Physics: Conference Series **656** (2015) 012048 doi:10.1088/1742-6596/656/1/012048

was observed. Therefore it became necessary to find the suitable flow velocity for the cavitation erosion test of polyethylene without heat melting. Cavitation tests of Z-LLDPE were carried out at a constant cavitation number of 0.025 as a function of flow velocity. The flow velocity of 150 m/s (upstream absolute pressure $p_u$=14.5MPa and downstream absolute pressure $p_d$=0.29MPa) and cavitation number 0.025 were found to be suitable for the polyethylene tests when using the cavitating liquid jet apparatus, because no heat melting occurs and the erosion proceed gradually.

After 24 hours, the eroded surface of the specimen was observed three dimensionally by digital microscopy. Figure 1 shows the surface profile through the maximum depth of erosion (MaxDE). This, plus the piling up height and the width are shown in the figure. MaxDER is the MaxDE divided by the exposure time of 24 hours.

Figure 2 shows the relationship between the 100% modulus of polyethylene and MaxDER. The dashed line is the MaxDER of austenitic stainless steel SUS304. Since the MaxDE of SUS304 was 370 μm after 24 hours, the MaxDER is 15μm/h, which is shown by a horizontal line. The MaxDER of Z-LLDPE and E-LDPE-30% is higher than that of SUS304 and the rest of the polyethylene has superior erosion resistance. Hattori et al. [3] reported that the erosion resistance of plastics is high because the bubble collapse impact load becomes small due to the low acoustic impedance. Since the material density of all polyethylene used in this study is almost the same, the 100% modulus is equivalent to the acoustic impedance $\rho c$, because $\rho c=\sqrt{E\rho}$ ($\rho$ is material density, $c$ is sound velocity and $E$ is Young's modulus). $E$ is proportional to the 100% modulus. The MaxDER was expected to enlarge due to the increase in 100% modulus, but the result in this study shows an opposite trend. This is because the range of 100% modulus in this study is narrow and because the mechanical property was changed, thereby overlapping the 100% modulus.

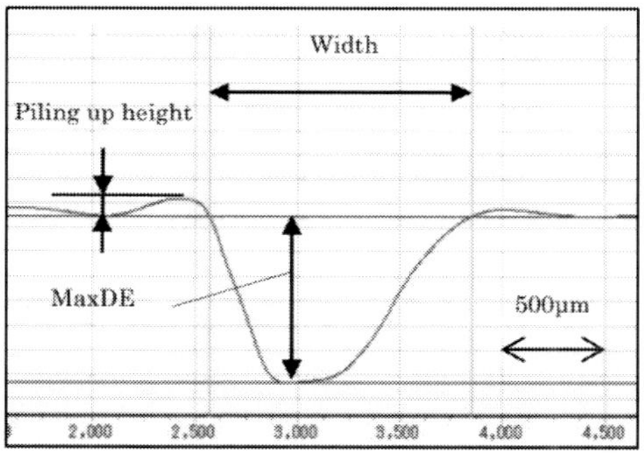


**Figure 1.** Surface profile (Z-LLDPE, 24h).

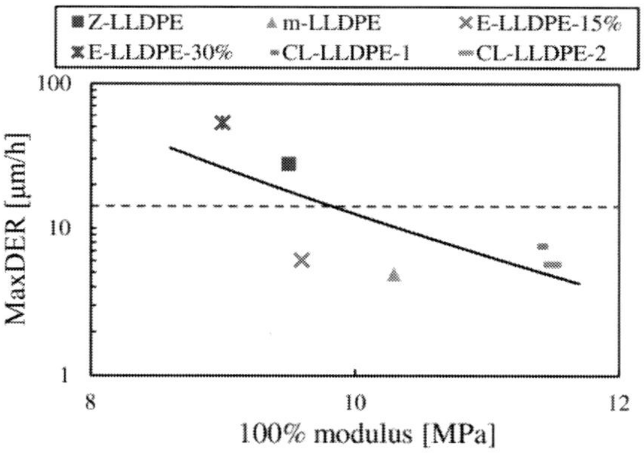


**Figure 2.** Relationship between 100% modulus and MaxDER with a trend curve.

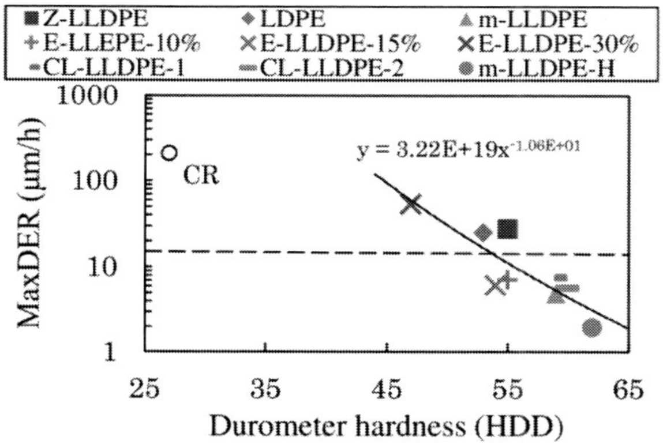


**Figure 3.** Relationship between durometer hardness and MaxDER.

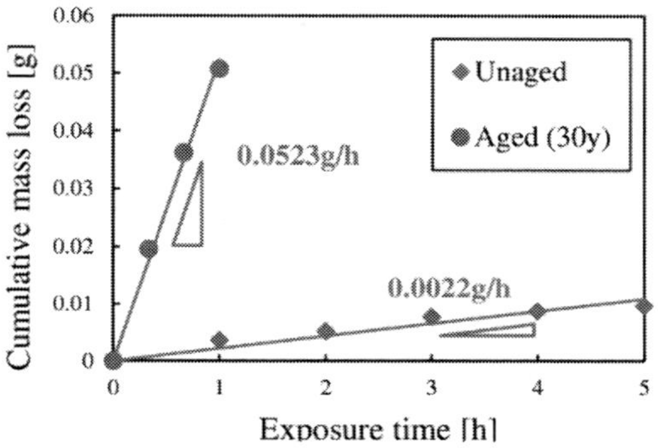


**Figure 4.** Cumulative mass loss-time curves for chloroprene rubber.

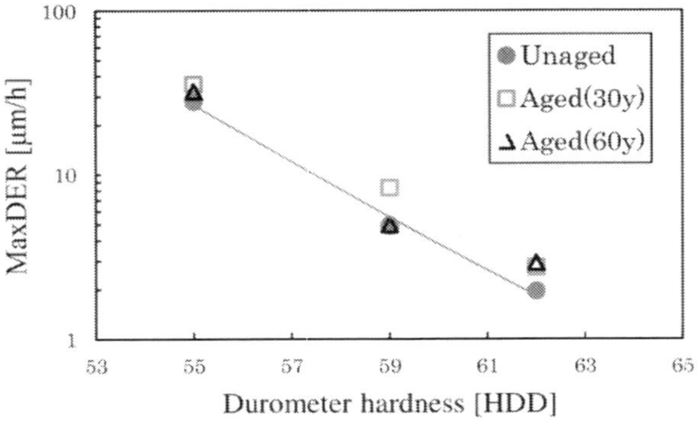


**Figure 5.** Relationship between durometer hardness and MaxDER for aged specimen.

Figure 3 shows the relationship between MaxDER and the durometer hardness of polyethylene and a chloroprene rubber (CR). The MaxDER of polyethylene decreases with durometer hardness. Thus, the erosion resistance of polyethylene is improved by an increase in hardness, as is the case with regular metals.

*3.2 .Cavitation erosion of aged polyethylene and chloroprene rubber*

Aged materials were made by the accelerated oxidation method which was obtained by a long immersion of lining material in tap water at 85℃. The average temperature of seawater in the real pipes is 20℃.

Figure 4 shows the cumulative mass loss-time curves for unaged and 30-year-aged specimen of chloroprene rubber. The mass loss rate is defined as the slope of the cumulative mass loss-time curves. The mass loss rate of the 30-year-aged specimen is about 24 times higher than that of the unaged specimen. Thus, the erosion resistance of chloroprene rubber decreases with aging.

Figure 5 shows the relationship between durometer hardness and the MaxDER for unaged and aged polyethylene. The hardness of aged polyethylene is almost the same as that of unaged polyethylene. The increasing ratio of the MaxDER for aged polyethylene is at most twice as high as with the unaged polyethylene. Thus, it was concluded that the effect of aging on the erosion rate is very small.

## 4. Conclusions

In this study, nine kinds of polyethylene, and a chloroprene rubber, were fabricated and used for cavitation erosion tests using a cavitating liquid jet apparatus. The erosion rate was evaluated in terms of the mechanical properties of the materials.

(1) A flow velocity of 150 m/s (upstream absolute pressure $p_u$=14.5MPa and downstream absolute pressure $p_d$=0.29MPa) and cavitation number 0.025, were suitable for the erosion tests of the polyethylene using the cavitating liquid jet apparatus.

(2) The MaxDER does not decrease with decreased 100% modulus. The MaxDER decreases with durometer hardness. Thus, the erosion resistance of polyethylene is improved by the increase in hardness, as is the case with regular metals.

(3) The erosion resistance of chloroprene rubber decreases with aging, because the hardness of the chloroprene was reduced when the material was aged. The increasing ratio of the MaxDER for aged polyethylene was at most twice as high as that for the unaged polyethylene. Thus, the effect of aging on the erosion rate is very small. It was thought that the oxidation rate of polyethylene was low as compared with that of chloroprene rubber.

## References

[1] Barletta A and Ball A 1983 *Proc. of 6th Int. Conf. on Erosion by Liquid and Solid Impact* 1-1
[2] Kobayashi R, Iwamoto S, Nemoto M and Akaike S 1997 *Piping Engineering* 101 (in Japanese)
[3] Hattori S, Itoh T and Mori H 2011 *Wear* **271** 1103
[4] Deplancke T, Lame O, Cavaille J-Y, Fivel M, Franc J-P and Riondet M 2015 *Wear* **328-329** 301

# Effect of Tensile Stress on Cavitation Erosion and Damage of Polymer

**M Hibi[1], K Inaba[1], K Takahashi[1], K Kishimoto[1], and K Hayabusa[2]**
[1] Tokyo Institute of Technology, 2-12-1 Ookayama, Meguro-ku, Tokyo 152-8550, Japan
[2] Ebara Corporation, 4-2-1, Honfujisawa, Fujisawa, Kanagawa 251-8502, Japan

E-mail: hibi.m.aa@m.titech.ac.jp, inaba@mech.titech.ac.jp, ktakahashi@mech.titech.ac.jp, kkishimo@mep.titech.ac.jp, hayabusa.keisuke@ebara.com

**Abstract.** Cavitation erosion tests for epoxy, unsaturated polyester, polycarbonate, and acrylic resin were conducted under various tensile stress conditions (Tensile-Cavitation test). A new testing device was designed to conduct the Tensile-Cavitation test and observe specimen surface during the experiment based on ASTM G32. When tensile stress of 1.31 MPa was loaded on epoxy resin, cracks occurred on the specimen after 0.5 hours during cavitation erosion. When no tensile stress was loaded on the epoxy resin, the damage was general cavitation erosion only. As well as the epoxy resin, unsaturated polyester resin applied tensile stress of 1.31 MPa and polycarbonate resin of 6.54 MPa indicated erosion damages and cracks. When tensile stress of 6.54 MPa was loaded on acrylic resin, the erosion damage was almost the same as the results without tensile stress. We confirmed that anti-cavitation property of epoxy resin was higher than those of acrylic and polycarbonate without tensile stress while the damage of epoxy resin was much serious than that of acrylic resins under tensile stress loadings.

## 1. Introduction

Cavitation erosion occurs in fluid machinery like pumps and valves. Repairing of the machineries have been paid attention for extending life and decreasing maintenance cost by polymer materials. So far, the evaluation method of the repairing part has not been well established yet. There are many studies to examine anti-cavitation properties of materials [1], [2], [3], while anti-cavitation properties of coating materials in pumps may be affected by tensile stress. Our previous study [4] reported that cracks were generated in epoxy resins under tensile stress loads in cavitation erosion tests. In this study, cavitation erosion test was conducted for various polymer materials (epoxy, unsaturated polyester, polycarbonate, and acrylic resins) by applying tensile stress (Tensile-Cavitation test) in order to examine the mechanism of cavitation damages.

## 2. Experimental procedure

New testing equipment was designed. It can monitor specimen condition during Tensile-Cavitation test. Figure 1(a) shows schematic of the device. Tests were carried out based on ASTM G32 [5].

Magnetrostrictive vibrating device generates the ultrasonic wave that is amplified by horn and results in vibration of the disk made of stainless steel. Cavitation is generated around the stainless disk. The magnetrostrictive vibrating device's frequency was 19.5 kHz, and the half amplitude is 25 μm. Specimen was placed at the distance of 1mm from the disk. The specimen was fixed by chucks and applied tensile stress through the stainless wire by weight. Chucks were fixed on two linear guides, and it can move in a tensile direction. If creep deformation occurs on specimen, specimen can be applied a constant tensile force and specimen position can be adjusted easily. Mirror placed in water tank was inclined to observe the specimen condition by video camera placed in front of the equipment outside of the tank. Testing liquid was distilled water whose temperature was kept at 25°C by the circulation tank. Figure 1(b) shows closed-up view of the test section. Specimen and stainless disk can be observed in the mirror.

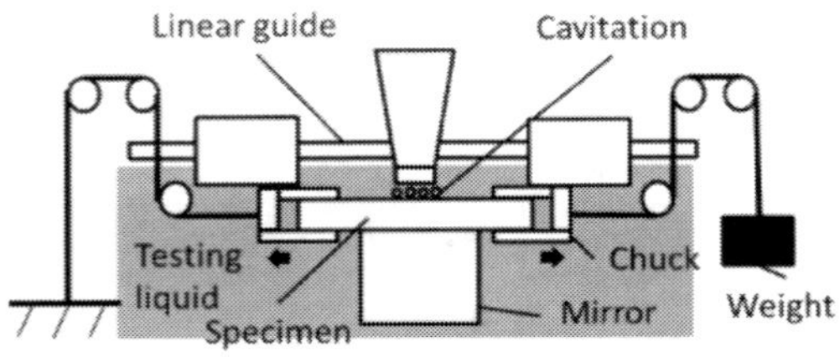


(a) Schematic of the test section

(b) Enlargement of test section

Fig. 1 Testing device

To confirm applied tensile load by the weight, strain gages were put on both sides of epoxy resin. Stress was calculated by the strain signals, Young's modulus and cross sectional area of the specimen. It was found that the applied load was about 70% of the weight. Therefore, we use the magnitude of tensile stress that is 0.7 times smaller of the weight.

Specimens were epoxy (EP), unsaturated polyester (UP), polycarbonate (PC), and acrylic resin (PMMA). EP and UP specimens were made by mixing main agent and curing agent to form the dog bone shape of the general specimen. PC and PMMA specimens were readymade and shaped by machining.

## 3. Results and Discussion

### *3.1. Epoxy resin (EP)*

Figure 2 shows extracted pictures from the movie in the Tensile-Cavitation test for epoxy resin. Figure 2(a) shows epoxy surface after 28 hours without tensile stress. Only cavitation erosion was observed after 2.5 hours without tensile stress. Figure 2 (b), (c) shows epoxy specimen applied tensile stress of 1.31 MPa. We can also see the stainless disc of the tip of the horn behind the specimen. After 0.5 hours, crack was generated. After 2 hours, several cracks were generated and gradually developed. Some cracks were combined into one large crack. Figure 2 (d) shows epoxy surface after test. Epoxy resin was broken after 7.5 hours. Cavitation erosion was also observed as well as the case without tensile stress. We confirmed that cracks were generated in epoxy resin by tensile stress under cavitation loading. In this case, crack was observed earlier than the occurrence of cavitation-erosion without tensile stress. By microscope observations, it was revealed that micro cracks were generated around the voids after the test. The voids were created due to the entrainment of the air when the specimens were prepared. We considered that the stress concentration occurred around the voids due to the cavitation impact under tensile stress loading and micro cracks were generated and gradually developed into the macroscopic cracks.

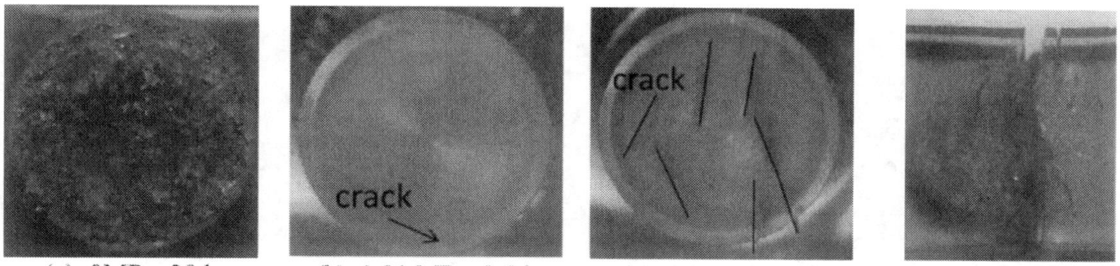


(a) 0MPa, 28 h (b) 1.31 MPa, 0.5 h (c) 1.31 MPa, 2 h (d) 1.31 MPa, 7.5 h

Fig. 2 Damage of epoxy resin

### *3.2. Fractography*

Fracture surface of EP ruptured by Tensile-Cavitation tests was observed. The specimen was applied tensile stress of 0.5 MPa and ruptured after 25.2 h. Figure 3 (a) is a monitoring image after 4 h when a crack was observed. Figure 3 (b) shows fracture surface near starting point of crack observed by laser microscope. Left side of this image is erosion surface and near the origin of fracture. Near the surface, mirror area and beach marks were observed. Around the middle of this figure, many small cracks were observed (Scaly pattern). From these results, cracks were expected to be developed by fatigue from the erosion surface side.

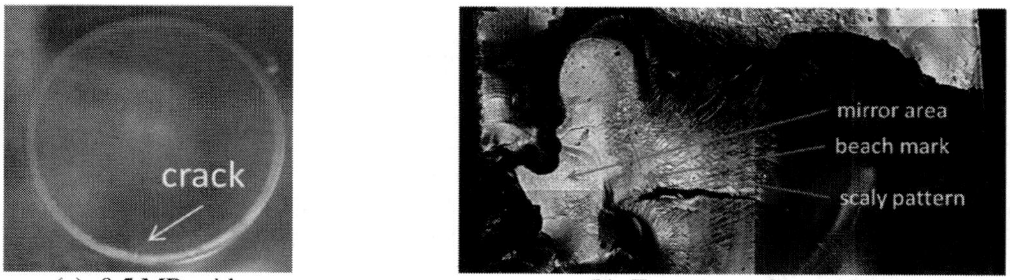


(a) 0.5 MPa, 4 h (b) Fracture surface of EP

Fig. 3 Monitoring image and fracture surface

### *3.3. Results of Four resins*

Tensile-Cavitation tests for UP, PC and PMMA were carried out as well as epoxy resin. Figure 4 (a) shows relation of stress and crack initiation time about EP, UP and PC. Closed symbol means crack initiation time. Opened symbol means finishing time of the test without crack in the specimen. This graph shows that the higher tensile stress is loaded on EP, the earlier crack is generated. Cracks were not observed when EP was applied tensile stress of less than 0.24 MPa. Cracks were not observed when UP was applied tensile stress under 0.62 MPa. It is believed that the threshold tensile stress exists. Cracks were observed on PC plane specimen applied tensile stress of 6.54 MPa. However, crack was not observed on PC applied tensile stress of 6.54 MPa when PC specimen had two holes to make stress concentration areas. The damages with stress concentration specimen by holes can be different from the specimen without stress concentration with tensile stress 6.54 MPa. Crack was not observed on PMMA applied tensile stress of 6.54 MPa. We are planning to conduct Tensile-Cavitation test for PMMA plane specimen applied higher stress and confirm whether crack is generated. Figure 4 (b) shows relation of stress and erosion initiation time about EP, UP, PC and PMMA. Although erosion was generated in all materials, we cannot find the relation between stress and erosion initiation time so far.

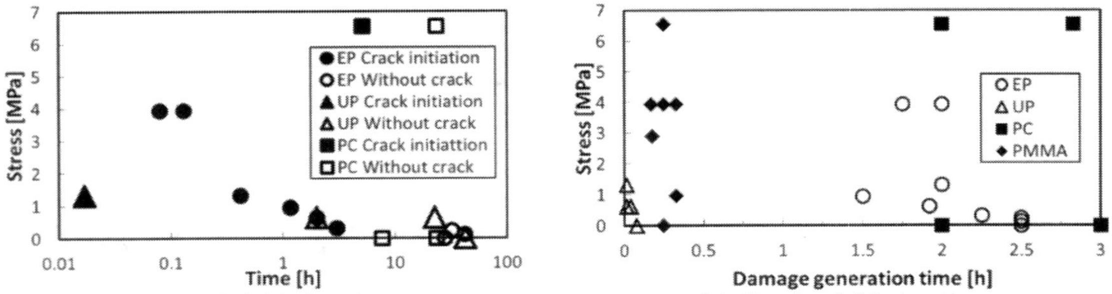


(a) Crack initiation time (b) Erosion initiation time

Fig. 4 Stress-Time diagrams

*3.4 Discussion*

It is suspected that crack is generated on EP, UP and PC applied tensile stress of more than threshold value. Although cracks were observed on EP and UP earlier than erosion, on PC later than erosion. EP and UP specimen included many voids mixed in preparing (molding) process. PC specimen included no voids. It seemed that mixed voids and micro cracks by cavitation erosion resulted in macroscopic cracks. Voids and micro cracks are necessary to be observed by microscope when cracks begin to generate. We will investigate the relation of crack generation and voids size in the future works. From observation of fracture surface, cracks were expected to be generated and developed by fatigue. Moreover, an EP specimen was ruptured only by applying tensile load after cracks were well observed. This implies that once crack length exceeds the threshold, stable fracture occurs. We will further investigate the initiation mechanism of cracks and fracture mechanics.

## 4. Conclusion

We made a new testing device that enabled us to monitor specimen condition in order to reveal the damage behavior of polymer materials applied tensile stress and cavitation erosion. Tensile-Cavitation tests for EP, UP, PC and PMMA were carried out. On EP without tensile stress, erosion was generated after 28 hours. However, on epoxy applied tensile stress of 1.31 MPa, crack started to be observed after 0.5 hours and specimen was broken after 7.5 hours. Cracks were generated on UP and PC under tensile stress. Tensile stress enhanced damages on EP, UP and PC. It is suspected that crack is generated on EP, UP and PC applied tensile stress of more than threshold value. Anti-cavitation property of EP without tensile stress showed stronger than that of acrylic. However, crack was generated on EP applied tensile stress of 1.31 MPa. Crack was not generated on PMMA applied tensile stress of 6.54 MPa. Anti-cavitation property under tensile stress was reversed. When polymer materials are used to coat for repairing, anti-cavitation property considering cavitation erosion and tensile stress will be required for its reliability. We observed fracture surface of EP specimens and confirmed beach marks near the erosion surface.

## References

[1] Hattori S, Itoh T and Mori H 2005 Cavitation erosion resistance of high polymer materials *Transactions of the JSME* A **71** 705
[2] Soyama H, Kumano H and Saka M 2001 A new parameter to predict cavitation erosion *CAV2001* A3 002
[3] Richman R and McNaughton W 1990 Correlation of cavitation erosion behavior with mechanical properties of metals *Wear* **140** 63-82
[4] Takahashi K, Arai D, Inaba K, Kishimoto K, Nakamoto H and Hayabusa K 2014 Evaluation of anti-cavitation property of coating materials for structural repair *Turbo kikai* **42** 12
[5] 2005 ASTM Designation *Annual Book of ASTM Standards* 106

Journal of Physics: Conference Series **656** (2015) 012050 doi:10.1088/1742-6596/656/1/012050

# Numerical simulation of cavitation erosion on a NACA0015 hydrofoil based on bubble collapse strength

**V Hidalgo[1,3], X Luo[1], X Escaler[2], R Huang[1] and E Valencia[3]**

[1] State Key Laboratory of Hydro Science & Engineering, Tsinghua University, Beijing, China
[2] Centre for Industrial Diagnostics and Fluid Dynamics, UPC, Barcelona, Spain
[3] Department of Mechanical Engineering, Escuela Politecnica Nacional, Quito, Ecuador

E-mail: `luoxw@mail.tsinghua.edu.cn`

**Abstract.** The prediction of erosion under unsteady cavitation is crucial to prevent damage in hydraulic machinery. The present investigation deals with the numerical simulation of erosive partial cavitation around a NACA0015 hydrofoil. The study presents the calculation of the bubble collapse strength, *Sb*, based on the bubble potential energy to identify the surface areas with highest risk of damage. The results are obtained with a numerical scheme assuming homogeneous mixture flow, implicit LES and Zwart cavitation model. The 3D unsteady flow simulation has been solved using OpenFOAM. Python language and OpenFOAM calculator (foamCalcEx) have been used to obtain and represent *Sb*. The obtained results clearly show the instants of erosive bubble collapse and the affected surface areas.

## 1. Introduction

Cavitation erosion has been a subject of deep study for long time using different techniques [1]. The main methods, which have permitted to understand the mechanisms of the phenomenon and the damage generation, are experimental correlations with noise, vibrations and material properties [2, 3]; analytical methods, such as pit distribution quantification by image processing [4, 5]; and numerical simulations using special turbulence models to reproduce cavitation morphology and dynamic behavior [6]. For the latter, Kubota et al. [7] introduced the idea of homogeneous mixture flow with the bubble two-phase flow (BTF) model. Then Coutier et al. [8] showed that the condensation and vaporization processes are controlled by a barotropic state law and simulated the cavity shedding in a venturi. The research works on horse-shoe vortex structures shed around a twisted hydrofoil [9] and on the transient sheet-cloud cavitating flows [10] also corroborate the validity of the homogeneous mixture flow models.

Regarding the cavitation erosion models, Van et al. [11] revised some models. In those models, the potential energy of the bubble, $E_p$, is described as main erosion factor when a group of bubbles collapse simultaneously in cascade and the gap between the wall and the cloud of bubbles is very thin. However, Dular et al. [4] indicated that erosion is also due to the re-energizing of bubbles close to the walls and the generation of high speed jets, which showed speeds close to 100 m/s. In this context, Li [6] proposed the pressure gradient, $\partial p/\partial t$, as a new erosion index based on $E_p$. Moreover, Bergeles et al. [12] presented the cavitation aggressiveness index(CAI) based on total derivatives of pressure, vapor fraction and the Rayleigh-Plesset equations for unsteady and quasi steady flow calculation. However, the CAI equation is not directly correlated with the quantity of energy liberated in the bubble collapse. Consequently, the quantification of this energy is still a challenge.

Bearing in mind these aspects, the present investigation presents a numerical simulation method based on homogeneous mixture flow assumption using implicit LES and the Zwart-Gerber-Belamri cavitation model to predict the cavitation erosion at high Reynolds numbers [13, 14]. The proposed collapse strength, $S_b$, is based on the erosion index proposed by Li [6]. Additionally, the pressure implicit with splitting of operators (PISO) algorithm [13] is used to solve the Navier-Stokes equations to simulate the behavior of unsteady flow on a NACA0015 hydrofoil.

## 2. Model description

### 2.1. Homogeneous mixture flows

The two-phase flow is modeled in each control as homogeneous mixture with vapor volume fraction, $\alpha_i$, as indicated in equations (1) and (2)

$$\alpha_i = \frac{\forall_{Vi}}{\forall_i}, \tag{1}$$

$$\frac{\partial(\alpha_i \rho_V)}{\partial t} + \nabla(\alpha_i \rho_V U) = \dot{m}, \tag{2}$$

where $\forall_{Vi}$ and $\forall_i$ are the vapor and total volume respectively, $\rho_V$ is the vapor density, $U$ is the velocity and $\dot{m}$ is the inter-phase mass transfer rate per unit volume. Moreover, $\forall_i$ is considerer as constant value, and $\dot{m}$ is a source term, which is decided based on the Zwart-Gerber-Belamri cavitation model and it can be calculated using the equation (3).

$$\dot{m} = \begin{cases} \dot{m}^{+} = F_V \dfrac{3r_{nuc}(1-\alpha)\rho_V}{R_B}\sqrt{\dfrac{2}{3}\left(\dfrac{p_V - p}{\rho_l}\right)} & if\ p < p_V \\ \dot{m}^{-} = -F_C \dfrac{3\alpha\rho_V}{R_B}\sqrt{\dfrac{2}{3}\left(\dfrac{p - p_V}{\rho_l}\right)} & if\ p > p_V \end{cases}, \tag{3}$$

where $p$ and $p_V$ are the pressure and vapor saturation pressure respectively, $F_V = 300$ and $F_C = 0.03$ are the selected calibration constants for vaporization and condensation, $r_{nuc} = 5.0 \times 10^{-6}$ is the nucleation site volume fraction, $R_B = 1.0 \times 10^{-6}$ m is the typical bubble size in water [14].

### 2.2. Bubble collapse strength

The total energy of one bubble in a control volume, $E_{bi}$, is the sum of potential energy, $E_{pi}$, and kinetic energy, $E_{ki}$. However, the potential term is more important than the kinetic term ($E_{pi} >> E_{ki}$) because before the collapse the bubble velocity is low. Therefore, the bubble collapse strength, $S_b$, calculated with equation (4) can be considered as an updated model to evaluate cavitation aggressiveness.

$$S_b = \sum_{i=1}^{n}\left[(p - p_V)\frac{\partial \alpha_i}{\partial t} + \alpha_i \frac{\partial p}{\partial t}\right], \tag{4}$$

where $n$ is the control volume number that presents cavitation in the computational domain. To calculate $S_b$, an algorithm based on foamCalcEx and Python language has been developed.

Obviously, to produce plastic deformation and erosion on solid parts, the $S_b$ must be higher than the material threshold resistance, $S_M$, which quantifies the material resistance to cavitation erosion and should be evaluated with previous experimental works [2]. Furthermore, the accumulative bubble collapse strength, $S_{bT}$, that causes damage can be calculated within one cycle of cavitation shedding process with equation (5):

$$S_{bT} = \frac{1}{N}\sum_{t_o}^{t_f} S_b \left\{ \begin{array}{l} S_b \neq 0,\ S_b > S_M \\ S_b = 0,\ S_b < S_M \end{array} \right. , \tag{5}$$

where $N$ is the number of the given results in one cycle, $t_0$ and $t_f$ are the initial and final time.

### 2.3. Implicit LES formulation

The implicit LES formulation, ILES, is indicated in equation (6) with the filtered operations [13, 14].

$$\frac{\partial(\rho\overline{u_i})}{\partial t} + \frac{\partial(\rho\overline{u}_i\overline{u}_j)}{\partial x_j} = -\frac{\partial\overline{p}}{\partial x_i} + \frac{\partial(\overline{\tau}_{ij} - \tau'_{ij})}{\partial x_j}. \tag{6}$$

The subgrid stress tensor $\tau'_{ij}$ is a nonlinear term, so it is separated as indicated in equation (7).

$$\tau'_{ij} = \rho(\overline{\overline{u}_i\overline{u}_j} - \overline{\overline{u}}_i\overline{\overline{u}}_j + \tilde{\tau}'_{ij}), \tag{7}$$

where, the tensor $\tilde{\tau}'_{ij}$ is modeled by using the truncation error to act as a dissipative action to avoid the explicit subgrid scale model (SGS) of LES [14].

## 3. Computational domain and boundary conditions

The computational domain and the main boundary conditions are indicated in figure 1. The free stream inlet velocity, $U_\infty$, is 17.3 m/s, the outlet pressure, $p_\infty$, is 302 kPa and the cavitation number, $\sigma$, is 2.01. The chord length of the hydrofoil NACA0015, $c$, is 60 mm with an angle of attack, $\widehat{AOA}$, of 8°. A structured mesh with 878280 hexahedron elements suitable for the ILES conditions of $y^+$ equal to 2.0 has been built. The quality of mesh has been checked by the analysis of the diagonals and Jacobians of the hexahedron [15].

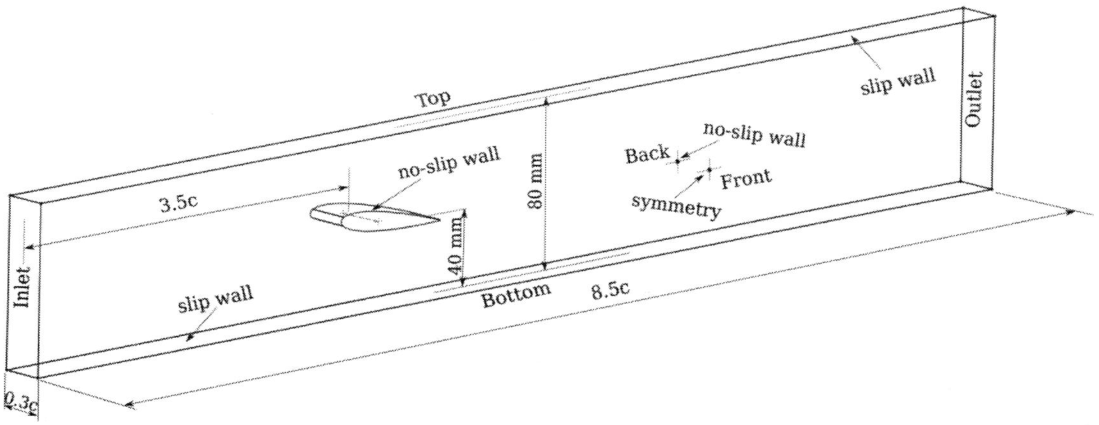


Figure 1: Computational domain and boundary conditions.

## 4. Results and Discussion

Figure 2 shows the simulated unsteady partial cavitation on the hydrofoil suction side at three different times within one shedding cycle from 0.01 s to 0.0175 s. The images show: (a) the cavity sheet, (b) the shedding of cavitation clouds and (c) the formation of a horse-shoe cavity convected downstream by the main flow. The hydrofoil surface areas exposed to possible erosive bubble collapses have been identified from the control volumes with highest $Sb$ values. From the results, it is confirmed that the highest aggressiveness takes place during the collapses as shown in figure 2(b). On the contrary, during the reentrant jet formation and the collapse of horse-shoe cloud shown in figure 2(c) only small areas are slightly affected by collapses. This might be due to a large distance between the bubbles and the hydrofoil surface. Results are similar to analogous risk assessment models and experiments [6].

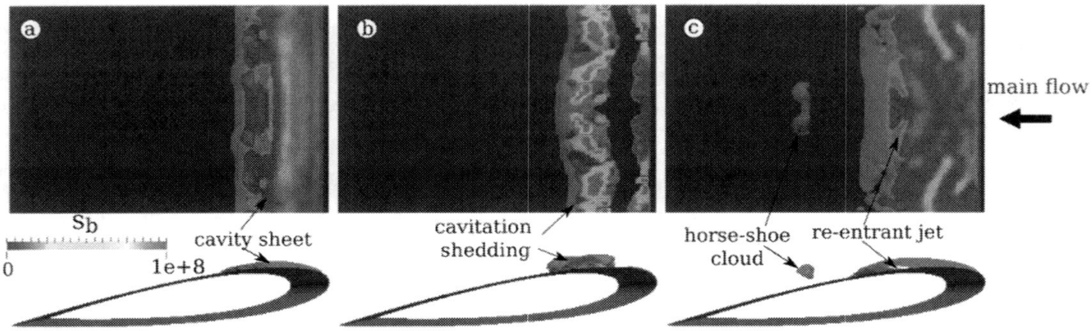


Figure 2: Erosion prediction for a typical cavitation cycle divided in three times a, b and c.

## 5. Conclusions

A numerical simulation of cavitation erosion on a NACA0015 has been carried out with a free OpenFOAM software package and an algorithm based on foamCalcEx and Python language. The bubble collapse strength has been calculated based on the potential energy and homogeneous mixture flow assumption using implicit LES and Zwart-Gerber-Belamri cavitation model. Results confirm the suitability of using $Sb$ and the developed algorithm for cavitation erosion prediction.

## Acknowledgments

This work was financially supported by the National Natural Science Foundation of China (Project Nos. 51306018, 51206087 and 51179091) and State key Laboratory of Hydroscience and Engineering, Tsinghua University (Project Nos. 2014-KY-05 and 2015-E-03).

## References

[1] Berchiche N, Franc J P and Michel J M 2002 *Journal of Fluids Engineering* **124** 601–606
[2] Escaler X, Farhat M, Avellan F and Egusquiza E 2003 *Wear* **254** 441–449
[3] Franc J P 2009 *Journal of Fluids Engineering* **131** 021303–021303
[4] Dular M, Stoffel B and Sirok B 2006 *Wear* **261** 642–655
[5] Petkovek M and Dular M 2013 *Wear* **300**
[6] Li Z 2012 *Assessment of Cavitation Erosion with a Multiphase Reynolds-Averaged Navier-Stokes Method* Ph.D. thesis Delft University of Technology
[7] Kubota A, Kato H and Yamaguchi H 1992 *Journal of fluid Mechanics* **240** 59 – 96
[8] Coutier-Delgosha O, Reboud J L and Fortes-Patella R 2003 *Journal of Fluids Engineering* **125** 38–45
[9] JI B, LUO X w, PENG X x and WU Y l 2013 *Journal of Hydrodynamics, Ser. B* **25** 510–519 ISSN 1001-6058
[10] Huang B, Zhao Y and Wang G 2014 *Computers & Fluids* **92** 113–124
[11] Van T, Fitzsimmons P, Li Z and Foeth J 2009 *Cav2009* (Michigan, USA) pp 1–13
[12] Bergeles G, Koukouvinis P and Gavaises M 2014 *ICHD 2014* (Singapore: N.T.U.) pp 619–627
[13] Hidalgo V, Luo X, Ji B and Aguinaga A 2014 *Chinese Science Bulletin* 1–7
[14] Hidalgo V, Luo X, Escaler X, Ji J and Aguinaga A 2014 *IOP Conference Series* **22** 052013
[15] Stimpson C, Ernst C, Knupp P, Pébay P and Thompson D 2006 *The verdict geometric quality library.* (SNL)

# Impact load measurements with a PVDF pressure sensor in an erosive cavitating flow

**J Hujer[1], JB Carrat[2], M Müller[1], M Riondet[2]**

[1] Technical University of Liberec, Studentska 2, 461 17 Liberec, Czech Republic
[2] LEGI, Univ. Grenoble Alpes, CNRS, F-38000 Grenoble, France

E-mail : JanHujer@seznam.cz

**Abstract**. A PVDF pressure sensor was used to measure the pressure peaks due to the collapse of cavitation bubbles in the high-speed tunnel of the LEGI laboratory. It was flush mounted on a stainless steel disk in the most erosive area of the test section. The recorded data were post-processed in order to get the impact load spectra for different velocities at constant cavitation number. The results are presented as cumulative histograms of peak rate and maximal impact load for different flow velocities of the high-speed tunnel.

## 1. Introduction

Cavitation is one of the main issues in turbomachinery since it may be responsible for efficiency losses, vibration and erosion. When a cavitation bubble collapses close to a surface, a microjet and a pressure wave are usually created, which impact the surface on a very small area and can deform and/or erode it. It has been shown from pitting tests that the impact area is of the order of tens or hundreds of micrometers in diameter (see for example, Franc et al [1]).

Moreover, the pressure pulse resulting from the collapse of a cavitation bubble is characterized by a large amplitude, of the order of 100 MPa (Wang et al [2]) or more (> GPa) (Momma et al [3]), and a very short duration measured typically in microseconds or even nanoseconds (Wang et al [4]).

Because of these extreme characteristics, it is very difficult to measure such pressure pulses whereas an accurate knowledge of their features is essential in view of damage prediction. So experimenters have developed special pressure sensors able to resist to cavitation erosion and with a high resonance frequency suited to the small duration time of the phenomenon.

Franc et al [5] measured the pressure pulse signal in a cavitation loop with a commercial pressure sensor (PCB 108A02) and showed that their sensor was not optimal for such measurements because of its limited resonance frequency.

Several authors use piezoelectric polyvinylidene fluorine (PVDF) films because of their high mechanical resistance and high resonance frequency. Momma et al [5] developed a PVDF pressure sensor and measured impact loads in a cavitating jet apparatus. Their PVDF transducer was mounted on a stainless steel support and protected on top by polyamide tapes from water and cavitation damage. Wang et al [1] recorded the impulsive pressure generated by cavitation bubble collapse with a single PVDF sensor and an array of PVDF sensors.

The purpose of the present work is to measure the impact loads due to cavitation bubble collapses in a cavitation loop with a PVDF sensor flush mounted in the region of maximum damage. Measurements were conducted at different flow velocities in order to investigate the effect of velocity on the flow aggressiveness at constant cavitation number i.e. for geometrically similar cavitating flows. Details on the facility and the pressure sensor are given in section 2. Section 3 is devoted to presentation of the results and the conclusion.

## 2. Experiments

### *2.1. Experimental Facility*

The experiment was carried out in the high-speed tunnel of the LEGI laboratory, described in [5], generally used for cavitation erosion tests on samples made from different materials. In cavitating conditions, a cavity develops in the test section, which generates small vapor structures that collapse and may erode the sample. Erosion appears as a ring whose dimensions, minimum radius ($R_{min}$) and maximum radius ($R_{max}$), depend on the cavitation number. Present tests were carried out at a constant value $\sigma = 0.9$ of the cavitation number that leads *to* $R_{min} \approx 21\ mm$ *and* $R_{max} \approx 26\ mm$ on average. The same value of cavitation number as in [1] was chosen with the aim of comparing pitting tests and pressure pulse measurements.

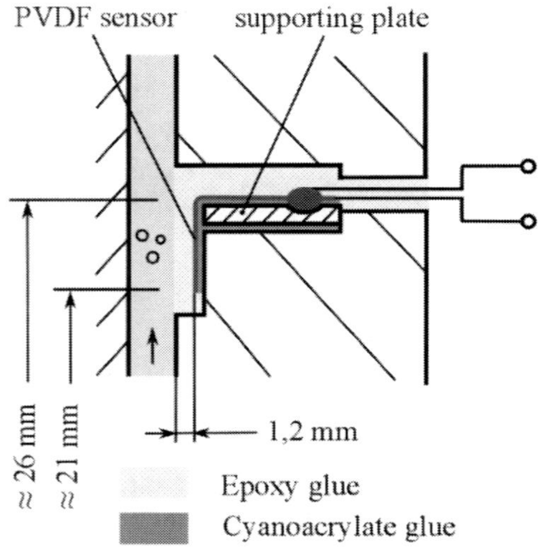


**Figure 1.** Detail of the mounted PVDF Sensor.

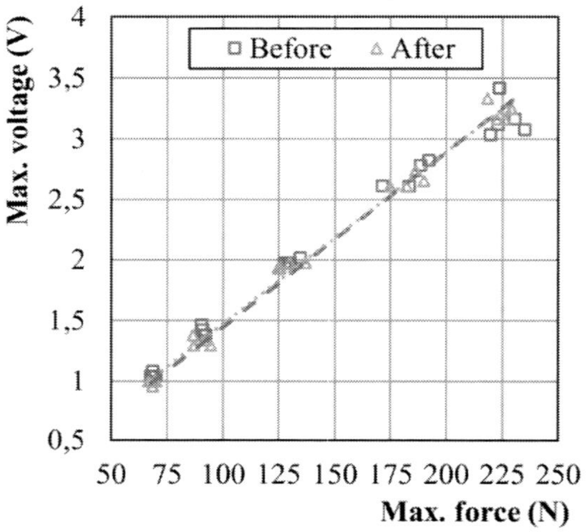


**Figure 2.** Calibration curve of the mounted PVDF sensor before (blue square and dashed line) and after (orange diamond and dotted line) experimentation.

Cavitation impacts are concentrated between the minimum and the maximum radius. A 28 μm thick commercial PVDF (DTI-028K/L – Measurement Specialties, Inc.) sensor was placed in this area and mounted on a cavitation stainless steel disk (AISI 304 L), as shown in figure 1. The active area of the PVDF sensor was cut from 12 mm x 30 mm to 12 mm x 5 mm approximately. At first, the PVDF sensor was fixed on an adjusted supporting plate made from polymethyl methacrylate (PMMA) and then glued on the disk by cyanoacrylate glue (Loctite Super Attak Power Flex Gel). Finally, the hole was filled with epoxy glue (Bison Epoxy 5 minutes) and the 1.2 mm thick upper layer was used to protect the sensor from cavitation damage.

### *2.2. Calibration of the sensor*

The piezoelectric PVDF sensor delivers a voltage $V$ proportional to the force $F$ or pressure $P$ applied on its active surface. So a calibration curve $F=f(V)$ or $P=f(V)$ is necessary for estimating the force or the pressure applied on the sensor from the measured voltage. In the present work, force is preferred since pressure is far from being uniformly applied on the sensitive surface of the sensor.

A ball drop test technique was used for calibration. A stainless steel ball is dropped on the homemade sensor, including its protective layer on top, from a known height $h_1$ and the height $h_2$ of the first rebound is measured. On the same time, the voltage is recorded as a function of time using the 8-bit Bus-Powered USB digitizer NI USB-5132.

The mean force on the transducer is calculated from a momentum balance:

$$F = \frac{m(V_1 + V_2)}{\tau}$$

where $V_1$ is the velocity just before the impact on the sensor, $V_2$ the rebound velocity just after impact, m the mass of the stainless steel ball and $\tau$ the duration of the impact. Both velocities $V_1$ and $V_2$ can be expressed as $V_i = \sqrt{2gh_i}$ ($g$ gravity acceleration). The impact duration $\tau$ was read on time axis of the calibration signal. The ratio between the maximum force and mean force is determined from the shape of the calibration signal.

The calibration curve in figure 2 was plotted from five calibrations settings characterised by different heights and stainless steel balls diameters, each one including four runs. It showed a good reproducibility. Furthermore, the performance of the sensor before and after the experiments is not changed significantly, which proves that the sensor is not damaged by the cavitating flow in spite of its high aggressiveness.

### *2.3. Acquisition and processing of the data*

The impulses generated by the cavitation were registered by the PVDF sensor connected directly to a digitizer (NI-USB 5132) and the signal acquisition was made by a LabVIEW virtual instrument control at the highest sampling rate of 50 MS/s. The high-speed cavitation tunnel, disc, digitizer and computer were grounded to limit noise.

The acquisition time was ajusted according to the cavitation erosion potential, which depends on the flow velocity (upstream pressure) of the cavitation tunnel. As a consequence, for a flow velocity of 22 and 34 m/s the acquisition time was 50 s and for 44.7, 54.8, 63.2 and 70,7 m/s it was 30 s.

The recorded data were post-processed by using two parameters: a threshold, depending on noise of each signal, and a locking time of 10 µs. So if one peak is detected, any other fluctuation of the signal during the following 10 µs is ignored in order to avoid multiple counting for one impact.

## 3. Results and conclusion

The results of the post-processing are presented in figure 3 as cumulative histograms. For the different flow velocities, cumulative peak rates in peaks per unit surface area of the PVDF sensor and per unit time are displayed as a function of the maximal impact load. The maximal impact load in newtons was determined from the maximal voltage in volts using the calibration.

The results show that the impact load increases as flow velocity increases. Also the peak rate increases with flow velocity. Concave parts of the spectra at high amplitude for each flow velocity suggests a limit value for the maximal impact load. Results at low amplitude are dependent upon the threshold used for filtering the noise.

As a conclusion, the use of PVDF pressure sensors appears to be an accurate technique for measuring the aggressiveness of a cavitating flow. In the present work, measurements have been made up to a maximum flow velocity of 71 m/s (25 bar) which corresponds to a flow of high aggressiveness. The final objective is to correlate these measurements with cavitation damage in order to propose a method of prediction of damage based on impact load measurements.

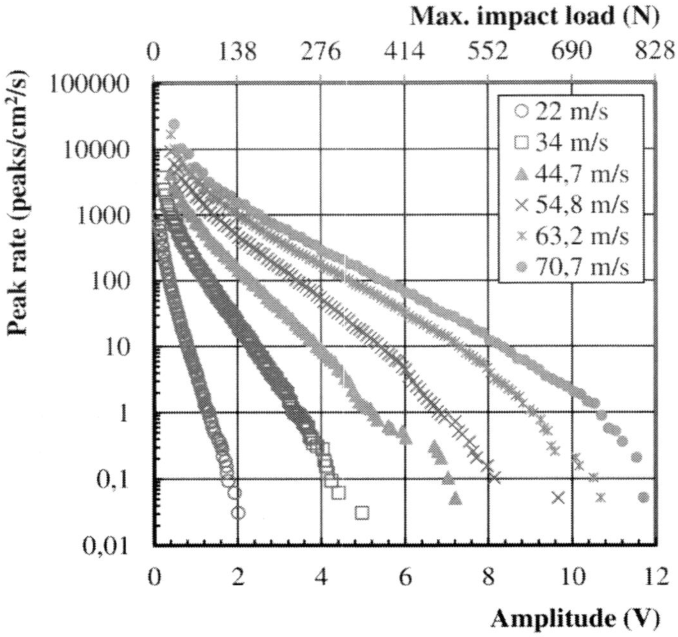


**Figure 3.** Impact load spectra at different flow velocities.

**Acknowledgments**
We gratefully acknowledge the supports of the project TUL No. SGS 21000, the project OP VK reg. no. CZ.1.07/2.3.00/20.0139 and the Foundation Grenoble INP though the industrial chair ALSTOM "HYDRO'LIKE".

**References**
[1] Franc, Jean-Pierre, Michel Riondet, Ayat Karimi, and Georges L. Chahine. "Material and Velocity Effects on Cavitation Erosion Pitting." Wear 274–275 (January 2012): 248–59.
[2] Wang, Yi-Chun, and Yu-Wen Chen. "Application of Piezoelectric PVDF Film to the Measurement of Impulsive Forces Generated by Cavitation Bubble Collapse near a Solid Boundary." Experimental Thermal and Fluid Science 32, no. 2 (November 2007): 403–14.
[3] Momma, T., and A. Lichtarowicz. "A Study of Pressures and Erosion Produced by Collapsing Cavitation." Wear 186–187 (August 1995): 425–36.
[4] Wang, Yi-Chun, Ching-Hung Huang, Yung-Chun Lee, and Ho-Hsun Tsai. "Development of a PVDF Sensor Array for Measurement of the Impulsive Pressure Generated by Cavitation Bubble Collapse." Experiments in Fluids 41, no. 3 (August 24, 2006): 365–73.
[5] Franc, Jean-Pierre, Michel Riondet, Ayat Karimi, and Georges L. Chahine. "Impact Load Measurements in an Erosive Cavitating Flow." Journal of Fluids Engineering 133, no. 12 (2011): 121301. doi:10.1115/1.4005342.
[6] Momma, T., and A. Lichtarowicz. "A Study of Pressures and Erosion Produced by Collapsing Cavitation." Wear 186–187 (August 1995): 425–36.

# Numerical 3D flow simulation of attached cavitation structures at ultrasonic horn tips and statistical evaluation of flow aggressiveness via load collectives

**S Mottyll and R Skoda**

Chair of Hydraulic Fluid Machinery, Ruhr-Universität Bochum, Universitätsstr. 150, 44801 Bochum, Germany

E-mail: stephan.mottyll@ruhr-uni-bochum.de

**Abstract.** A compressible inviscid flow solver with barotropic cavitation model is applied to two different ultrasonic horn set-ups and compared to hydrophone, shadowgraphy as well as erosion test data. The statistical analysis of single collapse events in wall-adjacent flow regions allows the determination of the flow aggressiveness via load collectives (cumulative event rate vs collapse pressure), which show an exponential decrease in agreement to studies on hydrodynamic cavitation [1]. A post-processing projection of event rate and collapse pressure on a reference grid reduces the grid dependency significantly. In order to evaluate the erosion-sensitive areas a statistical analysis of transient wall loads is utilised. Predicted erosion sensitive areas as well as temporal pressure and vapour volume evolution are in good agreement to the experimental data.

## 1. Introduction

Ultrasonic horn transducers are widely used as a standard test case for the investigation of the material resistance to cavitation erosion (mass loss curves) [2]. If the horn tip diameter is sufficiently small and driven at high amplitude with ultrasonic frequency, large cavitation clouds are generated, which cover the complete horn tip surface. These clouds oscillate at their own frequency falling into a subharmonic range referred to as the acoustic excitation frequency [3]. Znidarcic et al. [3] denominate this cavitation type as 'acoustic supercavitation' as it exhibits acoustic as well as hydrodynamic characteristics.

Different authors (Dular et al. [4] and Znidarcic et al. [5]) have figured out that standard hydrodynamic cavitation models (mass transfer models based on the Rayleigh equation) - due to the simplifications of the Rayleigh equation - cannot correctly predict the vapour volume and subharmonic oscillation frequency compared to experiments. Therefore, Znidarcic et al. [5] have developed a new mass transfer model especially for the simulation of 'acoustic supercavitation', which takes into account the second derivative term of the Rayleigh-Plesset equation. Applying this model for the simulation of acoustic supercavitation, accurate results have been achieved in terms of attached cavity oscillation frequency, vapour volume and pressure pulsations.

In contrast to Znidarcic et al. [5] we apply a compressible density-based CFD flow algorithm for the simulation of the ultrasonic horn flow. This method does accurately capture compressible wave dynamics, which is directly associated with the occurrence of cavitation erosion [6]. For the investigation of the erosion sensitive areas a statistical analysis of transient wall loads is used, which has been successfully applied in the past to hydrodynamic cavitation in the case of micro channels [7] and hydrofoils [8]. Furthermore, a statistical analysis of single collapse events in wall-near flow regions is utilised to evaluate flow aggressiveness, which has been developed by Mihatsch et al. [1] and applied to axis-symmetric and planar nozzles [1, 9] (hydrodynamic cavitation).

In the present study the simulation results of the density-based flow solver and the erosion analysis procedure are compared to measurement data (vapour amount, pressure pulsations, cavity oscillation frequency and eroded sample surfaces) in order to:

- validate our in-house solver hydRUB for 'acoustic supercavitation',
- verify the erosion analysis procedure for application in acoustically induced cavitation.

## 2. Test cases

Two horn tip set-ups are investigated:

- **case A**: tip diameter of 16 mm with counter sample in 500 µm distance; driving frequency: 19.82 kHz; peak-to-peak amplitude: 35.8 µm.
- **case B**: tip diameter of 3 mm; driving frequency: 20 kHz; peak-to-peak amplitude: 164 µm.

Case A corresponds to the standard erosion test case where a counter sample is added below the horn tip sample [2]. The tip amplitude and the driving frequency are defined from in-house far-field microscopy imaging. Erosion sensitive areas are determined from surface pictures and confocal microscopy images of the horn tip sample [10]. For the assessment of erosion sensitive areas from our simulation results, we assume that the region of highest erosion probability corresponds to the location with the highest material removal. From previous numerical studies [10] it can be deduced that the total gap is filled with a large vapour cloud, which pulsates on subharmonic time scales. Accordingly, we assume case A - despite the fact that it has a larger tip diameter and is driven at medium amplitude - to induce 'acoustic supercavitation'.

Case B is based on the set-up of Znidarcic et al. [5] who have investigated the vapour volume structure and vapour dynamic with high-speed shadowgraphy images as well as the acoustic pressure propagation with a hydrophone (7 mm distance to the horn tip).

## 3. Method

For a detailed description of the numerical and physical model as well as the erosion analysis procedure we refer to previous publications [10, 11].

### *3.1. Numerical and physical model*

Our in-house solver hydRUB [7, 8] is applied which has been extended by a compressible inviscid low Mach-number consistent Godunov-type flux formulation [12] and an optimised explicit Runge-Kutta scheme [12]. Velocity and density are reconstructed by the SMART [13] and MINMOD [14] scheme, respectively. A homogeneous mixture of liquid and vapour in equilibrium (un-dissolvable gas is neglected) is assumed. A barotropic equation of state with an isentropic phase change [15] is utilised.

### *3.2. Erosion analysis procedure*

For the evaluation of erosion-sensitive areas we use a statistical analysis of transient wall loads for two combined erosion indicators (wall pressure and condensation rate) referred to as erosion probability [10].

Furthermore, the 'Collapse Detector' by Mihatsch et al. [9] is applied, which is based on a set of physical criteria to detect collapses of isolated vapour clouds and characterises the strength of the collapse pressure. In order to reduce the grid dependency of the collapse pressure and the event rate, scaling laws are utilised, which are based on the linear decay law of spherical waves [1]. Collapse pressure and event rate are scaled by the ratio of characteristic grid spacing and a uniform reference length scale $x_{ref}$, and the event rate is further corrected by a power law with exponent $\kappa = 2/3$. In [1], $\kappa = 2/3$ is proposed based on a hydrodynamic cavitation test case.

### *3.3. Numerical set-up and verification*

Slip wall boundary conditions are applied, and the horn tip surface is moving with an oscillating grid kinematics. The free surface is modelled as symmetry (a small part is modelled as Dirichlet pressure boundary and therefore fixes the pressure level to 1 bar) and accounts for pressure wave reflection. Water at room temperature is used. The flow region around the ultrasonic horn is meshed with a double O-grid topology. For test case A a grid sensitivity study on three successively refined grids (factor 2 in each direction) is performed. The total number of cells per gap are: 21600 (coarse), 172800 (middle), 1382000 (fine). We find that the erosion probability on all grids features the same erosion sensitive areas on the tip: minimum erosion probability at the edge, maximal erosion probability at a ring surface and nearly continuous erosion probability in the middle of the sample, which is in excellent agreement with the material removal at the surface pictures (cf. figure 1). Furthermore, the same vapour dynamic - subharmonic oscillation of the vapour cloud inside the gap at every 10th driving cycle - is predicted on all grids (not shown).

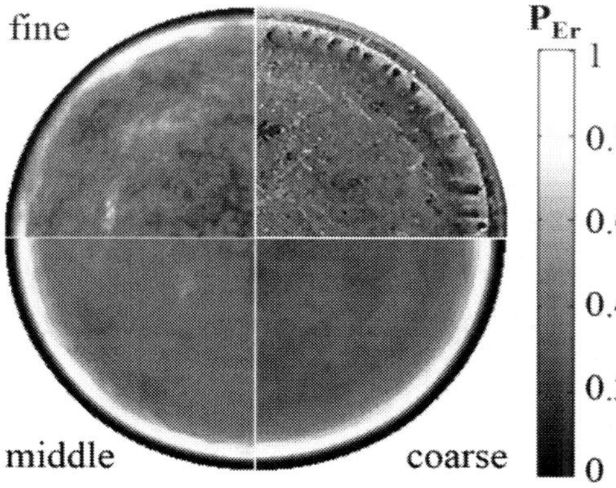


**Figure 1.** Erosion probability at the horn tip for different grid resolutions compared to an erosion surface picture.

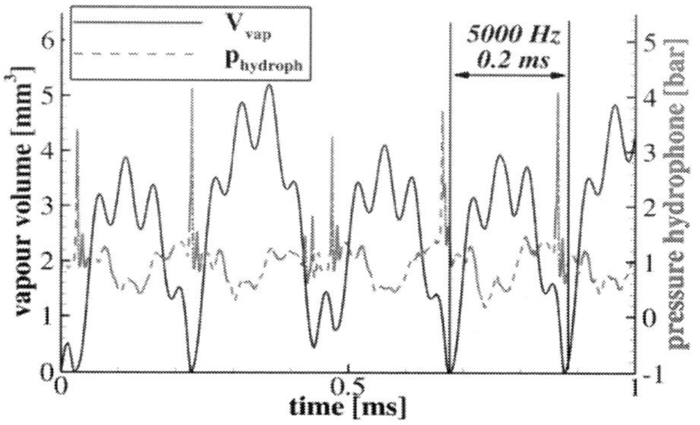


**Figure 2.** Temporal evolution of the virtual hydrophone pressure $p_{hydroph}$ and vapour volume $V_{vap}$ from the CFD simulation of test case B.

In order to evaluate temporal convergence behaviour the maximal deviation as well as the standard deviation of the erosion probability $P_{Er}$ (perimeter averaged) are compared between different analysis time intervals (not shown here). We conclude, that erosion sensitive areas can be reliably predicted after very short simulation time intervals (horn tip: $T_{analyse} = 2\,\text{ms} \,\hat{=}\, 40$ driving cycles) and that a temporally converged solution is reached at the horn tip sample six times faster than at the counter sample. Reason is the more frequent loading at the horn tip sample, as it is stressed at every driving cycle (partial collapses at the edge of the vapour cloud), whereas the counter sample is mainly stressed after each subharmonic cycle (every 10th cycle).

## 4. Results

### *4.1. Flow aggressiveness (case A)*

The 'Collapse Detector' is evaluated for wall-adjacent (distance $\leq 100\,\mu\text{m}$) flow regions at the entire horn tip and counter sample (cf. figure 3). The load collectives show an exponential decrease. For a scaling exponent of $\kappa = 2/3$ grid dependence of the load collective at the horn tip can essentially be eliminated (figure 3a), whereas at the counter sample, a considerable grid dependence remains, in particular for high collapse pressure (not shown). A significant reduction of grid dependence at the counter sample for high loads is obtained by a $\kappa = 3/2$ (figure 3b). However, although not significant for erosion prediction, the results for low loads remain grid-dependent for $\kappa = 3/2$. Furthermore, a comparison of loads between tip and counter sample is no longer possible due to the different values of $\kappa$. In particular, the extrapolation to even higher loads (trend lines in figure 3) suggest a higher load and therefore higher erosion sensitivity at the counter sample than on the tip which is a contradiction to experimental findings. We conclude that there is no unique value of $\kappa$, and it may even vary within the counter sample area.

### *4.2. Cavitating flow data (case B)*

In agreement to the flow measurements from Znidarcic et al. [3], we observe a large attached vapour cloud oscillating on subharmonic time scales and changing its shape from mushroom

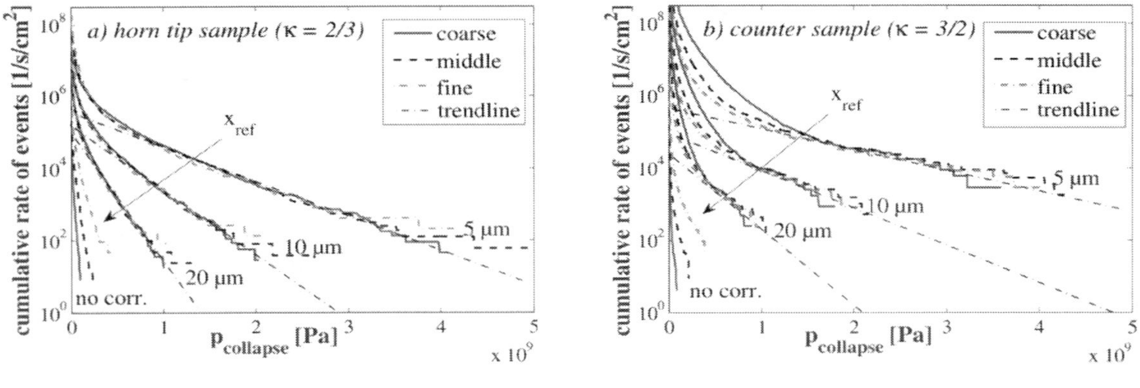


**Figure 3.** Load collectives (cumulative event rate vs collapse pressure) for the near wall region (100 µm) of the horn tip (a) and the counter sample (b). Un-corrected and corrected (according to [1]) load collectives as well as high-load trend lines are included.

to cone shape until collapse. The measured subharmonic frequency of 5050 Hz (1/4 of driving frequency) can be well reproduced (cf. figure 2). The peak pressure at the virtual hydrophone is predicted in the same range as in the experiment (figure 2, $2\,\text{bar} \leq p_{hydroph} \leq 4\,\text{bar}$ [3]). However, the produced vapour volume is predicted too low (ca. 30 %) compared with experimental data. We suppose the discrepancy appears due to the experimental procedure of determining vapour volume using shadowgraphy images (fixed brightness threshold; assume rotational symmetric vapour volume; neglect of inside liquid phase). This assumpton is confirmed, as the analogue numerical procedure via silhouettes of iso-surfaces (vapour volume fraction threshold: 10 %) predicts the vapour volume in better agreement to the experiment (not shown here).

## 5. Conclusions

We conclude that the compressible flow solver with barotropic cavitation model can well reproduce the main features of 'acoustic supercavitation' (attached cavity oscillation frequency, vapour amount and structure as well as pressure pulsations). Erosion sensitive areas can reliably be predicted. An exponential decrease of the load collective is also found for 'acoustic supercavitation'. For grid independent load collectives, different scaling exponents $\kappa$ have been figured out for the horn tip and counter sample, so that the choice of $\kappa$ seems to be case dependent and needs a deeper analysis which is on-going.

## References

[1] Mihatsch M S, Schmidt S J, Thalhamer M and Adams N A 2012 *Proc. 8th Int. Symp. on Cavitation* (Singapore)
[2] Pohl M and Stella J 2002 *Wear* **252** pp 501-11
[3] Znidarcic A, Mettin R, Cairos C and Dular M 2014 *Physics of Fluids* **26** 023304
[4] Dular M, Mettin R, Znidarcic A and Truong V A 2012 *Proc. 8th Int. Symp. on Cavitation* (Singapore)
[5] Znidarcic A, Mettin R and Dular M 2015 *Ultrasonics Sonochemistry* **22** pp 482-92
[6] Schmidt S J, Sezal I H, Schnerr G H and Thalhamer M 2007 *Proc. 8th Int. Symp. Experimental and Computational Aerothermodynamics of Internal Flows* (Lyon)
[7] Skoda R, Iben U, Morozov A, Mihatsch M S, Schmidt S J and Adams N A 2011 *WIMRC Proc. 3rd Int. Cavitation Forum* (Warwick)
[8] Skoda R, Iben U, Güntner M and Schilling R 2012 *Proc. 8th Int. Symp. on Cavitation* (Singapore)
[9] Mihatsch M S, Schmidt S J, Thalhamer M, Adams N A, Skoda R and Iben U 2011 *WIMRC Proc. 3rd Int. Cavitation Forum* (Warwick)
[10] Mottyll S, Mueller S, Niederhofer P, Hussong J, Huth S and Skoda R 2014 *EPJ Web of Conferences* **67** 02078
[11] Deimel C, Mottyll S, Skoda R, Kiermeir J, Güntner M and Schilling R 2014 *Proc. 8th Int. Conf. Computational Fluid Dynamics* (Chengdu)
[12] Schnerr G H, Sezal I H and Schmidt S J 2008 *Physics of Fluids* **20** 040703
[13] Gaskell P H and Lau A K C 1988 *Int. J. Numerical Methods in Fluids* **8** pp 617-41
[14] Harten A 1983 *J. Computational Physics* **49** pp 357-93
[15] Iben U 2002 *Systems Analysis Modelling Simulation* **42** pp 1283-307

# Modeling Fluid-Structure Interaction in Cavitation Erosion: Preliminary Results

**Yves Paquette[1], Christian Pellone[1], Marc Fivel [2], Giovanni Ghigliotti[1], Eric Johnsen[3], Jean-Pierre Franc[1]**

[1]LEGI, Univ. Grenoble Alpes, CNRS, F-38000 Grenoble, France
[2]SIMAP, Univ. Grenoble Alpes, CNRS, F-38000 Grenoble, France
[3]Mechanical Engineering Department, University of Michigan, Ann Arbor, MI 48109, USA

E-mail: Yves.Paquette@legi.grenoble-inp.fr

**Abstract.** This paper is devoted to cavitation erosion modeling. It presents some recent numerical developments made in the code initially developed in collaboration with E.Johnsen and collaborators at University of Michigan [1] in order to account for fluid-structure interaction. The considered test case is that of a single air bubble collapsing near a wall due to an incident shock wave in the surrounding water. In our investigation, we focus on the code implementation and optimization and the bubble implosion mechanism. The paper is focused on the various events occurring during bubble collapse and the computation of the time evolution of the pressure distribution. The influence of the amplitude of the incident wave and the distance of the bubble to the wall are investigated.

## 1. Introduction

Cavitation phenomena and erosion prediction are not totally understood. Recent efforts have been made in order to compute as accurately as possible the collapse of a single bubble [1], [2] or of a cloud of bubbles that is supposed to generate cavitation erosion on neighboring walls [3]. Experimental results are also available and new developments in measurements techniques provide experimental data about single bubble dynamics and the pressure field due to bubble collapse [4], [5]. To improve the prediction technique, 2D simulations using a high-accuracy CFD code have been made, that are able to capture shock waves with high reliability in particular due to bubble implosion. These simulations are able to compute the wall load with great accuracy that will be passed to a solid mechanics solver to investigate the effect of the fluid / structure interaction on the cavitation erosion process. To achieve this goal, we have developed a 2D CFD code for a single bubble collapse jointly with University of Michigan. This code allows us to compute the pressure signal due to bubble collapse, that is presented and discussed in the present paper.

## 2. CFD Code FSI4CAV

The CFD two-dimensional code FSI4CAV (Fluid Structure Interaction for CAVitation) is developed jointly by University of Michigan [1] and LEGI and SIMAP Laboratories of Grenoble University. The main goal is to model a single cavitation bubble that collapses close to a wall including the shock waves generated during the collapse.

### *2.1. Physical Modeling*

The code is based on four conservation equations [1]: mass, momentum (2 equations in 2D) and energy, completed by the Fourier law for heat transfer, the expression of the viscous stress and an equation of state for both gas and liquid [6]. To solve these equations, a fifth-order accurate finite volume WENO reconstruction is used for spatial discretization [7], and the HLL Rieman solver [8] is used to compute the fluxes at each cell interface for each conservation equation. Viscous stress and thermal conductivity are computed directly from the respective equations. Time marching is made with a fourth-order accurate Runge-Kutta scheme.

In the initial version of the code [1], the mesh was cartesian with uniform spacing. Cell nodes could not be changed during simulation, which makes it challenging to account for the deformation of the wall.

The code was modified in order to accommodate the deformation of the wall so that cell size is no longer constant. Then cells, and particularly cells close to the wall, can be deformed, and in our case this is due to the high pressure caused by the bubble collapse applied to the compliant wall. This space dependency required many modifications in the WENO process since all formulas will now depend on cell size.

The code simulates the dynamics of two fluids: water and air, without mass transfer and phase change. The same system of equations is solved in each cell for both water and air. The calculation starts with a single air bubble in water near a wall at atmospheric pressure. To make the bubble collapse, a shock wave is generated at infinity that moves towards the bubble and is reflected by the wall. A typical behavior of the bubble collapse is shown in figure 1.

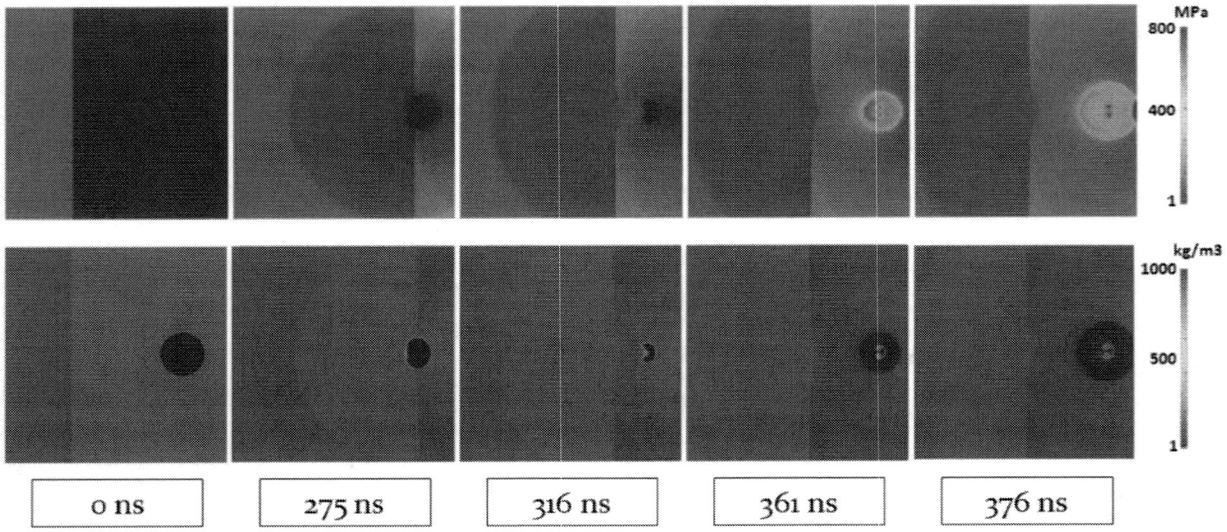


**Figure 1.** Bubble collapse during simulation. Above: pressure field, below: density.

After being hit by the shock wave, the bubble starts collapsing around 275ns. As expected, a micro-jet develops in direction of the wall. As a result, the bubble is split into two parts and the micro-jet hits the wall. In this paper, we will focus specially on the pressure increase on the wall due to the jet impact.

### *2.2. Pressure Signal*

We consider here the case of a bubble of $50\mu m$ in initial radius, the bubble interface is at $50\mu m$ from the wall and there is a resolution of $0.25\mu m$ per cell. Figure 2 presents the time evolution of pressure on the wall, where the shock wave due to the bubble collapse hits the wall, compared to the time evolution of the maximum pressure in the whole domain.

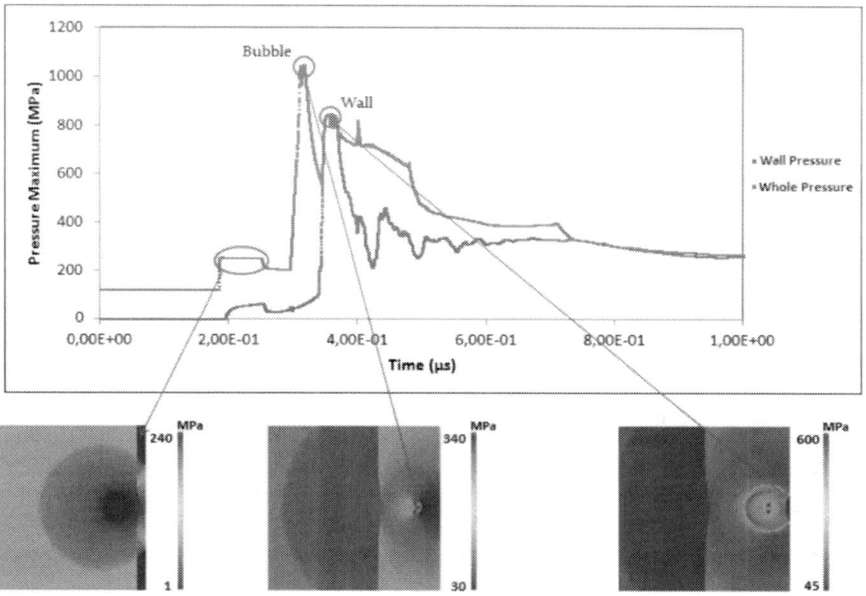


**Figure 2.** Pressure signal during bubble collapse.

Three peaks are visible on the maximum pressure in the whole domain (blue curve). The first one is due to the incident wave. This wave hits first the top-right and bottom-right borders of the domain. At the top and the bottom of the domain, the incident wave crosses only water, while in the middle part of the domain, it crosses air where propagation speed is lower. That is why, top right and bottom right borders are hit first by the incident wave. The maximum pressure is twice the pressure of the incident wave because of cumulation of the incident and the reflected wave. There is also a pressure increase in the middle part of the right border (red signal) which is however attenuated by the crossing of the bubble. The second peak is related to the microjet that develops during the collapse of the bubble. Just before the bubble collapses, the microjet hits the most right interface of the bubble and generates a shock wave associated to the second pressure peak. This shock wave will further hit the wall and cause the third peak visible on both pressure signals in Figure 2. This last peak is the most interesting for coupling, it corresponds to the maximum load applied to the material during the bubble collapse.

### *2.3. Incident Shock Wave dependence*

Various simulations were conducted in order to investigate the influence of (i) the magnitude of the incident pressure shock wave and (ii) the distance between the bubble right interface and the wall. The following results were obtained with a resolution of $0.50\mu m$ per cell.

(i) The pressure magnitude of the incident wave varies between 100 atm and 1200 atm. We focus on the collapse duration and the overpressure ratio between the maximum pressure on the wall and the pressure of the incident wave.

| Incident Wave (MPa) | 100 | 200 | 600 | 1200 |
|---|---|---|---|---|
| Collapse Time (ns) | 772 | 405 | 289 | 203 |
| Overpressure ratio | 6.0 | 6.2 | 5.4 | 4.8 |

**Table 1.** Dependence on incident wave.

As expected, the collapse time decreases while the pressure of the incident wave increases. It is due to the fact that the velocity of the microjet increase with the amplitude of the incident wave. As for the overpressure ratio, Table 1 shows that this ratio decreases when the incident wave pressure increases. However this ratio is still in the same order of magnitude: between 5 and 6 times the incident pressure.

(ii)Then, considering an incident wave of 1200 atm, the distance of the bubble to the wall was changed between 1 bubble radius and 0. The closer the bubble is to the wall, the higher the impact pressure on the wall. This is partly due to the fact that the attenuation of the shock wave becomes more important when the interface moves away from the wall.

| Distance interface-wall (bubble radius) | 0 | 0.25 | 0.5 | 1 |
|---|---|---|---|---|
| Pressure maximum on wall (MPa) | 1430 | 1100 | 916 | 822 |

**Table 2.** Dependence on distance interface-wall.

## 3. Conclusion and perspectives

The code 2D FSI4CAV was developed in order to model the fluid-structure interaction for cavitation erosion prediction purposes. This code allows us to compute the different events occurring during a single bubble collapse. It was shown that the maximum pressure in the whole domain presents three successive peaks. The first peak is due to the impact of the incident shock wave on the wall. The second one is due to the impact of the microjet on the bubble interface. It appears within the fluid. The shock wave that results from this impact propagates to the wall and generates a third peak when impacting the wall. The amplitude of this peak is significantly larger than the magnitude of the incident way and the ratio varies between 5 and 6 for the different cases investigated in this paper. The pressure distribution will now be passed to a solid code in order to model the wall deformation and an iterative procedure will be developed in order to update the computational domain after deformation.

## Acknowledgments

This work has been partially supported by the LabEx Tec 21 (Investissements d'Avenir - grant agreement nANR-11-LABX-0030) and ONR grant N00014-12-1-0751.

## References

[1] S. A. Beig, E. Johnsen, Maintaining interface equilibrium conditions in compressible multiphase flows using interface capturing 2015, submitted *J. of Comp. Phys.*

[2] E. Johnsen, T. Colonius, Numerical simulations of non-spherical bubble collapse 2009 *J. Fluid. Mech.* **629** 231-262

[3] K.-H. Kim, G. Chahine, J.-P. Franc,A. Karimi, 2014 *Advanced Experimental and Numerical Techniques for Cavitation Erosion Prediction*,Vol. 106, Springer, Netherlands, 345-372

[4] Y.-C. Wang, Y.-W. Chen, Application of piezoelectric PVDF film to the measurement of impulsive forces generated by cavitation bubble collapse near a solid boundary 2007 *Exp. Thermal and Fluid Science* **32** 403-414

[5] A. Jayaprakash, C.-H. Hsiao, G. Chahine, Numerical and Experimental Study of the Interaction of a Spark-Generated Bubble and a Vertical Wall 2012 *J. of Fluids Eng.* **134**

[6] O. Le Métayer, J. Massoni, R. Saurel, Modelling evaporation fronts with reactive Riemann solvers 2005 *J. of Comp. Phys.* **205** 567-610

[7] X. D. Liu, S. Osher, T. Chan, Weighted Essentially Non-Oscillatory Schemes 1994 *J. of Comp. Phys.* **115** 200-212

[8] E.F. Toro, M. Spruce and W. Speares, Restoration of the contact surface in the HLL-Riemann solver 1994 *Shock Waves* **4** 25-34

# Numerical Modelling and Prediction of Erosion Induced by Hydrodynamic Cavitation

**A. Peters**[*1], **U. Lantermann**[*], **O. el Moctar**[*]

[*] Institute of Ship Technology, Ocean Engineering and Transport Systems, University of Duisburg-Essen, Bismarckstr. 69, 47057 Duisburg, DE

E-mail: andreas.peters@uni-due.de

**Abstract.** The present work aims to predict cavitation erosion using a numerical flow solver together with a new developed erosion model. The erosion model is based on the hypothesis that collapses of single cavitation bubbles near solid boundaries form high velocity microjets, which cause sonic impacts with high pressure amplitudes damaging the surface. The erosion model uses information from a numerical Euler-Euler flow simulation to predict erosion sensitive areas and assess the erosion aggressiveness of the flow. The obtained numerical results were compared to experimental results from tests of an axisymmetric nozzle.

## 1. Introduction

In ship technology and maritime environments high flow velocities cause regions of low fluid pressure, which lead to the generation of vapour structures. Once these vapour structures reach regions of higher pressure, they start to rapidly collapse and are able to cause efficiency losses, vibrations and erosion.

An approach to predict cavitation erosion by [1] refers to the hypothesis of potential energy of a macroscopic vapour structure being converted into acoustic energy of pressure waves, which damage a surface directly. Krumenacker et al. [2] supposed that high values of $\partial p/\partial t$ initiate erosive bubble collapses. They state that erosion depends on the acoustic energy of pressure waves. This energy is calculated using a hybrid approach of a Reynolds-Averaged Navier-Stokes (RANS) method combined with a solver to calculate the dynamics and acoustic energies of single bubbles. Nohmi et al. [3] suggested a formula to numerically predict the aggressiveness of a flow depending on the local vapour volume, pressure and their temporal derivatives. Li [4] states that the driving erosion mechanism is the collapse of a cloud cavitation and developed a numerical erosion model, where erosion potential is qualitatively assessed by looking at the time derivative of pressure.

In a different hypothesis it is supposed that only cavitation bubbles in the direct vicinity of a surface lead to erosion. Kato et al. [5] stated an erosion model, which refers to the hypothesis, that single bubbles, which are shed from sheet cavitation cause erosion, when collapsing near a surface. A similar model was developed by Dular et al. [6]. The authors suppose that single bubbles oscillate and collapse due to the radiation of pressure waves through the fluid. It is suggested that erosion is caused once these bubbles collapse near a surface and form a liquid

[1] Corresponding author

water jet of high flow velocity, the so called *microjet*. At impact on the surface, high pressures are generated by shock impacts that are able to damage the surface. Peters et al. [7] have done further development of this erosion model to predict erosion based on the near wall behaviour of cavitation. An erosion model could be developed, which pedicts erosion based on the amount of impacts in a certain area, as well as the impact intensities.

## 2. Numerical Method

The flow is simulated using an implicit, pressure-based flow solver from the open source CFD package OpenFOAM. A two phase Euler-Euler flow is simulated with the Volume of Fluid (VoF) method to track the phase's interfaces. The equations of conservation of mass and momentum are solved for the homogeneous mixture of the two continuous phases. The mixture in each numerical control volume is obtained from the vapour volume fraction $\alpha$, which relates the vapour volume in a control volume to the total volume of the cell. The transport equation for the volume fraction reads:

$$\frac{\partial \alpha}{\partial t} + \frac{\partial \alpha u_i}{\partial x_i} = S_\mathrm{e} - S_\mathrm{v} \,. \tag{1}$$

$u_i$ is the velocity in coordinate direction $x_i$ and $t$ is the time. $S_\mathrm{e}$, $S_\mathrm{v}$ are the source terms due to condensation and evaporation processes, respectively, which are given by a cavitation model. In this work, the cavitation model by Schnerr and Sauer [8] was used:

$$\begin{aligned} S_\mathrm{e} &= \frac{\rho_\mathrm{v}\rho_\mathrm{l}}{\rho}\frac{3\alpha(1-\alpha)}{R_\mathrm{b}}\sqrt{\frac{2}{3}\frac{p_\mathrm{b}-p}{\rho_\mathrm{l}}}\,, \quad \text{for } p < p_\mathrm{v}\,, \\ S_\mathrm{c} &= \frac{\rho_\mathrm{v}\rho_\mathrm{l}}{\rho}\frac{3\alpha(1-\alpha)}{R_\mathrm{b}}\sqrt{\frac{2}{3}\frac{p-p_\mathrm{b}}{\rho_\mathrm{l}}}\,, \quad \text{for } p \geq p_\mathrm{v}\,. \end{aligned} \tag{2}$$

$\rho_\mathrm{v}$ and $\rho_\mathrm{l}$ are the densities of vapour and liquid, respectively, and $\rho$ is the density of the mixture. $R_\mathrm{b}$ is the radius of a cavitation bubble and $p_\mathrm{b}$ the inner bubble pressure. $p_\mathrm{v}$ is the vapour pressure. The cavitation model is based on a simplified Rayleigh-Plesset equation for bubble dynamics. To enable the simulation of unsteady sheet and cloud cavitation, a turbulence correction is applied, which considers the reduction of turbulent kinetic energy due to high compressibility in the mixture region.

## 3. Erosion Model

The developed erosion model is based on the microjet hypothesis [6] and was derived as explained in [7]. An erosion is supposed to happen, when two conditions are fulfilled: 1. It is possible that a microjet process takes place near the surface. 2. The velocity of the microjet needs to exceed a critical velocity, to damage a given material. The microjet velocity can be related to the pressure caused by a microjet, while the critical velocity depends on the stress needed to plastically deform the regarded material. For a dimensionless stand-off distance of $\gamma = 1.1$ the local jet velocity can be estimated by the following formula, as written in [7]:

$$v_\mathrm{jet} \approx 10.8\sqrt{\frac{p-p_\mathrm{v}}{\rho_\mathrm{l}}}\,. \tag{3}$$

An expression for the critical velocity velocity can be derived from a one-dimensional impact of a liquid mass onto a solid boundary [9]. The Tait equation was used to relate pressure and density of liquid water. The formula for the critical velocity is given as:

$$v_\mathrm{crit} = \sqrt{\frac{p_\mathrm{y}}{\rho_\mathrm{l}}\left(1-\left(1+\frac{p_\mathrm{y}}{B}\right)^{-1/n}\right)}\,. \tag{4}$$

Journal of Physics: Conference Series **656** (2015) 012054 doi:10.1088/1742-6596/656/1/012054

$B = 300$ MPa and $n = 7$ are standard coefficients for liquid water in the Tait equation of state. To evaluate the generation of microjets on a surface, the presence of single cavitation bubbles in the vicinity of the solid boundary needs to be predicted. Therefore, the control volumes near a face of the surface are checked for a minimum volume of vapour. Afterwards, it is evaluated, whether the local jet velocity is higher than the critical velocity. When both conditions are fulfilled, an erosion impact for this time step on the regarded face is supposed to happen. For each impact, an intensity coefficient is calculated:

$$c_{\text{intensity}} = \frac{v_{\text{jet}}}{v_{\text{crit}}} . \tag{5}$$

With the given method, it is difficult to quantify the amount of bubbles of different shapes and to estimate, which bubbles are able to form erosive microjets. A dimensionless deformation coefficient is therefore introduced to predict erosion qualitatively. It is assumed that the number of impacts, as well as their intensities, are the main factors to assess erosion potential. For each face of a regarded surface, the number of impacts, weighted with their intensity coefficients, is summarized and related to the total erosion impacts. The deformation coefficient is introduced, which is a measure of the erosion taking place in one face, compared to the total erosion:

$$c_{\text{def}} = \frac{\sum_t^T c_{\text{intensity},t}}{\sum_n^N (\sum_t^T c_{\text{intensity},t})_n} . \tag{6}$$

Here, $t$ is the time step index and $T$ the total calculated time of the erosion model. $n$ is the index of a face and $N$ the total number of all eroded faces.

## 4. Results

The flow through an axisymmetric nozzle was investigated by Franc et al. [10, 11]. Water is flowing through a cylinder of 8 mm radius from top to bottom. The cylinder is connected to a radial divergent outflow part via a 1 mm radius. Cavitation is generated at the radius and travels further downstream. The cavitation structures mostly collapse in specific regions, which leads to a characteristical ring of erosion on the bottom target plate, within a given distance from the rotational axis. Figures 1 and 2 show a momentary snapshot of the numerical simulation of internal flow through the nozzle. Cloud and sheet cavitation collapse within a characteristic distance from the radial axis. This leads to the high erosion potential predicted on the bottom boundary in this region.

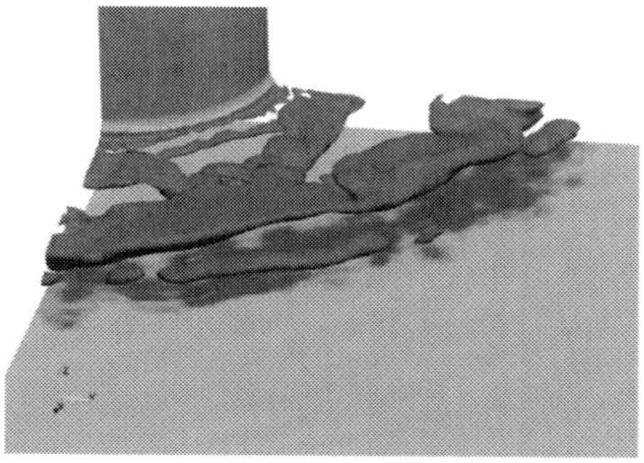

**Figure 1.** Cavitation structures travelling downstream from the radius. The predicted erosion on the bottom boundary is marked in red.

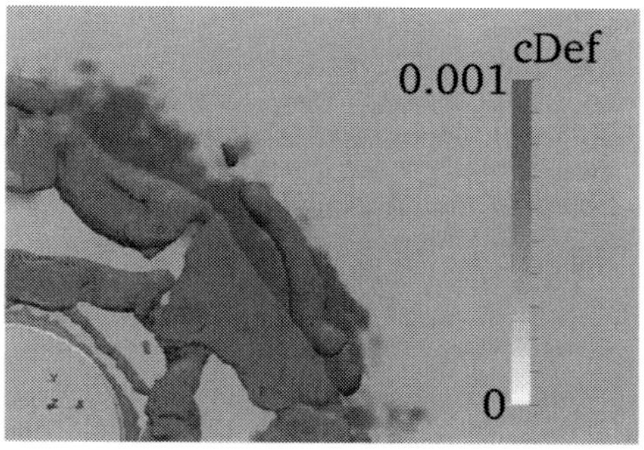


**Figure 2.** Top view of the correlation between cavitation and predicted erosion.

The numerical erosion prediction is compared to the experimental results from top view in Figure 3. The eroded target sample from the experiment is shown on the left and the numerical erosion prediction on the bottom boundary is shown on the right. The black circle marks the area, where the highest erosion damage was measured during experiments [10, 11]. It is apparent that the numerical method is well able to identify the area of highest erosion potential.

A statistical comparison of experiment and simulation is shown in Figure 4. To enable a qualitative comparison, the dimensionless damage distributions are shown, which were normalized by the maximum values of predicted erosion (simulation) and surface deformation (experiment), respectively. The numerical erosion model is able to give a good prediction of the position of maximum erosion potential, but predicts a larger total area to be eroded.

## 5. Conclusions

The developed erosion model predicts erosion based on the microjet hypothesis stated by Dular et al. [6]. The numerical method is able to simulate cavitating flows and predict erosion based on the present flow conditions. A comparison of the numerical prediction and the experimental results shows good agreement.

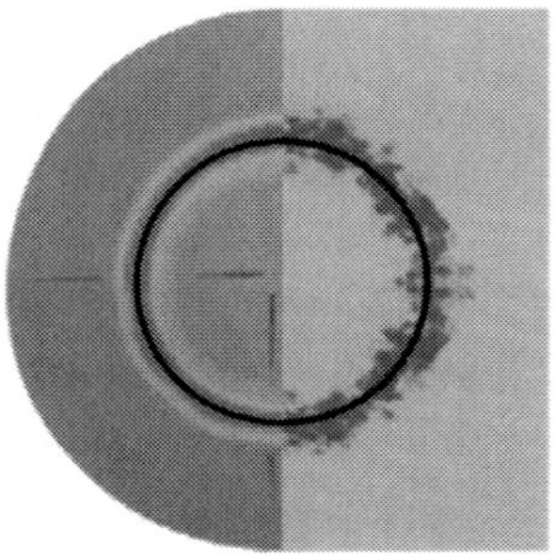

**Figure 3.** Top view of eroded target sample from experiment (left) and the numerical erosion prediction (right).

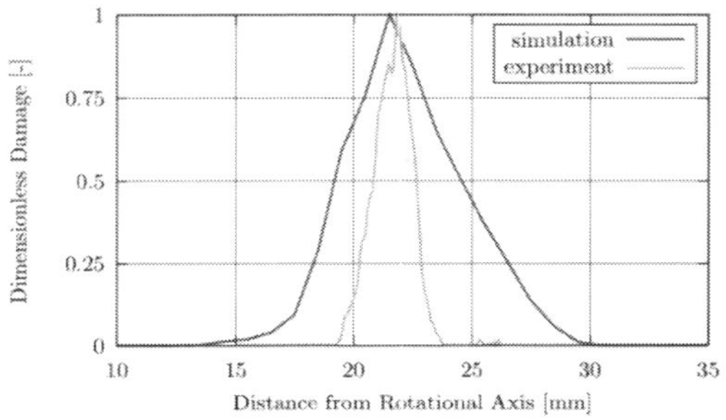


**Figure 4.** Comparison of dimensionless damage distributions from experiment and simulation.

## 6. Acknowledgement

The present study originates from the project KonKav III, which is funded by the development program "Shipping and Marine Technology for the 21st Century" of the Federal Ministry of Economic Affairs and Energy.

[1] Fortes-Patella R, Reboud J L and Briancon-Marjollet L 2004 *EROCAV Workshop, Val de Reuil, France*
[2] Krumenacker L, Fortes-Patella R and Archer A 2014 *27th IAHR Symposium on Hydraulic Machinery and Systems*
[3] Nohmi M, Iga Y and Ikohagi T 2008 *Proceedings of FEDSM 2008, Jacksonville, Florida, USA*
[4] Li Z R 2012 *Assessment of Cavitation Erosion with a Multiphase Reynolds-Averaged Navier-Stokes Method* PhD Thesis Delft University of Technology Delft, Netherlands
[5] Kato H, Konno A, Maeda M and Yamaguchi H 1996 *Journal of Fluids Engineering* **118** 582–588
[6] Dular M and Coutier-Delgosha O 2009 *International Journal for Numerical Methods in Fluids* **61** 1388–1410
[7] Peters A, Sagar H, Lantermann U and el Moctar O 2015 *Wear* **338–339** 189–201
[8] Sauer J and Schnerr G H 2000 *Proceedings of FEDSM 2000, 4th Fluids Engineering Summer Conference, New York, NY, USA* pp 1073–1079
[9] Lush P A 1983 *Journal of Fluid Mechanics* **135** 373–387
[10] Franc J P and Riondet M 2006 *Proceedings of the 6th International Symposium on Cavitation, CAV2006, Wageningen, Netherlands*
[11] Franc J P, Riondet M, Karimi A and Chahine G 2011 *Journal of Fluids Engineering* vol 133

# DES Prediction of Cavitation Erosion and Its Validation for a Ship Scale Propeller

**Dr Dmitriy Ponkratov**

Senior Consultant, Lloyd's Register, Technical Investigation Department, London, UK

E-mail: Dmitriy.Ponkratov@lr.org

**Abstract**. Lloyd's Register Technical Investigation Department (LR TID) have developed numerical functions for the prediction of cavitation erosion aggressiveness within Computational Fluid Dynamics (CFD) simulations. These functions were previously validated for a model scale hydrofoil and ship scale rudder [1]. For the current study the functions were applied to a cargo ship's full scale propeller, on which the severe cavitation erosion was reported. The performed Detach Eddy Simulation (DES) required a fine computational mesh (approximately 22 million cells), together with a very small time step (2.0E-4 s). As the cavitation for this type of vessel is primarily caused by a highly non-uniform wake, the hull was also included in the simulation. The applied method under predicted the cavitation extent and did not fully resolve the tip vortex; however, the areas of cavitation collapse were captured successfully. Consequently, the developed functions showed a very good prediction of erosion areas, as confirmed by comparison with underwater propeller inspection results.

## 1. Introduction

Undoubtedly, the identification of cavitation aggressiveness at the design stage has significant value to ship-owners, as the list of potential solutions at this point is much wider, compared to a limited list of actions when the ship has already been built. The treatment of the problem at the design stage is considerably cheaper, compared to "when built" troubleshooting.

Apart from traditional model experiments [2], it has now become practical to use Computational Fluid Dynamic (CFD) methods to predict cavitation and erosion due to the development in computational power. As summarised in [1], two main approaches can be used for the numerical prediction of cavitation aggressiveness.

One of these can be regarded as complete numerical resolution of the cavitation problem [3], [4], [5] and [6]. As the simulation of shock waves induced by cavity collapses requires very fine mesh, together with a time step in the order of 1 micro second, this approach is initially applicable for model scale objects. When the simulation of marine hydrodynamics is required, such as ship propellers and marine structures, this approach would be impractical, as very fine meshes around these objects would be required, resulting in cell counts in the order of hundreds of millions.

As a practical solution to address these problems, numerical erosion functions have recently been introduced by various researchers. One of the first studies was the development of erosive indexes in [7] and [8]. The validation of these indexes showed good results for a propeller in model scale. Further investigations in the development of erosive functions were completed in [9], [10] and [11], where analysis of existing functions was also performed.

Erosion functions have also been developed by LR TID in [12] and validated for a ship rudder in model scale. As cavitation simulations require a fine mesh, some simplifications had to be introduced in [12]; the real hull and propeller geometry were not included in the calculation and were taken into account by modifying the flow upstream of the rudder.

Results presented by LR TID in [1] were shown to be more accurate, as the realistic geometry of the hull and rotating propeller in ship scale was directly taken into account. A Detach Eddy Simulation (DES) was used, which required a fine mesh consisting of 22 million cells.

LR TID is aware of many cases where ship rudders suffered from cavitation erosion, however, the main concern of designers and ship-owners is aggressive cavitation on propellers, inducing extensive vibration and erosion problems, which cannot be treated easily. This concern has motivated the further study of developed erosion functions for a ship scale propeller completed in the current paper. A cargo vessel was selected for the subject of the study. After approximately 6 months of service, extensive erosion damage on all propeller blades was reported, as shown in **Figures 3** and **6**. LR TID performed borescopic cavitation observations on this vessel and recorded the development of cavitation extent on the propeller. Some results of the observations are presented in **Figure 2**.

**Figure 1**. Full scale vessel simulated by DES and the surface mesh on propeller, rudder and hull.

## 2. Case set up

As the cavitation on the propeller depends on the wake induced by the hull, it was decided to include the hull in the propeller calculations. The hull and propeller geometries were obtained in a high quality IGES file format. The rudder was built separately in STAR-CCM+ using drawings. **Figure 1** shows the full scale geometry simulated in this study.

Since the vessel stern wave had a significant height, which will influence the hydrodynamic pressure on the propeller blades, it was necessary to capture the shape of the stern wave in the modelling. In order to reduce the number of cells, the whole analysis was split into two stages. In the first stage of the analysis, the vessel flow was calculated with the free surface, in order to determine the shape of the waves. The cavitation model in this stage was not active. Once the calculation was converged, the deformed free surface was exported as a geometrical entity. This free surface geometry was used in the second stage of the calculation and set to an upper symmetry boundary. In this stage, the cavitation model was active.

The overall mesh for cavitation simulation was approximately 22 million cells; 18 million of which were in the rotating mesh region. Local volume refinements behind the blade tips and propeller hub were carried out, in order to capture the cavitation extent with high resolution.

A DES was employed to capture the dynamics of the highly turbulent flow. Propeller rotation was modelled using rigid body rotation of a cylindrical mesh region around the propeller, with sliding interfaces between the static and rotating mesh regions. For the propeller rotation, a second-order temporal differencing scheme was employed for the unsteady term.

Since the vessel speed, shaft speed and vessel draughts were recorded during the cavitation observations carried out on the vessel, the cavitation analysis was performed in identical conditions.

An indication of the stability of the solution was confirmed by the evolution of thrust, torque and vapor volume; these quantities were stable over the period of 10 sec and a periodic solution was achieved for all parameters.

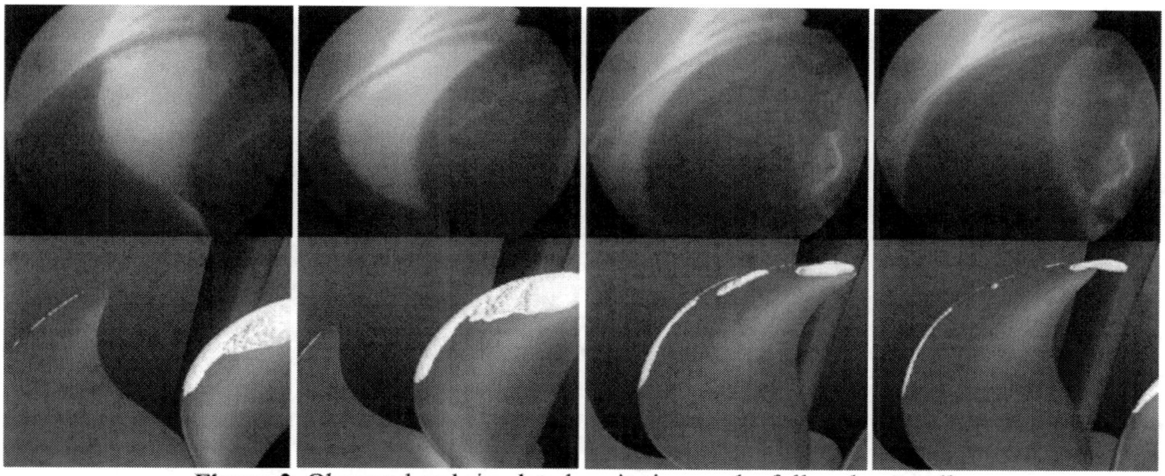

**Figure 2**. Observed and simulated cavitation on the full scale propeller.

DES was able to correctly resolve the sheet cavity and its rapid collapse, as shown in **Figure 2**. However, the development of tip and hub vortexes was significantly under predicted despite the fact that a fine mesh was built behind the blade tip and in the hub region. This could be due to modelling parameters such as the mesh not being fine enough in these regions or limitations of the cavitation model which is based on the simplified Rayleigh–Plesset equation implemented in STAR-CCM+.

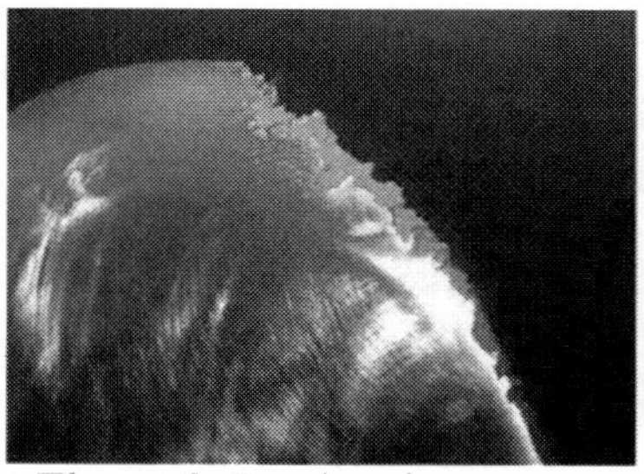

**Figure 3**. Erosion damage on the propeller.

**Figure 4**. Erosion damage predicted by Function 5.

**Figure 5**. Erosion damage predicted by Function 8.

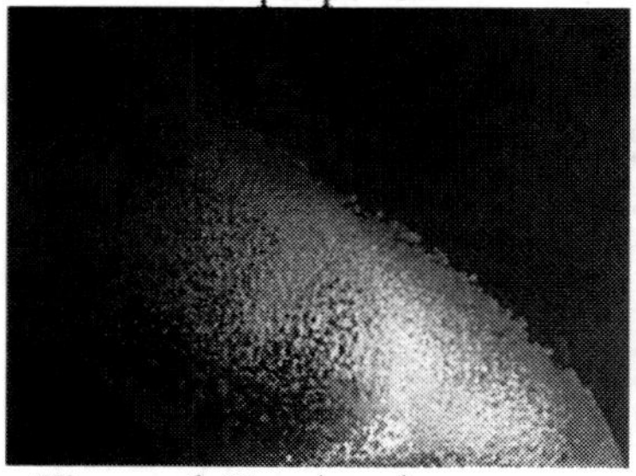

**Figure 6**. Erosion damage on the propeller.

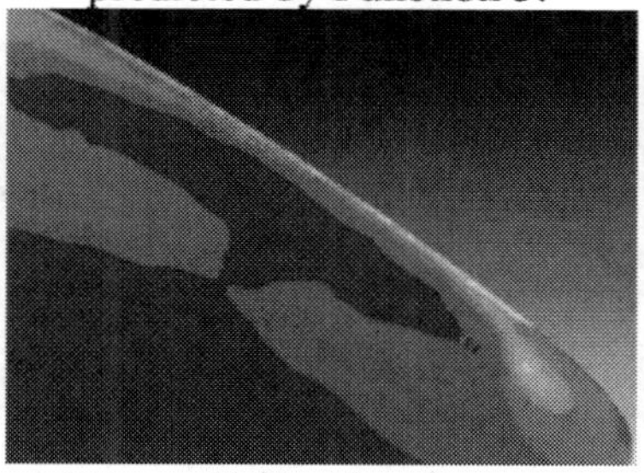

**Figure 7**. Erosion damage predicted by Function 5.

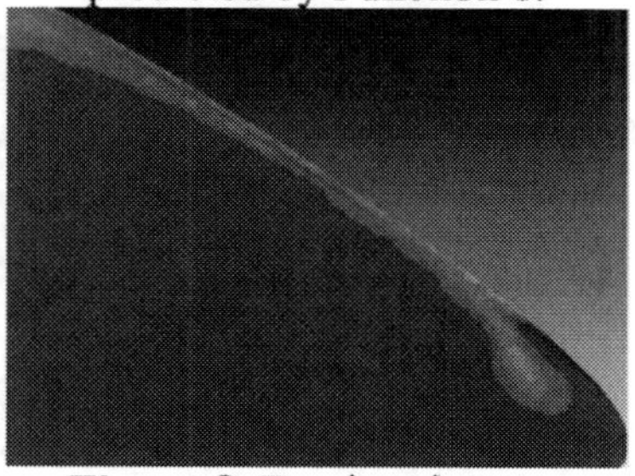

**Figure 8**. Erosion damage predicted by Function 8.

As discussed in [1] twelve functions have been developed in order to numerically predict cavitation erosion. The functions defined as Function 5 and 8 showed the best results for model scale hydrofoil and ship scale rudder [1]. For the current study these functions again showed promising results, mainly due to the correct prediction of sheet cavity collapse. The figures present the erosion damage on the propeller blade in the tip region closer to the leading edge (**Figures 4** and **5**) and trailing edge (**Figures 7** and **8**).

The presented figures were obtained from one propeller revolution and the predicted areas were similar for all blades. Further calculations for three propeller revolutions showed the same areas of potential erosion, which confirms high stability of cavitation prediction.

## 3. Conclusions

The erosive functions developed in LR TID and previously reported in [1] were applied for the ship scale propeller behind the hull. The borescopic observations and underwater propeller inspection revealed significant erosion damage on the blade tip, due to a rapid transformation of sheet cavity in the tip vortex. The DES calculation was performed for this vessel in the same conditions as recorded during the borescopic observations. CFD under predicted development of the tip vortex, possibly due to the mesh not being fine enough or due to a simplified formulation of the Rayleigh–Plesset equation; however, the sheet cavity collapse was captured accurately. As a consequence, the developed functions showed very good prediction of erosion areas, as confirmed by comparison with underwater propeller inspection results.

## 4. Acknowledgement

The current research project was funded under the internal LR R&D program. The author would like to thank his colleagues C. Craddock, A. Caldas and C. Fetherstonhaugh for their significant support.

## 5. References

[1] Ponkratov D and Caldas A 2015 Numerical Prediction of Cavitation Erosion and its Validation against Model and Ship Scale Results *Proc. of the 4th Int. Symp. on Marine Propulsors, SMP'15*, Austin, USA.

[2] ITTC recommended procedure 7.5-02-03-03.

[3] Schmidt S, Sezal I, Schnerr G and Thalhamer M 2007 Shock Waves as Driving Mechanism for Cavitation Erosion, *Proc. of the 8th Int. Symp. on Experimental and Computational Aerothermodynamics of Internal Flows*, Lyon.

[4] Schmidt S, Sezal I, Schnerr G and Thalhamer M 2008 Numerical Analysis of Shock Dynamics for Detection of Erosion Sensitive Areas in Complex 3-D Flows *Cavitation: Turbomachinery & Medical Applications, WIMRC FORUM 2008*, Warwick University, UK.

[5] Schmidt S, Schnerr G and Thalhamer M 2009 Inertia Controlled Instability and Small Scale Structures of Sheet and Cloud Cavitation *Proc. of the 7th Int. Symp. on Cavitation CAV2009* – Paper No. 17, Ann Arbor, Michigan, USA.

[6] Mihatsch M, Schmidt S and Adams N 2013 Estimation of incubation times through numerical simulation of 3-D unsteady Cavitating flows *SHF Conf. Hydraulic Machines and Cavitation /Air in Water Pipes*, Grenoble, France.

[7] Nohmi M, Ikohagi T and Iga Y 2008 Numerical Prediction Method of Cavitation Erosion, *Proc. of FEDSM2008*, ASME Fluids Engineering Conference, Jacksonville, Florida USA.

[8] Hasuike N, Yamasaki S and Ando J 2009 Numerical Study on Cavitation Erosion Risk of Marine Propellers Operation in Wake Flow, *Proc. of the 7th Int. Symp. on Cavitation CAV2009* – Paper No. 30, Michigan, USA.

[9] Li Z 2012 Assessment of Cavitation Erosion with a Multiphase Reynolds-Averaged Navier-Stokes Method *PhD thesis*, Delft University of Technology.

[10] Li Z and Terwisga T 2012 On the capability of a RANS method to assess the cavitation erosion risk on a hydrofoil *Proc. of the 8$^{th}$ Int. Symp. on Cavitation CAV2012*, Singapore.

[11] Eskilsson C and Bensow R 2015 Estimation of Cavitation Erosion Intensity Using CFD: Numerical Comparison of Three Different Methods Results *Proc. of the 4th Int. Symp. on Marine Propulsors, SMP'15*, Austin, USA.

[12] Boorsma A and Whitworth S 2011 Understanding Details of Cavitation, *Proc. of the 2nd Int. Symp. on Marine Propulsors, SMP'11*, Hamburg, Germany.

# Francis turbine draft tube modelling for prediction of pressure fluctuations on prototype

**S Alligné[1,3], C Landry[2], A Favrel[2], C Nicolet[1] and F Avellan[2]**

[1] Powervision Engineering sàrl, Chemin des champs courbes 1, 1024 Ecublens, Switzerland
[2] EPFL Laboratory for Hydraulic Machines, Avenue de Cour 33 bis, 1007 Lausanne, Switzerland

E-mail: sebastien.alligne@powervision-eng.ch

**Abstract.** The prediction of pressure fluctuations amplitudes on Francis turbine prototype is a challenge for hydro-equipment industry since it is subjected to guarantees to ensure smooth and reliable operation of the hydro units. The European FP7 research project Hyperbole aims to setup a methodology to transpose the pressure fluctuations measured on the reduced scale model to the prototype generating units. This paper presents this methodology which relies on an advanced modelling of the draft tube cavitation flow, and focuses on the transposition to the prototype of the draft tube model parameters identified on the reduced scale model. Different modelling assumptions of the draft tube are considered and their influence on the eigenmodes and the forced response of the system are presented.

## 1. Introduction

In the framework of the European FP7 research project Hyperbole, a methodology is setup to predict pressure fluctuations on prototype based on experimental measurements on the reduced scale model. This methodology relies on an advanced modelling of the draft tube cavitation flow. Previous investigations aimed to identify experimentally the model parameters for prediction of prototype stability such as Dörfler in 1982 [1] and Jacob et al. in 1993 [2]. In 2014, the draft tube model has been improved and dissipation induced by phase change during cavitation volume fluctuations has been introduced in the model [3]. Specific measurements to quantify this dissipation with the remaining parameters are required [4]. This paper presents the methodology and focus on the transposition to the prototype of the draft tube model parameters identified on the reduced scale model.

## 2. Methodology for prediction of pressure fluctuations on prototype

The methodology to predict the pressure fluctuations on prototype from measurements on reduced scale model is illustrated in figure 1. This paper is focused on the transposition process for part load conditions.

[3] To whom any correspondence should be addressed.

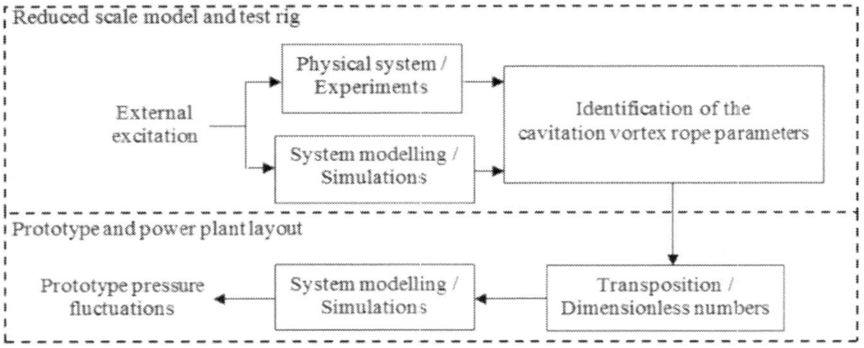


**Figure 1.** Methodology for prediction of pressure fluctuations on prototype.

The modelling of the draft tube cavitation flow is described by continuity and momentum equations including the convective terms and the divergent geometry [3]. Three cavitation vortex rope parameters describing this model have been identified experimentally at the reduced scale model:

- the local wave speed $a$ defined implicitly by the cavitation compliance $C_c$ (the mass flow gain factor $\chi$ is not considered in this study);
- the bulk viscosity $\mu''$ induced by the phase change during cavitation volume fluctuations;
- the excitation source $S_h$ induced by the interaction of the helical swirling flow and the elbow.

For transposition purpose of the wave speed and the bulk viscosity, two dimensionless numbers have been derived by applying the Buckingham-$\Pi$ theorem [4], see equation (1).

$$\Pi = \frac{\rho_w a^2}{p_{out} - p_v} = g(\beta) \quad ; \quad M'' = \frac{\mu'' f_{natural}}{p_{out} - p_v} = \Pi^2 \left(1-\beta^2\right) \frac{\rho_c}{\rho_w} \tag{1}$$

Regarding the excitation source, the amplitude identified at the reduced scale model is transposed by assuming both Thoma similitude and Froude similitude conditions. The influence of the Reynolds number on the transposition is assumed to be negligible. The prototype amplitude is defined by the ratio of heads between model and prototype, see equation (2).

$$A^P = A^M \left( \frac{H^P_{ref}}{H^M_{ref}} \right) \tag{2}$$

## 3. Modelling of the power plant

The power plant of interest features four 435MW generating units, each equipped with a Francis type turbine rated at power output of 444 MW under the 171m rated net head. Two part load operating points are investigated : PL1 and PL2 defined at the same nED but different QED [4]. A numerical model of the power plant is set up with the SIMSEN software. Distributed and lumped draft tube models are compared. The distributed model is characterized by several control volumes along the draft tube length where continuity and momentum equations are applied, whereas the lumped model features only one continuity control volume. In table 1, the predicted first eigenmode characterized by its frequency $f_1$ and its damping $\alpha_1$ is given depending on the draft tube modelling approach for operating point PL1. Moreover, it is mentioned if higher order eigenmodes are predicted unstable or not. In figure 2, the shapes of pressure and discharge fluctuations of the first eigenmode in the hydraulic system are compared between draft tube models with constant and variable parameters.

**Table 1.** Influence of wave speed and bulk viscosity distribution on frequency and damping of eigenmodes for PL1.

| | Distributed Model | | Lumped Model |
|---|---|---|---|
| | Constant Para. | Variable Para. | |
| σ (–) | 0.11 | | 0.108 |
| Void Frac.β (–) | 0.0124 | | 0.0128 |
| a (m/s) | 76.9 | 55 – 100 | 75.7 |
| μ" (Pa.s) | 3.06E+05 | 1.5E+05 – 5.7E+05 | 3.12E+05 |
| α1 ($s^{-1}$) | -0.40 | -0.38 | -0.40 |
| f1 (Hz) | 0.65 | 0.63 | 0.6 |
| Unstable eigenmodes | No | No | Yes |

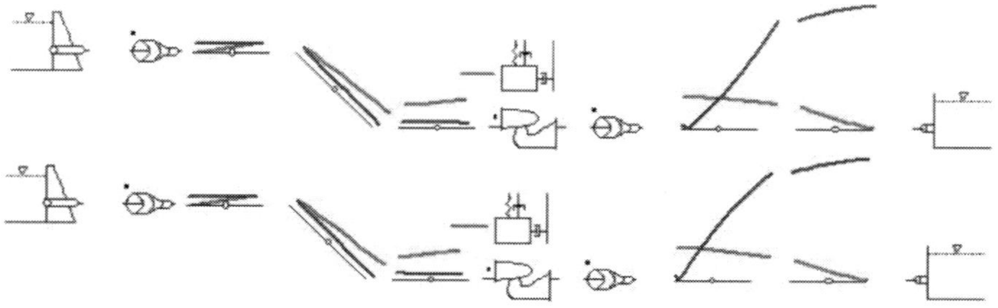

**Figure 2.** Influence of wave speed and bulk viscosity distribution on the first eigenmode shape for PL1 (Top: constant parameters – Bottom: variable parameters).

Two types of distributed models are compared: one with constant wave speed and constant bulk viscosity in all the control volumes and the other one with variable parameters along the draft tube length. In table 1, the set of transposed parameters is available for each model and the resulting predicted eigenmodes are given. It is shown that distributed model with variable parameters does not change the prediction of the first eigenfrequency compared to constant parameters. However, if a lumped model is used, unstable high order eigenmodes with positive dampings are predicted. The shape of pressure mode (red lines) is unchanged whereas the discharge mode (blue lines) is slightly modified. Distributed model with constant parameters is recommended for simplicity.

## 4. Prediction of pressure fluctuations amplitudes and discussions

Table 2 shows the influence of the operating point on the prototype draft tube parameters and on the resulting first eigenfrequency of the power plant. Distributed model with constant parameters is used. By changing the operating point from PL1 to PL2, the amount of cavitation volume is increased. Hence, the first eigenfrequency value $f_1$ is decreased from 0.65Hz to 0.39Hz and the eigendamping value $\alpha_1$ is increased towards positive values.

**Table 2.** Influence of the operating point on the draft tube parameters and on the resulting first eigen frequency.

| | PL1 | PL2 |
|---|---|---|
| σ (–) | 0.11 | 0.11 |
| β (–) | 0.0124 | 0.0421 |
| a (m/s) | 76.9 | 45.3 |
| μ" (Pa.s) | 3.06E+05 | 6.14E+04 |
| α1 ($s^{-1}$) | –0.40 | –0.10 |
| f1 (Hz) | 0.65 | 0.39 |

The source amplitude identified at PL2 at the reduced scale model features higher amplitude than at PL1 at the vortex rope frequency. By applying the transposition law defined by equation (2), the system response is computed and pressure fluctuations in the draft tube are represented in figure 3. If the excitation frequency matches with the first eigenfrequency $f_1$, then maximum amplitude is observed. At PL2 operating point, since the source features higher amplitude and the damping value $\alpha_1$ is higher, then the system response is higher and reaches 8.2% of the net head (~14m) compared to 1.2% (~2m) for PL1. However, the vortex rope frequency being between 0.2 and 0.4 times the runner frequency (yellow area in figure 3), a matching with the first eigenfrequency is only feasible at PL1 where amplitudes are rather small for resonance conditions.

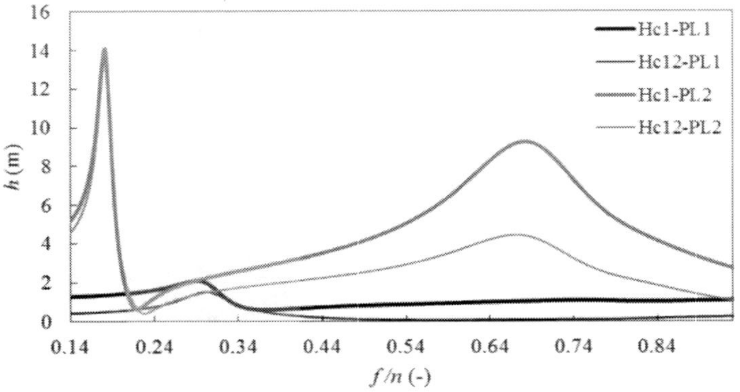


**Figure 3.** Prediction of pressure fluctuations in the draft tube at PL1 and PL2 depending on frequency of the vortex rope.

## 5. Conclusions

The methodology developed in the framework of the European Hyperbole project has been applied to a 435MW generating units of Francis type turbine. Parameters of the reduced scale draft tube model have been transposed to the prototype for two operating points at part load. A matching between the vortex rope frequency and the first eigenmode is predicted for one of the operating point, but the resulting pressure fluctuations amplitudes are rather small for resonance conditions. Measurements on prototype are foreseen for validation.

## 6. References


[1] Dörfler P 1982 System dynamics of the francis turbine half load surge. In *Proc. of Operating Problems of PumpStations and Powerplants,* Amsterdam, Netherlands.

[2] Jacob T and Angelico G 1994 Fonctionnement instable d'un modèle réduit de turbine Francis à débit partiel : évaluation de l'intensité de l'excitation, In *Proc. of the 17th IAHR Symposium on Hydraulic Machinery and Systems,* Beijing, China.

[3] Alligné S, Nicolet C, Tsujimoto Y and Avellan F 2014 Cavitation surge modelling in Francis turbine draft tube *J. of Hydraulic Research.* **52** (3).

[4] Landry C, Favrel A, Müller A, Nicolet C and Avellan F 2014 Experimental identification of the local wave speed and the second viscosity in cavitating draft tube flow *J. of Hydraulic Research* In reviewing process.



## Acknowledgments

The research leading to the results published in this paper is part of the HYPERBOLE research project, granted by the European Commission (ERC/FP7-ENERGY-2013-1-Grant 608532). The authors would like to thank BC Hydro for making available the reduced scale model and the prototype for validation, in particular Dany Burggraeve and Jacob Iosfin.

# Identification of the wave speed and the second viscosity in cavitating flow with 2D RANS computations – Part II

**S Alligné[1,4], J Decaix[2], C Nicolet[1], F Avellan[3] and C Münch[2]**

[1] Power Vision Engineering Sàrl, Chemin des Champs Courbes 1, 1024 Ecublens, Switzerland
[2] HES SO Valais-Wallis, Route du Rawyl 47, 1950 Sion 2, Switzerland
[3] EPFL Laboratory for Hydraulic Machines, Avenue de Cour 33 bis, 1007 Lausanne, Switzerland

E-mail: sebastien.alligne@powervision-eng.ch

**Abstract.** The 1D modelling of cavitation vortex rope dynamics in Francis turbine draft tube is decisive for prediction of pressure fluctuations in the system. However, models are defined with parameters which values must be quantified either experimentally or numerically. In this paper a methodology based on CFD simulations is setup to identify these parameters by exciting the flow through outlet boundary condition. A simplified test case is considered to assess if 1D cavitation model parameters can be identified from CFD simulations. It is shown that a low wave speed and a second viscosity due to the cavitating flow can be identified.

## 1. Introduction

Francis turbines develop a cavitation vortex rope at the runner outlet which induces pressure fluctuations propagating in the whole hydraulic system. Interaction between this excitation source and the system may lead to resonance phenomena. To investigate such problematic, a one-dimensional (1D) model of the system is setup including a model of the cavitation volume dynamics which is decisive to predict potential interactions. In the framework of the European FP7 research project Hyperbole, experimental investigations [1] identified the wave speed and the second viscosity parameters of the cavitation model. This paper is aiming to identify them numerically from two phase flow CFD simulations. The developed methodology is applied to a simplified test case: a venturi geometry [2].

## 2. Methodology for identification of cavitation dynamics parameters

The case study is a venturi geometry [2] featuring a divergent angle of 4 degrees downstream the throat. Due to the increase of velocity in the throat region and the downstream divergent, a cavitation sheet develops for low cavitation numbers. A CFD model [3] and a one-dimensional (1D) compressible model of the venturi geometry are setup. The 1D model is based on continuity and momentum equations including the convective terms and the divergent geometry [4]. In the throat region where cavitation develops a 1D cavitation model is used which parameters are the following:

- the local wave speed $a$ defined implicitly by the cavitation compliance $C_c = (gAL)/a^2$ ;

[4] To whom any correspondence should be addressed.

- the second viscosity $\mu''$ induced by the phase change during cavitation volume fluctuations.

In the cavitation free region, the wave speed value is set to 1000 m.s$^{-1}$. The evolution of the dimensionless static pressure head $h^* = h/\left(C_{in}^2/(2g)\right)$ along the venturi is in agreement with the CFD model. To achieve this matching between the two models, the Darcy-Weisbach value is set to $\lambda = 0.0085$. In figure 1, the methodology to identify the two cavitation parameters $a$ and $\mu''$ is presented. The system response to a fluctuating outlet static pressure of the CFD model and the 1D model are compared. The set of parameters of the 1D model is optimized to get the best fitting between the two system responses.

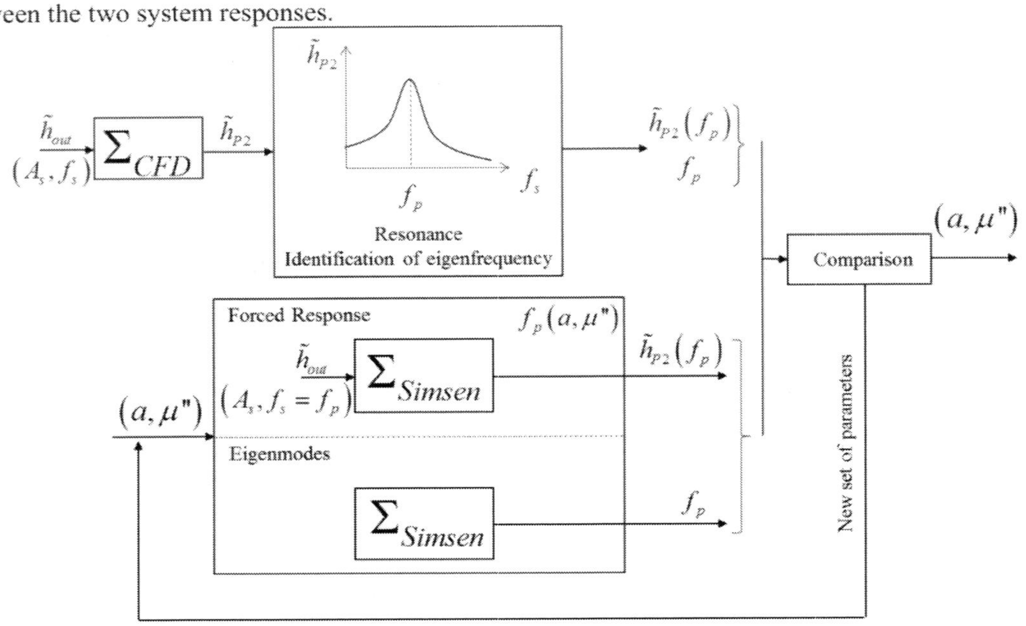


**Figure 1.** Methodology for identification of 1D cavitation model parameters.

On the one hand, the CFD model is excited in a given range of frequency to find a resonance of the cavitation volume fluctuations. The resonance found, the amplitude of pressure fluctuations in Plane2 (see the plane definition in [3]) $\tilde{h}_{P2}$ and the related resonance frequency $f_p$ are the two objectives for the optimization process. On the other hand with the 1D SIMSEN model, the set of cavitation parameters is found by minimizing the error for the two objectives. For the amplitude of pressure fluctuations in Plane2 a forced response analysis is performed in the frequency domain at the frequency $f_p$ and for the resonance frequency, the eigenfrequency of the 1D model is computed.

## 3. Results and discussions

The CFD model features a resonance frequency value of $f_p = 2$ Hz where maximum of pressure and cavitation volume fluctuations are experienced [3]. Time history of dimensionless pressure fluctuations in Plane2 and in Plane4 are represented and compared to the excitation outlet pressure set to a frequency value of $f_s = 6$ Hz in figure 2 and to the resonance frequency $f_s = 2$ Hz in figure 3. For out of resonance conditions, the amplitude of pressure fluctuations decreases from the outlet excitation source to the cavitation sheet which behaves like a pressure node for the system. However, at resonance conditions the amplitude near the cavitation sheet is higher than the outlet excitation source. Moreover, pressure fluctuations do not feature sinusoidal shape anymore. These typical shapes can be observed in Francis turbine draft tubes at resonance conditions due to system non-linearities.

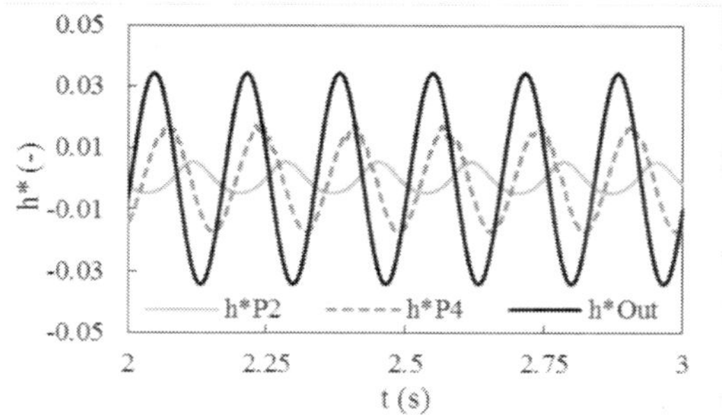


**Figure 2.** Pressure fluctuations for out of resonance conditions ( $f_s = 6Hz$ ).

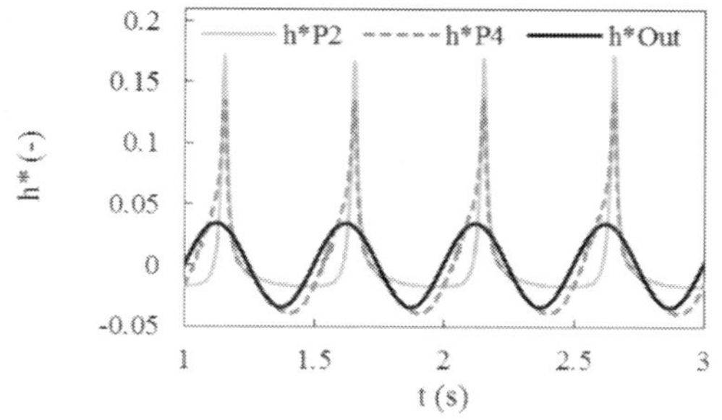


**Figure 3.** Pressure fluctuations for resonance conditions ( $f_s = 2Hz$ ).

In the methodology, the eigenmodes are computed with the 1D SIMSEN model as function of the set of cavitation parameters. Since the two parameters have both influence on damping and frequency of the first eigenmode, they must be considered simultaneously in the optimization process as presented in figure 1. In figure 4 and figure 5, the forced response of the system is computed respectively for two second viscosity values and the resulting pressure fluctuations along the venturi abscissa is plotted. The black solid line corresponds to the outlet excitation frequency set at the first eigenfrequency of the 1D system. The two remaining grey lines represent an excitation frequency out of resonance conditions. The behavior observed with the CFD model is reproduced with the 1D model. Indeed, for excitation frequency different to the eigenfrequency, the cavitation volume behaves like a pressure node for the system. However, with an excitation at the first eigenfrequency, the amplitude near the cavitation volume is higher than the outlet excitation source. Moreover, it is shown that the second viscosity value influences the amplitude of pressure fluctuations near the cavitation volume. Finally, by running the optimization process, the set of 1D cavitation model parameters leading to a matching of the eigenfrequency and the amplitude of pressure fluctuations in Plane2 is given in table 1. Moreover, the dimensionless cavitation compliance $C_c^* = C_c \cdot h_{in} / V_{throat}^3$ is computed.

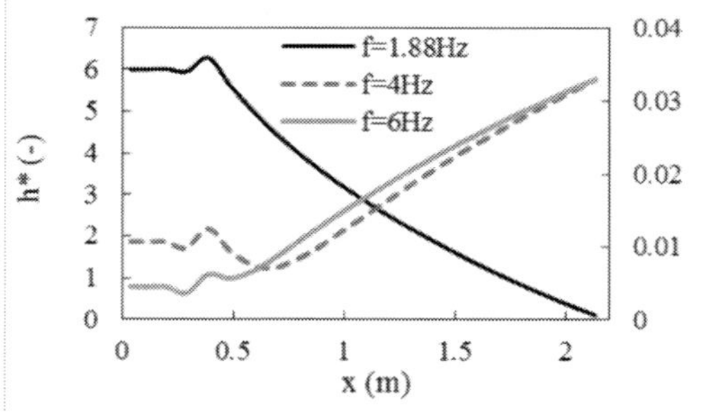


**Figure 4.** Forced response of the system with the set of cavitation parameters $a = 4$ m.s$^{-1}$ and $\mu'' = 280$ Pa.s .

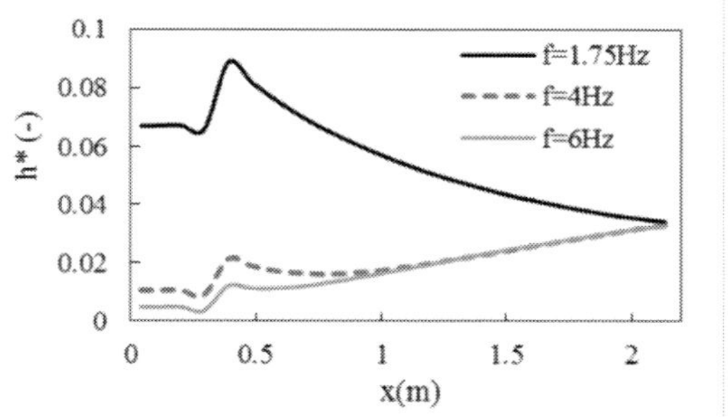


**Figure 5.** Forced response of the system with the set of cavitation parameters $a = 4$ m.s$^{-1}$ and $\mu'' = 1000$ Pa.s .

**Table 1.** Set of cavitation 1D model parameters.

| a | $C_c^*$ | μ" |
|---|---|---|
| (m/s) | (–) | (Pa.s) |
| 4.3 | 3.1 | 790 |

In the methodology, the 1D cavitation model parameters are considered as constant. However, the typical shape of pressure fluctuations observed in resonance conditions in figure 3 suggests non-linear parameters depending on pressure. By considering constant parameters the simulated system response features sinusoidal pressure and discharge fluctuations as shown in figure 6 and in figure 7.

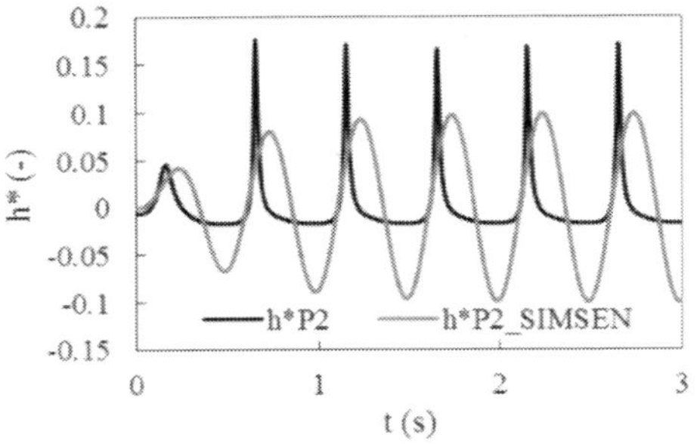


**Figure 6.** Comparison of the simulated pressure fluctuations in Plane2 between the CFD model and the 1D model.

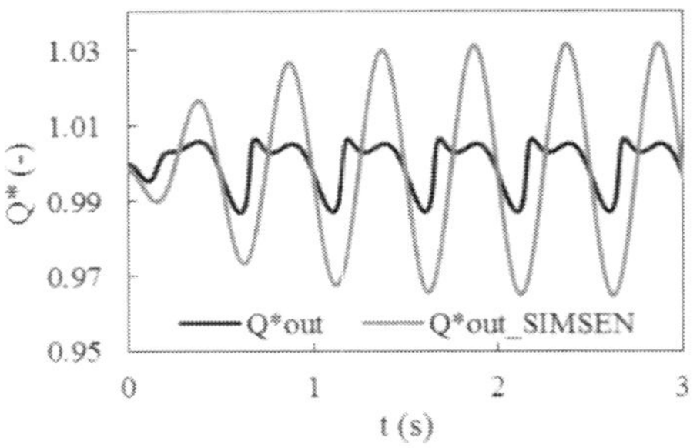


**Figure 7.** Comparison of the simulated downstream discharge fluctuations between the CFD model and the 1D model.

## 4. Conclusions

A methodology based on CFD simulations is setup to identify 1D cavitation model parameters. A venturi geometry is considered as case study and it is shown that a low wave speed and a second viscosity due to the cavitating flow can be identified. Moreover, the 1D model and the CFD model have the same behaviour to an outlet excitation source depending on its frequency. However, parameters are considered as constants and therefore the simulated system response does not reproduce the non-linear CFD simulation results. Further investigations will be carried out by taking into account non-linear parameters depending on pressure. Then, the methodology will be applied to identify parameters of the cavitation vortex rope modelling in the draft tube which involves one additional parameter: the mass flow gain factor. Results will be compared to an identification performed experimentally.

## 5. References

[1] Landry C, Favrel A, Müller A, Nicolet C and Avellan F 2014 Experimental identification of the local wave speed and the second viscosity in cavitating draft tube flow *J. of Hydraulic Research* In reviewing process.

[2] S. Barre J, Rolland G, Boitel E, Goncalves and Patella RF 2009 Experiments and modeling of cavitating flows in venturi: attached sheet cavitation *Eur. J. Mech. - B/Fluids* **28** (3).

[3] Decaix J, Alligné S, Nicolet C, Avellan F and Münch C 2015 Identification of the wave speed and the bulk viscosity of cavitation flows with 2D RANS computations – Part I. In *Proc. of the 9th International Symposium on Cavitation,* Lausanne, Switzerland.

[4] Alligné S, Nicolet C, Tsujimoto Y and Avellan F 2014 Cavitation surge modelling in Francis turbine draft tube *J. of Hydraulic Research.* **52** (3).

**Acknowledgments**

The research leading to the results published in this paper is part of the HYPERBOLE research project, granted by the European Commission (ERC/FP7-ENERGY-2013-1-Grant 608532).

Journal of Physics: Conference Series **656** (2015) 012058 doi:10.1088/1742-6596/656/1/012058

# Estimations of scale effects on blade cavitation

**Eduard Amromin**

Mechmath LLC, Federal Way, WA98003, USA

E-mail: amromin@aol.com

**Abstract**. Estimations of scale effects on blade cavitation require consideration of multiple models for both water flows and cavities. In particular, distinction of laminar and turbulent boundary layers is very important. A qualitative impact of selection of models is manifested for blade sheet cavitation. Its quantitative impact is shown for vortex cavitation inception.

## 1. Introduction

Naval hydrodynamics considers three different kinds of flows. The first kind is created in laboratories; these flows are convenient for measurements and can be repeatable. The second kind exists in full-scale conditions, where measurements are difficult and expensive. The third kind consists of imaginary CFD flows. Up to now model (laboratory) tests are considered as the ultimate tool for prediction of ship hydrodynamic characteristics, but the difference of model test and full scale flows forces employ CFD tools at least for estimations and comprehension of scale effects. Confidence of such estimations critically depends on selection of the employed CFD models.

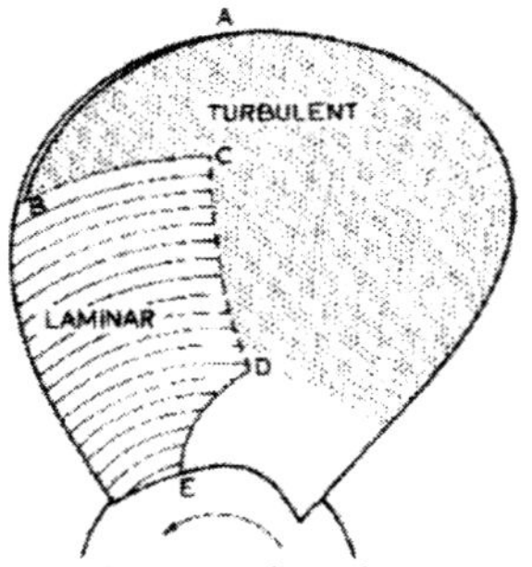


**Figure 1.** Boundary layers on a blade [4]

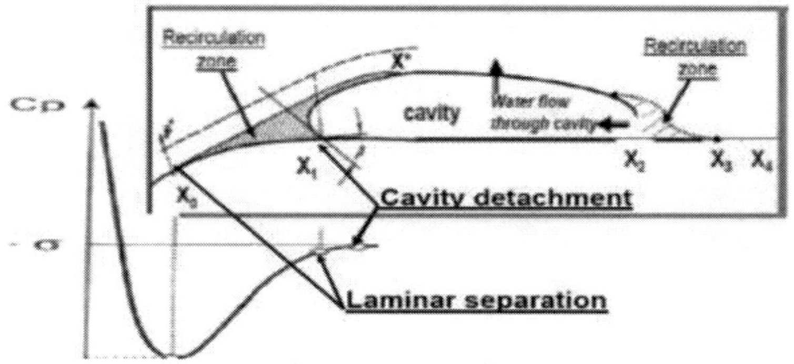


**Figure 2.** Scheme [5] of the flow around a sheet cavity

As already noted [1], single-model approaches (like ideal fluid cavitation or fully-turbulent multi-phase flows) are convenient for computations and can satisfactory describe some kinds of cavitating flows, but such models are insufficient for estimations of scale effects of cavitation that, as noted in [2]-[3], may be significant. This insufficiency takes place because, first, two kinds of boundary layer exist on blades (as on a blade model shown in [4] and in Fig.1 here) and, second, a sheet cavity is usually not a uniform mix of small bubbles, but it is more similar to a single gaseous volume (as presented in Fig.2). As shown below for several examples, multi-model (multi-zone) computations make it possible to understand and predict scale effects on various forms of blade cavitation.

## 2. Scale effect on pressure side cavitation

Blades rotation in the ship wake can lead to periodical appearances and disappearances of cavities. This periodical phenomenon occurs due to variations of the angles of attack $\alpha$ of the blade sections crossing the wake. The corresponding variations of the cavity volume can substantially contribute to pressure pulsations. As usual, pulsation predictions have been based on the model test results, but the marine propeller models have shown significant differences of pressure harmonics in comparison with the full scale propellers. The inflow velocities at the propeller disc experience a substantial scale effect just in its top sector, where cavitation is expected. The local inflow similarity can be achieved in this sector with increasing the blade advance ratio [6], but the cavities tend then to rise on the blade pressure side and the similarity of cavitation form becomes lost for another sector of the propeller disk.

The presented estimations were carried out for a 3D hydrofoil with distribution of the load, section thicknesses and cambers corresponding to typical blade characteristics (this is similar to the approach of [7]). The non-uniform incoming flow was fitted to a typical ship wake (like measured in [8]). Then viscous-inviscid interaction approach [5] considering boundary layer separation around the cavity and capillarity on its surfaces was used in these estimations (as in Fig.2). The outer inviscid flow has been analyzed here with including the unsteady vortex sheets behind the sections. The leading edge flows are laminar over both sides of the blade. Though in some ranges of the azimuth $\theta$ the pressure coefficients drops below $-\sigma$ there, the cavity appearance is mitigated by the capillarity impact and this impact is much greater in the model test conditions. On the contrary, the scale effect on the cavity location and length is small for the blade suction side due to great pressure gradients at its leading edge (as in Fig.3, where C is the section chord) and even ideal fluid computations may be acceptable, as noted in [3]. The effect of the cavity detachment on the pressure side cavitation, as shown in Fg.4, is in a significant difference of cavity lengths.

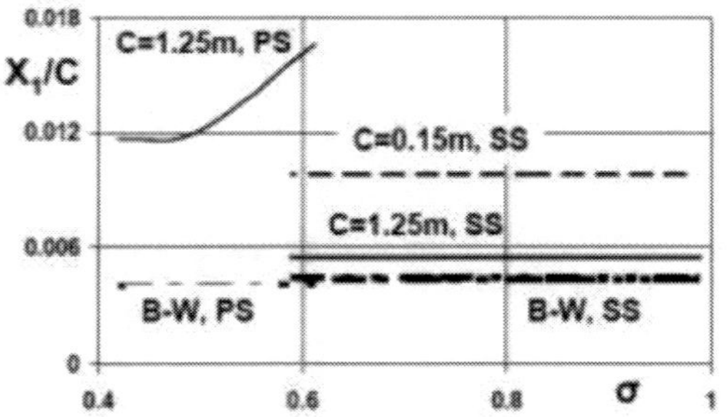


**Figure 3.** Cavity detachment on section suction side (SS) and pressure side (PS) of a blade. B-W relates to solution with Brillouin-Villat condition.

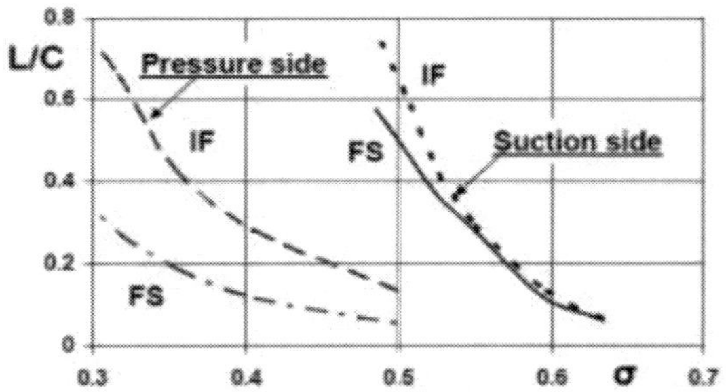


**Figure 4.** Cavity lengths on the same section; B-W marks the location corresponding to Brillouin-Villat condition, IF – ideal fluid, FS – full scale

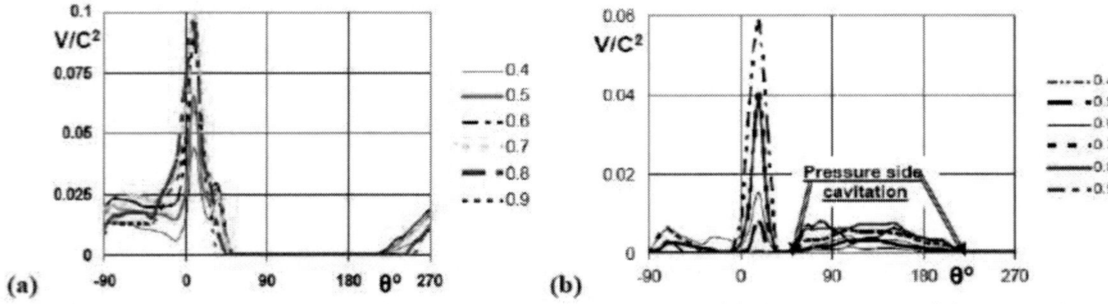


**Figure 5.** Normalized cavity section areas for model test (**a**) and full-scale (**b**) conditions; number show ratios of the section radius to the blade radius.

As seen in Fig.5 due to this difference, a cavity exists on the full-scale blade pressure side during a big part of the period and the cavity volume dependencies on θ experience the qualitative scale effect. So, the high pressure harmonics should not be similar.

## 3. Scale effects on vortex cavitation inception

The dream of engineers to link model test and full-scale results by simple scaling laws in the form of smooth functions of Reynolds number is rarely attainable for cavitating flows. In particular, the attempts to elaborate the universal scaling low $\sigma_I \sim \mathrm{Re}^{\varepsilon}$ with Re-independent power ε for cavitation inception in vortices have lead to failures since the pioneer study [9]. As shown in Fig.6, the experimental data [9] on cavitation inception number for tip vortices do not collapse into a single trend, but there are two different trends for high-Re band and low-Re band there. Obtaining an acceptable approximation for the low-Re band with two mistakable assumptions (on the fully turbulent flow at any Re and on the proportionality of the vortex core radius to the boundary layer thickness), McCormick even supposed [9] that the high-Re data were inaccurate.

Later trends $\sigma_I \sim \mathrm{Re}^{\varepsilon}$ with close ε for turbulent flows were found via a similarity analysis [10] (ε=0.24), RANS computations [11] (ε=0.22÷0.23) and an approximation of the model test data at Re>4.7×10$^6$ [12] (ε=0.21). Let us point out that cavitation inception of ships and their self-driven models has to be determined by a jump of hydrodynamic noise and it is important to register just this jump. It is incorrect to determine the cavitation inception speed as the highest speed of the low noise level (the speed of calm operations). In particular in [13], the speed of calm operations was 8.8 knots for the ship and 15 knots for its self-driven model. Its use gave ε =0.3, but this value indeed relates to cavitation-free flows. Using the experimental points [13] just with the jumps of noise level due to cavitation, one got ε ≈0.21. Nevertheless, this scaling problem cannot be considered as the solved problem yet because there is an issue to determine the point of trend change. As easy to find from Fig.6 with data of [9] (fig.14), an early transition from a low-Re trend to a high-Re trend will underestimate full-scale $\sigma_I$, On the other hand, one will overestimate it assuming that flow is laminar.

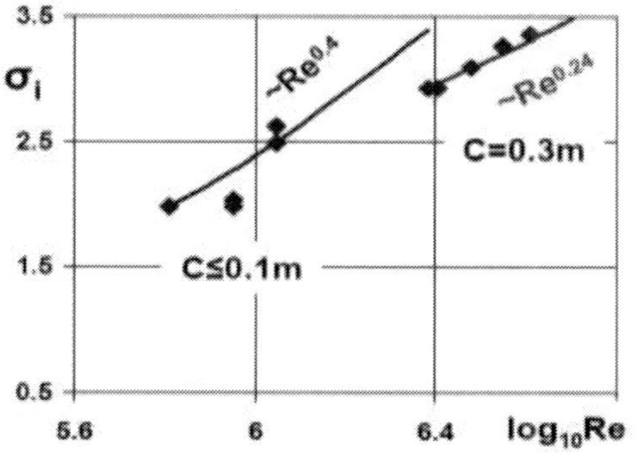


**Figure 6.** Approximations of experimental data [9] for hydrofoil tip vortex cavitation inception

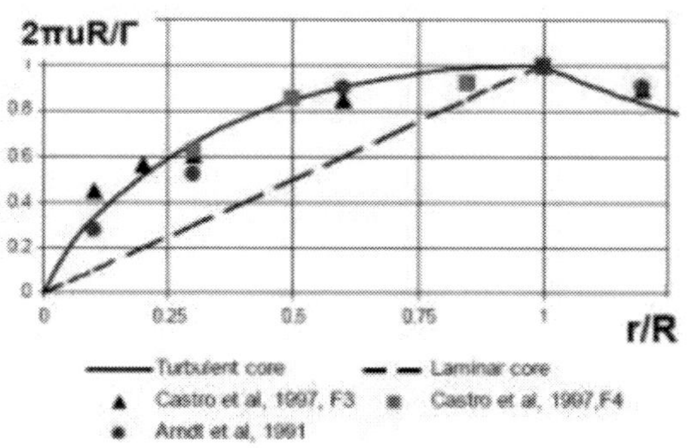


**Figure 7.** Comparison of measured (symbols) velocities with Eq.(1) and Rankine vortex

An inverse mistakable assumption affects prediction and scaling of cavitation inception in vortices located in wakes. It requires analyze pressure in vortex cores. The common assumption is the existence of a laminar vortex core (Rankine vortex) there. However, as derived by Amromin [14], indeed the vortex core velocity profile in turbulent flows is described by formula

$$u(\xi) = \Gamma \xi (1 - \ln \xi)/(2\pi R), \tag{1}$$

where $\xi = r/R$, Γ is vortex intensity, and the core radius $R = \sqrt{-\Gamma \nu/(\pi < u'v' >)}$. Eq.(1) is the exact solution of Reynolds momentum equation for axisymmetric flows. It can be applied to situations, where R is much smaller than the scale of a substantial variation of the Reynolds stress $< u'v' >$. One can see Fig.7 with data for quite high *Re* that Eq.(1) gives the much better approximation of the measured velocity profiles. Besides, this equation gives the substantially lower pressure within the core (shown in Fig.8) and, at the vortex axis

$$\min(C_P)\Big|_{Turbulent} = 1.5\min(C_P)\Big|_{Rankine} - 0.5 \qquad (2)$$

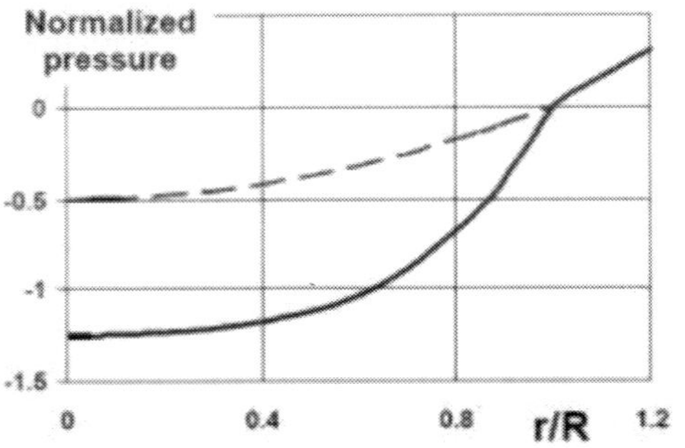


**Figure 8.** Pressure drop within laminar (dashed curve) and turbulent (solid curve) cores [14]

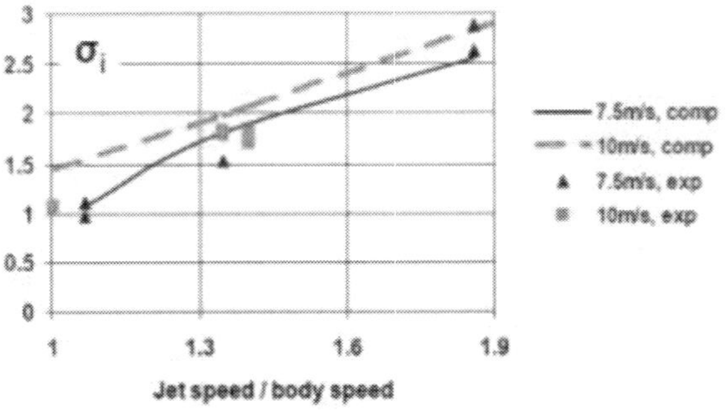


**Figure 9.** Computed and measured $\sigma_I$ for vortices in the wake of a self-driven body [15]

The use of Eq.(1).(2) made it possible a satisfactory correspondence of the dependencies computed with the viscous-inviscid interaction approach to experimental data, as shown in Fig.9. The corresponding axisymmetric flow was computed with consideration of the laminar boundary layer, transition, attached turbulent boundary layer and the zero-momentum wake [15], whereas the cavities were considered as spherical bubbles. However, the author supposes that one may improve some results of RANS computation for vortex cavitation simply including Eqs.(1) and (2).

## Conclusions

The advantages of multi-model CFD analysis of scale effect on cavitation are manifested in this paper. The proven multi-model viscous-inviscid approach was employed in presented computations, but it would not be realistic to expect an extensive spread and amelioration of this approach in the near future. It would be more realistic to seek modifications of RANS or LES solvers or their merging with other solvers allowing for the multi-model analysis of blade cavitation and its scale effects.

## References

[1] Amromin E L 2014 *ASME J. Fluids Eng.* **136** 081303
[2] Arndt R E A 1981 *Ann. Rev. Fluid Mech.* **13** 273-328
[3] Salvatore F, Streckwall H., van Terwisga T 2009 *1st Int. Symp. Marine Prop., Trondheim*
[4] Kuiper G 1981 *Cavitation inception on ship propeller models*, *Neth. Ship Model Basin*
[5] Amromin E L 2007 *ASME J. Fluids Eng.* **129** 1105-1111
[6] Bosschers J, Vaz G, Starke A R, van Wijngaarden E 2008 *RINA conf. MARINE CFD2008, UK*
[7] Foeth E-J, van Doorne C, van Terwisga T 2008 *ASME J. Fluids Eng.* **130**, 071303
[8] Jessup S D and Wang H C 1997 *Propeller/Shafting'97 Symp., Virginia-Beach*
[9] McCormick B W 1962 *J. Basic Eng.* **84** 369-386
[10] Amromin E L 2006 *Ocean Eng.* **v33** 530-534
[11] Hsiao C-T, Chahine C L 2008 *27th Symp. Naval Hydr., Seoul*
[12] Chesnacas C J and Jessup C D 2003 ASME *paper FEDSM2003-45320*
[13] Lee J-H, Jung J-K, Lee K-J, Han J-M, Park H-G, Seo J-S 2012 *J. Hydrodyn.* **24**, 658-667
[14] Amromin E L 2007 *Phys. Fluids* **19** 118108
[15] Arndt R E A, Amromin EL, Hambleton W T 2009 *ASME J. Fluids. Eng.* **131** 111302

# OpenFOAM investigations of cavitation in a flushed water-jet inlet

**A Gattoronchieri[1] and R Bensow[2]**

[1] Termomeccanica Pompe, La Spezia, Italy
[2] Department of Shipping and Marine Technology, Chalmers University of Technology, Gothenburg, Sweden

E-mail: a.gattoronchieri@gmail.com

**Abstract**. The cavitation on the lip of a flushed water-jet inlet has been simulated with a transient RANS model and the results has been validated against experiments. The k-ω SST turbulence model has been adopted together with the cavitation correction proposed by Reboud. The defined setup shows promising results and the vortex shedding has been qualitatively predicted. Moreover, the importance of the sufficient spatial resolution to capture the cavity closure and its extension has been studied and proved to be crucial.

## 1. Introduction

In the 20th century and even more in the last two decades, water-jets have been widely used for high speed crafts propulsion. The typical speed range determines relatively low free stream cavitation numbers which may easily lead to cavitation problems in different parts of the system. Beside the pump cavitation, which has been deeply investigated [1] [2], the cavitation in the inlet duct has been pointed out by various authors [3], [4]. The water-jet inlet is fundamental for the overall propulsion performance and for the pump operative life, therefore the study of the cavitation that may occur in different parts of the duct, namely the ramp, the bend and the inlet lip, is of particular interest. Previous authors pointed out that, for a given geometry, the cavitation at the inlet lip is considerably influenced by the waterjet inlet velocity ratio [4], [5] and by the hull boundary layer thickness [5]. Moreover, Bulten [4] highlights that cavitation tunnel measurements of the cavitation inception may be considerably influenced by tunnel walls. Consequently, reliable numerical simulations for the prediction of the cavitation on the inlet lip would be a valuable tool in the design process.

The here reported study is part of an intense simulation campaign aimed at the understanding of the capability of the open source CFD code OpenFOAM to predict the flow features in a flushed water-jet inlet through the comparison of the numerical results with a well-documented experimental campaign [5]. In the first part of this research, good results in simulation of non-cavitating conditions have been obtained, considering the minor uncertainties on the prediction of the influence of boundary layer thickness on the flow separation occurring along the ramp at the highest inlet velocity ratio (IVR) that have been pointed out [6]. The following presents the preliminary results obtained with a setup modified in order to predict the cavitation at the inlet lip documented in the same experimental campaign.

## 2. Experimental setup

The experiments have been carried out in the AMC Tom Fink Cavitation Tunnel, a closed-circuit variable-pressure water tunnel, shown in Figure 1. Detailed information on the tunnel and water-jet inlet test loop are given in Brandner and Walker [7]. The experiments have been carried out with the natural test section boundary layer and with a thickened boundary layer of 20 and 50 mm respectively, maintaining a turbulence and velocity distribution similar to those generated on a flat plate. The model has been instrumented to measure the principal flow features: boundary layer development, ramp pressure distribution, lip incidence and the pressure and velocity distributions at the pump face. Details of the instrumentation, not reported here for the sake of shortness, are available in the reference paper. The cavitation in this area is strongly influenced by the value of the IVR which determines the incidence on the lip and consequently the low pressure zone. For IVR smaller than that corresponding to the ideal incidence, the stagnation point moves towards the hull bottom, while for higher value it moves inside the duct, thus the IVR determines if cavitation may occur above or below the lip. The authors pointed out that also the boundary layer thickness influences the cavitation inception, as shown in Figure 2, where the inception cavitation number is reported as a function of the IVR and boundary layer thickness. The cavitation number has been defined as:

$$\sigma_\infty = \frac{p_{\infty L} - p_v}{1/_2 \rho U^2}$$

where $U$ is the free stream velocity when the cavitation occurs below the lip and the mean exit velocity in the other condition; $p_{\infty L}$ is the free stream static pressure at lip elevation and $p_v$ is the vapour pressure. Furthermore, the reference article reports four picture of the cavitation occurring on the outside of the lip at different IVR and Reynolds number in the thickened boundary layer configuration.

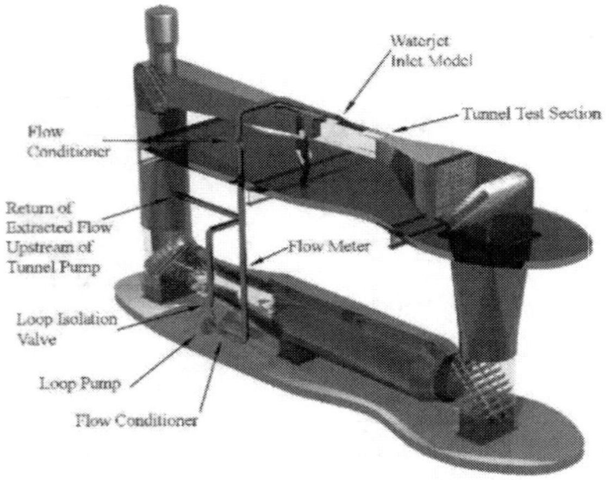


**Figure 1.** Water-jet test loop.

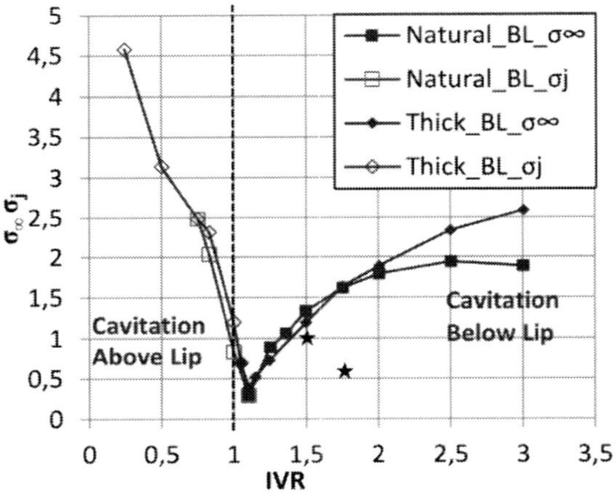


**Figure 2.** Lip cavitation inception and analysed operative condition.

## 3. Numerical model

Consistently with the first part of this research, [6], a computational domain coincident with the test section has been chosen for the simulations and it has been discretised using hexahedral elements in the sweepable parts and tetrahedral in the remaining. The prism layer has been defined in order to satisfy the requirements of the low-Reynolds wall treatment. The grid obtained from the mesh dependency study carried out for the wet-flow simulations has been refined several times in correspondence of the of the lip, varying the average edge sizes from about 3 mm used in the original grid to 0.6 mm used for the results presented in the following. This modification increases the total number of elements from about 1.3 to 5 millions. A mesh dependency study is currently being performed with a further level of refinement. No more details about the grids are reported here due to the short paper format.

The two phases coexisting in the cavitation have been modelled as a mixture of two incompressible fluids using the transient solver interPhaseChangeFoam, introducing a transport equation for the vapour volume fraction and the widely used Schnerr-Sauer mass transfer model [8]. The effects of the turbulence closures on the simulations of unsteady sheet cavitation have been pointed out in previous work [9] [10]. These publications highlight that the implicit LES model is capable in predicting the unsteady nature of this type of cavitating flows. However, in order to limit the computational time and to be consistent with the previous simulations of the waterjet inlet duct, the RANS approach has been adopted. The k-ω SST turbulence model available in OpenFOAM has been modified in order to include the Reboud correction [11]. This ad hoc correction reduces the turbulent viscosity in the mixture which is responsible for the inability of the RANS models to produce the unsteady cavitation behaviour, as it has been stressed in several publications [11] [12]. The time step has been fixed in order to ensure the maximum Courant number equal one, necessary due to the explicit implementation of the solver in the OpenFOAM version used.

## 4. Results

As mentioned before, the reference paper documents the cavitation under the inlet lip in four different operative conditions. Two of these, identified by the "star" marker in Figure 2, have been chosen to verify if the numerical simulations are able to correctly predict the cavitation behaviour.

In the first condition, characterized by $Rn = 1 * 10^6$, IVR=1.5 and cavitation number $\sigma_\infty = 1$, a sheet cavity with highly unstable closure and cavitating vortex filaments in the turbulent shear layer is documented in the reference paper. Figure 3 shows the picture taken in the model test in comparison with two isosurfaces computed from the numerical solution with different threshold of vapour volume fraction. The shape and extent of the sheet cavitation are well captured in proximity of the leading edge and in the centre of the duct, while its closure in the external part is visible only with the higher isovalue of the vapour volume fraction. The cloudy structure shed from the previous sheet and the cavitating vortex filaments visible in the experiments have not been observed in the numerical solution.

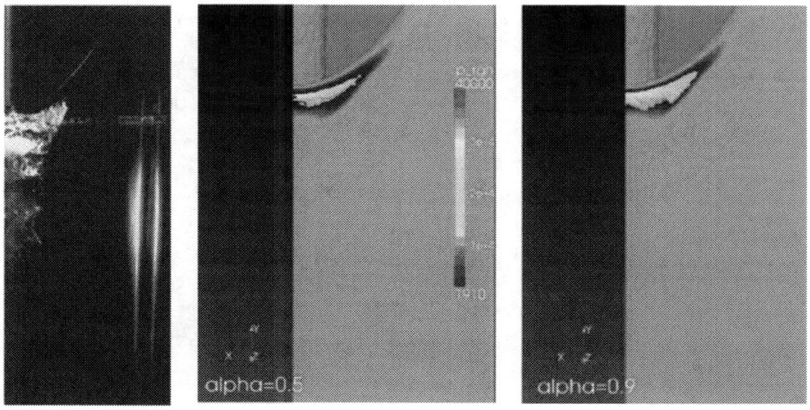


**Figure 3.** External lip cavitation detected in the experiments and visualized in the numerical simulation with $\sigma_\infty = 1$ and IVR = 1.5.

In the second simulated condition, characterized by the same Reynolds number, IVR=1.75 and cavitation number $\sigma_\infty = 0.6$, the cavity grows in base width and length and a more coherent closure mechanism cause the shedding of horseshoe vortices. The comparison between the model test and the numerical results, reported in Figure 4, confirms what was been pointed out for the first operative condition: the initial part of the cavity is visible with both threshold values while, the downstream cloudy structure is not predicted by the numerical solution; however, alpha equal to 0.9 highlights the shedding of cavitating vortices. It has been shown through experimental measurements that the vapour volume fraction could well be around 10 percent in relatively large regions in this type of cloudy cavitation [13]. Nevertheless, the difference between the iso-surface computed with the two threshold value indicate that further grid refinements are still necessary. It should be noted that the discrepancy between the numerical results and experimental data in prediction of the cavity size is also dependent on the accuracy of capturing the separated vortex. Having appropriate mesh resolution is vital for preserving the vortex core strength while transporting it downstream. Lack of spatial resolution will lead to smearing the pressure gradients especially at the core of the separated vortex, and therefore over prediction of the pressure in the vortex region. Consequently over prediction of the pressure will lead to over prediction of vapour to liquid mass transfer.

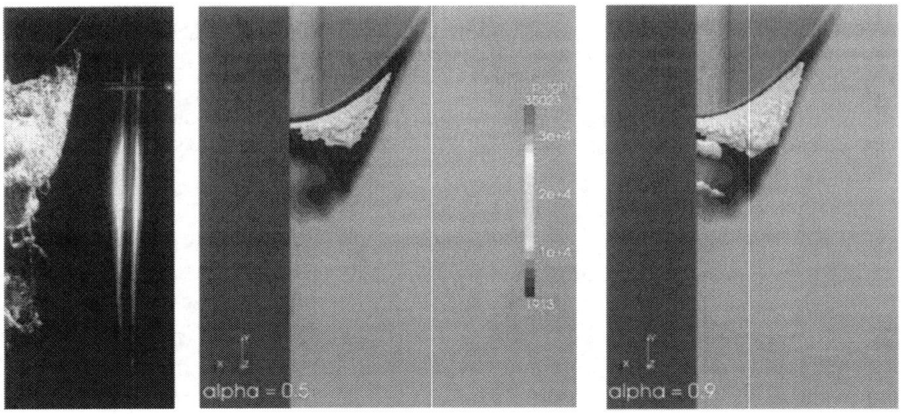


**Figure 4.** External lip cavitation detected in the experiments and visualized in the numerical simulation with $\sigma_\infty = 0.6$ and IVR = 1.75.

## 5. Conclusions

The defined setup provides promising results, the vortex shedding has been qualitatively predicted and the importance of the spatial resolution to capture the cavity closure has been pointed out. Further analysis are still required to refine the simulation setup; for this purpose more information about the experiments would be valuable for the validation of the numerical model.

## References

[1] Dang J, Liu R and Pouw C 2013 Waterjet System *Performance and Cavitation Test Procedures Third Int. Symp. on Marine Propulsors* smp'13 Tasmania Australia
[2] Gulich J F 2010. *Centrifugal Pumps* Springer
[3] Faltinsen O M 2005 *Hydrodynamics of High-Speed Marine Vehicles* 70-73
[4] Bulten N W 2006 *Numerical Analysis of a Waterjet Propulsion System* PhD Thesis 101-105
[5] Brandner P A and Walker G J 2007 An Experimental Investigation Into the Performance of a Flush Water-Jet Inlet *J. of ship research*
[6] Gattoronchieri A and Cravero C 2014 OpenFOAM investigations of a flushed water-jet inlet performance *proc. NuTTS'14* Marstrand Sweden
[7] Brandner P A and Walker G J 2001 A waterjet test loop for the Tom Fink Cavitation Tunnel *proc. of Waterjet Propulsion Conference III, RINA*, February 20 Gothenburg Sweden
[8] Schnerr G H and Sauer J 2001. Physical and Numerical Modeling of Unsteady Cavitation Dynamics. *Proc. 4th Int. Conf. on Multiphase Flow* New Orleans U.S.A.
[9] Bensow R 2011. Simulation of the unsteady cavitation on the Delft Twist11 foil using RANS, DES and LES. Sec. Int. Symp. on Marine Propulsors. Hamburg Germany
[10] Asnaghi A, Feymark A and Bensow R 2013 Effect of Turbulence Closure on the Simulation of the Cavitating Flow on the Delft Twist11 Foil. *NuTTS13*
[11] Reboud J L, Stuntz B and Coutier O 1998 Two-Phase Flow Structure of Cavitation: Experiment and Modeling of Unsteady Effects. *3rd Int. Symp. on cavitation. Grenoble France*
[12] Coutier-Delgosha O et al 2003 Evaluation of the Turbulence Model Influence on the Numerical Simulations of Unsteady Cavitation
[13] Stutz B and Reboud J L 1997 Two-phase Flow Structure of Sheet Cavitation Phys. Fluids 9 (12)

# Identification of the wave speed and the second viscosity of cavitation flows with 2D RANS computations – Part I

**J. Decaix**[a,1], **S. Alligné**[b], **C. Nicolet**[b], **F. Avellan**[c] **and C. Münch**[a]

[a]University of Applied Sciences and Arts, Western Switzerland, Switzerland.
[b]Power Vision Engineering Sàrl, Ecublens, Switzerland.
[c]Laboratory for Hydraulic Machines, Ecole Polytechnique Fédérale de Lausanne, Switzerland.

Email : jean.decaix@hevs.ch

**Abstract.** 1D hydro-electric models are useful to predict dynamic behaviour of hydro-power plants. Regarding vortex rope and cavitation surge in Francis turbines, the 1D models require some inputs that can be provided by numerical simulations. In this paper, a 2D cavitating Venturi is considered. URANS computations are performed to investigate the dynamic behaviour of the cavitation sheet depending on the frequency variation of the outlet pressure. The results are used to calibrate and to assess the reliability of the 1D models.

## 1. Introduction

1D hydro-electric models are used to investigate the dynamic behaviour of hydro-power plants. Particularly, the behaviour at off-design operating points of Francis turbines is of main interest since hydraulic power plants are more and more used to stabilize the electrical power network. To increase the reliability of the 1D model, the parameters involved in the models have to be calibrated.

This paper deals with the development of a methodology to calibrate 1D cavitation models by using results provided by CFD computations. A 2D Venturi geometry [1] is considered. An unsteady pressure condition is set at the outlet. Depending on the frequency of the outlet pressure, the response of the cavitation sheet is analysed.

## 2. Case study

The test case consists of a Venturi geometry [1] characterized by a divergent angle of 4 degrees (see Figure 1). The inlet velocity is equal to $C_{in}$ = 10.8 m/s ($Q_{in}$ = 0.024 m$^3$/s). At the experimental point, the cavitation sheet develops from the throat and extends approximately 0.08 meter downstream of the throat.

## 3. 1D modelling of cavitation dynamics

The 1D model is the one developed in the SIMSEN software [2]. For a cavitating flow, the mass conservation equation and the momentum equation write:

[1] To whom any correspondence should be addressed.

$$-\frac{dV_c}{dt} = \chi \frac{dQ_1}{dt} + C_c \frac{dh}{dt} \tag{1}$$

$$\frac{1}{gA}\frac{\partial Q}{\partial t} + \frac{Q}{gA^2}\frac{\partial Q}{\partial x} - \frac{Q^2}{gA^3}K_x + \frac{\partial h}{\partial x} + \frac{\tau_0 \pi D}{\rho g A} - \frac{\mu''}{\rho g A}\frac{\partial^2 Q}{\partial x^2} = 0 \tag{2}$$

With $Vc$ the cavity volume (m$^3$), $Q$ the discharge (m$^3$ s$^{-1}$), $\chi$ the mass flow gain factor (s), $C_c$ the cavitation compliance (m$^2$), $g$ the gravitational acceleration (m s$^{-2}$), $A$ the cross section (m$^2$), $K_x$ the gradient of the pipe section (m), $h$ the piezometric head (m), $\tau_0$ the tangential component of the stress tensor (Pa), $D$ the pipe diameter (m) and $\mu''$ is the second viscosity (Pa s). The cavitation compliance $C_c$ is related to the local wave speed whereas $\mu''$ introduces dissipation due the phase change.

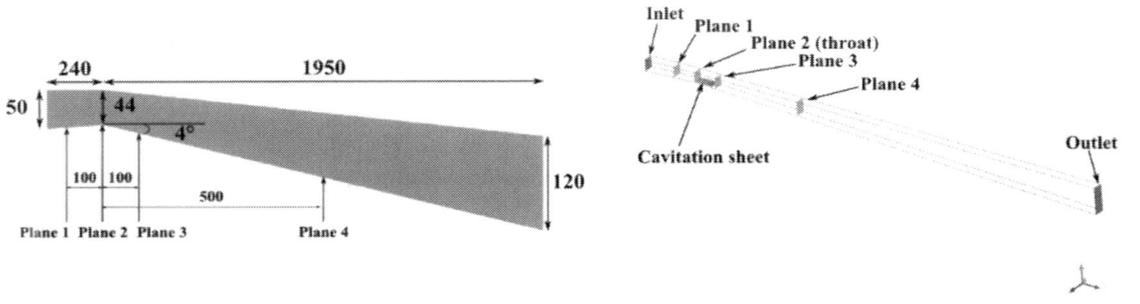


**Figure 1: 2D (left) and 3D (right) views of the Venturi.**

## 4. Flow numerical simulation set up

Unsteady cavitating RANS computations are performed with ANSYS CFX 15.0. The two-phase flow is computed using a homogeneous approach. Turbulence is modeled using the k-ω SST model, whereas cavitation is modeled using a transport equation for the vapour volume fraction with the source term based on the simplified Rayleigh-Plesset equation. The pressure vapour is set to $p_{vap}$ = 2’300 Pa. The two parameters of the cavitation model are set to their default value: $F_v$ = 50 and $F_c$ = 0.01. The computations are performed on a 2D computational domain discretized with a structural mesh composed of almost 40’000 nodes. A node is set in the depth since ANSYS CFX is based on the finite volume method. The depth is arbitrary taken equal to 0.044 m.

The discharge is fixed at the inlet to the experimental value $Q_{in}$ = 0.024 m$^3$/s. Non-slip walls are considered and a symmetry condition is imposed on the side walls. The outlet pressure is provided using a sinusoidal profile. The dimensionless outlet piezometric head $h^*_{out}$ is:

$$h^*_{out} = h^*_{ref} + h_a \sin(2\pi f t) \tag{3}$$

With: $h^*_{ref}$ = 1.22 and $h_a$ = 0.034 ($h^* = p/(\rho C^2_{ref}/2g)$). Two frequencies are investigated: $f$ = 2 Hz and $f$ = 8 Hz. The time step is set to $\Delta t = 10^{-4}$ s and the time duration of each simulation is equal to 3 seconds.

## 5. Dynamic behaviour of the cavity

The cavity volume variation depends on the outlet pressure frequency. Figure 2 (left) shows the outlet and Plane 3 pressure time history. For the highest frequency, the pressure amplitude is almost the same at the outlet and at the plane 3. On the contrary, for the lowest frequency, the amplitude at plane 3 is ten times larger, which corresponds to resonance. Figure 2 (right) shows the cavitation volume time history. For the highest frequency, the cavity volume fluctuates slightly, whereas for the lowest frequency, the cavity volume fluctuates strongly. The low frequency corresponds approximately to the resonance frequency.

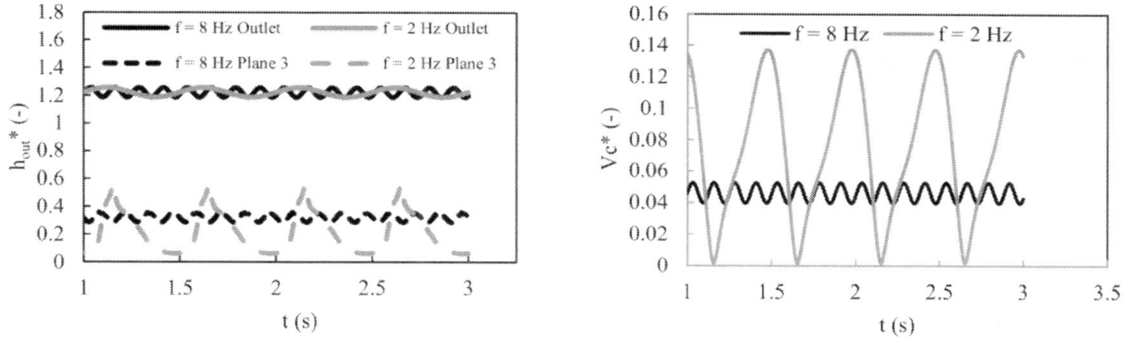


**Figure 2: Left: Dimensionless piezometric head time history at the plane 3 and the outlet. Right: Dimensionless volume of vapour time history. Unsteady RANS computations.**

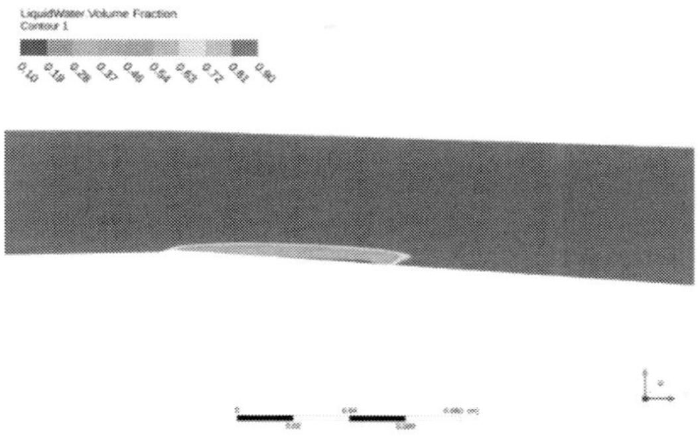

*f* = 8 Hz (maximum cavity volume)

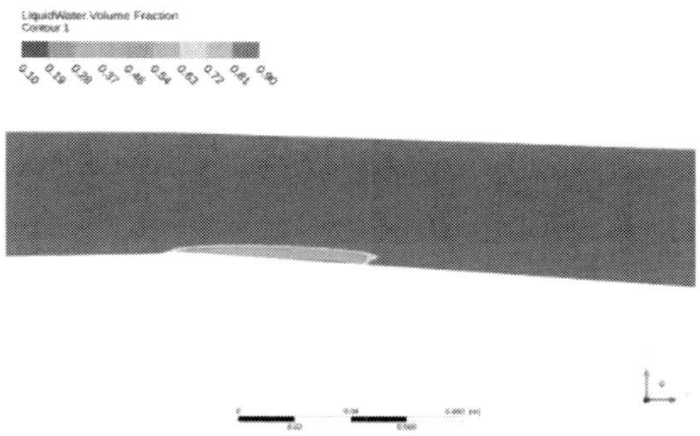

*f* = 8 Hz (minimum cavity volume)

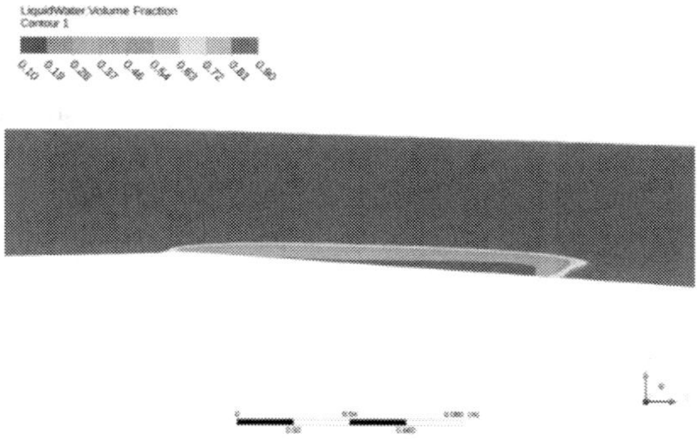

*f* = 2 Hz (maximum cavity volume)

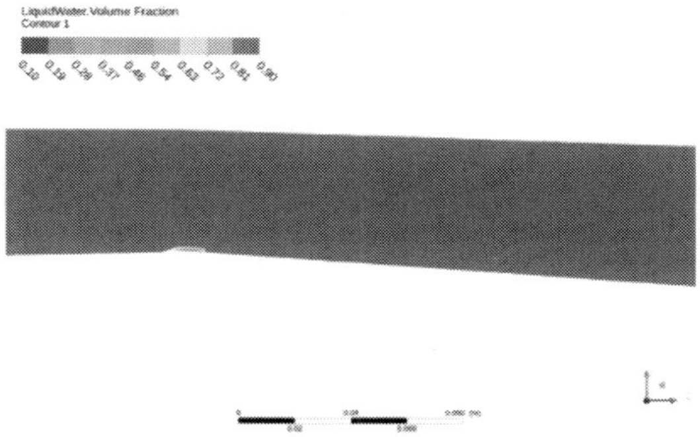

*f* = 2 Hz (minimum cavity volume)

**Figure 3: Contours of the liquid volume fraction at the minimum and maximum cavity volume for the two frequencies investigated.**

The contours of the volume liquid fraction are displayed on Figure 3 for the two frequencies at the minimum and maximal cavity volume. For the highest frequency, the size and the shape of the cavity are rather the same. For the lowest frequency, the minimum of the cavity volume correspond to a small cavitation sheet close to collapse.

The dimensionless streamwise velocity contours are shown on Figure 4. A re-entrant jet is noticeable at the cavity closure except at the minimum cavity volume for the frequency of 2 Hz. For unforced unsteady configurations, the re-entrant jet flows upstream until the cavitation sheet collapses, which can lead to the resonance. The unsteady outlet condition allows controlling and reproducing such a behavior.

Journal of Physics: Conference Series **656** (2015) 012060 doi:10.1088/1742-6596/656/1/012060

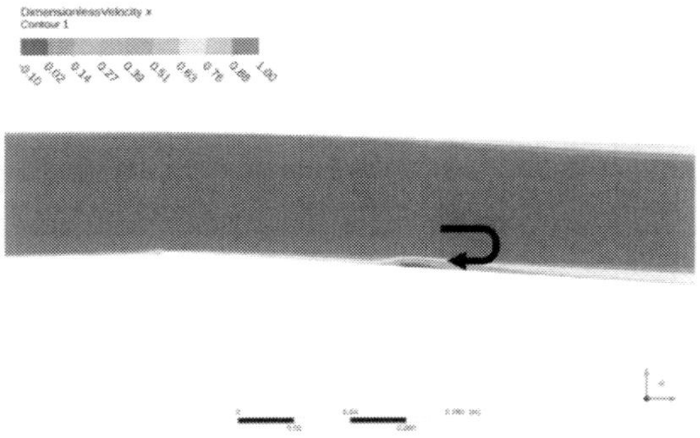

$f = 8$ Hz (maximum cavity volume)

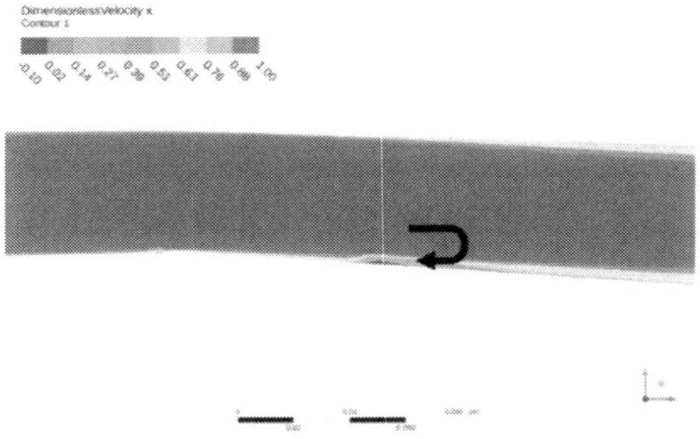

$f = 8$ Hz (minimum cavity volume)

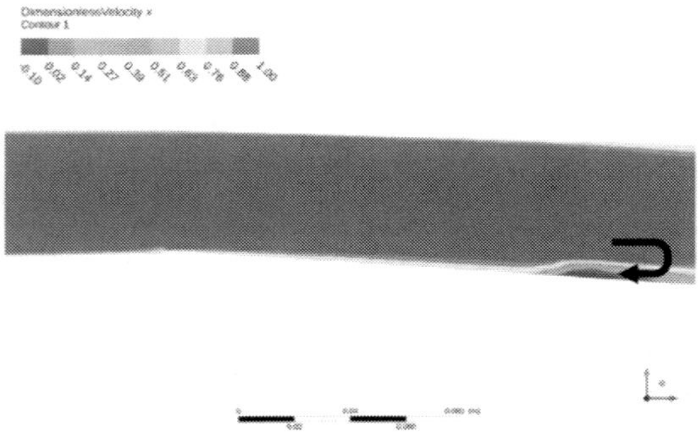

$f = 2$ Hz (maximum cavity volume)

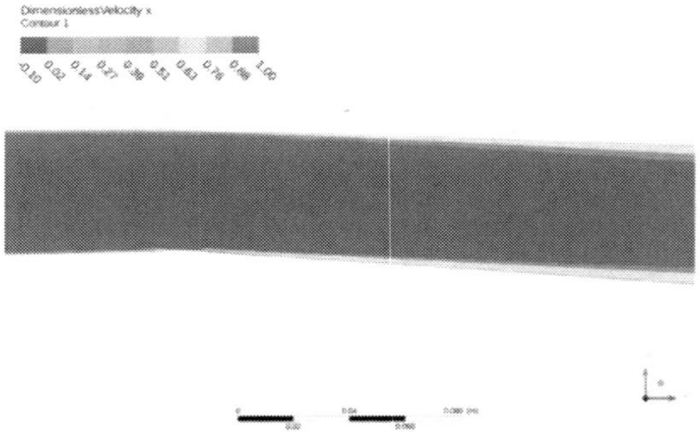

$f = 2$ Hz (minimum cavity volume)

**Figure 4: Contours of the dimensionless streamwise velocity at the minimum and maximum cavity volume for the two frequencies investigated. The black arrow refers to the re-entrant jet.**

## 6. Conclusion

The cavitating URANS computations capture the resonance of a cavitation sheet by setting a periodic outlet pressure. The dynamic behaviour of the cavitation sheet can then be used to determine the cavitation compliance $C_c$ and the second viscosity $\mu''$ by matching the cavitation volume fluctuations predicted by the 1D model with the ones provided by the RANS computations [3].

## 7. References


[1] Barre S., Rolland J., Boitel G., Goncalves E., and Patella R. F., "Experiments and modeling of cavitating flows in venturi: attached sheet cavitation", *Eur. J. Mech. - B/Fluids*, vol. 28, no. 3, pp. 444–464, May 2009.

[2] Alligné S., Nicolet C., Tsujimoto Y., and Avellan F., "Cavitation Surge Modelling in Francis Turbine Draft Tube", *J. Hydraul. Res.*, vol. 52, no. February 2015, pp. 1–13, 2014.

[3] Alligné S., Decaix J., Nicollet C., Avellan F. and Münch C., "Identification of the wave speed and the second viscosity in cavitating flow with 2D RANS computations – Part II". In *Proc of the 9th International Symposium on Cavitation CAV 2015*, 6-10 Dec 2015, Lausanne, Switzerland.



## Acknowledgments

The research leading to the results published in this paper is part of the HYPERBOLE research project, granted by the European Commission (ERC/FP7-ENERGY-2013-1-Grant 608532).

# Analysis of the part load helical vortex rope of a Francis turbine using on-board sensors

**F Duparchy[1], A Favrel[2], P-Y Lowys[1], C Landry[2], A Müller[2], K Yamamoto[2], F Avellan[2]**

[1]Alstom Hydro France, 82 Av. Léon Blum – BP75, 38041 Grenoble Cedex 9, France
[2] EPFL Laboratory for Hydraulic Machines, Avenue de Cour 33 bis, 1007 Lausanne, Switzerland

E-mail: florian.duparchy@power.alstom.com

**Abstract**. The cavitation helical vortex rope arising at Francis turbine partial load induces pressure fluctuations at the rope precession frequency, which can be decomposed into two components; the convective one and the synchronous one. The latter acts as an excitation source for the hydraulic system and is amplified in case of resonance. The present paper aims to highlight the impact of both components on the mechanical behaviour of the runner by performing on-board measurements. It is shown that the use of on-board pressure sensors enables to naturally separate the components of the pressure fluctuations. In addition, the synchronous component has little impact on the mechanical behaviour of the runner, even for the case in which hydro-acoustic resonance occurs.

## 1. Introduction

In the case of Francis turbine part load condition, a cavitation vortex rope is formed in the draft tube cone (figure 1), with a frequency of precession of about $0.3 \times f_0$, where $f_0$ is the rotational frequency of the runner. It induces pressure fluctuations in the draft tube cone at the same frequency, which can be decomposed into two different components [1]. The convective or asynchronous one corresponds to the rotation of the pressure pattern with the vortex core. The second one, called synchronous component, is known to propagate into the hydraulic system, acting consequently as an excitation source for the hydraulic system [2].

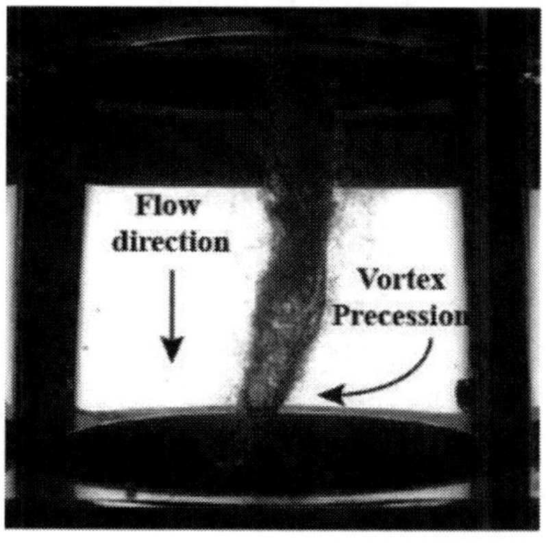


**Figure 1.** Visualization of the cavitation precessing vortex rope in the draft tube cone.

In case of resonance, the latter is amplified, leading to a dynamic pressure and power surge which may jeopardize the system stability [3]. The pressure fluctuations observed at part load condition are expected to impact the mechanical behaviour of the runner and therefore its lifetime. However, the exact impact of both synchronous and convective components is sparsely documented, notably in case of hydro-acoustic resonance between the hydraulic system and the excitation source. For this purpose, an experimental investigation with an instrumented runner is carried out, including on-board pressure sensors and strain gauges. The impact of resonance on both dynamics pressure and strain measured on the runner blade is then analysed.

## 2. Experimental setup

The test case is a reduced scale physical model of a 16 blades Francis turbine with a specific speed $\nu = 0.27$. The model is installed on EPFL test rig PF3 operating in a close-looped configuration. Dynamic pressure is measured in the draft tube cone by 4 flush-mounted piezo-resistive sensors installed in one cross-section of the cone. On-board sensors are installed on the runner blades, including pressure sensors and strain gauges (figure 2). The investigation is focused on a particular operating point at partial load. The speed factor $n_{ED}$ is equal to 0.288 (nominal value) while the discharge factor is fixed at $Q_{ED} = 0.128$, corresponding to 64% of the value at the best efficiency point.

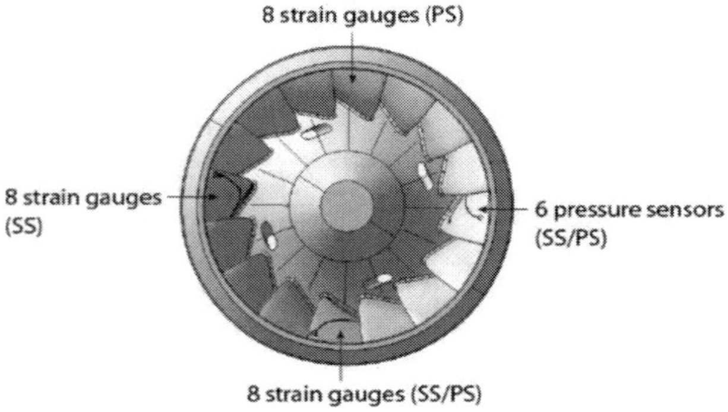


**Figure 2.** Instrumented runner with the location and number of pressure sensors and strain gauges (SS: suction side, PS: pressure side)

## 3. Pressure sensors in the draft tube cone

Pressure sensors in the draft tube cone are traditionally used during model tests to analyse the pressure fluctuations generated by the helical vortex rope. Fourier analysis of the recorded time signal evidences that the dominant frequency is equal to $f_s = 0.31 \times f_0$ (figure 3). This corresponds to the classic hallmark of the helical vortex rope seen in the non-rotating frame. The second harmonic is also observed, but at a significantly lower level.

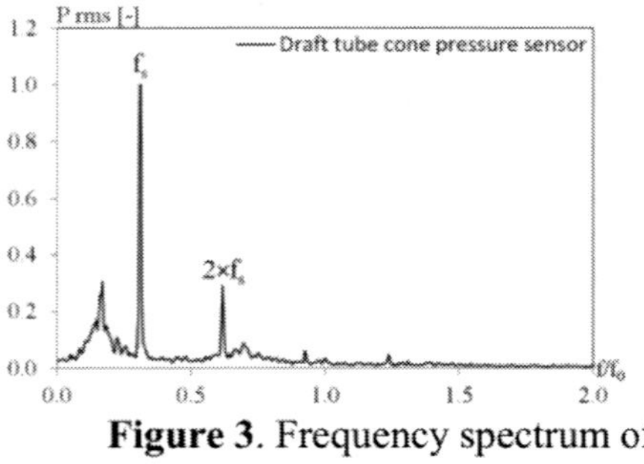


| Frequency | Level (norm. pressure) |
|---|---|
| $f_s$ | 1.00 |
| $2 \times f_s$ | 0.30 |

**Figure 3**. Frequency spectrum of a draft tube cone pressure sensor ($\sigma = 0.11$).

The synchronous and the convective components of the helical vortex rope are then observed at the same frequency. It is possible to separate them by using a set of sensors distributed around the draft tube cone at the same elevation and to apply the relevant mathematical method called Spatial Harmonic Decomposition [4]. It should be noted that the eigen frequency identified in [5] is observed in figure 3 at $0.17 \times f_0$.

## 4. Pressure sensors on the runner

By definition, the frequency of the synchronous part of the helical vortex does not depend on the reference frame of the sensor, whereas the frequency of the asynchronous component depends whether or not the sensor rotates. Consequently, the frequency spectra of the on-board pressure sensors show two distinctive peaks: a first one at $f_s = 0.31 \times f_0$ and a second one at $f_a = f_0 - f_s = 0.69 \times f_0$ (figure 4). The latter corresponds to the rotating component of the vortex rope seen from the rotating frame (figure 5). Using on-board pressure sensors rotating with the runner enables to separate naturally the components of the part load helical vortex rope and to analyse them individually. Moreover, the peak at $0.17 \times f_0$ is observed at the same frequency in the rotating frame, confirming the synchronous nature of this pressure fluctuation.

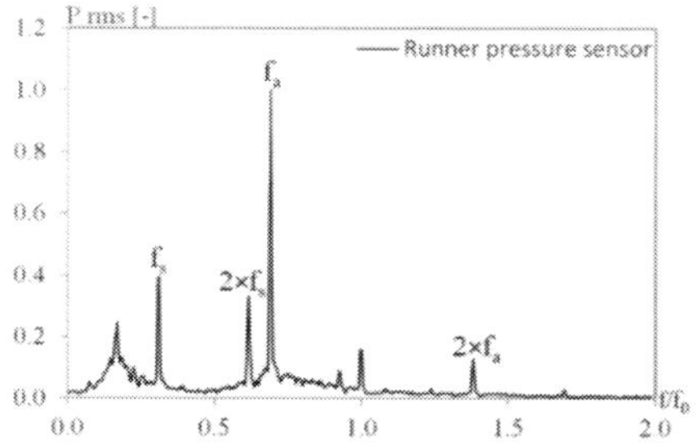


| Frequency | Level (norm. pressure) |
|---|---|
| $f_s$ | 0.46 |
| $2 \times f_s$ | 0.33 |
| $f_a$ | 1.00 |

**Figure 4**. Frequency spectrum of a runner pressure sensor ($\sigma = 0.11$).

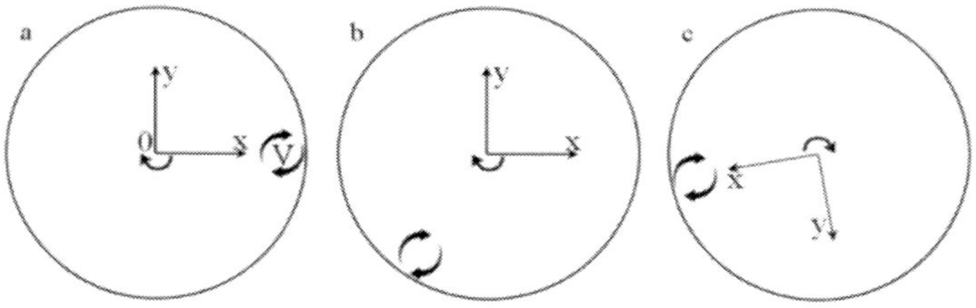


**Figure 5.** Relative position of the runner rotating frame (Oxy) and the part load vortex rope (V). At t = 0 (a), the runner frame and the vortex rope are aligned. At $t = 1 / f_0$ (b), the rotating frame has made one revolution while the vortex has made $f_s / f_0$ revolution. At $t = 1 / (f_0 - f_s)$ (c), the runner frame and the vortex are aligned again.

## 5. Strain gauges on the runner

Strain gauges are sensitive to the mechanical behaviour of the material on which they are located. By placing them at the surface of the runner, they give the opportunity to observe how the pressure fluctuations generated by the part load vortex rope impact the structure. As these sensors are in the rotating frame, the frequency spectra shows two peaks similar to those observed with the on-board pressure sensors. However, it appears clearly that the synchronous component of the vortex rope, at $f_s$, has a low impact on the runner compared to the asynchronous one (figure 6). This means that it is the convective component of the vortex rope that is the most likely to generate fluctuating stresses and potential fatigue damage to the runner.

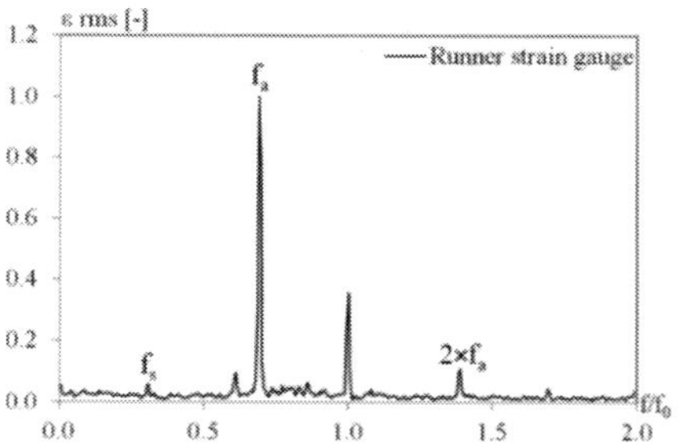


| Frequency | Level (norm. strain) |
|---|---|
| $f_s$ | 0.05 |
| $f_a$ | 1.00 |

**Figure 6.** Frequency spectrum of a runner strain gauge ($\sigma = 0.11$).

## 6. Impact of resonance

The cavitation number, $\sigma$, is modified step by step from $\sigma = 0.11$ (prototype condition) to $\sigma = 0.37$ (cavitation-free condition) by changing the pressure over the free surface of the downstream reservoir. By increasing the Thoma number, the cavitation volume in the draft tube cone decreases, inducing a rise of the hydraulic system's eigen frequencies [6]. The resonance conditions are met when the first eigen frequency of the system matches with the precession frequency [5], which occurs for this test case at $\sigma = 0.17$.

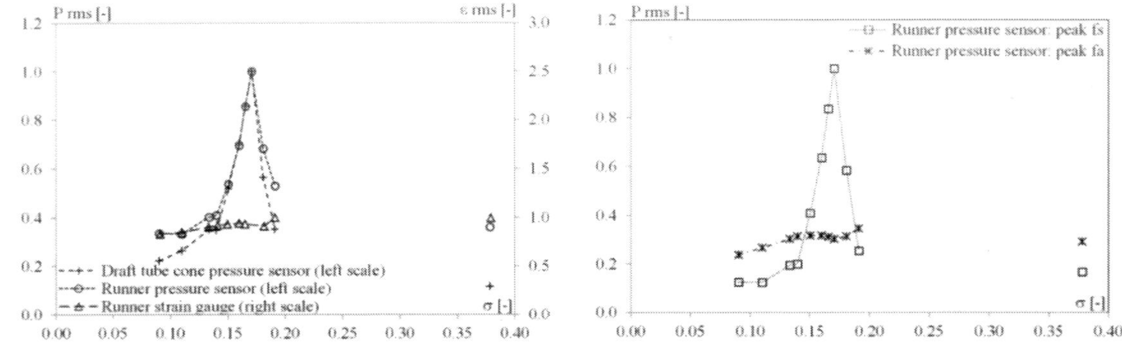


**Figure 7.** Evolution of the dynamic pressures and strain plotted against $\sigma$ (left) and evolution of the amplitude of the peaks $f_s$ and $f_a$ of the runner pressure sensor plotted against $\sigma$ (right).

Pressures in the draft tube cone and in the runner show a similar evolution, with a significant increase in resonance conditions (figure 7). On the contrary, the dynamic strain is almost not affected by the hydro-acoustic resonance. By separating the synchronous component from the asynchronous one using a pressure sensor on the runner, it can be observed that only the former is sensitive to the resonance. Moreover, the asynchronous component, $f_a$, has the same evolution as the dynamic strain.

## 7. Conclusion

On-board strain gauges located at the surface of the runner reveal that the asynchronous component of the part load helical vortex rope is a predominant source of mechanical excitation. On the contrary, the synchronous component of the vortex has a lower impact on the mechanical excitation. In the case of hydro-acoustic resonance, it is principally the synchronous component which is affected. As a consequence, the dynamic pressure significantly increases, while the dynamic strain is almost not affected. These observations show that the usual peak-to-peak pressure fluctuations measured by the draft tube cone pressure sensors are not sufficient to estimate the impact of the cavitation part load vortex rope on the mechanical stress of the runner. A more physical approach based on the decomposition of the components of the part load vortex rope should be preferred.

## Acknowledgments

The research leading to the results published in this paper is part of the HYPERBOLE research project, granted by the European Commission (ERC/FP7- ENERGY-2013-1-Grant 608532).

## References

[1] Nishi M and Liu S 2013 *International Journal of Fluid Machinery and Systems* **6** 33
[2] Dörfler P 1982 Systems Dynamics of the Francis turbine half load surge *Proc. 11th Symposium on Hydraulic Machinery and Systems* (Amsterdam)
[3] Rheingans W J 1940 *Transactions of the ASME* **62** 171
[4] Duparchy A, Guillozet J, De Colombel T and Bornard L 2014 *IOP Conference Series: Earth and Environmental Science* **22** 032016 (IOP Publishing)
[5] Favrel A, Landry C, Müller A, Yamamoto K and Avellan F 2014 *IOP Conference Series: Earth and Environmental Science* **22** 032035 (IOP Publishing)
[6] Landry C, Favrel A, Müller A, Nicolet C, Yamamoto K and Avellan F 2014 *IOP Conference Series: Earth and Environmental Science* **22** 032037 (IOP Publishing)

# Study on cavitation influence for pump head in an axial flow pump

**K Hosono[1], Y Kajie[1], S Saito[2] and K Miyagawa[2]**

[1] Graduate School, Department of Applied Mechanics, WASEDA University, 3-4-1 Okubo, Shinjuku-ku, Tokyo 168-8555, Japan
[2] Department of Applied Mechanics and Aerospace Engineering, WASEDA University, 3-4-1 Okubo, Shinjuku-ku, Tokyo 168-8555, Japan

E-mail: hoppi-kazu@fuji.waseda.jp

**Abstract.** The size of axial flow pumps used in drainage pump stations has recently decreased, and their rotation speeds have increased, causing an increase in the risk of cavitation. Therefore, to provide highly reliable pumps, it is important to understand the internal flow of pumps under cavitating conditions. In this study, high-speed camera measurements and computational fluid dynamics analysis were performed to understand the cavitation performance of an axial flow pump. The mechanism that causes the head to change as a result of cavitation under low net positive suction head values is shown to be the balance between the increasing angular momentum and the loss indicated by the changing streamlines.

## 1. Introduction

Recently the size of axial flow pumps has decreased, and their rotation speeds have increased. This has caused an increase in the risk of cavitation, which causes performance degradation. Therefore, to provide highly reliable pumps, it is important to understand the internal flow under cavitating conditions.

Many studies have proposed methods of determining the internal flow of axial flow pumps under cavitating conditions. Using experimental measurements of an axial flow pump under cavitating conditions, Saito [1] showed that impeller outlet flow is affected by the occurrence of cavitation at the impeller. Goltz *et al* [2] examined the performance of an axial flow pump and showed a difference in cavitation due to the flow rate. Zheng *et al* [3] discussed the tip leakage flow of an axial flow pump based on their experiments and computational fluid dynamics (CFD) analysis and showed that the tip leakage vortex is influenced by the blade loading and operation conditions.

Thus, the influence of cavitation on pump performance has already been studied, meaning the internal flow in an axial flow pump has been measured. However, the mechanisms that cause the change in performance have not yet been clarified. Through the results of experiments and CFD analysis, this paper reveals the mechanisms that change the performance of an axial flow pump when cavitation occurs.

## 2. Experimental apparatus and computational fluid dynamics analysis

### 2.1. *Experimental apparatus and procedures*

Figure 1(a) shows the axial flow pump test apparatus in the closed-loop water tunnel. The diameter of the pipes at the test section is 145 mm, and that elsewhere is 155.4 mm. The test impeller is shown in figure 2. The number of impeller blades is 4, and the specific speed of this impeller at the best efficiency point is $N_s$ = 1400 [m, m$^3$/min, rpm]. The diameter of the impeller is 144.4 mm, and the clearance between the impeller tip and the casing is 0.3 mm. The casing of the test section, which is shown in figure 1(b), is made of acrylic glass, allowing the internal flow and cavitation to be visualized. The net positive suction head (NPSH) can be changed by the vacuum pump connected to the upstream tank for cavitation tests.

The flow rate is changed by the valve at the discharge side of the pump and the flow rate is measured by a magnetic flowmeter. The head is calculated from the differences in the static and dynamic pressures caused by changing the cross sectional area. Moreover, the shaft power is computed by a torque meter at the impeller shaft. The internal flow and cavitation are imaged by a high-speed digital video camera. The frame rate of this camera is set to 1000 fps.

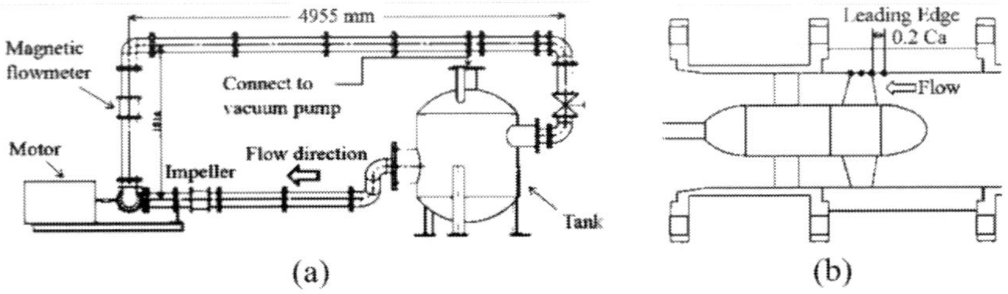


**Figure 1.** Axial flow pump test apparatus. (a) Overview of apparatus, (b) test section of axial flow pump.

**Figure 2.** Test impeller.

*2.2. Pump performance and cavitation characteristics*

Figure 3(a) shows the efficiency, the head coefficient $\Psi$, and the shaft power coefficient $\nu$ with respect to the nondimensionalized flow rate $Q/Q_d$ at the design point; these graphs represent the performance curves of the axial flow pump. There is a monotonic decrease in the head coefficient from the shut point to $Q/Q_d = 0.7$, followed by an increase in the range $Q/Q_d = 0.7$-$0.85$ as a positive-slope characteristic. Figure 3(b) shows cavitation characteristics of the head coefficient $\Psi$ with respect to NPSH. At the design point flow rate, the head coefficient achieves a maximum value at nearly NPSH = 2.2 m. Conversely, there are no peaks for flow rates less than the design point flow rate.

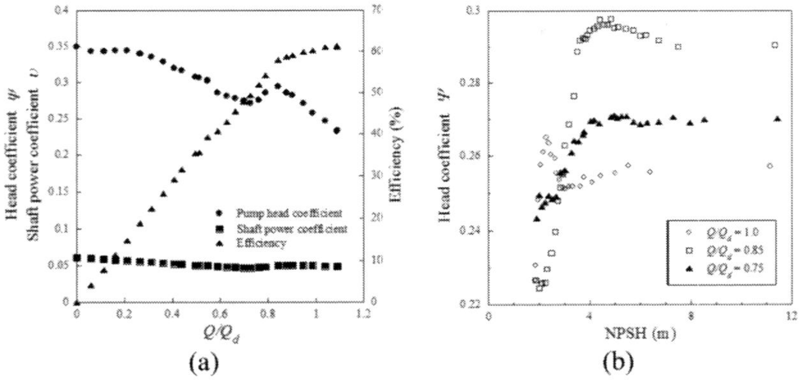


**Figure 3.** (a) Experimentally obtained performance curves and (b) cavitation characteristics of the axial flow pump.

*2.3. Computational fluid dynamics analysis and procedures*

In this study, the commercial CFD code ANSYS CFX 14.5 is used. The computational grid consists of approximately 4.2 million hexa-elements and 4.3 million nodes. There are 20 layers in the tip clearance. In this CFD analysis, the standard k-ε model is used as the turbulence model, and the simplified Rayleigh-Plesset based model in ANSYS CFX is used as the cavitation model. The boundary conditions of constant total pressure and constant mass flow rate are set at the inlet and outlet, respectively.

## 3. Result and discussion

### *3.1. Comparison of computational fluid dynamics analysis and experimental results*

Figure 4 shows the cavitation characteristics of the axial flow pump obtained experimentally and by CFD analysis at the design point. The CFD analysis and experiments are in good agreement and both have a maximum head coefficient point at NPSH = 2.2 m. Therefore, the internal flow under cavitating conditions can be simulated by CFD analysis with high accuracy.

The photographs on the left of figure 5 show the cavitation at the design point for NPSH values of 6.3 m and 4.2 m. Vortex cavitation occurs at NPSH = 6.3 m, and both vortex and sheet-type cavitation occur at NPSH = 4.2 m. The images on the right in figure 5 show the visualization of the cavitation area (void fraction = 0.1) obtained by CFD analysis. In the case of NPSH = 4.2 m, the sheet-type cavitation can be simulated, but the vortex cavitation cannot.

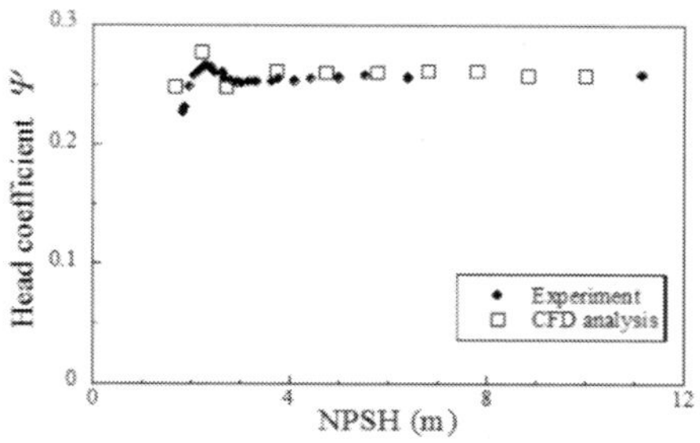


**Figure 4.** Cavitation characteristics obtained experimentally and by CFD analysis at the design point.

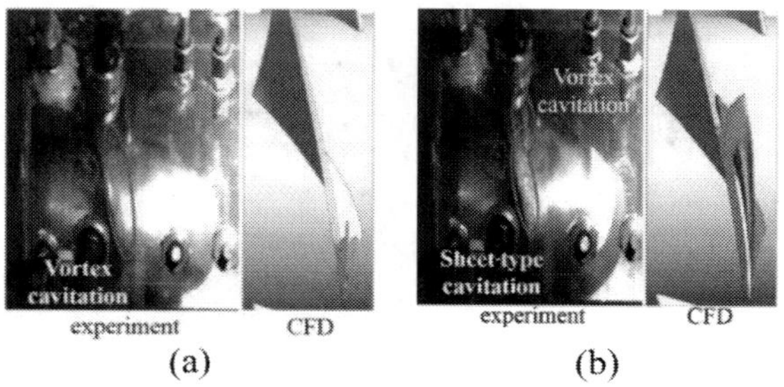


**Figure 5.** Visualization of cavitation at the design point for NPSH = (a) 6.3 m (b) 4.2 m.

### *3.2. Head coefficient changing under low net positive suction head*

Figure 6 shows the distributions of the pressure coefficient and the angular momentum from the front to the back of the impeller obtained by CFD analysis at the design point. In the pressure coefficient distribution, it is linked to the head coefficient changing. However, the angular momentum increases with decreasing NPSH, especially in the shroud. Under these conditions, the angular momentum is highest at NPSH = 1.8 m, but the pressure coefficient is highest at NPSH = 2.2 m.

The impeller blade loading distribution from leading edge (L.E.) to trailing edge (T.E.) at the hub and the shroud are shown in figure 7. At the hub, CFD analysis shows that the blade loading is largest at NPSH = 1.8 m. By contrast, the blade loading values at NPSH = 1.8 m and 2.2 m in the shroud are almost equal. In spite of this, the head coefficient is highest at NPSH = 2.2 m. This demonstrates that changes in the head coefficient do not correspond to changes in the impeller blade loading distribution.

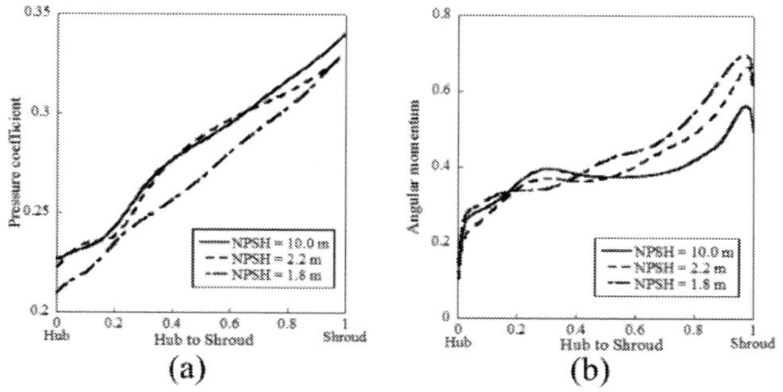


**Figure 6.** Distributions of (a) the pressure coefficient and (b) the angular momentum at the design point.

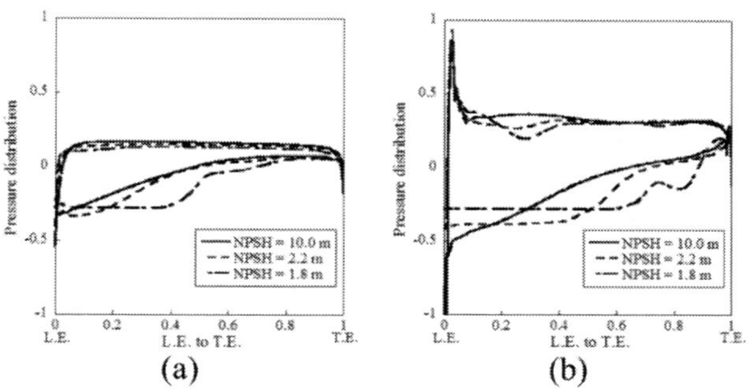


**Figure 7.** Impeller blade loading at (a) the hub and (b) the shroud at the design point.

### *3.3. Mechanisms of head coefficient changing under low net positive suction head*

Figure 8 shows the average loss coefficient at the pump meridional plane obtained by CFD analysis at the design point. The loss coefficient (L.C.) is defined as the nondimensionalized relative total pressure decrease from the inlet value. The high loss coefficient area increases as the NPSH becomes smaller. The loss caused by the tip leakage flow is high near the shroud area. Moreover, near the hub area, the loss caused by the cross sectional flow due to cavitation is higher at low NPSH values.

Figure 9 shows the streamlines obtained by CFD analysis on the suction side of the blade. Comparing the streamlines for NPSH = 2.2 and 1.8 m indicates that the position of the cross sectional flow moves downstream with decreasing NPSH. The pump head changing due to cavitation is the result of the balance between the increasing angular momentum and the loss indicated by the streamlines changing.

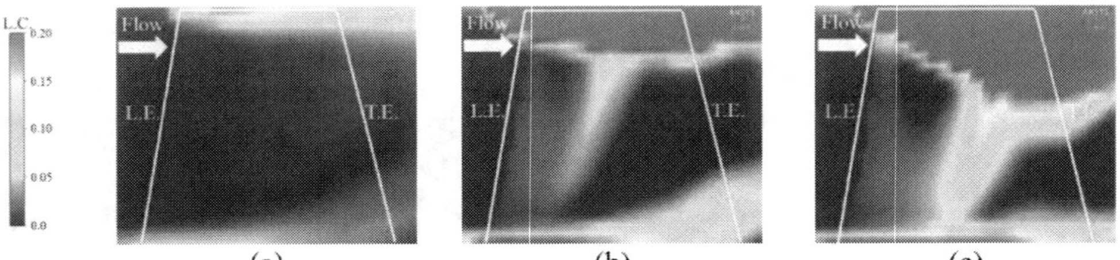


**Figure 8.** Loss coefficient contour at the meridional plane at the design point for NPSH = (a) 10 m, (b) 2.2 m, (c) 1.8 m.

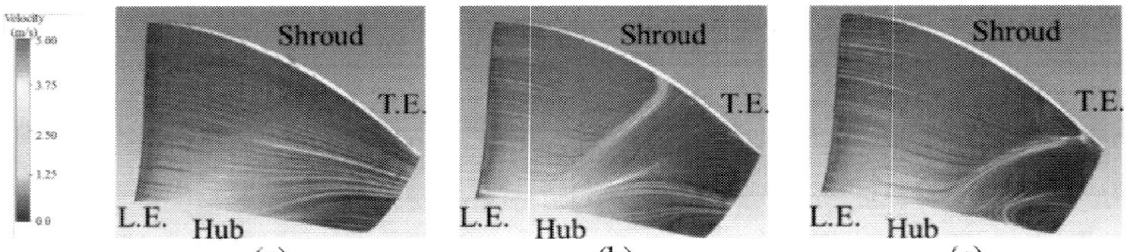


**Figure 9.** Streamlines at the blade surface suction side at the design point for NPSH = (a) 10 m, (b) 2.2 m, (c) 1.8 m.

## 4. Conclusion

The results obtained from this study are as follows.

(1) The angular momentum increases with decreasing NPSH under cavitating conditions. However, changes in the pressure and head coefficients do not correspond to increasing angular momentum.
(2) With decreasing NPSH, the high loss coefficient area caused by the tip leakage and cross sectional flows due to cavitation increases.
(3) The changes in the pump head due to cavitation are the result of the balance between the increasing angular momentum and the loss indicated by the changing streamlines.

## References

[1] Saito S 1987 Cavitation Aspect and Flow Pattern in an Axial-Flow Impeller *T. JSME B* **53-492** pp 3682-90
[2] Gortz I, Kosyna G, Stark U, Saathoff H and Bross S 2003 Stall inception phenomena in a single-stage axial-flow pump *Proc. IMechE* (Braunschweig, Germany) pp 471-79
[3] Zhang D, Shi W, Esch B, Shi L and Dubuisson M 2015 Numerical and experimental investigation of tip leakage vortex trajectory and dynamics in an axial flow pump *Computers & Fluids* **112** pp 61-71

Journal of Physics: Conference Series **656** (2015) 012063 doi:10.1088/1742-6596/656/1/012063

# Experimental comparison of backflow-vortex cavitation on pump inducer between cryogen and water

**Y Ito, A Tsunoda, and T Nagasaki**
Department of Energy Sciences, Tokyo Institute of Technology, 4259-G3-33-402 Nagatsuta-cho, Midori-ku, Yokohama, Kanagawa 226-8502, Japan

E-mail: itoyu110@00.alumni.u-tokyo.ac.jp

**Abstract.** The backflow-vortex cavitation on a rotating inducer was experimentally observed in both liquid nitrogen and water. The cavitation appeared "foggy" in nitrogen and "foamy" in water. This is explained by the critical Weber number theory. The number of backflow-vortex cavitation columns decreased as the head coefficient increased. This tendency is explained by a theoretical analysis in which backflow vortices are regarded as an array of vortex filaments rotating along the casing around the inducer center. The number of backflow-vortex cavitation columns is determined by the fluid dynamics and is independent of thermodynamic properties.

## 1. Introduction

Turbopumps for liquid-fueled rocket engines are required to be small for lightness, to suck in propellants at the minimum possible pressure to allow thin tank walls and a high payload, and to discharge propellants at the maximum possible pressure for a large thrust and a high specific impulse (ISP; with a higher ISP, a lower propellant flow rate is required for the same amount of thrust). Accordingly, a small impeller must operate at a high rotational rate; thus, cavitation is inevitable in turbopumps. The combination of an inducer (which has a low pressure ratio but resists cavitation) and a centrifugal pump (which breaks down owing to cavitation but has a high pressure ratio) is the best choice for rocket turbopumps. The inducer preliminarily compresses the working fluid and crushes the

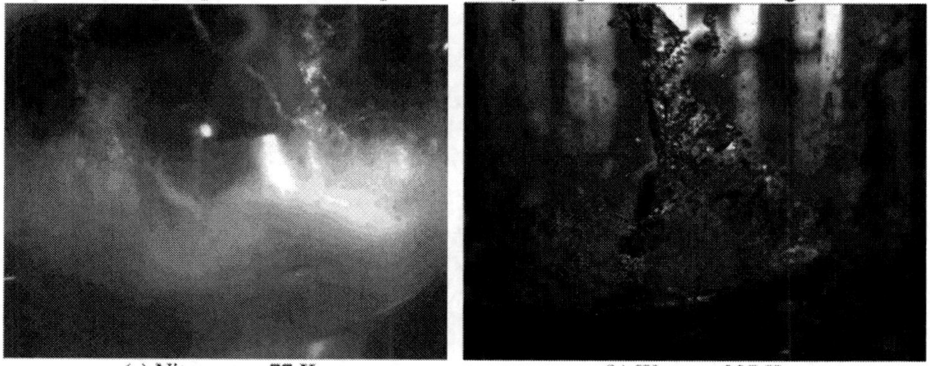

(a) Nitrogen at 77 K
$\sigma = 0.059$, $\phi/\phi_{max} = 0.47$, $\psi/\psi_{max} = 0.53$
$\omega/\omega_{ind} = 0.28$, $d/d_{ind} = 0.72$, $\delta/d_{ind} = 0.0046$

(b) Water at 337 K
$\sigma = 0.058$, $\phi/\phi_{max} = 0.45$, $\psi/\psi_{max} = 0.55$
$\omega/\omega_{ind} = 0.29$, $d/d_{ind} = 0.69$, $\delta/d_{ind} = 0.10$

**Figure 1.** Comparison of backflow-vortex cavitation on inducer between liquid nitrogen and water.

cavitation. Then, the centrifugal pump boosts the pressure in one shot. However, the inducer pump head abruptly fails at a very low cavitation number, $\sigma$. To effectively use the inducer, it is necessary to allow controllable cavitation to form.

Although many types of cavitation have been observed on inducers, this study focused on backflow-vortex cavitation, which is shown in Figures 1 and 2. Backflow-vortex cavitation often occurs in practical rocket turbopumps. It not only causes oscillating flows on the inducer but also affects the inducer performance. The backflow-vortex cavitation on an inducer inlet was experimentally observed by Acosta [1]. Subsequently, Aoki and Yamamoto [2], Tsujimoto et al. [3], and Yokota et al. [4] focused their research on water backflow-vortex cavitation.

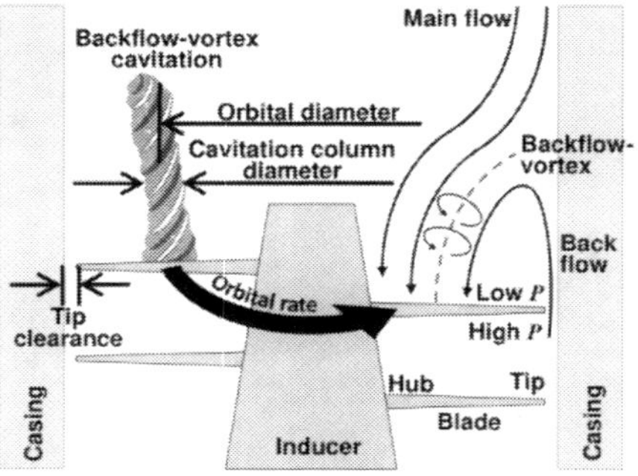

**Figure 2.** Schematic of backflow-vortex cavitation.

High-performance rocket engines use cryogens as propellants. Cryogens have "thermodynamic effects" whereby the temperature around the cavitation decreases because of latent heat removal, followed by a decrease in the saturation pressure and the suppression of additional cavitation. These effects decrease $\sigma$ at the cavitation inception, decrease $\sigma$ at the cavitation breakdown, and improve the pump head at low $\sigma$ compared with room-temperature water. The consequences of the thermodynamic effects were reported by Stahl and Stepanoff [5] for hot water. Later, they were confirmed by Salemann [6], Stepanoff [7], Spraker [8], Meng [9], Ball et al. [10], Ruggeri and Moore [11], Franc et al. [12], Watanabe et al. [13], and Ito et al. [14, 15]. Ball et al. [10], Watanabe et al. [13] and Ito et al. [14, 15] visualized the cryogenic cavitation on a rotating inducer. However, there have been very few studies of the thermodynamic effects on backflow-vortex cavitation, although such cavitation may cause phenomena whereby the thermodynamic effects improve the pump performance.

Ito et al. [16] used a visualization facility to compare the backflow-vortex cavitation on a rotating inducer in liquid nitrogen with that in water. They reported that the orbital rotational rate, orbital rotational diameter, and diameter of each backflow-vortex cavitation column were determined by the head coefficient, and that among these, only the diameter of each backflow-vortex cavitation column was affected by thermodynamic effects. In the present paper, we report quantitative measurements of the number of backflow-vortex cavitation columns using the same facility as [16] and investigate whether the thermodynamic properties affect the number of backflow-vortex cavitation columns.

## 2. Experimental Results and Discussion

### *2.1. Comparison of backflow-vortex cavitation on inducer between liquid nitrogen and water*

Figure 1 compares the backflow-vortex cavitation in liquid nitrogen at 77 K with that in water at 337 K under the same condition as Ito et al. [16] but at a different timing. The cavitation in nitrogen exhibits finer bubbles, i.e., cavities, compared with that in water. According to Sarosdy and Acosta [17], the cavitation behind a disk in water was "clear and well defined," and that in Freon was "indistinct and frothy." Freon, like cryogen, has a strong thermodynamic effect; thus, nitrogen cavities are similar to Freon cavities. Later, in 1965, Gadd and Grant [18] visualized the cavitation behind a disk in water, reporting that a very sharp edge yielded a "glassy clear" cavity and a dull edge yielded a "rough and striated frosted-glassy" cavity. In 1967, Ball et al. [10] visualized the cavitation around a rotating inducer in liquid hydrogen. Unfortunately, although a "foggy" cavitation is observed in their visualization photos, they reported nothing about the appearance of the cavitation in hydrogen. In 2004, Franc et al. [12] visualized the cavitation around a rotating inducer in Freon (R114). In their photos, the sheet cavitation due to the leading edge on the inducer surface is white. This probably means that the sheet cavitation is a cavity with a rough and striated surface or that it comprises small bubbles. In photos obtained using reflected light, the rough and curved surfaces are whiter, and the smooth and planar surfaces are blacker. Thus, the cavitation in Freon observed by Franc et al. exhibited similar features to that observed by Sarosdy and Acosta [17]. Yoshida et al. [19] compared the cavitation around a rotating inducer in liquid nitrogen observed by Watanabe et al. [13] with that in water at the same cavitation number, rotational rate, and flow rate. They described the cavitation in nitrogen and water as "creamy" and "icy," respectively. Nonetheless, according to Figure 1, the cavitation in nitrogen and water is better described as "foggy" and "foamy," respectively. Cavitation bubbles are generated by shears between the backflow and blade tip and are transported to the backflow-vortex region, as shown in Figure 2. To describe the stable sizes of the cavitation bubbles, $D_{Bmax}$, Ito et al. [16] introduced the critical Weber number, $\mathrm{We}_c$, as

$$D_{Bmax}=\mathrm{We}_c\,\frac{s_L}{U^2\rho_L}, \tag{1}$$

where $s_L$ is the surface tension, $U$ is the local velocity, and $\rho_L$ is the density. Here $D_{Bmax}$ for nitrogen is 1/4 of that for water. The smaller bubbles have a larger area per volume; thus, they allow better heat transfer during growth and extinction. This is one of the reasons why nitrogen exhibits stronger thermodynamic effects than water. Furthermore, the nitrogen cavitation easily vanishes where the pressure recovers; consequently, the nitrogen pump exhibits better performance than water under low-$\sigma$ operation.

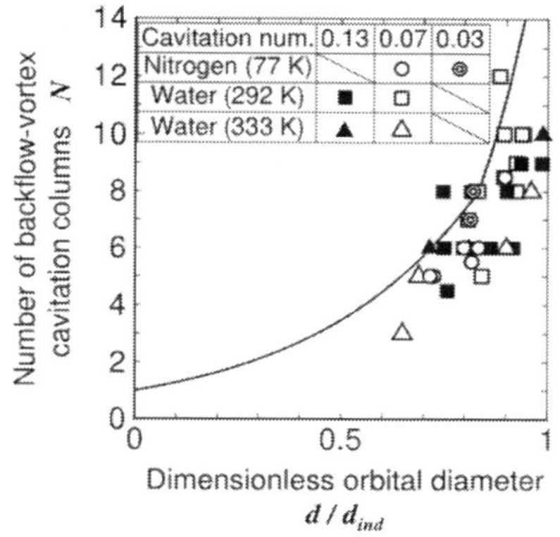


**Figure 3.** Number of backflow-vortex cavitation columns versus orbital diameter.

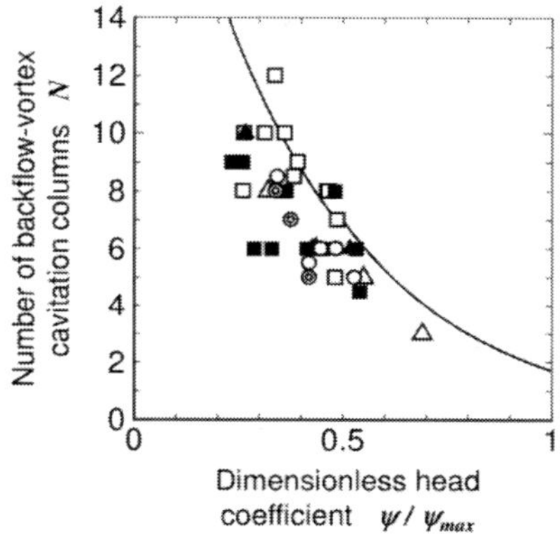


**Figure 4.** Number of backflow-vortex cavitation columns versus head coefficient.

### 2.2. Number of backflow-vortex cavitation columns

At the tip region, the backflow occurred because of the pressure difference between the inducer blade surfaces, as shown in Figure 2. The backflow made a deep inroad into the main flow and was pushed back to the inducer surface by the main flow. The backflow contained a rotational-velocity component that was affected by the rotating inducer. On the other hand, the main flow had little rotational velocity. This flow was similar to the Taylor-Couette flow. At the contact surface between the backflow and the main flow, a shear flow was formed that generated a backflow-vortex. The backflow-vortex rotated around itself in the same rotational direction as the inducer, and the orbital rotation also had the same direction as the inducer because the outer flow was faster than the inner flow.

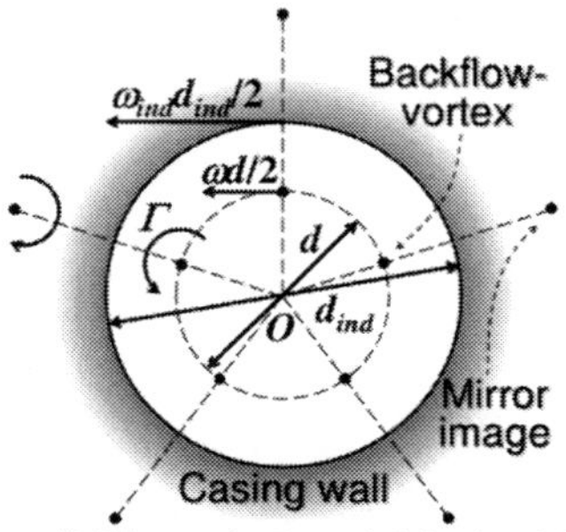


**Figure 5.** Theoretical model for backflow-vortices (e.g. $N = 5$)

The number of backflow-vortex cavitation columns was investigated. Figure 3 shows the relation between the number of backflow-vortex cavitation columns, $N$, and the dimensionless orbital diameter, $d$. $N$ increased as $d$ approached unity. On the other hand, Figure 4 shows the relation between $N$ and the dimensionless head coefficient, $\psi$. Here, to clarify how to determine $N$, a theoretical analysis is conducted, as shown in Figure 5. The backflow-vortices are regarded as an array of vortex filaments rotating along the casing wall around the center of the inducer. Yokota et al. [4] conducted a stability analysis of this system; however, we did not. This is because each backflow-vortex is always generated by a shear of the Taylor-Couette flow that is different from that of the Karman vortex flow. Therefore, an array of vortex filaments can exist if the flow is physically admissible. The circulation $\Gamma$ of each backflow-vortex around itself was assumed as

$$\Gamma=\pi\,\omega_{ind}d_{ind}^2\left\{1-(\omega/\omega_{ind})(d/d_{ind})\right\}\left\{1-(d/d_{ind})\right\}/8, \tag{2}$$

where $\omega$ is the experimental rotational rate of the backflow-vortices around $O$. According to [16], the relation among $\omega$, $d$, the inducer rotational rate, $\omega_{ind}$, and the inducer diameter, $d_{ind}$, is

$$(\omega/\omega_{ind})=1.36\{1-(d/d_{ind})\}^{1.49}+0.119\,. \tag{3}$$

In Figure 5, the rotational rate, $\omega'$, of each backflow-vortex around $O$ by all of the other vortices is theoretically calculated as follows:

$$\left(\frac{\omega'}{\omega_{ind}}\right)=\frac{1}{4(d/d_{ind})}\left\{1-\left(\frac{\omega}{\omega_{ind}}\right)\left(\frac{d}{d_{ind}}\right)\right\}\left\{1-\left(\frac{d}{d_{ind}}\right)\right\}\left\{\frac{N-1}{2}\frac{1}{(d/d_{ind})}+\frac{(d/d_{ind})}{1-(d/d_{ind})^2}-\sum_{n=1}^{N-1}\frac{(d/d_{ind})^3-(d/d_{ind})\cos(2\pi n/N)}{(d/d_{ind})^4-2(d/d_{ind})^2\cos(2\pi n/N)+1}\right\} \tag{4}$$

In cases where the $\omega'$ indicated by Equation (4) equals the $\omega$ indicated by Equation (3), an array of backflow-vortices is physically admissible. The curves in Figures 3 and 4 show splines of these discrete solutions. According to Equation (3), as $d$ increases, $\omega$ decreases; thus, a larger $N$ is derived at a larger $d$ from Equation (4), as shown by the curve in Figure 3. As indicated by [16], as $\psi$ increases, $d$ decreases and $\omega$ increases; therefore, a smaller $N$ is derived at a higher $\psi$, as shown by the curve in Figure 4. Both figures indicate that the cavitation number and the type of fluid hardly affected $N$. This suggests that $N$ is determined by the fluid dynamics and independent of thermodynamic properties.

## 3. Conclusions

The backflow-vortex cavitation on a rotating inducer was experimentally observed in both liquid nitrogen and water. The results are as follows:

– The cavitation appeared "foggy" in nitrogen and "foamy" in water. This is explained by the critical Weber number theory.

– The number of backflow-vortex cavitation columns decreased as the head coefficient increased.

– The number of backflow-vortex cavitation columns was independent of thermodynamic properties. This is explained by a theoretical analysis of the backflow-vortex cavitation.

## 4. References

[1] Acosta A J 1958 *2nd Symposium on Naval Hydrodynamics*, 533
[2] Aoki M and Yamamoto K 1992 *ASME J. of Turbomachinery* **114** 392
[3] Tsujimoto Y, Yoshida Y, Maekawa Y, Watanabe S and Hashimoto T 1997 *ASME J. of Fluids Engineering* **119** 775
[4] Yokota K, Kurahara K, Kataoka D, Tsujimoto Y and Acosta A J 1999 *JSME International J. Series B* **42** 451
[5] Stahl H A, Stepanoff A J and Phillipsburg N J 1956 *Transactions of ASME* 1691
[6] Salemann V 1959 *ASME J. of Basic Engineering* **81** 167
[7] Stepanoff A J 1961 *ASME J. of Engineering for Power* **83** 79
[8] Spraker W A 1965 *ASME J. of Engineering for Power* **87** 309
[9] Meng P R 1968 NASA TN D-4423
[10] Ball C L, Meng P R and Reid L 1969 NASA TM X-1360
[11] Ruggeri R S and Moore R D 1969 NASA TN D-5292
[12] Franc J P, Rebattet C and Coulon A 2004 *ASME J. of Fluids Engineering* **126** 716
[13] Watanabe M, Nagaura K, Hasegawa S, Niiyama K, Yoshida Y and Sugita E 2010 JAXA RM-09-010
[14] Kurishita Y, Ito Y and Nagasaki T 2012 *Asian Joint Conference on Propulsion and Power 2012*, AJCPP2012-093
[15] Ito Y, Kurishita Y, Kitano S and Nagasaki T 2012 *The 8th International Symposium on Cavitation*, CAV2012 No.104
[16] Ito Y, Tsunoda A, Kurishita Y, Kitano S and Nagasaki T 2015 "Experimental Visualization of Cryogenic Backflow Vortex Cavitation with Thermodynamic Effects" *AIAA J. of Propulsion and Power.*
[17] Sarosdy L R and Acosta A J 1961 *ASME J. of Basic Engineering* 399
[18] Gadd G E and Grant S 1964 *J. of Fluid Mechanics* **23** 645
[19] Yoshida Y, Kikuta K, Niiyama K and Watanabe S 2012 *ASME Fluids Engineering Summer Meeting*, FEDSM2012-72212

## Acknowledgments

This work was financially supported by the Iwatani Naoji Foundation and the JAXA's Engineering Digital Innovation (JEDI) Center. The authors thank Professor Emeritus Toshio Nagashima at the University of Tokyo, who provided advice. The authors also thank Mr. Ichiro Takaki at Tokyo Institute of Technology, who selected additional data from experimental movies.

# Progress in Numerical Simulation of Turbulent Cavitating Flow in an Axial Waterjet Pump

**Sung-Eun Kim**[1]

Naval Surface Warfare Center Carderock Division, 9500 McArthur Blvd., West Bethesda, MD 20817, U.S.A.

E-mail: sungeun.kim@navy.mil

**Abstract**. A progress made in numerical simulation of turbulent cavitating flow in an axial waterjet pump (ONR AxWJ-2) is presented. The computational approach is designed to resolve the interaction between the rotor and the stator both in time and in space comprising the full 360° sector of the pump, using a scalable finite-volume-based Navier-Stokes solver. Unsteady Reynold-Averaged Navier-Stokes (URANS) and a hybrid of URANS and Large Eddy Simulation (LES) are employed for turbulence modeling and also for directly capturing some of large-scale coherent structures in the flow. A continuum mixture approach based on phasic volume-fraction is adopted along with a finite-rate mass-transfer model. It is shown that both the major quantities of interest (QOI) of waterjet pumps and the salient features of the flow such as tip-leakage vortex (TLV) cavitation can be predicted with a commendable accuracy.

## 1. Introduction

There have been attempts to numerically predict performance of axial waterjet pumps, with more recent ones employing URANS [1,2,3]. Still, the complex physics of three-dimensional turbulent cavitating flow in axial waterjet pumps has evaded an explication of various flow phenomena such as thrust-breakdown.

This article concerns the turbulent cavitating flow in an axial waterjet pump, AxWJ-2, that has a 6-bladed rotor and a 8-bladed stator. The rotor-stator assembly is housed in a cylindrical shroud with the rotor-tip clearance of 0.7 mm. The pump was tested in the 36-inch cavitation tunnel at the Naval Surface Warfare Center Carderock Division (NSWCCD) [4,5]. It was also tested separately in the Rolls-Royce pump loop facility [6]. The AxWJ-2 model of 12-inch diameter was used by the group at the Johns Hopkins University (JHU) for detailed measurements and visualizations of the flow-fields [7]. In tandem with the experimental works, a high-fidelity computational capability has been developed aiming at simulating the entire pump and resolving the interaction between the rotor and the stator both in time and in space comprising the full 360° sector of the pump. The goal was to be able to predict not only the global quantities of interest (QOI) such as thrust, torque and head but also to capture the salient features of the turbulent cavitating flow including different types of cavitation (e.g., sheet cavity, tip-leakage vortex cavitation) and their impacts on the pump performance.

In the following, the key elements of the computational approach are summarized first. The progress made to date will be discussed along with the challenges, lessons learned, and future work.

## 2. Computational Approach – Finite-Volume Method with Continuum Approximation

A homogeneous equilibrium mixture model is adopted to describe two phases (water liquid and vapor). The individual phases can be modeled as incompressible or compressible fluid, which allows us to resolve shock and acoustic waves generated by cavitating flows. The phase compositions are represented by volume-fraction, with the phase change modeled using, among others available, a finite-rate mass transfer model derived from a modified Rayleigh-Plesset's equation [8]. An implicit, finite-volume-based projection algorithm is employed for a coupled system of equations for velocity, pressure and volume-fraction. A sliding-grids method [1] is used to pursue time-accurate solutions for rotor-stator interaction. The details of the computational method and validations can be found in the references [1,8].

For turbulence modeling, URANS approach is usually sufficient to predict the global quantities and to capture the gross features of the flow. The URANS computations for the AxWJ-2 pump have been made to date using the family of linear k-ω models, although more sophisticated models such as nonlinear or explicit algebraic k-ω models and differential Reynolds-stress models can potentially improve the accuracy of URANS predictions. Capturing large-scale coherent structures requires a "scale-resolving" approach. To that end, we chose to employ a hybrid URANS-LES approach based on the Improved Delayed Detached Eddy Simulation (IDDES)[9].

## 3. Results

### *3.1. Computational Details*

URANS computations have been carried out to simulate the NSWCCD's 36-inch tunnel experiment which covered a wide range of cavitation number and flow-rate including thrust-breakdown at the rotor speed of 2000 RPM and the inlet flow velocity around 11 m/s. Three computational grids with 2.2M-element, 4.3M-element and 11M-element grids were used for the computations. All the computations were made using the second-order spatial and temporal discretization schemes. An implicit time-marching algorithm with nonlinear iterations was employed to advance the solutions with the time-step

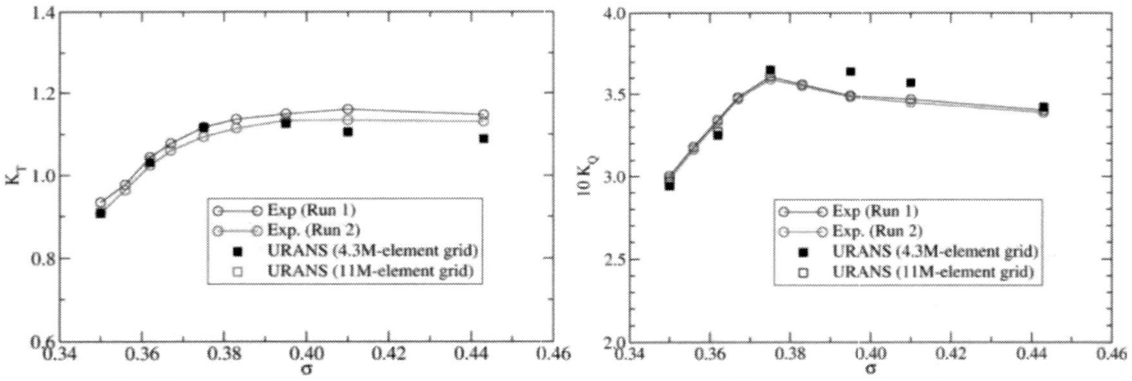


(a) Thrust coefficient (b) Torque coefficient

Figure 1. URANS predictions of thrust and torque coefficients as a function of cavitation number for the AxWJ-2 pump

size of $2 \times 10^{-5}$ second. With this time step size, one revolution of the rotor is resolved by 1500 time steps (0.24° per time step). For RANS turbulence modeling and near-wall treatment, we started with a pragmatic approach, namely, a combination of an isotropic eddy-viscosity model (Menter's SST k-ω) and wall functions.

Additional computations were made using the IDDES on the 11M-element grid in order to capture some of the large-scale turbulent structures in the flow. A smaller time-step size of $5 \times 10^{-6}$ was used for the IDDES computations.

### *3.2. Thrust and Torque on the Rotor*
Figure 1 shows the time-averaged $K_T$ and $K_Q$ obtained from the URANS computations with the 4.3M-element grid. In addition, the results from the 11M-element grid are shown for the two lowest cavitation (Thomas) numbers ($\sigma$ =0.350, 0.362). It was found that the overall difference between the predictions 2.2M-cell and the 4.3-cell grid was roughly 4%, with the difference further decreasing between the 4.3M-cell and the 11M-cell grids. The scatter of all the URANS predictions done so far using different turbulence models and numerical schemes seems to be around 6% from the tunnel data, which is fairly reasonable in view of the complexity of the flow. More importantly, the figure shows that the URANS closely captures the trend associated with the thrust-breakdown.

### *3.3. Cavitation Pattern and Dynamics*
The tests conducted at the NSWCCD, Rolls-Royce and JHU, albeit with all different rotor speeds (N = 2000, 1200, 900 RPM, respectively), have revealed the occurrence of the TLV and the sheet cavitation on the suction side of the rotor. The TLV cavitation is illustrated in Figure 2(a) by the photograph taken in the NSWWCD's 36-inch tunnel at $N^* = (p_{t3} - p_v)/\rho n^2 D^2 = 0.994$ and $Q^* = Q_J/nD^3 = 0.803$, where $p_{t3}$, $p_v$ and $Q_J$ are the inlet static pressure, the saturation pressure and the flow rate, respectively. The figure, depicting an instantaneous snapshot, portrays the characteristic shape of the TLV cavity

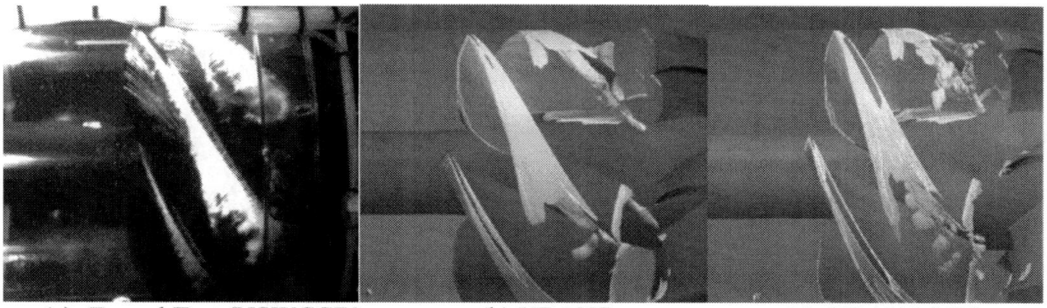

(a) Tunnel Test (NSWCCD) (b) URANS (c) URANS-LES
Figure 2. The tip-gap leakage vortex cavitation predicted using URANS and IDDES with 11M-cell grid N*= 0.994, Q*=0.803

expected in the near and post-breakdown regime – the "delta"-like overall shape and the shedding of cloud cavity, which is believed to reflect the interaction between the TLV and the vortices emanating from the suction side of the rotor. The shedding of vortices in the wake of the TLV cavity, which are oriented largely perpendicular to the blade surface, was also observed in the JHU's experiment [7]. They dubbed it "perpendicular cavitating vortex" (PCV). Both the URANS and the IDDES predictions, shown in Figures 2(b) and 2(c) using two iso-surfaces of volume fraction at 0.5 (mixture of 50% liquid and 50% vapor) and 0.9 (mixture of 90% liquid and10% vapor), are seen to reproduce the main features of the TLV cavitation. In particular, the IDDES captures finer cavitating structures associated with the vortices shed from the TLV cavity and the trailing edge of the rotor. Interestingly, cavitating vortices resembling the PCVs observed and discussed at length by the JHU group seem to be captured by the IDDES. Among other features captured by the URANS and IDDES are the small pockets of cavitation near the tips of the stator blades (seen in the figures) and the cavitating hub vortex (not shown here), which were also observed in the experiment [6].

Although not shown here due to space limitation, the URANS and IDDES results provide detailed space-time maps of the pressure distribution on the blade surfaces. The change in the surface pressure distribution with the progress of thrust-breakdown offers a crucial information that sheds light on what causes thrust-breakdown. The question comes down to how different types of cavitation in the pump contribute to unloading the pressure side of the rotor blades. Based on the numerical results, it is surmised that the elliptic nature of pressure plays an important role in unloading the pressure side, as discussed in [1] using the Poisson's equation for pressure. The computational results show that the pressure unloading mainly starts from the leading edge of the rotor blades rather than around the trailing edge. In view of this, the elliptic nature of pressure appears to work in a way that the suction side's low pressure (saturation pressure) is impressed upon the adjacent (incoming) pressure side, causing the unloading of the pressure side.

## 4. Summary and Conclusion

The URANS predictions were shown to predict the pump characteristics including the thrust-breakdown with a decent accuracy. In view of the complex nature of the flow such as the large streamline curvature and swirl, the accuracy URANS predictions may well be improved by resorting to more advanced turbulence models. The hybrid URANS-LES approach (IDDES) seems to be able to capture finer details of the turbulent cavitating flow without incurring an excessive computational cost. The elliptic nature of pressure appears to be responsible for thrust-breakdown, inasmuch as it can lead to unloading the pressure side of the rotor blades by the suction side's low pressure "impressed" on to the neighboring pressure side. Among our future work are computations with high-order RANS turbulence models, hybrid URANS-LES computation on a finer grid, and comparison with the experimental data in terms of the flow-fields such as the mean velocity and Reynolds' stresses.

## 5. Acknowledgement

This work was supported by the Office of Naval Research S &T program monitored by Dr. Ki-Han Kim and partly by the NSWCCD's ILIR program.

## References

[1] Kim S E and Schroeder S 2010 Numerical study of thrust breakdown due to cavitation on a hydrofol, a proepller, and a waterjet *28th International Conference on Naval Ship Hydrodynamics*, Pasadena, California September 12-17

[2] Lindau J W Pena C Baker W J Dreyer J J Moody W L Kunz R F and Paterson E G 2012 Modeling of cavitating flow through waterjet propulsors. *International Journal of Rotating Machinery 2012*

[3] Ahn S J and Kwon O J 2013 Numerical investigation of cavitating flows for marine propulsors using an unstructured mesh technique *International Journal of Heat and Fluid Flow **43*** 259-267

[4] Michael T Schroeder S and Becnel A 2009 Design of the ONR AxWJ-2 axial flow water jet pump", *Hydromechanics Department Report NSWCCD-50-TR-2008/066*

[5] Chesnakas C J, Donnelly M J Pfitsch D W Becnel A J Schroeder S D 2009 Performance evaluation of the ONR axial waterjet 2 (AxWJ-2) *Hydromechanics Department Report NSWCCD-50-TR-2009/089*

[6] Andersson L 2010 ONR AxWJ-2 Pump Loop Test *Report PR-1223* Rolls-Royce Hydrodynamic Research Ceneter

[7] Tan D, Li Y, Miorini R, Vagnoni E, Wilkes I and Katz J 2012 Role of large-scale vortical structures in the rotor passage of an axial waterjet pump in performance breakdown *30th Symposium on Naval Ship Hydrodynamics*, Hobart, Tasmania Australia November 2-7 2012

[8] Kim S E 2009 A numerical study of unsteady cavitation on a hydrofoil *7th International Symposium on Cavitation CAV2009* Ann Arbor Michigan August 17 – 22

[9] Travin A K, Shur, M L, Spalart, P R and Strelets, M K 2006 Improvement of delayed detached-eddy simulation for LES with wall modelling. In *ECCOMAS CFD 2006: Proceedings of the European Conference on Computational Fluid Dynamics,* Egmond aan Zee, The Netherlands September 5-8

# Application of a compressible flow solver and barotropic cavitation model for the evaluation of the suction head in a low specific speed centrifugal pump impeller channel

**P Limbach, T Müller and R Skoda**

Chair of Hydraulic Fluid Machinery, Ruhr-Universität Bochum, Universitätsstr. 150, 44801 Bochum, Germany

E-mail: phillip.limbach@ruhr-uni-bochum.de

**Abstract.** Commonly, for the simulation of cavitation in centrifugal pumps incompressible flow solvers with VOF kind cavitation models are applied. Since the source/sink terms of the void fraction transport equation are based on simplified bubble dynamics, empirical parameters may need to be adjusted to the particular pump operating point. In the present study a barotropic cavitation model, which is based solely on thermodynamic fluid properties and does not include any empirical parameters, is applied on a single flow channel of a pump impeller in combination with a time-explicit viscous compressible flow solver. The suction head curves (head drop) are compared to the results of an incompressible implicit standard industrial CFD tool and are predicted qualitatively correct by the barotropic model.

## 1. Introduction

For the complex cavitating flow in centrifugal pumps, commonly, incompressible flow solvers with an additional transport equation for the void fraction are used (e.g. [1]). Mass transfer is calculated by a source/sink term based on a simplified bubble dynamics equation, such as the Rayleigh equation (e.g. [2, 3]). Due to the simplification of the physics, empirical parameters are implemented in the source/sink terms, which may need to be calibrated to the particular flow situation. In contrast, a barotropic equation of state is used by several authors (i.e. [4, 5, 6]). While [4] use an isothermal barotropic model function for the mixture and [5] use a barotropic relation, which accounts for effects of thermal cavitation, the model of [6] is based on an isentropic path in the temperature-entropy diagram. This model have been applied e.g. on micro throttles [7], on hydrofoils [8] (companion paper) and ultrasonic horn flows [9], [10] (companion paper), because of their accurate resolution of collapse-induced wave dynamics, which lead to erosion damages. Moreover, the structure and measured shedding frequency of cloud cavitation can be captured with a good accuracy although the grid sensitivity and computational effort are high [8]. In the present study the barotropic cavitation model of [6] is applied on a single flow channel of a centrifugal pump with low specific speed, in order to evaluate the potential of this model to predict the pump head in cavitating flow conditions. The investigated pump is a single-stage glanded centrifugal pump of low specific speed ($n_q = 12\,\mathrm{min}^{-1}$), which contains a closed impeller with seven blades and a volute. The geometry and operating conditions of the pump are documented in [11].

Journal of Physics: Conference Series **656** (2015) 012065 doi:10.1088/1742-6596/656/1/012065

## 2. Method

### 2.1. Numerical and physical model

On the one hand the commercial flow solver Ansys CFX 15.0 is used to solve the incompressible Reynolds-averaged Navier-Stokes equations. Second order discretisation is used for time and space and the SST turbulence model [12] in combination with an automatic wall-function is applied. The two-phase flow is treated as a homogeneous mixture and described by an additional transport equation for the void fraction. The mass transfer model by Zwart et al. [3] is used and based on a simplified Rayleigh-Plesset equation [13, 14], i.e. Rayleigh equation, which governs the dynamics of a single, spherical bubble, a mass transfer term is included in the transport equation of the void fraction. For the present study parameters of the cavitation model are defined ($R_B = 10^{-6}\,\mathrm{m}, \alpha_{nuc} = 5e^{-4}, F_{vap} = 50, F_{cond} = 0.01$) as recommended by Zwart et al. [3]. This method is referred to as *incompressible* in the following.

On the other hand, our block-structured inhouse-code hydRUB contains a compressible flow solver with an explicit optimised four-stage Runge-Kutta time integration scheme [15]. The explicit time integration leads to very small time steps, hence, Reynolds-averaging of the compressible Navier-Stokes equations is not valid. As a consequence, viscous fluid is used but no explicit turbulence model is applied. The convective fluxes are modelled using a low Mach-number consistent Godunov-type flux formulation [15]. Second order discretisation in space is achieved by the reconstruction of density and velocity with the high resolution schemes MINMOD [16] and SMART [17], respectively. The fluid is treated as a homogeneous mixture of vapour and liquid assuming both phases to be in thermodynamic equilibrium. The energy equation is neglected and a barotropic cavitation model, i.e. equation of state is used [6] that assumes an isentropic phase change. No model parameters are available in contrast to the mass transfer model [3]. This method is referred to as *compressible* in the following.

### 2.2. Numerical set-up

In the present first application of the compressible method to the flow in a low specific speed centrifugal pump impeller, we assess the ability of the compressible method to predict the onset of head drop if the pressure level is successively decreased. A simplified geometry is used, i.e. a single flow channel of the impeller ($\approx 51°$ segment). The volute and sidewall gaps are neglected. The suction pipe and the outlet of the impeller channel are extended to reduce the influence of boundary conditions on internal flow. However, the reflection of pressure waves particular at the inlet is unavoidable for the compressible flow solver by our reflecting boundary conditions. Nevertheless, we cope with the pressure wave reflections at the present first study, since we are interested in time averaged head values. A confuser is added at the radial outlet section to avoid positive pressure gradients in the outlet region. A relatively coarse grid is used (60,000 nodes). A preliminary grid study showed (120,000; 240,000 nodes) that for these grids the time-averaged head varies by less than 2 % for single phase (both solvers) and cavitating flow conditions (only incompressible tested). A schematic view of the numerical set-up is shown in figure 1.

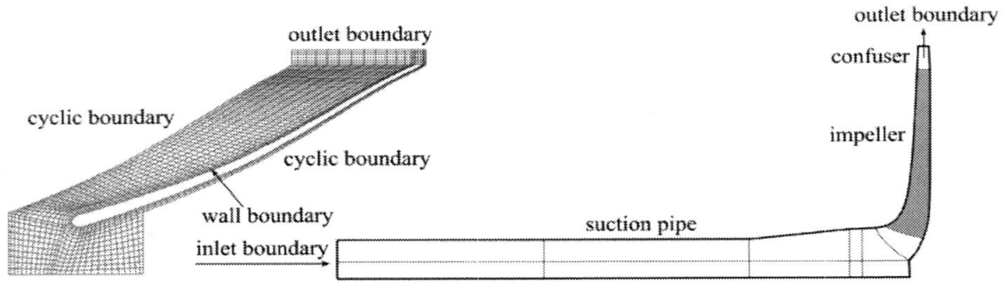


**Figure 1.** Numerical set-up of the impeller in conformal mapping and numerical model in a schematic meridian view.

A constant velocity is specified at the inlet of the suction pipe corresponding to the best efficiency point and for part- (−30 %) and overload (+30 %). Static pressure is set at the outlet, periodic boundary conditions are used in circumferential direction. Instationary simulations are performed in relative frame with frozen rotor. The convergence criterion is defined as $R_{max} \leq 1 \times 10^{-4}$ and the time step size is set to a value corresponding to a rotation of 1° of the impeller for the incompressible solver. The time-explicit scheme of the compressible solver leads to time steps of approximately $2.8 \times 10^{-8}$ s ($CFL \leq 0.95$). Concerning single phase flow simulations the measured pump head is overpredicted even with a full model containing a 360° geometry of the impeller, sidewall gaps and the volute [11]. The single channel flow model contains more geometry simplifications, hence, we do not focus on achieving quantitative agreement between simulation results and measured data, but rather on a first assessment of the barotropic model to capture the head drop curve.

## 3. Results

The pressure level at the outlet is defined high enough to ensure non-cavitating flow conditions (3 bar). For the prediction of the characteristics of the Net Positive Suction Head (NPSH) the pressure level is decreased successively. As a consequence, cavitation zones appear and grow, leading to a head drop. The computed head over time is presented exemplarily in figure 2.

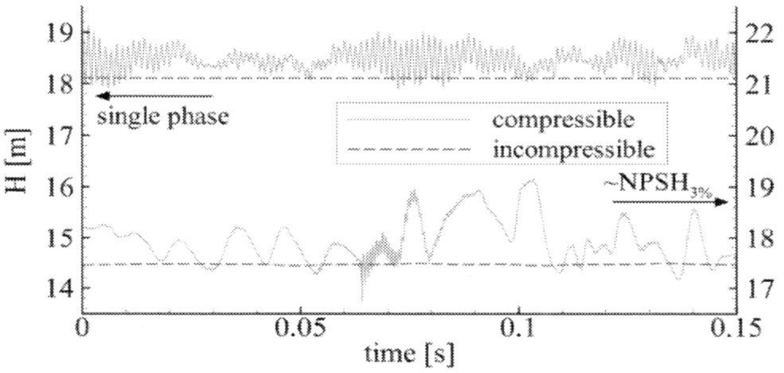


**Figure 2.** Pump head in dependence on time for overload single phase and cavitating flow conditions

For single phase flow conditions a constant head is predicted using the incompressible flow solver. In contrast, oscillations occur in the results of the compressible flow solver, which correlate with the length of the flow path and the speed of sound (~ 1480 m/s) since pressure waves are reflected at the inlet and outlet boundaries. Regarding cavitating flow conditions only minor fluctuations are visible for the incompressible flow solver. In the compressible solution the void regions oscillate with a relatively low frequency and lead to corresponding head oscillations. Occasionally occuring vapour region collapses close to the leading edge or suction pipe section generate pressure waves, which are propagated to the boundaries and lead to a sudden head drop (e.g. figure 2, 0.06 s). Although the occurence of cloud cavitation is likely for low NPSH [19], only sheet cavitation is observed in our simulation results, which may be attributed to the rather coarse grid.

Figure 3 and 4 presents the results of the head for single phase and cavitating flow conditions, respectively, which is averaged over a sufficiently long time interval. Both solvers predict essentially the same head characteristics for single phase flow conditions, while slight deviations in the overload operating point can be observed. Due to only small head deviations between both methods, the influence of the turbulence modelling (eddy viscosity versus no explicit turbulence model) seems to be minor for the evaluation of the head: Regarding the compressible solver we assume that the numerical diffusion acts as a subgrid-scale eddy viscosity. We expect that the local flow field in cavitating flow conditions is certainly influenced by a grid-induced arbitrary subgrid-scale viscosity magnitude, but time averaged head values do not seem to be dominated by turbulence effects, a conclusion which is also supported by [18]. In comparison to [11] an essentially linear slope of the head curve and higher deviation between simulation results and measured data (not shown here) is obtained due to the neglect of the volute and side chambers.

The head drop due to blockage of the flow channel with vapour can be captured by both methods. For all operating points, a sharp drop of the head is predicted. In nominal load

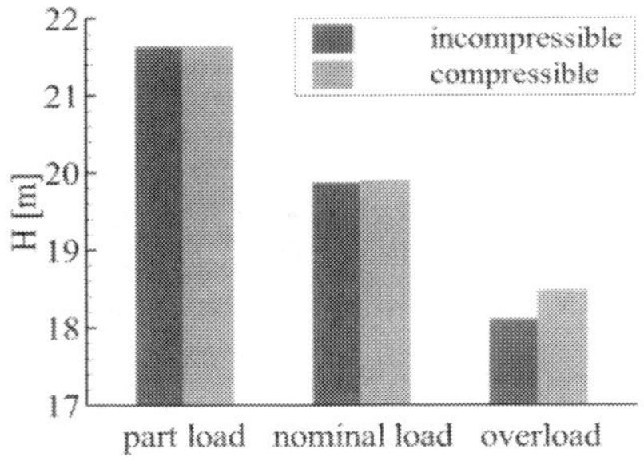


**Figure 3.** Predicted pump head for single phase flow conditions.

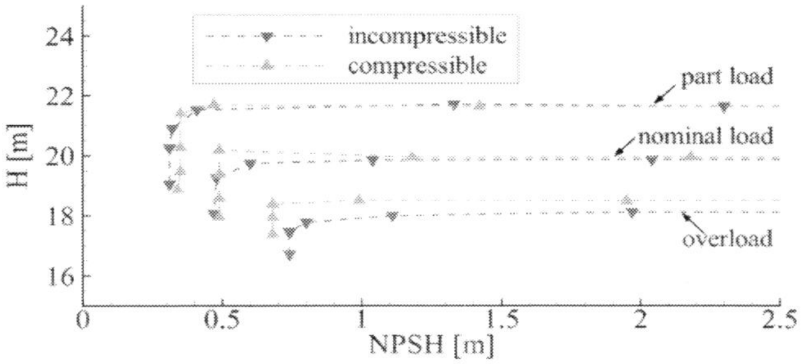


**Figure 4.** Head drop curve for three flow rates.

a local increase of the head is observed ($NPSH \sim 0.5\,\mathrm{m}$) using the compressible flow solver. Concerning this operating point, the flow around the blade leading edge is influenced by the vapour zones and attached flow on the pressure side is achieved downstream, whereas backflow is observed in the corresponding results of the incompressible flow solver. This qualitatively different result of both solvers needs to be verified on finer grids. Differences in the head already present in figure 3 (overload) are also present at lower NPSH values (figure 4, overload). The resulting $\mathrm{NPSH}_{3\%}$ (NPSH at a head drop of three percent) increases with increasing flow rate as expected from common theoretical and experimental findings (e.g. [19]).

## 4. Conclusions

The suction head curves (head drop) are predicted qualitatively correct by the barotropic model, although no empirical model parameters have been adjusted. This promising performance needs to be verified on finer grids and more realistic pump models including sidewall gaps and volute casing, which is done in ongoing work by the authors.

## Acknowledgments

The authors would like to express their gratitude to the *Verband Deutscher Maschinen- und Anlagenbau (VDMA)* for financial support. Further, the authors are grateful to *University of Cologne* for the grant of computing time on the Cheops Compute Cluster and gratefully acknowledge the company *WILO SE* for the supply of the pump data.

## References

[1] Iiu D, Iiu H, Wang Y, Zhuang S, Wang J and Hui D 2012 *Proc. 8th Int. Symp. on Cavitation* (Singapore)
[2] Schnerr G H and Sauer J 2001 *Proc. 4th Int. Conf. on Multiphase Flow* (New Orleans)
[3] Zwart P J, Gerber A G and Belamri T 2004 *Proc. 5th Int. Conf. on Multiphase Flow* (Yokohama)
[4] Goncalves E and Patella R F 2009 *Computers and Fluids* **38** (9) pp 1682-96
[5] Rapposelli E and d'Agostino L 2003 *Proc. 5th Int. Symp. on Cavitation* (Osaka)
[6] Iben U 2002 *Systems Analysis Modelling Simulation* **42** pp 1283-307
[7] Skoda R, Iben U, Morozov A, Mihatsch M, Schmidt S J and Adams N A 2011 *WIMRC Proc. 3rd Int. Cavitation Forum* (Warwick)
[8] Blume M and Skoda R 2015 *Proc. 9th Int. Symp. on Cavitation* (Lausanne)
[9] Mottyll S, Mueller S, Niederhofer P, Hussong J, Huth S and Skoda R 2013 *EPJ Web of Conferences* **67**
[10] Mottyll S and Skoda R 2015 *Proc. 9th Int. Symp. on Cavitation* (Lausanne)
[11] Limbach P, Kimoto M, Deimel C and Skoda R 2014 *Proc. of ASME Turbo Expo* (Düsseldorf)
[12] Menter F R 1994 *AIAA Journal* **32** pp 1598-605
[13] Rayleigh L 1917 *Phil. Mag.* **34** pp 94-8
[14] Plesset M S 1949 *ASME J. of Appl. Mech.* **16** pp 228-31
[15] Schnerr G H, Sezal I H and Schmidt S J 2008 *Physics of Fluids* **20** 040703
[16] Harten A 1983 J. Computational Physics **49** pp 357-93
[17] Gaskell P H and Lau A K C 1988 Int. J. Numerical Methods in Fluids **8** pp 617-41
[18] Wursthorn S and Schnerr G H 2002 *Proc. in Applied Mathematics and Mechanics* **1** pp 375-6
[19] Gülich J F 2014 *Centrifugal pumps* (Springer, Berlin Heidelberg)

Journal of Physics: Conference Series **656** (2015) 012066 doi:10.1088/1742-6596/656/1/012066

# Numerical predictions of the turbulent cavitating flow around a marine propeller and an axial turbine

**M Morgut[1], D Jošt[2], E Nobile[3] and A Škerlavaj[2]**

[1] Turboinštitut d.d., Ljubljana, Slovenia
[2] Turboinštitut d.d., Ljubljana, Slovenia, presently at University of Trieste, Department of Engineering and Architecture, Italy
[3] University of Trieste, Department of Engineering and Architecture, Italy

E-mail: mitja.morgut@kolektor.com

**Abstract.** The numerical predictions of cavitating flow around a marine propeller working in non-uniform inflow and an axial turbine are presented. The cavitating flow is modelled using the homogeneous (mixture) model. Time-dependent simulations are performed for the marine propeller case using OpenFOAM. Three calibrated mass transfer models are alternatively used to model the mass transfer rate due to cavitation and the two-equation SST (Shear Stress Transport) turbulence model is employed to close the system of the governing equations. The predictions of the cavitating flow in an axial turbine are carried out with ANSYS-CFX, where only the native mass transfer model with tuned parameters is used. Steady-state simulations are performed in combination with the SST turbulence model, while time-dependent results are obtained with the more advanced SAS (Scale Adaptive Simulation) SST model. The numerical results agree well with the available experimental measurements, and the simulations performed with the three different calibrated mass transfer models are close to each other for the propeller flow. Regarding the axial turbine the effect of the cavitation on the machine efficiency is well reproduced only by the time dependent simulations.

## 1. Introduction

Recently the University of Trieste (Italy) and Turboinštitut d.d. from Ljubljana (Slovenia) joined in the ACCUSIM-EU project that aims, primarily, to develop reliable, high fidelity methods for the accurate predictions, and optimization, of the performances of hydro-machinery and marine propellers.

In reference to this project, in this work selected results obtained from this successful collaboration are presented for a marine propeller and an axial turbine. In particular, the numerical investigations of the CNR-INSEAN E779A model propeller are presented, and the simulations performed considering a 6-blade Kaplan turbine are discussed.

In current predictions the turbulent cavitating flow was simulated using the so called homogeneous model. This model treats the working fluid as a homogeneous mixture of two fluids, i.e water and vapour, behaving as a single one, and the mass transfer rate due to cavitation is regulated by the mass transfer model. In the literature it is possible to find several mass transfer models. In this work the models originally proposed by Kunz et al. [1], Singhal et al. [2], Zwart et al. [3] and calibrated as described in [4] were considered.

In the case of the marine propeller working in non-homogeous (in wake) inflow conditions, the simulations were performed using OpenFOAM, an open source CFD (Computational Fluid Dynamics)

toolbox [5]. Time-dependent simulations were carried out using alternatively all the three different calibrated mass transfer models in combination with the SST turbulence model.
As far as the Kaplan case is concerned, the predictions were performed using ANSYS-CFX 15, a commercial CFD code. Steady-state and time-dependent predictions were carried out using the SST turbulence model and the SST-SAS method, respectively. In both cases the curvature correction and Kato-Launder limiter, for the production in the turbulent kinetic energy equation, were included following previous studies [6]. The mass transfer rate due to cavitation was evaluated using the native Zwart et al. mass transfer model with tuned empirical coefficients [3, 4].

## 2. Meshing

The meshes used in the current simulations were generated using ANSYS-ICEM CFD 15, a commercial meshing tool. In figure 1 two snapshots of the grids generated for the propeller and axial turbine are presented.

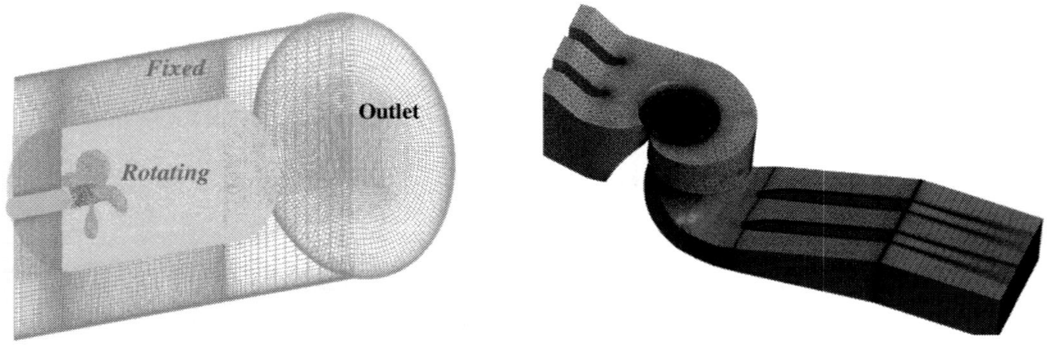


**Figure 1.** Computational grids for the marine propeller (left) and Kaplan turbine (right).

The mesh for the propeller simulations was generated using two different approaches. The rotating region (in yellow in figure 1) was created following the hybrid approach (tetrahedral + prisms layers at the walls) while the fixed region was discretized by the hexa-structured grid. The overall mesh had about 8.1 million cells. Regarding the Kaplan turbine a hybrid mesh was generated for the semi-spiral casing with stay vanes, while for the guide vane cascade, runner and elbow draft tube hexa-structured meshes were generated. The overall mesh had about 8.3 million nodes. All the grids were refined at the walls according to the recommended values of $y^+$.

## 3. Numerical simulations

### *3.1 E779A propeller*

The numerical simulations were carried out following the experimental/numerical setup described in [7]. In particular, here we point out that on the domain inlet the non-homogeneous inflow (nominal wake), kindly provided by CNR-INSEAN (private communication) was set. On outlet boundary a fixed value of static pressure was imposed. On the solid surfaces the no-slip wall condition was enforced. A second order upwind scheme was adopted for the discretisation of the convective terms while a first order implicit scheme was used for the time discretization.
From figure 2 it is possible to note that a similar cavity evolution was predicted using alternatively the three different mass transfer models. As can be seen, the overall numerical predictions agree well with the available experimental data.

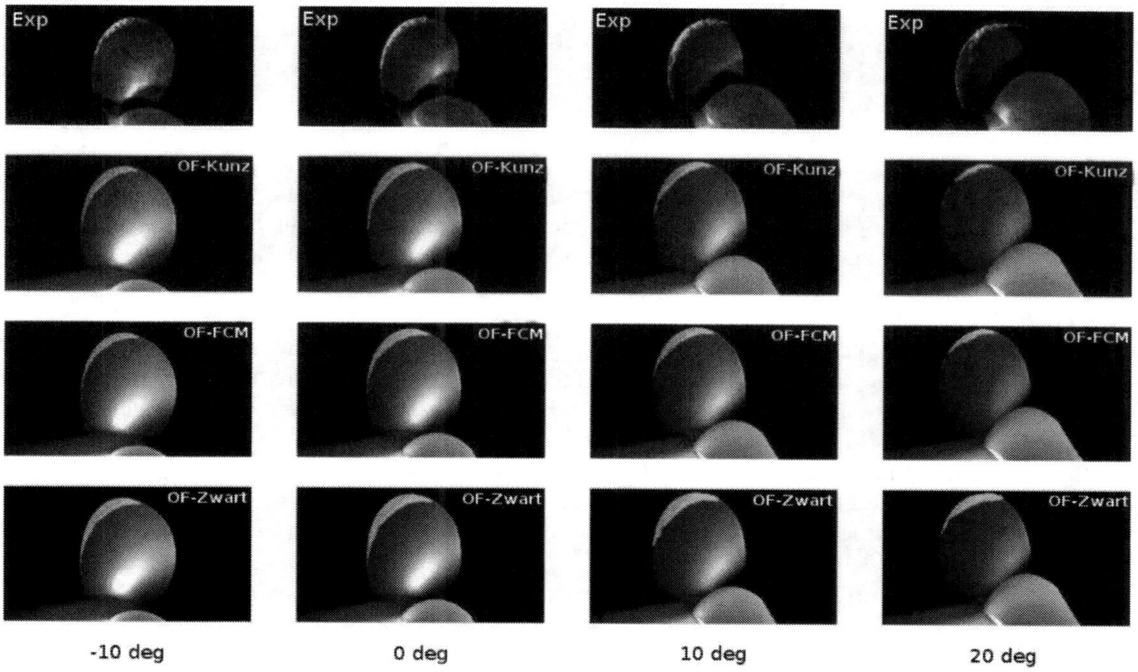


**Figure 2.** Cavity evolution during propeller rotation. Numerical cavitation patterns depicted using isosurfaces of vapour volume fraction equal to 0.1. In the above figures OF stays for OpenFOAM, while FCM stays for Full Cavitation Model [2].

*3.2 Kaplan turbine*

In order to determine the effect of cavitation on turbine efficiency (sigma-break curve) successive simulations were performed lowering the reference pressure. Steady state simulations were performed using the MRF (Multiple Reference Frame) approach in combination with the *frozen rotor* frame/change mixing model. Time-dependent simulations were performed using *sliding grids*. In both cases the robust high resolution scheme was used for the discretization of the convective terms.

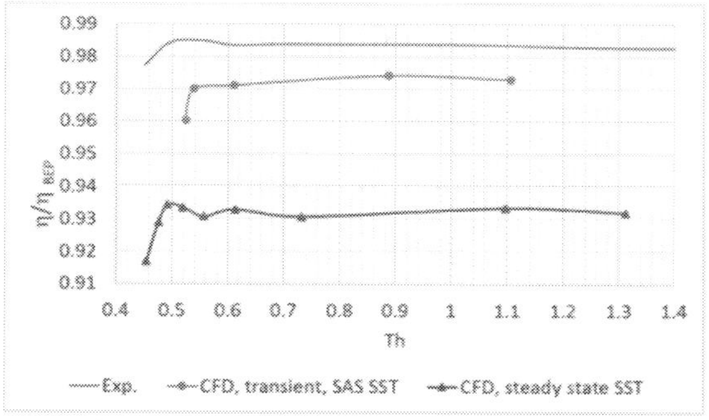


**Figure 3.** Sigma break curve

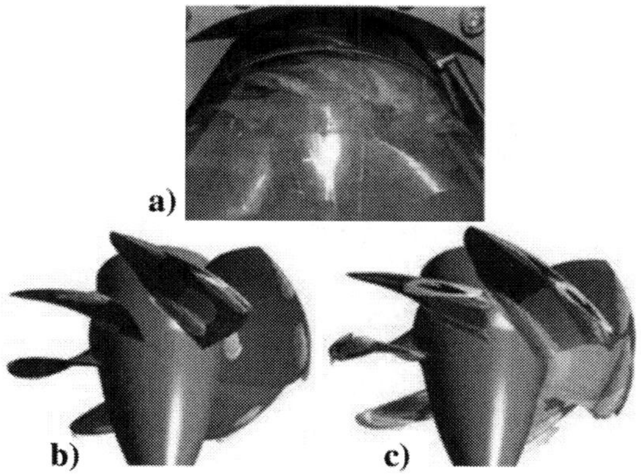


**Figure 4.** Cavity on runner blades: a) observation on the test rig, b) steady state simulation, c) time dependent simulation ; Th=0.52

Numerical results were compared with the observation of cavity size on the test rig and with the measured sigma-break curve. Steady-state simulations did not predict the same amount of cavitation on all blades due to frozen rotor assumption, which preserved differences in circumferential direction. Besides, the extent of cavitation was too small compared to the experimental one, as illustrated in figure 4.
With transient simulations the same amount of cavitation on all runner blades was obtained and the shape and extent of sheet cavitation agreed well with the cavitation observed on the test rig (figure 4). Steady-state simulations significantly underpredicted the efficiency, as illustrated in figure 3. Transient simulations predicted the efficiency more accurately, althought the Thoma number, Th, where the efficiency dropped for 1%, was slightly overpredicted (see figure 3).

## 4. Conclusions

In this paper the numerical predictions of the turbulent cavitating flow around a marine model propeller and a Kaplan model turbine were presented. The numerical predictions were carried out using both a commercial and an open source CFD code, solving the governing equations of the homogeneous (mixture) model.
In the case of a marine propeller three different calibrated mass transfer models were compared. The numerical results agreed well with the available experimental data and the three different mass transfer models ensured similar results, thus proving the importance of proper calibration of the models.
In the case of the Kaplan turbine the steady-state and time-dependent approaches were compared for the evaluation of the effect of the cavitation on turbine efficiency. The time dependent simulations, with the SAS SST turbulence model, better predicted the amount of cavitation and efficiency level, even though a slightly premature drop of the efficiency was observed.

## 5. Acknowledgements

The research leading to these results has received funding from the People Programme (Marie Curie Actions) of the European Union's Seventh Framework Programme FP7/2007-2013/ under REA grant agreement n°612279.

## 6. References

[1] Kunz R F, Boger D A, Stinebring D R, Chyczewski T S, Lindau J W, Gibeling H J, Venkateswaran S and Govindan T R 2000 A preconditioned Navier-Stokes method for two-phase flows with application to cavitation prediction *Computers and Fluids* **29**(8), 849-875
[2] Singhal A K, Athavale M M, Li H and Jiang Y 2002 Mathematical Basis and Validation of the Full Cavitation Model *Journal of Fluids Engineering* **124**(3), 617-624
[3] Zwart P, Gerber A G and Belamri T 2004 A Two-Phase Model for Predicting Cavitation Dynamics *ICMF 2004 International Conference on Multiphase Flow, Yokohama, Japan*
[4] Morgut M, Nobile E and Biluš I 2011 Comparison of mass transfer models for the numerical prediction of sheet cavitation around a hydrofoil *Int. Journal of Multiphase Flow* **37**(6), 620-626
[5] www.openfoam.com
[6] Jošt D, Škerlavaj A and Lipej A 2014 Improvement of Efficiency Prediction for a Kaplan Turbine with Advanced Turbulence Models, *Journal of Mechanical Engineering* **60**(2), 124-134
[7] Salvatore F, Streckwall H and van Terwisga T. 2009 Propeller Cavitation Modelling by CFD-Results from the VIRTUE 2008 Rome Workshop *Proc. of the First International Symposium on Marine Propulsors, smp'09, Trondheim, Norway*

# Numerical investigation of cavitation flow inside spool valve with large pressure drop

**Jian Deng, Dingyi Pan[1], Fangfang Xie and Xueming Shao**

State Key Laboratory of Fluid Power Transmission and Control, Department of Engineering Mechanics, Zhejiang University, Hangzhou 310027, China

**Abstract.** Spool valves play an important role in fluid power system. Cavitation phenomena happen frequently inside the spool valves, which cause structure damages, noise and lower down hydrodynamic performance. A numerical tools incorporating the cavitation model, are developed to predict the flow structure and cavitation pattern in the spool valve. Two major flow states in the spool valve chamber, i.e. flow-in and flow-out, are studies. The pressure distributions along the spool wall are first investigated, and the results agree well with the experimental data. For the flow-in cases, the local pressure at the throttling area drops much deeper than the pressure in flow-out cases. Meanwhile, the bubbles are more stable in flow-in cases than those in flow-out cases, which are ruptured and shed into the downstream.

## 1. Introduction

Spool valves are widely used in fluid power system. However, due to the complicated internal structure and high speed flow stream, cavitation phenomena are found in these valves [1]. It is normally harmful that cavitation may cause structure damages, produce noise [2], and lower down system hydrodynamic performance. Experimental and numerical endeavors has been given to predict the cavitation inception and effects of cavitation, and correlate the hydraulic parameters, valve structure and cavitation phenomena [3, 4].

In current study, concerning typical throttling valves with pairs of U-shaped grooves, a fully numerical simulation tool has been developed, and it has been applied to predict the viscous heating phenomenon and spool deformation [5, 6]. By incorporating the cavitation model, we are aiming to investigate the correlation between cavitation phenomena, e.g. cavitation inception, evolution of cavitation, and relevant hydraulic parameters, e.g. pressure drop, local structure, valve opening, etc. Two major classes of cases are investigated, flow-in cases and flow-out cases, corresponding to different local structures. Comparison with the experiment are given, which convinced the reliability of the numerical tool. In the following, the physical model and numerical method are introduced, and then, numerical results and discussion are presented.

## 2. Physical model and numerical method

In current study, each spool consists two pairs of symmetric U-shaped grooves on both sides of the neck. The corresponding schematic slices of the valve section of two typical cases are shown in Fig. 1. For the flow-in cases, the left pair of grooves and the sleeve forms the throttling area, and for flow-out cases, the right pair of grooves take places. The geometry of the groove

[1] To whom any corresponding should be addressed. Email: dpan@zju.edu.cn

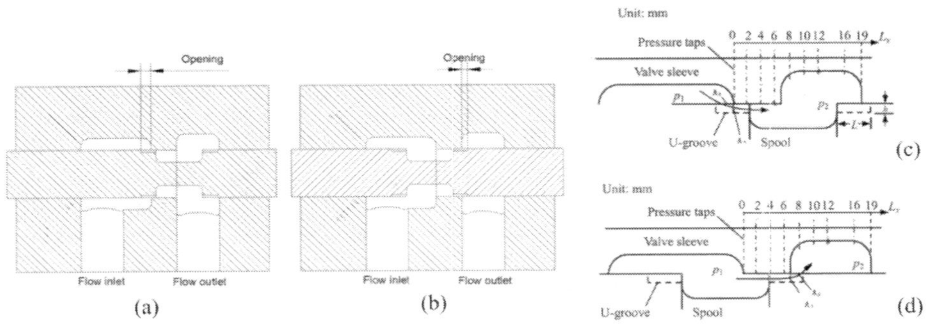


**Figure 1.** Schematics of the valve structure and working condition. Cut-view for flow-in (a) and flow-out (b) cases. Locally enlarged view for flow-in (c) and flow-out (d) cases.

is described by the following parameters, i.e. groove length ($L$), end radius ($r$) and depth (h). These parameters are fixed as $L$ = 5mm, $r$ = 1mm and $h$ = 1.5mm. The leakage between the spool and the sleeve is neglected. It is worthy to mention that there are two possible narrowest areas in the chamber, i.e. $A_1$ and $A_2$. With the variation of the opening, $A_2$ increases monotonously while $A_1$ remains constant.

The hydraulic oil in the system, is assumed to be incompressible, viscous and isothermal. The Navier-Stokes equations are solved numerically with commercial software FLUENT. The finite volume method (FVM) and the SIMPLEC algorithm are adopted [7]. The structured meshes are generated, yielding precise numerical accuracy. There are 290000 − 346000 cells and the throttling area is refined along the short edge of the groove. The Renormalization-group $k-\epsilon$ model is employed for turbulence modelling, which has a good adaptability for flow separation. The fluid is assumed as a mixture of liquid and vapor. The transport equation of vapor mass fraction is

$$\frac{\partial}{\partial t}(\rho f) + \nabla \cdot (\rho \mathbf{u} f) = \nabla \cdot (\mu \nabla f) + R_e - R_c, \tag{1}$$

where $\mathbf{u}$ is the velocity, $\mu$ the effective exchange coefficient, $f$ the vapor mass faction, and $R_e$ and $R_c$ the vapor generation and condensation rate, derived from the Rayleigh-Plesset equation, taking account of the effects of turbulence-induced fluctuation and non-condensable gas [8]

$$R_e = C_e \frac{\sqrt{k}}{\sigma} \rho_l \rho_v \left[\frac{2}{3}\frac{P_v - P}{\rho_l}\right]^{\frac{1}{2}} (1-f) \quad (P < P_v); \quad R_c = C_c \frac{\sqrt{k}}{\sigma} \rho_l \rho_l \left[\frac{2}{3}\frac{P - P_v}{\rho_l}\right]^{\frac{1}{2}} f \quad (P > P_v).$$

Here, $\sigma$ represents the surface tension, $P_v$ is the saturated vapor pressure, and the values of the empirical constants of $C_e$ and $C_c$ are 0.02 and 0.01, respectively. To validate our numerical method, a comparison with the experiment result [4] is given in Fig. 2. The bubble interfaces are demonstrated as $\rho$ = 700kg/m$^3$, and it is shown that the bubble sizes at different outlet pressures are quite close to the experiment results.

## 3. Results and discussions

The fluid properties are: liquid density $\rho_l = 889\text{kg/m}^3$ and kinetic viscosity $\nu_l = 4 \times 10^{-5}\text{m}^2/\text{s}$, vapor density $\rho_v = 10.95\text{kg/m}^3$ and kinetic viscosity $\nu_v = 7 \times 10^{-6}\text{m}^2/\text{s}$, and saturated vapor pressure $P_v$ = 50kPa. The inlet pressure is set to $P_1$ = 3MPa, and the outlet pressure ($P_2$) varies, leading to different pressure drops. The turbulence intensity at inlet is set to 5%, and the hydraulic diameter is set to 0.02m. Standard wall function is applied, corresponding to a fully developed turbulent flow in a circular tube. For each major class, i.e. the flow-in or flow-out cases, results of two typical pressure drops are compared.

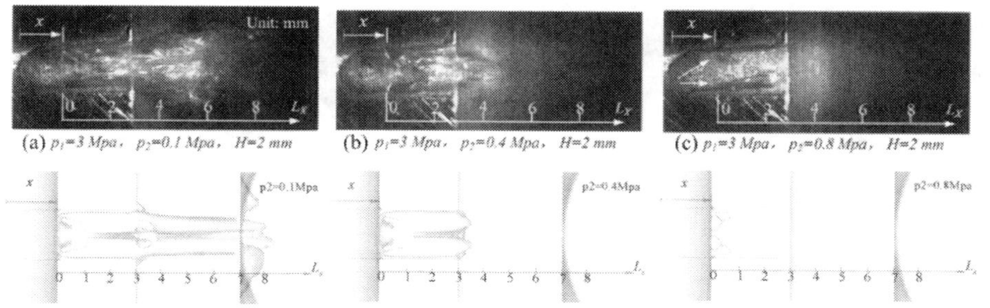


**Figure 2.** Comparison of the bubbles patterns of the experimental [4] and numerical results at different outlet pressures.

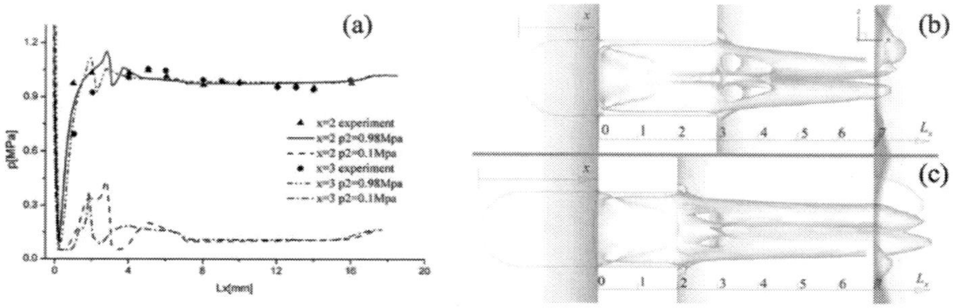


**Figure 3.** Pressure distribution along the spool direction ($L_x$) of flow-in cases with two typical outlet pressure constants, $P_2 = 0.98$MPa and $P_2 = 0.1$MPa (a), and the cavitation bubbles pattern for low outlet pressure case ($P_2 = 0.1$MPa) with different openings (b) and (c).

### 3.1. *Flow-in cases*

Fig. 3 (a) shows the local pressure along spool wall in throttling area. $x$ represents opening as shown in Fig. 1. The pressure drops quickly at $L_x = 0$, where the cross section area of the flow chamber turns to be minimal ($A_1$ in left sub-figure of Fig. 1). After then, as the cross section area increases, the pressure recovers quickly to the background pressure (outlet pressure) with fluctuation. Relevant experimental data are also given as solid symbols in Fig. 3 (a), and the results agree well with the experiment.

The patterns of the cavitation bubbles at $P_2 = 0.1$MPa are given in Fig. 3 (b) and (c), corresponding to different openings. It is shown that cavitation bubbles develop at the minimal pressure point, and are elongated at downstream. The larger opening the longer bubble.

### 3.2. *Flow-out cases*

The pressure distribution of flow-out cases are given in Fig. 4. For all cases, the pressure first keeps constant close to the inlet pressure, and then, decreases at $L_x = 4$, where the cross section area becomes minimal. Compared to flow-in cases, the pressure only decreases to the back pressure, without a recovery process from a low pressure to the back pressure. The decrease slopes are much smoother the those of the flow-in cases. It is believed that, the local structures around the throttling area of the major classes are quite different with each others, and hence, resulting in different flow structures with different pressure distributions.

According to the pressure results, we conjecture that it is difficult to induce the cavitation inception in flow-out cases, and the numerical results confirm this. It is difficult to capture the cavitation bubble structures if the interface is set to $\rho = 700\text{kg/m}^3$, indicating that there is only

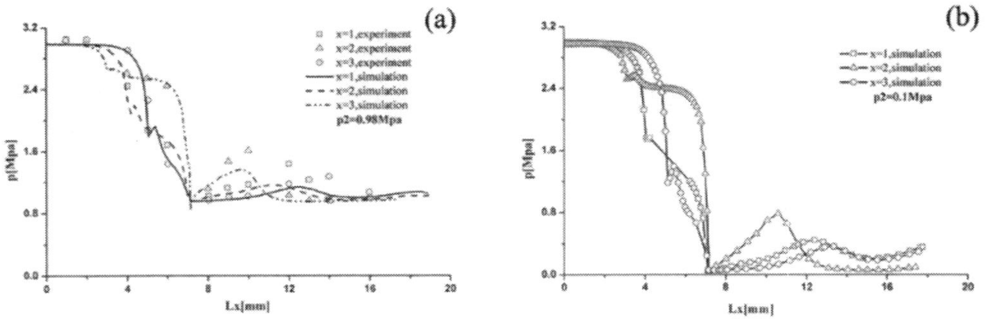


**Figure 4.** Pressure distribution along the spool direction ($L_x$) of flow-out cases with two typical outlet pressure constants, (a) $P_2 = 0.98$MPa and (b) $P_2 = 0.1$MPa.

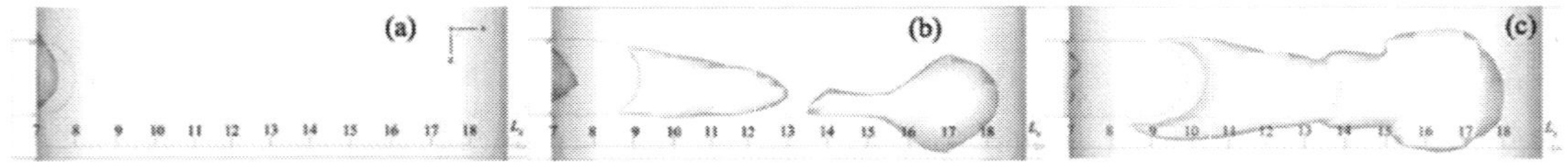


**Figure 5.** Cavitation bubbles pattern for low outlet pressure ($P_2 = 0.1$MPa) of flow-out cases with different openings.

little amount of liquid has phase changed. Instead, the interface with $\rho = 200\text{kg/m}^3$ are given in Fig. 5. Snapshots with three different openings are presented. The larger opening the bigger bubble. In the downstream the bubbles are first narrowed and then expanded, and are ruptured at the narrow neck and then shed into the downstream. We does not capture this phenomenon in flow-in cases, in which the cavitation bubbles are more stable than those in flow-out cases.

## 4. Conclusions

Two major flow states in the spool valve are studies numerically. The cavitation model is applied to predict the evolution of cavitation bubbles. The pressure distributions along the spool wall are investigated, and the results agree well with the experiment data. For the flow-in cases, the local pressure at the throttling area drops much deeper than the pressure in flow-out cases, inducing the cavitation inception more easily. The bubbles are more stable in flow-in cases than those in flow-out cases, which are ruptured and shed into downstream. Correlations between local flow structures and cavitation phenomena are required deep investigation in the future.

## Acknowledgments

This work is supported by the National Natural Science Foundation of China (Nos: 11272283, 11402230), Zhejiang Provincial Natural Science Foundation of China (No: LY12A02006).

## References

[1] McCloy D and Beck A 1967 *Proc. Inst. Mech. Eng.* **182** 163–174
[2] Martin C S, Medlarz H, Wiggert D C and Brennen C 1981 *J. Fluids Eng.* **103** 564–576
[3] Oshima S, Leino T, Linjama M, Koskinen K T and Vilenius M J 2001 *Int. J. Fluid Power* **2** 5–13
[4] Fu X, Du X, Zou J, Ji H, Ryu S and Ochiai M 2007 *Int. J. Fluid Power* **8** 29–37
[5] Deng J, Shao X M and Fu X 2009 *Proc. Inst. Mech. Eng., C: J. Mech. Eng. Sci.* **223** 2571–2581
[6] Deng J, Shao X M, Fu X and Zheng Y 2009 *Energy Conversion and Management* **50** 947 – 954
[7] Patankar S V 1980 *Numerical heat transfer and fluid flow* (Taylor and Francis)
[8] Singhal A K, Athavale M M, Li H Y and Jiang Y 2002 *J. Fluids Eng.* **124** 617–624

# Hydraulic Performance Comparison for Axial Flow Impeller and Mixed Flow Impeller with Same Specific Speed

**Zhongyong Pan[1], Yongyan Ni[2], Jianping Yuan[1] and Pei Ji[1]**

[1] National Research Centre for Pumps, Jiangsu University, Zhenjiang 212013, China
[2] School of Naval Architecture and Ocean Engineering, Jiangsu University of Science and Technology, Zhenjiang 212003, China

E-mail: pzy@mail.ujs.edu.cn

**Abstract**. An axial flow impeller and a mixed flow impeller with same specific speed were experimentally investigated, and the suction performance was studied with the help of CFD simulations. The results show that the axial impeller is roughly better than the mixed flow one. Especially under the design condition and a low flow rate condition range near the designed one, the axial flow impeller is more stable and therefore more suitable to be used in a water jet propulsion, while under these conditions the mixed flow impeller displays significant discrepancies. On the other hand, though its efficiency at the best efficiency point is lower than that of the axial flow one, the mixed flow impeller has a larger range of high efficiency conditions and is more convenient to be controlled to satisfy the irrigation and drainage systems that ought to be adjusted to varied flow rate conditions under a fixed head. In addition, the numerical investigation at the rated point shows that the axial impeller has a much better suction performance than the mixed flow impeller, which contradicts with the experience knowledge and therefore details need to be further studied.

## 1. Introduction

Heuristic materials show that an axial flow impeller is adopted when the specific speed $n_s$ of the pump is larger than 500, while a mixed flow one is more reasonable if the specific speed $n_s$ is lower than 500 but greater than 300. However, recently more and more mixed flow impellers with vane diffusers are used in the electric power plants, condensation systems and so on, which shows aggressive applications at even higher specific speed that axial flow impeller are conventionally selected [1]. As both mixed flow pumps and axial flow pumps with vane diffusers have too much long length in the axial directions and their guide bearings are submerge in the liquid medium, more recently mixed flow impellers with volute casings are also under consideration as alternative configurations. In addition, in the field of water jet propulsion, whether axial flow impellers or mixed flow ones are more convenient is always of discussion due to the limitation of the vessel hull [2]. Subsequently, more and more efforts are performed to investigate the details of the pumps, especially the flow structures such as tip leakage vortex, wake characteristics and so on. However, rarely comparative study has yet been carried out to examine their difference on overall performance. In this paper, both an axial flow impeller and a mixed flow impeller, which have the same specific speed $n_s$ of 576, are experimentally investigated to examine their inherence difference.

Individuals argued that commercial CFD code, though it is used to tackle many engineering problems, is so time consuming and leads to complex consequences, and is not possible, as yet, to draw general conclusions over a wide parameter range [3]. In addition, Ni insisted that the CFD code plays a certain negative role on the development of hydraulic machinery as the design process for them receives less and less attention it deserves [4]. Nevertheless, a CFD technology was performed to evaluate the suction performance of the impellers as the gauges of the test rig in this research were not accurate enough to measure the cavitation phenomenon.

## 2. Design parameters and test apparatus

The axial flow impeller with vane diffuser and the mixed flow one with volute are shown in Fig. 1(a) and Fig. 1(b). According to the model test requirements, the normal diameters of the impellers are 300mm. The rotation speed is $n$=1450r/min, the designed flow rate is $Q$=0.347m$^3$/s, the designed head is $H$=9.5m, and therefore the specific speed is $n_s=3.65nQ^{0.5}/H^{0.75}=576$. The obvious features of the two structures are that the axial flow pump has a larger axial length and a compact radial size, while that of the mixed flow one has a pancake type and therefore a robust configuration.

(a) Axial flow pump

(b) Mixed flow pump

(c) Schematic of closed test rig

**Figure 1.** Pumps and test rig.

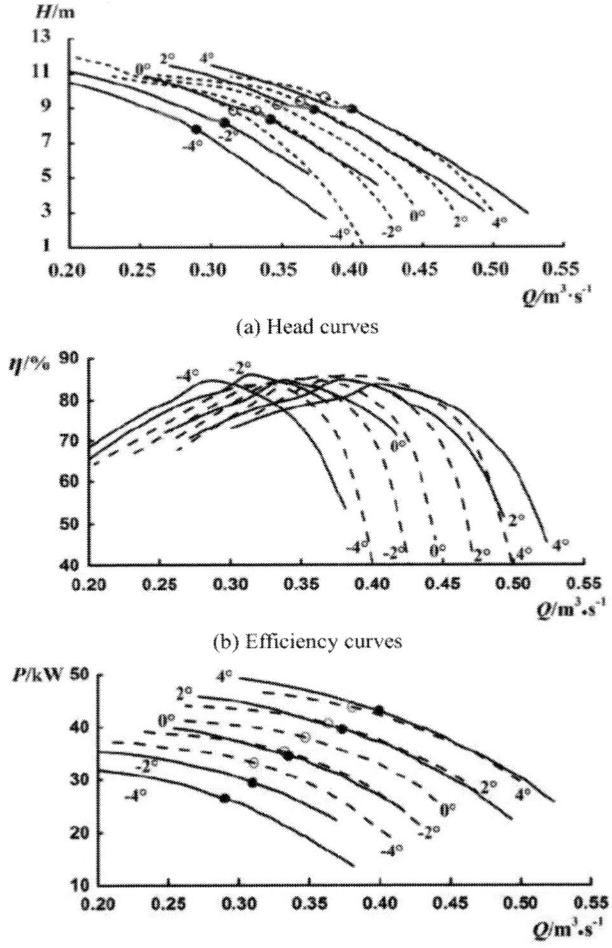


(a) Head curves

(b) Efficiency curves

(c) Power curves

**Figure 2.** Tested performance of pumps.

The experimental investigation was conducted at the closed loop rigs of the National Research Center of Pumps, Jiangsu University. As the configurations of the two pumps are of significant difference, the measurements of the two pumps were performed on two different test beds that of the same experimental methods as schematically shown in Fig. 1(c). The connection parts aft and prior the pump section should be replaced according to the pump flanges or other junctions.

## 3. Analysis on tested data

Figure 2 shows the tested performance curves in which the dashed lines indicate the data for the axial flow pump and the solid lines for the mixed flow one. Some open and closed circles shown in both Fig. 2(a) and Fig. 2(c) are the best efficiency points. The head-capacity curves of +4° and +2° show good coincidence with each other in a certain large flow rate range from the best efficiency point, while the curves of 0°, -2° and -4° for the mixed flow pump significantly drift to the left. Consequently, the adjustable range for flow rate of the mixed flow impeller is rather larger than that of the axial flow one, therefore the mixed flow impeller is more suitable for applications of flow rate large change with head nearly fixed. Nevertheless, insignificant discrepancies were found at the region that lower and near the best efficiency points on every head-capacity curves of the mixed flow pump, and with the blade angle increases, the discrepancies are more and more easily recognized.

For a mixed flow pump, a kind of discrepancy may appear on the head-capacity curves, which occurs under the conditions of about 0.5-0.7 times of the design flow rate and is related with an instability so-called rotating stall. In this case as shown in Fig. 2(a)., the left limit flow rates of the discrepancy regions are roughly 0.904, 0.958, 0.938, 0.949 and 0.964 times of the flow rate under best efficiency points for the blade angles -4°, -2°, 0°, +2° and +4°, respectively. More recently, Liu displayed similar discrepancies near the design points on a pump-turbine under pumping mode, who insisted that the phenomenon was contributed by low cavitation number and showed some variation of the event with the cavitation number changing [5]. Obviously, the discrepancies displayed here are not the same as that of the rotating stall related, which is somewhat similar with the rotating choke. Further investigations in detail will be conducted in the future.

Returning now to the comparison of the axial flow impeller and the mixed flow impeller, the mixed flow one studied here is not capable of being used to a water jet propulsion as the latter is usually working under the condition that appreciably lower than the design point to achieve the largest thrust. In general, the axial flow impeller has higher efficiency under a narrow range of flow rates near the best efficiency point, while the mixed flow impeller has higher efficiency under conditions that higher than the best efficiency point and has a wider high efficiency zone. Moreover, according to the power curves of +4° and +2°, the start-up power of the axial flow impeller is smaller.

## 4. Suction Characteristics

Figure 3 shows the curves of *NPSHR* vs. flow rate at the blade angle of 0°, in which the critical value of *NPSHR* was evaluated at the point that the efficiency decreased 1 per cent. The axial flow impeller has a better suction performance than the mixed flow one under a flow rate range near the design point. The test instruments shown in Fig. 1(c) and the results in Fig. 3 cannot depict the suction performance in detail. Therefore the numerical simulated data were displayed to discuss some of the difference. The best efficiency points at blade angle of 0° were taken into consideration, and the calculated critical points were that the efficiency decreased 1 per cent as the tested ones.

Commercial CFD code Fluent was performed to calculate the flow field in the impellers, in which the cavitating turbulent flow was simulated using the Partially-Averaged Navier–Stokes (PANS) method and a mass transfer cavitation model with the maximum density ratio effect between the liquid and the vapor, and Fig. 4 shows their suction performance under the best efficiency points. The best efficiency for the axial flow pump and the mixed flow pump are 85.19% and 85.27%, respectively. As shown in Fig. 4, the cavitation number under which the performance degradation occurs due to cavitation for the axial flow pump ($\sigma$=0.22) is much lower than that for the mixed flow pump ($\sigma$=0.41). Meanwhile, the critical cavitation number for the axial flow pump ($\sigma$=0.16) is also lower than that of the mixed flow one ($\sigma$=0.26). The results partly demonstrated the conclusion that leading edge cavitation for Kaplan turbines is less likely to occur than that for Francis turbines. While the conventional belief is that a mixed flow hydraulic machine has better suction performance than an axial flow one. Cavitation volume fraction profiles for six conditions marked by open or solid circles in Fig. 4 were also attached, the cavitation fraction is depend upon not only the cavitation number but also the geometry of the blade, details need more investigations.

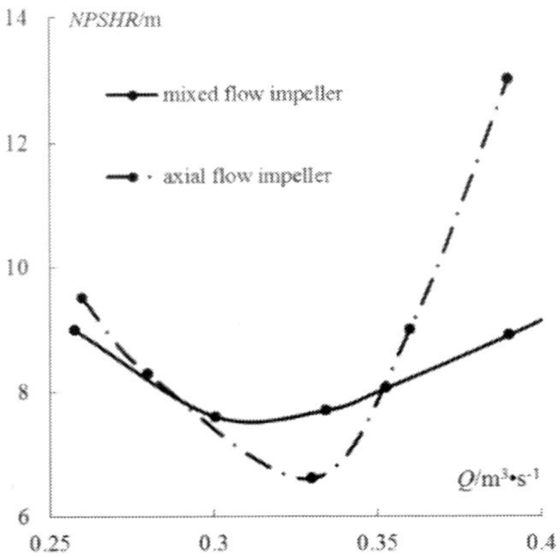


**Figure 3.** *NPSHR* curves tested.

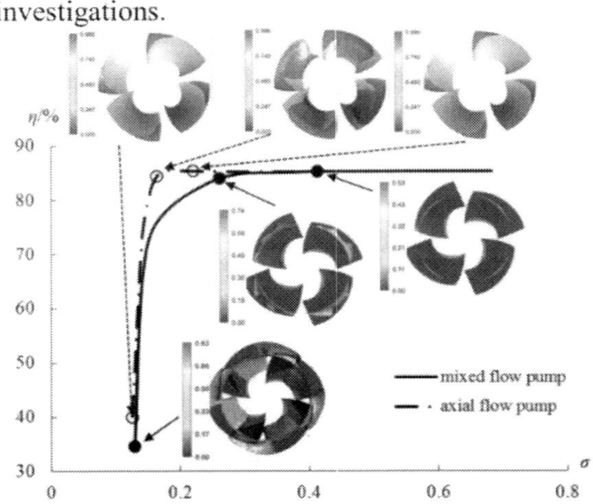


**Figure 4.** Suction performance calculated.

## 5. Conclusions

The axial flow impeller with a vane diffuser has a long axial length and compact radial configuration. Therefore it is more conveniently horizontal installed in a thin hull of high performance vessel to be a water jet propulsion. The mixed flow impeller with a volute collector has a robust pancake profile and is of vertical reliability.

Discrepancies near the design points were found on the performance curves of the mixed flow pump. Further study need to be done to investigate the detail as there have rarely materials concerning this phenomenon. The axial flow impeller has a good performance under the design condition while the mixed flow one has a larger region of high efficiency. In addition, the axial flow impeller also has a better suction performance under the design point, which contradicts with the experience knowledge and therefore details need to be further studied.

## 6. Acknowledgements

This work was financially supported by the National Natural Science Foundation of China (Grant No. 51209108, Grant No. 51239005), and Jiangsu Provincial Natural Science Foundation of China (Grant No. SBK2015020780).

## References

[1] Pan Zhongyong, Ni Yongyan, Yuan Shouqi, et al. Overview for research on mixed flow pumps with vaned diffuser[J]. *Fluid Machinery*, 2009, 37(9): 37–40. (in Chinese)

[2] Ni Yongyan, Wu Taotao. Analysis and improved design for the water-jet propulsor[J]. *Ship and Ocean Engineering*, 2012, 41(5): 61–63. (in Chinese)

[3] Brennen C E. *Fundamentals of multiphase flows*[M]. Cambridge: Cambridge University Press, 2009.

[4] Ni Yongyan, Liu Weimin. Overview on research of water-jet propulsion[J]. *Ship and Ocean Engineering*, 2013, 42(5): 1–5. (in Chinese)

[5] Liu D M, Zhao Y Z, Liu X B, et al. Pump hump characteristic research based on mass transfer equation[C]//*International Symposium of Cavitation and Multiphase Flow*, Beijing, China, October 18-21, 2014.

# Pressurized air injection in an axial hydro-turbine model for the mitigation of tip leakage cavitation

**A Rivetti[1], M Angulo[1], C Lucino[1] and S Liscia[1]**

UNLP Laboratory of Hydromechanics, 47 Street N° 200, La Plata, Argentina.

E-mail: arturorivetti@gmail.com

**Abstract**. Tip leakage vortex cavitation in axial hydro-turbines may cause erosion, noise and vibration. Damage due to cavitation can be found at the tip of the runner blades on the low pressure side and the discharge ring. In some cases, the erosion follows an oscillatory pattern that is related to the number of guide vanes. That might suggest that a relationship exists between the flow through the guide vanes and the tip vortex cavitating core that induces this kind of erosion.
On the other hand, it is known that air injection has a beneficial effect on reducing the damage by cavitation. In this paper, a methodology to identify the interaction between guide vanes and tip vortex cavitation is presented and the effect of air injection in reducing this particular kind of erosion was studied over a range of operating conditions on a Kaplan scale model.
It was found that air injection, at the expense of slightly reducing the efficiency of the turbine, mitigates the erosive potential of tip leakage cavitation, attenuates the interaction between the flow through the guide vanes and the tip vortex and decreases the level of vibration of the structural components.

## 1. Introduction

Axial hydro-turbines present a clearance between the runner blades and the discharge ring. This feature in combination with the high pressure gradient among both sides of the blade, induce a secondary flow known as tip vortex, which has been studied by several authors [1].
The high intensity of the vortex and the low pressure at the outlet of the runner may generate cavitation of the vortex core [2]. As a consequence, erosion issues at the runner blades and discharge ring can occur at the prototype machine. Moreover, the tip vortex can interact with the wakes generated by the flow leaving the guide vanes giving rise to a phenomenon of interaction that is similar to well-known rotor stator interaction (RSI). The existence of this interaction is suggested in the prototype machine by the presence of as many erosion patches at the discharge ring (on a horizontal plane) as guide vanes [3]. Some mitigation methods, such as anti-cavitation lips, have been tried but their effectiveness has not always been ensured. The use of air injection is an interesting option for cavitation problems, as it has been used in many industrial applications. Arndt et al. [4] examined the effect of air injection on NACA profiles for the mitigation of cloud cavitation, and found that it was an effective method of minimizing the potential of erosion.
In this work, experimental investigation of pressurized air injection in a physical model of a Kaplan turbine is studied. The main objectives are the mitigation of tip vortex cavitation and the suppression of the RSI. In the first part, the experimental setup and the test procedure are presented. Then, the trajectory of the air in the flow is characterized with the aid of high speed visualization and the main results are summarized in terms of efficiency and vibrations. Finally, the RSI pattern is analyzed.

## 2. Methods and materials

Experiments were carried out at the test rig facility in the Hydromechanics Laboratory of La Plata, Argentina. The circuit is equipped with a feeding pump with variable rotational speed, allowing for the regulation of the specific energy (E) and the flow rate (Q). Experimental investigations on the test rig were based on IEC 60193-standard [5]. The Kaplan model is located between the high and low pressure tanks. The diameter of the runner (D) was 0.34 m and the rotational speed was 1000 rpm, reaching a Reynolds number of 6.05E6 based on the blade tip velocity. The discharge ring was manufactured with a transparent material for the sake of visualization.

The model was instrumented with two accelerometers located at the discharge ring, one hydrophone at the draft tube inlet, and a phase sensor at the runner shaft (Figure 1). The sampling frequency of the signal was 50 kHz. An air compressor was connected, through a pressurized manifold, to 20 evenly spaced 3-mm-diameter holes located on a horizontal plane above the runner centerline. A ball air flow-meter was placed between the compressor and the manifold.

In Figure 1b the operating points tested over the hill chart are shown. Air was injected at each operating point for 8 different air flow rates ($Q_a$=0.2, 0.3, 0.4, 0.8, 1.3, 1.7 and 2.1‰) in combination with 11 $\sigma$ numbers ranging from cavitation-free conditions to a fully developed cavitating regime. Finally, a high-speed camera (PHOTRON SA4) was used to analyze the air trajectory.

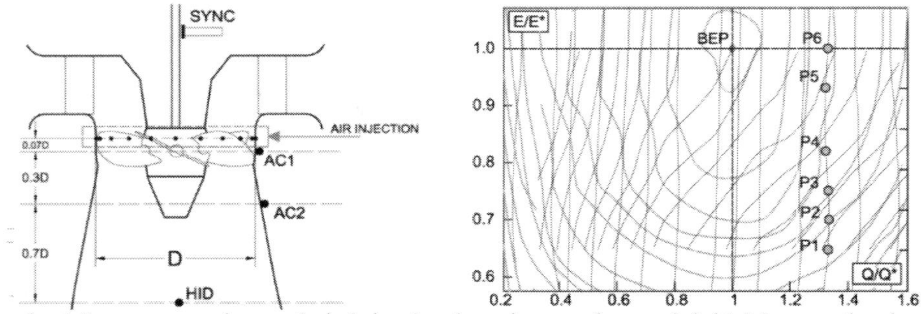


**Figure 1.** a) Instrumentation and air injection location on the model. b) Measured points in the hill chart. AC1, AC2 and HID denote the position of the accelerometers and the hydrophone, respectively. E* and Q* are the nominal value for specific energy and flow rate.

## 3. Results

### *3.1. Flow visualization*

Figure 2 presents high-speed visualization snapshots at different positions of the runner blade with air injection for a free-cavitation regime. Air entrance occurs when the pressure at the manifold is greater than the fluctuating fluid pressure (Figure 2a). Fluid pressure is seen to rise due to the high gradient between both sides of the blade as the leading edge interferes with the air jet (snapshot 2). Air travels along the edge of the blade (snapshot 3) until getting immersed into the vortex core (snapshot 4).

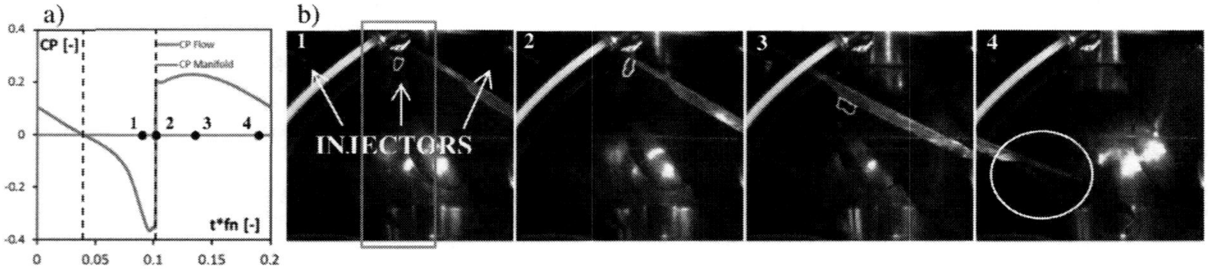


**Figure 2.** a) Fluid pressure fluctuations at the location of one injector and in the manifold for one blade passage, as a function of the pressure coefficient (CP) and time (*t*fn*), where *fn* is the runner rotational frequency. b) Snapshots of air injection for the blade position (1, 2, 3 and 4) indicated in 2.a) for a cavitation-free regime.

### *3.2. Hydrophone measurements*

For a cavitation-free regime, the ensemble average of the hydrophone signal remains essentially flat (Figure 3a), which implies a random behavior. For $\sigma$ = 1.40, there is an incipient tip leakage cavitation. The oscillatory pattern is driven by a frequency of 25 *fn* (Figure 2b) and remains the same as the amplitude increases with decreasing values of $\sigma$ (Figure 2c). Simultaneously, an increase of the random component is observed, which is in agreement with the higher development of tip cavitation. When air is injected, the frequency of 25 *fn* vanishes, giving rise to components of 20 and 5 *fn*.

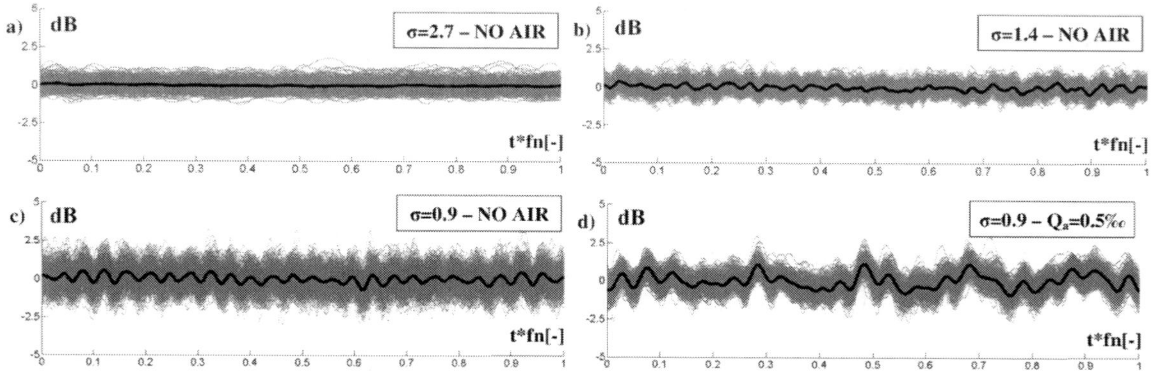


**Figure 3.** Hydrophone signal for the operating condition P2: ensemble average (black line) and instant values (gray dots) for different $\sigma$ values with and without air.

### *3.3. Global results*

The intensity of vibration rises as $\sigma$ decreases, reaching a local maximum for $\sigma \sim 1.2$. A further decrease in $\sigma$ results in levels of vibration going down until $\sigma \sim 0.9$ upon which a new rise is observed, in coincidence with full cavitation conditions (Figure 4a). The level of vibration is also seen to decrease as the rate of air injection increases for a constant $\sigma$ regardless of the operating conditions (Figure 4b). Beyond approximately $Q_a$= 0.5‰, air injection has no further influence on the level of vibration and the efficiency drop becomes greater than 1%. The evolution of the amplitude of the hydrophone signal corresponding to a frequency of 25 *fn* rises as the guide vane opening ($\alpha$) increases (Figure 4c). The trend is positive except for the maximum guide vane opening conditions (P1).

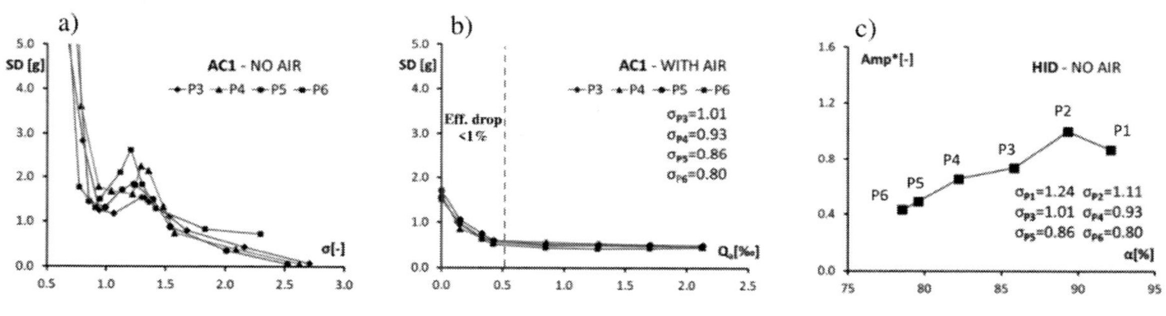


**Figure 4.** a) Influence of the development of cavitation on the vibration level as a function of the standard deviation of the acceleration signal (SD) and $\sigma$; b) Influence of air injection rates on the vibration level for constant $\sigma$ values; c) Amplitude of the hydrophone signal corresponding to a frequency of *25 fn* as a function of the guide vane opening. Pi (i = 1,2,…,6) denote the operating conditions shown in Figure. 1.

## 4. Discussion

### *4.1. RSI analysis*

As σ decreases, cavitation develops mainly as tip leakage cavitation. The influence of the non-uniform flow leaving the guide vanes can be identified by the analysis of the modulation process generated by the stationary flow field of the 24 guide vanes ($Z_0$), and the rotating flow field due to the 5 runner blades ($Z_b$), as proposed by Ruchonnet et al. [6]. Table 1a shows the RSI patterns for several orders of harmonics for the first diametrical mode number $k_1$. The highest amplitude expected is for the first order of harmonics of flow through the guide vanes ($n$ = 1) and the fifth order of harmonics of flow through the runner ($m$ = 5), since $k_1$ is the minimum. The characteristic frequency of 25 $fn$, given by the expression $f/fn = mZ_b$, has to be present if the interaction exists. When air is injected, there is a novel interaction between the 20 injectors ($Z_i$) and the blade passage, which is summarized in table 1b. In this case, $k1$ is minimum for $m$= 4 and $n$=1, and the characteristic frequency is 20 $fn$, where $f/fn = mZi$.

**Table 1.** First diametrical mode number $k_1$ according to harmonic order $m$ for the runner blades ($Z_b$=5) and $m$. a) For the guide vanes ($Z_0$=24); b) For the injectors ($Z_i$=20).

a)

| | $k1=mZ_b-nZ_0$ | | | | |
|---|---|---|---|---|---|
| **n/m** | 1 | 2 | 3 | 4 | 5 |
| 1 | -19 | -14 | -9 | -4 | 1 |
| 2 | -43 | -38 | -33 | -28 | |
| 3 | -67 | -62 | -57 | -52 | -47 |

b)

| | $k1=mZb-nZ_i$ | | | | |
|---|---|---|---|---|---|
| **n/m** | 1 | 2 | 3 | 4 | 5 |
| 1 | -15 | -10 | -5 | 0 | 5 |
| 2 | -35 | -30 | -25 | -20 | -15 |
| 3 | -55 | -50 | -45 | -40 | -35 |

## 5. Conclusions

The tip vortex cavitation that develops at lower cavitation numbers was singled out as one of the main causes of the increase of vibration levels.

The analysis of the signal of the hydrophone reveals the presence of RSI-related phenomena which becomes apparent when tip vortex cavitation develops.

Air injection was found to fluctuate in correspondence with periodic oscillations of fluid pressure due to the blade passage. Regardless, it can be an effective tool both in eradicating the influence of the flow driven by the guide vanes and in reducing the level of vibration. This may have a beneficial effect on the mitigation of the erosion due to its related tip vortex cavitation.

Vibration levels were reduced 50 % for an air flow rate of $Q_a$ = 0.5 ‰ at the expense of an efficiency drop of less than 1 %.

## References

[1] Dreyer M, Decaix J, Munch-Alligne C, Farhat M, 2014: Mind the gap: a new insight into the tip leakage vortex using stereo-PIV, Exp Fluids 55:1849, DOI 10.1007/s00348-014-1849-7

[2] Nilsson H and Davidson L 2000: A numerical comparison of four operating conditions in a Kaplan water turbine, focusing on tip clearance flow, 20th IAHR Symposium (Charlotte North Carolina, U.S.A.)

[3] Nennemann B and Vu T C, 2007:Kaplan turbine blade and discharge ring cavitation prediction using unsteady CFD 2nd IAHR Int. Meeting of the Workgroup on Cavitation and Dynamic Problems in Hydraulic Machinery and Systems (Timisoara, Romania)

[4] Arnt R, C Ellis, S Paul, 1995: Preliminary Investigation of the Use of Air Injection to Mitigate Cavitation Erosion. Journal of Fluids Engineering, Vol. 117, pp. 498-504.

[5] IEC standars. Hydraulic turbines, starages pumps and pumps-turbines-model acceptance tests; IEC60193-1999.

[6] Ruchonnet N, Nicolet C, Avellan F, 2006: Hydroacoustic Modeling of Rotor Stator Interaction in Francis Pump-Turbine, IAHR Meeting of WG on Cavitation and Dynamic problems in Hydraulics Machinery and Systems, Barcelona.

# Detailed investigation of cavitation and supercavitation around different geometries using various turbulence and mass transfer models

**M R Pendar**[1], **E Roohi**[2]
[1,2]High Performance Computing (HPC) Laboratory, Department of Mechanical Engineering, Ferdowsi University of Mashhad, Mashhad, Iran

E-mail: e.roohi@ferdowsi.um.ac.ir

**Abstract.** In this paper cavitation around 3D hemispherical head-form body as well as disk and conical cavitators were studied numerically with particular emphasis on detailed comparisons of the various turbulence and mass transfer models. Dynamic and unsteady behaviors of cavitation were simulated using the large eddy simulation (LES) and k-ω SST turbulence models together with Kunz, Sauer and Zwart mass transfer models. In addition, the compressive volume of fluid (VOF) method is used to track the cavity interface. Simulation is performed under the framework of the OpenFOAM package. Discussions on the boundary layer separation, re-entrant jet, cavity cloud, vortices and pressure/volume fraction contours over a broad range of cavitation numbers, especially in very low cavitation numbers, are provided. Our numerical results compared fairly well with experimental data and a wide set of analytical relations. Our results indicate that the most accurate solutions will be obtained with the LES/ Kunz approach. Moreover, for the first time, we present a correlation between the cavity length and diameter for hemispherical head-form bodies.

## 1. Introduction

Cavitation is a multi-phase, unsteady, three-dimensional, discontinuous or periodic phenomenon which occurs when the local liquid pressure becomes lower than its saturated vapor pressure. This process is characterized by a dimensionless number; i.e., $\sigma=(P_\infty-P_g)/0.5\rho U_\infty^2$ called cavitation number. To increase the performance of submarines and reduce the viscous drag, underwater vehicles usually operate in the cavitating conditions. Challenging issues to be considered in the numerical simulation of a 3-D cavitation are: selection of an appropriate mass transfer model, solution technique for the advection equation of the free surface and application of an appropriate turbulence model. There were different investigations of 2D and 3D cavitating flows over various geometries, i.e., see Refs. [1-4]. Compared to previous studies, the current work reports flow at very low cavitation numbers, i.e., σ=0.02 and 0.035, and reports a correlation between the cavity length and diameter for hemispherical head-form bodies.

## 2. Governing equations

The governing equations are the incompressible Navier-Stokes equations. The interface between the liquid and vapor phases is captured by using the volume of fluid (VOF) method [5]. Kunz et al. [6] and Schnerr and Sauer [7] mass transfer models were used. Large eddy simulation (LES) turbulence approach is based on computing large, energy-containing eddies that are resolved on the computational grid, whereas the smaller, more isotropic, sub-grid structures are modelled. In the current study, sub-grid scale terms are modeled using "one equation eddy viscosity" model. In addition to the LES, the shear stress transport k-ω SST model [8] is utilized for the turbulence modelling.

## 3. Numerical method

Figure 1 shows the 3D computational domain of the disk, conical cavitator and a hemispherical head-form body with the diameter of D=2.54 cm, 0.8284 m, 0.2 m respectively. $U_\infty$, which is the free stream velocity, is imposed as 20 m/s.

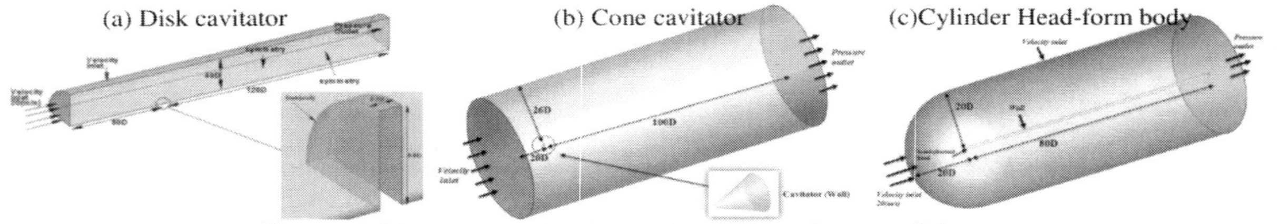


**Figure 1.** The computational domain and boundary conditions

The effect of using three different grid sizes for the conical cavitator is reported in table 1. From table 1, it is observed that Grid 2 provides close solutions for predicting different characteristic of the cavity. In a similar way, we performed our simulations using $1\times10^6$ cells for $\sigma > 0.07$ and $2.4\times10^6$ cells for $\sigma=0.02$ for the cone, and $1.5\times10^6$ cells for $\sigma>0.05$ and $2.4\times10^6$ cells for $\sigma=0.02$ for the hemispherical head-form cylinder and also $3\times10^6$ cells for the disk at $\sigma=0.035$. Two non-dimensional number: Reynolds and cavitation numbers are considered. In all cases, we considered the Reynolds number as $Re=5\times10^5$. We use a fine grids to obtain appropriate $y^+$, i.e., the maximum values of $y^+$ were 241, 275 and 236, and the mean values of $y^+$ were 8.59, 7.9 and 7.225 for $\sigma=0.02$ in conical cavitator case, hemispherical head-form cylinder case and $\sigma=0.035$ disk case, respectively. For $30 < y+ < 300$ at most positions, i.e., adjacent to the body region, OpenFOAM uses wall functions. A blending function is employed in the OpenFOAM's LES to manage different values of $y^+$ for the first grid cell and its influence in near-wall function selection. The blending function is given by Spalding Law.

Figure 2 shows the residuals convergence history of the liquid volume fraction (alpha) and pressure (p) for the five last time steps at $\sigma=0.07$ for cone cavitator.

**Table 1.** Effect of different grid sizes (conical cavitator)

| Mesh for $\sigma=0.02$ | Cell number ($\times10^6$) | $L_{cavity}/d_{cavitator}$ | | $D_{cavity}/d_{cavitator}$ | |
|---|---|---|---|---|---|
| | | Simulation | Reichardt | Simulation | Reichardt |
| Grid1 | 1.8 | 67.25 | 55.53 | 4.30 | 3.96 |
| Grid2 | 2.4 | 61.46 | 55.53 | 3.91 | 3.96 |
| Grid3 | 3 | 59.42 | 55.53 | 3.92 | 3.96 |

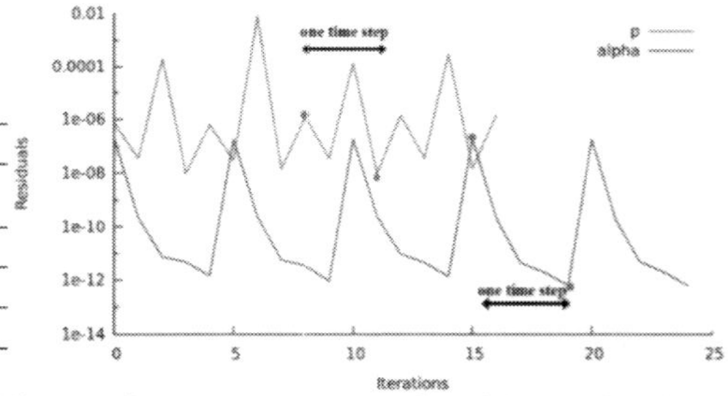


**Figure 2.** Convergence of residuals for five last time steps at $\sigma=0.07$ (LES, Kunz).

## 4. Results and Discussion

Figure 3 illustrates 3D views of the of the cavitating flow around different geometries at $\sigma= 0.035$ and 0.02. For the disk cavitator, we showed cavity at $\sigma= 0.035$ as the cavity length is approx the same with cone cavitator at $\sigma= 0.02$.The cavity has a roughly pulsating surface in the k-ω SST/Sauer solution while LES/Sauer predicts a smooth and regular shape. LES/Kunz shows a wavy surface for the disk cavity. All models show a steady behavior with a very long cloud length at the lowest cavitation number.

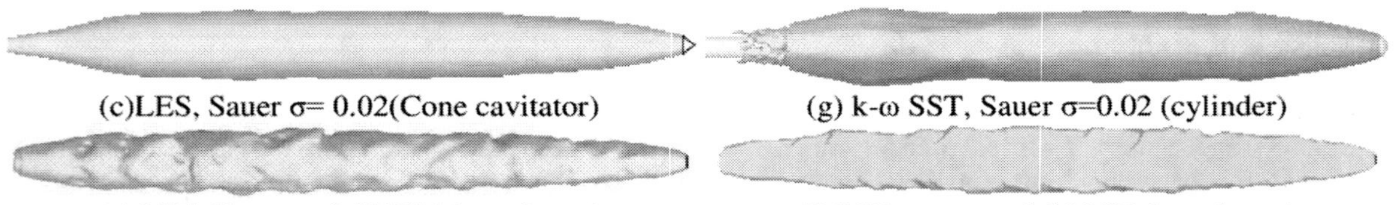


**Figure 3.** 3D cavitation around different cases

Journal of Physics: Conference Series **656** (2015) 012070 doi:10.1088/1742-6596/656/1/012070

Figure 4 frames (a-d) illustrates maximum cavity length and diameter of the cavity at various cavitation numbers from different models. We evaluate our results with an extensive set of analytical relations [9] and experimental data, and generally a suitable agreement was observed. Kunz model with the LES turbulence approach predicts the most accurate L/d, also Kunz and Sauer models with the LES predict relatively accurate solutions for the cavity diameter in comparison of other result. Based on the current numerical solutions in frames (e-f) of figure 4, we deduce inverse power law correlations for the cavity length and diameter over the hemispherical body as follows:

$$\frac{L}{D}=\frac{0.14}{\sigma^{1.60}} \qquad (0.02 \le \sigma \le 0.5) \qquad (1) \qquad\qquad \frac{d}{D}=\frac{0.4}{\sigma^{0.4}} \qquad (0.02 \le \sigma \le 0.5) \qquad (2)$$

The cavity length and diameter exponentially increase with decreasing the cavitation number. The results of different models are quite close to each other.

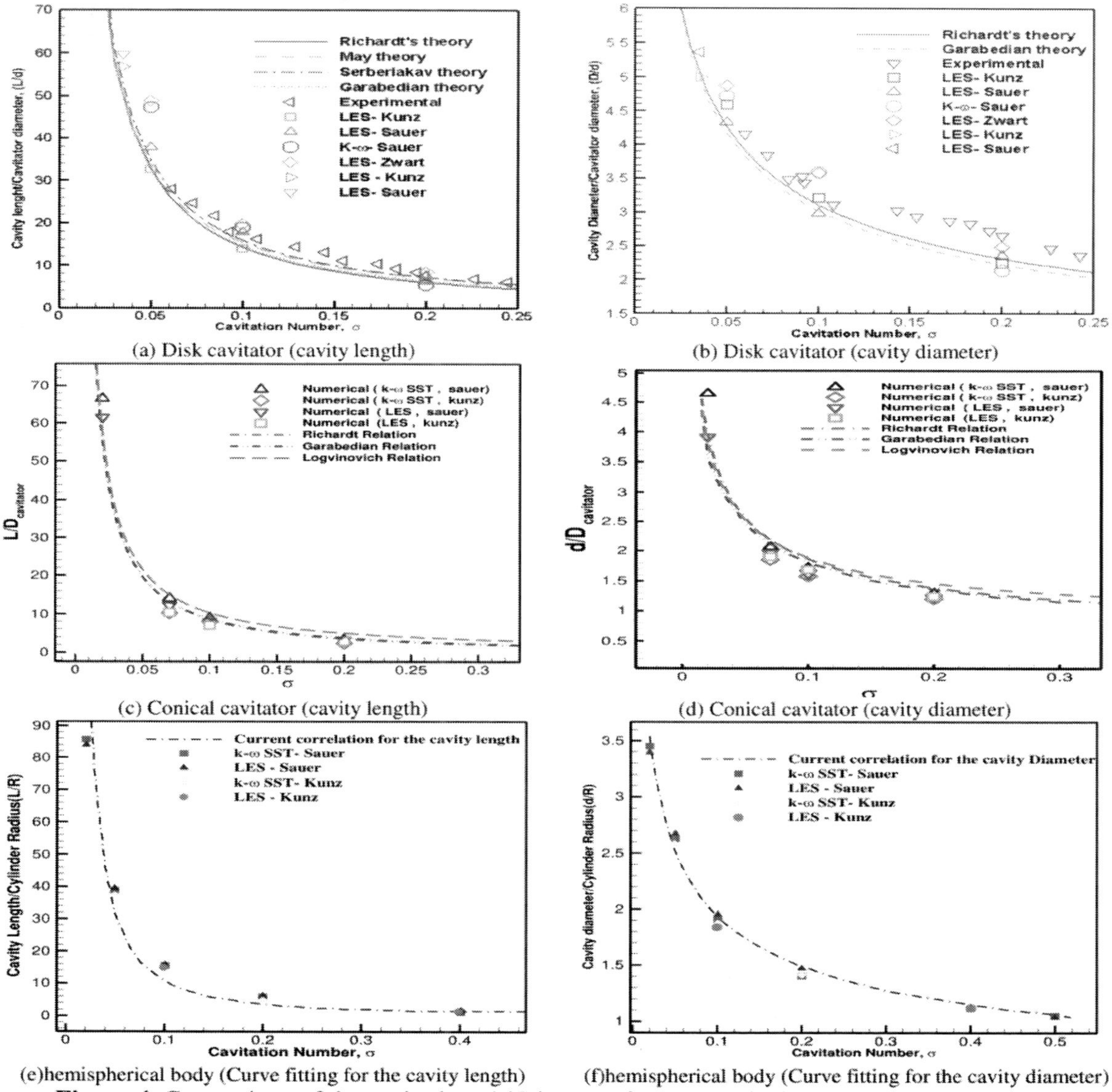


(a) Disk cavitator (cavity length) (b) Disk cavitator (cavity diameter)

(c) Conical cavitator (cavity length) (d) Conical cavitator (cavity diameter)

(e)hemispherical body (Curve fitting for the cavity length) (f)hemispherical body (Curve fitting for the cavity diameter)

**Figure 4**. Comparison of the cavity length/ Diameter from numerical and analytical solutions.

Figure 5 depicts the volume fraction of the liquid phase. The re-entrant jet is generated at the cavity closure. LES predicts a longer length for the re-entrant jet, but K-ω SST model predicts a weak re-entrant jet. As the cavitation number decreases, re-entrant jet becomes weaker. The separation point location is indicated by point C. For instance, the separation point position is located at (X/R) =12.88 in LES/Sauer solution. This figure illustrates that K-ω SST turbulence model is unable to model the vapor separation point. Figure 6 shows the pressure coefficient on the hemispherical body surface from various cavitation models and compared with experimental data reported by Rouse and McNown suitably [10].

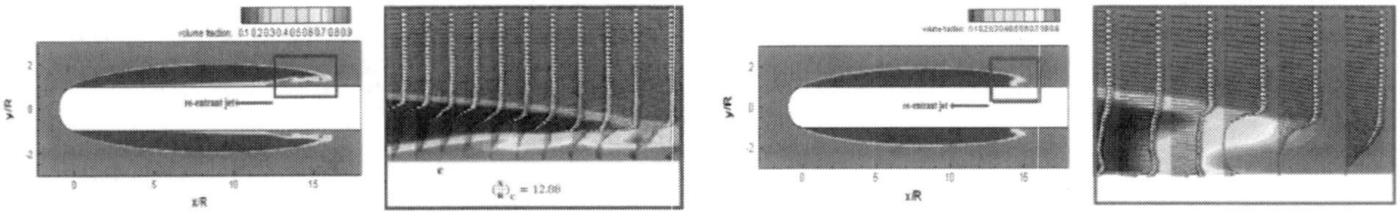


LES, Sauer K-ω SST, Kunz

**Figure 5**. Volume fraction contours at σ=0.1 over hemispherical head-form cylinder

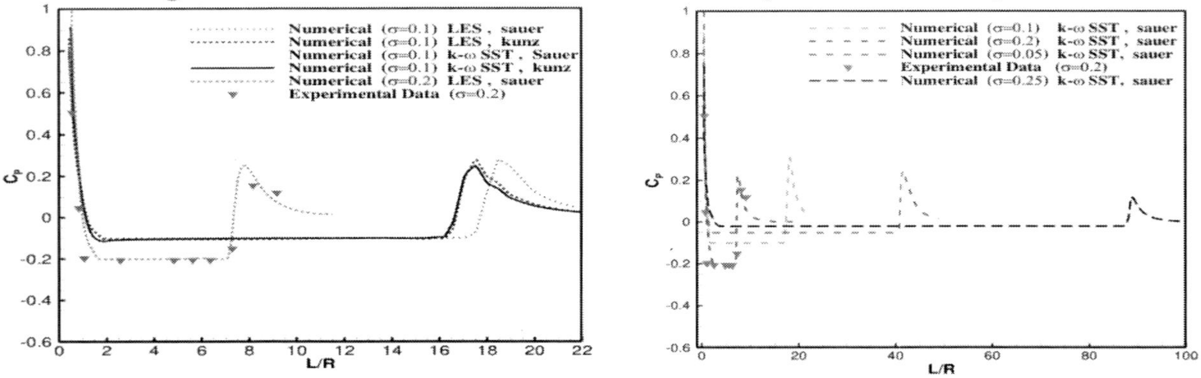


**Figure 6**. pressure coefitiont from different cavitations models over hemispherical head-form cylinder

## 5. Conclusion

In the present study, unsteady cavitation and supercavitation flow around 3-D disk/ conical cavitator and hemispherical head-form body were considered. Simulations were performed under the framework of the OpenFOAM package in very low cavitation number σ= 0.07, 0.035, 0.05, 0.02. Benefiting from VOF method, large eddy simulation (LES) and k-ω SST turbulence models as well as Kunz and Sauer mass transfer models, we simulated cavitation accurately. The principal innovation of this work is to provide a detailed comparison of the various sets of turbulence and mass transfer models in all cases and present a correlation between the cavity length and diameter for the hemispherical head-form body. Our numerical results were validated using experimental data and an extensive set of analytical relations for the cavity length and diameter, showing suitable accuracy. It is found that the most accurate solutions for the cavity characteristics are obtained by applying the LES turbulence with the Kunz model.

## 6. References

[1] Guo J, Lu C and Chen Y 2011 *J. Shanghai Jiaotong University (Science)* **16**(2) 236-241
[2] Park S and Hyung S 2012 *J. Computers & Fluids* **70** 73–85
[3] Shang Z 2013 *J.Applied Mathematical Modeling* **37** 8836-45
[4] Chen Y, Lu C, Chen X and Cao J 2015 *J. European Mechanics B/Fluids* **49** 153–170
[5] Hirt F H and Nichols B D 1981 *J. Comput. Phys.* **39** 201
[6] Kunz R F, Chyczewski T, Boger D, Stinebring D, Gibeling H and Govindan T R 1999 *Pro. FEDSM 99, 3rd9 ASME/JSME Joint Fluids Engineering Conference*
[7] Schnerr G H and Sauer J 2001 *Pro. 4th International Conference on Multiphase Flow*
[8] Menter F R 1994 *J. AIAA* **32**(8) 1598–1605
[9] May A 1975 *Water Entry and the Cavity-Running Behavior of Missiles, National Technical Information Service Springfield* **2** *3-9*
[10] Rouse H, McNown JS 1948 *Cavitation and pressure distribution in engineering Bulletin* **32**

# Cavitation measurements on a pump-turbine model

**H Schmidt[1], O Kirschner[1], S Riedelbauch[1]**

[1] Institute of Fluid Mechanics and Hydraulic Machinery; University of Stuttgart, Pfaffenwaldring 10, 70569 Stuttgart, Germany

hschmidt@ihs.uni-stuttgart.de

**Abstract**. When a hydraulic turbine is operated at off-design conditions, cavitation on the runner and other machine parts can occur. Vibration, noise and erosion caused by cavitation can damage the turbine and lead to a limitation of the operational range. To avoid damage of the turbine, it is fundamental to get knowledge of the presence of cavitation. In this paper, the acoustic emissions at a pump-turbine model at different operating conditions with and without the presence of cavitation were recorded and analysed. High speed video recordings were carried out simultaneously to validate the acoustic measurements. The main goal of the investigation was to compare the acoustic emissions with the visual observations at operating conditions with cavitation on the leading edge of the turbine runner. The analysis of the recorded signals and the visual observations are in good accordance for the investigated operating points.

## 1. Introduction

Cavitation at the runner and guide vanes can occur when a turbine is operated at off design conditions. If cavitation gets fully developed, parts of the turbine are prone to erosion damage. The turbine manufacturer permits an incipient state of cavitation, which is observed and evaluated during model tests and is which is supposed to be not erosive. Cavitation detection in model turbines based on acoustic emissions, vibrations and optical methods are state of the art [1] [2] [3] [4] [5]. When cavitation is measured in a model turbine by a physical quantity like acoustic emissions, it is important to get knowledge about the occurrence and state of cavitation visually. Unlike in prototype turbines the visual accessibility is widely possible in axial model turbines [1] [2] [3]. In radial model turbines the view onto the leading edge is hindered due to limited illumination and the design of the runner but observation through the transparent draft tube cone onto the runner outlet is a current method [4][5].

In this paper the acoustic measurement of the pressure side leading edge cavitation at a model pump-turbine with the visual evaluation via high speed video recording is presented. Therefore a window and an illumination system were installed in the lower wicked gate ring. Several operating conditions with different states of cavitation were measured and high speed video recordings were captured. The goal of the present experimental investigation is the detection of cavitation at the leading edge based on the acoustic emissions and get knowledge of the relation between measured signals and visual observations.

## 2. Experimental setup

The measurements were carried out at the closed loop test rig (fig. 1) at the Institute of Fluid Mechanics and Hydraulic Machinery, University of Stuttgart. The investigated pump-turbine model

Journal of Physics: Conference Series **656** (2015) 012071 doi:10.1088/1742-6596/656/1/012071

was equipped with an acoustic emission sensor (AE-sensor). The sensor was installed at the upper cover. The position on the upper cover was chosen to get as close as possible to the cavitation structures on the leading edge of the runner. The sensor was glued directly to the machine surface.

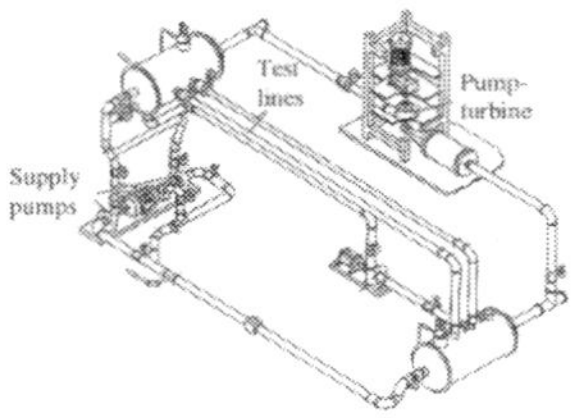


**Figure 1.** Closed loop test rig.

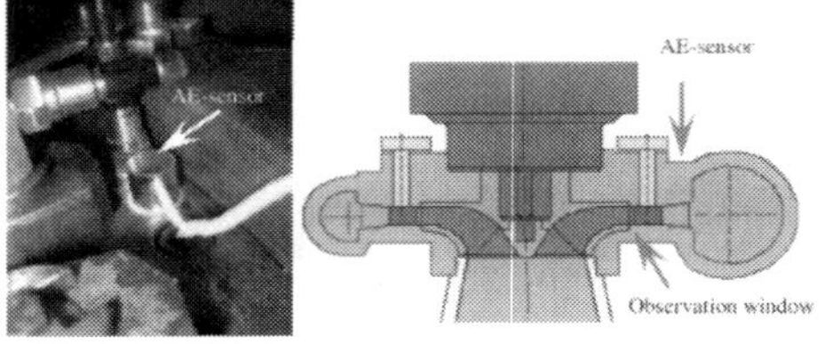


**Figure 2.** Acoustic emission sensor on upper cover (left) and cross section of the pump-turbine model (right).

Figure 2 shows the installed acoustic emissions sensor. The measured signals are filtered, amplified and recorded by a data acquisition system with a sampling rate of 6 MS/s.

To get visual access to the leading edge of the runner in turbine mode, observation windows and illumination were installed in the lower distributor ring of the pump-turbine model. The window with a diameter of 30 mm permits the view on the leading edge using a high speed camera. To obtain sufficient illumination for high speed recording, four high power light emitting diodes with a maximum output of 1665 lumens were installed in milled-out portions besides the window. Figure 3 shows the modifications of the lower distributor ring viewed from the runner-side. The high speed video records were captured with 6500 fps and analyzed by rating the extent of the visible cavitation structures.

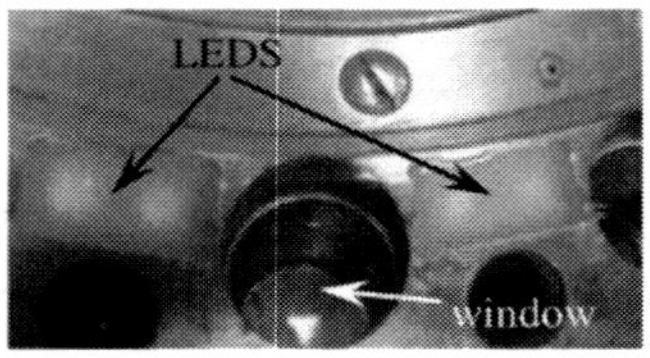


**Figure 3.** Observation window and illumination

## 3. Measurement

To induce cavitation at the leading edge of the runner, one operating point at low head was adjusted for the measurement. For the examined operating point the specific flow Q1' was 83% of $Q1'_{BEP}$ and the specific speed n1' was 132% of $n1'_{BEP}$. The head, discharge and machine speed were kept constant and the cavitation number σ was gradually lowered by lowering the system pressure of the closed loop test rig to obtain operating conditions with increasing cavitation. The cavitation number σ was constant during the measurement of one operating point. The measurement and the high-speed video were synchronously triggered by a key-phaser located on the machine shaft.

## 4. Analysis of the measured signals

The acoustic emissions caused by cavitation are broadband signals reaching the ultrasonic range. To avoid mechanical background noise from the turbine, the analysis of the signals was focused on the frequency range above f = 100 kHz. The signals were conditioned and wavelet-denoised. The time series at each σ-value were divided into several sections over one machine revolution to get an angular resolution and the power spectrum was calculated for each section. As cavitation sets in, the distribution of the power spectrum in the analysed frequency range is characterised by a distinct increase. By summing up, normalising and weighting the signal power amplitudes exceeding a set threshold in the frequency domain, characteristic values were derived to characterise the cavitation state of the turbine runner. A detailed description of the method can be found in [6].

The development of the characteristic curve for the threshold value at the examined sensor position is plotted versus the cavitation number, see figure 4. The development shows a nonuniform increase with a significant step around σ = 0.4. Four of the examined operating conditions, numbered 0 to 3, are exemplarily presented.

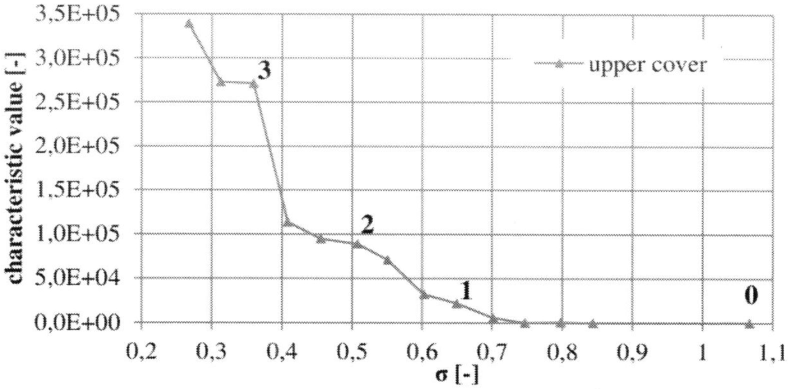


**Figure 4.** Development of characteristic value

Figure 5 shows a reference picture at σ = 1.06 without cavitation. The viewing perspective is upwards from the lower wicket gate ring on the guide vane and leading edge of the runner towards the turbine head cover. At operating condition 1 first cavitation structures at the leading edge are visible (fig. 6). Not all runner blades are affected, which is referred to the inequality of the leading edges of the runner blades.

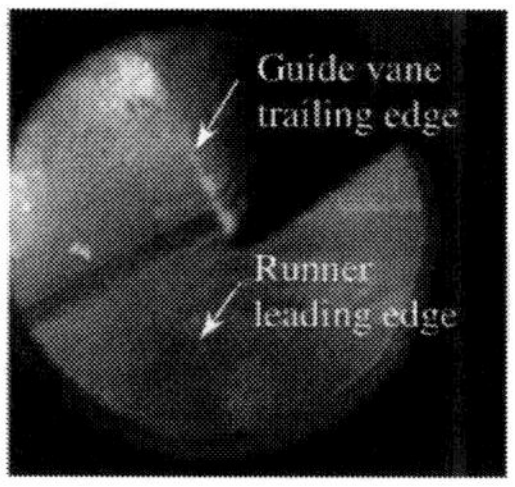


**Figure 5.** Operating condition 0, no cavitation.

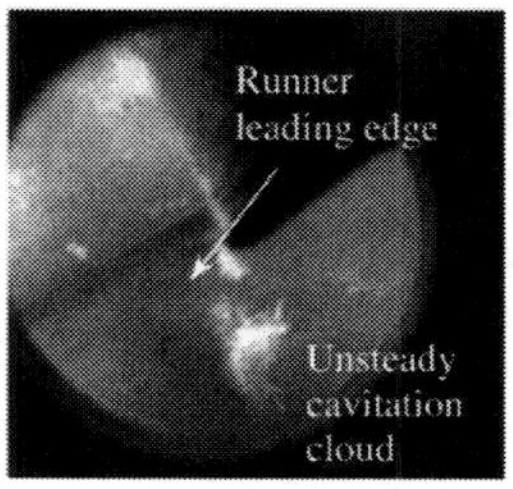


**Figure 6.** Operating condition 1

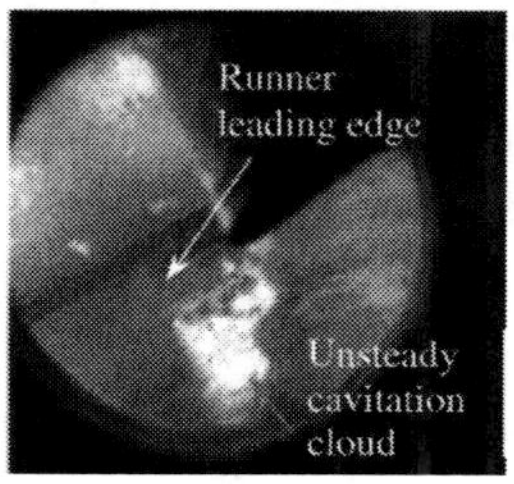


**Figure 7.** Operating condition 2

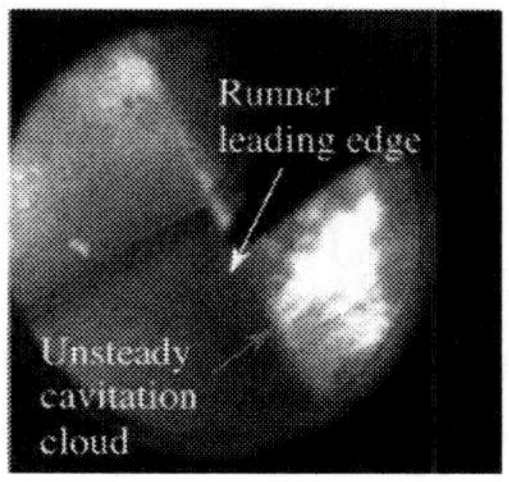


**Figure 8.** Operating condition 3

Operating condition 2 (fig. 7) is characterised by distinct cavitation structures at the leading edge but not all runner blades are affected over one revolution. Further a fluctuation of the volume of the cavitation structures while passing the guide vane is observed. This indicates a changing pressure field at the runner leading edge caused by an interaction of runner and guide vanes. At operating condition 3, severe cavitation is visible with large detaching structures on all runner blades (fig. 8).

Regarding the occurrence and extent of the observed cavitation structures, a correlation to the development of the characteristic curve determined from the acoustic emissions can be found. At the operating condition 0 cavitation was neither detected with the analysis of the recorded acoustic emissions signals nor visually. The appearance of sporadic cavitation structures lead to an increase of the characteristic value. The step in the development of the characteristic value correlates with the observation that all runner blades are affected by cavitation.

## 5. Results and Conclusions

The comparison of a characteristic value derived from measured acoustic emissions and the visual observations at operating points with pressure side leading edge cavitation show a good accordance at the examined operating points. The observed increase of the extent of the cavitation structures by lowering the cavitation number corresponds to the increase of the characteristic value. The results of the presented investigation show, that the analysis of acoustic emissions is an appropriate method to detect leading edge cavitation in the presented pump-turbine model. The visual access to the leading edge of the runner facilitates the matching of a cavitation indicator based on measured quantities.

## 6. Outlook

The purpose of the investigation presented here is to validate the results of the analysis of the acoustic emissions signals for cavitation detection at the leading edge of a radial runner. The results provide contribution in the development of a universal cavitation monitoring system. The next steps to be done are the analysis and validation of different operating points and compare the results with the analysis of axial model turbines and prototype turbines.

## Acknowledgement

The research leading to the results presented in this paper is part of a common research project of the University of Stuttgart and Voith Hydro Holding GmbH. The authors would like to thank Voith Hydro Holding GmbH for the financial support of this project.

## References

[1] Grekula M and Bark G 2001 "Experimental Study of Cavitation in a Kaplan Model Turbine", *Proc. of the 4th international Symposium on Cavitation (Pasadena, USA, June 20-23, 2001)*

[2] Širok B, Hočevar M, Kern I and Novak M 1999 "Monitoring of the Cavitation in the Kaplan Turbine", *Proc. of the IEEE International Symposium on Industrial Electronics (Bled, Slovenia July 12-16, 1999)*

[3] Rus T, Dular M, Širok B, Hočevar M and Kern I 2007 "An Investigation of the Reationship Between Acoustic Emission, Vibration, Noise, and Cavitation Structures on a Kaplan Turbine" *ASME* **129** 1112-22

[4] Escaler X, Egusquiza E, Farhat M, Avellan F and Coussirat M 2006 "Detection of cavitation in hydraulic turbines", *Mechanical Systems and Signal Processing* **20** 983–1007

[5] Gruber P, Odermatt P, Etterlin M, Lerch T and Farhat M 2013 "Cavitation Detection via Ultrasonic Signal Characteristics", *Proc. of the 5th International Workshop on Cavitation and Dynamic Problems in Hydraulic Machinery (Lausanne, Switzerland, Sept. 9-11, 2013)*

[6] Schmidt H, Kirschner O., Riedelbauch S., Necker J. and Arantes G. 2015, "Cavitation detection in hydraulic turbines", *Hydro 2015 (Bordeaux, France, Oct. 26-28, 2015), paper submitted and accepted.*

# On Rayleigh-Plesset based cavitation modelling of fluid film bearings using the Reynolds equation

**Troy A. Snyder, Minel J. Braun[1], and Kristopher Pierson**
Department of Mechanical Engineering
University of Akron, Akron, OH 44325

**Abstract.** In the 'universe' of the general cavitation phenomena the issue of cavitation in bearings, due to its particular application and the mostly non-homogeneous working fluids associated with it, has presented a rather specialized challenge. The present paper models the phenomenon of pseudo-cavitation in fluid film bearings and offers a physics-based approach that conserves mass while solving the Reynolds (RE) and Rayleigh-Plesset (RP) equations in a coupled, fully transient environment. The RP solution calculates a time dependent void fraction synchronized with the RE transient solution, where density and viscosity are (re)calculated at every grid point of this homogeneous two-phase fluid. The growth and evolution of the cavitation zone expanse is physics-based and thus can accommodate evaporation, diffusion, or pseudo-cavitation as separate processes. This is a step beyond the present available cavitation models both for the RE and the Navier-Stokes equations.

## 1. Introduction

Following the early non-mass-conserving bearing cavitation models of Gümbel, and Swift-Stieber, mass conserving models have emerged. Coyne and Elrod [1], Mori et al. [2], Birkhoff and Hays [3] accounted for fluid transport above and beneath the bubble and used mass conservation to define the cavity interface. Floberg [4, 5], and Jakobsson and Floberg [5], assumed striated flow with the liquid being transported in-between gas fingers that extended across the clearance, implemented a mass conservation model and subcavity pressures. Many authors have offered analytical and numerical solutions, the list of excellent contributions being too long to do it justice here. The interested reader is further referred to review publications of Dowson et al. [6], Brewe et al. [7], and Braun and Hannon [8]. Zuber and Dougherty [9] in 1982 were the first to propose a model for a homogeneous two-phase lubricating film mixture in conjunction with the RE. Natsumeda and Someya [10] in 1987 followed the same concept of a homogeneous two-phase flow RE by introducing the RP equation as the source of void production for the homogeneous two-phase lubricant film. The authors proved experimentally the existence of tensile subcavity pressures, (as high as 1.2MPa), and used a modified RP equation that retained the surface tension, $(\frac{2S}{\rho_l R_B})$, viscous, $(4\nu_l \frac{\dot{R}_B}{R_B})$, and viscous surface dilatation effects $(\frac{4\kappa^s}{\rho_l R_B^2}\dot{R}_B)$. The latter term importance in modeling and sustenance of the subcavity pressures was clearly defined and recognized both by [10] and Gehannin et al. [11]. Solving for the growth in the bubble radius through the offices of RP provides the mechanism for the calculation of the void fraction, α, as well as the calculation of the transport properties for the two-phase homogeneous RE.

[1] Corresponding author: mjbraun@uakron.edu

## 2. Scope of Work

The present work addresses specifically the development of pseudo-cavitation in a fluid film bearing, that is the formation and evolution of the cavitated region in the divergent section of the bearing through the growth of a collection of kernel bubbles due to depressurization, but without mass addition; this situation is particular to lubrication oils which contain large numbers of air kernel bubbles and have rather low hydrocarbon components evaporation pressure thresholds. The model presented herein is a step forward compared to other models available in that the RP retains all physics-significant terms (viscosity and surface dilatational viscosity) and calculates at every grid point a time-dependent void fraction that allows discrete (re)calculation of local density and viscosity in synchronicity of the RE transient solution for this homogeneous two-phase fluid.

## 3. On the Surface Dilatational Viscosity

For most flows, the thermodynamic and mechanical pressures are sufficiently close, to cause the bulk viscosity $\kappa$ to be negligible in magnitude, thus validating the Stokes's hypothesis. Donnan in his preface to McBain [12] defined an interface as "*two-dimensional molecular world the dynamics of which is analogous to that of ordinary three dimensional world of homogeneous phases in bulk*". Therein lies the analogy between $\kappa$ and $\kappa^s$ ,whereby the latter provides the source for the dynamic interfacial stress (e.g. oil and vapor, or oil and gas) similar to the one provided by the three-dimensional bulk viscosity in an isotropic fluid that expands, contracts, or goes through a shock wave. The dynamic equation for the thin spherical surface between the fluids inside and outside the bubble reduces to a dynamic stress jump condition (body forces and inertia are neglected)

$$\Delta\sigma_{fluid-bubble} = \frac{2\sigma}{R_B} + \frac{4\kappa^s}{R_B^2}\dot{R}_B \tag{1}$$

The surface dilatational viscosity stress is an excess stress and its action offers a dissipative resistance, acting towards restoring equilibrium temporally. In a moving (deforming) situation when the dynamic tension at the interface exceeds the equilibrium tension, $\frac{2\sigma}{R_B}$, by the quantity $\frac{4\kappa^s}{R_B^2}\dot{R}_B$, the latter through its derivative $\dot{R}_B$, always exerts an action *opposing the motion of the bubble wall* resulting into an interface stress jump that causes the deviation from the static case.

## 4. The Model Governing equations

The cavitation model consists of a simultaneous solution of the transient RE and a simplified RP equations (inertia terms are neglected based on an order of magnitude analysis). Within a single global time step, the field of initial bubble radii is advanced through 10 sub-iterative steps of the RP equation

$$\frac{4\nu_L}{R}\frac{DR}{Dt} + \frac{2\sigma}{\rho_L R} + \frac{4\kappa^s}{\rho_L R^2}\frac{DR}{Dt} = \frac{P_{g0}}{\rho_L}\left(\frac{R_0}{R}\right)^{3k} - \frac{P}{\rho_L} \tag{2}$$

The advective velocities within the substantial derivative of bubble radius are directly obtained from the emerging pressure field and averaged across the film thickness. Following the update of the bubble radii, the void field is recalculated as the ratio $\alpha = \frac{\forall_{bubble}}{\forall_{cell}}$. Then the homogeneous bulk flow properties of the lubricant film are updated using the newly found void field

$$\rho = \alpha\rho_B + (1-\alpha)\rho_L \qquad \mu = \alpha\mu_B + (1-\alpha)\mu_L \tag{3a, b}$$

Finally, the new pressure field is computed using the new $\rho(\alpha)$ and $\mu(\alpha)$ in the transient RE

$$\frac{\partial}{\partial x}\left(\frac{\rho h^3}{12\mu}\frac{\partial P}{\partial x}\right) + \frac{\partial}{\partial z}\left(\frac{\rho h^3}{12\mu}\frac{\partial P}{\partial z}\right) = \frac{\partial}{\partial x}\left(\frac{\rho h U}{x}\right) + \frac{\partial(\rho h)}{\partial t} \tag{4}$$

The RE is discretized using the finite-difference method. It is solved using a marching technique which begins at an axial boundary of the bearing and solves an implicit tri-diagonal system at each subsequent axial station.

## 5. Results and Discussion

Figure 1 presents graphically a comparison between 1000 and 5000 [rpm] cases with the surface dilatational viscosity held constant at a baseline value of $\kappa^S$=7.85E-3 [N.s/m], [10] and the relative eccentricity held constant at ε=0.4. The results show that when the $\kappa^S$ used in the RP equation is kept constant, the pressure curve exhibits a subcavity pressure region which varies moderately as rotational speed increases, Figure 1b. This new model also shows that the cavitation zone may extend considerably from the divergent zone into the convergent zone as rotational speed increases. The contour graphs in Figure 1a, show clearly by comparison with each other, the advance of the cavitation zone in the circumferential direction as the rotational speed increases. It is important to note that the contour of this zone was obtained 'organically' through the coupling of the physics of the RP and two-phase homogeneous RE (RP-RE model). No film rupture boundary conditions were necessary. The color bar indicates the pressure magnitudes. Figure 1b presents the pressure and void evolution around the circumference, as they are calculated by the model presented herein, superimposed on a baseline case provided by the RE using the Gümbel cavitation condition (RE-G model). While the RE-G model considers the convergent region containing only liquid of $\rho_L$ density and $\mu_L$ viscosity, the RP-RE model allows for the existence of compressed gas in the liquid, contributing to weighted average of ρ and μ in the two phase mixture throughout the bearing. This fact ('softer fluid'), and the advection of void through the $\dot{R}_B = \frac{\partial R_B}{\partial t} + u_i \frac{\partial R_B}{\partial x_i}$ term, lowers the maximum pressure and pushes the cavitation zone further into the divergent zone, and even convergent zone, respectively. This differentiates the RP-RE from the RE-G models both qualitatively and quantitatively.

## 6. Conclusions

The paper presents a coupled RP and homogenous two-phase RE equations where the RP equation is used as a source of void in the computation of the ρ and μ of the mixture. The core of the present approach eliminates the need for film rupture boundary conditions as used in [3, 4, 5, 1, 2], allowing cavitation to build up 'organically' on the basis of local void advection and growth. The weighted density influences significantly the magnitude of the pressure in the convergent zone causing the maximum pressure to be lower than the one that would be computed by a RE-G model. The present model allows also for the development of subcavity pressure zones based on the magnitude of the surface dilatational viscosity. As the latter increases, the capability of the fluid to sustain subcavity pressures (tensile stresses) increases. Based on the numerical results one can infer that the Sommerfeld and half-Sommerfeld (Gümbel-Swift-Stieber) curves represent limiting cases between a high and low surface dilatational viscosity values. Finally these differentiated results have potentially a significant effect on the static (load, attitude angle, torque) and dynamic performance (stiffness, damping, stability) of the fluid film bearings.

## 7. References

[1] J. C. Coyne and H. G. Elrod Jr, "Conditions for the rupture of a lubricating film, part I: theoretical model," *J. of Lubric.Technol.,* vol. 92, no. July, pp. 451-456, 1970.

[2] H. Mori, H. Yabe and Y. Fujita, "On the separation boundary conditions for fluid lubrication theories of journal bearing," *ASLE Transactions,* vol. 11, pp. 196-203, 1968.

[3] G. Birkhoff and D. Hays, "Free boundaries in partial lubrication," *J. Math. Phys., 1963, 42(2), 126-138,* vol. 42, no. 2, pp. 126-138, 1963.

[4] L. Floberg, "On journal bearing lubrication considering the tensile strength of the liquid lubricant," *Transactions of the Machine Elements Division, Lund Technical Uuiversity, Lund, Sweden,,* pp. 1-26, 1973.

[5] B. Jakobsson and L. Floberg, " The finite journal bearing considering vaporization.," *Trans,actions of Chalmers University Technology, Goteborg, Sweden,* vol. 190, pp. 1-119, 1957.

[6] D. Dowson, M. Godet and C. M. Taylor, "Cavitation and related phenomena in lubrication," in *In Proceedings of the 1st Leeds-Lyon Symposium on Tribology (Eds D. Dowson, M. Godet, and C. M. Taylor), pp 1-238*, University of Leeds, Leeds, UK (Mechanical Engineering Publications, Ltd, London), 1974.

[7] D. E. Brewe, "Theoretical modeling of the vapor cavitation in dynamically loaded journal bearings," *J. of Tribology.*, Vols. 1986, , no. 4, pp. 628-638., 1986.

[8] M. Braun and W. Hannon, "Cavitation Formation and Modelling for Fluid Film Bearings: a Review," *J. of Engineering Tribology, Proc. IMechE,* vol. 224, no. J9, pp. 839-863, 2010.

[9] N. Zuber and D. Dougherty, "The Field Equations for Two-Phase Reynolds Film Flow with Change of Phase," *ASLE Transactions,* vol. 25, pp. 108-116, 1982.

[10] S. Natsumeda and T. Someya, "Negative Pressures in Statically and Dynamically Loaded Journal Bearings, Paper III(ii)," in *Proc. of the 13th Leeds-Lyon Symp. of Tribology, Leeds, UK, pp 498-505,* 1987.

[11] J. Gehannin, M. Arghir and O. Bonneau, "Evaluationof the Rayleigh-Plesset Equation Based Cavitatin Models for Squeeze Film Dampers," *J. of Tribology,* vol. 131, pp. 15-25, 2009.

[12] J. McBain, An Introduction to Surface Chemistry, Cambridge: Cambridge University Press, 1926.

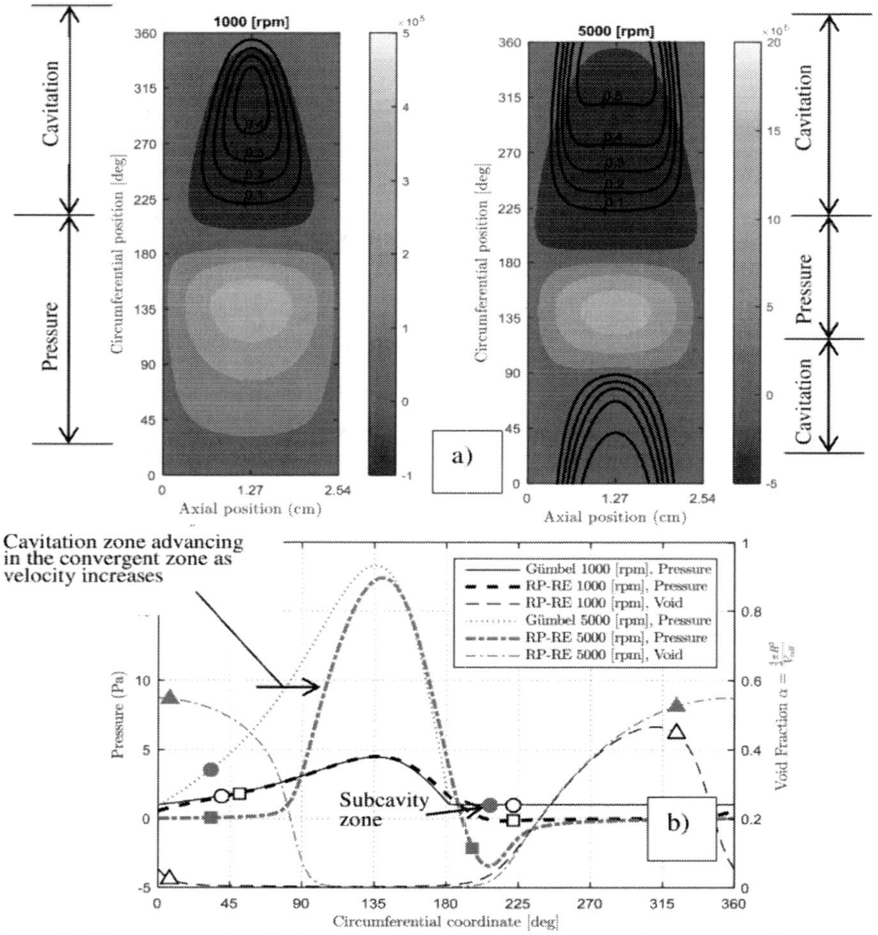


Figure 1. Pressure and void fraction at bearing midspan and corresponding contour plots of pressure and void fields at 1 and 5 krpm and $\varepsilon$=0.4. c/R=0.001; $\mu_L$=7.1E-3 Pa.s; $\mu_B$=1.81E-5 Pa.s; $\rho_L$=854 kg/m$^3$; $\rho_B$=1 kg/m$^3$; $\sigma$=3.5E-2 N/m$^2$; $\kappa^s$=7.85E-3 Ns/m; $R_{g0}$=3.85e-7 m

# Prediction of Performance of a Cavitating Propeller in Oblique Inflow

**Ye Tian[1] and Spyros A. Kinnas[2]**

[1,2] Ocean Engineering Group, Department of Civil, Architectural and Environmental Engineering, The University of Texas at Austin, Austin, TX 78712, USA

[1]tianye@utexas.edu, [2]kinnas@mail.utexas.edu

**Abstract**. A cavitating propeller subject to an oblique inflow in a cavitating tunnel is simulated using potential flow methods coupled with a Reynolds-averaged Navier-Stokes (RANS) solver. The propeller is mainly modelled using a panel method, while the inflow to the propeller is evaluated by coupling a Vortex-Lattice Method (VLM) with the RANS solver. The effects of the tunnel wall are incorporated into the calculated effective inflow to the propeller. The predicted propeller forces and cavity pattern are correlated with experiment. The fully wetted open water characteristics of the propeller predicted by the panel method are presented as well.

## 1. Introduction

Potential flow methods are intensively used in the design stage of marine propellers. In general the methods are sufficiently accurate for axisymmetric inflows. In the real design practice, the methods are commonly used with a measured wake behind a hull. Nonetheless the methods have not been carefully validated in the cases of non-axisymmetric inflows.

In order to provide computational fluid dynamics (CFD) groups with experimental data of a propeller in non-axisymmetric inflow with clearly defined boundary conditions, a series of experiments on a propeller in oblique inflow was conducted by SVA Potsdam GmbH in the workshop of the fourth international symposium on Marine Propulsors 2015 (SMP' 15) [1]. The test case is referred to the acronym PPTC'15 (Potsdam Propeller Test Case 2015)[1].

In this paper PPTC'15 will be simulated using mainly a panel method with effective wake calculated by coupling a Vortex-Lattice Method (VLM) with a RANS solver. The results will be correlated with experimental data.

## 2. Methodology

### *2.1. The panel method and the cavity model*

A panel method code, PROPCAV is adopted to predict propeller performance. Details of PROPCAV can be found in [2]. A potential flow model which combines a constant pressure condition with a tangency condition to determine the cavity extent and shape is used to simulate the cavity. This cavitation model was proposed by Fine and Kinnas [3]

### *2.2. Effective wake calculation*

In PPTC'15, the cavitating test was conducted in a cavitation tunnel. In order to consider the effects of the tunnel wall but without model it in the panel method, the effective wake concept is adopted to prepare the inflow to the propeller. The effective wake is defined as the time averaged difference between the total velocity and the propeller induced velocity, thus contains the information of the surrounding boundaries and the ambient flow of the propeller. The effective wake can be determined by coupling a potential flow solver with a RANS solver. The effective wake approach was proposed by Kerwin et al. [4] as an intuitive extension of the actuator disk concept to model ducted propellers. The approach was later improved by the Ocean Engineering Group (OEG) at the University of Texas at Austin and has been successfully applied in the case of a ducted propeller in axisymmetric inflow [5] and hull/thruster interaction [6]. In this paper, a VLM code, MPUF-3A is coupled with ANSYS Fluent to determine the effective wake.

## 3. Experimental Setup

In the experiment, the propeller is operating in a pull configuration with the hub cap pointing upstream. The propeller axis is inclined by $12^{\circ}$. More details of the experiment can be found on the website of SMP'15 [1]. The same propeller in axisymmetric uniform inflow was tested in the workshop of the second international symposium on marine propulsors 2011 (SMP'11) [7].

## 4. Numerical implementation and results

### *4.1. Wetted open water characteristics in axisymmetric uniform inflow*

Before simulating the propeller in oblique inflow, the results from the panel method are first validated against experiments in the case of axisymmetric uniform inflow. The propeller is simulated using $80\times 30$ (chordwise $\times$ spanwise) panels. Two trailing edge wake models of the propeller are adopted: the full wake alignment (FWA) model proposed by Tian and Kinnas [8] and a simplified wake model, namely PSF-2 type alignment [9]. Figure 1 shows the discretization of the blades and the trailing edge wake from FWA at J=0.8. Figure 2 correlates the predicted open water characteristics with experiments. Clearly the FWA model significantly improves the results at low advance ratios.

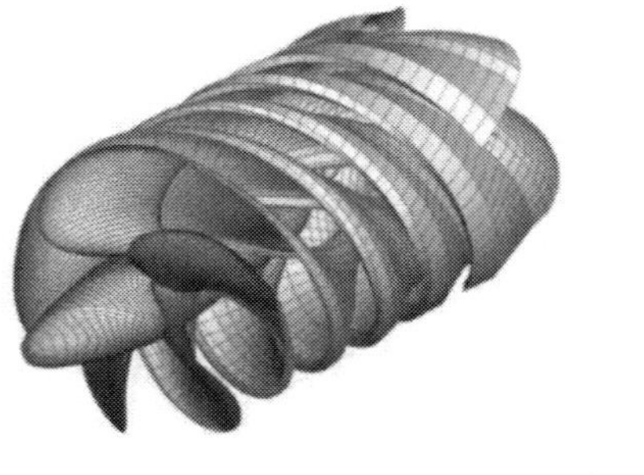

**Figure 1.** Surface panels of Propeller VP 1304 in axisymmetric uniform inflow with FWA, $J = 0.8$

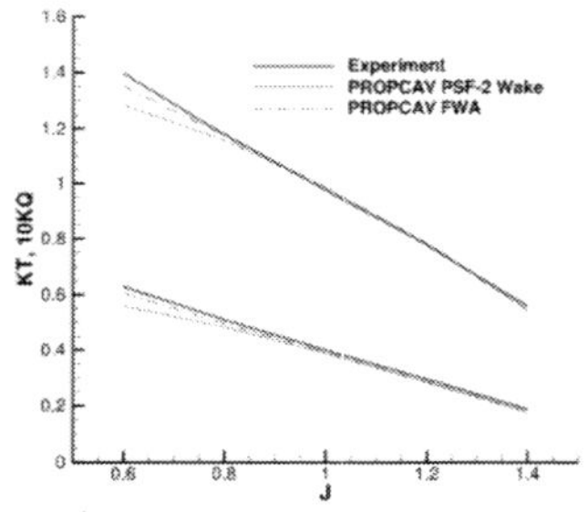


**Figure 2.** Predicted performance of Propeller VP 1304 in axisymmetric uniform inflow, comparing with experiment.

### *4.2. Wetted open water characteristics in oblique uniform inflow*

The wetted open water characteristics of the propeller in oblique uniform inflow are then investigated. The propeller is simulated using $60\times 20$ (chordwise $\times$ spanwise) panels and $80\times 30$ panels. Unfortunately the FWA has not been extended into non-axisymmetric cases yet. Only the PSF-2 type alignment is adopted in the following simulations. The numerical solutions are convergent, as shown in figure 3. Figure 4 compares the predicted mean performance of the propeller against the experiment. Reasonable correlations can be observed, whereas the KT and KQ at low advance ratios are under-estimated. The FWA model should improve the low J results as what has already been observed in the axisymmetric inflow case.

### *4.3. Cavitating propeller in oblique inflow*

In order to take account of the effects of the tunnel wall, the effective wake is calculated by coupling MPUF-3A with ANSYS Fluent. The RANS grid has about 1.3 million cells, as shown in figure 5. Standard $k$-$\epsilon$ turbulence model is applied, along with QUICK scheme for spatial discretization. The total velocity field from RANS is shown in figure 6.

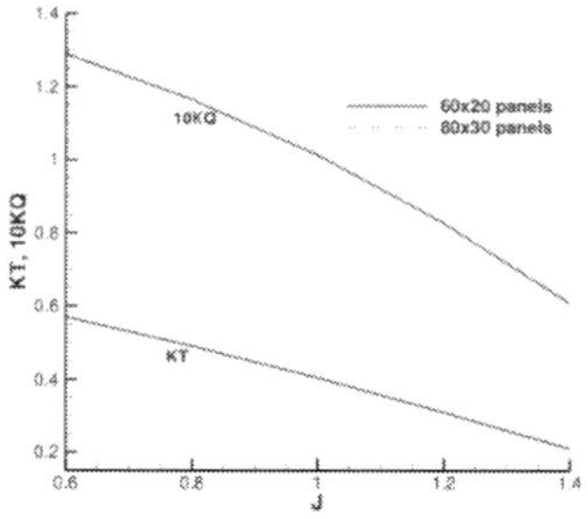


**Figure 3.** Predicted mean performance of Propeller VP 1304 in oblique uniform inflow from different spatial resolution.

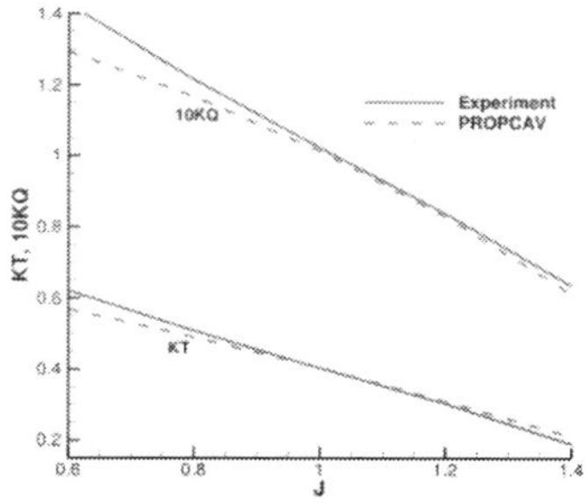


**Figure 4.** Predicted mean performance of Propeller VP 1304 in oblique uniform inflow, comparing with experiment.

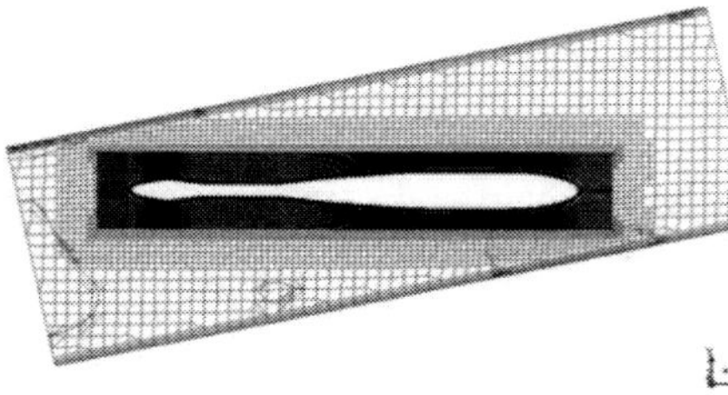

**Figure 5.** The RANS grid for the effective calculation

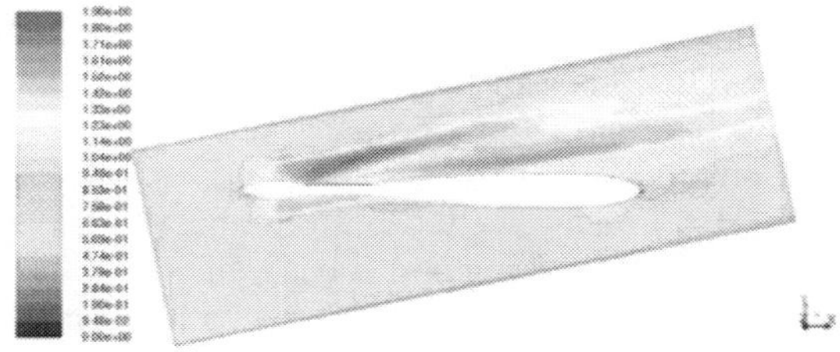

**Figure 6.** Total velocity field predicted by RANS/VLM coupling.

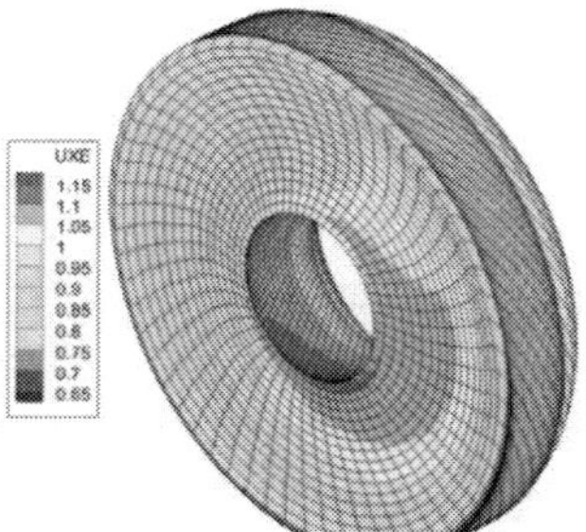


**Figure 7.** Predicted axial effective wake, normalized by $U_\infty$

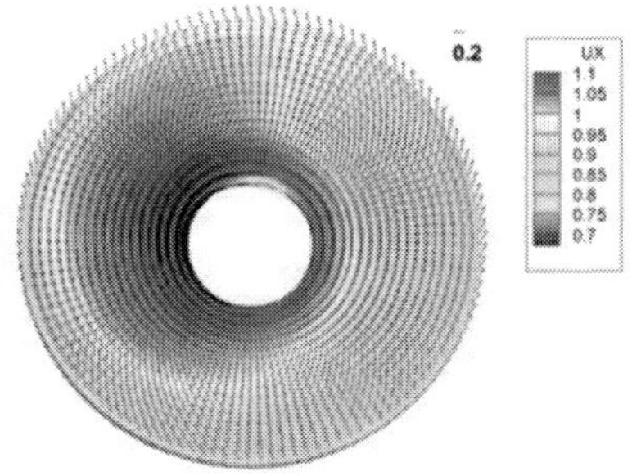


**Figure 8.** Predicted effective wake on the blade mid-chord surface, normalized by $U_\infty$

Figure 7 and 8 plot the effective wake. The transverse components of the velocity due to the inclination of the shaft are largely retained. In the meantime, the upper half of the blade zone is subject to higher axial velocity than the lower half, because of the presence of the upper tunnel wall.

Figure 9 and 10 show the cavity pattern from both the numerical simulation and the experiment. The predicted cavity appears largely at correct blade angles. However, the tip vortex at blade angle 216° is missing from the numerical results and thus needs further investigation.

Figure 11 plots the KT from PROPCAV on top of the experimental measurement. Clearly thrust breakdown is well predicted.

**Figure 9.** Predicted cavity pattern from the numerical simulation.

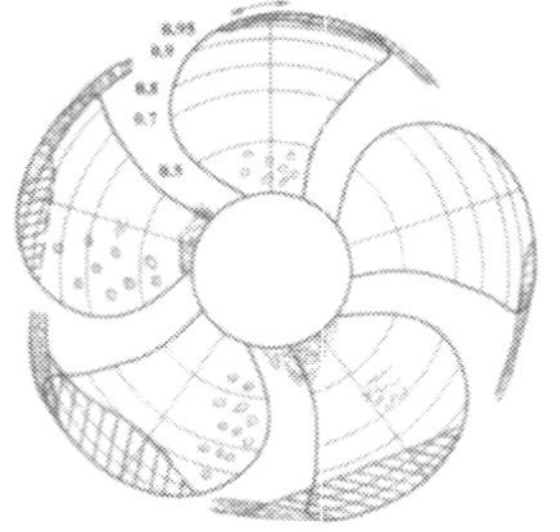

**Figure 10.** Observed cavity pattern in the experiment.

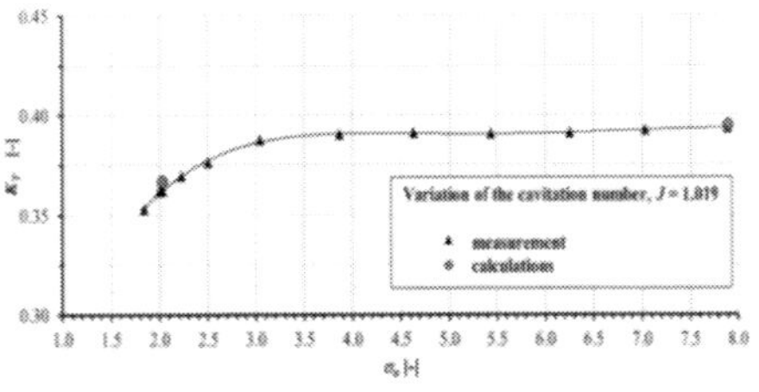


**Figure 11.** Predicted KT vs. experiment.

## 5. Conclusions and future work

The wetted and cavitating performance of a propeller in oblique inflows are simulated using a panel method code, PROPCAV. The numerical results are validated against the experimental data in PPTC'15. For the wetted performance of the propeller in oblique inflow, the predicted performance recovers the experimental measurement at the design advance ratio, but slightly deviates from the experimental data at low advance ratios. The discrepancy could be reduced if the FWA model was used. For the cavitating performance of the propeller in the tunnel, a coupled VLM-RANS method is adopted to calculate the effective wake to the propeller. Reasonable cavity pattern and good KT value are predicted by the panel method. Nonetheless the tip vortex at certain blade angles is missing.

In the future, the FWA model will be extended in the case of non-axisymmetric inflow. Other J and $\sigma_n$ combinations tested in PPTC'15 will be simulated as well. The time history of the blade forces will be also investigated.

## References

[1] SMP'15: http://www.caee.utexas.edu/smp15/smp_workshop_2015.html
PPTC'15: http://www.sva-potsdam.de/pptc_data_15.html

[2] Hsin C 1989 *Development and Analysis of Panel Methods for Propellers in Unsteady Flow.* (PhD dissertation, M.I.T., Department of Ocean Engineering)

[3] Fine N E and Kinnas S 1993 A boundary element method for the analysis of the flow around 3-d cavitation hydrofoils *J. Ship.Res.* **37** 213-224

[4] Kerwin J E, Keenan D, Black S, and Diggs J 1994 A Coupled Viscous/potential Flow Design Method for Wake-adapted, Multi-stage, Ducted Propulsors using Generalized Geometry. Discussion. Authors' Closure. *Trans. SNAME.* **102** 23-56

[5] Tian Y, Jeon C-H and Kinnas S 2014 On the Accurate Calculation of Effective Wake/Application to Ducted Propellers. J. Ship. Res. **58** 70-82

[6] Tian Y and Kinnas S 2015 Thruster and Hull Interaction *J. Offshore Mech. Arct. Eng.* **137** 4

[7] SMP'11:http://www.sva-potsdam.de/pptc_data.html#workshop

[8] Tian Y and Kinnas S 2012 A Wake Model for the Prediction of Propeller Performance at Low Advance Ratios *Intl. J. Rotating. Mach.* **2012** Article Id 372364

[9] Greeley D S and Kerwin J E 1982 *Trans. SNAME.* **90** 415-453

# Numerical simulations of the cavitation phenomena in a Francis turbine at deep part load conditions

**J Wack and S Riedelbauch**

Institute of Fluid Mechanics and Hydraulic Machinery, University of Stuttgart, Pfaffenwaldring 10, 70555 Stuttgart, Germany

E-mail: jonas.wack@ihs.uni-stuttgart.de

**Abstract**. In recent years, the operating range of hydraulic machines has been more and more extended. As a consequence, the turbines are facing off-design conditions with highly complex flow phenomena like cavitation. In the present study, the occurrences of cavitating inter blade vortices at deep part load conditions in a Francis turbine are investigated using two-phase flow simulations. The numerical simulations require small time steps and fine meshes to reproduce the required flow characteristics and resolve the minimum pressure in the vortex core. Furthermore, the treatment of the outlet boundary condition is important, as this operating point is facing severe backflow in one diffusor channel in the draft tube. The simulation results indicate that the inter blade vortices can be reproduced.

## 1. Introduction

Cavitation is a key factor in the hydraulic design of turbines. For a Francis turbine different types of cavitation can occur, depending on the head and discharge [1]. In the present study, the focus is on the inter blade vortices that occur at deep part load conditions. Deep part load conditions are characterized by a low discharge. The small guide vane opening causes a misaligned flow to the runner tip which results in diverse unsteady flow phenomena like inter blade vortices. These vortices can cause severe mechanical vibrations over the whole hydro power plant. Penstock vibrations that can be traced back to this cavitation phenomenon are reported by Dörfler et al. [2].

Only little research has been performed on deep part load in the past. Investigations performing single-phase numerical simulations show that deep part load conditions are accompanied by severe pressure pulsations with possibly higher amplitudes compared to full load and part load conditions [3]. First predictions of the inter blade vortex limits with numerical simulations have been carried out by Zuo et al. [4]. There, two-phase simulations showed a better accuracy compared to single-phase simulations. An experimental study has been carried out by Yamamoto et al. [5]. It is reported that the speed factor $n_{ED}$ has a strong impact on the occurrence of the inter blade vortices.

## 2. Modelling

Several different concepts are applicable for two-phase modeling [6]. In the present study, the simulations are performed based on the homogenous model [7]. It allows considering the mixture as a pseudo-fluid. Thus, the governing equations consist of the mass conservation, one set of momentum equations and a transport equation for the volume fraction.

The phase change due to cavitation is considered as source term in the transport equation for the volume fraction and has to be modeled by a cavitation model. In the present study, the Zwart model is applied. It contains four parameters that are set according to the recommendation by Zwart et al. [8].

## 3. Numerical setup

The numerical simulations are performed on a reduced scale model Francis turbine with the flow domain ranging from the inlet of the spiral case to the outlet of the draft tube as displayed in figure 1. For the investigation of the treatment at the outlet boundary condition, one additional setup contains a tank that is attached to the draft tube outlet. In the present study, structured meshes are used and three different mesh sizes are investigated. The smallest mesh consists of approximately 11 million (11M) elements. The 26M mesh is refined in all domains but with the focus on the runner and the draft tube to resolve regions of strong vortex flow and cavitation occurrence. The 35M mesh has a strong refinement in the runner to account for resolving the core of the inter blade vortices.

All simulations are carried out with Ansys CFX. The calculations are performed with a transient rotor stator approach. At the inlet boundary condition, a constant mass flow is specified according to the investigated operating point. At the outlet, the static pressure is specified and thus the cavitation number $\sigma$ is set. For spatial discretization a high resolution scheme and for temporal discretization a second order backward Euler scheme is used. The SST model is applied as RANS turbulence model.

## 4. Numerical results

In the present study, an operating point at deep part load conditions, which is specified in table 1, is investigated. It is characterized by a discharge below 30% of the discharge at best efficiency point $Q_{BEP}$. At these operating conditions, the experiments observed that no vortex rope is present any longer and the flow is dominated by a huge stagnant region in the center of the draft tube cone [5].

**Table 1.** Specification of the investigated operating point

| $Q_{ED}$ (-) | $Q/Q_{BEP}$ (-) | $n_{ED}$ (-) | n (rpm) | $\sigma$ (-) |
|---|---|---|---|---|
| 0.055 | 0.275 | 0.317 | 880 | 0.09 |

First, an appropriate time step is selected. Single-phase simulations with the 11M mesh are performed for four different time steps. For the torque, the deviation to the measurement results is investigated. The results are listed in table 2. For huge time steps, the torque significantly depends on the time step size. Regarding the results for a time step of 0.26° and 1° shows only minor differences and thus indicate a time step independency. As a consequence, for the following investigations, a time step according to 1° of runner revolution is selected.

**Table 2.** Deviation of the torque to experiments depending on the time step

| Time step size (° of runner revolution) | 10.56 | 6.60 | 1.056 | 0.264 |
|---|---|---|---|---|
| Torque – Deviation to experiment (%) | -3.9 | -0.5 | 2.4 | 2.5 |

In the second step, a suitable treatment of the boundary condition at the outlet of the draft tube is chosen that takes into account the present flow regime with low discharge. Three different single-phase simulation setups are investigated. The outlet boundary type permits outflow and prohibits inflow while the opening boundary condition allows both, outflow and inflow. In the third setup, a tank which is connected to the draft tube outlet is added. It corresponds to the setup of the test rig. The simulation results for the three setups are presented in table 3.

A comparison of the torque shows similar results for all boundary treatments, whereas the evaluation of the head indicates differences. When assessing the head deviations, it has to be taken into account that the calculation for the setups outlet and opening might be affected by setting the outlet boundary condition just at this location. In terms of accuracy, a difference of 0.7% to the tank simulation seems to be acceptable as the accurate prediction of head is not a priority objective of this study. The analysis of the discharge in one channel of the draft tube diffusor indicates that the investigated operating point is characterized by strong backflow in one channel. This cannot be modeled by the outlet treatment. The results for the opening and tank treatment are in the same range. All in all, the setups opening and tank are suitable for further analyses. Due to the fact that the opening treatment needs approximately 20% less mesh elements, it is selected for the following analysis.

**Table 3.** Comparison of head, torque and discharge in one draft tube diffusor channel for three different treatments at the outlet boundary condition

| Boundary treatment | Outlet | Opening | Tank |
|---|---|---|---|
| Torque – Deviation to experiment (%) | 2.42 | 2.36 | 2.37 |
| Head – Deviation to experiment (%) | -7.6 | -6.4 | -6.9 |
| $Q_{DT,Channel1}/Q_{SC,In}$ (%) | 0.1 | -41.8 | -39.1 |

The results of two-phase simulations with the 11M mesh indicate that the core of the inter blade vortices is not resolved properly, as no cavitation of the vortices occurs. Only a small cavity volume arises that is located at the trailing edge of the runner close to the hub. Thus, further investigations are carried out with the 26M and 35M mesh. Figure 2 displays the course of the vapor volume over time for different meshes. It is noticeable that in terms of vapor volume a mesh dependency is still present. The 35M mesh resolves the channel vortex better and thus the best resolution of the minimum pressure in the vortex core is obtained. Consequently, the biggest vapor volume is achieved there. Furthermore, it is conspicuous that the vapor volume is oscillating and the corresponding frequency is changing with the mesh refinement. The oscillation might be a result of differences in the discharge through the different guide vane channels. A comparison of head and torque shows a deviation of less than 0.1% between the 26M and 35M simulation. Thus, it can be stated that a mesh dependency is still present but only for resolving the vortex core. This is in good agreement with the investigations carried out by Stein et al. [9]. For entire mesh independence, a finer mesh is necessary which results in huge computational efforts.

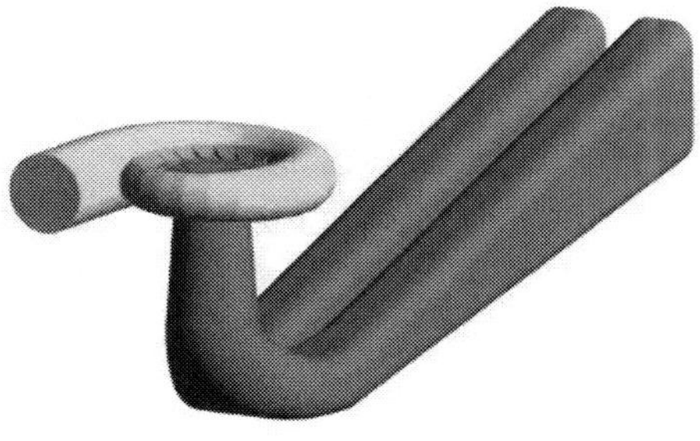

**Figure 1.** Simulation domain

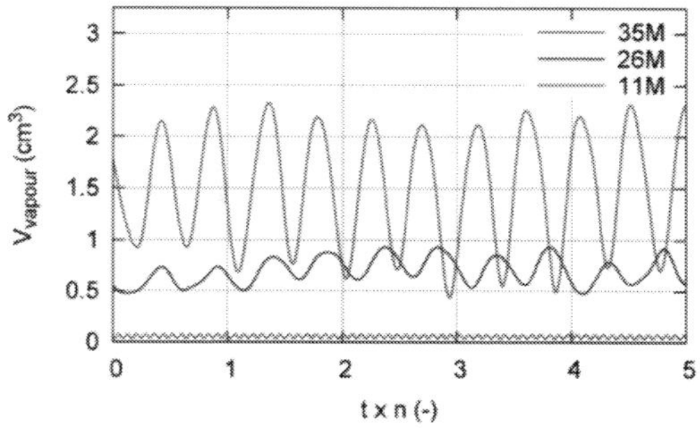


**Figure 2.** Vapor volume for different mesh sizes

Figure 3 shows a visualization of the inter blade vortices. The simulation results are displayed for the 35M mesh. In figure 3(a) the vortices are visualized with an isosurface of the velocity invariant. The coloring is related to the static pressure, blue being the lowest pressure. The simulation can reproduce the inter blade vortices which can also be visualized with an isosurface of the volume fraction of the liquid phase (see figure 3(b)). Both ways of visualization indicate that the vortices differ in strength in the different runner channels. In some runner channels the simulation does not even show cavitating vortices. High speed visualization recording of the experiments also shows differences in the cavity size in different runner channels (see figure 3(c)). However, the measurements indicate that the inter blade vortices are cavitating in every channel of the runner. As the simulated vapor volume is lower compared to the experiments, a deviation in the pressure level in the runner might be present between simulation and measurements. Further experimental and numerical analyses are required and will be carried out in the future for a better understanding of the processes at these extreme off-design conditions.

**Figure 3.** Visualization of the inter blade vortices – (a) Isosurface of the velocity invariant – (b) Isosurface of the volume fraction of the liquid phase – (c) Experiment [5]

## 5. Conclusion and outlook

In the current study, cavitation simulations are carried out on a Francis turbine at model scale for an operating point at deep part load. Preliminary investigations with single-phase simulations indicate that the time step size should be at most 1° of runner revolution and that the treatment of the boundary condition at the outlet has an impact on the flow characteristics.

The results of the two-phase simulations highlight the necessity of fine grids in the runner to resolve the vortex core of the inter blade vortices. An optically comparison with high speed visualization from experiments shows the existence of inter blade vortices in both, simulation and measurement. Nevertheless, the simulation seems to underestimate the occurrence of cavitation.

In future work, further investigations have to be carried out to better understand the formation of inter blade vortices. An additional measurement campaign will focus on the pressure on the runner blades. That will make it possible to compare the pressure level between simulation and measurement at a location close to the cavity region. Furthermore, existing pressure fluctuations can be analysed in detail. As a main objective, the simulation results will be used as input for a finite element analysis to predict the stresses in the runner. This enables an assessment of the structural integrity for operating turbines at deep part load conditions.

## Acknowledgments

The research leading to the results published in this paper is part of the HYPERBOLE research project, granted by the European Commission (ERC/FP7-ENERGY-2013-1-Grant 608532). Special thanks to EPFL, in particular Keita Yamamoto, Andres Müller, Arthur Favrel, Christian Landry and François Avellan, for providing all experimental data enabling comparisons presented here.

## References

[1] Avellan F 2004 *6th Int. Conf. on Hydraulic Machinery and Hydrodynamics* Timisoara
[2] Dörfler P, Sick M and Coutu A 2013 *Flow Induced Pulsation and Vibration in Hydroelectric Machinery* (London: Springer)
[3] Magnoli M V 2014 *Ph.D. Thesis Technische Universität München*
[4] Zuo Z, Liu S, Liu D, Qin D, Wu Y 2013 *Adv. Mech. Eng.* **2013** Article ID 397583
[5] Yamamoto K, Müller A, Favrel A, Landry C and Avellan F 2014 *IOP Conf. Series: Earth and Env. Sci.* **22** issue 3, p. 022011
[6] Koop A H 2008 *Ph.D. Thesis University of Twente*
[7] Bouziad Y A 2006 *Ph.D. Thesis EPFL Lausanne*
[8] Zwart P J, Gerber A G, Belamri T 2004 *Int. Conf. on Multiphase Flow* Yokohama paper no.152
[9] Stein P, Sick M, Dörfler P, White P, Braune A 2006 *Proc. 23rd IAHR Symp. on Hydraulic Machinery and Systems* Yokohama paper F228

# Numerical Investigation of Cavitation Improvement for a Francis Turbine

**Zhifeng Yao[1,2], Ruofu Xiao[1,2], Fujun Wang[1,2*], Wei Yang[1,2]**

[1]College of Water Resources and Civil Engineering, China Agricultural University, Beijing 100083, China
[2]Beijing Engineering Research Center of Safety and Energy Saving Technology for Water Supply Network System, Beijing 100083, China
E-mail: wangfj@cau.edu.cn

**Abstract**. Cavitation in hydraulic machine is undesired due to its negative effects on performances. To improve cavitation performance of a Francis turbine without the change of the best efficiency point, a model runner geometry optimization was carried out. Firstly, the runner outlet diameter was appropriately increased to reduce the flow velocity at runner outlet region. Then, to avoid the change of the flow rate at the best efficiency point, the blade shapes were carefully adjusted by decreasing the blade outlet angles and increasing the blade wrap angles. A large number of the modified runners were tested by computational fluid dynamic (CFD) method. Finally the most appropriate one was selected, which has the runner outlet diameter 10% larger, the blade outlet angles 3 degrees smaller and the blade wrap angles 5 degrees larger. The results showed that the critical cavitation coefficient of the model runner decreased at every unit rotational speed after the optimization, and the effect was much remarkable at relative high flow rate. Besides, by analysing the internal flow field, it was found that the zone of the low pressure on pressure surface of the optimized turbine blades was reduced, the backflow and vortex rope in draft tube were reduced, and the cavitation zone was reduced obviously.

## 1. Introduction

Cavitation can occur within the entire operating conditions of hydro turbines, once the absolute static pressure at a certain location falls below the saturated vapor pressure of the fluid at the prevailing temperature conditions [1]. The bubbles condense when they reach a region of higher pressure and finally collapse or implode. Cavitation in hydraulic machinery may lead to flow instabilities, excessive vibration, damage to material surfaces and degradation of machine performance [2]. Nowadays, cavitation issue is becoming more and more important, because the operating speeds are increasing as the results of the output power growth and reduction of geometry dimensions to decrease the cost of its components. Also it is subjected to operate modern turbines in the conditions far from the best efficiency point, and the cavitation phenomena are prone to occur at off design operating conditions [3].

For a Francis turbine, one of the well-known cavitation phenomena is the draft tube whirl, which is a cavitation vortex-core flow formed just below the runner cone in the center of the draft tube [2]. Müller et al.[4] gave an introduction to the cavitation in hydraulic machinery. He pointed out that the whirl cavitation developed from the hub of the runner to the center axis of the draft tube in the bulk flow. The vortex rotates in the same direction as the runner at part load and in the opposite direction at overload. In low flow rate regimes, the vortex core takes a helical shape and rotates at a speed between 0.25 - 0.40 times the runner rotational frequency. The whirl development usually is the main source of pressure fluctuations in hydraulic installation [5]. Alligne et al [6] predicted the cavitation surge phenomenon in the draft tube. It is shown that convective terms have a stabilizing influence modifying stability limit prediction driven by the divergent geometry modelling of draft tube. Wu et al. [7] applied numerical simulation and model test to investigate the cavitation flow in Francis turbines. Simulation results showed reasonable agreement with the experiment data by using the *SST k- ω* turbulence model.

In present paper, flow passage geometry of a Francis turbine was optimized to improve the cavitation performance. CFD method was employed to uncover the detail of the flow field inside the runner and draft tube. And the cavitation characteristics before and after optimization were compared.

## 2. Model runner and optimization procedure

The original runner was a reduced scale model having 16 stay vanes and 24 guide vanes. The height of the guide vane was 54mm. The runner inlet diameter $D_1$ was 410 mm and the outlet diameter $D_2$ was 280 mm. The runner had 15 long blades and 15 short blades. Figure 1 shows the three dimensional model of original runner.

The optimization procedure mainly included two steps. Firstly, the runner outlet diameter was increased by 3% to 12% of the original outlet diameter. The cavitation performance may improve due to the flow velocity decrease at the runner outlet region. But the flow rate at the best efficiency point will definitely changed, and also the efficiency performance may decline. In order to avoid such negative effects, the blade shapes were carefully adjusted by decreasing the blade outlet angles around 2 to 4 degrees and increasing the blade wrap angles 2 to 7 degrees.

A large number of the modified runners were tested by computational fluid dynamic (CFD) method. Finally the most appropriate one was selected. Figure 2 shows the comparison of meridional shapes of original and optimized runners. The blue line shown in Figure 2 is the optimized runner, and the black line is the original one. The outlet diameter of the optimized runner was 1.1 times of the original one, while the inlet diameter $D_1$ was kept the same. The blade of the runner should be adjusted properly due to the increase of outlet, and the best efficiency point was also kept the same with the original one. The final blade outlet angles of the runner after optimized were 3 degrees smaller than the original one. The blade wrap angles were 5 degrees larger than the original one. The inlet of draft tube was increased to match the outlet of optimized runner with the shape lines unchanged.

**Figure 1.** Original runner

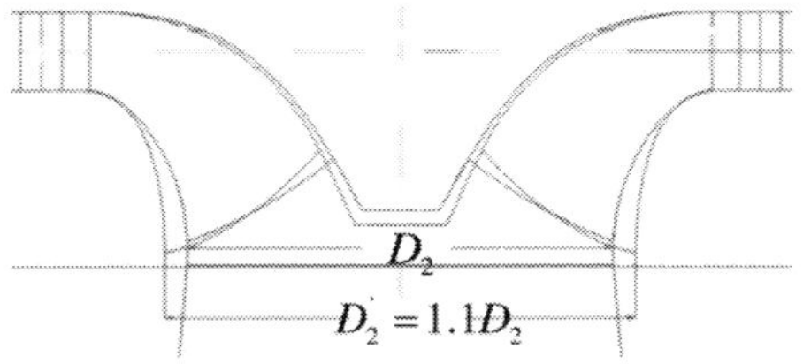


**Figure 2.** Comparison of meridional shapes of original and optimized runners

## 3. Numerical simulation method and meshing

The Francis turbine model used in CFD mainly consisted of five components, such as spiral volute, stay vane, guide vane, runner and draft tube. The flow domains were meshed in ICEM. The total mesh element quantity was 1,270,700 after grid-independent analysis. Structured meshes were generated to define the daft tube, stay vane, guide vane, runner and spiral volute around 138,200, 99,300, 168,200 and 572,500, respectively. For the spiral volute, unstructured mesh was applied with 292,500 elements, and a mesh refinement was used near the tongue zone.

The flow in Francis turbine was considered to be incompressible, viscous and fully turbulent. The Reynolds averaged Navier–Stokes (RANS) approach with *SST k-ε* turbulent model was used to simulate the steady flow by using the commercial code CFX. The Rayleigh-Plesset cavitation model was used to capture the cavitation phenomena. The wall functions based on the logarithmic law were adopted to model the boundary layers. The PISO algorithm was used to calculate pressure velocity coupling. High resolution scheme was used for advection terms and turbulence numeric. Frozen rotor method was adopted to simulate the relative motion of the runner grid with respect to the suction and volute. For a specific operating condition, the total pressure boundary condition was established at the inlet, and the average pressure boundary was imposed at the outlet of the computational domain. The volute and draft tube walls were set in stationary frame and modeled using no-slip boundary conditions.

## 4. Results and Discussions

### *4.1. Comparison of efficiency*

Figure 3 shows the comparison of efficiency between the optimized and original runners. At relative high unit rotational speed, the efficiency of optimized one is a little bit higher than the original one at the same guide vane opening. While at relative small unit rotational speed, the efficiency of optimized one is lower than the original one. However, the largest difference is lower than 1%. Figure 4 shows the comparison of the pressure distribution along the blade surface flow line at 90% span near bottom ring (Guide vane opening is 18° , $n_{11}$=62.5 r/min, $P_{tout}$=20 kPa.). The pressure imposed on the blade surface close to runner outlet for the optimized one is much larger than the original one.

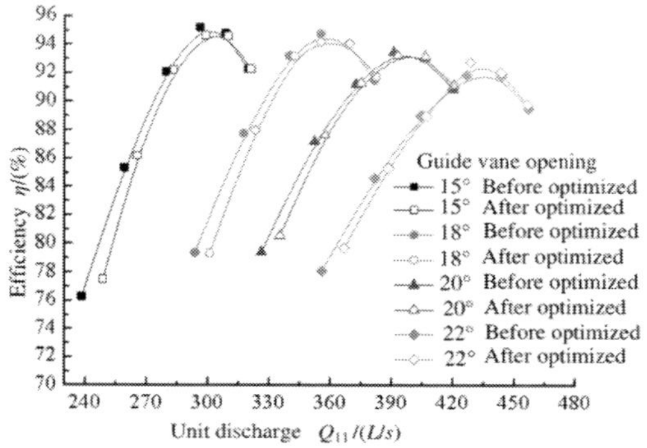


**Figure 3.** Comparison of efficiency

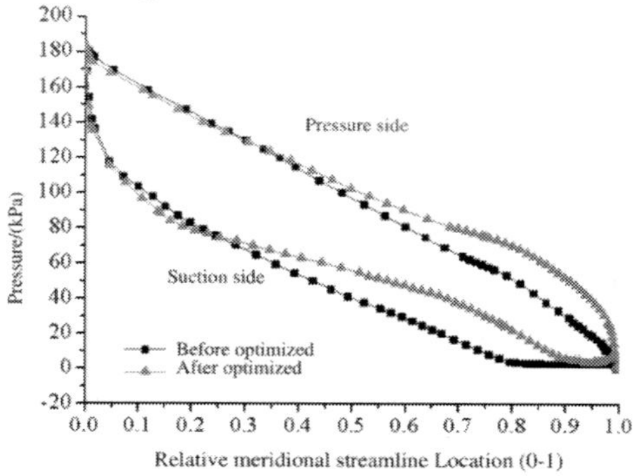


**Figure 4.** Comparison of the pressure distribution on blade surfaces

### *4.2. Comparison of cavitation*

The cavitation coefficient was defined as $\sigma = {P_{tout} - P_v}/{\rho g H}$ . where $P_{tout}$ was the total pressure of the outlet of draft tube, Pa. $P_v$ was the saturated vapor pressure at specific temperature, Pa. $\rho$ was the density of water, kg/m$^3$. $g$ was the acceleration of gravity, m/s$^2$. $H$ was the head of the turbine, m. The specific cavitation coefficient corresponding to the 1% drop of efficiency was defined as the critical cavitation coefficient $\sigma_c$ . Figure 5 presents the comparison of critical cavitation coefficient when the unit rotational speeds were 52.5 *r/min* and 67.5 *r/min*. It is quite obvious that runner optimization has greatly decreased the critical cavitation coefficient of original runner. And the effect is much remarkable in relative high flow rate.

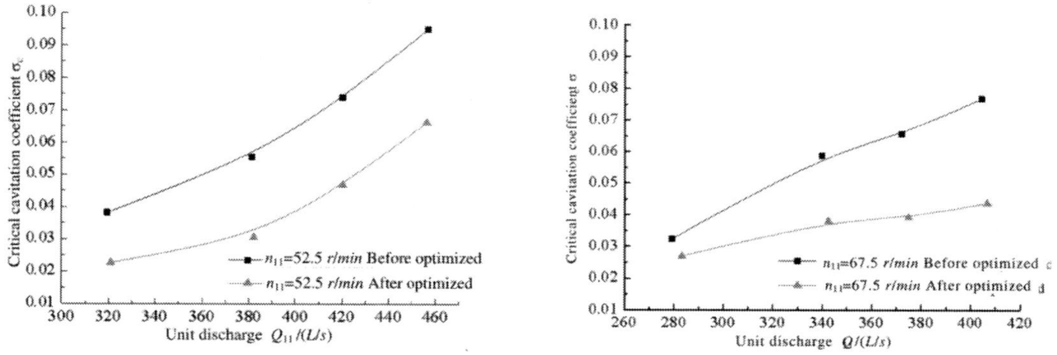


(a) $n_{11}$=52.5r/min (b) $n_{11}$=67.5r/min

**Figure 5.** Comparison of critical cavitation coefficient

Figure 6 shows the comparison of gas volume fraction in runner and draft tube with the isosurface value equaling to 10%. The large cavitation regions appear in runner outlet and draft tube for the original one. While only small cavitation regions on the blade close to outlet are discovered in the optimized runner. The backflow and vortex rope in draft tube are reduced. Increasing runner outlet diameter can reduce the flow discharge velocity and thus increases the static pressure at the runner outlet contributing to positive effect on cavitation. Therefore, the cavitation performance is greatly improved after optimization.

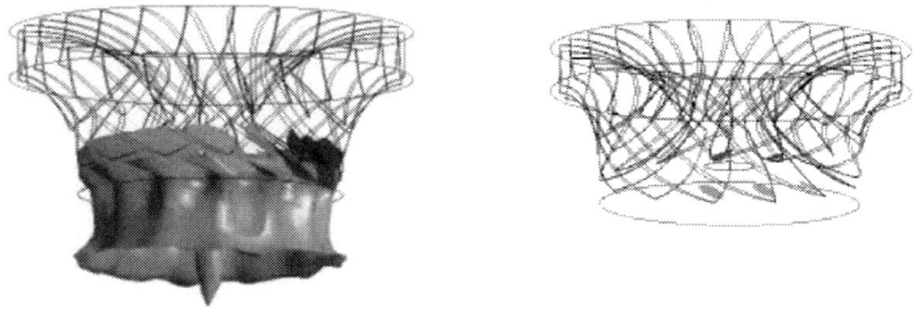

（a） Before optimization （b） After optimization

**Figure 6.** Comparison of vapour volume fraction in runner and draft tube (Isosurface value=10%, guide vane opening: 18° , unit rotational speed n11=62.5 r/min)

## 5. Conclusions

To reduce the flow velocity at the runner outlet region of the original model runner, the diameter of runner outlet was enlarged. Combining with the adjustments of the blade shapes by decreasing the blade outlet angles and increasing the blade wrap angles, the best efficiency point was kept the same. After a large number of modification runner test by CFD method, the final optimized one was selected. The optimized runner had the outlet diameter 10% larger than the original one. The blade outlet angles were 3 degrees smaller, and the blade wrap angles were 5 degrees larger than the original ones. The critical cavitation coefficients of the optimized runner decreased at every tested unit rotational speed comparing with the original one, especially at the relative large guide vane opening conditions. The zone of the low pressure on pressure surface of the optimized runner blades was reduced, which had the positive effect on cavitation. Proper increase the outlet of the Francis runner combined with adjustments of the blade shapes may have benefits on improving cavitation performance without changing the efficiency.

## Acknowledgement

The authors would like to acknowledge the financial support given by the National Natural Science Foundation of China (No. 51139007, 51409247) and the National Science & Technology Support Project of China (No. 2012BAD08B03, 2015BAD20B01).

## References

[1] Kumar P, Saini R P. Study of cavitation in hydro turbines--A review. Renewable & Sustainable Energy Reviews, 2010, 14(1): 374-383.

[2] Escaler X, Egusquiza E, Farhat M, Avellan F, Coussirat M. Detection of cavitation in hydraulic turbines. Mechanical Systems and Signal Processing, 2006, 20(2006): 983-1007.

[3] Hasmatuchi V, Farhat M, Roth S, Botero F, Avellan F. Experimevidence of rotating stall ina pump-turbine at off-design conditions in generating mode. Journal of Fluids Engineering, 2011, 133(5): 623-635.

[4] Müller A, Dreyer M, Andreini N, et al. Draft tube discharge fluctuation during self-sustained pressure surge: fluorescent particle image velocimetry in two-phase flow. Experiments in fluids, 2013, 54(4): 1-11.

[5] Müller A, Favrel A, Landry C, et al. On the physical mechanisms governing self-excited pressure surge in Francis turbines //IOP Conference Series: Earth and Environmental Science. IOP Publishing, 2014, 22(3): 032034.

[6] Alligne S, Nicolet C, Tsujimoto Y, Avellan F. Cavitation surge modelling in Francis turbine draft tube. Journal of Hydraulic Research, 2014, 52(3): 399-411.

[7] Wu Y, Liu J, Sun Y, Liu S, and Zuo Z. Numerical analysis of flow in a Francis turbine on an equal critical cavitation coefficient line. Journal of Mechanical Science and Technology, 2013, 27(6): 1635-1641.

# The numerical research of runner cavitation effects on spiral vortex rope in draft tube of Francis turbine

**J Yang[1], L J Zhou[2], Z W Wang[*1]**

1 Department of Thermal Engineering, Tsinghua University, Beijing, 100084, China

2 College of Water Conservancy and Civil Engineering, China Agricultural University, Beijing, 100083, China

**Abstract:** The spiral cavitating vortex rope developed in the draft tube of Francis turbine under part load condition maybe causes serious pressure fluctuations and power swings, which threatens the safety and stability of the power plant operations. Many works have been performed to explore the mechanisms of it. In this paper, the runner cavitation and spiral vortex rope under part load conditions were studied to investigate the relations of runner cavitation and the spiral vortex rope. The results proved the existence of obvious interaction between them. The swirl flow at the runner outlet plays an important role in the formation of vortex rope. And the periodic procession of vortex rope in turn intensifies the uneven pressure distribution near the runner outlet and causes the asymmetric cavitation on the runner blades, which then give rise to the modification of swirl flow at the runner blades and thereby affects the characteristics of vortex rope.

## 1. Introduction

The stability of Francis turbine caused by the cavitating flow at off-design conditions has always been the hot topic that people focus on. Many interesting investigations about draft tube cavitation have been done to explore the formation and evolution mechanisms of cavitating vortex rope. Jacob [1] studied the pressure surge under the entire range of operating conditions by both model experiments and site measurements. Nishi [2] investigated the effect of $\sigma$ on the flow field inside draft tube and the pressure pulsations by experiment. Alligne [3] also conducted detailed experimental and numerical research on cavitation in draft tube of Francis turbine. According to these results, it has been widely accepted that the precession of vortex rope at part load conditions may cause the serious instability problems. Since the main reason for the vortex rope formation is the swirl flow at the draft tube inlet, which may be modified by the periodic variations of cavitation presented on runner blades. That is to say, the impact of runner cavitation on the vortex rope also worth more attention to during the cavitating flow instability analysis. In this paper, the 3D two-phase cavitating flows in Francis turbine from inception cavitation to critical cavitation under part load conditions were simulated. The detailed

*1 wzw@mail.tsinghua.edu.cn

evolutions of runner cavitation during the vortex rope precession were discussed. The interaction of runner cavitation and vortex rope was analysed, which provides a helpful view on cavitating flow instability analysis.

## 2. Analysis Model and Boundary Conditions

**Figure 1.** Flow passage model, runner mesh and the pressure recording points

A model Francis turbine with runner diameter of 0.42m was used to simulate the cavitating flow, see Figure 1. After the mesh independence check, the total mesh of 1,666,000 nodes and 7,140,000 elements were chosen. The unsteady two phase cavitating flow from inception to critical cavitation at part load condition ($n_{11}$=$n_{opt}$, $Q_{11}/Q_{opt}$=0.75) were simulated. The LES turbulence model and ZGB cavitation model, which were verified through hydrofoil cavitation in former paper [4] and [5], were used to perform the unsteady cavitation flow simulation. The "total pressure" boundary condition was imposed at the inlet boundary. The outlet boundary condition was set as "static pressure" and the pressure value was computed by net head and the $\sigma=(P_{out}/\rho g- h_z- P_v/\rho g)/H$. Here, $P_{out}$ is the outlet pressure of the draft tube, the saturated pressure $P_v$ =3540*pa*, $h_z$ is the setting level, H is the net head. The time step was set as 0.000167s in unsteady simulation and the total computation time was 10s.

## 3. Results and Analysis

The runner cavitation during one cycle of vortex rope procession was detailed analysed for four cavitation numbers, which represented the four stages from inception cavitation ($\sigma$=0.025) to critical cavitation ($\sigma$=0.001) as shown in Figure 2. The green colour areas on the blades surface stand for the cavity zone. Five typical moments during one vortex rope's rotation period were chosen to analyse the evolution of runner cavitation. Under the inception cavitation condition, the vortex ropes is tiny, and there are only small bunches of cavity clustered at the trailing edge of blades near the shroud. And the locations of cavity zone vary with time. As the cavitation gradually developed, a significant asymmetrical cavitation can be observed on the suction side of opposite blades at this condition. The reason for asymmetrical cavitation may be the eccentric motion of cavitation vortex rope under part load condition, which cause the uneven pressure distributions in the circumference direction of flow field, thus induce the asymmetry pressure distribution at the runner outlet. However, as the $\sigma$ further reduced, the asymmetrical phenomena seem mitigated a little. That's because the range of runner cavitation extended due to the diameter of vortex rope became larger.

As is known to all, the swirl flow at the runner outlet is the key reason for the formation of vortex rope. The cyclic changes of runner cavitation inevitably modify the angle of swirl flow, and thus affect the characteristics of vortex rope. To verify this, the axial and circumference velocity profiles at the runner outlet in a cycle of vortex rope procession were plotted in Figure 3. The velocity profiles keep changing with the variation of runner cavitation. That may add an effect on the evolution of the vortex rope shape. Besides, the vortex rope in turn disturbs the pressure distribution and finally causes the asymmetric runner cavitation. The results give an evidence for the interaction of the runner cavitation and vortex rope, which implies that the runner cavitation should be considered during the instability analysis about the cavitation in draft tube.

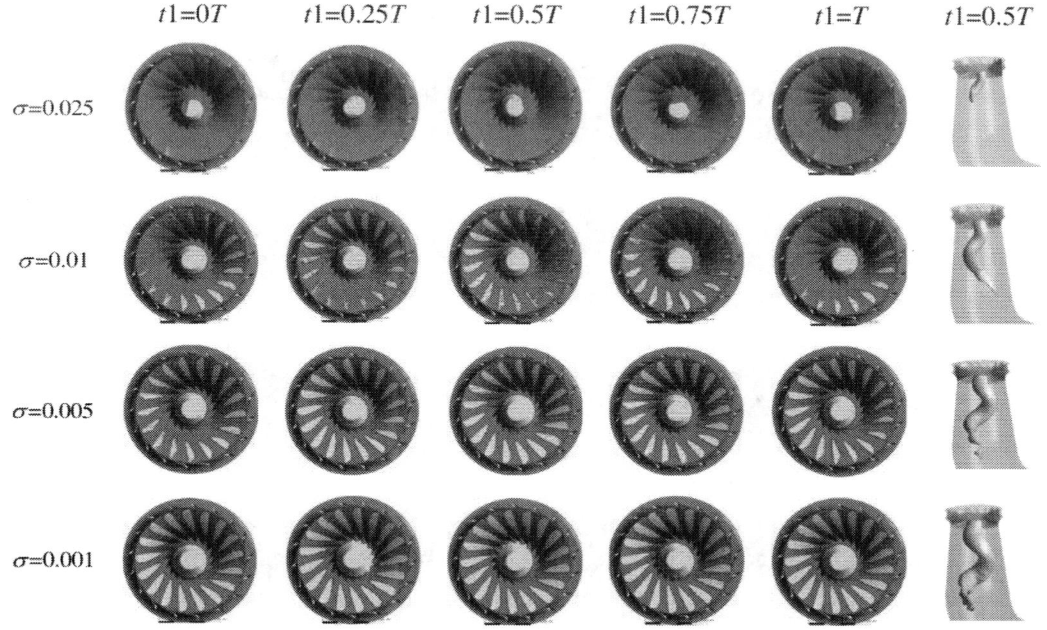


**Figure 2.** Runner cavitation evolution in one vortex rope cycle at different conditions

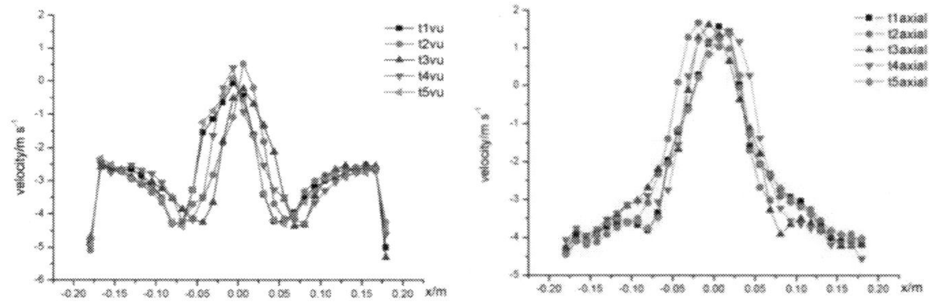


**Figure 3.** Circumference velocity and axial velocity of one vortex rope cycle at runner outlet ( $\sigma$=0.01 )

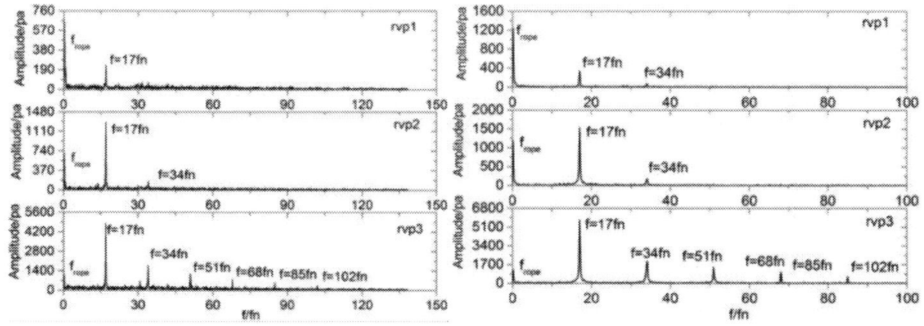


**Figure 4.** Frequency of pressure record point at runner outlet (left: $\sigma$=0.005 , right: $\sigma$=0.01 )

The cavitation evolutions on blades also reveal the pressure change in runner. Three pressure monitoring points were set along a blade trailing edge (see Figure 1) and the frequencies of the pressure points were analysed, as shown in Figure 4. The main frequency of pressure point near the crown (rvp1) is vortex rope rotational frequency of the vortex rope, which also provides an explanation for periodic blade cavitation variations. Although the influences of vortex rope still exist for the other two points, the main frequency changed to blade passing frequency. The amplitude of pressure pulsations is the highest at the pressure point near the band (rvp3). What's more, the frequency multiplication also appeared at this point. And from the comparing of the results, we can see that the amplitude at each point under $\sigma$=0.01 is higher than that from $\sigma$=0.005, which is in accordance with the blade cavitation evolution. The asymmetric phenomena under $\sigma$=0.01 is the most obvious, thus the uneven flow field lead to the serious pressure pulsations.

## 4. Conclusions

Based on the results of spiral vortex rope cavitating flow simulation under part load conditions performed on a Francis turbine, the cavitation on the surface of runner blades were analysed. During the process in which the cavitation developed from inception to critical stage, the vortex rope gradually grows in thickness and length. The cavitation on suction side of blade also presented periodic changes with the vortex rope procession. The cyclic changes of runner cavitation causes would bring a modification for swirl flow angle, which have impacts on the formation of vortex rope. The vortex rope intensifies the uneven pressure of flow field which in turn further affects the runner blades cavitation developments and leads to the asymmetric cavitation on runner blades. The asymmetric extent was the most obviously between the inception and critical cavitation conditions.The presented results show significant interaction between runner cavitation and vortex rope. The evolution of runner cavitation should also be considered on the instability analysis induced by vortex rope in the draft tube of Francis turbine.

## Acknowledgements

The authors thank the China Postdoctoral Science Foundation (No. 2015M571046) for supporting present work.

## References

[1] Jacob T, and Prenat J E, 1996, Francis turbine surge: discussion and data base, *Proceedings of 18th IAHR Symposium*, Valencia, Spain.

[2] Nishi M, and Liu S H, 2012, An outlook on the draft-tube-surge study, *Proceedings of the 26th IAHR Symposium on Hydraulic Machinery and Systems*, Beijing, China.

[3] Alligné S, Maruzewski P, Dinh T, 2010, Prediction of Francis turbine prototype full load instability from investigations on a reduced scale model. *Proceedings of the 25th IAHR Symposium on Hydraulic Machinery and Systems*, Timisoara, Romania.

[4] Luo X W, Ji B, Peng X X, Xu H Y, Nishi M, 2012, Numerical Simulation of Cavity Shedding from a Three-Dimensional Twisted Hydrofoil and Induced Pressure Fluctuation by Large-Eddy Simulation, *Journal of Fluid Engineering, Transactions of the ASME,* 134(1): 041202:1-10.

[5] Yang J, Zhou L J, Wang Z W, 2011, Numerical Simulation of 3-D Cavitation around a Hydrofoil. *Journal of Fluid Engineering, Transactions of the ASME,* 133(1): 081301:1-9.

Journal of Physics: Conference Series **656** (2015) 012077 doi:10.1088/1742-6596/656/1/012077

# Predicting the Inception Cavitation of a Reversible Pump-Turbine in Pump Mode

**Ran Tao[1], Ruofu Xiao[1], Di Zhu[1], Weichao Liu[2]**

[1]College of Water Resources and Civil Engineering, China Agricultural University, Beijing 100083, China

[2]Dongfang Electric Machinery Co., LTD, Deyang, Sichuan Province 618000, China

E-mail: xrf@cau.edu.cn

**Abstract**. Inception cavitation is a crucial indicator for reversible pump-turbines especially in pump mode. In actual applications, it is difficult to use CFD for the inception cavitation character. In this study, CFD simulation is conducted to find a proper way to evaluate the inception cavitation, different levels of vapor volume fraction in the impeller is predicted based on the tested results. Results show that the prediction of the location and scale of cavitation is accurate. The predicted cavitation number also matches the experimental data well. The vapor volume fraction levels from 0.0001% to 0.001% are recommended as the criterion of inception cavitation.

## 1. Introduction

Cavitation is crucial for hydraulic turbo-machinery because it will cause noise, vibration and damage to the components. Conducting a specific study on the inception cavitation would be meaningful for the operation stability and security. For a reversible pump-turbine, cavitation is more serious when it works as a pump [1]. Compared with common pumps, it is necessary to use the inception cavitation standard in reversible pump-turbines due to the larger cavitation scale. In the pump mode, cavitation usually incepts on the leading edge when the fluid separates here with an abrupt pressure drop [2]. Thus, it is important to study the inception of leading edge cavitation of pump-turbine.

There are already a lot of researches on the cavitation in hydraulic turbo-machineries. Escaler et al [3] conducted an experimental investigation on the cavitation in hydraulic turbine. They provided a good guidance for researchers to detect the inception cavitation in different types of turbine in actual applications. Cencîc et al [4] also conducted experiments to detect the leading edge cavitation in pump mode. It is helpful for researchers to investigate the inception cavitation in a pump-turbine. However, experimental studies are difficult to obtain all the flow details in the reversible pump-turbine. Thus, numerical simulation is used as an effective supplement for experiments. Based on the experiment, Anciger et al [5] predicted the rotating stall and cavitation inception in a pump-turbine in pump mode. Even the numerical simulation results were consistent with the experimental results, the criterion of inception cavitation are not clear. Jese et al [6] also gave an experimental-verified simulation for the inception cavitation in pump-turbine, they used the void ratio in the first cavitating cell as the criterion of inception cavitation. The numerical results also matched well with the experimental data, but there were only the simulations under off-design conditions. Liu et al [7] and Zhang et al [8] also conducted simulations for cavitation in pump-turbine in pump mode. These numerical simulations performed well in the prediction of the location of cavitation. It shows that the numerical simulation is applicable for the inception cavitation.

In our previous research, numerical simulation was already used for the optimization of a reversible pump-turbine [9]. In this study, a detailed study for the inception cavitation in pump-turbine in pump mode is conducted. Based on the former works, CFD simulation is used for the predicting of the inception cavitation. Different levels of vapor volume fraction is predicted. Both the design condition and off-design conditions are taken into consideration. On one hand, the locations of cavitation are indicated. On the other hand, the proper way of criterion of inception cavitation for reversible pump-turbine in pump-mode are discussed.

## 2. Research Object and Methods

The research object in this study is a Francis pump-turbine. It mainly consists of five components including the volute, stay-vane, guide-vane, impeller and draft tube. The flow domains for numerical simulation were modeled as shown in Fig. 1. In the pump mode, fluid flows into the draft tube first, then impeller, guide-vane, stay-vane and volute. In this study, both the design condition ($Q_{md}$ = 360 kg/s) and 4 off-design conditions (0.8$Q_{md}$, 0.9$Q_{md}$, 1.1$Q_{md}$ and 1.2$Q_{md}$) were discussed. The relationship between flow rate $Q_m$ and guide-vane opening angle $\alpha$ is shown in Fig. 2.

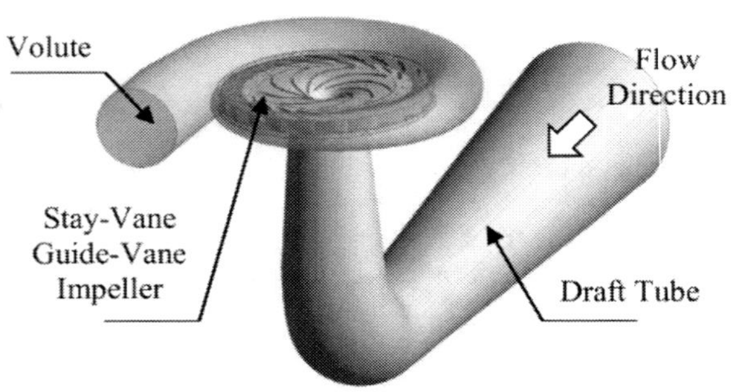


**Figure 1.** The flow domain of the pump-turbine model

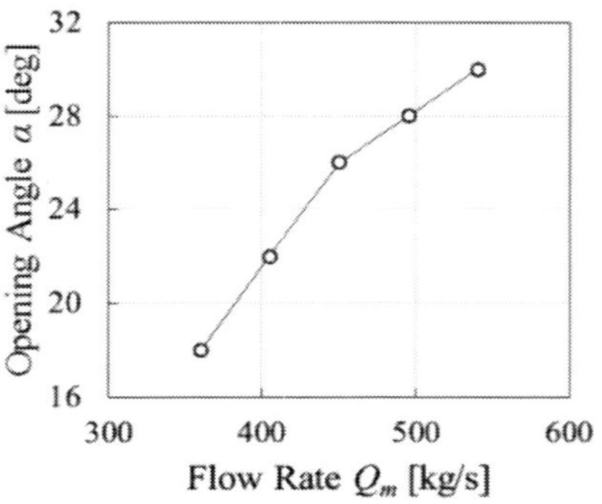


**Figure 2.** The relationship between flow rate and guide-vane opening angle

In this study, RANS simulation was conducted by using commercial CFD code in ANSYS CFX. Before the simulation, the flow domains were meshed using ICEMCFD. Structural elements were used for draft tube, impeller, guide-vane and stay-vane. Unstructured elements were used for the volute. A proper mesh scheme was determined with an independence check. By making sure the variation of simulated pump head less than 1%, the mesh scheme was finalized with altogether 2’863’660 nodes.

In the RANS simulation for turbulent flow, the SST $k$-$\omega$ model was used to close the averaged equations. The Zwart Model was used for cavitation. The multiple reference model was used with the impeller domain rotating in 1200 r/min and other domains keeping stationary. In the boundary condition setup, a mass flow inlet was given at the draft tube inflow and a pressure outlet was given at the volute outflow. All the solid walls were set as no-slip wall boundary. Transient sliding interfaces were set to connect the flow field of the domains. The environment pressure was set as 1 Atm.

## 3. Results and Discussions

### *3.1. Calculation of Vapor Volume Fraction*

Vapor volume fraction in the impeller is used to evaluate the inception cavitation. By modifying the pressure on the outlet boundary, it is able to control the domain pressure and monitor the variation of vapor volume fraction in the impeller. Figure 3 shows the process of finding the vapor volume fraction $f_v$ = 0.001%. For clearance, Fig. 3 shows only part of the data sets. The $p_{out}$-$f_v$ curves drop abruptly then become flat. Looking conversely at the horizontal axis, when the pressure is high enough, cavitation would not happen. With the pressure decreasing, water turns into vapor in somewhere, usually on the blade leading edge, in the impeller. If the pressure continue decreasing, the vapor volume fraction grows up rapidly.

### *3.2. Predicting the Leading Edge Cavitation*

In Fig. 3, we can find the point that the fraction curve rises up from flat, and define it as the "turning point". Even the saturation pressure are the same for these conditions, the "turning point" are different due to the different flow rate. Under the design flow rate of $1.0Q_{md}$ (450 kg/s), the "turning point" occurs when the outlet pressure $p_{out}$ is about 500 kPa. Under the off-design condition of $0.9Q_{md}$ (405 kg/s) and $1.1Q_{md}$ (495 kg/s), it occurs when $p_{out}$ is about 550 kPa. Under the $0.8Q_{md}$ (360 kg/s) and $1.2Q_{md}$ (540 kg/s), it occurs when $p_{out}$ is about 650 kPa. It means that the condition departs further from the design point, the cavitation happens earlier. Figure 4 shows the reason of the inception of leading edge cavitation. When operation goes away from the design point, it will cause the leading edge separation. Pressure drops in the separation region and induces the leading edge cavitation.

CFD prediction is very useful to indicate the location of the inception cavitation on the leading edge. As shown in Fig. 5, under lower off-design flow rate ($0.8Q_{md}$), leading edge cavitation occurs on the suction surface. On the contrary, leading edge cavitation is found on the pressure surface under higher off-design flow rate ($1.2Q_{md}$).

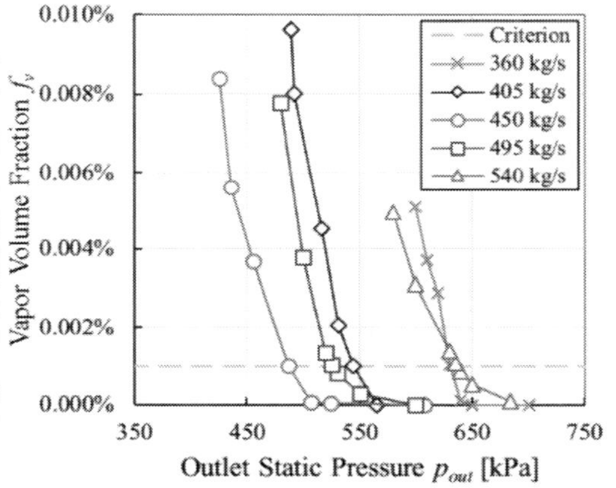


**Figure 3.** The process of finding the vapor volume fraction of 0.001%

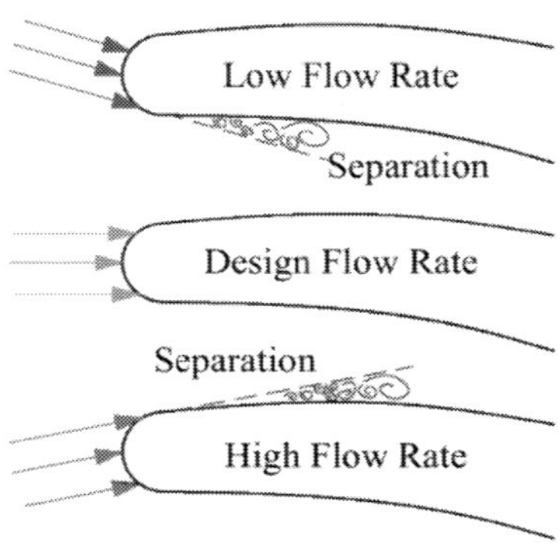


**Figure 4.** The reason of the inception of leading edge cavitation

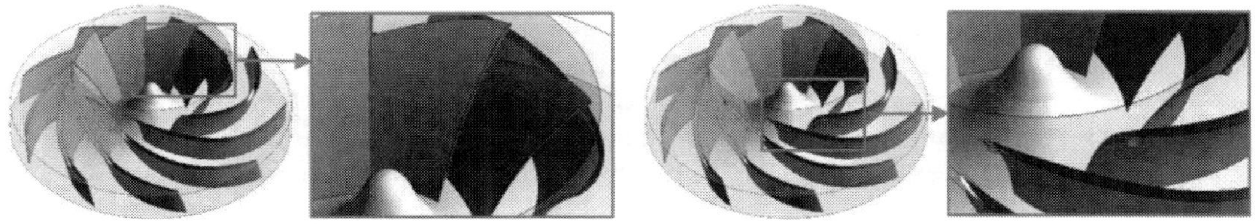

Low Flow Rate: $0.8Q_{md}$ High Flow Rate: $1.2Q_{md}$

**Figure 5.** The position of inception cavitation under different flow rates

### *3.3. Criterion of Inception Cavitation*

Even the vapor volume fraction can indicate the cavitation, an exact criterion of volume fraction is still necessary for the inception cavitation. For this purpose, different vapor volume fraction levels were simulated and compared with experimental data. Figure 6 is the test rig and facilities. In the experiment, the first three bubbles were captured and defined as the inception cavitation. Figure 7 shows the cavitation coefficient $C_\sigma$ by both the experiment and the simulation, where $C_\sigma$ is defined as:

$$C_\sigma = \left( \frac{p_s}{\rho g} + \frac{V_s^2}{2g} - \frac{p_v}{\rho g} \right) / H \quad (1)$$

where $\rho$ is the density of water, $g$ is the acceleration of gravity, $p_s$ and $V_s$ is the pressure and velocity at the reference position (here is the draft tube inlet), $p_v$ is the saturation pressure, $H$ is the pump head.

The $C_\sigma$ value of the vapor volume fractions that $f_v$ = 0.0001%, $f_v$ = 0.0005%, $f_v$ = 0.001% and $f_v$ = 0.002% are simulated. Comparing with the experimental data, the $C_\sigma$ curve from $f_v$ = 0.0005% matches the experimental curve best. However, the range from fraction 0.0001% to 0.001% is recommended as the criterion of inception cavitation for reversible pump-turbines.

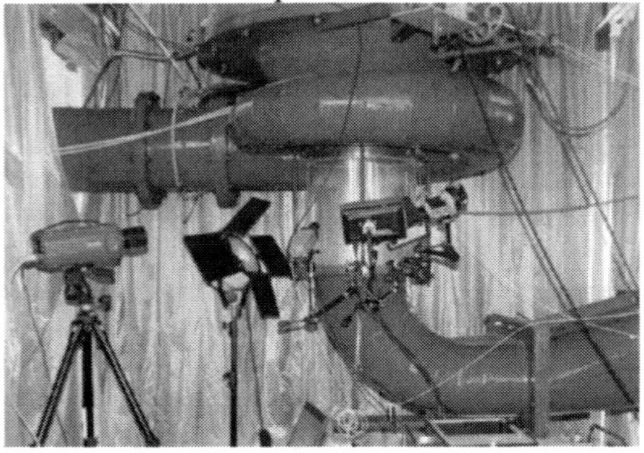

**Figure 6.** The test rig and facilities for inception cavitation measurement.

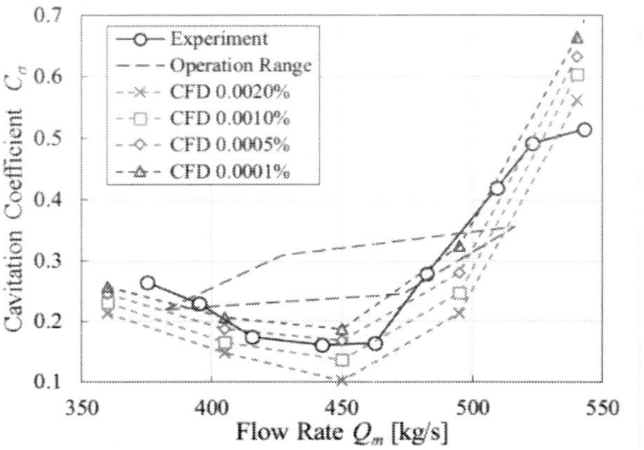


**Figure 7.** The cavitation coefficient $C_\sigma$ from both the experiment and the simulation.

## 4. Conclusions

CFD is available and effective for the prediction of inception cavitation in reversible pump-turbine. Firstly, it is effective for the calculation of vapor volume fraction which can be used as the criterion of inception cavitation. Secondly, CFD is available to visualize the location of the leading edge cavitation. By conducting CFD simulation and comparing it with experiment data, it is found that the vapor volume fraction level from 0.0001% to 0.001% is a proper criterion for inception cavitation. This rule can be applied to the engineering applications of reversible pump-turbines.

## References

[1] Zhou Y W. Draft head options for pump turbine of pumped storage plant. *Mechanical & Electrical Technique of Hydropower Station*, 2006, 29, 12-15.

[2] Pan Z Y, Yuan S Q. Fundamentals of cavitation in pumps. *Jiangsu University Press*, Zhenjiang, China, 2013.

[3] Escaler X, Egusquiza E, Farhat M, et al. Detection of cavitation in hydraulic turbines. *Mechanical Systems and Signal Processing*, 2006, 20: 983-1007.

[4] Cencîc T, Hoĉevar M, Ŝirok B. Study of erosive cavitation detection in pump mode of pump-storage hydropower plant prototype. *Journal of Fluids Engineering*, 136(5): 051301.

[5] Anciger D, Jung A, Aschenbrenner T. Prediction of rotating stall and cavitation inception in pump turbines. *25th IAHR Symposium on Hydraulic Machinery and Systems*, Timisoara, Romania, 2010.

[6] Jese U, Fortes-Patella R, Antheaume S. High head pump-turbine: Pump mode numerical simulations with a cavitation model for off-design conditions. *27th IAHR Symposium on Hydraulic Machinery and Systems*, Montreal, Canada, 2014.

[7] Liu J T, Wu Y L, Liu S H. Study of unsteady cavitation flow of a pump-turbine at pump mode. *IOP Conference Series: Materials Science and Engineering*, 2013, 52(6): 062021.

[8] Zhang L, Jing X, Wang Z, et al. Analysis of Francis pump-turbine runner cavitation flows in pump mode. *ASME 2009 Fluids Engineering Division Summer Meeting*, Colorado, USA, 2009.

[9] Tao R, Xiao R, Yang W, et al. Optimization for cavitation inception performance of pump-turbine in pump mode based on genetic Algorithm. *Mathematical Problems in Engineering*, 2014: 234615.

# Blade Section Design of Marine Propellers with Minimum Cavitation Induced Pressure Fluctuations

Zhibo Zeng [1]
China Ship Scientific Research Center, Wuxi214082, China
Gert Kuiper
Soetendaalseweg 106721 XB, Bennekon, The Netherlands

E-mail: zbzeng80@gmail.com

**Abstract.** To minimize cavitation induced pressure fluctuations by marine propellers with minimum efficiency loss, the paper presents a new design and optimization method using a blade section design method. The sheet cavity volume variation on a two-dimensional blade section in quasi-steady condition has been simplified to a relation with only a limited number of non-dimensional parameters. This results in a fast prediction method of the cavity volume of a blade section passing a wake peak, using a pre-calculated database. This makes optimization feasible. The optimization method was applied to the propeller of a container ship. Extensive tests in a towing tank and a cavitation channel validated the reduction of pressure fluctuations: 33% reduction in the first blade frequency amplitude and 18% reduction in the second blade frequency amplitude, with the same open water efficiency.

## 1. Introduction

Traditionally the objective of propeller design was to obtain maximum efficiency with a more or less shock-free pressure distribution on the blade sections in the tangentially averaged wake distribution. Nowadays we are able to calculate the pressure distribution on propeller blades in unsteady condition. However it has to be related to effects that are relevant, such as cavitation inception, induced pressure fluctuations or erosion. These phenomena are essentially unsteady and thus the wake has to be taken into account. And these also require that the cavitation can be calculated.

In case of cavitation inception calculation of the cavity can be avoided and an objective of the non-cavitating pressure distribution can be formulated: the minimum pressure on the blades during one revolution remains above the vapor pressure. This requires that all blade sections operate within the cavitation bucket of the section. This is essentially a two-dimensional method and the section designs have to be modified for three dimensional effects. Such a design method using the Eppler method was already developed by Kuiper and Jessup in 1993 [1]. This method was recently updated and automated by Zeng [2].

However, in most operating conditions of commercial ships it is not possible to avoid cavitation. One detrimental effect of cavitation is induced pressure fluctuations, resulting in hull vibrations. Then the design objective becomes: minimum cavitation induced pressure fluctuations. Efforts have been made to relate the cavitation induced pressures directly with properties of the pressure distribution on the blades. Johnsson [3] used a new type of section thickness distribution with two ellipses at the

[1] To whom any correspondence should be addressed.

leading edge to obtain a favorable pressure distribution from cavitation point of view. Yamaguchi et al [4] tried to relate the type of pressure distribution in steady condition with the cavitation induced pressure fluctuations. Wang et al [5] showed that a shift of the locations of maximum camber and thickness towards the trailing edge decreased calculated induced pressure fluctuations, but this improvement was not confirmed by experiments. Dang [6] used much wider cavitation buckets than that of a NACA section to reduce the total cavity volume. So the efforts to replace the criterion of minimum pressure fluctuations by pressure distributions have not been successful. A pressure source is proportional to acceleration of the cavity volume, and the cavity acceleration is very sensitive to the (unsteady) pressure distribution in the wake. This relation is strongly non-linear, so a prediction independent of the wake distribution and without cavitation might not be possible.

The design method in this paper includes the cavitation behavior in a wake. The objective of minimum induced pressure fluctuations was obtained using an optimization procedure. This approach is a continuation of the optimization procedure of the authors for maximum inception speed, so some elements are re-used.

## 2. Cavity volume prediction

The prediction of the cavity volume on a propeller blade is simplified by considering two dimensional cavitation on each blade section. The propeller blade sections are transformed in a linearized way by replacing the geometrical camber line with the effective camberline from the chordwise loading distribution [1]. When a section passes a wake peak the cavity on the section will experience inception, development and disappearance with the rotating propeller blade position θ. The cavity area: $V_c$ on a section can be calculated using CFD as proposed by Singhal [7]. But to carry out such a calculation on each section in a range of blade positions requires considerable computing time and this makes optimization impossible. Therefore an approximate method was developed.

The cavity volume on such blade section is directly related with the fully wetted pressure distribution at the leading part of the suction side which is now related with parameters from the Eppler method. Figure 1 shows typical pressure distributions from the Eppler method at two angles of attack. There is a constant pressure distribution $-C_{p1}$ in the middle segment at angle of attack $\alpha_1$ and another constant pressure distribution $-C_{p2}$ on the first segment at another angle of attack $\alpha_2$. The difference $\Delta C_p = -C_{p2} - (-C_{p1})$ and the location $x_1$ between the two segments are the dominant parameters controlling this change in pressure distribution.

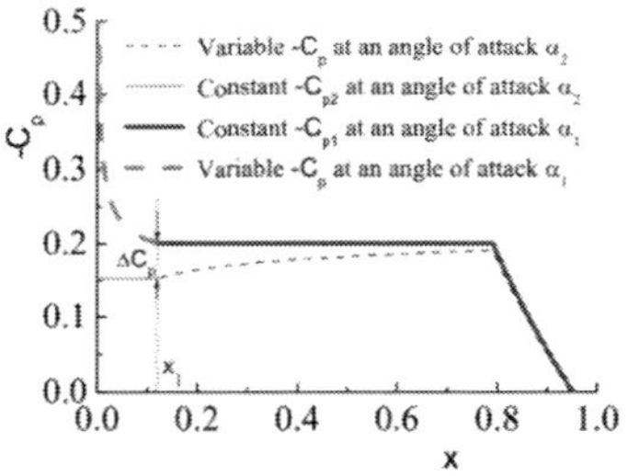


Figure 1. Parametric representation of pressure distribution at the leading part of the suction side of a section in fully wetted condition

To make optimization possible a range of sections were generated using the Eppler parameters and varying the parameters $x_1$ and $\Delta C_p$. For each case, the cavity area $V_c$ was calculated using Singhal full cavitation model [7] with a 10% volume fraction of vapor phase as cavity boundary. First the cavitation number $\sigma$ was varied with fixed angle of attack $\alpha$ and then $\alpha$ was varied with fixed $\sigma$. When $V_c/(\alpha - \alpha_i)$ is plotted versus the non-dimensional condition parameter $\sigma^* = (\sigma + C_{p1})/(\alpha - \alpha_i)$ a unique relationship between $V_c$ and ( $\sigma$ , $\alpha$ ) is found, as shown in figure 2. Here $\alpha_i(\sigma)$ is the inception angle of attack at $\sigma$, a parameter originally proposed by Le, Franc & Michel [8]. The unique relationship for each case can be fitted by a power function also shown in the figure

$$\frac{V_c}{\alpha-\alpha_i}=\frac{B}{((\sigma+C_{p1})/(\alpha-\alpha_i))^k} \tag{1}$$

where B, k are the coefficients utilized to describe the variation of cavity area for a section.

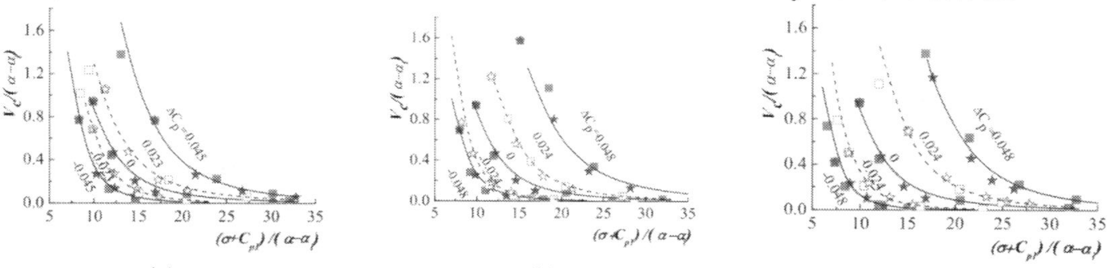


(a) (b) (c)

Figure 2. The database of pre-calculated cavity area characteristic of section in equation (1) related with a series of pressure distributions: (a) $x_1$=0.017; (b) $x_1$=0.067; (c) $x_1$=0.110 ("■ and □" denote a variation of $\sigma$ with $\alpha$ unchanged; "★ and ☆"denote a variation of $\alpha$ with $\sigma$ constant ; The dotted and solid lines are their fittings)

To further reduce the amount of calculations the cavity area characteristic described by [B, k] was related with the pressure distribution parameters ( $x_1$ , $\Delta C_p$ ) using the above pre-calculated database. Through statistic analysis of the database, a regression formula is obtained,

$$[B, \; k]= f(\Delta Cp, \; x_1) \tag{2}$$

When a section, its geometry prescribed by the Eppler parameters, passes a wake peak, the non-dimensional condition parameter $\sigma^*$ can be calculated (the operating curve of a section). On each section the development of the cavity area $V_c$ with the blade position $\theta$ can then be determined from the cavity area characteristic of the section in equation (1) with [B, k] calculated by formula (2).

## 3. Blade section optimization and validation

The induced pressure fluctuations are periodical with blade passages and section optimization is therefore done in the frequency domain. The cavitation induced pressure fluctuations amplitude of a certain blade frequency $[p_c]_{(mZ)}$ (m is blade frequency, Z is blade number) thus can be derived as,

$$[p_c]_{(mZ)} \cong \frac{\pi\rho n^2}{r_c}\left[\frac{\partial^2 V_c}{\partial \theta^2}\right]_{(mZ)} = \frac{\pi\rho n^2}{r_c} Z^2 m^2 [V_c]_{(mZ)}, \text{ when } V_c = V_{c0}+\sum_{m=1}^{\infty}(a_m \cos mZ\theta + b_m \sin mZ\theta) \tag{3}$$

where water density $\rho$, propeller rotation speed $n$, and the distance of the field point $r_c$. The focus in this paper is on the first three blade frequency harmonics, the objectives can be formulated as $\underset{X\in\Omega}{Min}[C_1(X), \; C_2(X), \; C_3(X)]$, where X is combination of design parameters in Eppler method, $\Omega$ is a design space of these parameters, and $C_1=[V_c]_{(Z)}$, $C_2=4[V_c]_{(2Z)}$, $C_3=9[V_c]_{(3Z)}$ by neglecting the influence of the change of $r_c$ . The optimization procedure follows a similar path as used in the section optimization for maximum inception speed [2].

The method was applied for the propeller of a container ship, where only one section was optimized. Figure 3 shows the objectives of an optimized section at r/R=0.8 in behind condition. It can be seen that the amplitudes of the first and second blade frequency of induced pressure fluctuations are significantly reduced compared with the conventional NACA section; the third one increases, however both of them are small. Figure 4 shows that the optimized section has much smaller acceleration of the cavity area with the blade position θ, which is responsible for the much lower blade frequency amplitudes.

With the optimized two-dimensional section, a propeller blade has been redesigned in a similar way as was done for the optimization of the maximum inception speed[2]. The open water performance test and the cavitation test in behind condition for both the optimized propeller and the NACA propeller were carried out in the towing tank and in the large cavitation channel of CSSRC in Wuxi, China. The test results show that the loads and efficiency are very close. The first blade frequency amplitude of the pressure fluctuations of the optimized propeller has a reduction of 33%, and 18% for the second one, and the third one is close to each other. Figure 5 shows the cavity extent of the optimized propeller model is much smaller which indicates that the optimization also significantly reduced the cavity extent (the volume can only be calculated).

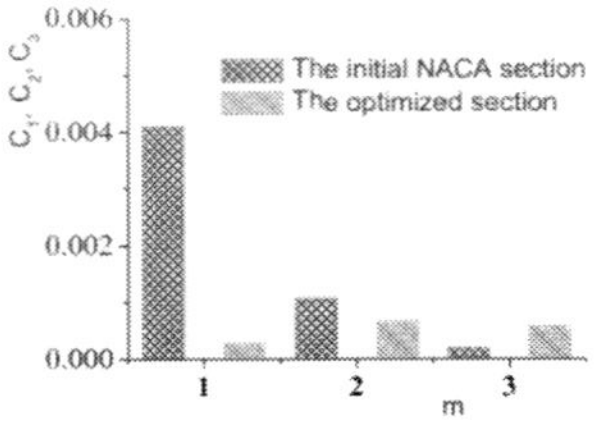


Figure 3. The optimized objectives

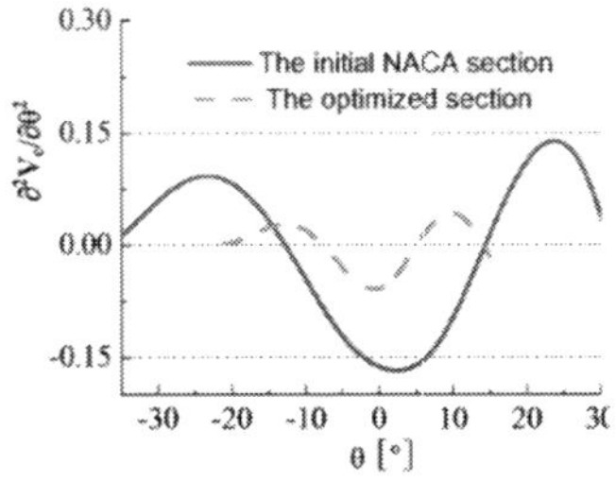


Figure 4. The variation of cavity area acceleration

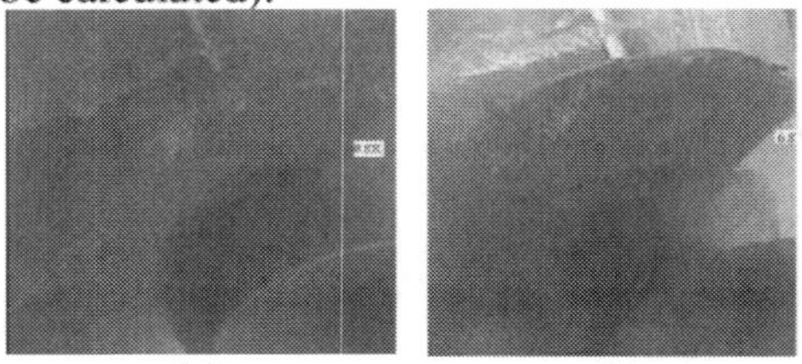

(a) (b)

Figure 5. Comparison of cavity extent: (a)the NACA propeller model, (b) the optimized propeller model (at the 12 o'clock position in behind condition)

## 4. Conclusion

A design technique using blade sections optimized for minimizing cavitation induced pressure fluctuations is shown to be effective. The developed method has the following main features: 1) A non-dimensional parameter $\sigma^* = (\sigma + C_{p1})/(\alpha - \alpha_i)$ is used to represent the working condition of two-dimensional blade sections beyond inception. 2) The calculation of two-dimensional cavity area on a blade section during a revolution was made so efficient that it could be used in an optimization program. This was done by making a database of pre-calculated cases. 3) It was shown that the amplitude of the blade frequencies of the pressure fluctuations was proportional to the amplitude of the cavity volume variations. This omitted the calculation of higher derivatives of the cavity volume and made the calculation stable. It also reduced the calculation time.

There are many simplifications in the method. Experimental and numerical validation of the optimized cavity volume variation is recommendable. After engineering practice, the method proposed can be further developed to be a design tool to improve propeller with favorable ship vibration.

## 5. Acknowlegements

This work was carried out in CSSRC. Special thanks to Prof. Shitang Dong and Prof. Denghai Tang for the valuable discussions and being very supportive throughout the development of the propeller design method in this paper.

## References

[1] Kuiper G and Jessup S D 1993. *SNAME Transactions,* Vol.101 247–273.
[2] Zeng Z and Kuiper G *Proceedings of the 7th International Symposium on Cavitation: CAV2009* (Ann Arbor, 17-22 August 2009) No.114.
[3] Johnsson C A 1980 *The naval Architect* No.3 113-115.
[4] Yamaguchi H *et al.* 1988 *Journal of the Society of Naval Architects of Japan* No.164 28-42.
[5] Wang D *et al. International Conference on Propeller Cavitation (*University of Newcastle, 16-18 May 1995) 73-89.
[6] Dang J. 2004 *Int. Shipbuilding Progress* Vol. 51 No. 4 353-376.
[7] Singhal A K *et al.* 2002 *Journal of Fluids Engineering* Vol. 124, 617-624.
[8] Le Q Franc J P and Michel J M 1993 Journal of Fluid Engineering Vol. 115 243-248.

# On instationary mechanisms in cavitating micro throttles

**B Beban[1], S Legat[1], S J Schmidt[1] and N A Adams[1]**

[1] Technische Universität München, Institute of Aerodynamics and Fluid Mechanics, D-85746 Garching, Germany

E-mail: bruno.beban@tum.de

**Abstract.** The current investigation presents numerical simulations of cavitating flows in a simplified model of a mushroom valve chamber of a piezo common rail injection system. Two discharge throttles with different step diameters are investigated. The developed models are able to predict relevant features of cavitating flow in fuel injectors. Special attention is put on the investigation of wave dynamics and related instationary mechanisms in the discharge throttle and the valve chamber. To this respect, a compressible flow solver with a homogeneous mixture model and barotropic description of the diesel-like-fluid is utilized. Highly unsteady phenomena are observed in both investigated designs. The structure of the cavitating flow is further analyzed with an emphasis on the interaction between collapsing vapor clouds in the throttle step and reentrant motion in the discharge throttle. Furthermore, numerical simulations reveal significant influence of the throttle step diameter on the cavity dynamics.

## 1. Introduction

Modern common rail injection systems exceed injection pressures of more than 2000 bar, with the aim to improve fuel injection efficiency and to meet demanding emission standards. However, huge pressure drops through the throttles cause the working fluid to cavitate. For the valve chamber, which regulates fuel injection sequencing, cavitating throttles in choked regime are used to ensure constant and pressure-drop independent mass flow rates. Formation of the instantaneous pressure peaks due to the collapse of cavitating clouds, accompanied by the high loads on the material surface, may therefore cause a failure of system components.

In this study, we investigate the cavitating flow in the valve chamber of a piezo common rail injection system. Since a realistic, three-dimensional model of the chamber – which consists of the discharge throttle, mushroom valve, return spring, valve piston and outlet passage – is expensive from the computational viewpoint, a simplified model is developed. It enables a detailed investigation of the complex flow physics at reduced geometrical complexity. Experimental investigation of micro channels is usually limited to planar geometries [1, 2, 3], or large-scaled orifices [4], since they are accessible by optical measurement techniques. Recent x-ray radiography measurements show promising results of averaged vapor volume prediction in axis-symmetric micro channels [5]. However, experimental measuraments of highly instationary cavitating flows in micro channels still represent a challenge and hence numerical simulation can provide valuable information about the flow dynamics of small scale structures.

## 2. Numerical method

For the numerical simulations, we apply our in-house code CATUM (**Ca**vitation **T**echnische **U**niversität **M**ünchen), which is a finite volume, density-based compressible flow solver. The compressible formulation captures the formation of pressure peaks during the collapse stages of vapor clouds, and consecutive development of shock fronts, which are a driving mechanism for cavitation erosion. Under the assumption that all phases are in thermodynamic equillibrium, phase transition is modeled with a homogeneus mixture approach. We neglect viscosity, since the focus of this study is put on the effects dominated by intertia. Detailed description of the underlying numerics can be found in [6]. The system of governing equations is closed with the barotropic equation of state for diesel-like ISO4113 fluid. Fitting parameters for the equation of state are taken from [7] and [8].

## 3. Computational grids and numerical setup

The present numerical study includes two different designs of the discharge throttle that connects the high pressure chamber with the valve chamber. The setup, together with an overview of the whole numerical domain, is sketched in Fig. 1. The geometries differ only in the throttle design, keeping the valve chamber and the distance between the throttle outlet and front side of the mushroom valve constant. Both designs feature a cylindrical throttle with length $l = 0.5$ mm, diameter $d = 0.2$ mm and a sharp inlet edge. Designs A and B have step diameters $D_A = 0.5$ mm and $D_B = 1$ mm, respectively. The step length $L = 0.5$ mm is equal for both cases. Block-structured decomposition of the computational domain is utilized. For each design, three grid levels have been created: coarse, medium and fine. The fine grid consists of approximately $2.2 \cdot 10^6$ hexahedral cells. On the fine grid, the discharge throttle is resolved with 49 cells in diameter and 48 along its axis, with a cell size of approximately 4 $\mu$m across the diameter.

At inlet and outlet, asymptotic boundary conditions for static pressure are prescribed ($p_{in} = 2000$ *bar*, $p_{out} = 60$ *bar*). For velocities, von Neumann boundary conditions are imposed. Walls are modelled as adiabatic, inviscid walls.

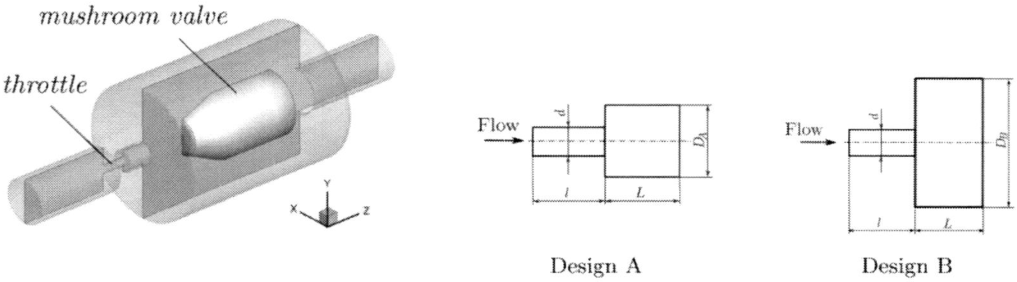


**Figure 1.** Computational model of the simplified mushroom valve chamber (left) and sketches of the utilized designs (right).

Explicit time discretization of the governing equations results in an order of magnitude $\mathcal{O}(1 \cdot 10^{-10})$ for the time step. Since the time step size is directly proportional to the smallest cell length, a grid sequencing technique is imployed to reduce the computational effort: flow field is first established on the coarser grid and successively interpolated onto the finer. In the following, results obtained on the finest grid level are presented.

## 4. Results and discussion

*Design A*

A transient behaviour of the cavitating flow in the valve chamber and throttle is observed for design A. Figure 2 depicts a series of six instants in time with a time interval $\Delta t = 1.38 \cdot 10^{-6}$ s between consecutive pictures. The entire interval $t_{1-6}$ spans $6.4 \cdot 10^{-6}$ s. At the first time instant, a cavitating sheet in the throttle separates from the stable cavity that begins from the throttle entrance. A high speed cavitating jet discharges from the throttle into the valve chamber with velocity of approximately 670 m/s. Low pressure cores in the shear layer around the high speed jet and the surrounding liquid initiate vortex cavitation and form toroidal vapor clouds. Such clouds grow until time instant (2), while the sheet cavity in the throttle continuously sheds further vapor clouds. At time instant (3), a part of the vapor cloud is convected downstream, while the other part remains entrapped by the recirculation eddy that forms within the step. At time instant (4), a supercavitating sheet that spans through the complete throttle is observed. The convected cloud separates and the chamber pressure is imposed on the vapor cloud in the step. This pressure forces the cavitating structure to recondensate and finally to collapse (frames (4)-(6)). Between time instants (5) and (6), the vapor fragment marked by a circle has collapsed, causing a significant pressure rise in the step. At time instant (6), the sheet cavity breakes up near the throttle exit and reduces in length. Once annular shaped vapor structures form in the throttle step, the cycle starts anew.

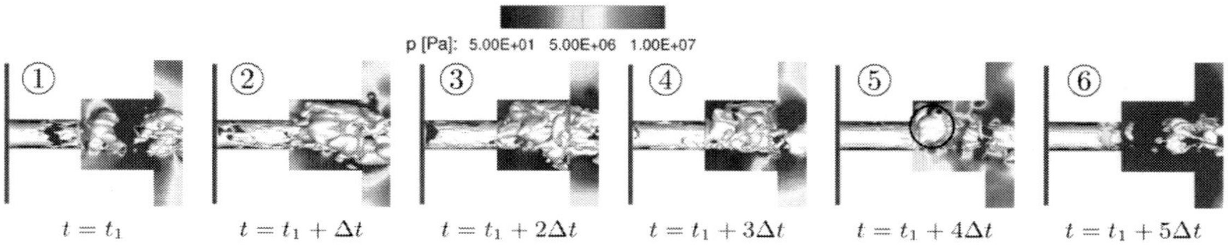


**Figure 2.** Pressure contours and isosurfaces of vapor volume fraction $\alpha = 0.1$ showing the instationary process in the throttle step. Time interval $t_{1-6} = 6.4 \cdot 10^{-6}$ s.

The onset of reentrant flow is induced before the final collapse stage of the vapor cloud. This leads us to the conclusion that not the shock wave formation after the cloud collapse, but the inertial effect of the collapsing vapor cloud triggers the beginning of the reentrant motion.

*Design B*

For this configuration, fluid from the chamber discharges into the valve chamber and two counter-rotating eddies form in the step. These vortices interact with the high speed jet. The pressure drop in the shear layer forms toroidal vapor clouds similar to those in design A. However, the highly instationary mechanism that initiates reentrant flow in the throttle differes from that in design A, since no vapor clouds are enforced by the recirculation eddies to stay in the step and to collapse there. Instead, vapor clouds partially collapse at the exit of the throttle step, emitting shock waves that lead to a pressure rise in the step. This mechanism triggers a reentrant motion in the throttle. Figure 3 depicts the process of shock wave propagation in the throttle step at four successive instants in time. The entire time interval $t_{1-4}$ is $4.66 \cdot 10^{-7}$ s long, with $\Delta t = 1.165 \cdot 10^{-7}$ s between each frame. In frame (4), an increase of the pressure in the throttle is clearly visible and can be connected directly to the collapse of the vapor fragments at the throttle exit.

In order to investigate the periodic behaviour of the cavitation flow, we analyse the mass flow at the throttle outlet for both designs, which is depicted in Fig. 4. A periodic drop in mass flow is caused by the reentrant flow at the throttle exit. For design A, frequency spectra reveal one dominant frequency $f_{1A} \approx 135$ kHz, which has also been identified by the visual inspection as the frequency of the collapsing vapor cloud in the throttle step. At least two harmonic modes

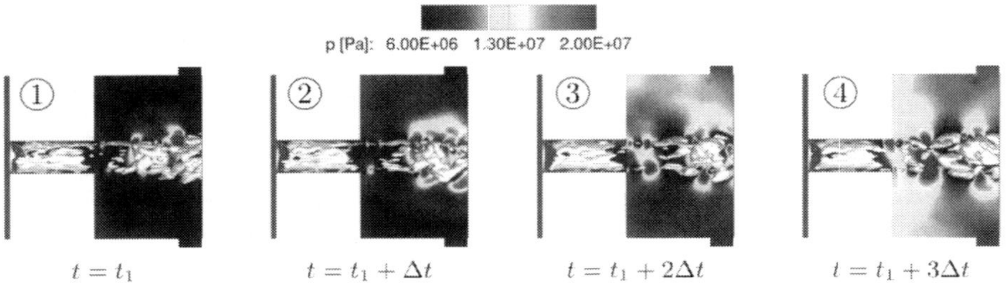


**Figure 3.** Formation of a shock wave after the collapse of a vapor fragment near the exit of the throttle step, visualized by pressure contours and isosurfaces of vapor volume fraction $\alpha = 0.1$.

are captured by the fast Fourier algorithm. Design B features also one dominant frequency, $f_{1B} \approx 190$ kHz.

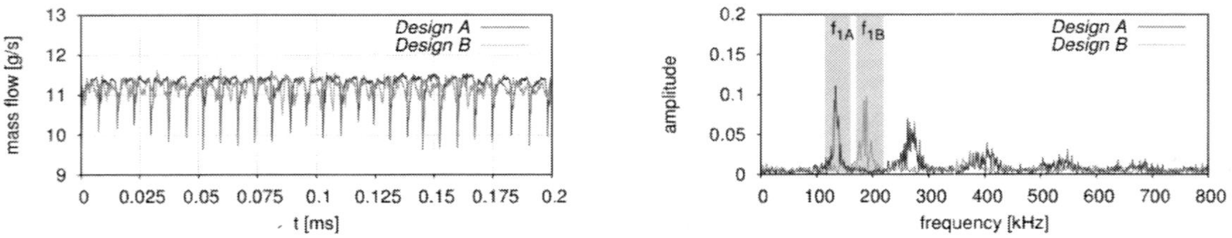


**Figure 4.** Time history of the mass flow at the exit of the discharge throttle (left); Fast Fourier analysis of the mass flows (right). Dominant frequencies highlighted with $f_{1A}$ and $f_{1B}$.

We therefore conclude that two different mechanisms triggering the reentrant motion in the discharge throttle have been detected. Owing to the shrinking vapor cloud, an 'inertial mechanism' is found for the smaller step diameter. However, for the larger step diameter, a 'shock wave mechanism' is identified due to the collapse of the vapor fragments at its exit.

## References

[1] Mauger C, Méès L, Michard M, Azouzi A and Valette S 2012 *Experiments in Fluids* **53** 1895–1913
[2] Iben U, Morozov A, Winklhofer E and Skoda R 2011 *WIMRC 3rd International Cavitation Forum 2011, University of Warwick*
[3] Iben U, Morozov A, Winklhofer E and Wolf F 2011 *Experiments in Fluids* **50** 597–611
[4] Stanley C, Barber T, Milton B and Rosengarten G 2011 *Experiments in Fluids* **51** 1189–1200
[5] Duke D, Kastengren A, Tilocco F Z and Powell C 2013 *25th Annual Conference on Liquid Atomization and Spray Systems, ILASS-Americas, Pittsburgh, PA, May* pp 5–8
[6] Schmidt S J, Sezal I H, Schnerr G H and Thalhamer M 2008 *46th AIAA Aerospace Sciences Meeting and Exhibit* pp 7–10
[7] Chorażewski M, Dergal F, Sawaya T, Mokbel I, Grolier J P E and Jose J 2013 *Fuel* **105** 440–450
[8] Ndiaye E H I, Bazile J P, Nasri D, Boned C and Daridon J L 2012 *Fuel* **98** 288–294

# Dynamic cavitation inside a high performance diesel injector – an experimental and CFD investigation

**Daniel Bush, Celia Soteriou, Mark Winterbourn & Christian Daveau**

Delphi Diesel Systems, Gillingham, UK

daniel.bush@delphi.com

**Abstract**. A combination of simulation and special experimental techniques has been used to investigate the transient flow and cavitation phenomena of a control device inside a high performance diesel injector. Dynamic cavitation behaviour was captured on a large scale transparent model, which was then used to develop and validate an advanced turbulence CFD model with Large Eddy Simulation. These techniques are used within Delphi to gain insight and optimise injector performance at real-size.

## 1. Introduction

The diesel injector is arguably the most important element in achieving premium engine performance with low emissions. It therefore follows that injectors for high performance engines must be controlled with precision. As a result, the hydraulic control valves and orifices must be designed and manufactured with high accuracy, and understanding their fundamental behaviour will provide an important contribution to the optimisation and fine-tuning process. Delphi uses a combination of simulation and special experimental techniques to investigate the flow characteristics of these devices.

As with all new injectors, a High Performance Diesel Injector under development at Delphi has been the subject of much simulation activity. Of particular interest to this paper, are the CFD investigations of the hydraulic control functions.

Flow cavitation is a familiar topic in diesel injection equipment; previous experimental experience showed that the industry standard Reynolds-Averaged Navier-Stokes (RANS) modelling was unable to give accurate results for dynamic cavitation behaviour. Hence in addition to the CFD work, large scale models (LSM) of different parts of the hydraulic control functions were commissioned. The use of transparent models of control valves and orifices at greatly increased scales enables a developed understanding of hydraulic performance [1] that is not easily achievable with other techniques. The hydraulic performance is sufficiently independent across all scales to be useful as a development tool [2].

In CFD simulation, resolving large scale turbulent structures using Large Eddy Simulation (LES) based models shows transient modelling benefit over RANS simulations [3] [4] [5]. The serious disadvantage of LES based CFD in an industrial environment is the huge computing resource required, of the order of several hundred dedicated cores. Without such a resource, the run time for a single case can take months, making serious investigations impossible in a sensible time frame.

Significant developments have been made within CFD modelling which attempt to combine the benefits of both LES and RANS turbulence modelling [6]. These models are often termed hybrid models and give greater potential for LES to be utilised within an industrial environment.

A hybrid LES-RANS CFD model, using a mixture multi-phase method for cavitation was created to simulate the performance of the control orifice. Turbulence modelling using a hybrid LES-RANS method allows significant computational gains over full LES, especially in the high Reynolds number (Re) wall-bounded flow geometries within a diesel injector.

## 2. Experimental Work

Transparent models were created of hydraulic control parts from within a diesel injector, including a high pressure control orifice, between the needle top and control valve. Due to the small scale of real injectors the model was created at 40 times real-size, as it has been shown in previous studies that orifice flow performance is reasonably independent of scale [1]. This increased scale allowed much greater ease of visualisation using a high speed camera, a Vision Research Phantom V1210. Such a camera with appropriate lighting captures detail of the flow not visible to the human eye, and allows the hydraulic behaviour to be tracked via the cavitation process.

The orifice diameter is around 100μm and tapers slightly before a larger conical section. Flow is from bottom to top through the orifice (Figure 1). Pressure measurements are taken upstream and downstream. Scaling calculations from accurate 1D simulation results allow the LSMs to be run under conditions which are equivalent to those of the real-scale component in normal operation [7]. The LSM was setup with scaled running conditions to match those of an injector operating at 1000, 1500 & 2000 bar rail pressure. As well as the large void and small bubble cavitation, as seen in previous orifice studies, the imaging revealed a new effect not previously seen, where the cavitation showed a pulsing motion along the orifice.

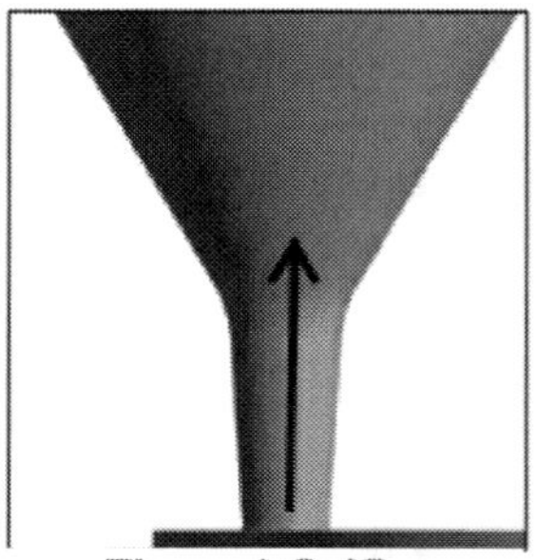

**Figure 1 Orifice geometry and flow**

Figure 2 shows a complete pulse of the cavitation in 8 images. The frequency was consistent over a significant number of events. The three injection pressures gave results which were visually similar but had different frequencies. For the 1000bar case the frequency was 160Hz and the Coefficient of Discharge calculated from the flow rate and pressure drop was 0.71.

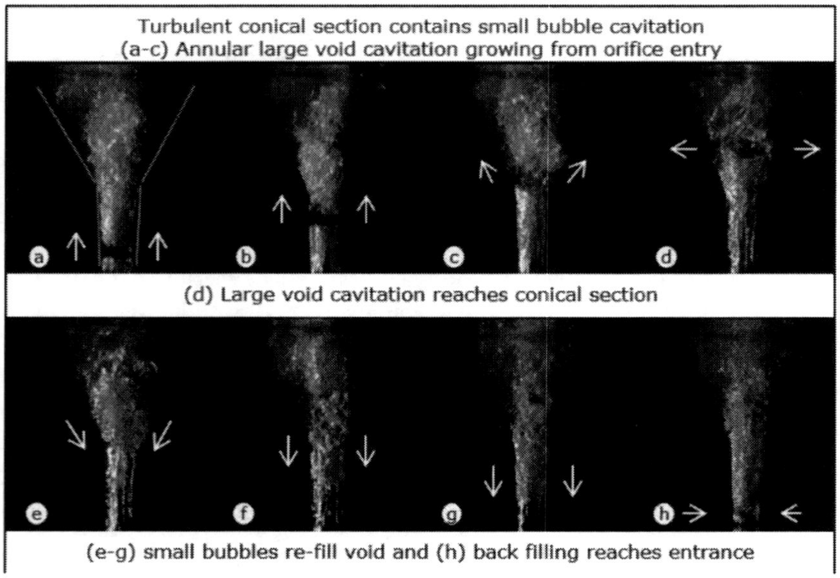


**Figure 2 Transient effect of the pulsing cavitation seen on LSM - 1000bar equivalent injection pressure**

## 3. CFD Simulation

The standard CFD modelling uses RANS turbulence modelling due to the low computational requirements and accuracy of Cd prediction. It was obvious that the RANS modelling could not predict the transient behaviour seen on the Large Scale Model from nominally steady pressure conditions.

A turbulence simulation method was chosen which combined both the LES and RANS techniques. This was set up using the commercial code ANSYS Fluent v15, with the intention of developing a simulation technique that could be used on a wide range of geometries. The hybrid LES-RANS technique was chosen to allow the model to run in RANS mode on a coarser mesh in areas of lesser interest, while still achieving the required detail within the orifice.

The 1000Bar equivalent case was simulated using the same scale pressure drop as calculated for the Large Scale Model. For the cavitation, a mixture model was used. This assumes that both liquid and vapour travel in a homogeneous multiphase flow; although not without limitations, it is a valid assumption for cavitating flows. Two incompressible phases are modelled, which is reasonable at the low pressure drops used in the large scale modelling.

The simulation of the 1000Bar equivalent injection pressure gave transient behaviour which matched the experimental results, as shown in Figure 3 with the eight experimental images from Figure 2. The frequency is 156Hz, giving excellent correlation with the experimental value of 160Hz.

Movies of both the experimental flow and the CFD results show a striking similarity.

The average value of Cd from the hybrid LES-RANS model is 0.71. This gives excellent agreement with the value measured on the Large Scale Model.

These results give confidence in the modelling and strong validation for the combination of turbulence and multiphase CFD models. Full details are given in [8].

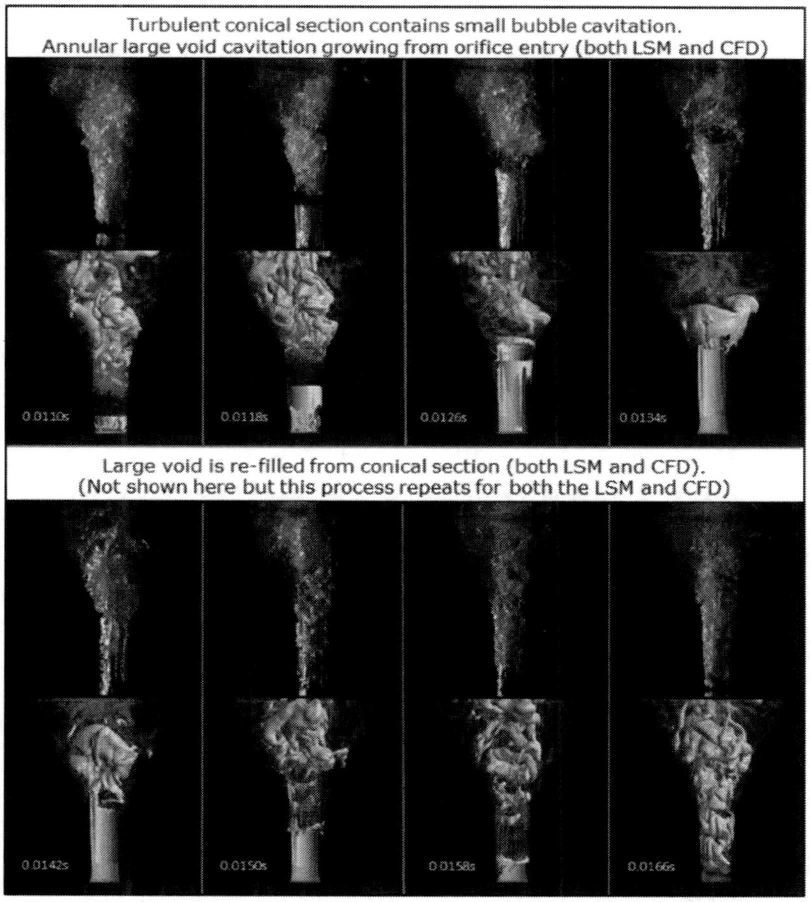


**Figure 3 CFD results correlate well with LSM experimental results. CFD (rows 2 & 4) cavitation (iso-surface of 90% liquid fraction - white) and vortex structures (iso-surface Q = 1e$^7$ – green)**

## 4. Real-size CFD Modelling

The CFD model was run for the real size case and produced similar pulsation close to 150 kHz. The expected frequency from the scaling calculations is 160 kHz, which further confirms the method.

## 5. Discussion

A realistic method has been demonstrated which enables the use of LES CFD modelling in product improvement. Some validation of the method has also been demonstrated, which is not common for LES CFD in this field.

In summary:

- High speed imaging of cavitation within a control orifice revealed a pulsing hydraulic effect not seen in previous studies. Large void cavitation was created along the orifice wall and once the conical section was reached, backfilling occurred upstream along the orifice. When the backfilling reached the orifice entrance, there was a flow structure transition, large void cavitation reformed and the process repeated
- Advanced turbulence modelling in the CFD simulations produced results that are hugely more accurate than standard models when compared with experimental results.
- The CFD matched the pulsing large void cavitation along the orifice and the back-filling event.
- The frequency of this behaviour on the CFD results also agreed with that of the experiment.
- The CFD revealed that the flow rate was also pulsating, linked with the cavitation behaviour described above. The fact that the average Cd in the CFD agreed with that measured in the experiment gave confidence that the CFD flow rate pulsation was accurate.
- Using LSMs enabled the development of the LES CFD models in a manageable time-frame. This saved considerable time in developing the models for real-size components.
- In the case of the real injector, the frequency of the pulsing cavitation was too high to influence the engine performance. The awareness of the phenomenon provided by this work will enable any potentially adverse effects to be avoided in future design levels.

## References

[1] C. Soteriou, M. Smith and R. Andrews, "Cavitation hydraulic flip and atomization in direct injection diesel sprays," in *IMechE C465/051/93*, 1993.

[2] C. Soteriou, M. Smith and R. Andrews, "Diesel injection - laser light sheet illumination of the development of cavitation in orifices," in *IMechE C529/018/98*, 1998.

[3] B. Befrui, P. Spiekermann, M. A. Shost and M.-C. Lai, "VoF-LES Studies of GDi Multi-Hole Nozzle Plume Primary Breakup and Comparison with Imaging Data," in *ILASS Europe Conference on Liquid Atomization and Spray Systems*, Chania, Greece, 2013.

[4] R. E. Bensow, "Simulation of the unsteady cavitation on the Delft Twist11 foil using RANS, DES and LES," in *Second International Symposium on Marine Propulsors*, Hamburg, Germany, 2011.

[5] C. Egerer, S. Hickel, S. Schmidt and N. Adams, "Analysis of turbulent cavitating flow in a micro-channel," in *SHF Conference on Hydraulic Machines and Cavitation / Air in Water Pipes*, Grenoble, France, 2013.

[6] M. L. Shur, P. R. Spalart, M. K. Strelets and A. K. Travin, "A hybrid RANS-LES approach with delayed-DES and wall-modelled LES capabilities," in *International Journal of Heat and Fluid Flow*, 2008.

[7] C. Arcoumanis, J. M. Nouri and R. J. Andrews, "Application of Refractive Index Matching to a Diesel Nozzle Internal Flow," in *6th International Symposium on Applications of Laser Techniques to Fluid Mechanics*, 1992.

[8] D. Bush, C. C. E.Soteriou, M. Winterbourn and C Daveau, "Investigating hydraulic control components in high performance injectors" in *Fuel Systems for IC Engines, IMechE* 2015.

# Large Eddy Simulation of a Cavitating Multiphase Flow for Liquid Injection

**M Cailloux[1], J Helie[2], J Reveillon[1] and F X Demoulin[1]**

[1]CNRS UMR 6614 CORIA, Rouen, France
[2]Continental Automotive SAS, Toulouse, France

E-mail : demoulin@coria.fr

**Abstract**. This paper presents a numerical method for modelling a compressible multiphase flow that involves phase transition between liquid and vapour in the context of gasoline injection. A discontinuous compressible two fluid mixture based on the Volume of Fluid (VOF) implementation is employed to represent the phases of liquid, vapour and air. The mass transfer between phases is modelled by standard models such as Kunz or Schnerr-Sauer but including the presence of air in the gas phase. Turbulence is modelled using a Large Eddy Simulation (LES) approach to catch instationnarities and coherent structures. Eventually the modelling approach matches favourably experimental data concerning the effect of cavitation on atomisation process.

## 1. Introduction

In internal combustion engines, injectors are used to deliver the fuel to the engine in the most efficient way in order to trigger the combustion. This mixing between air and gasoline is improved by the formation of a spray with fine droplets at very high injection pressure. The injected liquid flows through the nozzle at very high speeds, usually in the order of hundreds meters per second, it is compressible and highly transient. In addition, the strong acceleration during the passage of the fluid throughout the nozzle, generate a pressure drop that can be less than the fuel vaporization pressure, triggering cavitation. Cavitation can help in promoting the primary atomization, however it also forces the liquid inside the injector nozzle to be distributed unevenly and results in unpredictable behavior of the spray shape. Therefore in order to obtain a precise control of the injection and for the improvement of engine performance, it is crucial to study the formation and development of cavitation.

OpenFOAM® [1] libraries used in this work include originally solvers for resolving a compressible multiphase flow as well as models for cavitation such as Kunz [2] and Schnerr-Sauer [3] for instance. However, both models hold generally for a flow composed only by the couple liquid/vapour, which does not represents the reality of fuel injection, where the ambient gas is not vapour but for instance air. The aim of this work is to complete the standard OpenFOAM® libraries for simulating multiphase flows with an appropriate physical cavitation model to represent the engine injectors.

The structure of the paper is as follow: first of all the modelling approaches are described together with their implementations. Then the models are validated against Sou's experimental data [4] for a liquid-vapour flow. The final part is dedicated to the extension to a gas mixture that can include air in addition to the vapour.

## 2. Mathematical and physical model

### *2.1. Transport equations of volume fraction*

The present method includes a transport equation for the volume fractions $\alpha_i$ ("i" stands for vapour, liquid or air), defined as the ratio between fluid i volume to the total volume. $\alpha_i$ ranges between 0 and 1, where 0 means that there is no fluid i in this cell and 1 corresponds to a cell full of this fluid.

Then we have the following equations for the volume fractions:

$$\begin{aligned} \partial_t(\alpha_l\rho_l) + \nabla.(\alpha_l\rho_l U) &= \dot{\boldsymbol{m}} \\ \partial_t(\alpha_v\rho_v) + \nabla.(\alpha_v\rho_v U) &= -\dot{\boldsymbol{m}} \\ \partial_t(\alpha_{nc}\rho_{nc}) + \nabla.(\alpha_{nc}\rho_{nc} U) &= 0 \end{aligned} \quad (1)$$

The source term added in bold for liquid and vapour equations traduces the exchanges between this two fluids induced by the cavitation. This term has been added to the compressible multiphase solver "compressibleMultiphaseInterFoam" of OpenFOAM®. The subscript *nc* is used to described the air phase which is a non cavitating fluid in this flow. The subscripts *l* and *v* stand for liquid and vapour respectively, leading to $\alpha_l + \alpha_v + \alpha_{nc} = 1$.

Bulk densities for liquid, vapour and air follow a linear relation to the pressure to include compressible phenomena:

$$\rho_l = \rho_{l0} + \psi_l p\,, \quad \rho_v = \rho_{v0} + \psi_v p, \quad \rho_{nc} = \rho_{nc0} + \psi_{nc} p \quad (2)$$

$\psi_i$ represents the compressibility coefficient of specie "i". Rewriting the equations (1) by substituting the densities with equations (2) leads to:

$$\begin{aligned} \partial_t(\alpha_l) + \nabla.(\alpha_l U) &= -\frac{\alpha_l\psi_l}{\rho_l} D_t p + \frac{\dot{m}}{\rho_l} \\ \partial_t(\alpha_v) + \nabla.(\alpha_v U) &= -\frac{\alpha_v\psi_v}{\rho_v} D_t p - \frac{\dot{m}}{\rho_v} \\ \partial_t(\alpha_{nc}) + \nabla.(\alpha_{nc} U) &= -\frac{\alpha_{nc}\psi_{nc}}{\rho_{nc}} D_t p \end{aligned} \quad (3)$$

Then, by combining these equations the velocity divergence is obtained:

$$\nabla.U = -\left(\frac{\alpha_l\psi_l}{\rho_l} + \frac{\alpha_v\psi_v}{\rho_v} + \frac{\alpha_{nc}\psi_{nc}}{\rho_{nc}}\right) D_t p + \dot{m}\left(\frac{1}{\rho_l} - \frac{1}{\rho_v}\right) \quad (4)$$

### *2.2. Cavitation model*

In the literature, among existing models for the cavitation we start with approaches proposed by Kunz [2] and then Schnerr-Sauer [3] is considered (not presented in this abstract).

Kunz's approach decomposes the source term, $\dot{\boldsymbol{m}}$, into 2 parts. One part describes the liquid to vapour phase change (cavitation), which occurs when the local pressure is less than the saturation pressure: $\dot{m}^-$ . On the other hand, when the pressure is higher than the saturation pressure, the vapour to liquid phase change is activated (condensation): $\dot{m}^+$.

$$\dot{m}^- = \frac{C_{\mathrm{dest}}\rho_v\alpha_l \min(0, p - p_{\mathrm{sat}})}{(0.5\rho_l U_\infty^2)t_\infty} \text{ and } \dot{m}^+ = \frac{C_{\mathrm{prod}}\rho_g {\alpha_l}^2(1-\alpha_l)}{t_\infty} \quad (5)$$

$C_{\mathrm{prod}}$, $C_{\mathrm{dest}}$, $U_\infty$ and $t_\infty$ are constants to be specified.

## 3. Validation against Sou's experiments

### *3.1. Description of Sou's experiments*

Work of Sou [4] aims to show different cavitation regimes, based on an experimental set up. He pointed out 4 main types of cavitation, depending on the cavitation number as well as the Reynolds numbers.

The cavitation number is defined as follow, where $P_a$ is the atmospheric pressure, $P_s$ the saturation vapour pressure and $V_N$ the bulk liquid velocity in the nozzle.

$$\sigma = \frac{P_a - P_v}{0.5\rho_l V_N^2} \tag{6}$$

Sou's experiments point out that for cavitation number higher than 1.2, no cavitation bubbles are observed. Then for $0.75 \leq \sigma \leq 1.2$, cavitation cloud appeared in the upper half of the nozzle. For a cavitation number between 0.55 and 0.75, cavitation zone extended from the inlet to the injector exit. This is the super cavitation regime. In this condition, the liquid jet atomization is improved as droplets and ligaments appear quickly and the spray angle is enhanced. For lower cavitation numbers, we can observe the formation of hydraulic flip, in which air goes downstream at one edge of the nozzle up to the inlet.

### *3.2. Validation of the approach for a liquid-vapour flows*

In the injection simulations presented in this section, it is considered that the liquid is injected to the chamber which is filled only with vapour. The cavitation phenomenon starts at the entrance of the injector nozzle where the composition of the ambient gas does not play any role. The simulations of Sou's experiments, based on the previous work of Lu [5], assuming only vapour give results presented on figure 1.

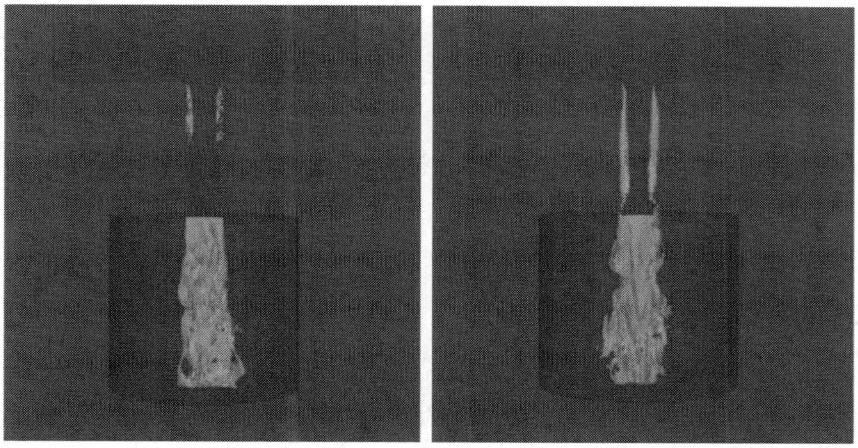

**Figure 1.** Simulations of the Sou's experiments. Cavitation number of 0.94 (left) and 0.65 (right)

The simulations are able to recover the cavitation regime proposed by Sou et al [4] more validations are proposed by Lu et al [5]. However, the ambient pressure is greater than the saturation pressure thus the model will consider condensation process acting on the spray jet because the simulated ambient gas is vapour instead of the actual air. The introduction of air in the composition of the gas phase has to be considered to avoid this bias.

## 4. Cavitation in multiphase flows

### *4.1. Method*

Generally cavitation modelling considers one specie with two phases liquid and vapour. Application to atomization requires to consider for the gas phase other species such as air. As a consequence, under the assumption of thermodynamic equilibrium at the surface, this is the partial pressure of vapour that is equal to the saturation pressure instead of the total pressure in the gas. This is included in the mass transfer source term by replacing the total pressure by the vapour partial pressure. The second issue is related to the treatment of the liquid surface if any interface capturing method is considered. Numerical method should be revisited to carefully avoid any diffusion between liquid and gas but allowing for diffusion between different species of the gas phase. Here a splitting method has been used separating the treatment of the convection term from the diffusion term in the context of a VOF approach. More details regarding this aspect will be given during the presentation due to size restriction of this abstract.

### *4.2. Results and enhancements*

The following picture shows results obtained for gas phase composed by vapour and air. A tank full of liquid at high pressure flows out to external environment composed of air. The nozzle geometry corresponds to a Sou's experiment. The Kunz's cavitation model is used in this case, see figure 2.

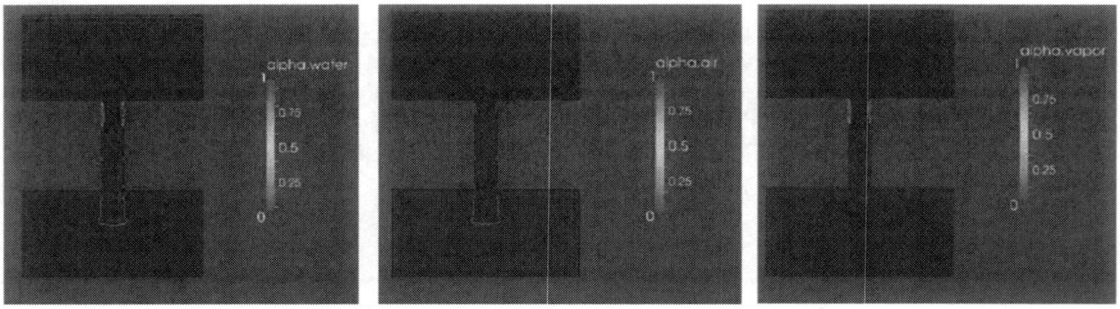


**Figure 2.** Simulation of cavitating injection: liquid, vapour and air volume fraction fields

We can see that now it is possible to have an injection of liquid into air with a production of vapour in the cavitation area. Additional comparisons between injection in vapour and in air will be presented.

## 5. Conclusions and perspectives

This work is an attempt to represent cavitation effect on atomisation processes with a solver based on interface capturing method for the two-phase flow, LES for the turbulence modelling and including the possibility various composition of the gas phase. Further improvements are expected to link the cavitation model to the vaporisation model. This should be done in particular by including the effect of the temperature. Additional improvement concerns the representation of subgrid scale wrinkling of the interface. A subgrid method should be applied, as the LES version of ELSA developed by Hecht et al [6] for the atomization.

## References

[1] www.openfoam.org

[2] Kunz R, Boger D, Stinebring D , Chyczewski T, Lindau J, Gibeling H, Venkateswaran S and Govindan T 2000 *A preconditioned Navier-Stokes method for two-phase flows with application to cavitation prediction* Computers and Fluids 29 (8): 849-875

[3] Sauer J, Winkler G and Schnerr G H 2000 *Cavitation and condensation - common aspects of physical modeling and numerical approach* Chem. Engng & Technol 23: 663-666

[4] Sou A, Hosokawa S and Tomiyama A 2007 *Effects of cavitation in a nozzle on liquid jet atomization* IJHT 50 : 3575-3582

[5] Lu N, Demoulin F X, Reveillon J and Chesnel J 2014 *Large eddy simulation of cavitation and atomization in injector flows using OpenFOAM* (Bremen : ILASS)

[6] Hecht N and Demoulin F X 2014 *Towards general purpose LES model of injection and atomization* (Bremen : ILASS)

# Cavitation Inside High-Pressure Optically Transparent Fuel Injector Nozzles

**Z. Falgout, M. Linne**

Department of Applied Mechanics, Chalmers University
Gothenburg, SE, 41296

Email: falgout@chalmers.se

**Abstract**. Nozzle-orifice flow and cavitation have an important effect on primary breakup of sprays. For this reason, a number of studies in recent years have used injectors with optically transparent nozzles so that orifice flow cavitation can be examined directly. Many of these studies use injection pressures scaled down from realistic injection pressures used in modern fuel injectors, and so the geometry must be scaled up so that the Reynolds number can be matched with the industrial applications of interest. A relatively small number of studies have shown results at or near the injection pressures used in real systems. Unfortunately, neither the specifics of the design of the optical nozzle nor the design methodology used is explained in detail in these papers. Here, a methodology demonstrating how to prevent failure of a finished design made from commonly used optically transparent materials will be explained in detail, and a description of a new design for transparent nozzles which minimizes size and cost will be shown. The design methodology combines Finite Element Analysis with relevant materials science to evaluate the potential for failure of the finished assembly. Finally, test results imaging a cavitating flow at elevated pressures are presented.

## 1. Introduction

Elevated injection pressures and microscopic nozzle orifices are used in injection systems in modern direct-injection engines to enhance atomization and deliver sufficient fuel mass quickly. The flow inside the injector must turn abruptly to enter the nozzle orifice, which in many cases causes cavitation to appear on the downstream side of the orifice inlet corner. Cavitation is generally considered a hazard, because collapsing cavitation bubbles can damage the surfaces of durable metals such as hardened steels. In addition, while some forms of cavitation have been shown to be beneficial for atomization, even removing carbon deposits in gasoline injectors, hydraulic flip suppresses atomization completely[1]. Until cavitation is better understood and easier to control, it seems logical to design the injector nozzle in such a way as to prevent cavitation.

The most effective way to study nozzle orifice flow cavitation is to construct an injector with an optically accessible nozzle so that the orifice flow can be visualized directly. This is not a simple task. Optically transparent engineering materials are most often brittle at room temperature. While brittle materials often have higher elastic moduli and stronger theoretical strengths, flaws left from the processing required to create the desired geometry can significantly reduce the load capacity of the finished component.

An important issue with optical tips is to find a match between the indices of refraction for the nozzle material and the fluid used. If they are not matched, then the edges of the passage will appear dark in a shadowgram. That would make it difficult to image wall cavitation. Acrylic and quartz have an index that comes close to that of most fuels (and water) but unfortunately they have low tensile strength due to processing flaws. They usually break well below realistic fuel pressures. Sapphire, on the other hand, has thermal and strength properties similar to steel, but it is poorly index matched to the fluids of interest.

Optically accessible (OA) injector nozzles have been used in several past studies of interior flows of injector nozzle geometries found in direct-injection compression-ignition engines. Many of these past studies were performed at injection pressures significantly lower than those of the related industrial injector, due to the challenge of designing an OA injector which can withstand high injection pressures. To compensate, some used scaled-up geometries to match Reynolds number. Studies which attempted to reach realistic injection pressures will be summarized here, since these are the most relevant to the design presented in this paper, and ultimately are the most accurate when it comes to reproducing the flow of interest.

Journal of Physics: Conference Series **656** (2015) 012082 doi:10.1088/1742-6596/656/1/012082

Arcoumanis et al. examined the effect of scaling geometries while matching Reynolds and Cavitation numbers in the real-size and scaled up geometry[2]. They found that the cavitation structures were different in the two geometries at similar cavitation numbers. Both geometry scales showed geometric cavitation on the downstream side of the upper orifice edge(with the injector pointing downward), and string cavitation existing on the opposite side of the orifice hole. The main difference observed by the authors was attributed to the difference in bubble residence time in the flow due to the different orifice lengths. The large scale injector was made of acrylic glass, while the real scale injector was made of a quartz rod attached to a Bosch common-rail injector body and surrounded by an acrylic support gallery. The acrylic piece was significantly larger than the injector's original steel nozzle. Reid et al. used sapphire to create a simplified version of the geometry found in valve-covered-orifice (VCO) injectors[3]. They achieved injection pressures of 2000 bar with the design. It seems that the orifice was constructed by layering sapphire plates in the axial direction with a series of thru holes which formed the final geometry. This was a clever way to simplify the processing of the parts. How the device was sealed, and how the plates were held in place, was not included in the published results. Two forms of string cavitation were observed: one which formed in the vortices attached to the part of the geometry designed to mimic the needle, and another which formed in a vortex that connected two orifice holes.

Badock et al. achieved elevated injection pressures using acrylic nozzles, although much like Arcoumanis et al, the piece was significantly larger than the original injector nozzle[4]. Studying primary breakup of a spray in such a nozzle is impossible due to the lensing effect of the surface of the drilled reliefs which fix the length of the nozzle orifice passages. Blessing et al. achieved injection pressures of 800 bar using an acrylic injector nozzle with a single off-axis orifice[5]. The design was compact enough to allow the primary breakup of the spray emitting from the nozzle orifice to be studied simultaneously with the in-nozzle orifice cavitation. Unfortunately, no further information regarding the design, such as dimensions or sealing mechanism, was provided. Another effort which deserves mention is the recent study by Butcher et al[6]. The group constructed a quartz nozzle in order to study flash boiling sprays relevant to spark-ignition engines and even installed a modified needle to reduce the stress concentration in the nozzle from the force of the needle seating. Their injection pressures were limited to 40 bar.

The most recent study that achieved elevated injection pressures is that of Mitroglou et al[7]. The authors examined the cavitation structures inside an optically transparent VCO geometry with six orifice holes. They reached 400 bar, but due to slow flaw growth and the desire to collect enough data to build ensemble averaged images of cavitation structures, the injection pressure was limited to 300 bar. Cavitation was observed on both sides of the orifice, as in the earlier study of Arcoumanis. In this study, string cavitation was observed only during the opening phase of the needle valve, and corresponded to the location of a vortex attached to the injector needle face.

To sum up, the three most common materials used in OA injector nozzle designs are acrylic glass, sapphire, and quartz/fused silica. There are more studies which successfully reach elevated pressures with acrylic than the other materials. In most of these studies, a large piece was used and that made it difficult, if not impossible, to simultaneously visualize the spray formation region and the orifice flow. This was necessary to reach elevated injection pressures with these designs.

## 2. Injector Design

The real injector has orifices approx. 0.75 mm in diameter, which have a large angle of separation between their central axes and the central axis of the injector. The engine belonging to the real injector is a two-stroke marine diesel engine, with a large exhaust valve occupying the center of the cylinder. This causes the injectors to be mounted in the periphery of the cylinder head, and as a result, the sprays are aimed inwards toward the middle of the cylinder. The orifices in the real injector are grouped in a line and oriented so that the sprays create a fan-like pattern. Unlike most road vehicle Diesel engine fuel injectors, these injectors use a hollow stop shaft to control injection, which has radial holes that allow the fuel to through its hollow center and enter the sac volume.

The transparent injector used in this study was half-scale. The injector had only two orifices to simplify observation of primary breakup of the isolated sprays and to allow for observation of potential cavitation between the two orifices. During the design process a commercial Finite Element Analysis (FEA) tool was used to estimate the stresses in the finished design. The nozzle has a symmetry plane that passes through the central axes of the two orifice holes, and this was used to reduce the computational domain in the FEA's. For the analysis, the injection pressure of the real device (800 bar) was applied as a boundary condition to the interior surfaces of the acrylic nozzle piece. Given that brittle materials most often fail in tension, sapphire clamps were added to the design to compress the nozzle from the outside and minimize the tensile forces at the inner surfaces. Since the clamps were on the outside of the nozzle, an optically transparent material had to be used for their construction. Sapphire was chosen due to its high durability. In a sense, this made the sapphire clamps the load bearing components of the nozzle, even though the piece to which the pressure was applied directly was made of acrylic.

The final design with 3 sapphire clamps, fabricated and assembled, is shown in Figure 1. The outer dimensions of the nozzle are 21.5 mm in the viewing direction in Figure 1 and 6.75mm in the vertical direction. O-ring seals are used on either side of the nozzle.

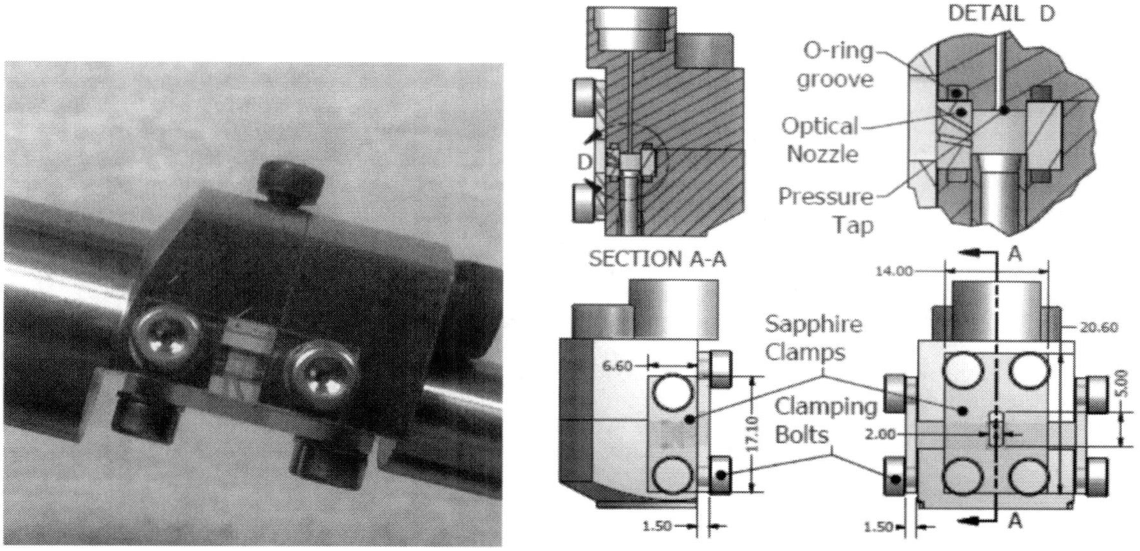


**Figure 1:** Design for high-pressure optically accessible injector nozzle with 3 sapphire clamps.

## 3. Results and Discussion

This design was tested with diesel fuel supplied by a commercial reciprocating piston pump with 3 stages, driven by a large electric motor. The original injection duration was set to 250ms, but the sac volume pressure measurements indicated that the pump was not able to sustain the commanded pressure for the entire period. The resulting injection period was approximately 100ms, and the injection pressure tapered off slowly during this period.

**Figure 2:** Representative still shot of the flow during steady state in the optically accessible nozzle when the injection pressure is at 350 bar.

The flow inside the nozzle was captured in a transillumination configuration with a Phantom V1210 high-speed camera. Because the ambient was at atmospheric back pressure, the orifice flow cavitated easily. A representative image with an injection pressure of 350 bar is shown in Figure 2. A full video of the interior flow as well as of the orifice flow and spray formation region at this flow condition has also been provided. This flow condition is at a Reynolds number of approximately 85,000. During the entire injection period, wall cavities appear on the side of the orifices furthest away from the injector body, where the inlet corner is sharper. The cavities extend almost all the way to the orifice exits, where they shed clouds that perhaps survive outside the injector body. No evidence of string cavitation in the sac volume can be seen, although vorticity is evident in the high-speed videos. Even without string cavitation, transient cavitation cloud shedding appears on the side of the orifice opposite the standing wall cavity in the orifice with the larger inclination angle from the central axis of the injector. The standing wall cavity in this orifice exists along the entire orifice inlet edge, so perhaps these clouds are shed from the wall cavity, in which case they would be a result of geometric cavitation. An important difference between this injector and other injectors in past studies is the absence of a needle. The stationary stop shaft is hollow, and so the largest vortex in the sac volume actually exists at the farthest axial position in the sac volume, where the flow turns back upstream to enter the nozzle orifices. String cavitation between the orifices was not observed in these studies.

When the nozzle failed, the sac volume pressure would drop suddenly, and this was seen in the sac volume pressure measurement. The ultimate pressure that the final design could withstand was 600 bar. While this injection pressure isn't as high as some past studies, the nozzle orifices of this design are unusually large.

## 4. Conclusions

An optically transparent fuel injector nozzle was designed using an approach which evaluated more carefully the nozzle's propensity for failure due to the materials used in its construction. This approach considered the most common failure mechanisms for brittle materials, and used Finite Element Analysis to resolve the operating loads on the nozzle in detail. The design used sapphire clamps with simple geometries to compress an acrylic nozzle piece in order to minimize the tensile stresses on the inner surfaces of the piece where the pressure from the fuel was applied. For a given pressure capacity, this design is less expensive than one constructed exclusively from sapphire and more durable than one constructed exclusively from acrylic.

The cavitation in the orifices showed dependence on geometry, specifically on the inclination angle of the central axis of the orifice. While cavitation was found on both side of the orifice with the larger inclination angle, it was not due to string cavitation, which is what cavitation in this area has been most commonly attributed to in past studies.

Future work will remove the clearance holes from the sapphire clamps, and instead use simpler sapphire pieces mounted inside stainless steel clamps. This change is expected to reduce the stress concentration in the sapphire pieces and allow greater clamping forces than were possible with the previous design. The new pieces will be tested to confirm their performance and then experiments will be conducted over a range of back pressures (cavitation umbers) in a pressurized chamber. This will produce a range of cavitating flows for further study.

## Acknowledgements

The authors would like to thank the EU for supporting this work under the HERCULES-C project, and also to thank the Swedish Energy Agency for their support.

## References

[1] Soteriou, C., Andrews, R., and Smith, M., 1999, "Further Studies of Cavitation and Atomization in Diesel Injection," (724).

[2] Arcoumanis, C., Flora, H., and Gavaises, M., 2000, "Cavitation in Real-Size Multi-Hole Diesel Injector Nozzles," (724).

[3] Reid, B. a., Hargrave, G. K., Garner, C. P., and Wigley, G., 2010, "An investigation of string cavitation in a true-scale fuel injector flow geometry at high pressure," Phys. Fluids, **22**(3), p. 031703.

[4] Badock, C., Wirth, R., Fath, A., and Leipertz, A., 1999, "Investigation of cavitation in real size diesel injection nozzles," Int. J. Heat Fluid Flow, **20**(5), pp. 538–544.

[5] Blessing, M., König, G., Krüger, C., Michels, U., and Schwarz, V., 2003, "Analysis of Flow and Cavitation Phenomena in Diesel Injection Nozzles and I ts Effects on Spray and Mixture Formation."

[6] Butcher, a. J., Aleiferis, P. G., and Richardson, D., 2013, "Development of a real-size optical injector nozzle for studies of cavitation, spray formation and flash-boiling at conditions relevant to direct-injection spark-ignition engines," Int. J. Engine Res., **14**(6), pp. 557–577.

[7] Mitroglou, N., McLorn, M., Gavaises, M., Soteriou, C., and Winterbourne, M., 2014, "Instantaneous and ensemble average cavitation structures in Diesel micro-channel flow orifices," Fuel, **116**, pp. 736–742.

# Evaluation of friction heating in cavitating high pressure Diesel injector nozzles

**R. Salemi[2], P. Koukouvinis[1], G. Strotos[1], R. McDavid[3] , Lifeng Wang[3], Jason Li[3], M. Marengo[4,5] and M Gavaises[1,*]**

[1] City University London, Northampton Square EC1V 0HB, United Kingdom
[2] Caterpillar UK Engines Co. Ltd, Peterborough, PE1 5FQ U
[3] Caterpillar Inc., Mossville, IL 61552, USA
[4]Dept. of Engineering and Applied Sciences, University of Bergamo, Viale Marconi 5, 24044 Dalmine (BG)
[5]School of Computing, Engineering and Mathematics, University of Brighton, Cockcroft Building, Lewes Road, BN2 4GJ, Brighton, UK

m.gavaises@city.ac.uk

**Abstract.** Variation of fuel properties occurring during extreme fuel pressurisation in Diesel fuel injectors relative to those under atmospheric pressure and room temperature conditions may affect significantly fuel delivery, fuel injection temperature, injector durability and thus engine performance. Indicative results of flow simulations during the full injection event of a Diesel injector are presented. In addition to the Navier-Stokes equations, the enthalpy conservation equation is considered for predicting the fuel temperature. Cavitation is simulated using an Eulerian-Lagrangian cavitation model fully coupled with the flow equations. Compressible bubble dynamics based on the R-P equation also consider thermal effects. Variable fuel properties function of the local pressure and temperature are taken from literature and correspond to a reference so-called summer Diesel fuel. Fuel pressurisation up to 3000bar pressure is considered while various wall temperature boundary conditions are tested in order to compare their effect relative to those of the fuel heating caused during the depressurisation of the fuel as it passes through the injection orifices. The results indicate formation of strong temperature gradients inside the fuel injector while heating resulting from the extreme friction may result to local temperatures above the fuel's boiling point. Predictions indicate bulk fuel temperature increase of more than 100°C during the opening phase of the needle valve. Overall, it is concluded that such effects are significant for the injector performance and should be considered in relevant simulation tools.

## 1. Introduction

Cavitation phenomena are known to affect the injection performance and nozzle durability [1]. High injection pressures reaching 3000bar are needed for more efficient heavy duty Diesel engines. Computational methodologies have a strong potential in understanding the nature of cavitation under real operating conditions. Several methodologies have been proposed such as treating the cavitating fluid as a single mixture [2], the two-fluid method in which conservation equations are solved for both

* Corresponding author.

phases separately [3], bubble based cavitation models [4], modified VOF methodologies [5], models which account for compressibility effects [6, 7], homogeneous equilibrium models which estimate the cavitation growth by using a barotropic equation [8], as also Eulerian-Lagrangian models which implement the full Rayleigh-Plesset equation including thermal effects [9]. The challenging task of quantifying the effect of needle motion in fuel injectors has attracted interest and various computational techniques have been used to simulate the needle movement. Due to the small timescales involved with the whole phenomenon, the majority of the aforementioned studies are based on the assumption of isothermal flow, but under extreme pressurisation conditions, the flow is characterized by strong friction forces which induce fuel heating and this may even lead to fuel boiling. The present work aims to shed light on these thermal phenomena occurring in high pressure Diesel nozzles by solving the energy equation and including the heating induced by friction forces. In the following sections, the simulation model and the results obtained are presented.

## 2. Numerical model and fuel properties

The flow solver used has been developed by the authors' group and solves the Navier-Stokes equations in an unstructured mesh. With regards to cavitation, the Eulerian-Lagrangian model used is based on the CFD code developed by [10]; this work is an extension of [11, 12], quantifying the effect of the motion of the injector's needle valve. The presence of the cavitating phase is taken into account through the calculation of the liquid volume fraction in a computational cell; the corresponding source terms that can be found in [13] and account for the interaction between the two phases. These source terms include the energy exchange due to mass transfer, the interfacial heat transfer and the work of viscous interfacial forces. The turbulent viscosity $\mu_{turb}$ is calculated from the k-ε turbulence model and the turbulent Prandtl number $Pr_{turb}$, is taken equal to 0.85. It has to be noted that the solved equations contain both the reversible and the irreversible work of viscous forces; the latter is commonly known as viscous heating and represents the heating induced by the friction forces. The fuel used is the so-called "summer diesel" and its properties were taken from [14] as function of temperature and pressure. For the purposes of the present work, they have been extrapolated up to 3000bar and 400°C for all cases simulated. The extrapolation method adopted, uses the functions given in [14] but extents the limits of pressure and temperature up to the point at which the property under consideration reaches a local minimum or maximum; beyond this point, each property is assumed to be equal to the corresponding value of the local minimum or maximum.

## 3. Diesel injector geometry and simulation results

### *3.1. Injector geometry*

The geometry has been provided by Caterpillar; it is a 5-hole tapered injector (see Fig. 1a). The computational domain was split in a set of moving, deforming and stationary zones. The current production injector operates at an average inlet pressure level of ~2000bar and outlet pressure of 50bar; however, as industrial developments target much higher pressures, simulations have been performed for pressures as high as 3000bar. The pressure pulse and the corresponding needle motion were calculated using 1-D tools (Fig 1b) and were used as boundary conditions for the simulation. The needle motion (see Fig. 1b) is assumed to be in the axial direction only, so any eccentricity effects were omitted. In reality eccentricity effects are significant during the early opening and late closing phases, but introduction of such effects will impose the simulation of the complete injector geometry, which will be much more complicated. In the current simulation, only $1/5^{th}$ of the domain was solved and periodic boundary conditions have been employed at the sides of the domain. The computational mesh used is block-structured, with the exception of a zone in the sac prior to the orifice entrance, which is unstructured tetrahedral. Inflation layers were introduced in the area of interest, such as sac and orifice to capture the boundary layers. Mesh motion was introduced by a smoothing algorithm at low lifts and the mesh layer addition at higher lifts. The mesh size was approximately 1 million cells at the beginning of the simulation; the needle motion introduces additional cell layers inside the volume denoted with blue colour in fig. 1a, so the actual cell count increases over time.

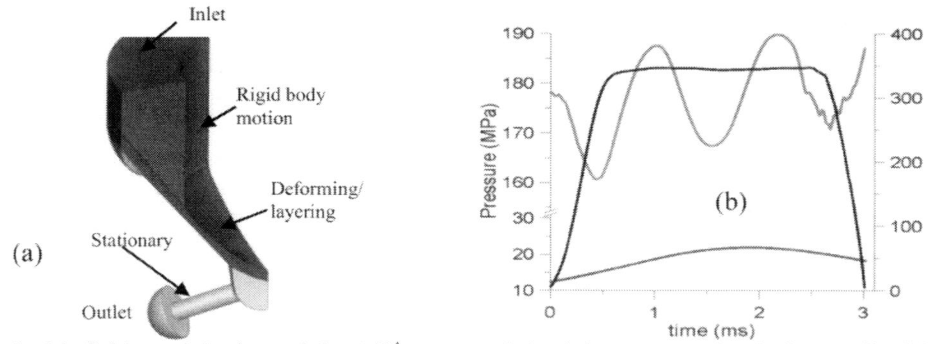


Fig. 1. (a) Grid particioning of the 1/5$^{th}$ sector of the injector examined, (b) needle lift, injection and back pressure profiles considered in the simulations

### *3.2. Simulation results*

Moving needle simulations which resemble a realistic injection event are examined in this section. In these cases the lift law plays important role, due to the timescales involved. Generally an injection event has a small duration and the flow may not reach a quasi-steady state, while the closing phase is usually much shorter than the opening phase. As a result, different behaviour of the fuel heating is observed in the opening and closing phase. In Fig. 2a the fuel heating for two inlet pressures (2000 and 3000bar); both single-phase and two-phase flow simulations are presented. In Fig. 2b the fuel heating and the nozzle's discharge coefficient are presented for both cases.

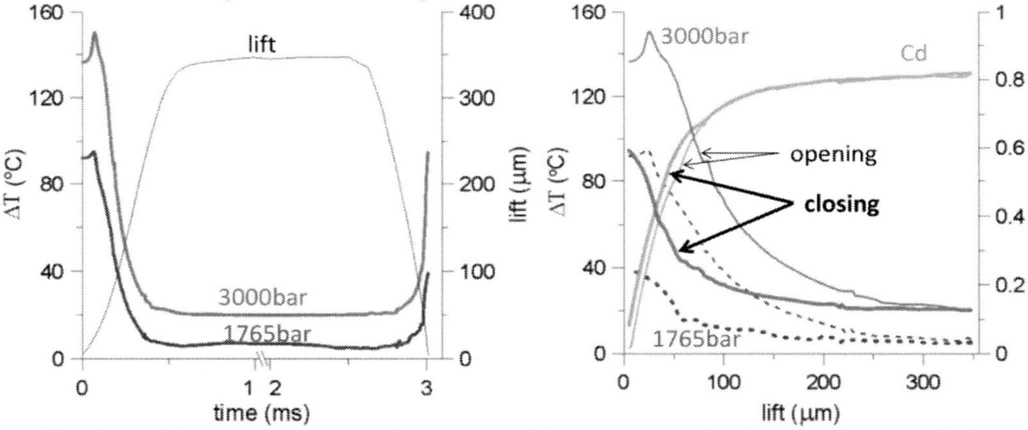


Fig. 2. (a) Fuel heating versus time for two inlet pressures. (b) Fuel heating and discharge coefficient versus lift

The nozzle's discharge coefficient shown in Fig. 2b varies at low lifts, while it approaches an asymptotic value for lifts higher than 80μm. Furthermore, at lifts lower than 80μm the amount of liquid injected in the opening and the closing phases is different. This hysteresis is ought to the presence of the sac volume. In the opening phase mass is accumulated to the nozzle's sac volume, while in the closing phase the already accumulated mass is pushed from the downwards needle motion towards the nozzle hole. Thus, a higher flow rate is observed in the closing phase. Regarding the fuel heating, the fuel is abruptly heated in the opening phase reaching its maximum value at 20μm and then the fuel is gradually cooled tending to reach the steady state heating predicted from the constant lift simulations. During the needle's closing phase, the fuel starts to heat-up again but lower values are predicted compared to those of the opening phase. This is attributed to the fact that closing is faster than opening and the flow has not enough time to reach a "steady" state, as also to the enhanced cavitation during the closing phase. In Fig. 2a the fuel heating for the two inlet pressures simulated is also presented. It is obvious that increasing inlet pressure results in more heating. The effect of cavitation on the mean liquid temperature becomes appreciable for needle lift values exceeding 40μm. This was deduced by performing single-phase simulations at the same conditions (not shown here) and it is attributed to the reduction of viscous forces due to the presence of the bubbles [12].

## 4. Conclusion

The thermal phenomena appearing in the flow inside a high pressure Diesel injector nozzle have been studied with a CFD model. The model includes variable fuel properties as a function of pressure and temperature. A significant temperature increase is predicted with increasing injection pressure. The moving needle simulations are shedding light on the differences observed in opening and closing phase respectively. The temperature field is affected by the needle motion, while two-phase flow simulations show decreased mass flow rate and reduced fuel heating.

## Acknowledgements

The research leading to these results has received funding from the People Programme (IAPP Marie Curie Actions) of the European Union's Seventh Framework Programme FP7/2007-2013/ under REA grant agreement n. 324313.

## References

1. Arcoumanis, C., et al., *Cavitation in Real-Size Multi-Hole Diesel Injector Nozzles.* SAE Paper 2000-01-1249, 2000.
2. Chen, Y.L. and S.D. Heister, *Two-Phase Modeling of Cavitated Flows.* Computers & Fluids, 1995. **24**(7): p. 799-809.
3. Alajbegovic, A., H.A. Grogger, and H. Philipp. *Calculation of Transient Cavitation in Nozzle Using the Two-Fluid Model.* in *Proc. 12th ILASS.* 1999. Indianapolis, Indiana, USA.
4. Yuan, W. and G.H. Schnerr. *Cavitation in Injection Nozzles - Effect of Injection Pressure Fluctuations.* in *Proc. CAV2001, Fourth International Symposium on Cavitation.* 2001. Pasadena, California, USA.
5. Singhal, A.K., et al. *Mathematical basis and validation of the full cavitation model.* in *Proc. of FEDSM'01 - 2001 ASME Fluid Engineering Division Summer Meeting.* 2001. New Orleans, Louisiana, USA.
6. Ando, K., T. Colonius, and C. Brennen, *Numerical simulation of shock propagation in a polydisperse bubbly liquid.* International Journal of Multiphase Flow, 2011. **37**: p. 596–608.
7. Zeravcic, Z., D. Lohse, and W. Saarloos, *Collective oscillations in bubble clouds.* Journal of Fluid Mechanics, 2011. **680**: p. 114–149.
8. Salvador, F.J., et al., *Computational study of the cavitation phenomenon and its interaction with the turbulence developed in diesel injector nozzles by Large Eddy Simulation (LES).* Mathematical and Computer Modelling, 2013. **57**(7–8): p. 1656-1662.
9. Keller, J. and M. Miksis, *Bubble oscillations of large amplitude.* Journal of the Acoustical Society of America, 1980. **68**(2): p. 628–633.
10. Giannadakis, E., M. Gavaises, and C. Arcoumanis, *Modelling of cavitation in diesel injector nozzles.* Journal of Fluid Mechanics, 2008. **616**(1): p. 153–193.
11. Theodorakakos, A., et al., *Friction-induced heating in nozzle hole micro-channels under extreme fuel pressurisation.* FUEL, 2014. **123**: p. 143-150.
12. Strotos, G., et al., *Transient heating effects in high pressure Diesel injector nozzles.* Int J heat and Fluid Flow 2015. **51**: p. 257–267.
13. Städtke, *Gasdynamic Aspects of Two-Phase Flow.* 2007: Wiley-VCH Verlag GmbH and Co. KGaA.
14. Kolev, N., *Multiphase Flow Dynamics 3: Turbulence, Gas Absorption and Release, Diesel Fuel Properties.* 2002: Springer Verlag Berlin Heidelberg.

# A comprehensive Two-Fluid Model for Cavitation and Primary Atomization Modelling of liquid jets - *Application to a large marine Diesel injector*

**Chawki Habchi[1,2], Julien Bohbot[1,2], Andreas Schmid[3] and Kai Herrmann[4]**

[1] IFP Energies nouvelles,1 et 4 avenue de Bois-Préau, 92852 Rueil-Malmaison, France
[2] Institut Carnot IFPEN Transports Energie
[3] Winterthur Gas & Diesel Ltd., CH-8401 Winterthur, Switzerland
[4] Paul Scherrer Institute, CH-5232 Villigen PSI, Switzerland

E-mail: Chawki.HABCHI@ifpen

**Abstract.** In this paper, a comprehensive two-fluid model is suggested in order to compute the in-nozzle cavitating flow and the primary atomization of liquid jets, simultaneously. This model has been applied to the computation of a typical large marine Diesel injector. The numerical results have shown a strong correlation between the in-nozzle cavitating flow and the ensuing spray orientation and atomization. Indeed, the results have confirmed the existence of an off-axis liquid core. This asymmetry is likely to be at the origin of the spray deviation observed experimentally. In addition, the primary atomization begins very close to the orifice exit as in the experiments, and the smallest droplets are generated due to cavitation pocket shape oscillations located at the same side, inside the orifice.

## 1. Introduction

Efficient transportation engines are required worldwide in order to decrease greenhouse emissions due to mobility not only by automobiles and airplanes, but also by ships. High-fidelity experimental data together with reliable two-phase flow simulation tools can help us better understand the physics governing the pollutants formation and consequently design more efficient engines in the future. Pollutant features and amounts from engines combustion are known to be influenced by in-nozzle flow and spray characteristics. But, currently, Computation Fluid Dynamic's (CFD) software cannot properly predict this key behaviour because of two main reasons: (1) internal injector scales can differ from internal combustion scales by three orders of magnitude; (2) a lack of understanding of the different physical phenomena (cavitation, primary atomization ...) and therefore their modelling. Some recent studies have focused on coupling the nozzle flow and ensuing spray in a "static" fashion [1] [2] [3]. In a static coupling approach the nozzle flow and spray simulations are performed separately. The nozzle flow simulations provide the boundary conditions at the nozzle exit to initialize a classical Eulerian-Lagrangian spray calculation. This kind of adhoc coupling at the nozzle exit boundary is not predictive in nature. Also, not all flow features can be captured in a static coupling approach since the spray results do not feedback to affect the nozzle flow development. In addition, transient injection periods (such as the opening and closing periods) can hardly be modelled using such static approach.

[1] Corresponding author (Dr C. Habchi, IFPEN).

Journal of Physics: Conference Series **656** (2015) 012084 doi:10.1088/1742-6596/656/1/012084

Besides, high spatial accuracy in the spray can be ensured if the simulation of the dense region is performed using an Eulerian-Eulerian (EE) approach rather than the classical Eulerian-Lagrangian (EL) framework. Indeed, in the near nozzle region, the proportion of liquid phase is very high and there is experimental evidence that there are very few discrete droplets in this region. The liquid phase may also comprise of voids or bubbles due to absorption of the ambient gas and cavitation from inside the nozzle. In this case, the liquid will be the continuous phase and void fractions would be the discrete phase. For these reasons, a comprehensive two-fluid model has been recently developed at IFPEN. In this work, this model is applied to the numerical simulation of a typical large marine injector, for which experimental results are available [4] [5] [6]. The results of these high fidelity simulations are presented. In the following section, the main sub-models will be described briefly since they are available elsewhere [7] [8]. Next, the computational configuration of the nozzle will be defined. Then, the results discussion section presents the main findings of the study, before the conclusions.

## 2. Description of the Two-fluid model

In this work, a comprehensive highly compressible two-fluid multi-species model is used. This model has been recently developed in order to compute in-nozzle cavitating flow and the primary atomization of actual liquid jets, simultaneously. It involves an equation for the transport of the liquid volume fraction in addition to two different sets of partial differential equations for the gas and the liquid phase. The multicomponent gas species phase is governed by an ideal gas equation of state (EOS) while the stiffened gas EOS is specified to the single-component liquid phase as a first attempt to compute highly compressible flows using a thermodynamic consistent model. Instead of the Reynolds Averaged Navier-Stokes (RANS) formulation adopted in [7] and [8], a standard Smagorinsky model is adopted for sub-grid scale turbulence modelling. Besides, the Gibbs free energy relaxation model (*GERM*) cavitation model described in detail in [7] is used. Unlike previous models, the *GERM* model may simulate two different cavitation regimes. The first so-called "gaseous cavitation" may appear in regions in which the static pressure is close to but above the liquid saturation pressure. In this case, not only the disequilibrium of pressure but also the disequilibrium of temperature between the liquid and the gas are responsible for the growth or collapse of the bubbles without the help of any phase change. The second cavitation regime may happen when the static pressure goes below the liquid saturation pressure. In this case, the liquid becomes superheated and leads to a "vaporous cavitation" regime. In addition, the *GERM* model is able to reproduce the appearance and collapse of cavitation without using adjustable parameters. Present simulations also adopt the *TwoSD* atomization model developed recently by Devassy et al. [8]. This model uses a two-surface density approach within the framework of the same Eulerian two-phase approach of the *GERM* cavitation model. The method followed distinguishes the primary and secondary atomization processes using separate equations of surface density: One for the liquid core and the other for the dispersed phase (the spray droplets). As the phenomenon of primary atomization is different from droplet breakup, a two surface density model may facilitate the modelling stages for both the droplets and liquid core. Thereby, by separating the liquid phase into two sub-phases (droplets and liquid core); it is possible to deal with the atomization and breakup phenomena more precisely. For instance, with this approach the droplets generated due to the primary atomization and secondary breakup can be filtered and studied separately for determining the droplets size PDFs and can later be vaporized using appropriate evaporation models.

## 3. Computational configuration

In this study, a simplified generic single-hole nozzle is used (Figure 1) [6]. This specimen has been designed in order to mimic the injection process from a typical five holes nozzle of large marine Diesel engines [4] [5]. It consisted of an elongated tip with a single injection hole. The direction of the fuel flow, the location of the spray orifice and the resulting main injection direction are indicated on the photograph of injection configuration inside the spray combustion chamber (SCC) on left of Figure 1. Two cross-sections (in the middle and on the right side of Figure 1) show the details of the nozzle configuration in lateral (A-A) and axial (B-B) direction of the nozzle bore. Important to mention here is the role of the drilled hole with diameter **D**, which bypasses the flow of four additional orifices – as they would exist in a serial injector – and has hence four times the area of the spray orifice with diameter **d**. This second hole leads the bypassed fuel amount into the discharge fuel system as indicated in Figure 1 (left). More details may be found in [9].

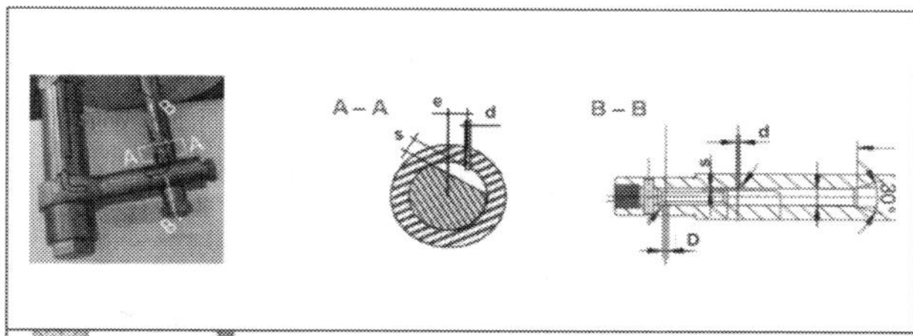


**Figure 1.** Generic injector configuration. Photograph inside the SCC (left). Cross-section A-A through the injection orifice (d=0.75 mm and e=1.5 mm for the eccentric configuration) (middle). Cross-section B-B through the injector bore with s=0.75 mm and D=1.3 mm (right).

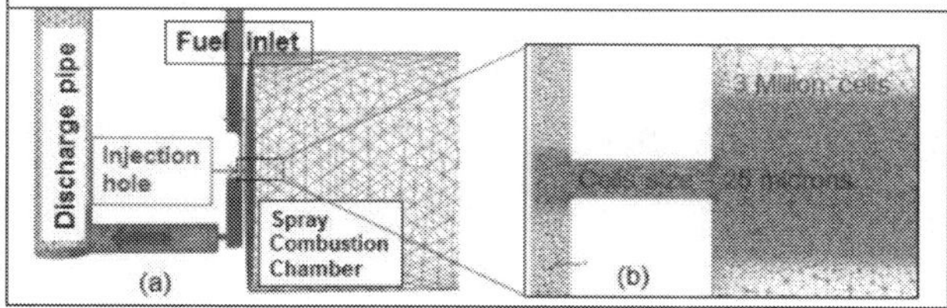


**Figure 2.** The 3D mesh. (a) Global view of the injector including the discharge system. (b) Meshing in the cross-section B-B highlighting the refinement extent inside the hole and in the SCC.

In addition, only a small cylindrical part of the SCC has been considered for the simulation of the liquid jet. Figure 2 shows the corresponding 3D unstructured mesh used in this study. It consists of ten million cells (tetra, prism, pyramid ...) with a smallest characteristic size in the range of 25 to 50 microns inside the hole and in the first 20 mm downstream in the SCC (see Figure 2b). This refined grid is used to be able to compute the cavitation inception and the primary atomization of the liquid jet as explained in [7] [8]. Initially, the whole injector is full of liquid (n-dodecane) while the combustion chamber is full of gas (Nitrogen). As such, the liquid-gas interface is assumed to be located initially at the middle of the orifice. Since the transient needle lift of the injector is not simulated, the injection is initiated using "shock tube" like conditions upstream of the channel of width **s** (see Figure 1). The injection pressure value (70 MPa) is initialized in the inlet part of the injector while the rest of the configuration pressure is initialized equal to 4 MPa, which is the same pressure than the gas phase in the combustion chamber. The temperature of the gas in the combustion chamber is set to 400 K; while the injected fuel temperature is assumed equal to ambient, 295 K. This temperature is also assumed to be that of the walls. In addition, wall laws derived for the liquid-gas mixture, are applied for the computation of the velocity and heat transfer at the walls as the grid (although relatively refined around 25 μm, see Figure 2) does not generally capture properly the near wall effects. Finally, the computations are assumed to start with a zero velocity.

## 4. Results and Discussion

The numerical simulations have been carried out using the newly developed two-fluid model implemented in the in-house IFPEN code [10]. The cavitation appears classically close to the sharp inlet edge, whether in the A-A or B-B section (Figure 3). The spray plume is deflected in the same side as the cavitation pockets, as shown in the cross-section A-A ($t > 100$ μs). The liquid flow bypasses the cavitation pocket, and then leaves the orifice with a small deviation to the bottom. This result is in good correlation with the experiments obtained recently by Schmid et al. [6]. Besides, pressure oscillations due to the opening of the needle (not shown here) have led to a very transient behaviour of the liquid flow and cavitation. The maximum velocity is up to 400 $ms^{-1}$ inside the orifice. It is also important to notice that smaller droplets are produced in the side of the cavitation pocket, especially at the end of the transient injection, as shown in the cross-section B-B in Figure 4 for ($t > 100$ μs). This is the main results of this study. But, validation of the *TwoSD* atomization model is needed in future work.

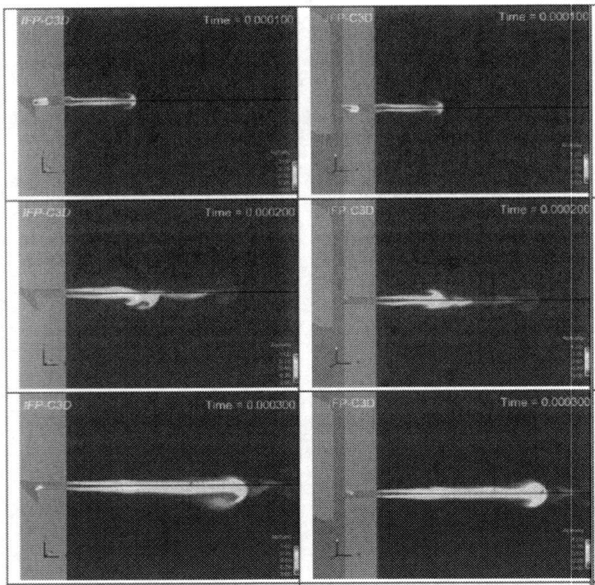

(a) Cross-section A-A

(b) Cross-section B-B (Movie available online)

**Figure 3.** (color online) Liquid volume fraction in the cross-sections A-A and B-B (see Figure 1).

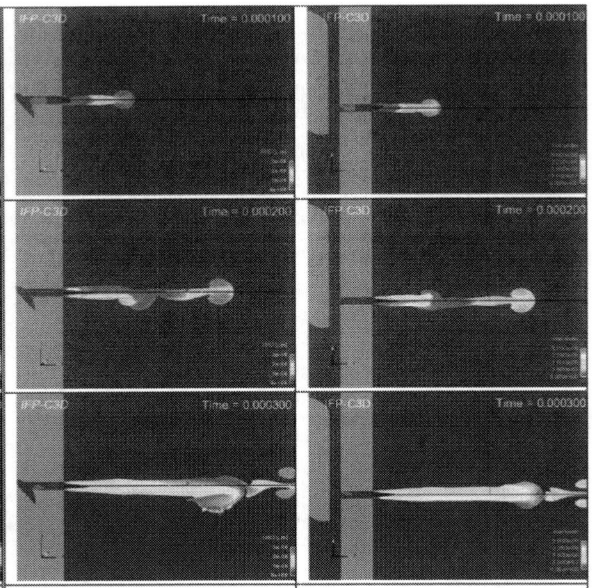

(c) Cross-section A-A

(d) Cross-section B-B (Movie available online)

**Figure 4.** (color online) Droplets size distribution in the cross-sections A-A and B-B (see Figure 1).

## 5. Conclusions

In this article, a strong correlation of cavitation and primary atomization of a liquid jet emerging from a large marine injector was demonstrated using a comprehensive two-fluid model that includes the *GERM* cavitation model and the *TwoSD* atomization model newly developed at IFPEN. The numerical results seem to indicate that cavitation is at the origin of the spray deviation off-axis observed experimentally for an eccentric nozzle configuration. In addition, this liquid spray deviation has been triggered by the cavitation collapse in the injection starting period. Later, the deviation is maintained at the cavitation side, by a highly asymmetric and transient velocity profile inside the orifice. This strong correlation between the in-nozzle cavitating flow and the liquid jet morphology (injection orientation and atomization processes) makes very difficult the simulations using a "static" coupling methodology.

**Acknowledgments**

The present work has been partly conducted during the HERCULES-C project within EC's 7th Framework Program, Contract SCP1-GA-2011-284354.

## 6. References

[1] Arcoumanis, C., Gavaises, M., and French, B. *SAE Technical Paper 970799* (1997).
[2] Bianchi, G. et al. *ASME J. of Engineering for Gas Turbines and Power* **123**, 419–427 (2001).
[3] Som, S. and Aggarwal, S. *Combustion and Flame* **157**(6), 1179 – 1193 (2010).
[4] Herrmann, K. et al. In *Proceedings of ISME, Paper-ISME111, Kobe, Japan, Oct. 17-21,* (2011).
[5] Hensel, S. et al. In *Proceedings of the COMODIA conference*, (2012).
[6] Schmid, A. et al. In *Proceedings of ILASS-Europe, 8-10 Sep. 2014, Bremen, GERMany*, (2014).
[7] Habchi C., C. *Atomization and Sprays,* **25 (4)**, 317–334 (2015).
[8] Devassy, B. M., Habchi, C., and Daniel, E. *Atomization and Sprays,* **25(1)**, 47–80 (2015).
[9] Herrmann, K. Technical report, PSI on behalf of WCH (2015).
[10] Velghe, A., Gillet, N., and Bohbot, J. *Computers & Fluids* **45**(1), 116 – 121 (2011).

# An Investigation of Flow in Nozzle Hole of Dimethyl Ether

**M Kato[1,2], T Yokota[2] , J Weber[2], D Gill[3],**

[2]DENSO CORPORATION, 1-1, Showa-cho, Kariya-shi, 448-8661, Aichi, Japan

Masaaki_kato@denso.co.jp

[3]AVL List GmbH, Hans List Platz 1, 8020 Graz, Austria

**Abstract.** For over twenty years, DME has shown itself to be a most promising fuel for diesel combustion. DME is produced by simple synthesis of such common sources as coal, natural gas, biomass, and waste feedstock. DME is a flammable, thermally-stable liquid similar to liquefied petroleum gas (LPG) and can be handled like LPG. However, the physical properties of DME such as its low viscosity, lubricity and bulk modulus have negative effects for the fuel injection system, which have both limited the achievable injection pressures to about 500 bar and DME's introduction into the market. To overcome some of these effects, a common rail fuel injection system was adapted to operate with DME and produce injection pressures of up to 1000 bar. To understand the effect of the high injection pressure, tests were carried out using 2D optically accessed nozzles. This allowed the impact of the high vapour pressure of DME on the onset of cavitation in the nozzle hole to be assessed and improve the flow characteristics.

## 1. Introduction

Dimethyl Ether (DME) is an excellent fuel for diesel combustion [1, 2, 3]. It burns completely soot free and produces less NOx than the equivalent diesel. However, it is a liquefied gas of lower bulk modulus at normal temperatures. In addition, the energy content of DME by volume is also about half of that of diesel, so that nearly twice as much DME has to be injected to produce the same power. This combination of low injection pressure and large injection nozzle flow area for DME has meant that the fuel spray from the injection nozzle has been better atomised, but has shorter penetration than the equivalent diesel [4]. Thus the objective of the work is to improve the flow characteristic of the nozzle hole without detriment to the spray ultimately the combustion.

## 2. Flow Analysis in Spray Holes

The flow characteristics of diesel injection nozzle spray holes have been improved by such measures as inlet edge rounding, by extrusion honing or hydraulic grinding [5, 6]. In addition, tapering the shape of the holes has also improved the flow characteristics by reducing the cavitation tendency of the fluid. As DME has a higher vapour pressure than diesel, it will tend to cavitate more readily. Therefore, it was decided to compare the effect of these measures with DME to see how the flow characteristics and cavitation tendency of DME was affected by the hole shape and the effects on the spray break up [7, 8, 9, 10] .

---

[1] Corresponding author.

### *2.1. Flow and optical tests on different throttle shapes*

In order to study the flow differences between the various spray hole shapes, two dimensional models were cut out of 0.3mm thick plate with the geometries required. The test set up was designed to enable optical access to the throttles and also measure the throttle flow. The setup is shown in Figure 1. The plates were sandwiched between two optical quality quartz glass windows which were then clamped together between steel plates. Fuel inlet and outlet connections and windows were provided in the steel plates so that the fuel flow and flow regimes could be observed. The fuel temperature and pressure were measured before and after the throttle. Steady state flow tests were carried out with different throttle shapes to simulate the flow through injector nozzle spray holes. The flow is adjusted by changing the backpressure after the throttle. The fuel is fed at different pressures through the apparatus and the geometry of the throttle is cut out of a thin metal plate sandwiched between two glass discs. The start and development of cavitation in the orifice are of particular interest as these affect the spray break up after the throttle. In order to investigate the specific nozzle hole geometries given in the test they had to be approximated by rectangular flow channels. For this purpose, three different hole shapes were prepared for assessment, namely a straight, sharp edged hole and a plus & minus tapered hole with rounded inlet edges. A straight and tapered hole shapes are typical of today's injection nozzles for diesel fuel. These geometries are shown in Figure 1.

Type 2 has tapered value k4 of k-factor. Type 4 has k-5 and longer hole length than Type 1 & 2. The k-factor is defined as percent of hydraulic taper ratio (HTR) shown following equation.

$$k=(Di-Do)\cdot 100/L=100HTR$$

Where, Di= inlet diameter of hole (inlet width of throttle)
Do= outlet diameter of hole (outlet width of throttle)
L= hole length (throttle length)

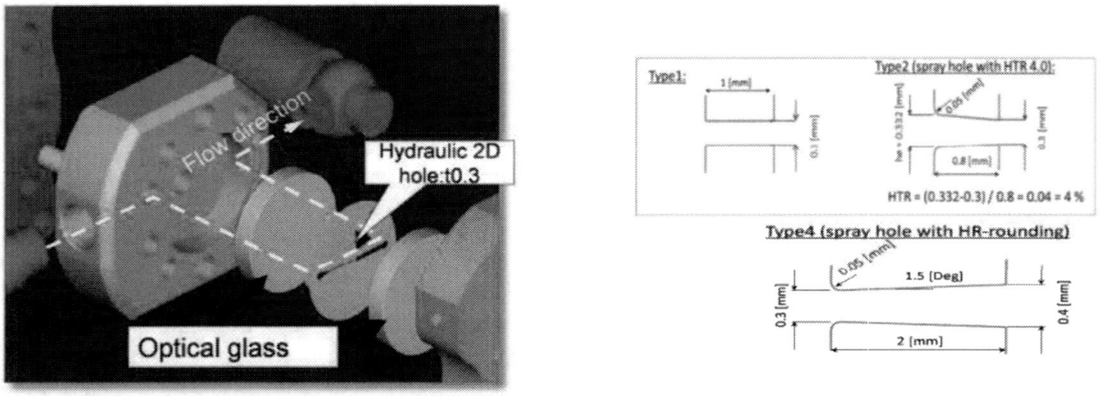


Fig. 1. Test set up to observe flow through throttles, and throttle type 1, 2, & 4

### *2.2. Flow characteristics and cavitation with optical access*

The flow tests through the throttles were carried out with inlet pressures of 200, 400, and 600 bar. The pressure difference was changed by varying the back pressure. The characteristic of the Type 2 nozzle, compared to the Type 1 shown in Figure 2 in case of inlet pressure 200bar. The curves are less inclined and the choked 2-phase area is less distinctive. As shown in Figure 2, the flow with throttle type 2 is 30% higher than for type 1 which would result in correspondingly shorter injection duration. By analyzing the images over 50 shots, a statistical picture can be formed which helps to understand how and where the cavitation is formed and develops. The red colored areas show purely liquid and the blue areas are purely vapor. The color change via yellow and green to blue are regions where the cavitation occurrence is increasing. This flow improvement of Type 2 is a result of the inlet rounding and the tapering effect of the throttle shape. And the maximum flow through the Type 2 throttle occurs at a higher pressure difference than for Type 1. Any further increase in pressure difference does not increase the flow rate. The flow rate can only be increased by increasing the inlet pressure.

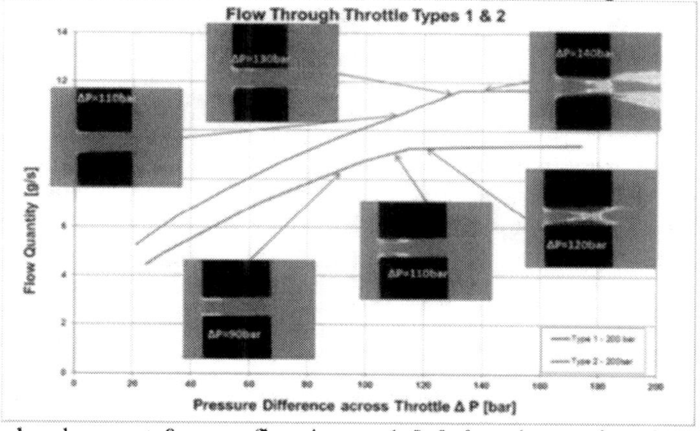


Fig. 2. Cavitation development & mass flow in type1 & 2 throttle as a function of pressure difference

These measurements were made in a way that the pressure upstream the nozzle was kept constant and the back pressure was varied. Thus, characteristic curve progressions are achieved which clearly distinguish different flow regimes. The hydraulic flow follows the Bernoulli relationship. It may either consist of a pure liquid flow regime or show local cavitation zones:

$$m_{hyd} = C_{hyd}\, A\sqrt{2\Delta P \cdot \rho}$$

Where, $m_{hyd}$ = mass flow, $C_{hyd}$ = flow coefficient, $A$ = cross sectional area of throttle, $\Delta P$ = pressure drop across the throttle=$P_{in}$-$P_{out}$, $\rho$ = density of the fluid

At specific conditions, the cavitation zones extend over the whole cross section which causes a choking of the flow due to an almost homogeneous 2-phase flow regime. This regime is compressible like a gas but it has a considerably high density due to the liquid inclusions. As a consequence, the local speed of sound drops which leads to the choked flow characteristics. Fig. 2 shows that in this regime the mass flow rates remain constant if the upstream pressure is kept constant. In other words, the flow characteristic is dependent on the throttle back pressure until a pressure difference is reached at which the flow no longer increases and remains constant independent of the backpressure. The flow characteristic of the choked flow regime is given by:

$$m_{2ph} = C_{2ph}A\sqrt{2(P_{in} - P_v) \cdot \rho} \cong C_{2ph}A\sqrt{2P_{in} \cdot \rho}$$

Where, $m_{2ph}$=mass flow, $C_{2ph}$= flow coefficient at choked flow condition, $P_v$=vapor pressure. Since $P_v$ is negligible small, $P_v$ is able to be assumed zero. According to these considerations the flow characteristics of different throttles can be comprehensively defined by the two flow coefficients $C_{hyd}$ and $C_{2ph}$. In this context it is emphasized that the combination of these two coefficients even defines the particular flow regime of any individual state. This can be explained by the following simple consideration. At the transition between hydraulic flow and choked 2-phase flow the mass flow rates from the above formulas can be equalized. This leads to the following identity

$$C_{hyd}A\sqrt{2\Delta P \cdot \rho} = C_{2ph}A\sqrt{2P_{in} \cdot \rho}$$

or re-arranged: $\frac{P_{in}-P_{out}}{P_{in}} = (\frac{C_{2ph}}{C_{hyd}})^2$

This simple formula indicates that the transition point can be clearly defined by the flow coefficients and that hydraulic flow appears at $\frac{P_{in}-P_{out}}{P_{in}} < (\frac{C_{2ph}}{C_{hyd}})^2$, and choked 2-phase flow at $\frac{P_{in}-P_{out}}{P_{in}} > (\frac{C_{2ph}}{C_{hyd}})^2$.

In this paper the transition point to the choked 2-phase flow is denominated as $K_{lim}$ [8].

$$K_{lim} = (\frac{C_{2ph}}{C_{hyd}})^2$$

The flow coefficients, $C_{hyd}$, $C_{2ph}$, & $K_{lim}$ of the two optical nozzles are shown in Table 1.

It should be noted that $K_{lim}$ shows the tendency to enter the choked 2-phase flow regime.

Table1 Flow and cavitation coefficients of the tested throttles

| Nozzle | $C_{hyd}$ | $C_{2ph}$ | $K_{lim}$ |
|---|---|---|---|
| Type 1 | 0.830 | 0.660 | 0.58 |
| Type 2 | 0.924 | 0.798 | 0.72 |

The flow test in the Type 4 throttle was also carried out. The Type 4 throttle with expanding flow area after the throat has $C_{hyd}$ values greater than 1. This indicates that the recovering the pressure after the narrowest section in Type 4 is better than the straight and converging throttles. The expanding throttle has a higher total flow capacity than the converging ones.
Decreasing $P_{out}$ will not affect the flow once cavitation has begun in the throttles.

## 3. Conclusion

The optical throttles indicated that similar effects to those seen with diesel sprays could be seen with DME: i.e. increasing the radius of the inlet to the spray hole and tapering the form of the hole, the intensity of the cavitation in the fluid at the exit from the spray hole could be reduced. As a consequence it would be expected that the spray would be better atomized and have higher penetration as well as an increase in flow rate. The results showed that different nozzle hole shapes changed the flow and the cavitation onset and tendency. These characteristics should be utilized for an optimization of the spray and combustion.

## References

[1] Mc Carthy C., et al: A New Clean Diesel Technology: Demonstration of ULEV Emissions on a Navistar Diesel Engine using a New Alternative Fuel; SAE Paper 950061.
[2] Sorensen S, et al: Performance and Emissions of a 0.273 Liter DI Diesel Engine with a New Alternative Fuel: SAE Paper 950064.
[3] Ofner H., Gill D., Krotschek C.,: DiMethyl Ether as a Fuel for CI Engines – A New Technology and its Environmental Potential, SAE 981158.
[4] Ocktaeck Lim, Sejun Lee, Sedoo Oh,: An Investigation on the Spray Characteristics of DME with variation of Nozzle Holes Diameter using the Common Rail Fuel Injection System, ICLASS, 2012,Heidellberg, Germany.
[5] Trengrove S.,: Using abrasive flow machining (AFM) to (a) increase the fatigue strength of diesel injection equipment components and (b) to generate high discharge coefficient diesel injection nozzles. C610/002/2003 I. Mech. E. Fuel Systems Conference.
[6] Blessing M., König G., Krüger C. Michels U., Schwarz V.,: Analysis of flow and cavitation in diesel injection nozzles and its effect on spray and mixture formation. C610/006/2003 I. Mech. E. Fuel Systems Conference.
[7] Kato, M., Takeuchi, H., Koie, K., Sekijima, H., Kajitani, S., Chen, L. Z., Hashimoto, S.,: A Study of Dimethyl Ether Flow in Diesel Nozzle, SAE Paper 2004-01-0081.
[8] Gill, D., Ofner, H., Stoewe, C., Wieser, K., Winklhofer, E., Kato, M., Yokota, T., Weber, J.,: An Investigation into the Effect of Fuel Injection System Improvements on the Injection and Combustion of DiMethyl Ether in a Diesel Cycle Engine, SAE Paper 2014-01-2658.
[9] Konno, M., Chiba, K., and Okamoto, T.,: Experimental and Numerical Analysis of High Pressure DME Spray, SAE Paper 2010-01-0880, 2010.
[10] Sasaki, S., Kato, M., et al,:AN experimental Study of Injection and COmbustion with Dimethyl Ether, SAE Paper 2015-01-0932.

# Simulation of throttle flow with two phase and single phase homogenous equilibrium model

**P Koukouvinis**[1,2], **M Gavaises**[2]

[2]City University London, Northampton Square EC1V 0HB, United Kingdom

foivos.koukouvinis.1@city.ac.uk

**Abstract.** This paper aims to compare the results of two commonly used methods for the simulation of cavitating flows; one is the two phase mass transfer approach and the other is a homogenous equilibrium model. Both methodologies are compared in a shock tube and a throttle flow, which resembles the constrictions in Diesel injector passages. The mass transfer rate in the two phase model plays the fundamental role in affecting how close to equilibrium the model is; by increasing the mass transfer the two phase model comes close to the homogenous equilibrium model.

## 1. Introduction

Over time, several methodologies have been developed for the simulation of cavitation effects, depending on the assumptions on the thermodynamic equilibrium, mechanical and thermal equilibrium, interface capturing and compressibility. Interface capturing methodologies are applicable in practical cases only for simulating a small number of bubbles. For industrial applications, involving potentially billions of interacting bubbles, such approaches are not feasible, so one has to resort instead either to mixture approaches, homogenous [1-3] or inhomogenous [4], depending on the mechanical and thermal equilibrium of the two phases, or to bubble tracking Eulerian Lagrangian methods [5]. In the present work, the focus is on homogenous mixture model, that is both phases have the same velocity, pressure and temperature. The main difference that will be examined is the influence of the thermodynamic equilibrium assumption; one of the examined methodologies assumes a finite mass transfer rate between vapour and liquid. The other methodology assumes thermodynamic equilibrium, by employing an equation of state that resembles the phase change. In practice both methods are equivalent, when the mass transfer of the two phase model is large enough.

## 2. Numerical background

Numerical simulations presented in this work are based on a the solution of the Navier Stokes equations, using a commercial pressure-based solver, Fluent [6]. The equations solved consist of the continuity and momentum equations, while the energy equation has been omitted since heat effects were ignored. Depending on the cavitation treatment, an additional transport equation is solved for tracking the vapour phase (two phase model with mass transfer), or a barotropic equation of state, resembling the phase change, is prescribed (Homogenous Equilibrium Model).

[1] Corresponding author.

### *2.1. The two phase model and mass transfer*

The two phase model, assumes mechanical equilibrium between the two phases, i.e. both liquid and vapour phase share the same pressure and velocity fields. An additional advection equation corresponding to the vapour fraction is solved with source terms describing the mass transfer. These terms are commonly associated with semi-empirical bubble dynamic models, based on the simplified, asymptotic Rayleigh-Plesset equation, but include additional user calibrated terms. In fact, the two phase model could be treated as a non-thermodynamic equilibrium model and an increase of the mass transfer rates towards infinity will push the model towards thermodynamic equilibrium.

For applying the two phase model, one needs to specify the properties and equation of state for the different materials. Here, for the liquid phase the Tait equation of state is used. On the other hand, the gas/vapour phase is assumed to be incompressible; it must be highlighted here that even if the pure vapour phase is incompressible, the mixture is not, since mass transfer is involved. The bulk modulus of the liquid is $B$=181MPa (which corresponds to a characteristic speed of sound for Diesel of ~1250m/s), the liquid density at saturation $\rho_{L,sat}$=828kg/m$^3$, the vapour density is $\rho_V$=0.058kg/m$^3$, the saturation pressure $p_{sat}$=892Pa and the viscosities are 2.1mPa.s and 7.5μPa.s respectively.

### *2.2. The homogenous equilibrium model (HEM)*

For the homogenous equilibrium model additional phase field variables are not needed, since the mass transfer occurs instantaneously, linking pressure to density only. So, one needs an appropriate equation of state that corresponds to the phase change of the liquid to the liquid/vapour mixture.

In this work, for simplicity the influence of thermal effects have been omitted and a barotropic equation of state is constructed as a piecewise function of the Tait equation of state for the liquid, the isentropic gas equation of state for the gas and a transitional function for the mixture, based on the Wallis speed of sound formula [7]. Eventually, the complete equation of state is the following:

$$p(\rho)=\begin{cases} B\left[\left(\dfrac{\rho}{\rho_l}\right)^n-1\right]+p_{sat,L} & \rho\geq\rho_l \\ \dfrac{c_v^2c_l^2\rho_l\rho_v(\rho_v-\rho_l)}{c_v^2\rho_v^2-c_l^2\rho_l^2}\ln\left(\dfrac{\rho}{c_l^2\rho_l(\rho_l-\rho)+c_v^2\rho_v(\rho-\rho_v)}\right)+p_{ref} & \rho_v\geq\rho>\rho_l \\ C\rho^\gamma & \rho<\rho_v \end{cases} \tag{1}$$

where $C$ is the constant of the isentropic gas and $c$ is the speed of sound, while $l$, $v$ correspond to liquid and vapour respectively. In the above equation $p_{ref}$ and $p_{sat,L}$ are reference values in order to make sure that the pressure is a continuous function of density, thus $\lim_{\rho\to\rho_v^+}p(\rho)=\lim_{\rho\to\rho_v^-}p(\rho)$ and $\lim_{\rho\to\rho_l^+}p(\rho)=\lim_{\rho\to\rho_l^-}p(\rho)$. It becomes obvious from the formulation of the equation, that during the phase change there is a small pressure difference equal to $\Delta p=p_{sat,L}-p_{sat,V}$. In practice, this difference is small in comparison with the pressure levels involved in the simulation, e.g. for the present case, the difference is around 1700Pa, whereas the pressure level in the current simulation is of the order of ~100 bar. Moreover, while it is true that the equation of state is not perfectly accurate for the sharp change of pressure in the saturation dome, it has the advantage of having a continuous speed of sound, which helps achieving convergence with the pressure-based solver utilized.

For the barotropic model, the bulk modulus of the liquid is $B$= 181MPa, the saturation pressure for the liquid is $p_{sat,L}$=1730Pa and for the vapour $p_{sat,V}$=56Pa, the liquid density at saturation is $\rho_{L,sat}$=828kg/m$^3$, the vapour density is $\rho_{V,sat}$=0.0037kg/m$^3$ and the speed of sound for vapour is $c_{V,sat}$=125m/s, calculated assuming ideal gas at 40°C. Viscosities are the same as the two phase model, i.e. 2.1mPa.s and 7.5μPa.s respectively.

### 2.3. *Comparison in a shock tube case*

As a first step for comparing the two modelling methodologies, we have used the shock tube case. The shock tube set-up resembles the conditions under consideration of the throttle case examined, thus the configuration involves interaction of pressurized liquid at ~100bar (left state) with vapour/liquid mixture at ~892Pa i.e. saturation conditions (right state). For the barotropic model the pressure levels correspond to the density of the liquid/vapour phases. On the other hand, for the 2 phase model the vapour fraction has to be specified explicitly in the mixture region. In order to be consistent with the barotropic model, a vapour fraction of 0.99 was used. Velocity is initially zero everywhere. The resolution employed is 1000 equispaced finite volumes, while the domain extends from -2m to 2m and the solution is taken at the time instant of 1ms.

In fig. 1 the shock tube solution is shown, for the cavitation models examined, along with the exact solution of the Riemann problem with the barotropic equation of state. It becomes apparent that the pressure-based solver can replicate the correct wave pattern that is predicted from the exact solution of the Riemann problem for the barotropic case. Also, in practice the two phase model results are practically indistinguishable from the barotropic model results. The only discrepancy found is the overshoot in the velocity field, that is associated with the shock wave passing. Still, this does not prevent the proper wave structure to be replicated by both cavitation models

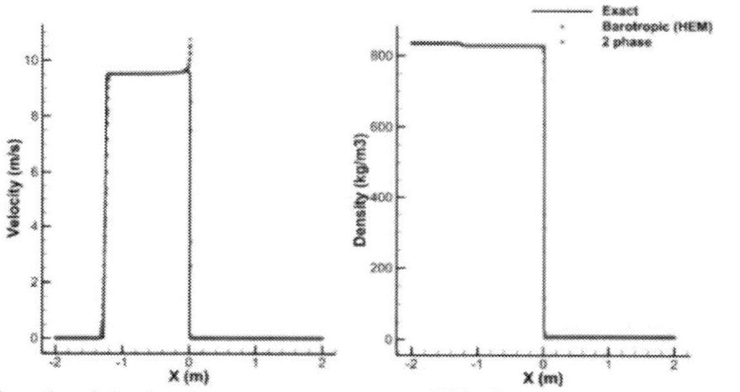


Fig. 1. Shock tube case solved with the homogenous equilibrium barotropic model and with the two phase mass transfer model. Exact solution of the Riemann problem is shown with black line.

## 3. Throttle case

The main comparison of the two models is in a throttle flow, used commonly as a benchmark for cavitation in Diesel injector passages. The inlet pressure is 100bar and the outlet pressure is 40bar. Simulation is performed with an advanced LES model, termed as Coherent Structure Model or CSM [8, 9], which has the advantage of diminishing at the near wall region while adapting to the vortical structures, since it is based on the second invariant of the velocity gradient.

The throttle consists of a rectangular channel of 3mm height and 0.3mm wide, which is constricted to a square passage of 0.3mm x 0.3mm, located at the middle of the channel; at the exit of the throttle the dimensions are the same of the upstream rectangular channel. The throttle passage has a length of 1mm and sharp angles at entrance and exit; for more information refer to the work of Perkovic et al. [10]. The working fluid is diesel liquid with saturation pressure at 892Pa, density 828kg/m$^3$, viscosity 2.1mPa.s at saturation conditions. The diesel vapor is assumed to have a density of 0.06kg/m$^3$ corresponding to the ideal gas density at the saturation pressure and temperature of 40°C. The inlet pressure is 100bar whereas the outlet pressure is 40bar.

Simulation with the LES CSM model, under the configuration outlined above, gives an average mass flow rate of 7.23g/s for the 2 phase model and 7.27g/s for the barotropic model, whereas the experimental flow rate is 7.3g/s. In fig. 3a the average distribution of vapor fraction is shown and in fig. 3b the maximum pressure recorded on the wall of the channel. As can be seen from the results, a similar pattern is found from the two cavitation models both in terms of average density distribution and maximum wall pressure. It has to be highlighted that with the tuning of the mass transfer coefficients the minimum pressure in the two phase model is -3bar; if no tuning was used, then the minimum pressure is of the order of -20bar.

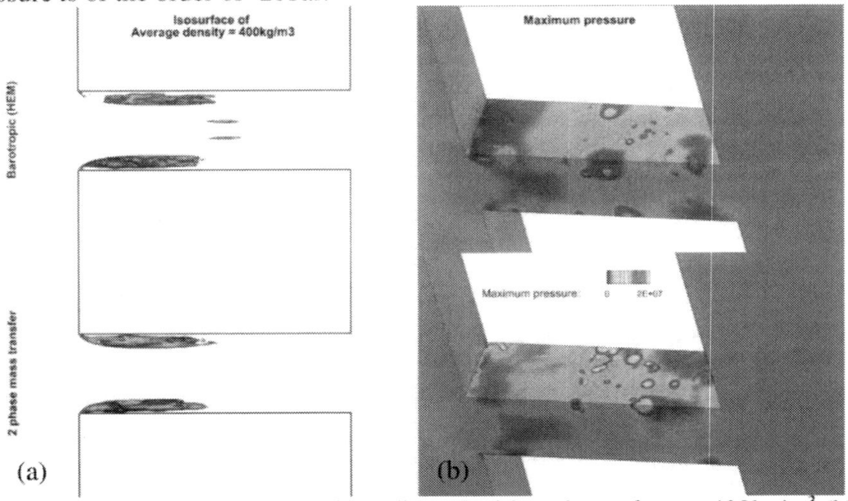


Fig. 2. (a) Average density inside the throttle, indicated with an isosurface at 400kg/m$^3$ (b) Maximum pressure distribution on the throttle wall, pressures may exceed 500bar in white zones.

## 4. Conclusion

In this paper, the performance of two cavitation models has been analyzed in a fundamental shock tube case and in a throttle case. From the shock tube case, it is evident that both the barotropic and the two phase cavitation models are equivalent, provided that the mass transfer of the latter is high enough to tend towards thermodynamic equilibrium. The throttle case confirms this finding, since both models show a similar cavitation zone, while also predict pressure peaks indicating the collapse of cavitation structures.

## References

[1] Dular M and Coutier-Delgosha O 2009 Numerical modelling of cavitation erosion. *Int J Numer Methods Fluids*, **61**: p. 1388-1410.

[2] Goncalves E and Patella R F 2009 Numerical simulation of cavitating flows with homogeneous models. *Computers & Fluids*, **38**: p. 1682–1696.

[3] Egerer C, et al. 2014 Large-eddy simulation of turbulent cavitating flow in a micro channel. *Phys Fluids*, **26**: p. 30.

[4] Pelanti M and Shyue K-M 2014 A mixture-energy-consistent six-equation two-phase numerical model for fluids with interfaces, cavitation and evaporation waves. *J Comput Phys*, **259**: p. 331-357.

[5] Gavaises M, et al. 2015 Visualisation and les simulation of cavitation cloud formation and collapse in an axisymmetric geometry. *Int J Multiphase Flow*, **68**: p. 14-26.

[6] ANSYS 2013 ANSYS Fluent 15.07.

[7] Brennen C 1995 Cavitation and Bubble Dynamics. (Oxford University Press).

[8] Kobayashi H and Wu X 2006 Application of a local subgrid model based on coherent structures to complex geometries, Center for Turbulence Research. p. 69-77.

[9] Edelbauer W, Strucl J, and Morozov A 2014 Large Eddy Simulation of cavitating throttle flow, in *SimHydro 2014:Modelling of rapid transitory flows* (Sophia Antipolis).

[10] Perkovic I, et al. 2008 3D CFD Calculation of Injector Nozzle Model Flow for Standard and Alternative Fuels, in *HEAT 2008, Fifth International Conference on Transport Phenomena In Multiphase Systems* (Bialystock, Poland).

Journal of Physics: Conference Series **656** (2015) 012087 doi:10.1088/1742-6596/656/1/012087

# Compressible bubble dynamic simulations with central-upwind schemes

**P Koukouvinis**[1,2], **M Gavaises**[2], **A Georgoulas**[3], **M Marengo**[3,4]

[2]City University London, Northampton Square EC1V 0HB, United Kingdom
[3]Dept. of Engineering and Applied Sciences, University of Bergamo, Viale Marconi 5, 24044 Dalmine (BG)
[4]School of Computing, Engineering and Mathematics, University of Brighton, Cockcroft Building, Lewes Road, BN2 4GJ, Brighton, UK

foivos.koukouvinis.1@city.ac.uk

**Abstract.** This paper discusses the implementation of an explicit density-based solver, based on the central-upwind schemes originally suggested by Kurganov, for the simulation of cavitating bubble dynamic flows. Explicit density based solvers are suited for highly dynamic, violent flows, involving large density ratios, as is rather common in cavitating flows. Moreover, the central-upwind schemes have the advantage of avoiding direct evaluation of the Jacobian matrix or estimation of the wave pattern emerging from Euler equations. Second order accuracy can be achieved with TVD MUSCL schemes. Basic comparison with the predicted wave pattern of the central-upwind schemes is performed with the exact solution of the Riemann problem showing an excellent agreement. Then several different bubble configurations were tested, similar to the work of Lauer et al. (2012). The central-upwind schemes prove to be able to handle the large pressure and density ratios appearing in cavitating flows, giving similar predictions in the evolution of the bubble shape.

## 1. Introduction

The subject of cavitation bubble dynamics has been studied traditionally using the Rayleigh-Plesset equation [1, 2]. However, either the original Rayleigh-Plesset equation, or its more complex variants, assume that the bubble shape is perfectly spherical. In practice this is not the case, since many works, both experimental [3] and numerical [4-6], suggest that the bubble shape may be strongly deformed in the presence of pressure fields (e.g. due to gravity, due to passing sound waves), or due to the presence of boundaries (walls, free surfaces, etc.).

Thus, if one wishes to predict the asymmetric bubble collapse near a wall, then it is necessary to do so by properly integrating the Navier Stokes equations. The complexity of the involved flow pattern is significant, since the flow is multiphase, involving a strongly deforming liquid/gas interface and very high velocities. In the past, there have been efforts to perform such simulations; one of the first was the pioneering work of Plesset and Chapman [4].

In the present investigation, a method similar to the one used by Adams et al. [7] and Pohl et al. [8] is employed; the cavitation bubble is described as a density difference of a single fluid, governed by a complex equation of state which represents the isentropic phase change due to cavitation. As it will be

[1] Corresponding author.

explained in the next section, a central-upwind scheme is applied, since it has some attractive features. The aim is to predict the outcome of several bubble collapse events near wall configurations, in order to determine the performance of the scheme employed.

## 2. Numerical method

The Euler equations are resolved with an in-house code in conservation form, considering the influence of cylindrical symmetry, see Toro [9], to reduce the computational cost. A piecewise barotropic equation of state is used, which is a combination of the Tait equation of state above saturation and a formula that resembles the isentropic transformation within the saturation dome [10]:

$$p(\rho)=\begin{cases} B\left[\left(\dfrac{\rho}{\rho_{sat,L}}\right)^{n}-1\right]+p_{sat} & \rho \geq \rho_{sat,L} \\ p_{sat}+C_{sat,L}\left(\dfrac{1}{\rho_{sat,L}}-\dfrac{1}{\rho}\right) & \rho < \rho_{sat,L} \end{cases} \tag{1}$$

The values used for the equation of state are: density of saturated liquid $\rho_{sat,L}$=998.2kg/m$^3$, speed of sound of the saturated liquid $C_{sat,L}$=1450m/s, saturation pressure $p_{sat}$=2339Pa, bulk modulus $B$=294MPa and the exponent $n$=7.15, commonly used for weakly compressible liquids.
In order to evaluate the flux at the interface of the finite volumes, the central-upwind scheme of Kurganov is used [11]. Linear interpolations are utilized at the cell interfaces, handled with the MUSCL scheme with a SuperBee limiter [9]. This scheme has the advantage of being universal, in the sense that it does not need the tuning of the AUSM+up scheme coefficients, while it does not require the identification of the wave structure of the solution of the Riemann problem, as e.g. the HLLC or Roe solvers do. On the other hand, the 2nd (or higher) order extension ensures low numerical diffusion. Boundaries are handled either as transmissive or as slip walls, depending on the configuration (see Toro [9]).
Time integration is performed in an explicit manner, with a splitting scheme [9], i.e. initially for the homogenous part of eq. 1 and then for the source term. In this preliminary work, 1$^{st}$ order Euler integration is used, with a CFL of 0.5, whereas in the future higher order Runge Kutta integration will be implemented.

## 3. Validation of the numerical method

In Fig. 1a the solution of the Kurganov scheme for the Riemann problem with initial conditions $\rho_L$=1002.88kg/m$^3$ and $\rho_R$=9.99kg/m$^3$, u=0m/s everywhere is shown; also the Lax-Friedrichs and exact solution are shown for reference. It is of interest that the central-upwind scheme is successfully capturing the correct wave pattern, with the same spatial resolution as the Lax-Friedrichs scheme without smearing of the shock, due to the inherent numerical dissipation of the latter. It should be highlighted here that obtaining the exact solution of the Riemann problem is not trivial for an arbitrary EOS, such as the one in eq. 1, due to the nature of the Riemann invariants in the rarefaction zones.
In order to validate the 2D axis-symmetric solver, which will be used later, a comparison with the 1D solver with spherical symmetry was performed. The initial conditions for this comparison are similar to the above, ρ=9.99kg/m$^3$ for R<1m and ρ=1002.88kg/m$^3$ for R>1m, resembling an implosion configuration. In Fig. 1b the comparison between the 2D axis-symmetric and the 1D spherical symmetric case is shown, showing a perfect agreement.

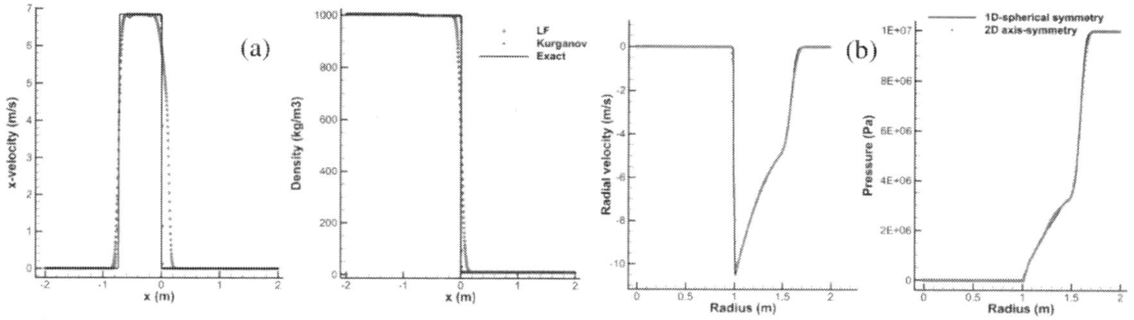


Fig. 1. (a) Shock tube test; comparison with exact, Lax-Friedrichs (LF) and Kurganov schemes for cavitating case at 0.5ms (b) Cavitation implosion simulation with 2D axis-symmetric and 1D with spherical symmetry at 0.4ms.

## 4. Results

The case of interest is the collapse of a water vapor bubble in the vicinity of a wall, in the same arrangement as used in Lauer et al. work [5], using the framework analyzed in section 2. The bubble has a radius of 400μm and its center is placed 416, 140 and -140 μm from a wall. The surrounding fluid has a pressure of 100bar, whereas the pressure within the bubble is equal to the saturation i.e. 2340Pa. In Fig. 2 indicative instances of the bubble deformation during the collapse are shown for the three configurations. The influence of the wall distance is apparent: when the initial bubble centre is above the wall, then the jet effect is formed on the axis of symmetry, causing the bubble to take a heart like shape or a torus in the cases of 416μm and 140μm respectively. On the other hand, in the case where the initial bubble centre is positioned below the wall, the jet forms in the $x$ direction at the wall, deforming the bubble in a pin-like shape.

In all cases, the formation of the jet, either on the symmetry axis, or in the $x$-direction causes the development of very high pressures due to momentum focusing, see Fig. 2. At the late stages of the bubble collapse, the jet will eventually impact on the wall, causing pressures of the order of 10000bar; such pressures are well beyond the yield stress of many common materials (e.g. SS316L has a yield stress of the order of $2\text{-}4 \cdot 10^3$bar), implying that such bubble collapse configurations will contribute to the erosion damage of the underlying solid material. From a numerical point of view, the employed scheme performed well, in the sense that it is able to handle pressure ratios of almost $5 \cdot 10^5$ and density ratios of 1000, without serious problems. High accuracy enabled a clear capturing of the interface within 1-2 cells, without oscillations, thanks to the TVD properties of the MUSCL scheme.

## 5. Conclusion

This paper outlines the development of an explicit density based solver for cavitating flows, based on the central upwind schemes of Kurganov et al. and the application on bubble collapse using 2D axis-symmetry conditions. The schemes have been tested in comparison with the exact solver of the Riemann problem, showing excellent accuracy and robustness. Application of the schemes on the bubble collapse cases showed a similar collapse pattern with the one that has been reported by Laeur et al. and similar pressure levels on the wall, even though a Homogenous Equilibrium assumption is used for the thermodynamic model. Overall, the performance of the central-upwind scheme was very good, considering the violent nature of the collapse and the involved pressure and density ratios. One of the main targets in the future is the implementation of higher accuracy in the time marching and possibly inclusion of thermal effects, with a potential application the simulation of bubble clusters.

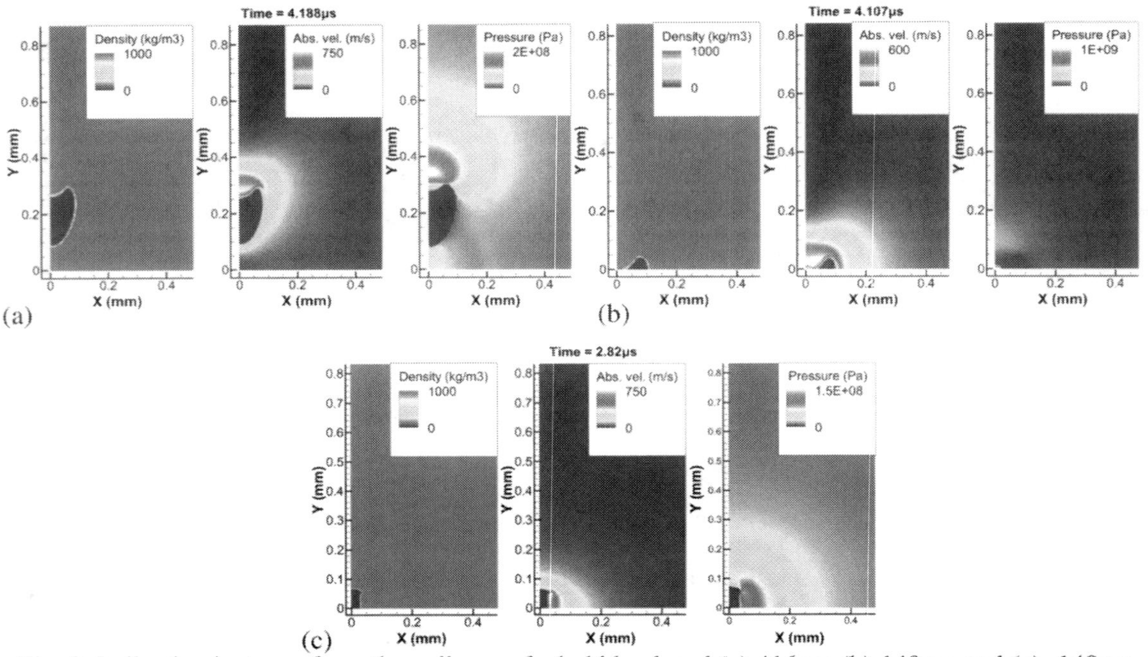


Fig. 2. Indicative instance from the collapse of a bubble placed (a) 416μm (b) 140μm and (c) -140μm from a wall; *y*-axis is axis of symmetry, *x*-axis is wall.

**Acknowledgements**
The research leading to these results has received funding from the People Programme (IAPP Marie Curie Actions) of the European Union's Seventh Framework Programme FP7/2007-2013/ under REA grant agreement n. 324313.

**References**

[1] Franc J-P and Michel J-M 2005 Fundamentals of Cavitation. (Kluwer Academic Publishers).

[2] Brennen C 1995 Cavitation and Bubble Dynamics. (Oxford University Press).

[3] Obreschkow D, et al. 2013 The Quest for the Most Spherical Bubble. *Exp Fluids*, **54**: p. 18.

[4] Plesset M S and Chapman R B 1970 Collapse of an initially spherical vapor cavity in the neighborhood of a solid boundary, California Institute of Technology. p. 42.

[5] Lauer E, et al. 2012 Numerical modelling and investigation of symmetric and asymmetric cavitation bubble dynamics. *Computers & Fluids*, **69**: p. 1-19.

[6] Hawker N A and Ventikos Y 2012 Interaction of a strong shockwave with a gas bubble in a liquid medium: a numerical study. *J Fluid Mech*, **701**: p. 55-97.

[7] Adams N A and Schmidt S J 2013 Shocks in cavitating flows, in *Bubble dynamics & Shock waves*, C.F. Delale, Editor. (Berlin Heidelberg, Springer-Verlag). p. 235-256.

[8] Pohl F, et al. 2014 Evaluation of cavitation-induced pressure loads applied to material surfaces by finite-element-assisted pit analysis and numerical investigation of the elasto-plastic deformation of metalic materials (article in press). *Wear*.

[9] Toro E 2009 Riemann Solvers and Numerical Methods for Fluid Dynamics: A Practical Introduction. (Springer-Verlag Berlin Heidelberg).

[10] Egerer C, et al. 2014 Large-eddy simulation of turbulent cavitating flow in a micro channel. *Phys Fluids*, **26**: p. 30.

[11] Kurganov A, Noelle S, and Petrova G 2001 Semidiscrete central-upwind schemes for hyperbolic conservations laws and Hamilton-Jacobi equations. *SIAM Journal on Scientific Computing* **23**: p. 707-740.

# Large Eddy Simulation of diesel injector opening with a two phase cavitation model

**P Koukouvinis**[1,2], **M Gavaises**[2], **J Li**[3], **L Wang**[3]

[2] City University London, Northampton Square EC1V 0HB, United Kingdom
[3] Caterpillar Inc., Mossville, IL 61602, US

foivos.koukouvinis.1@city.ac.uk

**Abstract.** In the current paper, indicative results of the flow simulation during the opening phase of a Diesel injector are presented. In order to capture the complex flow field and cavitation structures forming in the injector, Large Eddy Simulation has been employed, whereas compressibility of the liquid was included. For taking into account cavitation effects, a two phase homogenous mixture model was employed. The mass transfer rate of the mixture model was adjusted to limit as much as possible the occurrence of negative pressures. During the simulation, pressure peaks have been found in areas of vapour collapse, with magnitude beyond 4000bar, which is higher that the yield stress of common materials. The locations of such pressure peaks corresponds well with the actual erosion location as found from X ray scans.

## 1. Introduction

Diesel injection systems play a fundamental role in internal combustion engines since they affect the formation of the fuel spray, atomization and combustion, the formed emissions and the engine efficiency. The jet velocities formed are of the order of 500m/s, with upstream pressures around 2000bar. Current trends show injection pressures to even rise to 3000bar, in order to meet the future EU legislations in emissions. However, higher pressure levels causes very high velocities through the tight passages in the Diesel injector and strong accelerations in sharp direction changes (corners, fillets etc.), which lead to local static pressure dropping below the saturation pressure and causing cavitation. Furthermore, cavitation may lead to cavitation erosion damage and serious degradation of the injector performance, even catastrophic injector failure, which could damage the engine, if the injector tip breaks off.

There have been many efforts for the prediction of the Diesel injector erosion, see e.g. the work of Koukouvinis et al. [1]. The aim of the current work is to simulate the flow inside a Diesel injector in a more fundamental level, including compressibility effects of the liquid phase and also using a Large Eddy Simulation for describing turbulence. The reason for employing compressibility effects is the large density difference in the injector, which may reach 10% [2], not to mention the high liquid velocities that can reach a Mach number of 0.5. Furthermore, resorting to Large Eddy Simulation techniques is because RANS/URANS are inadequate for capturing the complicate vortex patterns which affect cavitation formation [3], while even modified RANS turbulence models are situational

[1] Corresponding author.

[4]. To the authors knowledge, a similar work does not exist in literature; Sezal et al. [5] and Salvador et al. [6] numerical work on injectors includes compressibility effects, but not the needle motion of an actual injection cycle.

## 2. Numerical background

Numerical simulations presented in this work are based on a the solution of the Navier Stokes equations, using a commercial pressure-based solver, Fluent [7]. The equations solved consist of the continuity and momentum equations, while the energy equation has been omitted since heat effects were ignored. However, since Diesel properties vary significantly with the pressure levels in the injection systems, both liquid phase viscosity and density are assumed variable, as functions of pressure. For density, the Tait equation of state was used:

$$p = B\left[\left(\frac{\rho}{\rho_{sat,L}}\right)^{n} - 1\right] + p_{sat} \tag{1}$$

where $\rho_{sat,L}$= 747.65kg/m$^3$ is the density at saturation pressure $p_{sat}$=1.2bar and temperature of 395K. The temperature was estimated through simplified 1D analysis for the pressure levels in the injector [8] and is an estimated average during the injection event. The bulk modulus is $B$=110MPa and $n$=7.15. The Tait equation of state is a compromise in the sense that is known to predict accurately the behaviour of liquids, whereas more accurate data are not available in the pressure ranges involved.

For Diesel liquid kinematic viscosity, $v$, a relation provided by N. Kolev [2] is used, applied for the same temperature level as above:

$$\log_{10}(10^6 v) = 0.035065275 - 0.000234373 \cdot p/10^5 \tag{2}$$

For simulation of cavitation effects, an additional transport equation is solved for tracking the vapour phase, of the form:

$$\frac{\partial(a\rho_v)}{\partial t} + \nabla(a\rho_v \mathbf{u}) = R_e - R_c \tag{3}$$

where $a$ is the vapour fraction, $\rho_v$ is the vapour density, $\mathbf{u}$ is the velocity field and $R_e$, $R_c$ are the mass transfer rates for condensation (c) and evaporation (e), prescribed by the Zwart-Gerber-Belamri model [9]. The vapour properties used, are density $\rho_V$=6.5kg/m3, the dynamic viscosity $\mu_v$=7.5μPa.s. The two phase model, assumes mechanical equilibrium between the two phases, i.e. both liquid and vapour phase share the same pressure and velocity fields. The mass transfer terms have been tuned in order to limit the existence of negative pressures inside the computational domain as much as possible; after the tuning the minimum pressure inside the injector is approximately -20bar. The aim of the two phase model employed is to replicate the growth and, more importantly, Rayleigh collapse of vapour structures which leads to high pressures during cavitation collapse and to cavitation erosion.

In order to capture the complicated turbulent structures inside the sac volume and orifice, the Wall Adapted Local Eddy-viscosity (WALE) LES model was employed [10], since it has much better performance in wall-bounded flows, as the one developing inside the injector.

## 3. Diesel injector simulation

### *3.1. Case set-up*

The Diesel injector tip geometry is shown in fig. 1a; it has 5 tapered orifice holes. The injector tip inlet and outlet pressure boundary conditions were calculated using 1D CFD tools. Note also that at the end of the orifice of the injector an additional hemispherical volume was added (fig. 1b), in order to move the influence of the outlet boundary further away from the orifice, especially considering that cavitation structures may reach or even exit the orifice, as will be shown later. The configuration resembles the injection test benches where fuel is squirted into a collector filled with liquid.

The needle motion is assumed to be in the axial direction only, so any eccentricity effects were omitted. In reality eccentricity effects are significant during the early opening and late closing phases, but introduction of such effects will impose the simulation of the complete injector geometry, which will be much more complicated. In the current simulation, only 1/5$^{th}$ of the domain was solved and periodic boundary conditions have been employed at the sides of the domain. The computational domain was split in a set of moving, deforming and stationary zones, as shown in fig. 1b.

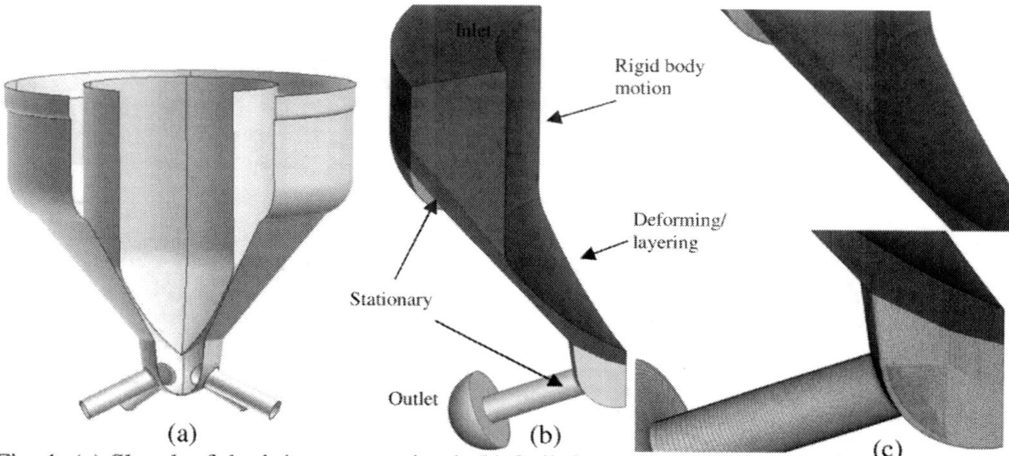


Fig. 1. (a) Sketch of the injector examined, (b) Splitting of the geometry to accommodate mesh motion (c) details of the mesh at the needle seat passage and sac volume.

The computational mesh used is block-structured, with the exception of a zone in the sac prior to the orifice entrance, which is unstructured tetrahedral (see fig. 1c). Inflation layers were introduced in the area of interest, such as sac and orifice to capture the boundary layers. Mesh motion was introduced by a smoothing algorithm at low lifts and the mesh layer addition at higher lifts. The mesh size was approximately 1 million cells at the beginning of the simulation; the needle motion introduces additional cell layers inside the volume denoted with blue colour in fig. 1b, so the actual cell count increases over time. A bounded central scheme (hybrid between central and second order upwind) was used for momentum, while second order upwind for density and QUICK for volume fraction. Time advancement was performed with an implicit, second order, backward differentiation with a time step of 5ns, in order to be able to capture the complicated turbulent patterns. The implicit integration avoids the time step restriction due to compressibility effects, which would further limit the time step to even lower values.

### *3.2. Simulation results*

In fig. 2a the complicated cavitation patterns inside the sac volume and the injector orifice are shown. Also in fig. 2b the evolution of the maximum pressure on the wall is shown. From the simulation results, cavitation starts to develop at the edges of the jet formed at low lifts between the needle and needle seat, due to shear layer instability. Later on, cavitation extends inside the sac volume, due to the vortex formation. At the same time, cavitation collapses on the needle produce strong shock waves, with pressure levels exceeding 4000bar and even peaking at 10000bar. When the needle opening is high enough, e.g. at ~39μm pressure starts to rise inside the sac volume and cavitation structures collapse, causing sporadic pressure peaks on the sac wall. Eventually, cavitation is limited only inside the orifice, at the upper orifice surface. Note that cavitation may extend outside the injector orifice.

## 4. Conclusion

This paper outlines the methodology and results from Large Eddy Simulation inside a Diesel injector, including liquid compressibility effects. The turbulence model used is capable of capturing the complicated vortex patterns inside the injector sac and orifice, which in turn affect the cavitation formation location and collapse. Cavitation collapses produce strong shockwaves, with pressure levels exceeding 4000bar. The locations of the pressure peaks correspond well with erosion patterns from X-ray imaging of the actual tested injectors.

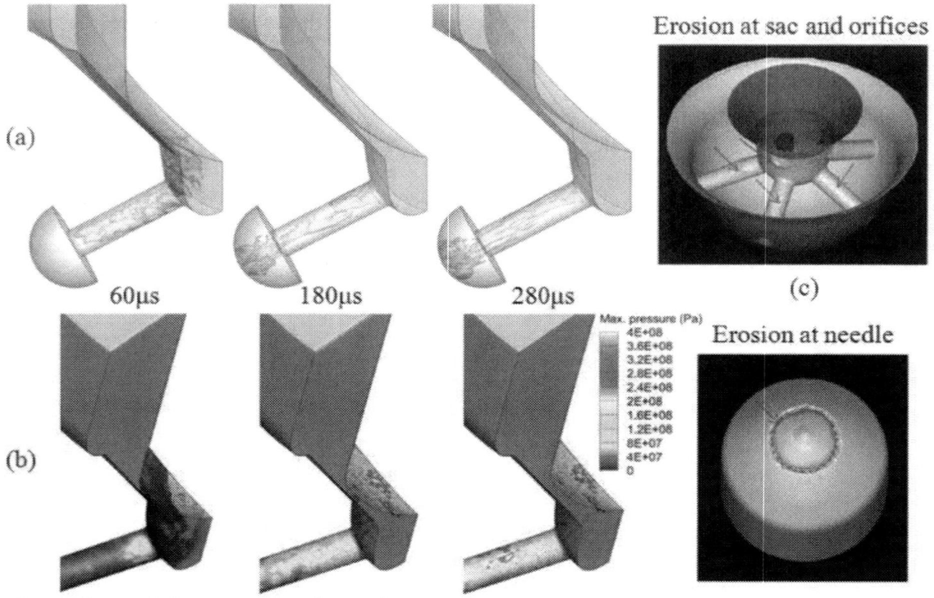


Fig. 2. (a) Isosurface of the vapour volume fraction, value 0.5, (b) Accumulation of maximum pressure peaks on the injector walls for three time instances (c) CT scans showing erosion locations.

**Acknowledgements**
The research leading to these results has received funding from the People Programme (IAPP Marie Curie Actions) of the European Union's Seventh Framework Programme FP7/2007-2013/ under REA grant agreement n. 324313.

**References**
[1] Koukouvinis P, et al. 2014 Simulation of cavitation inside diesel injectors, including erosion modelling, in *Fuel Systems for IC Engines*Institution of Mechanical Engineers (IMechE, London). p. 125-135.
[2] Kolev N 2007 Multiphase Flow Dynamics 3. (Springer Berlin Heidelberg).
[3] Coutier-Delgosha O, Reboud J L, and Delannoy Y 2003 Numerical simulation of the unsteady behaviour of cavitating flows. *Int J Numer Methods Fluids*, **42**: p. 527-548.
[4] Edelbauer W, Strucl J, and Morozov A 2014 Large Eddy Simulation of cavitating throttle flow, in *SimHydro 2014:Modelling of rapid transitory flows* (Sophia Antipolis).
[5] Sezal I H, et al. 2009 Shock and wave dynamics in cavitating compressible liquid flows
in injection nozzles. *Shock Waves* **19**: p. 49-58.
[6] Salvador F J, et al. 2013 Computational study of the cavitation phenomenon and its interaction with the turbulence developed in diesel injector nozzles by Large Eddy Simulation (LES). *Mathematical and Computer Modelling*, **57**: p. 1656-1662.
[7] ANSYS 2013 ANSYS Fluent 15.07.
[8] Strotos G, et al. 2015 Transient heating effects in high pressure Diesel injector nozzles. *Int J Heat Fluid Flow*, **51**: p. 257-267.
[9] Ji B, et al. 2013 Numerical analysis of unsteady cavitating turbulent flow and shedding horse-shoe vortex structure around a twisted hydrofoil. *Int J Multiphase Flow*, **51**: p. 33-43.
[10] Nicoud F and Ducros F 1999 Subgrid-scale stress modelling based on the square of the velocity gradient tensor. *Flow, Turbulence and Combustion*, **62**: p. 183-200.

# Visualisation of diesel injector with neutron imaging

**E Lehmann[1], C Grünzweig[1], S Jollet[1], M Kaiser[2], H Hansen[2] and F Dinkelacker[2]**

[1] Paul Scherrer Institut, CH-5232 –Villigen PSI, Switzerland
[2] Institut für Technische Verbrennung, Leibniz Universität Hannover, Welfengarten 1A, Deutschland

E-Mail: eberhard.lehmann@psi.ch

**Abstract**. The injection process of diesel engines influences the pollutant emissions. The spray formation is significantly influenced by the internal flow of the injector. One of the key parameters here is the generation of cavitation caused by the geometry and the needle lift.

In modern diesel engines the injection pressure is established up to 3000 bar. The details of the flow and phase change processes inside the injector are of increasing importance for such injectors. With these experimental measurements the validation of multiphase and cavitation models is possible for the high pressure range. Here, for instance, cavitation effects can occur. Cavitation effects in the injection port area destabilize the emergent fuel jet and improve the jet break-up.

The design of the injection system in direct-injection diesel engines is an important challenge, as the jet breakup, the atomization and the mixture formation in the combustion chamber are closely linked. These factors have a direct impact on emissions, fuel consumption and performance of an engine. The shape of the spray at the outlet is determined by the internal flow of the nozzle. Here, geometrical parameters, the injection pressure, the injection duration and the cavitation phenomena play a major role.

In this work, the flow dependency in the nozzles are analysed with the Neutron-Imaging. The great advantage of this method is the penetrability of the steel structure while a high contrast to the fuel is given due to the interaction of the neutrons with the hydrogen amount. Compared to other methods (optical with glass structures) we can apply real components under highest pressure conditions. During the steady state phase of the injection various cavitation phenomena are visible in the injector, being influenced by the nozzle geometry and the fuel pressure. Different characteristics of cavitation in the sac and spray hole can be detected, and the spray formation in the primary breakup zone is influenced.

## 1. Introduction

The relationship of the internal nozzle flow and the primary break up is important for combustion process. These processes of injection are very complex. There have to many physical processes, such as the momentum exchange between liquid phase and gas phase turbulence in the liquid phase and in the gas phase, transient heat transfer, boiling behaviour, surface tension effects, cavitation etc. are included (Figure 1) [1].

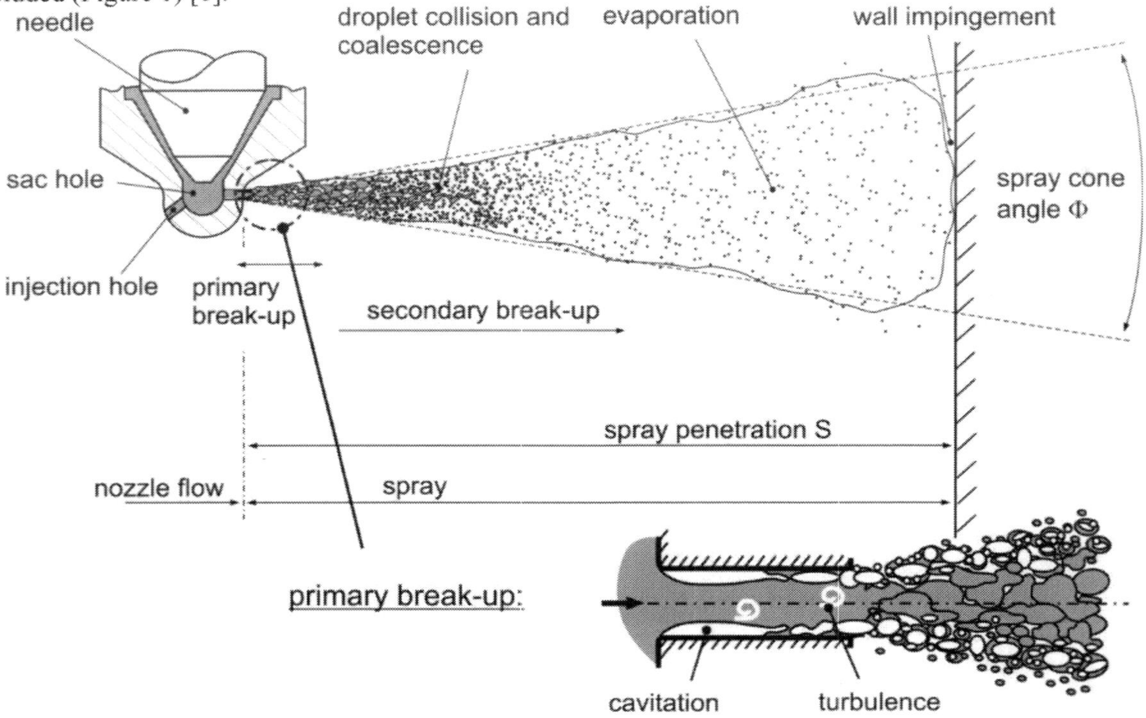


**Figure 1.** Features during the spray formation in the Diesel injection nozzle [1]

In the nozzle flow injection pressures are used up to 3000 bar. The liquid fuel is through the injection ports with speeds exceeding 500 m / s is injected into the combustion chamber. In the spray holes the cavitation zones and prevailing turbulence are the main departure mechanisms of the jet break-up at spray hole outlet [2]. Immediately after leaving the spray hole breaks the liquid jet into drops and ligaments, which form a dense spray in the vicinity of the nozzle. This first section of the spray break-up is called primary break-up. During the subsequent disintegration process, the so-called secondary break-up, existing droplets fall into more smaller droplets. In this areas, drops are collides and associates. The secondary break-up is caused due to aerodynamic forces. The relative speed between droplets and gas phase decelerated the droplets. Droplets on the spray holes are delayed through the strong resistance forces than the droplets which follow these drops. Thus, the jet is always replaced by new droplets and thus reflects the spray spread. The fuel jet is characterized in this case by the penetration depth and the spray cone angle. Also found in the area the spray cone a massive evaporation instead of the liquid fuel [3],[4],[5].

## 2. Neutron Imaging approach

As shown in Figure 2, the Diesel injection nozzles are relatively transparent for thermal neutrons while the small amounts of Diesel fuel gives considerable contrast. This advantage of the method can be used favorably for dedicated studies. On the one hand, the spray formation close to the nozzle outlet in near filed until the density limit is reached can be observed in stationary mode or in pulsed operation where the neutron imaging detector is triggered via the injection signal [6].

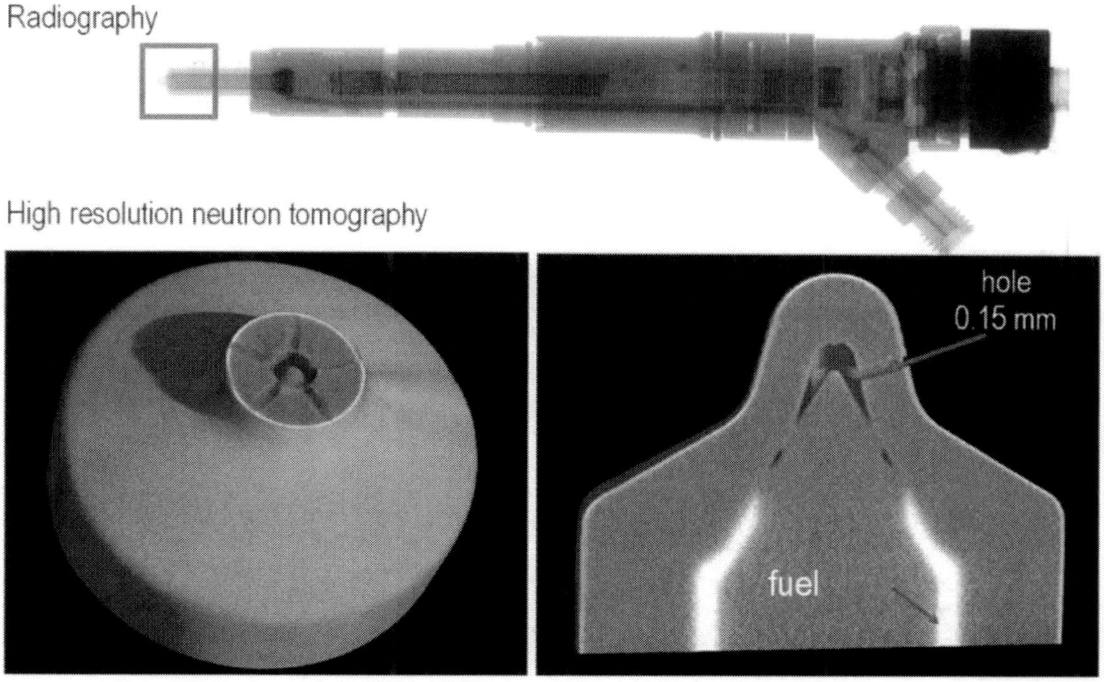


**Figure 2.** Neutron transmission image of a Diesel injection nozzle (top) showing the fuel distribution (dark lines) inside the steel structure with high transparency. The lower part is the neutron tomography of the nozzle outlet: horizontal slice through the holes (left), vertical slice through the middle (right), where remaining fuel is indicated in white contrast

Further, the structure of the nozzle outlet holes and their alterations can be investigated non-destructively by means of neutron tomography, as it is shown in Figure 2. The spatial resolution is on the order of 10 um and will be improved to about 5 um soon.

We have in mind, to investigate the running injection process with the highest possible spatial resolution and in dynamic mode in order to see the processes like cavitation and gas formation in the fuel directly. We will use a gated and intensified CCD detection system with high sensitivity for neutrons. This technique will be tried at first in a stationary process for averaged values.

Neutron imaging is a well-established technique at PSI, based on the operation of more than two facilities with high-end performance [7].

## 3. Appraoch

It is the intention to combine in this project the already established knowledge and experience (simulation and optical tests) with the new approach of neutron imaging in the best manner.

## 4. Results and Discussion

For the moment, we can only show independently achieved results: studies with optically transparent "dummy" cells and the related process simulations and neutron imaging data with static components. In addition, we can give results of dynamic processes of the spray formation near the nozzle exit, obtained with coarse spatial resolution for optical measurements. For the neutron imaging the

Journal of Physics: Conference Series **656** (2015) 012089 doi:10.1088/1742-6596/656/1/012089

combination of the high resolution option with the high contrast for the fuel will enable to see the stationary averaged processes in the nozzle outlet in the best possible way.

## 5. Conclusions

We described a new approach for the investigation of cavitation effects in the fuel exits of Diesel injection nozzles by means of neutron imaging. Using the high penetration power through steel, the high contrast for small amounts of fuel and the high spatial resolution of the newly developed neutron imaging detectors it will be possible to perform dedicated studies of the cavitation effects under realistic conditions.

## References

[1] Baumgarten C 2006 *Mixture Formation in Internal Combustion Engines* (Berlin Heidelberg New York: Springer)

[2] Jollet S 2014 *Experimentelle und numerische Untersuchungen der instationären dynamischen Innenströmung in Dieselinjektoren* (Leibniz Universität Hannover: Dissertation)

[3] Walther J 2002 *Quantitative Untersuchung der Innenströmung in kavitierenden Dieseleinspritzdüsen* (TU Darmstadt: Dissertation)

[4] Blessing M 2004 *Untersuchung und Charakterisierung von Zerstäubung, Strahlausbreitung und Gemischbildung aktueller Dieseldirekteinspritzsysteme* (Universität Stuttgart: Dissertation)

[5] Heilig A 2013 *Untersuchung des Primärzerfalls von Hochdruckdieselsprays mittels optischer Messtechnik* (Leibniz Universität Hannover: Dissertation)

[6] E.H. Lehmann E H, Tremsin A, Grünzweig C, Johnson I, Boillat P and Josic L 2011 *Neutron imaging — Detector options in progress*, JINST 6 C01050

[7] http://www.psi.ch/niag/niag

# Cavitation Inception in Immersed Jet Shear Flows

**R.D. Lockett**[1], **N. Ndamuso**[2], **R. Price**[3]

[1,2]Dept of Mech. Eng. & Aero., The City University, London, EC1V 0HB, UK
[3]Shell Global Solutions, Brabazon House, Threapwood Rd, Mancs, M22 0RR, UK

r.d.lockett@city.ac.uk

**Abstract**. Cavitation inception occurring in immersed jets was investigated in a purpose-built mechanical flow rig. The rig utilized custom-built cylindrical and conical nozzles to direct high-velocity jets of variable concentration n-octane-hexadecane mixtures into a fused silica optically accessible receiver. The fluid pressure upstream and down-stream of the nozzles were manually controlled. The study employed a variety of acrylic and metal nozzles. The results show that the critical upstream pressure to downstream pressure ratio for incipient cavitation decreases with increasing n-octane concentration for the cylindrical nozzles, and increases with increasing n-octane concentration for the conical nozzle.

## 1. Introduction

Cavitation inception in immersed liquid jet flows into liquid has been investigated extensively over the years. The results obtained have been inconsistent and ambiguous, and largely dependent on the particular experimental configuration [1]. It is thought that cavitation inception in shear jet flows is specified by the behaviour of the turbulent vortices that develop in the liquid surrounding the jet [2]. The transition of a homogeneous liquid flow to a multi-phase cavitating flow is important in a number of engineering and scientific applications, ranging from cavitating flows in hydraulic pumps [3], in the neighbourhood of, and on the surface of marine propeller blades [4], to cavitating flows occurring inside diesel injector nozzles [5], and diesel fuel injection equipment (FIE) spill-return valves [6].

This paper is focused on the determination of the conditions necessary for cavitation to occur in diesel pumps and nozzles using diesel-like mixtures. Cavitation occurring in these devices may result in surface erosion and/or carbonaceous deposits [6]. This may lead to declining performance, and in extreme cases, equipment failure.

## 2. Experimental

### 2.1. Experimental Setup

A custom mechanical flow rig was designed and manufactured at City University in order to investigate and determine the variables affecting cavitation inception occurring in high-speed liquid jets into stationary liquid. The different components of the rig are shown in Figure 1 below. The high pressure bottles and control valve enabled the variation of absolute fluid pressure upstream of the nozzle from 0.2 MPa to 10.0 MPa, while the fused silica receiver and downstream control valve enabled the variation of absolute fluid pressure downstream of the nozzle from 0.1 MPa to 2.0 MPa. The fused silica receiver was 150 mm long, with an inner diameter of 60 mm inner diameter and an outer diameter of 70 mm. It provided optical access to the cavitating jet flow and was capable of supporting a fluid pressure of up to 2.7 MPa. Upstream and downstream fluid pressures were measured using pressure gauges. Custom cylindrical and conical nozzles with 0.14 mm and 0.25 mm nozzle exit diameters with sharp internal edges were manufactured in the City University School of Engineering workshop using acrylic (for optical access), brass and aluminium. Figure 2 shows the design profile of the three nozzle types employed in this study.

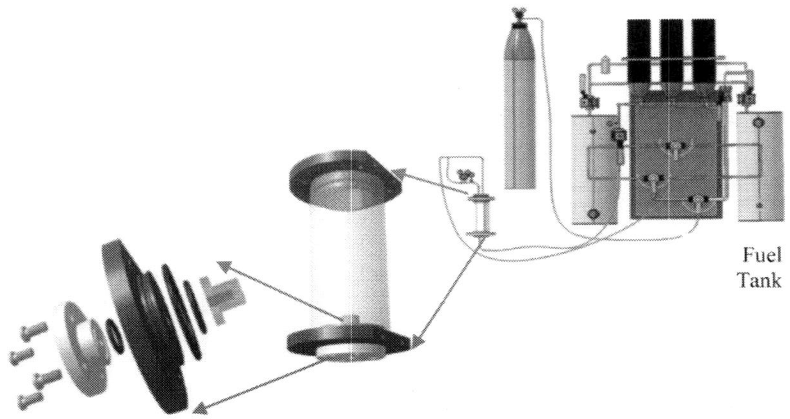


**Figure 1:** Detailed design of the mechanical flow rig used for the study of incipient cavitation.

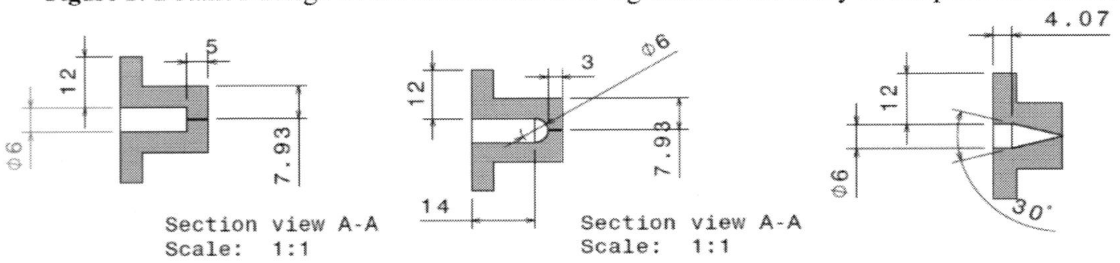


**Figure 2:** Nozzle design layout for cylindrical, hemispherical, and conical nozzles.

*2.2 Liquid Samples*

various mixtures of n-octane in n-hexadecane were prepared for the incipient cavitation experiment (0.1%, 1.0%, and 10.0% v/v). The hexadecane and n-octane were both 99% pure, supplied by Alfa-Aesar and Sigma-Aldrich respectively. The mixture properties are specified in Table 1 below.

| Mixture (% n-octane- hexa-decane v/v) | 0.10% | 1.0% | 10.0% |
|---|---|---|---|
| Density @ 20 °C (kg/l) | 0.7734 | 0.7727 | 0.7664 |
| Estimated Viscosity @ 20 °C (Ns) | 0.003462 | 0.003361 | 0.002548 |

**Table 1:** Properties of n-octane-hexadecane mixtures.

*2.3 Experimental Methodology*

The high pressure bottles containing approx. 10.5 l of fuel mixture in total were filled in sequence using 2 atm pressure obtained from a nitrogen bottle, which was connected to the tank through a priming valve. Once the high-pressure bottles were filled and isolated, the high-pressure valve was opened to admit high pressure nitrogen, which was employed to force the mixture through the nozzle into the fused silica receiver, and then on to the receiving tank at ambient pressure.

In order to obtain incipient cavitation in the nozzle or receiver, the downstream fluid pressure was initially set to ambient. The upstream fluid pressure was adjusted and set using the upstream needle valve. This caused the mixture to cavitate in the receiver. The downstream receiver mixture pressure was then increased until the cavitation stopped. Fine adjustment of the downstream mixture pressure enabled the determination of the critical downstream pressure corresponding to the point of cavitation

 IOP Publishing
doi:10.1088/1742-6596/656/1/012090

inception. The measurements were repeated 25 times each in order to obtain a mean and standard deviation of the downstream pressure for incipient cavitation.

## 3. Results

Cavitation inception could always be observed to occur in the receiver a short distance downstream of the nozzle exit. On occasion, cavitation inception was also observed to occur at the entrance to the nozzle hole. The incipient cavitation in the receiver was always coincident with cavitation being initiated at the entrance to the nozzle. The incipient cavitation was also observed to produce sound noise, which could be heard.

**Figure 3:** Near incipient cavitation in receiver downstream of nozzle.

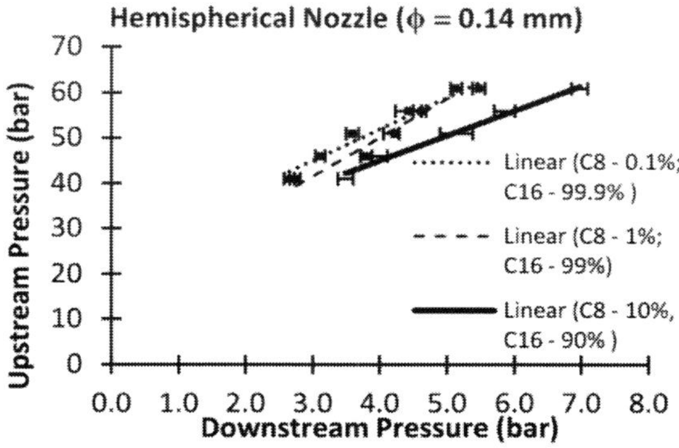


**Figure 4:** Upstream-downstream incipient cavitation pressure ratio for hemispherical bore nozzle.

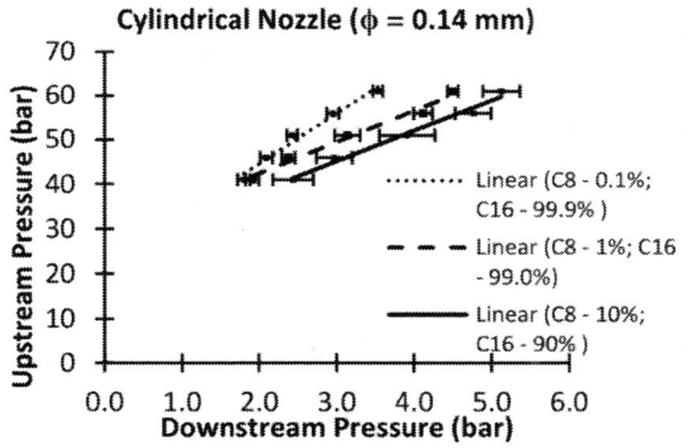


**Figure 5:** Upstream-downstream incipient cavit-pressure ratio for cylindrical bore nozzle.

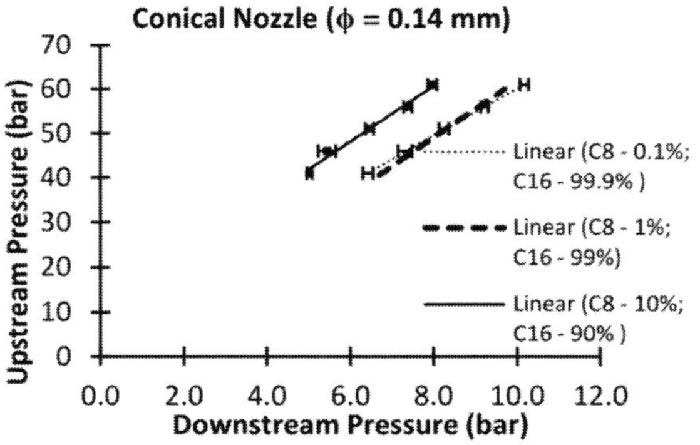


**Figure 6:** Upstream-downstream incipient cavitation pressure ratio for conical nozzle.

Figure 3 shows a photograph of near incipient cavitation occurring in the receiver, using a conical nozzle. The results of the incipient cavitation measurements are shown in the graphs contained in Figures 4 to 6. Figure 4 shows the upstream to downstream pressure ratio for incipient cavitation for the three n-octane-hexadecane mixtures (0.1%, 1.0%, 10.0% v/v), using a 3 mm long, 0.14 mm diameter cylindrical-hemispherical nozzle. Figure 5 shows the upstream to downstream pressure ratio for incipient cavitation for the three n-octane-hexadecane mixtures, using a 5 mm long, 0.14 mm diameter cylindrical-cylindrical nozzle. Figure 6 shows the upstream to downstream pressure ratio for incipient cavitation for the three n-octane-hexadecane mixtures, using a conical nozzle.

## 4. Discussion

The graphs presented in Figures 4 and 5 show that the pressure ratio required for incipient cavitation decreased as a function of n-octane concentration in the n-octane-hexadecane mixture, for both the cylindrical nozzles. This was to be expected, as the saturated vapour pressure of the mixture increased with increased n-octane concentration. However, the conical nozzle appeared to produce the opposite result: the pressure ratio for incipient cavitation increased with increasing n-octane concentration: a counter-intuitive finding.

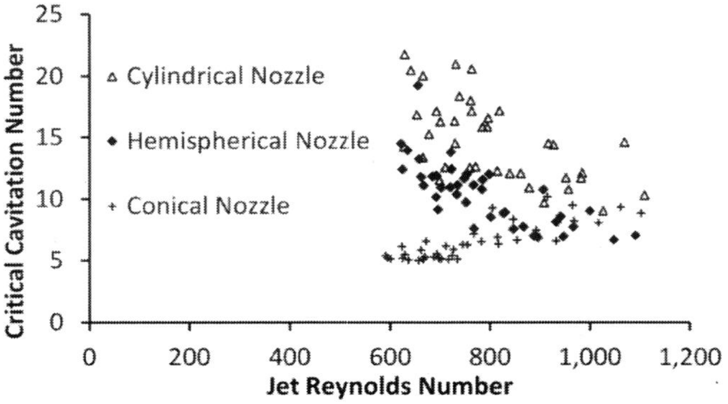


**Figure 7:** Scatter-graph of Critical Cavitation Number for Inception versus Jet Reynolds Number

Figure 7 shows the critical cavitation number for inception plotted against the jet Reynolds number for the jets developing from the three nozzles. The two cylindrical hole nozzles produce a decreasing critical cavitation number with jet Reynolds number, while the conical nozzle produces an increasing critical cavitation number with jet Reynolds number. This suggests that the character of the flow producing cavitation may be different for the conical nozzle compared with the cylindrical nozzles.

It is thought that cavitation inception in shear jet flows was caused by the fluid pressure in the centre of turbulent vortices adjacent to the jet dropping to below the critical pressure for cavitation to occur. In the case of the cylindrical nozzles, the internal flow was laminar, and the velocity profile was fully developed. This was in contrast to the flow exiting the conical nozzle, which had the character of a plug flow, or even a flow with maximum velocity near the edge of the jet, decreasing towards the central axis [7]. This difference in flow velocity profile appeared to gives rise to the difference in the upstream to downstream pressure ratio required for cavitation inception.

## 5. Conclusion

Cavitation inception has been investigated in mixtures of n-octane in hexadecane in a purpose designed and built mechanical flow rig. Two nozzles with 0.14 mm diameter cylindrical holes and a conical hole nozzle have been employed in the investigation. The flow in the cylindrical nozzle holes was fully developed, laminar flow, leading to a decrease in the upstream to downstream pressure ratio as a function of increasing n-octane concentration in hexadecane (increasing saturation vapour pressure). This is in contrast to the results obtained from the conical nozzle, which showed an increasing upstream to downstream pressure ratio with increasing n-octane concentration in hexadecane. The difference for cavitation inception obtained from the two nozzle types is thought to be caused by the variation in flow velocity profile and resultant turbulent shear mixing in the receiver.

## References

[1] Franc J-P., Michel J-M. (Eds) 2004 *Fundamentals of Cavitation* (Kluwer Academic Publishers).
[2] Xing T., Frankel S. 2002 *AIAA J.* **40:11** 2266 - 2276.
[3] Tullis J. 1989 *Hydraulics of Pipelines - Pumps, Valves, Cavitation, Transients* (John Wiley & Sons).
[4] Carlton J. 2012 *Marine Propellers and Propulsion* (Butterworth-Heinemann).
[5] Arcoumanis C., Flora H., Gavaises M., Badami M. 2000 *SAE* 2000-01-1249.
[6] Price R., Blazina D., Smith G., Davies T., 2015, *Fuel* **156**: 30 - 39.
[7] Rolon J., Veynante D., Martin J., 1991 *Exp. Fluids* **11**: 1 - 12.

**Acknowledgements:** The authors would like to acknowledge Shell Global Solutions for funding this work, and Mr G. Clow of City University for his assistance in building the rig.

# An Investigation into the Effect of Hydrodynamic Cavitation on Diesel using Optical Extinction

**R.D. Lockett[1], Z. Fatmi[2], O. Kuti[3], R. Price[4]**

[1,2,3]Department of Mechanical Engineering & Aeronautics, The City University, Northampton Square, London EC1V 0HB
[4]Shell Global Solutions, Brabazon House, Concord Business Park, Threapwood Road, Manchester, M22 0RR

r.d.lockett@city.ac.uk

**Abstract**. A conventional diesel and paraffinic-rich model diesel fuel were subjected to sustained cavitation in a custom-built high-pressure recirculation flow rig. Changes to the spectral extinction coefficient at 405 nm were measured using a simple optical arrangement. The spectral extinction coefficient at 405 nm for the conventional diesel sample was observed to increase to a maximum value and then asymptotically decrease to a steady-state value, while that for the paraffinic-rich model diesel was observed to progressively decrease. It is suggested that this is caused by the sonochemical pyrolysis of mono-aromatics to form primary soot-like carbonaceous particles, which then coagulate to form larger particles, which are then trapped by the filter, leading to a steady-state spectral absorbance.

## 1. Introduction

Modern diesel passenger car engines employ common rail fuel systems that typically operate at pressures in the range of 1,600 – 2,500 bar. The diesel fuel is supplied to a high-pressure pump, which compresses the diesel and provides it at high pressure to a common rail (pressure accumulator), which is connected to the fuel injectors. Metering of the diesel fuel flow in the pump is managed in different ways by different pump manufacturers. Some manufacturers meter the fuel at the inlet of their common rail pumps, while another utilizes a high-pressure pump spill return. However, there is a tendency for convergence on inlet metering for energy saving purposes. Both methods of metering create the opportunity for cavitation to occur during fluid compression, across the piston supply valves, and in the spill return. The injectors can be either solenoid or piezo actuated designs. The actuation of the injector, and the corresponding supply of fuel to the engine relies on the equalization of fuel pressure inside the diesel injector body through a spill path created by the lift of the actuator pin. The actuation of the spill port results in a pressure difference between the fluid pressure at the base of the injector (adjacent to the needle) and the fluid pressure above the needle, which induces the needle lift. Hence a significant proportion of the high pressure diesel fuel supplied to the pump and the injectors is returned to the tank at low pressure.

It is suggested that the fuel flow across the cylinder supply valves and spill ports in the pump, and the spill ports in the injectors result in cavitating flow, which may result in significant fuel degradation and ageing [1 - 8]. Indeed, cavitating fuel flows in diesel fuel injection equipment may be responsible for deposit formation in the injectors, leading to poor fuel-air mixing in the engine, and ultimately injector failure [1].

This consideration has led to a systematic investigation into the effect of sustained high pressure cavitation on diesel. Two cavitation recirculation flow rigs have been designed and manufactured, one of which was described in the paper of Lockett and Jeshani [9]. The experimental rig described in this paper is based on a similar design, using a high-pressure diesel pump.

## 2. Experimental

A schematic of the high-pressure cavitation flow rig is shown in Figure 1. It consists of a small tank, a low pressure priming pump, a fuel filter, a Denso Mk 2 diesel pump which is driven by a 7 kW electric motor, a common rail feeding a custom injector, which consists of a standard Delphi injector body housing an injector nozzle with a single 213 μm hole located on the symmetry axis. The diesel nozzle was provided with a 20% honed cylindrical nozzle hole. This nozzle was calibrated to provide a diesel mass flow rate of 0.3 l/min with 100 bar upstream pressure, ambient downstream pressure. A rail pressure of 1,650 bar is estimated to produce a diesel mass flow rate of approximately 1.2 l/min. The rig and nozzle are capable of producing nozzles flows with user-selectable cavitation numbers of up to 1,650, and Reynolds numbers of up to 50,000.

The nozzle discharges the cavitated diesel into an aluminium receiver. The diesel then passes from the receiver to a shell-and-tube heat exchanger, and then back to the tank. The rig is able to hold a maximum volume of approximately 4.0 litres. The volume of diesel used during a test is normally 3.1 ± 0.10 litres. After the fuel filter, an extra pipe passes a fraction of the diesel to a optically accessible cell. A 405 nm diode laser is directed through a 45° angled 90-10 TR beam-splitter, through the cell and onto a laser power meter. The 10 % reflected from the beam splitter is directed onto a second power meter (reference power meter). The optical arrangement and its calibration is described in detail in the IJER paper of Lockett and Jeshani [9].

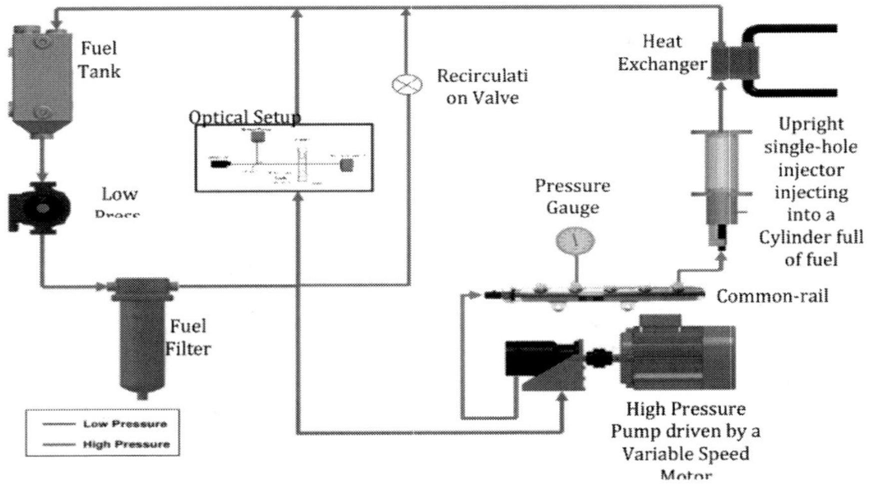


**Figure 1:** Schematic Diagram of High Pressure Diesel Cavitation Flow Rig

The rig was flushed twice with diesel prior to the cavitation flow experiment, in order to reduce contamination from the previous test to less than 1 %. In these experiments the diesel was subjected to 40 hours discontinuous cavitation flow at 550 bar, 1,100 bar and 1,650 bar (10 hours per day for 4 days). This was followed by subjecting diesel fuel and paraffin-rich model diesel (> 98 % paraffin content) to 130 hours of discontinuous cavitation (10 hours per day for 13 days).

## 3. Results

Figure 2 shows the variation in normalized spectral transmission at 405 nm with time for diesel samples cavitated at 550 bar, 1,100 bar and 1,650 bar. The spectral transmission coefficient is observed to decrease more quickly as a function of fuel pressure. The minimum spectral transmission coefficient achieved appears to be approximately the same for all three pressures.

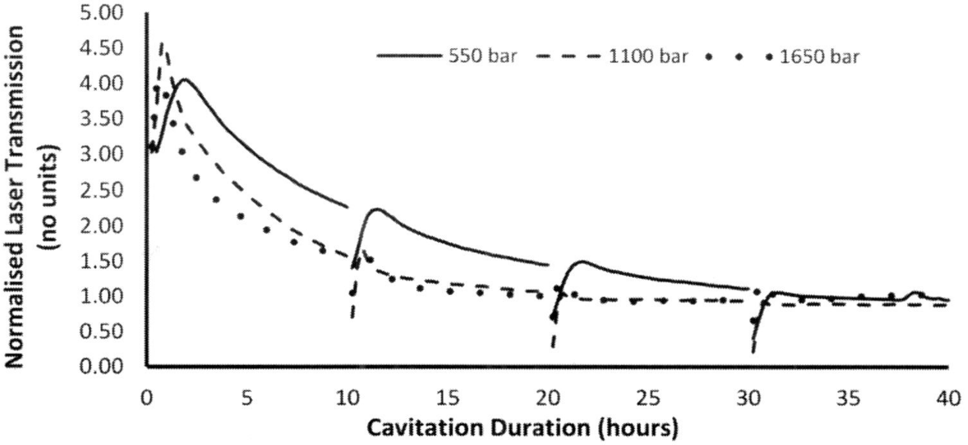


**Figure 2:** Graph of Normalised Transmission Coefficient of Diesel as a Function of Cavitation Time

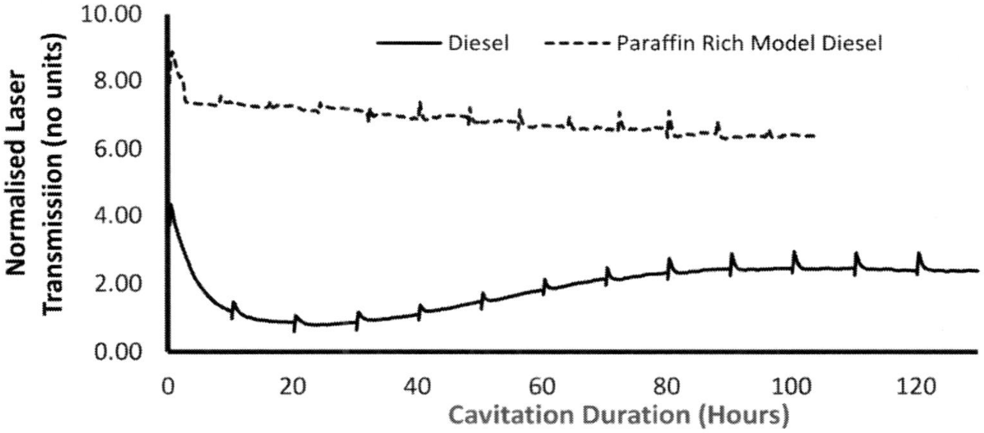


**Figure 3:** Graph of Normalised Transmission Coefficient of Diesel and Paraffin-Rich Model Diesel as a Function of Cavitation Time.

Figure 3 shows the variation in spectral transmission coefficient at 405 nm with time for diesel and paraffin-rich model diesel when the samples were subjected to cavitation flow at 1,650 bar for 130 hours. The spectral transmission coefficient for the diesel reaches a minimum value, and then begins to increase asymptotically to a steady-state value. On the other hand the spectral transmission coefficient for the paraffin-rich model diesel continues to decrease.

## 4. Discussion

The high-pressure cavitation flow rig employs a nozzle with an approximate mass flow rate of 1.2 l/min from a rail pressure of 1,650 bar. The volume of diesel in the rig is approximately 3.0 l. Therefore it takes approximately 2.5 minutes to cycle an entire volume of diesel through the injector, resulting in a net cycling rate of approximately 24 cycles/hour.

Figure 2 shows the difference in spectral transmission as a function of cavitation time for three rail pressures (550 bar, 1,100 bar and 1,650 bar). The characteristic timescales for the relative decrease in each curve (maximum to minimum) is approximately 23 hours (1,650 bar curve), 31 hours (1,100 bar curve), and 40 hours (550 bar curve). These results suggest that the rate of decrease of transmission coefficient is principally dependent on the cycle time (determined by the mass flow rate). Taking into account the effect of the cycle time on the spectral transmission profile suggests that the cavitation intensity (measured by its impact on the diesel sample) is relatively independent of rail pressure.

It is suggested that the difference between the recorded variations in normalised transmission coefficient between conventional diesel and the paraffin-rich model diesel is due to the impact of cavitation on major chemical components present in diesel, but not present in the paraffin-rich model diesel. Conventional diesel consists of approximately 70 % - 75 % paraffins, 20 % - 25 % mono-aromatics, 3 % di-aromatics, 1 % tri-aromatics, and other trace components (< 1 % additives, detergents and cleaning agents) [10]. Given that the mono-aromatics were the dominant components present in diesel that were not present in the paraffin-rich model diesel, it is suggested that the hydrodynamic cavitation induced in the high pressure cavitation flow rig produced a significant sonochemical pyrolysis effect on the mono-aromatics present in the diesel. 2-Column gas chromatography measurements and laser particle size and concentration measurements have been conducted on a number of cavitated diesel samples, which show that the concentration of mono-aromatics present in the diesel decreases during high-pressure cavitation, leading to the formation of a diesel particle suspension. This is to be published elsewhere [11].

The spectral transmission profile shown in Figure 3 is explained by the following argument. The hydrodynamic cavitation of the diesel results in sonochemical pyrolysis of the mono-aromatics, leading to primary soot-like particle formation (responsible for the decrease in spectral transmission coefficient). The formation of particles in this manner and their observation has been identified in R. Price et al. and G. Price et al. using ultrasound cavitation [1, 4]. The coagulation of primary particles to form larger particles (through Brownian motion and coagulation kinetics) [1], together with large particle filtration trapping produces the later increase in spectral transmission to a steady-state quasi-equilibrium value.

## 5. Conclusion

A conventional diesel fuel and a paraffin-rich model diesel have been subjected to long term cavitation tests in a new high-pressure cavitation flow rig. Optical absorption measurements conducted during the cavitation have demonstrated the effect of pressure on the spectral transmission coefficient. The results suggest that the profiles are affected principally by the cycle time, and that the cavitation intensity is independent of the rail pressure. Long term cavitation results in the spectral transmission coefficient decreasing to a minimum, followed by a rise to a steady-state quasi-equilibrium value. A sonochemistry pyrolysis mechanism, together with particle coagulation kinetics and large particle filtration, is suggested to explain the observed spectral transmission coefficient profile with time.

### References

[1] Price R., Blazina D., Smith G., Davies T., 2015, *Fuel* **156**: 30 - 39.
[2] Suslick KS, Gawienowski JJ, Schubert PF and Wang HH., 1983 *J Phys Chem* **87(13)**: 2299–2301.
[3] Price GJ and McCollom M. 1995, *Ultrason Sonochem* **2(2)**: S67–S70.
[4] Price GJ and McCollom M., 1995, *Fuel* **74(9)**: 1394–1397.
[5] Ullman J, Geduldig M, Stutzenberger H, Caprotti R, Balfour G. SAE Paper 2008-01-0926, 2008.
[6] Pehan S, Jerman MS, Kegl M and Kegl B. 2009 *Fuel* **88(6)**: 970–979.
[7] Caprotti R, Breakspear A, Graupner O, Klaua T and Kohnen O. SAE Paper 2006-01-3359, 2006.
[8] DI Common Rail Diesel Engine Nozzle Coking Test, 2008, CEC F98-08 (S) DW-10.
[9] Lockett R., Jeshani M. 2013 *Int. J. Eng. Res.* **14:6** 606 - 621.
[10] Erwin J., Division of Fuel Chemistry, American Chemical Society 1992 *Symposium on Processing and Product Selectivity of Synthetic Fuels*, 1915 - 1923.
[11] Lockett R., Jeshani M., 2015 to be published.

**Acknowledgements:**
The authors would like to acknowledge Shell Global Solutions for funding this work, and the technical assistance of G. Clow and J. Ford of City University.

Journal of Physics: Conference Series **656** (2015) 012092 doi:10.1088/1742-6596/656/1/012092

# Numerical Simulation of the Effect of 3D Needle Movement on Cavitation and Spray Formation in a Diesel Injector

**B Mandumpala Devassy[1], W Edelbauer[1], D Greif[2]**

[1]AVL List GmbH, Hans List Platz 1, 8020 Graz, Austria
[2]AVL-AST d.o.o., Ul. kneza Koclja 22, 2000 Maribor, Slovenia

E-mail: Bejoy.MandumpalaDevassy@avl.com

**Abstract**. Cavitation and its effect on spray formation and its dispersion play a crucial role in proper engine combustion and controlled emission. This study focuses on these effects in a typical common rail 6-hole diesel injector accounting for 3D needle movement and flow compressibility effects. Coupled numerical simulations using 1D and 3D CFD codes are used for this investigation. Previous studies in this direction have already presented a detailed structure of the adopted methodology. Compared to the previous analysis, the present study investigates the effect of 3D needle movement and cavitation on the spray formation for pilot and main injection events for a typical diesel engine operating point. The present setup performs a 3D compressible multiphase simulation coupled with a standalone 1D high pressure flow simulation. The simulation proceeds by the mutual communication between 1D and 3D solvers. In this work a typical common rail injector with a mini-sac nozzle is studied. The lateral and radial movement of the needle and its effect on the cavitation generation and the subsequent spray penetration are analyzed. The result indicates the effect of compressibility of the liquid on damping the needle forces, and also the difference in the spray penetration levels due to the asymmetrical flow field. Therefore, this work intends to provide an efficient and user-friendly engineering tool for simulating a complete fuel injector including spray propagation.

## 1. Introduction

Fuel injectors are one of the essential components in an internal combustion engine system. High pressure injection systems employ adequate atomization and mixture formation which are required for improved engine performance and reduced emissions. The present work addresses high pressure injection events including fuel injection and sprays by using a coupled 1D/3D nozzle flow simulation technique. The method has already been discussed in previous studies, as shown in [1] and [2]. The present approach introduces the co-simulation technique for pilot and main injections, and it is extended in such a way that the 3D motion of the needle, with 3 degrees-of-freedom, can be simulated in the 1D code: AVL BOOST- HYDSIM [3], and the corresponding mesh movement is represented in the 3D code: AVL FIRE® [4]. The longitudinal and transverse motion of the needle is coupled with the cavitating 3D nozzle flow. At selected time instants the flow field at the nozzle exit is written to separate data files, called nozzle-files. These files provide the boundary conditions for the sub-sequent Euler-Lagrangian spray simulation of the combustion chamber. The high pressure liquid medium is treated compressible [1], and the effect of needle damping forces is studied thoroughly. Cavitation and spray models have already been described in previous works, and the main approaches are based on the Euler-Eulerian, Euler-Lagrangian formulation, or their combination, as discussed in [5] and [6] for

the spray and in [7] and [8] for the cavitation and cavitation erosion models. The authors would like to refer to their previous works in [1] for detailing the applied 1D/2D/3D coupling interface, the governing equations and different models employed for the current simulation. The present study intends to investigate the spray injected from an eroded-asymmetrical and a non-eroded-symmetrical nozzle, which is already described in [1].
The paper is organized as follows: the second section presents the details of the test case with the computational mesh, and initial and boundary conditions. Section 3 provides a brief analysis of the results with a detailed discussion, and Section 4 contains conclusions and discusses future work.

## 2. Six-hole Fuel Injection Test Setup

The fuel injector with the combustion chamber geometry is shown in Figure 1. The motivation of this study is to identify the effect of 3D needle movement, compressiblity and nozzle asymmetry on the spray formation and its characteristics. Figure 1 illustrates the mesh of the eroded 6-hole injector in a zoom-in view, and the piston bowl geometry of the combustion chamber. Typical diesel engine conditions at a full-load operation point, as listed in Table 1, are applied for the spray simulation. The computational mesh of the injector flow simulation contains the nozzle sac, the needle shaft, the needle seat and the nozzle holes. Here, the asymmetric geometry is supposed to mimic the production inconsistency and/or possible erosion effects. For the actual injector the eroded regions in the nozzle holes have been randomly chosen. The full needle movement with three degrees of freedom is studied. Consequently a full 360 degree geometry is necessary for the current investigation.
In the present study the 3D-1D coupled simulation runs for 57 hrs using 16 Intel(R) Xeon(R) CPU X5620 @ 2.40GHz processors.

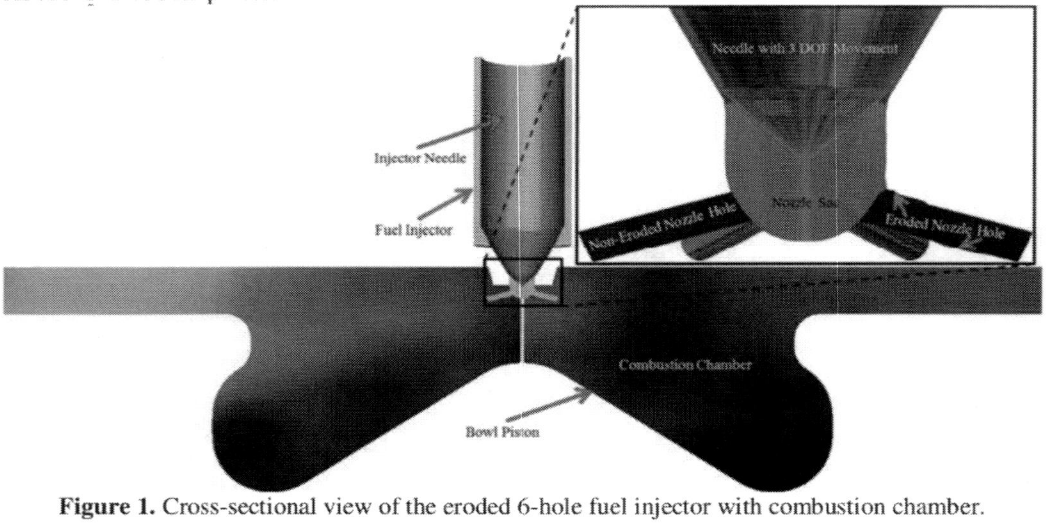


**Figure 1.** Cross-sectional view of the eroded 6-hole fuel injector with combustion chamber.

**Table 1.** Operating Conditions.

| | |
|---|---|
| Liquid density | 830 kg/m$^3$ |
| Temperature of the fuel inlet | 333.15 K |
| Each nozzle hole diameter, D | 218.5 μm |
| Injection pressure | 160 MPa |
| Pressure in the combustion chamber | 6.7 MPa |
| Temperature in the combustion chamber | 900 K |
| Piston, head and liner wall boundaries | 550, 500 & 400 K |

Table 1 shows the operating conditions of the simulation. The calculation is carried out for 2.3 ms with a rail pressure of 160 MPa. Pilot and main injection events are taken into account and cavitation as well as spray formation is studied for the symmetrical and the asymmetrical nozzle condition.

## 3. Results and Discussion

For both cases a considerable quantity of vapor has been detected. For the symmetrical mesh, the vapor volume fraction appears to be symmetric, while for the asymmetric mesh, owing to the eroded nozzle-hole regions, the vapor regions are clearly different from hole to hole. Note, that different holes exhibit different cavitation patterns. In several holes the cavitation cloud reaches the nozzle orifice. Figure 2 and Figure 3 present the liquid mass flow rates, and the cumulative quantity of liquid fuel mass injected for the entire injection cycle. The difference in the curves is caused by the differences in the vapor generation, where symmetrical geometry produces more vapor than the asymmetrical geometry. This behavior is also visible in the zoom-in view in Figure 4.

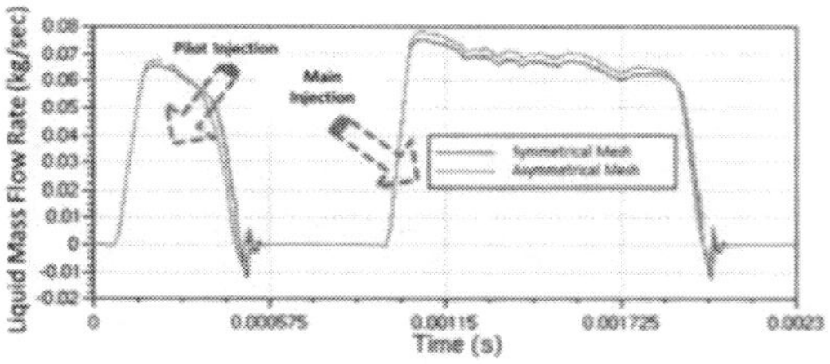


**Figure 2.** Liquid mass flow rate for symmetrical and asymmetrical geometries.

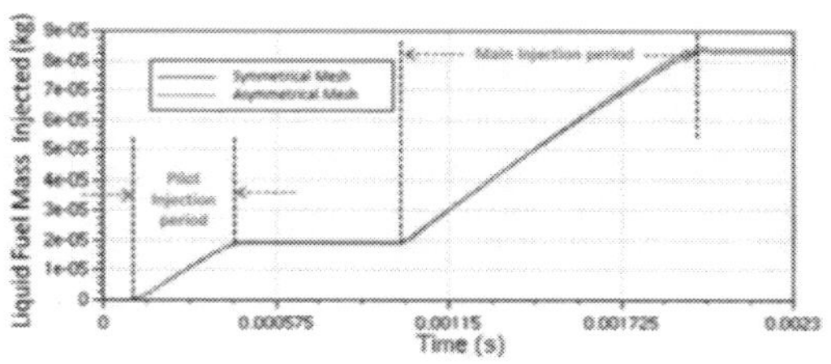


**Figure 3.** Liquid and vapor fuel mass injected for symmetrical and asymmetrical geometries.

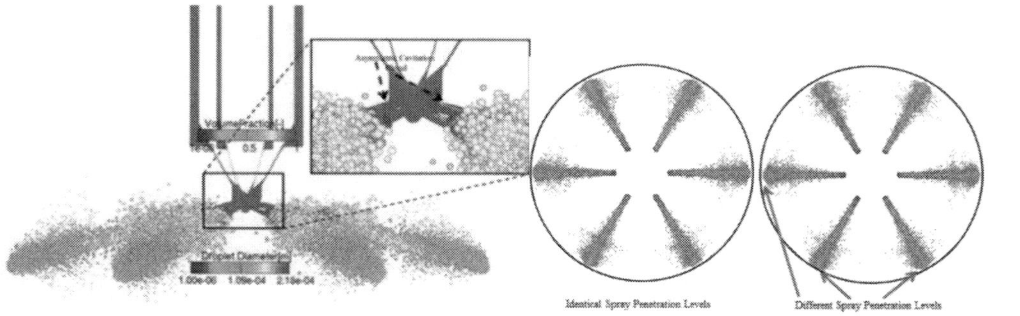

**Figure 4. Left**: Fuel injection and sprays from an eroded 6-hole asymmetrical nozzle at 0.16ms after start of injection. **Right**: Different spray penetration levels from a six-hole symmetrical (left) and asymmetrical nozzle holes (right).

In the applied workflow two separate calculations, using the Eulerian approach as a two-fluid model inside the nozzle tip, and the Lagrangian Discrete Droplet Model (DDM) approach outside the injector in the spray regions, are performed. The flow data in the nozzle orifice are transferred to the DDM model [5] as initial conditions. Additionally, a primary break-up model is applied within the DDM.
The results show that the asymmetrical flow field generated from an eroded nozzle (also shown in Figure 4) develops an asymmetrical spray mainly due to the variation in cavitation fields and the oscillatory needle movement. This is clearly notable by visualizing the spray characteristics, especially the spray tip penetration, as shown in Figure 4 (right). There one can observe slight differences in the spray contour and clear differences in the spray tip penetration between eroded and non-eroded nozzle

holes. Figure 5 (left) gives detailed insight of the eroded nozzle-hole positions. The induced asymmetricity here is supposed to mimic the production inconsistency and the erosion effects. For the actual injector the eroded regions in the nozzle holes have been randomly chosen: denoted as 'E' in Figure 5 (left). Due to the difference in the flow field (see Figure 4), the injected fuel mass varies between the nozzle holes, as shown in the bar chart of Figure 5 (right). This difference is mainly due to the uneven distribution of cavitation regions causing the difference in the spray penetration length.

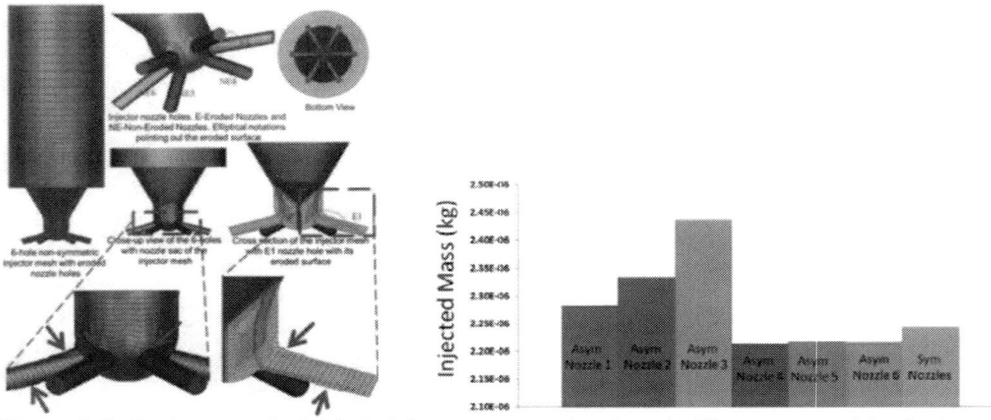


**Figure 5. Left**: Asymmetrical 6-hole injector computational mesh. The assymmetricity is due to the eroded nozzle holes sketched as "E". **Right**: Comparison of the injected fuel mass for different nozzle holes of the asymmetric and the symmetric injector nozzle holes until 0.3ms after start of the pilot injection.

## 4. Conclusions

This work intends to provide a novel flow simulation technique for studying nozzle cavitation and spray formation for pilot and main injection events in a multi-hole fuel injector. The results indicate clear differences in terms of cavitation generation and flow fields for both, the symmetric and the asymmetric, test case geometries. Owing to the asymmetrical flow field in the injector, differences in the spray tip penetration are clearly visible. This study combines AVL FIRE® and AVL BOOST-HYDSIM tools in order to simulate the flow in a high-pressure fuel injector and the subsequent spray formation. Liquid compressibility, cavitation and 3D needle movement are taken into account.

## References

[1] Mandumpala Devassy B, Caika V, Sampl P, Edelbauer W, and Greif D 2015 *Proc. Institution of Mechanical Engineers Conference on Fuel Systems for IC Engines (London, March 10-11).*

[2] Caika V, Sampl P, and Greif D 2013 Integrated 1D/2D/3D Simulation of Fuel Injection and Nozzle Cavitation *SAE Int. J. Engines* **6** (3) (doi:10.4271/2013-24-0006).

[3] AVL BOOST HYDSIM 2013 User's Guide, Graz.

[4] AVL FIRE®, Eulerian Multiphase 2014 Graz.

[5] Von Berg E, Edelbauer W, Alajbegovic A, Tatschl R, Volmajer M, Kegl B and Ganippa L C 2005 Coupled Simulations of Nozzle Flow, Primary Fuel Jet Breakup, and Spray Formation *Journal of Engineering for Gas Turbines and Power* **127** (4) (ASME, New York) pp 897-907.

[6] Edelbauer W 2014 Coupling of 3D Eulerian and Lagrangian Spray Approaches in Industrial Combustion Engine Simulation *Journal of Energy and Power Engineering* **8** pp 190-200.

[7] Alajbegovic A, Grogger H and Phillip H 1999 Calculation of Transient Cavitation in Nozzle using Two-Fluid Method *Int. Conf. on Liquid Atomization and Spray Systems* (Indianapolis, IN, USA, 209-213).

[8] Greif D and Wang D M 2007 Modeling of Cavitation Erosion within Diesel Injector and its Effect on Spray Propagation *Int. Conference Towards Clean Diesel Engines* (Ischia, Italy).

Journal of Physics: Conference Series **656** (2015) 012093 doi:10.1088/1742-6596/656/1/012093

# Influence of cavitation on near nozzle exit spray

**M Mirshahi[1,2], Y Yan[2], JM Nouri[2]**

[2] City University London, Northampton Square EC1V 0HB, United Kingdom

Milad.Mirshahi.1@city.ac.uk

**Abstract.** The importance of cavitation inside multi-hole injectors for direct injection internal combustion (IC) engineshas been addressed in many previous investigations. Still, the effect of cavitation on jet spray, its stability and liquid breakup and atomisation is not yet fully understood. The current experimental work aims to address some of these issues. It focuses on the initiation and development of cavitation inside a 7x enlarged transparent model of a symmetric 6-hole spark ignition direct injection (SIDI) injector and quantifies the effect of cavitation on near-nozzle spray cone angle and stability utilising high speed Mie scattering visualisation. The regions studied include the full length of the nozzle and its exitjet spray wherethe primary breakup takes place.

## 1. Introduction

Multi-hole injectors for IC engines are wildly used in both diesel and gasoline engines and have many advantageous such as their flexibility in terms of number of holes and their arrangements which can be fitted to different combustion cylinder head, their ability to produce stable spray which is critical for spray guided combustion concept and their ability to be used at high injection pressure to ensure enhanced atomization and evaporation [1, 2]. Experimental studies of in-nozzle flow characteristics, in particular, cavitation in enlarged transparent models of multi-hole injectors and spray characteristics of real size ones have been extensively investigated by City research group [3-10]. These studies revealed the formation of three different types of cavitation, namely needle, geometric and string cavitation. Cavitation within the nozzle holes of multi-hole SIDI injectors can lead to significant spray instabilities and lack of targeting which can cause problems in combustion when operating in stratified mode. The simultaneous presence of these types of cavitation regimes creates a complex two-phase flow structure in the nozzle holes which seems to be responsible for hole-to-hole and cycle-to-cycle spray variations. Investigations [7-10] revealed the cavitation sites, their frequency of formation, and the locations of erosion within the nozzle. In addition, they showed how to minimise surface erosion using tapered converging holes. Large variations in the instantaneous fuel injection quantity of individual injection holes have been recorded when a cavitation string is observed inside them. Combination with model predictions has revealed that the observed reduction in the individual hole flow rate is partially attributed to the increased vapour fraction inside the hole when a string is present; the vortex flow developing upstream of the hole entry is the main reason for the observed trend [9]. Despite available information, the true dynamics and mechanisms of cavitation development and its link with near nozzle liquid break-up at the exit of the nozzle hole are not fully understood. Therefore deeper understanding of in-nozzle flow characteristics and the dynamic link between cavitation and the emerging jet spray is essential. This calls for high temporal simultaneous imaging of the in-nozzle

[1] Corresponding author

cavitation and the exiting liquid jet spray in order to establish how cavitation influence spray shape, instability and break up. Thus the focus of this study is to establish these links and their mechanisms in a 7x 3-D transparent models of a six-hole mini-sac-type nozzle utilised with gasoline direct injection; a high speed camera has been used to obtain simultaneous images at a rate of 50,000fps.

## 2. Experimental test-rig

The experimental set up is shown in fig. 1a. A multistage centrifugal pump provides the upstream pressure in the transparent enlarged model. Working fluid is water at 25°C. A long telescopic extension tube was used on the CCD camera to maximise the magnification ratio of the screening area. The injection pressure was varied from 0.5 bar to 3bar and the sprays from each nozzle hole were injected into the atmosphere and were collected back into the supply tank. A 100mm diameter focusing lens with the focal length of 300mm was used to increase the light intensity around the nozzle area to more than 2 million Lux. Three set of 45° hot mirrors were used to filter out infrared wavelength in order to protect the Perspex nozzle from high temperatures. To ensure capturing high quality images, 2 set of cooling fans were used to decrease the temperature further around the nozzle and also to remove the water mists of the spray away from the near nozzle towards the suction collector to avoid fouling the imaging window. The injector assembly has a needle lift mechanism adjusted by a micrometer to set the exact needle height. The flow inside the nozzles is continuous, i.e. steady state flow condition, which means that transient nature of needle during its opening and closing processes is absent. The enlarged transparent model is geometry similar to the real-size injector and operates at similar Reynolds numbers.

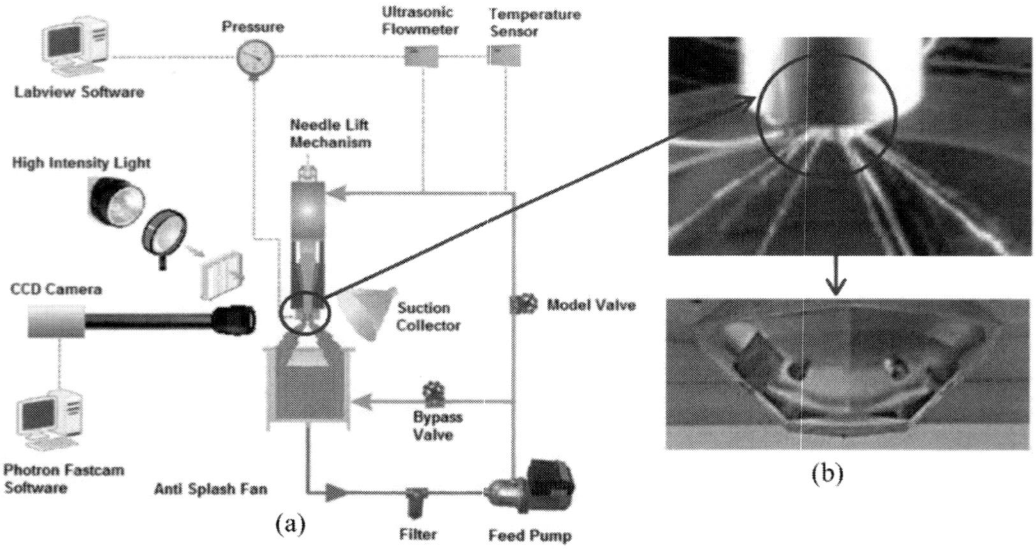


**Fig. 1.**(a) The closed loop steady-state flow rig of the transparent 7 times enlarged injector with the schematic diagram of flow circuit (left); (b) Close-up of transparent nozzle injector assembly.

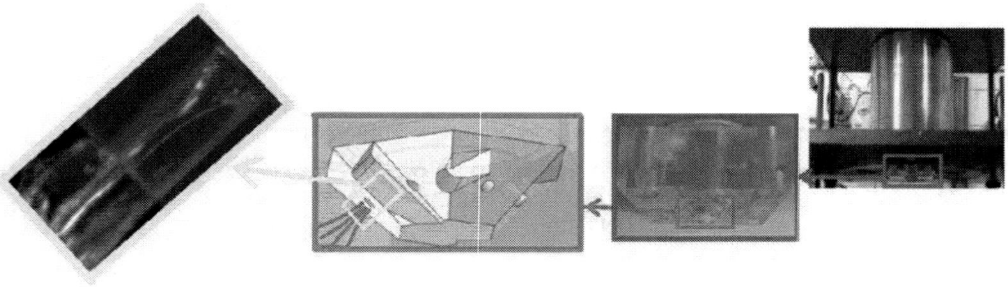

**Fig. 2.**Front view of the 7-times enlarged model installed on the test rig (red box) and schematic of the imaging area and the near-nozzle jet(yellow box).

## 3. Results

Sample images of the results are shown below. Fig. 3 shows a sequence of 6 images of the in-nozzle flow and the emerging spray at full lift (1.05mm) at cavitation numbers of CN = 0.75, 1, 1.5, 1.75, 2, and 2.5. Cavitation number is defined here as CN= $(P_{inj} - P_{back})/(P_{back} - P_{vap})$. Cavitation was not present at CN range between 0.75 to 1.0. Increasing cavitation above 1 induces cavitation in the nozzle and resulted to full film cavitation from CN=2 and above. It can be claimed that as the cavitation number rises from 1.0 to 1.5, the upper near-nozzle spray angle increases due to cavitation as can be seen from the images shown in fig. 3.

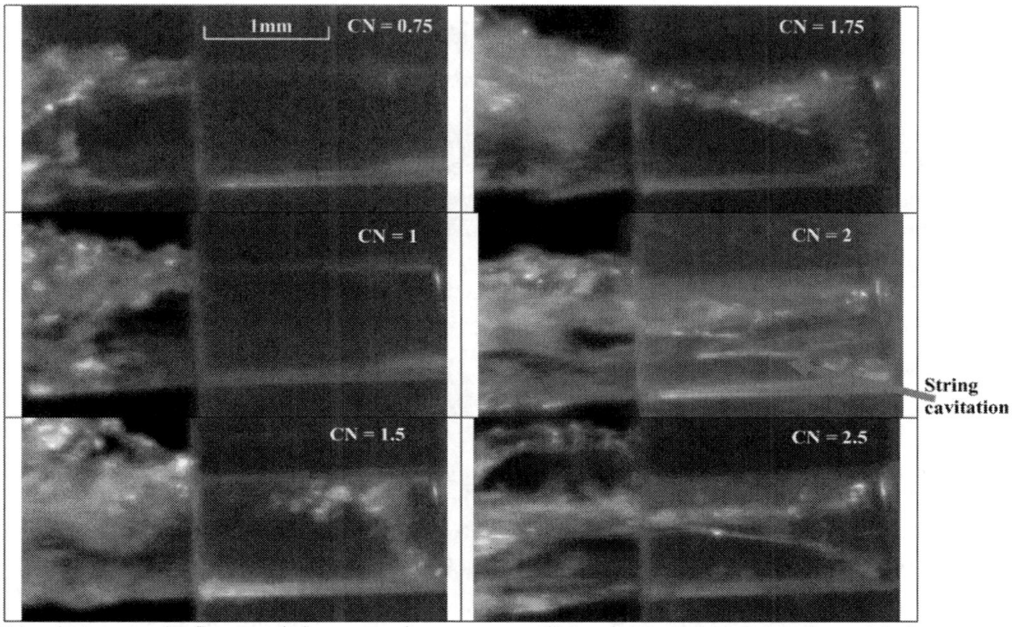


**Fig. 3.** In-nozzle flow and the emerging spray at CN = 0.75, 1, 1.5, 1.75, 2, and 2.5.

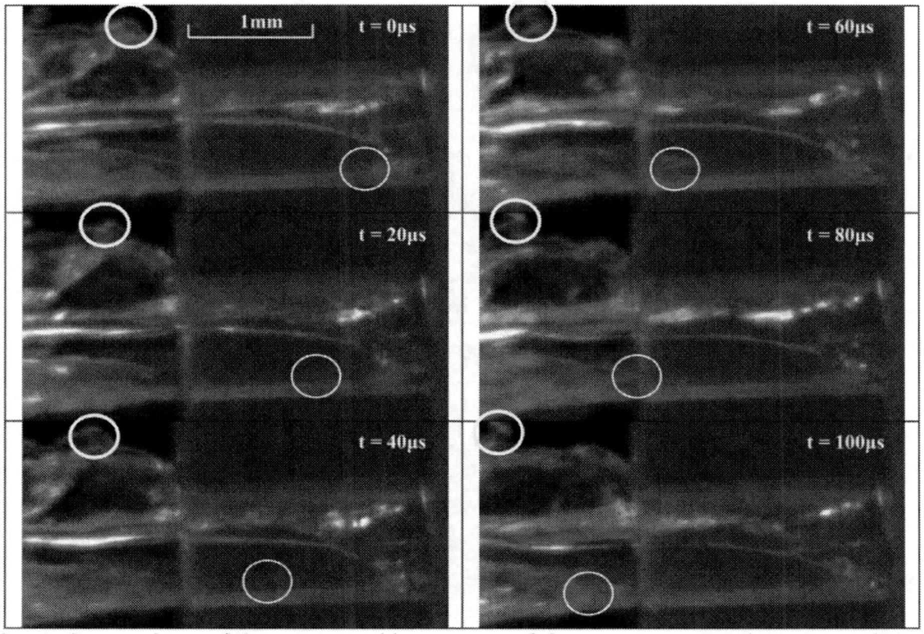


**Fig. 4.** Comparison of the upper and lower part of the very near-nozzle spray at CN=2.

As cavitation number gradually increases from 1.5 to 2.0, the geometric cavitation becomes more intense and stretches down the nozzle towards the exit. Another observation is that the structure of the geometric cavitation on top of the nozzle undergoes a transition with increase in CN from a cluster of bubbly cloud vapour into a smoothly horseshoe film cavitation as evident from images at CN=2 and CN=2.5 respectively; this had impact on the structure of the upper part of the near exit spray. Fig. 4 shows a sequence of 6 consecutive images of the in-nozzle flow and the emerging spray at CN = 2.0 during 100μs. The sequence depicts the near-nozzle structure of the spray with the extension of the smooth curved line that is stretched out of the nozzle. At t = 0, a pocket of bubbles was observed at the lower part of the nozzle inlet. This moves down and reaches the nozzle exit at t=100μs. This shows that the average velocity of this pocket is around 20m/s. However the upper part of the spray has much lower velocity (around 7m/s; tracked by the bold yellow circles) compared to the bottom part of the nozzle where the bulk of the liquid exists.

## 4. Conclusion

Simultaneous imaging of in-nozzle cavitation and emerging spray have been obtained in a 7 times enlarged model of SIDI injector at different injection pressures and needle lifts under steady state conditions. Observations showed that bubble cloud cavitation influence the structure and the angle of emerging spray, in particular, on the upper part of the nozzle which could lead to spray instabilities. It has also been observed that once the geometric cavitation reaches the exit of the nozzle, its structure transform from a bubbly cloud into a smoothly curved lines film cavitation, which greatly influences the exiting spray structure. Further image analysis showed lower spray velocities on the upper part of the nozzle, where the spray is dominated by in-nozzle cavitating flow, than the lower part.

## 5. References

[1] Fraidl G,et al.2003 The potential of next generation gasoline direct injection technologies. AssociazioneTecnicadell'Automobile, **56**: p. 5-17.
[2] Wirth M. et al.2004A Cost Optimised Gasoline Spray Guided Direct Injection System for Improved Fuel Economy, Seminar on Fuel Economy and Engine Downsizing. IMechE symposium, 13 May 2004London.
[3] Nouri J.M., et al.2012Effect of viscosity, temperature and nozzle length-to-diameter ratio on internal flow and cavitation in a multi-hole injector.In Proc. IMechE Seminar on Fuel Injection. 14-15March 2012, London.
[4] Arcoumanis et al, 1998 Analysis of the flow in the nozzle of a vertical multi-hole diesel engine injector. Transactions Journal of Engines, SAE paper 98-08-11.
[5] Soteriou, et al. 1993 Cavitation Hydraulic Flip and Atomization in Direct Injection Diesel Sprays, IMechE 1993. London.
[6] Roth H, et al.2002 Cavitation Initiation, Its Development and Link with Flow Turbulence in Diesel Injector Nozzles. Transactions Journal of Engines, SAE paper 2002-01-0214, SAE Transactions, **111**(3): p. 561-580.
[7] Gavaises, M. et al. 2009 Characterization of string cavitation in large-scale Diesel nozzles with tapered holes. Int J Physics of Fluids, **21**: p. 052-107.
[8] Gavaises, M., 2007 Link Between Cavitation Development and Erosion Damage in Diesel Injector Nozzles. SAE paper 2007-01-0246.
[9] Andriotis, A., Gavaises, M. & Arcoumanis, C. 2008. Vortex flow and cavitation in diesel injector nozzles. Journal of Fluid Mechanics, 610, pp. 195-215
[10] M Gavaises et al. 2008. Flow in valve covered orifice nozzles with cylindrical and tapered holes and link to cavitation erosionand engine exhaust emissions. International Journal of Engine Research **9**: 43-52.

# Application of cone-beam micro-CT on high-speed Diesel flows and quantitative cavitation measurements

**N. Mitroglou[1], M. Lorenzi[1], M. Santini[2], M. Gavaises[1], D. Assanis[3]**

[1] School of Mathematics Computer Science and Engineering, City University London, UK
[2]Department of Engineering and Applied Sciences, University of Bergamo, Italy
[3]Department of Mechanical Engineering, StonyBrook University, USA

n.mitroglou@city.ac.uk

**Abstract**. X-ray computed tomography (CT) is well-known and widely used in the medical sector for diagnosis of various illnesses. The technique is based on the absorption (i.e. attenuation) of the ionising electromagnetic radiation by the object. The amount of energy to be absorbed depends on the density and its thickness; the transmitted radiation through the object is then compared to the incident radiation that leads to a reconstruction of attenuation coefficients versus spatial position in the object. Thus, the resulting three-dimensional slices of the object are used (a) to identify internal geometric features of objects, and (b) to distinguish between media of different densities, i.e. liquid and air/vapour. In this study, the geometry extraction capability has been applied on time-averaged cavitation pocket shapes, as well as, the capability of density differentiation measurements on Diesel fuel flows. Results appear promising and pose a challenge in providing quantitative measurements of cavitation vapour fraction inside an injection hole.

## 1. Introduction

Fuel injection systems for modern direct-injection Diesel engines play a distinct and paramount role that is well recognised as a means of controlling their performance and meeting the ever more stringent emission regulations. Cavitation phenomena become the dominant and frequently uncontrolled flow characteristics that affect the durability of the fuel system and the properties of the near-nozzle spray. Investigations over the last few years have demonstrated that Diesel injector nozzles cavitate [1, 2] under typical operating conditions, a fact that complicates further the already complex design of high-pressure Diesel injection systems. Apart from the manufacturers' quest for environmentally friendly engines, the fuel industry has recently joined this race by providing increased quality Diesel fuel, compared to previous years. The latter is primarily achieved by standardised procedures at the refinery, however, in most cases it is cost-efficient to use fuel additives in order to achieve Diesel of higher standards. Almost all diesel fuel sold commercially contains additives that are necessary to maintain efficient FIE operation, engine performance and reduce emissions. Due to their proprietary nature and relatively complex chemical structures, additives and mechanisms by which they affect injector flow and/or spray formation are not yet well-known. Detergents are one of the most important components in a diesel additive package and are used for deposit control purposes [3]. Their mechanisms include keeping new injectors clean by generating a protective film on metal surfaces and by preventing agglomeration of deposit precursors, as well as cleaning up fouled injectors by deposit removal and dispersion [4].

Chemistries used for deposit control applications are not unlike household detergents in their basic structure. Recently, a significant and unexpected increase in power was reported with the use of fuel containing specific quaternary ammonium salts [5]. PIB-based quaternary ammonium salts at a treat rate as low as 500 ppm and 1000 ppm resulted in a significant power gain compared to un-additised fuel. The mechanism by which the increase in power was achieved is not known; however, there is evidence that the flow rate through the injector is altered. Based on these findings, the present study attempts to investigate the tendency in Diesel fuel to cavitate when enriched with the above type of additives.
The success of such an experimental investigation depends on the measurement technique and the accuracy of the acquired results. X-ray computed tomography is widely spread in the medical sector, as means of measuring density variations in organs of the human body. When multiple projections are gathered at various angles, then, reconstruction of the single radiographies results in a 3-D representation of the shape of the object by means of absorption coefficients. This capability is exploited in this study in order to measure differences in density between liquid and vapour Diesel.

## 2. Experimental set-up

Multi-hole Diesel injector nozzles present strong flow interactions among the several orifices [2, 6]. The developed complicated flow phenomena render it impossible to study the fundamentals of cavitation inception and development and compare subtle differences for a single orifice. Therefore, a prototype nozzle was designed with certain features that enable unobstructed access to the full length of the region of interest. In detail, as illustrated in Figure 1, the nozzle comprises a single injection hole, whose axis is at an offset to the main axis of symmetry of the device. The latter, and the non-symmetric needle, ensure a flow asymmetry that causes cavitation mainly on one side of the orifice entry, enabling unobstructed view of the cavitation pocket created from any angle.

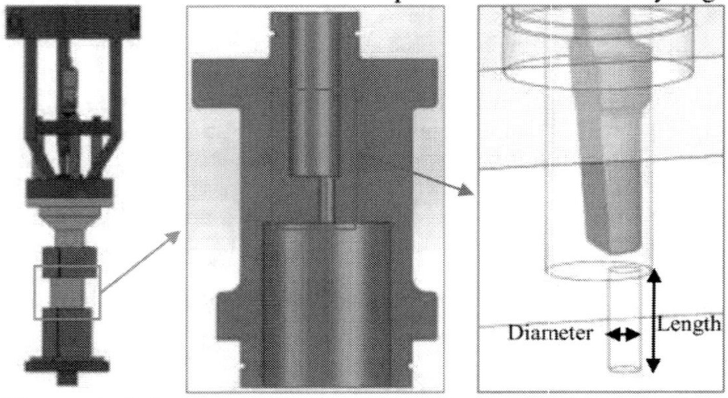


**Figure 1.** Schematic representation of assembled nozzle (left-most) and cut-out of the nozzle specific design parameters.

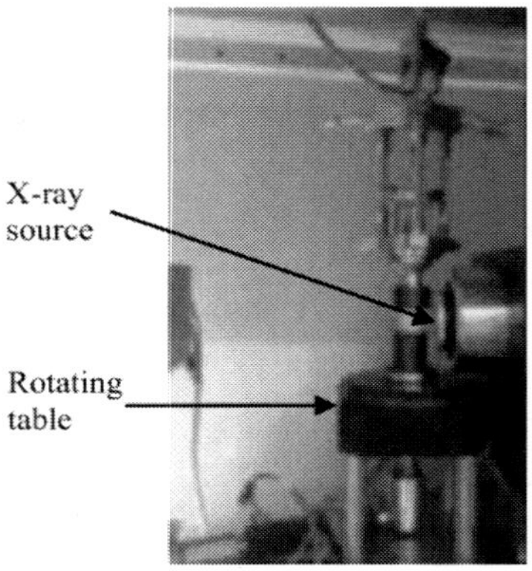


**Figure 2.** Test-rig inside the mCT scanner

The test-rig has been designed specifically for use inside a micro-CT scanner developed by University of Bergamo (Figure 2). The limitations in x-ray detector technology (acquisition freq. ~30Hz) and x-ray energy available from a polychromatic source, dictated the design of a steady state experiment. Therefore, the needle present in the nozzle is set at a fixed position and does not change during a complete 360° acquisition. Moreover, the hydraulic unit responsible for pressurised fuel delivery provides a maximum pressure of 140bar. The latter is not representative fuel injection pressure for common rail Diesel systems, however, the nozzle material that had to be transparent to x-rays is selected to be Peek and its strength is limited; additionally, it has been our intention to study the fundamentals of cavitation inception and development. Nevertheless, the test-rig is equipped with flow control valves, pressure transducers and a temperature control system able of controlling fuel temperature at ±1° around the set point. Finally, the nozzle has an orifice diameter of 3mm and a length of 10mm. The selection of large scale dimensions served the purpose of increased spatial resolution; the combination of a cone-beam CT scanner and the given nozzle size resulted in a voxel size of 8x8x8μm.

### *2.1. Measurement technique and post-processing*
A cone-beam CT scanner comprises a point source that emits cone beam shaped x-rays that penetrate through the object of interest, they attenuate and the remaining energy is captured by a detector, as illustrated in *Figure 3a*. In the same *Figure 3*, images b, c and d illustrate sample single projections (radiographies) of the flow orifice under different operating conditions. The difference in radiation absorption between air and Diesel is immediately visible; the latter secures enough contrast for the reconstructed volume to enable differentiation between Diesel liquid and vapour phases.

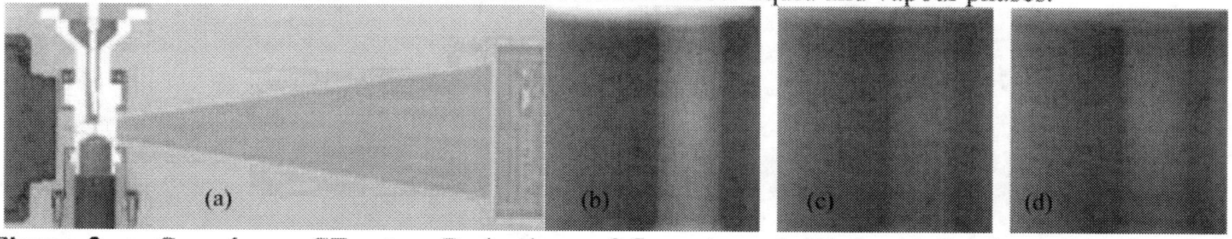


**Figure 3. a.** Cone beam CT setup. Projections of flow channel filled with **b.** air, **c.** Diesel and **d.** cavitating flow.

The reconstruction algorithm applied on cone-beam CT applications is the so-called back-projection algorithm of Feldkamp, Davies and Kress [7]. The algorithm is approximate but has useful properties, including errors that are relatively small in many practical instances and a form that leads to convenient computation. According to this technique, following the logarithmic normalisation and the application of a ramp filter on the raw radiographies, the image captured on the detector is back-projected to the Cartesian axis system of the object of interest and a three-dimensional density function is reconstructed. Effectively, the density function is created in terms of density slices of the object of interest normal to the rotational axis. Since the technique is based on radiation absorption comparison, it is vital for the incident cone beam to have equal energy over the whole area of interest. This is rather impossible due to the polychromatic x-ray source, a fact that potentially leads to increased calculation error. As seen in *Figure 3b,c*, along the axis of the orifice the captured intensity is not constant; it is evident that the projection is brighter at the middle. Therefore, in order to reconstruct the density map without including the error from radiation intensity differences, we devised a calibration procedure, which calibrates every voxel of the reconstructed volume against the corresponding voxel of two calibration sets, with Diesel liquid and air. Finally, the assumption that Diesel liquid and air share similar absorption properties to pressurised Diesel (at 60 bar) and Diesel vapour (at ~20 kPa) introduce relative errors of 1.2% and 2.5% for Diesel liquid and vapour respectively.

## Results
The followed calibration and post-processing procedures are expected to provide results that are not affected from the spatial intensity variations of the source; an error analysis showed a relative error of ~1%. As illustrated in Figure 4, the matrix of absorption coefficients is created for every slice of the orifice for both cases, with air and Diesel. The corresponding histogram then sets the range of absorption coefficient values that represent any possible mixture of Diesel and vapour for this particular slice. Finally, the reconstructed slices of any operating condition are calibrated against these ranges, per slice.

In detail, the micro-CT technique requires approximately 2 seconds to acquire enough signal for a single angular step. This causes the acquired data to be averaged over this time period; thus, showing the mean cavitation vapour fraction of the flow. Figure 5 illustrates two slices from two different locations along the orifice length, where vapour appears darker. According to the calibration mentioned above and the corresponding colour bar one can easily extract the cavitation vapour fraction for each pixel of the slice.

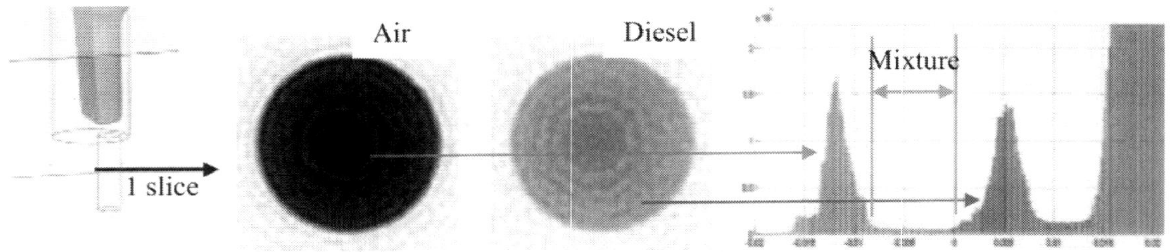


**Figure 4.** Calibration procedure. The super-imposed histograms of air and Diesel absorption coefficients illustrates the range of pixel values for this particular slice that represent any possible mixture other than 100% Diesel or air.

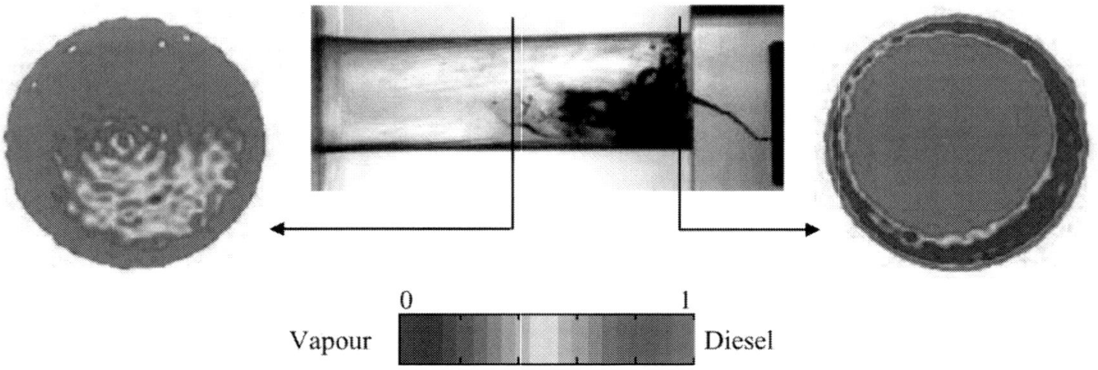


**Figure 5.** Quantitative measurements of vapour fraction at two different locations along the orifice.

## Conclusion
The possibility of applying the micro-CT technique on highly transient cavitating Diesel fuel flows has been investigated and results appear promising in terms of time-averaged quantitative fuel vapour measurements. Since this is an ongoing investigation, we intend to keep the scientific community updated with our future findings.

## Acknowledgements
The research leading to these results has received funding from the People Programme (IOF Marie Curie Actions) of the European Union's Seventh Framework Programme FP7/2007-2013/ under REA grant agreement n. 300410.

## References
[1] Arcoumanis, C., et al., *Cavitation in Real-Size Multi-Hole Diesel Injector Nozzles.* SAE Transactions, Journal of Engines, 2000-01-1249, 2000. 109-3.
[2] Mitroglou, N., et al., *Instantaneous and ensemble average cavitation structures in Diesel micro-channel flow orifices.* Fuel, 2014. 116(0): p. 736-742.
[3] Owen, K., *Gasoline and diesel fuel additives*. 1989: New York, NY; John Wiley and Sons Inc.
[4] Owen, K. and T. Coley, *Automotive fuels reference book*. 1995.
[5] Barbour, R., R. Quigley, and A. Panesar, *Investigations into Fuel Additive Induced Power Gain in the CEC F-98-08 DW10B Injector Fouling Engine Test.* SAE Technical Paper 2014-01-2721, 2014. SAE Technical Paper 2014-01-2721.
[6] Reid, B.A., et al., *On the formation of string cavitation inside fuel injectors.* Experiments in Fluids, 2014. 55(1): p. 1-8.
[7] Feldkamp, L.A., L.C. Davis, and J.W. Kress, *Practical cone-beam algorithm.* Journal of the Optical Society of America A, 1984. 1(6): p. 612-619.

# Evaluation of Turbulence Models Performance in Predicting Incipient Cavitation in an Enlarged Step-Nozzle

**H Naseri[1, 2], P Koukouvinis[2] and M Gavaises[2]**
[2]School of Mathematics, Computer Science & Engineering, City University London

E-mail: homa.naseri.2@city.ac.uk

**Abstract.** Predictive capability of RANS and LES models to calculate incipient cavitation of water in a step nozzle is assessed. The RANS models namely, Realizable k-ε, SST k-ω and Reynolds Stress Model did not predict any cavitation, due to the limitation of RANS models to predict the low pressure vortex cores. LES WALE model was able to predict the cavitation by capturing the shear layer instability and vortex shedding. The performance of a barotropic cavitation model and Rayleigh-Plesset-based cavitation models was compared using WALE model. Although the phase change formulation is different in these models, the predicted cavitation and flow field were not significantly different.

## 1. Introduction

Formation of vapour due to cavitation affects the fluid flow and a two-way interaction exists between the bubbles and the turbulent oscillations. Chahine [1] has developed a cavitation model to account for the two-way interaction of bubbles and the flow field. Okabayashi and Kajishima [2] investigated the two-way interaction between cavitation and turbulence using Direct Numerical Simulation (DNS). They reported the modulation of turbulence by cavitation which can form a basis for a Sub Grid Scale (SGS) model for cavitation in Large Eddy Simulation (LES). Iyer and Ceccio [3] used Fluorescent Particle Image Velocimetry (PIV) to assess the effect of cavitation on shear layer instabilities and flow turbulence downstream the shear layer, using a sharp edged plate in a cavitation channel. The importance of accurately capturing the turbulence pressure fluctuations in cavitating flows is highlighted in Edelbauer *et al.* [4]. They have compared Reynolds Averaged Navier-Stokes (RANS) and LES simulations of a cavitating throttle flow and show the situational applicability of the RANS model. They conclude that RANS can predict cavitation with a reasonably acceptable accuracy in an operating condition with high pressure difference, whereas it fails to predict the cavitation at a low pressure difference. Grid requirements and therefore the computational cost of RANS simulations is lower than LES, and in some studies [5] they proved to predict cavitation accurately, however their applicability varies from case to case.

Cavitation can be modelled using the Rayleigh-Plesset equation for bubble dynamics [6], which has been implemented in cavitation models of Singhal *et al.* [7], Schnerr and Sauer [8] and Zwart *et al.* [9]. Another approach to model cavitation is based on the assumption of thermodynamic equilibrium using an equation of state, such as the cavitation model of Schmidt *et al.* [10].

---

[1] Corresponding author.

In the current study CFD solver FLUENT V15.0 is used to compare the performance of a number of turbulence models and cavitation models for modelling incipient cavitation in a step nozzle geometry, based on the experimental study of Sou *et al.* [11].

## 2. Test Cases and Simulation Setup

The experimental setup is extensively reported in the reference study[11], so in the present report the operating conditions and geometry are only briefly presented. Water is discharged into a rectangular nozzle with 4.8 ml/s flow rate and the outlet is subjected to atmospheric pressure. Schematic of the nozzle is shown in figure 1.

A coarse mesh is systematically refined to perform a grid independency test. Simulations are done using Shear Stress Transport (SST) k-ω model. Inlet pressure of 2.42 bar and atmospheric outlet pressure is set as boundary conditions. After the grid independency test, performance of three RANS models, namely the Realizable k-ε, SST k-ω and Reynolds Stress Model (RSM) is assessed and compared to Wall Adaptive Local Eddy Viscosity (WALE) LES simulation for incipient cavitation. WALE is used to compare three different cavitation models, namely a barotropic model (see Koukouvinis *et al.* paper [12]), Zwart-Gerber-Belamri (ZGB) and Schnerr-Sauer (SS) cavitation models, and one simulation is set-up using ZGB model with modification. The ZGB model is modified to push the model towards thermodynamic equilibrium by increasing the evaporation and condensation rates. The multiphase model for all the cases is a homogeneous mixture model with both phases sharing the same velocity/pressure field.

## 3. Results and Discussion

Grid independency test results are reported in table 1 along with effect of grids on the flow rate. A consistent trend is achieved in the predicted velocities and it can be concluded that no significant improvement is reached by further refinement of the grid.

**Table 1.** Grid parameters and their effect on flowrate.

| Grid | Cells | Max $y^+$ | Min $y^+$ | Flowrate (ml/s) |
|---|---|---|---|---|
| **Coarse** | 1M | 55 | 1 | 4.7 |
| **Medium** | 2.3M | 45 | 0.5 | 4.7 |
| **Fine** | 6.8M | 37 | 0.2 | 4.8 |

Mean streamwise velocities calculated by Realizable k-ε, SST k-ω and RSM model and the LES WALE model at z = 1.5 mm are presented in figure 2. The minimum $y^+$ for the RANS models is ~1 and for the LES it is ~ 0.2. The Realizable k-ε and SST k-ω have less than 1% difference in the majority of the flow section. The most significant disparity between this two models is at the region between the recirculating flow and the bulk flow moving above it, where the SST k-ω over-predicts the velocity by ~24% and Realizable k-ε by ~20%. On average the two-equation models have ~30% discrepancy with experiment in mean streamwise velocity at z = 1.5 mm. The RSM results over-predicts the velocity at the recirculation region and on average RSM results have ~20% more discrepancy with experiment velocities compared to two-equation models.

No cavitation is predicted by the RANS models and the shear layer vortices are not predicted. RANS models are not designed to capture the shedding of small vortices, instead the effect of these vortices is modelled by artificially increasing the local viscosity. LES can predict vortex shedding and pressure fluctuations in the shear layer. LES can predict the cavitation and the results significantly improve in the near wall and the recirculation region. On average the LES velocity profile was ~10% closer to experiment result compared to two-equation models predictions.

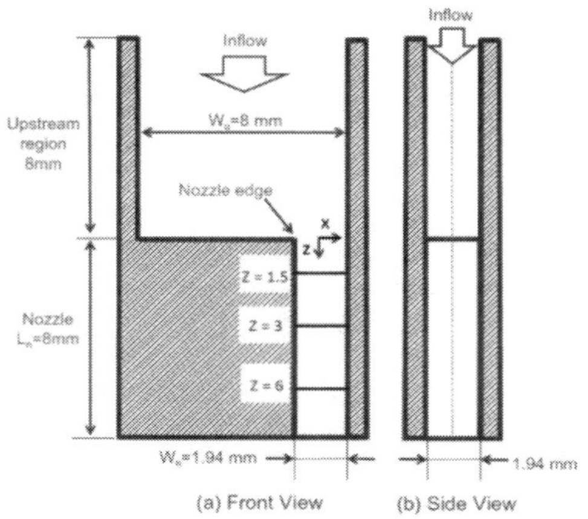


**Figure 1.** Nozzle geometry [11]

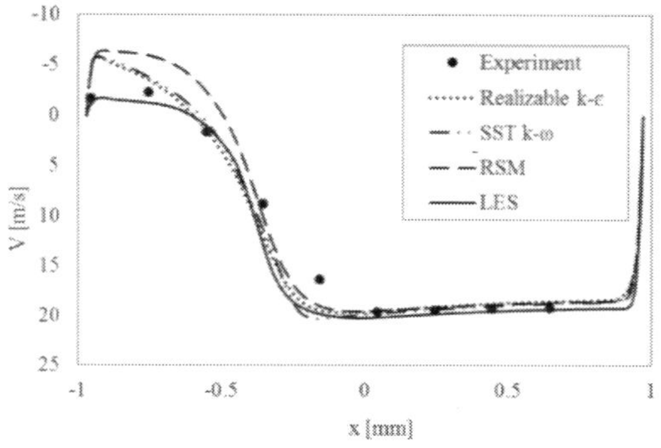


**Figure 2.** Mean streamwise velocity at z = 1.5 mm calculated by different turbulence models

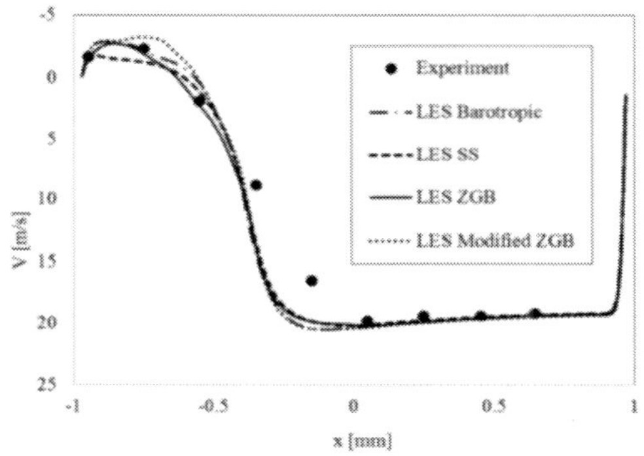


**Figure 3.** Mean streamwise velocity at z = 1.5 mm calculated using different cavitation models

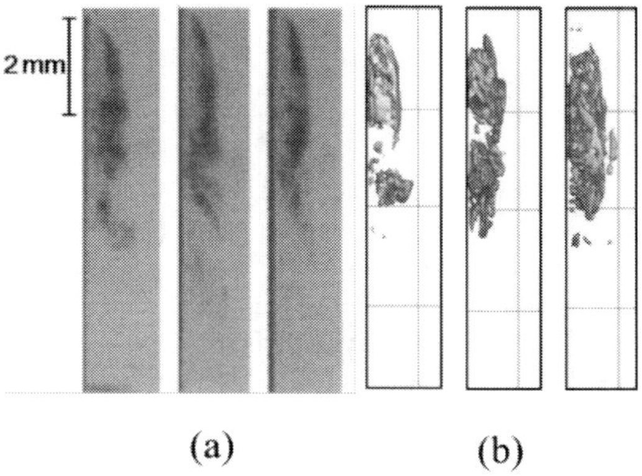


**Figure 4.** Cavitation in experiment [11] (a) and CFD results showing iso-surface of 50% vapour volume fraction (b)

It is evident from figure 3 that changing the cavitation model, only affects the streamwise velocity in the region of cavitation and the bulk flow remains unaffected. Furthermore the velocity predictions by the barotropic and the ZGB model are nearly identical, and have the best match with experiment velocity measurements. The minimum mean pressure in the flow field for the barotropic and the modified ZGB model are ~10,000 Pa and for the default SS and ZGB models it is ~8,000 Pa. The default models predict regions of negative pressure as low as -13,000 pa which can be the reason of lower minimum pressure calculated by these models. The minimum pressure calculated with modified ZGB is close to zero, and the barotropic model does not predict any region of negative pressure. Negative pressures have been observed in delicate experimental conditions only and in such states liquid is said to be in tension[13]. However existence of negative pressure regions in this experiment is questionable. In industrial cases, such as water in cavitation tunnels, the pressure inside the cavity have been reported to match the saturation pressure at a given temperature.

## 4. Conclusion

This paper evaluates the predictive capability of two-equation and 7-equation RANS models to simulate incipient cavitation in a rectangular step nozzle, and compares the results with WALE model

predictions. The LES model is then used to further investigate the performance of barotropic and non-equilibrium cavitation models.
This test case shows the situational applicability of RANS model for predicting cavitation. All the RANS models used for this study, i.e. the Realizable k-ε, SST k-ω and RSM model failed to predict pressures below the saturation pressure. RANS is a useful tool for many cavitation problems as seen in the literature [5], but its limited capability has also been reported for cases with small amounts of cavitation[4]. For problems such as incipient cavitation in a nozzle where the pressure drop from inlet to outlet is low, small vortices are formed that act as nucleation cites for bubbles. In order to capture these flow structures, more rigorous turbulence models such as LES are required.
The average minimum pressure predicted by the barotropic and the non-equilibrium cavitation models is above the saturation pressure of water. This result further justifies the minimum pressure predicted by RANS models, which is above saturation pressure. Furthermore, changing the cavitation model did not significantly affect the streamwise velocity outside the cavitation region. The predicted shape of the cavity was in agreement with experimental images, however quantitative measurements inside the vapour volume is required to judge the accuracy of the calculated cavitation.

**Acknowledgements**
The research leading to these results has received partial funding from the People Programme (IAPP Marie Curie Actions) of the European Union's Seventh Framework Programme FP7/2007-2013/ under REA grant agreement n. 324313.

## 5. References

[1] Chahine G L 2009 Numerical simulation of bubble flow interactions *J. Hydrodyn. Ser. B* **21** 316–32
[2] Okabayashi K and Kajishima T 2009 Investigation of turbulent modulation by cavitation for subgrid-scale modeling in LES *Proc. Of 7th Int. Symp. On Cavitation* vol 89 (Ann Harbor, Michigan, USA, 17-22 August) )
[3] Iyer C O and Ceccio S L 2002 The influence of developed cavitation on the flow of a turbulent shear layer *Phys. Fluids* **14** 3414–31
[4] Edelbauer W, Strucl J and Morozov A 2014 large eddy simulation of cavitating throttle flow *SimHydro Int. Conf. Model. rapid transitory flows*
[5] Andriotis A, Gavaises M and Arcoumanis C 2008 Vortex flow and cavitation in diesel injector nozzles *J. Fluid Mech.* **610** 195–215
[6] Plesset M S 1949 The dynamics of cavitation bubbles *J. Appl. Mech.* **16** 277–82
[7] Singhal A K, Athavale M M, Li H and Jiang Y 2002 Mathematical basis and validation of the full cavitation model *J. Fluids Eng.* **124** 617–24
[8] Schnerr G H and Sauer J 2001 Physical and numerical modeling of unsteady cavitation dynamics *Fourth Int. Conf. Multiph. Flow*
[9] Zwart P J, Gerber A G and Belamri T 2004 A two-phase flow model for predicting cavitation dynamics *Proc. Of 5th International Conference on Multiphase Flow*
[10] Schmidt S J, Mihatsch M, Thalhamer M and Adams N A 2011 Assessment of the prediction capability of a thermodynamic cavitation model for the collapse characteristics of a vapor-bubble cloud *Proc. Of WIMRC Cavitation Forum* pp 4–6
[11] Sou A, Biçer B and Tomiyama A 2014 Numerical simulation of incipient cavitation flow in a nozzle of fuel injector *Comput. Fluids* **103** 42–8
[12] Koukouvinis P and Gavaises M 2015 Simulation of throttle flow with two phase and single phase homogenous equilibrium model *CAV2015 Conf.*
[13] Brennen C E 2013 *Cavitation and Bubble Dynamics* (Cambridge University Press)

# Large-eddy simulation of cavitating nozzle and jet flows

**F Örley[1], T Trummler[1], S Hickel[1,2], M S Mihatsch[1], S J Schmidt[1], and N A Adams[1]**

[1] Institute of Aerodynamics and Fluid Mechanics, Department of Mechanical Engineering, Technische Universität München
[2] Chair of Computational Aerodynamics, Faculty of Aerospace Engineering, TU Delft

E-mail: felix.oerley@aer.mw.tum.de

**Abstract.** We present implicit large-eddy simulations (LES) to study the primary break-up of cavitating liquid jets. The considered configuration, which consists of a rectangular nozzle geometry, adopts the setup of a reference experiment for validation. The setup is a generic reproduction of a scaled-up automotive fuel injector. Modelling of all components (i.e. gas, liquid, and vapor) is based on a barotropic two-fluid two-phase model and employs a homogenous mixture approach. The cavitating liquid model assumes thermodynamic-equilibrium. Compressibility of all phases is considered in order to capture pressure wave dynamics of collapse events. Since development of cavitation significantly affects jet break-up characteristics, we study three different operating points. We identify three main mechanisms which induce primary jet break-up: amplification of turbulent fluctuations, gas entrainment, and collapse events near the liquid-gas interface.

## 1. Introduction

Modern fuel injection systems play a key role for the optimization of the air-fuel mixing process in the combustion chamber. This is a key requirement for a more efficient combustion process in order to meet future emission standards. Recent developments aim towards increasing injection rail pressures, which enhances the jet break-up and mixing and hence improves the combustion. Higher flow acceleration goes hand in hand with thermo-hydrodynamic effects, such as cavitation, which occurs when the pressure locally drops below saturation conditions. The collapse of vapor pockets in regions of higher pressure causes strong shock-waves and high-velocity liquid jets directed towards the wall surface. This can generate high local stresses inside the surrounding structure. Loads induced by such phenomena are employed to clean nozzles from surface deposits, and can promote primary jet break-up. Also, cavitation can lead to choked conditions in a duct and hence maintains a mass flow rate independent of the pressure-drop. However, collapse events may cause material erosion and failure of the component.

A particular challenge in the context of injection of liquid jets into a combustion chamber is the mutual interaction of cavitating liquids and non-condensable gas. Sou *et al* [1, 2] assume that primary break-up of liquid jets is promoted by enhanced turbulent fluctuations caused by collapse events of cavitation structures near the nozzle outlet.

In this work we apply a simple, closed-form barotropic two-fluid cavitation model including a non-condensable gas component proposed by Örley *et al* [3]. The thermodynamic model is an extension to our compressible framework for performing implicit large-eddy simulation of turbulent, cavitating flows of Egerer *et al* [4]. The model is used to simulate an experimental reference configuration of Sou *et al* [1, 2] of a cavitating water jet injected into air.

## 2. Thermodynamic model

The fluid consists of the three phases: liquid water ($W$), water-vapor-mixture ($M$) and non-condensable gas ($G$). The volume averaged density $\rho$ inside a computational cell is

$$\rho = \sum_{\Phi} \beta_{\Phi} \rho_{\Phi}, \tag{1}$$

where $\beta_{\Phi}$ denotes the volume fraction and $\rho_{\Phi}$ the mean density of phase $\Phi$. For a detailed description of the thermodynamic model refer to Örley *et al* [3].

The actual vapor-liquid interface of cavitation structures is not reconstructed with sub-cell resolution. Surface tension thus is neglected. The cavitation model has been extensively validated for LES of turbulent wall-bounded flows by Egerer *et al* [4] and Hickel *et al* [5]. By using corresponding thermodynamic closure relations for each phase in Eqn. (1), the equation of state can be formulated in a suitable way to solve for the cell-averaged pressure $p = p(\rho)$.

Liquid water and liquid-vapor mixtures are modelled as barotropic mixture fluid. Integration of the isentropic speed of sound leads to

$$\rho = \rho_{s,liq} + (p - p_s)/c^2, \tag{2}$$

where $p_s$ is the saturation pressure and $\rho_{s,liq}$ is the saturation density for liquid water. The formation of vapor is modelled by a homogenous mixture model. For purely liquid water, i.e. $p > p_s$, the speed of sound is $c = c_{liq} = 1482.35$ m/s at ambient conditions. Comparison to more accurate models, such as the Tait equation, shows negligible deviation for pressures up to 200 bar. For liquid-vapor mixtures, i.e. $p < p_s$, the same equation of state is used, but a different speed of sound is employed. Here, we consider an average of the speed of sound between a frozen and an equilibrium isentropic phase change in the two-phase region. We use a numerical value of $c = c_M = 1$ m/s, which corresponds to a conservative upper limit.

The non-condensible gas phase is modelled as an ideal, isothermal gas at reference temperature $T_{ref} = 293.15$ K

$$\rho_G = \frac{p}{\mathrm{R}_G T_{ref}}. \tag{3}$$

## 3. Numerical Method

We employ an implicit LES approach based on the Adaptive Local Deconvolution Method (ALDM) by Adams *et al* [6] and Hickel *et al* [7, 8]. In contrast to explicit SGS models, implicit LES merges turbulence modelling and numerical discretization. ALDM is a nonlinear finite volume method and incorporates free parameters that control the implicit SGS model. An SGS model that is consistent with turbulence theory is obtained through parameter calibration [7]. The compressible version of ALDM [8] can capture shock waves while smooth pressure waves and turbulence are propagated without excessive numerical dissipation. More details on the validation of ALDM for cavitating flows are discussed by Egerer *et al* [4].

## 4. Setup

We adopt the setup presented by Sou *et al* [1, 2], see Fig. 1, who experimentally investigated different flow regimes in a nozzle flow. We conduct LES for a set of cavitation numbers that lead to different cavitation characteristics: Case 1: $\sigma = 1.27$ – no cavitation inside the nozzle; Case 2: $\sigma = 0.78$ – developing cavitation; and Case 3: $\sigma = 0.65$ – supercavitation. The

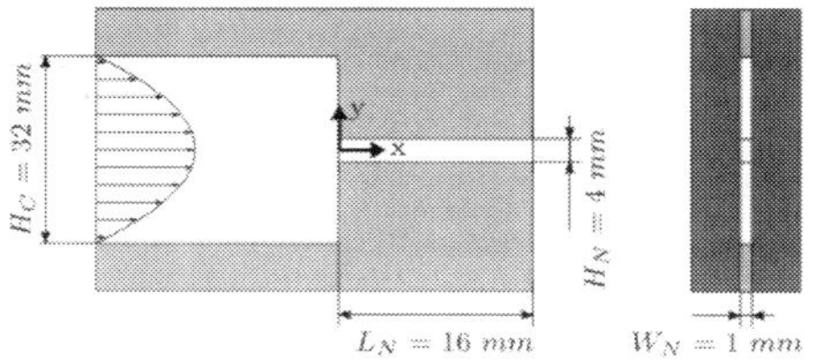


**Figure 1.** Setup of a cavitating, rectangular nozzle flow.

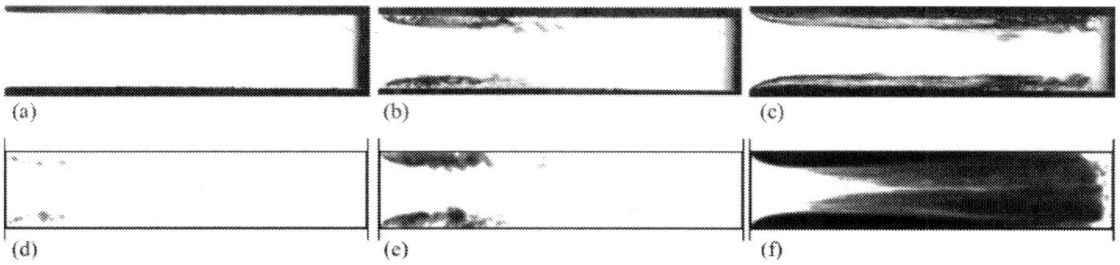


**Figure 2.** Side view of vapor structures inside the nozzle for $\sigma = 1.27$ (a/d), $\sigma = 0.78$ (b/e), and $\sigma = 0.65$ (c/f). Top row: instantaneous vapor structures observed in experiments (Reprinted from Sou *et al* [2], with permission from Elsevier); bottom row: contours of depth averaged vapor volume fraction $0.01 < \langle\alpha\rangle_z < 1.0$ (adopted from Örley *et al* [3]).

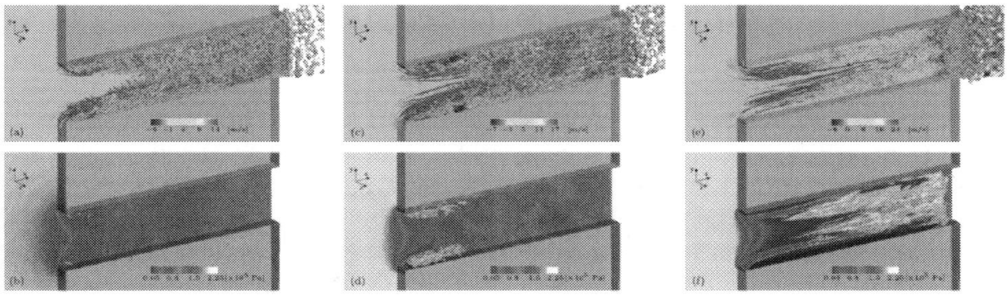

**Figure 3.** Snapshots of iso-surfaces of $\lambda_2 = -1 \times 10^8\ 1/s^2$ colored by streamwise velocity $u$ (top row) and iso-surfaces of vapor volume fraction $\alpha = 0.1$ and wall pressure (bottom row) for $\sigma = 1.27$, $\sigma = 0.78$, and $\sigma = 0.65$ (from left to right). Figures adopted from Örley *et al* [3].

domain is discretized on a Cartesian mesh with grid sequencing and $43 \times 10^6$ cells on the finest level. The smallest cell size has a dimension of $3.91\mu m$, which imposes a time step size of $\Delta t = 7.92 \times 10^{-10}$ s based on the CFL condition. Local grid refinement is applied at the nozzle wall and in the outflow region. A laminar duct flow profile is prescribed at the inlet to meet the specific cavitation number. The outflow region is initially filled with gas.

## 5. Results

Figure 2 shows transmitted light images of instantaneous vapor structures observed in the experiment in comparison to our LES data. Cases 1 and 2 compare well, whereas the numerical results for case 3 show vapor generation inside large vortical structures in the nozzle center. Turbulent structures inside the duct as iso-surfaces of the $\lambda_2$-criterion coloured by axial velocity, together with cavitation structures as iso-surfaces of the vapor volume fraction $\alpha$ and wall pressure are visualized in Fig. 3. Turbulence is damped in regions of high vapor content.

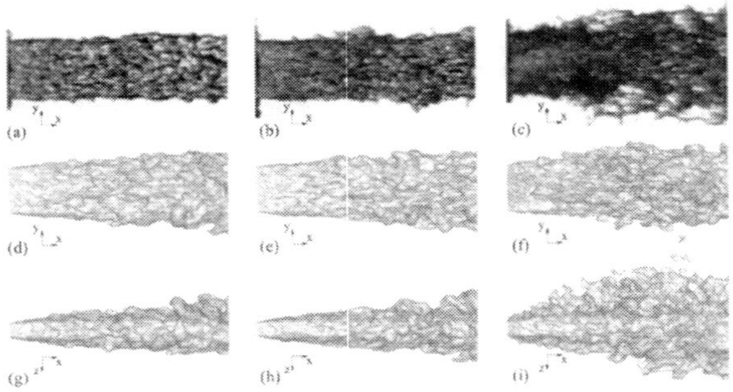


**Figure 4.** Experimental transmitted light images (top row, reprinted from Sou *et al* [2], with permission from Elsevier) and LES snapshots of $x$-$y$ view (middle row) and $x$-$z$ view (bottom row) showing iso-surfaces of gas volume fraction $\beta_G = 0.99$ in the range 16 mm $< x <$ 32 mm (adopted from Örley *et al* [3]) for $\sigma = 1.27$, $\sigma = 0.78$, and $\sigma = 0.65$ (left to right).

Stable, cavitating corner vortices are observed for low cavitation numbers.

Finally, the effect of the cavitating nozzle flow on primary liquid jet break-up inside the gas domain is shown in Fig. 4. For cavitation numbers $\sigma = \{1.27, 0.78\}$, see Fig. 4(a/b), the jet structure in the experiment is similar. For $\sigma = 0.65$, Fig. 4(c), in contrast, spray formation is observed. Small droplets and ligaments of liquid detach from the surface and cause an increased jet angle. Our numerical results, see Fig. 4(d/e/f), show only little effect of the cavitation number on the jet angle in the $x$-$y$ plane. In contrast, significant differences between the higher cavitation numbers $\sigma = \{1.27, 0.78\}$, and the low cavitation number $\sigma = 0.65$ are found in the $x$-$z$ plane, see Fig. 4(g/h/i). Here, we clearly notice a widening of the jet and a detachment of liquid structures from its surface, which closely resembles the observations in the experiments.

From an analysis of the transient data we identified three main mechanisms that lead to distortions of the jet surface and, ultimately, to a widening and break-up of the jet. First, turbulent fluctuations, which are induced by collapse events in the proximity of the exit plane of the nozzle, add to the momentum in wall-normal direction. This observation confirms the hypothesis of Sou *et al* [2]. Second, low pressure vapor regions near the nozzle exit and the gas filled plenum form a pressure gradient, which enables entrainment of gas from the outlet region into the nozzle. When the gas is being ejected back out, the water is accelerated towards the side walls and creates large scale bulges of liquid. Third, collapse events of cavitation structures inside the jet near the liquid-gas interface induce high velocity liquid jets directed towards the interface. This effect resembles the findings of Kobel *et al* [9], and causes small, needle-like structures of liquid in our simulation. A detailed discussion is provided in Örley *et al* [3].

## References

[1] Sou A, Tomiyama A, Hosokawa S, Nigorikawa S and Maeda T 2006 *JSME International Journal. Series B, Fluids and Thermal Engineering* **49** 1253–1259
[2] Sou A, Hosokawa S and Tomiyama A 2007 *International Journal of Heat and Mass Transfer* **50** 3575–3582
[3] Örley F, Trummler T, Hickel S, Mihatsch M S, Schmidt S J and Adams N A 2015 *Physics of Fluids* **27** 086101
[4] Egerer C P, Hickel S, Schmidt S J and Adams N A 2014 *Physics of Fluids* **26** 085102
[5] Hickel S, Mihatsch M and Schmidt S J 2011 Implicit Large Eddy Simulation of Cavitation in Micro Channel Flows *WIMRC 3rd International Cavitation Forum, University of Warwick, UK, 2011* ed Li S C
[6] Adams N A, Hickel S and Franz S 2004 *Journal of Computational Physics* **200** 20–20
[7] Hickel S, Adams N A and Domaradzki J A 2006 *Journal of Computational Physics* **213** 413–436
[8] Hickel S, Egerer C and Larsson J 2014 *Physics of Fluids* **26** 106101
[9] Kobel P, Obreschkow D, Dorsaz N, de Bosset A, Nicollier C and Farhat M 2005 *Phys Rev Lett* **97**

Journal of Physics: Conference Series **656** (2015) 012097 doi:10.1088/1742-6596/656/1/012097

# LES of cavitating flow inside a Diesel injector including dynamic needle movement

**F Örley[1], S Hickel[1,2], S J Schmidt[1], and N A Adams[1]**

[1] Institute of Aerodynamics and Fluid Mechanics, Department of Mechanical Engineering, Technische Universität München

[2] Chair of Computational Aerodynamics, Faculty of Aerospace Engineering, TU Delft

E-mail: felix.oerley@aer.mw.tum.de

**Abstract.** We perform large-eddy simulations (LES) of the turbulent, cavitating flow inside a 9-hole solenoid common-rail injector including jet injection into gas during a full injection cycle. The liquid fuel, vapor, and gas phases are modelled by a homogeneous mixture approach. The cavitation model is based on a thermodynamic equilibrium assumption. The geometry of the injector is represented on a Cartesian grid by a conservative cut-element immersed boundary method. The strategy allows for the simulation of complex, moving geometries with sub-cell resolution. We evaluate the effects of needle movement on the cavitation characteristics in the needle seat and tip region during opening and closing of the injector. Moreover, we study the effect of cavitation inside the injector nozzles on primary jet break-up.

## 1. Introduction

Recent developments of fuel injection systems aim towards increasing rail pressures for improving the mixing and combustion to meet future emission standards. A detailed understanding of the flow is necessary to control hydrodynamic effects, such as cavitation, that go hand in hand with higher rail pressures, and their influence on jet and spray characteristics. Typical dimensions of fuel nozzles inside Diesel injectors are in the range of a few hundred micrometers. This makes the instrumentation with diagnostic equipment for an experimental flow characterization a challenging task. Moreover, short intrinsic timescales, imposed by inherent flow dynamics or by functional components such as the opening or closing of the injector needle, complicates time-accurate measurements. Experimental analysis of cavitation erosion can thus supply information on incubation time, position, and the progress of erosion damage, but does not necessarily reveal the relevant flow physics. Computational Fluid Dynamics, on the other hand, can provide time-resolved information on flow structures in arbitrary small geometries. Numerical simulations thus have become an important tool in the design process. Recent work was presented, e.g., by Andriotis *et al* [1] for stationary, and by Battistoni *et al* [2] for dynamic needle positions.

In this work, we investigate the flow inside a 9-hole solenoid common-rail (CR) injector in terms of fluid flow turbulence, cavitation characteristics, and primary jet break-up. Dynamic needle movement is handled by a conservative cut-element based immersed boundary method. We employ a simple, closed form barotropic single-fluid cavitation model for ISO4113 air-free test oil including non-condensable gas effects.

## 2. Thermodynamic model

We apply a homogenous-mixture single-fluid model for cavitation, which is extended by a component of free gas to a two-fluid mixture model. A validation of the model has recently been published in Örley *et al* [3]. The single-fluid cavitation model has been extensively validated in previous studies, e.g., for LES of cavitating, turbulent, wall-bounded fuel flows by Egerer *et al* [4], and closing control valves by Örley *et al* [5].

The fluid consists of the three phases: liquid Diesel fuel surrogate (ISO4113 air-free test oil, $D$), fuel-vapor-mixture ($M$), and non-condensable gas ($G$). The volume averaged density $\rho$ inside a computational cell is

$$\rho = \sum_{\Phi} \beta_{\Phi} \rho_{\Phi}, \tag{1}$$

where $\beta_{\Phi}$ denotes the volume fraction and $\rho_{\Phi}$ the mean density of phase $\Phi$. The actual vapor-liquid interface of discrete vapor structures inside a computational cell is not reconstructed, unlike with sharp-interface methods, see e.g., Lauer *et al* [6]. Surface tension is thus neglected.

Liquid fuel and liquid-vapor mixtures are modelled as a mixture fluid in thermodynamic equilibrium. For purely liquid fuel, i.e. if the pressure $p$ is higher than the saturation pressure $p_s$, we employ a Tait equation of state

$$\rho_D = \rho_{ref} \left( \frac{p + B}{p_{ref} + B} \right)^{1/N}. \tag{2}$$

Here, $B = 1.19 \times 10^8$ Pa and $N = 11.5$ are calibrated parameters. By choosing a set of $\{B, N\}$, the fuel can be modelled as either isothermal or isentropic. We use the saturation conditions for liquid Diesel fuel as reference conditions, i.e. $p_{ref} = p_s$ and $\rho_{ref} = \rho_{s,liq}$. The Tait equation is able to model the liquid phase in the full pressure range of interest.

For liquid-vapor mixtures, i.e. $p < p_s$, we use a linearized equation of state based on the integration of the isentropic speed of sound, which leads to

$$\rho_M = \rho_{s,liq} + (p - p_s)/c_M^2. \tag{3}$$

The mixture speed of sound is set to $c_M = \sqrt{p_s/\rho_s}$.

The non-condensable gas phase is modelled as an isothermal ideal gas

$$\rho_G = p/\mathrm{R}_G T_{ref} \tag{4}$$

at reference temperature $T_{ref} = 293.15$ K.

Equations (1)-(4) are combined to a closed equation of state that employs the non-condensable gas mass fraction $\beta_G$, which is transported as a scalar. The equation of state is solved iteratively for the cell average pressure $p = p(\rho, \beta_G)$.

## 3. Numerical Method

We use an implicit LES approach based on the Adaptive Local Deconvolution (ALDM) proposed by Adams *et al* [7] and Hickel *et al* [8, 9]. ALDM merges turbulence modelling and numerical discretization of the conservation equations. A subgrid-scale (SGS) model that is consistent with turbulence theory was obtained through parameter calibration, see Hickel *et al* [8] for details. The compressible version of ALDM [9] can capture shock waves, while smooth pressure waves and turbulence are propagated without excessive numerical dissipation. More details on the validation of ALDM for cavitating flows are given in Egerer *et al* [4]. We employ Cartesian multi-block grids with hanging nodes. Complex moving geometries are represented with the conservative immersed boundary method recently proposed by Örley *et al* [5]. This so-called cut-element method is an extension of Ref. [10] and enables the highly accurate and oscillation free simulation of weakly compressible fluid flows in moving geometries.

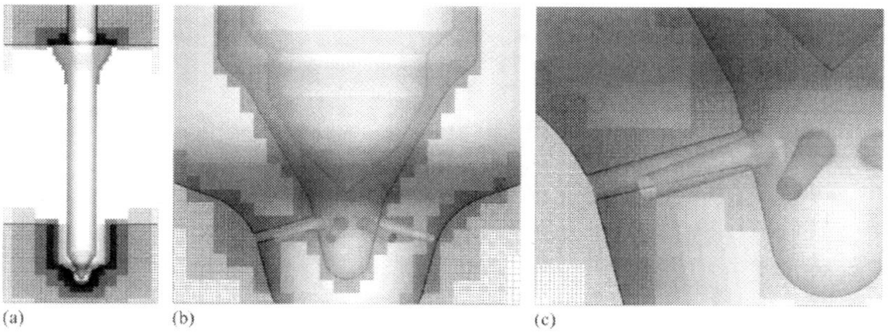


**Figure 1.** Medium grid and surface geometry of the 9-hole solenoid CR injector (geometry provided by Huber and Ulbrich [11]).

## 4. Setup

We investigate the flow field during a full injection cycle at different rail pressures $p_R$, and include the full 360 degree injector geometry to capture circumferential asymmetries in the flow field and communication between the nozzles. The prescribed needle lift is obtained from a multi-domain simulation model by Huber and Ulbrich [11] for specific injection pressures, see Fig. 2. We initialize the domain with liquid fuel at $p_R = \{1500, 2000\}$ bar above the closed needle seat, and gas at chamber pressure $p_c = 10$ bar, otherwise.

Geometry data was provided by Huber and Ulbrich [11]. Multiple grid resolutions are tested. The medium grid resolution including the injector geometry and the needle is shown in Fig. 1. The nozzle holes have a mean diameter of $d_H \sim 200$ $\mu$m. The grid resolution in the proximity of the nozzle holes is on the order of $\Delta x = \{5, 10, 20\}$ $\mu$m. for the fine, medium, and coarse mesh with cubic cells. The fluid domain is resolved with approx. 20 cells over the nozzle diameter. Grid coarsening is employed in regions far away from the injector. The medium grid consists of approx. $50 \times 10^6$ cells. Owing to the small timestep size of 1.75 ns, the real time for one cycle is on the order of 5 days on 4000 CPUs, whereas the recomputation of geometrical parameters in each time step consumes 50% of the computational effort due to the large number of cut-cells. The computations are run on the Leibnitz Supercomputing Centre, Munich.

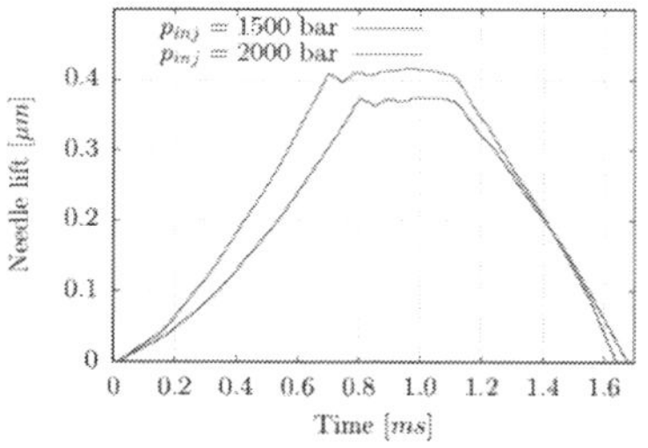


**Figure 2.** Needle lift computed from a multi-domain simulation model by Huber and Ulbrich [11] for rail pressures $p_R = \{1500, 2000\}$ bar.

## 5. Results

Figure 3 shows coherent turbulent structures and cavitation structures during the initial phase of the injection cycle. While the interior flow field in the sac volume quickly undergoes transition to fully turbulent flow, larger vortical structures in the jet are generated in the shear layer of the liquid-gas interface inside the combustion chamber, see Fig. 3(a). Cavitation in the needle seat due to a strong expansion of the liquid, visualized in Fig. 3(b), is already found in the early stages of the injection. The liquid flow inside the nozzle holes shown in Fig. 4 does not yet cavitate, but already reveals characteristic vortical structures.

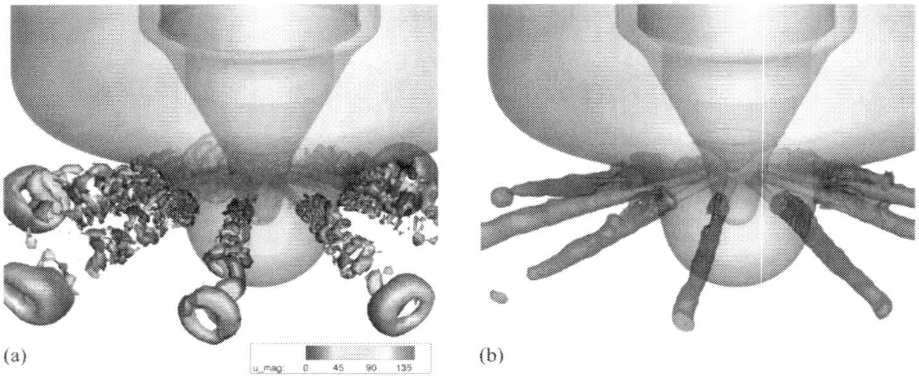


**Figure 3.** Initial phase of the injection process at $t = 0.11\ ms$. Left: iso-surfaces of $\lambda_2 = -1 \times 10^{10}$ 1/s colored by absolute velocity $u_{mag}$; right: iso-surfaces of void fraction $\alpha = 0.1$ (blue) and gas volume fraction $\beta_G = 0.9$ (grey).

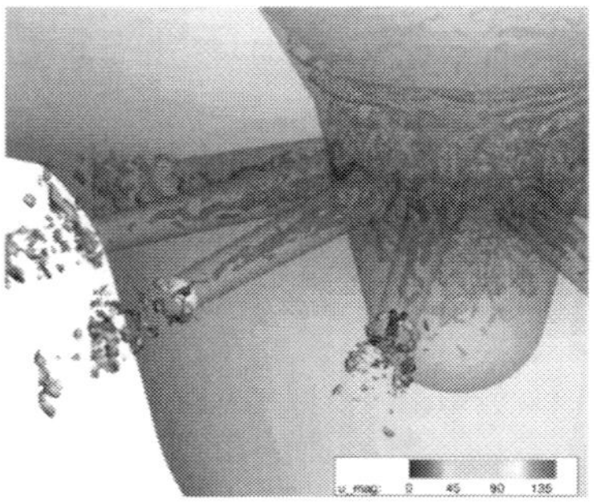


**Figure 4.** Vortical structures inside the injection holes at $t = 0.11\ ms$ visualized by iso-surfaces of $\lambda_2 = -4 \times 10^{11}$ 1/s colored by absolute velocity $u_{mag}$.

We will present a detailed analysis of the flow field during the injection using isothermal and isentropic flow, multiple rail pressures, and different injection times.

## Acknowledgements
We would like to thank B. Huber and H. Ulbrich (Institute of Applied Mechanics, Technische Universität München) for providing the injector geometry data and curves for needle lift. Computing time was granted by the Leibnitz Supercomputing Centre (LRZ) of the Bavarian Academy of Sciences and Humanities.

## References
[1] Andriotis A, Gavaises M and Arcoumanis C 2008 *Journal of Fluid Mechanics* **610** 195–215
[2] Battistoni M, Xue Q, Som S and Pomraning E 2014 Effect of Off-Axis Needle Motion on Internal Nozzle and Near Exit Flow in a Multi-Hole Diesel Injector Tech. rep. Warrendale, PA
[3] Örley F, Trummler T, Hickel S, Mihatsch M S, Schmidt S J and Adams N A 2015 *Physics of Fluids* **27** 086101
[4] Egerer C P, Hickel S, Schmidt S J and Adams N A 2014 *Physics of Fluids* **26** 085102
[5] Örley F, Pasquariello V, Hickel S and Adams N A 2015 *Journal of Computational Physics* **283** 1–22
[6] Lauer E, Hu X Y, Hickel S and Adams N A 2012 *Physics of Fluids* **24** 052104
[7] Adams N A, Hickel S and Franz S 2004 *Journal of Computational Physics* **200** 20–20
[8] Hickel S, Adams N A and Domaradzki J A 2006 *Journal of Computational Physics* **213** 413–436
[9] Hickel S, Egerer C and Larsson J 2014 *Physics of Fluids* **26** 106101
[10] Meyer M, Devesa A, Hickel S, Hu X Y and Adams N A 2010 *Journal of Computational Physics* **229** 6300–6317
[11] Huber B and Ulbrich H 2014 Simulation and Validation of a Common Rail Diesel Injector *Proceedings of the Ninth International Conference on Engineering Computational Technology*

# Modelling of single bubble-dynamics and thermal effects

**D Papoulias**[1], **M Gavaises**[2]

[1] CD-adapco, 200 Shepherds Bush Road, London W6 7NL, UK
[2] Department of Mechanical Engineering & Aeronautics, City University London, Northampton Square, London EC1V 0HB, UK

dimitrios.papoulias@cd-adapco.com

**Abstract**. This paper evaluates the solution effects of different Rayleigh-Plesset models (R-P) for simulating the growth/collapse dynamics and thermal behaviour of homogeneous gas bubbles. The flow inputs used for the discrete cavitation bubble calculations are obtained from Reynolds-averaged Navier-Stokes simulations (RANS), performed in high-pressure nozzle holes. Parametric 1-D results are presented for the classical thermal R-P equation [1] as well as for refined models which incorporated compressibility corrections and thermal effects [2, 3]. The thermal bubble model is coupled with the energy equation, which provides the temperature of the bubble as a function of conduction/convection and radiation heat-transfer mechanisms. For approximating gas pressure variations a high-order virial equation of state (EOS) was used, based on Helmholtz free energy principle [4]. The coded thermal R-P model was validated against experimental measurements [5] and model predictions [6] reported in single-bubble sonoluminescence (SBSL).

## 1. Introduction

Based on potential flow theory, Rayleigh and Plesset [7, 8] derived and tested the so-called Rayleigh-Plesset model for predicting the growth/collapse dynamics of cavitation bubbles. Since its inception, the R-P model has been further developed in order to compensate for the inherited assumptions of the method, which prohibit the physical realization of the violent bubble collapse phase. Inhomogeneous pressure effects developing upon the collapse of bubbles were addressed by Moss et al. [2]. To treat this model limitation, Moss introduced a time-derivative term of the gas-pressure in the classical R-P model, which effectively reproduced the rapid damping of the bubble collapse rebounds, as observed in SBSL experiments. Compressibility correction terms were formulated in the work of Keller and Miksis [3], and Lohse and Hilgenfeldt [5], as a function of the bubble's Mach number. The solution effects of the aforementioned compressible R-P models were examined by Prosperetti and Hao [9], assuming homogeneous bubbles and perfect-gas behavior. The Keller-Miksis R-P model was further extended in the work of Yasui [6], in order to also account for evaporation/condensation mass-transfer effects and chemical kinetics. For the calculation of the internal bubble gas-pressure, most of the aforementioned studies relied on idealized equations of state (EOS) or improved equations designed for real-gases, such as the second-order van der Waals model. Model limitations associated with the development of high gas-pressures, precursors of shock-waves in bubble sonoluminescence, were raised in the paper of Löfstedt et al. [10].

The thermal-bubble model proposed in this paper combines in an explicit way a multiparameter Helmholtz-type EOS, the energy equation and the R-P models presented in the reviewed studies. Details regarding the formulation of the implemented models are available in the following section.

## 2. Thermal bubble-dynamic R-P model

In the following analysis the Keller-Kiksis R-P equation (K-M) is presented, alongside to the suggested supplementary energy model for coupling the dynamic and thermal behavior of bubbles.

$$\left(1-\frac{\dot{R}}{c_L}\right)R\ddot{R}+\frac{3}{2}\left(1-\frac{\dot{R}}{3c_L}\right)\dot{R}^2=\frac{1}{\rho_L}\left(1+\frac{\dot{R}}{c_L}+\frac{R}{c_L}\frac{\mathrm{d}}{\mathrm{dt}}\right)\left[p_B-p_\infty-p_s(t)\right] \quad (1)$$

where R is the radius of the spherical bubble and the overdot symbol denotes time-derivative terms, i.e. bubble-wall velocity and acceleration. The bubble pressure $p_B$, the ambient static-pressure $p_\infty$ and the imposed acoustic pressure-field $p_s$ are collected inside the square brackets. Closure of equation (1) requires an additional expression for the dynamic boundary condition at the bubble-wall:

$$p_B=p_g-\frac{2\sigma}{R}-\frac{4v_L}{R}\dot{R} \quad (2)$$

equation (2) provides a relationship between the internal bubble gas-pressure $p_g$ and the pressure acting on the external side of the bubble-wall. The remaining variables participating in equations (1) and (2) refer to the properties of the surrounding liquid-fluid, i.e. density $\rho_L$, speed of sound $c_L$, viscosity $v_L$, and surface tension σ. These parameters are extrapolated from available correlations [11], for fixed pressure and temperature conditions.

The gas temperature of cavitating bubbles $T_B$ is approximated by solving the energy equation:

$$\Delta U=m_B C_v \Delta T_B=-p_g\Delta V_B-kA_B\nabla T_B-hA_B\left(T_B-T_\infty\right)-\varepsilon\sigma A_B\left(T_B^4-T_\infty^4\right) \quad (3)$$

equation (3) encapsulates variations of the internal energy ΔU of bubbles due to the growth/collapse pressure-volume work and the modeled heat-transfer mechanisms, i.e. convection, conduction and radiation. The variables $m_B$, $V_B$ and $A_B$ stand for the mass, volume and surface area of the bubble. The thermal conductivity k and heat-transfer coefficient h of the internal gas are calculated as functions of the Nusselt number [12]. In the thermal radiation term, σ corresponds to the Stefan-Boltzman constant and ε is the emissivity of the bubble, which is modeled as a black-body.

The gas pressure and temperature variables are coupled by integrating a Helmholtz virial EOS:

$$Z=\frac{p_g}{\rho_g R_g T_B}=1+\delta\frac{\partial}{\partial\delta}a^r(\delta,\tau) \quad (4)$$

the left-hand side in equation (4) is the ideal gas law, where $R_g$ is the gas constant and $\rho_g$ is the density. The right-hand side includes the residual Helmholtz energy $\alpha^r$ contribution to the ideal part, also known as compressibility coefficient Z, which is expressed as a multiparameter function of the reduced density δ and temperature τ. Similar virial expansion models are also used for estimating the heat-capacity $C_v$ in equation (3).

Numerical integration of the formulated system of equations is performed iteratively, using the embedded fifth-order Runge-Kutta-Fehlberg algorithm and an adaptive step-size control with a minimum step of 1.0e-15 s. The bubble gas pressure and density unknowns in equation (4) are calculated by integrating successive Newton-Raphson iterations. The solution performance of the described thermal R-P models is examined in the forthcoming discussion, for sonoluminescence and cavitation bubble-dynamics.

## 3. Sonoluminescence and cavitation bubble-dynamics

Validation results of the implemented thermal R-P model are plotted in figure 1. The scatter dots in these graphs correspond to a) light-scattering measurements of the bubble-radius [5] and b) calculations of the gas temperature [6], while the solid lines are the equivalent predictions of the coded K-M bubble model. The reproduced SBSL case assumed a gas nucleus with initial size $R_0$ = 4.5 μm, which is perturbed from its equilibrium ($T_\infty$ = 20 C and $p_{B,0}$ = $p_\infty$ = 1 bar) by a pressure-wave (dashed line). The amplitude and frequency of the acoustic-wave are 1.35 bar and 20 kHz, respectively. As shown in these results, the gas bubble grows asymptotically during the rarefaction phase of the wave, driven by the low ambient pressure. Gradually, the pressure-field recovers to positive values causing the bubble to collapse. The bubble collapse-rate is decelerated by the internal gas, which is compressed to high pressures and temperatures. Eventually, inertial effects force the bubble into repeatable growth/collapse rebound oscillations. Model predictions seem to be aligned with this sequence of dynamic events, both in terms of the measured bubble-radius and the corresponding prediction for the gas temperature. Löfstedt [5] calculated the peak temperature of this SBSL event to be ≈ 8500 K. Differences in the comparison of these results are noticed at the bubble's rebounds, mainly due to neglected physical mechanisms, e.g. mass-transfer and shock-waves effects [10].

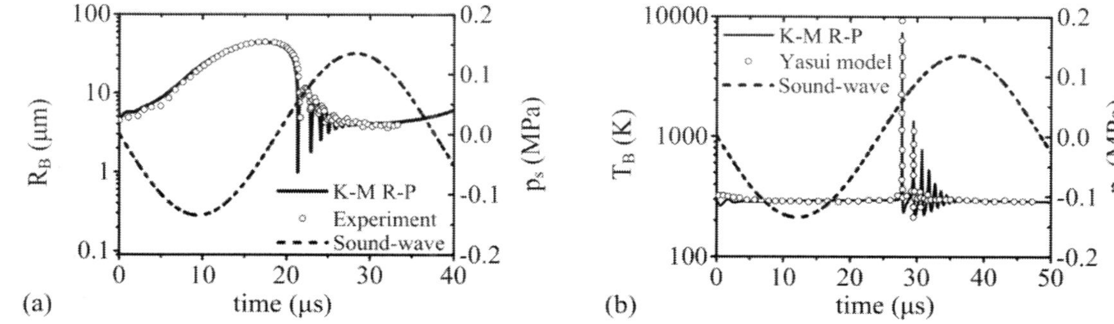


**Figure 1**. Validation of the thermal K-M model against reported a) experiments and b) predictions.

Additional single-bubble calculations are performed using the K-M and the classical thermal R-P equations as well as the compressible adiabatic model proposed by Moss. In these tests, the prescribed pressure boundary at the bubble-wall is extracted from incompressible single-phase RANS flow simulations inside a micro-sized nozzle (D = 200 μm). The pressure-drop applied to the water-liquid ($T_\infty$ = 353 K) across the hole varied from 200 MPa at the inlet to 5 MPa at the outlet. The simulated cavitation nucleus is considered to be in equilibrium with the upstream flow, i.e. $p_{B,0}$ = 200 MPa and $T_{B,0}$ = 353 K, at an initial size equal to $R_0$ = 1 μm. Model predictions of the growth/collapse bubble-radius are plotted in figure 2 (a), as a function of the normalized spatial location x/L inside the hole where L is the length of the channel. In figure 2 (b) velocity and pressure contours of the internal flow are shown.

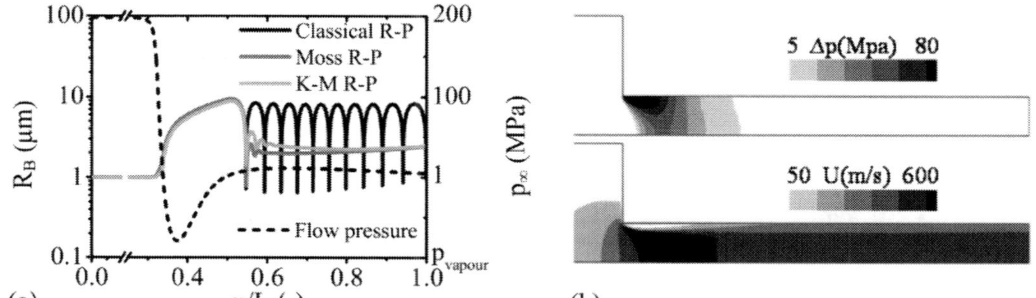


**Figure 2**. a) Cavitation growth/collapse dynamics calculated inside b) a high-pressure sharp nozzle.

Starting its journey from the upstream inlet, the gas nucleus arrives at the sharp hole step (x/L≈0.4) where it begins to grow due to the formed flow tension-site. The inception of the cavitation bubble is followed by a slow-rate asymptotic growth, which as it appears is equivalently predicted by the tested R-P models. This similarity implies that compressibility and thermal effects are insignificant during the initial growth phase. Further downstream (x/L≈ 0.55), the pressure is recovered (5 MPa) and the bubble is forced to implode. The bubble-dynamics ensuing after the termination of the growth phase are interpreted differently, depending on the utilized R-P equation. The compressible Moss and K-M models indicated a rapid decay of the cavitation oscillations, shortly after the collapse episode. On the contrary, the incompressible thermal R-P equation predicted repeatable rebounds with equally large amplitudes and consistent frequency. Damping in this case arises only from the viscous diffusion term.

Differences are also noticed in the predicted minimum bubble-radius, attained upon the initial collapse event. The thermodynamic behaviour of the cavitating bubble is depicted in figure 3.

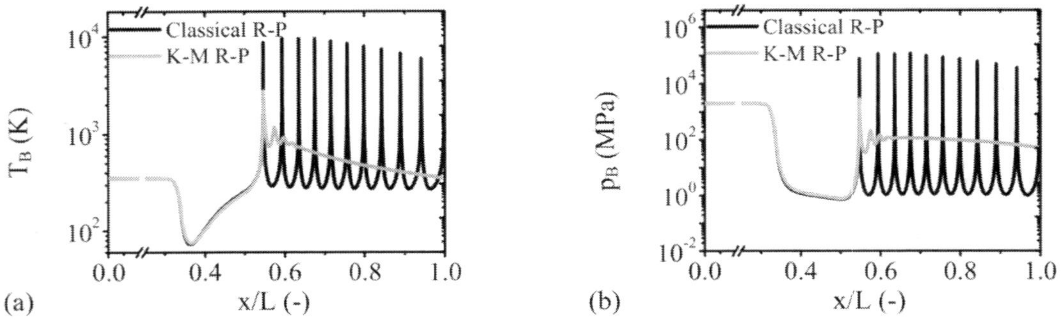


**Figure 3**. Homogeneous gas a) temperature and b) pressure predictions of the cavitation bubble.

Temperature and pressure variations of the cavitating gas-bubble are plotted for the thermal R-P models, i.e. the classical and K-M R-P equations. As seen in these graphs, during the initial asymptotic bubble-growth the internal gas is expanded to subatmospheric pressures and low temperatures. These conditions are promoted due to the absence of a mass-transfer model, i.e. evaporation effects. The development of hot-spots and high gas-pressures coincided with the locations where the bubble collapsed. The duration of these instances lasted only a few picoseconds, while their frequency of occurrence is a function of the damping terms integrated in each model. In absolute numbers, the classical thermal R-P model predicted temperatures and pressures in the order of 10000 K and 100 GPa, respectively. These values are considerably lower when the thermal K-M equation is used, ranging close to 3500 K and 0.5GPa.

## 4. Conclusions

In this paper a high-order Helmholtz EOS was coupled with different thermal R-P models. The method was validated against SBSL experiments and equivalent model predictions. Single-bubble cavitation dynamics were also calculated in high-pressure nozzle channels. The presented results indicated that during the violent implosion of bubbles the internal gas is elevated to high temperatures followed by pressures of the order of GPa. To proceed with the integration of bubble-dynamics within these physical limits, high-fidelity models are required for the characterization of the thermodynamic state variables and gas properties, as well as for addressing the role of shock-waves [13].

## References

[1] Brennen C E 1995 *Cavitation and Bubble Dynamics* (New York: Oxford University Press)
[2] Moss W S, Levatin J L and Szeri A J 2000 *Proc. R. Soc. Lond.* A **456** 2983-94
[3] Keller J B and Miksis M 1980 *J. Acoust. Soc. Am.* **68**(2) 628-33
[4] Lemmon E W, Jacobsen R T, Penoncello S G and Friend D G 2000 *J. Phys. Chem. Ref. Data* **29** (3) 331-85
[5] Lohse D and Hilgenfeldt S 1997 *J. Chem. Phys.* **107**(17) 1359-62
[6] Yasui K 1997 *Phys. Rev.* E **56** (6) 6750-60
[7] Lord Rayleigh 1917 *Phil. Mag.* **34** 94-8
[8] Plesset M S 1949 *J. Appl. Mech.* **16** 277-82
[9] Prosperetti A and Hao Y 1999 *Phil. Trans. R. Soc. Lond.* A **357** 203-23
[10] Löfstedt R, Barber B P and Putterman S J 1993 *Phys. Fluids* A **5**(11) 2911-28
[11] Perry R H 1999 *Perry's Chemical Enginners' Handbook* (New York: McGraw-Hill)
[12] Incropera F P, Dewitt D P, Bergman T L and Lavine A S 2013 *Principles of Heat and Mass Transfer* (Singapore: John Wiley & Sons)
[13] Löfstedt R, Barber B P, R Hiller and Putterman S J 1992 *J. Acoust. Soc. Am.* **91**(1) 2331

Journal of Physics: Conference Series **656** (2015) 012099 doi:10.1088/1742-6596/656/1/012099

# String cavitation formation inside fuel injectors

**B. A. Reid**[1], **M. Gavaises**[2], **N. Mitroglou**[2], **G. K. Hargrave**[1], **C.P. Garner**[1], **and R. M. McDavid**[3]

1 Wolfson School of Mechanical and Manufacturing Engineering, Loughborough University, Loughborough, Leicestershire LE11 3TU

2 School of Engineering and Mathematical Sciences, City University London, 10 Northampton
Square, London EC1V 0HB, UK

3 Caterpillar Inc., Product Development & Global Technology Division, Mossville IL USA

B.A.Reid@lboro.ac.uk

**Abstract.** The formation of vortex or 'string' cavitation has been visualised at pressures up to 2000 bar in an automotive-sized optical diesel fuel injector nozzle. The multi-hole nozzle geometry studied allowed observation of the hole-to-hole vortex interaction and, in particular, that of a bridging vortex in the sac region between the holes. Above a threshold Reynolds number, their formation and appearance during a 2 ms injection event was repeatable and independent of upstream pressure and cavitation number. In addition, two different hole layouts and three-dimensional flow simulations have been employed to describe how, the relative positions of adjacent holes influenced the formation and hole-to-hole interaction of the observed string cavitation vortices, with good agreement between the experimental and simulation results being achieved.

## 1. Introduction

Vortex or string cavitation structures have often been observed in the sac volume of enlarged nozzle replicas at fluid pressures of only a few bar (selectively [1-4] and only briefly observed in real-scale nozzle-holes operating at several hundred bar [5, 6]. The present work reports the presence of cavitation strings in a real-scale nozzle to pressures up to 2000bar and explores the influences on the formation of string cavitation at such conditions. In particular, the hole-to-hole interactions of the cavitation strings are investigated with respect to the position of the holes within the injector geometry and the prevailing fluid parameters. This work explores the hypothesis that, by better understanding the formation of string cavitation, the phenomenon can be ultimately be controlled.

## 2. Experimental set-up

Visualisation of the formation of string cavitation was performed in the near-hole sac volume region of a true-scale optical multi-hole fuel injector. The static nozzle geometry was designed to reproduce the asymmetric hole-inlet flow characteristics of a VCO or mini-sac type injector, where the flow in the sac volume, between the conical needle tip and the injector casing, must rapidly change direction to exit through the nozzle hole. Figure 2 illustrates the construction of the optical nozzle assembly, which consisted of three separate plates, one made of polished sapphire and two made of steel, clamped together to form the desired geometry and allow optical access to the flow volume upstream of the nozzle-hole entrances. Two different multi-hole plates were used, having a different hole spacing distance between the two rearward, covered, holes of the triangle. A 'standard' spacing of 0.7 mm was complimented by a 'wide' spacing of 0.9 mm to investigate the effect of hole separation distance.

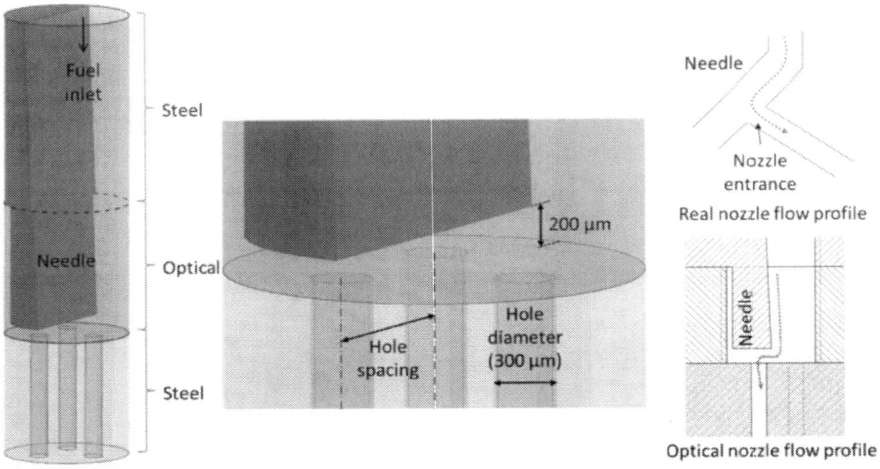


Fig 2. Layout and fluid flow path of the manufactured transparent nozzle.

Steady-state and 2 ms duration transient injections were delivered to the nozzle at pressures ($P_{rail}$) up to 2000 bar. Nozzle back pressure, $P_{back}$, was created by injecting into a pressurised nitrogen chamber downstream of the nozzle. Imaging of the internal nozzle flow was carried out using high speed imaging and single-shot photography, both utilising pulsed laser illumination of the nozzle flow to produce short-exposure images that 'froze' the flow features being imaged.

## 3. CFD Simulations

In the absence of quantitative measurements of the pressure and velocity fields in the areas where cavitation forms and develops, the computational fluid dynamics (CFD) code of [7] was utilised; Use of CFD was considered essential not only for assisting in the interpretation of the acquired images but also during the design process of the specific test nozzles utilised here.

## 4. Results

The first stage of the investigation used the rig in its steady-state configuration running at a 'low' pressure of $P_{rail} = 150$ bar.

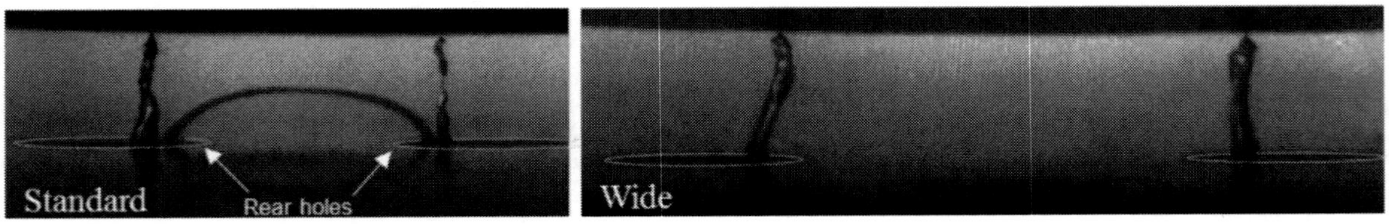


Fig. 3: Representative flow images obtained for the two different nozzles tested. [$P_{rail} = 150$ bar, $P_{back} = 20$ bar]

Two distinct two-phase flow structures were observed: the 'bridging' string that connects the two holes and two vertical strings, one in each hole, appearing to the side of the bridging string. The absence of the bridging string in the wide case, indicated that the formation of the vortex depends on hole spacing and the subsequent interaction of vortices from both holes. It was noted that the absence of the bridging string in the wide case resulted in the increased stability of the vertical string formation, where the two vertical strings were observed in both holes in almost every frame of the high speed video contrasting with their frequent but intermittent appearance in the standard case.

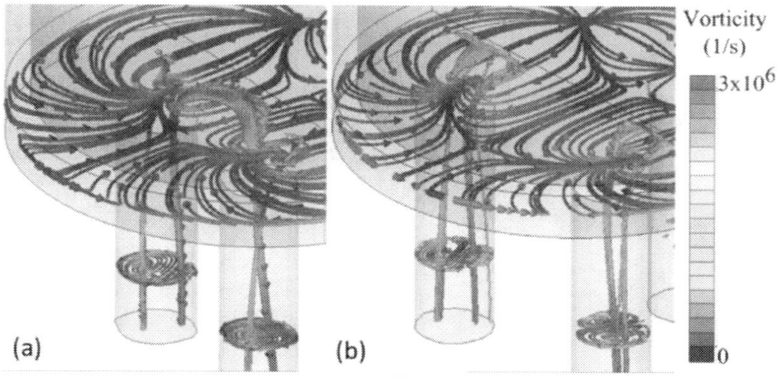


Fig. 4. CFD simulations of the flow inside and upstream of the injection hole inlet (a) vortex core illustration for standard spacing and (b) vortex core illustration for wide spacing nozzles. [$P_{rail}$ = 150 bar, $P_{back}$= 20 bar]

Increasing the injection pressure to c.2000 bar and running under transient injection conditions resulted in no change in the appearance of the cavitation strings for the standard hole spacing, as shown in figure 5, with the two vertical strings and the bridging string still present and of similar dimensions; changes to back pressure, between 1 and 90 bar, at these elevated inlet pressures also had no apparent effect on the appearance of the cavitation.

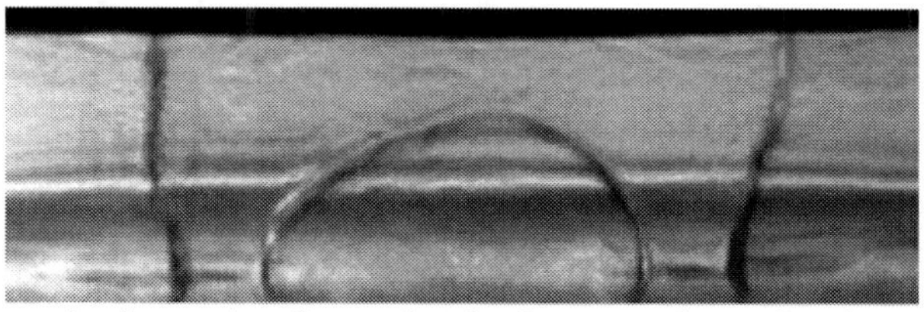

Fig. 5. Image of the cavitation string formation in the sac volume for the standard hole-spacing geometry for $P_{rail}$ = 2050 bar. $P_{back}$ = 90 bar. t = 1 ms ASOI

However, for the wide spacing case, when the injection pressure reached 2000 bar the formation of the bridging string was also observed to occur, as shown by the image in Figure 6. The formation of the bridging string at this injection pressure was also successfully captured by the CFD simulations.

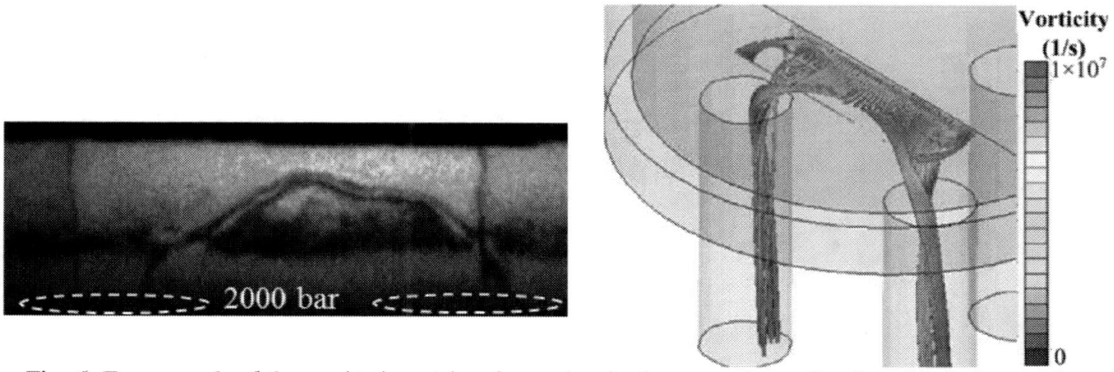


Fig. 6. Frame grab of the cavitation string formation in the sac volume for the wide hole-spacing geometry for $P_{rail}$ = 2000 bar. $P_{back}$ = 1 bar. t = 1 ms ASOI, and the corresponding CFD simulation result for the same condition

## 5. Conclusions

1. Cavitating flow structures have been visualised at a range of pressures up to 2050 bar in a real-scale flow geometry.
2. The occurrence of the hole-to-hole interaction of cavitation strings, observed here as the bridging string, was found to be affected by the spacing between adjacent holes and the respective flow rate through a particular geometry. As hole spacing was increased the flow rate, or injection pressure, required to form the bridging string was significantly increased.
3. Once formed, increases to the fluid pressure, or reductions in back pressure, have no influence on the appearance of cavitation strings.

## Acknowledgements

This work has been technically and financially supported by Caterpillar Inc. and the UK Technology Strategy Board. Project No, TP/3/DSM/6/I/15289. This project was co-funded by the Technology Strategy Board's Collaborative Research and Development programme, following an open competition.

The Technology Strategy Board is an executive body established by the Government to drive innovation. It promotes and invests in research, development and the exploitation of science, technology and new ideas for the benefit of business - increasing sustainable economic growth in the UK and improving quality of life.

## References

[1]Soteriou, C., R. Andrews, et al. (1995). "Direct injection diesel sprays and the effect of cavitation and hydraulic flip on atomization." SAE technical paper **950080**.

[2]Arcoumanis, C., M. Gavaises, et al. (1998). "Analysis of the flow in the nozzle of a vertical multi-hole diesel engine injector." Transactions Journal of Engines, SAE paper 980811, SAE Transactions **107**(3).

[3]Soteriou, C., M. Smith, et al. (1998). Diesel injection: laser light sheet illumination of the development of cavitation in orifices. IMechE Paper C529/018/98.

[4]Arcoumanis, C., H. Flora, et al. (1999). "Investigation of cavitation in a vertical multi-hole injector." SAE transactions **108**(3): 661-678.

[5]Walther, J. (2002) Quantitative Untersuchungen der Innenströmung in kavitierenden Dieseleinspritzdüsen, PhD. Thesis, TU Darmstadt http://tuprints.ulb.tu-darmstadt.de/241/ accessed Dec 2013

[6]Roth, H., E. Giannadakis, et al. (2005). "Effect of multi-injection strategy on cavitation development in diesel injector nozzle holes." SAE transactions **114**(3): 1029-1045.

[7]Giannadakis, E., M. Gavaises, et al. (2008). "Modelling of cavitation in diesel injector nozzles." Journal of Fluid Mechanics **616**(1): 153-193.

[8]Theodorakakos A., Mitroglou N., et al. (2012). Simulation of heating effects caused by extreme fuel pressurisation in cavitating flows through Diesel fuel injectors. CAV2012, Singapore.

# Numerical simulation of internal and near-nozzle flow of a gasoline direct injection fuel injector

**Kaushik Saha**[1] , **Sibendu Som**[1], **Michele Battistoni**[2], **Yanheng Li**[3], **Shaoping Quan**[3], **Peter Kelly Senecal**[3]

[1]Argonne National Laboratory Lemont, Illinois, USA Emails:ksaha@anl.gov,ssom@anl.gov

[2]University of Perugia
Perugia, Italy Email:michele.battistoni@unipg.it

[3]Convergent Science Inc.
Madison, Wisconsin, USA Email:yanheng.li@convergecfd.com

**Abstract.** *A numerical study of two-phase flow inside the nozzle holes and the issuing spray jets for a multi-hole direct injection gasoline injector has been presented in this work. The injector geometry is representative of the Spray G nozzle, an eight-hole counterbore injector, from the Engine Combustion Network (ECN). Simulations have been carried out for the fixed needle lift. Effects of turbulence, compressibility and non-condensable gases have been considered in this work. Standard* k$-\epsilon$ *turbulence model has been used to model the turbulence. Homogeneous Relaxation Model (HRM) coupled with Volume of Fluid (VOF) approach has been utilized to capture the phase change phenomena inside and outside the injector nozzle. Three different boundary conditions for the outlet domain have been imposed to examine non-flashing and evaporative, non-flashing and non-evaporative, and flashing conditions. Inside the nozzle holes mild cavitation-like and in the near-nozzle region flash boiling phenomena have been predicted in this study when liquid fuel is subjected to superheated ambiance. Noticeable hole to hole variation has been also observed in terms of mass flow rates for all the holes under both flashing and non-flashing conditions.*

## 1. Introduction

In the field of fuel injection systems for internal combustion engine applications cavitation and flash boiling are two common phase change phenomena. In the fuel injection systems when the local pressure drops below the saturation pressure corresponding to the fuel temperature, there is a strong potential for the liquid fuel to transform to vapor. Cavitation is prevalent for high pressure diesel injectors, while flash boiling generally occurs for direct injection gasoline injectors with elevated fuel temperatures. Cavitation is pressure-driven vaporization occurring at low fuel temperatures, where heat transfer is almost instantaneous; therefore, cavitation is basically inertia-driven. At elevated fuel temperatures saturation vapor density is going to be much higher compared to the one at low fuel temperatures and consequently, considerably more energy is required for phase-transition per unit volume of vapor, compared to low fuel temperatures. Flash boiling is thus a thermal non-equilibrium process unlike cavitation. This phenomenon can be explained better with the help of non-dimensional Jakob number ($Ja$), $Ja = \frac{\rho_l C_P \Delta T}{\rho_v h_{l_v}}$ [1]. Jakob number is the ratio of sensible heat energy available to energy required for vaporization. Higher $Ja$ indicates process is close to equilibrium, since heat transfer time-scale is much lower than flow time-scale, and hence cavitating. In case of lower $Ja$ process is more likely to be in non-equilibrium i.e. flash boiling, as the heat transfer time-scale will not be negligible compared to flow time-scale. The vapor bubbles, forming inside the injector holes, will expand rapidly at the nozzle outlet where the chamber pressure is relatively low. Such phase transformations internally and in the near-nozzle region affect the spray formation and overall charge formation for in-cylinder combustion.

Several numerical studies on flash boiling for GDI injectors are available in the literature. The modeling studies that resorted to bubble based models looked in to mono/poly-disperse bubbles, bubble-to-bubble interaction (collision, breakup etc.) and effect of different turbulence modeling approaches [2, 3]. The modeling studies that utilize empirical coefficients mainly utilizes the Homogeneous Relaxation Model (HRM) [1, 4–8] which represents the phase transition by one equation by estimating the time scale of phase change. The time scale provides the estimate of the deviation from thermal equilibrium. HRM lies in between the two extremes of thermodynamic two-phase models - homogeneous equilibrium model (HEM) and homogeneous frozen model (HFM). In case of HEM, the two-phases are assumed to be mixed perfectly homogeneous and heat transfer occurring contiguously. In real world scenario of two-pahse flows such as bubbly flows instantaneous heat transfer is not feasible. The other extreme, HFM assumes zero heat transfer or heat transfer time-scale to be infinitely long. HRM manages to capture the in-between practical two-phase flow scenarios. Experimental studies available in the literature adopted different optical and laser-based diagnostics techniques to provide insights of the flash boiling phenomenon [9–11]. The experimental investigations revealed - chances of high levels of spray plume interactions under the influence of intense flashing, possibility of considerable flashing at or above 20 K of superheat etc.

The current study wants to focus on development of high fidelity flash boiling models inside and the near-nozzle regions of multi-hole GDI fuel injector. The CONVERGE CFD code has been used [12]. HRM has been coupled with Volume of Fluid (VOF) approach. The previous published works have successfully demonstrated the potential of the CONVERGE code to predict the cavitating two-phase flow in diesel injectors, operating under high injection pressure e.g. Battistoni et al. (2014) [13]. The present work will investigate HRM performance for different thermodynamic conditions - flashing, non-flashing and non-evaporating.

## 2. MODEL FORMULATION

The problem considered in this study is a 8-hole counter-bored GDI fuel injector, which is denoted as Spray G nozzle in the Engine Combustion Network (ECN)[14]. The internal geometry of Spray G nozzle is not symmetric since it has 8 holes and 5 dimples which results in uneven flow passages. Therefore for realistic predictions full nozzle geometry has been simulated in this study. Details of the cases studied in this work are summarized in Table 1. The conservation equations of mass, momentum, species, energy, turbulence (standard $k - \epsilon$) are solved. Trace amounts of $N_2$ and $O_2$ are assumed to be present within the liquid fuel. The source term in species equation of vaporizing species (iso-octane) is modeled by HRM. HRM provides estimate of rate of change of vapor mass fraction i.e. $\frac{Dx}{Dt}$ [12].

$$\frac{Dx}{Dt} = \frac{\overline{x} - x}{\theta} \tag{1}$$

Journal of Physics: Conference Series **656** (2015) 012100 doi:10.1088/1742-6596/656/1/012100

The time-scale $\theta$ is calculated as $\theta = \theta_0 \alpha^{-0.54} \psi^{-1.76}$ where $\alpha = void(gas + vapor) fraction$; $\theta_0 = 3.84 \times 10^{-7}$; $\psi = \frac{p_{sat} - p}{p_{crit} - p_{sat}}$.

**Table 1.** Cases Studied

| Parameters | Spray G | Spray G2 | Spray GCool |
|---|---|---|---|
| Inj. Press.($MPa$) | 20 | 20 | 20 |
| Chamber Press.($kPa$) | 600 | 53 | 600 |
| Chamber Temp.($K$) | 573 | 293 | 293 |
| Fuel Temp.($K$) | 363 | 363 | 363 |
| Chamber Fluid | Pure $N_2$ | Pure $N_2$ | Pure $N_2$ |

## 3. Results and Discussions

In this work results using 180 $\mu$m and 140 $\mu$m as base grid sizes have been presented. There are 1.4 million cells for 180 $\mu$m base grid size while 2.8 million cells are available for 140 $\mu$m base grid size. Smallest cell sizes are 22.5 $\mu$m and 17.5 $\mu$m respectively. These resolutions are relatively finer than the ones used by Moulai et al. [8]. From thermodynamic considerations it can be assessed that Spray G is non-flash boiling and Spray G2 is moderately flashing, since for Spray G2 iso-octane fuel is subjected to 12 K of superheat and Ja number equal to 24.

For Spray G considerable vapor formation has been observed in the outlet domain, which can attributed to evaporation due to high chamber temperature. SprayGcool has been studied to isolate this issue by making the chamber temperature much lower. Figure 1 unravels the

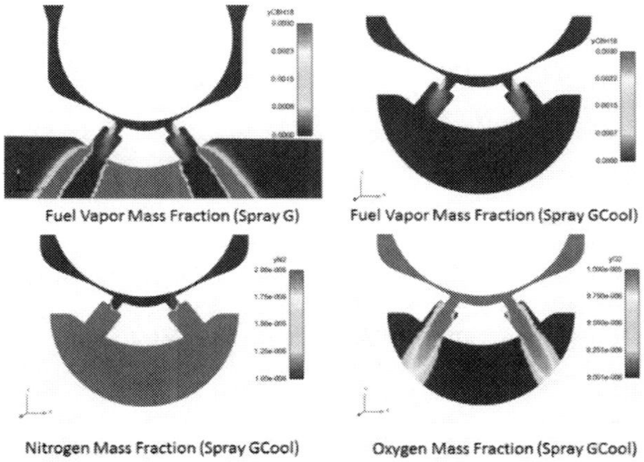


**Figure 1.** Contours of Spray G and Spray GCool

that the vapor formation for Spray G case is indeed due to the high temperature evaporative environment. The color codings in the contours have been carefully adjusted to analyze the nature of the two-phase composition inside the holes and the counter-bores. It is evident that there is mild vapor formation i.e. cavitation-like phenomenon occurring inside the holes. $N_2$ and $O_2$ concentrations are at 1.0e-05 mass fraction inside the holes. However in the counter-bores there is relatively higher presence of $N_2$ gas. Considering the variation of $O_2$ mass fraction it can be postulated that counter-bore region is subjected to back-flow of $N_2$ from the chamber. Spray GCool have been carried out with a smaller outlet domain to save computational time.

For Spray G2 the vapor mass fraction contours on a vertical cut-plane are provided in Fig. 2 and moderate flash boiling is observed in the near-nozzle regions. The left image shows the full

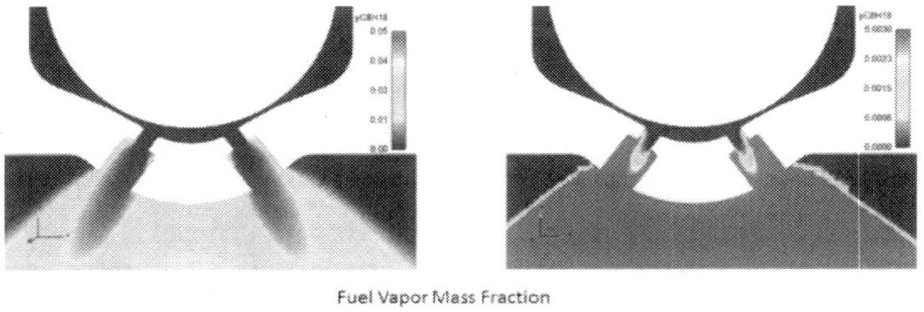


**Figure 2.** Vapor mass fraction contours of Spray G2

range of variation of the iso-octane vapor mass fraction. Moulai et al. [8] predicted modest level of flashing for Spray G2 condition with the peak mass fraction being 0.03, which is close to the peak mass fraction estimate (= 0.05) from the present study. The right image has an adjusted color coding to unravel the mild vapor formation inside the holes. The chamber pressure for Spray G is 600 kPa and that for Spray G2 is 53 kPa. Spray G2 being a higher pressure differential case with same fuel temperature and nozzle geometry, vapor formation inside the holes is obvious, since Spray G case had vaporization inside the holes. The full-range contour of mass fraction also demonstrates the possibility of plume-to-plume interaction as time progresses.

For both Spray G and Spray G2 hole-to-hole variations have been noticed in terms of mass flow rates. The internal geometry is very asymmetric because of five dimples that narrow down the internal flow passage. More in-depth analysis of hole-to-hole variations is necessary and will be investigated at a later time.

## 4. Summary

As a summary of the numerical findings it can be mentioned that Homogeneous Relaxation Model coupled with Volume of Fluid approach, implemented in CONVERGE, has been successfully applied to analyze flash boiling and cavitation-like phenomena in a Gasoline Direct Injection Spray G nozzle. Further studies with Large Eddy Simulations and moving needle problems will be undertaken in future.

## References

[1] Neroorkar K D 2011 *Modeling of Flash Boiling Flows in Injectors with Gasoline-Ethanol Fuel Blends* PhD Thesis University of MassachusettsAmherst
[2] Zeng Y and Lee C F 2001 *Combustion Science and Technology* **169** 45–67
[3] Zuo B, Gomes A M and Rutland C J 2000 *International Journal of Engine Research* **1** 321–336
[4] Bilicki Z and Kestin J 1990 *Proceedings of Royal Society London A* **428** 379–397
[5] Downar-Zapolski P, Bilicki Z, Bolle L and Franco J 1996 *International Journal of Multiphase Flow* **22** 473–483
[6] Schmidt D P, Gopalakrishnan S and Jasak H 2010 *International Journal of Multiphase Flow* **36** 284–292
[7] Brennen C E 1995 *Cavitation and Bubble Dynamics* (Oxford University Press)
[8] Moulai M, Grover R, Parrish S and Schmidt D 2015 Internal and near-nozzle flow in a multi-hole gasoline injector under flashing and non-flashing conditions SAE Paper 2015-01-0944 (Society of Automobile Engineering Conference)
[9] Zeng W, Xu M, Zhang G, Zhang Y and Cleary D J 2012 *Fuel* **95** 287–297
[10] Xu M, Zhang Y, Zeng W, Zhang G and Zhang M 2013 Flash boiling: Easy and better way to generate ideal sprays than the high injection pressure SAE Paper 2013-01-1614 (Society of Automobile Engineering Conference)
[11] Weber D and Leick P 2014 Structure and velocity field of individual plumes of flashing gasoline direct injection sprays (ILASS-Europe 2014, $26^{th}$ Annual Conference on Liquid Atomization and Spray Systems)
[12] Richards K J, Senecal P K and Pomraning E *CONVERGE v2.2 Documentation* Convergent Sciences Inc.
[13] Battistoni M, Xue Q, Som S and Pomraning E 2014 *SAE International Journal of Fuels and Lubricants* **7** 167–182
[14] `Sandia National~Laboratory` 2014 Spray G operating condition url `http://www.sandia.gov/ecn/`

# On the relation between tonal and broadband content of hull pressure spectra due to cavitating ship propellers

**J Bosschers**
MARIN, Haagsteeg 2, 6700 AA, Wageningen, the Netherlands

E-mail: j.bosschers@marin.nl

**Abstract.** Cavitating propellers generate pressure fluctuations on the hull of the ship. These pressure fluctuations are usually analyzed in the frequency domain using FFTs and the spectrum is composed of tonals at multiples of the blade passage frequency and a broadband part. The two are often considered separately but a relation between the two exists which has been investigated by theoretical signal analysis. It will be shown that the broadband part is related to the variability of the signal between blade passages and a simple procedure is proposed to quantify the variability in terms of amplitude and phase angle. The procedure has been applied to a data set obtained at sea trials.

## 1. Introduction

Cavitating propellers may generate inboard noise and vibration hindrance for crew and passengers on-board ships. Often, it is the hull pressure spectrum that is analyzed to evaluate the hull-excitation force of the cavitating propeller. This spectrum is composed of tonal components at harmonics of the blade passage frequency and a broadband part. The broadband part is known to be caused by the variability of the pressure signal [1][2], but a quantitative description is missing in literature. The present paper presents a formulation that can fill this gap although mathematical proof is still to be completed. Before presenting the formulation some basic aspects of the spectrum are discussed.

## 2. Stylised spectra of a single blade passage

Stylised hull pressure time traces and corresponding spectra of a cavity collapse during a single blade passage are shown in figure 1. The signal was generated using the absolute value of a sine for a number of cycles. Signal 1a and 1b have no damping and the width of the broadband hump in the spectrum, with a centre frequency at 10 Hz which is the repeat frequency of the pressure pulse, is inversely proportional to the number of cycles. For signal 2a and 2b an exponential damping was added to the time trace which leads to a smooth variation of the spectrum from frequency zero to the broadband hump.

## 3. Stylised spectra of multiple blade passages

Analytical formulations for the effect of amplitude and phase angle modulation on spectra of pulse trains, referred to as blade passages in the present paper, have been presented by McFarlane [3] and the theory is briefly restated here. Consider that the amplitude density spectrum of a single blade passage at time zero is given by $G(\omega)$.

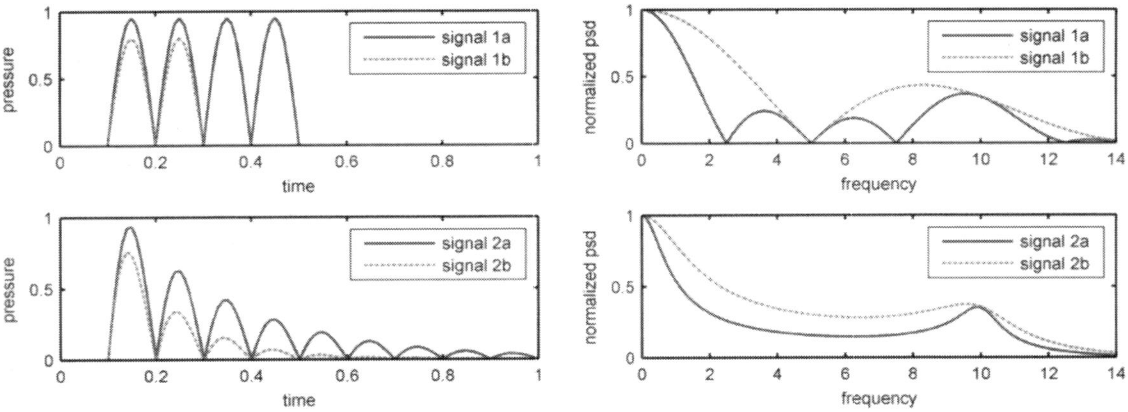


a) b)

**Figure 1.** Stylised pressure time traces (left) and normalised power spectral densities (right) due to a cavity collapse and rebounds during a single blade passage

The spectrum, $S$, of a sum of $2N+1$ (non-overlapping) blade passages spaced T apart, with amplitude $a_n$ and relative time shift $t_n$, is then given by

$$S(\omega)=G(\omega)\sum_{n=-N}^{N} a_n \exp\left[i\omega\left(nT+t_n\right)\right] \tag{1}$$

and the power spectral density (psd) is given by $R(\omega)=\left|S(\omega)\right|^2/(2N+1)T$.

With amplitude modulation the values for $a_n$ are statistically described by the mean $\bar{a}$ and standard deviation $\sigma_a$ while there is no relative time shift $(t_n=0)$. After some algebraic manipulations in which an infinite series of blade passages is assumed, we find the psd for amplitude modulation,

$$R(\omega)=\left[\sigma_a^2+\bar{a}^2\omega_r\Delta_{\omega_r}(\omega)\right]\frac{\left|G(\omega)\right|^2}{T} \tag{2}$$

where $\omega_r=2\pi/T$ corresponds to the blade passage frequency (bpf) and $\Delta_{\omega_r}(\omega)$ to a Dirac comb with frequency spacing $\omega_r$. The spectrum is composed of a broadband part with amplitude $\sigma_a^2$ summed with a tonal part with amplitude $\bar{a}^2$ weighted with the frequency bin width $\omega_r$.

Phase modulation is defined here as an arbitrary variation in arrival time $nT+t_n$ of the pressure signal for each blade passage $n$ while the amplitude does not vary in time $(a_n=a)$. The probability of $t_n$ is given by a Gaussian distribution $q$ with zero mean and standard deviation $\sigma_p$,

$$q(t_n)=\frac{1}{\sigma_p\sqrt{2\pi}}\exp\left(-t_n^2/2\sigma_p^2\right). \tag{3}$$

After lengthy algebraic manipulations, the following expression for the psd for phase angle modulation is obtained,

$$R(\omega)=a^2\left[1-\exp\left(-\sigma_p^2\omega^2\right)+\exp\left(-\sigma_p^2\omega^2\right)\omega_r\Delta_{\omega_r}(\omega)\right]\frac{\left|G(\omega)\right|^2}{T} \tag{4}$$

The effect of the amplitude and phase angle modulation is illustrated in figure 2, following [2]. A stylised spectrum is sketched of a single blade passage together with the spectrum of a series of blade passages. The frequency is non-dimensionalized by the bpf. If the signal repeat perfectly there is no modulation and the spectrum consists of tonals at harmonics of the bpf. Random variations in amplitude alone between different blade passages hardly affect the amplitudes of the tonals. However, the spectrum now does contain a broadband part of which the magnitude is proportional to the standard deviation of the amplitude variations. Random changes in time of arrival of the signal lead to random phase angle variations of which the magnitude increases linearly with frequency. This random phase modulation decreases the amplitude of the tonals and redistributes the power over a broadband region. The spreading is most pronounced at higher frequencies and may cause the complete disappearance of the tonal.

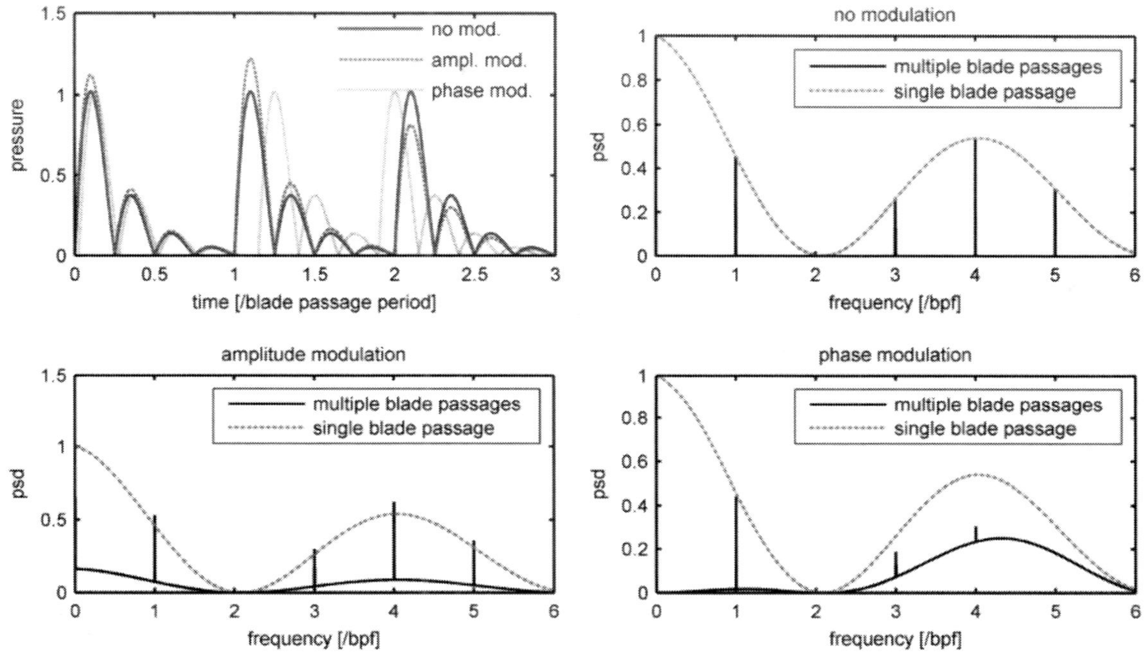


**Figure 2.** Stylised pressure time traces and resulting spectra for multiple blade passages with amplitude and phase modulation

## 4. Practical example of combined amplitude and phase angle modulation

For the combination of amplitude and phase angle modulation, the equations (2) and (4) are combined in a heuristic manner with mathematical proof still to be completed. The formulation is given by

$$R(\omega)=\left\{\sigma_a^2+\left[1-\exp\left(-\sigma_p^2\omega^2\right)+\exp\left(-\sigma_p^2\omega^2\right)\omega_r\Delta_{\omega_r}(\omega)\right]\bar{a}^2\right\}\frac{\left|G(\omega)\right|^2}{T} \tag{5}$$

In reality, the shape of the spectrum also varies with the blade passages and the standard deviation for amplitude and phase is frequency dependent. The formulation has been applied to the hull pressure signal generated by a four bladed cavitating propeller during a sea trial.

Data sets are analyzed by subdividing the time trace into 'shaft revolutions' with a shaft revolution defined by a constant number of samples which does not change in time, see figure 3a. This implies that variations of the shaft revolution rate in time due to sea state are interpreted as phase angle modulation. Each shaft revolution is analyzed with an FFT, after which the mean and standard deviation of both amplitude and phase angle can be computed for each harmonic of the shaft rate frequency. The variation of standard deviation of amplitude and phase angle with frequency is given in figure 3c and 3d. The phase angle was 'unwrapped' which explains values for the standard deviation of the phase larger than 360 deg.

The spectrum that is generated from the amplitude information alone by using equation (2) is referred to as AM and the result is given in figure 3b. To simplify the graph, the tonals are given for the harmonics of the bpf only. Using equation (5), we can reconstruct the spectrum with amplitude and phase modulation (designated APM in figure 3b) and compare this with the reference spectrum that is obtained by taking an FFT over a large number of revolutions. For simplicity, a linear fit of the standard deviation of the phase angles at harmonics of the bpf is taken except for the first bpf. The pressure amplitude at the first bpf is also due to the passage of the non-cavitating blade which repeats well as shown by the small values of the relative standard deviation. Figure 3b shows reasonable agreement between the two spectra with some differences at the 2nd and 3rd bpf suggesting that, after some further improvement, the proposed formulation and analysis procedure may be useful for quantifying the relation between higher order tonals and broadband content of hull pressure spectra.

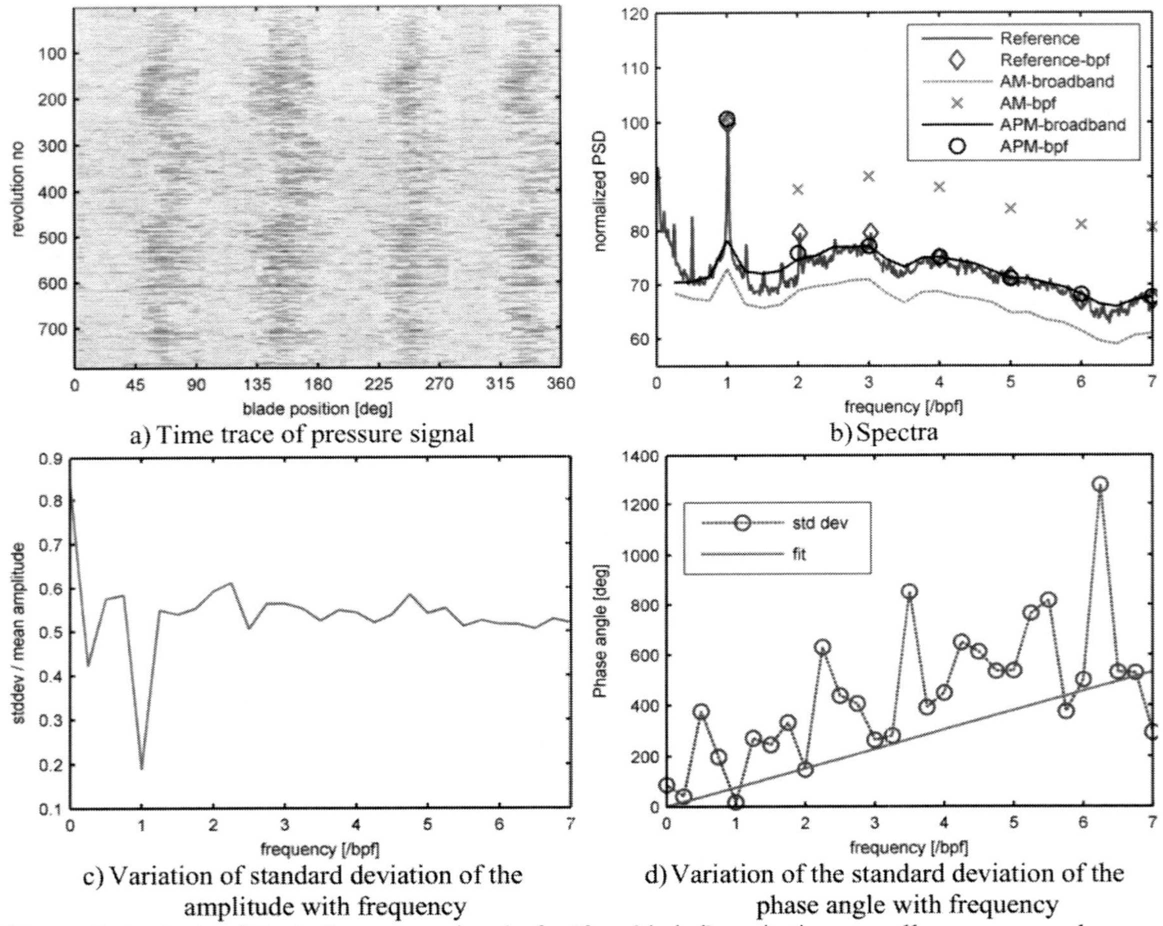


a) Time trace of pressure signal

b) Spectra

c) Variation of standard deviation of the amplitude with frequency

d) Variation of the standard deviation of the phase angle with frequency

**Figure 3.** Analysis of the hull pressure signal of a (four bladed) cavitating propeller as measured on a sea trial

## References

[1] Bark G 1988 *On the mechanisms of propeller cavitation noise* (Ph.D. thesis, Chalmers University of Technology, Göteborg, Sweden)

[2] Bosschers J 2009 Investigation of hull pressure fluctuations generated by cavitating vortices, *First International Symposium on Marine Propulsors, smp'09*, Trondheim, Norway

[3] MacFarlane G G 1949 On the energy-spectrum of an almost periodic succession of pulses, *Proc. of the Inst. of Radio Eng.*, Vol. 37, No. 10, pp. 1139-1142

# Research on the noise induced by cavitation under the asymmetric cavitation condition in a centrifugal pump

**J X Lu [1], S Q Yuan [1], J P Yuan [1], X D Ren [2], J Pei [1], Q R Si [1]**
[1]National Research Center of Pumps, Jiangsu University
[2]School of Mechanical Engineering, Jiangsu University
Zhenjiang, 212013, Jiangsu, China
E-mail address: shouqiy@ujs.edu.cn

**Abstract:** An experimental investigation has been carried out to research the noise induced by cavitation under the asymmetric cavitation (AC) condition in a centrifugal pump. The acoustic pressure signals at the pump inlet and outlet were measured respectively during the development of cavitation in a closed hydraulic test rig. It could be found that both the pump inlet and outlet acoustic pressures changed obviously with the development of cavitation. The time domain and the power spectrum density of the pump inlet and outlet acoustic pressure pulsations were analyzed. The broadband pulses of the acoustic pressure pulsations were found and the reasons for the phenomenon were given.

## 1. Introduction

Most hydraulic machineries including turbopump inducer, turbine, hydrofoil, propeller and pump, etc. were exposed to cavity [1]. Cavitation was responsible for the noise, vibration and instability generation, for the erosive damage with the surface of flow channels, and also for the loss of performance in pumps [2-6]. The structural vibration, acoustic emissions and hydrodynamic pressures have been selected by Escaler to evaluate the detection of cavitation in actual hydraulic turbines [7]. With the pressure decreasing, the process of cavitation in a centrifugal pump could be summarized as incipient cavitation, quasi-steady cavity, and unsteady cavitation [8, 9]. While various kinds of inducer cavitation instabilities have been reported, the major ones were cavitation surge (CS), rotating cavitation (RC), asymmetric cavitation (AC) and high-order instabilities [10]. The noise generated in a centrifugal pump not only depended on its geometry structure and the operating conditions, but also induced by the flow instability which may be caused by stall, surge, and especially the cavitation in the pump [11]. Some severe instabilities were caused by cavitation under the AC condition, which would produce lots of noise. In this paper, we mainly focused on the noise induced by cavitation under the AC condition in a centrifugal pump.

## 2. Experimental methods

The acoustic pressure signals at the pump inlet and outlet were measured during the development of cavitation in a closed hydraulic test rig. The centrifugal pump in the rig was driven at a constant speed of 2900 r/min by an electromotor, with a blade passing frequency $f_d$=290 Hz. The acoustic pressure pulsations were measured by four hydrophones which were installed at the pump inlet and outlet respectively. The hydrophones with the range of 0.1Hz～180kHz were produced by B&K Company, whose type was 8103. The sensitivity of the hydrophones was -211dB/1VμPa. A PXI 4472B data acquisition module made by National Instruments Company was applied to capture the electric signals and converted them to digital signals. The test rig could meet the secondary precision of national specifications of China.

## 3. Results and discussion

### *3.1 Asymmetric cavitation*

According to the reports of Y. Tsujimoto and T. Kimura, et al [9, 12, 13], in this paper it could be considered that the values of *NPSHa*=1.74 to 1.44 could be regarded as the condition of AC in the centrifugal pump.

### *3.2 Pump inlet noise signals*

It could be found that the acoustic pressure presented a periodic fluctuations over the time under the non-cavitation condition that *NPSHa*=9.87m in figure 1. However, the periodic variational regularity of the pump inlet noise signals gradually weakened under the AC condition with the development of cavitation. The periodic trends almost disappeared when *NPSHa*=1.44m. Not only had the amplitudes of the acoustic pressure pulsations under the AC condition become obviously lower than that under the non-cavitation condition, but also its intensity intensified. The explanation for this phenomenon was that on the one hand a larger cavity or sheet cavitation were produced by the single bubbles which were independent free in the pump while cavitation developed to a certain stage. The larger cavity collapsed in the downstream passages which was far away from the pump inlet. On the other hand, it was considered that the larger cavity absorbed part of the noise and shock pressure which were caused by the collapse of vapor bubbles under the severe cavitation condition [14]. The larger cavities showed a strong flexibility and no fixed shape characteristics due to the interaction between the larger cavities and the surrounding high pressure liquid which made the bubbles collapse or condense.

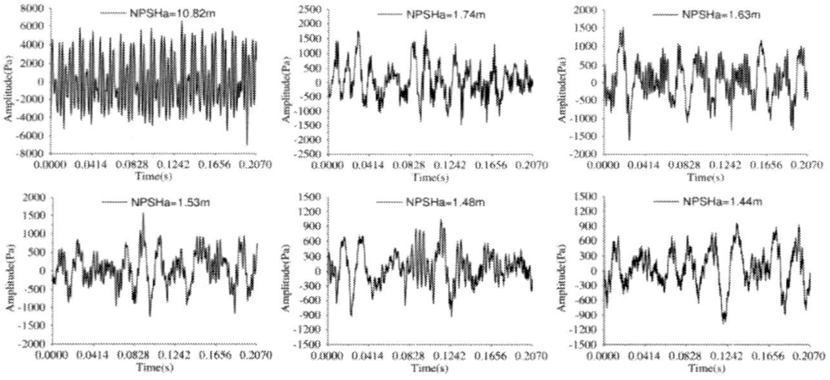


**Figure 1** Time domain signals of pump inlet acoustic pressure pulsations at $Q_d$.

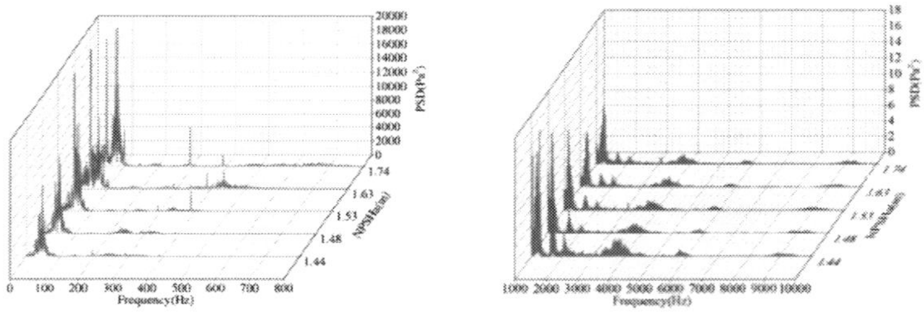


(a) (b)

**Figure 2** PSD signals of pump inlet acoustic pressure pulsations at $Q_d$.

The power spectrum density (PSD) signals of the pump inlet acoustic pressure pulsations under the AC condition was shown in figure 2. It could be found that the main frequency of the pump inlet acoustic pressure was the shaft frequency and its amplitudes were much higher than that of other frequency bands. The amplitudes of the main frequency decreased gradually with the development of cavitation. It was found that the broadband pulsations of the acoustic pressure distributed between 100 and 800Hz. With the decreasing of *NPSHa*, the range of broadband frequency gradually moved to low frequency and its intensity weakened. The broadband pulses could be apparently found between 1000 and 5000Hz and the frequency of the broadband presented no migration with the development of cavitation. Not only were the amplitudes of the broadband pulses at 1200Hz higher and its intensity relatively more severe than that of other frequency, but also the amplitude gradually increased.

### *3.3 Pump outlet noise signals*

It was apparent that the periodic trends of the pump outlet acoustic pressure pulsations over the time could be found under both the non-cavitation and AC conditions in figure 3. The amplitudes of the pump outlet acoustic pressure pulsations increased gradually with the development of cavitation under the AC condition. The explanation for the phenomenon was that either the free type of cavitation or the sheeting cavitation would be flowed to the downstream of the pump with the fluid, then the cavities would collapsed near the pump outlet. So the noise produced by the vortexes and collapse of cavitation were relatively high near the hydrophones which were installed at the outlet of the pump. On the other hand, a large number of cavitation bubbles were converged and dilated as the further decreasing of *NPSHa*, which would increase of the intensity of the noise induced by the collapse of vapor bubbles.

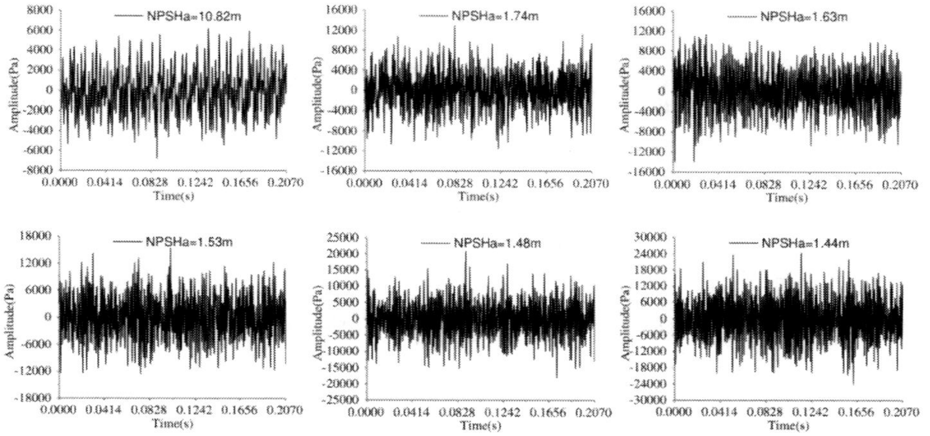


**Figure 3** Time domain signals of pump outlet acoustic pressure pulsations at $Q_d$.

The PSD signals of the pump outlet acoustic pressure pulsations in the AC condition were shown in figure 4. It could be found that its main frequency was 2 times of the blade passing frequency. The broadband pulses distributed near the 4 and 10 times of the blade passing frequency, and also between the 700 and 900Hz. The intensity of the broadband pulses became severe with the development of cavitation. Some weaker broadband pulses distributed between 4000 and 9000Hz could be found with the further development of cavitation. It indicated that the high frequency acoustic pressure pulsations were produced by the collapse of cavitation bubbles under the AC condition. Due to the convergence and dilatation of the cavities, the intensity of the collapse was strengthened which could lead to the intensification of the noise. It could be found that the intensity of the pump outlet acoustic pressure pulsations was clearly much stronger than that of the pump inlet acoustic pressure pulsations. The main reason for the phenomenon was that the cavities moved to the downstream and the location of the collapse was closer to the hydrophones which were installed at the outlet of the pump.

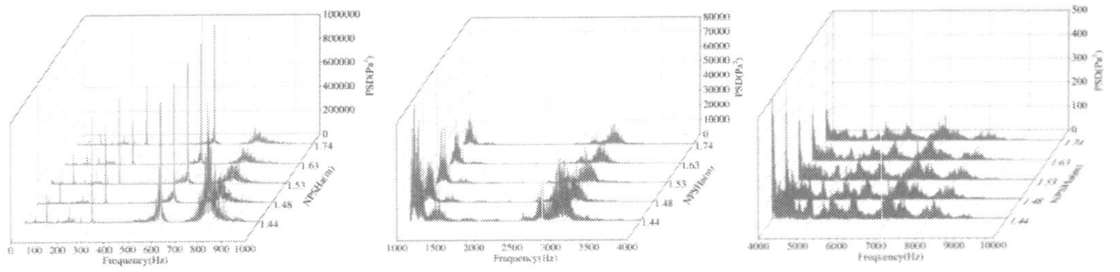

(a) (b) (c)

**Figure 4** PSD signals of pump outlet acoustic pressure pulsations at $Q_d$.

## 4. Conclusion

An experimental investigation has been carried out to research the noise at the pump inlet and outlet induced by cavitation under the AC condition in a centrifugal pump. The results indicate that:

- Both the pump inlet and outlet acoustic pressures changed obviously with the development of cavitation under the AC condition.
- The high frequency acoustic pressure pulsations were radiated by the collapse of cavitation bubbles under the AC condition. Due to the convergence and dilatation of the cavities, the intensity of the collapse were strengthened which would lead to the intensification of the noise.
- Compared with the pump inlet acoustic pressure pulsations, the intensity of the pump outlet acoustic pressure pulsations was clearly much stronger than that of the pump inlet acoustic pressure owing to the cavities moved to the downstream.

## Acknowledgments

This research was supported by the Natural Science Foundations of China (Grant No. 51239005 and 51479082) and the Priority Academic Program Development of Jiangsu Higher Education Institutions.

## References

[1] Yu X X, Huang, C G, Du T Z, et al. 2014 *J Fluids Eng.* **136** 051303-1-8
[2] Ni Y Y, Yuan S Q, Pan Z Y, et al. 2008 *Chinese Journal of Mechanical Eng.* **21** 46-49
[3] Chudina, M 2003 *Acoustical Physics* **49** 463-474
[4] Cernetic J 2009 *Mechanical Eng. Science* **223** 1645-1655
[5] Horiguchi H, Watanabe S, Tsujimoto Y 2000 *J. Fluids Eng.* **122** 798-805
[6] Semenov Y, Fujii A, Tsujimoto Y 2004 *J. Fluids Eng.* **126** 87-93
[7] Escaler X, Egusquiza E, Farhat M, et al. 2006 *Mechanical Systems and Signal Processing* **20** 983-1007
[8] Ravi B, Eugene S, Simon B 2012 *Proceedings of the Second Middle East Turbomachinery Symposium* Doha Qatar
[9] Kimura T, Yoshida Y, Hashimoto T, et al. 2008 *J. Fluids Eng.* **130** 051104-1-9
[10] Tsujimoto Y, Yoshida Y, Maekawa Y, et al. 1997 *J. Fluids Eng.* **119** 775-781
[11] Chudina M 2003 *Acoustical Physics* **49** 463-474
[12] Lu J X, Yuan S Q, Zhou B L, et al. 2014 *The 6th International Symposium on Fluid Machinery and Fluids Eng.* Paper No.90047. Wuhan China
[13] Tani N, Yamanishi N, Tsujimoto Y 2012 *J. Fluids Eng.* **134** 021302-1-13
[14] Xu C H, Xu D H, Wu Y L, et al. 2003 *J. Agricultural Machinery* **34** 139-142

# Dynamics and acoustics of a cavitating Venturi flow using a homogeneous air-propylene glycol mixture

**M Navarrete[1], J Naude[2,3], F Mendez[2], FA Godínez[1]**

[1] Laboratorio de Pruebas No Destructivas, Instituto de Ingeniería, Universidad Nacional Autónoma de México, UNAM. Av. Universidad 300, C. U. Distrito Federal 04510, México.
2 Departamento de Termofluidos, Facultad de Ingeniería, Universidad Nacional Autonoma de México, UNAM.

E-mail: jorge_naude@hotmail.com

**Abstract**. Dynamics and acoustics generated in a cavitating Venturi tube are followed up as a function of the input power of a centrifugal pump. The pump of 5 hp with a modified impeller to produce uniform bubbly flow, pumps 70 liters of propylene glycol in a closed loop (with a water cooling system), in which the Venturi is arranged. The goal was to obtain correlations among acoustical emission, dynamics of the shock waves and the light emission from cavitation bubbles. The instrumentation includes: two piezoelectric transducers, a digital camera, a high-speed video camera, and photomultipliers. As results, we show the cavitation patterns as function of the pump power, and a graphical template of the distribution of the Venturi conditions as a function of the cavitation parameter. Our observations show for the first time the sudden formation of bubble clouds in the straight portion of the pipe after the diverging section of the Venturi. We assume that this is due to pre-existing of nuclei-cloud structures which suddenly grow up by the tensile tails of propagating shock waves (producing a sudden drop in pressure).

## 1. Introduction

The behavior of cavitating Venturi nozzles has been investigated with great interest over the past 60 years because of the wide range of phenomena associated with their flow [1]. Accordingly, in [2] a theoretical development for bubbly flows together with experimental measurements emphasizing the role of the void fraction is found. In the same line of research, but including a theoretical-experimental treatment of shock emissions is found in [3]. Another theoretical analysis showed in [4], provides a robust set of expressions for the flow and shock waves generated in the nozzle. Another related topic is the design of commercial Venturi flow meters/controllers by ASME standards [5-6]. Nevertheless, these codes do not take into account the formation of complex flow patterns that may affect the proper operation and/or calibration of the device. Thus, a robust design requires studying the complexity of the phenomena associated with the flow including: two-phase bubbly flow, shock waves, luminescence, nucleation-expansion-compression of bubbles as single entities or cooperative-interacting clusters, formation of turbulent structures (gas-liquid-vapor), noise, and pulsating jets, among others as stated in [7-11]. Regarding numerical simulations, above complexities are addressed in [12].Our observations show the complexity of the above mentioned local bubble phenomena.

Experiments also reveal the noise generated, and the light pulses emitted as well as how the flow stream is affected by turbulence, vorticity, and shock waves.

## 2. Experimental setup

A schematic of the test rig is shown in Fig.1; the connecting PVC pipes are 2.5 inches in diameter. The Venturi is manufactured in transparent PMMA with 33 cm long, 64 mm of inner diameter and 16.5 mm diameter in throat section. The pump has a modified impeller (with a reduction of 36% in the impeller output area), which acts as a generator of nuclei, increasing the bubbles formation; besides the circuit has a regulating valve, and an auxiliary tank. A compact variable frequency drive Yaskawa J1000, is used to change the output speed of the pump; a photomultiplier (PMT) Hamamatsu Photonics R5783-04 for luminescence detection; a dynamic force sensor PCB 200B05 and a piezoelectric transducer 5 MHz Olympus V310-SU are used for sensing the acoustics. All signals are acquired with a digital Lecroy LC584AM oscilloscope. High-speed video is obtained from a Phantom v9.1 camera in the interval of 25000-40000 fps.

## 3. Experimental procedure

The experimental run begins when the fluid is recycled in order to obtain a homogeneous air-liquid mixture (dissolved oxygen = 20) with the cooling device activated until the mixture reaches a uniform temperature of 15°C; the power output of the pump is maintained at 10% of full power. Later on, the mixture flow rate is changed by varying the rotation frequency of the pump, and consequently its total power percentage, %. Acoustic and luminescence measurements as well as high speed video are acquired for each flow rate stage. The Venturi is used also as a flow rate meter; in which a U-tube mercury manometer is installed. The test rig and its instrumentation are tested with tap water; these tests showed that the visible choked condition in the Venturi throat is reached at 35% of pump power.

## 4. Results

Fig 1 shows the pump power (%) versus cavitation parameter $\sigma$ at different temperatures. The shock condition in the Venturi is attained at 35% of the pump power which corresponds at $\sigma = 0.5 \pm 0.2$, and the light emission is detected starting from $\sigma = 0.7$. The place where the shock wave and light flash occur, shifts as a function of flow rate; e.g. for $\sigma \sim 0.7$ it occurs in the exit zone of the throat and for $\sigma \sim 0.01$ it occurs in the transparent pipe line, see Fig 1.

General observations of the flow dynamics as a function of pump power are summarized in Table 1. At 20%; incipient cavitation is visible, the Kelvin-Helmholtz instabilities generate swirls at the interphases along the horizontal line [11]. As we know, inside the swirls the pressure is less than in the outside giving place to evaporation, leading to the formation of a mixture cavitation layer. At 30 %; the implosion of bubbles takes place at 48 mm from the throat, where non periodical bubble filaments appear, the flow rate in the wall reaches 6.8 m/s and 14 m/s in the "middle". At 35%; the implosion of the filaments occurs at 29 mm away from the throat, while the flow rate in the middle reaches 28 m/s. At this level, the emission of shock waves becomes observable. At 36%; a cloud stagnates and goes growing in an ordered fashion at 13.6 m/s until the diverging section of the Venturi is completely filled. This cloud is formed and destroyed periodically in the same place. At 38%; the cloud undergoes spallation and the clusters generated are dragged by the flow towards the straight section of the pipe. Some of these clusters become into nuclei-cluster structures that are practically imperceptible to the naked eye. Further increase in the pump power (41%) causes the cloud to be "*reborn*" in the straight section of the pipe. We think that this is partially due to pre-existence of nuclei-cluster structures which suddenly grow up by a sudden drop in pressure by the tensile tail of a shock wave. At this level the clouds in the diverging zone and at the straight section of the pipe emit shock waves into liquid phase at approximately 2450 m/s. The shock waves are reflected and returned up (identified with high density zones which are seen as darken regions in images) reducing the zone of remaining bubbles and clouds to some nuclei-cloud structures which are difficult to discern at a glance. At 42%; two clouds are clearly observed in the corresponding image, one filling the divergent zone of the Venturi and the other emerging from accumulation of the nuclei-cluster structures after the propagation of a shock wave. From the analysis of the acoustical noise the wideband lies between (6 - 10) kHz; while the light pulse emission is approximately between (10 to 20) ns in width and is accompanied systematically by shock waves with a bandwidth of 2 MHz approximately.

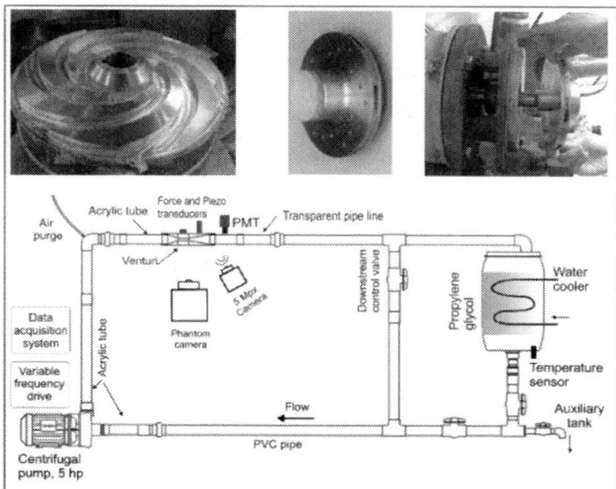


**Fig 1.** Block diagram of closed loop propylene glycol. The upper images show the special design of the pump impeller.

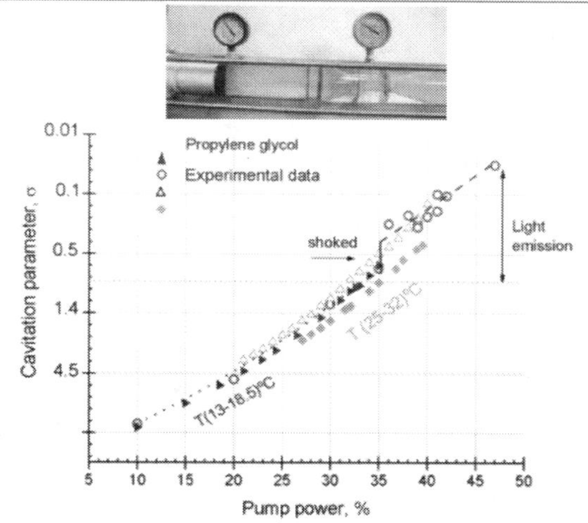


**Fig 2.** Pump power versus cavitation parameter at different temperatures. The upper image show the Venturi device.

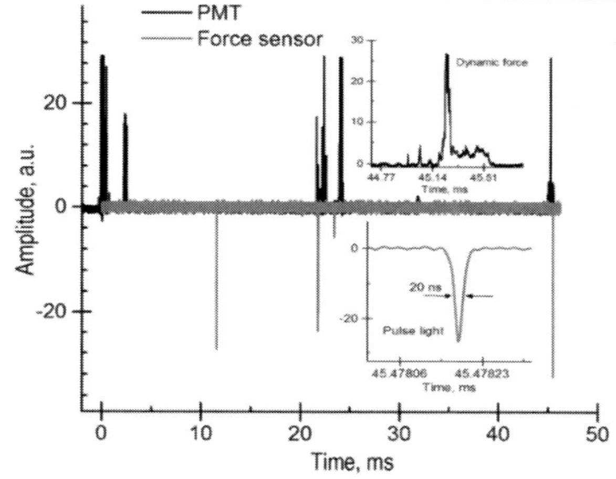


Fig 3. Signals from photomultiplier and force sensor. The inserted graphics are a zoom of the pulse force and light emission

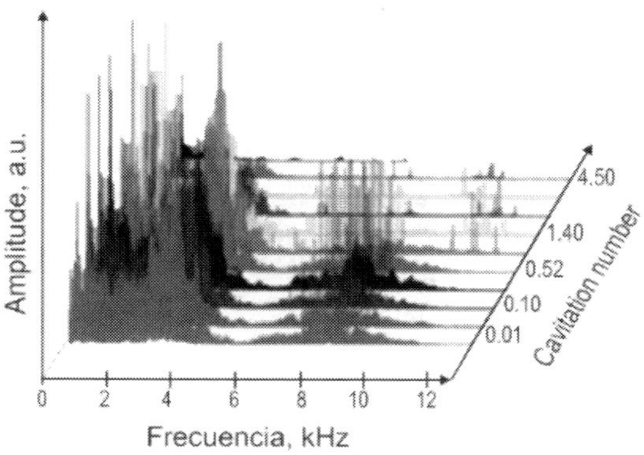


Fig 4. FFT-Spectrum amplitude of the noise signals as a function of power pump and the corresponding σ.

## 5. Concluding remarks

Dynamics and acoustics of a cavitating Venturi using a homogeneous air-propylene glycol mixture have been experimentally studied. We show the cavitation patterns as function of the pump power, and a graphical template of the distribution of the Venturi conditions as a function of the cavitation parameter. Our observations show for the first time the sudden formation of bubble clouds in the straight portion of the pipe after the diverging section of the Venturi. We hypothesized that this is due to pre-existence of nuclei-cloud structures which suddenly grow up by a sudden drop in pressure by the tensile tails of shockwaves.

**Acknowledgments**
The authors gratefully acknowledge financial support from UNAM-DGAPA-PAPPIT under Grant IN106515, IN12515, and IISGBAS-114-2013.

**Table 1**. An example of the evolution of the cavitation in Venturi device using propylene glycol.

| % | Representative frame | two-dimensional scheme | Description |
|---|---|---|---|
| 20 |  | 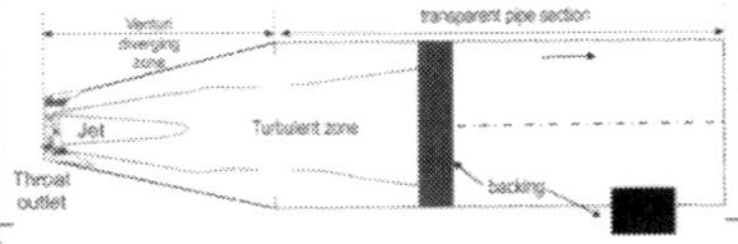 | A jet and turbulent zone (mixing zone inside the straight pipe) immersed in a homogeneous liquid. |
| 30 |  | 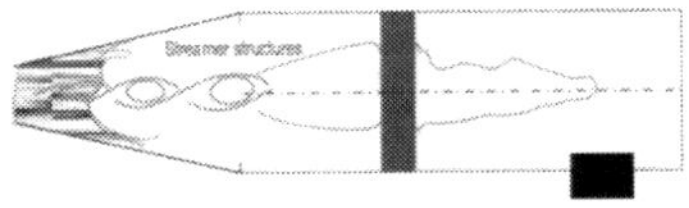 | Onset of streamer structures and turbulent zone. |
| 35 |  | 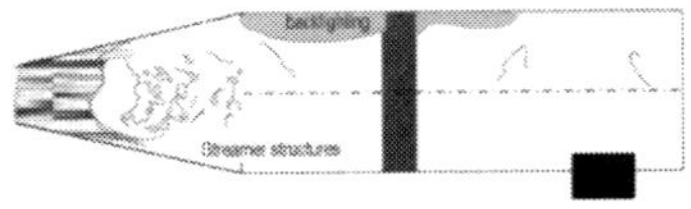 | Mixed cavitation in the axial layers and the onset of the emission of shock waves. |
| 36 |  | 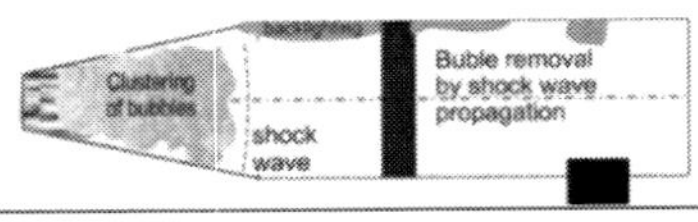 | Phase change in the throat and the divergent zone of Venturi. |
| 38 |  |  | The stagnation cloud undergoes spallation and the clusters are dragged by the flow towards the straight section of the pipe. |
| 40 | 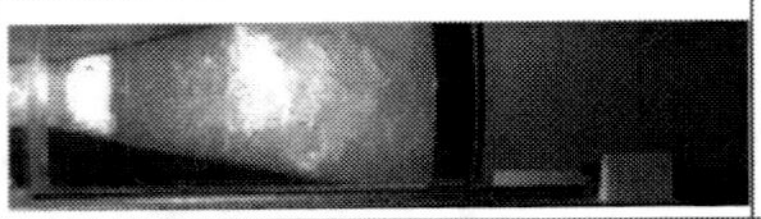 | 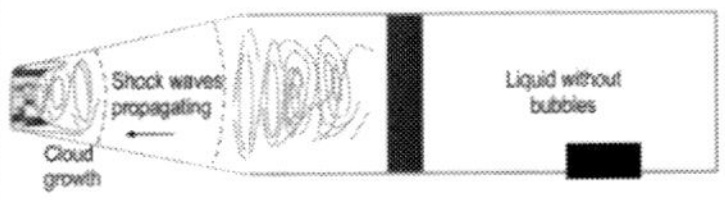 | New cloud formation and dissolution of the former. |
| 41 | 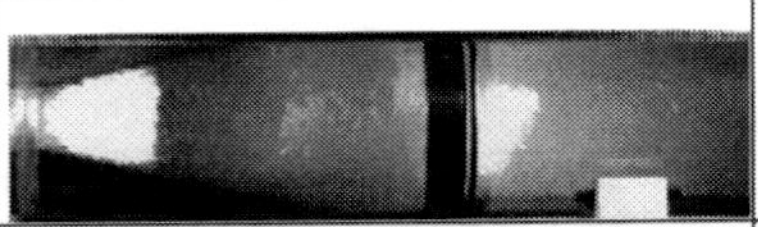 | 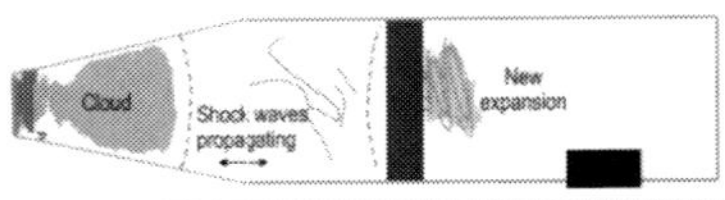 | The former cloud "reborn" in the straight section of the pipe by a sudden drop in pressure by the tensile tail of a shockwave. |
| 42 |  | 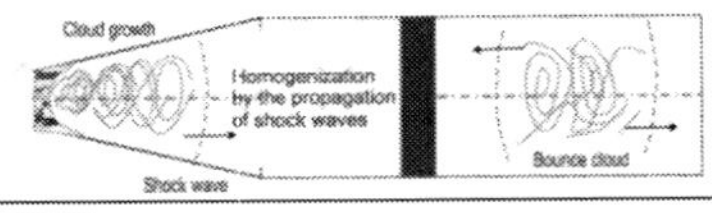 | Two clouds, one fills the divergent zone and the other emerges from the nuclei-cluster structures after propagation of a shock |

## 6. References

[1] Jarman P, and Taylor K 1965 *Brit. J. Appl. Phys.,* **16** 675
[2] Ishii R, Yoshikuni U, Shigeaki M and Norihiki S 1993 *Phys. Fluids A:Fluid. Dyn.* **5** 1630
[3] Ando K, Sanada T, Inaba K, Damazo J, Shepherd J, Colonius T and Brennen C 2011 *J. Fluid Mech.***671** 339
[4] Pasinlioglu S, Delale C and Schnerr G 2009 *IMA J. of Appl. Math.***74** 230
[5] ASME MFC-7M, 1987
[6] http://www.optimal-systems.demon.co.uk/flow-orifice.htm, © 1999 Optimal Systems Limited
[7] Reshetnikov A, et.al. 1987 *Fluid Mechanics: Sov. Res.* **16** 61
[8] Amini I and Owen I 1995 *Flow Meas. and Instr.***6** 41
[9] Hui T, Peng Z, Nanjia Y and Guopbiao Cai 2014 *Flow Meas. and Instr.* **38** 21
[10] Goncalves E and Fortes Patella R 2009, *Comput. Fluids* **38** 1682
[11] Lecoffre Y, Cavitation Bubble Trackers, 1999 A Balkema
[12] Brennen C, 2013 *J. Fluids Engineering* **135** 061301

Journal of Physics: Conference Series **656** (2015) 012104 doi:10.1088/1742-6596/656/1/012104

# Mechanism of noise generation by cavitation in hydraulic relief valve

**K Okita[1], Y Miyamoto[2], T Kataoka[2], S Takagi[3] and H Kato[3]**

[1] Department of Mechanical Engineering, College of Industrial Technology, Nihon University, 1-2-1 Izumi-cho, Narashino-shi, Chiba 275-8575, Japan
[2] KOMATSU, 400 Yokokura Shinden, Oyama-shi, Tochigi 323-8558, Japan
[3] Department of Mechanical Engineering, School of Engineering, The University of Tokyo, 7-3-1 Hongo, Bunkyo-ku, Tokyo 113-8654, Japan

E-mail: okita.kohei@nihon-u.ac.jp

**Abstract.** In order to clarify the mechanism of noise generation in a hydraulic relief valve, oil cavitating flows in a half cut model of the valve were observed by means of a high-speed camera and were simulated numerically. As the result of image analysis, the fluctuation of cavitation volume is corresponding to the pressure fluctuation of downstream, and the both fluctuations take peaks at frequencies from 1.5 to 2.5 kHz depending on the back pressure. In addition, as the back pressure increases, the frequency of the pressure fluctuation increases and the peak value decreases. These phenomena were also qualitatively reproduced in the numerical simulation.

## 1. Introduction

The noise of kHz order in a hydraulic relief valve has been often reported in construction machinery. In the tests of using an actual hydraulic relief valves, the noise has been confirmed in the particular conditions of the flow rate and back pressure, and considered to be owing to cavitation. However, the mechanism of the noise generation has not been clarified yet, because the noise occurred unpredictably. Although the operation of the hydraulic relief valve is not disturbed by the cavitation, a problem is the discomfort of construction machine operators by the noise. Early resolution of the noise generation is required to the development of the future hydraulic relief valves. In the present study, oil cavitating flows in a half cut model of the hydraulic relief valve are observed by means of a high-speed camera and are simulated numerically to clarify the mechanism of the noise generation owing to cavitation.

## 2. Visualization of the oil cavitating flow in the hydraulic relief valve

### *2.1. Visualization experimental apparatus*

The half cut model based on an actual hydraulic relief valve and the outline of a test section are shown in Fig.1. As shown in Fig. 1(a), the half cut model is composed of the sheet, valve and case. In the actual hydraulic relief valve, the pressure rise between the seat and the valve moves the valve to the right. Then, hydraulic fluids pass through a gap between the seat and the valve and flow outside of the case. But, in the half cut model employed in the present study, the valve is fixed. And the gap between

(a) Half cut model (b) Test section

**Figure 1.** Visualization experimental apparatus.

(a) 0.07MPa (b) 0.5MPa (c) 2MPa

**Figure 2.** Instantaneous images of cavitation in the downstream of the gap

the seat and the valve is adjusted as 100μm by spacers. On the other hand, as shown in Fig. 1(b), the test section is a sturdy structure to avoid the leak flows between an acrylic plate and the half cut model at operating pressure of several tens MPa.

### *2.2. Results and discussions of the visualization experiment*

The oil cavitating flows in the half cut model were captured by the high-speed camera with a long-distance microscope. Capture conditions were respectively the frame rate of 50,000 fps and the shutter speed of 1/330,000. The pump pressure was 20 MPa and the flow speed through the gap was about 100 m/s. The back pressure was changed from 0.07 to 2 MPa in the present study.

Instantaneous images of cavitation in the downstream of the gap for back pressure of 0.07, 0.5 and 2 MPa are shown in Fig. 2. As shown in Fig. 2(a), cavitation develops to the hole of the case by the jetting flow from the gap. As the back pressure increases, the cavitation region decreases as shown in from Fig. 2(a) to (c).

The cavitation region is defined by binarization processing on the basis of the luminance value, and the cavitation volume is calculated in the partial region surrounded by the broken line in Fig. 2(a). A time-series data of the cavitation volume is then obtained from all captured images. Figure 3 shows the results of the fast Fourier transform for the time development of both the cavitation volume and the pressure at the case exit of the half cut model. The fluctuation of cavitation volume is corresponding to the pressure fluctuation. The both fluctuations take peaks at 1.5 and 2.5 kHz for the back pressure of 0.5 and 1.0 MPa, respectively. In the actual hydraulic relief valve, it is observed that the noise occurs depending on the back pressure. Therefore, cavitation is considered to be a cause of the noise generation in the hydraulic relief valve.

## 3. Numerical simulation of the oil cavitating flow in the hydraulic relief valve

### *3.1. Numerical model and numerical settings*

According to the experiment as described above, the oil cavitating flow in the hydraulic relief valve was simulated by means of the numerical code, in which cavitation is modeled as the source/sink of vapor phase and is represented by the pressure difference from a vapor pressure[1][2][3]. Figure 4 shows the two-dimensional numerical model of the valve with axial symmetry. The oil flows from the gap between the sheet and valve to outside of the case. Boundary conditions are uniform flow at the inlet,

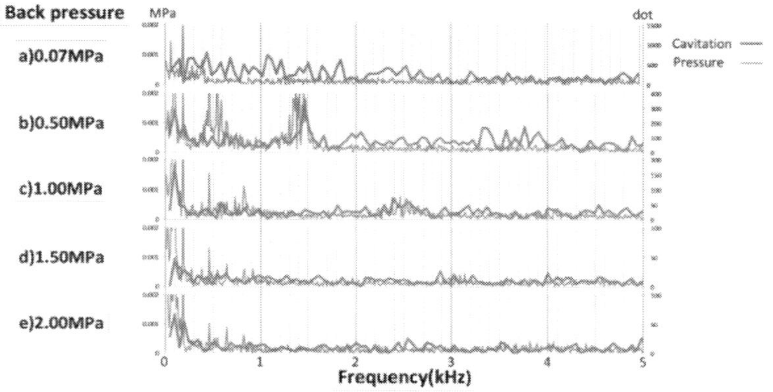


**Figure 3.** Results of the fast Fourier transform of the cavitation volume and the pressure at case exit for various back pressure conditions

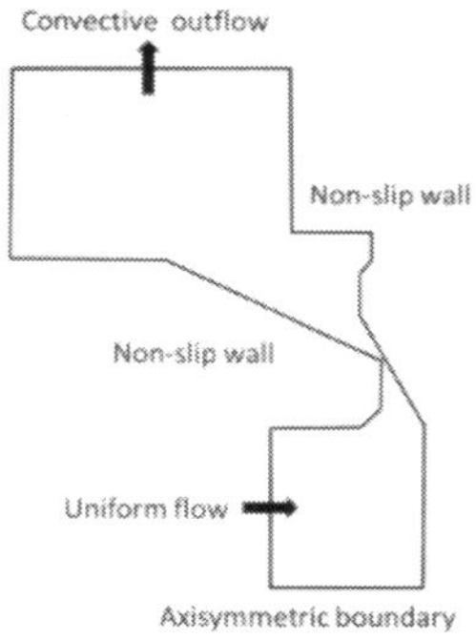


**Figure 4.** Numerical model

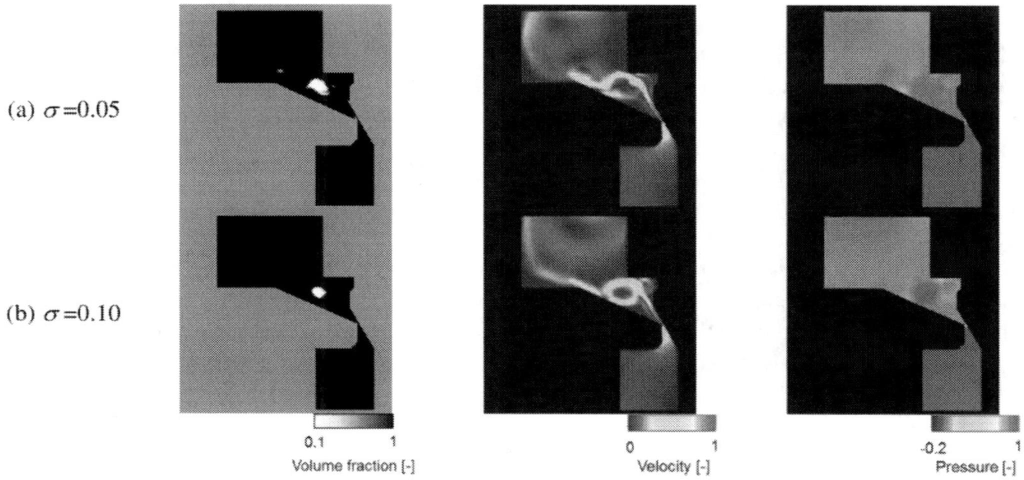


**Figure 5.** Instantaneous distribution of the volume fraction of oil, velocity magnitude and pressure.

convective outflow at the outlet and non-slip condition on the wall of the sheet, valve and case. Two-dimensional boundary fitted structured mesh is employed and discretized in space by the grid number of 450x120. Reynolds number is defined as Re = $v_g\ \delta / \nu$, where $v_g$, $\delta$ and $\nu$ are the velocity through the gap, the width of the gap and the kinetic viscosity, respectively. The cavitation number is also defined as $\sigma = (p_0 - p_v)/(\rho\, v_g^2/2)$. In the present study, flow conditions are Re = 100 and $\sigma$ = 0.04~0.10.

### *3.2. Numerical results and discussions*

Figure 5 shows instantaneous distribution of the volume fraction of oil, velocity magnitude and pressure for $\sigma$ = 0.05 and 0.10. In the both cavitation number cases, cavitation is observed in the low pressure region in the large vortex which is generated by the high speed flow from the gap to the wall of the case. The size of the cavitation becomes large with decreasing cavitation number. In addition, time-series of the results show that the size of cavitation in the large vortex fluctuates corresponding to the pressure fluctuation.

Figure 6 shows the results of the fast Fourier transform for the pressure averaged in space on the outlet boundary for various cavitation numbers. The pressure fluctuations take peaks at 2, 3 and 4 kHz for $\sigma$ = 0.04, 0.06 and 0.08, respectively. As the cavitation number decreases, the peak values of the pressure fluctuation decrease with increasing peak frequency. These phenomena were observed in the

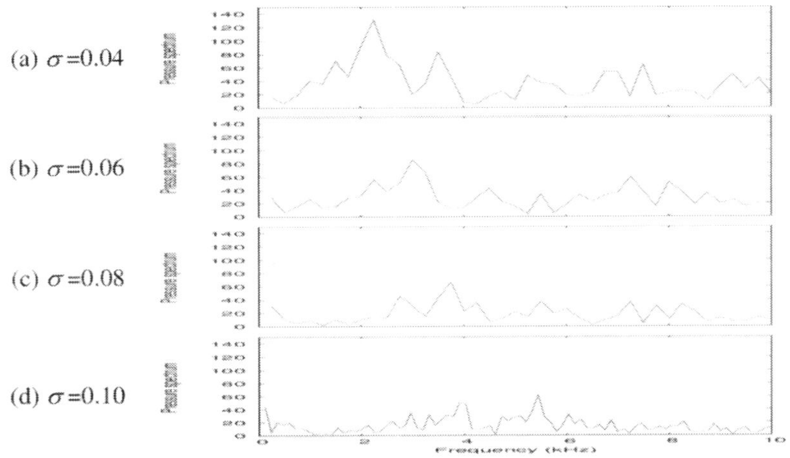


**Figure 6.** Results of the fast Fourier transform of the pressure averaged on outlet boundary for various cavitation numbers.

experiment results as shown in Fig. 3. Thus, it is considered that the present numerical simulation could reproduce the oil cavitating flows in the relief valve qualitatively.

Since the vapor pressure of the hydraulic fluid is low, the simulation for $\sigma = 0.10$ corresponds to the experiment for the back pressure of 0.5MPa. However, the peak frequency of 4 kHz in the simulation is higher than the peak frequency of 1.5 kHz in the experiment. Here, comparing the cavitation region in the experiment and simulation results (Fig. 2(b) and Fig. 5(b)), the cavitation region in the simulation is reproduced smaller than that in the experiment. As shown in the Fig. 6(a), the pressure fluctuation in the simulation result for $\sigma = 0.04$, in which the cavitation grows larger, takes peak at the frequency of 2.5 kHz. This peak frequency is corresponding to the peak frequency in the experiment for the back pressure of 1.0 MPa, as shown in Fig. 3(c). Thus, it is necessary to examine the correspondence of the calculation conditions and the experiment conditions.

## 4. Conclusions

To clarify the mechanism of the noise generation owing to cavitation in the hydraulic relief valve, the oil cavitating flows were experimentally visualized by means of the half cut model and were numerically reproduced. As the results of the image analysis, it was confirmed that the fluctuation of the cavitation volume is corresponding to the pressure fluctuation. Moreover, it was found that the pressure fluctuations take peaks at the frequency of the order of kHz in the particular back pressure conditions. Therefore, cavitation is considered to be a cause of the noise generation in the hydraulic relief valve. On the other hand, in the numerical simulation, as the cavitation number decreases, the peak values of the pressure fluctuation decrease with increasing peak frequency. Although the peak frequency of the pressure fluctuation in the simulation was higher than that in the experiments, the present numerical simulation could reproduce the oil cavitating flows in the relief valve qualitatively.

## References

[1] Okita K, Kajishima T, 2000 Numerical Investigation of Unsteady Cavitating Flow around a Rectangular Prism *Proc. 4th JSME-KSME Thermal Engineering Conference* **2** 571.

[2] Okita K, Takagi S, Matsumoto Y, Yagi Y, Murase M, 2004 Numerical Investigation on Mechanism of Cavitation Erosion at the Exit of a Long Orifice *Proc. 5th International Conference on Multiphase Flow, ICMF'04*, Paper No.466.

[3] Okita K, Ugajin H, Matsumoto Y, 2009 Numerical Analysis of the influence of the tip clearance flows on the Unsteady Cavitating Flows in a Three-dimensional Inducer *J. Hydrodynamics Ser. B* **21**(1) 34.

Journal of Physics: Conference Series **656** (2015) 012105 doi:10.1088/1742-6596/656/1/012105

# Experimental investigation of the flow-induced vibration of hydrofoils in cavitating flows

**Guoyu Wang, Qin Wu, Biao Huang[1], Yuan Gao**
Beijing Institute of Technology, 5 South Zhongguancun Street, Beijing 100081, China

E-mail: huangbiao@bit.edu.cn

**Abstract.** The objective of this paper is to investigate the correlation between fluid induced vibration and unsteady cavitation behaviours. Experimental results are presented for a modified NACA66 hydrofoil, which is fixed at $\alpha$=8°. The high-speed camera is synchronized with a single point Laser Doppler Vibrometer to analyze the transient cavitating flow structures and the corresponding structural vibration characteristics. The results showed that, with the decreasing of the cavitation number, the cavitating flows in a water tunnel display several types of cavitation patterns, such as incipient cavitation, sheet cavitation and cloud cavitation. The cavity shedding frequency reduces with the decrease of the cavitation number. As for the cloud cavitation regime, the trend of the vibration velocity goes up with the growth of the attached cavity, accompanied with small amplitude fluctuations. Then the collapse and shedding of the large-scale cloud cavities leads to substantial increase of the vibration velocity fluctuations.

## 1. Introduction

Flow induced vibration is one of the major issues in a wide range of fields, such as fluid mechanics, structural mechanics, computational fluid dynamics and acoustics. Discussions about these subjects can be found in the references [1-4]. Yao et al. [5] experimentally investigated the flow induced vibration on a hydrofoil with Donaldson trailing edge shape. The results showed that the hydrodynamic damping is significantly increased with the Donaldson cut and the vibration is significantly reduced with a slight modification of the trailing edge without alternation of the hydrodynamic performances. Ducoin et al. [6] experimentally characterized that the vibrations of hydrofoil were induced by the laminar to turbulent boundary layer transition and depended on the vortex shedding frequency.

Since the cavitation involves complex interaction between phase-change and vortex structures, the unsteady breakdown and shedding of the cavities will induce strong transient loads and lead to further hydrodynamic instabilities, even structure failures. Hence, the flow induced vibration under cavitating flows has attracted considerable interests. Torre [7] conducted a series of experiments to investigate the influence of the sheet cavitation and supercavitation on the added mass effects experienced by a modified NACA0009 hydrofoil. The results showed that the added mass decreases when cavitation appears because of the increased cavity length. Amromin and Kovinskaya [8] analyzed the vibration of an elastic hydrofoil with an attached cavity in periodically perturbed flow, where a beam equation was used to describe the hydrofoil vibration. The results showed that the structural vibration increased significantly due to the cavitation. The high-frequency band was attributed to the hydrofoil resonance

[1] To whom any correspondence should be addressed.

with the cavity adding a certain damping and the low-frequency band was corresponding to the cavity volume fluctuations.

Although the phenomenon of flow induced vibration has been widely investigated, the correlation between fluid induced vibration and unsteady cavitation behaviours are still not obvious, since it's difficult to measure the unsteady structural vibration and moreover synchronize the observation of cavitation behaviours with the measurement of vibration fluctuations. The objective of this work is to shed light on the unsteady cavitating flow and corresponding vibration characteristics, and to offer a basis for future numerical validation studies.

## 2. Experimental set-up

Experimental studies are conducted in a closed-loop cavitation tunnel at Beijing Institute of Technology [9], which is shown in figure 1. The test section is 0.7m long and has a rectangular section with width of 0.07m and height of 0.19m. As shown in figure 2, the modified-NACA66 [10] is placed with $\alpha$=8°, which is subjected to a nominal free stream velocity of $U_\infty$=8m/s, yielding a moderate Reynolds number of $Re=U_\infty c/\nu=6.0\times10^5$. The outlet pressure is set to vary according to the cavitation number $\sigma=(p_\infty-p_v)/(0.5\rho_l U_\infty^2)$. It should be noted that a small gap (1mm) is existed between the front wall of the test section and the free end of the hydrofoil, so that the tip section is free to enable the vibration of the hydrofoil.

The cavitation patterns are observed with a high-speed digital camera (HG-LE, by Redlake) [10] and 2500 fps is used in this study. The vibration velocities are measured on the pressure side of the hydrofoil with a single point Laser Doppler Vibrometer (LDV) Polytec PSV-100 [10]. The smallest possible measurement range (20mm/s as its peak) is used to optimize the signal-to-noise ratio and the sample frequency is 22kHz, with the measure point located at the $x/c$=0.2 in the mid-plane of the hydrofoil. Then the velocity signal can be fed into digital inputs of a data acquisition system [9].

Figure 3 shows the layout of simultaneous sampling system, the high-speed visualization and unsteady vibration measurement setups are combined by a controller. When the controller is triggered, the cavitation images and unsteady vibration velocities will be captured simultaneously. Even though the sampling rate of high-speed camera is far less than that of LDV system, each cavity image has its own corresponding value of the vibration velocity according to the actual time. The uncertainty of simultaneous sampling is negligible compared with periodic time of the unsteady cloud cavitation [9].

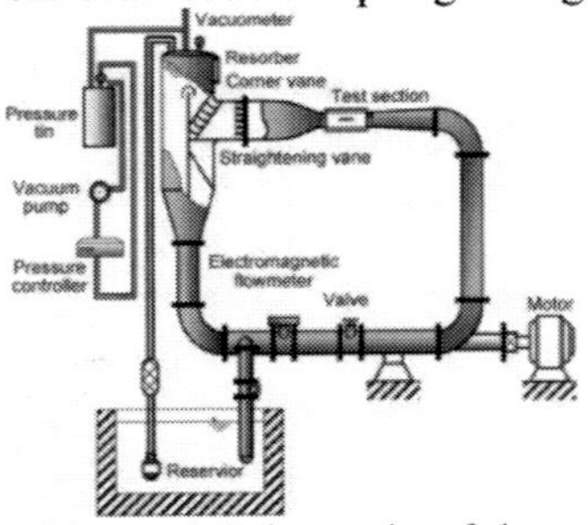


Figure 1 Schematic of the cavitation tunnel

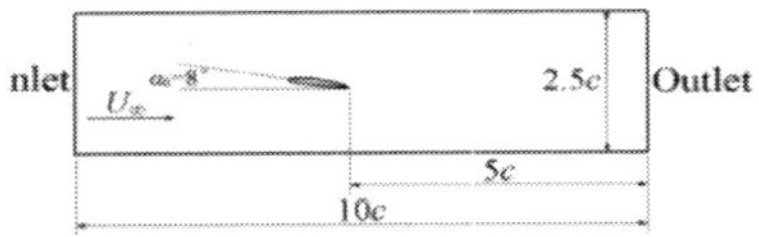


Figure 2 Sketch of the foil's position in the test section

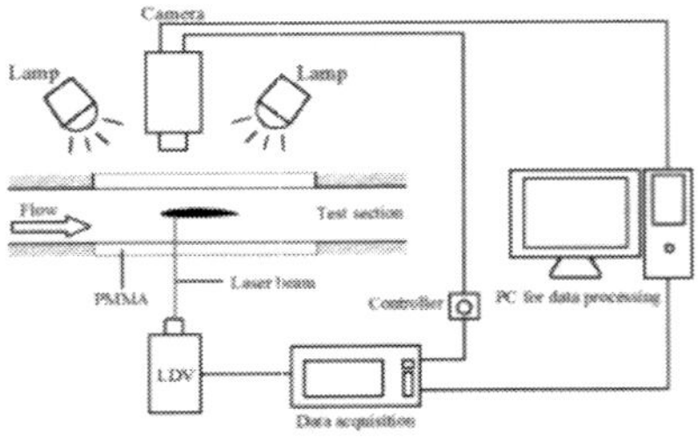


Figure 3 Layout of simultaneous sampling system

## 3. Results and discussion

### *3.1. The spectral analysis of the vibration velocities in different cavitation regimes*

Figure 4 shows the time evolution of the typical cavity shapes observed in the experimental visualization for different cavitation regimes with $\sigma$=2.3, 1.9, 1.2 and 0.8. It is shown that the cavitation inception is observed at about $\sigma$=1.9. In this regime, the cavitating flow appears in white at the leading edge of the foil, indicating that it contains large number of micro-sized vapor bubbles. Reducing the cavitation number, the cavitation changes from the incipient cavity to sheet cavity at about $\sigma$=1.2. Although the cavity is attached at the leading edge of the foil, the rear portion of the sheet cavity is unsteady and rolls into small bubbly vortices that shed intermittently. Further lowing the cavitation number to about σ=0.80, the sheet cavity grows and the trailing edge becomes increasingly unsteady accompanied with cloud cavities shedding massively.

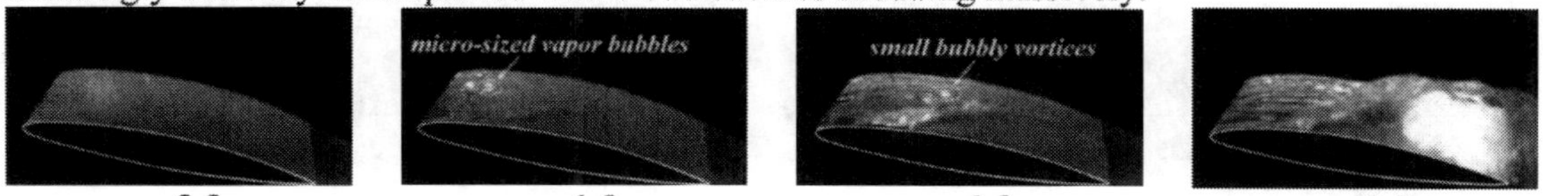


σ=2.3 σ=1.9 σ=1.2 σ=0.8

Figure 4 The typical cavitation patterns for different cavitation number

To investigate the effects of cavitation on hydroelastic response, figure 5 shows spectral characteristics of the measured vibration velocity of the hydrofoil, which are obtained by the Fast Fourier Transform of the vibration velocity. One of the main flow-induced frequencies, as shown in the amplitude peaks in figure 5, are 165Hz, 90Hz and 18Hz for the cases with σ=1.9, 1.2 and 0.8 respectively, are accordance with the cavity shedding frequency. This is confirmed by comparing the spectra to the cases of fully turbulent flow, where such components are not visible. The cavity shedding frequency decreases with the decrease of the cavitation number because the cavity increases in length and requires more time to evolve between the subsequent cavitational cycles.

The Strouhal number $Str=fc/U$ against $\sigma/2\alpha$ is plotted in figure 6. Reasonable agreement can be obtained between the present results and pervious experimental measurements [11-13] for different hydrofoils. Although some discrepancies exited between the foils at a constant value of $\sigma/2\alpha$, all cases show similar trends. The Strouhal number $Str$ remains almost constant when the values of $\sigma/2\alpha$ less than approximately 4, and increases with $\sigma/2\alpha$ for larger $\sigma/2\alpha$ values.

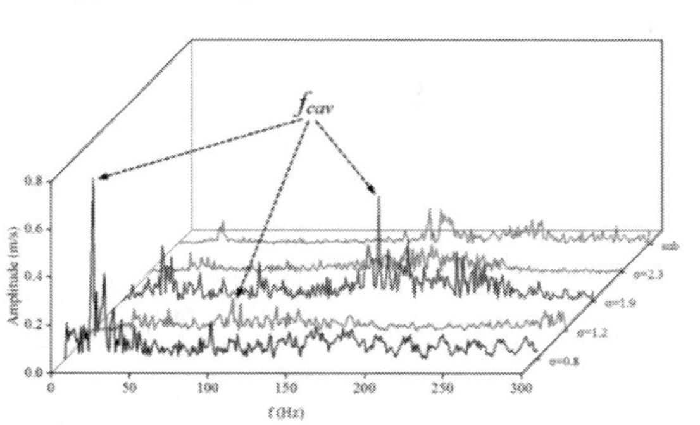


Figure 5 The effect of σ on the frequency spectrum of vibration velocity

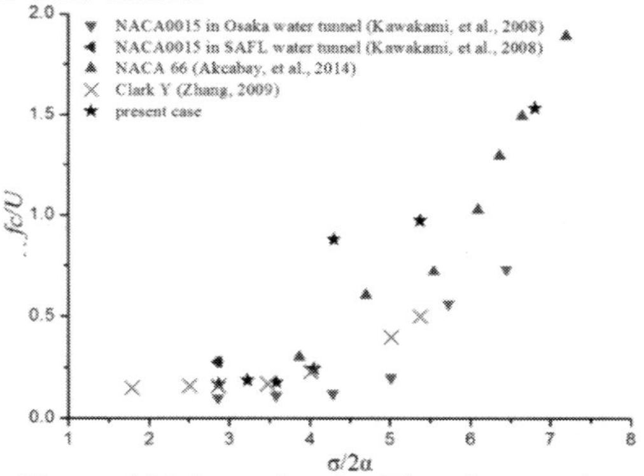


Figure 6 Main cavity shedding frequencies compared with results in Refs.

*3.2. The vibration velocity associated with the unsteady cloud cavitation behaviors*

As to the regime of cloud cavitation, the unsteady breakdown and shedding of the cavity occurs violently, which leads to highly unsteady hydrodynamic forces on the hydrofoil, and hence may result to strong dynamic instabilities [14-15]. In the present study, when the cavitation number reduces to σ=0.8, the transient cloud cavitation develops, which has a distinctly quasi-periodic pattern. Figure 7 shows the time evolution of the vibration velocity within several cycles, followed by the experimental observation of the cavity shapes at representative times in one cycle, as shown in figure 8. The period of the cloud cavitation $T_{cycle}$ is about 55ms, which is corresponding to $5.8c/U_\infty$. Combined figures 7 and 8, the evolution of the transient cavity shapes and the corresponding vibration characteristics could be divided into two stages: (I) development of the attached cavity (from $t_1$ to $t_3$) and (II) shedding and collapse of the attached cavity (from $t_3$ to $t_{12}$). The principal characteristics is that the trend of the vibration velocity goes up with the growth of the attached cavity (seen the red dash line shown in figure 7), accompanied with small amplitude fluctuations. Then the vibration velocity tends to decrease with large amplitude fluctuations, which is corresponding to the shedding and collapse of the cloud cavity. The detailed development of the flow structures in this stage can be characterized as follows: From $t_4$ and $t_6$, the re-entrant jet begins to form at the rear part of the cavity, leading to the formation of the cloud cavity. As seen in figures 7 and 8(f), the moment ($t_6$) the cloud cavity breaks up from the primary attached cavity, the vibration velocity of the hydrofoil hits the peak. After that, from $t_7$ and $t_{10}$, the collapse of the large-scale cloud cavity leads to the substantial increase of vibration velocity fluctuations, as shown in figure 7, because the propagated pressure pulse and the energy released by the cavity collapse aggravate the vibration of the foil. This is also confirmed by Chen et al [9]. Finally, due to the unsteady shedding of the cavity, the amplitude of the vibration velocity fluctuations decrease.

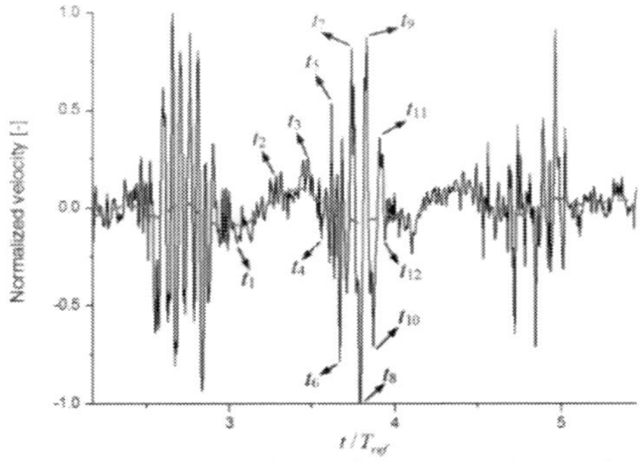


Figure 7 Time evolution of the vibration velocity.

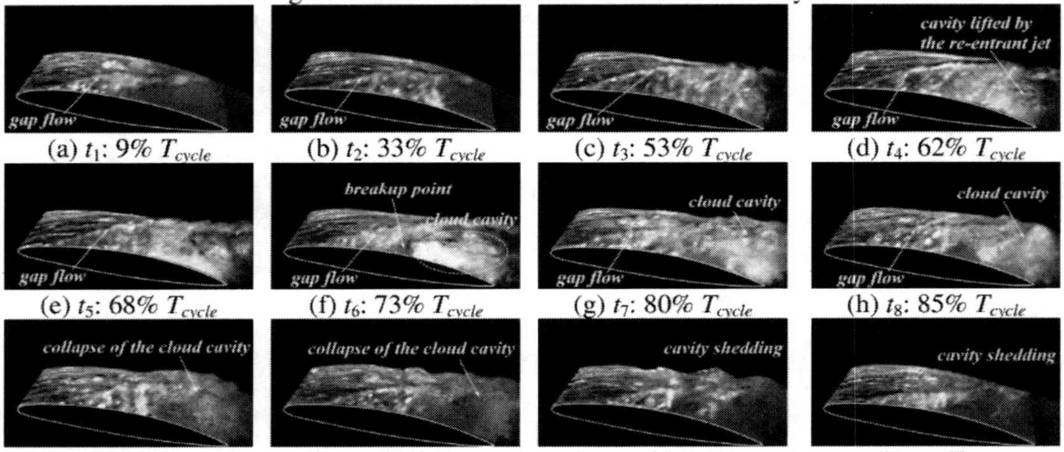


(a) $t_1$: 9% $T_{cycle}$ (b) $t_2$: 33% $T_{cycle}$ (c) $t_3$: 53% $T_{cycle}$ (d) $t_4$: 62% $T_{cycle}$
(e) $t_5$: 68% $T_{cycle}$ (f) $t_6$: 73% $T_{cycle}$ (g) $t_7$: 80% $T_{cycle}$ (h) $t_8$: 85% $T_{cycle}$
(i) $t_9$: 90% $T_{cycle}$ (j) $t_{10}$: 94% $T_{cycle}$ (k) $t_{11}$: 98% $T_{cycle}$ (l) $t_{12}$: $T_{cycle}$

Figure 8 Time evolution of the cloud cavitation patterns.

## 4. Conclusions

In this paper, experimental studies are presented for a NACA66 hydrofoil with the angle of attack $\alpha$=8° at Re=600,000 for different cavitation regimes ($\sigma$=2.3, 1.9, 1.2 and 0.8). The primary findings include:

(1) With the decreasing of the cavitation number, the cavitating flows in a water tunnel display several types of cavitation patterns, such as incipient cavitation, sheet cavitation and cloud cavitation. The cavity shedding frequency reduces with the decrease of the cavitation number because the cavity increases in length and requires more time to evolve between the subsequent cavitational cycles. And the vortex shedding frequency remains almost constant.

(2) For the cloud cavitation regime, the evolution of the transient cavity shapes and the corresponding vibration characteristics could be divided into two stages: (I) development of the attached cavity and (II) shedding and collapse of the attached cavity. During the first stage, the trend of the vibration velocity goes up with the growth of the attached cavity, accompanied with small

amplitude fluctuations. During the latter stage, the vibration velocity tends to decrease with large amplitude fluctuations, which is corresponding to the shedding and collapse of the cloud cavity.

(3) As for the correlation between fluid induced vibration and unsteady cavitation behaviours, the moment the cloud cavity breaks up from the primary attached cavity, the vibration velocity of the hydrofoil hits the peak. The collapse of the large-scale cloud cavity leads to the substantial increase of vibration velocity fluctuations, because the propagated pressure pulse and the energy released by the cavity collapse aggravate the vibration of the foil. Due to the unsteady shedding of the cavity, the magnitude of the vibration velocity fluctuation decreases.

**Acknowledgements**
This work was supported by the National Science Foundation of China (Grant No.51306020) and the Beijing Natural Science Foundation (Grant No.3144034).

**References**

[1] Ausoni P, Farhat M, Escaler X, Egusquiza E and Avellan F 2007 *ASME J. Fluids Eng.* **129** 966-73.
[2] Donaldson M E 1956 *J. Eng. Power* **78** 1141-47.
[3] Howe M S 1988 *J. Sound Vibration* **126**(3) 503-23.
[4] Seo J H and Lele S K 2009 *Proc. 7th Int. Symp. Cav.* Ann Arbor, Michigan, USA.
[5] Yao Z F, Wang F J, Dreyer M and Farhat M 2014 *J. Fluids Struct.* **51** 189-98.
[6] Ducoin A, Astolfi J A and Gobert M-L 2012 *J. Fluids Struct.* **32** 37-51.
[7] Torre D L, Escaler X, Egusquiza E and Farhat M 2013 *J. Fluids Struct.* **39** 173-87.
[8] Amromin E and Kovinskaya S 2000 *J. Fluids Struct.* **14**(5) 735-51.
[9] Chen G H, Wang G Y, Hu C L, Huang B and Zhang M D 2015 *Exp. Fluids* **56**(32) 1-11.
[10] Wu Q, Huang B, Wang G Y and Gao Y 2015 *Int. J. Multi. Flow* doi: http://dx.doi.org/10.1016/j.ijmultiphaseflow. 2015.03.023
[11] Akcabay D T, Chae E J, Young Y L, Ducoin A and Astolfi J A 2014 *J. Fluids Struct.* **49** 463-84.
[12] Zhang B 2009 Thesis of PhD, Beijing Institute of Technology.
[13] Kawakami D T, Fuji A, Tsujimoto Y and Arndt R E A 2008 *ASME J. Fluids Eng.* **130** 1-8.
[14] Arndt R E A 1981 *Ann. Rev. Fluid Mech.* **13** 273-328.
[15] Luo X and Ji B. 2012 *ASME J. Fluids Eng.* **134** 041202 1-10.

# Study on Prediction of Underwater Radiated Noise from Propeller Tip Vortex Cavitation

**Takuyoshi Yamada, Kei Sato, Chiharu Kawakita, Akira Oshima**

Research &Innovation Center, Mitsubishi Heavy Industries Ltd.
5-717-1, Fukahori-Machi, Nagasaki, Japan

E-mail takuyoshi_yamada@mhi.co.jp

**Abstract**. The method to predict underwater radiated noise from tip vortex cavitation was studied. The growth of a single cavitation bubble in tip vortex was estimated by substituting the tip vortex to Rankine combined vortex. The ideal spectrum function for the sound pressure generated by a single cavitation bubble was used, also the empirical factor for the number of collapsed bubbles per unit time was introduced. The estimated noise data were compared with measured ship's ones and it was found out that this method can estimate noise data within 3dB difference.

## 1. Introduction

Propeller cavitation is known as one of measure sources of vibration and noise of ships. Especially, it is important to reduce cavitation noise in designing propellers of the ships such as oceanographic research vessels because they use acoustic instruments and cavitation noise are harmful in using them. Therefore, it is indispensable to quantitative prediction of propeller cavitation noise. Among various types of propeller cavitation, it is very difficult to predict the noise of tip vortex cavitation, which occurs at a distance propeller blade, as well as to predict inception of tip vortex cavitation. In this study, the practical method in order to adapt at the design stage of propeller was investigated. The simple calculation model was combined with experimental data and the method to predict underwater radiated noise from propeller tip vortex cavitation was developed. Finally, this method was verified in comparison with measured ship's noise data.

## 2. Prediction method

The prediction method is constructed, as follows.

### 2.1. Prediction of tip vortex cavitation

Analogous to McCormick's assumptions [1], propeller tip vortex is substituted to a Rankine combined vortex, in which the pressure P is shown in Eq.(1).

$$P = P_0 + \frac{1}{2}\left(\frac{\Gamma r}{2\pi a^2}\right)^2 - \left(\frac{\Gamma}{2\pi a}\right)^2 (r \le a), P = P_0 - \frac{1}{2}\left(\frac{\Gamma}{2\pi r}\right)^2 (r \ge a) \qquad (1)$$

It is necessary to know the strength of tip vortex and the radius of vortex core. The circulation of propeller at r/R=0.95 calculated by propeller lifting surface theory[2] was adopted as the strength of tip vortex. And also, it was assumed that the radius of vortex core is proportional to the displacement thickness of boundary layer on propeller blade surface at r/R=0.95 with using some empirical correction factor K [3] as shown in Eq. (2)

$$a = K\left(\delta^*{}_B + \delta^*{}_F\right) \qquad (2)$$

It was considered that the inception of a cavitation bubble occurred at center of a Rankine combined vortex. To calculate the growth of a cavitation bubble, Rayleigh-Pleset equation (3) was used.

$$R\frac{d^2R}{dt^2}+\frac{3}{2}\left(\frac{dR}{dt}\right)^2=\frac{P_v-P_0}{\rho} \tag{3}$$

At that time, it was assumed that the center of bubble placed at center of the vortex, so, the pressure at the bubble surface was equal to the pressure of the vortex at bubble radius. Finally, the maximum radius of bubble $R_M$ and the pressure $P_C$ at $R_M$ were gotten.

### 2.2. Prediction of noise levels

A cavitation bubble with radius $R_M$ collapsed under the pressure $P_M$. Ideal spectrum function $S_P(\omega)$ for the sound pressure generated by a single cavitation bubble was introduced as shown in Fig.1 [4]. In Fig.1, the angular frequency becomes non-dimensional by multiplying $\tau_C$ as shown in Eq. (4).

$$\tau_C = R_M\sqrt{\rho/P_C} \tag{4}$$

In Fig.1, SP ($\omega$) is divided to 3 parts. Each part is expressed in Eq. (5), (6), (7).

Part 1 $$\frac{S_P(\omega)}{R_M{}^4\rho P_C}=\frac{\pi}{18}(\omega\tau_C)^4\left[\frac{\cos(\omega\tau_C)}{(\pi/2)^2-(\omega\tau_C)^2}\right]^2 \tag{5}$$

Part 2 $$\frac{S_P(\omega)}{R_M{}^4\rho P_C}=\frac{2.4}{9\pi}(\omega\tau_C)^{-0.4} \tag{6}$$

Part 3 $$\frac{S_P(\omega)}{R_M{}^4\rho P_C}=\frac{2.4}{9\pi}(\omega_S\tau_C)^{-0.4}\left(\frac{\omega}{\varpi_S}\right)^2,\ \omega_S=\frac{1}{9\tau_C}\left(\frac{\rho c_m}{P_C}\right)^{5/6},\ c_m=\frac{c}{\sqrt{1+\beta_g\rho c/\rho_g c_g}} \tag{7}$$

And also, to consider the size of cavitation bubbles and noise level measured at sea, the numbers of collapsed bubbles per unit time were determined

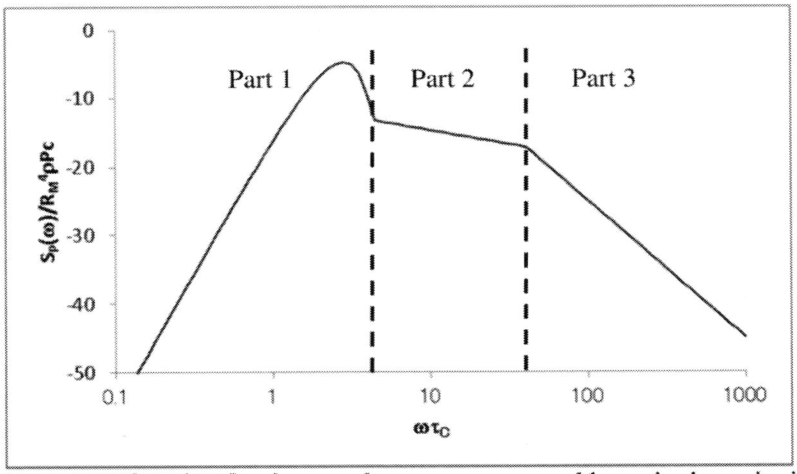


Fig.1 Ideal spectrum function for the sound pressure generated by a single cavitation bubble

## 3. Results of prediction

The predicted noise data were compared with measured ones, as follows.

### 3.1. Subject ships

The prediction method was adapted to 3 ships as shown in Table 1.

Ship A and B are oceanographic research vessels and Ship C is a training ship. Because the adopted case of the developed prediction method is limited to tip vortex cavitation noise, the ships and operating conditions which only tip vortex cavitation was observed in model tests were chosen.

The radiated noise from subject ships were measured by using a hydrophone which was suspended from measuring boat. The waters depth was about 1000m and subject ships were running at the point within 100m from measuring boat. So, noise propagation was assumed as spherical diffusion. The source level Ls was calculated from the measured one Lp using the distance r between hydrophone and ship, considering of the sound reflection at sea surface, as follows,

$$L_S = L_P + 20\log(r) - 3 \tag{8}$$

. Table 1 Principal Particulars of Ship

| Ship | A | B | C |
|---|---|---|---|
| Length (m) | 100.0 | 116.0 | 87.6 |
| Breadth (m) | 16.2 | 19.0 | 13.6 |
| Depth (m) | 8.9 | 13.2 | 8.8 |
| Displacement (ton) | 3991 | 8600 | 3515 |
| Engine Output (kW) | 5590 | 7355 | 3900 |
| Speed at Research (kn) | 16 | 16 | 8 |
| Propeller Diameter (m) | 3.7 | 4.1 | 3.3 |

### *3.2. Results of prediction*

Predicted source levels are compared with measured ones as shown in Fig.2~4.

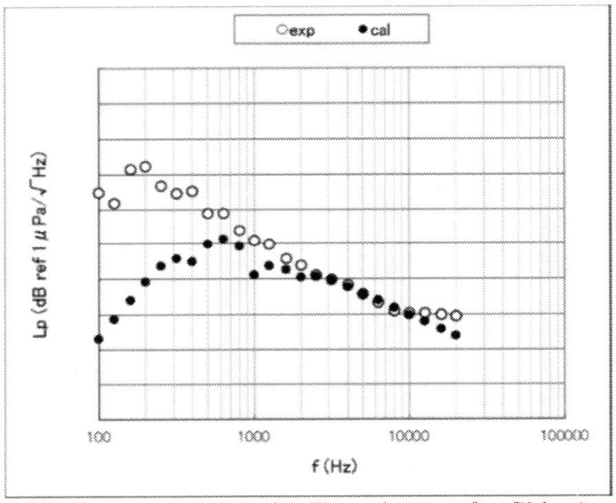


Fig.2 Comparison with Experiments for Ship A

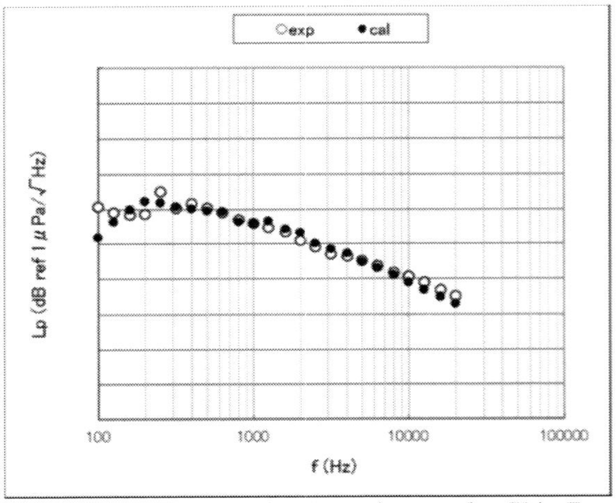


Fig.3 Comparison with Experiments for Ship B

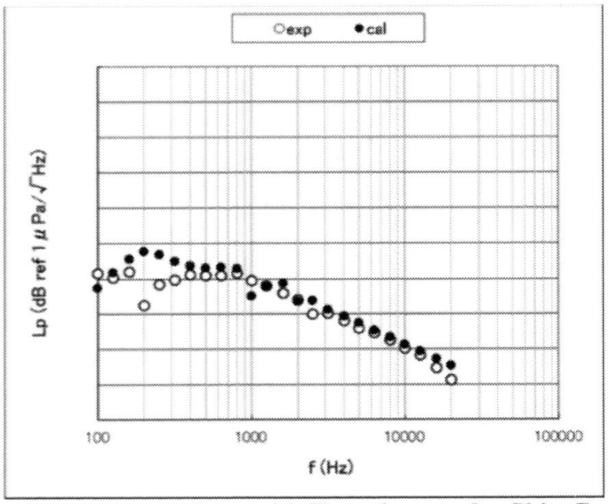


Fig.4 Comparison with Experiments for Ship C

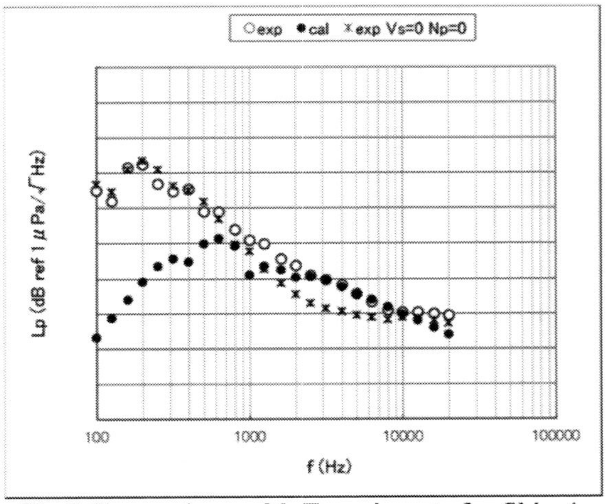


Fig.5 Comparison with Experiments for Ship A

In the case of Ship A, measured noise levels were larger than those in prediction. At full-scale, measured noise data included not only propeller cavitation noise but also machinery ones. Therefore, it was considered that machinery noise were larger than those of cavitation with a range below 1 kHz. In the case of ship A, noise were measured at that the ship speed and propeller shaft speed were kept 0. The results are shown in Fig.5. It is clearly shown that the machinery noise is dominant with a range below 1 kHz. In the case of ship B and C, predicted data show good agreement with measured ones with a range above 100Hz. Calculated noise levels at 2, 5, and 10 kHz are compared with measured ones as shown in Fig.6. The difference of noise levels between calculation and experiment is less than 3dB. So, the developed prediction method can be evaluated that it is useful in design of propeller.

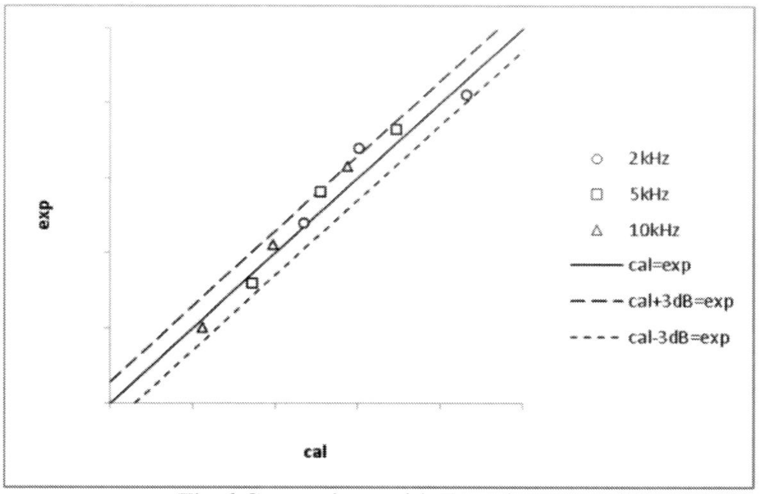


Fig.6 Comparison with Experiments

## 4. Conclusion

The method to predict the radiated noise from propeller tip vortex cavitation was developed. Tip vortex was substituted to Rankine combined vortex. The growth of a cavitation bubble was calculated by using Rayleigh-Pleset equation. Ideal spectrum function for the sound pressure generated by a single cavitation bubble was introduced. And also, to consider the size of cavitation bubbles and noise level measured at sea, the number of collapsed bubbles per unit time was determined. This method was verified in comparison with measured noise data at sea. The difference noise level between calculation and experiment was less than 3dB, so, it can be said that present method is useful at design stage of propeller.

At present, many researchers are working on the analysis of flow field near tip vortex by using CFD and detailed measurement. It can be expected that suitable model of phenomenon is developed and the accuracy of prediction increases. Furthermore, it is necessary to measure propeller noise at full-scale and to separate cavitation noise.

## References

[1] McCormick, B.W., "On Cavitation Produced by a Vortex Trailing From Liftng Surface," Joruna1 of Basic Engineering Trans. ASME, Vol.84, No.3, 1962, pp369-379
[2] Hoshino, T., "Application of Quasi-Continuos Method to Unsteady Propeller Lifting Surface Problems," Soc. of Naval Architects of Japan, Vol.158, 1985, pp51-71
[3] Oshima, A., "Study on Tip Vortex Cavitation Inception of Propeller," The 2nd International Symposium on Cavitation, 1994, pp367-372
[4] Blake, W., Mechnics of Flow-Induced Sound and Vibration Volume I 6.4.1 Dependence of Sound on Stages of Bubble History, Accademic Press, 1986, pp404-413

# Effect of cavitation on removal of alkali elements from coal

**Srivalli.H**[1], **Nirmal.L**[1], **Nagarajan.R**[1]*
[1]Department of Chemical Engineering, Indian Institute of Technology-Madras, Chennai-600036, India

*Corresponding author
Telephone: +91-44-22578070; E-mail: nag@iitm.ac.in

**Abstract.** The main impurities in coal are sulphur, ash and alkali. On combustion, the volatile forms of these impurities are either condensed on the boilers, or emitted in the form of potentially hazardous gases. The alkali elements present in coal help the fly ash particles adhere to boiler surfaces by providing a wet surface on which collection of these particles can take place. Use of ultrasonic techniques in cleaning of coal has stirred interest among researchers in recent times. Extraction of alkali elements by cavitation effect using low-frequency ultrasound, in the presence of reagents ($HNO_3$ and $H_2O_2$) is reported in this paper. Powdered coal was dissolved with the reagent and exposed to ultrasonic fields of various frequencies at different time intervals. The treated solution is filtered and tested for alkali levels.

## 1. Introduction

Coal accounts for around 40% of the world's source of power. Although harnessing clean, green sources of energy has been gathering momentum in recent years, nearly 66% of power in India still comes from coal [1]. Coal is projected to dominate the world's power sector in the decades to come considering the abundant reserves available, and the high costs involved in using alternate forms of energy. Most coals contain ash and sulphur as impurities. On combustion, ash particles get deposited on heat transfer surfaces and reduce efficiencies, while sulphur is hazardous to the eco-system on the whole. The third form of impurities which affect heat transfer surfaces indirectly are alkali elements. Sodium and potassium are known to enhance fouling tendencies in heat transfer equipment. These elements in their volatile form act as 'glue', stick to the ash particles that impact the surface of heat transfer equipment, and hold them together [2, 3]. The high adhesion forces provided by the molten alkali salts make it increasingly difficult for the deposits to leave the surface. It is therefore important to remove these alkali elements from coal.

Ultrasound is an oscillating pressure wave with frequency higher than the human hearing range. Ultrasonic technology has found tremendous applications in medicine, imaging, non-destructive testing, etc. This technology can be used to remove sulphur and ash from coal prior to combustion. Though Indian coals have a high ash content of 35-45% [1], ultrasonic coal washing is not commonplace in India yet. Application of ultrasound on liquids has two main effects- acoustic cavitation and acoustic streaming. At higher frequencies, acoustic streaming is prominent. Acoustic streaming refers to the unidirectional flow that is induced in the liquid when a high-frequency ultrasound is coupled to it.

Cavitation is the implosion of bubbles resulting in a shock wave. This phenomenon is observed at lower frequencies. If the ambient pressure is sufficiently reduced, a liquid boils without heating. This technique is put to use on a small scale in ultrasonication by reducing the pressure at a local level. When the pressure is reduced, small bubbles of vapor form. They subsequently implode, releasing energy as a shock wave in the compression cycle [5]. Thus, cavitation effects are stronger at lower frequencies.

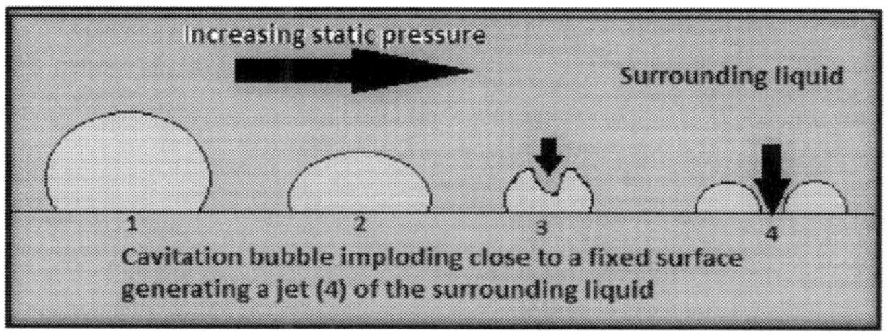


**Figure 1.** Formation of cavitation bubble and implosion close to the surface

Ambedkar et al. [5] used $HNO_3$ and $H_2O_2$ as reagents to carry out experiments on de-sulphurisation of coal. They concluded that at low frequencies, the cavitation phenomenon, being dominant, leads to breakage of coal particles and results in better contact with the oxidising agents, resulting in higher removal of sulphur. Balakrishnan et al. [4] used two ultrasonic frequencies, viz. 25 *kHz* and 430 *kHz*, in sequence to remove alkali elements from coal.

From a brief review of literature, it is apparent that some experiments have been conducted based on ultrasonic technology that are aimed at removing ash and sulphur from coal. However, very little data are available for removal of alkali from coal using ultrasonic methods. In this paper, the effects of low-frequency ultrasound and two reagents, hydrogen peroxide and nitric acid - on alkali removal from coal are studied.

## 2. Procedure

### *2.1 Preparation of coal sample*

Grade-G Coal from Orissa coal mines was used in as-received condition for this study. Lumps of coal were ground in a roll mill and the size was reduced. Next, the coal was placed in hard-metal tungsten carbide grinding bowls of 250ml capacity, and powdered in a planetary mill of FRITSCH make and model Pulverisette 5 for 30 minutes at 100 *rpm* with stainless steel balls, and with a sample-to-ball ratio of 1:7 [2]. The powdered coal was sieved, and coal particles of size less than $210 \mu m$ were used for the experiments after drying in an oven at 110°C to remove excess moisture and dissolving the dried coal in acid. The initial sodium and potassium levels, estimated using the standard method (ASTM D6349-09), were found to be 31.7 mg/l and 47.2 mg/l, respectively.

### *2.2 Ultrasonic method*

The reagents used were 2N $HNO_3$ and 3 volume percent $H_2O_2$, both of analytical grades. 10g of the coal sample was added to 100ml of the reagent, and the resulting liquid was placed in the ultrasonic tank. Tanks of four different frequencies were used: 25 *kHz*, 40 *kHz*, 58 *kHz* and 68 *kHz*. The sample was removed after 10 minutes and filtered. The clear liquid was collected and tested for sodium and potassium levels using Inductively Coupled Plasma (ICP) spectrometer of Perkin Elmer make and model Optima 5300DV [4]. This experiment was repeated at intervals of 10 minutes for one hour and for different frequencies and reagents. An aluminium foil test was performed to check the efficiency of the ultrasonic tanks.

## 3. Results and discussion

### *3.1. Effect of time*

From Figure 2 and Figure 3, it may be inferred that the initial leaching of both potassium and sodium remains low, and increases only when the sample is exposed to ultrasound for a longer period of time. Sodium shows a sudden jump in leaching tendency, while the amount of potassium leached with time shows a more gradual increase. This indicates that sodium is bonded to the coal differently from potassium, as corroborated later by the study on effect of solvent used.

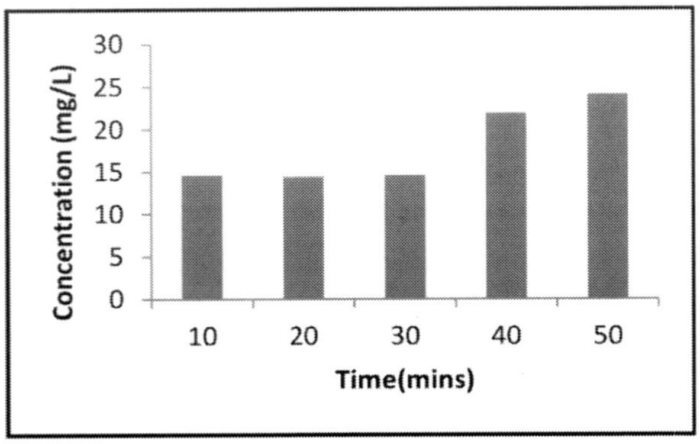


**Figure 2.** Sodium leached at 68 *kHz* using $HNO_3$

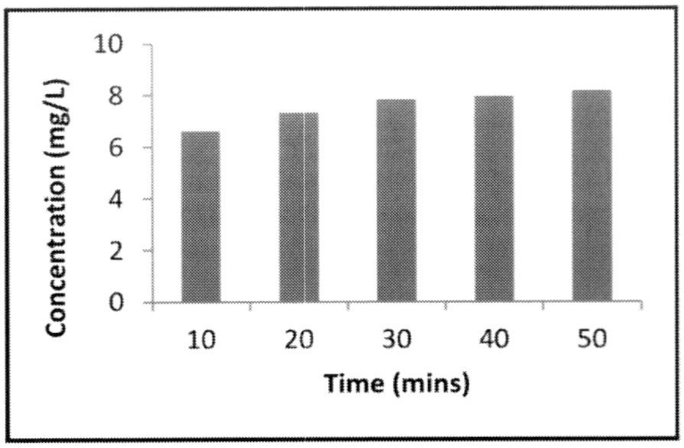


**Figure 3.** Potassium leached at 68 *kHz* using $H_2O_2$

### *3.2 Effect of frequency*

The result of effect of frequencies on extraction of sodium is reported in Figure 4 and Figure 5. 58 *kHz* appears to be the most effective frequency for removal of both the alkali metals. This is likely to be due to the fact that the 58 *kHz* displays the highest cavitation intensity among the systems employed. Although cavitation should scale inversely with frequency, this trend can be influenced by the design and operating conditions of the specific system.

The results of the aluminium foil test (exposure, followed by mass loss measurement) to measure cavitation intensity were analysed quantitatively. 25 *kHz* appears to be most eroded, while extent of erosion decreases progressively as the frequency increases. However, leaching behaviour is strongest at 58 *kHz*. This may be due to an optimum combination of cavitation and streaming effects being simultaneously present at 58 *kHz*, making it ideal for removal of alkali.

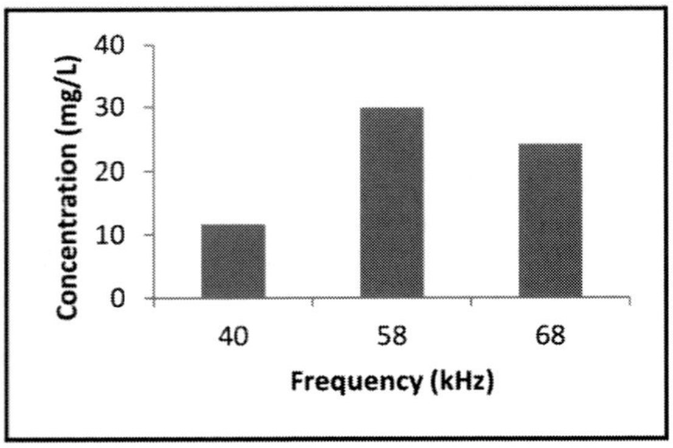


**Figure 4.** Sodium leached for various frequencies- 20 mins using $H_2O_2$ as solvent

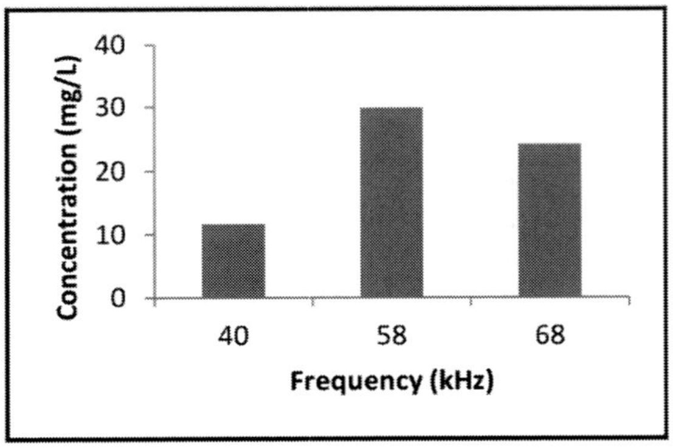


**Figure 5.** Potassium leached for various frequencies-30 mins using $HNO_3$ as solvent

*3.3 Effect of solvent*

When a comparison is made of the solvents used, nitric acid appears to be only slightly better compared to hydrogen peroxide as far as removal of sodium is concerned. However, for potassium removal, nitric acid seems to be far better than hydrogen peroxide as a reagent (Refer Figure 6 and Figure 7). This may be due to the fact that potassium is in the form of silicates and aluminates in coal, and requires a stronger solvent such as nitric acid to achieve significant leaching behaviour.

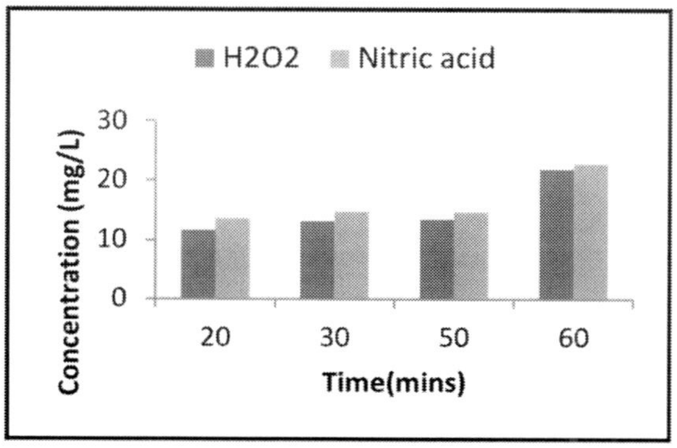


**Figure 6.** Sodium leached at 40*kHz* for $HNO_3$ and $H_2O_2$ as solvents

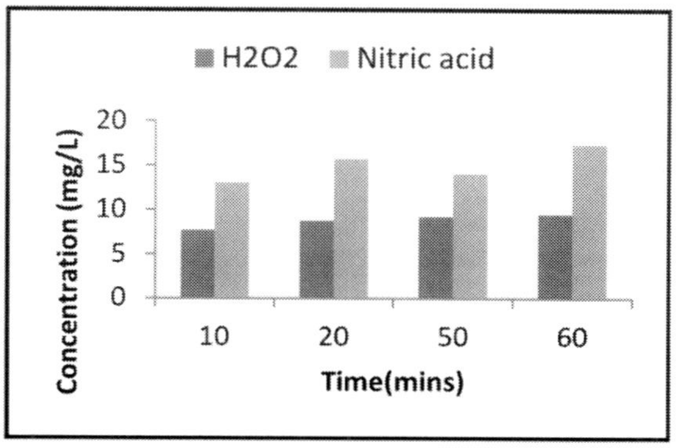


**Figure 7.** Potassium leached at 25*kHz* for $HNO_3$ and $H_2O_2$ as solvents

## 4. Conclusions

Exposure of coal to low-frequency ultrasound leads to the following interesting results:

- The effects of frequency, time and solvents on the removal of sodium and potassium from coal were studied. It is notable that percent extraction is consistently low, indicating that the samples need to be exposed additionally to high-frequency ultrasound in order to optimize the leaching effect.
- Further analysis needs to be done on effectiveness of the solvents used in leaching of sulphur from coal. A comprehensive process optimisation study is required to identify the ideal sonication time, best frequency to be used, and the best reagent to leach sodium, sulphur and ash from coal. These results will be presented at the Conference.
- It has been observed that reagents such as nitric acid, when used at a high concentration, can break the potassium-silica bonds [5] and lead to significant removal of potassium.

## 5. References

[1] Sarath.K.Guttikunda, Puja Jawahar 2014 Atmospheric emissions and pollution from the coal-fired thermal power plants in India . *Atmospheric Environment* **92** 449-460.

[2] D.E. Rosner, R. Nagarajan 1987 Toward a mechanistic theory of net deposit growth from ash-laden flowing combustion gases: self-regulated sticking of impacting particles and deposit mechanism in the presence of vapor deposited - or submicron mist - "glue," in: R. w. Lyczkowski (Ed.), *AIChE Symposium Series Heat Transfer* 289-296.

[3] J.S. Ross, R.J. Anderson, R. Nagarajan 1988 Effect of sodium on deposition in a simulated combustion gas turbine environment, *Energy & Fuels* **2** 282.

[4] Ambedkar B, Nagarajan R, Jayanti S 2011 Ultrasonic coal-wash for de-sulfurization *Ult. Son.* **18** 718-726.

[5] Balakrishnan R 2014 Modeling and simulation of fly ash deposition on boiler heat-transfer surfaces and its mitigation strategies for Indian coals *Ph.d thesis* IIT Madras 114-122.

# Mechanical Surface Treatment of Duralumin Plate by Bubble Induced by Pulse Laser

**H Soyama**[1], **H Sasaki**[2], **S Endo**[3] **and Y Iga**[4]
[1] Professor, Department of Nanomechanics, Tohoku University, Sendai, Japan
[2] Ph. D. Student, Graduate School of Engineering, Tohoku University, Sendai, Japan
[3] Graduate Student, Graduate School of Engineering, Tohoku University, Sendai, Japan
[4] Associate Professor, Institute of Fluid Science, Tohoku University, Sendai, Japan

E-mail: soyama@mm.mech.tohoku.ac.jp

**Abstract.** Surface of duralumin plate can be treated by not only laser abrasion but also bubble induced by pulse laser. The pulse laser produced two kinds of shock waves related to laser abrasion and bubble collapse. In the case of laser peening, the shock wave induced by abrasion was normally used. In the present paper, the behaviour of bubble was observed and noise was detected by the hydrophone. It was revealed that the impact induced by the bubble collapse produced by the pulse laser was stronger than that of the laser abrasion. A numerical simulation was also carried out to investigate phenomenon of bubble collapse.

## 1. Introduction

Pulse laser is utilized for mechanical surface treatment, i.e., laser peening [1-3]. In the case of laser peening, shock wave induced by laser abrasion is used. As well known, bubble is also produced by pulse laser [4], and it develops and collapses as same as cavitation bubble. In the present paper, the bubble induced by pulse laser was called as laser cavitation. The shock wave induced by laser cavitation can be used for the mechanical surface treatment, as hydrodynamic cavitation was applied for mechanical surface treatment, i.e., cavitation peening [5, 6].

Laser peening methods can be classified two types. In order to confine shock wave induced by pulse laser, one type of them uses a water film on target with pulse laser, and the other type uses a water filled chamber. The bubble cannot be developed in a water film type, as the water film is broken by the laser abrasion. However, the laser cavitation can be developed in the water filled chamber type. Although it was reported that the shock wave induced laser abrasion was larger than that of the bubble collapse [7], it might be possible to enhance an impact at bubble collapse.

In the present paper, the laser cavitation was observed by a high speed video camera with noise which corresponded to the shock wave, comparing with shock wave generated by a laser abrasion. A numerical simulation was also carried out in order to investigate behaviour of bubble collapse.

## 2. Experimental apparatus and procedures

Figure 1 illustrates a schematic diagram of pulse laser system, in which Nd:YAG laser with Q-switch was used. The wave length, the maximum energy, the beam diameter, the pulse width and the repetition frequency were 532 nm, 0.2 J, 6 mm, 10 Hz and 5 ns, respectively. The pulse laser was irradiated to a target which was place in a water filled chamber made of quartz glass. The focal length of final convex lens was 100 mm. The distances from the final lens to the target $s_{Laser}$ was 102 mm,

Journal of Physics: Conference Series **656** (2015) 012108 doi:10.1088/1742-6596/656/1/012108

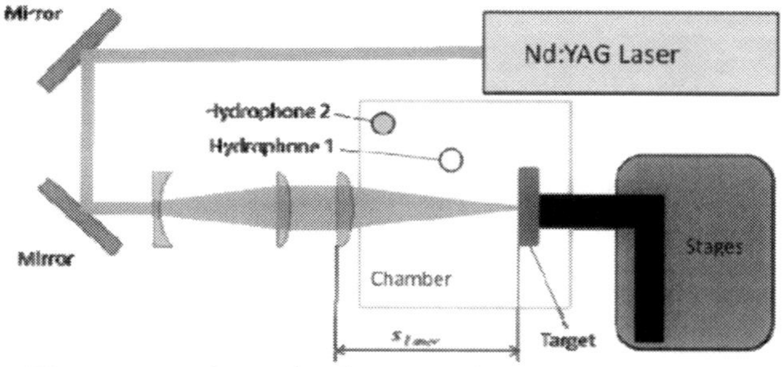


**Figure 1**. Schematic diagram of pulse laser system.

which was optimized by evaluating peening intensity changing with $s_{Laser}$. The above mentioned $s_{Laser}$ = 102 mm was chosen by measuring maximum arc height of duralumin plate changing with $s_{Laser}$.

The laser cavitation was observed by a high speed video camera whose maximum framing rate was 230,000 frames per second with noise detected by a hydrophone.

In order to demonstrate mechanical surface treatment by the present pulse laser system, duralumin plate A2017-T3 whose thickness was 3 mm was treated, and the arc height was measured, then the inverse of curvature was calculated to show the peening effect.

A numerical simulation was also carried out for the collapse stage of the hemispherical vapour bubble on a wall. A simple numerical method for gas-liquid two-phase / material coupling analysis was applied [8], in the fluid analysis, locally homogeneous model was used and the bubble interface was expressed by gradient of void fraction. And compressibility was considered in liquid phase and gas phase was assumed as ideal-gas. In the material analysis, homogeneous isotropic elastic body was assumed and propagation of stress wave was reproduced numerically.

## 3. Results

In order to reveal effect of mechanical surface treatment by the pulse laser system, Fig. 2 shows the arc height of duralumin plate changing with density of pulse laser. The pulse laser produced a curvature of the duralumin plate. That means the present pulse laser system can introduce compressive residual stress. Namely, the present system can be applied for mechanical surface treatment of metals.

Figure 3 reveals the noise level $p_N$ detected by the hydrophone changing with time $t$ after irradiated pulse laser. The 1st and 2nd peaks were observed at $t \approx 0$ ms and 0.68 ms. Regarding Fig. 4, which shows the aspect of laser induced bubble, it was revealed that the 1st peak was related to laser abrasion and the 2nd peak was produced by the collapse of the laser cavitation. Note that the amplitude of the 2nd peak $p_{N\,2\mathrm{nd}}$ was larger than that of 1st peak $p_{N\,1\mathrm{st}}$. Namely, the shock wave induced by laser cavitation was stronger than that of laser abrasion.

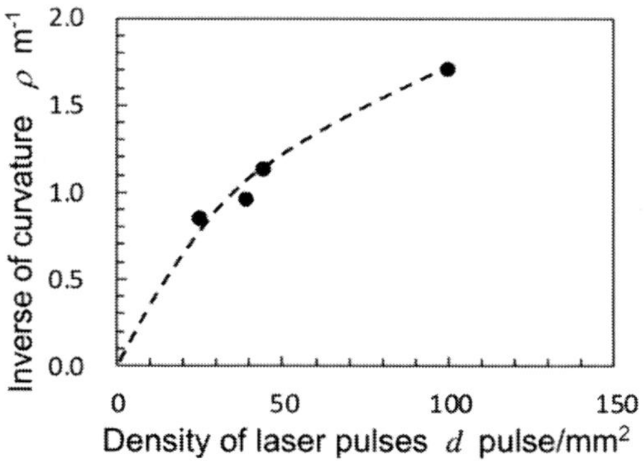


**Figure 2.** Peening effect changing with density of laser pulses.

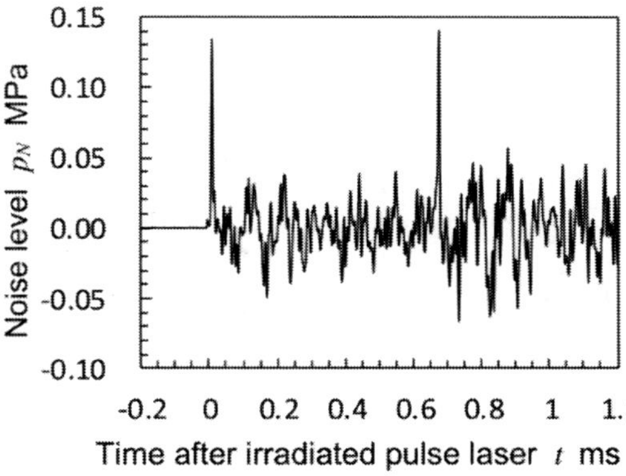


**Figure 3.** Noise induced by laser abrasion and laser cavitation.

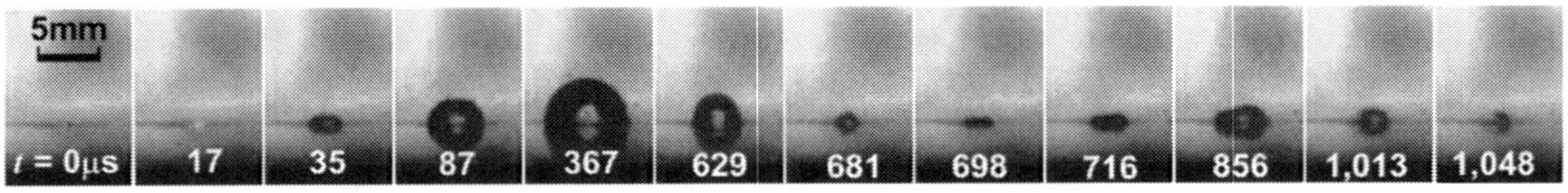


**Figure 4.** Aspect of laser cavitation (laser induced bubble).

In order to investigate key parameter of amplitude of 2nd peak of the noise, the bubble size and the noise level were measured changing with energy of pulse laser $E_{PL}$. Figure 5 reveals the noise level at 2nd peak $p_{N\,2nd}$ as a function of volume of bubble at the maximum size $V_{max}$. When $V_{max}$ was larger, $p_{N\,2nd}$ was larger. However, $p_{N\,2nd}$ seems to be saturated at large $V_{max}$ in the present condition. The correlation coefficient $r^2$ of the relation between $V_{max}$ and $p_{N\,2nd}$ was about 0.66 for 15 points.

Figure 6 shows the relation between the noise level at 1st peak $p_{N\,1st}$ and the noise level at 2nd peak $p_{N\,2nd}$. As shown in Fig. 6, $p_{N\,2nd}$ was increased with an increase of $p_{N\,1st}$. The $r^2$ of the relation between $p_{N\,1st}$ and $p_{N\,2nd}$ was 0.76, which was better than that of between $V_{max}$ and $p_{N\,2nd}$. It might be happed that the inside pressure of the bubble at large $p_{N\,1st}$ would be less than that at low $p_{N\,1st}$, as noise at 1st peak was generated by the laser abrasion. Namely, the laser abrasion produced negative pressure wave just after shock wave, and the negative pressure would be larger when the shock wave, i.e., $p_{N\,1st}$, was larger. Thus, pressure difference between inside and outside of bubble would be large at large $p_{N\,1st}$ at the time of maximum bubble radius because of drastic growth of the bubble. It can be concluded that the large pressure difference might produce large impact at bubble collapse. These experimental result was examined by following numerical simulation.

A numerical simulation was also carried out for the collapse stage of the hemispherical vapour bubble on a wall. A simple numerical method for gas-liquid two-phase / material coupling analysis was applied [8], in the fluid analysis, locally homogeneous model was used and the bubble interface was expressed by gradient of void fraction. And compressibility was considered in liquid phase and

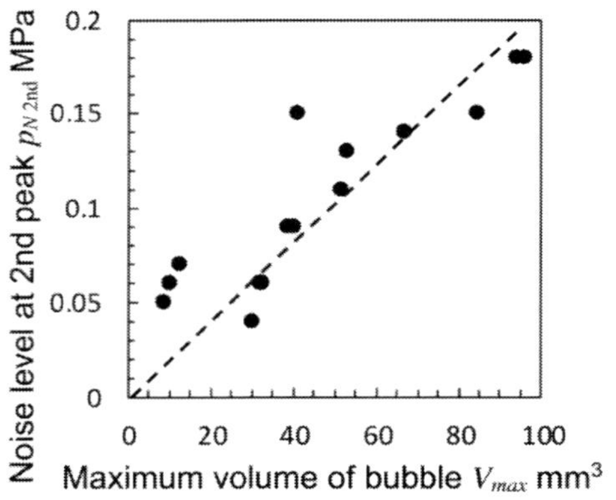


**Figure 5.** Noise level at 2nd peak as a function of maximum volume of bubble

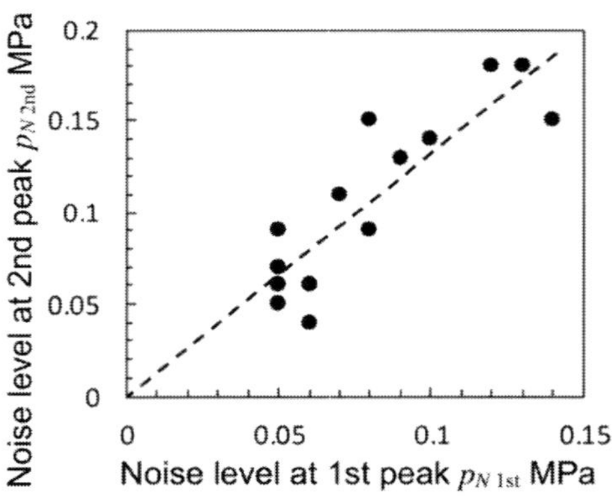


**Figure 6.** Noise level at 2nd peak as a function of noise level at 1st peak

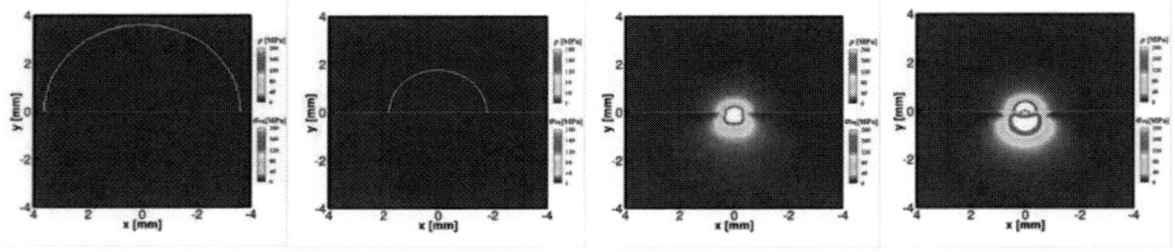

(a) $t = T_0$ (b) $t = T_0 + 300.96$ μs (c) $t = T_0 + 333.59$ μs (d) $t = T_0 + 333.72$ μs

**Figure 7.** Aspect of bubble collapse (Upper: fluid, lower: material, Initial condition: radius 3.6 mm, bubble pressure 2,300 Pa, ambient pressure 0.1 MPa, temperature 293.15 K)

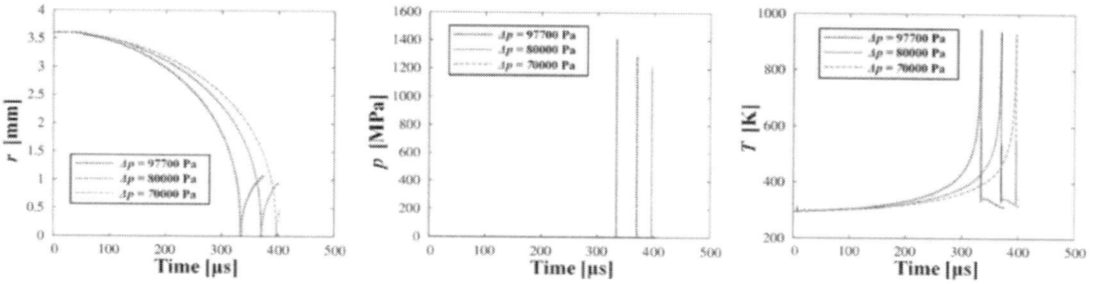


**Figure 8.** Time evolution of bubble radius, centre pressure and centre temperature inside a bubble at the collapse

gas phase was assumed as ideal-gas. In the material analysis, homogeneous isotropic elastic body was assumed and propagation of stress wave was reproduced numerically. In Fig. 7, time evolution of aspect of bubble collapse is shown in a case of initial bubble radius 3.6 mm. Pressure contour distribution and iso-void fraction line at 0.5 are shown in fluid (upper) and equivalent stress is shown in material (lower). In Fig. 7, (a) is the time of maximum radius, (b) is half radius, at the time (c) maximum collapse pressure occurs in fluid and maximum equivalent stress is yielded in material, after that, bubble starts rebound, pressure wave propagates in liquid and stress wave propagates in material at the time (d).

Next, time evolution of bubble radius, centre pressure and centre temperature inside a bubble at the collapse stage are shown in Fig. 8 with varying initial pressure difference. As show in the figure, higher pressure and higher temperature is caused by higher pressure difference condition although initial bubble radiuses are same. Besides, because ideal gas was assumed for vapour inside the bubble, the maximum values were not accurate at such a high pressure and high temperature condition at final stage of bubble collapse. But relative evaluation was considered to be effective. In a future work, influences will be analysed for ambient void fraction which corresponds residual bubble nuclei, temperature and flow field around a bubble. In addition, the tendency that higher pressure difference produced higher pressure at bubble collapse well corresponded to above mentioned experimental results.

## 4. Conclusions

In order to demonstrate the possibility of bubble induced by pulse laser, i.e., laser cavitation, for mechanical surface treatment, the high speed observation was carried out with detecting the noise, which corresponded to the shock wave, evaluating peening intensity by curvature of duralumin plate. The numerical simulation was also carried out for a behaviour of bubble collapse, pressure and temperature distribution inside the hemispherical vapour bubble, and equivalent stress distribution in the material were analysed. It was concluded that the pulse laser at the present condition can be applied for the peening and impact induced by the laser cavitation was larger than that of laser abrasion.

This work was partially supported by The Amada Foundation and The New Technology Development Foundation. The authors thank Mr. M. Mikami, technician, and Mr. Y. Ueno, graduate student, Tohoku University for their help with the experiment.

## References

[1] Sano Y, Obata M, Kubo T, Mukai N, Yoda M, Masaki K and Ochi Y 2006 *Mater. Sci. Eng.* A **417** 334
[2] Hatamleh O, Lyons J and Forman R 2007 Inter. J. Fatigue **29** 421
[3] Gill A, Telang A, Mannava SR, Qian D, Pyoun Y-S, Soyama H, Vasudevan VK 2013 *Mater. Sci. Eng.* A **576** 346
[4] Lauterborn W and Bolle H 1975 *J Fluid Mech.* **72** 391
[5] Soyama H and Sekine Y 2010 *Inter. J. Sustain. Eng.* **3** 25
[6] Soyama H 2014 *Mater. Sci. Appl.* **5** 430
[7] Sasoh A, Watanabe K, Sano Y and Mukai N 2005 *Appl. Phys.* A **80** 1497
[8] Sasaki H, Iga Y and Ikohagi T, 2010, *Proc. SNA+MC2010*, 10189

Journal of Physics: Conference Series **656** (2015) 012109 doi:10.1088/1742-6596/656/1/012109

# The acoustic signatures of cavitation erosion on grade DH36 steel

**I. Armakolas, J. Carlton, M. Vidakovic, T. Sun and K.T.V. Grattan**

City University London, Northampton Square EC1V 0HB, United Kingdom

Ioannis.Armakolas.1@city.ac.uk

**Abstract**. Cavitation can cause considerable erosion to adjacent materials. Erosion is accompanied by acoustic emissions, related to crack formation and propagation inside the material. In this study a piezoelectric acoustic sensor mounted on the back of a grade DH36 steel plate is used to identify the acoustic signatures of cavitation. Cavitation is induced near the plate by means of an ultrasonic transducer (sonotrode). Various 'non-erosive' and erosive test rig configurations are examined and an acoustic threshold value for the onset of cavitation erosion is identified and presented. The use of a fibre Bragg grating (FBG)-based acoustic sensor developed at City University London for acoustic monitoring purposes is also examined. Acoustic signals from both sensors are analysed, by means of a fast Fourier transform, showing a very good agreement in terms of captured frequencies.

## 1. Introduction

A series of phenomena are believed to be the cause of cavitation erosion, including bubble collapse and rebound, micro-jet formation, clouds of collapsing micro bubbles and cavitation vortices [1]. Studies with high speed photography have shown high pressure waves emitted from collapsing cavities [2, 3]. Recent studies have suggested that a very high proportion of the collapse energy of a large cavity can be focused into a small region of the solid surface and cause significant erosion, under specific geometries [4]. It has also been suggested that the potential power of macro cavities could be converted into collapsing clouds of micro bubbles, resulting into shock waves that can cause significant erosion on the solid surface [5]. The initiation of erosion can produce acoustic emissions which are essentially high frequency waves related to crack formation and propagation inside the material [6]. The present study is focused on the investigation and design of a monitoring system for cavitation erosion related acoustic emissions from metals, based on piezoelectric acoustic sensors and fibre Bragg grating (FBG) sensors.

## 2. FBG and piezoelectric acoustic sensors

Fibre Bragg gratings have become an important sensing element for the measurement of strain, temperature, and wide range of other parameters [7]. Their inherent advantages such as immunity to electromagnetic interference ease of multiplexing, small size and typically lightweight have made them suitable to be used in various environments [8]. An FBG is a piece of an optical fibre with its core refractive index being periodically modulated, which creates a narrow band reflection. The principle of operation of an FBG-based sensor system is based on monitoring of the wavelength shift of the reflected Bragg signal, as a function of a measurement, such as strain or temperature applied to the FBG. On the other hand piezoelectric acoustic sensors are operating on the principle of piezoelectricity, which can be defined as the linear electromechanical interaction between the mechanical and the electric state in crystalline materials [9]. Electrical charge is produced by the sensor in response to applied mechanical stress which can then be captured as an acoustic signal.

## 3. Experimental setup

Figure 1 shows an experimental setup for the cavitation test of a grade DH36 steel plate. The dimensions of this square shaped steel plate are 50 x 50 x 5 mm. The plastic bridge placed at the bottom of the water tank is used both to support the steel plate and to adjust the distance between the metal plate and the sonotrode. The sonotrode is a Hielsher UIP1000hd ultrasonic transducer, operating at a standard frequency of 19.5 kHz and it is placed above the steel plate surface at a predefined distance. The power output of the sonotrode is also adjustable and directly affects cavitation intensity.

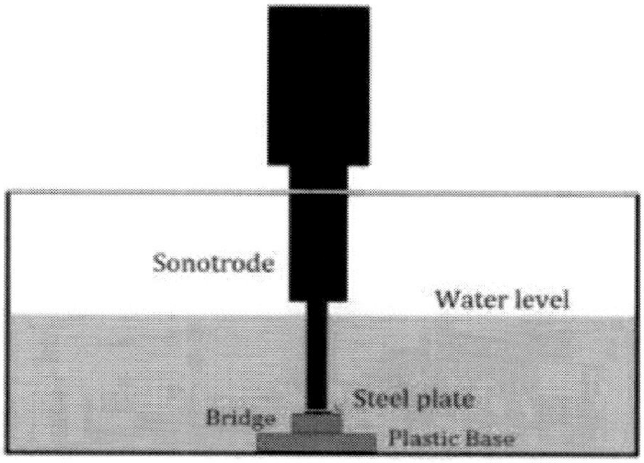


Fig. 1. Cavitation test rig

Figure 2 illustrates the detection systems for both the FBG and piezoelectric acoustic sensors. The signal detected by the photodiode is cross-compared with the signal detected by the piezoelectric sensor, placed adjacent to the FBG on the same metal plate. Prior to the recording of the piezoelectric sensor data, both pre-amplification and signal attenuation modules are included to optimize the signal output. Acoustic Signals from both sensors are captured through an Oscilloscope at a sampling rate of 31.3 MHz.

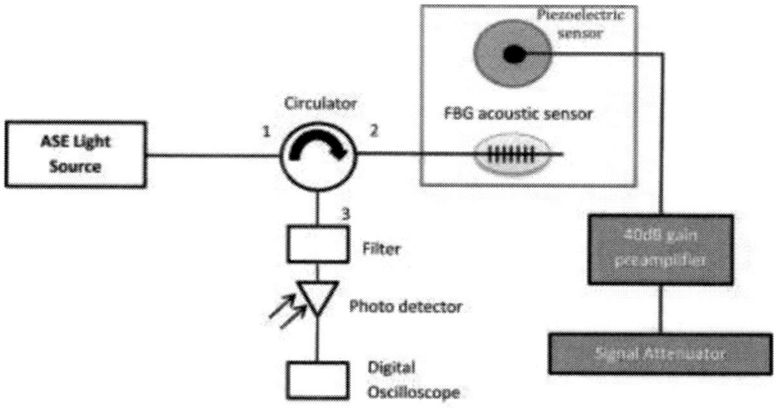


Fig. 2. Instrumentation of steel plate with both FBG and piezoelectric acoustic sensors

## 4. Experimental results

### *4.1. Acoustic threshold for cavitation erosion (piezoelectric acoustic sensor)*

Four different test rig configurations are presented, based on the impact of cavitation on the exposed steel samples in terms of erosion. These configurations were selected by measuring the cavitation erosion related mass loss of the examined steel samples over a period of 2 h. Adjustable parameters of the cavitation test rig include the gap between the sonotrode and the sample as well as the power output of the sonotrode. The distributions of the acoustic emission signals captured through the piezoelectric sensor are presented in Figure 3. Signals are presented in terms of acoustic power, corresponding to different cavitation intensities and consequently erosion of the steel plate. It should be noted that signals were captured under a sampling rate of 31.3 MHz and the average duration of the acoustic samples was 0.4 s containing approximately 1M acoustic data points.

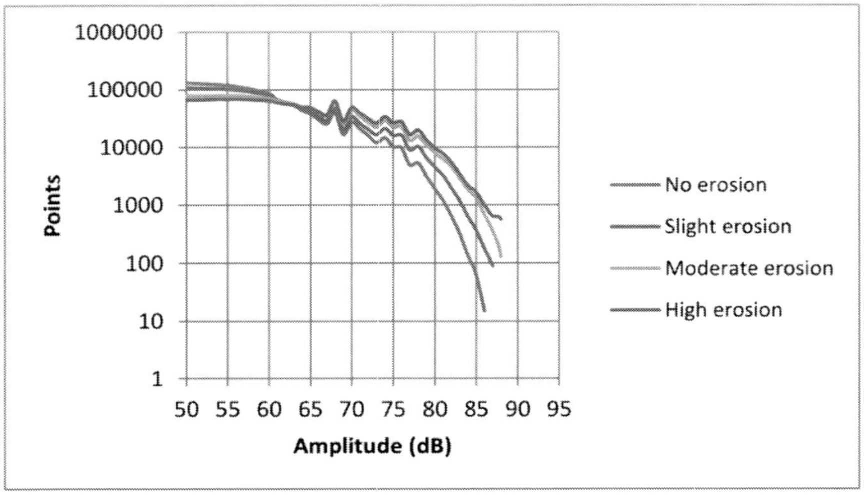


Fig. 3. Acoustic emissions distribution for various test rig configurations

The threshold acoustic value for the 'non-erosive' test rig configuration is noticeable and distinct and is of the order of 86 dB. The next 'slight erosion' setting, which indicates the onset of cavitation induced erosion has a maximum acoustic power value of 88 dB. Acoustics signals with even higher amplitudes were recorded for the 'moderate erosion' and 'high erosion' configurations.

### *4.2. Comparison between FBG and piezoelectric acoustic sensors*

The use of fibre Bragg grating (FBG)-based acoustic sensors for cavitation erosion monitoring was investigated through comparison with the piezoelectric sensor. Signals from both FBG and piezoelectric sensors were captured through an oscilloscope at a sampling frequency of 31.3 MHz using a 'normal erosion' test rig configuration. The average duration of signals from both sensors was 0.4 s, each one containing approximately 1 M acoustic data points. Signals from both sensors were further processed with the use of a fast Fourier transform (FFT) algorithm in order to investigate the frequency components of which the two signals are composed. Figure 4 shows the frequency-domain data obtained from both sensors for cross-comparison after the use of the fast Fourier transform (FFT). It is noticeable that the same frequency element of 19.5 KHz, which matches with the operating frequency of the sonotrode and is directly related to cavitation activity, has been captured by both sensors. Compared to the piezoelectric sensor, the FBG sensor has shown some additional second harmonic signals as well as some frequencies below 19.5 KHz. It should be noted that cross-comparison between FBG and piezoelectric sensors, in terms of captured frequencies could only be made in the range of 0 to 100 KHz due to the absence of any other resonances above this frequency for the FBG sensor.

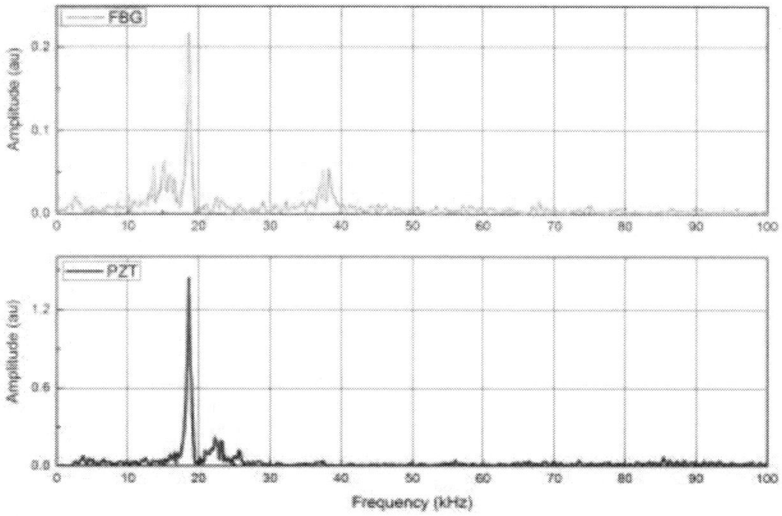


Fig. 4. Frequency response comparison between FBG and piezoelectric sensor

## 5. Conclusion

Acoustic emissions related to cavitation erosion were successfully identified by means of a piezoelectric acoustic sensor - based monitoring system. The acoustic threshold value related to the onset of cavitation erosion was successfully identified by testing various 'non-erosive' and erosive test rig configurations. Comparison between FBG and piezoelectric sensors, has shown a common frequency element which matches with the excitation frequency of the sonotrode and is directly related to cavitation activity. The good agreement made between both sensors has demonstrated the potential of FBG-based acoustic sensors that could be possibly used for a variety of industrial applications and has to be further investigated.

## References

[1] Franc, J.P. and Michel, J.M., *Fundamentals of Cavitation.*2004.

[2] Philipp, A. and Lauterborn, W., *Cavitation erosion by single laser-produced bubbles.* Journal of Fluid Mechanics, 1998. **361**: p. 75–116.

[3] Lindau, O. and Lauterborn, W., *Cinematographic observation of the collapse and rebound of a laser-produced cavitation bubble near a wall.* Journal of Fluid Mechanics, 2003.

[4] Bark, G., Berchiche, N. and Grekula, M., *Application of principles for observation and analysis of eroding cavitation.* The EROCAV observation handbook, Edition 3.1, 2004.

[5] Fortes-Patella, R., Reboud, J.L. and Briancon-Marjolett, *A phenomenological and numerical model for scaling the flow aggressiveness in cavitation erosion.* EROCAV Workshop, Val de Reuil, 2004.

[6] Gorman, M., Prosser, W.H., *AE orientation by plate wave analysis.* Journal of Acoustic Emission, 1991. **9(4)**: p 283-288

[7] Grattan, K.T.V. and Sun. T., *Fiber optic sensor thechnology: An overview.*Sens. Actuators A:Phys, 2000. **82**: p. 40-61.

[8] Kersey, A., Davis, M., Patrick, H., et al., *Fiber grating sensors.* Journal of Lightwave Technology, 1997. **15(8)**: p. 1442-1463.

[9] Gautschi, G., *Piezoelectric sensorics: Force, Strain, Pressure, Accelaration and Acoustic emission Sensors, Materials and Amplifier,*2002. Springer.

# X-ray Diagnostics for Cavitating Nozzle Flow

**Daniel J. Duke**[1], **Andrew B. Swantek**[1], **Alan L. Kastengren**[2] and **Christopher F. Powell**[1]

[1] Energy Systems Division, Argonne National Laboratory, Lemont, Illinois, USA
[2] X-ray Science Division, Argonne National Laboratory, Lemont, Illinois, USA

E-mail: `dduke@anl.gov`

**Abstract.** Cavitation plays a critical role in the internal flow of nozzles such as those used in direct fuel injection systems. However, quantifying the vapor fraction in the nozzle is difficult. The gas-liquid interfaces refract and multiply scatter visible light, making quantitative extinction measurements difficult. X-rays offer a solution to this problem, as they refract and scatter only weakly. In this paper, we report on current progress in the development of several x-ray diagnostics for cavitating nozzle flows. X-ray radiography experiments undertaken at the Advanced Photon Source at Argonne National Laboratory have provided measurements of total projected void fraction in a 500 $\mu$m submerged nozzle, which have been directly compared with numerical simulations. From this work, it has been shown that dissolved gases in the liquid also result in the formation of vapor regions, and it is difficult to separate these multiple phenomena. To address this problem, the liquid was doped with an x-ray fluorescent bromine tracer, and the dissolved air substituted with krypton. The fluorescent emission of Br and Kr at x-ray wavelengths provide a novel measurement of both the total void fraction and the dissolved gas component, allowing both cavitation and dissolved gas contributions to be measured independently. *[199/200 words]*

## 1. Introduction

Cavitation is known to play an important role in high-pressure direct fuel injection systems [1]. Although the fundamental physics of cavitation nucleation is reasonably well-understood [2], the combination of cavitation and turbulent fluid flow in a fuel injector nozzle can lead to complex multi-phase flow structures which are not fully understood and remain an area of active research . It has been established that cavitation inside the nozzle can affect atomization and combustion , and that cavitation leads to erosion and emissions control issues [3].

Visualization experiments have revealed a great deal about the morphology of nozzle cavitation [4]. However, it is difficult to obtain quantitative vapor fraction measurements from these experiments. Visible light refracts strongly at both the nozzle walls and larger gas-liquid interfaces, and will also be multiply scattered by clouds of small bubbles, making measurements of the void fraction difficult.

In recent years, x-ray radiography has been demonstrated as a useful diagnostic for studying dense fuel sprays [5], since x-rays refract and scatter only very weakly [6]. With an appropriate choice of wavelength, x-ray absorption can be easily related to the projected density of the flow field [7]. These techniques are equally applicable to cavitating nozzle flows. Void fraction measurements in a cavitating pipe have previously been demonstrated using x-ray CT scanning [8]. However, x-ray tube sources are limited in spatial and temporal resolution. Focused, monochromatic synchrotron x-rays provide orders of magnitude more flux, and can probe flow at scales relevant to fuel injection, with a spatial resolution on the order of 5 microns [9]. X-ray radiography can be directly compared to numerical simulations, since a line-of-sight projection of the void fraction can be easily computed from a 3D model.

Recent work has shown that non-condensable gas can also play an important role in determining the vapor distribution in a cavitating nozzle [10]. Since x-ray radiography cannot distinguish between vapor due to gas bubbles and vapor due to cavitation, we have developed an x-ray fluorescence experiment which uses separate tracers in the fuel and gaseous phases to separate these effects [11]. Unlike visible light, x-rays are sufficiently energetic to interact with core shell electrons, such that elemental species can be tracked without regard for the valence state. The fluorescence photons are themselves x-rays [12];

they also interact weakly with the sample. We report on progress in the development of x-ray radiography and fluorescence techniques for the study of cavitation in a submerged model nozzle.

## 2. Methods

The test geometry was a submerged plastic nozzle 500 $\mu$m in diameter and 3.05 mm in length ($L/D = 6.1$), with an expansion/contraction diameter ratio of 5. The working fluid was a commercial gasoline surrogate (Viscor 16br) at 25°C with a vapor pressure of 3 torr and a density of 0.773 g/cc. A piston accumulator (Fig. 1) was used to supply a continuous flow of fuel to the nozzle, with a static inlet pressure of $P_1 = 10$ bar abs. A needle valve downstream of the experiment was used to regulate the back pressure on the nozzle to $P_2 = 1.0$ bar abs, with the fuel returned to a storage tank held under partial vacuum [9]. In this nozzle, $\mathrm{Re}_D = 1.5 \times 10^4$ and cavitation number $\sigma = 0.11$.

The experiments were performed at the 7-BM beamline of the Advanced Photon Source at Argonne National Laboratory [13]. A monochromatic x-ray beam was focused to a spot of $5 \times 6$ $\mu$m cross-section, and the nozzle was traversed through the fixed beam to build up a two-dimensional raster map of measurements. Fig. 2 shows the setup with a cross-section through the nozzle throat. For radiography, the beam energy was set at 8 keV (3.0% full width at half-maximum bandpass), and was passed through a reference intensity monitor ($I_0$) and collected by a PiN diode ($I_1$). The absorption of x-rays in the liquid is related to the projected void fraction $\alpha$ by the Beer-Lambert Law;

$$\int \alpha \, dz = \frac{-1}{\mu_l \rho_l} \ln\left(\frac{I_1/I_0}{I_1'/I_0'}\right), \tag{1}$$

where $I_1'/I_0'$ is a reference measurement with the nozzle full of liquid and no vapor, $\rho_l$ is the liquid density and $\mu_l$ is the attenuation coefficient for the liquid fuel, which is calibrated against a known standard [9].

For the fluorescence experiments, the same setup was used, but an energy-resolving silicon drift diode was placed at a 100-degree angle to the incident beam, capturing the fluorescent emission where the beam interacts with the flow. The liquid was doped with tetrabromomethane, which dissolves well in hydrocarbons. Dissolved air in the fuel was substituted with krypton, which has similar solubility to nitrogen at standard conditions, despite its higher density [14]. The beam energy was increased to 14.7 keV (1.0% FWHM bandpass), and the Br and Kr $\mathrm{K}_\alpha$ emission lines were used to track the elemental concentrations.

The projected fluid density is determined from the integrated count rate of Br $\mathrm{K}_\alpha$ emission $I_k = I_{\mathrm{Br}}$;

$$\int \rho_k \, dz = \frac{c_k f_s(x,y)}{I_0 f_a(x,y,I_1/I_0)} \frac{t_{\mathrm{live}}}{\Delta t} I_k, \tag{2}$$

where $c_k$ is a constant scaling coefficient which depends on the detector solid angle, fluorescence yield and species concentration, $f_a$ is a correction function for the absorption of the beam in the sample, and $t_{\mathrm{live}}/\Delta t$ is the fraction of time that the detector is active. The sample time per point was $\Delta t = 5$ s. $f_s$ is a signal trapping function which corrects for the increasing absorption of photons in the fluid and nozzle

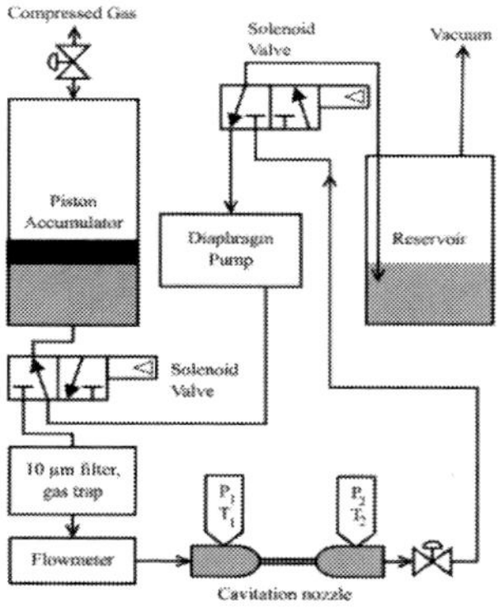


**Figure 1:** Fuel delivery system.

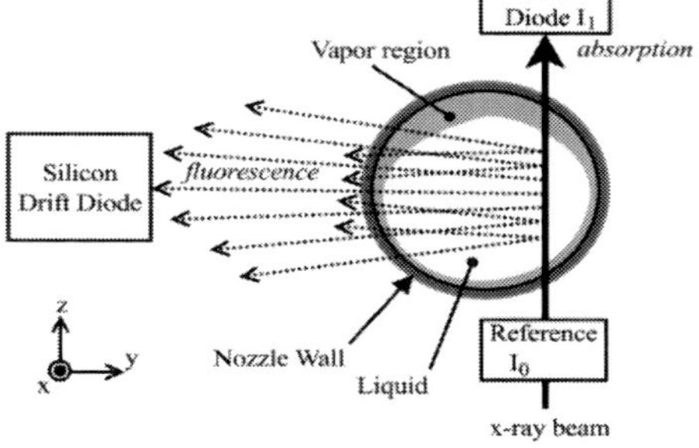


**Figure 2:** Schematic of x-ray experiment setup, showing a cut-plane of the nozzle.

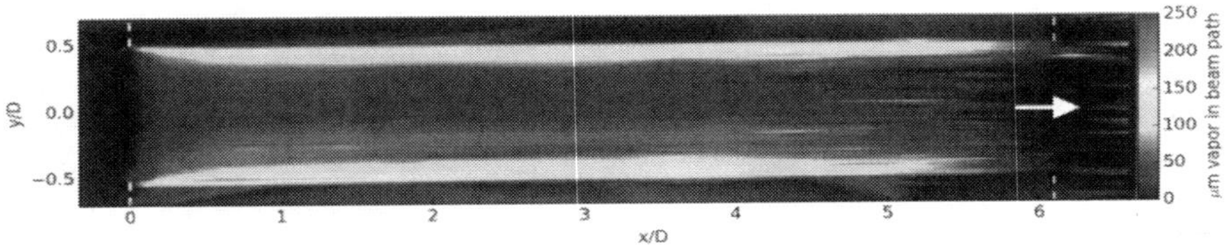


(a) Projected void fraction at 0 degrees.

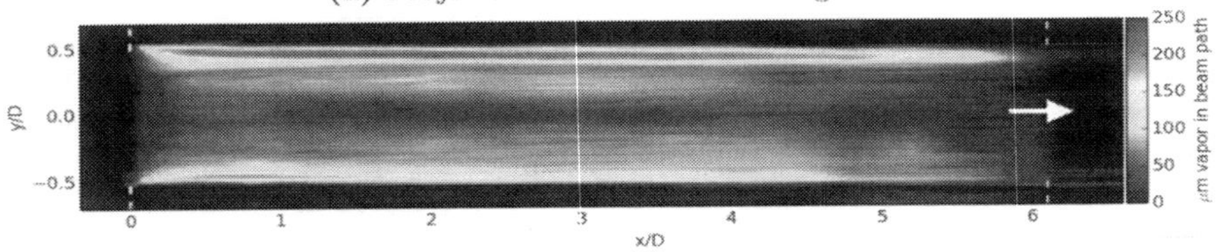


(b) Projected void fraction at 90 degrees.

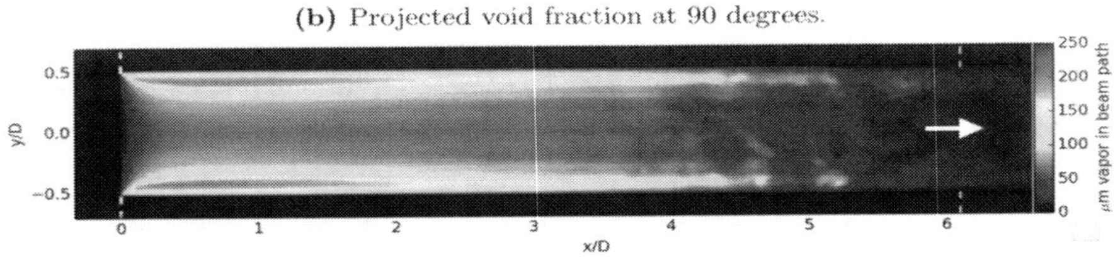


(c) Projected void fraction from LES.

**Figure 3:** Projected total void fraction $\int \alpha\, dz$ with degassed fuel at $P_1 = 10$ bar, $P_2 = 1$ bar from (a,b) two orthogonal views using x-ray radiography and (c) large eddy simulation. Flow is left to right.

wall as the distance between the beam and detector increases. It was iteratively solved by updating the estimates of the fluid density distribution in a plane, applying an Abel inversion, and computing the absorption toward the detector [11]. The path length of non-condensable gas is similarly solved using Eqn. 2 with $I_k = I_{\mathrm{Kr}}$. Given the known path length through the nozzle $\delta z$, the fluid density, and the partial density of the gas, it is possible to solve for the total projected vapor $\int \alpha\, dz = \delta z - \int \rho_{\mathrm{Br}}\, dz/\rho_l$ and the non-condensable gases $\int \alpha_g\, dz = \int \rho_{\mathrm{Kr}}\, dz/\rho_v$. When $\int \alpha\, dz > \int \alpha_g\, dz$, we assume that the additional vapor is due to cavitation; $\int \alpha_v\, dz = \int \alpha\, dz - \int \alpha_g\, dz$.

## 3. Results and Discussion

In the first set of experiments, we removed all dissolved air from the fuel and made a time-averaged radiography measurement of the total projected vapor fraction from two orthogonal views. Figs. 3(a-b) show that most of the vapor is attached to the nozzle wall and persists until about one nozzle diameter from the exit. The results compare favorably with three-dimensional large eddy simulations of the flow (Fig. 3c), which were performed using a homogeneous relaxation cavitation sub-model [15, 16].

During the experiments, we found that the introduction of dissolved air caused a large decrease in the fluid density [9]. Simulations suggested that this was due to gas in the fuel coming out of solution due to the sudden pressure drop at the nozzle inlet [10]. To investigate further, we turned to the x-ray fluorescence diagnostic. Kr was substituted for air at standard conditions, giving a mass ratio of $1.039 \times 10^{-4}$. The results are shown in Fig. 4. The total projected vapor and gas from Br fluorescence is markedly increased (Fig. 4a) compared to the degassed case (Fig. 3a). The projected volume fraction of Kr (Fig. 4b) shows a large amount of gas throughout the nozzle. An increase in the centerline Kr concentration with $x$ indicates addition gas coming out of solution. With a bulk velocity $\overline{U} \approx 58\ \mathrm{ms}^{-1}$, we note that the nozzle transit time is 53 $\mu$s. This process occurs orders of magnitude slower than cavitation inception, but its time scales are still relevant for diesel injection.

The difference between the Br and Kr measurements (Fig. 4c) reveals that the only region of true cavitation is an annular bubble which is asymmetric and disappears by $x/D = 4$. The asymmetry may be partially due to machining defects in the nozzle which differ between the two experiments. However, the reduction in true cavitation relative to the completely degassed case suggests that nucleating gas bubbles (which have orders of magnitude higher density and pressure than the cavitation vapor) are displacing the cavitation zones which would otherwise occur.

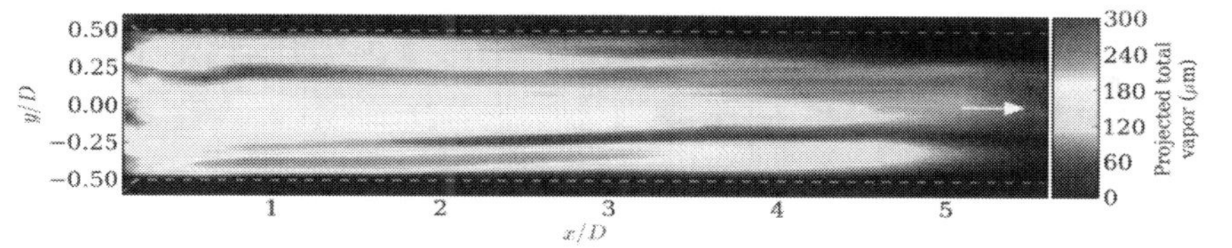


(a) Projected total vapor from bromine fluorescence.

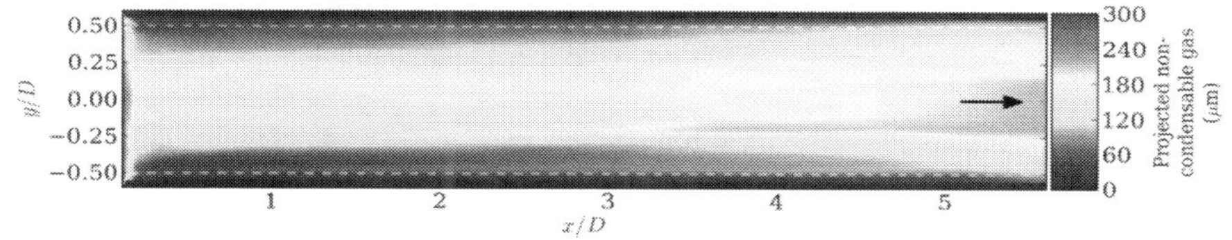


(b) Projected non-condensable gas distribution from krypton fluorescence.

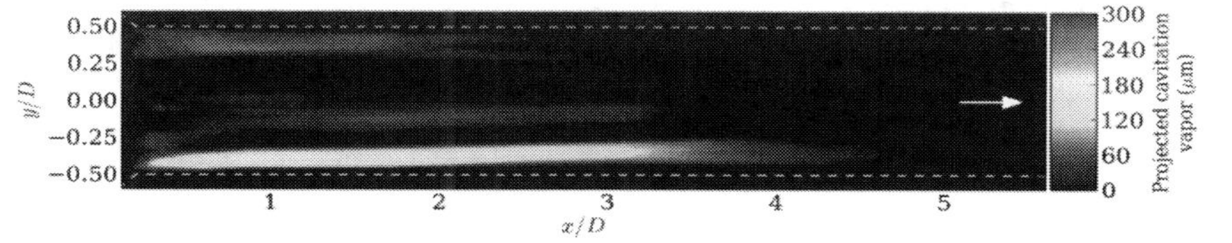


(c) Projected cavitation distribution.

**Figure 4:** X-ray fluorescence measurements of fuel with dissolved krypton gas at $P_1 = 10$ bar, $P_2 = 1$ bar. Fluorescence of (a) the liquid phase gives the total void fraction, from which (b) the non-condensable gas contribution can be removed to compute (c) the contribution due to cavitation. Flow is left to right.

## 4. Conclusions

These early results suggest that simulations of cavitation in fuel injection systems need to take into account the dynamics of dissolved gases in order to yield accurate predictions. Furthermore, when the fluid contains non-condensible gas, experimental studies must account for the fact that many of the observed voids may not be true cavitation, particularly in the scaled-up case where the transit time through the nozzle can be longer, allowing more opportunity for phase change.

X-ray radiography and fluorescence measurements permit useful quantitative analysis of cavitating flows, which in future work may be readily extended to spraying nozzles and transparent injectors.

## Acknowledgments

The authors wish to thank Professor David Schmidt (University of Massachusetts-Amherst) for the use of *HRMFoam* , which was used to perform the simulation shown in Fig. 3c. This research was performed at the 7-BM beam line of the Advanced Photon Source at Argonne National Laboratory. Use of the APS is supported by the U.S. Department of Energy (DOE) under Contract No. DE-AC02-06CH11357.

## References

[1] Gavaises M 2006 *SAE Technical Paper* **2006-01-1114**
[2] Brennen C 1995 *Cavitation and Bubble Dynamics* (Oxford University Press)
[3] Hiroyasu H, Arai M and Shimizu M 1996 *Recent Advances in Spray Combustion* (AIAA)
[4] Arcoumanis C, Gavaises M and Flora H 2001 *Mécanique & industries* **2** 375–381
[5] Kastengren A and Powell C F 2014 *Experiments in Fluids* **55** 1686
[6] Als-Nielsen J and McMorrow D 2001 *Elements of Modern X-Ray Physics* (Wiley, New York)
[7] Kastengren A L, Tilocco F, Duke D, Powell C, Zhang X and Moon S 2014 *Atom Sprays* **24** 251–272
[8] Bauer D, Chaves H and Arcoumanis C 2012 *Measurement Science and Technology* **23** 055302
[9] Duke D, Kastengren A L, Tilocco F, Swantek A B and Powell C 2013 *Atom Sprays* **23** 841–860
[10] Battistoni M, Duke D J, Swantek A B, Tilocco F Z, Powell C F and Som S 2015 *Atom Sprays* **25** 453–483
[11] Duke D J, Swantek A, Kastengren A, Fezzaa K and Powell C 2015 *SAE Int J Fuels & Lubricants* **8** 135–146
[12] Kastengren A, Powell C, Dufresne E and Walko D 2011 *J Synchrotron Rad* 811–815
[13] Kastengren A, Powell C F, Arms D, Dufresne E M *et al.* 2012 *J Synchrotron Rad* **19** 654–657
[14] Clever H L, Battino R, Saylor J H and Gross P M 1957 *The Journal of Physical Chemistry* **61** 1078–1082
[15] Neroorkar K, Shields B, Grover Jr R O, Torres Plazas A and Schmidt D P 2012 *SAE Journal* **2012-01-1269**
[16] Duke D J, Schmidt D P, Neroorkar K, Kastengren A L and Powell C 2013 *Int J Engine Research* **14** 578–589

Journal of Physics: Conference Series **656** (2015) 012111 doi:10.1088/1742-6596/656/1/012111

# Hydrodynamic Nuclei Concentration Technique in Cavitation Research and Comparison to Phase-Doppler Measurements

**Eric Ebert, Willfried Kröger, Nils Damaschke**
Institute for General Electrical Engineering, Photonics and Optoelectronics, Albert-Einstein-Straße 2, 18059 Rostock Germany

E-mail: eric.ebert@uni-rostock.de

**Abstract.** Small particles, especially bubbles in the micro-meter range, influence the cavitation of the propellers. The prediction of cavitation inception and water quality measurements are important in cavitation research. The Hydrodynamic Nuclei Concentration (HDNC) technique can be used for reliable bubble concentration measurements in fluid flows. The HDNC technique bases on the analysis of scattered light from the cavitation nuclei in the water. The HDNC technique can distinguish between bubbles and solid particles. The particle type classification is important, because the number concentration of solid particles is often much higher than the nuclei concentration in cavitation tunnels and in seawater. Verification experiments show, that the HDNC technique reaches similar capabilities in number concentration estimation as Phase Doppler (PD) technique in much shorter acquisition time.

## 1. Introduction and Motivation

Cavitation processes at naval vessels causes vibrations, erosion of the propeller and rudder and impairs the efficiency. Cavitation prediction models are therefore required during the design process of vessels. Current models do not represent the complexity of the cavitation process exact enough for a reliable cavitation prediction of a full scale vessel. Therefore ship model basins and propeller designers are interested in the improvement of their numerical cavitation prediction models to obtain valid results early in the design process.

The cavitation inception behaviour of a ship propeller is influenced by the water quality and especially the bubble count concentration. At low pressure regions near the propellers surface bubbles expand (fig. 1). The expanded bubbles are collapsing at high pressure regions and cause the cavitation effects. The prediction of cavitation inceptions and the determination of the scale effects include the experimental measurement of the nuclei count concentration and the nuclei size distribution in the wake flow of model and full scale. The results are applied as boundary condition for experimental and numerical prediction.

In the framework of the KonKav project, founded by the Federal Ministry of Economics and Technology (BMWi) in Germany, cavitation prediction with focus on the influence of water quality was investigated. Model scale and full scale measurements of velocity fields, cavitation observation and nuclei concentration were performed. Project partners are the Flensburger Schiffbau-Gesellschaft (FSG), Hamburg Ship Model Basin (HSVA), Technical University of Hamburg Harburg (TU-HH), Potsdam Model Basin (SVA) and the University of Rostock. First results of nuclei measurements by Phase-Doppler (PD) technique and the concept of full scale measurements were presented on the previous CAV conference in 2012 [1,2] and the further development was documented in [3].

Therefore a robust measurement technique is necessary to determine the cavitation nuclei properties in the wake flow of a ship. The developed Hydrodynamic Nuclei Concentration (HDNC) technique combines a robust, calibration free setup with moderate hardware effort and the ability to measure particle count concentrations with bubbles in the micro-meter range. Previous measurements by PD technique [3] show, that the estimation of concentration and nuclei spectra is strongly influenced by solid particles. The concentration of solid particles in navel applications is often 10 times higher than nuclei concentration. The separation of gas bubbles and solids is therefore essential for a reliable cavitation nuclei distribution statistics and is also solved by HDNC technique.

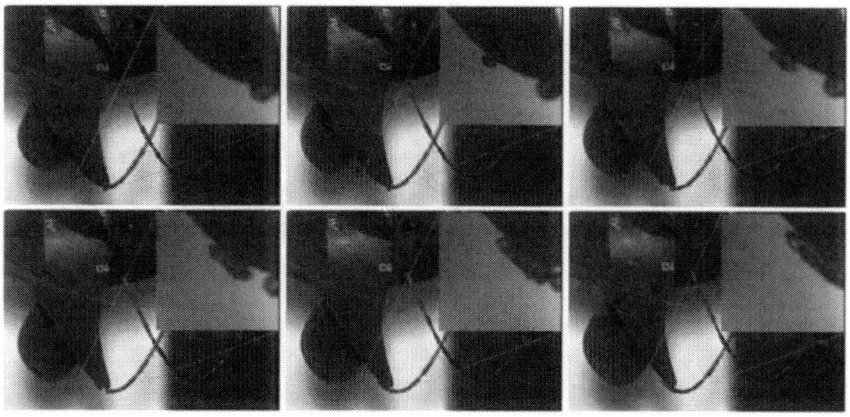

**Figure 1.** High speed recording of an expending cavitation nuclei near the propeller blade surface which causes tip vortex cavitation

## 2. Measurement principle, Setup and Image Processing

The HDNC technique uses a single laser beam which illuminates the nuclei in the wake flow and a camera for imaging the particles (fig. 2). We used a 532nm 300mW cw Nd:YAG DPSS laser.
The incident laser light is scattered by the particle. A spherical bubble has two glare points in the right scattering angle of about 90°. These are imaged out of focus. An interference fringe pattern occurs. The spacing between the fringes is proportional to the bubble diameter. The exact proportionality between fringe spacing and particle diameter has to be calculated for each individual optical setup, depending on observation distance, focal length, f-number and aperture of the optics.

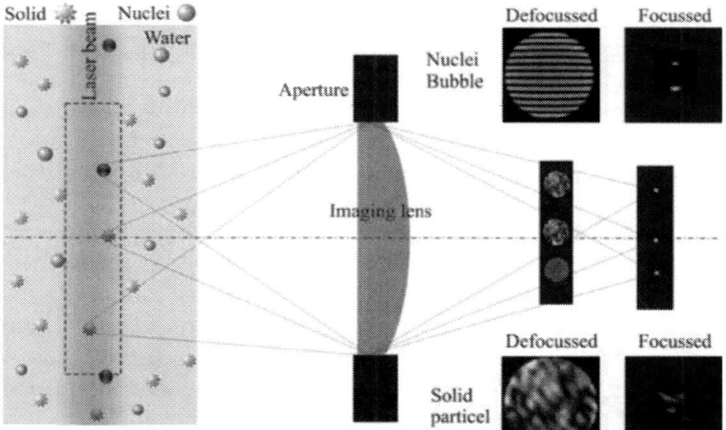


**Figure 2.** Schematic configuration and image generation of HDNC technique [4].

An elongated region of interest with reduced number of pixel can be used. The low required laser light energy (no pulse laser is required), the well separated and aligned defocussed particle images along the imaged laser beam, the reduced hardware effort, the moderate expense alignment and the straightforward detection volume definition (required for a corrected nuclei count concentration) are the advantages of the HDNC technique (compared to PD [5] and standard IPI/ILIDS [6, 7]). A disadvantage is the higher required frame rate to reach the same statistical confidence compared to the standard IPI. Because of the significant reduced number of pixels and simpler image processing the particle rate is increased and real-time processing is possible.
An important parameter of the HDNC setup is the scattering angle between the laser beam and the imaging optics. The scattering angle decides about the imaged scattering behaviour of the nuclei. The

calculations of the light scattering of spherical bubbles in water were performed for different scattering angels, two polarizations and different sizes of the receiving aperture by using the algorithms described in [5]. We found scattering angles between 80 deg and 102 deg are suitable for measurements [4]. A scattering angle of 90 deg is preferred, because of minimized optical aberrations.
The processing steps for nuclei concentration estimation are shown in Fig. 3. After pre-processing for background removal and noise suppression, the particle detection is based on template matching by different masks, based on the aperture shape. The interference structure can be used for classification of bubbles, because the spectral image characteristic of a solid particle differs from spectral characteristic of bubble images [8]. A speckle pattern appears by imaging non-spherical or inhomogeneous particles, whereas bubbles generate parallel interference fringes. After classification the nuclei or bubble diameter can be estimated from the fringe number by a spectral analysis. Classification and size calculation leads to the nuclei count concentration in advance. It is necessary to correct the obtained particle size distribution because of the inhomogeneous laser illumination and the nuclei size dependence of the scattered light intensity. Small particles are underestimated compared to larger ones.

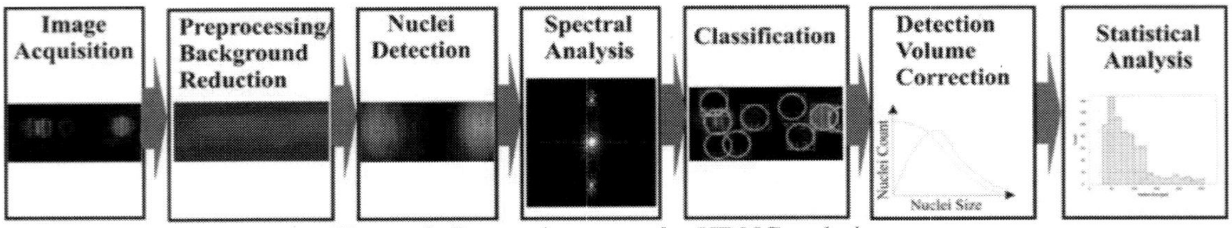


**Figure 3.** Processing steps for HDNC technique.

## 3. Measurement Results

Measurements in different cavitation tunnels (K15a SVA Potsdam and HYKAT HSVA) and full scale measurements (vessel "Amandine") were performed. The setups and the results for full scale were already presented in [8]. The following discussion focusses on the reliability of the techniques.
One measurement campaign at K15a was intended for a comparison of the HDNC and the PD techniques. Fig. 4 shows an overview of the results (right) from this campaign. The upper graph indicates the operating conditions of the cavitation tunnel K15 during the measurements. HDNC and PD techniques were used in parallel in front of the propeller at the same position. The lower graph in Fig. 4 shows the estimated nuclei concentration. Because of the small elliptical detection volume (< $0.016 mm^3$), the applied PD technique requires approx. 12 minutes (width of the green bars) for a reliable nuclei statistics. The HDNC technique measures comparable number of nuclei in significant shorter time (width of red bars corresponds of 10s), because the detection volume was $67 mm^3$. For each condition $10^4$ frames were acquired by HDNC with a frame rate of 10kfps at the beginning of each PD measurements. The inevitable uncertainty from the particle counts (Poisson distribution) is shown as error bars in Fig. 4. After processing and detection volume correction the results are comparable. The first measurements (#1…13) show a quantitative discrepancy between PD and HDNC results. The deviations indicate systematic errors in the measurement process. Because of the complexity of the measurement systems and the lack of quantitative reliable methods, the systematic errors cannot be assigned to PD or HDNC technique. The authors more believe in the accuracy of the HDNC technique, because HDNC has much less parameters (Laser power, in-coupling efficiency, secondary scattering, photodetector sensitivity and gain, estimation of detection volume size) and optical components in comparison to PD-technique [5]. Furthermore the detection of solid particles is more precise for HDNC as the statistical correction in PD-technique and PD technique has generally large uncertainty (about 50%) for concentration measurements. Final statements would require well-controlled reference experiments for nuclei concentration, which are not available. Nevertheless, all experiments (especially #13…50) show same tendencies for the different working conditions and agree well by considering the complexity of the measurements.

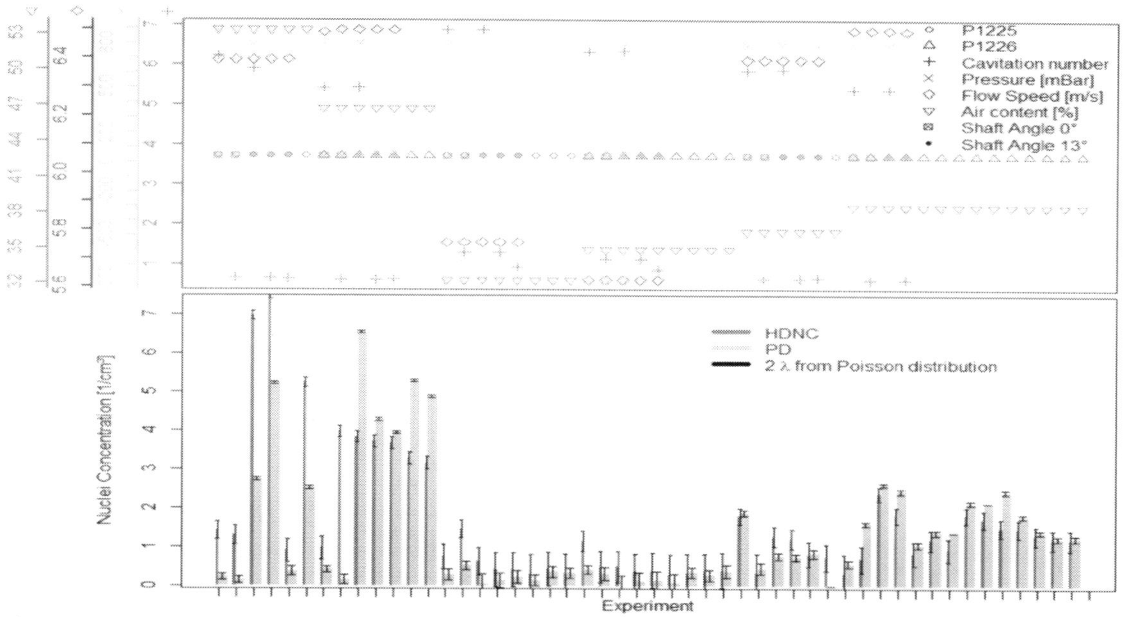


**Figure 4.** Operating conditions (upper graph) and comparison (lower graph) of PD and HDNC nuclei concentration measurements in K15.

## 4. Conclusion

In comparison to PD-technique, the HDNC technique is appropriate for the measurement of the cavitation nuclei concentration in the cavitation research. The reduced optical setup and higher data rate represent advantages. Quantitative deviations are in the order of the expected systematic errors of both techniques.

## 5. Acknowledgements

The authors thank the German Federal Ministry of Economics and Technology (BMWi) (grant number 03SX316D) and the project partners for the support of the KonKav-project.

## 6. References

[1] A. Kretschmer, S. Hohne, and N. Damaschke, "Particle characterization with the Time-Shift-Technique," presented at the Sensors, 2011 IEEE, 2011.

[2] A. Kleinwächter, E. Ebert, R. Kostbade, and N. Damaschke, "Concept for optical full-scale measurements on ship propeller inflow and bubble size distribution," presented at the 8th International Symposium on Cavitation (CAV2012), Singapore, 2012.

[3] M. Degner, N. Damaschke, M. Schaeper, and I. Menn, "Microarraybasiertes Ortsfilter," Germany Patent WO/2011/063876, 2011.

[4] E. Ebert, K. Domke, W. Kröger, and D. Nils, "Particle characterization system for industrial and naval applications," in *8th International Conference on Sensing Technology*, Liverpool, UK, 2014.

[5] H.-E. Albrecht, M. Borys, N. Damaschke, and C. Tropea, *Laser Doppler and Phase Doppler Measurement Techniques*. Berlin Heidelberg New York: Springer-Verlag, 2003.

[6] N. Damaschke, H. Nobach, T. I. Nonn, N. Semidetnov, and C. Tropea, "Multi-dimensional particle sizing techniques," *Experiments in Fluids,* vol. 39, pp. 336-350, 2005.

[7] T. Kawaguchi, Y. Akasaka, and M. Maeda, "Size measurements of droplets and bubbles by advanced interferometric laser imaging technique," *Measurement Science and Technology,* vol. 13, pp. 308-316, 2002.

[8] E. Ebert, A. Kleinwächter, R. Kostbade, and N. Damaschke, "Interferometric Particle Imaging for cavitation nuclei characterization in cavitation tunnels and in the wake flow," presented at the 17h Int Symp on Applications of Laser Techniques to Fluid Mechanics, Lisbon, Portugal, 2014.

# A novel ultrasonic cavitation enhancer

**D Fernandez Rivas[1,2], B Verhaagen[2], Andres Galdamez Perez[3], Elena Castro-Hernandez[3], Ralph van Zwieten[1,2,4], Karin Schroen[4]**

[1]Mesoscale Chemical Systems, University of Twente, 7500AE Enschede, The Netherlands
[2]BuBclean, Institutenweg 25, 7521 PH Enschede, The Netherlands
[2]Mecanica de Fluidos, Universidad de Sevilla, C/ S. Fernando, 4, C.P. 41004-Sevilla, Spain
[4]Food Process Engineering, Wageningen University, 6700AA Wageningen, The Netherlands

E-mail: [1]d.fernandezrivas@utwente.nl

**Abstract.** We introduce a Cavitation Intensifying Bag as a versatile tool for acoustic cavitation control. The cavitation activity is spatially controlled by the modification of the inner surface of the bag with patterned pits of microscopic dimensions. We report on different measurements such as the transmission of ultrasound, temperature increase inside the bag during sonication. Several applications of interest to other scientific activities are also demonstrated.

## 1. Introduction

Ultrasonic cleaning relies on the formation and collapse of bubbles in the vicinity of the contaminated substrate. Recently, the different physical cleaning mechanisms involved when bubbles collapse have been discussed in a Special Issue of Ultrasonics Sonochemistry [1], but cavitation in general remains a stochastic process [2]. Control over the cleaning process can be obtained by defining the location and instant that the bubbles appear, influencing the cavitation nucleation process. Artificially created microscopic pits act as source of ultrasonic cavitation nuclei [3], and can be used for cleaning [4]. We introduce a three-dimensional way to bring cavitation near an arbitrary object in a controlled way, with the potential to enhance processes such as cleaning, sonochemistry [5] and emulsification [6].

### *1.1. General description of the Cavitation Intensifying Bag*

Plastic containers (hereafter named Cavitation Intensifying Bag or CIB) were modified to include pits in its inner surfaces. These CIB's were designed and patented by two of the authors, and commercialised as BuBble Bag by BuBclean. When a liquid is poured inside (e.g. water, ethanol, acetone, surfactant solutions), a gas bubble can be trapped in the pits depending on the gas content of the liquid. Upon exposure to ultrasound, microbubbles are generated at the predefined locations and in large quantities. Using CIB reduces the amount of chemicals used for cleaning: typically 10–100 mL instead of 1–100 L for the entire ultrasonic bath, or other liquid containers (e.g. metal, glass). The bag furthermore assures that the contaminants removed during cleaning do not cross-contaminate other objects cleaned inside the ultrasonic bath, or contaminate the ultrasonic bath content itself. The thin plastic material transmits ultrasound better than metal or glass containers, which have a high reflection coefficient.

Journal of Physics: Conference Series **656** (2015) 012112 doi:10.1088/1742-6596/656/1/012112

## 2. Characterisation of the CIB

### *2.1. Thermal conduction through the CIB wall*

An ultrasonic bath (DT510H-RC, Bandelin) operating at 35 kHz was pre-heated to 53.0±1°C; three CIBs were filled with tap water at 21.5±1°C. One T-type thermocouple (Z2-T-2-MP, Labfacility) was immersed in the ultrasonic bath, additional thermocouples were fixed inside the CIBs. The CIBs were immersed in the bath and the thermocouple values were recorded during 4 minutes using a data logger (TC08, Pico Technology). With the ultrasound switched on, the liquid inside reaches the temperature of the ultrasonic bath in 2 minutes, whereas without ultrasound, the temperature was 5°C below the temperature of the ultrasonic bath at the end of the experiment (4 minutes). After 2 minutes, the temperature inside the CIB was 44°C.

### *2.2. Acoustic transmission through the CIB wall*

The acoustic attenuation of the CIB walls was determined using a needle hydrophone (HNR-500, Onda) at one position. The hydrophone (not calibrated for 35 kHz) was connected to an oscilloscope (DPO 4034, Tektronix) recording 0.1 s of acoustic data at a rate of 12.5 MS/s and allowed for estimating the relative attenuation by the plastic wall material. The geometry and acoustic boundary conditions of the ultrasonic bath were recreated digitally in CFD software (COMSOL Multiphysics v5.0, COMSOL). The acoustic driving was modeled by 4 circular sources, equally distributed around the bottom of the bath, and driven at an amplitude of 0.34 $\mu$m, and a nominal frequency of 35 kHz changed sinusoidally with ±1 kHz every 1 ms (i.e. 100 Hz)[7]. A CIB was modelled as a shell, located inside the bath above one of the transducers, with standard material properties for the plastic material (PP). The CIB and the bath itself were filled with water at 30°C. The maximum element size was 8.6 mm, with finer mesh near the CIB. Each ultrasound period was discretised with 30 time steps and the total simulation time was 1.2 ms. The rms-value of the pressure at a frequency of 35 kHz was attenuated with 7.7%. The numerical simulations predicted that the pressure in the center of the CIB was 0.7% lower than the pressure at the same location above one of the other transducers. No phase difference was observed between the two evaluation points in the numerical model.

### *2.3. Visualisation of the cavitation activity*

One transducer from an ultrasonic bath (P2000, Qtec; frequency: 42.7 kHz) was mounted inside a transparent bath (10 L of tap water at 21°C). A CIB was mounted in front of the transducer at a distance of approximately 5 cm. A high-speed camera (SA7, Photron) with a 50 mm lens (Zeiss) was positioned outside the tank to record the cavitation activity at a speed of 1000 fps, and light was provided using a flood light (Hella). For sonochemiluminescence imaging, a CIB was filled with a 2 mM Luminol and NaOH solution at pH 10 and fixed horizontally inside an ultrasonic bath (DT510H-RC, Bandelin) filled with tap water at 21°C. The ultrasonic bath was located inside a light-tight box, equipped with a remote-controlled photo camera (D300, Nikon; aperture f/1.8, shutter time 30s, ISO Hi1.0) and 50 mm lens (Nikkor). The sonochemiluminescence was recorded during 30 s of ultrasound exposure to the CIB. The high-speed recordings showed clusters of bubble originating from the pits inside the CIB (Figure 1a). Some of these bubble clusters were interacting and formed combined bubble clusters, similar to those reported elsewhere [4]. The long-exposure photographs showed sonochemiluminescence signal coming from within the CIB only, in a distinct pattern corresponding to the pits inside the CIB and interacting bubble clouds (Figure 1b).

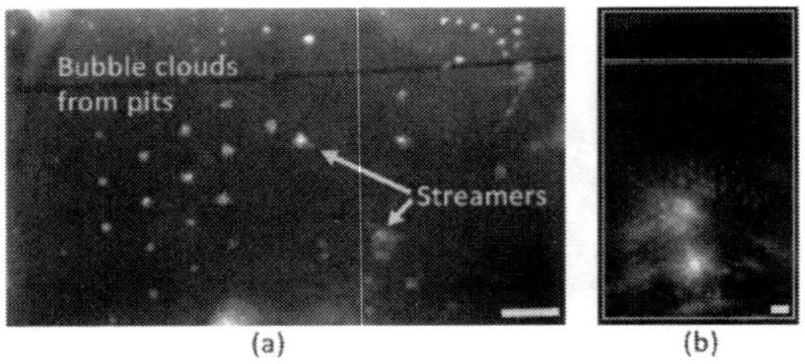


**Figure 1.** (a) High-speed image showing bubble streamers (white spots indicated with yellow arrows) originating from the regular pattern of pits inside the CIB. (b) Blue zones recorded as a result of the chemical reaction of radicals produced by the cavitation bubbles and the Luminol solution in the CIB (sonochemiluminescence). The white lines indicate the outer dimensions of the CIB. The yellow bars represent 5 mm.

## 3. Demonstration of functioning and potential applications

### *3.1. Cleaning indicators and 3D printed parts*

A Level 3 cleaning indicator (gke) was fixed inside a CIB to visualise the ultrasonic cleaning activity inside the ultrasonic bath filled with tap water at 21°C. The coloured side of the indicator, facing away from the transducer, was imaged with a camera (D5100, Nikon), and the removal of the cleaning indicator dye was monitored during ultrasound exposure. The dye removal occurred in a very specific 'honeycomb' pattern, corresponding to the locations of the pits in the opposite wall of the CIB. A Cleaning Challenge Device (CCD) [8] was printed in two-fold using a DLP 3D printer (S30L, RapidShape). The CCD represents a challenging 3D printed model from which support material needs to be removed. One CCD was cleaned using the standard protocol prescribed by the printer manufacturer (3 minutes in isopropylalcohol in the ultrasonic bath, followed by blowing with pressurised air and another 3 minutes of ultrasound). The second CCD was cleaned using the CIB filled with 50 mL of isopropylalcohol and placed for 1 minute in an ultrasonic bath filled with demiwater. It was found that the CCD treated in the CIB was cleaned in only 14% of the time needed for the first CCD.

### *3.2. Emulsion preparation*

Hexadecane emulsions were made by pipetting 2.25 mL of hexadecane into a CIB containing 10 mL of 1% SDS aqueous solution to obtain oil hexadecane concentrations of about 15%. The CIB was placed for 60 minutes inside an ultrasonic bath (P60H, Elma; frequency: 80 kHz, amplitude: 100%) filled with 1/3 tap water and 2/3 demi water at 21°C. The resulting emulsion droplet size distribution was measured using a particle sizer (Mastersizer Hydro 2000 SM, Malvern). Figure 2 shows the resulting droplet size distribution at three different ultrasound treatment times when using CIB and one blanc measurement with a non-treated bag. When comparing the CIBs with regular bags, the former showed a smaller droplet diameter than the latter; typically after one hour treatment the diameter is a factor of ten smaller. It can be seen that the CIB helps reducing the particle size over time, to a mean diameter of $d_{32} = 0.26$ $\mu$m. Compared to droplet mean diameters reported in literature [9], we find that the use of the CIB has resulted in a mean diameter that can be compared with classic emulsification techniques such as a high pressure homogeniser. The associated energy consumption based on the electrical power consumption (640W) is estimated to be on the order of $7 \cdot 10^3$ J/m$^3$.

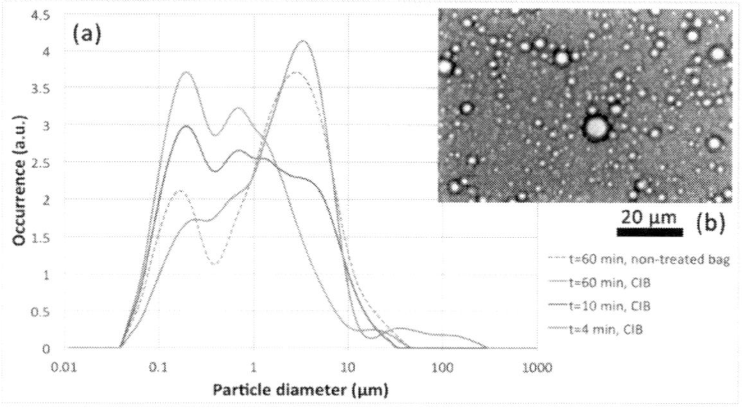


**Figure 2.** (a) Droplet size distribution at 4, 10 and 60 minutes ultrasound treatment times inside the CIB, and after 60 minutes in an untreated bag (red dashed line). (b) Microscope image of the obtained emulsion after 60 minutes in the CIB.

## 4. Discussion & Conclusions

Using low- and high-speed imaging techniques, we showed that cavitation activity increases significantly inside the CIB. The liquid temperature inside the CIB reaches the bath temperature in less than 2 min, and the CIB wall material does not attenuate considerably the ultrasound transmission. Numerical simulations predicted an ultrasound attenuation of 0.7%, whereas experiments showed an attenuation of 7.7 %; vibrations of the CIB may have led to a changed acoustic field and therefore a larger reduction in pressure. This difference in attenuation values is important as it refers to the loss in ultrasound pressure and bubble creation efficiency. From our simulations we found that the orientation of the bag does not cause such a large attenuation. The bubbles generated can perform surface modifications on different materials (cleaning indicator dye and removing 3D printed parts support materials with complex geometries). Sonochemical reactions can be performed inside the CIB, as illustrated by sonochemiluminescence imaging. We also presented promising preliminary results for using the CIB as emulsification device. We can conclude that the CIB is a versatile tool for performing acoustic cavitation studies, in which the cavitation is spatially controlled by the use of small pits inside the bag.

## 5. References

[1] Rivas DF, Verhaagen B, editors. Special Issue: Cleaning with bubbles. Ultrasonics Sonochemistry; 2015.
[2] Brennen CE. Cavitation and Bubble Dynamics. pp. 28-36, 84-95. Oxford, United Kingdom: Oxford University Press; 1995.
[3] Rivas DF, Prosperetti A, Zijlstra AG, Lohse D, Gardeniers JGE. Efficient Sonochemistry through Microbubbles Generated with Micromachined Surfaces. Angewandte Chemie International Edition. 2010;49(50):9699–9701.
[4] Rivas DF, Verhaagen B, Seddon J, Zijlstra AG, Jiang LM, Van der Sluis LWM, et al. Localized removal of layers of metal, polymer or biomaterial by ultrasound cavitation microbubbles. Biomicrofluidics. 2012;6:034114.
[5] Rivas DF, Cintas P, Gardeniers JGE. Merging microfluidics and sonochemistry: Towards greener and more efficient micro-sono-reactors. Chemical Communications. 2012;48(89):10935–10947.
[6] Berton-Carabin CC, Schroen K. Pickering Emulsions for Food Applications: Background, Trends and Challenges. Annual Review of Food Science and Technology. 2015;6(1).
[7] Bandelin. The ultrasonic peak output in Sonorex ultrasonic baths; 2012.
[8] Verhaagen B, Zanderink T, Rivas DF. Ultrasonic cleaning of 3D printed objects and Cleaning Challenge Devices. Applied Acoustics. 2015;p. – in press.
[9] Nazir A. Premix emulsification systems. University of Wageningen, The Netherlands; 2013.

# Air release measurements of V-oil 1404 downstream of a micro orifice at choked flow conditions

**H-A Freudigmann**[1], **U Iben**[1], **P F Pelz**[2]
[1]Robert Bosch GmbH, Robert-Bosch-Campus 1, D-71272 Renningen, Germany
[2]Chair of Fluid Systems, TU Darmstadt, D-64289 Darmstadt, Germany

E-mail: hans-arndt.freudigmann@de.bosch.com, uwe.iben@de.bosch.com, peter.pelz@fst.tu-darmstadt.de

**Abstract.** This study presents measurements on air release of V-oil 1404 in the back flow of a micro orifice at choked flow conditions using a shadowgraph imaging method. The released air was determined at three positions downstream of the orifice for different pressure conditions. It was found that more than 23 % of the initially dissolved air is released and appears downstream of the orifice in the form of bubbles.

## 1. Introduction

Dissolved air in hydraulic liquids often compromises the functionality of hydraulic systems as the air can be released in form of small bubbles. The resulting two phase system has an increased compressibility and a decreased density and viscosity. In hydraulic systems this can lead to reduced system stiffness, vibrations, noise, fluctuation of the mass flow rate and aging of the liquid due to oxidation and generation of carbon residues caused by micro-diesel effects [1–3]. As typical mineral-oil based hydraulic liquids can dissolve a high amount of air (7-12 Vol.-%, [1, 4]), degassing caused by lowering the static pressure of the liquid below the saturation pressure can lead to a considerable amount of free air. Degassing due to pressure reduction can therefore occur in the suction section of pumps [5], in hydraulic resistances such as valves or throttles [6] or in volume elements such as pipelines during pressure transients [7]. Moreover, experiments [4, 8] and theoretical considerations [9] demonstrate that hydrodynamic cavitation significantly affects the degassing dynamics of liquids. This has to be considered during the design process for robust hydraulic systems using appropriate models. In order to develop and validate these models experimental data are required.

To the authors knowledge and beside the previously mentioned work, this is the first systematic study investigating quantitatively the linkage of hydrodynamic cavitation and air release. In this work close attention is paid to the bubbly flow downstream of a cylindrical orifice at choked flow conditions. The flow is investigated in a channel of rectangular cross section applying a shadowgraph imaging method at three positions. The recorded images were processed using a developed bubble segmentation algorithm.

## 2. Materials and Methods

Shell V-oil 1404 (ISO 4113) is widely used as a testing oil for diesel injection systems and its liquid properties are well-known [10, 11]. Figure 1a) illustrates the experimental setup. A stirrer

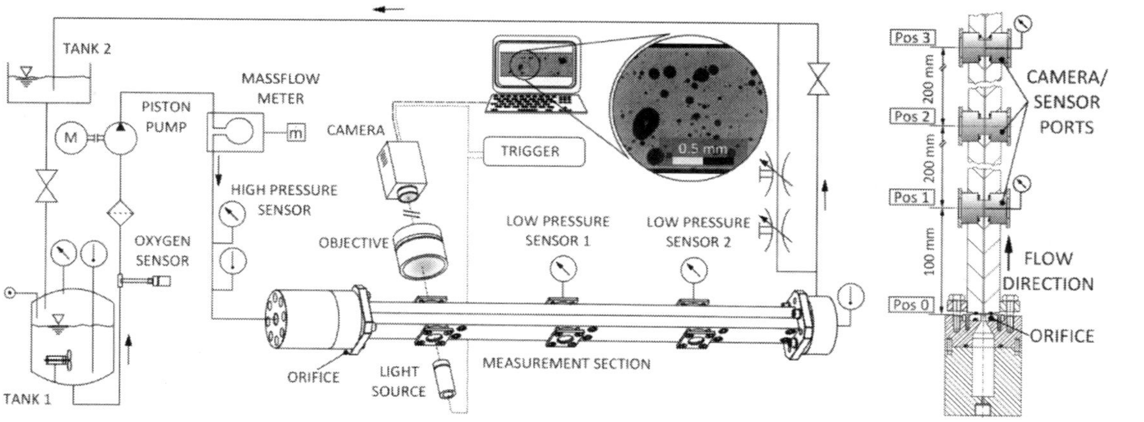


**(a)** Test rig scheme. **(b)** Measurement section.

**Figure 1:** Schematics of the experimental setup.

in the supply tank "TANK 1" ensures that the liquid is homogeneously saturated with air at atmospheric pressure conditions. During the experiment the gas volume above the liquid in "TANK 1" is increased by about 0.4 bar above atmospheric pressure to avoid degassing in the suction line of the speed-variable piston pump. Due to the long time-scales for the diffusion process, the increased gas pressure has virtually no influence on the initial concentration of the dissolved gas. An oxygen sensor (Hamilton, VisiFerm DO Arc 120) upstream of the pump is used for measuring the partial oxygen pressure allowing for monitor the saturation of the liquid with air. Rearward of the piston pump a mass flow meter (Siemens F C MASS 2100 DI 1.5), a high pressure sensor (Kistler 4065, 200 bar) and a temperature sensor are installed. The temperature of the liquid in "TANK 1" is between 23 °C and 25 °C.

Figure 1b) illustrates the measurement section in detail. The liquid passes a cylindrical micro orifice with 0.2 mm in diameter and 2 mm in length at "Pos 0". Rearward of the orifice the flow is released into a square flow channel with 1 mm × 1 mm in size. In this channel three ports are installed where the pressure is measured at two positions (Althen XPC10 35 bar) and the flow is observed with a camera at a third position. The positions of the pressure sensors and the camera are interchangeable. Two throttling valves are placed in series downstream of the square flow channel which enable an adjustment of the outlet pressure rearward of the orifice. Liquid which has passed the measurement section is collected in "TANK 2". Sensor signals are gathered by NI DAQ data acquisition cards (National Instruments) with a sampling rate of 1 kHz.

For observing the flow in the channel a shadowgraph technique is used with a spark flash (Impulsphysik GmbH, Nanolite EGG FX 800) as a light source, an objective (B. Halle, f=100 mm, k=2) and a camera (PCO 2000) similar to the setup described in [4]. The camera and the flash are synchronized; the applied exposure rate is 2 Hz with an exposure time of 1 µs of the camera. The scaling factor of the optical setup is 1.6 µm per pixel with a magnification of 4.6. For each operation point 100 images were evaluated.

Depending on the operation conditions, bubbles can be observed in the flow downstream of the orifice, cf. Figure 1a). In order to analyze the bubbles in the obtained shadowgraph images a segmentation algorithm was developed. The algorithm detects spherical bubbles using the Hough-Transformation and non-spherical bubbles using a histogram based binarization, cf. Figure 2a). The radii of the spherical bubbles are determined at the half irradiance level of the intensity profiles [12]. The radii of spherical out-of-focus bubbles are corrected using an empirical correction function applying the contrast of the bubble leading to sizing errors less

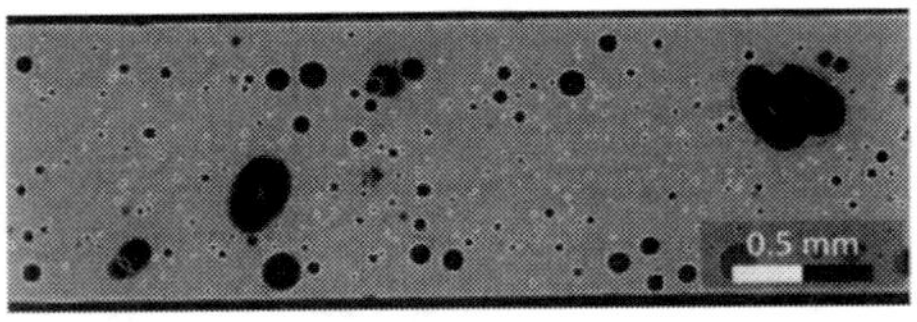


**(a)** Result of the bubble recognition.

**(b)** Cavitation plume.

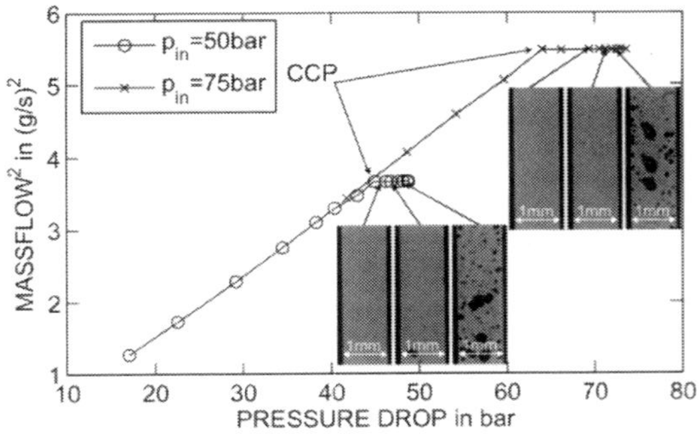


**(c)** Mass flow behavior of the cylindrical orifice.

**Figure 2:** a) Green circles indicate bubbles with an evaluated intensity profile. Due to overlaps, this was not possible for the red marked bubbles. Blue lines illustrate the best fitting ellipses for non-spherical bubbles. b) Exemplary image of the cavitation plume representing the average of 100 shadowgraph images at "Pos 0" recorded in a transparent channel located directly at the orifice exit. c) Mass flow behavior of the orifice and exemplary images of the bubbley flow for two fixed inlet pressures $p_{in}$. The outlet pressure was measured 101 mm downstream of the orifice.

than 10 %. The volume of a non-spherical bubble is approximated as two spheroids based on the best fitting ellipse. Errors due to mutually concealing bubbles were not taken into account. The detection algorithm was validated with manually evaluated bubble images.

## 3. Results

As the present study focuses on the appearance of bubbles in the backflow of an orifice at choked flow conditions, the mass flow characteristic of the used orifice is presented in Figure 2c). It is obtained using a pressure resistant flow channel with a cross-section area identical to the channel shown in Figure 1b). As expected, the square of the mass flow rate is proportional to the pressure drop over the orifice for operating points with no cavitation occurring. For a critical value of the pressure drop (Critical Cavitation Point CCP) significant cavitation occurs, cf. Figure 2b) and the mass flow is independent of the outlet pressure. For the region beyond the CCP bubbles can be observed in the backflow of the orifice e.g. at "Pos 1" as shown in Figure 2c). Images of the bubbly flow were recorded at three positions ("Pos 1"-"Pos 3") for two fixed inlet pressures, several outlet pressures and constant mass flow rates due to choked flow conditions. The observed bubbles downstream of the orifice tend to be spherical for smaller bubbles and non-spherical for larger bubbles. Bubble diameters range from around 7 µm up to 0.5 mm (spherical equivalent diameter). The bubbles are assumed to be completely filled with air due to the high liquid pressure in the channel (>1.3 bar) compared to the vapor pressure of the liquid (<10 Pa, [11]).

The mass fraction of the initially dissolved air $c_g$ in the suction line of the piston pump is determined by measuring the partial pressure of oxygen in the liquid, cf. Figure 1a). The ratio of the partial pressures of oxygen to air in the liquid is assumed to equal the ratio of the partial pressures of oxygen to air in the atmosphere. Hence, the pressure of dissolved air in the liquid can be determined. Knowing the Ostwald-Coefficient of the liquid for air ($L_V$=12 Vol.-% at 25 °C, [4]) the concentration of initially dissolved air (0.169 mass-‰± 1 %) is determined by employing Henry's law. To determine the amount of gas contained by the bubbles according to [4] the obtained images are analyzed using the presented bubble segmentation algorithm. Figures 3a) and 3b) show the mass fraction of the degassed air $\xi := \frac{m_g}{m_l \cdot c_g} \cdot 100\%$ for the stated conditions at three distances downstream of the orifice. $\xi$ is defined as the ratio of the degassed air mass $m_g$ to the initially dissolved air mass $m_l$ in a liquid element.

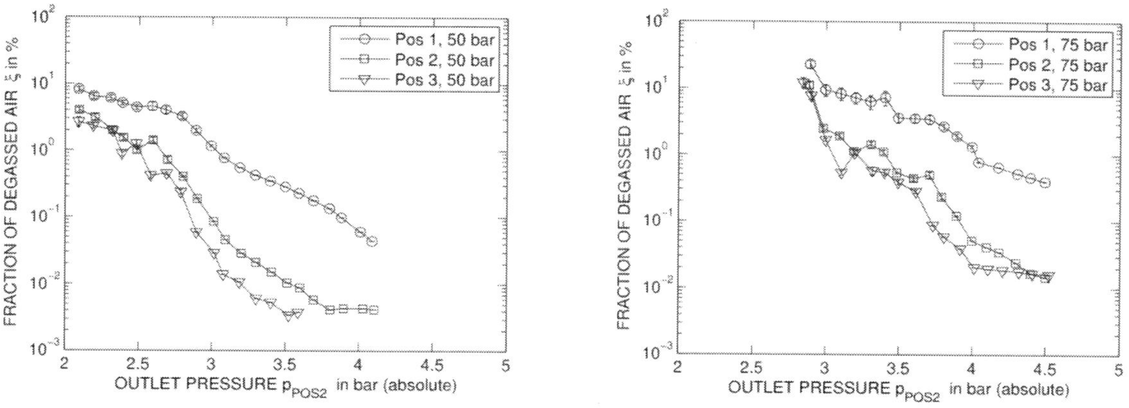


(a) $p_{in} = 50\,\text{bar}$ (b) $p_{in} = 75\,\text{bar}$

**Figure 3:** Mass fraction of released air $\xi$ for two fixed inlet pressures and several outlet pressures. Non-spherical bubbles are treated as spheroids by rotating the best fitting ellipsis of the shadowgraph image about its major or minor axis. The differences in the bubble volumes for each rotating axis are indicated by the error bars.

According to the results for a different orifice presented by Iben et al. [4] the release of air at choked flow conditions is very sensitive to the outlet pressure showing a nearly exponential relationship. The measurements at different positions for similar operating points show that the amount of free air is less at "Pos 2" than at "Pos 1". This is due to the reabsorption of the released air in the undersaturated liquid. However, the differences between "Pos 2" and "Pos 3" in the amount of free air are less significant.

## 4. Conclusion

The presented results show how hydrodynamic cavitation in an orifice is associated with a substantial increase of free air downstream of a cavitating regime. In addition the results reveal that a significant amount of the released air is reabsorbed into the liquid while traveling downstream and therefore extend the findings of Iben et al. The results can serve as validation data for modeling approaches of air release phenomena in hydraulics.

## References

[1] Bock W, Braun J, Puhl N and Heinemann H 2010 Air release properties of hydraulic fluids *IFK, Internationals Fluidtechnisches Kolloquium* vol 7 pp 327–340
[2] Ruan J and Burton R 2006 Bulk modulus of air content oil in a hydraulic cylinder *ASME 2006 International Mechanical Engineering Congress and Exposition* (American Society of Mechanical Engineers) pp 259–269
[3] Backé W and Lipphardt P 1976 *CI Mech. Eng. C97/76*
[4] Iben U, Wolf F, Freudigmann H A, Fröhlich J and Heller W 2015 *Experiments in Fluids* **56** 1–10
[5] Zhou J, Vacca A and Casoli P 2014 *Simulation Modelling Practice and Theory* **45** 35–49
[6] Lichtarowicz A and Pearce I D 1974 Cavitation and aeration effects in long orifices *Cavitation Conference, Institution Mechanical Engineers, Edinburgh* vol 19 p 74
[7] Wiggert D and Sundquist M 1979 *Journal of Fluids Engineering* **101** 79–86
[8] Sato K, Liu Z and Brennen C 1993 The micro-bubble distribution in the wake of a cavitating circular cylinder *ASME* 153 (American Society of Mechanical Engineers) pp 75–80
[9] Watanabe M and Prosparetti A 1994 The effect of gas diffusion on the nuclei population downstream of a cavitation zone vol 190 (ASME) pp 211–211
[10] Ndiaye E H I, Bazile J P, Nasri D, Boned C and Daridon J L 2012 *Fuel* **98** 288–294
[11] Chorążewski M, Dergal F, Sawaya T, Mokbel I, Grolier J P E and Jose J 2013 *Fuel* **105** 440–450
[12] Bongiovanni C, Chevaillier J P and Fabre J 1997 *Experiments in Fluids* **23** 209–216

# Laser diagnostics for characterization of sprays formed by a collapsing non-equilibrium bubble

**Y S Kannan, S Balusamy and B Karri**

Indian Institute of Technology, Hyderabad
Ordnance Factory Estate, Yeddumailaram, Telangana, INDIA 502205

E-mail: badarinath@iith.ac.in

**Abstract.** In this paper, we investigate the use of laser diagnostic tools for in-plane imaging of bubble induced spray using a laser sheet and Mie scattering technique. A perspex plate of thickness 10 mm with a hole of diameter 1 mm in the center is placed in the middle of a glass tank filled with water such that the top surface of the plate coincides with the water surface. A bubble is created just below the hole using a low-voltage spark circuit such that it expands against the hole. This leads to the formation of two jets which impact leading to a spray and break-up into droplets. The spray evolution is observed using a laser sheet directed in a plane through the center of the hole. The illuminated plane is imaged using a high-speed camera based on the Mie scattering from glass beads suspended in the liquid. Results show that Mie scattering technique has potential in studying bubble-induced sprays with applications such as in fuel sprays, drug-delivery etc, and also for validation of numerical codes. We present results from our ongoing experiments in this paper.

## 1. Introduction

An oscillating non-equilibrium bubble in the presence of a non-uniform pressure field can lead to the formation of a high-speed jet in the bubble during its collapse. This can occur in the presence of a free surface [1-2], a rigid boundary [3-6] or due to the interaction of the bubble with a shock-wave [7-9]. The jet while thought of as a mechanism for cavitation damage, can also be used to actuate flows at small scales. For example, the jet in a collapsing bubble has been shown to lead to pumping in both water [10], and viscous fluids [11] when there is a hole in the boundary at the location of jet impact. Recent work [12-14] has further shown that if the bubble expands nearly hemi-spherically against the hole in a plate positioned on the surface of a liquid, it can lead to the formation of two jets - an initial slow jet during the bubble expansion, and a faster second jet during the collapse of the bubble towards the wall. The interaction of the two jets can lead to the formation of fine sprays and droplets [12-13] and also high-speed jets at the micrometer length scale with potential application in needle-free drug delivery [14].

Study of the formation of fine spray is particularly relevant for studies on fuel atomization for better combustion and mixing. The spray characteristics such as droplet size distribution and jet/spray velocity give particularly useful information. One approach for such characterization is the Mie scattering technique where laser is used to illuminate a plane and the phenomena occurring in the plane are visualized using glass-beads (that scatter light) and collected by a camera through a narrow band-pass filter. In bubble related studies, the scattering of light by bubbles itself can give information on the bubble size [15-16]. The technique has also been used to study the droplet size in aerated sprays [17] and for collapse of a sonoluminescing bubble [18].

In this paper, we explore the possibility of using Mie scatter technique to characterize the spray formed due to bubble-induced jets. To create the spray, we use a similar experimental set-up as Karri et al. [12], but with an improved circuit for creating the spark bubble as presented by Goh et al. [19] which gives repeatable bubble sizes. A planar sheet of light using a laser is used to illuminate the plane of the hole. Micro-metric glass bead particles dispersed in the water scatter the incident radiation and these are then imaged using a high-speed camera. Preliminary results show that this technique can be applied to characterize the spray and thus give further insight into the potential of bubble-induced sprays in various applications.

## 2. Experimental set-up and procedure

The experimental set-up used is shown in Figure 1 (a). It comprises of a glass tank (250 mm × 250 mm × 250 mm) which is filled with water to nearly 75% volume. A perspex plate (80 mm × 80 mm × 10 mm) with a through-hole of 1 mm diameter in the center of the plate is positioned such that the top of the plate coincides with the free-surface. A low-voltage spark circuit [19] is replicated here and used to generate a bubble just below the hole. The circuit uses the principle of charging capacitors to a voltage of 60 V through a DC power supply, and then allowing them to be sparked through a pair of electrodes which are placed in the water. The charging, discharging and sparking circuits are connected to relays and a MOSFET, and the different parts of the circuit are controlled using an NI-DAQ (model USB-6008) through a LAB-view program. The bubble expansion and collapse below the plate leads to two separate jets which are then visualized using Mie scattering technique.

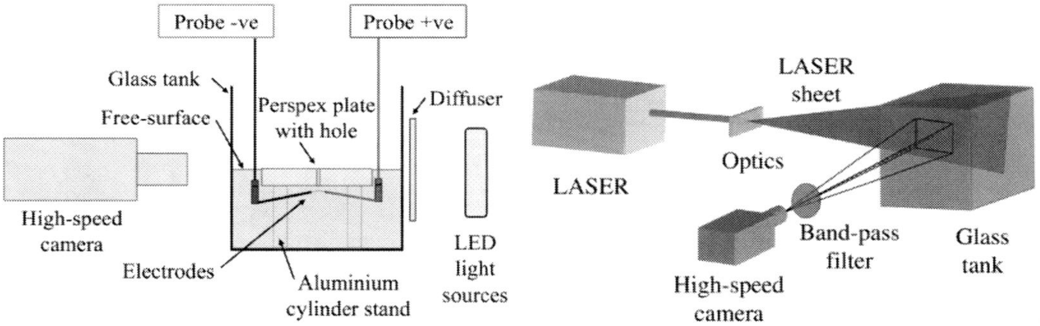


(a) Line-of-sight imaging (b) High-speed Mie scatter imaging

**Figure 1.** Experimental set-up.

We first tested the repeatability of the bubble sizes obtained using the circuit before studying the bubble under the hole. For these experiments, the sparking electrodes were placed in the tank such that they are at least 10 diameter lengths away from the nearest wall. The electrode length used were L = 10 mm. The capacitors were short-circuited through the electrodes to create a bubble and the bubble growth and collapse was recorded using the high-speed camera with diffused back-lit illumination. For the experiments involving the bubble near a holed plate, we used planar imaging using a laser. Here a continuous laser beam (532 nm wavelength) is projected through a cylindrical lens (f = -15 mm) to form a sheet of 1 mm thickness which passes through the central plane of the hole in the plate. The sheet is in a direction perpendicular to the camera as shown in the Figure 1 (b). Micro-metric glass beads (TSI, 8-12 $\mu$m diameter) are dispersed in the water as seeding particles to generate the Mie scatter. As the fluid is actuated due to the bubble growth and collapse through the hole, the glass beads also move along with the water (in the jets) and reflect the laser radiation. The Mie scatter light is captured by the camera through a narrow band-pass filter (wavelength: 532 nm ± 10 nm) to indicate the movement of the jet. Image processing (inversion) is then done on the images using MATLAB to get details of the flow.

## 3. Results and Discussion

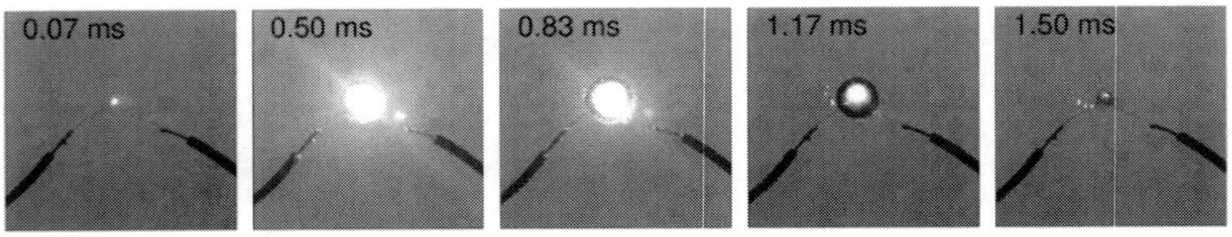


**Figure 2.** Evolution of a typical bubble using the spark circuit.

Figures 2 and 3 shows details about the bubble created by the spark-circuit. Figure 2 shows the evolution of a free-field bubble with time. The electrodes are short-circuited at t = 0.00 ms (when the spark begins), and the spark is shown here at t = 0.07 ms. The bubble grows, reaching a maximum size (around 8 mm diameter) around t = 0.83 ms, and collapses by about t = 1.50 ms. As shown in Figure 3, the maximum size obtained is slightly smaller than those by Goh et al.[19] possibly due to differences in the electrode material and wires used. The consistency of bubble sizes is necessary to get useful information from the Mie scatter technique in characterizing the spray.

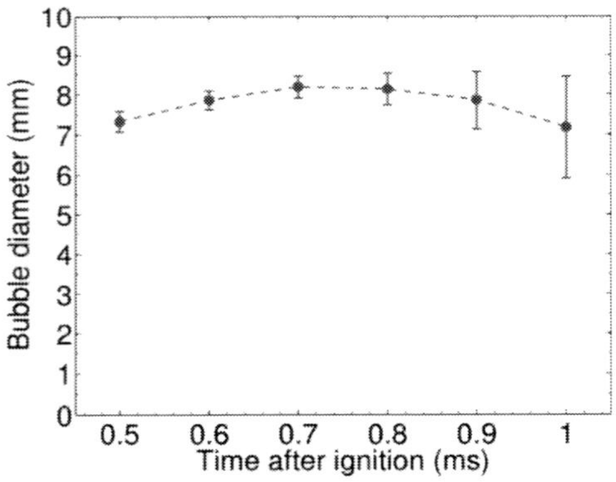


**Figure 3.** The variation of the bubble radius with time as obtained from five different experiments on a free-field bubble. The error bars show ± standard deviation indicating good repeatability of the bubble-sizes particularly at its maximum size.

Figure 4 shows the images obtained by the Mie scatter technique of the bubble expanding and collapsing below the hole in the perspex plate. The times after ignition (in milliseconds) are shown on top of each image. The position of the plate and the hole is shown in all the 6 images by two rectangles with a narrow gap(hole) in the middle. The bubble can be seen expanding at t = 0.24 ms. At t = 1.22 ms, the bubble has nearly collapsed and a faint first jet can be visualized on top of the plate. The next snapshot at t = 2.07 ms shows the situation after the second jet due to bubble collapse has impacted on the first jet. The evolution of the jet and its break-up into sprays (from the sides of the jet) can be seen from t= 2.44 ms to 3.66 ms. The images in this sequence were taken at 41,000 fps with an exposure time of 23 $\mu s$. The last two frames show 2 consecutive images of the spray at t = 3.63 ms and 3.66 ms respectively which can be processed further using a Particle Image Velocimetry (PIV) software, say to obtain the velocity information. This forms part of our ongoing study.

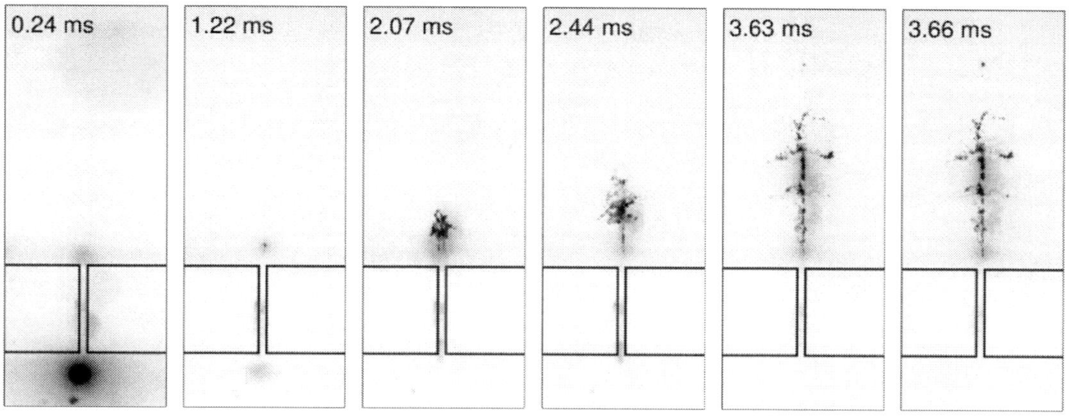


**Figure 4.** Mie scatter images of spray evolution formed by a collapsing non-equilibrium bubble. Images are normalized to the peak signal within the image sets.

## 4. Conclusions

An experimental set-up to use Mie scatter technique to visualize the sprays formed by bubble induced jets has been developed. The jets were produced by the collapse of a non-equilibrium oscillating bubble below a holed plate positioned on the surface of water, which leads to spray formation. A laser-based Mie scatter technique to visualize the sprays in the plane is demonstrated in this paper and will be expanded further to measure the flow-field data from the sprays.

## Acknowledgements

BK would like to thank his PhD supervisors Prof. Khoo Boo Cheong and Dr. Evert Klaseboer, and his colleague during PhD, Mr. Terence Goh. The authors also thank Ms. Jyothi, Mr. Mayur, Dr. Ashok Pandey, Dr. Ramji, Mr. Rakesh Kale and other IIT-Hyderabad students for the various ways in which they helped us in the experiment.

## References

[1] Chahine G L 1977 *Trans. ASME I J. Fluids Engg.* **99** 709
[2] Blake J R and Gibson D C 1981 *J. Fluid Mech.* **111** 124
[3] Kornfeld M and Suvorov L 1944 *J. Appl. Phys.* **15** 495
[4] Naudé C F and Ellis A T 1961 *Trans. ASME D J. Basic Engg.* **83** 648
[5] Benjamin T B and Ellis A T 1966 *Phil. Trans. Royal Soc. London* A **260(1110)** 221
[6] Plesset M S and Chapman R B 1971 *J. Fluid Mech.* **47** 283
[7] Delius M 2000 *Proc. 15th Int. Symp. on Nonlinear Acoustics* ISNA 15 **524** 23
[8] Zhu S L, Cocks F H, Preminger G M and Zhong P 2002 *Ultrasound Med. Bio.* **28** 661
[9] Sankin G N, Simmons W N, Zhu S L and Zhong P 2005 *Phys. Rev. Lett* **95** 034501
[10] Lew K S F, Klaseboer E and Khoo B C 2007 *Sens. Act.* A **133** 161
[11] Karri B, Pillai K S, Klaseboer E, Ohl S -W and Khoo B 2011 it Sens. Act. A **169** 151
[12] Karri B, Gonzalez-Avila S R, Loke Y C, O'Shea S J, Klaseboer E, Khoo B C and Ohl C -D 2012 *Phys. Rev.* E **85** 015303
[13] Karri B, Ohl S -W, Klaseboer E, Ohl C -D and Khoo B C 2012 *Phys. Rev.* E **86** 036309
[14] Avila S R G, Song C and Ohl C -D *J. Fluid Mech.* 2015 **767** 31
[15] He H, Li W, Zhang X, Xia M and Yang K 2012 *J. Quant. Spectros. Rad. Trans.* **113** 1467
[16] Zhan Y, Gu K, Yang F, Wu H and Yu M 2013 *Optik* **124** 4236
[17] Kristensson E, Berrocal E, Wellander R, Ritcher M, Alden M and Linne M 2011 *Proc. Comb. Inst.* **33** 855
[18] Weninger K R, Barber B P and Putterman S J 1997 *Phys. Rev. Lett* **78(9)** 1799
[19] Goh B H T, Oh Y D A, Klaseboer E, Ohl S -W and Khoo B C 2013 *Rev. Sci. Instru.* **84** 014705

# Visualization and image processing of spray structure under the effect of cavitation phenomenon

**M Ghorbani[(1)], G Alcan[(1)], D Yilmaz[(2)], M Unel[(1)], A Kosar[(1)*]**

[(1)]Mechatronics Engineering Program, Faculty of Engineering and Natural Sciences, Sabancı University, Orhanlı, Tuzla, Istanbul, Turkey
[(2)]Manufacturing Systems Program, Faculty of Engineering and Natural Sciences, Sabancı University, Orhanlı, Tuzla, Istanbul, Turkey

**Abstract.** This paper presents visualization and image processing of spray structure affected by cavitation bubbles and cavitating flow patterns. Experiments were conducted for a better understanding of cavitation and resulting flow regimes. Cavitation is generated with sudden pressure drop across a 4.5 *mm* long short micro-channel with an inner diameter of 152 *μm* connected to the setup using proper fittings. Generated cavitation bubbles and fluid flow patterns were observed by using a high speed camera. The spray structure was observed in four different segments and mainly the droplet evaluation in the lower segments for low upstream pressures was analyzed using several image processing techniques including contrast adjustments and morphological operators. Moreover, fluid flow regimes for different upstream pressures were investigated, and the flow patterns were analyzed in the separated regions of the spray.

## 1. Introduction

One of the most significant parameters affecting spray characteristics is cavitation with cavitation bubbles inside an orifice, which may extend to the outlet and impact the spray via bubble collapse. The cavitation bubbles and cavitating flow patterns were recently experimentally visualized in transparent nozzles [1-3]. Payri et al. [4] visualized existing cavitation bubbles at the outlet of the orifice using the special near-nozzle field visualization technique with the aid of a test rig pressurized with fuel. They attempted to investigate the effect of the nozzle geometry on cavitation patterns and the spray formation. They showed that cavitation inception and chocked flow conditions differed in terms of the pressure values, and also, the spray cone angle increased with cavitation. The flow regimes under cavitating flow conditions were studied by looking at the variations in the frequency of the cavitation shedding in the internal nozzles [5] and by analyzing the periodic shedding of the cavitation phenomenon using high speed cameras [6]. In this regard, De Giorgy et al. [7] studied flow regimes under the effect of cavitation phenomenon using a CCD camera and presented an analysis on pressure fluctuation frequency. Their experimental observations illustrated different flow regime maps (from inception to cavitating jets) under various working conditions.

In this study, the spray structure was visualized using a high speed imaging system in order to detect the details of the fluid flow regime and droplet discretization at the end of the spray. Acquired images were enhanced using several image processing algorithms such as contrast adjustment, morphological operators and connected component analysis (CCA). Enhanced images provide better visualization of bubbles and enable analysis of bubbles.

## 2. Experimental Setup and Procedure

### *2.1. Hydrodynamic Cavitation*

A schematic of the experimental setup is shown in Figure 1. The experimental setup consists of a high pressure pure nitrogen tank (Linde Gas, Gebze, Kocaeli), a liquid container (Swagelok, Erbusco BS, Italy), pressure sensors (Omega, USA.), fine control valves (Swagelok) at different locations, a micro

* To whom any correspondence should be addressed. Email: kosara@sabanciuniv.edu

filter (Swagelok), a turbinemeter (Omega, USA), a Phantom high speed camera (Phantom V320 high speed camera) with appropriate lenses, a workstation with visualization software (Phantom PCC 2.0 software), fittings (Swagelok), and microchannel (Small Parts, USA.) connected to the setup with appropriate fittings for cavitation formation. The tank was used as a container for de ionized (DI) water and high pressure nitrogen tank was connected to the liquid container in order to maintain high input pressure and propel the fluid to the microprobe. During the experiments, valves were fully opened and cavitation is generated with sudden pressure drop across a 4.5 *mm* long short microchannel with an inner diameter of 152 $\mu m$ connected to the setup using proper fittings.

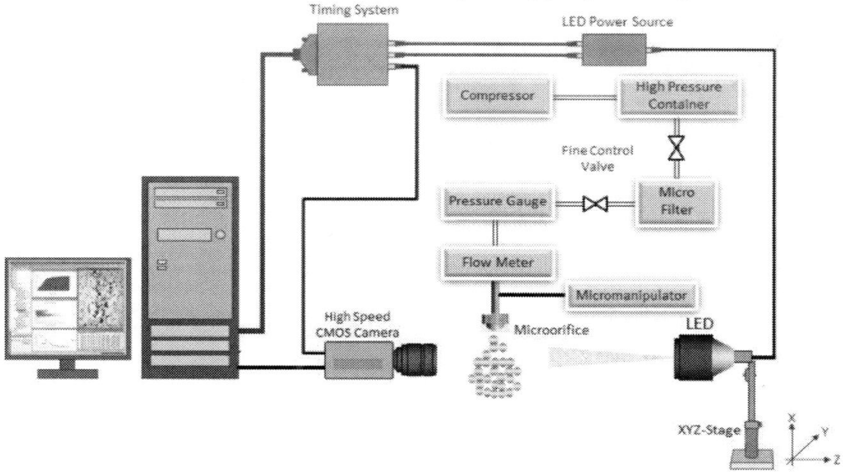


**Figure 1.** Experimental setup configuration with the orifice throat and exit area

*2.2. Experimental Procedure*

Experiments were conducted by adjusting the inlet pressures. Inlet pressures were varied from 10 *bars* to 120 *bars*, while outlet pressure at the microchannel was fixed to 1 *atm*. Volumetric flow rate was increased with the change in the inlet pressure. At a certain pressure level, cavitation inception was observed with the high speed camera. Bubbly cavitating flows generated in the microchannel was recorded at different regions (four segments, at distances of 0-3.5, 3.5-7.3, 7.3-11.1 and 11.1-14.9 *mm* from the microchannel outlet) downstream the microchannel outlet.

## 3. Results and Discussion

Emerging flow patterns from the micro orifice exit are displayed in Figure 2 in various segments for a constant upstream pressure of 30 *bars*. The variation in cavitation number is shown in Figure 3, which presents the effect of the upstream pressure on the spray shape. Cavitation number at lower injection pressures starts from 1.37 and decreases to 0.1108 at $P_i$= 110 *bars* ($P_i$ is upstream pressure) for segment 1, which indicates a highly vaporous cavitation pattern for higher upstream pressures. X/D indicates the ratio of the spray length with respect to the diameter of the microchannel. As shown in Figure 3, a highly vaporous bubbly flow is observed at cavitation numbers below 0.4, which corresponds to upstream pressures more than 80 *bars* in all segments and the first segment of the upstream pressure of 50 *bars*. Bubbly flow is visualized between cavitation numbers of approximately 0.8 and 0.4. Upstream pressures of 50 *bars* for all the segments and the first segment of 30 *bars* generate bubbly flow at the outlet of the microchannel. Jet cavitation is observed below the cavitation number of 1, which indicates the effect of generated cavitation bubbles inside the microchannel on the structure of the spray. Above this number, the spray contains just single liquid jet affected by the throat velocity. Above the cavitation number of 2, the droplets begin to appear and continuous flow changes to discrete droplet flow, which is shown in the third and fourth segments for the cases corresponding to upstream pressures below 20 *bars*.

Journal of Physics: Conference Series **656** (2015) 012115 doi:10.1088/1742-6596/656/1/012115

Before starting with the analysis of droplets in these segments, images need to be processed and purified from the noise to detect droplets in a correct way. Since high speed CMOS camera provides 8-bit grey-scale images, droplets could not be distinguished from background easily due to shadows, noise and undesired particles in unprocessed original images (Figure 4 (a)).

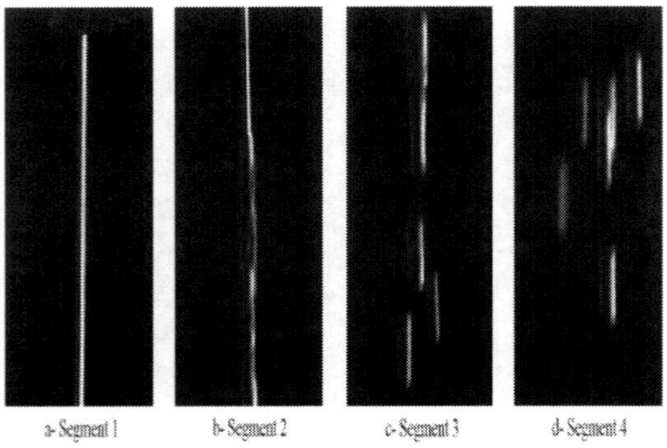


**Figure 2.** Bubble cloud emerging that is captured through increasing distance from the probe exit, respectively ($P_i$=30 *bars*)

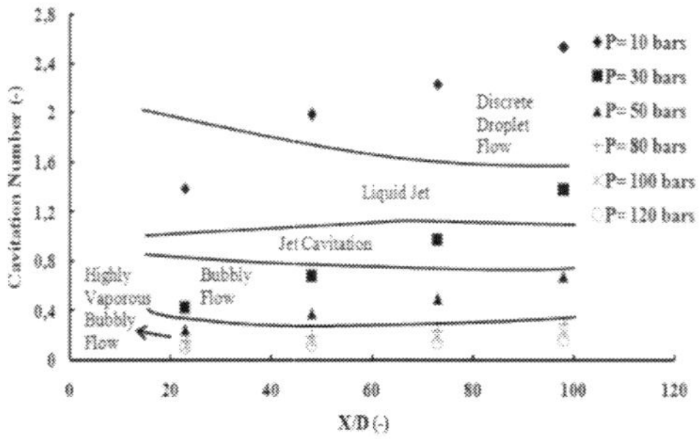


**Figure 3.** Cavitation number for different upstream pressure in various locations along the spray

Increasing the contrast of the images is the first step of visual processing. Image adjustment allows 1% of data to be saturated at low and high intensities of the image. This increases the contrast of the image and makes foreground distinguishable from background (Figure 4(b)). Histogram equalization enhances the contrast by transforming the values of the image, in the colormap of indexed image. It is generally applied to images with very close histogram values. To eliminate the noise which is generated by histogram equalization and to get smoother images, 'morphological opening' operation is applied (Figure 4(c)). Morphological opening is the dilation of the erosion of the image by a structuring element. In this case, structuring element was chosen as a 3x13 vertical rectangle because droplets were like rectangles, not circles. Morphological opening was the final step of noise removal process and purified images were thresholded by Otsu's method [8] to segment the droplet candidate regions. Finally binary images were formed to investigate the regions by Connected Component Analysis (CCA). CCA looks for the relationships between pixels, divides them into different groups and labels. In each group, pixels were located in the neighbourhood of each other. All labelled regions in processed binary images were candidates of droplets (Figure 4 (d)).

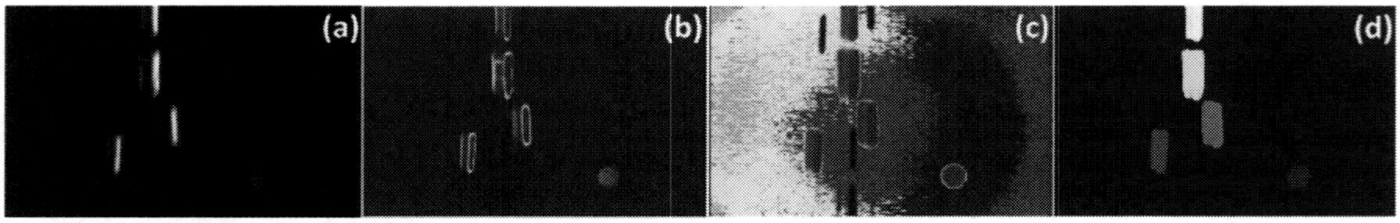


**Figure 4.** Image processing steps **(a)** Original raw image **(b)** Contrast adjustment **(c)** Histogram equalization and morphological opening **(d)** Connected component analysis ($P_i$=10 *bars*)

Until now, processing may not completely clean the noises. To eliminate the noise regions, region properties were utilized. Clearly, droplets have some considerable amount of pixel-wise area between 5000 and 15000. Thus, regions with less than a certain pixel-wise area can be treated as noise and were excluded. Apart from these areas, there were some circular regions which did not flow like other droplets. The shape of droplet was utilized in this case to get rid of circular noise (Figure 5). In geometry, eccentricity of a shape represents how close it is to a pure circle. This number is between 0 - 1 and it is close to 1 if the shape is pure circle. Obviously eccentricities of droplets and circular noises are far from each other.

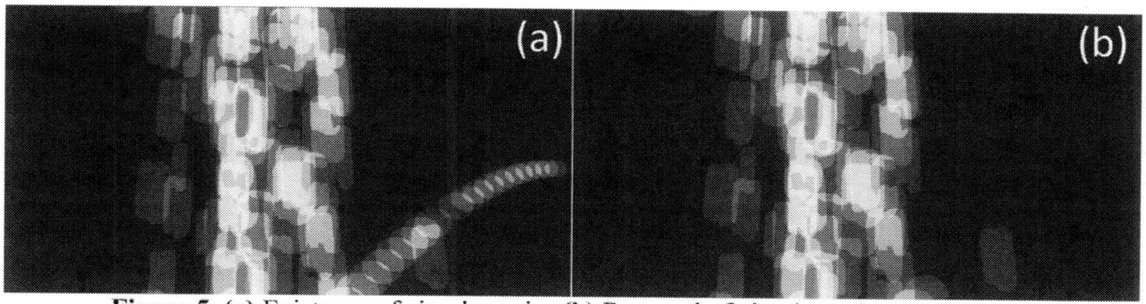


**Figure 5.** **(a)** Existence of circular noise **(b)** Removal of circular noise ($P_i$=10 *bars*)

In addition to discrete droplets visualization, scattered bubbles around jet spray were also determined by the same image processing method with a slight modification. After noise removal process steps were applied, CCA algorithm releases information about the properties of scattered bubbles. Figure 6 depicts the frequency of scattered bubbles in different segments of spray structure.

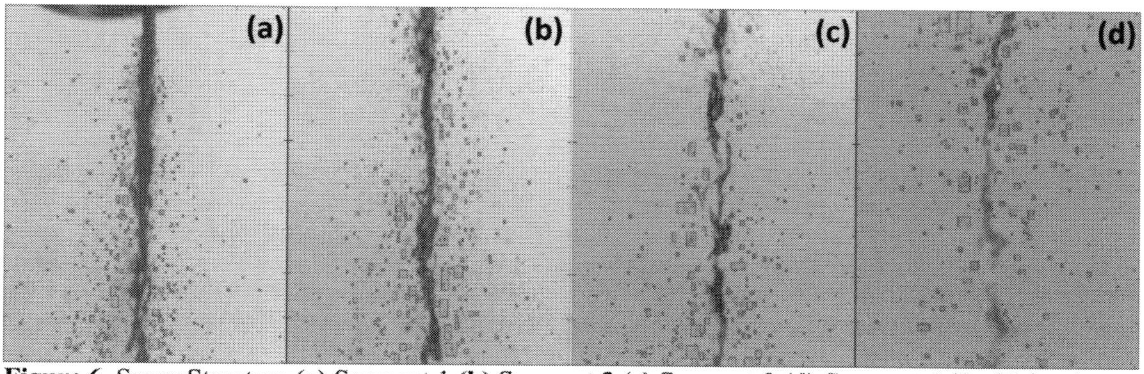


**Figure 6.** Spray Structure **(a)** Segment 1 **(b)** Segment 2 **(c)** Segment 3 **(d)** Segment 4 ($P_i$=80 *bars*)

**Conclusion**

Results show that there are different fluid flow regimes while the upstream pressure varies from 10 to 120 *bars*. Discrete droplet flow, liquid jet, cavitation jet, bubbly flow and highly vaporous bubbly flow were reported as the flow patterns for different cavitation numbers. Highly vaporous bubbly flow is observed at the outlet of the microchannel in the first segment due to very low cavitation number while the upstream pressure was more than 100 *bars*. Moreover, image processing methods were utilized to segment the discrete droplets accurately. Additionally, scattered bubbles around main flow were extracted.

## 4. References

[1] Andriotis A, Gavaises M, Arcoumanis C, 2008 *J. Fluid Mech.* **610** 195
[2] Suh H K, Lee CS, 2008 *I. J. Heat Fluid Flow* **29** 1001
[3] Mishra C, Peles Y, 2005 *Physics fluids* **17** 013601
[4] Payri R, Salvador F J, Gimeno J, De La Morena J, 2009 *I. J. Heat Fluid Flow* **30** 768
[5] Chandra B W, Collicott S H, 1999 In: 12th Annual Conference ILASS, Americas, 379
[6] Sato K, Saito Y, 2002 *JSME I. J. Ser. B Fluids Thermal* **45** 638
[7] De Giorgi M G, Ficarella A, Tarantino M, 2013 *I. J. Heat Fluid Flow* **39** 160
[8] Otsu N, 1975 *Automatica* **11** 23

# Multidimensional tensor array analysis of multiphase flow during a hydrodynamic ram event

**A Lingenfelter and D Liu**

2950 Hobson Way, Wright-Patterson AFB, OH 45433, USA

E-mail: andrew.lingenfelter@afit.edu

**Abstract.** Flow visualization is necessary to characterize the fluid flow properties during a hydrodynamic ram event. The multiphase flow during a hydrodynamic ram event can make traditional image processing techniques such as contrast feature detection and PIV difficult. By stacking the imagery to form a multidimensional tensor array, feature detection to determine flow field velocities are visualized.

## 1. Introduction

Fluid flow visualization is critical for quantitatively and qualitatively determining flow characteristics and properties in engineering and physics. The research community has made great strides in developing quantitative data acquisition techniques for cavitation and flow analysis such as high-speed cameras, particle image velocimetry (PIV), rapid pulse lasers for short exposure times, and increasing camera resolution [1]. During this hydrodynamic ram (HRAM) experiment, spherical steel projectiles were accelerated horizontally into a tank of water at velocities of 45 to 150 m/s. Imagery was collected at 41,025 frames per second with a 9.61 $\mu s$ exposure time utilizing a brightfield imaging technique [6]. Entrained flow fields inside the cavity were detected during testing. This effort attempts to measure velocities of the entrained flow field within the cavity. However, the cavity boundary masks the entrained flow field making contrast detection of the flow field difficult via traditional methods. This paper presents a new technique using multidimensional tensors to obtain velocity measurements of the entrained flow field.

## 2. Objective

During an HRAM event, a cavity forms behind the projectile consisting of fluid vapor from cavitation and entrained gasses from ambient air. When characterizing HRAM, it is necessary to detect and process items of interest in the flow field [2, 3]. Utilizing high-speed imagery, the spherical projectile is clearly discernible and detection of its position was completed in prior research [4]. However, the cavity's entrained flow is visually detected, but not as easily measured through traditional techniques such as PIV.

Edge detection of the flow field's front is difficult since the contrast is masked by the cavity's gas-liquid boundary layer. Predicting the contrast gradient to detect multiple flow fronts is computationally intensive due to contrast differences and the transient nature of the flow field. However, the formation of a multidimensional tensor enables flow field detection where traditional techniques have failed. Hence the development of the "image cube" for detection and analysis of multiphase flows.

An image, $A_n$ represented by Equation 1, is an $r \times c$ matrix composed of pixel values, $a$, from the specified image frame number, $n$. $A_n$ represents a typical image where vertical position is on the $y$-axis and horizontal position is on the $x$-axis. An image cube, Equation 2, is a multidimensional tensor of all the images collected with frame number residing on the $z$-axis shown in Figure 1(a). To determine the flow field's velocity, an image slice, $S_r$ via Equation 3 at the specified row, $r$, of interest is needed. The user must specify $r$ based on the row location of the flow field of interest in $A_n$. Row 130 was selected due to the proximity to the projectile shot line of which $S_{130}$ was generated. The box placed around the image cube in Figure 1(a) corresponds to row 130 for all the images composing $S_{130}$. Therefore, $S_{130}$ is a horizontal plane of the tensor and contains the frame number versus horizontal position data utilized for velocity analysis of the flow field for row 130.

$$\mathrm{A_n} = \begin{bmatrix} a_{1,1} & a_{1,2} & \cdots & a_{1,c} \\ \vdots & \vdots & \ddots & \vdots \\ a_{r,1} & a_{r,2} & \cdots & a_{r,c} \end{bmatrix} \quad (1)$$

$$\mathrm{IC} = [A_1 \ldots A_n] \quad (2)$$

$$S_r = \begin{bmatrix} \mathrm{A1}_{(r,1)} & \mathrm{A1}_{(r,2)} & \cdots & \mathrm{A1}_{(r,c)} \\ \mathrm{A2}_{(r,1)} & \mathrm{A2}_{(r,2)} & \cdots & \mathrm{A2}_{(r,c)} \\ \vdots & \vdots & \ddots & \vdots \\ \mathrm{An}_{(r,1)} & \mathrm{An}_{(r,2)} & \cdots & \mathrm{An}_{(r,c)} \end{bmatrix} \quad (3)$$

Vertical position and velocity analysis is possible by looking at the vertical plane of the tensor. To detect velocities in multiple directions, correlation of the flow field between the multiple tensor planes is necessary. However, the horizontal velocity is much larger than the vertical velocity in this experiment. Therefore, since the dominant flow is in the horizontal direction for this experiment, only the horizontal plane is utilized for the position and velocity data.

## 3. Methodology Verification

Although application of tensors for mathematical and statistical analysis is common [5], applying a multidimensional tensor for flow field analysis is unique. Therefore, utilizing a series of generated images of a target with known pixel locations, the image cube methodology was applied to verify the technique. The target had known pixel shifts of 1, 0, -3, and 10 pixels per frame in the horizontal direction. Applying Equation 3 generated the image slice seen in Figure 2. The slope of the lines located in regions 1 through 4 represent the respective 1, 0, -3, and 10 pixel per frame shifts of the target in the generated images, confirming the original known velocity of the target.

## 4. Results

Row 130 was selected to generate $S_{130}$ for HRAM data analysis due to its proximity to the projectile shot line. Using Figure 3(a), the first and darkest contrast line represents the projectile as it traverses the camera's field of view. The next contrast line is representative of the first entrained flow field and is highlighted by line 1. Line 1 begins at row 37, horizontal position 443 and is represented by $S_{130_{(37,443)}}$. $S_{130_{(37,443)}}$ corresponds to row 130, horizontal position 443 of frame number 37 represented by $A_{37_{(130,443)}}$ and shown in Figure 3(b). The contrast line is followed to $S_{130_{(51,235)}}$ which corresponds to $A_{51_{(130,235)}}$ shown in Figure 3(c). Comparison between the contrast line from the image cube to the raw imagery, it is apparent the image cube is sufficient at detecting the entrained flow field. Line 1's slope is indicative of the velocity and is equal to 14.9 pixels per frame–corresponding to a flow field entrainment velocity of 119.4 m/s.

Further analysis of Figure 3(a) indicates additional contrast lines for flow field analysis. The velocity of the entrained flow fields are summarized in Table 1. These additional contrast lines

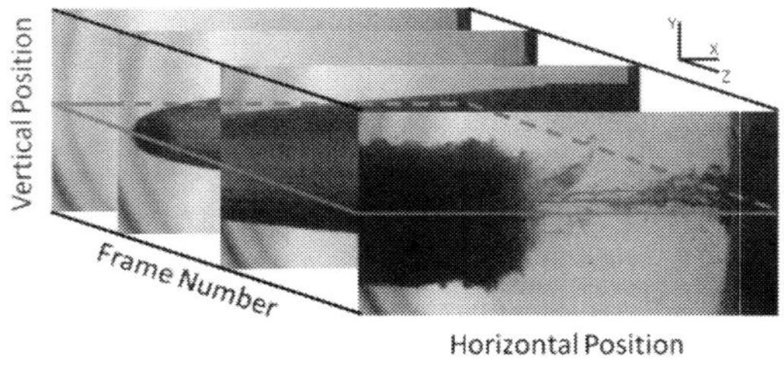


**(a)** Image Cube

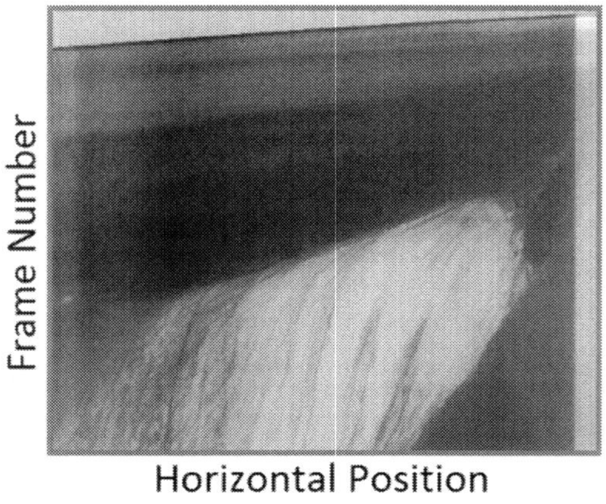


**(b)** Slice 130 ($S_{130}$) of Image Cube

**Figure 1:** A visual example of how the tensor is formed to create the image cube for flow analysis. Notice, instead of positional depth on the the $z$ axis, Frame Number is on the $z$ axis. This subtle difference is not instinctive but is important to conduct position and velocity analysis at the desired $S_r$. $S_{130}$ corresponds to row 130 of all images collected and formed together via Equation 3 to obtain Frame Number vs. Horizontal Position information. Time is obtained by using the frame number and the known camera's frame rate.

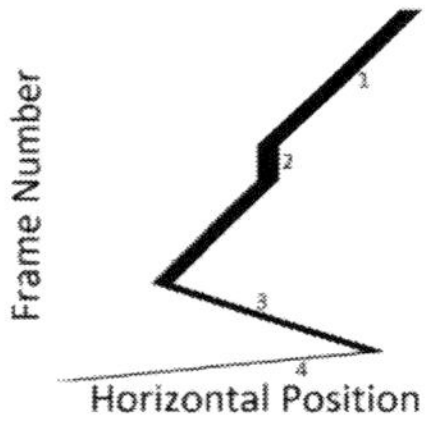


**Figure 2:** An example of an image slice from generated imagery of a target with known pixel positions. Regions 1, 2, 3, and 4 correspond to the square moving at 1, 0, -3, and 10 pixels per frame, respectively.

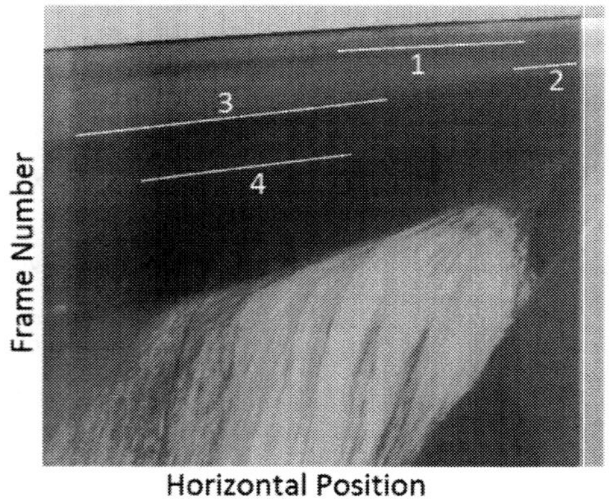


**(a)** Slice 130 ($S_{130}$)

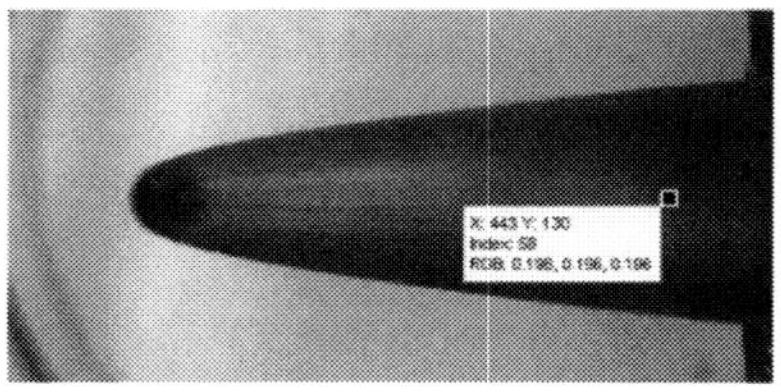

**(b)** Point 1 - Frame 37

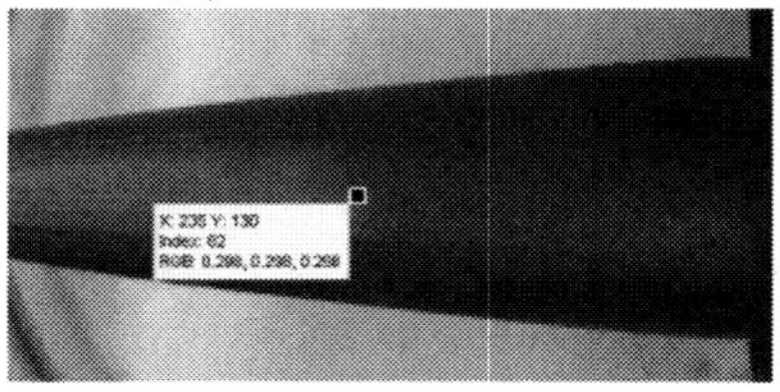

**(c)** Point 2 - Frame 51

**Figure 3:** Slice 130 of the image cube used to detect and characterize the entrained flow field velocity. Figure 3(b) and Figure 3(c) correspond to line 1 in Figure 3(a).

were chosen for a variety of reasons. Line 3 is representative of detecting a flow field with a lower velocity than the flow field represented by line 1. Line 2 represents calculating a flow field entrainment very close to the projectile penetration orifice. Lines 2 and 4 also exemplify calculating a flow field velocity in a heavily masked region.

In comparison, the PIV software utilized an adaptive correlation technique to determine the pixel shift between frames in a specified subset region of the image. Since the flow is not seeded, correlation between frames utilizes the pixel contrast to determine the shift between the subset regions. When the PIV software did detect the flow field, the average of the PIV vectors in the subset region is within the same order of magnitude as the calculated velocity from the image cube. In some cases, the entrained flow field was not detected by the PIV software even though the flow field is visually observed in the raw imagery. As stated earlier, the cavity boundary masks the entrained flow field making the contrast difficult to detect and correlate between frames for the PIV software. Lines 2 and 4 are examples when the PIV software was not able to detect the flow field velocity, yet application of the imagery cube methodology at $S_{130}$ was able to provide these measurements.

**Table 1:** Measured entrainment velocities derived from Figure 3(a) for a projectile impact velocity of 146 $m/s$.

| Reference Lines from Figure 3(a) | Calculated Velocity from image cube ($m/s$) | Average Velocity from PIV software ($m/s$) |
|---|---|---|
| 1 | 119.4 | 91.0 |
| 2 | 109.8 | Software did not detect |
| 3 | 59.9 | 45.5 |
| 4 | 53.2 | Software did not detect |

## 5. Conclusion

The methods presented here detail the ability of processing flow data utilizing a multidimensional tensor array. Application of this technique to calculate flow velocities in multiple directions requires additional development. However, where traditional techniques have failed, utilization of this technique was able to provide useful velocity calculations for flow field characterization in a multiphase flow during an HRAM event.

## 6. Acknowledgments

The authors thank Dr. Matthew Fickus for the helpful discussion about image processing techniques, and for collaborating on the method used here. We also would like to thank Eric Brickson, Scott Wacker, Patrick O'Connell, and all the $96^{th}$ Test Group, AVSF for their experimental assistance.

## References

[1] Noel T. Clemens. Flow imaging. *Encyclopedia of Imaging Science and Technology*, 2002.

[2] Peter J Disimile, Luke A Swanson, and Norman Toy. The hydrodynamic ram pressure generated by spherical projectiles. *International Journal of Impact Engineering*, 36(6):821–829, 2009.

[3] Nicolas Lecysyn, Aurlia Bony-Dandrieux, Laurent Aprin, Frdric Heymes, Pierre Slangen, Gilles Dusserre, Laurent Munier, and Christian Le Gallic. Experimental study of hydraulic ram effects on a liquid storage tank: Analysis of overpressure and cavitation induced by a high-speed projectile. *Journal of Hazardous Materials*, 178(13):635 – 643, 2010.

[4] Andrew J. Lingenfelter and David Liu. Development of methods for characterizing hydrodynamic ram cavity dynamics. AIAA, 2015.

[5] Douglas C Montgomery. *Design and analysis of experiments*, volume 7. Wiley New York, 1984.

[6] Gary S. Settles. Schlieren and shadowgraph techniques for fluid physics experiments - a brief tutorial. 2008.

Journal of Physics: Conference Series **656** (2015) 012117 doi:10.1088/1742-6596/656/1/012117

# A novel particle separation technique using 20-kHz-order ultrasound irradiation in water

**Hiroya Muramatsu[1], Sayuri Yanai[2], Yuki Mizushima[1] and Takayuki Saito[3]**

[1]Graduate School of Science and Technology, [2]Graduate School of Engineering, and [3]Research Institute of Green Science and Technology, Shizuoka University, 3-5-1 Johoku, Naka-ku, Hamamatsu, Shizuoka 432-8561, Japan

E-mail: saito.takayuki@shizuoka.ac.jp

**Abstract.** Ultrasound techniques such as washing, fine-particle manipulation and mixing have been investigated. MHz-band ultrasound was usually used in the previous work, and studies of kHz-order ultrasound are very rare. In the usual manipulation technique, μm-order particles are targeted due to wavelength limitations. We discovered an interesting phenomenon that holds promise for a novel particle separation technique using kHz-order ultrasound. Here, particles with sub-mm- or mm-order diameters were flocculated into a swarm in water irradiated by 20-kHz ultrasound. To develop a practical separation process, we investigated the stationary position and dia. of the particle swarms and the sound-pressure profiles in a vessel, as well as the flocculation mechanism, by varying the irradiation frequency, water level, particle diameter and particle amount. The primary stationary position corresponded to the wavelength calculated from the resonant frequency regardless of the particle diameter. Subtle changes in the frequency and water level resulted in a significant change in the stationary position. Based on these results, we propose a new separation process based on the particle diameter for sub-mm- or mm-order particles.

## 1. Introduction

Many research groups have been devoted to the investigation of MHz-band ultrasound techniques such as ultrasonic cleaning [1], fine-particle manipulation [2] and ultrasonic atomization [3]. In contrast, studies of kHz-order ultrasound techniques are very rare. In addition, regarding fine-particle manipulation, only μm-order particles are targeted due to limitations of the wavelength.

We discovered an interesting phenomenon when particles with diameters ranging from sub-mm to mm were flocculated into a spherical swarm in water irradiated by 20-kHz-order ultrasound. The present study's authors Mizushima and Saito [4] reported that the particle flocculation of a spherical swarm was caused by acoustic cavitation-oriented bubbles adhering to particles. The stationary position of the swarm shifted with a change in the particle concentration and particle diameter. Higher particle concentrations and larger diameters resulted in a lower stationary position.

In the present study, as part of our attempt to develop a new separation process, we investigated the stationary position and diameter of the particle swarm and the sound-pressure profile in a vessel, as well as the flocculation mechanism. Based on the experimental results obtained from careful control of the

[3]Tel/Fax: +81-53-478-1601, E-mail address: saito.takayuki@shizuoka.ac.jp

irradiation frequency, water level, particle diameter and particle amount, we propose a novel, practical and diametrical particle separation process against equivalent density and different-diameter particles, using kHz-order ultrasound.

## 2. Experimental Setup

Figure 1 is a schematic of the experimental setup used in this study. A sine wave voltage signal (frequency 19.7–20.6 kHz) was supplied to a bolt-clamped Langevin-type transducer (HEC-45254M, Honda Electronics, Tokyo) through a power amplifier circuit. Ultrasound was irradiated upwardly/vertically from the bottom of an acrylic water vessel (inner size: 54 mm × 54 mm). The coordinate origin was defined as the intersection point at the bottom of the vessel and the centre of the transducer. The $z$-axis was defined as the vertical direction, and the $x$-$y$ plane was horizontal. The water temperature in the vessel was controlled at approx. 20°C.

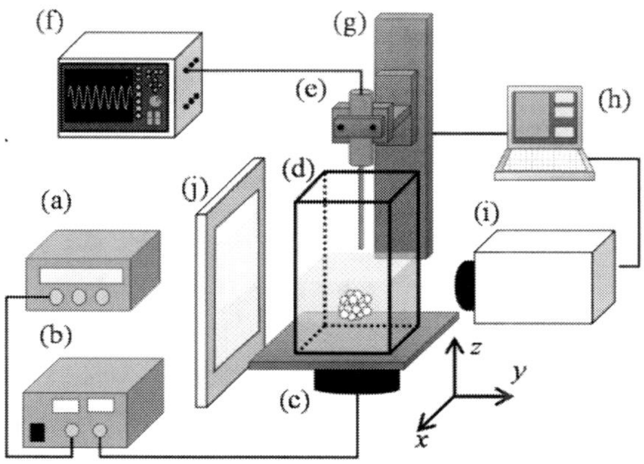


(a) Function Generator
(b) Power Amplifier Circuit
(c) BLT Transducer
(d) Acrylic Vessel
(e) Needle Hydrophone
(f) Recorder
(g) Servo Actuator
(h) PC
(i) Camera
(j) Light Source

Snapshots:
Upper (C): bubble adhering to the surface of a particle.
Lower (B): particle flocculation in the vessel.

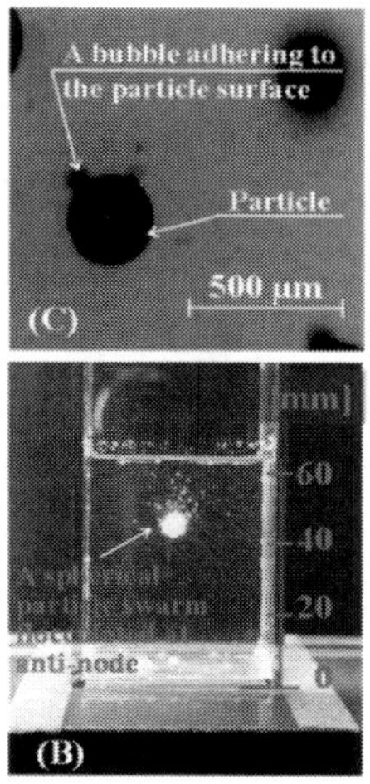


**Fig. 1.** Schematic of the experimental setup used in the present study.

### 2.1. *Experiment 1: Measurement of the sound-pressure profile*

Sound-pressure profiles in the vessel were measured through a hydrophone (HPM1/1 and HP1, tip dia. 1.5 mm, sensor dia. 1.0 mm; Precision Acoustics, Dorchester, UK) under several irradiation frequencies and water levels. The vessel was filled with degassed ultrapure water. The degassed water was made with a vacuum pump at −0.1 MPa for 24 hours. The water level was set at 65, 90 or 120 mm. The measurement range of the $z$-direction was set from 2 mm of $z$ axis to water surface. The hydrophone was moved at a rate of 4 mm/s in the $z$-direction by the servo actuator. The sound pressure was stored in a digital recorder (8861-50, Hioki, Nagano, Japan).

### 2.2. *Experiment 2: Visualization of particle flocculation*

To examine the new flocculation phenomenon in a practical separation process using high-speed visualization, we measured the stationary position and diameter of the spherical particle swarm under various parameters: water level 65 mm, 90 mm and 120 mm; irradiation frequency 19.7–20.6 kHz; particle dia. 400 μm and 800 μm, and the load amount of the particles. The density of the particles was 1.05 g/cm$^3$. The particle flocculation was filmed by a high-speed video camera (Phantom V9, Vision Research, Bedford, UK). The image size was 720 × 1200 pixels (resolution 81.8 μm/pixel), and the frame rate was 500 fps (exposure time 400 μs).

**Table 1.** Particle diameters and load amounts of particles

| Condition | Dia. [μm] | Mass [mg] | Dia. [μm] | Mass [mg] |
|---|---|---|---|---|
| Single-400 | 400 | 400 | – | – |
| Single-800 | – | – | 800 | 400 |
| Mix | 400 | 400 | 800 | 400 |

## 3. Results and discussion

### *3.1. Results of Experiment 1: Measurement of the sound-pressure profile*

Figure 2 shows the sound-pressure profile results for the central axis of the vessel under the irradiation frequency of 20.0 kHz and three water levels. With the increase in the water level, a standing wave formed from a traveling wave from the vessel bottom, and a reflected wave from the water surface grew to a clearer pattern with nodes and anti-nodes. The sound-pressure profile gradually became a suitable profile for trapping the particle swarm. Figure 3 shows the sound-pressure profile for the central axis under the water level of 120 mm and irradiation frequencies of 19.7–20.6 kHz. With the increase in irradiation frequency, the positions of the anti-nodes gradually shifted upward. Based on these results, we discuss the relationship between stationary positions of the particle swarm and the pressure profile of the standing wave below in Section 3.2.

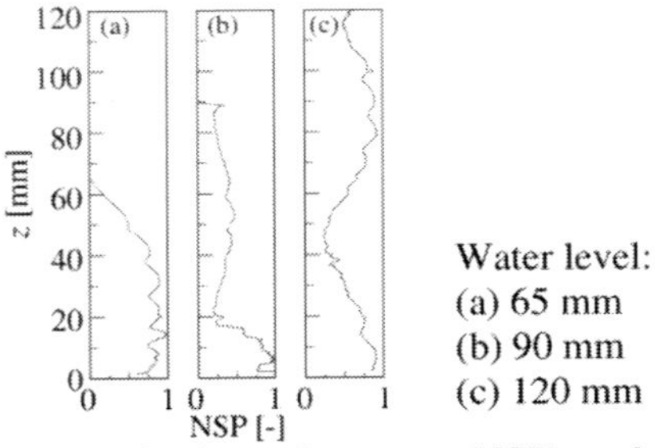


**Fig. 2.** Normalized sound pressure (NSP) on the central axis under several water levels, frequency 20.0 kHz.

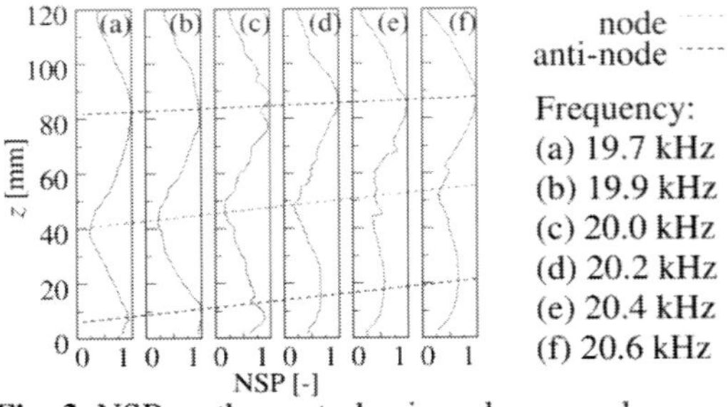


**Fig. 3.** NSP on the central axis under several irradiation frequencies at a water level of 120 mm.

### *3.2. Results of Experiment 2: Visualization of particle flocculation*

The stationary positions and diameters of the spherical particle swarm in the examined water levels are plotted in Figures 4 and 5, respectively. The swarm was trapped in the node and anti-node of the standing wave mentioned above. Under the condition of Single-400 and 19.7 kHz, the swarms were flocculated at two positions: the upper anti-node and the lower node. At the water levels of 65 and 90 mm, the Single-800 results were similar to the results obtained with the other conditions. Interestingly, the Single-800 results at the water level of 120 mm were significantly different from the Single-400 results and Mix results. The Single-800 swarm was shrinking in size under the frequencies of 19.7 and 20.2 kHz (black dots plotted in Figs. 4c and 5c). These phenomena occurred in case of transitional conditions between the cases particle flocculation trapped in the upper position or lower position. With the increase in the irradiation frequency to 20.2 kHz, the stationary position shifted toward the anti-node, and the Single-800 swarm flocculated at the lower anti-node was growing in size again. Although we observed the partial composition of the same-diameter particles, we speculate that this difference in the stationary position between Mix and Single-800 is the result of influences of the smaller particles on the behaviour of the larger particles. These phenomena must be related to the bubbles adhering to the particle surface (Fig. 1c) [4] and the fine structure of the sound-pressure profile; however, the mechanism of this particle flocculation is still unknown. In order to develop a precise separation process for sub-mm- or mm-order particles, further investigations to reveal as-yet unidentified phenomena are needed.

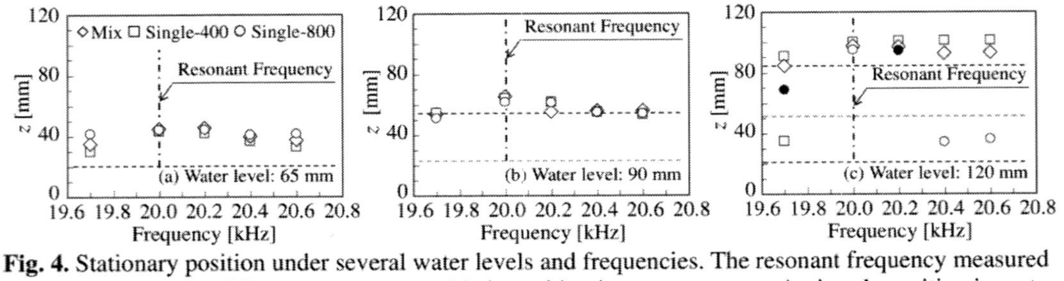


**Fig. 4.** Stationary position under several water levels and frequencies. The resonant frequency measured through a lock-in amplifier. ----- Node position in water ----- Anti-node position in water

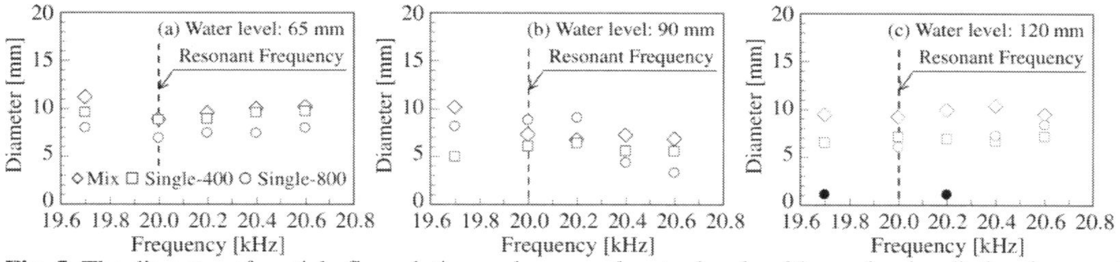


**Fig. 5.** The diameter of particle flocculation under several water levels with varying irradiation frequencies.

### *3.3. A proposed novel particle separation process, and a demonstration*

Based on the above results, we propose a novel particle separation process based on particle diameter, and we demonstrate its performance herein. Figure 6 shows the process flow and typical behaviour of the particle swarms during processing. Mixture particles were flocculated in the upper region in the initial phase (Fig. 6a). With the slight reduction of the irradiation frequency, the particles were immediately sinking (Fig. 6b–d). After that, with the return of the irradiation frequency to the initial value, the smaller particles re-flocculated in the upper region (Fig. 6e,f). The larger particles re-flocculated in the lower region. The entire process took approx. 10 sec. This separation mechanism is closely related to the sound-pressure profile; therefore under a water level of 120 mm (at which a clear sound-pressure profile was formed), a small transition in the sound-pressure profile played an important role against the particle separation. The stationary position of the particle swarm differed by particle diameter (*see* [4], Fig. 6b). Based on these results, the minimum differential diameter in this separation technique might be 200 μm when the particle mixture condition is set at 400 μm or 600 μm.

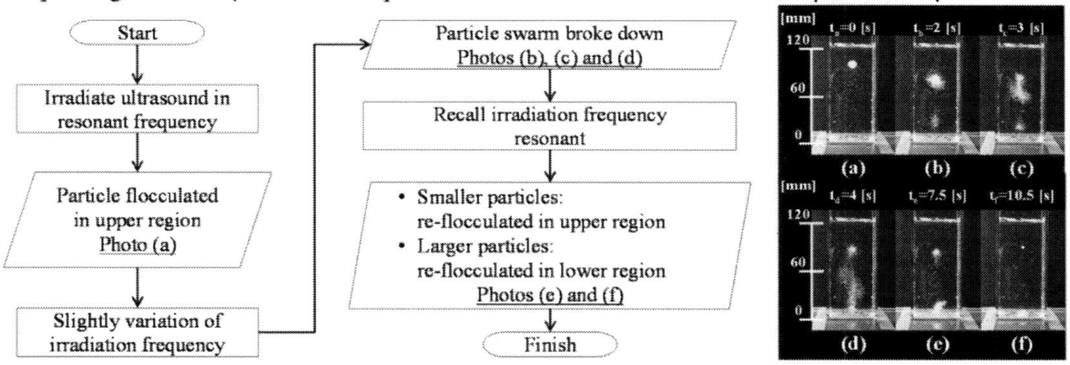


**Fig. 6.** Proposed separation process and particle behaviour.
Yellow particles: 800-μm dia.; white particles: 400-μm dia.; density of both: 1.05 g/cm$^3$.

## 4. Conclusion

Dispersed particles with a sub-mm diameter in water were flocculated into a spherical swarm through irradiation by kHz-order ultrasound. The stationary position and diameter of the particle swarm were investigated under various irradiation frequencies, water levels and particle conditions. We observed that the stationary position and the swarm diameter were well controlled through subtle changes in the frequency and water level. We also discovered that the stationary position for each particle diameter was significantly shifted during the transition state brought about by the changes in the frequency and water level. Based on these results, we propose a novel particle separation process based on the particle diameter, and we have demonstrated its performance.

## References

[1] Li J and Sanderson R D 2002 *J. Membr. Sci.* **205** 247–257
[2] Kozuka T 2005 *J. ASJ* **61** 154–159
[3] Julianna C S *et al* 2012 *Phys. Med. Biol.* **57** 8061
[4] Mizushima Y, Nagami Y, Nakamura Y and Saito T 2013 *Chem. Eng. Sci.* **98** 395–400

# Cavitation for improved sludge conversion into biogas

**A H Stoop[1], T W Bakker[1] and H J M Kramer[2]**
[1]Whirlwind International BV, Toldijk 17-19, 7901 TA Hoogeveen, The Netherlands
[2]Delft University of Technology, Leeghwaterstraat 39, 2628 CB Delft, The Netherlands

E-mail: albert.stoop@whirlwindi.nl

**Abstract.**
In several studies the beneficial influence of pre-treatment of waste activated sludge with cavitation on the biogas production was demonstrated. It is however, still not fully certain whether this effect should be mainly contributed to an increase in conversion rate of organics into biogas by anaerobic bacteria, and how much cavitation increases the total biogas yield. An increase in yield is only the case if cavitation can further disrupt otherwise inaccessible cell membrane structures and long chain organic molecules. In this study the influence of hydrodynamic cavitation on sludge that was already digested for 30 days was investigated. The total biogas yield could indeed be increased. The effect of the backpressure behind the venturi tube on the yield could not yet be established.

## 1. Introduction

Cavitation bubbles have the tendency to adhere to surfaces like cell membranes as well as to long chain molecules, and the stress exerted upon these surfaces during implosion of the bubbles leads to their disrupture. Cavitation can therefore be applied to improve the extraction of valuable ingredients from cells, as in the extraction of carotene from alginate cells [1]. Cavitation can also be used to accelerate and enhance biogas production from municipal waste activated sludge by anaerobic bacteria [2-3]. Waste activated sludge is also called secondary sludge and comes from domestic waste after treatment with aerobic bacteria. There are several methods to induce cavitation. The most frequently applied techniques are ultrasound [4-5], and hydrodynamic cavitation [6-7]. For these methods the energies needed to achieve an improvement of about 20% in biogas production were compared [2]. Hydrodynamic cavitation requires a substantially lower energy input. In these studies however, where the beneficial effect of cavitation was tested, it became not fully clear whether cavitation increases the total biogas yield or only the digestion rate. The aim of this study is to demonstrate whether cavitation indeed increases the biogas yield. For this proof cavitation was applied on sludge after long-term digestion by anaerobic bacteria. The most easily digestible organics are by then converted, and further production of biogas should come from conversion of membrane structures disrupted by cavitation and from disrupted long chain organic molecules. A second objective is to investigate whether an increase in pressure at the outlet of the hydrodynamic cavitation tool, here chosen to be a venturi tube, could increase the biogas production at the same pressure drop over the tool. Calculations have shown that an increase in backpressure behind a cavitation tool, leads to a higher collapse pressure [8], and thus a stronger disruption of membrane structures is expected.

Journal of Physics: Conference Series **656** (2015) 012118 doi:10.1088/1742-6596/656/1/012118

## 2. 50 Bar test circuit

A closed loop stainless test circuit with a total volume of 0.26 $m^3$ was constructed that could stand a working pressure of 50 bar. A venturi tube with a convergent inlet (31°) with a diameter decreasing from 46 mm to a 5 mm and a divergent outlet (90°) with a diameter increasing up to 12 mm was hanging in a test chamber with a diameter of 300 mm. The outlet of the tube in the test chamber was illuminated and could be watched through a viewing window. A flow was circulated through the venturi tube where cavitation could be induced by a pressure drop over the tube maintained by two centrifugal pumps (Grundfos, CRN 10-21SF, CRN 10-20) lined up in series. The backpressure in the test chamber could be increased up to 10 bar by use of a pressure accumulator at the top of the chamber. The maximum flow rate through the venturi tube was 6 $m^3$/h at a pressure drop of 40 bar.

## 3. Results and Discussion

First of all the zone was established where at various backpressures the flow rate sufficed to induce low pressure cavitation bubbles in the venturi tube. The presence cavitation and the size of the plume of cavitation bubbles leaving the tube outlet could only be determined by visual observation through the viewing window.
This cavitation zone was measured for water, because digested sludge is not transparent. Sludge has about the same density as water and only a slightly higher viscosity at a solid weight percentage of 2.5. So no large shift in position of the cavitation zone is expected for sludge.
In figure 1 the flow rate is plotted against the backpressure for experiments where the flow rate was gradually increased at backpressures between 2.5 and 9 barg. At flow rates above the dotted line cavitation was observed. The higher the flow rate the denser and longer the observed cavitation plume at a given backpressure became. This can be seen by comparing figure 1b and 1c both at a backpressure of 2.9 barg. In figure 1b the flow rate is 4 $m^3$/hr which lies about 0.5 $m^3$/hr above the dotted cavitation line, and in figure 1c the flow rate is 5 $m^3$/hr. In figure 1d the flow rate is 6 $m^3$/hr at a backpressure of 8.7 barg, which is also about 0.5 $m^3$/hr above the cavitation line, and the observed plume resembles the plume in figure 1b.

Digested sludge from the wastewater plant Houtrust in the Netherlands was used for the subsequent experiments. This sludge resulted from digestion by anaerobic bacteria for 30 days. Before use this digested sludge was filtered over a 4 mm size filter to prevent blockage of the venturi tube that is 5 mm wide at its narrowest cross-section. This sludge was circulated 50, 75 or 100 times over the venturi tube at a flow rate that was sufficiently high above the dotted cavitation line for water at the selected backpressure to induce cavitation. Although no cavitation could be visually observed, its occurrence was audible. After treatment of the digested sludge with cavitation the biochemical methane potential (BMP) was determined for comparison with the BMP of untreated digested sludge (the blanc). The results for two backpressure and flow rate conditions of respectively 2.9 barg and 5 $m^3$/hr and 8.5 barg and 6 $m^3$/hr, indicated by the red triangles in figure 1a, are presented in figure 2a and b. For the two conditions with flow rates of 5 and 6 $m^3$/hr similar cavitation plumes are expected as seen in figure 1c and 1d.

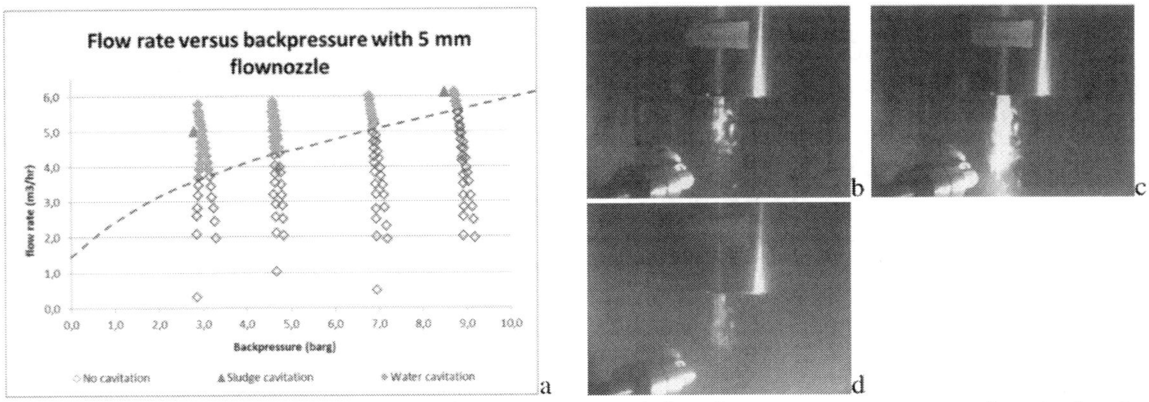


**Figure 1.** a) plot of flow rates versus backpressure and the absence or occurrence of cavitation in water; b and c) cavitation plumes in water at a backpressure of 2.9 barg and flow rates of respectively 4 $m^3$/hr and 5 $m^3$/hr; d) cavitation plume at a backpressure of 8.7 barg and a flow rate of 6 $m^3$/hr.

Before the methane production was measured 100 g sludge sample was mixed with 100 g untreated digested sludge as inoculum before it was fed into the AMPTS vessel. This inoculation was added in order to have enough anaerobic bacteria available for conversion of the organics. The mixture was kept at $36^0$ C and the methane production was measured over time for the digested sludge treated by cavitation under two operating conditions at three recirculation times as well as for the blanc. These results are presented in figure 2 after subtraction of the methane produced from the added inoculum itself. Comparison of the methane production during the first day most clearly shows that treatment of the sludge by cavitation strongly accelerates the conversion into methane. In figure 2a it is shown that the methane production increased about a 100% from 42 to 85 ml during the first day for a 100 times recirculation, while the total methane production after 11 days increased by about 50% from 153 to 232 ml. The effect of recirculation on the methane yield is also evident. When the recirculation is lowered to 75x or 50x the production rate and total production after 11 days are lowered proportionally.

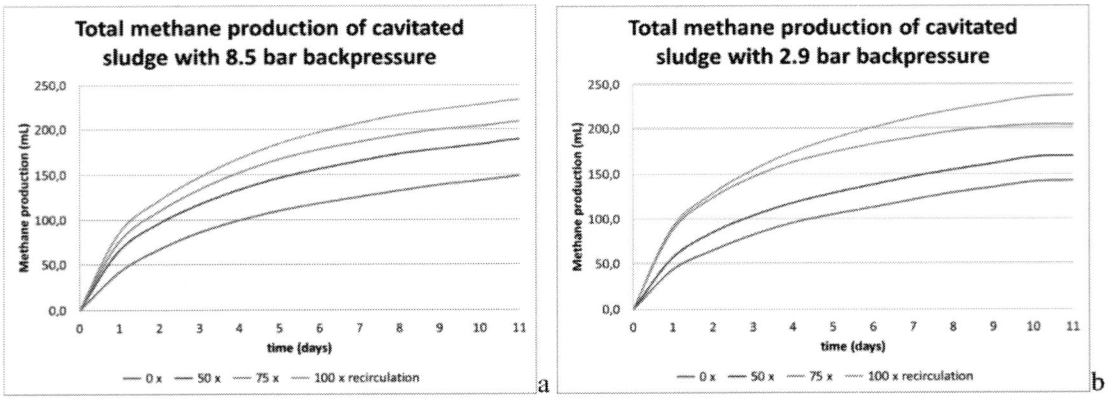


**Figure 2.** Methane production of digested sludge with cavitation at a backpressure of (a) 8.5 barg and a flow rate of 6 $m^3$/hand of (b) 2.9 barg and 5 $m^3$/h. 0x implies that the sludge is not cavitated as in the blanc.

With a backpressure of 2.9 bar and 100 times recirculation in figure 2b a more or less similar acceleration in methane production after the first day (110%) and increase in production after 11 days of about 66% can be seen as for the backpressure of 8.5 bar. But now the production rate and total yield after 11 days are lowered more at 75x and 50x recirculation than predicted by the proportional reduction in recirculation times.
Although a comparison between the results for both backpressures is not fully allowed because of the difference in flow rate, it can be observed that the total methane yield after 11 days is about equal for a 75 and 100 times recirculation. For 50 times recirculation the higher backpressure gives a better yield. At a higher backpressure, however the required energy input for an open circuit would become higher.

## 4. Conclusions

From these experiments with high recirculation times it can be concluded that treatment with cavitation further disrupts cell membranes and long chain molecules and makes more organic material accessible for conversion by anaerobic bacteria. It, however, remains undecided if for digested sludge and high recirculation times a higher backpressure is beneficial. Especially because more energy is required for cavitation in an open circuit at a higher back pressure.

## Acknowledgments

This study was co-financed by Topconsortia for Knowledge & Innovation of the Dutch ministry of economic affairs and was performed in cooperation with Delfluent Services BV and KWR Watercycle Research Institute. The authors thank Prof. Dr. ir. G.M. van Rosmalen for her contributions to this study.

## References

[1] Christaki E, Bonos E, Giannenas I, Florou-Paneri P. Functional properties of carotenoids originating from algae. *J Sci Food Agric*. Wiley-Blackwell; 2012 Oct 9;**93**(1):5–11.
[2] Whitlock D, Sandino J, Novak J, Johnson B, Fillmore L. Evaluation Methodology Framework for Processes to Reduce Waste Activated Sludge. *Proceedings of the Water Environment Federation* Water Environment Federation; 2010 Jan 1;**2010**(12):4815–49.
[3] Cano R, Pérez-Elvira SI, Fdz-Polanco F. Energy feasibility study of sludge pretreatments: A review. *Applied Energy*. Elsevier BV; 2015 Jul;**149**:176–85.
[4] Bougrier C, Albasi C, Delgenès JP, Carrère H. Effect of ultrasonic, thermal and ozone pre-treatments on waste activated sludge solubilisation and anaerobic biodegradability. *Chemical Engineering and Processing: Process Intensification*. Elsevier BV; 2006 Aug;**45**(8):711–8.
[5] Tiehm A, Nickel K, Zellhorn M, Neis U. Ultrasonic waste activated sludge disintegration for improving anaerobic stabilization. *Water Research*. Elsevier BV; 2001 Jun;**35**(8):2003–9.
[6] Petkovšek M, Mlakar M, Levstek M, Stražar M, Širok B, Dular M. A novel rotation generator of hydrodynamic cavitation for waste-activated sludge disintegration. *Ultrasonics Sonochemistry* Elsevier BV; 2015 Sep;**26**:408–14.
[7] Biogest. CROWN-disintegration system, Available at: http://www.biogest.com/products/sludge-treatment/crown-disintegration-system (accessed 12 august 2015)
[8] Gogate PR, Pandit AB. A review and assessment of hydrodynamic cavitation as a technology for the future. *Ultrasonics Sonochemistry*. Elsevier BV; 2005 Jan;**12**(1-2):21–7.

# Optical Measurements of Cavitation in Tribological Contacts

**Tian Tang, Nick Morris and Jeremy Coupland**

Department of Mechanical and Manufacturing Engineering, Loughborough University, UK

E-mail: T.Tang2@lboro.ac.uk

**Abstract.** The paper describes the use of a white light interformeter to measure the cavitation bubble and oil film thickness in a tribological contact and compares the results to theory. It is found that oil film thickness is best predicted by the theory proposed by Coyne and Elrod.

## 1. Introduction

The role of cavitation in tribological contacts has been discussed by many authors with regard to load capacity, lubricant thickness, frictional losses and wear in sliding or rolling contact bearings [1-15]. The pioneering work of Osborne Reynolds in 1886 showed that the lubricant pressure distribution along a smooth sliding bearing can be expressed by the equation that now bears his name,

$$\frac{\partial}{\partial x}\left(h^3\frac{\partial p}{\partial x}\right) = 6\eta U\frac{\partial h}{\partial x} \tag{1}$$

where $h(x)$ is the height profile, $p(x)$ is the oil film pressure, $\eta$ is the lubricant viscosity and $U$ is the bearing sliding velocity [1]. For the case of a convergent/divergent tribological contact the solution to this equation predicts a significant increase in gauge pressure within the converging section followed by a decrease in pressure (within the diverging section) that often falls below the saturation or vapor pressure which signifies the onset of cavitation or film rupture. The cavitation, however, clearly modifies the flow and must be properly accounted for in the theory using appropriate boundary conditions.

The first cavitation model was introduced by Gümbel [2] who assumed that film rupture occurred at the beginning of the divergent region. Swift [3] and Stieber [4] introduced an improved model assuming the pressure gradient equal to zero at the film rupture position. In order for the flow to exit the contact it must either pass around multiple cavitation bubbles or flow over (or under) a single cavity. These two theories have been studied extensively but separately. The assumption of flow around multiple cavitation bubbles has been studied by Floberg [5], Jacobson and Floberg [6], Olsson [7] and Floberg [8]. Taken together these works are referred to as JFO theory which not only satisfies mass flow continuity and predicts the film rupture position, but also locates the reformation position and is therefore capable of predicting the load carrying capacity. The theory has widely used after an efficient algorithm for calculation was introduced by Elrod [9, 10] and is now commonly referred to as the Elrod cavitation model.

Alternatively the lubricant can flow over (or indeed under) a single cavity. Prandtl's theory of boundary layers and flow separation was built upon by Birkhoff and Hays [11] to determine that the point at which flow separation will occur due to adverse pressure gradients causing reverse flow in the diverging region of the contact. Coyne and Elrod [12, 13] improved the model by introducing the influence of surface tension, gravity and fluid inertia. In essence, the inclusion of surface tension effects allows the local fluid pressure to drop below the cavity pressure and the point of film rupture moving further along the divergent section.

Both the assumption of flow around multiple cavitation bubbles and the assumption of flow over a single cavitation bubble are supported to a certain degree by experimental observation. Dowson [14] shows the flow between multiple cavitation bubbles generated in the lubricant flow between a sliding surface and a fixed convex surface. On the other hand, Heshmat [15] photographed the cavitation generated by a flat stationary surface and a rotating cylindrical surface showing flow between a single bubble and the rotating surface.

In a previous paper we have investigated the lubricant flow in a glass sliding bearing (modelling a ring-cylinder liner conjunction) using a digital holographic microscope [16]. In essence this transmission setup measures the phase and amplitude of the wavefront passing through the tribological contact. This information allowed the computation of high resolution images of the cavitation bubbles and, by comparing wavefronts before and after the onset of cavitation, high fidelity quantitative measurements of the bubble thickness are possible. Although the measurements clearly show multiple cavitation bubbles and suggest that the bubbles only contact the cylindrical surface, the transmission setup does not allow the oil film thickness to be measured. In this paper, we report an interferometric reflection geometry using white light to measure both the bubble and the film thickness. Measurement results are discussed and compared with the theoretical models of lubricant flows.

## 2. Experimental configuration

The contact consists of a plane glass surface and a cylindrical lens as shown in figure 1. The plane glass is driven by a bidirectional motorized translation stage with controlled speed up to 2.3 $mm/s$. The cylindrical lens (radius=13.25 $mm$, width=12.5 $mm$ and length=25 $mm$) is fixed on a stationary metal surface. The lubricant oil between the two bearing surface is additive-free, classic green gear oil with a viscosity $\eta = 1.508\ Pa.s$ measured at room temperature (20 C$^{o}$). The sliding plane glass is entirely supported by the hydrodynamic film with a constant load of $W = 2.65\ N$. In this experiment the bearing surfaces are constructed from high refractive index S-NF57 glass (n = 1.84) which, relative to the refractive index of the oil (n = 1.49), provides sufficient reflected light. The white light interferometer configuration is shown in figure 2.

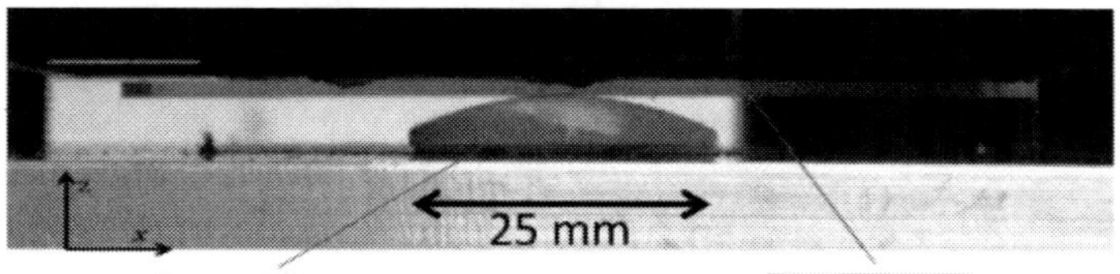


Figure 1. Glass bearing configuration.

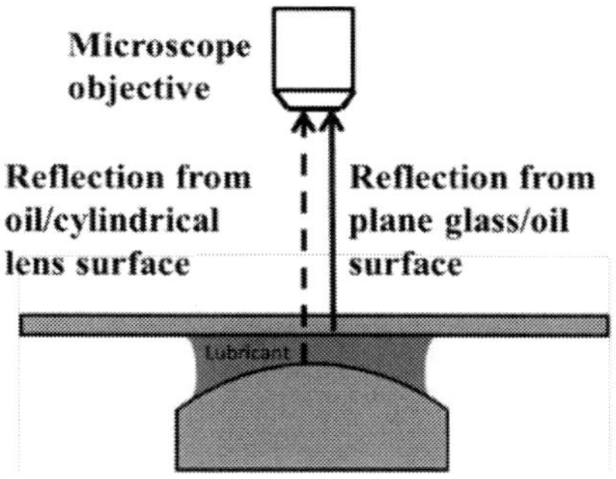


**Figure 2.** White light interferometer.

In this configuration as the illumination passes through the bearing, a reflection occurs at the interface between the plane glass and oil and another at the interface between the oil and cylindrical surface. Since the illumination is white light, interference is observed only if these surfaces are in close proximity as shown in figure 3. Using white light interferometry, the minimum thickness can be found unambiguously by using a theoretical model to fit the experiment results as follows.

## 3. Analysis

To a good approximation the intensity of two reflections interference, $\boldsymbol{I}$, can be expressed as a function of the thickness profile, $\boldsymbol{h(x)}$, such that,

$$I = I_1 - I_2 \cos\left(\frac{4\pi}{\lambda} h(x)\right) E\big(h(x)\big) \tag{2}$$

where $I_1$ and $I_2$ are constants, $\lambda$ is the average wavelength and $E$ is a function that describes the fringe envelope. Since the thickness profile is known, the thickness of the oil film $\boldsymbol{h(x)}$, can be expressed as a function of position $\boldsymbol{x}$,

$$h(x) = h_0 + R(1 - \cos(\arcsin(\tfrac{x}{R}))) \quad (3)$$

where $h_0$ is the minimum thickness and $R$ is the radius of cylindrical lens. Using this equation the values of the constants in equation 2 together with the fringe envelope function can be found by analyzing an inteferogram taken when the bearing is stationary. When driven, the oil film thickness can then be found by iteratively searching for $h_0$. Using data up to the point of film rupture the normalised fringe data ($I_{norm} = (I - I_1)/I_2$) is shown in figure 4. Note the origin in figure 4 is x = 336 $\mu m$ in figure 3. Here, the blue line represents the average normalized intensity while the red line is that predicted by equation 2. A similar method can be used to measure the thickness of the bubble

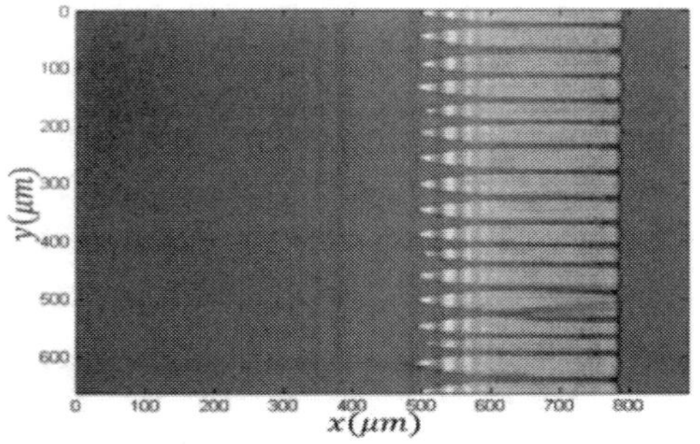


**Figure 3.** White light interferogram.

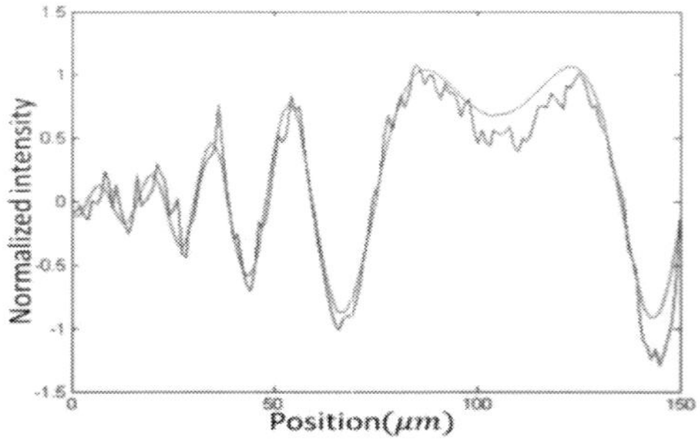


**Figure 4.** Average normalized intensity (blue) and theoretical fit (red).

## 4. Results

The reflection geometry has been used to measure the minimum thickness, bubble position and bubble thickness at various speeds (table 1) and has been compared with those predicted by the model discussed in section 1.

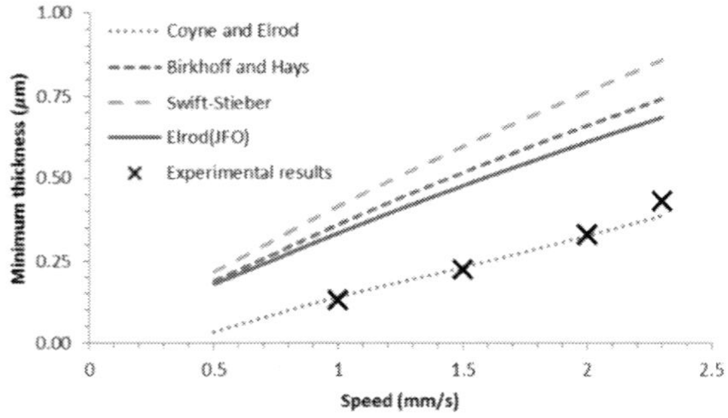


**Figure 5** Minimum thickness versus speed.

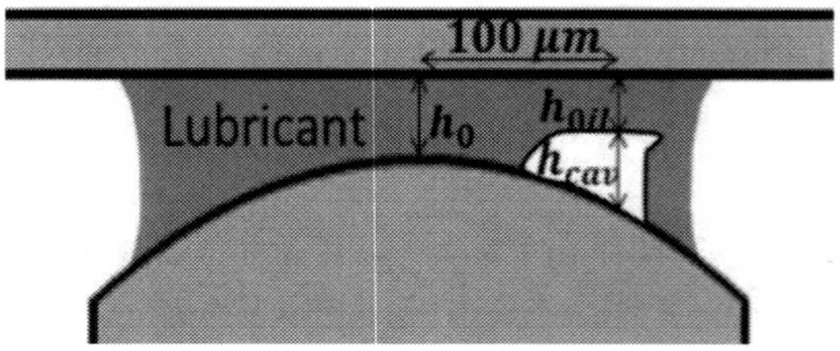


**Figure 6** Bubble geometry.

Figure 5 shows a comparison of the measured minimum film thickness as a function of entrainment speed with those predicted by the different theoretical models. It can be seen that although there are clearly multiple bubbles (as figure 3) the minimum film thickness is best predicted by the Coyne and Elrod model that assumes flow over a single bubble.

**Table 1** Bubble thickness and oil film thickness at 100 μm at different speeds

| Speed (mm/s) | $h_{cav}$ (μm) | $h_{oil}$ (μm) | $h_0$ (μm) |
|---|---|---|---|
| 1.0 | 0.2835 | 0.3092 | 0.149 |
| 1.5 | 0.3982 | 0.2902 | 0.256 |
| 2.0 | 0.506 | 0.285 | 0.359 |
| 2.3 | 0.5566 | 0.301 | 0.426 |

Since we measure both the position of the glass surfaces and the bubble thickness, the oil film thickness over the cavitation bubbles can also be calculated. With reference to figure 6 these quantities are tabulated in Table 1. It can be seen that with the increase of entrainment speed, the minimum film thickness and bubble thickness will increase; however, the oil film that flows over the bubbles is of almost constant thickness.

## 5. Conclusions

This paper has introduced a white light interferometer and associated analysis methods as an aid to understanding cavitation in tribological contacts. Using this approach we have been able to measure oil film and bubble thickness simultaneously and can compare these results with those predicted by theory. Although the results clearly show multiple bubbles (as figure 3) the minimum film thickness is best predicted by the Coyne and Elrod model that assumes flow over a single bubble. We have verified that the oil film does indeed flow over the bubbles and furthermore we have shown that the thickness of this component (approximately 300 nm) is almost independent of entrainment velocity.

It is evident that neither the multiple bubble (JFO, Elrod) nor single bubble (Elrod and Coyne) models fully explain real cavitation, at least in the lightly loaded flow conditions studied here. It is clear that aspects of both theories are necessary to properly predict the pressure boundary conditions and hence calculate the load conditions. Furthermore it is expected that surface roughness will strongly affect bearings performance and further work is planned to study its effect using interferometry.

## References

[1] Reynolds O 1886 On the theory of lubrication and its application to Beauchamp Tower's experiments including an experimental determination of the viscosity of olive oil *Philos. Trans. R. Soc. London Ser.* A 177, 157-233

[2] Gümbel L K R 1921 Vergleich der Ergebrusse der rectinerischen Behandling des Lagerschmierangsproblem mitneueren Versuchsergebrussen *Monatsbl. Berliner Bez Ver. Dtsch. Ing.* 125-128

[3] Swift W 1933 The stability of lubricating film in journal bearing *J. Inst. Civ. Eng.* 233 (1), 267-88

[4] Stieber W 1933 *Das Schwimmlager. Berlin: Ver. Dtsch. Ing*

[5] Floberg L 1973 On journal bearing lubrication consideringthe tensile strength of the liquid lubricant. *Transactions of the Machine Elements Division, Lund Technical University, Lund, Sweden* 1–26

[6] Jakobson B and Floberg L 1957 The finite journal bearingconsidering vaporization. *Transactions of Chalmers University Technology, Goteborg, Sweden* 190, 1–119

[7] Olsson K O 1965 Cavitation in dynamically loaded bearings *Transactions of Chalmers University Technology, Goteborg, Sweden* 308

[8] Floberg L 1974 Cavitation boundary conditions with regard to the number of streamers and tensile strength of the liquid *Cavitation and related phenomena in lubrication, Proceedings of the 1st Leeds–Lyon Symposium on Tribology, University of Leeds, Leeds, UK* 31–36

[9] Elrod H G and Adams M 1974 A computer program for cavitation and starvation problems *Cavitation and related phenomena in lubrication* 37-41

[10] Elrod H G 1981 A cavitation algorithm *Journal of Tribology* 103(3), 350-354

[11] Birkhoff G and Hays D F 1963 Free boundaries in partial lubrication *Journal of Mathematics and Physics* 42(2), 126

[12] Coyne J C and Elrod H G 1970 Conditions for the rupture of a lubricating film. Part I: theoretical model *Journal of Tribology* 92(3), 451-456

[13] Coyne J and Elrod H G 1971 Condition for the Rupture of a Lubricating Film-Part II: New Boundary condition for Reynolds Equation *Journal of Lubrication Technology* 93, 156

[14] Dowson D 1957 Cavitation in lubricating films supporting small loads *Proc. Inst. Mech. Eng. Conf. Lubr. Wear.* 93-99

[15] Heshmat H 1991 The mechanism of cavitation in hydrodynamic lubrication. *Tribology transactions* 34(2), 177-186

[16] Tang T, Morris N, Coupland J, and Arevalo L 2015 Cavitation Bubble Measurement in Tribological Contacts Using Digital Holographic Microscopy *Tribology Letters* 58(1), 1-10

# Comparison of cavitation intensity in water and in molten aluminium using a high-temperature cavitometer

**I Tzanakis[1], G S B Lebon[2], D G Eskin[1], K Pericleous[2]**
[1] Brunel Centre for Advanced Solidification Technology, Brunel University, Uxbridge, Middlesex, UB8 3PH, United Kingdom
[2] Centre for Numerical Modelling and Process Analysis, The University of Greenwich, London, SE10 9ET, United Kingdom

E-mail: iakovos.tzanakis@brunel.ac.uk

**Abstract.** The application of ultrasound to industrial casting processes has attracted research interest during the last 50 years. However, the transfer and scale-up of this advanced and promising technology to the industry have been hindered by difficulties in treating large volumes of liquid metal due to the lack of understanding of certain fundamentals. In the current study, experimental results on ultrasonic processing in deionised water and in liquid aluminium (Al) are reported. Cavitation activity was determined in both liquid environments using an advanced high-temperature cavitometer sensor. In water, the highest cavitation activity is obtained for the lowest sonotrode tip amplitudes. Below the sonotrode, the cavitation intensity in liquid aluminium is found to be four times higher than in water.

## 1. Introduction

In recent years, there has been an increased interest in fundamental and applied investigations on metal solidification with the use of ultrasound [1]. Ultrasonic treatment of metallic alloys is driven by cavitation and bubble dynamics and it has been proven to be effective and promising for degassing and refining grain structure [2, 3]. However, its industrial application remains rather limited. Commercial-scale ultrasonic melt treatment is hindered by the lack of fundamental knowledge and practical models needed to optimize the ultrasonic treatment conditions, particularly those concerning, (i) the interaction between ultrasound-induced cavitation and the processed volume (ii) the characteristics of cavitation zone in molten metals, which currently limits this advanced processing to laboratory scale. The development of such knowledge is required for any major technological breakthrough.

In the current study, two different approaches were used. The first one is related to the characterisation of cavitation activity within a particular water vessel; while the second one investigates cavitation activity within molten Al. Ultrasonic excitation for both cases is achieved by utilising an ultrasonic transducer with a sonotrode submerged into the melt. Measurements at various locations across the vessels revealed noticeable spatial variations in the cavitation activity. The local cavitation phenomena were explained based on the spectral characteristics of acoustic emission.

## 2. Methodology

In this study, two different experimental setups for water and Al were used, as shown in Figure 1. In the case where water was treated by sonication, a titanium sonotrode with a 40-mm tip diameter was driven by a 1-kW piezoelectric transducer which oscillates at a frequency of 20 kHz producing maximum peak-to-peak amplitude of 34 μm (Hielscher/Germany) at the tip. The tip of the sonotrode was submerged 20 mm below the water surface. A 3 litre cylindrical vessel with diameter of 16 cm and height of 15 cm filled with 2 litres of deionised water was used. The reason for choosing this geometry is that the geometrical features of this water vessel are very close to the crucible's geometry (dimension 15 cm/height 21 cm) where Al melt is treated. Water temperature was maintained at 22±1 °C. Experiments were performed at various distances across the central axis of the vessel, with the sonotrode placed at two different positions i) in the centre of the vessel and ii) near to the vessel's wall in order to investigate the effect of positioning on the cavitation intensity. All experiments were carried out with sonication amplitude adjusted at 50% (p-p 17 μm) and 100% (p-p 34 μm).

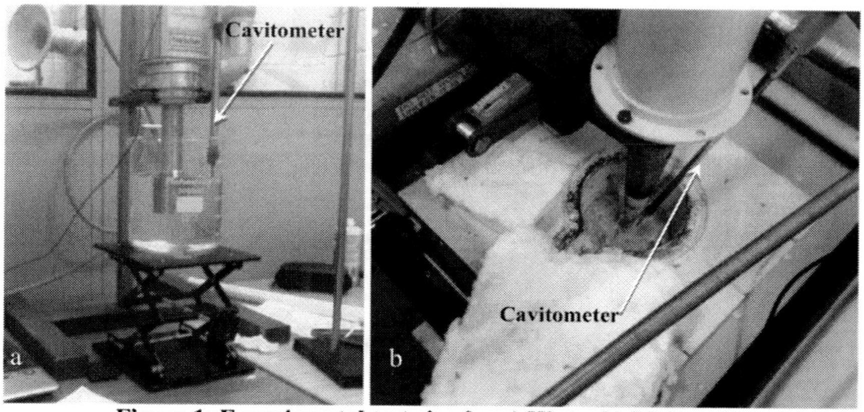


**Figure 1: Experimental test rigs for a) Water b) Molten Al.**

Pure Al was selected because it has been extensively studied and widely used in metallurgical, automotive and aerospace industry as an alloy base. Additionally, liquid Al and water have very close kinematic viscosities while their Newtonian behaviour is similar [4] making water a frequently used physical model of liquid Al. The charge of 5.2 kg (equal to 2 litres) of Al was melted in a clay–graphite crucible coated with boron nitride (BN) from inside at a temperature of 760 °C. The ultrasonic equipment consisted of a 5 kW magnetostrictive transducer (Reltec, Russia) with a Ti sonotrode of 20 mm in diameter submerged to a depth of 5 mm in the melt. Experiments were performed at a driving frequency of 17.5 kHz only for the case of 3.5 kW (p-p amplitude 39 μm inside Al melt). The tip was preheated and the melt temperature was controlled during the process. There was no controlled atmosphere.

The intensity of cavitation was directly measured using a calibrated high-temperature cavitometer (Belorussian State University of Informatics and Radioelectronics) (Figure 1). Cavitometer consists of a tungsten probe with a diameter of 4 mm and a length of 500 mm connected to a piezoelectric receiver mounted within a metallic enclosure. A full description of cavitometer can be found in [5]. The signal acquisition and processing was carried out using a dedicated external digital oscilloscope device "Picoscope" (Pico Technology/UK) that allowed real-time signal monitoring of the cavitometer sensor's data and ultrasonic parameters.

## 3. Experimental results and discussion

Figure 2 shows a typical acoustic spectrum for water and liquid Al as received by the cavitometer. The acoustic sources in water and in Al working at 20 kHz and 17.5 kHz, respectively, produce broadband signals well into the high frequency domain associated with the activity from cavitation bubbles. The general shape of the spectrum plot is similar, thus reinforcing further the opinion that water and Al can share similar behaviour under ultrasonic treatment. Specifically, the prominent fundamental frequency component ($f_0$) is clearly shown in both graphs with further contributions from sub- and ultra-harmonic frequencies. The difference is that harmonics in molten Al are less prominent, implying that cavitation activity has not been fully developed yet or that the acoustic signals received by the cavitometer in liquid Al are much lower than expected due to the attenuation effects from the liquid.

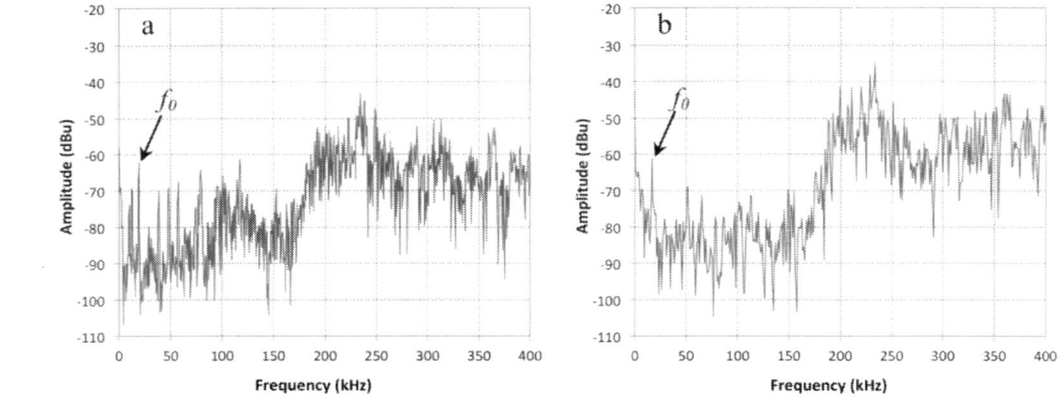


**Figure 2: Acoustic spectrum generated by ultrasonic transducer a) 20-kHz piezoelectric in water and b) 17.5-kHz magnetostrictive in Al. Measurements were taken below the sonotrode's surface**

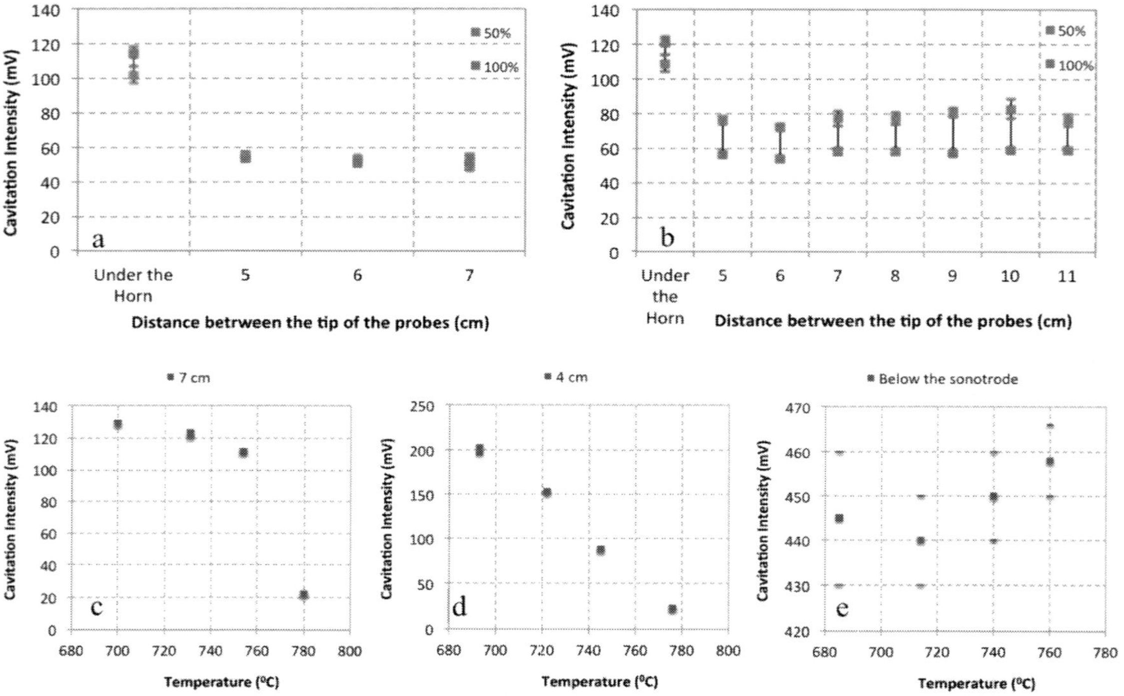


**Figure 3: Cavitation activity measured in: a-b) water for two positions of the sonotrode a) near to the side wall b) in the centre of the cylindrical vessel, as a function of distance between the centre tip of the cavitometer and the sonotrode under mechanical amplitudes of 50% and 100% and c-e) liquid Al with respect to temperature drop at various distances from centre tip of sonotrode with the sonotrode placed in the centre of the crucible.**

One should take into account that molten Al does not contain as many nuclei or pre-existed bubbles as in water, making the inception and development of powerful cavitation more difficult [1]. Although, the overall broadband component (>200 kHz), which is generated by the collapsing bubbles of a wide range of sizes with their shock emissions and liquid jets contributing further, is slightly higher in liquid Al than that of the water implying a more prominent activity from cavitation bubbles. Specifically, at higher frequencies i.e. in the range of 200–250 kHz, prominent peaks are shown, suggesting nonlinear activity from numerous cavitation bubbles. This leads us to a preliminary conclusion that in the studied melt the activity of bubbles with resonant sizes of 10–15 μm (according to Minnaert's equation [6]) could possibly prevail in the cavitation regime. This is in agreement with our recent in-situ study of cavitation in Al melts, showing that the majority of cavitation bubbles are indeed in that particular range [7]. Note that the hump at 160 kHz is attributable to the variation in the sensitivity response of the cavitometer.

Figure 3 shows the spatial variation data of the cavitation activity for different positions of the sonotrode and the cavitometer probe within the water vessel and the crucible. It can be clearly seen in Figures 3a, b that less power is more effective in producing higher cavitation activity in water. In the case when the sonotrode was placed in the centre of the cylindrical vessel, 50% power was producing slightly higher cavitation activity levels than 100%. However, when the sonotrode was moved near the side wall, the 50% output produces about 30% more cavitation activity compared with 100%. These amplitude differences regarding the power output and the positioning of the sonotrode are attributable to two factors; a) to the shielding and scattering effects; where the higher output level, i.e. 100%, produces an intense region of cavitation very close to the sonotrode tip surface, which prevents the further propagation of ultrasound into the liquid, thus attenuating the signal received by the cavitometer sensor and b) to the fact that activity can be higher near to the wall as bubbles can collapse easier due to the solid interface and also cavitation nuclei or many microscopic air bubbles are more available due to the surface roughness of the glass wall. Results are in a good agreement with the works of Rozenberg [8] and Hodnett et al. [9]. Additionally, when the cavitometer was placed below the sonotrode, intensity was measured to be similar to that at 100%.

Another interesting feature is that cavitation activity significantly increases with temperature drop in liquid Al as shown in Figures 3c-3e. This clearly appears in the areas away from the main cavitation zone. Below the sonotrode, the intense region of cavitation maintains high levels of cavitation activity regardless of temperature variation. The measurements of raw acoustic signal in liquid Al indicate that acoustic cavitation is much higher than in water e.g. 4 times higher in locations below the sonotrode. Thus, cavitation activity is expected to be much higher in liquid melts.

## 4. Conclusions

1) Ultrasonic processing of molten Al produces an acoustic spectrum comparable to water. The analysis of the broadband spectra showed that cavitation intensity is higher in liquid Al than in water implying a more prominent activity from cavitation bubbles.
2) Lower power outputs i.e. 50%, seem to be the best setting for producing the higher cavitation activity throughout the treated volume, suggesting nonlinearity in energy transfer to the liquid, while the location of the sonotrode is seen to affect cavitation activity within the liquid.
3) Cavitation intensity significantly increases with temperature drop in liquid aluminium, except for the region below the sonotrode where intense cavitation is maintained regardless of temperature variations.

**Acknowledgments:** This work is performed within the Ultramelt Project financially supported by the EPSRC grants EP/K005804/1 and EP/K00588X/1.

## References:

[1] Eskin G I and Eskin D G 2014 Ultrasonic Treatment of Light Alloy Melts (CRC Press)
[2] D.G Eskin et al. J. Mater. Process. Technol. 222 (2015) 148–154
[3] T.V. Atamanenko et al. Metall. Mater. Trans. A, 41A (2010) 2056–2066.
[4] D. Xu et al. Metall. Mater. Trans. B, 29B (1998) 1281-1288.
[5] I. Tzanakis et al. "Calibration and performance assessment of an innovative high-temperature cavitometer" Ultrason. Sonochem. (Under Review)
[6] M. Minnaert, Phil. Mag., 16 (1933) 235–248.
[7] W. Wu et al. in TMS Supplemental Proceedings, Wiley & Sons, Hoboken, NJ, USA, 2015, pp. 61–66.
[8] L.D. Rozenberg, in Powerful Ultrasonic Fields, Nauka, Moscow, 1968, pp. 221-266
[9] M. Hodnett et al. Ultrason. Sonochem. 11 (2004) 441–454

# Numerical Simulation of Supercavitating Flows using a Viscous-Potential Method

**Ji-Hye Kim and Byoung-Kwon Ahn**

Dept. of Naval Architecture and Ocean Engineering, Cellege of Engineering, Chungnam National University, 99 Daehakro(St), Yuseong-gu, Daejon, Korea

E-mail:bkahn@cnu.ac.kr

**Abstract**. A numerical method was developed to predict the supercavity around axi-symmetric bodies. Employing potential flow, the proposed method computes the cavity shape and drag force, which are the important features of practical concern for supercavitating objects. A method to calculate the frictional drag acting on the wetted body surface was implemented, which is called the viscous-potential method. The results revealed details of the drag curve appearing in the course of an increase in speed and cavity growth. In addition, the supercavity and drag features of the actual shape of the supercavitating torpedo were investigated according to the different depth conditions.

## 1. Introduction

When the speed of a submerged body increases and the cavity is sufficient to cover the entire body, it is called a supercavity. The cavity reduces the drag forces acting on the body and can help an underwater vehicle moves faster. This study focused on the cavitator located in front of the body, which generates a cavity and develops it until the natural supercavity covers the entire body. In the field of marine hydrodynamic applications, traditional methods to predict the extent and behavior of a cavity on the surface of a propeller blade are based mostly on the linearized lifting surface theory [1]. The surface panel method was developed to improve the accuracy near the high curvature section [3], [2]. Following that studies, present method distributed the normal dipoles and sources on the cavitator and the cavity surface to solve the supercavitating flow problem generated by various types of cavitators. The normal dipoles and sources on the cavity surface are moved to the newly computed cavity surface, where the boundary conditions are satisfied again. This iteration process is repeated until the results converge. All the results were validated by a comparison with existing analytic and empirical solutions through the previous research [4]. In addition, a viscous-potential method was introduced to compensate for the effects of viscosity.

## 2. Numerical Formulation

Geometric definition of cavitator and generated supercavity are shown in Figure 1 and positive x-axis direction is the direction in the wake. $S_B$ is the cavitator surface, $S_C$ is the supercavity surface and $\hat{n}$ is the unit normal vector from the surface to the outer flow. The induced velocity potential, $\phi$ , which is satisfied with the governing equation in the infinite fluid domain surrounded by boundary surface is also satisfied with the Green theorem. We divided the whole body into lots of panels. And we considered the piecewise constant distribution of the normal dipoles on the whole surface and the sources only on the cavity surface. The pressure and strength of the source distributions of the axi-symmetric cavity along the cavity surface were assumed to be the same as that of the two dimensional ones.

**Figure 1.** Geometric definition of body and supercavity

Then the total potential, $\Phi$, on the ith panel can be expressed by

$$\Phi_i = \underline{U}_\infty \cdot \underline{x}_i + \sum_{j=0}^{N^D-1} \frac{\mu_j}{2\pi} \int_{S_B \cup S_C} \frac{\partial \ln r}{\partial n} dS + \sum_{k=0}^{N^S-1} \frac{q_k}{2\pi} \int_{S_C} \ln r \, dS \tag{1}$$

where $N^D$, $N^S$ and $N^P$ is the number of normal dipoles, sources, and dipoles only on the cavitator. And $i, j$ are the index numbers from zero to $N^D - 1$ and $k$ is the index number from zero to $N^S - 1$. $\underline{U}_\infty$ is the oncoming velocity, $\underline{x}$ is the position vector and $\mu_j, q_k$ is the strength of the normal dipoles and sources. To know the effect of the existence of the supercavity on the flow around it, first we assume the length of the cavity and then find the tangential velocity on the cavity surface. Cavitation number and tangential velocity are constant on the cavity surface. And with the thickness correction factor, we make new cavity shapes. On this stage, we check that the new cavity surface is satisfied with the cavity closure condition and if it is not, code return to the first stage and give the new correction cavity as an initial geometry on the first stage. After iteration process, if the cavity closure condition is satisfied, then code computes the force and pressure distribution around the cavity surface. In this paper, we did not consider the influence of the wake behind the supercavity, because the lift force can be ignored in this case as compared with the drag force acting on the cavitators. And we considered a dummy body behind the cavitator in order to investigate more practical cases. Before the cavity covers the entire body, some parts of body surface are wet with water. In the case of turbulent flow, the coefficient of the friction force is strongly affected by the surface roughness [5],

$$C_f = \left(1.89 - 1.62 \log_{10} \frac{\varepsilon}{L}\right)^{-2.5} \tag{2}$$

where $\varepsilon$ is the surface roughness and $L$ is the length of the plate. Eq. (2) was used to calculate the water friction force on the surface and the value was fixed to 0.0032 in this paper. Until now, the term is introduced only in the pure liquid part and the value of the coefficient is dependent on each practical cases. We validated our method by comparisons with the analytic solutions, empirical values, and the results of the simulations by using ANSYS-FLUENT.

## 3. Results and Discussion

The model is close to the real-size practical model, which is similar to the supercavitating torpedo, called Shkval, which was developed in Russia. The body length is 8.2 m and the maximum body diameter is 533 mm. The maximum speed is known to be 200 knots or greater [6]. The diameter of the cavitator, $D_C$, of the proposed analysis model is 250 mm, the maximum diameter of the body, $D_B$ is 533 mm, and the length of a dummy body is 8 m. The model is assumed to move 10 m below sea level. The water density is 998.2 $kg/m^3$ and the vapor pressure is 2,300 $Pa$ according to the water temperature, 20° C.

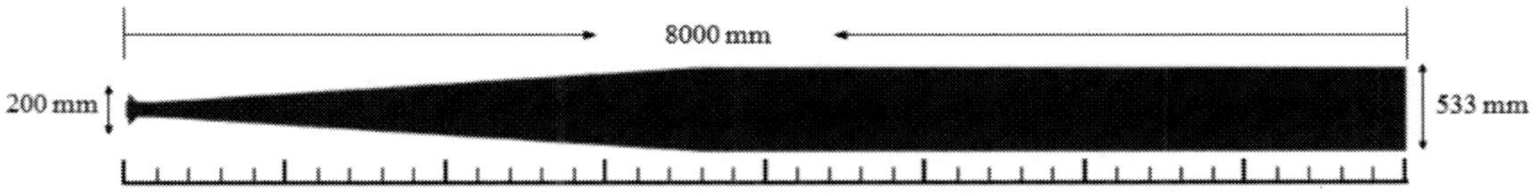


**Figure 2.** Computational model

**Table 1.** Predicted supercavity and hydrodynamic force

| V (m/s) | σ | Potential Drag(kN) | Viscous-Potential Drag(kN) | Ls [m] | Ds [m] |
|---|---|---|---|---|---|
| 27.80 | 0.511 | 14.40 | 27.55 | 0.4 | 0.33 |
| 39.67 | 0.251 | 24.09 | 50.11 | 1.0 | 0.43 |
| 52.58 | 0.143 | 38.56 | 80.78 | 2.0 | 0.54 |
| 63.11 | 0.099 | 53.39 | 107.89 | 3.0 | 0.62 |
| 71.11 | 0.078 | 66.44 | 126.63 | 4.0 | 0.69 |
| 78.06 | 0.065 | 79.06 | 135.24 | 5.0 | 0.75 |
| 85.89 | 0.053 | 94.67 | 142.06 | 6.0 | 0.80 |
| 91.68 | 0.047 | 107.19 | 138.43 | 7.0 | 0.85 |
| 97.04 | 0.042 | 119.50 | 129.06 | 8.0 | 0.89 |
| 106.78 | 0.035 | 143.66 | 143.66 | 10.0 | 0.97 |
| 115.53 | 0.030 | 167.32 | 167.32 | 12.0 | 1.05 |

The characteristic features of the supercavity developed by the cavitator according to the different speed or cavitation number can then be predicted, and more detail information is shown in Table 1. $L_S$ and $D_S$ is the predicted cavity length and maximum diameter. Figure 3 shows the drag forces according to each velocity conditions. The speed and acting forces can be determined according to different depth conditions.

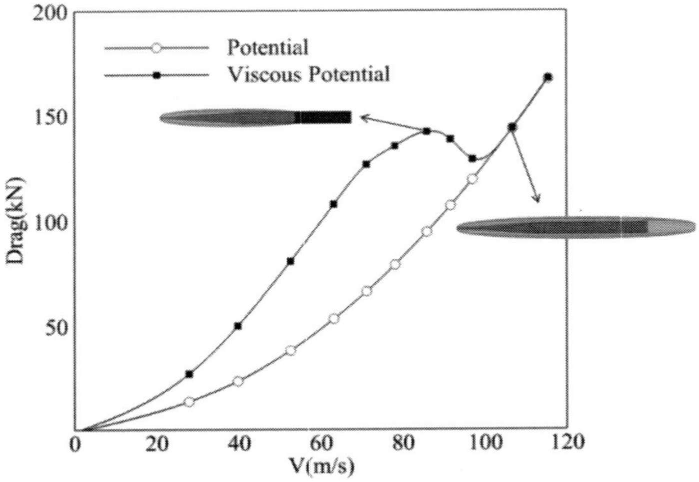


**Figure 3.** Predicted drag force; comparison of pressure and friction drag

Figure 4 shows the speed and drag forces below each sea level conditions when the predicted cavity length, which are non-dimensionalized with the diameter of the cavitator, are 30, 40, 50, and 60.

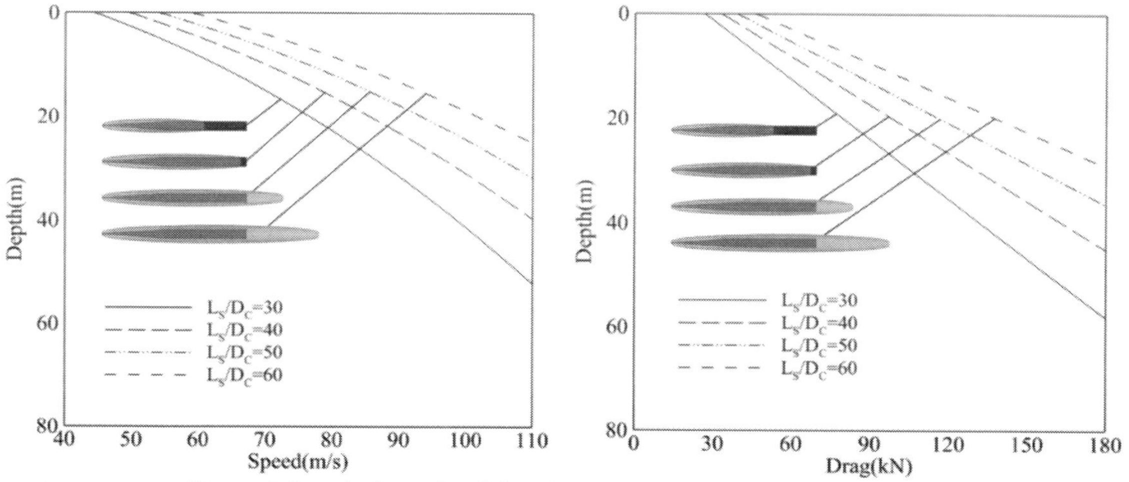


**Figure 4.** Required speed and drag force according to the depth conditions

## 4. Conclusions

In recent years, the practicality of supercavitation has been growing with the increased engineering applicability of the high speed underwater objects. The important features of a supercavity are the cavity shape and drag force acting on the body according to the operating conditions. In this study, a potential based numerical method was developed to predict the supercavity around various shapes of the cavitator. Based on the reliability of the two-dimensional analysis, a three-dimensional analysis was developed. The numerical results were compared with the existing analytic and empirical solutions, and showed reasonable agreement. By employing the coefficient of the friction force, a viscous-potential method was suggested and the drag forces for a certain model of a specific shape were calculated. The characteristics of drag forces appearing in the course of the cavity growth with increasing speed were determined. The supercavity of a practical model considering the influence of the depth conditions was also investigated.

## Acknowledgement

This work was carried out with the support of National Research Foundation of Korea(NRF-2014M3C1A9060785).

## References

[1] Lee C S 1979 *Prediction of steady and unsteady performance of marine propellers with or without cavitation by numerical lifting surface theory* (Cambridge Mass)

[2] Kim Y G and Lee C S 1990 Super-Cavitating Flow Problems about Two-Dimensional Symmetric Strut *J Soc Naval Archi Korea* **27** 4 p 15-26.

[3] Lee C S, Kim Y G and Lee J T 1992 A potential-based panel method for the analysis of a two-dimensional super- or partially-cavitating hydrofoil *J Ship Res.* **36** 2 p 168-81.

[4] Ahn B K, Jang H G, Kim H T and Lee C S 2012 Numerical Analysis of Axisymmetric Supercavitating Flows *Proceedings of the 8th International Symposium on Cavitation (Singapore, 14-16 August 2012)*

[5] Schilichting H 1979 *Boundary Layer Theory* (New York, McGraw-Hill)

[6] Polmar N and Moore K J 2004 *Cold War Submarines : The Design and Construction of U.S. and Soviet Submarines* (Potomac Books, Dulles)

Journal of Physics: Conference Series **656** (2015) 012122 doi:10.1088/1742-6596/656/1/012122

# Numerical investigation of shedding partial cavities over a sharp wedge

**B Budich**[1], **S Neuner**[1], **S J Schmidt**[1] **and N A Adams**[1]

[1] Institute of Aerodynamics and Fluid Mechanics,
Department of Mechanical Engineering, Technische Universität München,
Boltzmannstr. 15, D-85748 Garching b. München

E-mail: bernd.budich@tum.de

**Abstract.** In this contribution, we examine transient dynamics and cavitation patterns of periodically shedding partial cavities by numerical simulations. The investigation reproduces reference experiments of the cavitating flow over a sharp wedge. Utilizing a homogeneous mixture model, full compressibility of the two-phase flow of water and water vapor is taken into account by the numerical method. We focus on inertia-dominated mechanisms, thus modeling the flow as inviscid. Based on the assumptions of thermodynamic equilibrium and barotropic flow, the thermodynamic properties are computed from closed-form analytical relations. Emphasis is put on a validation of the employed numerical approach. We demonstrate that computed shedding dynamics are in agreement with the references. Complex flow features observed in the experiments, including cavitating hairpin and horse-shoe vortices, are also predicted by the simulations. Furthermore, a condensation discontinuity occurring during the collapse phase at the trailing portion of the partial cavity is equally obtained.

## Introduction

The understanding and assessment of transient cavity dynamics play a key role for a broad range of applications, such as hydraulic machinery or naval propulsors. Cavitation decreases the efficiency of fluid machinery, causes pressure oscillations in conduits and leads to noise and structural vibrations. Moreover, the occurrence of transient cavitating structures is frequently connected to material erosion, thereby limiting the lifespan of the affected components. Thus, an accurate prediction of transient cavitation phenomena by numerical methods is desirable in order to complement experimental work in the design process.

For an investigation of the cavitating flow over a sharp wedge, Ganesh [1] recently used X-ray densitometry in order to acquire a spanwise-averaged, time-resolved void fraction field. For intermittently shedding and periodic shedding partial cavities, experimental evidence for an upstream-propagating condensation discontinuity is obtained.

With the present numerical study, we reproduce the experiments by Ganesh [1] at an upstream cavitation number $\sigma_0 = 1.94$, which exhibits periodic shedding. In a previous numerical investigation of a NACA0015 hydrofoil by Schmidt *et al.* [2], a homogeneous mixture approach was applied. Utilizing an inviscid flow model, it has been shown by the authors that the condensation discontinuity can be predicted numerically. Therefore, the same strategy is pursued for the current contribution. Focus is put on the validation of the employed numerical model by comparing cavitation patterns and flow dynamics.

Journal of Physics: Conference Series **656** (2015) 012122 doi:10.1088/1742-6596/656/1/012122

## Numerical Method

A density-based, finite volume method in combination with a homogeneous mixture model [2, 3] is applied to the simulation of the two-phase flow of liquid water and water vapor. It is assumed that the flow is predominantly inertia-driven, i.e. viscous effects are neglected. Furthermore, local mechanical and thermodynamic equilibrium is assumed. Solved and non-condensable gas content is neglected and the flow is considered as barotropic. Phase change in two-phase regions, treated as saturated mixtures, is modeled as isentropic.

The governing equations are the unsteady, compressible Euler equations, spatially discretized on structured, body-fitted grids. Time integration is performed by a 4-stage Runge-Kutta method. A modified Tait-equation [4] is utilized for computing the pressure $p(\rho)$ in the pure liquid. For water-vapor mixtures, the pressure is obtained by integrating the equilibrium speed of sound $c^2 = \frac{\partial p}{\partial \rho}|_s$. The void fraction $\alpha$ is computed using the mixture density $\rho$ and the densities for the saturated liquid $\rho_{\mathrm{l,sat}}$ and vapor $\rho_{\mathrm{v,sat}}$ as $\alpha = (\rho - \rho_{\mathrm{l,sat}})/(\rho_{\mathrm{v,sat}} - \rho_{\mathrm{l,sat}})$. The values of $\rho_{\mathrm{l,sat}}$ and $\rho_{\mathrm{v,sat}}$ are evaluated at a constant reference temperature $T_{\mathrm{ref}} = 20°\mathrm{C}$.

## Numerical Setup

The experimental facilities are described in detail by Ganesh [1]. A sketch of the test section geometry is depicted in Fig. 1. The computational domain encompasses the double contraction leading to the reduced test section with a $76.2 \times 76.2$ mm cross section and a sharp step in the downstream portion. The wedge, with a contraction angle $\varphi_1 = 22.1°$, diffuser angle $\varphi_2 = 8.13°$ and a height $h = 25.4$ mm, is mounted on the lower wall. The full domain is shown in Fig. 2. The inlet is located at 1.5 m upstream of the wedge apex. A homogeneous inflow velocity is prescribed such that the experimentally measured velocity at position ①, $v_1 = 8$ m/s, is matched. A large volume at the downstream boundary reduces wave reflections to the interior. An asymptotic outlet pressure boundary condition is imposed, such that the pressure at position ②, $p_2 = 52$ kPa, agrees with the experiments. Remaining mesh boundaries are modeled as slip-walls.

A grid study involving four grid levels is conducted, with a total of $45 \cdot 10^6$ cells on the finest level. In order to reduce the computational cost, a grid sequencing method is employed. In the vicinity of the wedge, as depicted in Fig. 3, the flow field is discretized on the finest level with an average minimum cell length of 0.3 mm in wall-normal and 0.07 mm in wall-tangential direction.

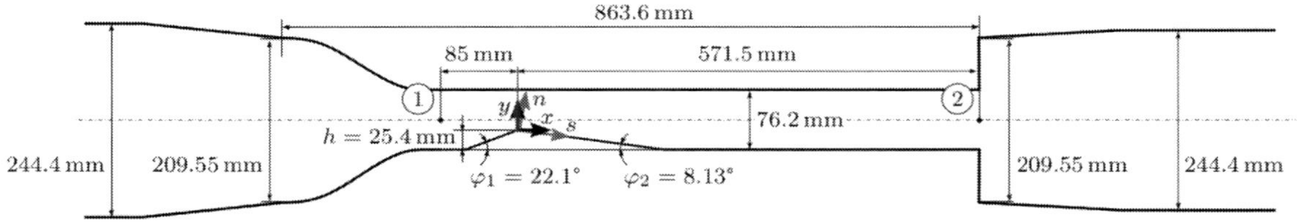


**Figure 1.** Test section geometry.

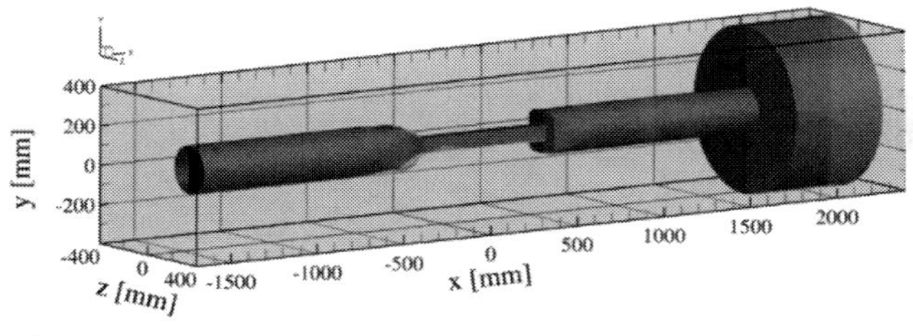


**Figure 2.** Overview of the computational domain, with upstream ducting and a double contraction, the test section with wedge, and the downstream diffuser.

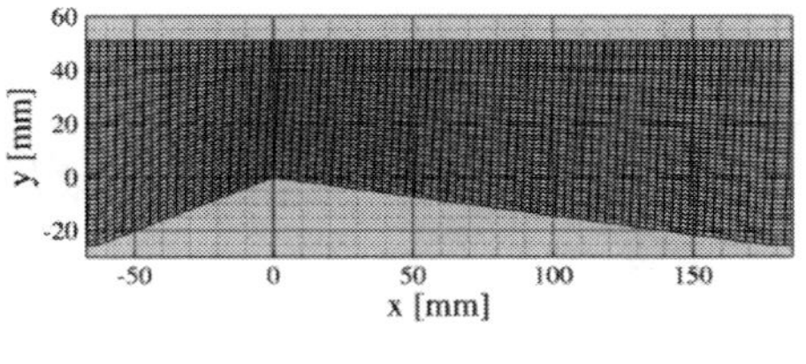


**Figure 3.** Numerical grid in the vicinity of the wedge (finest grid level, every $4^{\mathrm{th}}$ grid line shown).

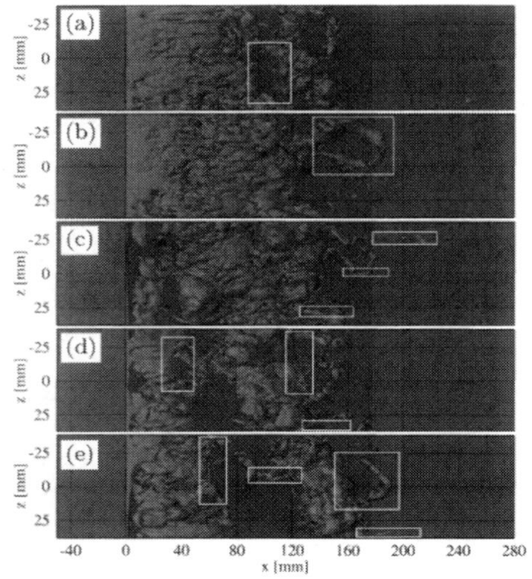


**Figure 4.** Time series of instantaneous vapor structures (iso-surfaces $\alpha = 0.1$), time interval between frames $\Delta t \approx 7.5 \cdot 10^{-3}$ s (from top to bottom).

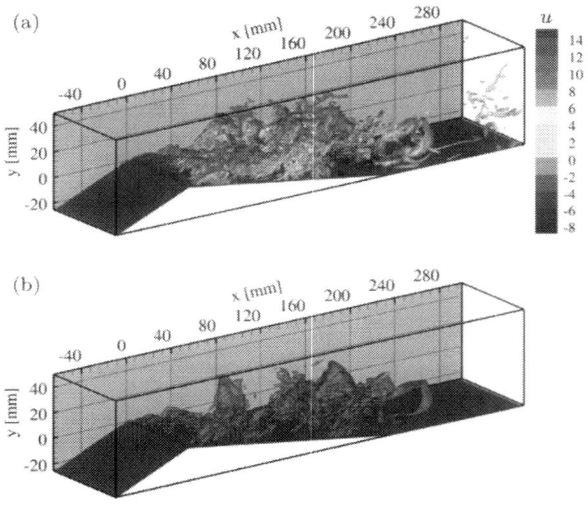


**Figure 5.** Comparison of instantaneous vortical structures (iso-surfaces $|\boldsymbol{\omega}| = 4000\,\mathrm{s}^{-1}$, colored by axial velocity $u$, top), and instantaneous vapor structures (iso-surfaces $\alpha = 0.1$, bottom)

## Results and Discussion

Onset of cavitation is observed at the sharp apex of the wedge. At the chosen operating conditions, a partial cavity develops and experiences periodic shedding. A characteristic Strouhal-number of $St = 0.28$ is obtained, which is in close agreement with experimentally reported values [1]. The shedding process is visualized in Figs. 4(a-e) by means of $\alpha = 0.1$ iso-surfaces, with a time interval of $\Delta t \approx 7.5 \cdot 10^{-3}$ s between selected frames. The maximum extent of the attached cavity sheet at $t = t_0$ is displayed in Fig. 4(a). While fragmented vapor structures from the previous cycle are convected further downstream, the attached part collapses towards the wedge apex. At $t = t_0 + 2\Delta t$, Fig. 4(c), the collapse front reaches the apex and the cavity completely detaches, leading to the generation of a new separated cloud. At $t = t_0 + 4\Delta t$, Fig. 4(e), a new partial cavity develops at the apex.

The simulation exhibits complex flow patterns that are also observed experimentally. Highlighted in Fig. 4, this includes cavitating horse-shoe vortices, Fig. 4(b,e), cavitation of streamwise-oriented hairpin vortices, Figs. 4(c,e), and crescent-shaped regions, Fig. 4(a,d,e). The latter are propagating shock wave structures occurring for localized collapses of vapor clouds [5]. In Fig. 5, vortical structures, visualized by iso-surfaces of vorticity magnitude $|\boldsymbol{\omega}| = 4000\,\mathrm{s}^{-1}$ (colored by axial velocity), are compared to iso-surfaces of 10% vapor volume fraction. Cavitating structures are closely connected to regions of increased vorticity.

In addition to a shedding mechanism involving a re-entrant flow, Ganesh [1] observes a condensation discontinuity propagating upstream towards the wedge apex. This mechanism is also predicted by the simulations. Figure 6 shows the spanwise-averaged instantaneous void

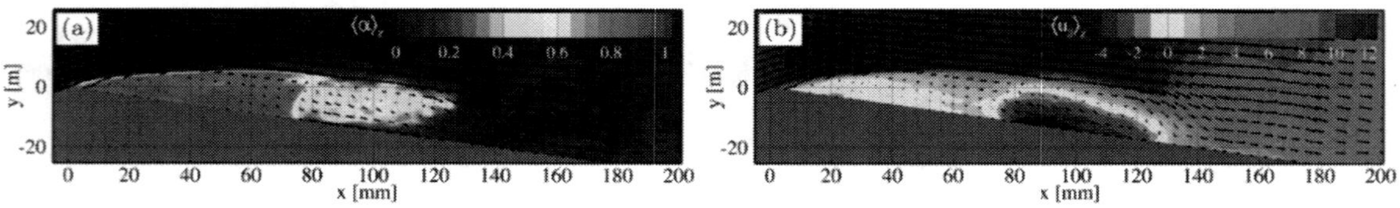


**Figure 6.** Spanwise-averaged instantaneous void fraction $\langle\alpha\rangle$ (left) and flow velocity tangential to the wedge surface $\langle u_{||}\rangle$ (right), with vectors of spanwise-averaged local flow velocity $\langle \boldsymbol{u} \rangle$.

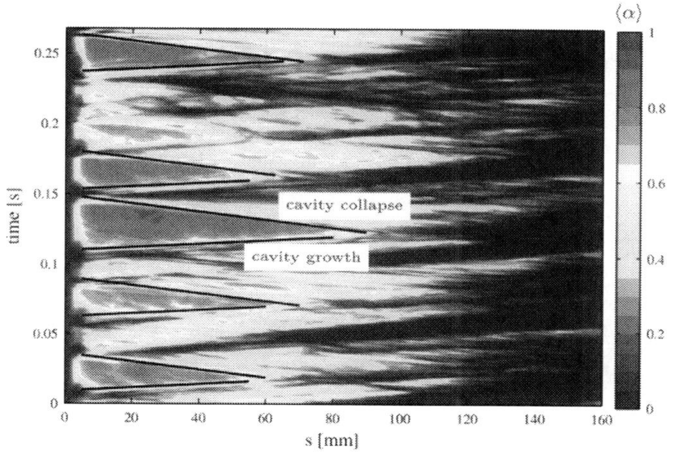


**Figure 7.** $s-t$-diagram of the spanwise-averaged void fraction $\langle\alpha\rangle$, recorded on a plane at distance $d = 2\,\mathrm{mm}$, parallel to the wedge surface. Lines with constant slopes of $7\,\mathrm{m/s}$ and $-3.5\,\mathrm{m/s}$ indicate cavity growth and collapse phases, respectively.

fraction $\langle\alpha\rangle$ and the flow velocity parallel to the wedge surface $\langle u_{||}\rangle$. A discontinuity in $\langle\alpha\rangle$ spans across the complete height of the cavity, separating the attached part from a detached cloud. Downstream of this discontinuity, the flow is directed towards the leading edge, with a maximum velocity of approximately -4 m/s. Behind the discontinuity, $\langle\alpha\rangle$ does not fully vanish. This can be accredited to the averaging procedure, as a noticeable variation in the location of the discontinuity in the spanwise direction is observed. On individual slices in the $x-y$ plane (omitted here for brevity), complete condensation to $\alpha=0$ is found.

Figure 7 shows the $s-t$-diagram of $\langle\alpha\rangle$, recorded on a plane at distance $d=2\,\mathrm{mm}$ parallel to the wedge surface. Cavity growth and collapse phases can easily be identified, as indicated by lines of positive and negative slope, respectively. For the cavity growth, a characteristic speed of $\approx 7\,\mathrm{m/s}$ is recorded for the observed shedding cycles. In close agreement with the experiments, the condensation discontinuity travels towards the apex with an average bulk velocity of -3.5 m/s, accelerating during the late collapse phase. The void fraction reaches maximum instantaneous values of $\langle\alpha\rangle \approx 0.95$, which is consistent with the experimental observations.

In summary, very good agreement between the experiments and the simulations is found. Experimentally observed flow and cavitation patterns as well as cavity dynamics are reproduced numerically, including the condensation discontinuity. It is concluded, that the chosen homogeneous mixture model is well suited for the simulation of shedding partial cavities.

## Acknowledgments

Support for this research was provided by the U.S. Office of Naval Research (ONR) and ONR Global through the NICOP project entitled "Numerical Investigation of Ship-Propeller Cavitation with Full Description of Shock-Wave Dynamics" (Contract No. N62909-12-1-7034). The authors are grateful for Dr. Ki-Han Kim (ONR HQs) and Dr. Woei-Min Lin (ONR Global) for their support and guidance throughout this project. The authors further greatly acknowledge Prof. Steven Ceccio and Dr. Harish Ganesh for sharing data from the experiments and for the fruitful discussions on the subject. Computing time was provided by the Gauss Centre for Supercomputing e.V. (www.gauss-centre.eu) on the GCS Supercomputer *SuperMUC* at the Leibniz Supercomputing Centre (LRZ, www.lrz.de).

## References

[1] Ganesh H 2015 *Bubbly Shock Propagation as a Cause of Sheet to Cloud Transition of Partial Cavitation and Stationary Cavitation Bubbles Forming on a Delta Wing Vortex* Ph.D. thesis University of Michigan
[2] Schmidt S J, Thalhamer M and Schnerr G H 2009 *7th International Symposium on Cavitation, CAV 2009* (Ann Arbor, Michigan, USA)
[3] Schmidt S J, Sezal I H, Schnerr G H and Thalhamer M 2008 *46th AIAA Aerospace Sciences Meeting and Exhibit, ASME 2008*
[4] Sezal I H 2009 *Compressible Dynamics of Cavitating 3-D Multi-Phase Flows* Ph.D. thesis Technische Universität München
[5] Reisman G E and Wang Y C 1998 *Journal of Fluid Mechanics* **355** 255–283

# The non-linear response of bubble clouds to pressure excitations

**D. Fuster and L. Bergamasco**

CNRS-UPMC (UMR 7190) Universite Pierre et Marie Curie, 4 Place Jussieu, 75005, Paris

E-mail: fuster@dalembert.upmc.fr

**Abstract.** In this article we numerically investigate the non-linear response of a bubble cloud against a periodic pressure excitation. By exciting a planar bubble curtain with an external acoustic pulse of given amplitude and frequency, we characterize the global dynamic response of the system using phase diagrams representing the void fraction against the excitation pressure. Even in the absence of mass transfer, the void fraction around which the system oscillates increases when increasing the excitation amplitude. We show how the maximum pressures reached during the collapse of bubbles are higher in polydisperse bubble clouds than in monodisperse clouds for strong pressure pulses.

## 1. Introduction

The response of bubble clouds to external pressure pulses is a difficult problem with application in various fields such as military applications, noise reduction or sea life protection. In this latter case, the presence of a bubble curtain around an acoustic source is used to attenuate the pressure waves and minimize the damage of the noise induced in its surroundings. Various authors have tried to model the response of bubble clusters using Direct Numerical Simulations. However, the range of applicability of such codes is very limited due to the lack of enough computational resources. Thus, similar to the development of LES models in turbulence, new lines of research have been devoted to propose simplified models able to capture the influence of thousands or millions of small bubbles in large scales. In this manuscript, we use the model proposed in [1] to investigate the influence of non-linear bubble cluster response and polydispersity. To this end, we explore the response of large bubble cloud curtains to non-linear pulses in order to gain new insight into the effect of bubbles on global and local variables.

## 2. Physical model and numerical code

In this work we use the model presented in Fuster & Colonius [1] implemented in a compressible flow solver that utilizes advanced WENO shock-capturing techniques [4]. This code has been validated against the theoretical solution of linear wave propagation in diluted bubbly liquids in Fuster et al [2]. The model treats bubbles as punctual sources of mass and momentum that depend on the local dynamic response of the bubbles at the subgrid scale. To obtain the bubble oscillating motion, the model resorts to a potential decomposition at the subgrid scale in order to derive a Rayleigh-Plesset like equation that allows us to integrate the bubble radius evolution in time. This model makes possible to capture the influence of bubble clouds containing a large number of bubbles in the pressure waves propagating through a bubble cloud. For more details about the model and the implementation in the code, the reader is referred to the publications mentioned above.

## 3. Numerical results

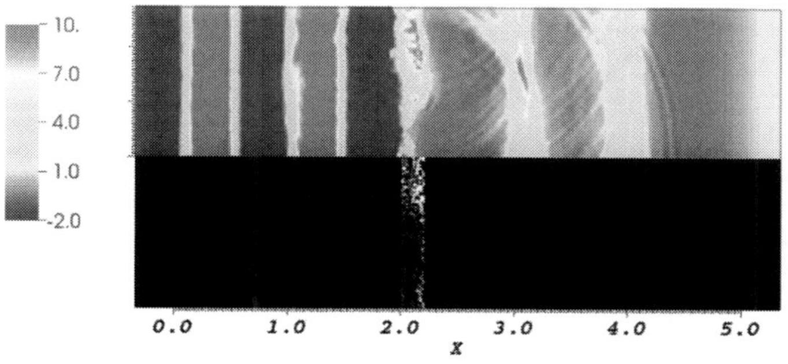


**Figure 1.** Snapshots of the pressure (top) and concentration contours (bottom) after some cycles. The presence of the bubble curtain significantly modifies the pressure field upstream, downstream and inside the cloud. Strong bubble collapses take place inside the cloud. The color scale for pressure is made dimensionless using the reference pressure.

In this manuscript we investigate the influence of polydispersity in the response of a bubble cloud excited with a sinusoidal planar wave of given amplitude, $\Delta P$, and frequency $f$ in a liquid with speed of sound $c_l$. The excitation is generated using a source in the energy equation far from the bubble cloud. We use the wavelength of the incoming wave $\lambda = c_l/f$ as a characteristic length of the problem. The liquid properties are those of water. The initial void fraction $\beta_0$ is set in all simulations equal to $10^{-4}$ vol/vol. The dimensionless width of the bubble curtain in the wave propagation direction is $\frac{L_x}{\lambda} = 0.4$. We consider a situation with infinite number of bubbles in the two transversal directions. The averaged dimensionless radius is set to $\frac{R_0}{\lambda} = 6.78 \cdot 10^{-4}$. The actual bubble cluster is generated with a random number function obeying a log-normal distribution with a given dispersion $\sigma$. The bubbles are randomly distributed in the domain. Heat transfer effects between the bubble and its surrounding are taken into account but in this case mass transfer effects are neglected.

The simulation domain size in dimensionless units is $6 \times 2 \times 0.1$. Periodic boundary conditions are used in the $y$ and $z$ directions in order to mimic the presence of an infinite bubble cloud. The grid size is set to $\frac{\Delta x}{\lambda} = 5 \cdot 10^{-3}$ except in the $z$ direction, where only one grid is used in order to save computational time.

Figure 1 shows two snapshots of typical pressure fields after the arrival of the incoming wave. The bubble curtain significantly disturbs the pressure wave inducing a large attenuation. Inside the bubble cluster, bubbles react to the pressure pulse expanding first and finally imploding violently and generating large local pressures.

In order to characterize the global response of the bubble cluster we use void fraction versus excitation pressure diagrams (top of Figure 2). As expected the bubble cluster describes periodic orbits around a fixed point. The location of this point strongly depends on the amplitude of the

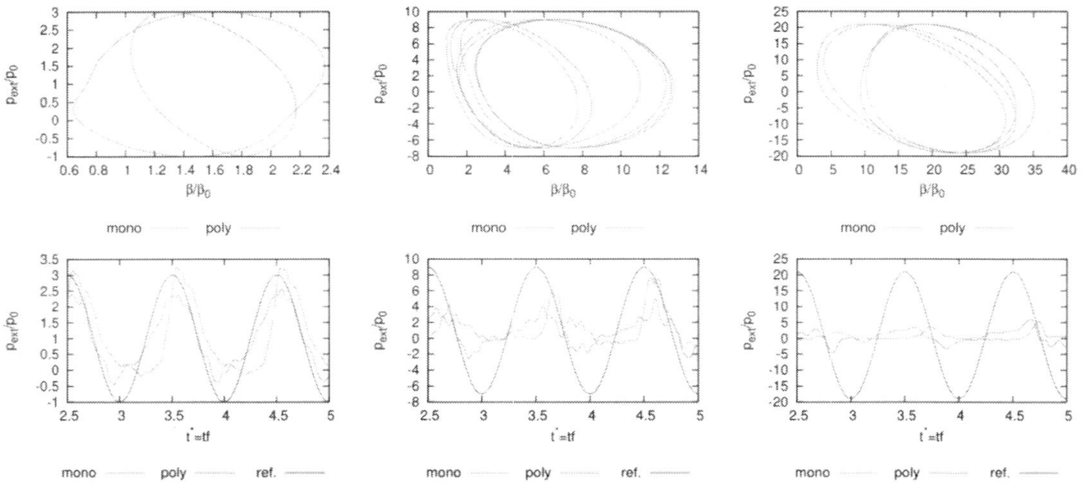


**Figure 2.** Top: Void fraction as a function of the excitation pressure for three different excitation amplitudes (from left to right: 2, 8, 20 atm). Bottom: Pressure sampled at a distance of 0.3 from the bubble cluster. Bubbles damp more effectively large amplitude waves due to the effective increase of the averaged void fraction during the excitation.

excitation. The larger the amplitude the higher the void fraction around which the cluster oscillates is. Note that we have neglected vaporization effects, so this effect cannot be attributed to a global change in the amount of gas/vapor along the simulation. Instead, this effect is attributed to the non-symmetry on the response of bubbles to pressure pulses in non-linear regimes. Bubbles tend to oscillate around an equilibrium point given by $R_0$ and $p_0$ only in the linear regime. For large amplitude excitations the increase on the void fraction induced during the bubble growth is more important than the compression taking place during the bubble collapse and we observe a *virtual* increase on the void fraction of the bubble cloud during the excitation of the cloud. This phenomenon can be interpreted in terms of potential energies [3], which show how the bubble expansion is energetically favored when the radius is larger than the Blake's radius ($R_c \approx \sqrt{3}R_0$). The increase on the effective void fraction attenuates more effectively the incoming pressure waves as the amplitude increases (bottom of Figure 2) validating the hypothesis that bubble curtains are especially suitable to damp large amplitude pressure waves. We can see how polydispersity does not play a major role on these quantities.

In addition to the global response of the cluster, the model allows us to access to local quantities such as the maximum local pressures reached during the bubble collapse. In particular we obtain statistics of the maximum pressures reached at the liquid interface during the collapse of every single bubble. Figure 3 shows the cumulative distribution of the maximum pressures reached during the bubble collapse for different conditions. As expected, on the left of Figure 3 we can see how the maximum pressures increase as the excitation amplitude increases. Remarkably, polydispersity plays a major role on these distributions (Figure 3 right). At low amplitudes (red line), the cumulative distribution is broader showing how it is possible to induce large pressure during the collapse only due to the bubble size variability in the cluster. In this case the median value remains approximately constant. For large amplitudes the situation is different and we clearly show how polydispersity effectively increases the maximum pressures reached during the collapse of bubbles. This effect is attributed to the non-synchronization of the bubble oscillation process enhanced by the presence of bubbles with different sizes. Note that these observations are consistent with the predictions in the low frequency limit obtained from the simplified theory presented in [3], although certainly dynamic effects play an important role in this situation.

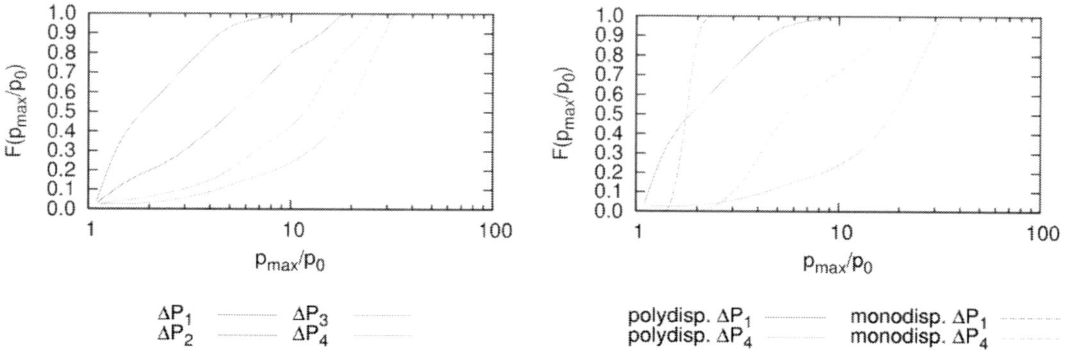


**Figure 3.** Cumulative distribution of the maximum pressures reached in the liquid along the simulation. Left: Results obtained for a polydisperse bubble cloud excited at various amplitudes ($\Delta P_1 = 2$, $\Delta P_2 = 4$, $\Delta P_3 = 8$ and $\Delta P_4 = 20$ atmospheres). Right: Comparison between the cumulative distribution obtained in a monodisperse and polydisperse bubble cloud for low amplitude excitations (2 atm) and large amplitude excitations (20 atm).

## 4. Conclusions

This paper presents numerical results used to investigate non-linear effects on the response of bubble clouds. For sinusoidal excitations we see how the system oscillates around a virtual fixed point with a *virtual* concentration larger than the equilibrium concentration without any excitation. This virtual concentration tends to increase when increasing the excitation amplitude for a fixed amount of gas and neglecting any mass transfer process. This effect is explained by the preference of bubbles to expand when they go beyond the critical Blake's radius.

The distribution of maximum pressures reached in the liquid have been shown to depend on the polydispersity of the bubble cloud. We systematically observe that the largest pressures are reached in the polydisperse case. Still, for low amplitudes, the median value remains constant irrespective of the polydispersity level. For large amplitudes, polydispersity significantly increases the maximum pressures reached during the bubble collapse.

## References

[1] D. Fuster and T. Colonius. Modeling bubble clusters in compressible liquids. *Journal of Fluid Mechanics*, 688:253–289, 2011.

[2] D. Fuster, J.M. Conoir, and T. Colonius. Effect of direct bubble-bubble interactions on linear-wave propagation in bubbly liquids. *Phys. Rev. E*, 90(6):063010, 2014.

[3] D. Fuster, K. Pham, and S. Zaleski. Stability of bubbly liquids and its connection to the process of cavitation inception. *Phys. Fluids*, 26:042002, 2014.

[4] E. Johnsen and T. Colonius. Implementation of WENO schemes in compressible multicomponent flow problems. *Journal of Computational Physics*, 219(2):715–732, 2006.

# How to control bubble nucleation from superhydrophobic surfaces

**Alberto Giacomello, Matteo Amabili and Carlo Massimo Casciola**
Dipartimento di Ingegneria Meccanica e Aerospaziale, Sapienza Università di Roma, via Eudossiana 18, 00184 Roma, Italy
E-mail: alberto.giacomello@uniroma1.it

**Abstract.** Superhydrophobicity is realized by entrapping gas bubbles inside surface roughness. While this strategy affords remarkable surface properties, it enhances the risk of cavitation from these gas nuclei at negative pressures. Here we use free energy molecular dynamics simulations and an extension of the classical nucleation theory to show that the relevant nucleation rates and barriers can be controlled by engineering the surface structure. Mimicking the re-entrant and chemically heterogeneous structure found in the leaves of the Salvinia molesta allows one both to stabilize the gas pockets against liquid intrusion and to reduce the risk of cavitation.

## 1. Introduction

Superhydrophobicity is emerging as a means to control the macroscopic properties of surfaces in contact with a liquid. Superhydrophobic properties include self-cleaning [1], anti-(bio)fouling [2], and drag reduction [3]. At the same time, however, the superhydrophobic state in surfaces under flow conditions may facilitate the nucleation of gaseous bubbles [4], starting the cavitation process. Here we discuss the issue of controlling such nucleation process by a suitable design of the nanopatterns [5].

Superhydrophobic properties are linked to the entrapment of gas pockets inside of surface roughness or patterns – the so-called Cassie state [6]. Capillary forces sustain this "suspended" state. The Cassie state can be achieved by combining surface roughness and hydrophobic coatings [7] or by special re-entrant structures [8]. Superhydrophobicity is lost if the liquid wets the surface realizing the transition to the so-called Wenzel state [9].

Similarly to nucleation, the Cassie-Wenzel transition is a thermally activated event over a broad range of pressures. For superhydrophobic surfaces, in the realm of negative pressures which is relevant to cavitation, at least three (meta)stable states can exist: Wenzel, Cassie, and pure vapor or nucleated state; free energy barriers make of the Wenzel-Cassie and of the Cassie-vapor transition two activated events, see Fig. 1. Special techniques are required in order to compute the path and the relevant free energy barriers for such transitions; these are usually referred to as *rare events methods* [10]. Among those, restrained molecular dynamics and the string method have proved useful in studying superhydrophobicity and nucleation [11, 12]; the atomistic approach is computationally expensive but yields a complete and assumptionless picture of molecular phenomena, non-classical effects, and nano-confinement.

On the other end, classical nucleation theory (CNT) [13, 14] is based on a macroscopic description of the fluid and works well in predicting the nucleation barriers relatively close to

Journal of Physics: Conference Series **656** (2015) 012124 doi:10.1088/1742-6596/656/1/012124

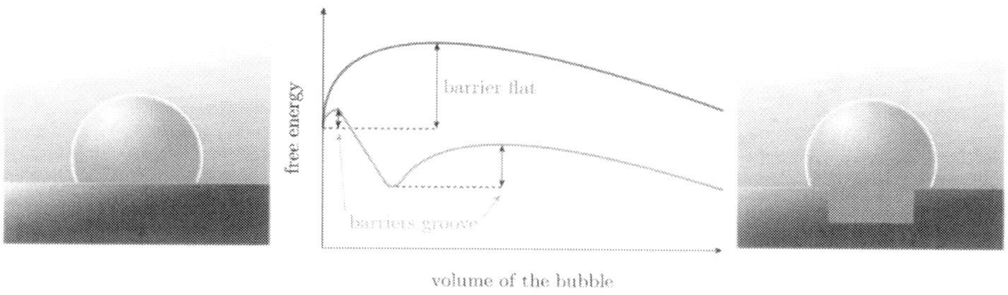


**Figure 1.** Catalytic effect of a rectangular groove on bubble nucleation as compared to a flat surface. For details on CREaM calculations for determining the free energy profiles see [4].

the liquid-vapor coexistence where the critical bubbles are of macroscopic dimensions; however, as the pressure decreases CNT fails to predict the nucleation rates, see, e.g., [15]. Another limitation of CNT is that it assumes the bubble radius $R$ as the order parameter to describe the nucleation process. While this approach is sufficient for simple cases such as homogeneous nucleation and nucleation from a flat surface, an extension of this macroscopic theory is needed in order to evaluate nucleation from generic surface geometries; a possible approach is the continuum rare events method (CREaM) proposed in Refs. [4, 16].

## 2. Results & Discussion

### *2.1. Catalytic effect on nucleation due to surface heterogeneities*

It is well known that the presence of topographical and chemical heterogeneities in walls and solid impurities may promote the nucleation of bubbles thus favoring the inception of cavitation. For instance, the crevice model by Atchley and Prosperetti predicts how the presence of pre-existing vapor bubbles entrapped in surface asperities influence the cavitation pressure [17].

One can extend the picture of the crevice model to consider also the Wenzel-Cassie transition which is part of the overall process of nucleating a bubble from the liquid. This has been investigated in [4] for various kinds of crevices. Figure 1 shows that the nucleation process of a gas bubble is divided into two activated sub-steps when it happens from a crevice: the Wenzel-Cassie and the Cassie-vapor transitions. The two free energy barriers are lower than the one needed to nucleate the same bubble from a flat surface with the same properties. In other words, the crevice acts as a catalyst, enormously increasing the nucleation rate from the surface [4] (see also the similar case of crystal nucleation in pores [18]).

### *2.2. Barriers in a typical case*

The picture emerging from the above discussion is that nucleation can be a multistep process, especially for superhydrophobic surfaces, and that the nucleation rate depends heavily on the status of the surface in contact with the liquid and in particular on the geometry and chemistry of the surface. A question of technological interest is whether cavitation can be controlled, by tuning such surface properties in such a way that the surface maintains its superhydrophobic character (Cassie state) and, at the same time, reduces the nucleation rates.

In order to answer to this question, in Fig. 2 we show the free energy barriers computed via restrained molecular dynamics for a re-entrant decoration of the surface with a heterogeneous chemistry: hydrophobic within the crevice and hydrophilic on the top layer. This peculiar configuration is inspired by the air-retaining properties of the water fern *Salvinia molesta* [19] which has been shown to have optimal stability properties of the Cassie state against nucleation and wetting [5].

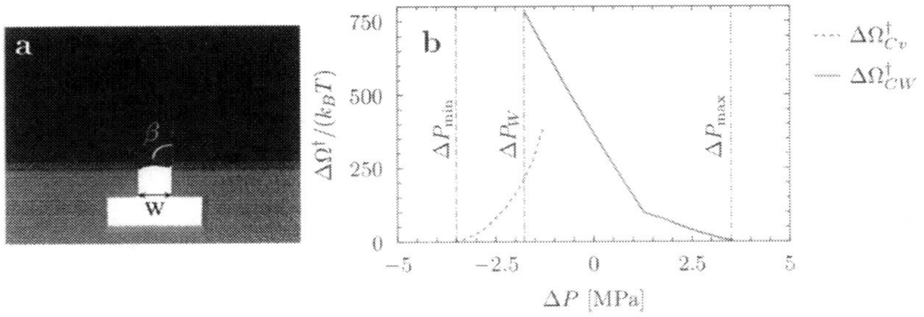


**Figure 2.** a) Salvinia-inspired geometry for the surface nanodecoration. The surface has an hydrophilic top (dark brown, contact angle $\theta_{Y,out} = 55°$), a hydrophobic interior (light brown, $\theta_{Y,in} = 110°$), and $w \approx 5$ nm. b) Nucleation ($\Delta\Omega^\dagger_{Cv}$) and wetting ($\Delta\Omega^\dagger_{CW}$) free energy barriers as a function of the pressure difference between the gas and the vapor phase $\Delta P \equiv P_l - P_g$. The vertical lines indicate the nucleation spinodal pressure $\Delta P_{\rm min}$, the intrusion one $\Delta P_{\rm max}$, and the (confined) liquid spinodal $\Delta P_{\rm W}$ where the Wenzel state becomes unstable (see text).

The free energy barriers for nucleation, $\Delta\Omega^\dagger_{Cv}$, diverge as the coexistence pressure $\Delta P \equiv P_l - P_g = 0$ is approached. What is more interesting is that for pressures below $\Delta P_{\rm min}$ $\Delta\Omega^\dagger_{Cv}$ vanishes: this is the spinodal for the Cassie-vapor transition. This spinodal occurs at very low pressures for the nanocavities ($\sim 5$ nm) here considered, $\Delta P_{\rm min} = -3.5$ MPa. The bulk spinodal of pure liquids is observed at even lower pressures ($\Delta P \approx -120$ MPa for water [20]). Below $\Delta P_{\rm min}$, therefore, the nucleation of bubbles is a deterministic event happening preferentially at the surface cavity. It is interesting that also the Wenzel state has a spinodal pressure $\Delta P_W$ at which it becomes unstable. In our case $\Delta P_W \approx -1.8$ MPa, which means that the Cassie state survives at lower pressures than the Wenzel one.

The wetting of the groove by the liquid is governed by the free energy barrier between the Cassie and the Wenzel state, $\Delta\Omega^\dagger_{CW}$; this barrier monotonically decreases with pressure, see Fig. 2. In this case the spinodal is found at positive pressures, $\Delta P_{\rm max} = 3.5$ MPa.

### *2.3. Pinning of the contact line*

Both the barriers and the spinodal pressures depend on the geometry and the chemistry of the surface, which can be used to control the stability of the Cassie state against bubble nucleation and liquid intrusion. In particular, it can be shown [5] that hydrophilic chemistry of the top layer dramatically increases $\Delta\Omega^\dagger_{Cv}$ while the hydrophobic interior increases $\Delta\Omega^\dagger_{CW}$.

Also the spinodals $\Delta P_{\rm min}$ and $\Delta P_{\rm max}$ can be controlled by tuning the geometry and chemistry of the surface decorations. For the re-entrant geometry shown in Fig. 2, this dependence can be made explicit with macroscopic models of capillarity as follows. The mechanical equilibrium of a liquid-gas meniscus is described by the Laplace law, which reads for the cavity in Fig. 2:

$$\Delta P = -2\gamma \cos\beta / w \,, \tag{1}$$

where $\gamma$ is the liquid-gas surface tension, $\beta$ is the angle formed by the vertical and the tangent to the liquid-vapor interface (Fig. 2), and $w$ is the width of the crevice mouth. The maximum and minimum of Eq. 1 represent the maximum and minimum pressures that can be balanced by capillary forces, i.e., the spinodals $\Delta P_{\rm max}$ and $\Delta P_{\rm min}$, respectively. The angle $\beta$ in Eq. 1 is determined by the pinning of the contact line at the crevice mouth, which is dictated on sharp corners or chemical contrasts by the Gibbs criterion [21]:

$$\theta_{Y,out} - 90° < \beta < 90° + \theta_{Y,in} \,, \tag{2}$$

where $\theta_Y$ denotes the Young contact angles ($\cos\theta_Y < 0$ for hydrophobic and $> 0$ for hydrophilic surfaces, respectively) for the top surface ($\theta_{Y,out}$) and the inner surface ($\theta_{Y,out}$). Combining Eq. 1 with the limits for $\beta$ imposed by Eq. 2 for the case under exam one obtains the spinodals $\Delta P_{\max} = 2\gamma/w$ and $\Delta P_{\min} = -2\gamma/w$, which is the maximum range of pressures in which the Cassie state is stable. In general, it is possible to write down the Laplace equation and the Gibbs criterion for a given surface and design its chemistry and geometry such that it minimizes the risk of cavitation, while maintaining the desired superhydrophobic properties.

## 3. Conclusions

We presented molecular dynamics simulations and continuum calculations of the free energy barriers and spinodals for the nucleation of bubbles from surfaces with hydrophobic cavities. Our results show that superhydrophobicity indeed increases the nucleation rates as compared to a perfectly smooth surface, but it is possible to decrease them substantially by a careful design of the surface microstructure. In particular, a hydrophilic top layer in combination with a hydrophobic cavity interior achieve the optimal results.

## Acknowledgments

The research leading to these results has received funding from the European Research Council under the European Union's Seventh Framework Programme (FP7/2007-2013)/ERC Grant agreement n. [339446]. We acknowledge PRACE for awarding us access to resource FERMI based in Italy at Casalecchio di Reno.

## References

[1] Barthlott W and Neinhuis C 1997 *Planta* **202** 1–8
[2] Ferrari M and Benedetti A 2015 *Adv. Colloid Interface Sci.* **222** 291–304
[3] Rothstein J P 2010 *Annu. Rev. Fluid. Mech.* **42** 89–109
[4] Giacomello A, Chinappi M, Meloni S and Casciola C M 2013 *Langmuir* **29** 14873–14884
[5] Amabili M, Giacomello A, Meloni S and Casciola C 2015 *Adv. Mater. Interf.*
[6] Cassie A B D and Baxter S 1944 *T. Faraday Soc.* **40** 546–551
[7] Lafuma A and Quéré D 2003 *Nat. Mater.* **2** 457–460
[8] Tuteja A, Choi W, Mabry J M, McKinley G H and Cohen R E 2008 *Proc. Natl. Acad. Sci. USA* **105** 18200–18205
[9] Wenzel R 1936 *Ind. Eng. Chem.* **28** 988–994
[10] Bonella S, Meloni S and Ciccotti G 2012 *Eur. Phys. J. B* **85** 97
[11] Giacomello A, Meloni S, Chinappi M and Casciola C M 2012 *Langmuir* **28** 10764–10772
[12] Giacomello A, Meloni S, Müller M and Casciola C M 2015 *J. Chem. Phys.* **142** 104701
[13] Volmer M 1939 *Kinetik der Phasenbildung* (Dresden: Theodor Steinkopff)
[14] Turnbull D 1950 *J. Chem. Phys.* **18** 198–203
[15] Talanquer V and Oxtoby D W 1996 *J. Chem. Phys.* **104** 1483–1492
[16] Giacomello A, Chinappi M, Meloni S and Casciola C M 2012 *Phys. Rev. Lett.* **109** 226102
[17] Atchley A A and Prosperetti A 1989 *J. Acoust. Soc. Am.* **86** 1065–1084
[18] Page A J and Sear R P 2006 *Phys. Rev. Lett.* **97** 065701
[19] Barthlott W, Schimmel T, Wiersch S, Koch K, Brede M, Barczewski M, Walheim S, Weis A, Kaltenmaier A, Leder A *et al.* 2010 *Adv. Mater.* **22** 2325–2328
[20] El Mekki Azouzi M, Ramboz C, Lenain J F and Caupin F 2013 *Nat. Phys.* **9** 38–41
[21] Oliver J, Huh C and Mason S 1977 *J. Colloid Interface. Sci.* **59** 568–581

# A four-way coupled Euler–Lagrange approach using a variational multiscale method for simulating cavitation

**Georg Hammerl, Wolfgang A. Wall**

Institute for Computational Mechanics, Technische Universität München, Boltzmannstraße 15, 85747 Garching b. München, Germany

E-mail: hammerl@lnm.mw.tum.de

**Abstract.** An Euler-Lagrange model is developed to simulate bubbly flow around an obstacle with the aim to resolve large and meso-scales of cavitation phenomena. The volume averaged Navier-Stokes equations are discretized using finite elements on an unstructured grid with a variational multiscale method. The trajectory of each bubble is tracked using Newton's second law. Furthermore, bubble interaction is modeled with a soft sphere contact model to obtain a four-way coupled approach. The new features presented in this work, besides using a variational multiscale method in an Euler-Lagrange framework, is an improved computation of the void fraction. A second order polynomial is used as filtering function and the volume integral is transformed by applying the divergence theorem twice, leading to line integrals which can be integrated analytically. Therefore, accuracy of void fraction computation is increased and discontinuities are avoided as is the case when the kernel touches a Gauss point across time steps. This integration technique is not limited to the chosen spatial discretization. The numerical test case considers flow in a channel with a cylindrical obstacle. Bubbles are released close to the inflow boundary and void fractions up to 30 % occur at the stagnation point of the obstacle.

## 1. Introduction

Cavitation is a very important phenomenon that can be highly detrimental, e.g. when occurring at ship propellers. Vapor bubbles, formed within the fluid phase due to a sudden drop of local pressure, can cause structural damage when collapsing or lead to power loss. Millions of small nuclei, which can be found in any liquid, strongly influence cavitation inception. Simplified approaches to simulate cavitation include homogeneous flow models with a proper equation of state or two-fluid Eulerian approaches considering the volume fractions of each phase. However, due to the recent developments in computational resources and methods, it becomes feasible to simulate each nucleus as a particle which is tracked throughout a simulation in the scope of an Euler–Lagrange model.

## 2. Numerical Model

The liquid phase is described by the volume-averaged incompressible Navier-Stokes equations. The continuity equation is reformulated as in [1] leading to a non-divergence free fluid field:

$$\nabla \cdot \mathbf{u} = -\frac{1}{\epsilon_l}\left(\frac{\partial \epsilon_l}{\partial t} + \mathbf{u} \cdot \nabla \epsilon_l\right) \tag{1}$$

where $\epsilon_l$ denotes the liquid-phase volume fraction and $\mathbf{u}$ is the volume averaged fluid-phase velocity. Following [7], the volume averaging procedure reads

$$\epsilon_l \rho_l \frac{\partial \mathbf{u}}{\partial t} + \epsilon_l \rho_l \mathbf{u} \cdot \nabla \mathbf{u} = -\nabla p + \nabla \cdot \boldsymbol{\tau} + \epsilon_l \rho_l \mathbf{g} + \mathbf{s} \tag{2}$$

using the pressure $p$, the temporal and spatial constant liquid-phase density $\rho_l$ and the viscous stress tensor $\boldsymbol{\tau} = \mu \left( (\nabla \mathbf{u}) + (\nabla \mathbf{u})^T \right) - \frac{2}{3}\mu (\nabla \cdot \mathbf{u})\, \mathbf{I}$ with the dynamic viscosity $\mu$ and the identity tensor $\mathbf{I}$. The gravity vector and the coupling term with the bubbles are denoted by $\mathbf{g}$ and $\mathbf{s}$, respectively.

The fluid is described in an Eulerian frame whereas the bubbles are handled as discrete points which are tracked individually in a Lagrangian frame using Newton's second law:

$$m_b \frac{\mathrm{d}}{\mathrm{d}t}\dot{\mathbf{v}} = \mathbf{F}_G + \underbrace{\underbrace{\mathbf{F}_D + \mathbf{F}_L + \mathbf{F}_{VM}}_{=-\mathbf{s}'} + \mathbf{F}_P + \mathbf{F}_{VISC}}_{=-\mathbf{s}} + \mathbf{F}_C \tag{3}$$

in which $m_b$ and $\mathbf{v}$ are the bubble mass and velocity, respectively. Gravity force, drag force, lift force, virtual mass force, and pressure force are chosen as in [2] using a constant lift and virtual mass coefficient $c_l = c_{vm} = 0.5$. The drag coefficient $c_d$ is adopted from [3]. Viscous forces

$$\mathbf{F}_{VISC} = V_b \nabla \cdot \tau \tag{4}$$

of a bubble with volume $V_b$ require the computation of second derivatives of the fluid velocity which are zero for linear finite elements as they are used in this work. Therefore, a reconstruction of the first derivatives using an L2-projection is applied. The collision force is computed using a soft sphere contact model with the normal gap $g_{n,ij}$, unit normal vector $\mathbf{n}_{ij}$ from bubble $i$ to bubble $j$, and the stiffness parameter $k_c$ according to $\mathbf{F}_C = k_c g_{n,ij} \mathbf{n}_{ij}$ if $g_{n,ij} \leq 0$. Bubble-wall collisions are modeled assuming infinite mass of the wall element. The coupling force $\mathbf{s}$ does not include gravity and collision forces when choosing the momentum equation as in (2). Furthermore, the pressure force and viscous force can be dropped from the coupling term $\mathbf{s}$ when combined with the respective terms in the momentum equation [4]. The coupling term is then denoted as $\mathbf{s}'$ and the momentum equation reads

$$\epsilon_l \rho_l \frac{\partial \mathbf{u}}{\partial t} + \epsilon_l \rho_l \mathbf{u} \cdot \nabla \mathbf{u} = -\epsilon_l \nabla p + \epsilon_l \nabla \cdot \boldsymbol{\tau} + \epsilon_l \rho_l \mathbf{g} + \mathbf{s}' \ . \tag{5}$$

The final formulation of the volume-averaged NS equations used in this work are obtained by dividing (5) by $\epsilon_l$, and the continuity equation as given in (1).

The Euler–Lagrange model is based on a variational multiscale method for the underlying fluid according to [5] and thus it is possible to distribute coupling forces from the bubbles using the shape functions evaluated at the position of the bubble within the element. Hence, a conservative coupling can be achieved because of the partition of unity property of finite element shape functions. Fluid velocity and acceleration are necessary to evaluate the bubble forces. Interpolation from fluid quantities such as velocity and acceleration is also performed using the shape functions evaluated at the bubble position. The computation of the volume fraction occupied by the bubbles cannot be handled by the finite element discretization directly. In order to avoid pulsating velocity fluctuations in the fluid field when assigning the full bubble volume to a single element as described in [6], a remedy are smooth kernel functions with zero tangents at their bounds of influence, see also [6]. This reduces oscillations due to Gaussian integration. In this work, a distribution using a quadratic filtering function

$$\Psi^d = \begin{cases} \frac{3}{4}\left(-\frac{\left(x^d - x_j^d\right)^2}{\sigma^3} + \frac{1}{\sigma}\right) & \text{, if} \quad -\sigma \le \left(x^d - x_j^d\right) \le +\sigma \\ 0 & \text{, else} \end{cases} \tag{6}$$

for direction $d$ around the bubble position $x_j^d$ together with an improved integration of the resulting volume integral is proposed. The radius of influence $\sigma = 1.2\,r_{b,j}$ is chosen to smear the volume fraction across the underlying fluid elements. The volume integral

$$\sum_{i=1}^{n_{bub}} V_{b,i} \int_{\Omega_j} \Psi_i^x \Psi_i^y \Psi_i^z \mathrm{dxdydz} \tag{7}$$

over the domain $\Omega_j$ of fluid element $j$ gives its partial volume that is occupied by all $n_{bub}$ bubbles. The bubble influence volume is assumed to be cubic, as proposed in [8], which allows for a special treatment of the volume integral that is usually solved using Gaussian integration. If the fluid elements are distorted and not enough Gauss points are chosen, loss of accuracy in the calculation of the void fraction can occur. Inspired by [9], the volume integral is transformed to line integrals via applying divergence theorem twice instead of using Gaussian integration. The evaluation of the resulting line integrals is as expensive as using four Gauss points in each direction but accuracy is highly increased. This integration scheme allows for an infinite bubble movement to result in an infinite change in void volume which is superior to Gaussian integration.

The fluid flow is solved on an unstructured grid using stabilized linear finite elements with implicit time integration with generalized-$\alpha$ time integration ($\alpha_m = \frac{5}{6}, \alpha_f = \gamma = \frac{2}{3}$) as applied in [5]. Larger time step can be chosen than for the explicit Velocity-Verlet algorithm for the bubbles. Hence, subcycling is applied and one fluid step is followed by 100 bubble steps.

## 3. Results

The following test case with flow around a cylindrical obstacle is chosen because it can be seen as a simplified geometry of a hydrofoil. The box like geometry of 5.0 m × 2.0 m × 0.4 m (length × height × depth) contains a cylindrical hole in width direction with a diameter of 0.5 m located in the height mid plane in a distance of 1.5 m from the inflow. Fluid inflow from the left is ramped from zero to 1.0 m/s within 0.05 s. This velocity is also prescribed on the top and bottom surface whereas a slip condition is prescribed on the sides in order to model an infinitely extended cylinder. On the outflow on the right a do nothing boundary condition is imposed. Bubble inflow of 2121 bubbles is started at time $t = 0.01$ s and repeated every 0.01 s within the first layer of fluid elements 0.01 m from the inflow. The inflow velocity is chosen equal to the underlying fluid velocity and the inflow position is determined by using a rectangular grid with 101 times 21 grid points in height and width direction, respectively. A random offset of up to 20 % of the bubble radius is added to each initial position to break symmetry. The bubble radius $r_b = 0.004$ m and its density $\rho_b = 0.001$ kg/m$^3$. The stiffness parameter $k_c$ is set to 0.021 N/m leading to a maximum overlap of less than 0.02 % of the bubble radius. The fluid with dynamic viscosity $\mu = 8.9 \cdot 10^{-4}$ Pa s and density $\rho_l = 1.0$ kg/m$^3$ is discretized with 53448 tri-linearly interpolated unstructured hexahedral elements without refinement in the near wall region and a time step size of $\Delta t = 0.01$ s. Gravity acceleration is neglected. In Figure 1 (left), the nuclei are depicted and colored with their velocity magnitude at $t = 5.0$ s. The boundary surfaces of the fluid are also included in the figure but the front and rear surface are removed. Vortex shedding is about to start in the wake of the cylindrical obstacle. Due to bubble Reynolds number smaller unity, the bubbles follow the fluid flow quite accurately and can be seen as marker in the fluid flow. The cross section of the channel is reduced at the obstacle position leading to an

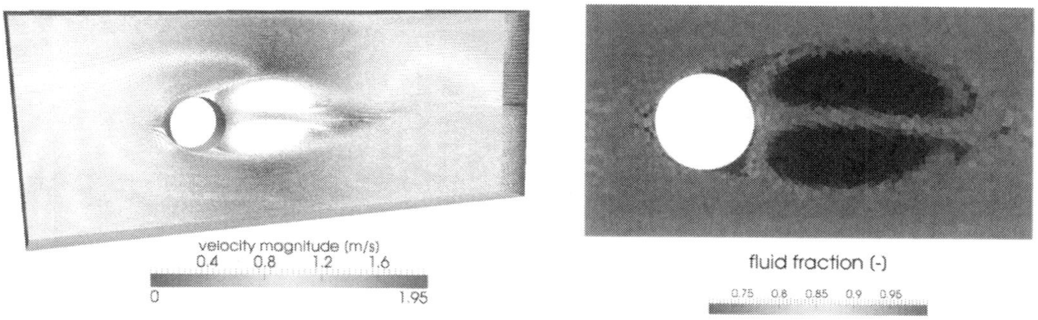


**Figure 1.** Bubble velocity (left) and close-up of fluid fraction in mid-plane (right) at $t = 5.0$ s.

increased fluid velocity which directly influences the bubble velocity. At the stagnation point of the obstacle bubbles are slowed down and extensive inter-particle contact as well as contact with the cylindrical surface occurs. In this region, fluid fraction goes down to about 70 % as depicted in Figure 1 (right).

## 4. Summary and Outlook

An Euler–Lagrange model based on a variational multiscale method for the fluid is presented in which bubble movement is driven by Newton's second law. The new feature of this contribution is the usage of the finite element model for the volume averaged Navier-Stokes equations in such a framework and an improved computation of the void fraction which is not limited to the discretization type. The volume integral is transformed into line integrals by applying divergence theorem twice. This obviates Gaussian integration and allows for analytical integration, which is more accurate. Further developments will focus on the investigation of damage due to cavitation. Therefore, the bubble radius, which is assumed constant so far, will be modeled based on the Rayleigh-Plesset equation. Thus, position and time of collapses can be detected dependent on the ambient conditions.

## References

[1] Shams E, Finn J and Apte S V 2011 A Numerical Scheme for Euler–Lagrange Simulation of Bubbly Flows in Complex Systems, *Int. J. Numer. Meth. Fluids* **67** 1865–1898

[2] Darmana D , Deen N G, Kuipers J A M 2006 Parallelization of an Euler–Lagrange model using mixed domain decomposition and a mirror domain technique: Application to dispersed gas-liquid two-phase flow *J. Comput. Phys.* **220** 216–248

[3] Deen N G, Van Sint Annaland M, Van der Hoef M A and Kuipers J A M 2007 Review of discrete particle modeling of fluidized beds *Chem. Eng. Sci.* **62** 28–44

[4] Di Renzo A and Di Maio F P 2007 Homogeneous and bubbling fluidization regimes in DEM–CFD simulations: Hydrodynamic stability of gas and liquid fluidized beds *Chem. Eng. Sci.* **63** 116–130

[5] Gravemeier V, Kronbichler M, Gee M W and Wall W A 2011 An algebraic variational multiscale-multigrid method for large-eddy simulation: generalized- time integration, Fourier analysis and application to turbulent flow past a square-section cylinder *Comput. Mech.* **47** 217–233

[6] Kitagawa A, Murai Y and Yamamoto F 2001 Two-way coupling of Eulerian–Lagrangian model for dispersed multiphase flows using filtering functions *Int. J. Multiphase Flow* **27** 2129–2153

[7] Crowe C T, Schwarzkopf J D, Sommerfeld M and Tsuji Y 2012 Multiphase Flows with Droplets and Particles (CRC Press)

[8] Tomiyama A, Zun I, Higaki H , Makino Y and Sakaguchi T 1997 A three-dimensional particle tracking method for bubbly flow simulation *Nuclear Engineering and Design* **175** 77–86

[9] Sudhakar Y and Wall W A 2013 Quadrature schemes for arbitrary convex/concave volumes and integration of weak form in enriched partition of unity methods *Comput. Methods Appl. Mech. Engrg* **258** 39–54

# The numerical study of the cavitation–structure interaction around 3D flexible hydrofoil

**Hu Shi-liang, Chen Ying and Lu Chuan-jing**

Department of Engineering Mechanics, Shanghai Jiao Tong University, No. 800 Dong chuan Road, Shanghai, China

E-mail: jerry542_hu@hotmail.com

**Abstract**. The closely coupled approach combined the Finite Volume Method (FVM) solver and the Finite Element Method (FEM) solver is applied to simulation the cavitation–structure interaction of a 3D cantilevered flexible hydrofoil in water tunnel. In the cavitating flow, the elastic hydrofoil would deform or vibrate in bending and twisting mode. And the motion of the foil would affect the characteristics of the cavity and the hydrodynamic load on the foil in turn. With smaller cavitation numbers ($\sigma_v$=2.15), the frequency spectrum of the lift on the foil would contain two frequencies which are associated to the cavity shedding and the first bend frequency of the hydrofoil. With larger cavitation number ($\sigma_v$=2.55), the frequency of the lift is completely dominated by the natural frequency of the foil.

## 1. Introduction

Recently, hydrofoil blades of marine propellers and turbomachinery are being designed with light weight and thin profile to increase the efficiency. With such flexible structures, the fluid load would induce structure oscillation and inversely the time-space distribution of the flow would also be influenced by the structure motion. When the pressure on the suction side of foil drops down under the saturating vapor pressure of liquid, the cavitation occurs and the interaction between the cavitation and structure could not be ignored.

Although many efforts were made in fluid-structure interaction (FSI) for aerodynamics application, relatively few works show concern for hydro-elasticity. Amromin and Kovinskaya[1] analyzed the vibration of an 2D elastic wing with an attached cavity in a periodically perturbed flow. Campbell and Paterson[2] developed and implemented a FSI solver to model a quasi-steady problem of a single 3D foil vibration in water. Torre et al.[3] experimentally investigated the influence of leading edge cavitation and supercavitation on the added mass effects experienced by a 2D truncated hydrofoil. Young[4] presented a coupled Boundary Element Model (BEM) and Finite Element Method (FEM) approach to study the fluid structure interaction of flexible composite propellers in subcavitating and cavitating flows. Ducoin et al.[5] analyzed the structural response of a rectangular cantilevered flexible hydrofoil submitted to various flow regimes through the experiment carried out in a hydrodynamic tunnel.

In the present work, the cavitation–structure interaction effects on a 3D cantilevered flexible hydrofoil in the water tunnel are numerical investigated. The Reynolds-averages Navier-Stokes equations (RANS) for fluid domain are solved by the Finite Volume Method (FLUENT). The turbulence effect is modeled using the RNG $k$-$\varepsilon$ model, and the cavitating flow is computed using the cavitation model suggested by Zwart et al.[6], in which a transport equation for the volume fraction of vapor phase is calculated. The hydrofoil is treated as elastic body with small deformation and the damping effect is neglected. The material of hydrofoil is considered as isotropic. The structure domain is dispersed by the Finite Element Method (ANSYS). The coupled method between the fluid and structure solver is based on the closely coupled approach[7].

## 2. Problem description

The fluid domain is 1m long and has a 0.192m square cross section, which is shown in figure 1. The hydrofoil is a NACA66 with a camber ratio of 2% and a relative thickness of 12%. The chord is c=0.15 m and the span is b=0.191 m. One end of the hydrofoil is fixed on the wall of the tunnel, and the other end is set free. The gap between the free end and the vertical wall is 1 mm. The angle between the chord and the horizontal coordinate X is defined as the attack angle α. Point A is the leading point at free end of the hydrofoil, and the central axis S is located at x/c=0.5. The flexible hydrofoil has a Polyacetate (POM) material with $\rho^s$=1480 kg/m$^{-3}$, $E$=3000 MPa, $\nu$=0.35.

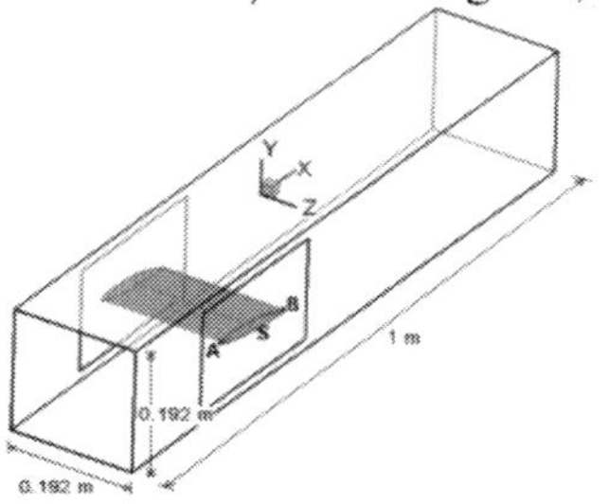


**Figure 1.** Geometric properties of fluid domain.

## 3. Result and discussion

The investigation of cavitation–structure interaction is carried out with the initial angle α=8° and Re=$\rho^f U_\infty c/\mu^f$=7.5×10$^5$. Through decreasing the pressure $p_\infty$ in the tunnel, five cases with the cavitation number $\sigma_v$=1.95, 2.15, 2.28, 2.55 and 2.80 are simulated, where the $\sigma_v$ is defined by:

$$\sigma_v = \frac{p_\infty - P_v}{1/2\,\rho^f\,U_\infty^{\,2}} \quad (1)$$

Table 1 compares the maximum lift coefficient and the maximum cavity length on the rigid and flexible hydrofoil with different cavitation numbers. It is shown that the elasticity of the structure would promote the maximum cavity length and increase the extreme value of lift on the foil. Figure 2 indicates that the numerical results of the maximum displacement of point A versus the maximum length of cavity have the same trend as the experimental ones. The free end displacement would increase with the enlargement of the cavity on the hydrofoil. Figure 3 illustrates the maximum twist angle of the section along the span and the maximum vertical displacement of the axis S. The results show that the deformation of the flexible foil increases with the reduction of cavitation number, whereas the cavitation number almost has no influence on the deformation mode in these cases.

**Table 1**. The comparison of the maximum lift coefficient and the maximum cavity length between the rigid and flexible foil.

| $\sigma_v$ | $Cl_{max}$ | | $(l_{cav})_{max}$ (m) | |
|---|---|---|---|---|
| | Rigid foil | Flexible foil | Rigid foil | Flexible foil |
| 1.95 | 1.310 | 1.391 | 0.0593 | 0.0615 |
| 2.15 | 1.271 | 1.343 | 0.0465 | 0.0486 |
| 2.28 | 1.262 | 1.304 | 0.0326 | 0.0347 |
| 2.55 | 1.151 | 1.179 | 0.0195 | 0.0214 |
| 2.80 | 1.098 | 1.101 | 0.0059 | 0.0059 |

### 3.1. $\sigma_v$=2.15

The time history of the lift coefficient and the displacement of point A for $\sigma_v$=2.15 are shown in figure 4. It is presented that the variation of the lift and tip displacement has the same phase. Figure 5 gives the frequency spectrum of the lift for both rigid and flexible foil. It reveals that the spectrum of the rigid foil only has one main frequency (about 20Hz) related to the cloudy cavity shedding, while the spectrum of the flexible foil has two peaks and one of the peaks (about 42Hz) is quite close to the first bending frequency of the hydrofoil in water ($f_1$=43Hz). Figure 6 and 7 compare the lift coefficient and the cavity length of the rigid and flexible foil in one period. In the ascent stage of lift (0~2/6$T_0$), the flexible foil moves upward synchronously (seen in figure 4). The motion of foil restrain the development of the cavity on the suction side which leads to the lift on the flexible foil is relatively small. In the descend stage of lift (2/6$T_0$~4/6$T_0$), the flexible foil moves downward. The growth rate of the cavity on the flexible foil is quicker than that of rigid foil and the lift on the flexible foil is much larger. At about 4/6$T_0$, the partial cavity on both rigid and flexible foil is cut by the reentrant jet and sheds from the leading edge.

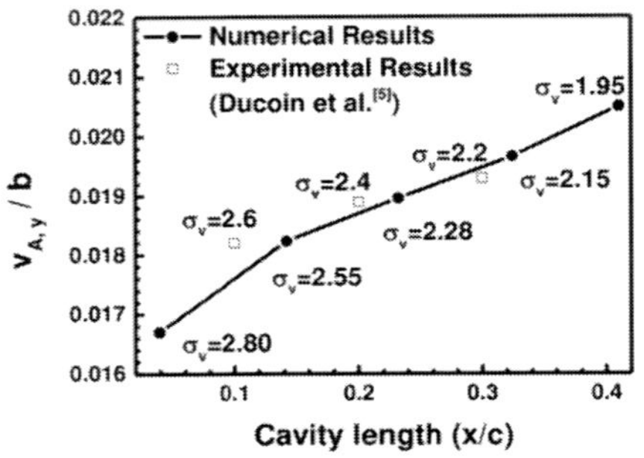


**Figure 2.** Maximum displacement of point A versus maximum cavity length.

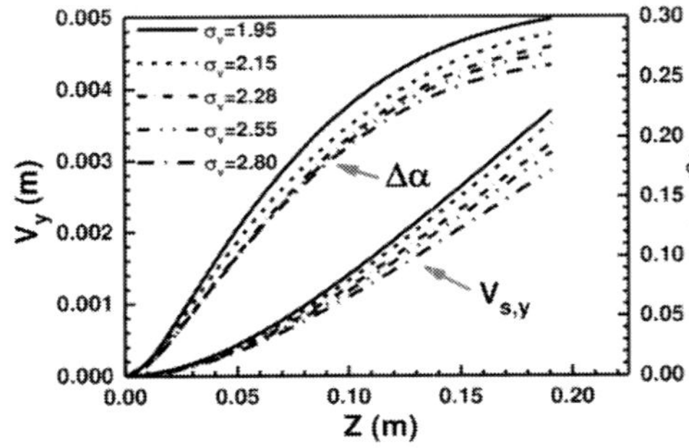


**Figure 3.** The maximum twist angle and the maximum displacement of axis S for different cavitation number.

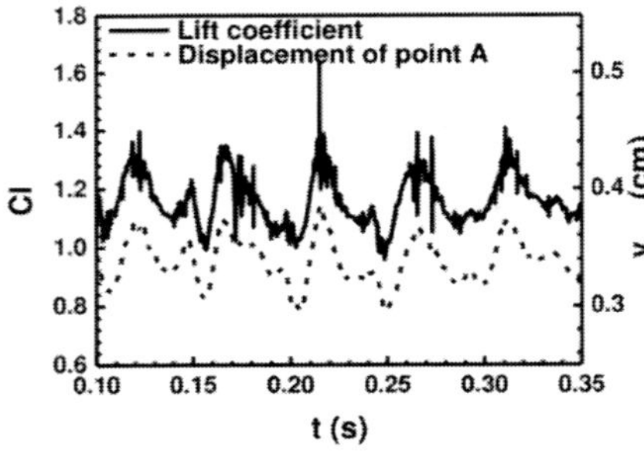


**Figure 4.** The time history of the lift coefficient on the hydrofoil and the displacement of point A for $\sigma_v$=2.15

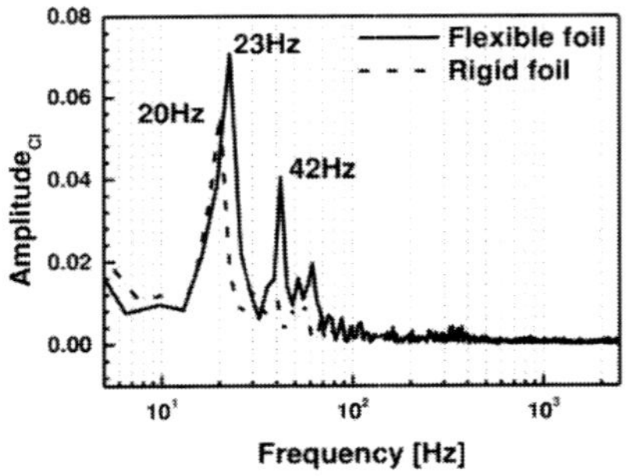


**Figure 5.** The comparison of the lift coefficient spectrum for $\sigma_v$=2.15.

### 3.2. $\sigma_v$=2.55

The time history of the lift coefficient on the rigid and flexible foil for $\sigma_v$=2.55(figure 8) displays that the vibration of the foil does change the regime of cavitating flow. The lift coefficient on the rigid foil has no obvious regularity and the amplitude of the fluctuation is narrow. However, on the deformable foil, the lift coefficient oscillates periodically with large amplitude. The spectrum (figure 9) shows that the single frequency peak of the lift coefficient on the flexible foil is 40Hz ($f_1$=43Hz), which indicates that the movement of the foil domains the unsteady characteristic of the flow field with $\sigma_v$=2.55. In figure 10, the vortex structure at the middle section of the fluid field (z/b=0.5) is visualized based on the Q-criterion. The influence of the structure motion on the flow regime could be clearly seen.

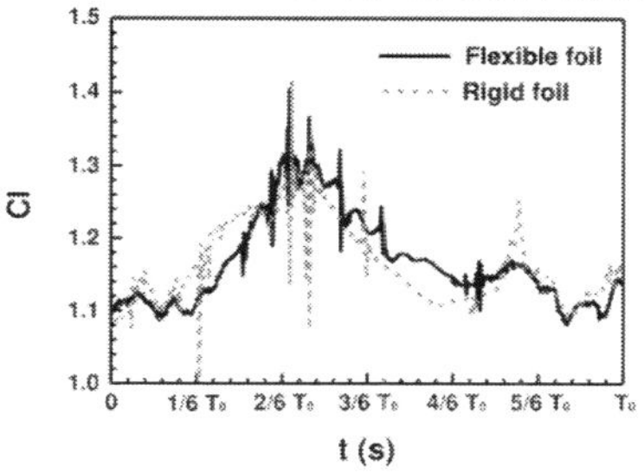


**Figure 6.** The lift coefficient in one cycle $T_0$.

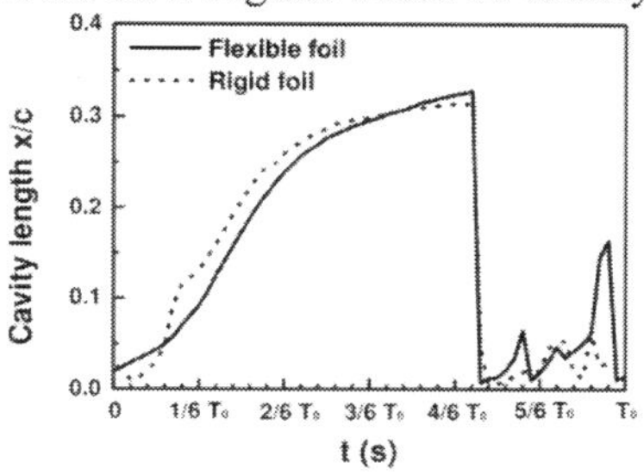


**Figure 7.** The cavity length in one cycle $T_0$.

## 4. Conclusion

In the present work, the closely coupling approach is selected as coupling strategy in the numerical modeling. And the cavitation–structure interaction for a cantilevered 3D flexible hydrofoil in the water tunnel is studied. The results show that the hydrodynamic loads would cause the bending and twisting deformation of the foil and the vibration of the foil would also affect the regime of the cavitating flow.

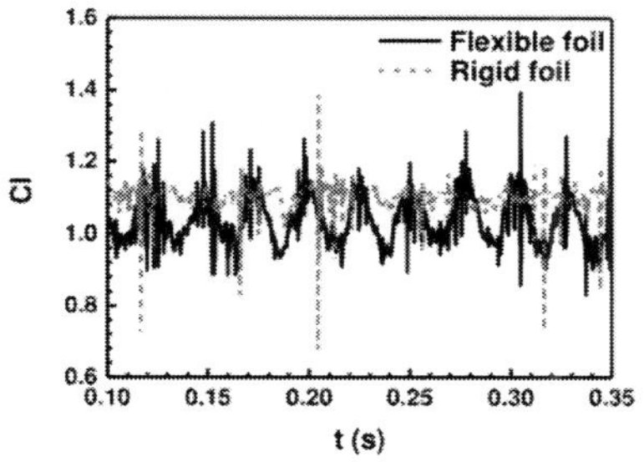


**Figure 8.** The lift coefficient of the flexible and rigid foil for $\sigma_v$=2.55.

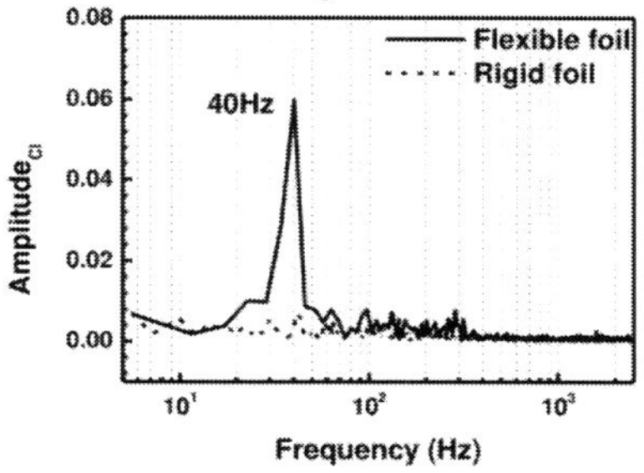


**Figure 9.** The comparison of the lift coefficient spectrum for $\sigma_v$=2.55.

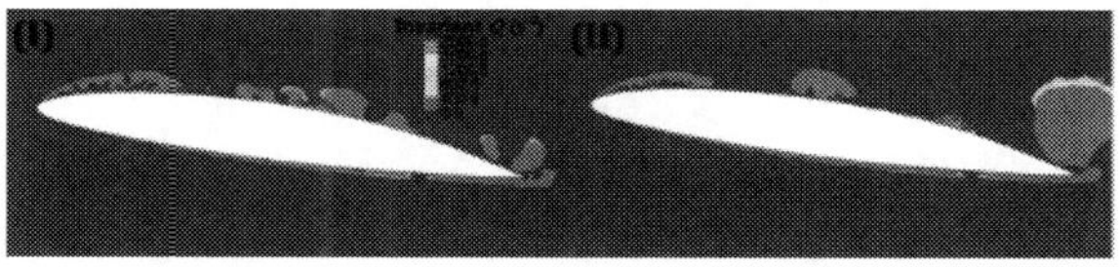


**Figure 10.** The vortex structure (based on the Q-criterion) at the middle section of the fluid field (z/b=0.5) with $\sigma_v$=2.55. (I) Rigid foil. (II) Flexible foil

**Acknowledgments**

This project is supported by the National Nature Science Foundation of China (Grant No. 11472174). Its financial support is gratefully acknowledged.

**References**

[1] Amromin E and Kovinskaya S 2000 *J. Fluid. Struct.* **14** 735–751.
[2] Campbell R L and Paterson E G 2011 *J. Fluid. Struct.* **27**, 1376–91.
[3] De La Torre O, Escaler X, Egusquiza E and Farhat M 2013 *J. Fluid. Struct.* **39**, 173–187
[4] Young Y L 2008 *J. Fluid. Struct.* **24**, 799–818.
[5] Ducoin A, Astolfi A J and Sigrist J F 2012 *Eur. J. Mech. B-Fluid* **36** 63–74.
[6] Zwart P J, Gerber A G and Belamri T 2004 *5th Int. Conf. Multiphase Flow (Japan: Yokohama).*
[7] Lian Y, Shyy W, Ifju P and Verron E 2002 *AIAA Paper* **2972** 2492–94.

# Numerical evaluation of cavitation shedding structure around 3D Hydrofoil: Comparison of PANS, LES and RANS results with experiments

**B Ji[1,2], X X Peng[3], X P Long[1,2], X W Luo[4], Y L Wu[4]**
1 School of Power and Mechanical Engineering, Wuhan University, Hubei 430072 China
2 Key Lab of jet theory and new technology of Hubei Province, Hubei 430072, China
3 China Ship Scientific Research Center, Wuxi 214082, China
4 State Key Laboratory of Hydroscience & Engineering, Tsinghua University, Beijing 100084, China

E-mail: jibin@whu.edu.cn

**Abstract.** Results of cavitating turbulent flow simulation around a twisted hydrofoil were presented in the paper using the Partially-Averaged Navier-Stokes (PANS) method (Ji et al. 2013a), Large-Eddy Simulation (LES) (Ji et al. 2013b) and Reynolds-Averaged Navier-Stokes (RANS). The results are compared with available experimental data (Foeth 2008). The PANS and LES reasonably reproduce the cavitation shedding patterns around the twisted hydrofoil with primary and secondary shedding, while the RANS model fails to simulate the unsteady cavitation shedding phenomenon and yields an almost steady flow with a constant cavity shape and vapor volume. Besides, it is noted that the predicted shedding vapor cavity by PANS is more turbulent and the shedding vortex is stronger than that by LES, which is more consistent with experimental photos.

## 1. Introduction

Cavitation shedding behavior is important in engineering interest since they seriously generate the hydro-machine vibration, noise and cavitation erosion. For example, the cavitating vortex rope in the draft tube can cause lower frequency pressure fluctuation and severe cavitation surge problems. This makes us realize the significance of understanding cavitation mechanism and may help to improve the performance and reliability of hydraulic systems.

In the past, due to various limitations in measurement techniques, noticeable efforts have been made to numerically simulate cavitating flows. Many cavitation models have been based on the assumption of the homogenous equilibrium medium, where the slip between the liquid and vapor interface is neglected and the liquid-vapor mixture is treated as a single fluid that satisfies the Navier-Stokes equations. A key point in this kind of model is how to define the mixture density. One approach is based on the state equation. Coutier-Delgosha et al. [1] used a barotropic state law that linked the mixture density to the static pressure to simulate cloud cavity shedding in a Venturi-type duct. Another method has used a multiphase mixture cavitation model based on the transport equation model (TEM) for the phase change, which introduced an additional equation for the vapor (or liquid) volume fraction including source terms for evaporation and condensation processes. Comparison studies of different TEM models have been shown by Morgut et al. [2].

Journal of Physics: Conference Series **656** (2015) 012127 doi:10.1088/1742-6596/656/1/012127

For simulation of cavitating flow, the turbulence model is crucial because the cavitation process is basically unsteady in nature and there must be strong interactions between the cavity interface and the boundary layer during the cavitation development. Though the current Reynolds Average Navier-Stokes (RANS) equation approach has been widely used to model turbulent flows in industry, the capability of RANS model with eddy viscosity turbulence models to simulate unsteady cavitating flows is limited and needs some modifications. There have also been attempts to predict the flow unsteadiness during cavitation using Large Eddy Simulation (LES) which are expected to more accurately predict larger-scale turbulent eddies [3-4], but most of their results are based on 2D LES solver due to very large computational resources needed. This leads to the requirement for efficient hybrid approached for application of cavitating flow, such as FBM, DES and PANS [5].
Inspired by their work, the aim of this paper is to investigate three dimensional cavitation structures and add knowledge on the impact of different turbulence models when implemented in the same code with the same mesh.

## 2. Description of the geometry and numerical setup

The twisted hydrofoil, studied by Foeth [6] is selected in present research. The hydrofoil is a wing of rectangular planform of a NACA0009 section with varying attack angle from 0° at the side section to 11° at the mid-section, which is symmetry with respect to its mid-span plane. This geometry can avoid the cavitation sidewall effect during the experimental test in water tunnel. So it is a very good case to resemble propeller cavitation with well defined experimental data that is easily studied.
In present paper, we compare the cavitation simulated results with a RANS approach, which is based on solving the time averaged equations, a LES approach (Ji et al. (2013)), based on the filtered equations, and a PANS approach (Ji et al. 2013), which is a bridging method from the RANS to DNS. We have previously evaluated capability of unsteady cavitation simulation with PANS model and LES model for 2D hydrofoil and 3D hydrofoil. In this paper, we briefly summarized these results and make some comparison of different models.

## 3. Results and discussion

The twisted design and the larger attack angle in the middle area cause the cavitation to mainly develop near the mid-span area close to the leading edge with a curved closure line. And the internal jet under below the cavitation is no longer purely with a straight upstream direction but also has a side component in the spanwise direction. The combined effect of these two components causes a very complex shedding process.
Figure 1 shows the time dependent total vapor volume $V_{cav}$ for RANS, PANS and LES. Here $V_{cav}$ was defined as:

$$V_{cav} = \sum_{i=1}^{N} \alpha_i V_i \tag{1}$$

where $N$ is the total number of control volumes in the computational domain, $\alpha_i$ is the vapor volume fraction in each control volume and $V_i$ is the volume of each cell.
The total vapor volume, $V_{cav}$, is a convenient parameter for understanding the transient behavior of cavitating flows. The total vapor volume calculated at each time step is shown in Fig. 1 with snapshots of three typical instants of cavitating flows with iso-surface values of $\alpha_v$=0.1 shown in Fig. 2. The choice of the vapor volume fraction value has a direct influence on the cavitation size. A higher value of the vapor volume fraction would result in less difference in terms of cavitation extent, but it tends to hide important details, i.e. interface distortion and cut-off. Based on previous work, we find the current selected vapor volume fraction is good to visualize the cavitation mechanisms.
As indicated in Fig. 1 and the left column of Fig. 2, the RANS model fails to simulate the unsteady cavitation shedding phenomenon and yields an almost steady flow with a constant cavity shape and vapor volume. As reported by many research work, this poor prediction of cavity shedding is due to the over-prediction of the turbulent viscosity in the rear part of the cavity, which suppressed the cavity shedding dynamics.

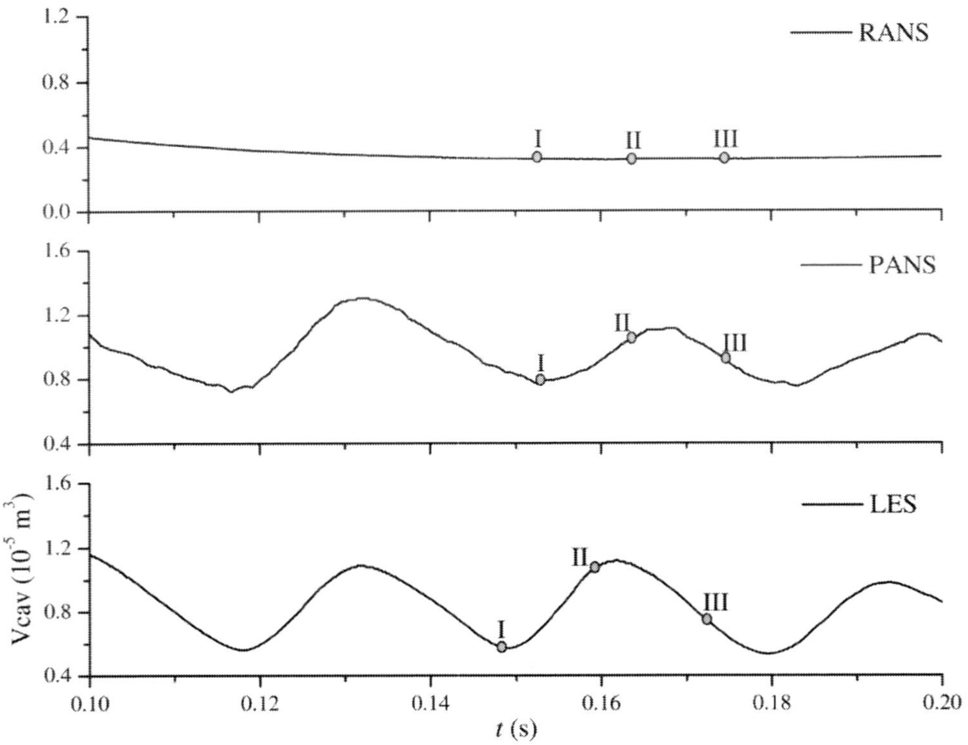


Fig. 1 Calculated vapor volumes for different turbulence model

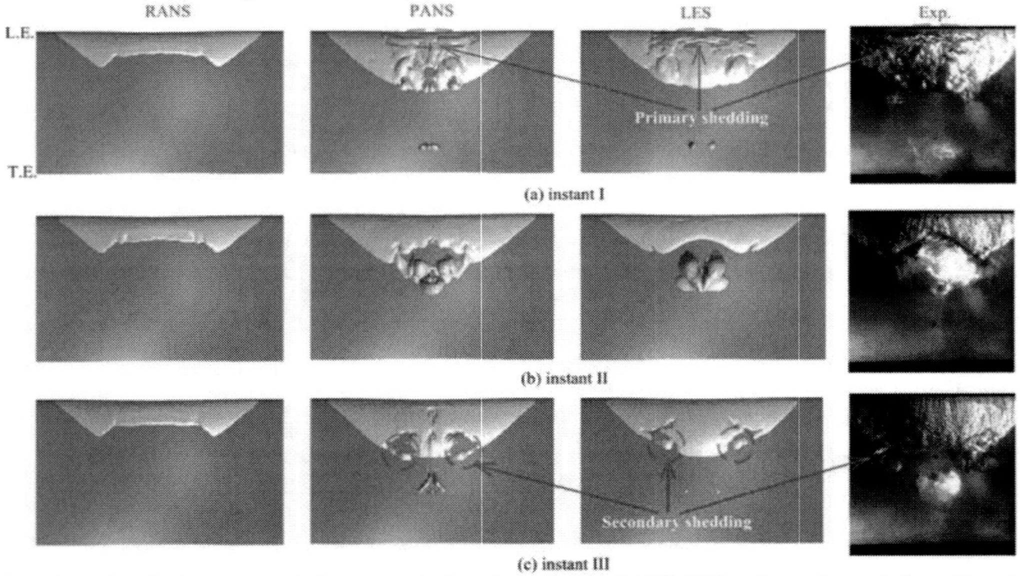


Fig. 2 Top view of cavitation patterns during one typical cycle for RANS, PANS, LES and experimental photo (three instants corresponding to the time marked as I, II and III in Fig. 1

If we used PANS or LES models, the cavitation shedding behavior occurs and the vapor volume variation exhibits periodically as shown in Fig. 1 The predicted shedding frequency by PANS is 31Hz, while that for LES is 31.7 Hz, which both agree fairly well with the measured frequency (32.2 Hz). The detailed analysis of the cavitation shedding structure by PANS and LES is discussed in the

following. In instant I, the total vapor volume is a minimum after the attached cavity in the center of the hydrofoil has shed from the leading edge due to the collision between the re-entrant flow and the cavity interface. This process is called the primary shedding. Then the shedding vapor cloud becomes more turbulent and is advected downstream by the main flow, as shown in instant II. Meanwhile, the tail of the attached cavity begins to curl into a concave shape and grows quickly from the leading edge, which explains the increase in the total vapor volume from instant I to instant II in Fig. 1 After that the attached cavity grows slowly and the shedding vapor cloud quickly shrinks and finally collapses, which caused the decrease of total vapor volume in Fig. 1 It should be noted that there is a secondary shedding of both downstream lobes of the remaining attached cavity in instant III. As is shown by experimental video and our detailed analysis based upon the simulated instantaneous cavitation animation, the primary shedding event (instant I) is rather periodic while the secondary shedding of cloud cavitation (instant III) is less regular.
Comparing the flow evolutions predicted by RANS, PANS and LES with experimental results in Fig. 2 the RANS is not able to simulate cavitation shedding dynamics correctly, which is due to over-prediction of eddy viscosity in the rear part of the cavity. Although the PANS and LES reasonably reproduce the cavitation shedding patterns around the twisted hydrofoil with primary and secondary sheddings in Fig. 2 the simulated shedding vapor cloud is a little smaller than that in experiment and do not survive as long as in the experiments. Whether it is due to the insufficient grid resolution, numerical scheme, the mass transfer cavitation model behavior or the combination of all is still under investigation. Besides, as indicated by Fig. 2 it is noted that the predicted shedding vapor cavity by PANS is more turbulent than that by LES, which is more consistent with experimental photos.

## 4. Conclusions
We have presented the unsteady cavitating turbulent flow simulation around a twisted hydrofoil with PANS, LES and RANS. The results show that the RANS model fails to simulate the unsteady cavitation shedding phenomenon and yields an almost steady flow with a constant cavity shape and vapor volume. And the PANS and LES can reasonably predicted the 3D complicated cavitation shedding patterns including the primary shedding and secondary shedding. Besides, it is noted that the predicted shedding vapor cavity by PANS is more turbulent and the shedding vortex is stronger than that by LES, which is more consistent with experimental photos.
The main advantage of the PANS simulation over LES applied on the present grid is clear in the cloud cavitation region, where very strong vortex shedding is observed. The LES applied on present grid resulted in a weak vortex. One of the possible reasons may be due to the fact that LES is highly dependent on the mesh resolution, while the PANS is not sensitive to grid resolution. Thus, the PANS is promising since it opens a possibility for using PANS for predicting the complicated cavitation phenomena and the mechanism for the interaction between the cavitation and turbulence.

## Acknowledgements
The authors would like to appreciate Dr. Foeth and Prof. & Dr. Tom van Terwisga, Delft University of Technology, for opening geometry of the Delft Twist-11 hydrofoil and the experimental data.
This work was financially supported by the National Natural Science Foundation of China (Project Nos. 51206087 and 51576143).

## References
[1] Coutier-Delgosha O, Reboud J and Delannoy Y 2003 *Int. J. Numer. Methods Fluids* **42** 527-548
[2] Morgut M, Nobile E and Bilus I 2011 *Int J Multiphase Flow* **37**(6) 620-626
[3] Roohi E, Zahiri A and Passandideh-Fard M 2013 *Appl Math Model* **37** 6469-6488
[4] Ji B, Luo X, Peng X and Wu Y 2013 *J Hydrodyn* **25**(4) 510-519
[5] Ji B, Luo X, Wu Y, Peng X and Duan Y 2013 *Int J Multiphase Flow* **51** 33-43
[6] Foeth E 2008 *Ph.D. Thesis*, Delft University of Technology, Wageningen, the Netherlands

# Shock-Induced Bubble Collapse versus Rayleigh Collapse

**Anil Kapahi**[1], **Chao-Tsung Hsiao, and Georges L. Chahine**
DYNAFLOW, INC.

10621-J Iron Bridge Road. Jessup, Maryland, USA.

E-mail: anil@dynaflow-inc.com

**Abstract.** This paper compares two physical mechanisms for the collapse of a bubble near a rigid wall: a traveling shock-induced collapse and a Rayleigh-like collapse due to a uniform rise of the pressure around the bubble. A multi-material compressible flow solver capable of handling material interfaces under high pressures is used to investigate these two scenarios for different levels of the driving pressure ranging from 1 *MPa* to 400 *MPa*. The two mechanisms are differentiated on the basis of the resulting bubble dynamics, the reentrant jet velocity, and the pressures imparted to the wall.

## 1. Introduction

Cavitation erosion is a major problem in fluid machinery such as propellers, pumps, turbines, .. etc. Microbubbles grow in low pressure regions then collapse when encountering high pressures to form near boundaries reentrant jets and shock waves [1, 2] leading to material damage and loss of performance. The collapse of these bubbles can also be used for useful purposes such as for cleaning [3] or erosion testing of materials [4]. Under various flow scenarios and applications, the collapse can be of a Rayleigh-type, where all parts of the bubble surface are exposed simultaneously to a high local pressure, or it could be shock-induced [5] where a shock wave (for example generated by a nearby bubble cloud collapse [6] or by a lithotripter) travels to the bubble and collapses it progressively. Depending on the distance of the centre generating the shock wave from the bubble surface, the shock wave can be spherical or planar. In this work we consider planar shock waves to simplify presentation.

## 2. Solution Method

The governing equations to study the bubble dynamics are Euler equations, described by a set of hyperbolic conservation laws [7]. These equations represent conservation of mass, momentum, and energy. The system of equations is closed with an equation of state for each phase. An ideal gas equation is used for air [8] and Tillotson's equation is used for water [9]. This model account for high compressibility effects including shock waves, but does not take into account effects due to viscosity, surface tension, and turbulence. These effects will be studied in future work. The details pertaining to the multi-material solver we have developed, 3DYNAFS_COMP©, with validation and verification examples are provided in [1]. The two collapse scenarios described

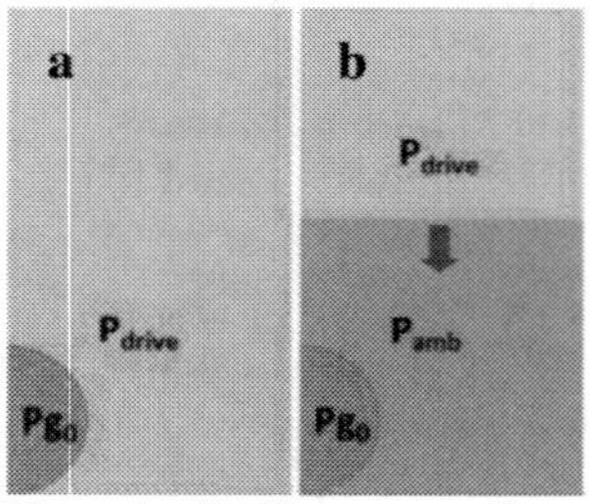


**Fig. 1** Different collapse scenarios a) Rayleigh collapse b) Shock induced collapse.

[1] To whom any correspondence should be addressed.

above and illustrated in Fig 1 are investigated using 3DYNAFS_COMP© for a set of driving pressures, $P_{drive}$. A domain size of 1cm by 1cm is used for all the cases. The bubble is resolved using a uniform grid of 10 μm. A reflection boundary condition was imposed on the axis of symmetry and at the bottom boundary, i.e. all physical variables such as density, pressure, velocities and energy are reflected from the axis, while transmission non-reflective boundary conditions (i.e. the flow variables are extrapolated along the characteristic wave direction) were imposed at the far-field boundaries.

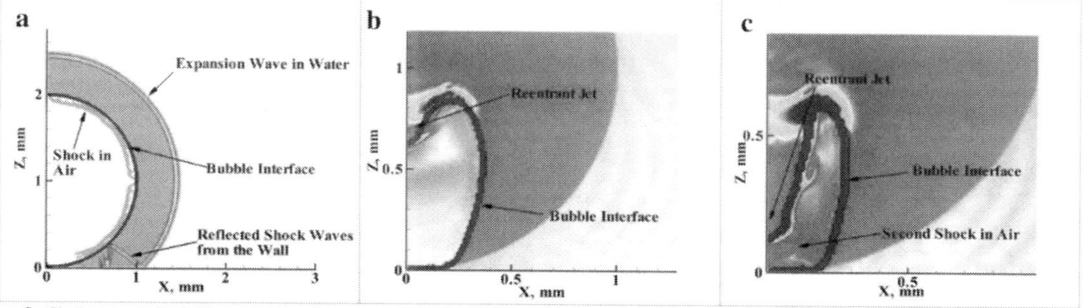


**Fig. 2** Coloured Schlieren shadowgraphs showing the evolution of the collapse for the Rayleigh-like scenario for a driving pressure of 10 *MPa*. Bubble initial radius and standoff distance to the wall of 1 mm, and initial bubble gas pressure of 0.1 *MPa*. (a) $t$=0.28 μs, (b) $t$=10 μs, (c) $t$=10.51 μs.

## 3. Rayleigh-like Collapse

In the case of a Rayleigh-like collapse scenario, the bubble is suddenly subjected to a high ambient pressure. As illustrated in Fig. 2, the sudden rise of the pressure (here from 0.1 *MPa* to 10 *MPa*) results in the generation and propagation of a shock wave inside the bubble, while an expansion wave travels outward in the water as illustrated by the Schlieren contours in Fig. 2a. The rarefaction wave reflects back at the rigid wall towards the bubble and interacts with the bubble rarefaction wave in an increasingly complex manner as the number of reflections increases. Also, the pressure difference between the inside and the outside of the bubble, compresses the bubble. However, due to presence of the wall, the bubble pole farthest from the wall moves toward the wall faster than the rest of the interface resulting in the formation of a reentrant jet, which is clearly defined in Fig. 2b and very developed in Fig. 2c. As the jet speed increases, the jet tip crosses the full bubble height and impacts the wall generating a high impulsive pressure. This can be seen in Fig. 3, which actually shows a multi peaked pressure. The following peaks are due to the collapse of the remaining toroidal bubble and following shock emissions and reflections [1, 2].

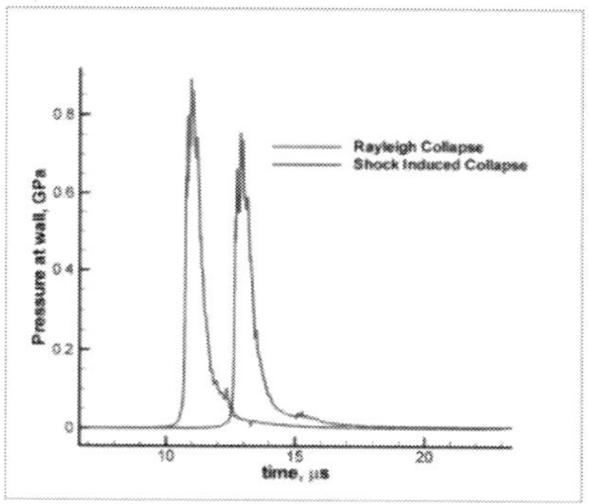


**Fig. 3** Pressures generated at the wall by collapse of bubbles in Fig. 2 and Fig. 4.

## 4. Shock-Induced Collapse

To present a shock-induced collapse, we consider here a very thick shock front (finite shock thickness is quite important and will be addressed in another publication) with a high pressure advancing toward the bubble. The bubble initial radius, the standoff to the wall, and the pressure magnitude are the same as in the previous section. The high pressure hits the air/water interface, sending a shock wave inside the bubble and an expansion wave in water as shown in Fig. 4a. The shock front is seen in this figure not having reached the wall. The bubble starts compressing in a gradual fashion starting from the top as the high pressure reaches lower and lower parts of the interface. A reentrant jet then forms and generates a second shock wave in front of it, which propagates inside the bubble (Fig. 4b). This shock is here stronger than the initial wave and catches it. In the liquid, the incident shock wave gets reflected from the wall, while the shock wave travelling inside the bubble also hits the wall and gets reflected. These details affect the history of the pole velocities presented in Fig. 5a. The figure shows almost no movement for the bottom pole for a long time. Finally, the combined shocks in the bubble hit the bottom pole (12.05 μs, pt. A) and reflect forcing its movement up into the bubble as shown in Fig. 5a (pt. B). The reflected wave from the wall inside the bubble hits the top pole, resulting in the slowing down of top pole (pt. C) before it collapses. Bubble contours showing the development of the reentrant jet and comparing Rayleigh-like collapse and shock-induced collapse are also shown in Fig. 5b and Fig. 5c. Both sets of contours are very much alike with the shock-induced reentrant jet occurring later and being slightly thinner. The motion up of the bottom pole is not shown to avoid overlapping contours.

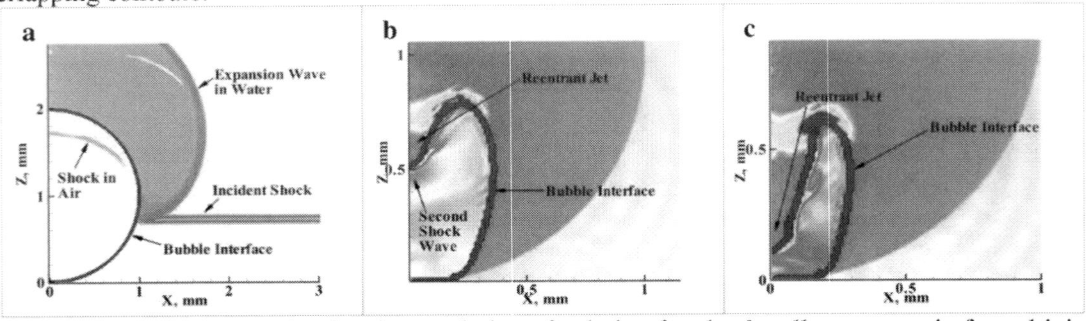


**Fig. 4** Coloured Shadowgraphs showing evolution of solution for shock collapse scenario for a driving pressure of 10 *MPa*. (a) $t$ =1.1 μs (b) $t$ =11.96 μs (c) $t$ =12.41 μs.

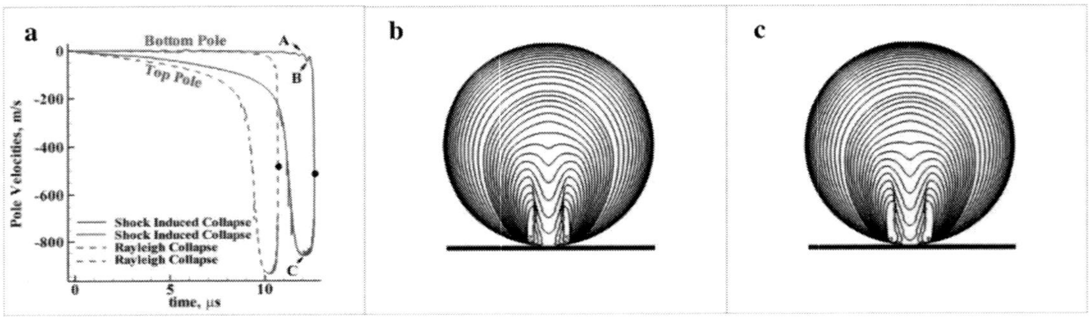


**Fig. 5** (a) Pole velocities for both Rayleigh-like collapse and shock-like collapse scenarios at a driving pressure of 10 MPa. (b) Bubble contours at selected times for the Rayleigh-like collapse. (c) Bubble contours at selected times for the shock-induced collapse. An adaptive time stepping is used. The contour plots are shown for every 100 time steps with time step varying from $10^{-8}$ s to $10^{-5}$ s.

## 5. Effects of the Driving Pressure

The two collapse scenarios are investigated further by analyzing the reentrant jet development (top pole position vs. time). The evolution of the top pole position with time for driving pressures ranging from 10 *MPa* to 400 *MPa* is shown in Fig. 6. As the driving pressure is increased the bubble period is reduced as expected. The Rayleigh-like collapse scenario results in a faster collapse compared to the shock-induced collapse scenario for the same driving pressure. The bubble collapse times are shown in Fig. 7a, and follow the expected $P_{\text{drive}}^{1/2}$ trend. The corresponding jet velocities are shown in Fig. 7c. The peak pressures monitored at the wall for both scenarios and for a whole set of driving pressures between 1 *MPa* and 400 *MPa* are shown in Fig. 7b. These peaks increase with increasing driving pressures. Moreover, the Rayleigh-like collapse results in a higher wall pressure compared to the shock-driven collapse. Finally the maximum jet velocities for both

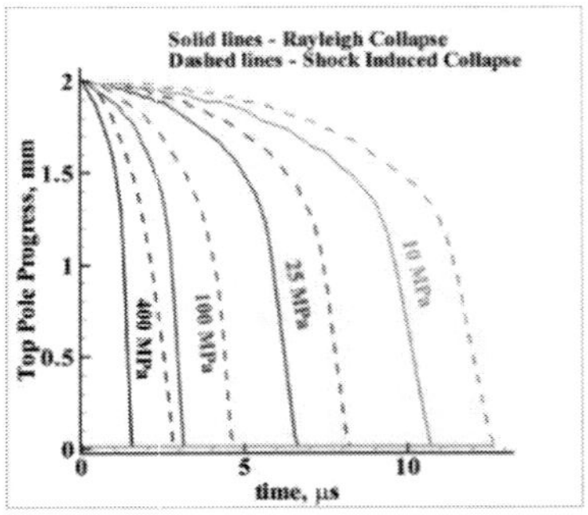


**Fig. 6** Evolution of top pole position with time for different pressure and different collapsing scenarios, uniform collapse (solid lines) and shock collapse (dashed lines).

scenarios are compared with the sound speed in the liquid and with the characteristic velocity of collapse of an empty sphere in an infinite medium or the Rayleigh time, $\sqrt{P_{drive}/\rho}$, for different driving pressures and are shown in Fig. 7c. The jet velocity for the Rayleigh-like collapse is much higher compared to the shock induced collapse. The lower speed in the shock induced cases is due to the slowing down of the top pole when the reflected shock from the wall hits the reentrant jet tip. This slowing down of jet is mostly observed at the highest driving pressures observed above 80 MPa, when the jet speed and the sound speed become of the same order. In the case of the Rayleigh-like collapse the jet speed is able to easily exceed the sound speed for the higher driving pressures.

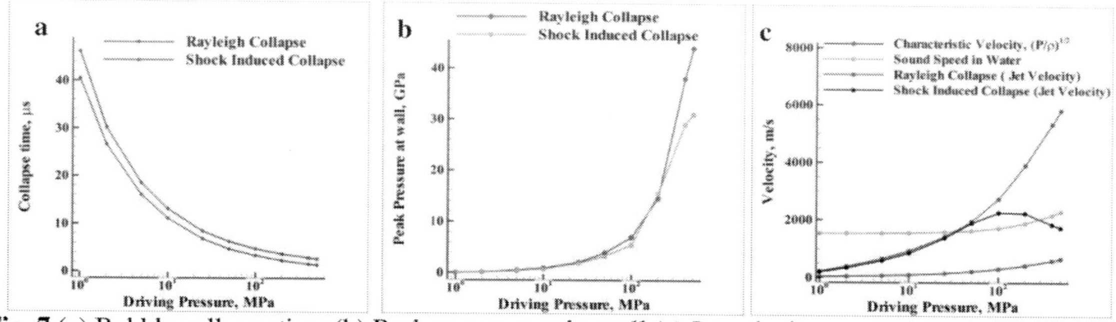


**Fig. 7** (a) Bubble collapse time (b) Peak pressure at the wall (c) Jet velocity comparison with sound speed and characteristic speed for both scenarios covering different driving pressures

## 6. Conclusion

In this work, we compared the characteristics of bubble collapse dynamics near a rigid wall for a Rayleigh-like bubble collapse and a shock-induced bubble collapse. The collapsing mechanisms were studied for different driving pressures and the resulting pole positions, reentrant jet velocities, and impact pressures at the wall were compared. The results show that, for the same driving pressure, the Rayleigh-like collapse is faster than the shock-induced collapse and results in higher impact pressures at wall. The reflected shock waves at the wall play an important role in the shock-induced collapse leading to slowing down of incoming jet for higher driving pressures. These effects are more amplified with thin shock waves, which we did not address here for sake of brevity.

## 7. Acknowledgements

This work was supported by a DOE SBIR Phase II (DE-SC0006443), by ONR under contract N00014-12-C-0382 monitored by Dr. Ki-Han Kim and by DYNAFLOW IR&D. We are grateful for this support.

## 8. References

[1] G. L. Chahine, "Modeling of Cavitation Dynamics and Interaction with Material," in *Advanced Experimental and Numerical Techniques for Cavitation Erosion Prediction*, A. Kim, K-H, Chahine, G.L., Franc, J-P., Karimi, Ed. Berlin: Springer, 2014, pp. 123–173.

[2] C.-T. Hsiao, A. Jayaprakash, A. Kapahi, J.-K. Choi, and G. L. Chahine, "Modelling of material pitting from cavitation bubble collapse," *J. Fluid Mech.*, vol. 755, pp. 142–175, Aug. 2014.

[3] G. L. Chahine, A. Kapahi, J.-K. Choi, and C.-T. Hsiao, "Modeling of Surface Cleaning by Cavitation Bubble Dynamics and Collapse," *Ultrason. Sonochem.*, vol. 10.1016/j., May 2015.

[4] G. L. Chahine and C.-T. Hsiao, "Modelling cavitation erosion using fluid–material interaction simulations," *Interface Focus*, no. http://dx.doi.org/10.1098/rsfs.2015.0016, 2015.

[5] C.-T. Ha, W.-G. Park, and C.-M. Jung, "Numerical simulations of compressible flows using multi-fluid models," *Int. J. Multiph. Flow*, vol. 74, pp. 5–18, Sep. 2015.

[6] M. R. Betney, B. Tully, N. A. Hawker, and Y. Ventikos, "Computational modelling of the interaction of shock waves with multiple gas-filled bubbles in a liquid," *Phys. Fluids*, vol. 27, no. 3, p. 036101, Mar. 2015.

[7] A. Kapahi, C.-T. Hsiao, and G. L. Chahine, "A multi-material flow solver for high speed compressible flow applications," *Comput. Fluids*, p. DOI: 10.1016/j.compfluid.2015.03.016, 2015.

[8] J. Anderson, *Computational Fluid Dynamics*. McGraw-Hill Education, 1995, p. 547.

[9] I. B. Zel'dovich, Y. B. Zel'dovich, and Y. P. Raizer, *Physics of Shock Waves and High-Temperature Hydrodynamic Phenomena*. Courier Corporation, 2002, p. 916.

[10] A. Wardlaw and J. A. Luton, "The Gemini Euler solver for the coupled simulation of underwater explosions," *NSWCIH - Tech. Rep.*, no. 2500, 2003.

[11] P. Colella, "A Direct Eulerian MUSCL Scheme for Gas Dynamics," *SIAM Journal on Scientific and Statistical Computing*, vol. 6, no. 1. pp. 104–117, 1985.

Journal of Physics: Conference Series **656** (2015) 012129 doi:10.1088/1742-6596/656/1/012129

# Consideration of a Phase Change Model Based on Apparent Phase Equilibrium

**S. Kashiwada**[1], **Y. Iga**[2]
[1] Graduate School of Engineering, Tohoku University, Sendai, Japan
[2] Associate Professor, Institute of Fluid Science, Tohoku University, Sendai, Japan

E-mail: kashiwada@cfs.ifs.tohoku.ac.jp

**Abstract**. It has been known that cavity volume is underestimated and there is a discrepancy between predicted and measured breakdown characteristics for the numerical simulation of unsteady cavitation around a hydrofoil at high angle of attack. Therefore, in this study, in order to predict the cavity volume with high accuracy, the phenomena that gas phase increases even at a pressure higher than saturated vapour pressure which is known as aeration is modelled, and applied to phase change term. It was assumed that the precipitation of dissolved air is promoted by mechanical stimulation such as Reynolds stress in unsteady flow. The effectivity of the proposed model is discussed through the comparison among some kinds of components of the pressure variation.

## 1. Introduction

In recent years, several cavitation models for numerical simulation have been proposed, and many commercial software which are implementing those cavitation models are now available. However, for the simulation of unsteady cavitating flow around a single hydrofoil at high angle of attack [1], it has been reported that the prediction accuracy of time averaged lift and drag is low, and the length of sheet cavity is prone to be underestimated. It can be thought that following two mechanisms contribute to the underestimation of lift coefficient. 1. Cavity volume is underestimated since pressure does not decrease. , and 2. Pressure does not decrease since cavity volume is underestimated. In the former case, cavity volume increases as pressure decreases even mass fraction of gas phase does not increase which means evaporation rate does not increase. In the latter case, cavity volume increases as evaporation increases and it leads to reduction of pressure. In the present study, the latter case is assumed. Hence, increase of evaporation rate is studied in unsteady cavitation.

In conventional models for the numerical analysis of cavitation, evaporation and condensation is estimated by whether the local pressure takes lower or higher value than the local saturated vapour pressure. On the other hand, it is well known that the cavity incepts on the condition that the local pressure is higher than the saturated vapour pressure [2]. The phenomenon depends on Reynolds number, which is called scale effect. The author also shown by an experiment that the gas phase increase even at a pressure higher than the local saturated vapour pressure if the liquid contains dissolved air and exposed to disturbance. Therefore new phase change model which is different from phase change models based on phase equilibrium at the saturated vapor pressure is necessary in order to express the increase of gas phase in actual flow field by homogeneous model which has difficulty in treating bubble nucleus.

Thus, in this study, a phase change model which was suggested by the authors [3] is used for numerical simulation of unsteady cavitation in which the situation can be reproduced where the gas phase increase even above saturated vapor pressure. In this model, precipitation of dissolved air is not taken into account explicitly since the model is based on homogeneous model, but the effect of dissolved air is taken into account implicitly for homogeneous model. The precipitation of dissolved air is assumed to be promoted by dynamic stimulation such as Reynolds stress or Baroclinic torque.

The flow field around single NACA0015 hydrofoil which is well known for benchmark simulation of cavitating flow, also known as the most difficult condition for prediction is selected. Through the comparison of time averaged pressure distribution on a hydrofoil between experimental data, the effectivity of apparent phase equilibrium model is considered.

## 2. Numerical Method

*2.1 Two -phase locally homogeneous medium model*

In this study, locally homogeneous model of gas-liquid two phase medium [4] is used. This model treats two phase medium as a pseudo single-phase medium which has locally homogeneous void fraction. Considering compressibility for both gas phase and liquid phase, governing equations are given as follows:

$$\frac{\partial \mathbf{Q}}{\partial t}+\frac{\partial(\mathbf{E}_j-\mathbf{E}_{vj})}{\partial x_j}=\mathbf{S} \quad \mathbf{Q}=\begin{bmatrix}\rho\\ \rho u_i\\ \rho Y\end{bmatrix},\quad \mathbf{E}_j=\begin{bmatrix}\rho u_i\\ \rho u_i u_j+\delta_{ij}p\\ \rho u_j Y\end{bmatrix},\quad \mathbf{E}_{vj}=\begin{bmatrix}0\\ \tau_{ij}\\ 0\end{bmatrix},\quad \mathbf{S}=\begin{bmatrix}0\\ 0\\ \Gamma\end{bmatrix} \tag{1}$$

where, $\rho$, $p$, $u$, are density, pressure, and velocity, respectively. $Y$ is mass fraction of gas phase, and $\Gamma$ in Eq. (1) is evaporation speed. In the present study, for stability of computation, phase equilibrium takes place instantaneously when pressure decreases under vapor pressure. In the model[5], there is no empirical constant.

Additionally, the working fluid in the present numerical study is room temperature water for which the temperature distribution is known to be almost uniform. Thus, energy conservation equation can be omitted from the governing equations given in Eq. (1). The equation of state of two phase medium is expressed by using $Y$ as follows:

$$\rho=\frac{p(p+p_c)}{K_l(1-Y)p(T+T_c)+R_gY(p+p_c)T} \tag{2}$$

where, $K_l$, $R_g$, $T_c$, $p_c$ are liquid constant, gas constant, liquid temperature constant, and liquid pressure constant, respectively. Void fraction $\alpha$ is calculated as:

$$\alpha=\frac{R_gY(p+p_c)T}{K_l(1-Y)p(T+T_c)+R_gY(p+p_c)T} \tag{3}$$

Finite difference method is used for discretization of governing equation, and the explicit TVD-MacCormack method with 2nd-order accuracy in time and space is used for time integration and evaluation of the convection and viscous terms. In addition, turbulence model is not used since the effectiveness of RANS model was not observed for the simulation of cavitating flow in a previous research [4]. However, large scale velocity variation is reproduced which is originated by oscillation of unsteady cavitation.

*2.2 Apparent phase equilibrium model*

In this study, APE (apparent phase equilibrium) model [4] is applied in which the situation that the gas phase increases even above saturated vapour pressure. In this model, saturated vapour pressure in phase change term $\Gamma$ is replaced by $p_v$* :apparent phase equilibrium pressure. The phase equilibrium pressure is considered to shift toward higher pressure side, $p_v$* is expressed as follows:

$$p_v^*=p_{sat}+f(\alpha)\,p'_{turb} \tag{4}$$

$$\text{Function i}: \quad f(\alpha)=\tanh\{(0.5-\alpha)2n\}/2\tanh(n)+0.5$$
$$\text{Function ii}: \quad f(\alpha)=\alpha(1-\alpha)/0.25 \tag{5}$$

where, $p'_{turb}$ is pressure variation term which changes according to local unsteady flow condition, and $f(\alpha)$ is a weighting function which is a function of void fraction $\alpha$. Two functions are compared in the present study. The Function i, which is previous function [3], takes a value 1 at void fraction $\alpha = 0$ and takes 0 *at* $\alpha = 1$. It represents the situation that apparent phase equilibrium takes place at start of evaporation and the phase change goes to real phase equilibrium ($p = p_{sat}$) at the end of evaporation. On the other hand, the Function ii takes 1 when void fraction $\alpha$ is 0.5 and takes 0 at $\alpha = 0$ and 1. It corresponds to the situation that there is no gas-liquid interface and phase change does not occur at $\alpha = 0$ and 1, then the effect of dissolved air does not appear at the condition. Both Function i and ii can represent the well-known fact that pressure inside fully developed attached sheet cavity becomes saturated vapour pressure [6].

$p'_{turb}$ is a component which contributes to the prediction of dissolved air, but we don't know what kind of component acts on the phenomenon. Therefore, the numerical experiment is conducted in the present study. The following three kinds of variation components were compared,

Model1: $$p'_{turb} = C\rho\overline{u_i' u_j'} \tag{6}$$

Model2: $$p'_{turb} = C\rho\left(L\frac{1}{\rho^2}(\nabla\rho\times\nabla p)\right)^2 \tag{7}$$

Model3: $$p'_{turb} = \begin{cases} -0.5C\rho u_{i,j}u_{j,i} & (u_{i,j}u_{j,i} < 0) \\ 0 & (u_{i,j}u_{j,i} > 0) \end{cases} \tag{8}$$

These components are based on Reynolds stress, Baroclinic torque, and Q value which is the second invariant of velocity gradient. $L$ is characteristic length which was used to adjust the dimension, and $C$ is control parameter. In the present study, the above values are roughly estimated not from turbulence model but from calculation result which is grid scale disturbance caused by homogeneously represented unsteady cavitation.

*2.3 Calculation condition*

Simulations of cavitating flow around NACA0015 hydrofoil were performed. Main flow velocity, temperature, and void fraction are 8[m/s], 293.15[K], and 1%, respectively. Angle of attack is 8[deg], and cavitation number $\sigma$ is set to be 1.4 since for this case the discrepancy between the predicted and measured breakdown characteristics is most prominent [1].

## 3. Results and Discussions

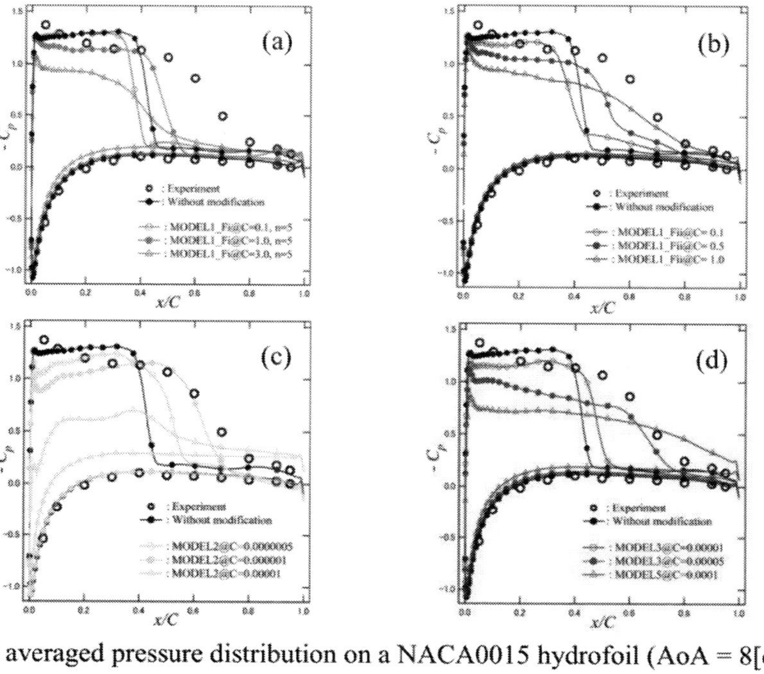


**Figure 1.** Time averaged pressure distribution on a NACA0015 hydrofoil (AoA = 8[deg], $\sigma = 1.4$)

The numerical results of time averaged pressure distribution on a NACA0015 hydrofoil by the present APE model are shown with that of experimental data in Fig.1. Also the numerical result that without the present modification [4] is shown, In the result without the modification, suction side pressure was fail to predict as much higher than that in experiment in the region after mid chord, which indicates that the cavity length was much shorter compared with the experiment. In this study, Function i was adopted for Model 1, and Function ii is adopted for Model 1, Model 2, and Model 3. From Model 1 with Function i in Fig. 1(a) and with Function ii in Fig. 1(b), by adopting Function ii, vapour pressure inside the cavity manages to remain at comparatively lower pressure when the pressure in the region after mid chord is decreased, although it is still insufficient. In Model 2 with Function ii in Fig. 1(c), the suction side pressure decreases sufficiently at mid-chord region, but the pressure inside the sheet cavity especially near the leading edge increased. In Model 3 with Function ii in Fig. 1(d), the pressure distribution showed similar trend to Model 2 using Function ii. In all Model with Function ii, the suction side pressure decreased in the region after the pressure recovery point to trailing edge which corresponds to the region where cloud cavity sheds, but the pressure inside the sheet cavity increased simultaneously in the region before the recovery point which corresponds to the region where sheet cavity develops.

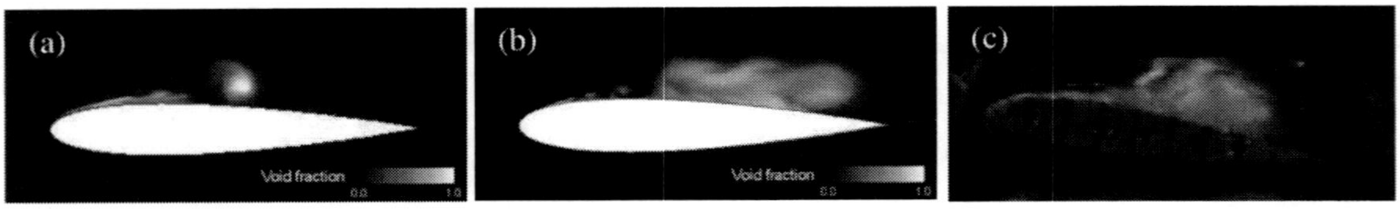


**Figure 2.** Comparison of the instantaneous aspect of cavitation between numerical results and experiment (NACA0015, AoA = 8[deg], $\sigma = 1.4$) (a) Without modification [4] (b) APE modification, Model 2(Function ii, C=1.0) (c) An experiment by The University of Tokyo

The instantaneous predicted void fraction distribution by Model 2 with Function ii and C = 1.0 in the present APE model and the snapshot of transient cavitation visualized by high-speed camera are shown in Fig. 2(b), and (c), respectively. By applying APE model, the cloud cavity volume is successfully improved compared with that without APE modification.

## 4. Conclusions

In order to improve the prediction accuracy of transient cavitating flow around NACA0015 hydrofoil at higher angle of attack, APE (apparent phase equilibrium) model was applied, and numerical simulation was carried out. The results are summarized as follows:

- By applying the APE model, the pressure distribution on suction side of a NACA0015 hydrofoil was slightly improved in the region after pressure recovery point to trailing edge which corresponds the region where cloud cavity sheds.
- The cavity volume and the aspect of cavity shedding were successfully improved by the APE model.
- For the component which affects the apparent phase equilibrium assumption, Reynolds stress and Q value are comparatively better.
- By applying a weighting function which represents the situation that there is no gas-liquid interface and phase change does not occur and the effect of dissolved air does not appear at void fraction 0 and 1, the pressure distribution improved compared with previous function.

## References

[1] Kato, C., *Proc. AJK2011-FED*, Hamamatsu, AJK2011-06084 (2011).
[2] T.J., O'HERN, *Journal of Fluid Mechanics* (1990) 215, pp. 365-391.
[3] Iga., *Earth and Environmental Science*, 22 (2014) No. 052010, pp. 1-11.
[4] Iga, Y., *Proc. The 8th International Symposium on Cavitation* (CAV2012), Singapore No. 160 (2012), pp. 691-695.
[5] Iga, Y., Hashizume, K., Yoshida, Y., Ikohagi, T., *Journal of Environment and Engineering*, Vol.4, No.3 (2009), pp.524-538.
[6] S., Kelly, *Journal of Propulsion and Power*, Vol. 29, No.2 (2013), pp. 410-416.

# Compressible Turbulent Flow Numerical Simulations of Tip Vortex Cavitation

**F. Khatami**[1], **E. van der Weide**[1] **and H. Hoeijmakers**[1]

[1]Mechanical Engineering Department, University of Twente
P.O. box 217, 7500 AE Enschede, The Netherlands
web page: http://www.utwente.nl/ctw/ts/

E-mail: f.khatami@utwente.nl, h.w.m.hoeijmakers@utwente.nl

**Abstract.**
For an elliptic Arndt's hydrofoil numerical simulations of vortex cavitation are presented. An equilibrium cavitation model is employed. This single-fluid model assumes local thermodynamic and mechanical equilibrium in the mixture region of the flow, is employed. Furthermore, for characterizing the thermodynamic state of the system, precomputed multiphase thermodynamic tables containing data for the appropriate equations of state for each of the phases are used and a fast, accurate, and efficient look-up approach is employed for interpolating the data. The numerical simulations are carried out using the Unsteady Reynolds-Averaged Navier-Stokes (URANS) equations for compressible flow. The URANS equations of motion are discretized using an finite volume method for unstructured grids. The numerical simulations clearly show the formation of the tip vortex cavitation in the flow about the elliptic hydrofoil.

## 1. Introduction

In most practical situations, cavitation can and typically does occur when the local static pressure of a liquid drops below the saturation pressure. During an unsteady process, vapor cavities appear in the low pressure zones and subsequently disappear in the areas with higher pressure. Cavitation may cause a number of problems in hydraulic systems, including noise, surface erosion and loss of turbomachinary efficiency. However, besides these undesirable effects, cavitation is used advantageously in medical applications such as e.g. ultrasound cavitation, and lithotripsy as well as in some industrial processes for cleaning of surfaces and the dispersion of particles in a liquid. To control effects of cavitation in these applications, it is essential to understand the structure and mechanisms of the cavitation phenomenon. Following Saurel *et al.* [1], Schnerr *et al.* [2], and Schmidt *et al.* [3] we employ an equilibrium cavitation model, a compressible single-fluid flow model which does not depend on empirical constants. A computational method is presented to numerically simulate an unsteady cavitating flow, see also [5]. The method is applied to predict the unsteady tip vortex cavitation occurring in the flow about Arndt's elliptic hydrofoil. The numerical simulations are performed by using the compressible flow URANS equations (and RANS equations for the case of fully wetted flow) which are closed with the thermodynamic state relations based on the employed equilibrium cavitation model.

## 2. Physical model

The Unsteady Reynolds Averaged Navier Stokes (URANS) equations are employed for the modeling of turbulence, which in integral conservation form are given by

$$\frac{\partial}{\partial t}\iiint_{\Omega} \mathbf{U}\, d\Omega + \iint_{\Gamma=\partial\Omega} \vec{\mathbf{F}}_j\left(\mathbf{U}\right) n_j\, d\Gamma = \iint_{\Gamma=\partial\Omega} \vec{\mathbf{F}}^v_j n_j\, d\Gamma. \tag{1}$$

Here, it is assumed that $\Omega$ is a bounded stationary polygon domain in $R^3$ with boundary $\partial\Omega$, $\mathbf{U} = [\rho, \rho u_i, \rho E]$ denotes the vector of conservative variables, with $\rho$ the density, $u_i$ a component of the velocity vector, and $E$ the specific total energy. The $\vec{\mathbf{F}}_j\left(\mathbf{U}\right) n_j$ and $\vec{\mathbf{F}}^v_j n_j$ are the normal components of the inviscid and viscous flux vectors, respectively, which in Cartesian coordinates can be written as:

$$\vec{\mathbf{F}}_j\left(\mathbf{U}\right) n_j = \begin{bmatrix} \rho\hat{u} \\ \rho\hat{u}u_i + pn_i \\ \rho\hat{u}H \end{bmatrix}, \quad \vec{\mathbf{F}}^v_j n_j = \begin{bmatrix} 0 \\ \hat{\tau}_{ij}n_j \\ \hat{\tau}_{jk}u_k n_j - \hat{q}_j n_j \end{bmatrix}, \tag{2}$$

where $\hat{u} = u_j n_j$ is the velocity normal to the surface $\Gamma = \partial\Omega$, $\hat{\tau}_{ij}$ and $\hat{q}_j$ are the components of the viscous stress tensor and the heat flux which can be written as $\hat{\tau}_{ij} = \mu_{tot}\left[\left(\frac{\partial u_i}{\partial x_j} + \frac{\partial u_j}{\partial x_i}\right) - \frac{2}{3}\delta_{ij}\frac{\partial u_k}{\partial x_k}\right]$, and $\hat{q}_j = -\mu^{\star}_{tot} C_p \frac{\partial T}{\partial x_j}$. Moreover, $\mu_{tot}$ is the total viscosity defined by $\mu_{tot} = \mu + \hat{\mu}$, with $\mu$ and $\hat{\mu}$ the dynamic and turbulent viscosities, respectively, $\mu^{\star}_{tot} = \frac{\mu}{Pr} + \frac{\hat{\mu}}{\hat{Pr}}$, with $Pr$ and $\hat{Pr}$ the dynamic and turbulent Prandtl numbers, respectively, and $C_p$ is the specific heat at constant pressure.

It is assumed that the transport properties are constant for both the liquid and vapor phases. The mixture transport properties are obtained via a blending of the properties of the liquid and vapor phase. The blending function used to compute the mixture values of $\mu$ and $Pr$ is the vapor void fraction $\alpha$, while the mass fraction of the vapor $f_v = \frac{\rho_v}{\rho}\alpha$ is used to compute the mixture values of $C_p$. For example, the mixture $\mu$ can be computed from $\mu_{mix} = \alpha\mu_v + (1-\alpha)\mu_l$. The turbulent viscosity $\hat{\mu}$ is computed from a turbulence closure model. In this paper we employ two-equation Menter's Shear Stress Transport (SST) model.

The URANS equations require additional thermodynamic relations in order to close the system of equations. The equilibrium cavitation model [1], [2], [3] is employed, which in the two-phase flow region assumes local thermodynamic and mechanical equilibrium. Note that in this model the phase transition does not depend on empirical constants, and a cavitation threshold at $p = p_{sat}$ is assumed. The computational method assumes Tait's equation of state for the liquid phase, perfect gas for the vapor phase, and an equilibrium model for the mixture phase (see e.g. Schmidt *et al.* [3] and Koop [4]). Khatami *et al.* [5] showed that these thermodynamic equations are highly computational-intensive. They replaced this approach by using precomputed multiphase thermodynamic tables containing the same thermodynamic information, and they developed a fast, accurate, and efficient method (POITC) [5] for the computational use of these tables. This computational approach is utilized in this paper.

## 3. Results

The computational method uses a cell-centered unstructured finite-volume method applied on a structured grid with the MUSCL-type reconstruction scheme (Van Leer [6]), and the AUSM+up for all speeds (Liou [7]) numerical flux scheme. To ensure the monotonicity of the solution, the Venkatakrishnan limiter [8] is employed. The temporal discretization is carried out using a third-order accurate three-stage TVD Runge-Kutta scheme (Shu [9]). The time step used in

Table 1: Conditions for the cavitating flow around Arndt's hydrofoil at 7 deg AOA.

| $U_\infty$ [m s$^{-1}$] | $p_\infty$ [$10^5$Pa] | $T_\infty$ [K] | $\rho_\infty$ [kg m$^{-3}$] | $c_\infty$ [m s$^{-1}$] | $\sigma$ [–] |
|---|---|---|---|---|---|
| 50 | 12.5 | 293 | 998.7 | 1540.0 | 1.0 |

the numerical simulations must be taken very small. This is because the equilibrium cavitation model takes into account the physics of both the incompressible (liquid) and compressible (vapor and mixture) flows, which are solved using the compressible flow approach. For such a flow the Mach number ranges from low values in the liquid (low-Mach number flow), with a large value of speed of sound $c$ (e.g. $c \approx 1500\,\mathrm{m\,s^{-1}}$), to relatively high values in the mixture, i.e. with low values of speed of sound (e.g. with $c < 1\,\mathrm{m\,s^{-1}}$). Therefore, in order to resolve the dynamics of the pressure waves, which are directly related to the quantity $\rho c$ (see e.g. [3]), and to satisfy the appropriate CFL condition for numerical stability, in particular for low-Mach number flow, the time steps that are needed are extremely small.

### *3.1. Tip vortex cavitation on elliptic hydrofoil*
The numerical simulation of vortex cavitation in the flow about the elliptic Arndt's hydrofoil (NACA$66_2$ – 415) [11], [12] at 7 deg AOA are presented. A cavitation number of $\sigma = 1$ is used, where $\sigma$ is defined by $\sigma = \frac{p_\infty - p_v}{1/2\rho_\infty U_\infty^2}$ with $p_v$ the vapor pressure, and $p_\infty$, $\rho_\infty$, and $U_\infty$ denoting the free-stream pressure, density, and velocity, respectively. The free-stream velocity and pressure are chosen at values higher than the standard conditions (keeping the cavitation number constant), in order to accelerate the cavitation cycles [4]. The initial solution for the unsteady compressible cavitating flow is the steady fully-converged fully-wetted flow solution. Prior to running the main unsteady simulations, it was checked whether or not the flow solution is indeed unsteady. Therefore, the numerical simulations were carried out using an approach for steady flow. It was observed that, when using this approach the simulations did not converge and the system showed an entirely unsteady-like behavior. Hence, an approach for time-accurate unsteady flow is pursued.

Results are presented in Figure 1, which are obtained using the grid shown in Figure 1a. For this grid, the value of $y^+$ at the tip of the hydrofoil has everywhere on the surface a value close to unity. However, it has a maximum value of $y^+ \approx 50$. Furthermore, Figure 1b (top) depicts isosurface of the negative of the pressure coefficient $C_p$, with $C_p \equiv \frac{p - p_\infty}{1/2\rho_\infty U_\infty^2}$, together with the streamlines on the suction side. Experimental studies [12] for similar flow conditions show that the vortex cavitates up to the farfield of the domain. However, from the plots in Figure 1, in particular from Figure 1c (top), it is observed that after a short distance downstream of the hydrofoil, the vorticity magnitude is decreased at a rather high rate. This results in a less compact vortex core with a static pressure that does not reach the low values required for cavitation to occur. The reason for this is mainly due to artificial dissipation that is introduced by the employed numerical approach, in particular the MUSCL reconstruction scheme. In order to improve the results one may use a finer grid, which is in particular refined (far enough from the hydrofoil) downstream of the hydrofoil in the areas in which the formation of the vortex core takes place and cavitation is expected to occur.

## 4. Conclusions
Results of compressible flow numerical simulations of the tip vortex cavitation in the flow about Arndt's elliptic hydrofoil have been presented. The URANS equations for compressible flow have been used which are closed with an equilibrium cavitation model. Furthermore, the

(a) (b) (c)

Figure 1: Results for numerical simulations of tip vortex cavitation about $NACA66_2 - 415$ for $\sigma = 1$: (a) computational grid consisting of approximately 2,000,000 cells; (b) top: isosurface of $C_p = -0.5$ and streamlines shown on suction side; bottom: cross-flow planes with pressure iso-contours; (c) top: cross-flow planes with iso-contours for vorticity magnitude and bottom: turbulent viscosity ratio $\frac{\hat{\mu}}{\mu}$.

thermodynamic state of the system has been characterized by using the precomputed multiphase thermodynamic tables containing data from the appropriate equations of state in the liquid, mixture, and vapor phases and the fast, accurate, and efficient look-up approach of POITC [5] has been adopted. The equations have been discretized using a cell-centered unstructured finite volume method on a structured grid. The numerical results clearly show the formation of tip vortex cavitation for the elliptic hydrofoil at $\sigma = 1$ and 7 deg AOA.

**Acknowledgment**
This research has been carried out under the sponsorship of AgentschapNL in the framework of the Maritime Innovation Program, MIP-IOP IMA10007.

## References

[1] Saurel R, Cocchi J and Butler J 1999 *J. Propulsion and power* **15**(4) 513-22
[2] Schnerr G, Schmidt S, Sezal I and Thalhamer M 2006 *Proceedings CAV2006 Sixth International Symposium on Cavitation* (Wageningen: The Netherlands)
[3] Schmidt S, Sezal I and Schnerr G 2006 *ECCOMAS CFD* (Egmond aan Zee: The Netherlands)
[4] Koop A 2009 *Numerical Simulation of Unsteady Three-Dimensional Sheet Cavitation* (PhD thesis, University of Twente: The Netherlands)
[5] Khatami F, van der Weide E and Hoeijmakers H 2015 *Heat Transfer Engineering (special topic issue)*, **36**(12) 1065-83
[6] Van Leer B 1979 *J. Comput. Phys.*, **32** 101-36
[7] Liou M 2006 *J. Comput. Phys.* **214** 137-70
[8] Venkatakrishnan V 1995 *J. Comput. Phys.* **118** 120-30
[9] Shu C -W 1988 *SIAM J. Sci. and Stat. Comput.* **9** 1073-84
[10] Roe P L 1981 *J. Comput. Phys.* **43** 357-72
[11] Maines B and Arndt R 1997 *J. Fluids Engineering* **119**(2) 413-19
[12] Maines B 1995 *Tip Vortex Formation and Cavitation* (PhD thesis, The University of Minnesota: USA)

# Numerical simulations of cryogenic cavitating flows

**Hyunji Kim, Hyeongjun Kim, Daeho Min and Chongam Kim**
Department of Aerospace Engineering, Seoul National University, Seoul 151-742, Korea
E-mail: chongam@snu.ac.kr

**Abstract.** The present study deals with a numerical method for cryogenic cavitating flows. Recently, we have developed an accurate and efficient baseline numerical scheme for all-speed water-gas two-phase flows. By extending such progress, we modify the numerical dissipations to be properly scaled so that it does not show any deficiencies in low Mach number regions. For dealing with cryogenic two-phase flows, previous EOS-dependent shock discontinuity sensing term is replaced with a newly designed EOS-free one. To validate the proposed numerical method, cryogenic cavitating flows around hydrofoil are computed and the pressure and temperature depression effect in cryogenic cavitation are demonstrated. Compared with Hord's experimental data, computed results are turned out to be satisfactory. Afterwards, numerical simulations of flow around KARI turbopump inducer in liquid rocket are carried out under various flow conditions with water and cryogenic fluids, and the difference in inducer flow physics depending on the working fluids are examined.

## 1. Introduction

Cryogenic cavitation occurs at turbopump inducer, one of the key components in liquid rocket propulsion system, which pressurizes oxidizer and fuel. The characteristics of cryogenic cavitation are quite different to those of water due to thermal effect and strong variations in fluid properties. Once cavitation occurs, not only caviation breakdown but also instabilities such as rotating cavitation or cavitation surge are frequently observed. As these phenomena give rise to degradation of turbopump performance, or even failure, understanding and quantifying characteristics of cryogenic cavitation are crucial for the design of turbopump.

The majority of CFD simulations for turbopump have been limited to the isothermal incompressible flow conditions (Athavale and Singhal [1], Dupont and Okamura [2], Coutier-Delgosha *et al.* [3], and Kiris *et al.* [4]). The effect of temperature variations in fluids is, by definition, not taken into account in these calculations. Recently, Hosangadi *et al.* [5], Utturkar *et al.* [6] and Goncalvés *et al.* [7] have computed compressible two-phase flows with cryogens as a working fluid.

Based on an accurate and efficient numerical scheme for all-speed two-phase flows (Refs. [8] and [9]), we first modify the numerical dissipations to enhance the numerical stability and accuracy in low Mach number regions. The extension to cryogenic fluids is achieved by removing the EOS(Equation of State)-dependent term. To validate the numerical code, Hord's experiments are simulated. Numerical simulations of KARI(Korea Aerospace Research Institute) turbopump inducer are then carried out under various flow conditions, and the inducer flow physics depending on the working fluids are examined.

Journal of Physics: Conference Series **656** (2015) 012131 doi:10.1088/1742-6596/656/1/012131

## 2. Governing equations

### 2.1. Homogeneous mixture equations

The homogeneous mixture model with mass fraction is adopted to describe two-phase flows by including fully compressible thermal effects. The governing equations consist of mixture mass, momentum, and energy conservation laws with one-phase mass conservation law. Then, system preconditioning is introduced for the convergence in low Mach number regions. The preconditioned form of two-phase governing equations are as follows:

$$\frac{\partial}{\partial t}\int_{\Omega}\boldsymbol{Q}d\Omega+\boldsymbol{\Gamma}\frac{\partial}{\partial\tau}\int_{\Omega}\boldsymbol{Q_p}d\Omega+\oint_{\partial\Omega}[\boldsymbol{F}-\boldsymbol{F_v}]\,dS=\int_{\Omega}[\boldsymbol{H}_{rot}+\boldsymbol{H}_{cav}]\,d\Omega. \tag{1}$$

The primitive variable vector $\boldsymbol{Q_p}$ reads

$$\boldsymbol{Q_p}=\begin{bmatrix} p & u & v & w & T & Y_1 \end{bmatrix}^T. \tag{2}$$

In equation (1), $\boldsymbol{\Gamma}$, $\boldsymbol{H}_{rot}$, $\boldsymbol{H}_{cav}$ indicates system preconditioning matrix from Ref. [10], rotating source term and cavitation source term, respectively. Since a pseudo-acoustic speed is defined as $c' = \min(c, \max(V, V_{ref}))$, the original governing equation is recalled at supersonic regions to achieve hyperbolic system of equations.

### 2.2. Equation of state

The definition of the mixture density $\rho$ plays the role of the mixture equation of state (EOS):

$$\frac{1}{\rho(p,T,Y_1)}=\frac{Y_1}{\rho_v\,(p,T)}+\frac{(1-Y_1)}{\rho_l\,(p,T)}. \tag{3}$$

In equation (3), $\rho_v$ and $\rho_l$ is the density of vapor and liquid phase respectively on each occupied volume within the computational mesh.

The properties of cryogenic fluids are sensitive to temperature variation due to their low latent heat compared to other fluids. Therefore, an accurate EOS is required to simulate cryogenic cavitations. Thermodynamic properties are generated from the National Institute of Standards and Technology (NIST) [11] for pure fluids.

### 2.3. Cavitation model

The cavitation source term is defined via simplified non-equilibrium finite rate form. Phase change rates are mainly determined by the difference between the local pressure and the vapor pressure. Here, cavitation models of Merkle [12], Kunz [13], FCM [14] and Mushy IDM [15] are employed.

## 3. Numerical methods

The numerical methods adopted in the study are based on the work by Kim *et al.* [9] and Ihm *et al.* [8].

## 4. Numerical results

### 4.1. Cryogenic cavitating flows around hydrofoil

As a validation case, numerical simulation of the experiments by Hord [16] for liquid nitrogen is demonstrated in figure 1. Flow conditions for run number 289C are presented in table 1.

**Table 1.** Flow conditions for validation problem.

| Run number | Geometry | Working fluid | T [K] | $V$ [m/s] | $\sigma$ |
|---|---|---|---|---|---|
| 289C | hydrofoil | liquid nitrogen | 88.64 | 23.5 | 1.55 |

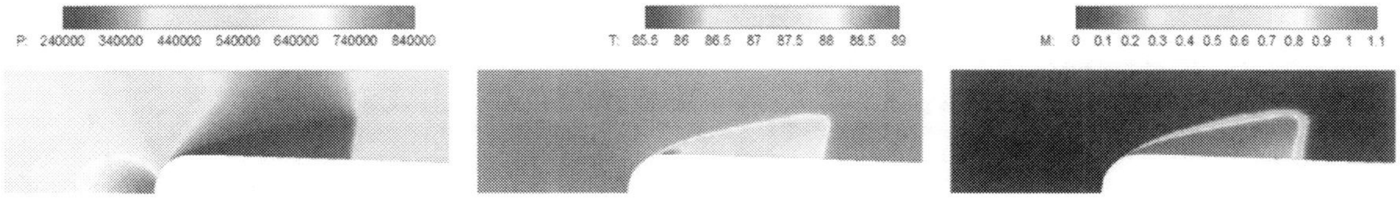


**Figure 1.** Numerical results of Hord's experiments: Run 289C.

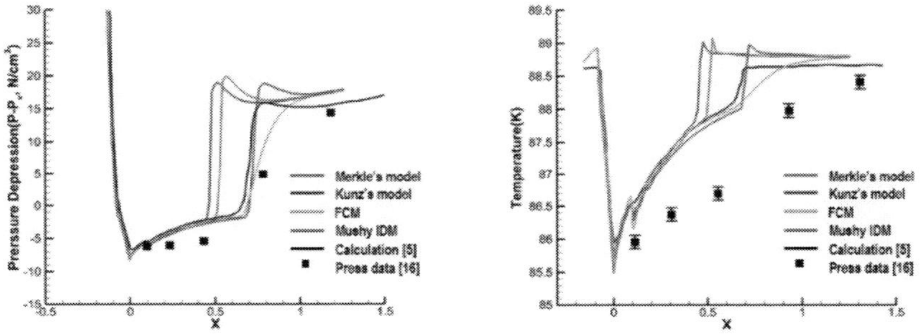


**Figure 2.** Pressure depression(left) and temperature variation(right).

As shown in figure 2, the calculated results qualitatively agree with the experimental ones despite the discrepancy in the cavity closure region. Further analysis on the empirical coefficients of the cavitation models and experimental uncertainty is expected to fill the gap between the computations and the experimental data in this region.

*4.2. Cavitating flows around turbopump inducer*
Numerical simulations of KARI turbopump inducer are performed in water, liquid oxygen and

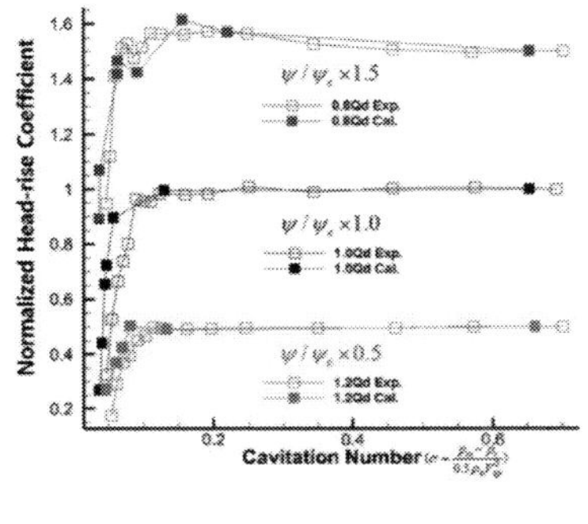


**Figure 3.** Normalized head-rise coefficient for water (3 flow rates).

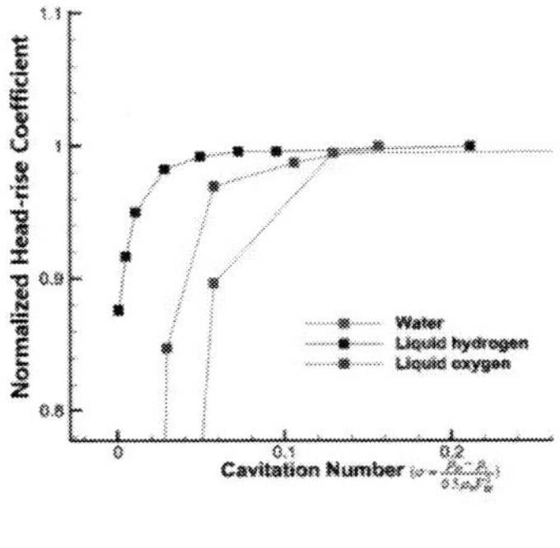


**Figure 4.** Normalized head-rise coefficient for cryogens.

Journal of Physics: Conference Series **656** (2015) 012131 doi:10.1088/1742-6596/656/1/012131

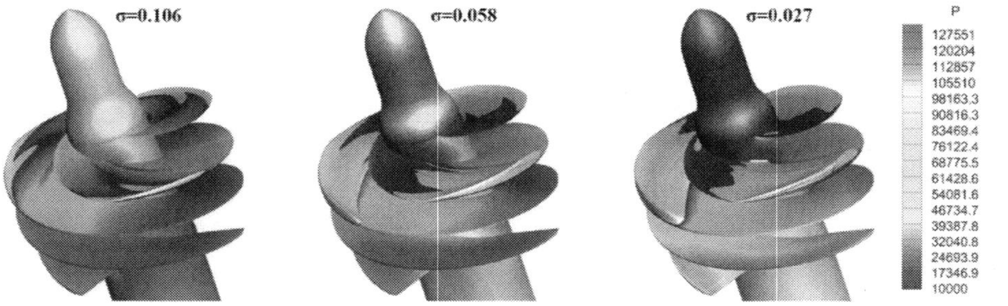


**Figure 5.** Pressure distribution on surface and volume fraction iso-surface at 0.2 (liquid oxygen).

liquid hydrogen with design flow rate and off-design flow rates in figures 3 to 5. Liquid hydrogen and liquid oxygen are simulated to compare thermal effect.

## 5. Conclusions

In the present study, simulations of cryogenic cavitation are performed. Two-phase numerical methods which have been developed for water-gas two-phase flows are extended to cryogenic two-phase flows by generalizing EOS. Scaling of numerical dissipations are also successfully applied so that the stability and accuracy in low Mach number regions are secured. After the validation of the numerical method with cryogenic two-phase flows around hydrofoil, numerical simulation of 3D KARI turbopump inducer is performed. Computed results with water at three different flow rates are reliable compared with experimental data. To examine thermal effect of cryogenic fluids, liquid oxygen and liquid hydrogen turbopump inducer are also successfully computed.

## Acknowledgments

The authors appreciate the financial support of the National Research Foundation of Korea funded by the Ministry of Science, ICT and Future Planning (NRF-2014M1A3A3A02034856) and Civil Military Technology Cooperation Center.

## References

[1] Athavale M M and Singhal A K 2001 *37th AIAA/ASME/SAE/ASEE Joint Propulsoin Conf. and Exhibit (Salt Lake City)* AIAA 2001-3400 (Virginia: AIAA)
[2] Dupont P and Okamura T 2003 *Int. J. Rotating Machinery* **9** 163
[3] Coutier-Delgosha O, Morel P, Fortes-Patella R and Reboud J L 2005 *Int. J. Rotating Machinery* **2005.2** 135
[4] Kiris C C, Kwak D C and Housman J A 2008 *Computers & Fluids* **37** 535
[5] Hosangadi A, Ahuja V, Ungewitter R J and Busby J 2007 *J. Propulsion and Power* **23** 1225
[6] Utturkar Y, Thakur S and Shyy W 2005 *43rd AIAA Aerospace Sciences Meeting and Exhibit (Reno)* AIAA 2005-1286 (Virginia: AIAA)
[7] Goncalvs E, Patella R F, Rolland J, Pouffary B and Challier G 2010 *J. Fluids Engineering* **132**
[8] Ihm S and Kim C 2008 *AIAA J.* **46** 3012
[9] Kim H, Min D and Kim C 2014 *Proc. ICCFD8 (Chengdu, China)*
[10] Weiss J M and Smith W A 1995 *AIAA J.* **33** 2050
[11] NIST Reference Fluid Thermodynamic and Transport Properties Database (REFPROP): Version 8.0, NIST standard Reference Database 23
[12] Merkle C L, Feng J Z and Buelow P E O 1998 *Proc. 3rd Int. Symp. on Cavitation (Grenoble, France)*
[13] Kunz R F, Boger D A, Stinebring D R, Chyczewski T S, Lindau J W, Gibeling H J, Venkateswaran S and Govindan T R 2000 *Computers & Fluids* **29** 849
[14] Singhal A K, Athavale M M, Li H and Jiang Y 2002 *J. Fluids Engineering* **124** 617
[15] Senocak I and Shyy W 2004 *Int. J. for Numerical Methods in Fluids* **44** 975
[16] Hord J 1973 *Cavitation in liquid cryogens 2: Hydrofoil* (Washington: NASA)

# Numerical Simulation of Cavitation Phenomena for Hybrid Contra-Rotating Shaft Propellers

**Sung-Hoon Kim[1], Sung-Han Choo[2], Jeong-Yong Park[2] and Gil-Hwan Choi[2]**

[1] Ship Performance Research Department, HHI, Seoul, South Korea
[2] Performance Evaluation Research Department, HHI, Ulsan, South Korea

E-mail: ksh3417@hhi.co.kr

**Abstract**. This paper deals with a numerical simulation of cavitation flow around a hybrid contra-rotating shaft propeller operating in wake field. The simulation for the cavitating flow is performed for straight operating and turning condition of podded propeller located behind the main propeller using unsteady Reynolds-Averaged Navier-Stokes. The behavior of the main propeller is almost similar regardless of the turning angle. In contrast, the cavitation behavior of the podded propeller depending on the turning angle appears to be entirely different due to the change of the load distribution on the podded propeller. At the large angle of the turning condition, the unstable cavity flow due to the large amount of cavitation and the hub vortex separated from the forward propeller as well as face cavitation is observed. Thus, a great caution on the cavitation phenomena is needed when designing and operating the HCRSP.

## 1. Introduction

Some EU projects show a possibility of a Hybrid Contra-Rotating Shaft Propeller (HCRSP, Figure 5) to reduce both operating cost and CO2 emission. In a TRIPOD project [1], an attempt which combined a CLT-type propeller and HCRSP has been tried and achieved energy savings of 9-11 percent at model tests. In an ENVIROPAX project [2], the HCRSP concept including various hydrodynamic issues like power split, propeller design, powering performance evaluation has been investigated. Along with a study of the HCRSP, a Japanese ROPAX Ferry has been equipped with an ABB hybrid CRP-POD system claiming energy savings of 13 percent in full scale [3].

Nonetheless of a success on energy savings, an application of the HCRSP is still reluctant to attempt due to the high cost of the system, the uncertainty of propulsive performance in full scale and uncertain cavitation phenomena, especially in turning condition of a podded propeller. In this study, therefore, cavitation flow around the HCRSP is numerically simulated in order to investigate a characteristic of the cavitation behavior in turning conditions as well as a straight operating condition.

## 2. Numerical Method

Governing equations for incompressible turbulent flow are an instantaneous conservation of mass (continuity equation) and momentum (Reynolds averaged Navier-Stokes equation, RANS). Reynolds stresses are important to predict turbulent flow, especially in high Reynolds number flow. To close the transport equations, a k-w SST turbulent model [4] is applied because the model is suitable for highly adverse pressure gradient flow such as propeller and turbine blades. For a calculation of cavity flow on propellers, a multi-phase mixture model assuming that a working medium is a single fluid with a homogeneous mixture of two phases (liquid and vapor) is used. Schnerr and Sauer model [5] is used for the cavitation model for the simulation.

A grid system is filled with unstructured mesh by STAR-CCM+. The meshes around leading edge and trailing edge are refined to capture steep pressure gradient. A size of the numerical domain for a numerical cavitation tunnel is determined to have the same size with a medium-sized cavitation tunnel (0.6 m × 0.6 m). In order to simulate wake field in the cavitation tunnel, an axial wake distribution which was numerically computed in full scale, is applied to a velocity inlet condition of the domain.

## 3. Cavitation Simulations of HCRSP

### *3.1. Design of HCRSP*

A HCRSP system has been designed for a 7,000 TEU container. Power ratio between a forward propeller and a podded propeller is divided equally and RPM ratio is determined to be 1:1.45. At all the speeds, the HCSRP requires about 10% of less power assumption than a conventional propulsion system. A tendency to save more energy at low speeds is observed through the results of the model tests. In order to perform cavitation simulations of the HCRSP, a thrust coefficient is calculated for the forward propeller from the results of the model tests. A rotation rate of the podded propeller is determined as a RPM ratio between two propellers.

The forward propeller rotates counter-clockwise while the podded propeller rotates clockwise when looking downstream with 0° of blade position defined at top position. The turning angle of the podded propeller is defined as the angle between the rotation axis of the forward and podded propeller.

### *3.2. Cavitation Simulation of HCRSP in straight operating condition (0° of turning condition)*

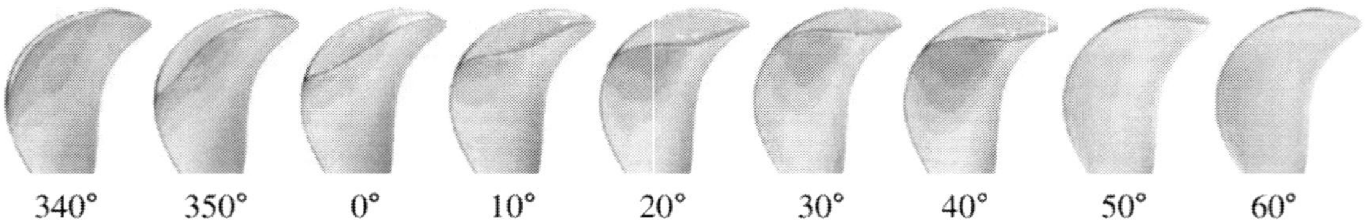


**Figure 1. Cavitation patterns of forward propeller in straight operating condition**

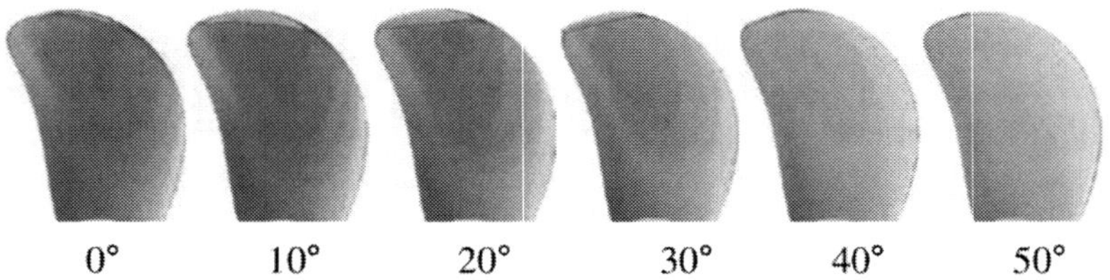


**Figure 2. Cavitation patterns of podded propeller in straight operating condition**

Figure 1 and 2 show the cavitation patterns of the forward propeller and podded propeller in the straight operating condition (0° of the turning condition). Back sheet cavitation starts appearing from 340° on a surface of the forward propeller. A maximum volume of cavitation appears on an upper side of the surface at 20° position and the cavitation disappears from 60°. The cavitation disappears smoothly around the tip region. Thus, there is no momentarily high value of the index indicating the erosion risk. From 70° to 330°, the propeller is wholly free from the sheet cavitation.

The cavitation volume of the podded propeller is relatively small even if the power density (power per unit area) of the podded propeller is larger than that of the forward propeller, because of a relatively low gradient wake field into the podded propeller. The back sheet cavitation starts appearing from 0° and disappearing from 50°. Likewise with the cavitation behavior of the forward propeller, the sheet cavitation stably behaves and shows no erosive cavitation.

### *3.3. Cavitation Simulation of HCRSP in turning condition*

**Table 1. Summary of cavitation observation results**

| Turning Angle | Back Cav. (Forward) | Face Cav. (Forward) | Back Cav. (Pod) | Cav. due to hub vortex (Pod) | Face Cav. (Pod) |
|---|---|---|---|---|---|
| -10 ° (Port) | 340° ~ 50° | none | 0°~10°/185°~325° | 210° ~ 275° | 355° ~ 95° |
| -8 ° (Port) | 340° ~ 55° | none | 0°~15°/185°~315° | 220° ~ 275° | 5° ~ 90° |
| -6 ° (Port) | 340° ~ 60° | none | 0°~25°/215°~290° | 235° ~ 260° | 65° ~ 80° |
| -4 ° (Port) | 340° ~ 60° | none | 0° ~ 25° | none | none |
| -2 ° (Port) | 340° ~ 60° | none | 0° ~ 35° | none | none |
| 0 ° (Straight) | 340° ~ 60° | none | 0° ~ 50° | none | none |
| 2 ° (Starboard) | 340° ~ 60° | none | 355° ~ 50° | none | none |
| 4 ° (Starboard) | 340° ~ 60° | none | 345° ~ 90° | none | none |
| 6 ° (Starboard) | 340° ~ 60° | none | 340° ~ 90° | 70° ~ 85° | 195° ~ 265° |
| 8 ° (Starboard) | 340° ~ 60° | none | 340° ~ 120° | 45° ~ 85° | 160° ~ 265° |
| 10 ° (Starboard) | 340° ~ 65° | none | 345° ~ 140° | 40° ~ 85° | 155° ~ 265° |

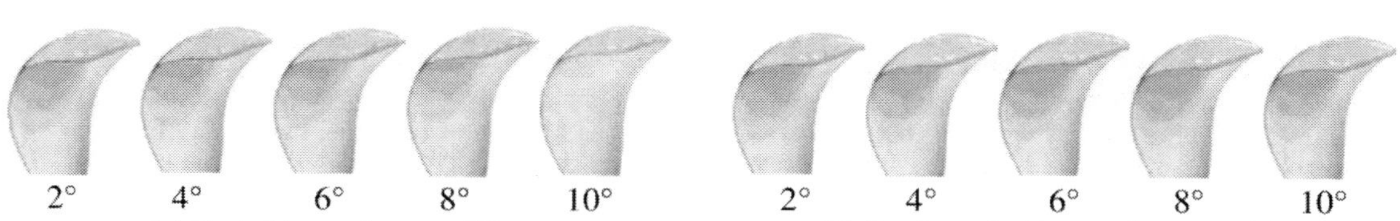


(a) Port-side turning condition (b) Starboard-side turning condition

**Figure 3. Patterns of sheet cavitation of forward propeller in turning condition (20° of blade position)**

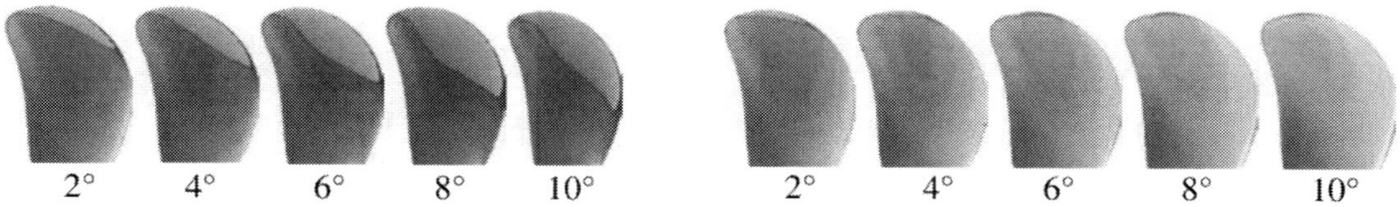


(a) Port-side turning condition (b) Starboard-side turning condition

**Figure 4. Patterns of sheet cavitation of podded propeller in turning condition (20° of blade position)**

The positions of cavitation occurrences of the HCRSP are summarized in table 1. The cavitation phenomena of the forward propeller are almost similar regardless of turning angle of the podded propeller. As depicted in figure 3, which shows the cavitation patterns at 20° of blade position for the forward propeller in turning condition of the podded propeller, the cavitation pattern and volume of the forward propeller maintains almost similarly regardless of the turning angle, except that the cavitation volume gets slightly bigger as turning angle goes towards starboard-side from port-side turning.

Unlike with the cavitation phenomena of the forward propeller, those of the podded propeller are entirely different depending on the turning angle of the podded propeller. The load of upper side becomes larger and that of lower side becomes smaller as the running angle gets bigger, which makes the cavitation volume get bigger at the upper side of the propeller because the actual pitch angle of the upper side becomes lager due to the turning angle. Directly contrary to the port-side turning condition, the cavitation of upper side gradually gets smaller and fades away in starboard-side turning condition due to the reduced actual pitch angle at the upper side.

In a contrary with the upper side, the actual pitch angle at the lower side gets to be larger in the starboard-side turning condition. Consequentially, the cavitation on back side of the lower side begins to occur and gradually expands, from 6° of the port-side turning angle. In addition, the change of load distribution due to the turning angle also results in an occurrence of face cavitation. The face cavitation on the lower side appears in the port-side turning condition, while the face cavitation appears on the upper side in the starboard-side turning condition. As the turning angle bigger, the occurrence range and volume of the cavitation gradually spreads out further and bigger.

From 6° of the port-side and starboard-side turning condition, moreover, the hub vortex separated from the forward propeller heats the surface of the podded propeller. The hub vortex induces the separated cavity on the blade surface near the root of the propeller. Since the behavior of the cavitation is quite unstable, the cavity flow could erode the blade surface by the hub vortex.

No sheet cavitation around the pod housing of the HCRSP at the turning condition is found regardless of the turning angle until 8°. In accordance with the turning angle increasing, it is found that the static pressure on the surface of the pod housing gradually goes down. At 10° of the port-side turning condition, eventually, the sheet cavitation is found on a fin located at the lower part of the pod housing.

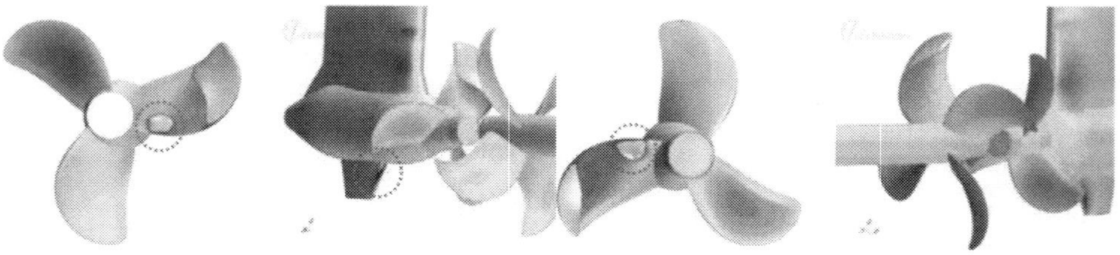

(a) Port-side turning condition (b) Starboard-side turning condition

**Figure 5. Cavitation pattern due to hub vortex and sheet cavitation of pod housing in turning condition**

## 4. Conclusion

The cavitating flow is simulated from -10° to 10° of the turning angles for the podded propeller behind the forward propeller. Irrespective of the turning angle, the characteristics of the forward propeller remain almost similar. However, the pattern and volume of cavitation of the podded propeller is completely different because the load distribution of the podded propeller is entirely changed according to the turning angle. At the large angle of the turning condition, the unstable cavity flow due to the large amount of cavitation and the hub vortex separated from the forward propeller as well as face cavitation is observed. Thus, a great caution on the cavitation phenomena is needed because the cavitation pattern of the podded propeller is wholly different according to the turning angle.

## Acknowledgement

The present work was carried out as a task of the Korean National R&D project, "Development of the key technology for a ship drag reduction and propulsion efficiency improvement (10040030)", supported by the Ministry of Trade, Industry and Energy.

## References

[1] A. Sánchez-Caja, M. Pérez-Sobrino, R. Quereda, M. Nijland, T. Veikonheimo, J. González-Adalid, I. Saisto and A. Auriarte 2013 Combination of Pod, CLT and CRP Propulsion for Improving Ship Efficiency: the TRIPOD project *Third International Symposium on Marine Propulsors smp'13 (Launceston, Australia)*

[2] Levander O., Sipilä H. and Pakaste R. 2005 ENVIROPAX ferries make promising progress *Marine News - Wärtsilä Customer Magazine* Nr *1*

[3] Ueda N., Oshima A., Uniseki T., Fujita S., Takeda S. and Kitamura T. 2004 The First Hybird CRP-POD Driven Fast ROPAX Ferry in the World *Mitsubishi Heavy Industries Ltd Techinal Review*

[4] Menter F. R. 1994 Two-Equation Eddy-Viscosity Turbulence Models for Engineering Apllications *AIAA Journal* **32**(8) 1598-1605

[5] Schnerr, G. H. and Sauer J. 2001 Physical and Numerical Modeling of Unsteady Cavitation Dynamics *Fourth International Conference on Multiphase Flow (New Orleans, USA)*

# Jet-Supercavity Interaction: Insights from CFD

**M Kinzel[1], M Moeny[1], M Krane[1], and I Kirschner[2]**
[1] Applied Research Laboratory, The Pennsylvania State University, University Park, PA, USA
[2] Applied Physical Sciences, Groton, CT, USA
E-mail: mpk176@psu.edu

**Abstract.** In this work, the interaction between a ventilated supercavity and a jet are examined using computational fluid dynamics (CFD). The CFD model is validated using experimental data, and shows to capture the correct trend in the bulk cavity behavior (qualitatively and quantitatively). Using these models, a number of novel insights into the physical characteristics of the interaction are developed. These interactions are described by: (1) the jet gas and ventilation gas poorly mix within the cavity, (2) the jet appears to cause additional gas leakage by transitioning the cavity from a recirculating flow to an axial flow, (3) the jet has the ability to lengthen the cavity, and (4) the jet invokes wake instabilities that drive cavity pulsation. These phenomena are documented and discussed in the following paper.

## 1. Introduction

Supercavitating vehicles operate within a supercavity that completely surrounds the vehicle in a gas to reduce skin friction. This is a concept that reduces power for high-speed underwater vehicles and mitigates adverse cavitation issues at high speed. Supercavitating vehicles often use ventilation gas to develop this cavity. Such cavities have been explored in detail without the coupling to propulsors.

A relatively unstudied interaction is that between a supercavity and a jet. Such an interaction is important when jets, or rockets, are used for propulsion. Previous studies in this topic are limited, and the work Paryshev [1] appears to be the most developed. Therein, the cavity-jet interaction was postulated to be dependent on the total pressure of the jet. When the total pressure of the jet is less than the free-stream total pressure of the liquid, the jet is relatively weak and acts to ventilate the cavity. On the contrary, as the total pressure of the jet increases above the total pressure in the liquid, the jet penetrates the cavity and increases gas leakage. Experimental observations tend to corroborate the theory [1].

This paper explores jets interacting with supercavities using CFD. The paper is organized as follows. A brief discussion of the experimental and numerical model setup is developed, which is followed by a benchmarking process. Results from the CFD model are then presented. This is followed by an interpretation of the solutions to provide new insight into cavity-jet interactions.

## 2. Methods

A validation effort is performed using a recent test campaign of cavity-jet interactions. Tests were performed in the 1.22m Garfield Thomas Water Tunnel at The Pennsylvania State University - Applied Research Laboratory [2]. The model consists of a sting-mounted, 30 deg, conical-shape cavitator (Fig. 1). A body extended from the cavitator to support a jet.

The CFD model is based on a three phase Eulerian multiphase model within Star-CCM+ [3]. The model uses 2$^{nd}$-order differencing (space/time), along with a volume of fluid method using High-Resolution

Interface Capturing (HRIC) for the phase-conservation convection terms. The model uses incompressible liquid and isothermal compressible ventilation gas and jet gases. In these efforts, we use the improved delayed detached eddy simulation (IDDES) turbulence model [4], which is consistent with modeling requirements established by Kinzel et al. [5]. All cases are run on a series of four mesh/time resolutions to ensure mesh/time asymptotic solutions, with a fine-mesh solution using roughly 3.8x10$^6$ hex-dominant, unstructured volumes. The validation effort indicated that the prediction is within the uncertainty of the experiment for nearly all conditions. The CFD domain represents the experiment and is shown in Fig. 1, which focuses on results for a 11mm diameter jet at 13.7 m/s. Boundary conditions are applied as follows: (1) the upstream boundary inlet is 1.22" upstream of the vehicle, (2) the outlet is 3.8m downstream of the vehicle, (3) slip walls are used to represent the test-section walls, (4) no-slip walls are used to represent the sting, cavitator, and body (as shown in the blow up in Fig. 1), (5) ventilation ports are just downstream of the cavitator, and (6) a jet gas inlet through a converging nozzle that exits through the aft-end of the body.

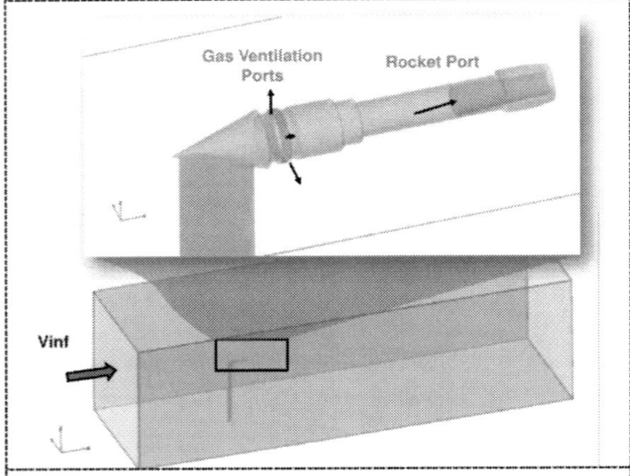


**Figure 1.** Diagram of the computational domain used to represent the experimental setup.

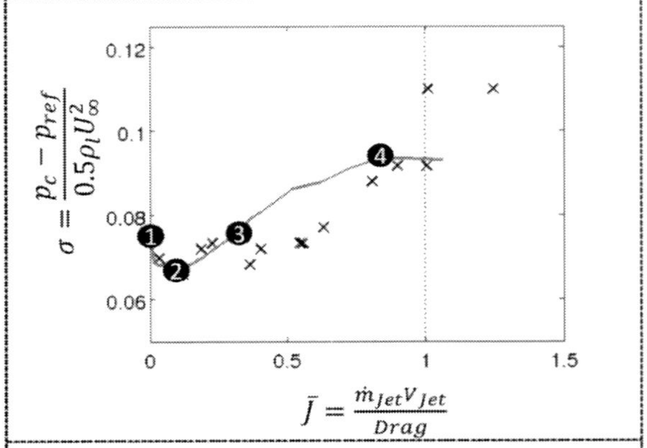


**Figure 2.** Benchmark of computational predictions (red line) as compared to the experimental measurements (X's).

## 3. Results

First, we review the validation of the CFD. Results from the studies are provided in Fig. 2 in the form of plots of $\sigma_c$ versus $\bar{J}$. Cavitation number, $\sigma_c = \frac{(p_c - p_\infty)}{0.5\rho_l U_\infty^2}$, relates to cavity size and $\bar{J}$, $\bar{J} = \frac{(\dot{m}_{Jet} V_{Jet})}{Drag}$, is similar to the thrust-to-drag ratio (introduced by Paryshev [1]). Note that $p_c$ is the cavity pressure, $p_\infty$ is the free stream pressure, $\rho_l$ is the liquid density, $U_\infty$ is the free stream speed, $\dot{m}_{Jet}$ is the jet mass flow, $V_{Jet}$ is the jet velocity, and *Drag* is the cavitator drag. For a $\bar{J}$ value of 0, there is no jet. As the jet is increases to a value of 1.0, thrust approximates self propulsion. And, for values greater than 1.0, the jet accelerates the vehicle. The CFD predictions (red line), predicted a range of conditions. The CFD and experiments indicate a similar trend and corroborate Paryshev [1]. This trend is defined as follows. Starting from a $\bar{J} = 0.0$, there is an initial drop in $\sigma_c$ with increasing $\bar{J}$ until a $\bar{J}$ value of 0.3. Thereafter, $\sigma_c$ increases with an increasing jet strength. In addition, the CFD predictions are in reasonable correlation with the experiments. With this, it is the opinion of the authors that the CFD is predicting the physical aspects of the interactions quite well and it is believed that insight from the CFD is relevant.

Flow visualization from the CFD results is provided in Fig. 3, which has four subfigures indicating:

(a) A representation of cavity shape (cyan-colored isosurface at a liquid volume fraction of 0.5) along with a corresponding pressure field contour plot on the center plane.
(b) Contour plot of the cavity ventilation gas indicating ventilation and jet gas mixing.

(c) The local air entrainment from the cavity as indicated by local air entrainment rate parameter, $C_{Q,loc}$, ($C_{Q,loc} = u/U_\infty \alpha_G$ , where $\alpha_G$ is the gas volume fraction and $u$ is the axial velocity). Positive values are plotted such that red indicates gas traversing from the cavity and blue indicates recirculating gas.
(d) The ratio of the local stagnation pressure to that in the free stream, plotted within the cavity. Red values indicate the gas has stagnation pressure equivalent to the liquid, blue indicates a weak jet. Note that stagnation pressure, $p_0$, is given as $0.5\rho U^2 + p$.

Each subfigure represents solutions from different conditions. These solutions are numbered (in the top-left corner), and correspond to the numbered points on the CFD solution results in Fig. 2. These results evaluate the solution at various regimes of the cavity-jet interaction.

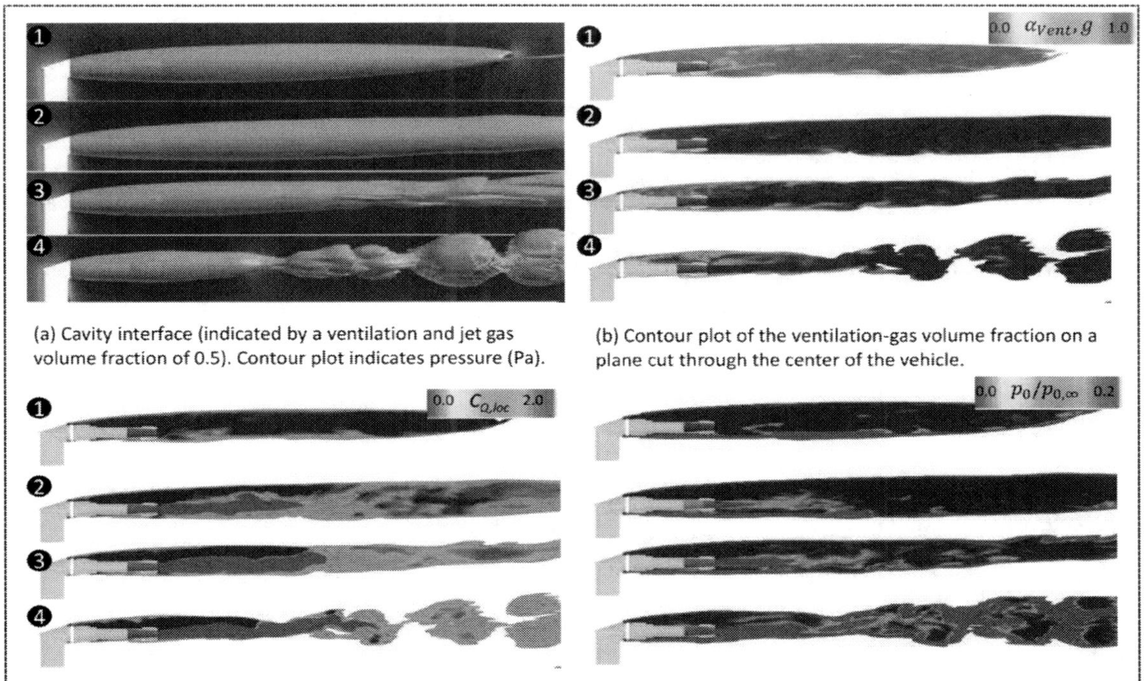


(a) Cavity interface (indicated by a ventilation and jet gas volume fraction of 0.5). Contour plot indicates pressure (Pa).

(b) Contour plot of the ventilation-gas volume fraction on a plane cut through the center of the vehicle.

(c) Contour plot of the local ventilation rate, $C_{Q,loc}$, on a plane cut through the center of the vehicle

(d) Contour plot of the stagnation pressure normalized by the free-stream stagnation pressure.

**Figure 3.** Flow field predictions. These plots are shown for four values of $\bar{J}$, corresponding to the numbered points in Fig. 2.

## 4. Discussion

### *4.1. Impact of Jet on the Cavity Shape*

First, consider the cavity character with increasing jet strength (or $\bar{J}$). As indicated in Fig. 2, as the jet activates, $\sigma_c$ sharply drops indicating the jets acts to ventilate. This is apparent in comparing the cavity sizes of the $\bar{J}$=0.0 and $\bar{J}$=0.11 cases in Fig. 3 (a), where the $\bar{J}$=0.11 case has a much larger cavity (note that cavity size ~ $1/\sigma_c$). As $\bar{J}$ increases to 0.37, the cavity diameter approaches the $\bar{J}$=0.0 case (note that $\sigma_c$ values are similar), but the cavity length approaches that of the $\bar{J}$=0.11 case. This longer cavity for $\bar{J}$=0.37 is a difference and, as suggested by Paryshev [1], shows that the jet has the ability to lengthen the cavity for a given $\sigma_c$ value. Lastly, for $\bar{J}$=0.84, a highly unsteady wake forms, and, as observed in animations, induces a pulsating cavity. In Fig. 3 (d), is the predicted $p_0/p_{0,\infty}$. The model of Paryshev [1] imposes a jet without decreases in $p_0$. CFD suggests the expansion process leads to rapid loss in $p_0$, and that the jet affects the cavity through different mechanisms.

### *4.2. Impact of Jet on the Cavity-Gas Composition*

The next observation focuses on the gas content within the cavity. The behavior is shown in Figs. 3 (b), indicating that, an active jet dominates the gas within the cavity. The ventilation gas adheres to the cavity shear layers, as suggested by Spurk [7] and Kinzel et al. [8], and weakly mixes with the jet gas.

### *4.3. Impact of Jet on the Cavity-Gas Entrainment Mechanisms*

Gas entrainment from the cavity is depicted in Fig. 3 (c). For the low $\bar{J}$ values, 0.0 and 0.11, gas moving out, shown by the non-blue $C_{Q,loc}$ regions, is dominated by cavity interface regions and the jet. With the weak jet ($\bar{J}$=0.11), the red shows the diffusing jet, with a blue region above the jet indicating that the recirculation process within the cavity is not disrupted. As $\bar{J}$ increases (to 0.37), the jet disrupts the re-circulatory region (when the cavity interface is scarred by the jet in Fig. 3 (a)). Aft of this point, is a transition from re-circulatory flow to axial flow. This axial flow appears to enable a more slender cavity. For $\bar{J}$=0.84, the re-circulatory region shortens and also develops into y axial flow. The difference in this condition, is that the axial flow region drives an oscillatory wake that could be driving cavity pulsation.

## 5. Summary and Conclusions

CFD simulations of cavity-jet interactions were performed and based on a new experimental data set. The CFD simulations tracked liquid, ventilation gas, and jet gas within the context of a IDDES simulation methodology. The CFD model was validated and found to be in reasonable agreement with experiment in terms of the correct trend and quantitative values. These simulation results are used to describe the physical characteristics of the cavity-jet interaction.

The cavity-jet interaction are defined as cavity size, character, gas composition, and gas entrainment. A regime of the jet acting to ventilate the cavity was observed, followed by a transition to the jet enhancing air entrainment. Observations indicated that re-circulatory cavity was altered by the jet and drives this breakdown process. For weak jets, where the jet was a ventilator, the re-circulatory process within the cavity was not disrupted. As the jet increased in strength, the re-circulatory flow was disrupted, and an axial cavity flow is developed leading to increased gas entrainment and cavity instabilities. This appears to dominate the cavity-jet interaction, which is not described in prior work. These fundamental processes observed within the CFD indicate new insight into the interaction of a jet with a supercavity.

## 6. Acknowledgments

Funding provided through ONR under contracts #N00014-13-1-0692 and #N00014-12-1-0489, with Dr. Ronald Joslin as the program manager.

## 7. References

[1] Paryshev E 2006 Approximate mathematical models in high-speed hydrodynamics *J. Eng. Math.* **55** 41–64

[2] Moeny M, Krane M, Kirschner I and Kinzel M 2015 Jet-Supercavity Interaction: Insights from Experiment *9th International Symposium on Cavitation (CAV2015)* (Lausanne, Switzerland)

[3] CD-Adapco 2014 Star-CCM+ 9.06

[4] Shur M L, Spalart P R, Strelets M K and Travin A K 2008 A hybrid RANS-LES approach with delayed-DES and wall-modelled LES capabilities *Int. J. Heat Fluid Flow* **29** 1638–49

[5] Kinzel M, Lindau J, Peltier L, Kunz R and Sankaran V 2007 Detached-Eddy Simulations for Cavitating Flows *18th AIAA Computational Fluid Dynamics Conference* Fluid Dynamics and Co-located Conferences (AIAA)

[6] Kinzel M P, Moeny M, Krane M, Kirschner I and Lindau J W 2015 A V&V Approach for a Parametric Investigation of Jet-Cavity Interactions for Supercavitation *ASME V&V2015* (Las Vegas, NV, USA)

[7] Spurk J H 2002 On the gas loss from ventilated supercavities *Acta Mech.* **155** 125–35

[8] Kinzel M P, Lindau J W and Kunz R F 2009 Air entrainment mechanisms from artificial supercavities: insight based on numerical simulations *7th International Symposium on Cavitation (CAV2009)* pp 1–14

# Comparison between low-order and high-order acoustic pressure solvers for bubbly media computations

**G S Bruno Lebon[1], I Tzanakis[2], K A Pericleous[1] and D G Eskin[2]**

[1] Centre for Numerical Modelling and Process Analysis, The University of Greenwich, London, SE10 9ET, United Kingdom
[2] Brunel Centre for Advanced Solidification Technology, Brunel University, Uxbridge, Middlesex, UB8 3PH, United Kingdom

E-mail: G.S.B.Lebon@gre.ac.uk

**Abstract.** Numerical modelling is a useful tool for the fundamental study of acoustic cavitation treatment in liquid metals. This treatment, also known as ultrasonic melt processing, significantly improves the properties and quality of metallic materials. However, the mechanisms leading to these observed improvements are still unclear and a fundamental study of cavitation treatment is required to understand this process. In this endeavour, this paper compares the use of high-order discretization schemes for solving acoustic pressures in cavitating liquids with its low-order counterpart. A fourth order scheme is shown to be more stable and accurate than a second order scheme when taking into account the acceleration of bubbles before their collapse, and is recommended for the full cavitation modelling of acoustic treatment of liquid metals.

## 1. Introduction

Ultrasonic treatment yields significant improvements in the quality and properties of metallic materials [1-3]. These improvements are attributed to acoustic cavitation [4]. However, this promising technology has not been successfully applied in the treatment of large volumes of melt, as is required by industrial processes like continuous casting. A fundamental study of the melt cavitation process is therefore required to circumvent difficulties in scaling up the ultrasonic treatment process [5]. The '*full cavitation model*' [6], originally developed for hydrodynamic cavitation in water, has been used by Nastac [7] and the present authors [10-11] to predict the cavitation treatment zone in a crucible and a launder respectively. Accurate prediction of these cavitation regions is required to optimize launder designs and minimize the power input, complexity, and time required to treat a certain volume of liquid metal. However, the pressure discretization models used so far [9-11] are at most second order in space and time, making the computations prone to numerical diffusion over long simulation run times. This paper investigates the use of higher-order methods in the discretization of the acoustic wave equations as applied to cases where the acoustic pressures are very large, to pave the way for longer simulations of ultrasonic melt treatment.

## 2. Theory

In the present model, the acoustic pressure $p$, acoustic wave velocities $v_i$, and bubble volume fraction $\phi$ are solved for in a segregated manner.

### *2.1. Governing equations*

Conservation of mass and momentum, and the identity $\partial p/\partial \rho \equiv c^2$, where $\rho$ is the medium's density, give the equations for the propagation of sound in a moving fluid medium:

$$\frac{\partial p}{\partial t} + \bar{v}_j \frac{\partial p}{\partial x_j} + \bar{p} c^2 \frac{\partial v_j}{\partial x_j} = S \tag{1}$$

$$\frac{\partial v_i}{\partial t} + \bar{v}_j \frac{\partial v_i}{\partial x_j} + \frac{1}{\bar{\rho}} \frac{\partial \mathrm{p}}{\partial x_i} = \frac{\partial \tau_{ij}}{\partial x_j} + F_i \tag{2}$$

$S$ contains mass sources such as sinusoidal signals representing the vibrating surface of a sonotrode or the contribution of the growth and collapse of cavitation bubbles to the acoustic pressure ($-\rho c^2 \partial\phi/\partial t$). The forcing terms $F_i$ contain velocity sources due to the vibrating surface. The material properties symbols are described in Table 1 and the overbar indicates mean flow variables. The speed of sound $c$ in the bubbly fluid is variable and depends on the resonant frequency of the oscillating bubbles. For frequencies much lower than the resonant frequency of the bubbles present in the liquid, $c = \sqrt{\frac{K_b K_l}{\{\phi K_l + (1-\phi) K_b\}[\phi \overline{\rho_b} + (1-\phi)\overline{\rho_l}]}}$ where $K$ represents the bulk modulus of the phase and the subscripts $b$ and $l$ denote bubble and liquid properties respectively [10]. The governing equation for the bubble volume fraction $\phi$ is

$$\frac{\partial(\overline{\rho_b}\phi)}{\partial t} + \frac{\partial(\overline{\rho_b}\overline{v_j}\phi)}{\partial x_j} = \Gamma \tag{3}$$

where $\Gamma$ is the net phase change. Following the derivation of the full cavitation model [6], by using the Keller-Miksis equation – which is numerically stable for the large forcing amplitudes that are encountered in ultrasonic melt treatment [11] – and keeping the acceleration term [12], the net phase change rate for bubble growth is obtained as

$$\Gamma = C_g \frac{v_{ch}}{\sigma} \rho_l \rho_b \left[ \frac{2}{3} \left( \frac{1 + v_{ch}/c}{1 - v_{ch}/3c} \right) \frac{p_v - P}{\rho_l} - \frac{2}{3} R \frac{d^2 R}{dt^2} \right]^{1/2} (1 - \phi) \tag{4}$$

applicable for $p_v > P$. For the collapse phase, i.e. $P > p_v$, the phase change rate is

$$\Gamma = -C_c \frac{v_{ch}}{\sigma} \rho_l \rho_b \left[ \frac{2}{3} \left( \frac{1 + \frac{v_{ch}}{c}}{1 - \frac{v_{ch}}{3c}} \right) \frac{P - p_v}{\rho_l} - \frac{2}{3} R \frac{d^2 R}{dt^2} \right]^{\frac{1}{2}} \phi \tag{5}$$

$v_{ch}$ is a characteristic bubble velocity that is estimated from an average bubble size [9], and $C_g$ and $C_c$ are empirical coefficients.

### *2.2. Wave equations discretization*

A fully staggered mesh is used in both differencing schemes presented: the scalar quantities acoustic pressure and bubble volume fraction are stored in cell centres at the end of each time step, and velocity components are stored at cell faces in the middle of each time step [13].

*2.2.1. Leapfrog scheme.* The leapfrog scheme is second order in space and time. The acoustic pressures and velocities are updated using the following relations:

$$p^k = p^{k-1} - \frac{\mathrm{c}\Delta\mathrm{t}}{\mathrm{x_i}} \left( u_{i,dwnd}^{k-\frac{1}{2}} - u_{i,upwnd}^{k-\frac{1}{2}} \right) + \mathrm{S}\Delta\mathrm{t} \tag{6}$$

$$u_i^{k+\frac{1}{2}} = u_i^{k-\frac{1}{2}} - \mathrm{c}\Delta\mathrm{t}\left(p_{dwnd}^k - p_{upwnd}^k\right) + \mathrm{F_i}\Delta\mathrm{t} \tag{7}$$

*2.2.2. Fourth-order scheme.* The dispersion-relation-preserving scheme [14] is used for the spatial differentiation and temporal integration steps of the convective terms in equations (1) and (2).

The 6-point stencil for spatial derivatives is given by

$$\frac{df}{dx} = \frac{1}{\Delta x}\sum_{j=-2}^{3} a_j f\left[x + \left(j - \frac{1}{2}\right)\Delta x\right] \tag{8}$$

The temporal integration is given by

$$\int_{-\frac{\Delta T}{2}}^{\frac{\Delta T}{2}} f(t)dt = \Delta t \sum_{m=0}^{3} b_m f(-m\Delta t) \tag{9}$$

with the coefficients $a_j$ and $b_m$ calculated so that the spatial and temporal schemes are fourth-order and third-order accurate respectively [13].

## 3. Problem description

For the example case presented in this paper, the geometry is the rectangular cuvette used by Žnidarčič *et al* [12]: the base is a 25 mm x 25 mm square, and liquid water with material properties given in Table 1 is filled up to a height of 50 mm. A 3 mm diameter sonotrode is immersed in the liquid at the axis of the cuvette, with the vibrating surface resting at 30 mm from the bottom. The sonotrode vibrates with an amplitude of 164 μm and frequency 20 kHz, with a power input $Q_{us}$ of 70 W. The resulting acoustic pressure is calculated as $p_{max} = \sqrt{2\rho c \, Q_{us}/A}$, where $A$ is the surface area of the sonotrode [15].

**Table 1: Material properties of aluminium and water [1]. Surface tension with hydrogen interface for aluminium and air interface for water.**

| Material Property | Aluminium (700 °C) | Water (20 °C) |
|---|---|---|
| **Sound speed $c$ (m s$^{-1}$)** | 4600. | 1482. |
| **Density $\rho_l$ (kg m$^{-3}$)** | 2375. | 1000. |
| **Dynamic viscosity $\mu_l$ (mPa s)** | 1.0 | 1.004 |
| **Surface tension $\sigma$ (N m$^{-1}$)** | 0.860 | 0.079 |
| **Vapour pressure $p_v$ (kPa)** | Negligible | 2.2 |
| **Bulk modulus $K$ (GPa)** | 41.2 | 2.15 |

## 4. Results and conclusions

Figure 1 shows the computed acoustic pressure at a probe position that is 7 mm horizontally from the vibrating surface of the sonotrode using the fourth-order solver. Figure 2 shows the predicted pressure profile in the cuvette. Three cases were run: with the fourth-order solver including the acceleration term in source terms (4) and (5), the fourth-order solver without the acceleration term, and the second order solver without the acceleration term. The same Courant number of 0.2 and a grid size of 0.5 mm were used. The fourth-order scheme converges with the acceleration term included while the second-order leapfrog scheme diverges. The second-order scheme converges to a wrong solution without the acceleration term. This instability arises in the second-order scheme because of the inaccurate prediction of acoustic pressure, which then results in an inaccurate bubble distribution profile: this is fed back to the pressure solver in the next iteration.

The fourth-order scheme is therefore more suitable for handling large acoustic pressures and bubble fraction variations; the higher-order method is both less prone to numerical diffusion and accurate with large Courant numbers of order 0.2. Since ultrasonic melt treatment processes are long – melt treatment in a typical crucible is of the order of minutes –, the numerical scheme for modelling acoustic cavitation will always suffer from numerical diffusion creeping in as the solution advances in time. Higher-order methods are therefore required in these cases so that numerical simulations can run to completion, without unrealistic prediction of acoustic pressures.

However, the method does not resolve the peaks arising from bubble collapses accurately: this is an inherent property of the homogeneous model. The fine details of the bubble dynamics are sacrificed to recover their overall effect on the domain acoustic pressure. This is to enable a sufficiently accurate simulation involving large time steps, which is a requirement for industrial-scale design of experiments simulations. This feature is crucial for the optimization of launder and crucible flow configurations, as would be required by industrial processes involving ultrasonic melt treatment. A numerical simulation employing baffles to increase the treatment time of liquid metal in cavitation zones will benefit from a higher-order scheme for the solution of acoustic pressure.

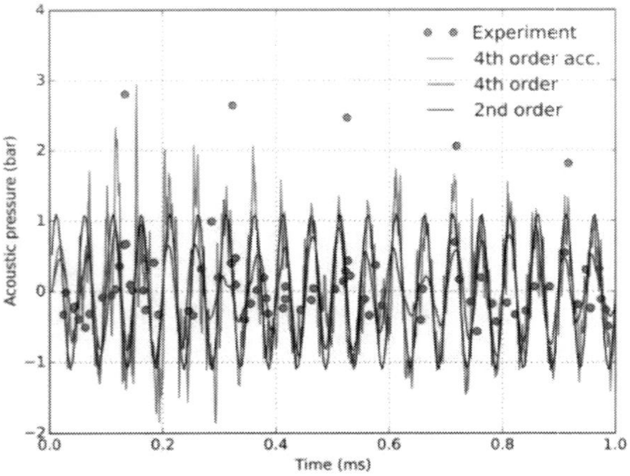


**Figure 1: Comparison between predicted acoustic pressure at probe position for different solvers and experimental data from [12].**

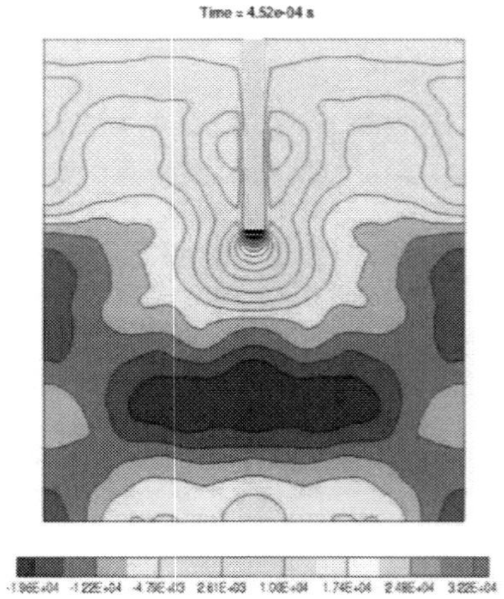


**Figure 2: Computed acoustic pressure contour with the fourth-order spatial scheme.**

**Acknowledgements**
The authors are grateful to the UK Engineering and Physical Sciences Research Council (EPSRC) for financial assistance for this research in contract numbers: EP/K00588X/1 and EP/K005804/1.

**References**

[1] Eskin G I and Eskin D G 2014 *Ultrasonic Treatment of Light Alloy Melts* (CRC Press)
[2] Campbell J 1981 *Int. Mater. Rev.* **26** 71
[3] Cao G *et al* 2008 *Mater. Sci. Eng. A-Struct.* **486** 357
[4] Eskin G I 1995 *Ultrason. Sonochem.* **2** 137
[5] Komarov *et al* 2013 *Ultrason. Sonochem.* **20** 754
[6] Singhal A K *et al* 2002 *J. Fluid Eng.* **124** 617
[7] Nastac L 2011 *Metall. Mater. Trans. B* **42** 1297
[8] Lebon G S B *et al* 2015 *IOP Conf. Ser.: Mater. Sci. Eng.* **72** 052050
[9] Lebon G S B *et al* 2015 *Proc. 144th TMS Annual Meeting & Exhibition (Orlando) Advances in the Science and Engineering of Casting Solidification* p 23
[10] Cheeke J D N 2012 *Fundamentals and applications of ultrasonic waves* (CRC Press)
[11] Keller J B and Miksis M 1980 *J. Acoust. Soc. Am.* **68** 628
[12] Žnidarčič A *et al* 2014 *Ultrason. Sonochem.* **22** 482
[13] Djambazov *et al* 2000 *AIAA Journal* **38** 16
[14] Tam C K W and Webb J C 1993 *J. Comput. Phys.* **107** 262
[15] Van Iersel *et al* 2008 *Ultrason. Sonochem.* **15** 294

# Large eddy simulation of turbulent cavitating flows

**A Gnanaskandan and K Mahesh**

Aerospace Engineering and Mechanics, University of Minnesota, USA

E-mail: kmahesh@aem.umn.edu

**Abstract.**

Large Eddy Simulation is employed to study two turbulent cavitating flows: over a cylinder and a wedge. A homogeneous mixture model is used to treat the mixture of water and water vapor as a compressible fluid. The governing equations are solved using a novel predictor–corrector method. The subgrid terms are modeled using the Dynamic Smagorinsky model. Cavitating flow over a cylinder at Reynolds number ($Re$) = 3900 and cavitation number ($\sigma$) = 1.0 is simulated and the wake characteristics are compared to the single phase results at the same Reynolds number. It is observed that cavitation suppresses turbulence in the near wake and delays three dimensional breakdown of the vortices. Next, cavitating flow over a wedge at $Re = 200,000$ and $\sigma = 2.0$ is presented. The mean void fraction profiles obtained are compared to experiment and good agreement is obtained. Cavity auto–oscillation is observed, where the sheet cavity breaks up into a cloud cavity periodically. The results suggest LES as an attractive approach for predicting turbulent cavitating flows.

## 1. Introduction

Turbulent cavitating problems pose unique challenges in that apart from resolving the wide range of length and time scales, steep gradients arising due to vapor cavity collapse have to be handled efficiently. A non–dissipative base scheme with dissipation added locally near the discontinuities is a very attractive candidate to address these challenges. Traditionally, RANS has been the preferred approach to simulate cavitating flows. However, RANS does not predict the unsteady nature of cavitation with great accuracy. LES can therefore be a good approach for highly unsteady cavitating flows. In this paper, we present LES of two turbulent cavitating problems using a numerical method developed by Gnanaskandan and Mahesh [1]: cyclic cavitation behind a circular cylinder and sheet to cloud cavitation over a wedge. We study cavitation over a cylinder at $Re = 3900$ and $\sigma = 1.0$ to understand the effect of cavitation on Karman vortex shedding and wake characteristics. While bluff body wakes have been studied extensively for single phase flows, only a few studies exist that shed light on the effect of cavitation on vortex shedding [2, 3, 4]. The second problem is sheet to cloud cavitation over a wedge at $Re = 200,000$ and $\sigma = 2.0$. Sheet to cloud cavitation causes severe damage to the underlying structure when the cloud collapses. Several experiments have been conducted to understand the sheet to cloud transition mechanism [5, 6, 7] and in this study, we numerically simulate sheet to cloud cavitation transition and compare to the experiments of Ganesh [8].

## 2. Physical model and numerical method

The physical model used is the homogeneous mixture model, which treats the mixture of water and water vapor as a single compressible fluid and assumes the constituent phases to be in thermal and mechanical equilibrium. The governing equations are the compressible Navier Stokes equation for the mixture of water and vapor along with a transport equation for mass fraction of vapor. The system is closed by a non–barotropic mixture equation of state obtained

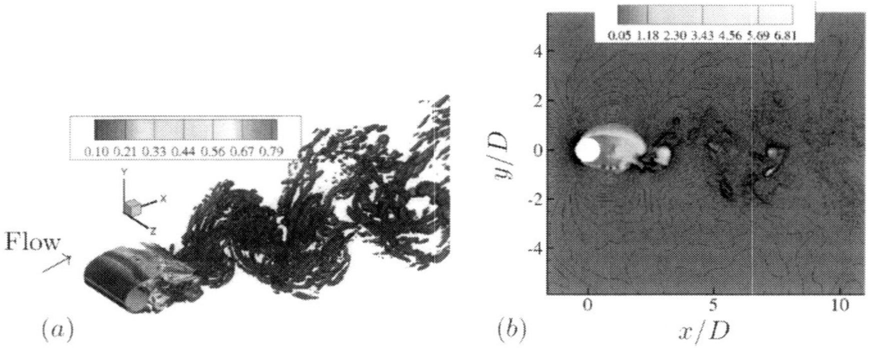


**Figure 1.** (a) Isocontours of Q-criterion colored by void fraction. (b) Instantaneous Mach number contours superimposed with lines of pressure.

from the stiffened equation of state for water and ideal gas equation of state for vapor. The density ratio of water to vapor in the simulations is 1000. The governing equations are Favre averaged and then spatially filtered to perform LES. The subgrid terms in the momentum and energy equation are modeled using the Dynamic Smagorinsky model (DSM), while the subgrid term in the mass fraction equation is neglected in the current study [1]. The numerical method uses a novel predictor–corrector algorithm. The predictor step uses a symmetric and non-dissipative scheme to obtain a predicted value for the conservative variables. The corrector step uses a characteristic-based filtering approach to add dissipation locally only in the vicinity of discontinuities. Further details on the predictor–corrector algorithm can be found in [1].

## 3. Turbulent cavitating flow over a circular cylinder

LES of turbulent cavitating flow at $Re = 3900$ and $\sigma = 1.0$ is performed. Figure 1(a) shows the instantaneous vortical structures in the form of Q-criterion colored by void fraction. Three dimensional flow structures of varying scales can be observed. The shear layer breaks up into smaller spanwise structures. The flow structures in the near wake, especially in the vapor cavity are larger when compared to the flow structures observed by Verma and Mahesh [9] in a single phase flow at the same Reynolds number. This is because of the effective lowering of Reynolds number by the presence of vapor. Further, small vapor pockets can be seen trapped in the smaller scale vortices downstream of the wake. Although the flow is three–dimensional inside the vapor cavity immediately downstream, breakdown to finer scales occurs downstream of the cavity closure. Figure 1(b) shows instantaneous contours of Mach number showing supersonic Mach numbers in the shear layer. Cavitation in wake vortices also causes locally supersonic Mach numbers there. The contour lines of pressure shows distorted fronts of pressure waves that are formed due to cavity collapse. Turbulence causes different points in the wave fronts to have different speeds resulting in their distortion. The Strouhal number corresponding to vortex shedding is 0.17 as compared to the single phase value of 0.2. Thus cavitation lowers the vortex shedding frequency.

The near–wake velocity profiles are compared to the single phase LES results of Verma and Mahesh [9] in Figure 2. The streamwise velocity profiles at all stations show a wider wake profile for the cavitating flow. The station $x/D = 3.0$ shows the maximum difference in the vertical velocity profile since it corresponds to the cavity closure region. Inside the cavity (except at $x/D = 1.06$), larger values of vertical velocity are obtained. Overall, cavitation yields a wider near–wake profile, representative of a low Reynolds number flow.

## 4. Turbulent cavitating flow over a wedge

LES of sheet to cloud cavitation over a wedge at $Re = 200,000$ and $\sigma = 2.0$ is performed. The averaged statistics presented in this section are obtained by performing a time average

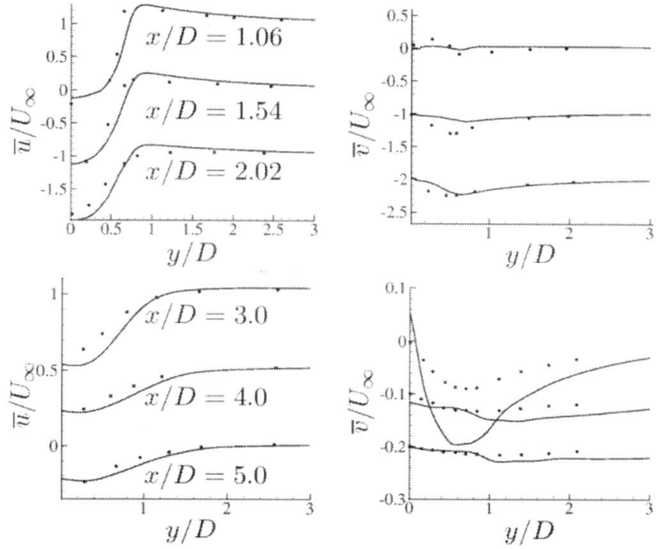


**Figure 2.** Comparison of velocity profiles at streamwise stations obtained from the present simulations to the single phase simulations at $Re = 3900$, • : Single phase results of Verma and Mahesh [9], —— : Cavitating flow (Present).

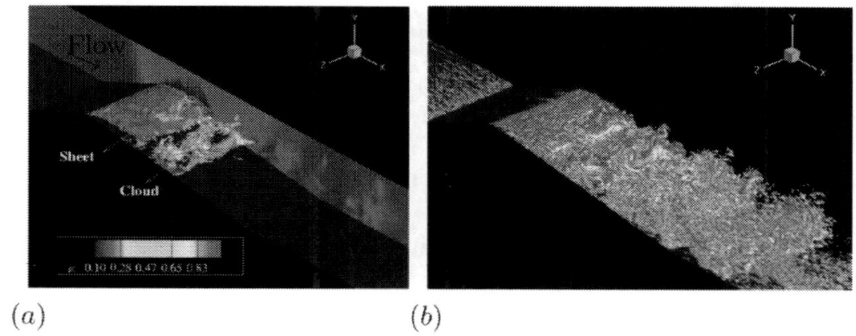


**Figure 3.** (a) Instantaneous snapshots of isocontours of void fraction showing sheet and cloud cavities, (b) Isocontours of Q-criterion colored by streamwise velocity.

over 40 $th/u_\infty$ and has approximately 30,000 samples, where $h$ is the height of the wedge and $u_\infty$ is the freestream velocity. The separation between the samples is 0.001 $th/u_\infty$ which captures all relevant high frequencies. The non–dimensional time step used in the simulation is $tu_\infty/h = 1 \times 10^{-5}$. The nature of the instantaneous solution is illustrated using isocontours of void fraction in Figure 3(a) which shows the presence of both sheet and cloud cavities. The sheet cavity grows to an average length of $x/h = 2.1$. The leading edge of the sheet cavity in Figure 3(a) varies in the spanwise direction and these spanwise variations grow larger as we move towards the trailing edge of the cavity. After attaining its maximum length, the sheet cavity pinches off and the trailing edge of the sheet cavity rolls up into a cloud. This cloud is highly three dimensional and it is the collapse of this cloud that causes noise, vibration and surface erosion. The vertical plane in Figure 3(a) shows pressure contours. It can be observed that pressure exhibits both wave–like behavior (close to leading edge if the cavity) and a highly intermittent behavior (downstream of the cloud). The pressure waves produced on cavity collapse impinges on the growing sheet cavity and causes it to break into smaller cavities. Figure 3(b) shows isocontours of Q-criterion colored with streamwise velocity. The intensely vortical nature of the flow and wide range of length scales are apparent in both sheet and cloud regions. Small secondary cloud shedding occurs just before and after the main cloud

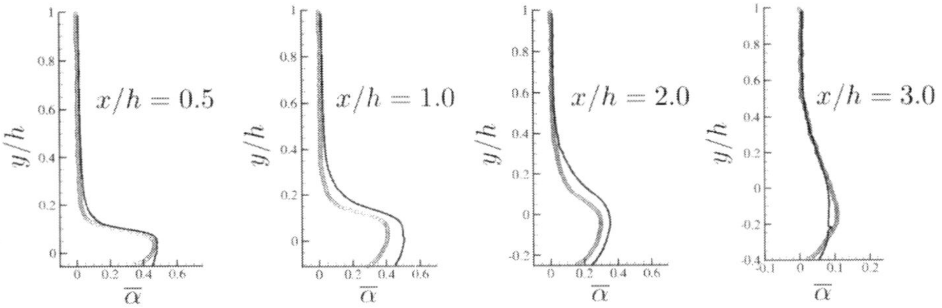


**Figure 4.** Comparison of mean volume fraction at different streamwise stations, —— : Present and ○ : Experiment [8].

sheds, which is normally not captured in time averaged simulations. Leroux *et al.* [6] have observed these secondary cloud shedding in their experiments on a hydrofoil and the LES is also able to capture this secondary shedding. The Strouhal number corresponding to the auto–oscillation (extracted from spectra as well as animations) is 0.28, which lies within the range of 0.25-0.30 obtained by the experiments of Ganesh [8].

The time–averaged and span–averaged values of void fraction at different streamwise locations on the wedge is compared to the experimental results of Ganesh [8] in Figure 4. The agreement in the mean void fraction is very good at $x/h = 0.5$ and $x/h = 3.0$ stations, while LES slightly over–predicts void fraction at the other two stations. While the length of the cavity predicted by the simulation agrees well with the experiment, the simulation slightly over–predicts the thickness.

## 5. Summary

LES is used to study two different turbulent cavitating problems. In the cavitating flow over a cylinder, it is observed that cavitation suppresses turbulence in the near wake and delays three dimensional breakdown of the Karman vortices. Cavitation is also seen to reduce the vortex shedding frequency. In sheet to cloud cavitation over wedge, a cavity auto–oscillation is observed, where the sheet cavity breaks up and forms a cloud cavity. The sheet cavity is mildly unsteady with larger spanwise fluctuations near its trailing edge, whereas the cloud cavity is highly unsteady and three dimensional. The simulation is also able to predict secondary cloud shedding that is normally not captured using RANS. Good agreement with experiment is observed for both mean void fraction inside the cavity and auto–oscillation frequency.

## 6. Acknowledgments

This work was supported by the United States Office of Naval Research under ONR Grant N00014-11-1-0497 with Dr. Ki-Han Kim as program manager. Computing resources were provided by the Minnesota Supercomputing Institute (MSI), the Texas Advanced Computing Center (TACC) and the Arctic Region Supercomputing Center of HPCMP.

## References

[1] Gnanaskandan A and Mahesh K 2015 *Int. J. Multiphase Flow* **70** 22-34
[2] Rao BC and Chandrashekhara DV 1975 *J. Fluids Eng.* **98** 461-66
[3] Ramamurthy AS and Bhaskaran P 1977 *J. Fluids Eng.* **99** 717-26
[4] Fry SA 1984 *J. Fluid Mech.* **142** 187-200
[5] Arndt R, Song CCS, Kjeldsen M, He J and Keller A 2000 *Proc. 23rd Int. Symp. on Cavitation*
[6] Leroux JB, Astolfi JA and Billard JY 2004 *J. Fluids Eng.* **126** 94-101
[7] Callenaere M, Franc JP, Michel J and Riondet M 2001 *J. Fluid Mech.* **444** 223-56
[8] Ganesh H 2015 Bubbly shock propagation as a cause of sheet to cloud transition of partial cavitation and Stationary cavitation bubble forming on a delta wing vortex, *PhD thesis*, Univ. of Michigan
[9] Verma A and Mahesh K 2012 *Phys. of Fluids.* **24** 1-24

# Evaluation of finite rate homogenous mixture model in cavitation bubble collapse

**M Bhatt, A Gnanaskandan and K Mahesh**
Department of Aerospace Engineering and Mechanics, University of Minnesota Twin Cities, Minneapolis, Minnesota- 55414, US
E-mail: kmahesh@umn.edu

**Abstract.** The objective of this study is to evaluate the performance of finite rate homogeneous mixture models in three key aspects: (a) the ability of the model to predict the dynamics of resolved small scale vapor regions, (b) the importance of finite rate mass transfer and (c) the impact of assuming $2D$ flow as is done in RANS simulations with a statistically homogeneous direction. We consider a bubble collapse problem to evaluate all these effects.

## 1. Introduction

Cavitation often causes erosion due to the high impact loading caused by vapor cavity collapse. Sheet to cloud cavitation is a classic phenomenon where the cloud cavity often collapses and causes severe noise, vibration and surface erosion. Homogeneous mixture modelling is the most commonly used physical model to study turbulent sheet to cloud cavitation and these models employ both finite rate (e.g. Singhal *et al.* [1]) and equilibrium mass transfer terms (e.g. Schnerr *et al.* [2]) to model mass transfer between water and vapor. Using the finite rate mass transfer model of Saito *et al.* [4] as a representative example, we make use of the numerical method of Gnanaskandan and Mahesh [3] to study three key effects: the performance of the model in predicting the dynamics of resolved small scale vapor regions, the effect of finite rate mass transfer and the consequence of two dimensional assumptions as is commonly done in RANS simulations.

## 2. Spherical bubble collapse

While it is well known that homogeneous mixture models are reasonably accurate in predicting large scale cavitation dynamics, the final stages of a cloud collapse involve vapor regions of smaller length scales. The performance of homogeneous models in predicting the dynamics of these smaller scale vapor regions is of interest. In this section, the finite rate homogeneous mixture model is applied to study a 0.4 mm spherical bubble collapse problem. Evolution of a spherical cavity in the absence of surface tension and fluid viscosity with the assumption of liquid incompressibilty is given by [5] ,

$$R\ddot{R} + \frac{3}{2}\ddot{R} = \frac{p(R) - p_\infty}{\rho}, \tag{1}$$

where $\rho$ is the liquid density, $p_\infty$ is the ambient pressure of liquid and $p$ is the pressure in the liquid at the interface. The collapse time ($t_{rayleigh}$) can be obtained by integrating the above

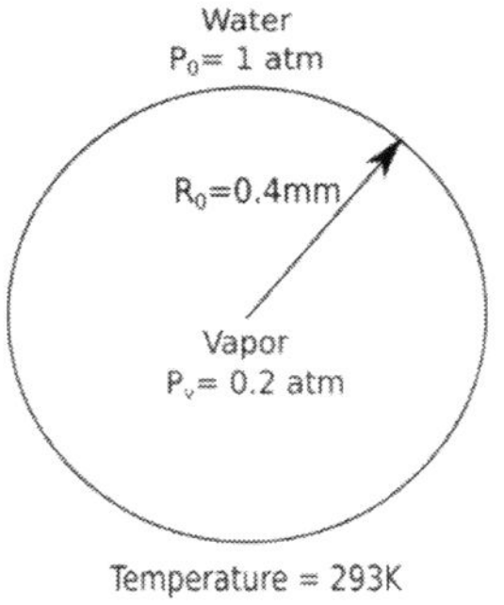


**Figure 1.** Schematic of the cylindrical bubble collapse.

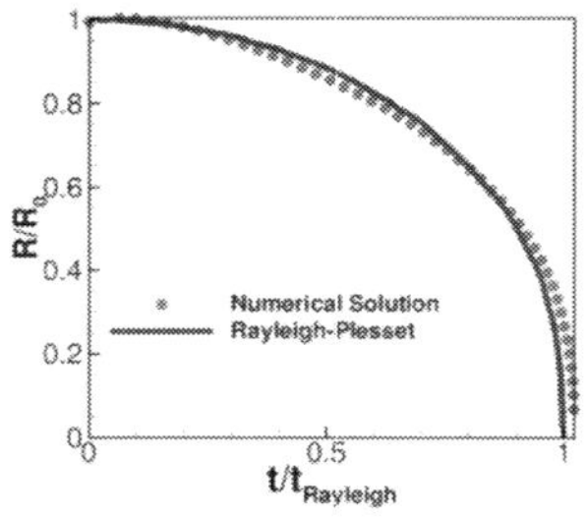


**Figure 2.** Evolution of radius with time.

equation. We consider a spherical vapor bubble of initial radius ($R_0$) in water as shown in Figure 1. The ambient pressure is $P_0$ and pressure inside the bubble is considered to be equal to vapor pressure ($P_v = 0.02$ atm at 293 K). Initial volume fraction of vapor ($\alpha_0$) is 0.99 inside the bubble and 0.01 outside the bubble. The reference values considered are $L_r = 0.4mm$, $U_r = 10m/s$ and $\rho_r = 1000kg/m^3$. We consider 75 equidistant grid points spanning the initial bubble radius. Reflections are minimized by placing boundaries at $25R_0$ from the the bubble region with absorbing sponge layer starting from $20R_0$ . The numerical result in Figure 2 shows good agreement with the analytical solution for evolution of bubble radius with time showing the suitability of the model in predicting the dynamics of small scale vapor regions as long as they are resolved on the computational grid.

## 3. Effect of finite rate mass transfer

Equilibrium models assume mass transfer between phases to occur at rates much faster than that of convection. In this section we evaluate this effect by first interpreting the condensation and convection time scales from the mass transfer terms. Then a series of numerical experiments are performed to corroborate our analysis and finally the interesting finite rate phenomenon of 'dynamic delay' is discussed.

### *3.1. Non-dimensional analysis*

The collapse process generates very high pressure and the magnitude of this collapse pressure depends on the relative rate of condensation and convection. In a cavity collapse problem, the evaporation term is zero. The non-dimensional mass fraction transport equation is thus given by

$$\frac{\partial \tilde{\rho} Y}{\partial \tilde{t}} = -\frac{\partial \tilde{\rho} Y \tilde{u_j}}{\partial \tilde{x}_j} - \tilde{S}_c, \tag{2}$$

where $\tilde{S}_c$ is the non-dimensional condensation source term given by

$$\tilde{S}_c = \frac{C_c \alpha^2 (1-\alpha)^2 \max(P - P_v, 0)}{\sqrt{2\Pi R_g T_s}} * (\frac{L_r}{\rho_r * U_r}), \tag{3}$$

where $\alpha$ is the void fraction, $C_c$ is an empirical constant, $P$ is the pressure, $P_v$ is the vapor pressure, $R_g$ is the characteristic gas constant of vapor and $T_s$ is the saturation temperature. Here $L_r$, $\rho_r$ and $U_r$ are the reference quantities. Defining $a_s$ as speed of sound in vapor at saturation temperature ($\sqrt{\gamma R_g T_s}$), $\tilde{S}_c$ can be rearranged as a product of three non–dimensional parameters,

$$\tilde{S}_c = (C_c L_r) * (\frac{P - P_v}{\rho_r U_r^2}) * (\frac{U_r}{a_s}) * [\alpha^2 (1 - \alpha^2)]. \tag{4}$$

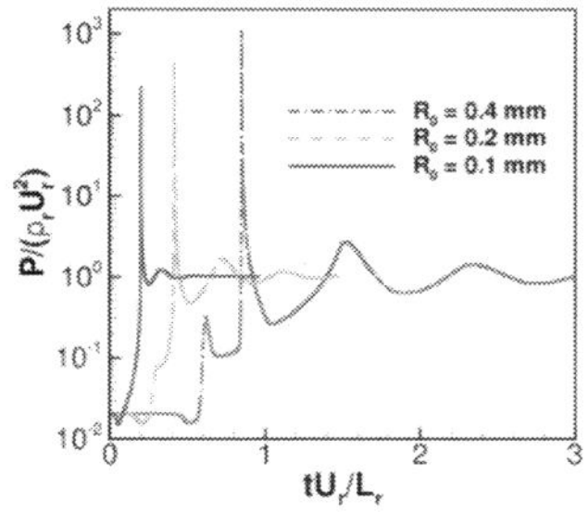


**Figure 3.** Temporal evolution of pressure at center for different initial bubble radius.

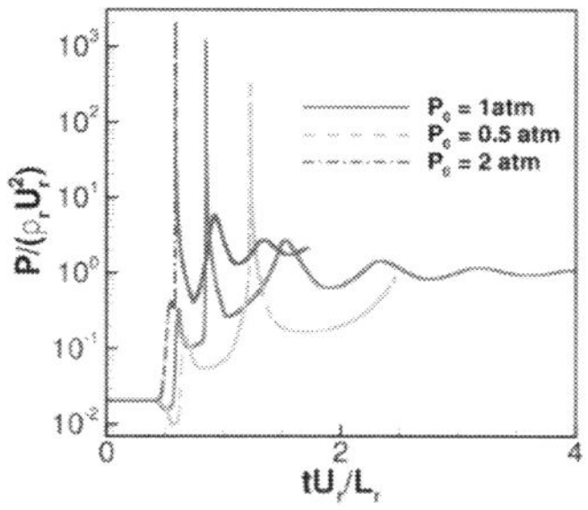


**Figure 4.** Temporal evolution of pressure at center for different ambient pressures.

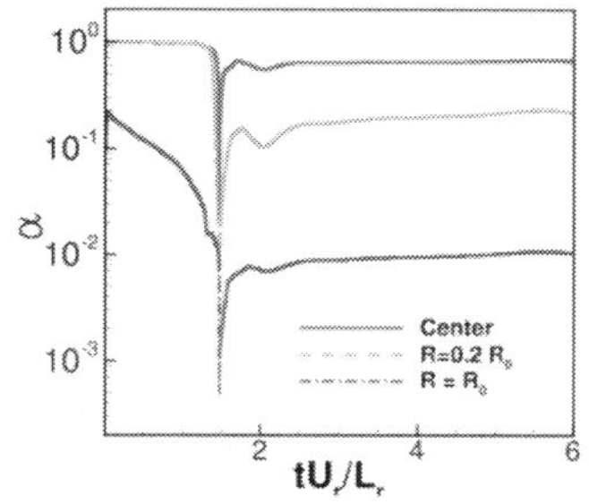


**Figure 5.** Volume fraction evolution at several locations showing residual vapor.

For the bubble collapse problem, the reference length is taken as the initial radius ($R_0$), and the reference velocity is the initial material front speed $\sqrt{(P_0 - P_v)/\rho_r}$. The non-dimensional source term can be considered as the relative strength of condensation with respect to convection. It can also be construed as the ratio of the convection and condensation time scales. Consequently, these non-dimensional parameters allow us to comment on the rate and strength of collapse under various flow conditions. As the bubble radius is increased, the non-dimensional parameter $C_c L_r$ increases, which increases the condensation strength. Thus collapse of larger bubbles should result in larger collapse pressure. Similarly larger driving pressure $P - P_v$ should also result in larger collapse pressures. Further, if we interpret this behavior in terms of time scales, increasing the driving pressure decreases the condensation time scale which allows rapid collapse of the bubble. These results are verified using numerical simulations in Section 3.2.

### *3.2. Effect of initial radius $R_0$ and liquid pressure $P_0$*

Parametric studies are conducted for $R_0 = 0.4, 0.2$ and 0.1mm, and $P_0 = 0.5, 1$ and 2 atm. Figure 3 shows the pressure evolution at the center of the bubble for $R_0 = 0.4, 0.2$ and 0.1 mm. It is observed that both the collapse pressure magnitude and the time taken for collapse ($\tau_c$) reduce as the initial radius is reduced. The larger bubble has more mass to be compressed and hence leads to larger collapse pressure. Figure 4 shows pressure evolution at the center for $P_0 = 0.5, 1$ and 2 atm. The main consequence of changing $P_0$ is in changing the initial velocity with which the material front moves ($U_0$). For the highest $P_0$, the difference in speed between material front and the compression wave is negligible and hence the initial pressure peak due to compression is not visible. Further maximum collapse pressure and minimum collapse time are obtained for $P_0 = 2$ atm. It is evident that higher the pressure gradient ($P_0 - P_v$), the faster the collapse will be and higher the collapse pressure.

### *3.3. Dynamic delay*

For $R_0 = 0.4$ mm and $P_0 = 1$ atm, an interesting phenomenon is observed. Once the material front reaches the center, all the vapor is expected to be converted to water since the pressure reaches values above vapor pressure. However, we observe that even long after collapse not all vapor gets destroyed. This is evident from the void fraction evolution at two locations close to the center in Figure 5. It is observed that the void fraction does not relax back to 0.01 and remains at a value as high as 0.7 at the center. The void fraction is significant even at $r = 0.2R_0$. This "dynamic delay" [6] phenomenon occurs because of the large difference in the values of time scales associated with condensation and convection. As the size of the vapor becomes very small, the condensation time scale becomes very large when compared to convection time scale and hence it takes a long amount of time to observe the effect of condensation. This phenomenon

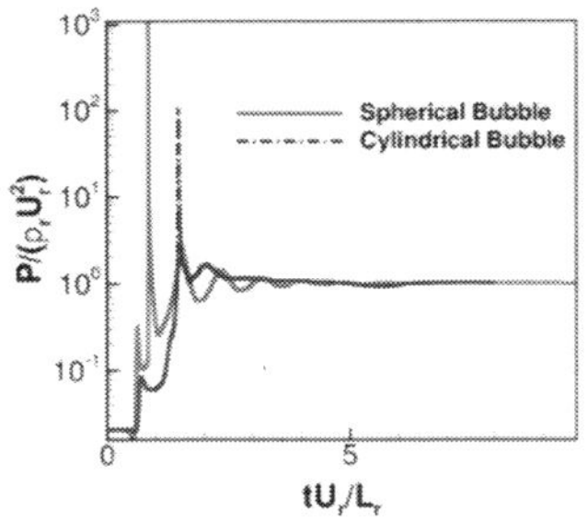


**Figure 6.** Comparison of temporal evolution of pressure at center for spherical and cylindrical bubbles.

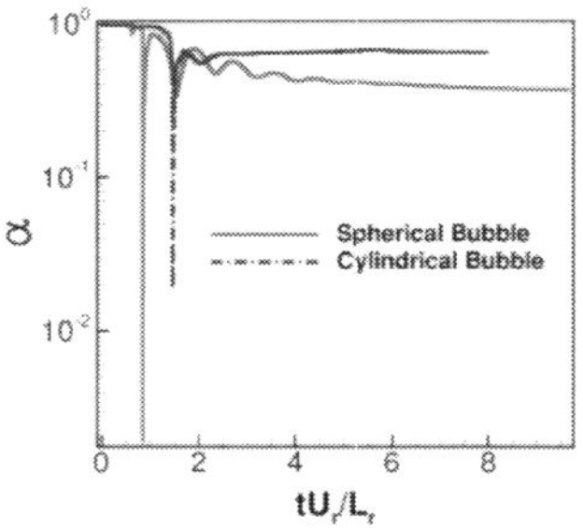


**Figure 7.** Comparison of temporal evolution of volume fraction at center for spherical and cylindrical bubbles.

is of practical significance since the undestroyed vapor can get convected by the flow and grow or collapse in other regions depending on the conditions. Equilibrium mass transfer models are not expected to predict this phenomenon.

## 4. Comparison of cylindrical and spherical bubble collapse

Here, we consider a case of collapsing cavitation bubble in 2D (i.e. a cylindrical shape bubble with infinite length in span wise direction) and compare it with spherical bubble collapse. All the parameters are same as those used for spherical bubble collapse in Section 3.2. As seen from Figure 6, a spherical bubble of same size under same flow parameters generates an order of magnitude higher collapse pressure compared to a cylindrical bubble. Further, from Figure 7, we can see that volume fraction of vapor remaining at the center is much lower for spherical bubble collapse compared to cylindrical bubble collapse, which shows that condensation strength is higher in the case of spherical bubble collapse. The practical significance of this result is that, 2D RANS simulations will under predict the condensation strength and collapse pressure.

## 5. Summary

A finite rate mass transfer homogeneous mixture model is used to study bubble collapse. It is observed that as long as the vapor region is sufficiently resolved, homogeneous models are accurate in predicting the dynamics of small vapor regions, making them a suitable approach to study sheet to cloud cavitation transition. The effect of finite rate mass transfer is analyzed using a non–dimensional analysis and a parametric study corroborates the non–dimensional analysis. A finite rate phenomenon - 'Dynamic delay' is also observed in the simulations. Finally the effect of considering $2D$ collapse as is commonly assumed in RANS is illustrated by comparing cylindrical and spherical bubble collapse. It is shown that cylindrical bubble collapse produces much smaller collapse pressure than a spherical bubble collapse which is an important design parameter.

## 6. Acknowledgments

This work is supported by the United States Office of Naval Research under Grant ONR N00014-11-1-0497 with Dr. Ki-Han Kim as the program manager.

## References

[1] Singhal AK, Athavale MM, Li H and Jiang Y 2002 *J. Fluid Engg.* **124** 617-624
[2] Schnerr GH, Sezal IH and Schmidt SJ 2008 *Phys. of Fluids* **20**
[3] Gnanaskandan A and Mahesh K 2015 *Int. J. Multiphase Flow* **70** 22-34
[4] Saito Y, Takami R, Nakamori I, Ikohagi T 2007 *Comp.Mech.* **40**85-96
[5] Plesset MS and Prosperetti A 1977 *Ann. Rev. Fluid Mech.* **9** 145-185
[6] Franc J-P and Michel J-M 2003 *Fundamentals of cavitation* (Kluwer Acad. Publishers) chapter 1 pp 1-14

# Prediction of cavitation erosion for marine applications

**T Maquil, S Yakubov and T Rung**

Institute for Fluid Dynamics and Ship Theory, Hamburg University of Technology, Am Schwarzenberg-Campus 4, 21073 Hamburg, Germany.

E-mail: `thierry.maquil@tuhh.de`

**Abstract.** The paper presents the development of a cavitation erosion prediction method. The approach is tailored to marine applications and embedded into a VoF-based procedure for the simulation of turbulent flows. Supplementary to the frequently employed Euler-Euler models, Euler-Lagrange approaches are employed to simulate cavitation. The study aims to convey the merits of an Euler-Lagrange approach for erosion simulations. Accordingly, the erosion model is able to separate different damage mechanisms, e.g. micro-jets, single and collective bubble collapse, and also quantifies their contribution to the total damage. Emphasis is devoted to the prediction of the cavitation extend, the influence of compressible effects and the performance of the material damage model in practical applications. Examples included refer to 2D validation test cases and reveal a fair predictive accuracy.

## 1. Introduction

Cavitation erosion is frequently observed in marine and hydraulic engineering, e.g. ship propellers and rudders, pumps or turbines. Due to the severe risks associated to operations under sustained erosive cavitation, the accurate prediction of cavitation erosion is of great importance from an industrial point of view. Such modelling approaches would allow replacing model tests by numerical simulations during the preliminary design phase and thereby yield a reduction of the expenditures and time frame. However, cavitation erosion predictions based on numerical models are a challenging problem for the computational fluid dynamics and computational engineering community. The challenges particularly refer to the wide range of involved scales of the compressible multiphase flow, but also the response of the solid materials and the influence of individual bubbles or the initial nuclei population.

When attention is given to the simulation of erosive cavitation, the erosion mechanisms are often masked by considerable modelling efforts inherent to the usually employed Euler-Euler modelling framework. Erosion is associated with strong pressure changes and travelling pressure waves, whose amplitude depends on the bubble sizes. It is thus almost inaccessible for Eulerian approaches which do not resolve down to the bubble scale. The present research aims to assess the predictive improvements offered by an Euler-Lagrange approach, which employs a dispersed Lagrangian vapour phase model and thereby provides access to vapour bubbles physics and realistic pressure-wave predictions.

## 2. Numerical Model

Results of the present study are obtained from the in-house solver FreSCo$^+$ [1]. The finite volume procedure uses a segregated algorithm based on the strong conservation form of the governing equations and employs a cell-centered, co-located storage arrangement for all transport properties. The solution is iterated to convergence using a modified SIMPLE method for pressure-velocity coupling [2]. The schemes takes into account for density variations due to cavitation using an additional source term in the compressible pressure-correction equation. A linearisation of this source term ensures a stable convergence for both steady and transient cases. Multiphase problems are addressed by an Euler-Euler VoF approach or an Euler-Lagrange technique.

The most popular technique to simulate cavitating flows refers to the VoF-based Euler-Euler approach [3]. It considers the two-phase flow as mixture of a liquid and a vapour phase which share the kinematic field. The vapour-volume fraction is computed from an additional transport equation that heuristically models the mass transfer between vapour and liquid using various empirical constants. The model parameters are unfortunately not universal and interact with the employed computational algorithm [4]. On contrary, the two-way coupled Euler-Lagrange approach allows to compute the evolution and motion of individual vapour bubbles [5]. Supplementary to an Eulerian mixture phase, separate equations for the size and the momentum are solved for each individual cavitation bubble/nuclei. The field properties follow from the Eulerian conservation equations, whereas the vapour part is governed by Newtonian motion of individual bubbles driven by the surrounding flow field. The present study employs a parallelised two-way-coupled strategy and accounts for inhomogeneous and transient water-quality effects such as nuclei spectra [6]. The resolution of yield loads (e.g. copper: 200 MPa) and the prediction experimentally observed pressure peaks in the range 0.8 to 2 GPa are crucial for a direct assessment of cavitation erosion, but pose a severe challenge for the computational model. Regarding Euler-simulations, reported values are limited to approximately 100 MPa, while for our Lagrange simulations realistic pressure peaks up to 1 GPa are seen. This yields the conclusion that for a quantitative prediction, Lagrange models offer benefits over Euler models.

## 3. Erosion assessment

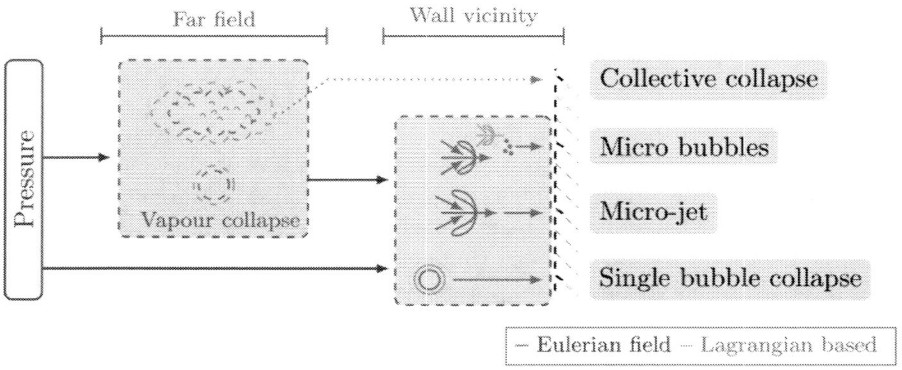


**Figure 1.** Erosion damage mechanism cascade using an Euler-Lagrange approach.

The study compares results obtained from Eulerian and Lagrangian modelling frameworks. Within the Eulerian approach an enhanced micro-jet model for erosion prediction is used [7]. The idealized mechanism focusses on (unresolved) small bubbles in the vicinity of the surface, that produce a micro-jet during their pressure induced collapse. The parametrised micro-jet is deemed responsible for the pitting of the solid surface. The model requires an assumption of the size and the population of small bubbles close to the wall - including their wall distance - and a threshold for an activation pressure. The corresponding mathematically framework can be coupled to the Euler-Euler cavitation approach without problem. However as it does not resolve any of the responsible mechanisms, it inheres a considerable amount of empiricism. An example refers to the ambiguity of the bubble state for a given mixture state, i.e. it is not clear what bubble spectrum composes a mixture.

Cavitation erosion has its origin in the collapse of cavitation structure whereby depending on the cavitation type the erosion aggressiveness varies. Hence a numerical prediction method needs to diversify cavitation type and rely on miscellaneous erosion definitions. The Lagrangian approach provides information about the cavitation type by evaluating the bubbles size and position and therewith overcoming the difficulties of the Eulerian mixture phase. Additionally these parameters allow to distinguish between the state of the art defined erosion mechanism. The overall Euler-Lagrange damage mechanism cascade is illustrated in Figure 1. Due to the Lagrangian treatment of the vapour phase (bubble parameters and position to the solid surface), the method covers the following range damage mechanisms:

(1) Symmetrical bubble collapse in the vicinity of the wall [8]
(2a) Micro-jet from asymmetrical collapse of bubbles very close to wall [9]
(2b) Erosion damage from micro bubbles created after asymmetrical bubble collapse [10]
(3) Collective collapse of multiple bubbles in cascade [11]

The erosion damage is calculated in Lagrangian manner. The compressible Rayleigh-Plesset equation accounting bubble interaction are solved for each bubble and the micro-jet model is used for bubbles in the vicinity of the wall. The choice of erosive mechanism is coupled to the wall distance of the respective bubble and the occurrence of neighbouring bubbles. The pressures are evaluated in the Lagrangian model. The commonly smaller Lagrangian time step compared to the one from the Eulerian background fluid field supports the resolution of pressure waves and enables an exact derivation of the erosion damage. On the other hand, Eulerian physics is retained and can be an important aspect, e.g. when considering the pressure waves of collapsing vapour structures away from the wall.

## 4. Validation

**Table 1.** Variation of the erosion rate with the ref. velocity for flow through a venturi nozzle.

| Case | Spec. material | Ref. vel. [m/s] | Cav. number | Exponent |
|---|---|---|---|---|
| Exp. He and Hammitt [12] | aluminium/steel | 30-50 | 0.76 | 4.0/1.1 |
| Exp. Knapp [12] | soft aluminium | 20-30 | ci. 0.3 | 6.4 |
| | | | 0.85 | 9.2 |
| Sim. FreSCo$^+$ | aluminium | 30 -50 | 0.76 | 6.4 |
| | | | 0.62 | 3.2 |

The first example is a simple two-dimensional venturi nozzle experimentally analysed by He and Hammitt [12]. Emphasis is given to the prediction of scale effects, i.e. the behaviour of the erosion rate $R$ in response to an increase of the reference velocity $V$ at constant cavitation number. Results are compared in terms of a corresponding exponent $n$, $R \sim V^n$. Table 1 summarises the experimentally reported and numerically obtained exponents.

The erosion of a circular leading edge hydrofoil has often been experimentally analysed, e.g. [7]. Figure 2 shows the erosion damage after 1 h of exposure. On the right a photography of the experimental damage illustrates the top view on the airfoil surface. The diagram on the left side compares the experimental and simulated (modified Euler-Euler approach) span-averaged surface damage. The numerically predicted damage is in reasonable agreement with experiments. Only around $x \sim 40\,\mathrm{mm}$ (i.e. $x/C = 0.37$) a major discrepancy is observed. Here, the cavitation clouds separate from the foil and the deficits of Euler-Euler approach, i.e. the activation pressure assignment to the wall, whip hand.

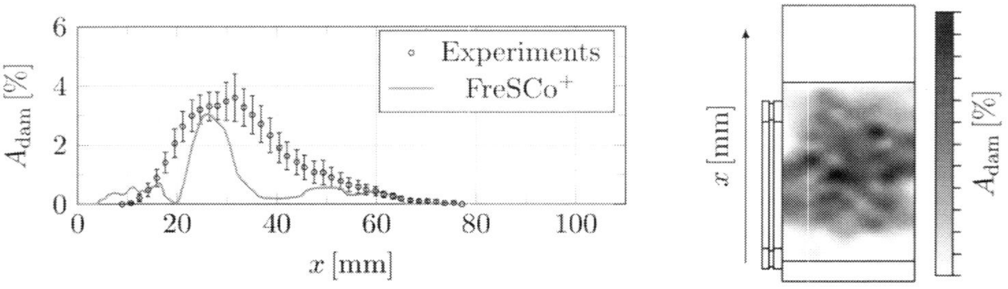


**Figure 2.** Erosion damage after 1 h exposure on the CLE hydrofoil (Exp. [7]).

## 5. Conclusions

In this study two approaches for assessing cavitation erosion are described. The well known Euler-Euler approach to simulated cavitating flows combined with an enhanced micro-jet erosion models shows satisfactorily results when appropriately adjusted to benchmark cases. However the empiricism of the technique remains the major deficit. The second approach employs using the Euler-Lagrange method and answers most of the weak spots per definition. The exact location, size and history of the bubbles are known and are used instead of empirical assumptions. Additionally cascade collapse and micro bubbles can be computed to account for further erosion mechanisms. However the computational cost increases with each feature and needs to be addressed. More results related to the above depicted Euler-Lagrange approach within the FreSCo$^+$ framework will be presented at the conference.

## Acknowledgments

The current work is supported by German Research Foundation (DFG, Grant No. RU 1575/2-1) and the Sutor-Stiftung. The simulations were performed at the North German Cooperation for High-Performance computing (HLRN). This support is gratefully acknowledged by the authors.

## References

[1] Rung T *et al.* 2009 *Jahrbuch der Schiffbautechnischen Gesellschaft* vol 103 pp 127–139
[2] Yakubov S, Rung T and Maquil T 2015 *Computers and Fluids* **111** 91–104
[3] Sauer J 2000 *Instationär kavitierende Strömungen - Ein neues Modell, basierend auf Front Capturing (VoF) und Blasendynamik* Ph.D. thesis Fakultät für Machinenbau, Universität Karlsruhe
[4] Maquil T *et al.* 2011 *WIMRC 3rd International Cavitation Forum* (University of Warwick, UK)
[5] Kim K, Chahine G, Franc J and Karimi A 2014 *Advanced Experimental and Numerical Techniques for Cavitation Erosion Prediction* (*Fluid Mechanics and Its Applications* vol 106) (Springer)
[6] Yakubov S *et al.* 2012 *8th Int. Symp. on Cavitation (CAV 2012)* 063 (Singapore)
[7] Dular M and Coutier-Delgosha 2009 *Int. Jour. for Numerical Methods in Fluids* **61** 1388–1410
[8] Fujikawa S and Akamatsu T 1980 *Journal of Fluid Mechanics* **97** 481–512
[9] Plesset M and Chapman R 1971 *Journal of Fluid Mechanics* **41** 283–290
[10] Isselin J, Alloncle A and Autric M 1998 *Journal of Applied Physics* **84** 5766–5771
[11] Ma J, Hsiao C and Chahine G 2015 *Journal of Fluids Engineering* **137** 041301–1–10
[12] He J and Hammitt F 1982 *Wear* **80** 43–58

# Numerical Optimization of converging diverging miniature cavitating nozzles

**Kanchan Chavan, B Bhingole[1], J Raut[2] and A B Pandit[1*]**
[1]Department of Chemical Engineering, Institute of Chemical Technology, Mumbai, India.
[2]Unilever R&D Bangalore, 64 Main Road, Whitefield, Bangalore, India.
[*]Corresponding author E-mail: ab.pandit@ictmumbai.edu.in

**Abstract:**
The work focuses on the numerical optimization of converging diverging cavitating nozzles through nozzle dimensions and wall shape. The objective is to develop design rules for the geometry of cavitating nozzles for desired end-use. Two main aspects of nozzle design which affects the cavitation have been studied i.e. end dimensions of the geometry (i.e. angle and/or curvature of the inlet, outlet and the throat and the lengths of the converging and diverging sections) and wall curvatures(concave or convex).
Angle of convergence at the inlet was found to control the cavity growth whereas angle of divergence of the exit controls the collapse of cavity. CFD simulations were carried out for the straight line converging and diverging sections by varying converging and diverging angles to study its effect on the collapse pressure generated by the cavity. Optimized geometry configurations were obtained on the basis of maximum Cavitational Efficacy Ratio (CER)i.e. cavity collapse pressure generated for a given permanent pressure drop across the system. With increasing capabilities in machining and fabrication, it is possible to exploit the effect of wall curvature to create nozzles with further increase in the CER. Effect of wall curvature has been studied for the straight, concave and convex shapes. Curvature has been varied and effect of concave and convex wall curvatures vis-à-vis straight walls studied for fixed converging and diverging angles.It is concluded that concave converging-diverging nozzles with converging angle of $20^{o}$ and diverging angle of $5^{o}$ with the radius of curvature 0.03 m and 0.1530 m respectively gives maximum CER.
Preliminary experiments using optimized geometry are indicating similar trends and are currently being carried out. Refinements of the CFD technique using two phase flow simulations are planned.
Keywords: Converging-diverging nozzle, cavitation, convex, concave.

## 1. Introduction:

Cavitation is the formation and subsequent implosion of bubbles in a liquid. Small bubbles form when the total tension applied to the liquid is greater than a critical pressure. Upon reaching subsequently high pressures they implode catastrophically leading to a generation of high pressures and temperatures. Hydrodynamic cavitation is produced by pressure variations in a flowing liquid due to the geometry of the system. The high pressures and temperatures can be used to break chemical bonds leading to accelerated chemical reactions, wastewater treatment and material synthesis, they can also be used to abrade and clean surfaces, in creating sprays and atomizationetc[2]. Hydrodynamic cavitation offers more versatility in terms of scale-up and is reportedly more efficient than acoustic cavitation[2]. Hydrodynamic cavitation can be created through the use of high-pressure/high-speed homogenization, orifice plates and converging-diverging nozzles. In the current study we report simulation studies on hydrodynamic cavitation in converging-diverging nozzles to arrive at design rules for creating optimal nozzle designs for various end applications.

There are three main aspects of the nozzle design that affect cavitation – (i) theend dimensions, angle and/or curvature of the inlet, outlet and the throat and the lengths of the converging($l_{converging}$)and diverging($l_{diverging}$)sections, (ii) the wall curvature in the converging and diverging sections and (iii) wall roughness. Studies in literature have focused on the optimization of the lengths and dimensions of the converging-diverging nozzles with straight walls[2]. Theoretical studies however have focused primarily on nozzles with straight walls or have not paid attention to the aspects of wall curvature. The effect of wall curvature of the converging and diverging sections on cavitating flows has never been reported in the literature to the best of our knowledge. The curvature which can broadly be classified as straight, convex or concave (Figure 1(a), (b) & (c)) influences the nature of the pressure profiles without affecting the absolute pressure values at the outlet and the throat (Figure 2). With increasing capabilities in machining and fabrication, it is possible to exploit the effect of wall curvature to create nozzles with maximum CER. Different wall curvatures were simulated using CFD software and bulk pressure and velocity history were obtained for further estimation of bubble dynamics. In the present study Rayleigh-Plesset equation is solved till the point where compressibility of liquid can be neglected (bubble wall velocity is less than the velocity of sound in liquid) after which Tomita-Shimaequation [3] is used when the compressibility of liquid medium is needs to be considered (bubblewall velocity is greater than the velocity of sound in liquid).

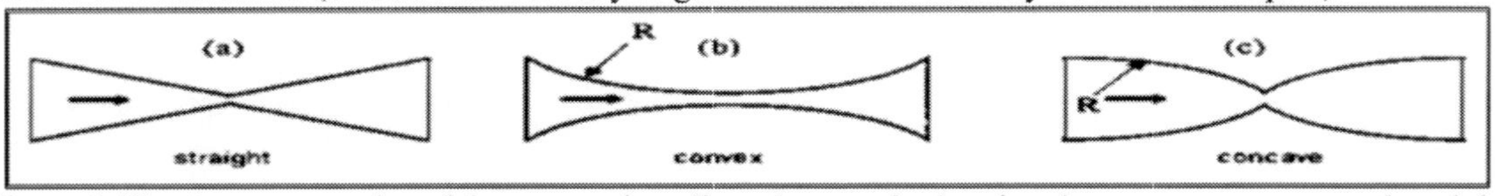


Figure 1: The three geometries of the converging–diverging nozzle showing the convex, concave and straight curvatures.

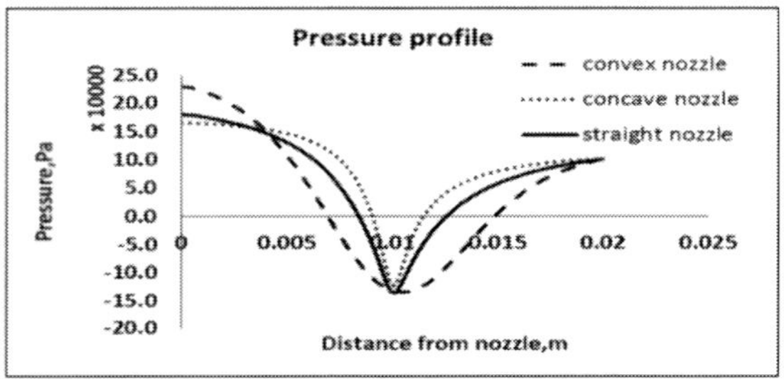


Figure 2: Pressure profiles along the nozzle length for the concave-concave, convex-convex and straight-straight nozzles for a constant flow rate ($l_{total}$= 0.02m, $r_{inlet}$=$r_{outet}$= 0.001m, $r_{throat}$ = 0.0005m, flowrate = 1.2 l/min).

## 2. Geometries selected for simulation:

CFD simulations were carried out for straight line converging-diverging nozzle by varying angle of convergence and divergence from 5° to 60° to study the effect of convergence and divergence angles on CER. CER in terms of collapse pressure can be defined as the ratio of cavity collapse pressure to permanent pressure drop in the system. Configuration which gives high collapse pressure and minimum permanent pressure drop has been selected as an optimum. CER is reported as $R_{efficiency}$ i.e. ratio of collapse pressure to permanent pressure drop. Inlet & outlet pipe diameters were taken as 0.002 m and throat diameter was taken as 0.001 m. Length of convergence and divergence were varied according to the angle of convergence and divergence (table 1).

Table 1: Parameters of simulated geometry.

| Angle of convergence / divergence | 5° | 10° | 20° | 30° | 40° | 60° |
|---|---|---|---|---|---|---|
| Length of convergence / divergence (m) | 0.010 | 0.0057 | 0.0028 | 0.0018 | 0.0013 | 0.0008 |

For Convex and concave converging-diverging nozzle, radius of curvature has been varied by keeping angles of convergence and divergence constant, to study the effect of curvature onCER.For concave converging-diverging nozzles,curvature of the converging section was varied from 0.0062 m to 0.0440 m and curvature of the diverging section was varied from 0.0430 m to 0.6420 m is discussed later.

## 3. Simulation Strategy:

Design module available in Ansys 13 was used to construct converging-diverging nozzle geometries. 2D geometry was sufficient for converging-diverging nozzle because of the axial symmetry. The meshing used was QUAD type meshing, and the mesh number was approximately 20,000.

Single phase steady state CFD simulations for the cavitational geometry are done in FLUENT 6.2. This involves solving momentum balance based on the geometry of the cavitational device, 2D axisymmetric simulations are done using Eulerian approach and k-ε realizable turbulence model was used. The simulations were carried out till the residual values dropped below $10^{-6}$.The cavity dynamics model developed by Mahulkar et. al. (2008)[1] for finding the cavity radius, the collapse pressure and cavity inner temperature of the collapsing cavity, themaximum pressure reached inside the collapsing cavity, the nature of the cavity dynamics (transient or oscillating) has been employed in this study.

## 4. Boundary Conditions:

The boundary conditions used for the simulations were, velocity inlet and pressure outlet. Generally, pipe velocity is kept in the range of 2-3 m/s for non-cavitating flows in pipe. To produce cavitating condition in the system, velocity was increased and chosen in the range of 6-9 m/s. Hence, the inlet velocity in the pipe was kept as 6.37 m/s arbitrarily and the outlet pressure was kept equal to atmospheric pressure. All simulations were carried for 1 μm of initial cavity size containing 75% non-condensable gases, after initial assessment of the effect (not reported here) of cavity diameter.

## 5. Results and discussion:

There is a significant pressure loss in the flow though the nozzle which can be attributed to shear loss as well as cavitation loss (phase change). Pressure lost in the nozzle is used for the generation of cavities which on collapse generates significantly high pressure which can be used for end-use, such as physico-chemical changes in the cavitating liquid.

### *5.1 Effect of converging-diverging angle*:

Effect of converging-diverging angle was studied by varying converging and diverging angle from 5° to 60°of straight line converging-diverging nozzle. It was observed that as the diverging angle increase CER decreases (Figure 4). It can be concluded from table 2 that converging angle of 20° and diverging angle 5° shows maximum ratio of collapse pressure to permanent pressure drop and can be said to be an optimum geometrical configuration in terms of highest CER.

Table 2: cavitational efficacy ratio (CER) of straight line converging diverging nozzles.

| Diverging angle → | 5° | 10° | 20° | 30° | 40° | 60° |
|---|---|---|---|---|---|---|
| Converging angle ↓ | CER | CER | CER | CER | CER | CER |
| 5° | 13056.43 | 12396.48 | 7726.77 | 4338.69 | - | - |
| 10° | 12062.49 | 11471.62 | 9292.53 | 4530.85 | - | - |
| 20° | 14454.05 | 12319.72 | 9956.28 | 4557.21 | - | - |
| 30° | 14148.70 | 12240.97 | 10956.80 | - | - | - |
| 40° | 13607.01 | 11850.38 | 7685.55 | - | - | 1831.29 |
| 60° | 11968.17 | 10835.17 | 6355.46 | - | - | 1454.95 |

CER - Ratio of collapse pressure to permanent pressure drop.
'-' indicates that cavity growth occurs in this configuration but do not exhibits cavitational activity such as cavity collapse and generation of pressure.

### *5.2 Effect of curvature*:

Effect of curvature was studied for the optimum converging and diverging nozzle configuration obtained earlier for the case of straight line converging-diverging simulations in the previous section by varying the radius of curvature (as defined in Figure 1(b,c) ). For concave converging-diverging nozzles, curvature radius of the converging section was varied from 0.0062 m to 0.0440 m andcurvature of the diverging section was varied from 0.0430 m to 0.6420 m. It was observed that CER in the case of concave curvature of converging- diverging nozzles is higher than straight line converging-diverging nozzles.Optimization of Concave shaped converging-diverging nozzles was doneby singleparameter optimization technique. Curvature radius 'R' of the converging section was varied from 0.0062 m to 0.0440m by keeping the radius of the curvature of the diverging section

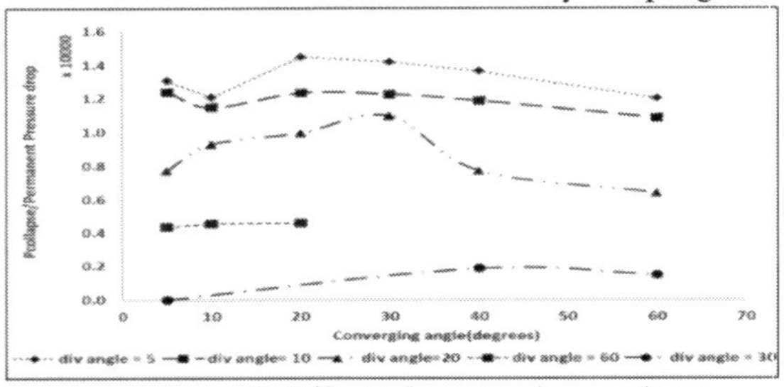


Figure 3: Effect of converging angle.

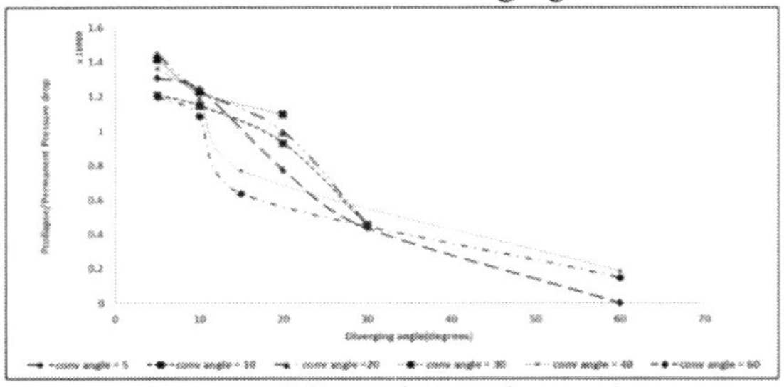

Figure 4: Effect of diverging angle.

0.1090 m constant and optimum converging section curvature was obtained as 0.03 m (Figure 5). Curvature radius of the diverging section was varied from 0.0430 m to 0.6420 m by keeping curvature radius of the converging section to 0.03 m constant and optimum radius curvature of the diverging section was obtained as 0.1530 m (Figure 6). As convex curvature of converging-diverging nozzle shows less CER than straight line nozzles, it is not discussed in details.

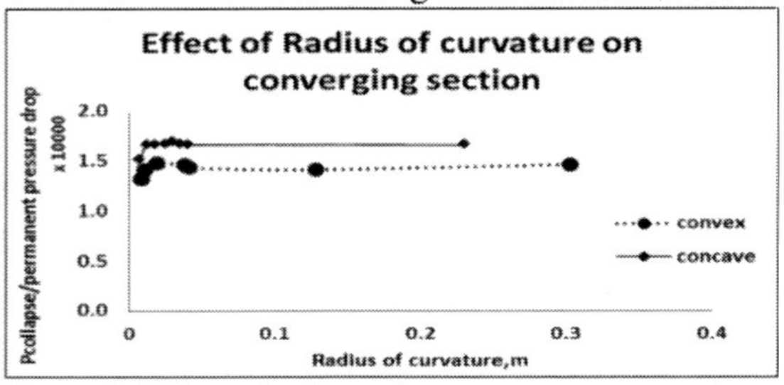


Figure 5: Effect of radius of curvature on converging section.

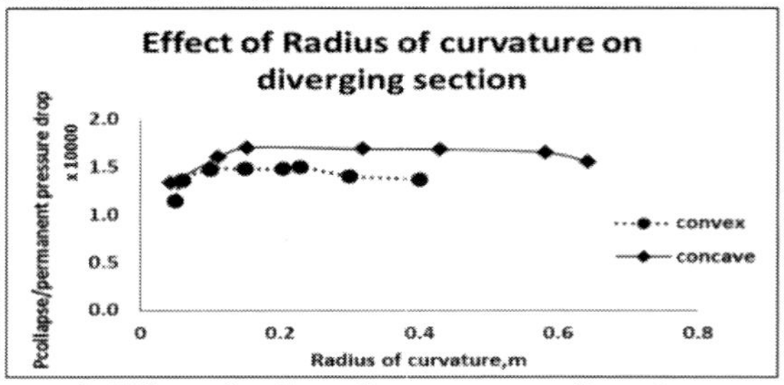


Figure 6: Effect of radius of curvature on diverging section.

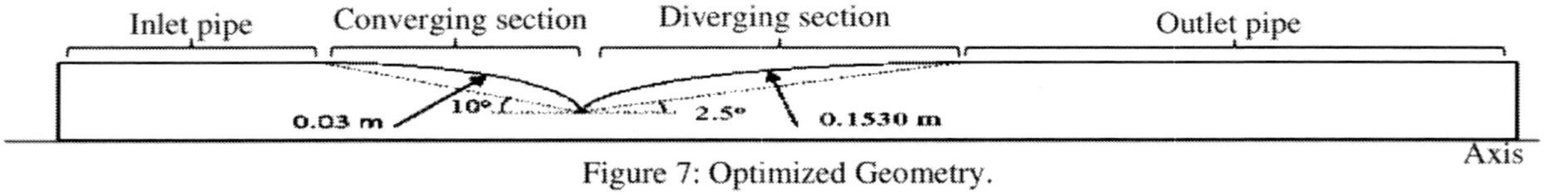


Figure 7: Optimized Geometry.

## 6. Conclusions:

Effect of converging diverging angles and the curvature (straight, concave and convex) on the converging-diverging nozzle has been studied and optimum configuration has been reported. Concave converging-diverging nozzles withconverging angle of 20° and diverging angle of 5° with the radius of curvature 0.03 m and 0.1530 m respectively gives maximum CERi.e. $1.7\times 10^4$, 1.18 times greater than straight line. However, further refinements in the CFD techniques using two phase flow simulations are planned.

## Reference:

[1] Mahulkar AV, Bapat PS, Pandit A B, Lewis F M, 2008.Steam Bubble Cavitation.*AIChe Journal* Vol. **54**, 7.

[2] Raut JS, Stoyanov SD, Duggal C, Pelan E G, Arnaudov L N, Naik VM,2012. Hydrodynamic cavitation: a bottom-up approach to liquid aeration.*Soft Matter*, **8**, 4562-4566.

[3] Shima A, Tomita Y, 1979. The behaviour of a spherical bubble in mercury. *Report 2. The Reports of the Institute of High Speed Mechanics*, **39** Sendai, Japan: Tohoku University,19–45.

Journal of Physics: Conference Series 656 (2015) 012139 doi:10.1088/1742-6596/656/1/012139

# CFD analysis of cloud cavitation on three tip-modified propellers with systematically varied tip geometry

**K W Shin**[1] **and P Andersen**[2]
[1]Propeller & Aftship R&D Department, MAN Diesel & Turbo, Frederikshavn, Denmark
[2]Department of Mechanical Engineering, Technical University of Denmark (DTU), Kgs.Lyngby, Denmark

E-mail: keun.shin@man.eu

**Abstract**. The blade tip loading is often reduced as an effort to restrain sheet and tip vortex cavitation in the design of marine propellers. This CFD analysis demonstrates that an excessive reduction of the tip loading can cause cloud cavitation responsible for much of noise and surface erosion. Detached eddy simulations (DES) are made for cavitating flows on three tip-modified propellers, of which one is a reference propeller having an experimental result from a cavitation tunnel test with a hull model, and the other two are modified from the reference propeller by altering the blade tip loading. DES results have been validated against the experiment in terms of sheet and cloud cavitation. In DES, non-uniform hull wake is modelled by using the inlet flow and momentum sources instead of including a hull model. A 4-bladed Kappel propeller with a smooth tip bending towards the suction side is used as the reference propeller. For the reference propeller, sheet cavitation extends over a whole chord length in the hull wake peak. As the blade gets out of the wake peak, the rear part of sheet cavity is detached in a form of cloud cavitation. For the reference propeller, the tip pitch reduction from the maximum is about 35%. When decreasing the tip pitch reduction to 10%, tip vortex cavitation is formed and cloud cavitation is significantly weakened. When increasing the tip pitch reduction to 60%, sheet cavitation slightly moves to inner radii and cloud cavitation grows larger.

## 1. Introduction

The mechanism of cloud cavitation on a hydrofoil is that the rear part of a sheet cavity is detached in a form of cloud cavitation by a re-entrant jet, when sheet cavitation is chordwisely extended more than a certain length [2]. Cloud cavitation can also be produced by a periodic flow disturbance on an oscillating hydrofoil [3]. Sheet cavitation builds up with an increase of the incident angle. As the incident angle decreases, a froth cloud is ejected from the downstream edge of the sheet cavity with the sheet cavity collapsing. The periodic inflow fluctuation on a hydrofoil resembles non-uniform hull wake on a ship propeller.

As the smooth interface of a sheet cavity grows instable by turbulent eddies, it breaks down into a microbubble cloud. As the cloud cavitation flows downstream, collapse of a single microbubble may initiate cascade implosions of collective microbubbles emitting shock waves with high energy concentration [4]. Cloud cavitation is associated with intense noise generation and surface erosion, hence the prevention of cloud cavitation is of major importance in the marine propeller design.

A marine propeller showing extensive sheet cavitation and cloud cavitation in a cavitation tunnel test is selected as a reference propeller for research. A cavitation simulation for the reference propeller with hull wake has been validated against the experimental result in terms of cavitation variation along the blade position [1]. Two other propellers are modified from the reference propeller by altering the blade tip loading. By comparing DES results for three propellers, the influence of the tip loading on cloud cavitation is investigated.

## 2. Propeller models

As a part of efforts to improve propulsive efficiency in ship operations, several kinds of tip-modified propellers have been introduced to reduce energy loss from tip vortices [5]. Due to tip bending, variation patterns from sheet cavitation to tip vortex cavitation on tip-modified propellers are different from conventional propellers. The reference propeller taken for research is a Kappel propeller which features a smooth tip bending towards the suction side. It is a 4-bladed propeller with a low area ratio of $Ae/Ao$=0.38, where $Ae$ is the expanded blade area and $Ao$ is the propeller disk area. A cavitation tunnel test conducted with a complete hull model at SSPA shows a large extent of sheet cavitation and detachment of sheet cavity in a form of cloud cavitation.

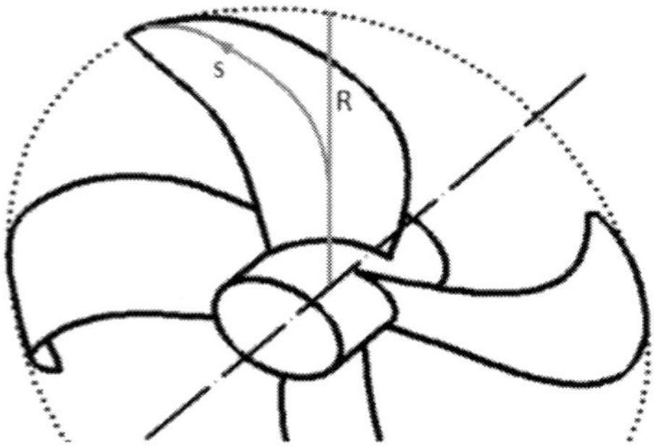


**Figure 1.** Orthogonal mid-chord locus of Kappel propeller

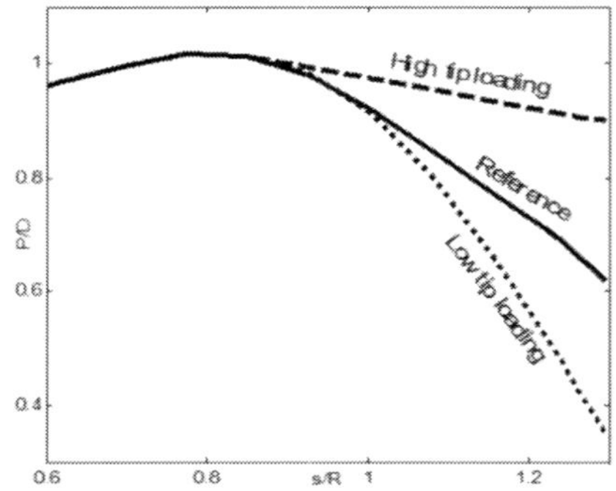


**Figure 2.** Pitch ratio $P/D$ as a function of relative mid-chord length $s/R$

**Figure 3.** Computational mesh around the reference propeller and rudder

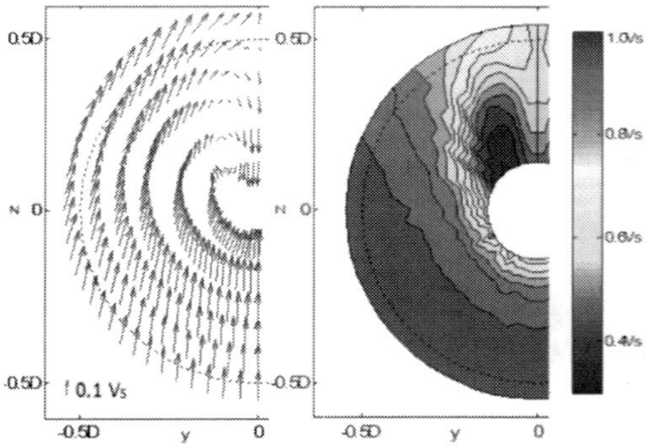


**Figure 4.** Transverse (left) and axial (right) velocity components of hull wake extracted from a bare hull simulation [1]

Two other propellers are modified from the reference by altering the blade tip pitch. For the reference propeller, the tip pitch reduction from the maximum is about 35%. The maximum pitch is at a relative mid-chord length of $s/R\approx0.80$, i.e. a relative radius of $r/R\approx0.75$. Two propellers are designed by modifying the tip pitch reduction to 10% and 60%, respectively. As shown in figure 2, the pitch drops from the same maximum with a different rate for three propellers and the one with the small reduction will have a higher loading than the reference propeller and the one with the larger reduction will have a lower loading. Since the lifting surface in the tip region of the tip-modified propeller is not on cylindrical surfaces, the angle of incidence to the blade sections is determined by the pitch angle and the vertical angle of aperture of the surface relative to the onset flow.

## 3. Cavitation simulation

Cavitation simulations are made by DES with the $k$-$\omega$ SST turbulence model. Unsteady computations are performed with a time step corresponding to 1° propeller rotation. A cylindrical domain is used with an inner rotating domain around a propeller model with a diameter of $D$=250 mm. The propeller rotation is simulated by rigid body motion with a sliding mesh.

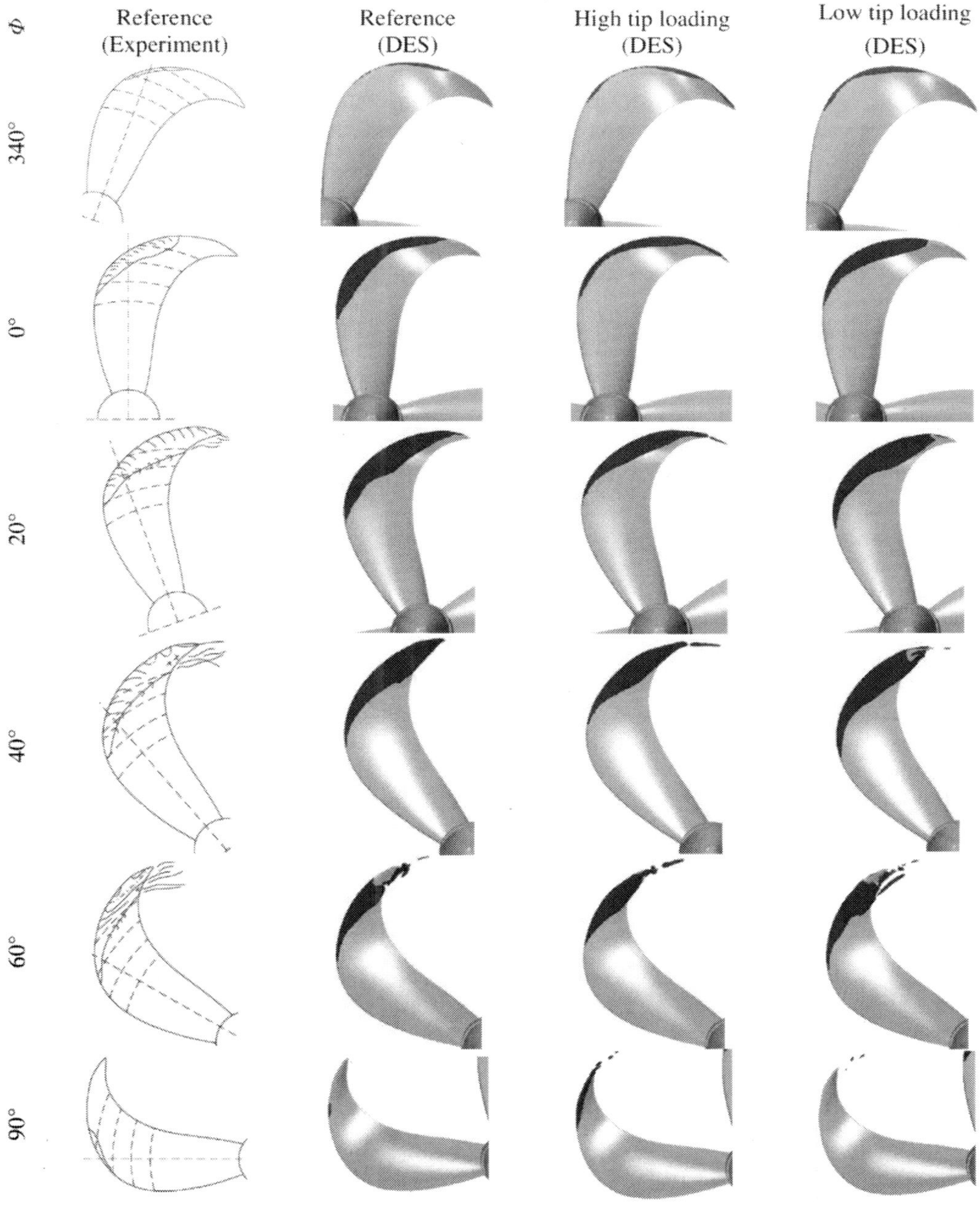


**Figure 5.** Cavitation for three propellers with different tip loadings

A hexahedral mesh trimmed by propeller and rudder models is generated with prism layers resulting mostly in a dimensionless first-cell height of $y^+<2$ (See figure 3). Mesh generation, computation and post-processing are made using the commercial CFD solver StarCCM+. Cavitation is modeled by the Volume-of-Fluid (VOF) and a vapor transport equation based on the Rayleigh-Plesset equation. The force of gravity is applied to take into account hydrostatic pressure.

Simulations are made in the same model-scale condition as the cavitation tunnel test: ship speed $V_S$=4.5 m/s, rotating speed $N$=24.0 rps and cavitation number $\sigma_N$=3.8. Instead of including a hull model, a wake field extracted from a bare hull simulation is applied as a propeller inflow by using the inlet boundary condition and momentum source [1]. Since hull wake of a single-screw ship is almost symmetric with respect to the vertical centerline, only half of the wake field is presented in figure 4. The rotating speed of two modified propellers is adjusted by $\Delta N \approx \pm 0.5$ rps to reach the same thrust $T$ with a difference of $\Delta T \leq 1\%$ as the reference. The propeller loadings are as follows: reference- $K_T$=0.213 (experiment), 0.215 (DES), high tip loading- $K_T$=0.225 (DES) and low tip loading- $K_T$=0.206 (DES), where $K_T = T/(\rho \cdot N^2 \cdot D^4)$. The efficiency $\eta$ of the high tip-loading propeller is 0.5% lower than the reference probably due to an energy loss from an increased tip vortex, where $\eta = T \cdot V_A/(2 \cdot \pi \cdot N \cdot Q)$ and $Q$ is the propeller torque and $V_A$ is the propeller advance speed. $\eta$ of the low tip-loading propeller is close to that of the reference with a difference of less than 0.1%.

In figure 5, the variation of cavitation for three propellers is summarized only for blade positions of $\varphi \approx 340$-90°, because cavitation on the reference propeller is suppressed at other blade positions. An isosurface of 10% vapor fraction is taken as a cavitation interface. As the blade enters a high wake region at $\varphi \approx 340°$, leading-edge sheet cavitation starts. Sheet cavitation is gradually developed at $\varphi \approx 340$-40°. At $\varphi \approx 40°$, it is extended over the whole chord length at $r/R \approx 0.9$-1.0. While the rear part of the sheet cavity is detached in a form of cloud cavitation in the experiment at $\varphi \approx 60°$, unstable sporadic large-scale structure of cloud cavitation is reproduced by DES, but microbubbles are not captured, because the computation grid of $\Delta x \approx 0.5$ mm is larger than microbubble scales.

DES for the high tip-loading propeller shows that sheet cavity is extended to the tip and tip vortex cavitation occurs without detachment of a sheet cavity. For the low tip-loading propeller, the sheet cavity is detached earlier and cloud cavitation is more intense than for the reference, as shown at $\varphi \approx 40°$ in figure 5. Since the blade loading is distributed more to the tip region for the high tip-loading propeller, overall sheet cavitation is lessened at $\varphi \approx 0$-40°. In contrast, the blade loading is concentrated on the maximum pitch region for the low tip-loading propeller and hence sheet cavitation becomes more extensive.

## 4. Conclusion

DES is made for cavitating flows on three propellers with different tip loadings. The DES result indicates that high unloading of a blade tip can cause or intensify cloud cavitation for propellers accompanying extensive sheet cavitation. Cloud cavitation can be prevented or reduced by increasing tip loading and so distributing blade loads more evenly from the maximum pitch region to the tip. Since high tip loading can, on the other hand, bring up tip vortex cavitation and propulsive efficiency loss from the tip vortex, an appropriate balance of the tip load distribution is required in the marine propeller design.

## References

[1] Shin K W, Regener P B and Andersen P 2015 *Proc. SMP'15* (Austin)
[2] de Lange D F, de Bruin G J and van Wijngaarden L 1994 *Bubble Dynamics and Interface Phenomena* (Netherlands: Springer) p 241-246
[3] Reisman G E, McKenney E A and Brennen C E 1994 *Proc. 21th Symp. on Naval Hydrodynamics* (Santa Barbara)
[4] van Terwisga T J C, Fitzsimmons P A, Ziru L and Foeth E J 2009 *Proc. CAV2009* (Ann Arbor)
[5] Andersen P, Friesch J, Kappel J J, Lundegaard L and Patience G 2005 *J. Marine Tech.* **42** 3
[6] Dang J 2004 *J. Int. Shipbuild. Progr.* **51** 4 353-376

# Application of a Panel Method in Performance Prediction of Cavitating and Non-cavitating Tip Loaded Propellers

**Yiran Su**[1] **and Spyros A. Kinnas**[2]

[1 2] Ocean Engineering Group, Department of Civil, Architectural and Environmental Engineering. The University of Texas at Austin, Austin, TX 78712, USA

Email: yiransu@utexas.edu

**Abstract**. A panel method is applied to predict the performance of tip loaded propeller with and without cavitation modeling. For fully wetted cases, the geometry of the wake is shown to be critical to the computational accuracy. To align the wake surface, two methods are proposed and the full wake alignment scheme behaves better then PSF2 alignment. The boundary layer correction can also improve the pressure distribution especially near the trailing edge region. Finally, for cavitating cases, the results based on full wake alignment tend to create larger cavitation region and higher cavitation thickness compared to results based on PSF2 alignment.

## 1. Introduction

Tip loaded propellers (TLPs) improves propeller efficiency by preventing circulation loss at the tip region[1]. To predict the performance of TLPs, panel method can be used because of its accuracy as well as efficiency. The panel method can also be coupled with various cavitation models to predict cavitation behavior.

To apply panel method to TLPs, some special treatments are needed.

First, the shape of the wake is critical to numerical accuracy. According to Tian and Kinnas[2], the pressure distribution on a blade is sensitive to the geometry of the wake surface shedding from the previous blade. This effect can be significant for TLPs because of the close proximity between the end plate to the wake surface shedding from the previous blade.

To determine the wake geometry in panel method, a wake alignment model is required. Greeley and Kerwin proposed a PSF2 wake alignment method[3], in which the wake geometry is determined by the solution of a vortex lattice method (VLM) calculation on a coarse gird. Tian and Kinnas proposed an iterative wake alignment scheme called the full wake alignment (FWA)[2]. In this method, the wake geometry is determined by the solution of the panel method itself.

Another treatment is the boundary layer correction. Yu and Kinnas treated the boundary layer effect as a blowing source on blade surfaces and wake surfaces[4]. The strength of the blowing source can be determined by solving the boundary layer equations[5]. By including the boundary layer effect, the pressure distribution near the trailing edge of the blades can be evidently improved.

In this paper, panel method is applied in P2504 tip loaded propeller. Both full wake alignment scheme and boundary layer correction scheme are tested. Results are compared to experiments data as well as RANS results. Finally, some cavitation solutions are shown and discussed.

## 2. Formulation

### *2.1. Governing Equation*

If we fix the system of coordinates onto the rotating propeller, then the total velocity can be decomposed into background velocity $U_\infty$, relative rotational velocity, and perturbation velocity u. The last component equals to the gradient of the flow potential $\phi$ which satisfys the Laplace's equation.

$$q = U_\infty + w \times r + u \tag{1}$$

$$\nabla^2 \phi = 0 \tag{2}$$

According to Green's second identity, equation (2) can be converted to the boundary integral form in which S includes propeller surfaces and cavity surfaces, $S_W$ represents the wake surfaces, $n_q$ is the normal vector and $G(p;q)$ is the three dimensional Green's function.

$$2\pi\phi_P = \iint_S \left[\phi_q \frac{\partial G(p;q)}{\partial n_q} - G(p;q)\frac{\partial \phi}{\partial n_q}\right] d_s + \iint_{S_W} \Delta\phi_W \frac{\partial G(p;q)}{\partial n_q} d_s \tag{3}$$

The kinematic boundary condition (equation 4) is applied on blade surfaces, hub surfaces, and cavity surfaces. For cavity surface the dynamic boundary condition, which describes the pressure value, should also be satisfied. This boundary condition can be converted to a Dirichlet type condition (equation 5) by applying Bernoulli equation and integrating the velocity along certain curves on the cavity surface.

$$\frac{\partial \phi}{\partial n} = -U_\infty \cdot n \tag{4}$$

$$\phi = \phi_c(s) \tag{5}$$

On the wake surface, a Kutta condition is applied where the wake dipole strength $\Delta\phi_W$ equals to the difference between the potential values on each side of the trailing edge.

$$\Delta\phi_W = \phi_+ - \phi_- \tag{6}$$

### *2.2. Wake Alignment Model*

In the full wake alignment scheme, a helical shape wake is used in the first step as an initial guess of the wake geometry. With the given wake geometry, we are able to solve the first iteration. Then, the wake surface is aligned to the downstream streamlines, which can be obtained by computing the induced velocities. The new wake geometry is then used for the subsequent iteration. It usually takes less than 5 iterations for the wake geometry to converge.

### *2.3. Boundary Layer Correction Model*

According to boundary layer theory, pressure within the boundary layer remains constant along the transverse direction and the flow outside the boundary layer can be treated as inviscid. Therefore, the influence of the boundary layer to the ambient inviscid flow can be seen as pushing the streamlines outward by some distance, which is called the displacement thickness. In the panel method, an additional blowing source can be used to simulate this effect and the strength of the source can be determined by the mass defect solved from boundary layer equation [5].

### *2.4. Cavitation Model*

For a given cavitation planform, we are able to obtain a solution by enforcing kinematic boundary condition on blade surfaces and dynamic boundary condition on cavity surfaces. Then, cavitation thickness can be computed from the normal velocity and the kinematic boundary condition on the cavity. In most cases, the cavitation thickness on the trailing edge of the cavity is not zero, which means the assumed cavitation planform is not compatible and therefore should be adjusted.

In the cavitation model, the cavitation planform is assumed in the first several attempts. For each attempt, the cavity thickness on the cavity trailing edge can be computed. Then the Newton Raphson method is adopted to find the converged cavitation planform which corresponds to a cavitation thickness distribution that equals to zero at cavity trailing edge. By this procedure, the cavitation distribution can be determined. Detailed information about this model can be found in Fine's paper[6].

## 3. Fully Wetted Results

### *3.1. Numerical Model*

In this chapter and the next chapter, panel method is used to predict the performance of P2504 tip loaded propeller[1]. Parameters of this propeller are shown in the table 1. To validate the fully wetted panel method, a RANS simulation is performed.

**Table 1:** Basic parameters of P2504 tip loaded propeller

| Propeller No. | Diameter | P/D at 0.7R | No. of blades |
|---|---|---|---|
| P2504 | 6.5m | 1.441 | 5 |

### *3.2. Pressure Distribution*

The pressure coefficients are plotted at two different blade stations, as shown in figure 1. The solution with PSF2 alignment scheme is shown to create larger errors when compared to the solution with full wake alignment scheme. Also, by adopting the boundary layer correction scheme, the pressure distribution on the trailing edge can be improved.

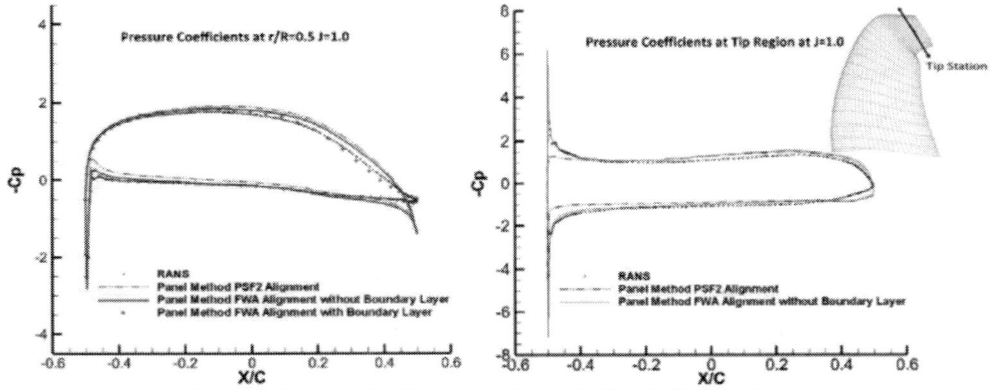


**Figure 1:** Pressure distribution at r/R=0.5 (left) and r/R=0.7 (right).

### *3.3. Thrust/Torque Coefficients*

Figure 2 shows the thrust and torque coefficients from RANS, from experiments[1], and from the panel method. In the panel method calculation, both full wake alignment scheme and the boundary layer correction model are both included.

Note that the geometry in the experiment is slightly different from the numerical models. This might cause some difference in the results.

Compared to the RANS results, the error of both coefficients maintains less than 5% for advance ratios between 0.4 and 1.2.

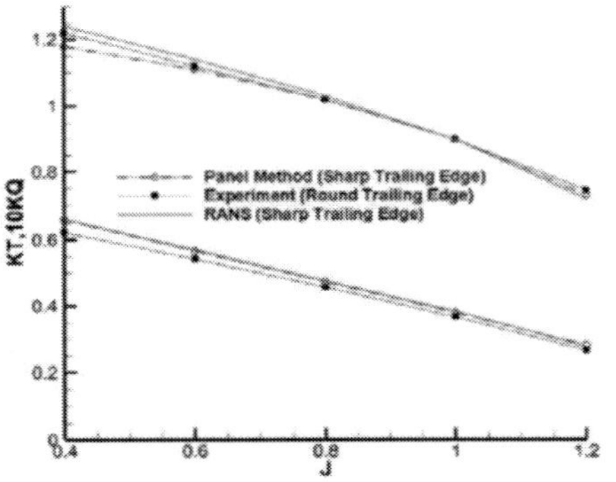


**Figure 2:** Thrust coefficients and torque coefficients.

## 4. Cavitating Solutions

In this chapter, panel method with cavitation model is applied to P2504 TLP. The advance ratio is set to 0.5 which is lower than design condition (J=0.9). Two cases with different cavitation numbers ($\sigma_*$=2 and $\sigma_*$=3) are tried.

Figure 3 shows circulation distribution on the blade. This figure shows the computational error between the two wake alignment models and also shows that cavitation effect can increase the circulation on the blade.

Figure 5 displays the cavitation region as well as the cavitation thickness for both cavitation numbers and both wake alignment models. The solution based on full wake alignment, which is proved to be more accurate in chapter 3, is shown to induce larger cavitation volume than the solution based on PSF2 alignment.

Figure 4 shows the pressure distribution at station r/R=0.8 and r/R=0.91 with cavitation number $\sigma_{\cdot}=3$. The cavitation region can be clearly identified by the constant pressure region. Only back cavitation occurs partially. The pressure distribution is not much affected by the cavitation effect at the other side of the blade and at the far field from the cavitation region.

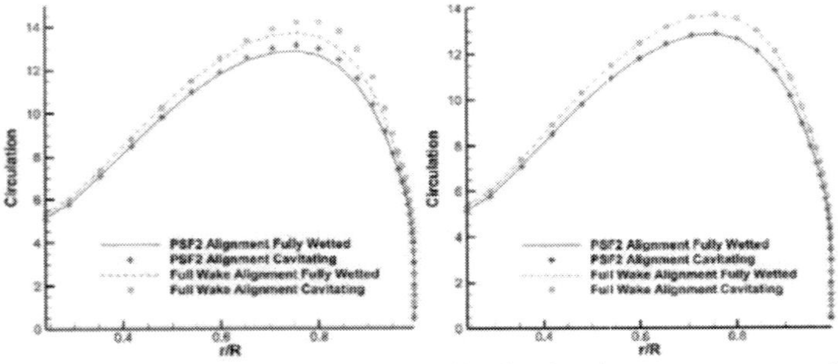


**Figure 3:** Circulation distribution for cavitation number $\sigma$=2 (left) and $\sigma$=3 (right).

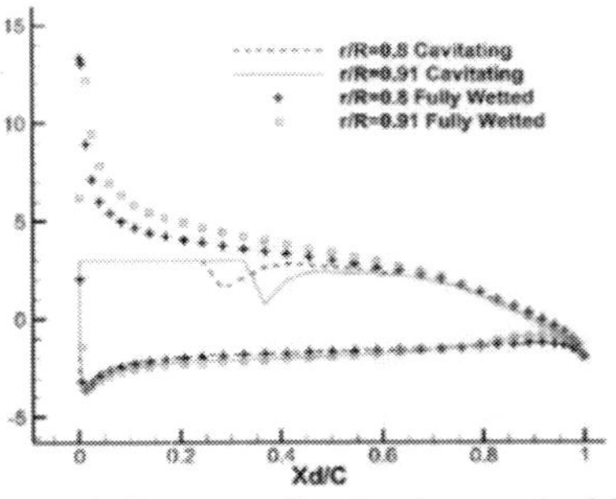


**Figure 4:** Pressure distribution on the blade

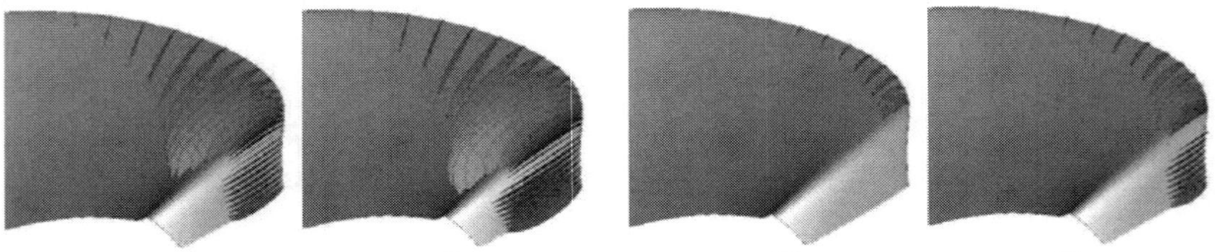

(1) (2) (3) (4)

**Figure 5:** Cavitation distribution for (1) $\sigma$=2 PSF2 alignment, (2) $\sigma$=2 FWA alignment, (3) $\sigma$=3 PSF2 alignment, (4) $\sigma$=3 FWA alignment.

**References**

[1] Brown M, Sánchez-Caja A, Adalid J G, Black S, et al 2014 Improving Propeller Efficiency through Tip Loading. *30th Symposium on Naval Hydrodynamics.*

[2] Tian Y, Kinnas S A 2012 A Wake Model for the Prediction of Propeller Performance at Low Advance Ratios. *International Journal of Rotating Machinery*, vol 2012, 372364.

[3] Greeley D S, Kerwin J E 1982 Numerical methods for propeller design and analysis in steady flow. *Transactions of Society of Naval Architects & Marine Engineers*, vol 90, p 415–453.

[4] Yu X M 2012 Three Dimensional Viscous/Inviscid Interactive Method and Its Application to Propeller Blades. Master Thesis, UT Austin.

[5] Drela M 1985 Two-dimensional Transonic Aero-dynamic Design and Analysis Using the Euler Equations. PhD Thesis, MIT.

[6] Kinnas S A, Fine N E 1993 A numerical nonlinear analysis of the flow around two- and three-dimensional partially cavitating hydrofoils. *Journal of Fluid Mechanics*, vol 254, p 151-181.

# Development of Numerical Simulation Method for Compressible Gas-Liquid Two-Phase Flows

**Y Tamura**
Information Sciences and Arts / Center for Computational Mechanics Research, Toyo University, Kujirai 2100, Kawagoe 350-8585 Japan

E-mail: tamura@toyo.jp

**Abstract.** A numerical simulation method of compressible gas-liquid two-phase flow is developed for analyses of a cavitation bubble. Thermodynamic state of both phases is described with stiffened gas equation of state. Interface of two phases is captured by Level-Set method. As internal energy jump between two phases is critical for the stability of computation, total energy equation is modified so that inviscid flux of energy is smoothly connected across the interface. Detail of governing equations as well as their discretization is described followed by the result of one-dimensional simple example computation.

## 1. Introduction
A strong pressure waves emitted from a collapsing cavitation bubble is very important issue in many engineering fields. However the numerical analysis of a collapsing bubble has some difficulties, e. g. moving interface between two phases exists, density ratio of two phases is large, compressibility of both phases must be considered to capture pressure waves (consequently energy equation / thermodynamics relation is inevitably required). There are a few researches in the literature. For example, Takahira et al. [1] developed improved ghost fluid method and Ochiai et al. [2] applied homogenization model used in cavitating flow simulations. Both of them successfully simulated bubble deformation and collapse. The former is based on two-fluid model and latter one-fluid model. In this research, two-fluid model is adopted because the interface should be clearly defined to capture the collapse of a bubble. Instead of (improved) ghost fluid method, which has rather complicated algorithm, the Level-Set method [3] is applied to capture interface and then the numerical algorithm is expected to be simpler and suitable for large-scale parallel computing, which should be required for detailed analysis of bubble collapse or bubble fission. The present author has tried to simulate two-dimensional / axisymmetric bubble collapse problems [4] which showed numerical instability in the case of practical density / pressure ratio of two phases. Detailed examination suggested that instability is not caused by the large density difference but by the internal energy difference given by the stiffened gas equation of state, also used in [1]. Hence the energy equation is modified in the present algorithm to overcome the numerical instability. In the next section, the detail of the present algorithm is described, followed by the result of simple one-dimensional problems to discuss its validity.

## 2. Governing equations and numerical method
Let the volume fraction of liquid be $\psi$. $\psi$ takes the value of 0 or 1 in almost whole region and the value between 0 and 1 in the vicinity of the interface. In such region, assuming that velocity and

pressure are the same in both phases, the following governing equations are derived in one-dimensional inviscid flow.

$$\begin{aligned}&\frac{\partial \rho}{\partial t}+\frac{\partial(\rho u)}{\partial x}=0\\&\frac{\partial(\rho u)}{\partial t}+\frac{\partial\left(\rho u^2+p\right)}{\partial x}=0\\&\frac{\partial e}{\partial t}+\frac{\partial\{(e+p)u\}}{\partial x}=0\end{aligned}\tag{1}$$

where

$$\begin{aligned}&\rho=\rho_L\psi+\rho_G(1-\psi)\\&e=\rho_L\varepsilon_L\psi+\rho_G\varepsilon_G(1-\psi)+\frac{1}{2}\rho u^2\\&p=(\gamma_L-1)\rho_L\varepsilon_L-\gamma_L\Pi_L=(\gamma_G-1)\rho_G\varepsilon_G\end{aligned}\tag{2}$$

The subscripts $L$ and $G$ denote liquid phase and gas phase, respectively. Thermodynamic relation is given by stiffened gas equation of state and $\gamma_L=2.8$, $\Pi_L=8.5\times10^8\,\mathrm{Pa}$, $\gamma_G=1.4$ [5]. As the atmospheric pressure has the order of $10^5$ Pa, large difference of internal energy in two phases exists obviously. Once $\psi$ is obtained, pressure is calculated from eq. (2).

The volume fraction of liquid $\psi$ is expressed with the so-called Level-Set function $\phi$, which indicates the signed distance from the interface as

$$\psi=0.5+H(\phi)\ .\tag{3}$$

$H$ is called the Heaviside function and expressed with the half width of the smoothed interface $\alpha$ as

$$H=\begin{cases}0.5, & \phi>\alpha\\ -0.5, & \phi<-\alpha\\ \dfrac{1}{2}\left\{\dfrac{\phi}{\alpha}+\dfrac{1}{\pi}\sin\left(\dfrac{\pi\phi}{\alpha}\right)\right\}, & \text{otherwise}\end{cases}\tag{4}$$

$\alpha$ is typically 2.5 times of the grid spacing. $\phi$ is advected by the fluid velocity and requires the re-initialization process after (each) advection step to maintain the function of distance from the interface. These are

$$\begin{aligned}&\frac{\partial\phi}{\partial t}+u\frac{\partial\phi}{\partial x}=0\\&\frac{\partial\phi}{\partial\tau}=S(\phi)\cdot\left(1-\left|\frac{\partial\phi}{\partial x}\right|\right)\quad\text{or}\quad\frac{\partial\phi}{\partial\tau}+\vec{W}_S\cdot\frac{\partial\phi}{\partial x}=S(\phi),\quad S(\phi)\equiv\frac{\phi}{\sqrt{\phi^2+\delta^2}},\quad\vec{W}_S=S(\phi)\cdot\frac{\partial\phi/\partial x}{|\partial\phi/\partial x|}\end{aligned}\tag{5}$$

Here $\tau$ in the lower equation is a virtual time and the equation is iteratively solved to obtain the steady state solution which recovers the signed distance.

Equations (1) – (5) give the complete set of governing equations. As all of the equations are advection ones, upwind schemes are suitable and Harten-Yee's upwind TVD scheme [6] is adopted here. The Roe's average is simply applied to calculate the interface values and the minimod limiter is used to obtain the second order accuracy. The speed of sound appearing in their scheme is described as

$$c=\sqrt{\frac{\gamma_m p}{\rho}+\frac{\gamma_m-1}{\gamma_L-1}\frac{\gamma_L\psi\Pi_L}{\rho}},\quad\gamma_m=\left(\frac{\psi}{\gamma_L-1}+\frac{1-\psi}{\gamma_G-1}\right)^{-1}+1\tag{6}$$

As described in the introduction, the computation tends to be unstable near the interface. Here the energy equation is modified as

$$\frac{\partial e'}{\partial t}+\frac{\partial\{(e'+p)u\}}{\partial x}=-\frac{\gamma_L \Pi_L}{\gamma_L-1}\left\{\frac{\partial\psi}{\partial t}+\frac{\partial(\psi u)}{\partial x}\right\}$$

$$e'=e-\frac{\gamma_L \Pi_L}{\gamma_L-1}\psi=\left(\rho_L\varepsilon_L-\frac{\gamma_L \Pi_L}{\gamma_L-1}\right)\psi+\rho_G\varepsilon_G(1-\psi)+\frac{1}{2}\rho u^2 \quad (7)$$

$$=\frac{p}{\gamma_L-1}\psi+\frac{p}{\gamma_G-1}(1-\psi)+\frac{1}{2}\rho u^2=\frac{p}{\gamma_m-1}+\frac{1}{2}\rho u^2 .$$

Formally this modification smoothly connects the internal energy of liquid phase and that of gas phase across the interface to avoid numerical instability. Note that equation (6) is still valid even though the above modification seems also modifying the characteristic speed of sound. The right-hand side of eq. (7) can be re-written using the following mass conservation equation of liquid under the assumption without phase change at the interface,

$$\frac{\partial(\rho_L\psi)}{\partial t}+\frac{\partial(\rho_L\psi u)}{\partial x}=0 , \quad (8)$$

and finally one obtains

$$-\frac{\gamma_L \Pi_L}{\gamma_L-1}\left\{\frac{\partial\psi}{\partial t}+\frac{\partial(\psi u)}{\partial x}\right\}=\frac{\gamma_L \Pi_L}{\gamma_L-1}\frac{1}{\rho_L}\frac{D\rho_L}{Dt} . \quad (9)$$

When the liquid density change is negligible (compared with the order of $\Pi_L$), the right-hand side of eq. (7) is found to be small.

**Table 1.** Computational conditions.

| | | |
|---|---|---|
| Liquid | density | 1100kg/m$^3$ |
| | pressure | 2 atm |
| | temperature | 15°C |
| | no. of grid points | 20 |
| Gas | density | 1.2kg/m$^3$ |
| | pressure | 1 atm |
| | temperature | 15°C |
| | no. of grid points | 50 |
| | time step | 4.41μs |

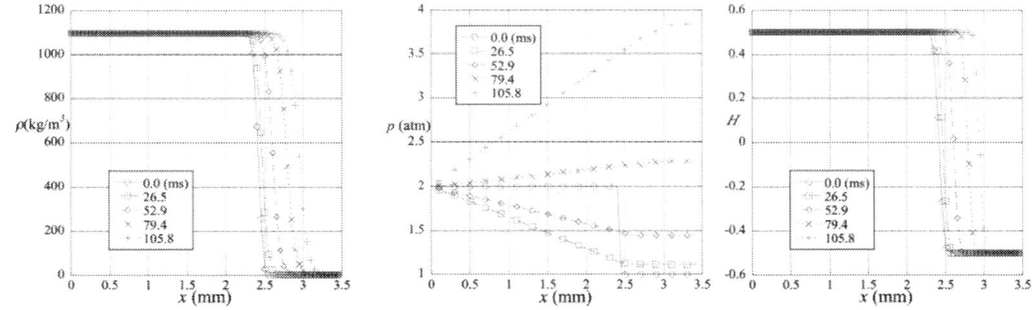


(a) (b) (c)

**Figure 1.** Compression process: (a) density, (b) pressure, (c) $H(\phi)$.

## 3. Numerical results

Results of one-dimensional two-phase flows are presented. The left side is liquid and the right side is gas. Liquid length is 2.5mm and gas length is 1.0mm. Pressure is fixed at the left boundary and the symmetric condition is posed on the right boundary. Detailed conditions are tabulated in Table 1. To stabilize the computation, the right-hand side of eq. (7) is set to be zero near the interface.

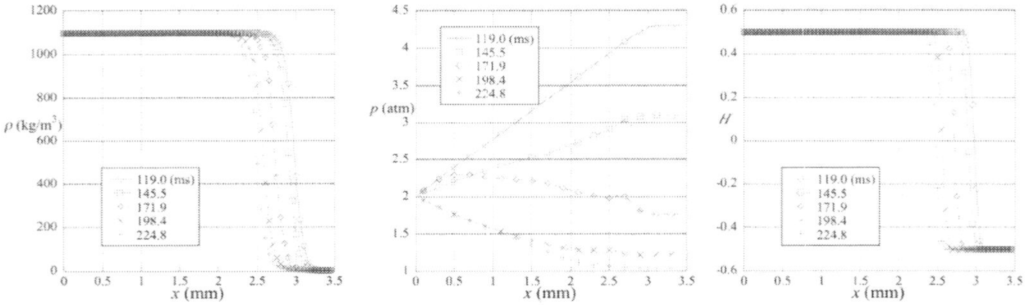


(a) (b) (c)

**Figure 2.** Expansion process: (a) density, (b) pressure, (c) $H(\phi)$.

Figure 1 shows density, pressure and Heaviside function changes during the compression process while Figure 2 shows those during the expansion process. The results of several time steps are plotted. The interface is clearly captured as shown in Fig. 1 (a), (c) and Fig. 2 (a), (c) (density and $H(\phi)$ are independently computed in the present method). Initial pressure discontinuity rapidly disappears as the pressure wave travels toward the liquid side with the liquid speed of sound. After that, liquid pressure gradually increases or decreases along with the compression or the expansion of the gas phase. The entire process is successfully and stably simulated with a small numerical oscillation of pressure observed in Fig. 2 (b). The case of larger pressure ratio of 10 (10 atm to 1 atm) is also examined. The numerical solution is again successfully obtained even though there is larger oscillation observed in pressure.

## 4. Conclusions

A new numerical method for compressible two-phase flow was proposed. Extension from the existing single-phase compressible flow code is straightforward. For the future works, the present method is extended to multi-dimensions and applied to bubble collapse or bubble fission problems.

## Acknowledgments

A part of the present research is supported by MEXT Japan of Supported Program for the Strategic Research Foundation at Private Universities.

## References

[1] Takahira H, Matsuno T and Shuto K 2008 *Fluid Dyn. Res.* **40** 510-20
[2] Ochiai N et al. 2011 *J. Fluid Sci. Tech.* **6** 860-74
[3] Sussmann M, Smereka and Osher S 1994 *J. Comp. Phys.* **114** 146-59
[4] Tamura Y, Hirano K and Tsurumi N 2010 Development of Simulation Method for Cavitation Bubble Collapse WCCM / APCOM 2010
[5] Robbins D J, et al. 2012 *World Academy of Sci. Eng. & Tech.* **68** 1143-9
[6] Yee H 1987 NASA TM 89464

# Prediction and analysis of jet pump cavitation using Large Eddy Simulation

**Hai Zi[1], Lingjiu Zhou[1], Long Meng [1]**

[1] College of Water Resources and Civil Engineering, China Agricultural University, Beijing 100083, China,
Email: zlj@cau.edu.cn

**Abstract**. 3D LES numerical simulations were performed to investigate cavitation performance inside a jet pump. The results were found to match the test data most closely. The cavitation characteristics of the jet pump were then analyzed using changes in the inlet and outlet pressure to isolate its effect on cavitation. Both results shows that the increase of the inlet pressure generally increases the Renolds number but decrease the cavitation number, thus aggravate cavitation. The closing of the outlet valve increase the outlet pressure but decrease the flowrate ratio, resulting in the increase of velocity difference and vorticity in the mixing layer. So the cavitation first declines and then grows. The cavities appear slender and extended longer in the throat with high flowrate ratio. Conversely, the cavities look short and located in the front part of the throat with low flowrate ratio. Flow analysis indicated that the turbulence behavior in the shear layer and the overall mean pressure has great influence on the local pressure in jet pump, which reveal the reason of different cavitation shape observed in experiment.

## 1. Introduction

A jet pump contains none of the moving mechanical components typical of other pumps. Rather, the propulsion is provided by a high velocity jet, which flows from the working inlet and is forced through a nozzle. The original jet flow and the entrained flow are mixed in the throat and both discharge from the outlet, as shown in Figure 1a. Many researchers have studied the internal flow of jet pumps using the 2-equation models based on the RANS method (e.g., '[1]', '[2]', '[3]'). The predicted 3-D cavitation evolutions based on LES, including the cavity growth, break-off and collapse downstream, and the shedding cycle as well as its frequency agree fairly well with experimental results (e.g., '[4]'). The use of LES to model cavitating flows is expected to give better prediction of the characteristic of cavity in jet pump.

In this paper, the LES method is used to calculate the 3D unsteady flow field. Cavitating flow field is predicted and compared with the experimental photos. The influences of different factors on the cavitation are analyzed. The performance parameters of the jet pump are defined in table.1, in which the $Q_n$, $Q_s$, $P_n$, $P_s$ and $P_c$ are shown in Fig.1a, $P_\infty$ and $v_n'$ is the mean pressure and velocity on the nozzle outlet section. $v_c'$ is the mean velocity of sucked fluid on the same section. All of them were calculated from the Bernoulli's equation.

## 2. Numerical models

Unsteady simulations of cavitating flow were performed with LES model with CFX. The sub-grid scale turbulent viscosity in SGS model is closed by LES wall-adapting local eddy-viscosity (WALE) model. A homogeneous multiphase model, namely the Zwart-Gerber-Belamri cavitation model, was employed to describe the evolution of the cavities. The basic geometry and calculation points are summarized in Fig. 1. A structured mesh was selected with about $2.43 \times 10^6$ nodes to meet the requirement of $y^+$, which is approximately 1 in the shear layer. The velocity was set on two inlets and static pressure was set on the outlet according to the experimental data. The time step was set to $10^{-5}$s. The residual values were below $10^{-5}$.

**Table 1.** Performance parameters of jet pump.

| Pressure ratio | Flowrate ratio | Efficiency | Cavitation number |
|---|---|---|---|
| $h=\dfrac{(\frac{p_c}{\rho g}+\frac{v_c^2}{2g}+z_c)-(\frac{p_s}{\rho g}+\frac{v_s^2}{2g}+z_s)}{(\frac{p_n}{\rho g}+\frac{v_n^2}{2g}+z_n)-(\frac{p_s}{\rho g}+\frac{v_s^2}{2g}+z_s)}$ | $q=\dfrac{Q_s}{Q_n}$ | $\eta=q\dfrac{h}{1-h}$ | $\sigma=(p_\infty-p_v)/(\rho v^2/2)$ <br> $v=v_n'-v_c'$ |

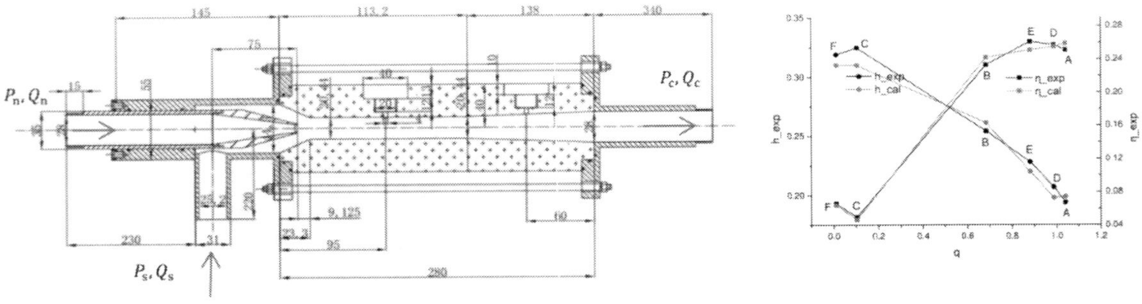


a) The structure diagram of the jet pump(mm) b) the calculation points

Figure 1. The structure diagram of the jet pump and the calculation points

## 3. Comparison of the calculation results and experimental data

The simulation results under 6 typical conditions （3 conditions with $P_n\approx$600kPa and 3 conditions with $P_n\approx$400kPa） were selected to compare with experimental photos. The predicted vapor fraction ISO surface is shown in Fig.2 and Fig.3. Fig.2 and Fig.3 shows the vapor fraction ISO surface is similar to the shape of the bubbles shown in experimental images. Their similarity on morphology suggests that the simulations are convincing. Fig. 1b shows that the predicted points on performance curves agree well with experimental data.

## 4. Analysis of the calculation results

### *4.1. The influence of different parameters on the cavitation performance*

The velocity in cavitation number is not the velocity at the reference point but the velocity difference before mixing $v=v_n'-v_c'$. The cavitation number is listed in Fig.2 and Fig.3 for different conditions. Generally the cavitation is severe with low cavitation number. However, the cavity vapor volume fraction does not increase linearly with the cavitation number. Since the jet pump has two inlets and one outlet, the pressures on each of the 3 section can influence both the overall pressure and mean velocity in the throat. This means that the pressures affected the cavitation in a more complicate way.

With the decrease of $P_n$, the outlet pressure and the flowrate on both inlets also decrease as expected. But generally the cavitation number is high and the Reynolds number is low, thus the cavitation on the throat is weak. In fact when $P_n$ is below 350kPa, no cavitation was observed during experiments since the outlet pressure was limited by the test system.

The outlet valve was closing from condition A to C or D to F with the pressure on the working inlet $P_n$ keeping almost constant. With the closing of the valve, the outlet pressure increased but the flowrate ratio decrease, which means the velocity difference decrease. Thus the cavitation number increase first then decrease. Both experimental and calculation results show that the cavitation first declined and then grew as expected. It is especially obvious for high pressure case (condition A to C,

$P_n \approx 600$kPa). Figure 2 and 3 show the cavitation variation in this process. Cai's experiment (Cai, 2005) also reported the same cavitation phenomenon in jet pump.

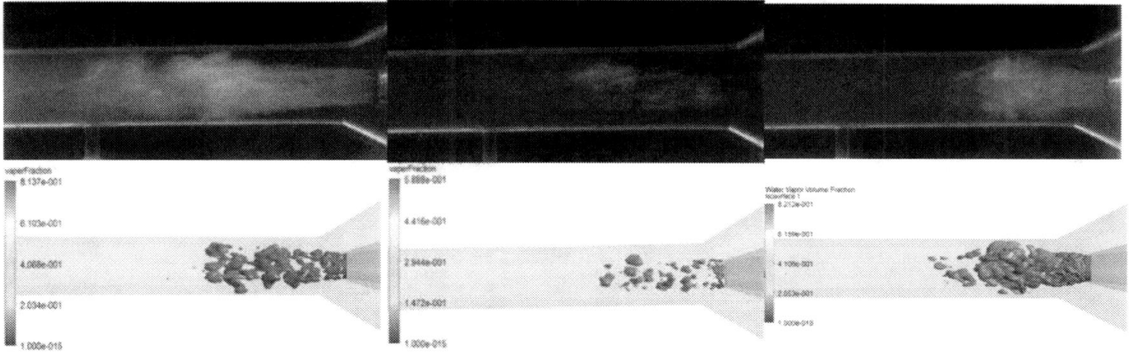

Condition A: q=1.04, $R_c = 2.46 \times 10^5$ $P_c$ =164kPa, $\sigma$ =0.171

Condition B: q=0.68, $R_c = 1.97 \times 10^5$ $P_c$ =208kPa, $\sigma$ =0.219

Condition C: q=0.10, $R_c = 1.30 \times 10^5$ $P_c$ =252kPa, $\sigma$ =0.191

Figure 2. Comparison of experimental photo (up) and the predicted vapor fraction ISO surface (bottom, Vapor fraction=0.1), $P_n \approx 600$kPa

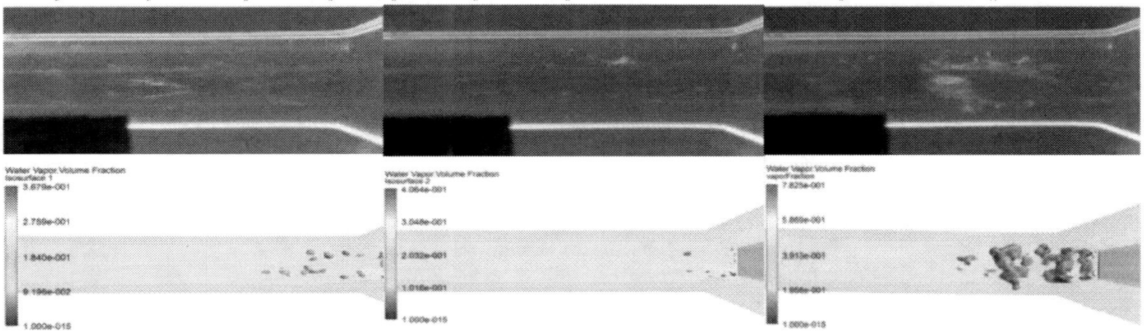

Condition D: q=0.99, $R_c = 1.85 \times 10^5$ $P_c$ =142kPa, $\sigma$ =0.378

Condition E: q=0.88, $R_c = 1.73 \times 10^5$ $P_c$ =151kPa, $\sigma$ =0.393

Condition F: q=0.01, $R_c = 0.899 \times 10^5$ $P_c$ =196kPa, $\sigma$ =0.332

Figure 2. Comparison of experimental photo (up) and the predicted vapor fraction ISO surface (bottom, Vapor fraction=0.1), $P_n \approx 400$kPa

### 4.2. *The vorticity in the mixing layer*

Above results show that the cavitation in jet pump is different from the wall attached cavitation since cavitation usually appear in the mixing layer. The turbulence behaviour in the shear layer has great influence on the local pressure, thus on cavitation. The pressure and contours of the vorticity in the throat are shown in Fig.4 for Condition A to C.

The ISO surfaces of pressure and vapor fraction form ring near the nozzle outlet. With the increase of the distance to the nozzle, both become fragmented and distributed in a random way. Obviously the high vorticity in the mixing region cause dramatic local pressure decrease. Meanwhile the overall mean pressure also influences local pressure level. Thus, with condition A, the overall mean pressure is the lowest in the throat because the outlet pressure is the lowest. The cavitation region extends much longer to the downstream than with Condition B and C. When the flowrate ratio decrease to condition B, the vorticity magnitude varies a little but the overall mean pressure level increases a lot. The cavitation thus decline. On the other hand, with condition C, even the overall mean pressure is the highest, local pressure decrease is much dramatic than with other 2 cases because the velocity difference and vorticity are high. Thus the cavity is short and the bubble is limited to the region near the nozzle. In Cai's experiments (Cai, 2005), the cavitation on condition A was referred as pressure-leading, while on Condition C velocity-leading .

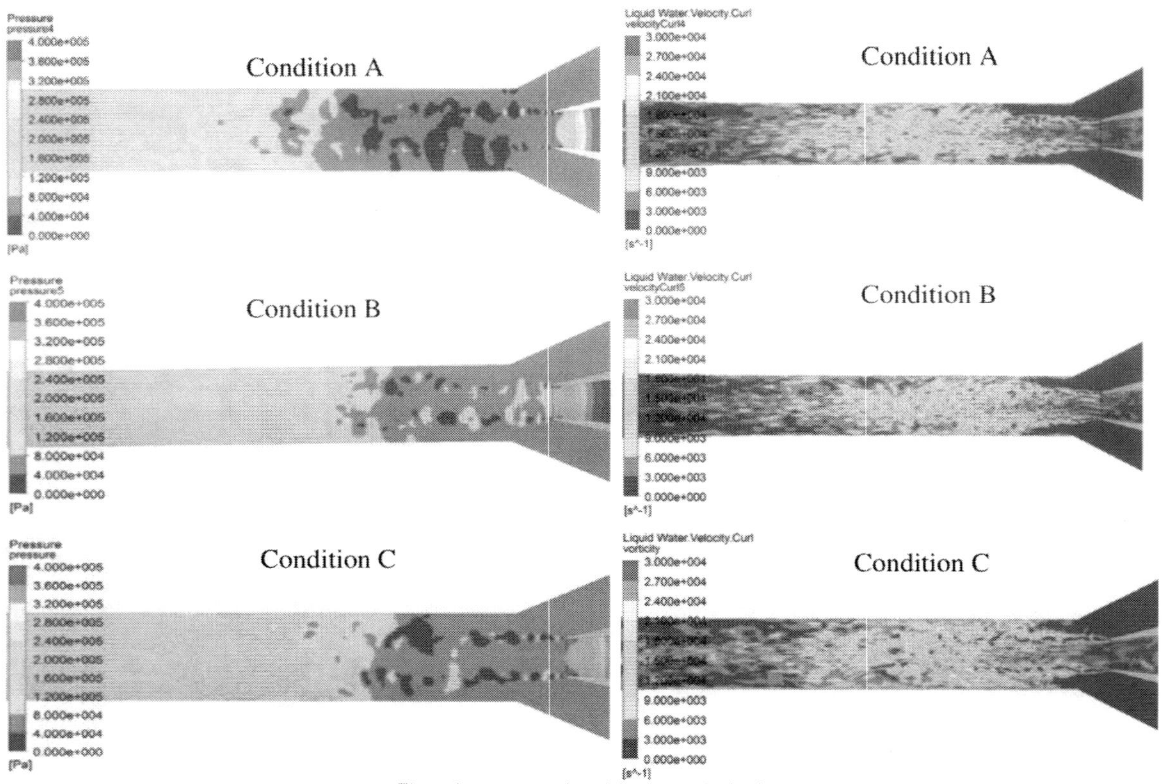


Figure 4. pressure and vorticity contour in the throat

## 5. Conclusions

In this work, the LES WALE mothed coupled with ZGB cavitation model has been employed to predict cavitation flows in jet pump. The results agreed well with experimental data. The numerical methods in this study are convincing. Both results shows that the increase of the inlet pressure generally increases the Renolds number but decrease the cavitation number, thus aggravate cavitation. The closing of the outlet valve increase the outlet pressure but decrease the flowrate ratio, resulting in the increase of velocity difference and vorticity in the mixing layer. So the cavitation first declines and then grows. The cavities appear slender and extended longer in the throat with high flowrate ratio. Conversely, the cavities look short and located in the front part of the throat with low flowrate ratio. Flow analysis indicated that the pressure field in the shearing layer is dependent on both overall mean pressure and local shear strain rate, which reveal the reason of different cavitation shape observed in experiment.

## Acknowledgments

The authors gratefully acknowledge the support from China Scholarship Council and National Natural Science Foundation of China (No. 51279205, 51279083).

## References

[1] Cai B H 2005 *Analysis and experimental study on incipient cavitation within jet pumps* (Wuhan: Wuhan University)
[2] Long X P, Han N, Chen Q 2008 *J. Mech. Scie. Tech.* **22** 1959-65
[3] Rafet Y, Aldaş K 2013 *Proc. IMechE Part A: J. Power and Energy.* **227** 438-49
[4] Ji B, Luo X W, Peng X X, Wu Y L 2013 *J. Hydr.* **25** 510-9
[5] ANSYS CFX Release11.0 2006 *ANSYS CFX-Solver Theory Guide* (Canonsburg: ANSYS, Inc.)

Journal of Physics: Conference Series **656** (2015) 012143 doi:10.1088/1742-6596/656/1/012143

# An algorithm for fast DNS cavitating flows simulations using homogeneous mixture approach

**A Žnidarčič[1], O Coutier-Delgosha[1], M Marquillie[2] and M Dular[3]**

[1] Laboratoire de Méchanique de Lille, Boulevard Paul Langevin, 59655 Villeneuve dAscq, France

[2] Le Centre de Ressources Informatiques (CRI), Université de Lille 1, Cité Scientifique, 59655 Villeneuve d'Ascq, France

[3] Laboratory for water and turbine machines (LVTS), Faculty of mechanical engineering, Aškerčeva cesta, 1000 Ljubljana, Slovenia

E-mail: anton.znidarcic@ensam.eu

**Abstract.** A new algorithm for fast DNS cavitating flows simulations is developed. The algorithm is based on Kim and Moin projection method form. Homogeneous mixture approach with transport equation for vapour volume fraction is used to model cavitation and various cavitation models can be used. Influence matrix and matrix diagonalisation technique enable fast parallel computations.

## 1. Introduction

Cavitation-turbulence interactions raise many questions in cavitation research. Numerous experiments were performed to understand them better, with numerical work following and adding to the discoveries [1, 2].

Over the years, cavitation simulations improved significantly. Today, they are able to describe unstable [1, 2] or even compressibility effects [3]. Presently, the mostly used cavitation models utilise homogeneous mixture approach with incompressibly treated phases. Additional transport equation for vapour volume fraction is often introduced. Because of many simplifications in development of such models, and as RANS turbulence models normally accompany and interact with them, the accuracy of simulations with them is limited. Consequently also the possible models' improvement and the knowledge about the mentioned interactions. To overcome this, cavitation-turbulence interactions in such simulations should be better analysed and DNS simulations seem a good tool for the task. There were some DNS simulations already performed, but their objectives were different [1, 2, 3] and their algorithms accepted only certain models [1, 2]. Therefore a new algorithm, capable of performing fast simulations with different models, is currently developed in our laboratory.

The aim of this paper is a brief description of the new algorithm. At first, the MFLOPS-3D code, representing a basis for our work, is described. Then, the new algorithm is presented through the equations, which up to now offered the most stable performance. At the end, some verification results are given.

Journal of Physics: Conference Series **656** (2015) 012143 doi:10.1088/1742-6596/656/1/012143

## 2. MFLOPS-3D code

As intention is to use various cavitation models in computationally heavy simulations, a code capable of fast DNS simulations is needed. The MFLOPS-3D code performs fast DNS incompressible flow simulations and was therefore taken as a basis for our development. Here, a brief description of it follows and more information can be found in [4].

Governing equations solved in this code are non-dimensional Navier-Stokes continuity and momentum equations. Non-incremental form of Kim and Moin pressure correction projection method is used to form the algorithm. Simulations are performed in parallel, where influence matrix technique ensures continuity of solutions over whole domain, while systems of equations in sub domains are solved with matrix diagonalisation (eigendecomposition). As a result, the code solves only Helmholtz or Poisson equations which have constant left hand side. Consequently the most time consuming procedures of matrix diagonalisation and influence matrix technique are done only once, at the start of a simulation. Only a diagonal system of equations remains to be solved in each time step. All of this combined enables faster DNS simulations.

Code utilises structured and collocated grids. Spatial discretization is done with compact finite differences, while time derivatives are described with $2^{nd}$ order backward difference scheme (BDS). $2^{nd}$ order Adams-Bashforth scheme is used for some explicitly treated variables.

## 3. New alogrithm for simulations of cavitating flow

### *3.1. The governing equations*

Description of cavitation with homogeneous mixture approach uses Navier-Stokes momentum (1) and continuity (2) equations and transport equation (3) for vapour volume fraction $\alpha$.

$$\rho\left(\frac{\partial\vec{u}}{\partial t}+(\vec{u}\cdot\nabla)\,\vec{u}\right)=-\nabla p+\nabla\cdot\left(\mu(\nabla\vec{u})+\mu(\nabla\vec{u})^T\right)-\frac{2}{3}\nabla\left(\mu(\nabla\cdot\vec{u})\right) \tag{1}$$

$$\frac{\partial\rho}{\partial t}+\nabla\cdot(\rho\vec{u})=0 \tag{2}$$

$$\frac{\partial\rho_v\alpha}{\partial t}+\nabla\cdot(\rho_v\alpha\vec{u})=S \tag{3}$$

$S$ in equation (3) is the source term, describing destruction or creation of $\alpha$. It depends on pressure $p$ and $\alpha$. The relationship between the density $\rho$ and viscosity $\mu$ of the mixture and both present phases is given in equation (4). Subscripts $v$ and $l$ depict vapour and liquid phase.

$$\rho=\alpha\rho_v+(1-\alpha)\rho_l\ ,\ \mu=\alpha\mu_v+(1-\alpha)\mu_l \tag{4}$$

### *3.2. Developed algorithm for solving presented system of governing equations*

Introduction of new governing equations demanded changes to the projection method in MFLOPS-3D code. The form of Kim and Moin method was retained, with the solution procedures for predictor velocity $\vec{u}^*$ and intermediate variable $\Phi$ heavily changed and additional step to obtain $\alpha$ added. In this paper, only the most stable algorithm is described.

Firstly, $\vec{u}^*$ is obtained. Equation for it is given in (5) and was, as in Kim and Moin method, developed from equation (1) by replacing $\vec{u}^{n+1}$ in time derivative and viscous terms with $\vec{u}^*$ [5].

$$\left(\Delta-\frac{3\rho_l}{2\Delta t\mu_l}\right)\vec{u}^*=\rho^n\frac{-4\vec{u}^n+\vec{u}^{n-1}}{2\Delta t\mu^n}+\frac{\rho^n}{\mu^n}((\vec{u}\cdot\nabla)\vec{u})^{n,n-1}-\frac{1}{\mu^n}(\nabla\mu^n)\cdot(\nabla\vec{u}^n)-\frac{1}{\mu^n}(\nabla\mu^n)\cdot(\nabla\vec{u}^n)^T$$
$$-\frac{1}{3}\nabla(\nabla\cdot\vec{u}^n)+\frac{2}{3\mu^n}\nabla\mu^n(\nabla\cdot\vec{u}^n)+\frac{\nabla p^n}{\mu^n}+\left(\frac{3\rho^n}{2\Delta t\mu^n}-\frac{3\rho_l}{2\Delta t\mu_l}\right)\vec{u}_e^* \tag{5}$$

Equation has many explicitly treated terms ($n$ denotes time level and $n, n-1$ use of Adams-Bashforth scheme). To lower their amount and provide constant left hand side, Concus and Golub method (CG) is used [6]. This gives the last term on the right hand side of equation (5) and demands iterative steps until $\vec{u}^*$ and its explicit value $\vec{u}_e^*$ converge. Pressure term is included as the most stable procedure is based on pressure incremental form of projection method.

$\Phi$ is solved next. Its equation is the difference between equations (1) and (5) and is given in equation (6) in most stable form (form, with $\rho$ included in $\Phi$, is also possible). Solving for $\Phi$ proved to be the most difficult part and was specially formed because of the code's demand for the constant left hand side. After applying divergence to equation (6), the unknown $\nabla \cdot \vec{u}$ is, as in [7], replaced with $S\left(1/\rho_v - 1/\rho_l\right)$, where $S$ is linearised. The best solutions are obtained if $S$ is linearised in regards to $p$ and $\alpha$, giving multiple terms including pressure change $dp$ (also used in $dS/d\alpha$). $dp$ is replaced with $\phi$ using $p - \Phi$ connection in equation (7). Finally, by using CG method again, this leads to more implicit and stable solving for $\Phi$. Equation for $\Phi$ solution is given in equation (8), where $\sigma$ is the constant introduced by the CG method. The equation is solved until $\Phi$ converges.

$$\frac{3\rho}{2\Delta t}\left(\vec{u}^{n+1} - \vec{u}^*\right) = -\nabla\Phi \tag{6}$$

$$p^{n+1} = \Phi + \mu\left(S\left(\frac{1}{\rho_v} - \frac{1}{\rho_l}\right) - \nabla\cdot\vec{u}^*\right) + p^n \tag{7}$$

$$(\Delta + \sigma)\,\Phi^{k+1} = -\frac{3}{2\Delta t}\left(\left(\vec{u}^k - \vec{u}^*\right)\cdot\nabla\rho^n + \left(\left(S^n + \frac{\partial S}{\partial p}dp + \frac{\partial S}{\partial \alpha}d\alpha\right)\left(\frac{1}{\rho_v} - \frac{1}{\rho_l}\right) - \nabla\cdot\vec{u}^*\right)\right) + \sigma\Phi^k \tag{8}$$

$k$ in equation (8) denotes iteration level. $\vec{u}^{n+1}$ is updated after each solution using equation (6). In linearised term $S$, only $\Phi$ values change between iterations, while $S$ as a whole is updated with converged $\Phi$. Next step is the solution of $\alpha$ equation. This is solved in each point separately, as it cannot be reshaped into Helmholtz equation. Equation (9), obtained from equation (3), is used for this. Iterative steps are performed, in which $\alpha^{n+1}$ and $S^{n+1}$ are updated, $k$ again denotes iteration level. Finally, $\rho$ and $\mu$ are updated using equation (4).

$$\alpha^{n+1,k} = \frac{\frac{S^{n+1,k-1}}{\rho_v} + \frac{\alpha^n}{\Delta t} - \left(\vec{u}^{n+1}\cdot\nabla\right)\alpha^{n+1,k-1}}{\frac{1}{\Delta t} + S^{n+1,k-1}\left(\frac{1}{\rho_v} - \frac{1}{\rho_l}\right)} \tag{9}$$

Boundary conditions, especially for $\Phi$, are an important factor. For $\Phi$, MFLOPS-3D code originally uses von Neumann boundary conditions. These are unsuitable for cavitation as they impose compatibility condition for $\Phi$. Mixed boundary conditions were implemented instead, in form of Dirichlet boundary condition as pressure value on the exit (using $\Phi$ gave unstable simulations) and von Neumann conditions elsewhere. Boundary conditions for $\vec{u}^*$ follow same logic as those in the original code, while $\alpha$ demanded use of Dirichlet conditions.

## 4. Verification results

The algorithm was mainly tested with a verification procedure using the Method of Manufactured Solutions [8]. Analytical equations describing the flow with variable density and corresponding source term $S$ were developed to see if the systems of governing equations are solved correctly. It was found that the algorithm indeed produces correct results. To test its limits, we compared it with an algorithm using Dirichlet boundary conditions for $\Phi$, enabling ideal stability. It was found out that both algorithms have same stability. The only difference was in the speed, with the ideal conditions making the code faster by roughly 30 %.

An example of solutions, obtained with the presented algorithm, and comparison with analytical values, is given on figure 1. In this case, the range of $\alpha$ was $\alpha = \{0; 1\}$ and $S$ was taken as $S = -\rho_v \alpha p$. Used domain was a cuboid with $x \in \{-0,2;0,2\}$ $m$, $y \in \{-0,2;0,2\}$ $m$ and $z \in \{-0,1;0,1\}$ $m$. It was split into four sub domains, two in $x$ and two in $y$ directions. Grid was non-uniform with 41 points in $x, y$ directions and 11 points in $z$. 1800 time steps with increment $\Delta t = 0,014$ $s$ were performed.

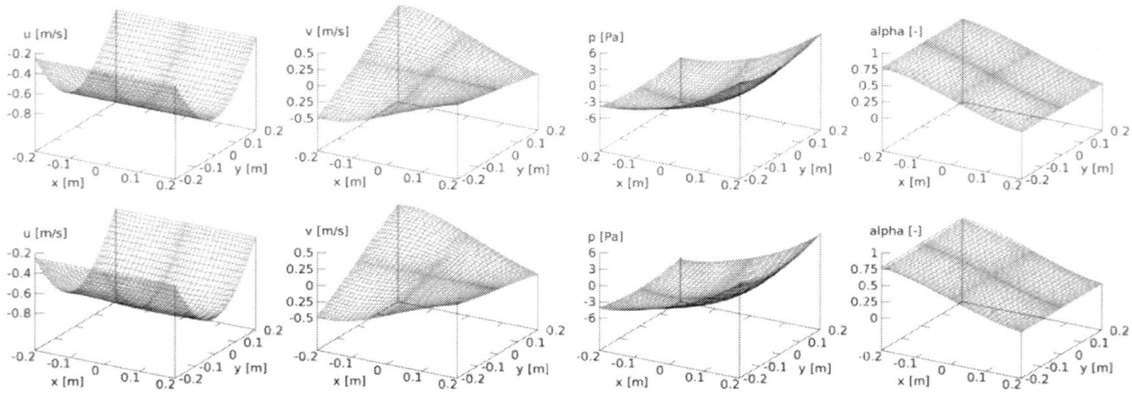


**Figure 1.** Comparison of analytical (above) and calculated (below) results for, from left to right, $u$,$v$ velocity, pressure $p$ and $\alpha$ at $z = 0$ $m$.

Results show very good agreement of analytical and calculated values. This case is actually limiting case still stably solved with described settings (coarser grid or longer time step result in unstable simulations). Errors were estimated using time averaged $L^2$ norm, where squares of the differences between analytical and computed values were divided (averaged) by number of points in the domain. Reported errors magnitudes are $10^{-3}$ for velocities, $10^{-1}$ for pressure and $10^{-2}$ for $\alpha$. Convergence tests show that MFLOPS-3D code with presented algorithm can reach $2^{nd}$ order accuracy. In case of incompressible flow tests (original code), a bit better convergence is achieved, but accuracy still tends to $2^{nd}$ order. This is mainly a consequence of using $2^{nd}$ order BDS for time derivatives. We therefore conclude the algorithm is ready for real flow simulations.

## 5. Conclusion

A new algorithm for fast DNS cavitating flow simulations is presented. The algorithm is constructed to enable use of different cavitation models and it has proven to be able to solve the system of governing equations correctly. The algorithm will next be tested in real flow cases, which has not yet been done because of high CPU demands.

## Acknowledgments

The work is funded by the DGA-Direction générale de l'armement - French government defence procurement agency and consortium CIRT-Consortium Industrie-Recherche en Turbomachine.

## References

[1] Kubota A, Kato H and Yamaguchi H 1992 *J. Fluid Mech.* **240** 59–96
[2] Xing T, Li Z and Frankel S H 2005 *J. Fluids Eng.* **127** 714–25
[3] Lu T, Samulyak R and Glimm J 2007 *J. Fluids Eng.* **129** 595–604
[4] Marquillie M, Laval J P and Dolganov R 2008 *J. Turb.* **9** 1–23
[5] Guermond J L, Minev P and Shen J 2006 *Comput. Methods Appl. Mech. Engrg.* **195** 6011–45
[6] Dimitropoulos C D and Beris A N 1997 *J. Compu. Phys.* **133** 186–91
[7] Yuan W, Sauer J and Schnerr G H 2001 *Mec.Ind.* **2** 383–394
[8] Eca L and Hoekstra M 2007 *Proc. 7th Int. Offshore and Polar Engineering Conf. (Lisbon)* vol 3 (Cupertino: International Society of Offshore and Polar Engineers (ISOPE)) p 2012

# A dynamic testbed for supercavitating vehicle control

**David Escobar, Devin Vollmer, and Roger Arndt**
St. Anthony Falls Laboratory, University of Minnesota, 2 Third Avenue SE, Minneapolis, USA
E-mail: descobar, voll0129, arndt001@umn.edu

**Abstract.** Supercavitating vehicles are capable of achieving unprecedented speeds due to the presence of a large gas cavity surrounding their bodies. The control design and validation of these vehicles is challenging because planing forces, oscillations, and instability arise when the vehicle pierces the supercavity. In this paper, we propose a methodology to test control algorithms for a supercavitating vehicle subject to planing. With a free-to-rotate scale vehicle in a high-speed water tunnel, we reproduce planing and are capable of assessing active control technologies experimentally. *Sponsored by the Office of Naval Research.*

## 1. Introduction

Supercavitation is a developed form of cavitation in which a large gas cavity is created behind an object that moves with respect to a fluid. An underwater vehicle surrounded by a gas cavity–supercavity–exhibits a decrease in skin friction drag; therefore, it is capable of achieving higher speed and lower power consumption than conventional vessels. High-speed transportation, ocean exploration, and defense are domains in which supercavitation provides unprecedented benefits. An example is the Ghost vessel [1] that travels above the water surface driven by two underwater supercavitating torpedoes.

We consider a vehicle that consists of a cylindrical body, a sharp disk cavitatior located at the vehicle front-end, and two lateral fins at the back-end. At the cavitator edges, the flow separates and the supercavity develops. Steering the vehicle is possible by rotating the cavtator and fins. A scale vehicle with this architecture is shown in Figure 1.

The advantages of supercavitation in terms of speed and power, come along with difficulties in the modeling, control, and experimental validation. In these three domains, the main challenge is the nonlinear interaction between the supercavity and vehicle body. When the vehicle body pierces the supercavity, suddenly a large force hits the vehicle back end. This force, referred to as *planing*, leads to oscillatory motion and sometimes instability. In this paper, we focus on an experimental method to validate control systems for the *longitudinal motion* of a vehicle subject to realistic flow conditions and planing.

Validating control schemes for a supercavitating vehicle in realistic flow conditions is particularly beneficial because the complex vehicle dynamics may not be fully described with a computational model. However, there is a lack of affordable small-scale experimental methods in the open literature to meaningfully validate control strategies for a supercavitating vehicle. In previous work [2], we developed a test method to evaluate control systems for the non-planing dynamics of a supercavitating vehicle. In this article, we propose an approach to validate control

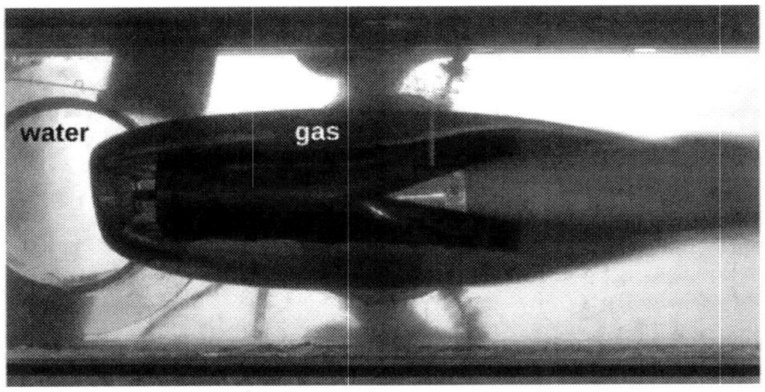


Figure 1: Experimental scale vehicle at the St. Falls Laboratory, University of Minnesota

algorithms subject to the most critical phenomena: planing, oscillations, and instability. This approach could alert engineers of possible pitfalls in the mathematical vehicle model and control algorithms before undersea testing.

## 2. Experimental methods

We conceived a methodology to conduct realistic control experiments in the high-speed water tunnel located at the St. Anthony Fall Laboratory. The main idea is that a free-to-rotate scale vehicle naturally achieves planing as the undersea counterpart. Conducting experiments with such a vehicle provides insight into the strengths and drawbacks of control approaches. Although the test method does not reproduce the exact unconstrained motion of a supercavitating vehicle traveling undersea, it captures with exactitude all the hydrodynamic forces. And most importantly, it exhibits the complex nonlinear interaction between the vehicle body and supercavity that originates planing, oscillations, and instability.

Figure 2a depicts our implementation of the proposed validation platform. A small scale vehicle is employed. It is equipped with a ventilation system that enables the formation of supercavities at speeds above 3 m/s. The test vehicle is a cylinder of 50 mm diameter and 148 mm length with an interchangeable disk cavitator and two interchangeable lateral fins. The fins and cavitator are capable of deflecting $\pm 20$ deg and $\pm 15$ deg respectively. The weight of the vehicle body is 1 Kg. The vehicle rotation, equal to the vehicle attack angle, is measured using a rotary encoder. By translating the vehicle along a slit plate, the vehicle center of rotation and dynamic behavior can be manually adjusted. Between the vehicle and encoder, a force transducer is attached. This transducer rotates together with the vehicle and therefore forces and moments are measured with respect to the vehicle axes.

Real-time measurements of forces, moments, and vehicle attack angle are continuously transmitted to a data acquisition and control computer ($DAQC^2$). The force and torque measurements are transmitted to the $DAQC^2$ via an Ethernet interface in a local area network. The encoder data are read by an NI PCI-6902 data acquisition card. The same card is used to generate Pulse Width Modulated (PWM) signals that command the positions of the actuators attached to the cavitator and fins. By commanding the cavitator and fins, the algorithms running onto the $DAQC^2$ control the vehicle motion.

A high-speed camera is used to characterize the supercavity dynamics and analyze the experiments. Synchronization between the $DAQC^2$ and video streams is achieved using an Light Emitter Diode (LED) connected to the NI PCI-6902 card. When the control experiment starts, the LED is turned on. The software running on the $DAQC^2$ is implemented under MATLAB/Simulink and Real-Time-Windows-Target (RTWT) with a sampling frequency of 100 Hz. Key features of the proposed platform include:

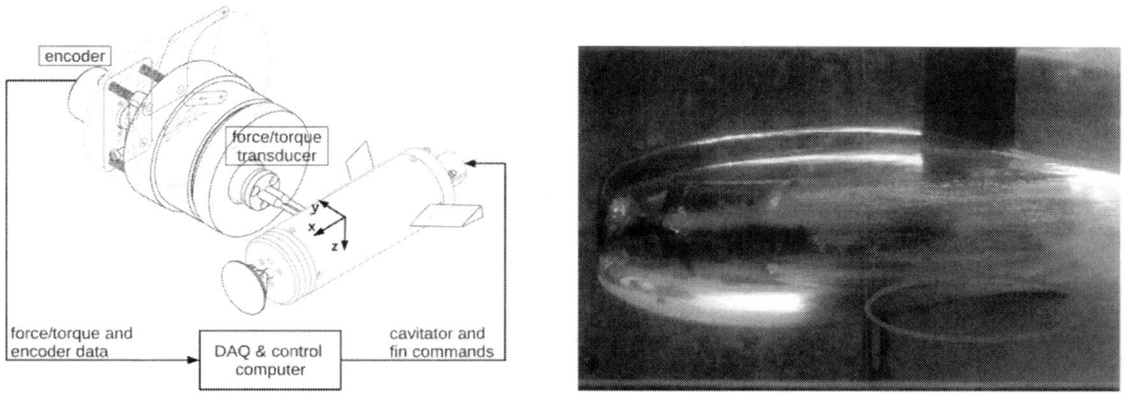


(a) Schematic (b) Induced lateral planing

Figure 2: Experimental Facility

- enables the modeling of hydrodynamic forces and supercavity dynamics
- provides an affordable solution to validate control schemes for supercavitating vehicles after design and before undersea testing
- recreates realistic flow conditions, oscillations, and instability

The main limitation of the platform is that the supercavity is pierced by the rotational shafts, leading to lateral planing. Figure 2b illustrates the induced planing regions. Lateral planing slightly damps the vehicle rotations and partially attenuate the effect of planing forces. Yet, the system exhibits oscillations and instability due to planing, as desired to assess controllers.

## 3. Vehicle dynamics and controls

The dynamics of both the unconstrained vehicle traveling undersea and the experimental vehicle in the tunnel share the same structure. Both dynamics are driven by the forces at the cavitator, fins, and planing regions, as well as the gravity. Figure 3 depicts a general scheme of the vehicle dynamics $G$ composed of the supercavity and body dynamics $G_s$ and $G_b$ respectively. $G$ is described by a nonlinear differential equation. According to the design method, the controller $K$ may be described by a static gain or a linear/nonlinear differential equation. Its main objective is to drive the vehicle to a desired state by using the cavitator and fin commands $u$. For the test vehicle in the tunnel, the goal of $K$ is to track attack angle commands. In this case the measurement $y$ is equal to the vehicle attack angle $\alpha$.

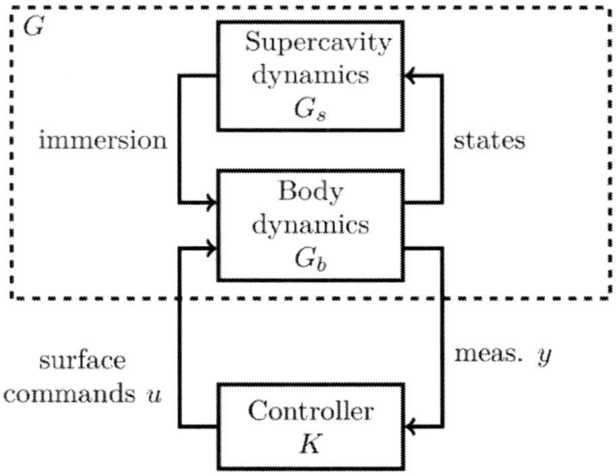


Figure 3: Vehicle and controller dynamics

## 4. Experiments

The main objective of the experiments we conducted was to asses diverse control schemes that we designed. An overall procedure to evaluate a controller consists of: (i) creating a mathematical model $G$ of the vehicle and supercavity motion, (ii) synthesizing a controller $K$, (iii) designing the attack angle commands to assess the controller, and (iv) conducting the experiment in the water tunnel. We carried out two experiments to evaluate how the control systems perform under realistic flow conditions. A controller $K_1$ was designed to track reference commands $r$ so that $r - \alpha$ is minimized at all times. This controller was designed with a model that assumes no planing. A controller $K_2$ was designed to track commands $r$ while guaranteeing performance in the face of planing. Both controllers were evaluated using step-up and step-down commands from -2 to 4 deg. When the vehicle rotates towards 4 deg, planing emerged.

## 5. Results

The results of the aforementioned experiment are presented in Figure 4. The top plot shows the responses of the vehicle attack angle to reference commands with controllers $K_1$ and $K_2$. The controller $K_1$ exhibits oscillations when the vehicle is subject to planing. These oscillations are minimized by $K_2$, designed to operate under planing conditions. In the middle plot, the cavitator deflections are presented. Interestingly, $K_2$ demands larger cavitator deflections than $K_1$, to move the supercavity and thereby minimize planing immersion. Since $K_2$ mostly uses the cavitator for control, it requires smaller fin deflections than $K_1$. The bottom plot illustrates the fin deflections. Our results suggest that including planing within the controller scheme significantly improves its performance.

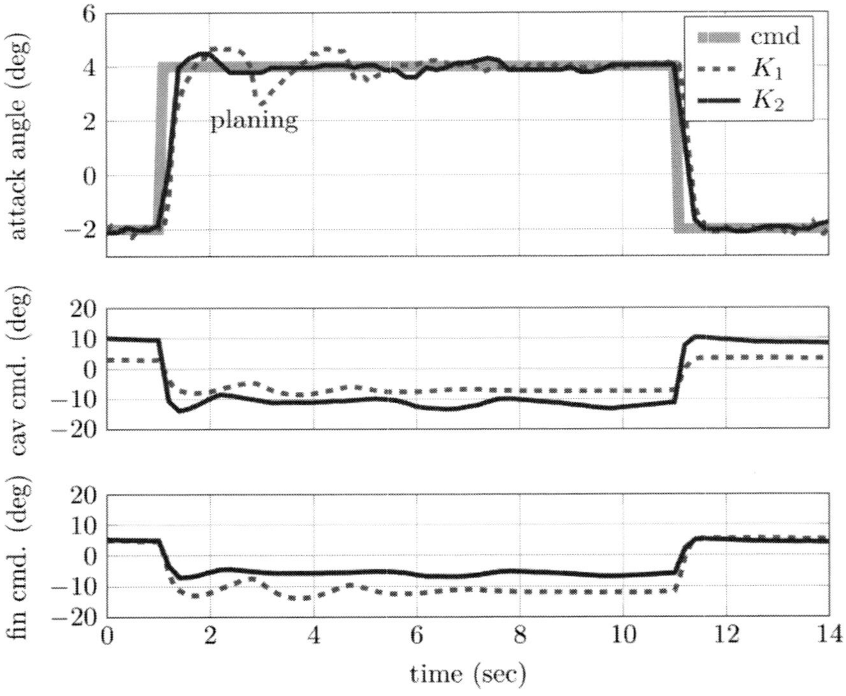


Figure 4: Experimental data with controllers $K_1$ and $K_2$

## 6. Summary and conclusions

We have shown that a free-to-rotate supercavitating vehicle in a high-speed water tunnel can be employed to assess control technologies for a supercavitating vehicle. This method is an affordable validation approach for research and development.

## References

[1] Inc J M S Ghost vessel description URL `http://www.julietmarine.com`
[2] Escobar D, Balas G and Arndt R E 2014 *Oceanic Engineering, IEEE Journal of* **PP** 1–12 ISSN 0364-9059

# Closure mechanisms of ventilated supercavities under steady and unsteady flows

**Ashish Karn[1,2], Rohan De[3], Jiarong Hong[1,2] and Roger E A Arndt[1,*]**,
[1] Saint Anthony Falls Laboratory, University of Minnesota, Minneapolis, MN , USA.
[2] Department of Mechanical Engineering, University of Minnesota, Minneapolis, MN, USA.
[3] Dept. of Aerospace Engineering and Mechanics, University of Minnesota, Minneapolis, MN
E-mail: arndt001@umn.edu

**Abstract.** The present work reports some interesting experimental results for ventilated supercavitation in steady and unsteady flows. First, a variety of closure modes obtained as a result of systematic variation in Froude number and air entrainment, are reported. The closure mechanisms were found to differ from the standard criterion reported in the literature. Further, the occurrence of a variety of stable and unstable closure mechanisms were discovered that have not been reported in the literature. Next, a hypothesis is presented to explain the cause behind these different closure mechanisms. The proposed hypothesis is then validated by synchronized high-speed imaging and pressure measurements inside and outside of the supercavity. These measurements show that the supercavity closure is a function of instantaneous cavitation number under unsteady flow conditions. (Research sponsored by *Office of Naval Research*, USA)

## 1. Introduction

Supercavitation is a technique for achieving high speeds underwater by enveloping a body in a gaseous cavity. It is of great practical interest owing to its advantages in the drag and noise reduction for high-speed underwater vehicles. A supercavity can be formed at lower speeds by injecting non-condensable gas, typically air downstream of the cavitator. This is commonly referred to as ventilated supercavitation. The parameters that characterize ventilated cavities are cavitation number, $\sigma = 2(P_\infty - P_C)/(\rho_w U^2)$; Froude number, $Fr = U/\sqrt{g d_c}$, and air entrainment coefficient, $C_Q = \dot{Q}/U d_c^2$, where $P_\infty$ and $P_C$ refers to the test-section pressure upstream of the cavitator and cavity pressure respectively, $\rho_w$, $U$ and $g$ correspond to liquid density, the free-stream velocity in the test-section and gravitational acceleration, respectively, $d_c$ denotes the cavitator diameter, and $\dot{Q}$ is the air ventilation rate. In understanding the mechanism of air entrainment, the study of supercavity closure, i.e. how a supercavity closes in the rear portion, is particularly relevant, since a majority of ventilated gas leaks from the closure region. The ventilation demand to form and sustain a supercavity is dependent upon the closure mode of supercavity under different flow conditions.

Four types of closure mechanisms have been reported in the literature. These include re-entrant jet, twin-vortex, quad-vortex, and pulsating modes. Three modes have been thoroughly reviewed by Franc and Michel (2005). Based on an empirical relation from Campbell and Hilborne (1958), the re-entrant jet mode should occur when the product $\sigma_c Fr > 1$ while the twin-vortex mode is expected to appear at $\sigma_c Fr < 1$. However, Kawakami and Arndt (2011) observed twin-vortex mode as the dominant closure mechanism in their experiments where $\sigma_c Fr$ was always much greater than one, where according to other authors the re-entrant jet regime would be expected. The third closure mechanism, i.e. quad-vortex mode, was first reported by Kapankin and Gusev (1982). This mode consists of four vortices situated in pairs, one pair above the other. According to Kapankin and Gusev (1982), this mode occurs when the cavitator angle of attack was less than some critical angle which is a function of $U_\infty$ (Tunnel

* To whom any correspondence should be addressed.

velocity), and the drag coefficient of the cavitator at zero cavitation number. Although the quad-vortex closure has been observed in the experiments at SAFL (Kawakami and Arndt 2011), the conditions of this closure did not match those predicted by Kapankin and Gusev (1982). Moreover, neither of the studies has provided clear physical explanations of the occurrence of quad-vortex closure mode. Finally, the pulsating mode refers to a particular unsteady closure phenomenon that occurs in a pulsating supercavity, which commonly occurs at extremely high ventilation flow rate (See details in §2.2.3).

Besides the existing inconsistency in the onset criteria for different closure mechanisms, a recent investigation on the closure mechanism at SAFL (see details in §2.3) reveals many new closure modes which are not readily interpreted and reconciled with prior literature based on existing empirical relations involving $\sigma_c$, $Fr$ and $C_Q$. The lack of agreement among various studies is commonly attributed to the differences in experimental procedure such as the tunnel type and blockage ratio. These limitations on our ability to explain the observed physical phenomena, including the change of flow conditions leading to transition between different closure modes, indicates a deficiency in our current understanding of the supercavity closure. The present work is aimed at bridging this gap in the current understanding of supercavity closure.

## 2. Experimental Methodology

In this work, the variation of closure modes of ventilated supercavitation is investigated by systematically varying $Fr$ and $C_Q$ for a range of cavitator sizes. This experiment uses a backward facing cavitator model (Figure 1). In the backward facing model, the strut is placed upstream of the cavitator to avoid the interaction between the formed cavity and the strut body which was shown to affect the supercavity closure (Kawakami and Arndt 2011). The experiments were conducted with 10, 20, 30 and 40mm-diameter cavitators to investigate the variation of closure modes over a wide range of parameters with $Fr$ varying in the range of 5-40 and $C_Q$ of 0.001-10 and $B$=5%, 9%, 14% and 19%. The experiments on unsteady ventilated supercavitation are conducted with the aid of a gust generator located upstream of the cavitator. The gust generator consists of two oscillating hydrofoils, which are placed upstream of the cavitator at a distance 180 mm. These two hydrofoils are oscillated in phase by the system's pivot arm to generate uniform gusts. This pivot arm is linked to a flywheel through a connecting arm, which extracts the periodic motion from the motor. An eccentric flywheel can allow for gusts of varying amplitudes. The detailed description of gust generator mechanism is provided in Kopriva *et al* (2008). High speed imaging is conducted and synchronized with pressure measurements inside the cavity and test section. The experiments were conducted at a fixed $Fr$ = 20 and $C_Q$ = 0.03 with a cavitator diameter of 10 mm to minimize blockage effects. The gust amplitude is kept fixed for all the experiments and the gust frequencies were varied between 1 – 10 Hz. The pressure measurements and high speed videos were captured at a sampling rate of 1500 Hz.

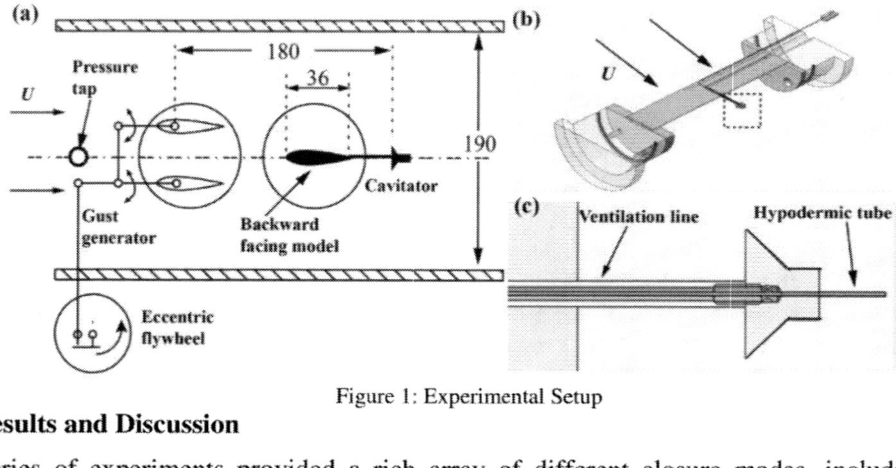


Figure 1: Experimental Setup

## 3. Results and Discussion

This series of experiments provided a rich array of different closure modes, including several modes that were not reported in the literature, as illustrated in Figure 2. In general, the closure modes are divided into stable modes and unstable modes. The four stable closure modes are Re-entrant Jet (RJ), Twin-vortex (TV), Quad-vortex (QV) and Foamy Cavity (FC) (Figure 2a - d). Note that FC is strictly not a closure mode, but it is one of the states in which a supercavity might exist. The unstable closure modes include Hybrid QV-RJ (QVRJ), Hybrid TV-QV (TVQV), Pulsating Twin vortex (PTV) and Interacting Vortex (IV). These unstable closure modes were either observed at the transition of two stable closure modes or at extremely high ventilation rates. QVRJ closure, formed at the transition of QV and RJ closure has the features of both QV and RJ modes and the cavity has a foamy end with four vortices. TVQV closure refers to the transition between QV and TV modes and may have one to four vortices. It was found that the bottom two vortices were more stable than the upper two, however sometimes even one of the two lower vortices disappears. Similarly, a TVRJ closure, characterized as the supercavity with foamy rear and two vortices, was also observed. Finally, at very high flow rates, PTV could be seen, where both the vortices pulsate in longitudinal direction, and closure may intermittently break into elongated pockets of air because of severe instabilities on the cavity surface. At similar flow conditions, but at a higher blockage, IV mode was observed, where the two vortices would interact with each other to form one single thick vortex at the closure. In our experiments, TV and RJ mode does not obey the Campbell-Hilborne criterion. Also, QV is not observed at the condition reported by Kapankin and Gusev. Clearly, no empirical criterion for other observed closure modes have been reported in the literature.

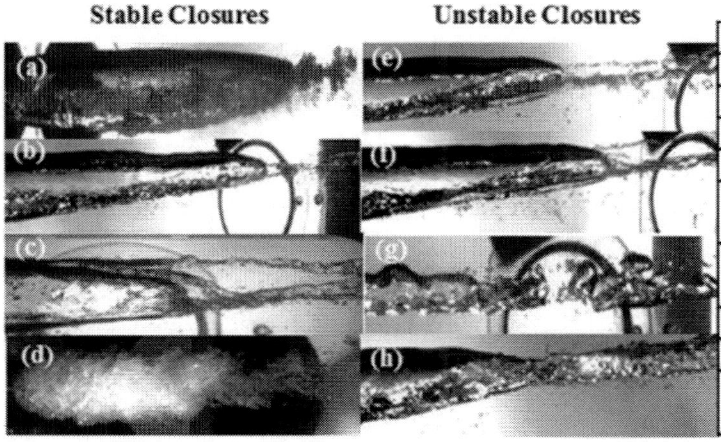


| Acronym | Name |
|---|---|
| RJ | Re-entrant Jet |
| TV | Twin vortex |
| QV | Quad vortex |
| FC | Foamy cavity |
| QVRJ | Hybrid quad vortex-re-entrant jet closure |
| TVQV | Hybrid twin vortex-Quad vortex closure |
| PTV | Pulsating twin vortex |
| IV | Interacting vortex |
| TVRJ | Hybrid twin vortex-re-entrant jet closure |

Figure 2: Assembly of different closure modes observed: (a) RJ (b) TV (c) QV (d) FC (e) TVRJ (f) QVRJ (g) PTV (h) IV.

A hypothesis on the physical mechanism that determines the closure modes of a ventilated supercavity is proposed as follows in order to interpret the above-mentioned observations. The hypothesis posits that the closure mechanism is mainly determined by the pressure difference ($\Delta\tilde{P}$) across the gas-liquid interface at the cavity closure. The pressures are normalized as follows: $\Delta\tilde{P} = (P_C - P_{TS}) / P_d = \tilde{P}_C - \tilde{P}_{TS}$, where $P_d$ ($= 0.5\rho U^2$) is the dynamic pressure in the test section of the water tunnel, $\tilde{P}_C$ is the pressure inside the cavity at the closure and $\tilde{P}_{TS}$ represents the pressure just outside the cavity closure. It is worth noting that $P_C$ is not necessarily equal to the mean cavity pressure due to the possibility of pressure gradients due to internal flow circulation within the supercavity. The formation of a stable closure is a result of a balance among pressure difference, gas/liquid phase momentum across the interface, and other factors such as surface tension, which could contribute to subtlety of different vortex-based closure mechanisms. We hypothesize that the change of $Fr$, $C_Q$, $B$ or type of experimental facility leads to $\Delta\tilde{P}$ resulting in the variation of different closures. Spurk (2002) notes that the opening of the cavity (and thereby, cavity closure) is determined by the change of streamline curvature and the free streamline at the cavity closure. Our experiments reveal that the radius of curvature ($R$) of the streamlines at the supercavity closure follows this general trend: $R_{TV} > R_{QV} \gg R_{RJ}$. Using this radius inequality, we hypothesize that the criterion to obtain different closure follows this general trend: $\Delta\tilde{P}_{RJ} \gg \Delta\tilde{P}_{QV} > \Delta\tilde{P}_{TV}$. RJ is formed when $\tilde{P}_{TS}$ is significantly higher than $\tilde{P}_C$. As $\Delta\tilde{P}$ reduces to a critical value when it is no longer sufficient to sustain a re-entrant jet, the closure region transitions from RJ characterized by chaotic mixing flow pattern to vortex-based closures, from QV to TV gradually as $\Delta\tilde{P}$ keeps decreasing. To get an insight into the physical mechanisms that cause the change in cavity shape and closure, we conducted simultaneous pressure measurements both inside and outside the cavity in synchronism with the high-speed imaging of cavity behavior under different gust frequencies. To quantify the effect of unsteadiness, the pressure data are recorded at steady state ($Fr$ = 20, $C_Q$ = 0.03), under which RJ closure is present. The flapping of the gust generator causes fluctuation of test-section pressure which in turn leads to an oscillation in cavity pressure. Figure 3 shows the frequency response analysis of cavity pressure data and it is observed that the cavity pressure oscillates at a frequency close to the gust frequency, although sometimes a minor frequency mode can be seen.

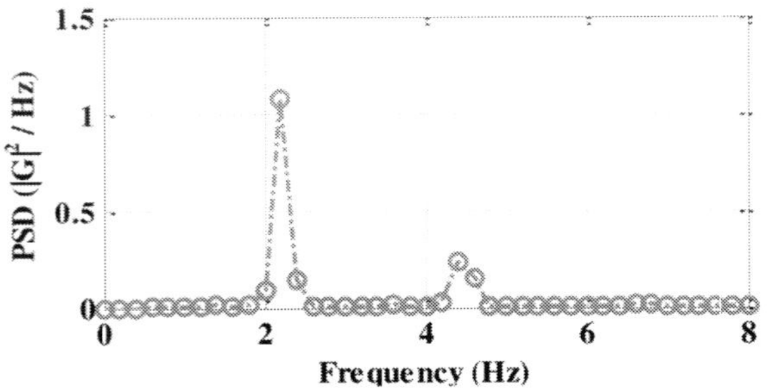


Figure 3: Power spectral density of cavity pressure data at a gust frequency of 2.2 Hz.

Figure 4 shows the periodic variation in normalized test-section pressure, normalized cavity pressure and their difference (i.e. cavitation number) over many gust cycles. Pressure information in the liquid and gas propagate at different speeds, resulting in a phase lag between the measurements of test-section and cavity pressure. This difference in phase of the test-section pressure and cavity pressure leads to the oscillating nature of cavitation number. Synchronized high-speed video and pressure measurements reveal that the cavity closure alternates between RJ and TV modes, with the RJ mode forming at instants when the cavitation number attains its maximum (0.2 - 0.22), while the TV mode is obtained when the cavitation number attains its minimum (~0.14). Thus, our experiments have demonstrated that cavity closure is dependent upon the pressure difference at the rear end of supercavity and can undergo transition even when none of the steady-state non-dimensional parameters, viz. $C_Q$, $Fr$ or blockage etc. in the flow has been altered. Simultaneous measurements of internal and external pressure at the rear portion of the cavity for different supercavity closure modes is required to further substantiate our proposed hypothesis.

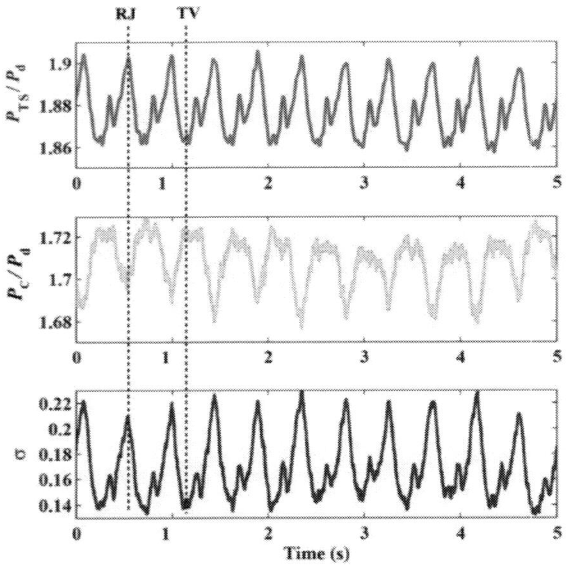


Figure 4: The periodic variation of normalized test-section pressure, cavity pressure and cavitation number at a gust frequency of 2.2 Hz.

## 4. Conclusions

In this paper, a variety of supercavity closures have been obtained by a systematic variation of Froude number and air entrainment. It is found out that the supercavity closure is also dependent upon the path taken. Further, a hypothesis is presented to explain the cause behind these different closure mechanisms. The proposed hypothesis is then validated by synchronized high-speed imaging and pressure measurements inside and outside of the supercavity. These measurements show that the supercavity closure is a function of instantaneous cavitation number under unsteady flow conditions.

## References

1. Campbell I J and Hilborne D V 1958 Air entrainment behind artificially inflated cavity," *Proceedings of the Second Symposium on Naval Hydrodynamics*, Washington.
2. Spurk J 2002 On the gas loss from ventilated supercavities *Acta Mechanica* **155** 125-135.
3. Kawakami E and Arndt R E A 2011 Investigation of the behavior of ventilated supercavities *J. Fluids Eng.* **133** (9) 091305-1 - 091305-11.
4. Kopriva J, Arndt R E A and Amromin E L 2008 Improvement of hydrofoil performance by partial ventilated cavitation in steady flow and periodic gusts *J. Fluids Eng.* **130** (3) 031301-1 – 031301-7.
5. Franc J P and Michel J M 2004 Fundamentals of Cavitation, Kluwer Academic Publishers, Dordrecht.
6. Kapankin Y N and Gusev A V 1984 *Proceedings of CAHI,* No. 2244, Moscow, Russia, pp. 19-28.

# Numerical 3D analysis of cloud cavitation shedding frequency on a circular leading edge hydrofoil with a barotropic cavitation model

**M Blume and R Skoda**

Chair of Hydraulic Fluid Machinery, Ruhr-Universität Bochum, Universitätsstr. 150, 44801 Bochum, Germany

E-mail: martin.blume@ruhr-uni-bochum.de

**Abstract.** A compressible density-based time-explicit low Mach number consistent viscous flow solver is utilised in combination with a barotropic cavitation model for the analysis of cloud cavitation on a circular leading edge (CLE) hydrofoil. For 5° angle of attack, cloud structure and shedding frequency for different cavitation numbers are compared to experimental data. A strong grid sensitivity is found in particular for high cavitation numbers. On a fine grid, a very good agreement with validation data is achieved even without explicit turbulence model. The neglect of viscous effects as well as a two-dimensional set-up lead to a less realistic prediction of cloud structures and frequencies. Comparative simulations with the Sauer-Schnerr cavitation model and modified pre-factors of the mass transfer terms underestimate the measured shedding frequency.

## 1. Introduction

Depending on geometry, fluid and flow conditions, different types of large-scale cavitation structures occur. One of the more aggressive types is cloud cavitation, which is characterised by a periodic separation of vapour sheets from a surface caused by a re-entrant jet and shedding of these vapour structures as clouds. This separation may also be influenced by shock waves from imploding vapour clouds [1]. Since cloud cavitation occurs in hydraulic machinery such as pumps, simple hydrofoil test cases are utilised to investigate cavitation models.

In the cavitation model of Schnerr and Sauer [2], referred to as Sauer-Schnerr model in the following, a mass transfer rate based on the Rayleigh equation is introduced as source/sink term in the void fraction transport equation and combined with an inviscid implicit flow solver. They determine a shedding frequency of 11 Hz for the NACA0015 hydrofoil at $\alpha = 6°$ and cavitation number $\sigma = 1.0$. For the same operating point and cavitation model, Li [3] and Oprea et al. [4] have found 11 Hz and 17 Hz as dominant frequency, respectively. Koop [5] and Sezal [6] apply a barotropic model and obtain 24 Hz and 9 Hz, respectively. Gosset et al. [7] compare different turbulence models in combination with the Sauer-Schnerr model for a wide range of operating points in 2D-simulations with one pre-factor $C_{cond} = 2$ in the mass transfer sink term. Frobenius [8] and Deimel et al. [9] have applied the Sauer-Schnerr model to a 2D circular leading edge (CLE) hydrofoil at $\alpha = 5°$ with a 2D computational set-up. Frobenius has restricted his investigations to cavitation numbers in a narrow range of $\sigma = 2.0$ to 2.2 and achieved a good agreement with measured shedding frequencies. Deimel et al. [9] have found that for higher cavitation numbers in contrast to the data no periodic shedding but steady sheet cavitation was predicted and introduced pre-factors in the source/sink term according to table 1 to obtain cloud cavitation in the entire range $\sigma = 2.0$ to 2.7. These exemplary cited and other CFD studies with mass transfer models based on a Rayleigh source/sink term show large scatter in shedding frequency. Dular et al. [10] apply a 2D implicit pressure-based solver on the CLE-hydrofoil and use a barotropic model with a smoothed density gradient to avoid too steep density variations and overcome numerical stability problems. A shedding frequency of about 60 Hz at $\sigma = 2.3$ is obtained. The present work aims at the assessment of a barotropic model in combination with a pressure wave resolving hyperbolic solver for prediction of shedding frequencies on the CLE-hydrofoil.

## 2. Method

For all simulations OpenFOAM® is utilised. A homogeneous mixture of liquid and vapour within each computational cell is assumed. For the Sauer-Schnerr model two empirical pre-factors $C_{cond}$ and $C_{vap}$ for the source/sink term in the vapour volume fraction equation enable calibration to specific flow problems, whose values have been adapted for CLE-hydrofoil simulations in a 2D numerical set-up in [9], as specified in table 1. The Sauer-Schnerr model is combined with an incompressible, time-implicit solver. The SST turbulence model [11] with a modification by Reboud [12] is used, which limits the eddy-viscosity in the two-phase region.

**Table 1.** Cavitation models.

| Model | Sauer-Schnerr [2] | barotropic [13] |
|---|---|---|
| Model type | mass transfer, Rayleigh equation | equilibrium, equation of state |
| Parameters | $C_{cond} = 4$, $C_{vap} = 16$ [9] | free of parameters |
| Num. solver | implicit, pressure-based | explicit, density-based [14] |
| Friction/turbulence | SST model [11] with limited eddy-viscosity [12] | without explicit turbulence model |

On the other hand a barotropic cavitation model [13] representing an isentropic path at thermodynamic equilibrium is applied as an equation of state $p(\rho)$ in combination with a compressible, density-based solver with low Mach number consistent flux functions and modified four-stage Runge-Kutta scheme with enlarged stability region, proposed by Schnerr et al. [14]. Density and velocity are reconstructed with the schemes MINMOD [15] and GammaV [16], respectively. This method accurately resolves the propagation of cavitation-induced pressure waves as shown by [5] for hydrodynamic cavitation and [17] for the prediction of erosion sensitive areas in acoustic applications. The OpenFOAM® implementation was done by [18].

## 3. CLE-hydrofoil

Böhm [19], Hofmann [20] and Bachert [21] have determined the shedding frequency of vapour clouds by means of high speed video analysis and FFT with a Hann-window of four pressure sensors on the suction side of the hydrofoil. The shedding frequency increases with cavitation number, with a low dependence on the Reynolds number, see figure 5. For our simulations we reproduce the test section geometry of [19] and choose a 5° angle of attack, see figure 1. Two grid resolutions, corresponding to O-Grid cell numbers of 196x35x50 and 392x70x100 in circumferential, wall-normal and spanwise direction, are investigated. The $y^+$ wall values are about 100 and 60, respectively. At the inlet a fixed velocity $u = 13\,\mathrm{m/s}$ corresponding to $Re = 1.3 \cdot 10^6$ is specified. The fluid is water.

A time step of $2 \cdot 10^{-6}$ s is chosen for the implicit time integration with the Sauer-Schnerr model and corresponds to a convective CFL number of 0.2 and a resolution of one shedding cycle with more than 3000 time steps. For the explicit time integration with the barotropic model the time step is limited to a maximum of $4 \cdot 10^{-8}$ s (acoustic CFL number $< 0.95$).

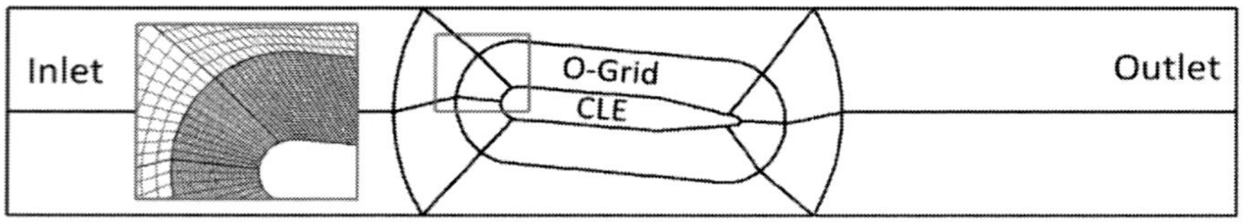


**Figure 1.** Block structure and zoom of coarse mesh (0.35 million cells).

Journal of Physics: Conference Series **656** (2015) 012146 doi:10.1088/1742-6596/656/1/012146

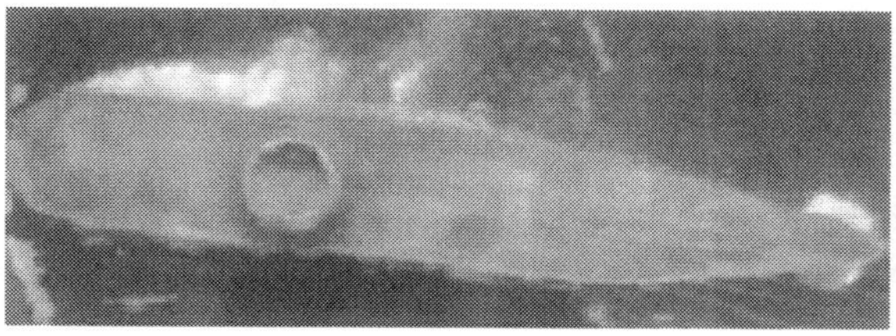

**Figure 2.** CCD image [21] of cloud structures at $\sigma = 2.5$.

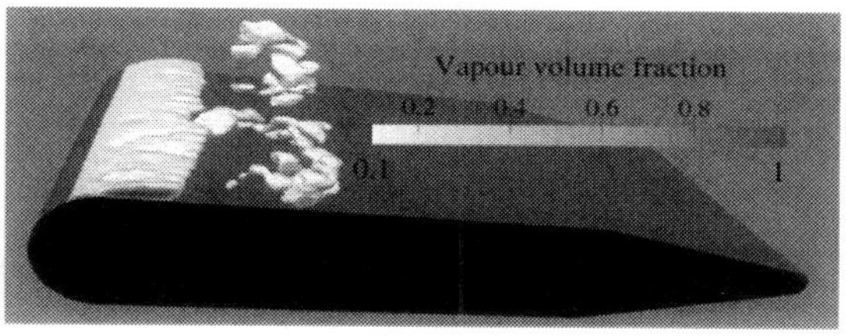


**Figure 3.** Exemplary time instant for a 3D Navier-Stokes simulation at $\sigma = 2.5$ with barotropic model on the fine mesh.

## 4. Results

For the assessment of the cloud structures in the simulation, iso-surfaces of void fraction are compared to integral data in spanwise direction from transmitted light images (see figure 2) as well as local data from light-section images [19]. The frequency is obtained in the same way as in the experiments applying a FFT to virtual pressure sensors [9], see figure 4. Since some ambiguity was found in particular for 2D results as already reported by [9], the FFT analysis is supported by a manual count of shedding structures.

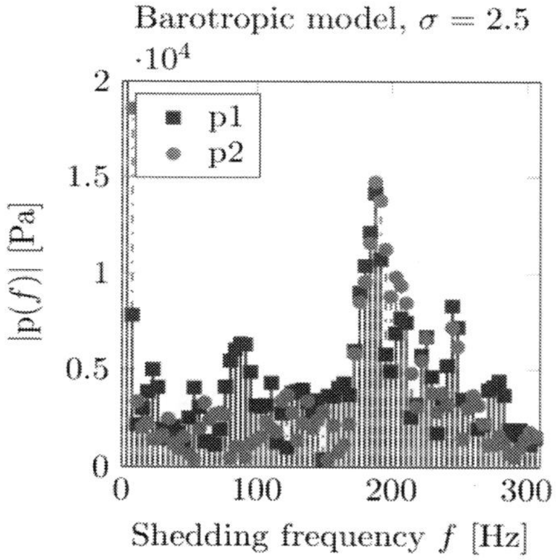


**Figure 4.** Exemplary FFT result of two virtual pressure sensors on the 3D coarse mesh.

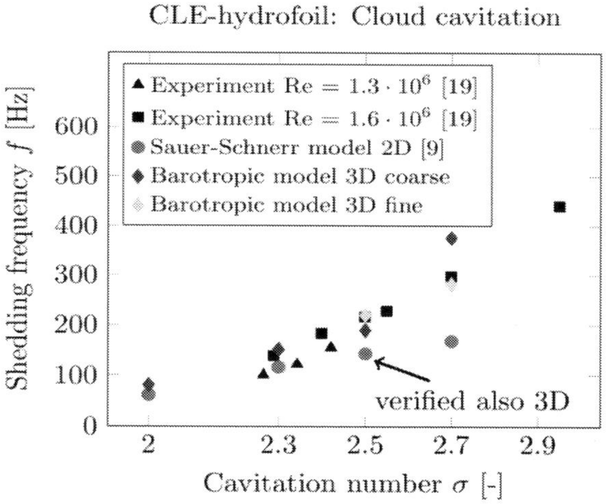


**Figure 5.** Frequency obtained with experiments and both cavitation models for different operating points.

For the barotropic model, after a relatively short initial-solution influenced time period, a quasi-periodic shedding occurs. By a time convergence study for $\sigma = 2.5$ on the coarse grid where up to 100 cycles have been calculated, a dependence of the shedding frequency on the number of evaluated cycles is found. Compared to the dominant frequency obtained after about 100 cycles, the deviation drops below 5 % for an analysis interval of 10 and more cycles. Due to the high calculation time on the fine grid, between 9 and 20 cycles are calculated for the barotropic model. We estimate that the frequency error is negligible on the coarse grid and lower than 7 % on the fine grid.

2D results by [9] are available for the Sauer-Schnerr model. Deimel et al. [9] have demonstrated that the fine grid of the present study yields grid-independent 2D results. In the present study 3D fine grid simulations for $\sigma = 2.5$ are performed to assess the influence of 3D effects. 3D structures develop in spite of the 2D geometry. However, the 3D-simulation yields only minor differences to the 2D set-up shedding frequency. On the other hand for 2D simulations with barotropic model and $\sigma = 2.5$, too low frequencies are found in combination with shedding clouds that are too small near the leading edge and too big near the trailing edge compared to measurements. For 3D-simulations, frequencies can be uniquely deduced from the FFT (exemplary 190 Hz for $\sigma = 2.5$ on the coarse grid, see figure 4), which are closer to the experiment. A similar cloud structure as observed in the experiment is found and shown for an exemplary time instant in figure 3.

By a comparison of the 3D Navier-Stokes with 3D Euler results for $\sigma = 2.5$, the effect of viscosity is assessed. In Euler simulations for the barotropic model, the re-entrant jet occurrence rate as well as shedding frequency are overestimated compared to data. The cloud structures are smaller due to the shorter periods between shedding events and in particular smaller than observed in the experiments. For Euler simulations with the Sauer-Schnerr model also a higher occurrence rate of the re-entrant is observed. However, bigger cloud structures develop, since clouds rotate without being advected and coalesce to bigger clouds which finally extend over more than one half (instead of one third as in the experiment) of the CLE-hydrofoil. Shedding of the bigger clouds leads to lower frequencies (e.g. 100 Hz at $\sigma = 2.5$).

Frequency vs cavitation number is shown in figure 5 for experiment and viscous simulations. Since 3D effects seem to have a minor effect for shedding frequencies obtained by the Sauer-Schnerr model, 2D results by [9] are included while as discussed above an exemplary 3D simulation reproduced the 2D shedding frequency for $\sigma = 2.5$. For the 3D barotropic model results, a serious grid dependence is observed, and no unique trend of the grid resolution is discernible for $\sigma = 2.5$ and $\sigma = 2.7$. The fine grid results are significant closer to experimental data, so that the coarse grid results are considered unreliable. Both cavitation models capture the increase of frequency with increasing cavitation number, while the Sauer-Schnerr model underestimates the data significantly for higher cavitation numbers. For these higher cavitation numbers, the barotropic model fine grid results are in very good agreement with the data.

## 5. Conclusions

Although empirical pre-factors in the URANS Sauer-Schnerr model have been modified to obtain the best agreement with data, the barotropic model shows particularly for high cavitation numbers a better shedding frequency prediction in spite of its simplifying assumption i.e. equilibrium thermodynamics and no use of an explicit turbulence model. However, the grid study needs to be extended to even finer grids. Due to the promising results of the barotropic model and since it does not contain empirical parameters, it is planned to further assess its potential to predict cloud cavitation also for more complex test cases, in spite of a strong grid dependence and high CPU times.

## References

[1] Arndt R E A 2011 *WIMRC Proc. 3rd Int. Cavitation Forum* (University of Warwick)
[2] Schnerr G H and Sauer J 2001 *ICMF Proc. 4th Int. Conf. on Multiphase Flow* (New Orleans)
[3] Li Z R 2012 *PhD Thesis* (TU Delft)
[4] Oprea O I and Bulten N 2011 *WIMRC Proc. 3rd Int. Cavitation Forum* (University of Warwick)
[5] Koop A H 2008 *PhD Thesis* (University of Twente)
[6] Sezal I H 2009 *PhD Thesis* (TU Munich)
[7] Gosset A, Lema M and López Peña F 2012 *Proc. 8th Int. Symp. on Cavitation* (Singapore)
[8] Frobenius M 2004 *PhD Thesis* (TU Munich)
[9] Deimel C, Günther M and Skoda R 2014 *EPJ Web of Conf.* **67** 02018
[10] Dular M and Coutier-Delgosha O 2009 *Int. J. for Numer. Meth. Fluids* **61** pp 1388-410
[11] Menter F R 1994 *AIAA Journal* **32** pp 1598-605
[12] Reboud J L, Stutz B and Coutier-Delgosha O 1998 *Proc. 3rd Int. Symp. on Cavitation* (Grenoble)
[13] Iben U 2002 *Systems Analysis Modelling Simulation* **42** pp 1283-307
[14] Schnerr G H, Sezal I H and Schmidt S J 2008 *Physics of Fluids* **20** 040703
[15] Harten A 1983 *J. Computational Physics* **49** pp 357-93
[16] Jasak H, Weller H G and Gosman A D 1999 *Int. J. for Numer. Meth. Fluids* **31** pp 431-49
[17] Mottyll S, Müller S, Niederhofer P, Hussong J, Huth S and Skoda R 2014 *EPJ Web of Conf.* **67** 02078
[18] Deimel C and Skoda R 2014 *9th OpenFOAM workshop* (Zagreb)
[19] Böhm R 1998 *PhD Thesis* (TU Darmstadt)
[20] Hofmann M 2000 *PhD Thesis* (TU Darmstadt)
[21] Bachert B 2004 *PhD Thesis* (TU Darmstadt)

# Comparative CFD Investigation on the Performance of a New Family of Super-Cavitating Hydrofoils

**S. Brizzolara**[1] **L. Bonfiglio**
MIT-iShip Innovative Ship Design Lab and MIT Sea Grant
Massachusetts Institute of Technology, 77 Mass ave, Cambridge, MA 02139

E-mail: stebriz@mit.edu

**Abstract**. We present a CFD characterization of a new type of super-cavitating hydrofoil section designed to have optimal performance both in super-cavitating conditions and in sub-cavitating conditions (including transitional regime). The basic concepts of the new profile family are first introduced. Lift, drag and cavity shapes at different cavitation numbers are calculated for a new foil and compared with those of conventional sub-cavitating and super-cavitating profiles. Numerical calculations confirm the superior characteristics of the new hydrofoil family, which is able to attain high lift and efficiency both in sub-cavitating and super-cavitating conditions. Numerical calculations are based on a multi-phase fully turbulent URANSE solver with a bubble dynamic cavitation model to follow the generation and evaporation of the vapor phase. The new profile family, initially devised for ultra-high speed hydrofoil crafts, may result useful for diverse applications such as super-cavitating or surface-piercing propellers or high-speed sailing boats.

## 1. Introduction

In high speed crafts, the option offered by hydrofoils in sustaining at least a portion of the vessel weight becomes an inevitable design choice to take when speed increases. The higher efficiency of hydrofoil supported crafts have been demonstrated in the 30-50 knots range (with few exceptions above) also in passenger ferries, a recent example reaching the broad public being the AC-72 catamaran hydrofoils of the 34th America's cup. Most hydrofoil applications including this last one are limited to the sub-cavitating or partially cavitating regime and yet few of them have been designed for speeds in the super-cavitating regime. At sailing speeds higher than approximately 50 knots traditional hydrofoils creating a meaningful lift force start to experience super-cavitation (i.e. the conditions at which the hydrofoil suction side is fully contained inside the vapor cavity that forms at its leading edge, see Figure 1). In this condition, conventional hydrofoils loose most of their ability to create lift and a different hydrofoil shape need to be designed. Conventional super-cavitating (SC) hydrofoils are designed to produce lift on the face and are optimized to operate in this regime: they have a pointed or sharp leading edge, a very slender entrance body to maintain the back inside the cavity and they recover strength (or inertia) by increasing their thickness at the trailing edge. The trailing edge

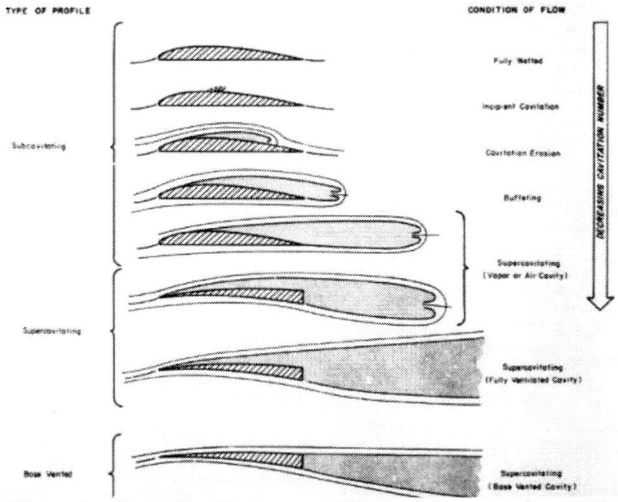


**Figure 1 – Cavitation Regimes and Hydrofoil types (after [10])**

[1] To whom any correspondence should be addressed.

usually carries the max thickness and it is sharp cut, to facilitate base cavitation. Different design by analysis methods have been proposed starting from the asymptotic theory of Tulin and Burkart [1], engineered by Johnson [2] and more recently optimization methods based on numerical solution of the potential flow with sheet cavitation, such as Mishima & Kinnas [3] or Pearce and Brandner [4].

## 2. SCSB hydrofoil family

We propose a new hydrofoil design that was initially developed [5] and patented [6] for application on ultra-high speed autonomous surface-piercing hydrofoil crafts (120 knots design speed). The design method of this new family was presented in [7]. The SCSB profile, in Figure 2, combines the shape of a SC profile with the pointed tail typical of conventional sub-cavitating profile, through two steps, which act as cavitation inception points (cavitators). The pressure and suction side of the entrance part are designed according [7], while the tail may be tuned (as a kind of flap) to give the required lift coefficient in sub-cavitating conditions. A member of the new family is identified by a code with two numbers SCSB-XX-YY: the first integer (XX) gives the design lift coefficient $100 \cdot C_L$ in supercavitating conditions; the second (YY) gives the design cavitation number $100 \cdot \sigma_0$.

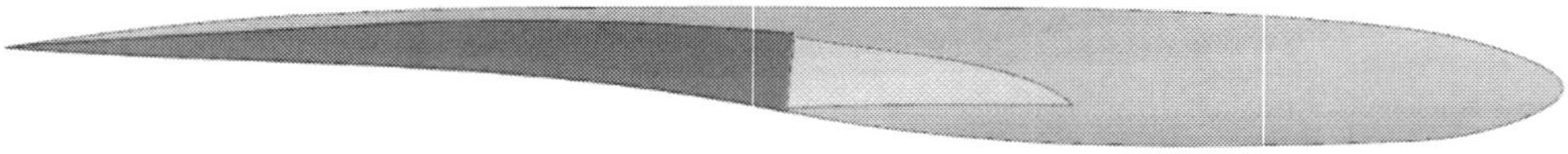

**Figure 2 – SBSC profile (patented in [6]) working in a super-cavity (blue). In red the entrance part, in yellow the tail.**

## 3. CFD Analysis of the SCSB-25-5 and equivalent NACA-65 profile

Figure 3 summarizes the global hydrodynamic characteristics of the SCSB profile, in terms of lift and efficiency versus the cavitation index $\sigma_0=(p-p_{vap})/(0.5\rho V^2)$, at the design angle of attack $\alpha$=5 deg. Its performances are compared to those of a good equivalent NACA profile, having the same maximum thickness to chord ratio, $(t/c)_{max}$=0.10, and the same effective chord of the SBSC profile in super-cavitating conditions. Furthermore the camber of the NACA profile is designed for the same lift at the same angle of attack in non-cavitating conditions. Figure 4 presents the cavity shapes predicted at different cavitation number (at the design angle of attack). Predictions were obtained by fully turbulent URANSE solver with Schnell-Sauer cavitation model, validated for instance in [8] [9] on different SC hydrofoils and with different solvers (StarCCM+ / OpenFoam). Full scale $Re=Vc/\nu$=O($10^7$) is considered.

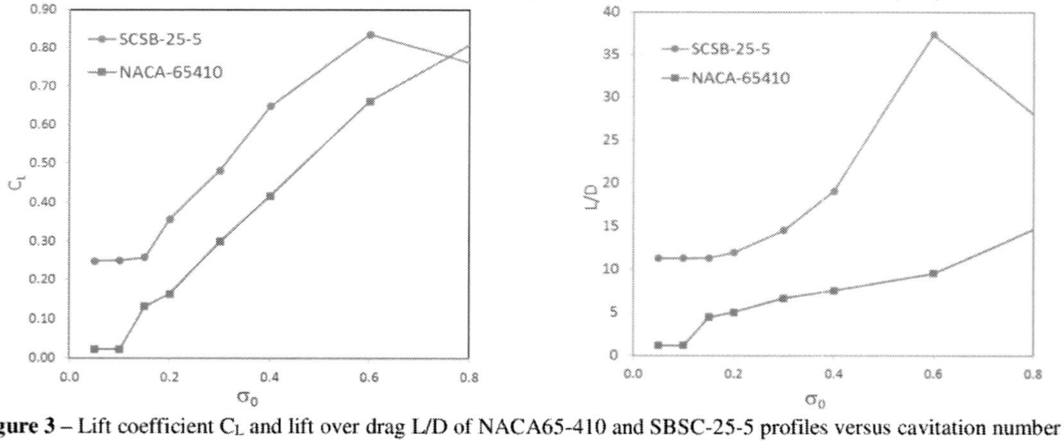


**Figure 3** – Lift coefficient $C_L$ and lift over drag L/D of NACA65-410 and SBSC-25-5 profiles versus cavitation number $\sigma_0$

As it can be observed, the NACA profile enters earlier into partial cavitation, at $\sigma_0$=0.8 the length of the cavity *l* is already larger than half the chord. As the cavity grows on the back ($0.4 \leq \sigma_0 \leq 0.8$), the lift rapidly drops with almost the same slope in both profiles. At $\sigma_0$=0.4 the cavity spans almost the full length of profile, although it is still highly unstable (buffeting regime). The instability of the cavity is evident from the evolution of the shape over time and it influences also the value of the hydrodynamic forces which show a periodic fluctuation around the average value (reported in Figure 3). In this regime, the difference of lift between the two profiles sets around $\Delta C_L$~0.25. The SCSB-25-5 enters first into the stable super-cavitation regime, at about $\sigma_0$=0.15, at which the NACA 65-410 still tends to develop an oscillating (buffeting) cavity. At the lowest investigated $\sigma_0<0.15$, the difference of lift coefficient and efficiency between the two profiles is dramatic: the SBSC sets onto $C_L$~0.25, its design value, while the NACA drops down to $C_L$~0.02. The reason is clearly the loss of the back side for the NACA profile, which remains with its non-optimal face shape to develop lift in these conditions. On the other hand, as the cavity gets shorter ($\sigma_0>0.8$) the NACA profile considerably reduces its drag (without increasing lift) and in non-cavitating conditions ($\sigma_0$ >~$1.5$) its efficiency rapidly climbs up to $L/D\approx90$, as opposed to the SCSB which stays on $L/D\approx35$. This is still a very attractive number, especially if compared to the efficiency of an equivalent conventional super-cavitating profile with blunt truncated trailing edge which typically stays around $L/D\approx6$ (as presented in [7]).

## 4. Conclusions

We have briefly introduced the design feature and hydrodynamic characteristics of a new family of super- cavitating hydrofoil, named SCSB: they maintain high lift and efficiency at both sub-cavitating and supercavitating regimes, at the contrary to optimal conventional sub-cavitating or conventional SC hydrofoils. In particular, the presented profile, SCSB-25-5 at its design angle of attack $\alpha$=5 degrees, produces about the same lift of a NACA-65-410 in non-cavitating conditions ($C_L$~0.8).

The earlier entrance in cavitating condition (with respect to $\sigma_0$) is what keeps the lift of the NACA profile lower than the SCSB, in the whole investigated range of cavitation indexes. This difference is considerable and arrives up to 100% at $\sigma_0$=0.15. Then the SCSB-25-5 enters earlier also in the stable supercavitating condition ($\sigma_0<0.15$) and there it stabilizes on an $L/D$=11.3, with an asymptotic $C_{Lmin}$=0.25, while the NACA-65410 confirms its inadequacy to work in this regime, practically losing the ability to develop lift ($C_{Lmin}$=0.02) and dragging the efficiency down to $L/D$=1. Due to its particular shape, the SCSB leaves the buffeting regime first and anticipates stable supercavitating conditions with respect to the NACA profile: this is an advantage for applications to high speed hydrofoils, since the dangerous range of unsteadiness which needs to be carefully avoided, is reduced. The advantages offered by the new profiles can be handy for other types of application, such as super-cavitating propellers or active control surfaces for ultra-high speed water craft. It would be useful to experimentally verify the numerical results presented here in a dedicated tests in a high speed, low pressure cavitation tunnel.

## Acknowledgments

This study received partial support from DARPA grant HR0011-14-1-0060 and ONR grant N000141310332.

## References

[1] Tulin, M. P., and Burkart, M. P. (1955). Linearized Theory for Flows About Lifting Foils at Zero Cavitation Number. Rep. C-638, David W. Taylor Model Basin, Navy Dept., Feb. 1955.
[2] Johnson V.E., Jr. (1961). Theoretical and Experimental Investigation of Supercavitating Hydrofoils Operating Near the Free Water Surface. NASA Technical Report, TR R93.
[3] Mishima, S. and Kinnas, S. (1996). A Numerical Optimization Technique Applied to the Design of Two-Dimensional Cavitating Hydrofoil Sections. Journal of Ship Research, 40, 1, March, pp. 28-38.
[4] Pearce B.W., Brandner P.A. (2007). Limitations on 2D Super-cavitating Hydrofoil Performance. Proceedings of 16th Australasian Fluid Mechanics Conference, Gold Coast, Australia, 2-7 December
[5] Brizzolara S. (2011). ONR USV HY2-SWATH. Hydrodynamic Design and Assessment by CFD Methods of HYbrid HYdrofoil SWATH hulls for a Super High Speed USV. Report ONRG-N62909-10-1-7116.
[6] Brizzolara S. (2014). "Watercraft Device". US patent US8820260. Italian Patent GE-2011-A000012 (2011).
[7] Brizzolara S. (2015) A New Family of Dual-Mode Super-Cavitating Hydrofoils. Fourth International Symposium on Marine Propulsors, smp'15, Austin, Texas, USA.
[8] Brizzolara S., Federici A (2011). Super-Cavitating Profiles for Ultra High Speed Hydrofoils: a Hybrid CFD Design Approach. Proceedings 9th Symposium on High Speed Marine Vehicles, HSMV 2011. Naples, March 2011, vol. 1, p. 1-13, ISBN/ISSN: 978-1-88906-112-0.
[9] Bofiglio L., Brizzolara S. (2015). A Numerical Investigation over Cavitating Flow Regime of Simple Super-Cavitating Hydrofoils. To appear in FAST 2015, 13th International Conference on Fast Sea Transportation, Sept. 1-4, Washington DC, organized by SNAME.

[10] Auslaender J. (1962). Low Drag Supercavitating Hydrofoil Sections. Technical report 001-7, Hydronautics Inc., Washington D.C.

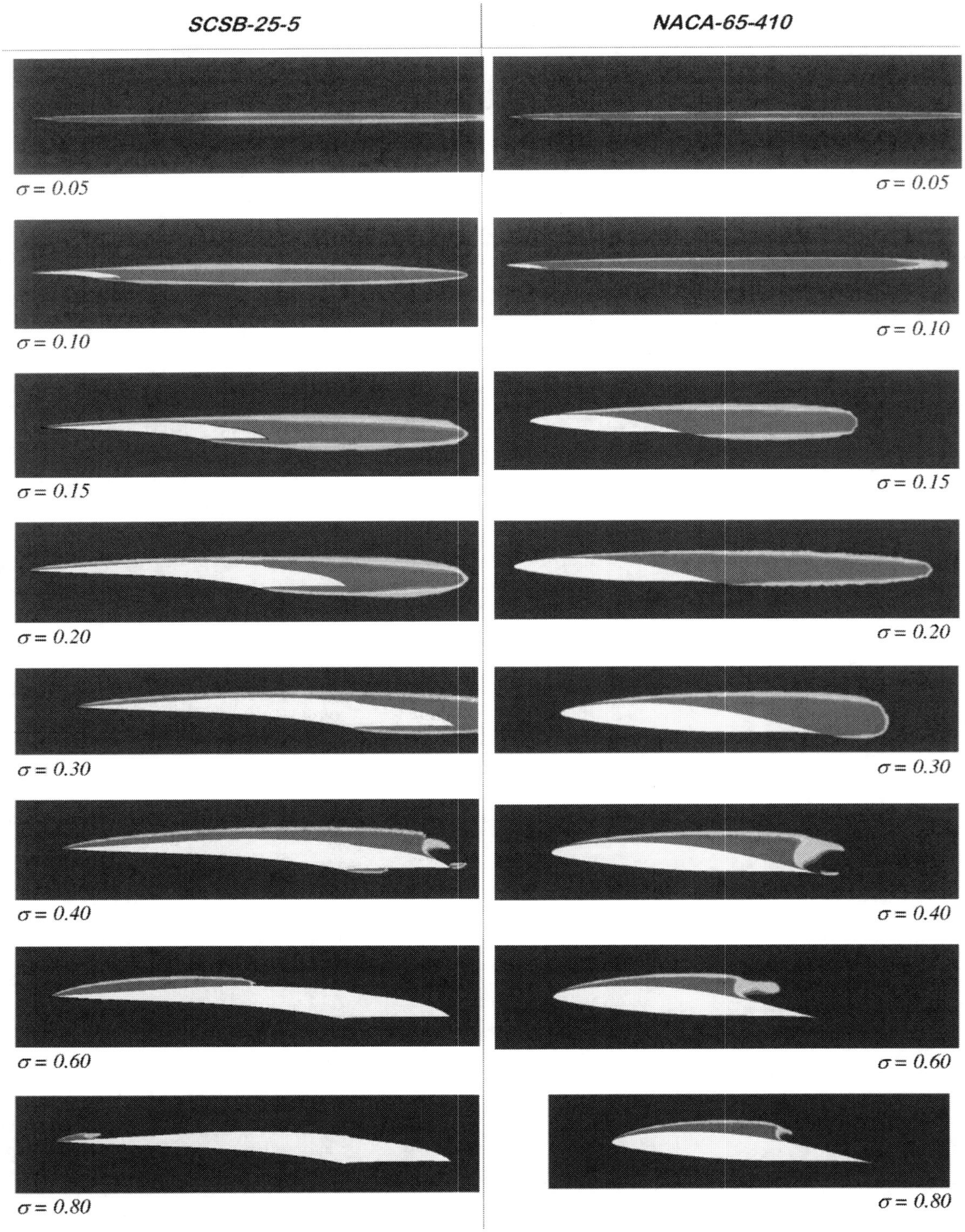


**Figure 4 - Predicted cavity shapes at different cavitation numbers (indicated at the bottom of each picture). SCSB-25-5 profile (left) and NACA-65-410 (right), angle of attack α=5 deg for both.**

# Study of the dynamics of a sheet cavitation

**Boris Charrière and Eric Goncalves**

University of Grenoble-Alpes, Laboratory LEGI, Grenoble, France
Department of Aeronautical Engineering ENSMA, Laboratory Pprime, Poitiers, France

E-mail: Boris.Charriere@legi.grenoble-inp.fr, Eric.Goncalves@ensma.fr

**Abstract.** A compressible, two-phase, one-fluid solver has been developed to investigate the behaviour of cavitation models. The model is based on a transport equation for the volume fraction of gas. Numerical simulations are performed on a cavitating Venturi flow where a self-sustained cavitation pocket develops. The sheet dynamics involving traveling pressure waves is investigated.

## 1. Introduction

The cavitation simulation is a challenging problem both in modelling of the physics and in developing robust numerical methodologies. Several physical and numerical models have been developed to investigate cavitating flows within the framework of the averaged approach. In its implementation, there are different approaches according to the assumptions made on the thermodynamic path during the phase transition and the slip condition between phases. A hierarchy of models exists, with the numbers of equations ranging from seven to three only.
In the present paper, we study an homogeneous, one-pressure and one-temperature model, composed by four equations: three mixture conservation laws and a void ratio transport equation. The mass transfer between phases is closed assuming its proportionality with the divergence of velocity [3, 4, 6]. This model is implemented in a compressible preconditioned Reynolds averaged solver with transport equation turbulence models. A self-sustained oscillating cavitation pocket developping on a Venturi geometry is investigated. The importance of the propagation of pressure waves in the physical mechanism is exhibited.

## 2. A four-equation model

The homogeneous mixture approach is used with the assumption of thermal and mechanical local equilibrium between pure phases, liquid and vapour. The model consists in three conservation laws for mixture quantities (mass, momentum and total energy) and an additional equation for the void ratio. The expression for the void ratio equation is:

$$\frac{\partial \alpha}{\partial t} + \operatorname{div}(\alpha \vec{V}) = \left( \frac{\rho_l c_l^2 - \rho_v c_v^2}{\frac{\rho_l c_l^2}{1-\alpha} + \frac{\rho_v c_v^2}{\alpha}} + \alpha \right) \operatorname{div} \vec{V} + \left( \frac{\frac{c_v^2}{\alpha} + \frac{c_l^2}{1-\alpha}}{\frac{\rho_l c_l^2}{1-\alpha} + \frac{\rho_v c_v^2}{\alpha}} \right) \dot{m} \tag{1}$$

where $\dot{m}$ is the mass transfer between phases, $\vec{V}$ is the velocity vector and $\rho_k$, $c_k$ are the pure phase density and speed of sound, respectively. Pure phases follow the stiffened gas equation of state. By assuming that the mass transfer is proportional to the divergence of the velocity, it is possible to build a family of models in which the mass transfer $\dot{m}$ is expressed as [3]

$$\dot{m} = \frac{\rho_l \rho_v}{\rho_l - \rho_v}\left(1 - \frac{c^2}{c^2_{wallis}}\right) \operatorname{div} \vec{V} \tag{2}$$

where $c_{wallis}$ is the propagation of acoustic waves without mass transfer. The system is closed by an equation of state for the mixture. Two formulations are considered to determine the pressure in the mixture. A sinusoidal law deduces the pressure from the void ratio:

$$p_m(\alpha) = P_{vap}(T_{ref}) + \left(\frac{\rho_l^{sat} - \rho_v^{sat}}{2}\right) c^2_{min} \operatorname{Arcsin}(1 - 2\alpha) \tag{3}$$

while the stiffened gas EOS, written with mixture quantities, proposes the following formulation:

$$p_m(\rho_m, e_m, \alpha) = (\gamma_m(\alpha) - 1)\rho_m(e_m - q_m(\alpha))) - \gamma_m(\alpha) p_{m,\infty}(\alpha) \tag{4}$$

The parameter $c_{min}$ refers to the minimal speed of sound in the mixture. Its influence and the EOS properties are detailed in [1]. Each closure expression has been studied on an expansion tube [3] and a Venturi flow [6].

## 3. The numerical tool

The numerical simulations are carried out using an implicit RANS solver for multi-domain structured meshes. This solver is based on a cell-centered finite-volume discretization. The Smith $k - \ell$ turbulence model is used. Unsteady simulations are performed using the dual time stepping method. Numerical schemes are second-order accurate in both time and space. All computations are performed on a H-type grid containing 174 nodes in the flow direction and 56 in the orthogonal direction. More details are given in [4, 6].

## 4. Computations and comparisons

### *4.1. Experimental conditions*

The test case is Venturi geometry constituted by an angle of the divergent part of 8°. In experiments, the pressure $P_{inlet}$ was lowered until the desired cavitation number. The lower wall was equipped with probe holes to evaluate the local pressure, void ratio and velocity. The selected operation point is characterized by the following parameters [2]:
$V_{inlet} = 7.04$ m/s, the inlet velocity; $T_{ref} = 293$ K, the reference temperature; $\sigma_{inlet} \simeq 2.15$, the cavitation parameter in the inlet section.
A typical self-oscillation behavior was observed with quasi-periodic vapour clouds shedding which frequency is measured closed to 45 Hz. Figure 1 shows an instantaneous photograph of the cavity and a scheme of the venturi profile.

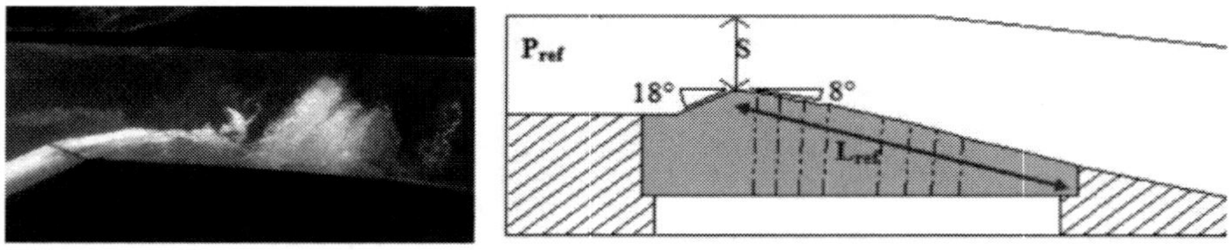


**Figure 1.** Photograph of the cavity and scheme of the geometry with probe locations.

### *4.2. 4-equation models comparison*

The objective of the study is to investigate the ability for the 4-equation model to correctly reproduce the cloud cavitation shedding. The infuence of the tunable parameter $c_{min}$ of the sinusoidal EOS (3) is evaluated. Two values are tested: 0.47 m/s and 1.36 m/s. Cavity pocket motions are illustrated on Figure 2 by representing the void ratio on a spatio-temporal graph $(x, t)$. The Venturi throat is located at the abscissa $x = 0$ m. The cavitating flow dynamic mainly depends on the considered equation of state. Moreover, variations of $c_{min}$ modify the pressure computation and influence the cyclic phenomenon frequency.

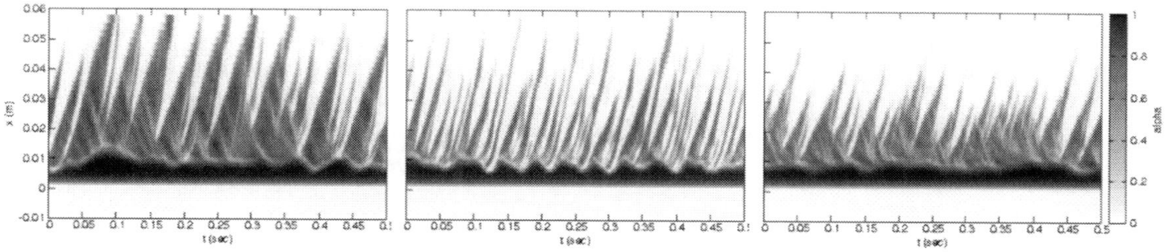


**Figure 2.** Sheet cavitation dynamic and pressure oscillation frequency using (from left to right): sinusoidal law $c_{min} = 0.47$ m/s (freq = 24.13 Hz); sinusoidal law $c_{min} = 1.36$ m/s (freq = 42.24 Hz) and stiffened gas law (freq = 40.94 Hz).

The physical mechanism of break-off cycles and cloud shedding are not clearly understood. The motion of a re-entrant liquid jet [2] is known as a main responsible of cavitation instability but recent experimental works [5] also report the presence of travelling pressure waves created by cloud collapses. In our simulations, the re-entrant jet is clearly observed by negative time-averaged velocity profiles near the wall. As presented on Figure 3, it remains identical for all computations and argue that the pressure and the speed of sound determine the periodic shedding frequency without changing the velocity field.

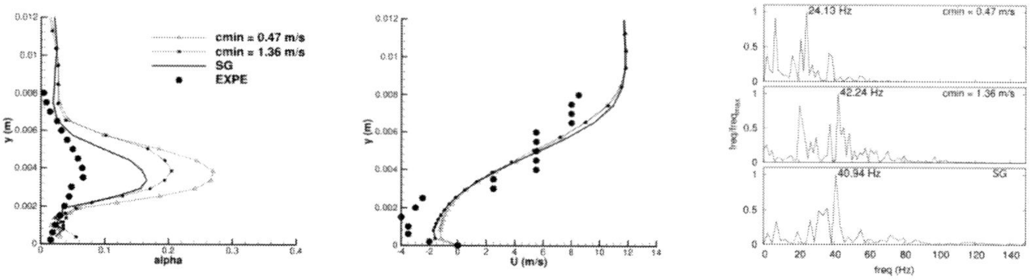


**Figure 3.** From left to right: comparison of void ratio and velocity profiles and pressure oscillation frequency for sinusoidal and stiffened gas law (probe number 2 : x = 0.0315 m).

In order to identify pressure waves, a numerical captor is located in the region of cloud condensation and the pressure signal is plotted on Figure 4. It reveals the presence of high pressure values between consecutive clouds cavitation which could disturb their advection.

### *4.3. Saturation versus metastable value for $\rho_v$*

The added transport equation allows to relax the local thermodynamic equilibrium and to introduce a metastable state for the vapour phase. Comparisons have been done for the sinusoidal EOS (3) between the two cases: equilibrium case with the saturation value $\rho_v = \rho_v^{sat}$ and metastable case according to the stiffened gas EOS. Figure 4 compares the pressure evolutions for both cases in the collapse region and demonstrates that a non-equilibrium state for $\rho_v$ increases the pressure magnitude. Similar results have been observed for speed of sound variations and consequently the Mach number computation in the shedding region. Thus, the pressure wave propagation caused by the vapour structure collapse is modified when the pure vapour density is no more constant.

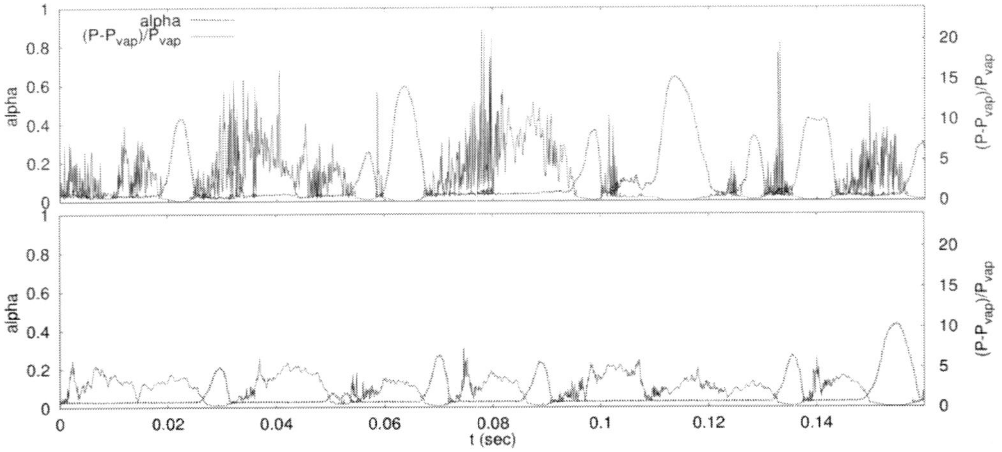


**Figure 4.** Comparison of pressure evolution in the collapse region (x = 0.0415 m; y = 0.0025 m) for metastable vapour density (top) and saturation value (bottom).

## 5. Conclusion

One-fluid RANS simulations based on a void ratio transport-equation model were performed to study an oscillating cavitation pocket developing along a Venturi geometry. The ability of simulation to reproduce the cloud shedding phenomenon is investigated by comparison with experimental data. Two equations of state are proposed (sinusoidal and stiffened gas). Results show that these relations and their dedicated parameters pilot the cavitating flow dynamic. Moreover, the introduction of a metastable state for the vapour phase modifies the speed of sound and increases pressure oscillations in the collapse region. These observations question the need of considering a non-constant vapour density for such cavitating flows.

## Acknowledgment

We thank Direction Générale de l'Armement (DGA) for a financial support to the PhD thesis "Modélisation et simulation d'écoulements turbulents cavitants avec un modèle de transport de taux de vide".

## References

[1] E. Goncalves and R. Fortes Patella, "Numerical simulation of cavitating flows with homogeneous models", *Computers & Fluids*, **38**, pp. 1682–1696, (2009).
[2] V. Aeschlimann, S. Barre and H. Djeridi, "Unsteady cavitation analysis using phase averaging and conditional approaches in a 2D Venturi". *Open J. of Fluids Dynamics*, **3**, pp.171–183, (2013).
[3] E. Goncalves, "Numerical study of expansion tube problems: Toward the simulation of cavitation", *Computers & Fluids*, **72**, pp.1–19, (2013).
[4] E. Goncalves and B. charriere, "Modelling for isothermal cavitation with a four-equation model", *Int. J. Multiphase Flow*, **59**, pp.54–72, (2014).
[5] C. Stanley, T. Barber and G. Rosengarten, "Re-entrant jet mechanism for periodic cavitation shedding in a cylindrical orifice". *Int. J. Heat and Fluid Flow*, **50**, pp.169–176, (2014).
[6] B. Charriere, J. Decaix and E. Goncalves, "A comparative study of cavitation models in Venturi flow", *Eur. J. of Mech. B/Fluids*, **49**, pp.287–297, (2015).

# On the non-equilibrium dynamics of cavitation around the underwater projectile in variable motion

**Y Chen, C J Lu, J Li, X Chen, Z X Gong**

School of Naval Architecture, Ocean and Civil Engineering, Shanghai Jiao Tong University, 800 Dongchuan Road, Shanghai, 200240, China

E-mail: cyofjs@sjtu.edu.cn

**Abstract**. In this work, the dynamic behavior of the non-equilibrium cavitation occurring around the underwater projectiles navigating with variable speed was numerically and theoretically investigated. The cavity collapse induced by the decelerating motion of the projectiles can be classified into two types: periodic oscillation and damped oscillation. In each type the evolution of the total mass of vapor in cavity are found to have strict correlation with the pressure oscillation in far field. By defining the equivalent radius of cavity, we introduce the specific kinetic energy of collapse and demonstrate that its change-rate is in good agreement with the pressure disturbance. We numerically investigated the influence of angle of attack on the collapse effect. The result shows that when the projectile decelerates, an asymmetric-focusing effect of the pressure induced by collapse occurs on its pressure side. We analytically explained such asymmetric-focusing effect.

## 1. Introduction

Partial cavitation usually appears around the shoulder the underwater projectiles navigating near sea surface. The deceleration of projectile caused by propulsion problem may result in the unstable evolution and collapse of the cavity, which will significantly affect the hydrodynamic loads acting on the projectile's body.

The primary objective of this work is to investigate the dynamic behavior of the cavity in non-equilibrium state around several underwater projectiles, which is induced by the instantaneously variable motion of the projectiles. The numerical works were carried out by solving the one-fluid compressible RANS equations using the commercial CFD package 'ANSYS FLUENT 14.0'. The cavitating flow is computed using the cavitation model suggested by Zwart et al. [1].

## 2. Problem statement

The axisymmetric underwater projectiles with three kinds of head form were studied. The projectiles move in an infinite stationary water area. A constant pressure is specified at the downstream boundary to adjust the initial cavitation number $\sigma$. The motion of the projectiles is achieved by dynamic mesh technique The initial velocity of the projectiles, $U_{b0}$, is specified as 30 m/s. Then the projectile is decelerated to an ultimate velocity $U_{bt}$ =20 m/s in 0.2 s, with the law: $U_{bt}+(1-t/0.2)^3(U_{b0}-U_{bt})$.

## 3. Results and discussion

### *3.1. Evolution of non-equilibrium cavity during variable motion of projectile*

Figure 1 presents the evolutions of the cavity during the decelerating motion and the following uniform speed period. The evolutions on the projectiles are basically similar but have some differences. The large-scale of separated flow around the blunt-nose projectile makes the secondary cavities to grow and collapse periodically and endlessly. However, the relatively smooth noses of the other two projectiles make the secondary cavities to gradually reach quasi-steady state after times of oscillation.

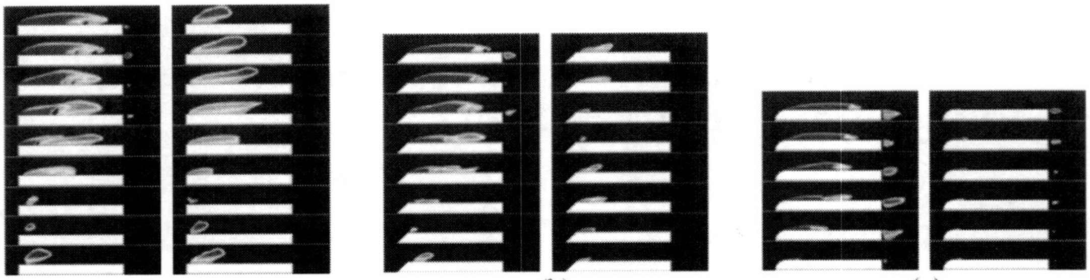

(a) (b) (c)

**Figure 1.** Evolution of the cavities around the projectiles: (a) blunt-nose projectile, $\sigma$ =0.3; (b) conic-nose projectile, $\sigma$ =0.25; (c) hemispheric-nose projectile, $\sigma$ =0.2.

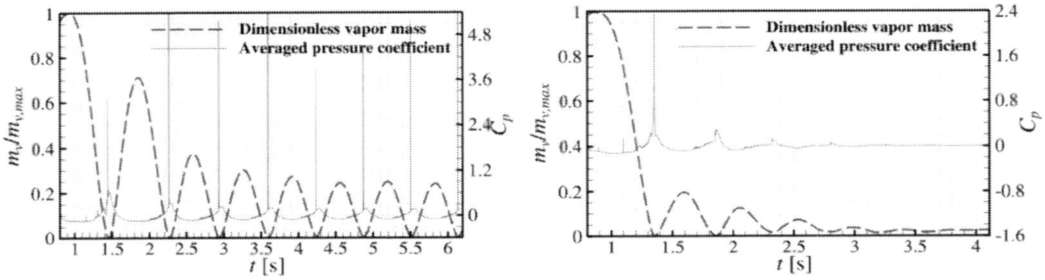


(a) (b)

**Figure 2.** The correlation between the total vapor mass and the averaged pressure in far field: (a) blunt-nose projectile; (b) conic-nose projectile.

We obtained the time histories of the total vapor mass and the averaged pressure coefficient on the upstream boundary, as shown in figure 2. In the case of the conic-nose projectile, a strong pressure peak appears around the first collapse, and each of the following valleys on the curve of vapor mass corresponds to an individual pressure peak. However, as for the blunt-nose projectile, the cavitation reaches a periodic state without damping after four or five times of oscillation.

For cavitating flows in dynamic equilibrium, the disturbed pressure $p'_\infty$ instead of $p_\infty$ should be related to the change-rate of the cavity. We assume that when the cavity has irregular shape, its collapse effect on the averaged pressure in far field is similar with that of a spherical cavity having the same volume. Here we propose the definition of the equivalent radius, $\tilde{R}$, using the instantaneous cavity volume $V_C(t)$, and establish a correlation between the averaged pressure coefficient and the change-rate of the specific kinetic energy of collapse:

$$\tilde{R}(t) = \left[\frac{3}{4}\frac{V_C(t)}{\pi}\right]^{\frac{1}{3}} \tag{1}$$

$$C_p \sim E(\tilde{R}) = \left[\tilde{R}\frac{d^2\tilde{R}}{dt^2} + \frac{3}{2}\left(\frac{d\tilde{R}}{dt}\right)^2\right] \Big/ \left(\frac{1}{2}U_{b0}^2\right) \tag{2}$$

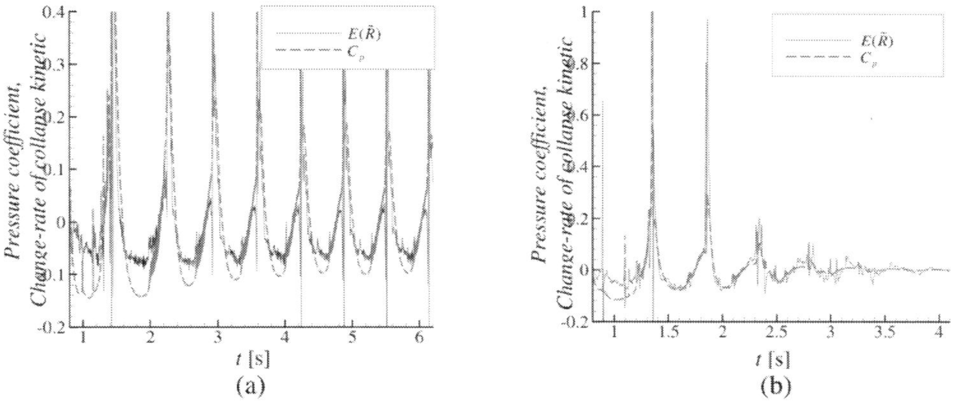


**Figure 3.** Comparison between the averaged pressure coefficient in far field and the change-rate of the specific kinetic energy: (a) blunt, $\sigma_0$=0.30; (b) conic, $\sigma_0$=0.25.

Figure 3 presents the comparison between these two dimensionless quantities. The two curves are in good agreement in the shape and position of their main peaks.

### *3.2. Asymmetric-focusing effect of the collapse-induced pressure*

We find if the projectile navigates with angle of attack, the collapse will entirely move to the pressure side of the projectile even if the angle is small, different with traditional cavitation theory [2].

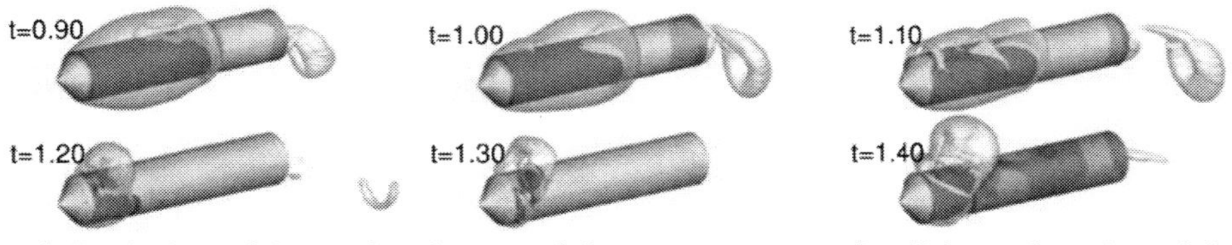


**Figure 4.** Evolution of the cavity shape and the pressure on projectile's surface ( $\sigma_0$=0.3, $\alpha = 2^\circ$ ).

Figure 4 shows the cavity evolution around the projectile navigating with $\alpha = 2^\circ$ . The filled contours of the pressure on projectile's surface are shown as well. The collapse produces a small block of high-pressure region on the pressure side. This phenomenon is unusual in common sense.

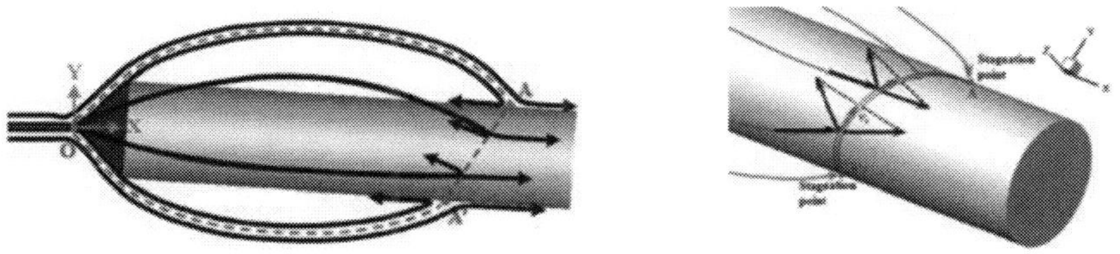

**Figure 5.** Diagram of the streamlines on an idealized cavity and the velocity along the closure.

We consider an ideal cavity whose shape is like ellipsoid [3]. It is well known that the re-entrant jet produced is reflected at the closure line [4]. By using the Cauchy-Lagrange integration in the reference frame fixed on the projectile, in the stage of the asymmetric focusing after the deceleration stops, the pressure difference between the stagnation points " A " and " A′ " can be written as:

$$\delta p = \rho_l \frac{D}{Dt} \delta\varphi = \rho_l \frac{D}{Dt} \int_{A'A} V_s ds \tag{4}$$

In section 3.1, it has been proven that the change-rate of the equivalent radius of cavity, $\tilde{R}$, has good correlation with the variation of pressure. It is considered that the length and the thickness of cavity tend to be isotropic when they are small enough. Since the annular cavity is converted into a spherical cavity of equivalent radius, $L_{OA}$ and $\tilde{R}$ are related as:

$$\frac{dL_{OA}}{dt} \sim O\left(\frac{\sqrt{\tilde{R}}}{\sqrt{R_n}}\frac{d\tilde{R}}{dt}\right) \tag{5}$$

The relation between the magnitude of the minimal axial component of the jet velocity $V_J$ and the corresponding maximal velocity on the closure line is:

$$V_{s,\max} = V_{J,\min}\frac{\cos(\theta-\alpha)}{\cos(\pi-2\theta+\alpha)} \tag{6}$$

This equation can also be expressed as $V_{s,\max} = C(\theta)V_{J,\min}$. The integration of the velocity along the closure line is in scale with the maximal velocity $V_{s,\max}$ :

$$\int_{A'A} V_s ds \sim O(\bar{V}_s R_n) \sim O(V_{s,\max} R_n) \tag{7}$$

The re-entrant jet reflects on the closure line "AA′", thus $V_{J,\min}$ should be greater than or at least not be lower than $dL_{OA}/dt$. The following relation exists:

$$V_{J,\min} \approx \left|\frac{dL_{OA}}{dt}\right| \tag{8}$$

After putting all the above-mentioned equations together, the pressure difference between "$A'$" and "$A$" is in proportional relation to a composite expression of the equivalent radius $\tilde{R}$ and its change-rates:

$$\delta p \sim \rho_l\sqrt{R_n}\frac{D}{Dt}\left(\sqrt{\tilde{R}}\left|\frac{d\tilde{R}}{dt}\right|\right) = -\rho_l\sqrt{R_n}\left[\sqrt{\tilde{R}}\frac{d^2\tilde{R}}{dt^2}+\frac{1}{2\sqrt{\tilde{R}}}\left(\frac{d\tilde{R}}{dt}\right)^2\right] \tag{9}$$

The $\tilde{R}$ satisfies the Rayleigh equation, and it's easy to obtain its change-rates in the above equation. Finally we have found that corresponding difference of pressure coefficient can be estimated as follows, given $\tilde{R}_0/\tilde{R} >> 1$:

$$\delta C_p \sim \frac{2}{3}\sigma\sqrt{\frac{R_n}{\tilde{R}}}\left(\frac{\tilde{R}_0}{\tilde{R}}\right)^3 \tag{10}$$

**Acknowledgments**
This project is supported by the National Nature Science Foundation of China (Grant No. 11472174). Its financial support is gratefully acknowledged.

**References**
[1] Zwart P J, Gerber A G and Belamri T 2004 *Proc. 5th Int. Conf. Mult. Flow* Yokohama, Japan.
[2] Callenaere M, Franc J P, Michel J M, et al. 2001 *J. Fluid Mech.* **444** 223.
[3] Savchenko Y N, Vlasenko Y D and Semenenko V N. 1999 *Int. J. Fluid Mech. Res.* **26** 365.
[4] Laberteaux K R and Ceccio S L 2001 *J. Fluid Mech.* **431** 43.

# Review of parameters influencing the structural response of a submerged body under cavitation conditions

**X Escaler**[1], **O De La Torre**[2] **and M Farhat**[3]

[1] CDIF, Universitat Politècnica de Catalunya, Barcelona, Spain
[2] Civil Engineering Department, National University of Ireland, Galway, Ireland
[3] LMH, Ecole Polytechnique Fédérale de Lausanne, Lausanne, Switzerland

E-mail: escaler@mf.upc.edu, dela.oscar@gmail.com, mohamed.farhat@epfl.ch

**Abstract**. Submerged structures that operate under extreme flows are prone to suffer large scale cavitation attached to their surfaces. Under such conditions the added mass effects differ from the expected ones in pure liquids. Moreover, the existence of small gaps between the structure and surrounding bodies filled with fluid also influence the dynamic response. A series of experiments and numerical simulations have been carried out with a truncated NACA0009 hydrofoil mounted as a cantilever beam at the LMH-EPFL cavitation tunnel. The three first modes of vibration have been determined and analysed under various hydrodynamic conditions ranging from air and still water to partial cavitation and supercavitation. A remote non-intrusive excitation system with piezoelectric patches has been used for the experiments. The effects of the cavity properties and the lateral gap size on the natural frequencies and mode shapes have been determined. As a result, the significance of several parameters in the design of such structures is discussed.

## 1. Introduction

The current interest among structural designers lies in seeking the limits of new materials or new structural configurations that offer improved and suitable qualities to create thinner, lighter and more flexible structures. At the same time, the concentration of power and the off-design operation of fluid machinery results in an increase of the unsteady hydrodynamic loads and the appearance of fluid instabilities. This general trend leads to more frequent material fatigue or resonance failures due to undesirable flow induced vibrations.

In this sense, a deep understanding of the fluid-structure interaction (FSI) of submerged bodies is fundamental for the design of a large variety of systems. This problem, although not new, has undergone a recent growth in popularity due to the large number of applications for fluids. In brief, the structural response of a submerged body depends significantly on its boundary conditions. However, the actual hydrodynamic conditions are often uncertain and it is difficult to determine them because the in-situ environment is not accessible for observation or measurement.

As a first approach, it is well known that a body in a dense and still fluid exhibits a different dynamic behavior than when surrounded by air due to the so-called "added mass effect" that is mainly translated to a reduction of the natural frequencies [1-4]. If the fluid flows around the body, additional uncertainties appear with a possible influence on the total damping and stiffness [5-6]. For example, the proximity of solid boundaries create small gaps filled with fluid that affect significantly the dynamic response [7-8]. And certainly, the presence of a two-phase flow due to hydrodynamic cavitation will add new boundary conditions that are not fully understood yet.

From a structural point of view, the interest in cavitation stems from the fact that many submerged bodies suffer from this phenomenon or they are prone to suffer it. The mixture of liquid and vapor water phases that forms the macroscopic hydrodynamic cavities can create averaged properties that vary from the expected effects of a pure liquid. Moreover, the complex structure and morphology of the two-phase flow enhances its scientific interest and opens new challenges to the design of submerged structures [9]. Consequently, evidence obtained from experiments and numerical simulations is summarized in the current paper to illustrate the significant parameters to take into account in such cases.

## 2. Experimental set-up

A series of experimental modal analysis were performed on an aluminum NACA0009 hydrofoil at the High Speed Cavitation Tunnel of the Laboratory for Hydraulic Machines (LMH-EPFL). A couple of Plumbum (lead) Zirconate Titanate (PZT) patches made of piezoelectric ceramic material were used as excitation and measuring systems perfectly embedded on the surface to avoid affecting the flow around the profile. The patch closer to the leading edge was used as exciter and the other one closer to the leading edge was used as sensor.

To measure structural vibration velocities, a single point Laser Doppler Vibrometer was used which is based on the "Doppler effect". Additional transducers consisted of two IEPE accelerometers, a force transducer and an instrumented hammer with a steel tip. For signal acquisition, recording and analyzing a sweep signal function generator, an analog to digital (A/D) converter with simultaneous channel sampling and a Fourier analyzer system were used.

The dynamic response of profile was studied under different flow conditions including leading edge attached sheet cavitation and supercavitation. Before filling the tunnel, a modal analysis was performed to obtain the reference natural frequencies ("Air" condition). Then, with the tunnel full of still water, the "Still water" conditions were also obtained. Finally, the "Cavitation" conditions were selected to ensure a series of stable attached cavities as shown on the left of Figure 1. The cavity lengths were varied from 2 to 75 % of the chord at 14 m/s free stream velocity and for two incidence angles of 1° and 2°.

## 3. Numerical model

A NACA0009 hydrofoil mounted inside the tunnel test section submerged with water was modelled with a FEM solver. A coupled Structural – Acoustic analysis was used to simulate the dynamic behaviour of the system. A sensibility analysis of the solid domain mesh was performed to determine its optimal size. Regarding the fluid domain, the dimensions were also selected to ensure that the lateral walls considered as fully reflective did not influence the numerical solution. Moreover, special attention was given to the small lateral gap between the hydrofoil tip section and the tunnel wall that has a size of approximately 0.12 mm as drawn in the centre of Figure 1. The final model can be seen on the right of Figure 1.

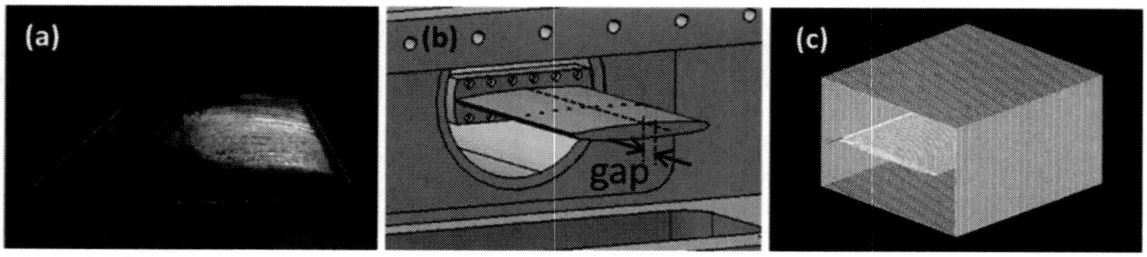


**Figure 1.** Photograph of partial cavitation (a), hydrofoil in the test section with lateral gap (b) and solid-fluid computational domain (c).

## 4. Results and discussion

To quantify the effects of various parameters on the modal behaviour of the hydrofoil, both the added mass coefficient, $C_M$, and the mode shapes have been determined and analysed. The $C_M$ is defined by Equation (1) where $f_{vacuum}$ and $f_{fluid}$ correspond to the natural frequencies for the hydrofoil surrounded by air and by water or cavitation, respectively.

$$C_M = \left(\frac{f_{vacuum_i}}{f_{fluid_i}}\right)^2 - 1 \qquad (1)$$

To characterize the attached cavity dimensions the *CSR* defined by Equation (2) has been used where $l$ is the cavity length and $c$ is the hydrofoil chord.

$$CSR = \frac{l}{2c} \qquad (2)$$

In particular, the three first modes of vibration named $f_1$, $f_2$ and $f_3$ corresponding to the first bending, the first torsion and the second bending ones, respectively, have been identified and considered for the studies.

### 4.1. Cavity properties

As observed on Figure 2, the added mass effects depend on the particular mode shape. They are maximum for $f_1$ and minimum for $f_2$. The *CSR* obviously determines the effect but no linear relationship is found with $C_M$ because other factors also influence the result. On the contrary, for the two angles of attack similar results are obtained.

A detailed analysis of the cavity morphology and dimensions [9] has demonstrated that a linear correlation can be found between $C_M$ and the so-called entrained mass which accounts for the average density of the cavity, the hydrofoil surface covered by the cavity and its relative displacement induced by the mode of vibration. Therefore, for an accurate prediction of the added mass effects under cavitation conditions, these variables must be taken into account which is not an easy task. In particular, it is necessity to improve the experimental and numerical tools to quantify the void ratio of the cavities.

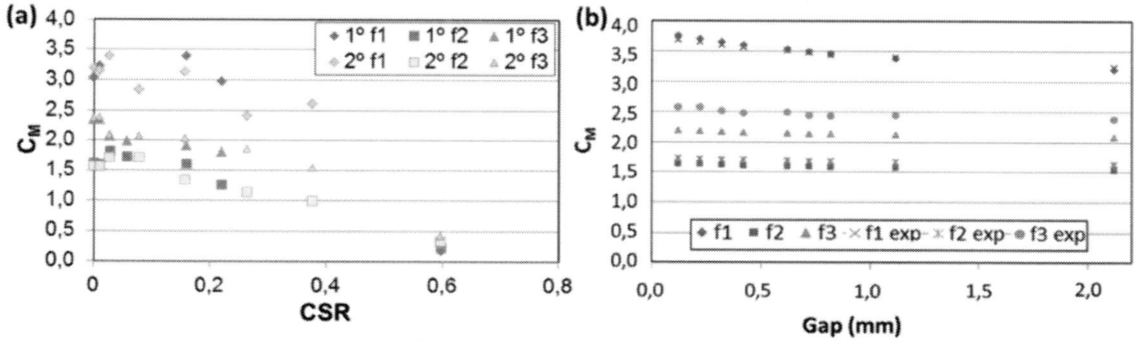


**Figure 2.** Added mass coefficient as a function of cavity surface ratio (a) and of lateral gap dimension without cavitation (b).

### 4.2. Lateral gap

A significant effect has been identified that is provoked by the narrow gap between the structure and the solid boundary. In our case, the experimental and numerical results plotted on the right of Figure 2 clearly show how the added mass effects increase following a quadratic law when the gap distance is reduced. This behaviour is observed in all the modes and the effect increases with mode order.

### 4.3. Mode shapes

Finally, a relevant mode shape alteration has been clearly measured regarding mainly the second bending mode, $f_3$, under cavitation conditions in comparison with the reference mode shape in air as shown in Figure 3. As observed, under partial cavitation the new mode loses the symmetry and it resembles a more complex bending-torsion coupled mode. The different properties of the fluids surrounding the hydrofoil -water and vapor-, their location and their size appear as the main factors determining the mode shape alteration.

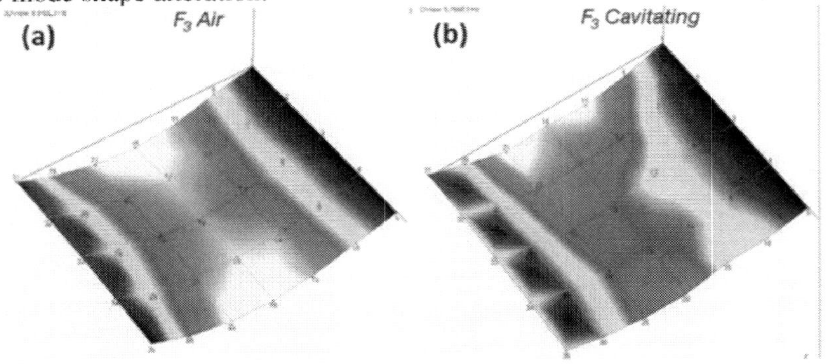


**Figure 3.** Mode shape of second bending mode of vibration obtained experimentally in air (a) and with attached partial cavitation (b).

## 5. Conclusions

The design of submerged structures that are prone to suffer macroscopic cavities attached to their surfaces requires the consideration of several parameters that have a significant effect on their dynamic response. The experimental and numerical studies demonstrate that the average properties of the cavity, the cavity size and the cavity location have a relevant effect on the added mass coefficients depending on the particular mode shape under consideration. Moreover, the existence of narrow gaps between the structure and the solid boundaries also has a significant influence on the overall behaviour. And finally, it must be noted that the new mode shapes might differ from the predicted ones in air conditions.

## References

[1] Lindholm U S, Kana D D, Chu W-C and Abramson H N 1965. Elastic vibration characteristics of cantilever plates in water *J. Ship Research* **9** 11-22

[2] Yadykin Y, Tenetov V, Levin D, 2003 The added mass of a flexible plate oscillating in a fluid. *J. Fluids and Structures* **17** 115-123

[3] Rodriguez C G, Egusquiza E, Escaler X, Liang Q W and Avellan F 2006 Experimental investigation of added mass effects on a Francis turbine runner in still water *J. Fluids and Structures* **22** 699-712

[4] Liang Q W, Rodriguez C G, Egusquiza E, Escaler X, Farhat M and Avellan F 2007 Numerical simulation of fluid added mass effect on a Francis turbine runner *Computers & Fluids* **36** 1106-1118

[5] Ducoin A, Young Y L and Sigrist J F 2010 Hydroelastic responses of a flexible hydrofoil in turbulent, cavitating flow *FEDSM-ICNM2010*, Montreal.

[6] Young Y L 2007 Time-dependent hydroelastic analysis of cavitating propulsors *J. Fluids and Structures* **23** 269-295

[7] Motley M R, Matthew R K and Young YL 2013 Free surface and solid boundary effects on the free vibration of cantilevered composites plates *Composite structures* **96** 365-375

[8] De La Torre O, Escaler X, Egusquiza E and Farhat M 2014 Numerical and experimental study of a nearby solid boundary and partial submergence effects on hydrofoil added mass *Computers & Fluids* **91** 1-9.

[9] De La Torre O, Escaler X, Egusquiza E and Farhat M 2013 Experimental investigation of added mass effects on a hydrofoil under cavitation conditions *J. Fluids and Structures* **39** 173-187

# Interaction of a Compressible Bubbly Flow With an Obstacle Placed Within a Shedding Partial Cavity

**Harish Ganesh, Simo A Mäkiharju, Steven L. Ceccio**

University of Michigan, Ann Arbor, MI 48109, USA

E-mail: gharish@umich.edu

**Abstract.** Bubbly shocks act as the dominant mechanism of shedding of partial cavities under certain conditions. The compressible nature of low void fraction regions in a partial cavity can provide conditions necessary for the existence of such bubbly shocks. Using X-ray densitometry as a flow visualisation mechanism, the dynamics of bubbly shock fronts with and without the presence of an obstacle are presented. Different flow features associated with the shock dynamics are identified and the associated flow physics explained.

## 1. Introduction

Hydrodynamic cavitation can occur in separated regions of flow behind objects resulting in partially filled vapor cavities. Once formed, partial cavities are generally stable, but can experience auto oscillations of cavity length resulting in shedding of vapor clouds, termed as cloud cavitation, carrying the vapor filled mixture that originally formed the cavity. Cloud cavitation and its onset are detrimental, and it is one of the principal agents of cavitation erosion. Measured values of void fraction in a partial cavity, as discussed in [1], [2], [3] and [4, 5] suggest that the cavity is not very dense under certain conditions. This means that the local speed of sound in such scenarios can be very small resulting in high local Mach numbers. A discussion of variation of speed of sound with void fraction can be found in [6]. Low values of speed of sound results in smaller propagation speed of disturbances than in liquids and lead to the presence of strong shock waves. [4] discusses the role of bubbly shocks in the shedding of partial cavities on a wedge and their relationship to the underlying cavitating flow field. The present study deals with the interaction of such propagating bubbly shocks with solid obstacles placed on the geometry used [4, 5]. Using X-ray densitometry as a flow visualization technique, the dynamics of the bubbly shock wave with and without an obstacle are studied. Different types of wave interactions that suggest the presence of complex flow physics involved are presented.

## 2. Experimental Setup

Experiments were carried out at the University of Michigan 9-Inch Water Tunnel. The tunnel has a 6:1 round contraction leading into a test section with a diameter of 23 cm (9 inches). The test section then transitions to a square cross-section that is 21 cm by 21 cm with chamfered corners. The flow velocity in the tunnel test section can be varied from $U_0 = 0$ to 18 m/s and the static pressure, $p_0$, from near vacuum to 200 kPa.

$$\sigma_0 = \frac{(p_0 - p_v)}{\frac{1}{2}\rho U_0^2} \qquad (1)$$

In the present experiments, the test section was further reduced in area to a conduit that had a 7.6 cm by 7.6 cm cross-section. This was done to reduce the baseline X-ray attenuation produced by the non-cavitating flow. The wedge geometry was chosen because it produces a nominally two-dimensional cavity with a well-defined line of cavity detachment. The wedge makes an angle of 22.1 degrees to the incoming flow, and has a downstream angle of 8.1 degrees. [4, 5] discuss the set-up and geometry in more detail. A schematic of the wedge in the test section is shown in Figure 1.

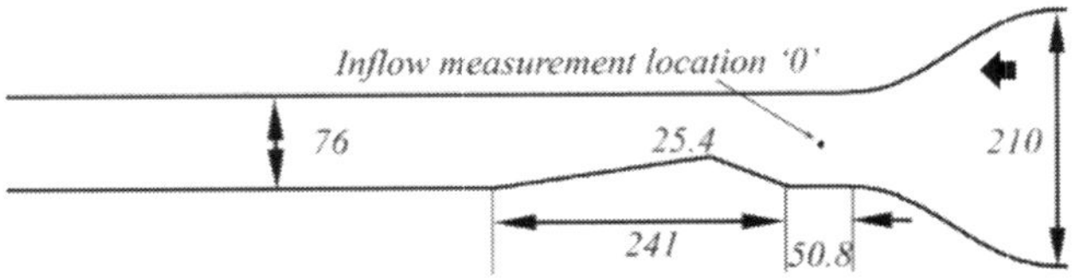


*Figure 1: Schematic of the wedge (without the obstacle) in the test section. All dimensions are in 'mm'.*

X-ray densitometry is based on the principle of Beer-Lambert's law of attenuation of ionizing radiation. The X-ray densitometry system used in the present study had a source capable of 433 mA at 150 kV, and the imaging system is comprised of an image intensifier coupled with a high-speed camera (Vision Research Phantom V9.0). By shooting an X-ray beam into the partial cavity with vapor, the void fraction flow field can be determined by analyzing the attenuated image of the cavity. From the void fraction field, one can track interfaces such as shock waves, cavities, and shed vapor clouds. A detailed description of the method can be found in [7].

## 3. Shedding Partial Cavity

The wedge geometry provides conditions for sustainable cavitation at the wedge apex. By systematically changing the inlet pressure $p_0$, the inlet cavitation number was varied to study different types of cavities. At higher cavitation numbers ($\sigma_0$ > 2.2), the cavities at the wedge apex were relatively stable and with reduction in inlet cavitation number, $\sigma_0$, the cavities began to exhibit significant length oscillations. When the inlet cavitation number was decreased to $\sigma_0$ <1.9 +/-0.1, periodically shedding cavities were observed.

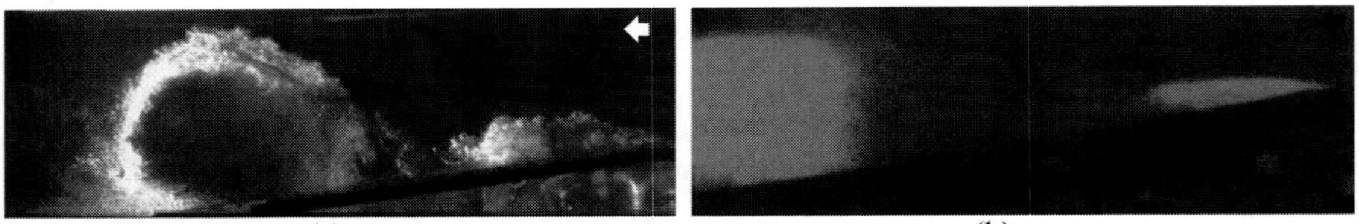

(a) (b)

*Figure 2: A shedding cavity at $U_0$ = 8 m/s (a) High speed video image at $\sigma_0$ =1.85 +/-0.1 (b) X-ray attenuation field image at $\sigma_0$ =1.81 +/-0.1.*

Figure 2(a) shows a snapshot of a periodically shedding cavity using high speed video. The X-ray densitometry snapshot image for periodically shedding cavity is shown in Figure 2(b). From the images, flow features like cavity closure and shed vapor cloud can be clearly seen. Figure 3 shows a time series of a shedding cavity with the associated flow features. The cycle has well defined growth and pinch-off with the propagating shock from clearly visible in Figures 3(c) and 3(d).

## 4. Interaction of shock wave with obstacle

A small obstacle of cross-sectional size 4mm x 4mm that spanned the wedge was placed with its leading edge 63mm from the wedge apex.

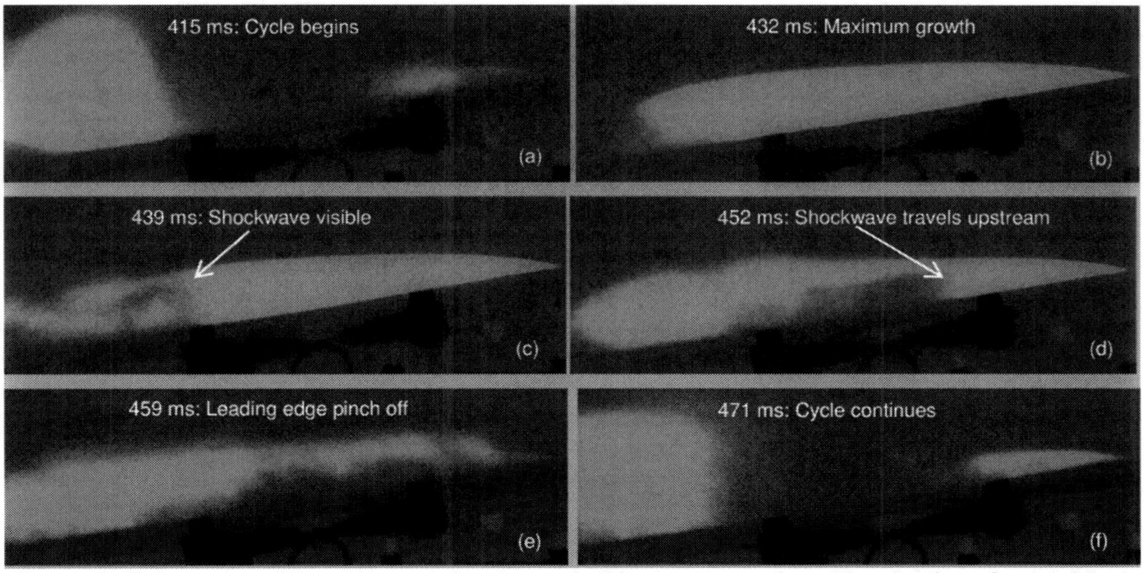


*Figure 3: Time series of X-ray images showing a shedding cavity cycle at $U_0 = 8$ m/s and $\sigma_0 = 1.85$ +/- 0.1.* In this case, no obstacle is present.

Time resolved X-ray attenuation videos were recorded to study the interaction of the shock wave with the obstacle as shown in Figures 4, 5 and 6. As the cavity grows, the void fraction in the cavity reduces resulting in low speed of sound values. Speed of sound for this particular flow is found to be around 3 to 5m/s [4, 5]. Once the cavity length crosses the obstacle the effect of low speed of sound is felt locally. Since the obstacle is in the region of recirculation, an oblique shock wave forms to aid the recirculating flow to cross the obstacle. This is different from the propagating normal shock wave that results in shedding. This is shown in Figure 4(a) and 4(b). Figure 4(b) also shows the interaction of the oblique shock wave with the expansion fan that exists in the cavity. It should be noted that this expansion is a result of the propagating shock being reflection back as an expansion from the wedge apex.

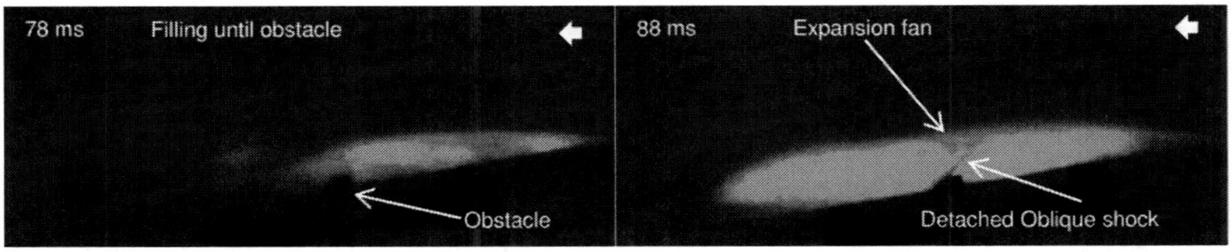


*Figure 4: Cavity dynamics in the presence of an obstacle in filling phase of the cycle for $U_0 = 8$ m/s and $\sigma_0 = 1.85$ +/-0.1*

Figures 5(a) and (b) show the interaction of the normal shock wave with the obstacle. Once the normal shock wave crosses the obstacle, the wake of the obstacle is filled with vapour. This could be due to the cavitation of the re-entering liquid drawn by the shock as it moves into a vapour filled cavity or due to the rarefactions waves produced by the passage of shock over a sharp edge. The recirculation zone present at the aft of the obstacle transitions into a stagnation mass of bubbly fluid once the void fraction values drop below a certain value. This results in the disappearance of the oblique shock (that was seen in 4b) because of the absence of the re-entrant flow.

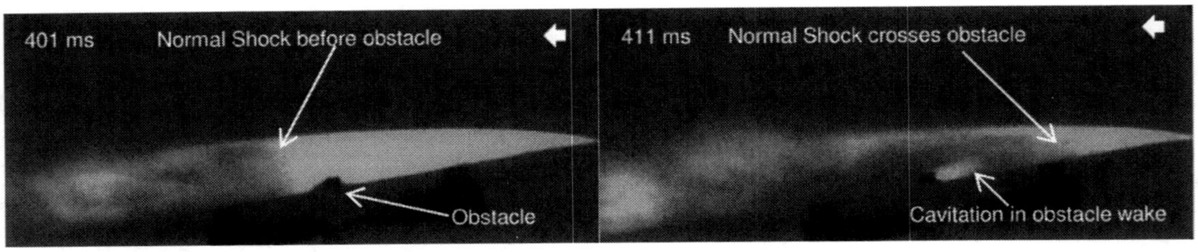


*Figure 5: Cavity dynamics in the presence of an obstacle in collapse phase of the shedding cycle for* $U_0 = 8$ *m/s and* $\sigma_0 = 1.85$ +/-0.1

For smaller cavity lengths and higher cavitation number, the interaction of the stationary oblique shock and the propagating normal shock is feasible. One such interaction is shown in Figure 6(a) and 6(b). For this case the cavity length was such that the detached shock occurrence due to changes in void fraction values near the obstacle and the initiation of a propagating bubbly shock occurred very close to each other in time. This can be visualised in Figures 6(b) where the complex interaction of the shock waves and the expansion fans can be seen. More information on the kind of interactions between the waves can be found in [8].

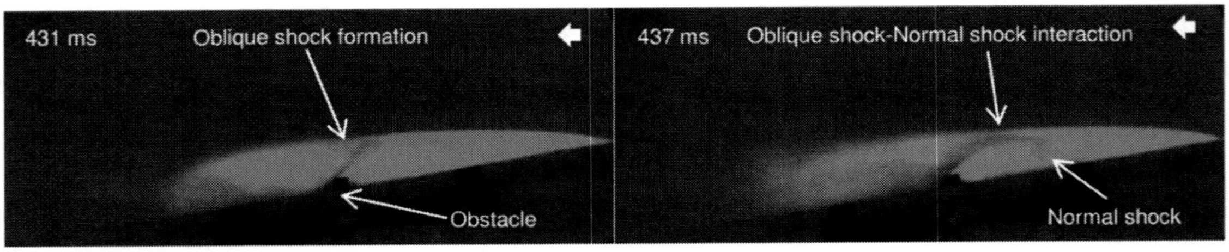


*Figure 6: Shock-Shock Interaction* $U_0 = 8$ *m/s and* $\sigma_0 = 1.94$ +/-0.1

**References**

[1] Coutier-Delgosha, O., Stutz, B., Vabre, V. & Legoupil, S. (2007) Analysis of cavitating flow structure by experimental and numerical investigations. *J. Fluid Mech.* **578**, 171-222

[2] Stutz, B & Reboud, J. L. (1997a) Experiements on unsteady cavitation. *Expt in Fluids.* **22**, 191-198Stutz, B & Reboud, J. L. (1997b) Two-phase flow structure of sheet cavitation. *Physics of fluids.* **9**, 12, 3678-3686.

[3] Stutz, B & Legoupil, S. (2003) X-ray measurements within unsteady cavitation. *Expt in Fluids.* **35**, 130-138

[4] Ganesh, H., Mäkiharju, S.A., & Ceccio, S.L. (2015) Bubbly shock propagation as a mechanism for sheet-to-cloud transition of partial cavities. *J. Fluid Mech.* (Submitted)

[5] Ganesh, H. (2015) Bubbly shock propagation as a cause of sheet to cloud transition of partial cavitation and stationary cavitation bubbles forming on a Delta wing vortex. *Ph.D Thesis, University of Michigan*

[6] Brennen, C, E. (2005) "Fundamentals of multiphase flows", *Cambridge University Press*

[7] Mäkiharju, S.A., Chang, N., Gabillet, C., Paik, B.-G., Perlin, M. & Ceccio, S.L. (2013) Time Resolved Two Dimensional X-Ray Densitometry of a Two Phase Flow Downstream of a Ventilated Cavity. *Experiments in Fluids*, **54**,

[8] Liepmann, H., W., & Roshko, A. (2001) "Elements of Gas Dynamics", *Dover Publications*

Journal of Physics: Conference Series **656** (2015) 012152 doi:10.1088/1742-6596/656/1/012152

# The performance of the water-entry cavity on a high-pressure gas-driven projectile

**Z X Gong, L Ma, Y L Fang, H Liu,Y Chen**

Department of Mechanics, MOE Key Laboratory of Hydrodynamics, Shanghai Jiao Tong University, Shanghai, China

E-mail: jackyff@sjtu.edu.cn

**Abstract**. We present the experimental results of water-entry cavity on a slender projectile, which moves in the opposite direction of gravity. The regime of the cavity including sheet cavity, supercavity and trailing cavity is exhibited and its relevance to Euler and Weber numbers are illuminated. Meanwhile, various pinch-off performances of the supercavity are analysed and the resulting re-entrance jet or cavity ripples are presented.

## 1. Introduction

Most existing works on water-entry cavity problems focus on a moving solid object impacting a free fluid surface and the cavity is formed by the air entrainment behind the moving solid. The first research can dates back to 1897, when Worthington & Cole investigated spheres impacting onto liquid surfaces using single-spark illumination photography. With the developments in experimental techniques and numerical simulation schemes, there are notable progresses on understanding the mechanism and characteristics of the water-entry cavity in the past decades.

Here, we pay attention to a different water-entry cavity. In our study, a slender projectile is driven by a high-pressure gas and moves in the opposite direction of gravity. The projectile launches from a gas-filled canister and enters the water with cavities around and/or behind it. A brief sketch for the comparison between two water-entry cavities are shown in Figure 1.

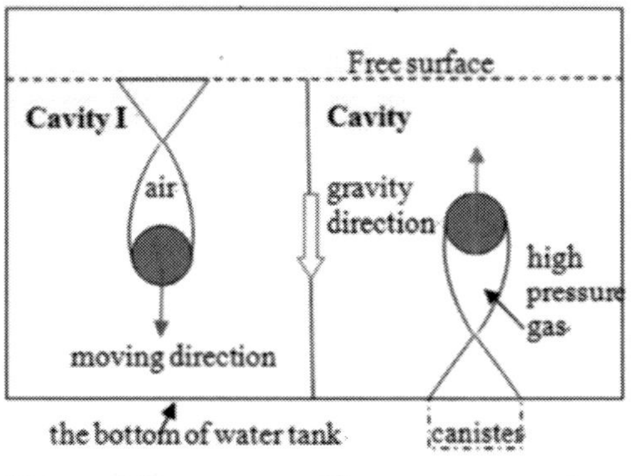


**Figure 1**. Sketch map of water-entry cavity.

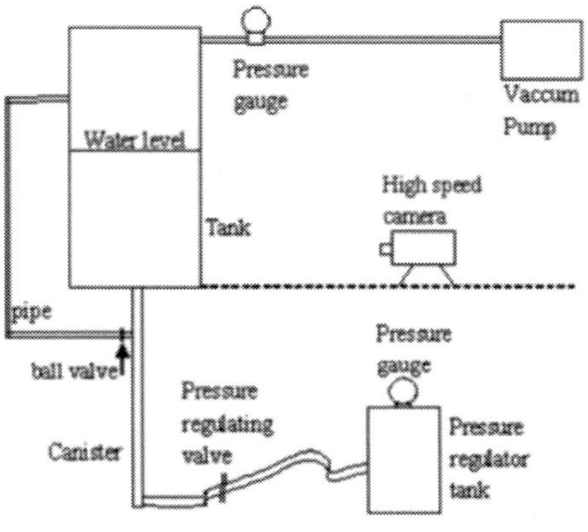


**Figure 2**. Schematic of the experimental set-up.

## 2. Experimental set-up

The experiment is conducted to observe the performance of the cavity on a high-pressure gas-driven projectile, which is a slender cylinder with hemispherical head. The experimental set-up is shown in Figure 2. The octagonal water tank is with 800mm diameter and 1500mm height and filled with 800mm height of water. The pressure on the top of the water level can be adjusted by a vacuum pump. The water tank contains three glass windows (40mm width and 100mm height) for the input of the light sources and the high-speed camera recording. The images are taken at 2000 frames/sec. The high-pressure gas from the pressure regulator tank runs into the inner of the canister to accelerate the projectile and the duration of ventilation is controlled by the pressure regulating valve. The preset-gas inside the canister is connected with the vacuumized air above the free surface inside the water tank and the connection is governed by a ball valve to avoid the gas at the front of the canister being over compressed. The canister and water tank are separated by a membrane which can sustain a pressure difference within 0.2atm. The projectile (28.6mm diameter and 170mm long) is initially placed at the bottom of the canister. The pressures at the bottom of the canister and at the top of the water tank are measured by the pressure gauges.

## 3. Results

Define the Froude number, Euler number and Weber number as $Fr = V^2/gD$, $Er = P_{wb}/(0.5\rho V^2)$, $We = \rho V^2 D/\sigma$ ,where $g$ is the acceleration of gravity, $D$ is the diameter of the projectile, $\rho = 998.2$ kg/m$^3$and $\sigma = 72.59$ mN/m is the density and surface tension of water, respectively. $P_{wb}$ is the static pressure at the bottom of the water tank, $V$ is the velocity when the bottom of the projectile gets out of the canister. In our experiments, $P_{wb}$ ranges from $1.4\times10^4$ to $4.8\times10^4$ Pa, and $V$ ranges from 3.17 to 10.31 m/s. Since the experimental projectile model does not change, the Bond numbers $Bo = \rho g D^2/\sigma$ maintain a constant value.

### *3.1. The regime of the cavity*

In our experiments, the shapes of the cavities vary with the flow conditions. Sheet cavity, supercavity and trailing cavity can be observed, as shown in Figure 3. The cavity performance corresponding to the Weber-Euler number is plotted in Figure 4, where the separated points are experiments results and the solid line is a fitted curve. It can be seen that with large Euler number and small Weber number, there is no cavity along the side of the projectile but a trailing cavity attached to its bottom; as the Euler number decreasing and Weber number increasing, sheet cavity is generated and the trailing cavity still exists; under small Euler number and large Weber number, a supercavity is formed and the trailing cavity disappears since the bottom of the projectile is inside the supercavity.

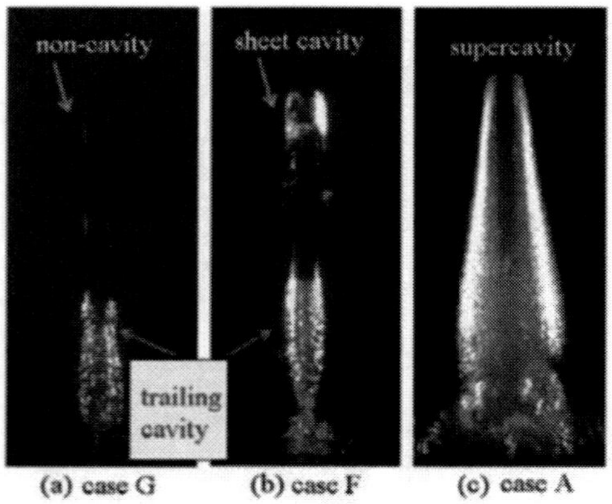


**Figure 3.** The shape of the cavities

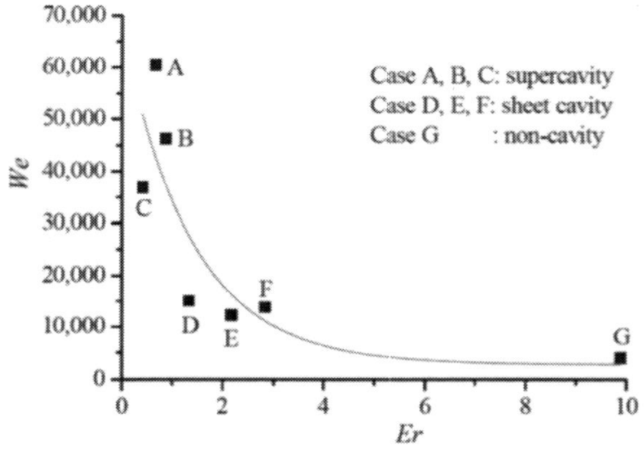


**Figure 4.** The cavity performance corresponding to the Euler-Weber number Variation.

We denote the pressure at the bottom of the canister as $P_{cd}$. Figure 5 exhibits the bottom pressure variation under case A, F and G, which is respectively corresponding to non-cavity, sheet cavity and supercavity situation. For each case, we denote $t_1$ as the time when the projectile start moving and $t_2$ as the time when the membrane gets broken. The time difference between $t_1$ and $t_2$ are almost identical, which stands that the mean velocity of projectile inside the canister is almost identical. Since the background pressure is approximately equal, the larger pressure inside the canister causes a higher launching velocity of the projectile. After $t_2$, $P_{cd}$ is greater than $P_{wb}$ in case A, approximately equal $P_{wb}$ in case B, and less than $P_{wb}$ in case C. The comparison among Figure 5 shows that the differences between $P_{cd}$ and $P_{wb}$ effect the amount of gas attached to the projectile. It can be concluded that the launching velocity and the pressure difference mainly determine the cavity performance.

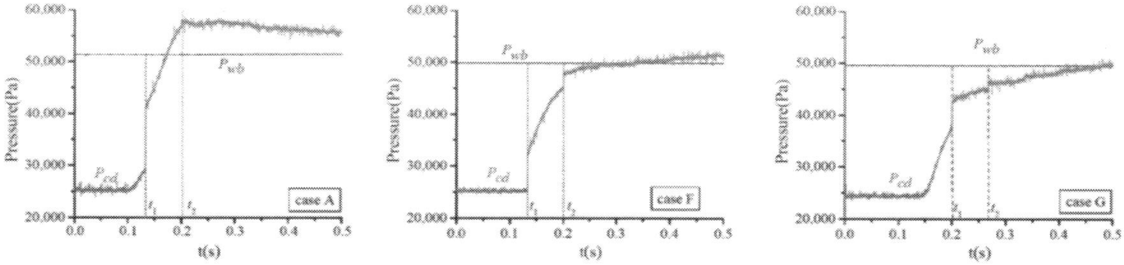


(a) case A (b) case F (c) case G

**Figure 5**. The monitored pressure variation of case A: $P_{wb}=5.1\times10^4 Pa$, $V=12.4m/s$; case F: $P_{wb}=5.0\times10^4 Pa$, $V=5.9m/s$; and case C: $P_{wb}=5.0\times10^4 Pa$, $V=3.2m/s$.

*3.2. The pinch-off of the supercavity*

In our experiments, three kinds of supercavity pinch-offs are observed and presented in Figure 6. For case A, the cavity surface is smooth and after the pinch-off a re-entrance jet is immediately produced and injects into the cavity. The trajectory of the re-entrance jet is very clear and its velocity is nearly twice of the supercavity' tail velocity. For case B, the end of the supercavity is not 'perfectly' closed when first pinch-off happens. A fraction of gas still moves downward along its surface to form a smooth closure area. Then the second pinch-off takes place and a re-entrance jet appears and impacts the inside of the supercavity. The extra gas are shed off downward. For case C, the closure style of the supercavity is particular. We regard the shrinking part of the cavity as the 'cavity neck', and its length here is considerable long. Therefore, the closure area is a line instead a point (in 2D view) and multi pinch-offs happen successively in a very short time. The corresponding pressure variation is given in Figure 7. The large pressure difference between $P_{cd}$ and $P_{wb}$ in case C makes more gas ejecting from the canister and such gas can continuously supply into the supercavity, which leads to the multi pinch-offs.

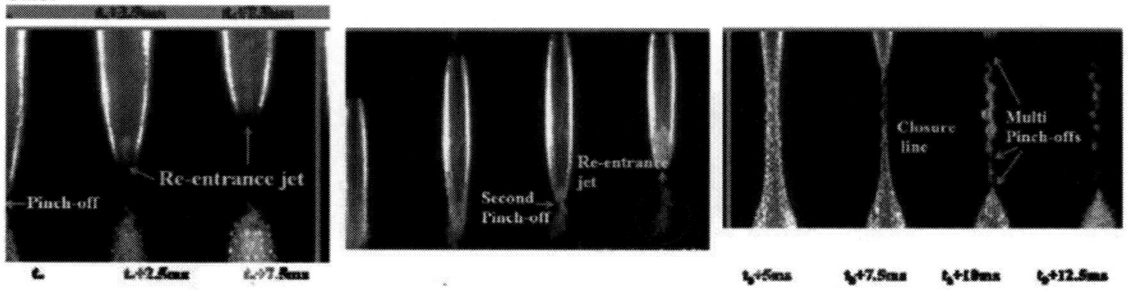


case A case B case C

**Figure 6**. The supercavity pinch-off of case A: $P_{wb}=5.1\times10^4 Pa$, $V=12.4m/s$; case B:, $P_{wb}=5.1\times10^4 Pa$, $V=10.8m/s$; and case C: $P_{wb}=1.9\times10^4 Pa$, $V=9.7m/s$.

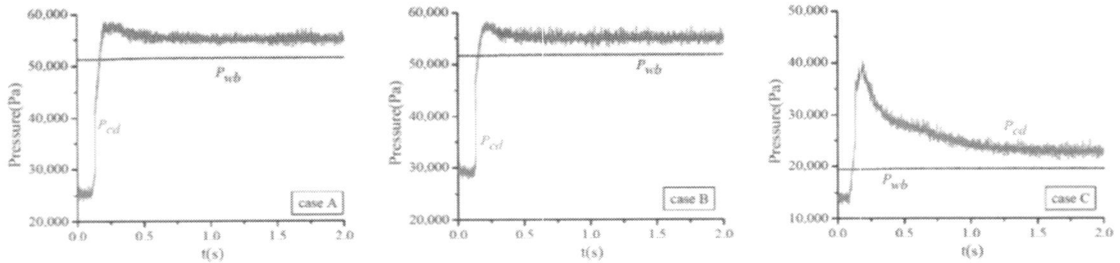


**Figure 7**. The monitored pressure variation of case A: $P_{wb}=5.1\times10^4 Pa$ , $V=12.4m/s$ ; case B:, $P_{wb}=5.1\times10^4 Pa$ , $V=10.8m/s$ ; and case C: $P_{wb}=1.9\times10^4 Pa$ , $V=9.7m/s$ .

### *3.3. The trailing cavity ripples*

The surface of the trailing cavity becomes fluctuation after pinch-off for all cases except the supercavity ones in our experiments. Such surface undulations are called as 'cavity ripples', which has been previously studied by Gekle et al(2008), Grumstrup et al.(2007), Bodily et al.(2014) and Mansoor et al. (2014). However, the mechanism still remains many unsolved questions.The evolution of the cavity ripples is shown in Figure 6. The flucuation is very obvious right after the pinch-off. The amplitude of the flucuation declines rapidly, however, the fuctuating frequecy nearly keeps constant, which can be concluded from the quasi-periodic shedding.

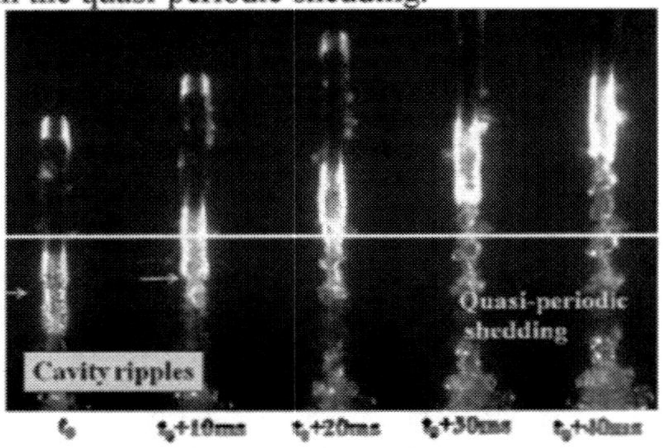


**Figure 8**. The evolution of the cavity ripples.
(case F: $p_{in}=4.0\times10^5 Pa$ , $P_{wb}=5.0\times10^4 Pa$ , $V=5.9m/s$ ;)

## 4. Conclusion

The water-entry cavity on a high-pressure gas-driven projectile is studied by experimental means. Its performance is a consequence of various flow conditions such as the launching velocity, the background pressure and the pressure inside the canister. Three kinds of supercavity regarding to the pinch-off behaviors are analyzed and the trailing cavity ripples are also observed.

## References

[1] Worthington A M, Cole R S 1897 Phil. Trans. R. Soc. Lond. 189, 137-148.
[2] Grumpstrup T, Keller J B and Belmonte A 2007 Phys. Rev. Lett. 99,114502.
[3] Gkele S, Bos A, Bergmann R, Meer D and Lohse D 2008 Phys. Rev. Lett. 100,084502.
[4] Bodily K G, Carlson S J and Truscott T T 2014 Phys. Fluids 26,072108.
[5] Mansoor M M, Marston J O, Vakarelski I U and Thoroddsen S T 2014 J. Fluid Mech. 743:295-326.
[6] Duclaux V, Caille F, Duez C, Ybert C, Bocquet L and Clanet C 2007 J. Fluid Mech. 591:1-19.
[7] Aristoff J M, Bush J W M 2009 J. Fluid Mech. 619:45-78.

Journal of Physics: Conference Series **656** (2015) 012153 doi:10.1088/1742-6596/656/1/012153

# Dynamics of a Bubble Cloud near a Rigid Wall

**J. Ma, C.-T. Hsiao[1], and G.L. Chahine**
DYNAFLOW, INC.
10621-J Iron-Bridge Road, Jesssup, MD 20794, USA

E-mail: *ctsung@dynaflow-inc.com*

**Abstract**. Simulations involving a large number of interacting bubbles in a bubble cloud are conducted to assess the effect of frequency and amplitude of a sinusoidal driving pressure on the bubble cloud behavior. As the pressure amplitude increases strong nonlinear bubble dynamics become more pronounced and higher pressures are generated at the wall. A resonance frequency, corresponding to the highest collective bubble behavior, results in a pressure peak orders of magnitudes higher than the excitation pressure. This frequency is significantly different from the natural frequency of a spherical cloud executing small amplitude oscillations.

## 1. Introduction

The collapse of a cloud of bubbles near a rigid boundary is known to be one of the most destructive forms of cavitation. It occurs in a variety of engineering applications including cavitation on propellers, ultrasonic cavitation, cavitating jets, Shock Wave Lithotripsy (SWL),.. etc. Numerical modeling of the dynamics is very challenging since it involves bubble-bubble, bubble-flow, and bubble-wall interactions [1]. We have examined in [2] several practical simulation models (i.e. not Direct Numerical Simulation) including a continuum homogeneous model, an Eulerian multi-component model, and an Eulerian-Lagrangian model. We found that, although all the models capture the average low-frequency dynamics, the microscale response of the discrete bubbles and the resultant high-frequency local fluctuations are only captured appropriately by the Eulerian-Lagrangian approach. This highlights that discrete bubble modeling is essential to a better understanding of the dynamics since it can resolve the physics down to the single bubble scale, while maintaining computation cost affordable to capture the overall large scale flow behavior. Using a generalized 3D two-way coupled Eulerian-Lagrangian model [3,4], this paper investigates the dynamics of a bubble cloud excited by a surrounding sinusoidal pressure field and computes the resulting pressures on a nearby rigid wall.

## 2. Numerical Method

The two-phase mixture in the cloud is treated as a continuum with unsteady continuity and momentum conservation equations expressed as follows:

$$\frac{\partial \rho_m}{\partial t} + \nabla \cdot (\rho_m \mathbf{u}) = 0, \qquad \rho_m \frac{D\mathbf{u}}{Dt} = -\nabla p + \mu_m \nabla^2 \mathbf{u}, \tag{1}$$

where $\rho_m$, $\mu_m$, **u**, and $p$ are respectively the mixture density, dynamic viscosity, velocity, and pressure. $\rho_m$ and $\mu_m$ are related to the liquid and gas properties and to the gas volume fraction, $\alpha$, by:

$$\rho_m = (1-\alpha)\rho_l + \alpha\rho_g, \qquad \mu_m = (1-\alpha)\mu_l + \alpha\mu_g. \tag{2}$$

[1] To whom any correspondence should be addressed.

We couple Eq. (1) with those describing the dynamics of the discrete individual bubbles in the cloud. Knowing at each instant all bubble radii and locations provides $\alpha$ (thus $\rho_m$ and $\mu_m$) as a function of space and substitutes for the need of a mixture equation of state to close the system of equations. Each bubble is treated as a source, which represents volume change, and a dipole to represent translation. The equivalent spherical bubble radius, $R(t)$, is obtained using a modified Rayleigh-Plesset-Keller-Herring equation [5] which accounts for the mixture compressibility and non-uniform pressure field:

$$\left(1-\frac{\dot{R}}{c_m}\right)R\ddot{R}+\frac{3}{2}\left(1-\frac{\dot{R}}{3c_m}\right)\dot{R}=\frac{\mathbf{u}_s^2}{4}+\frac{1}{\rho_m}\left(1+\frac{\dot{R}}{c_m}+\frac{R}{c_m}\frac{d}{dt}\right)\left[p_v+p_g-p_{enc}-\frac{2\gamma}{R}-4\mu_m\frac{\dot{R}}{R}\right]. \quad (3)$$

$c_m$ is the local sound speed in the mixture, $p_v$ is the liquid vapor pressure, $p_g$ is the bubble gas pressure, and $\gamma$ is the surface tension. The term $\mathbf{u}_s^2/4$, accounts for the pressure resulting from the difference between the host medium velocity, $\mathbf{u}_{enc}$, and the bubble velocity, $\mathbf{u}_b$, with $\mathbf{u}_s=\mathbf{u}_{enc}-\mathbf{u}_b$ [6]. $p_{enc}$ and $\mathbf{u}_{enc}$ are the encountered pressures and velocities averaged over the bubble surface to account for local non-uniform flow. The bubble trajectory is obtained using the following equation of motion [7]:

$$\frac{d\mathbf{u}_b}{dt}=-\frac{3}{\rho_l}\nabla p-2\mathbf{g}+\frac{3}{4}\frac{C_D}{R}\mathbf{u}_s|\mathbf{u}_s|+\frac{3}{2\pi}\frac{C_L}{R}\sqrt{\frac{\mu_l}{\rho_l}}\frac{\mathbf{u}_s\times\boldsymbol{\omega}}{\sqrt{|\boldsymbol{\omega}|}}+\frac{3\dot{R}}{R}\mathbf{u}_s, \quad (4)$$

where $C_L$ is a lift coefficient, $\boldsymbol{\omega}$ is the local vorticity, and $C_D$ is a drag coefficient. The last term is the Bjerknes force due to coupling between bubble volume rate and bubble motion.

## 3. Problem Setup

As illustrated in Fig. 1, an initially spherical bubble cloud with a radius, $A_0$, is driven by a pressure function $P(t)= P_0 - P_{amp}\sin(2\pi ft)$. The cloud center is initially at a distance $X_0$ from a rigid wall and is composed of small bubbles of initial radii $R_0$. The bubbles are randomly distributed within $A_0$, resulting in a quasi-uniform initial $\alpha_0$ within the cloud and all bubbles are initially at equilibrium with the initial pressure $P_0$.

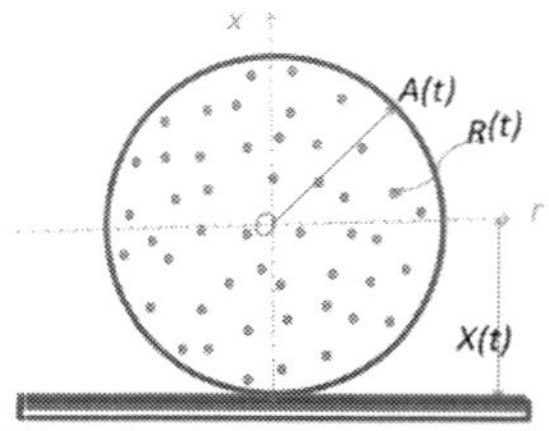


**Fig. 1:** Schematic of the problem of the dynamics of a bubble cloud near a rigid wall.

## 4. Driving Pressure Amplitude Effect

The effects of the driving pressure amplitude are illustrated in Fig. 2, which shows snapshots of the bubble cloud dynamics at different instants for $P_{amp}$ = 4 *atm* and 16 *atm*. The images from top to bottom are for increasing times during the first cycle of cloud oscillation. In both cases, the bubbles in the cloud grow first then collapse in response to the driving pressure. Overall, the cloud collapses in a cascade with the bubbles at the cloud top collapsing first and those on the bottom last. As the driving pressure amplitude increases weak bubble oscillations with small amplitudes transform into strong bubble growths and collapses as evidenced by the changes of the bubble sizes visualized by the colors in Fig. 2. This is made further clear in Fig. 4, which quantitatively compares, the variations of the radius of a bubble close to the wall for increasing amplitudes. For $P_{amp}$= 1 *atm*, the radius change is negligible. However, for $P_{amp}$ = 25 *atm*, the bubble grows to a maximum radius 2.5 times larger than $R_0$ and then collapses to a minimum of about

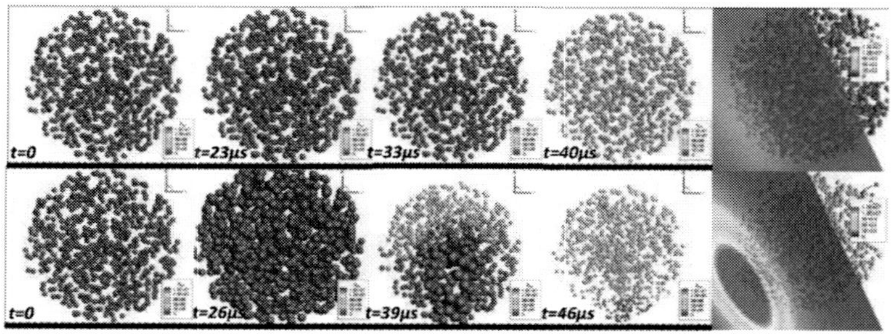


**Fig. 2.** Time sequence of bubble locations and pressures in a bubble cloud driven by pressure excitation with amplitude of 4 atm (top), and 16 atm (bottom). The last frame of each row is a side bottom view of the pressures at the rigid wall.

0.3 $R_0$. Fig. 3 compares the pressure histories monitored on the wall center and shows that the resulting peak increases from 10 *atm* to about 500 *atm* for increasing amplitudes.

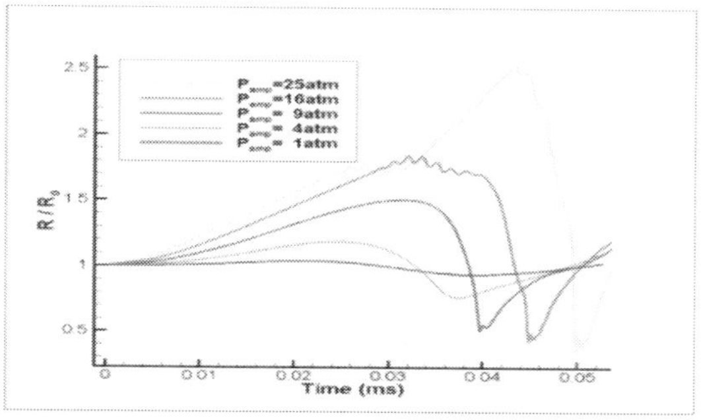


**Fig. 4**. Size variations of the bubble closest to the wall for different driving pressure amplitudes. $P_0$ =10 atm, $R_0$ =50 μm, $A_0$=1.5 mm, $\alpha_0$=5%, $f$=23 kHz and $X_0$=1.5

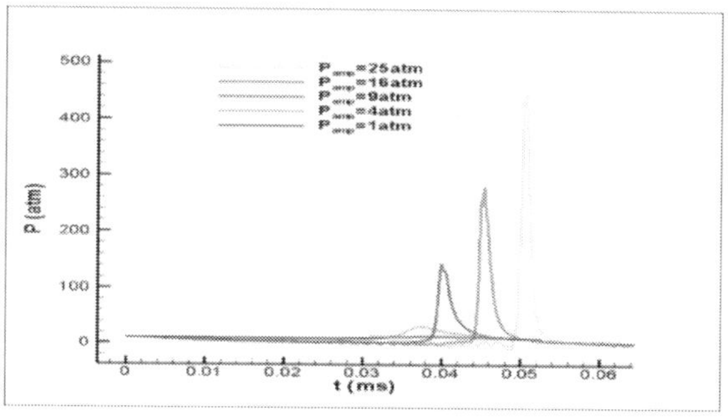


**Fig. 3:** Pressure at the wall center created by the cloud collapse for different excitation pressure amplitudes. $P_0$ =10 atm, $R_0$ =50 μm, $A_0$ =1.5 mm, $\alpha_0$=5%, $f$=23 kHz and $X_0$=1.5 mm.

## 5. Driving Frequency Effects

As illustrated in [8], tuning between the cloud characteristic frequency and the pressure field frequency is essential to generating very high collapse impulsive loads. When the driving pressure frequency matches the characteristic frequency of the bubble cloud, strong collective behavior occurs and very high pressures are generated at the wall. We compare in Fig. 5, for the same pressure amplitude of 9 *atm*, the effect of the driving frequency on the pressure at the wall. As the figure shows, the cloud behavior is very sensitive to the driving frequency. The highest collective effects appears to be at 8 *kHz* and result in a pressure peak of ~ 300 *atm*, two orders of magnitudes higher than the excitation pressure. As illustrated in Fig. 6, which shows bubble radii versus time for different bubbles in the cloud and the corresponding pressure variations, the optimum occurs when the driving pressure reaches its highest level while the individual bubbles are collapsing collectively. For lower frequencies, the driving pressure reaches its peak too early and drops before the bubbles start their collapse. Conversely, for higher frequencies, the driving pressure reaches its highest value too late and the bubble volumes change are not in phase. This optimum frequency of 8 *kHz*, is here much smaller than the natural frequency of a cloud, $f_{cloud}$ , derived for small amplitude oscillations and linear bubble dynamics theory [9,10].

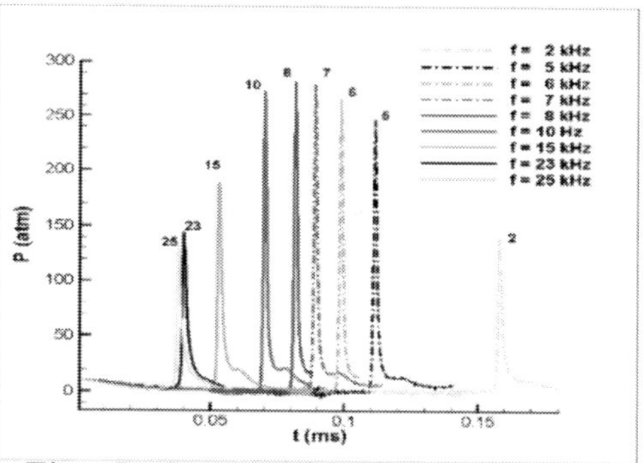


**Fig. 5:** Pressure at the wall center created by a cloud excited at different frequencies. $P_0$ =10 atm, $P_{amp}$ =9 atm, $R_0$=50μm, $\alpha_0$=5% and $A_0$=$X_0$=1.5 mm.

$$f_{cloud} = f_0\left(1+\frac{12}{\pi^2}\frac{A_0^2}{R_0^2}\frac{\alpha_0}{1-\alpha_0}\right)^{-1/2}, \text{ with } (2\pi f_0)^2 = \frac{P_{g0}}{\rho R_0^2}\left(3-\frac{2\gamma}{P_{g0}R_0}\right), \tag{5}$$

which gives a value of $f_{cloud}$ = 23 kHz. The discrepancy is due to the large amplitude oscillations and non-linear behavior of interest here.

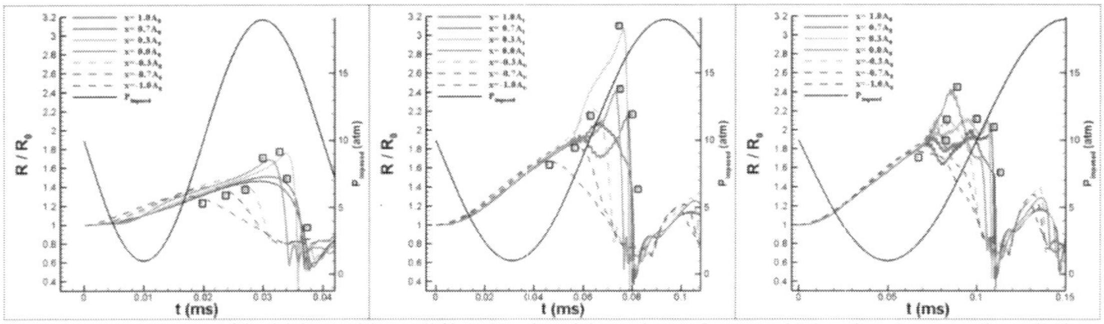


**Fig. 6:** Comparison of bubble radii at different cloud locations for a bubble cloud excited at different pressure frequencies: (a) $f$=25 kHz; (b) $f$=8 kHz, and (c) $f$=5 kHz. $P_0$=10 atm, $P_{amp}$=9 atm, $R_0$=50 μm, $A_0$=1.5 mm, $\alpha_0$=5%, $f$=23 kHz and $X_0$=1.5 mm.

## Conclusions

The dynamics of a bubble cloud subjected to an imposed pressure time variation and the resulting pressure loads on a nearby rigid wall was studied using a 3D two-way coupled Euler-Lagrange model. Strong nonlinear bubble dynamics occur as the amplitude increases and very high pressures are generated on the wall when the cloud collapses. At resonance frequency, a pressure peak orders of magnitudes higher than the excitation pressure occurs and corresponds to a matching between the driving pressure reaching its maximum and the collective individual bubble collapse. Off-resonance, the driving pressure reaches its peak amplitude too early or too late. The non-linear resonance frequency is significantly different from the classical bubble cloud natural frequency.

## Acknowledgements

This work was supported by a DOE SBIR Phase II (DE-SC0006443) and by ONR under contract N00014-12-C-0382 monitored by Dr. Ki-Han Kim. We are grateful for this support.

## References

[1] G. L. Chahine, "Numerical simulation of bubble flow interactions," *J. Hydrodyn. Ser. B*, vol. 21, no. 3, pp. 316–332, Jun. 2009.
[2] R. Raju, S. Singh, C.-T. Hsiao, and G. Chahine, "Study of Pressure Wave Propagation in a Two-Phase Bubbly Mixture," *ASME J. Fluids Eng.*, vol. 133, no. 12, pp. 121302–312, 2011.
[3] J. Ma, G. L. Chahine, and C.-T. Hsiao, "Spherical bubble dynamics in a bubbly medium using an Euler–Lagrange model," in *Chemical Engineering Science*, 2015, vol. 128, pp. 64–81.
[4] J. Ma, C.-T. Hsiao, and G. L. Chahine, "Euler-Lagrange Simulations of Bubble Cloud Dynamics Near a Wall," *ASME J. Fluids Eng.*, vol. 137, no. 4, pp. 041301–10, 2015.
[5] M. S. Plesset and A. Prosperetti, "Bubble Dynamics and Cavitation," *Annual Review of Fluid Mechanics*, vol. 9, no. 1. pp. 145–185, 1977.
[6] C.-T. Hsiao, G. L. Chahine, and H.-L. Liu, "Scaling effect on prediction of cavitation inception in a line vortex flow," *J. Fluids Eng.*, vol. 125, no. 1, pp. 53–60, 2003.
[7] V. E. Johnson and T. Hsieh, "The Influence of the Trajectories of Gas Nuclei on Cavitation Inception," in *Proc. 6th Symposium on Naval Hydrodynamics.*, 1966, pp. 163–179.
[8] G. L. Chahine, C.-. T. Hsiao, and R. Raju, "Scaling of Cavitation Bubble Cloud Dynamics on Propellers," in Kim K-H, Chahine G.L., Franc J-P, Karimi A. (eds): *Advanced Experimental and Numerical Techniques for Cavitation Erosion Prediction*, Chapter 15, pp 345-373, Springer, Berlin, 2014.
[9] L. D'Agostino and C. Brennen, "On the acoustical dynamics of bubble clouds," in *ASME Cavitation and Multiphase Flow Forum*, 1983, pp. 72–75.
[10] C. E. Brennen, *Cavitation and Bubble Dynamics*. New York: Oxford University Press, 1995.

# Observations and Measurements on Unsteady Cloud Cavitation Flow Structures

**L X Gu[1], G J Yan[1], B Huang[2]**
[1] Beijing Institute of Astronautical System Engineering, Beijing, 100076, China
[2] Beijing Institute of Technology, Beijing 100081, China

E-mail: huangbiao@bit.edu.cn

**Abstact** The objectives of this paper are to investigate the unsteady structures and hydrodynamics of cavitating flows. Experimental results are presented for a Clark-Y hydrofoil, which is fixed at $\alpha=0°$, $5°$ and $8°$. The high-speed video camera and Particle Image Velocimetry (PIV) are applied to investigate the transient flow structures. The dynamic measurement system is used to record the dynamic characteristics. The cloud cavitation exhibits noticeable unsteady characteristics. For the case of $\alpha=0°$, there exit strong interactions between the attached cavity and the re-entrant flow. While for the case of $\alpha=8°$, the re-entrant flow is relatively thin and the interaction between the cavity and re-entrant flow is limited. The results also present that the periodic collapse and shedding of the large-scale cloud cavitation, which leads to substantial increase of turbulent velocity fluctuations in the cavity region. Experimental evidence indicates that the hydrodynamics are clearly affected by the cavitating flow structures, the amplitude of load fluctuation are much higher for the cloud cavitating cases.

## 1. Introduction

It is well known that unsteady cavitating flow in turbomachineries and controlled surface would lead to many problems such as sudden changes in loads, pressure pulsation, vibrations and noise [1-2]. Among the various cavitation phenomenons, more attention has been paid to the cloud cavitation due to the unsteady characteristics [3-4]. Various experimental studies have been conducted to improve the understanding of the complex structures and unsteady hydrodynamics of cloud cavitating flows. Li et al. [5] employed a PIV method to examine the structures of developed cavitating flows. They showed that strong mass and momentum transfers occurs between the upper and lower flow layers, which leads to the lower velocity distribution in the core part of the cavitating region. Shridhar and Katz [6] used fluorescent tracer particles and an optical filter to cancel the incident laser light and measured the velocity filed in the wake of a cloud cavity. Reynolds shear stresses were found been increased by 25 to 40% due to cavitation.

Although much works have explored the effects of cavitation through experimental studies, the complex flow structure and hydrodynamic characteristics of unsteady cloud cavitation are still not well understood. The objective of the present study is to obtain a comprehensive set of flow field data, video recordings and load coefficients to investigate the structures and hydrodynamics of unsteady cloud cavitation with various angles of attack.

[2] To whom any correspondence should be addressed.

## 2. Experimental set-up

Experimental studies are conducted in a closed-loop cavitation tunnel at Beijing Institute of Technology [7]. The schematic of experimental setup is illustrated in figure 1. The cavitation phenomenon is documented by a high-speed digital camera (HG-LE, by Redlake), with 4000 fps used in this study. A two-component high-speed PIV system provided by Dantec is used. The recording rate of 2000 Hz (4000 fps) at 1024×512 pixel is used for the actual data acquisition. The lift and drag are measured by four strain gauges of a dynamic measurement device [7], as shown in figure 2. Signals are amplified, filtered and collected through an 8-channel 16-bit A/D digitizer at simultaneous sample with a maximum available sample frequency of 51.2 kHz. A Clark-Y hydrofoil with different angles of attack is adopted in the present study, and the chord length of the hydrofoil is c=0.07 m. The position of the hydrofoil inside the test section is shown in figure 3. In this experimental research, the reference velocity $U_\infty$ is fixed at 10m/s and the Reynolds number is $Re = U_\infty c/\nu = 7\times10^5$.

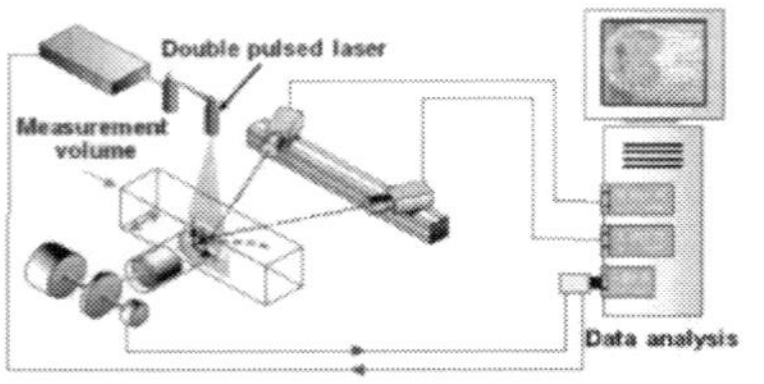


**Figure 1** Schematic of the cavitation tunnel

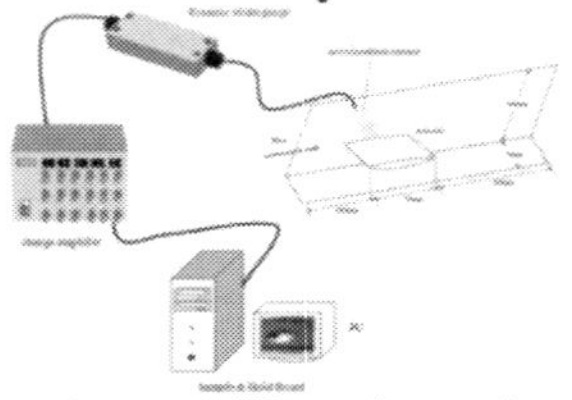

**Figure 2** The schematic diagram of the dynamic measurement system

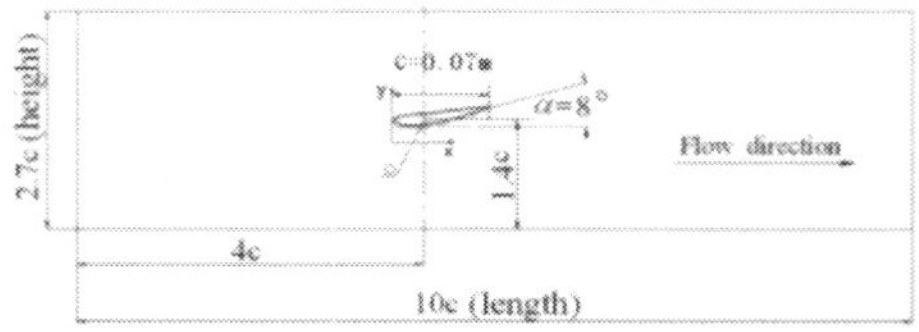


**Figure 3** Sketch of the test section

## 3. Results and discussion

### 3.1. The unsteady multiphase structures with cloud cavitating flows

Figure 4 shows the side views of the periodic cavity structures, within a single flow cycle for all the cases. Based on the experimental visualization, it can be observed that the development of cloud cavitation has a distinctly quasi-periodic pattern and the periods $T_{cycle}$ of the cloud cavitation is about 40ms, which is corresponding to $5.7c/U_\infty$. Figure 5 presents the measured time evolution of the nomarlized cavity area over several flow cycles at cloud cavitation regimes, which is drawn by an in-house feature-recognition software package [8]. The cavity areas change periodically and the measured mean value are $0.49S_c$, $0.59\ S_c$ and $0.72\ S_c$ respectively, where $S_c$ is the cross-section area of the hydrofoil.

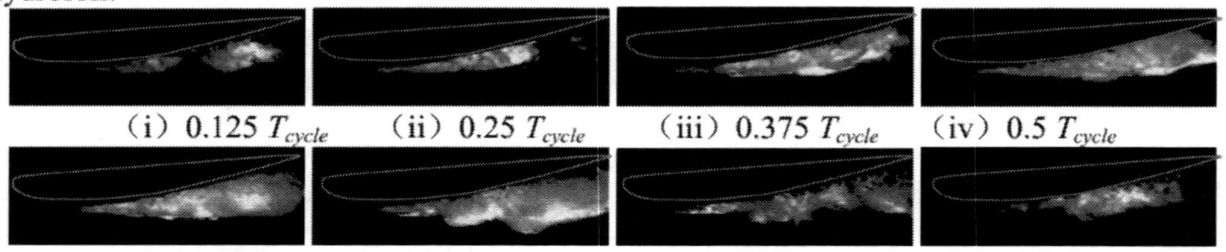

(i) 0.125 $T_{cycle}$ (ii) 0.25 $T_{cycle}$ (iii) 0.375 $T_{cycle}$ (iv) 0.5 $T_{cycle}$

(v) 0.625 $T_{cycle}$ (vi) 0.750 $T_{cycle}$ (vii) 0.875 $T_{cycle}$ (viii) 1.0 $T_{cycle}$

**Figure 4** Time evolution of cavity shapes in the typical cloud cavitation regimes (α=8°, σ=0.80).

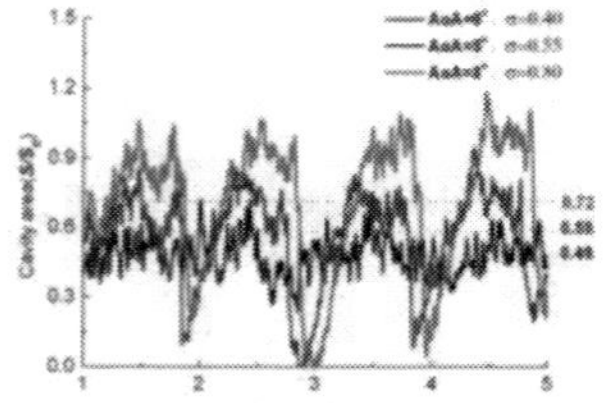

**Figure 5** The time evolution of the cavity area.

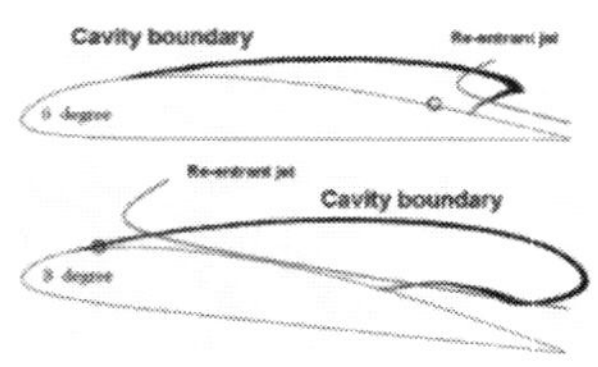


**Figure 6** The re-entrant flow in the cavity region

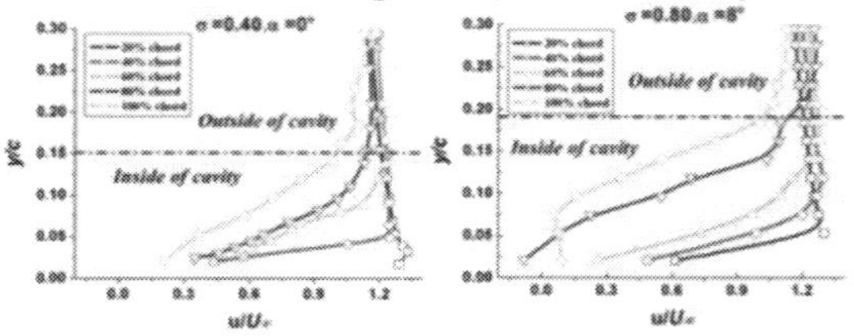


**Figure 7** Mean of the ensemble averaged axial velocity profiles at selected chordwise locations.

Figure 6 shows the schematic representation of the re-entrant flow in the region of an attached cavity in typical cloud cavitation regimes. It could be concluded that the cavity thickness is directly controlled by the angle of attack, and the ratio of the thickness of the cavity to that of the re-entrant flow is substantially important to estimate the unsteady behavior of the partial cavity. For $\sigma$=0.80 at $\alpha$=8°, the cavity is thick enough to limit the interaction between the re-entrant flow and the cavity interface, which allows the re-entrant flow to reach the cavity leading edge and the formation of a large scale vapor structures. However, for $\sigma$=0.40 at $\alpha$=0°, because of the smaller cavity thickness, there is a strong interaction between the cavity interface and the re-entrant flow. The re-entrant flow can only reach the rear part of the hydrofoil, contrary to the large scale cloud shed for $\sigma$=0.80 at $\alpha$=8°, the small vapor structures are formed.

### *3.2. The flow structures in cavitating region*

Figure 7 shows the mean values of the normalized ensemble averaged axial (u) velocity profiles at different chordwise locations for $\sigma$=0.80 at $\alpha$=8° and $\sigma$=0.40 at $\alpha$=0°. The Y axis represents the vertical distance from the suction side of the foil. For the case of $\sigma$=0.80 at $\alpha$=8°, recirculation breaks up and lifts the cavity upward, and the large scale shedding is formed, which significantly increases the thickness of the boundary layer and larger low velocity region can be observed compared to the case for $\sigma$=0.40 at $\alpha$=0°. Consequently, the gradients of the mean axial velocity profiles for the $\sigma$=0.40 at $\alpha$=0° case tend to be smaller than that for the $\sigma$=0.80 at $\alpha$=8° case.

With the fluorescent particles in the flow field (seen in figure 8(a), we can capture the turbulent velocity fields, figures 8(b) and (c) show the normalized amplitude of the averaged fluctuating velocities, $I = \sqrt{u'^2 + v'^2} / U_\infty$, where $u'$ and $v'$ are the horizontal and vertical components of the turbulent velocity fluctuations. For both cases, large velocity fluctuation occurs near the aft half-chord of the hydrofoil and continues into wake, indicating unsteady large-scale fluctuations due to the shedding of the cloud cavity, especially for the $\sigma$=0.80 at $\alpha$=8° case. Meanwhile, the turbulent fluctuations are confined to a thicker layer on the suction side.

(a) the fluorescent particles in the flow field (b) $\alpha$=0°,$\sigma$=0.40 (c) $\alpha$=8°, $\sigma$=0.80

**Figure 8** Normalized amplitude of the turbulent velocity fluctuations.

### *3.3. The hydrodynamics of unsteady cavitating flows*

Figure 9 shows the measured lift coefficient ($C_L$) and drag coefficient ($C_D$) over a range of non-dimensional parameter $\sigma/2\alpha$. When the flow is subcavitating, both lift and drag coefficients remain largely unchanged as $\sigma/2\alpha$ varies. In the incipient cavitation stage, the effect of cavitation on the lift and drag coefficients is very small. In the subcavitating and incipient cavitation stages, the hydrodynamic coefficients show little fluctuations due to the relatively stable cavity dynamics, as indicated by the error bars. Further decreasing $\sigma/2\alpha$, the development of cavitation increases the drag while the lift coefficient becomes lower. In the cloud cavitation stage, the large vortex shedding and related unsteady movement strongly affect the flow structures around the hydrofoil, leading to the relatively higher magnitude of drag and lower magnitude of lift. As indicated by the error bars of the hydrodynamic coefficients in the cloud cavitation stage, the measured values show large fluctuations, especially for the case of $\alpha$=8°, which is induced by the unstable cavity dynamics. Finally, the flow transitions from cloud cavitation to supercavitation, the cavities exhibits more steady characteristics and both drag and lift coefficients become lower.

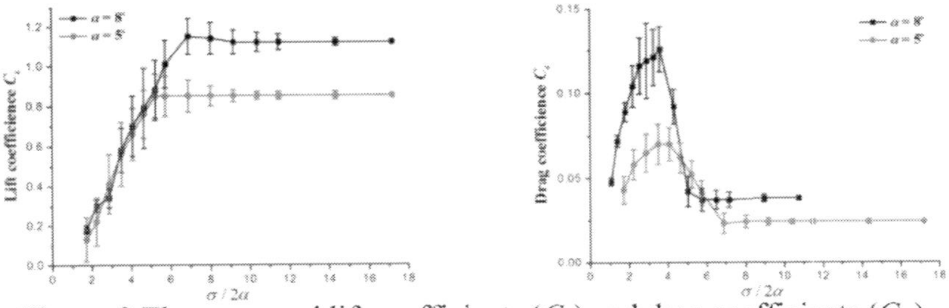


**Figure 9** The measured lift coefficients ($C_L$) and drag coefficients ($C_D$).

## 4. Conclusions

The objectives of this paper are to investigate the unsteady structures and hydrodynamic characteristics of the cavitating flows, and to provide experimental data for future numerical validating studies. Experimental studies are presented for a Clark-Y hydrofoil at the fixed angles of attack of α=0°, 5° and 8° with $Re$=7×10$^5$ for various cavitation numbers. Detailed analysis of the cavity behavior is applied to identify the re-entrant flow instability which leads to cloud cavitation at different angles of attack. The primary findings include:

(1) Cloud cavity is highly unsteady. For σ=0.40 at α=0°, the thickness of the attached cavity is comparable to the re-entrant flow, a relatively strong interaction exists between the cavity interface and the re-entrant flow throughout its upstream movement. The cavity in the rear part of the hydrofoil splits into small vapor structures. For σ=0.80 at α=8°, the attached cavity is thick enough to limit the interaction between the re-entrant flow and the cavity interface during its movement to the leading edge and the interface is cut, contrary to the small scale cloud shed for σ=0.40 at α=0°, the typical large vapor structures are formed.

(2) Compared to the case for σ=0.40 at α=0°, the velocities in the cavitating region are much lower, and the fluctuating velocities as well as turbulent intensities are much higher for the cloud cavitating case with σ=0.80 at α=8°, which lead to a much thicker turbulent boundary layer.

(3) The dynamic characteristics of the cavitation vary considerably with the angles of attack. The effect of the incipient cavity on the lift and drag coefficients is very small, little fluctuation of the dynamic is obseved. With the decrease of cavitation numbers, the lift and drag coefficients are affected by the cavitating flow structure. At cloud cavitation, the large scale shedding leads to the relatively higher magnitude of drag and decrease of the lift, and the dynamic coefficients exhibit substantial fluctuations. As supercavitation appears, the cavity shapes and dynamics seem more stable.

**Acknowledgements**

This work was supported by the National Science Foundation of China (Grant No.51306020) and Beijing Natural Science Foundation (Grant No.3144034).

**References**

[1] Arndt R E A 1981 *Ann. Rev. Fluid Mech.* **13** 273-328.
[2] Ausoni P, Farhat M, Escaler X, Equsquiza E and Avellan F 2007 *ASME J. Fluid Eng* **129** 966-73.
[3] Huang B, Ducoin A and Yong Y L 2013 *Phys. Fluids* **25** 102109.
[4] Wang G, Senocak I, Shyy W, Ikohagi T and Cao S 2001 *Prog. Aero. Sci.* **37** 551-81.
[5] Li X, Wang G, Zhang M and Shyy W 2008 *Int. J. Heat Fluid Flow* **47** 1263-75.
[6] Shridhar G and Katz J 2000 *Phys. Fluids* **12**(4) 895-911.
[7] Huang B 2012 *Thesis of PhD* Beijing Institute of Technology.
[8] Zhang M, Song X and Wang G 2006 *Trans. Beijing Inst. Technol.* **26** 983-6.

# Cavitation on hydrofoils with sinusoidal leading edge

**H Johari**

Mechanical Engineering, California State University, Northridge, CA 91330 USA

E-mail: hjohari@csun.edu

**Abstract**. Cavitation characteristics of hydrofoils with sinusoidal leading edge were examined experimentally at a Reynolds number of $7.2\times10^5$. The hydrofoils had an underlying NACA $63_4$-021 profile and an aspect ratio of 4.3. The sinusoidal leading edge geometries included three amplitudes of 2.5%, 5%, and 12% and two wavelengths of 25% and 50% of the mean chord length. Results revealed that cavitation on the leading edge-modified hydrofoils existed in pockets behind the troughs whereas the baseline hydrofoil produced cavitation along its entire span. Moreover, cavitation on the modified hydrofoils appeared at consistently lower angles of attack than on the baseline hydrofoil.

## 1. Introduction

Despite its large size (12 – 16 m) and rigid body, the humpback whale (*Megaptera novaeangliae*) is quite maneuverable compared to other species. This agility has been attributed to the humpback whales' use of its unique pectoral flippers [1]. The flippers have several distinct characteristics including a unique planform shape with an aspect ratio of about 6, a span of up to one-third of the body length, and large protuberances along the leading edge. The Reynolds number for the flipper ranges from one-half to one million. Tight turns during bubble netting are unique to the humpback whale, and the protuberances on the pectoral flipper leading edges are thought to be for flow control.

Idealized humpback whale flipper models with and without leading edge protuberances were tested in a wind tunnel by Miklosovic et al. [2]. The effects of a sinusoidal leading edge on the loading of full-span NACA $63_4$-021 foils were investigated in a water tunnel by Johari et al [3]. Depending on the geometry of the leading edge, lift on the modified foils was lower in the pre-stall regime and higher in the post-stall regime when compared to a baseline foil without any protuberances. No drag reduction was observed for any of the modified leading edge foils studied. Flow visualization revealed that the flow over the protuberance peaks remains attached well past the stall angle of attack of the baseline foil. Load measurements on wings with various planforms and sinusoidal leading edge have been reported by Custodio et al. [4]. Weber et al. [5] tested ship rudder geometries with and without leading edge protuberances in a water tunnel at Reynolds numbers up to $8.8\times10^5$. They indicated that the effects of leading edge modifications vanished at high Reynolds numbers. Yoon et al. [6] carried out numerical simulation of rectangular wings with a sharp tip and a semi-aspect ratio of 1.5 at Reynolds number of $10^6$ for ship rudder applications. The rudder had protuberance amplitude and wavelength of 2.5% and 20% of the chord length, respectively. Lift enhancement was observed only in the post-stall regime, similar to the lower Reynolds number full-span experiments.

The primary objective of this study was to establish the effect of various sinusoidal leading edge geometries on the cavitation characteristics of rectangular planform wings at a Reynolds numbers of $7.2\times10^5$ and angles of attack up to 24°.

## 2. Experimental Setup

To examine the cavitation characteristics of hydrofoils with leading edge protuberances, a qualitative flow visualization study was carried at the Naval Undersea Warfare Center closed-loop water tunnel facility in Newport, RI, USA. The water tunnel has a 0.30-m square test section, and the freestream velocity was set at 7.2 m/s. The freestream turbulent intensity on the tunnel centerline was 0.6%. The freestream velocity in the water tunnel was monitored at a point 0.66 m upstream of the models using a Laser Doppler Velocimeter. Monitoring and adjustment of the freestream velocity was required due to the variations caused by the different models and angles of attack. The angle of attack, $\alpha$, was varied from 12° to 24° in increments of 3°. Cavitation was absent at angles of attack below 12° at the freestream velocity of 7.2 m/s. Images were taken with a Nikon D200 Digital SLR camera and 200 μs strobe lighting to illuminate areas of cavitation; further details are available in [7]. The incipient cavitation criterion was the first visual cues of presence of vaporous cavitation near the leading edge.

Seven rectangular planform hydrofoils with mean chord length $c = 102$ mm and span of 219 mm were designed and CNC-fabricated from aluminum and are hydrodynamically smooth. To eliminate the sharp, flat edge at the free tip, rounded caps were attached to the free end of these models. Including the rounded cap, all hydrofoils had an aspect ratio of 4.3. All models had an underlying NACA $63_4$-021 profile, which is symmetric with a maximum thickness of 0.21 $c$ located at 0.4 $c$. This profile was chosen for its similarity to the humpback whale flipper cross-section. The Reynolds number based on the mean chord length and the freestream velocity was $7.2\times10^5$ for all cases.

The leading edge of the modified hydrofoils was defined by a sinusoidal pattern with specified amplitude and wavelength; the leading edge amplitude and wavelength were constant fractions of the mean chord length. Three amplitudes of 0.025 $c$, 0.05 $c$, and 0.12 $c$ along with two wavelengths of 0.25 $c$ and 0.50 $c$ were considered. These values were chosen as they are representative of those found on the humpback whale flippers. Each hydrofoil was designated by the wavelength (4 for 0.50 $c$ and 8 for 0.25 $c$) and amplitude (S for 0.025 $c$, M for 0.05 $c$, and L for 0.12 $c$) of the leading edge. The hydrofoils were designed so that the top and bottom surfaces were not wavy.

## 3. Results

The cavitation characteristics of modified hydrofoils with wavelengths of 0.50 $c$ and 0.25 $c$ are presented in Figs. 1 and 2, respectively. Cavitation can be distinguished by the white color of the vapor in the low pressure regions. Cavitation effects on the baseline hydrofoil can be seen in Figs. 1a, 2a. As with all the hydrofoils examined, cavitation was observed first at $\alpha = 12°$ in the core of the tip vortex. With increased angle of attack, cavitation in the tip vortex grew until $\alpha = 21°$, after which point tip vortex cavitation diminished as a result of stall at the higher angles. Figures 1b and 2b illustrate that the modified hydrofoils with protuberance amplitude of 0.025 $c$ (4S and 8S) had tip vortex cavitation characteristics similar to the baseline hydrofoil, with an increase in the vortex cavity size until $\alpha = 18°$. At angles of attack greater than $\alpha = 18°$, the size of the cavity generated by the tip vortex decreased gradually. The 4M hydrofoil in Fig. 1c produced a vortex cavity that diminished at an earlier angle than the baseline while the 8M hydrofoil in Fig. 2c revealed a trend similar to that of the baseline. The hydrofoils with the largest amplitude protuberances (4L and 8L) had tip vortex cavities that diminished at an earlier angle than the baseline. At the Reynolds number of the experiments, the baseline hydrofoil produced greater lift than the modified hydrofoils with large protuberances. This created a stronger tip vortex on the baseline hydrofoil, in turn generating a larger tip vortex cavity.

The rapid acceleration of flow near the leading edge of the hydrofoils leads to reduced pressure and cavitation along the leading edge. Similar to the cavitation that occurs in the tip vortex, cavitation at the leading edge was also modified by the presence of protuberances. Sheet cavitation developed on the leading edge of the baseline hydrofoil as angle of attack increased past $\alpha = 15°$, see Fig. 1a. Incipient cavitation occurred on the baseline hydrofoil at an angle of $\alpha = 15°$. At $\alpha = 18°$, sheet cavitation had developed along the majority of the baseline hydrofoil leading edge. As the angle of attack was further increased to $\alpha = 21°$, the cavity associated with sheet cavitation enlarged and remained essentially unchanged as the angle of attack reached $\alpha = 24°$.

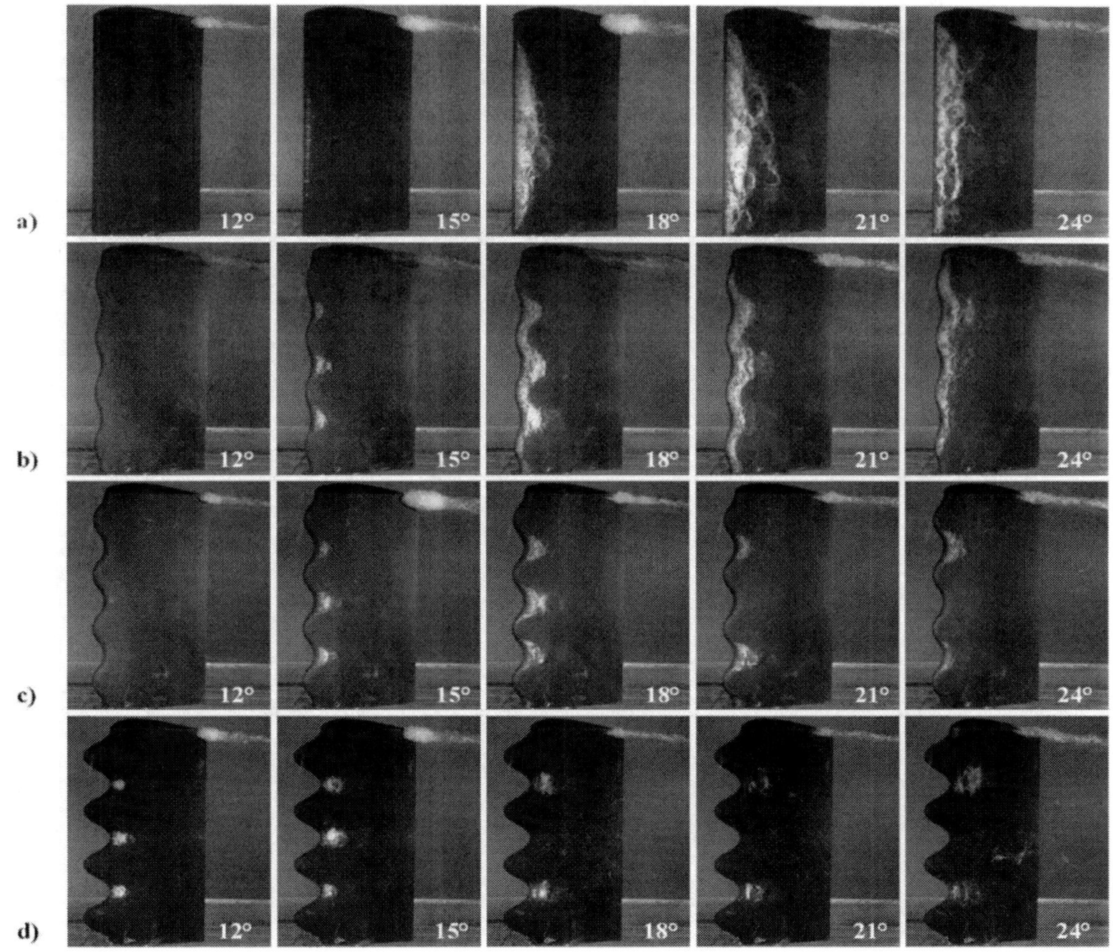


**Figure 1.** Cavitation on the 0.50 *c* wavelength hydrofoils: a) baseline, b) 4S, c) 4M, and d) 4L.

For all modified hydrofoils, incipient cavitation initiated in the troughs of the protuberances. This was due to the larger leading edge radius in the troughs of modified hydrofoils leading to the lowest local pressures. Modified hydrofoils with the largest protuberance amplitude of 0.12 *c* showed incipient cavitation at angles as low as $\alpha = 12°$ whereas all other modified hydrofoils showed the incipient condition at an angle of $\alpha = 15°$. Hydrofoils with the smallest protuberance amplitude of 0.025 *c* in Figs. 1b and 2b, developed sheet cavitation similar to the baseline hydrofoil, with increasing sheet size corresponding to an increase in angle of attack. However, sheet cavitation along the leading edge of the 4S and 8S hydrofoils was limited to a smaller fraction of the chord length.

At high angles of attack, unsteady flow on the surface of the modified hydrofoils led to transient cavitation cells behind the troughs. This was true for all modified hydrofoils at the higher angles of attack but especially true for the hydrofoils with the largest amplitude protuberances. The 4L and 8L modified hydrofoils cavitated only in the troughs and not along the peaks on the leading edge. Hydrofoils with medium (4M and 8M) and large amplitude (4L and 8L) protuberances had similar cavitation characteristics. Protuberance wavelength appeared to play only a minor role in establishing the cavitation patterns observed on the modified hydrofoils in comparison with amplitude. Generally, hydrofoils with the shorter wavelength developed transient cavitation cells at higher angles of attack than the hydrofoils with longer wavelengths.

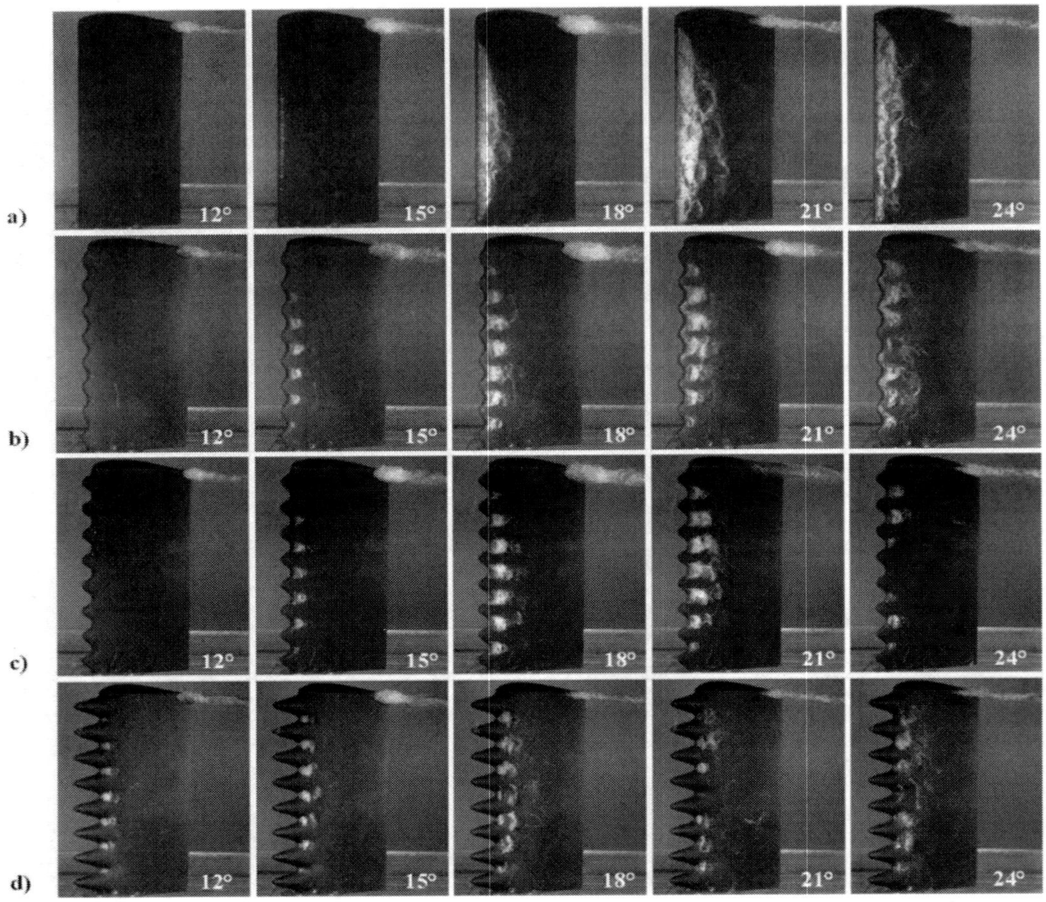


**Figure 2.** Cavitation on the 0.25 *c* wavelength hydrofoils: a) baseline, b) 8S, c) 8M, and d) 8L.

## 4. Conclusions

Cavitation characteristics of hydrofoils with various sinusoidal leading edge geometries were examined in a flow visualization experiment at a Reynolds number of $7.2\times10^5$. Still images revealed that cavitation on the modified hydrofoils was largely confined to the area directly behind the troughs of protuberances, whereas the baseline hydrofoil exhibited sheet cavitation over the entire span. The incipient cavitation on the baseline hydrofoil always occurred at an angle of attack greater than that for the modified hydrofoils. This implies that for a given local static pressure, leading edge-modified hydrofoils will always cavitate at a lower velocity than their straight leading edge counterpart.

## 5. References

[1] Fish F E and Battle J M 1995 *J. Morphol.* **225** 51–60
[2] Miklosovic D S, Murray M M, Howle L E and Fish F E 2004 *Phys. Fluids* **16** L39–42
[3] Johari H, Henoch C, Custodio D and Levshin A 2007 *AIAA J.* **45** 2634–42
[4] Custodio D, Henoch C and Johari H 2015 *AIAA J.* **53** (doi: 10.2514/1.J053568)
[5] Weber P W, Howle L E and Murray, M M 2010 *Marine Technol.* **47** 27–36
[6] Yoon H S, Hung P A, Jung J H and Kim M C 2011 *Comput. Fluids* **49** 276–89
[7] Custodio D 2012 Ph.D. dissertation, Worcester Polytechnic Institute, Worcester, MA, USA

## Acknowledgements

This project was supported by the Office of Naval Research-University Laboratory Initiative program under contract N00014-08-1-1043. Derrick Custodio and Charles Henoch carried out the experiments.

Journal of Physics: Conference Series **656** (2015) 012156 doi:10.1088/1742-6596/656/1/012156

# Jet-Supercavity Interaction: Insights from Physics Analysis

**IN Kirschner**[1], **MJ Moeny**[2], **MH Krane**[2], **MP Kinzel**[2]

[1] Applied Physical Sciences Corporation, Groton, CT, USA

[2] Applied Research Laboratory, The Pennsylvania State University, University Park, PA, USA

E-mail: IKirschner@APhySci.com

**Abstract.** Various closure conditions of a ventilated cavity enveloping all or part of a high-speed underwater body are introduced, including those involving a propulsion jet. The flow regimes for the latter are described based on Efros-Paryshev theory, which is extended to estimate the efficiency and fundamental limitations of a rocket-type propulsor.

## Introduction

The closure regions of nominally axisymmetric cavities have been the subject of extensive research. Emphasis has been placed on the closure of a supercavity. Since such a cavity closes downstream of the body that it envelopes, it is not subject to interference from a solid surface. For practical operation of self-propelled vehicles other closure conditions are of interest. Several closure conditions are shown in figure 1 as images acquired in experiments more fully described in Moeny, et al., (2015).

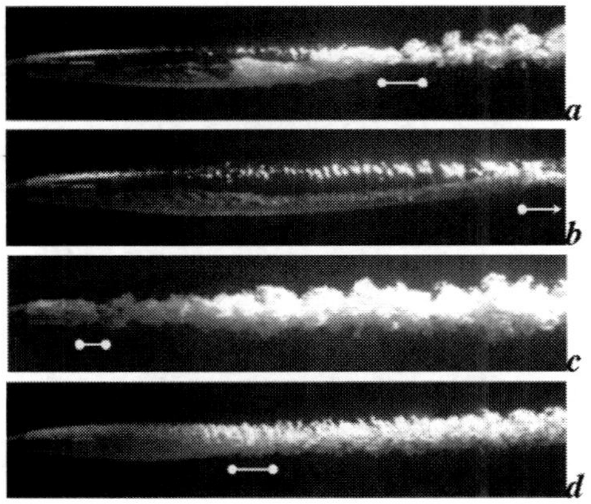


*Figure 1.* Some regimes of cavity closure: ***a*** – Re-entrant jet off body; ***b*** – Twin-vortex off body; ***c*** – Collapsed closure on body; ***d*** – Closure on propulsion jet. Approximate location of cavity closure (either in or downstream of the field of view) is indicated.

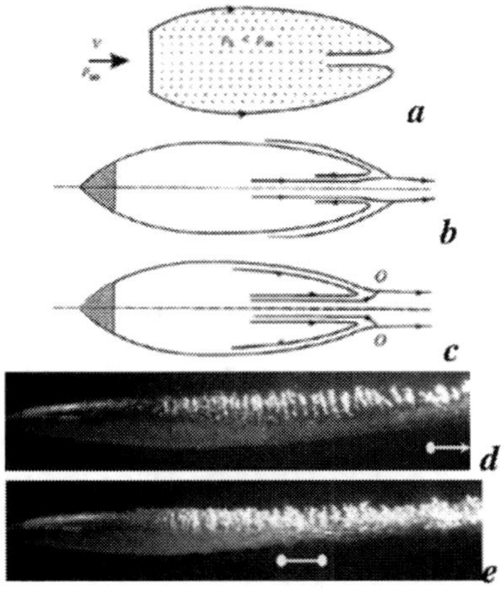


*Figure 2.* Efros and Paryshev closure models: ***a*** – Efros re-entrant jet model of supercavity closure; ***b*** – Extension of Efros model to closure on a "hard" jet; ***c*** – Paryshev's model for closure on a "soft" jet; ***d*** and ***e*** – Recent water tunnel results corresponding to ***b*** and ***c***, respectively. Images ***a-c*** from Paryshev (2006).

Other regimes of interest not shown in figure 1 include closure on a lifting or planing afterbody, which generates a vortex pair. Cavitator lift also generates trailing vortex structures. Also of interest are dynamical effects of cavity auto-oscillation. This article primarily addresses closure of nominally axisymmetric, steady cavities on a propulsion plume. That can have a strong effect on entrainment of ventilation gas from the cavity, and therefore direct consequences for design and operations.

**Methods**

As described in Paryshev (2006), classical re-entrant jet cavity closure theory (Efros, 1946) has been extended to cavity closure on a solid afterbody, and then later to cavity closure on a high-pressure gas jet. In the latter, the model is applied to flow systems for which the stagnation pressure of the jet is greater than that of the freestream flow of ambient liquid, a case referred to herein as a "hard" jet, since it is dynamically equivalent to closure on a solid body. In that case, high jet stagnation pressure forces ambient liquid to form re-entrant flow upstream along the jet into the cavity. A complimentary case is that of a "soft" jet, with stagnation pressure less than the ambient liquid flow. In that case, the central jet stagnates on the ambient liquid, forcing re-entrant flow of gas upstream along the cavity boundary. The original Efros model, its application to closure on a "hard" jet, and the case of cavity closure on a "soft" jet are depicted in figure 2, along with some high-speed video frames from recent experiments performed to explore these effects. The latter show subtle but important changes in the cavity-jet system. For the "hard" jet (figures 2b and c), the cavity remains convex well downstream of the field of view, and does so until the liquid just outside the cavity boundary stagnates on the central jet. For the "soft" jet, inflection occurs near cavity closure, due to rapid transition of cavity pressure from a nearly constant value, through a high pressure region near jet stagnation on the cavity boundary, followed by a rapid pressure drop to ambient conditions once the jet exits the cavity. In the short region of super-ambient pressure, the cavity boundary is concave. Downstream geometry is dominated by turbulent jet spreading after penetration of the liquid layer. Thus, the topology of closure on a jet is governed by the ratio of stagnation pressures of the ambient liquid and the jet: $P \doteq \rho V^2/\rho_j V_j^2$, where $\rho$ and $\rho_j$ are the densities of the liquid and the jet, respectively, and where $V$ and $V_j$ are the associated velocities.

A model for supercavity closure on a "soft" jet was offered by Paryshev (2006). The simplifying assumptions may be understood via figure 3a, and may be summarized as follows: (1) the flow system is slender; (2) the cavity sections obey the principle of independence of expansion (Logvinovich, 1969); such that, (3) standard approximations for cavity shape may be applied (for example, Logvinovich & Serebryakov, 1975); (4) the jet stagnates on the cavity boundary, such that all or part of the gas in the jet re-enters the cavity along the inner surface of the cavity boundary; and, (5) the region of high pressure associated with jet stagnation is confined to a short axial segment of the system as described above. Paryshev applies the model to deduce the condition required for part of the jet to penetrate the cavity boundary in terms of the momentum flux of the jet, $\rho_j V_j^2 S_j$ (where $S_j$ is the nominal area of the propulsion jet), made dimensionless with respect to the cavitator drag, $D_n$, resulting in the second dimensionless parameter of importance: $\bar{J} \doteq \rho_j V_j^2 S_j / D_n$. Paryshev's model results in a condition for penetration that $\bar{J} > 1/2$: for lower values, the "soft" jet is "weak," none of the gas penetrates the cavity, and it only supplements the ventilation gas; for values increasing above the cut-off, the "soft" jet is "strong," and an increasing fraction penetrates the cavity.

The three regimes demarcated by Efros-Paryshev theory are depicted in figure 3b. Neither model accounts for turbulent mixing and entrainment as depicted in figure 3c, which shows a budget for the volumetric flow of gas through the system in the Paryshev regime. In that figure, $q_{v0}$ represents the basic flow of ventilation gas supplied to the cavity independent of the propulsion gas, whereas $q_{j0}$ represents the flow of gas through the rocket nozzle. The Efros-Paryshev model captures the re-entrant flow of propulsion gas into the cavity, $q_3$, and any direct penetration of propulsion gas through the cavity boundary, $q_2$; but it does not account for entrainment due to turbulent mixing or other diffusive processes of the ventilation gas by the jet, $q_{ej}$, nor entrainment by the ambient liquid, $q_{ec}$. Paryshev's model estimates the supplemental ventilation of the cavity by the propulsion gas in the "soft jet" regime of closure. This is plotted as a function of both the stagnation pressure ratio and the dimensionless jet momentum flux in figure 4, and is complementary to the flow of gas exiting the cavity: the two flow rates sum to the total volumetric flow rate from the propulsion nozzle.

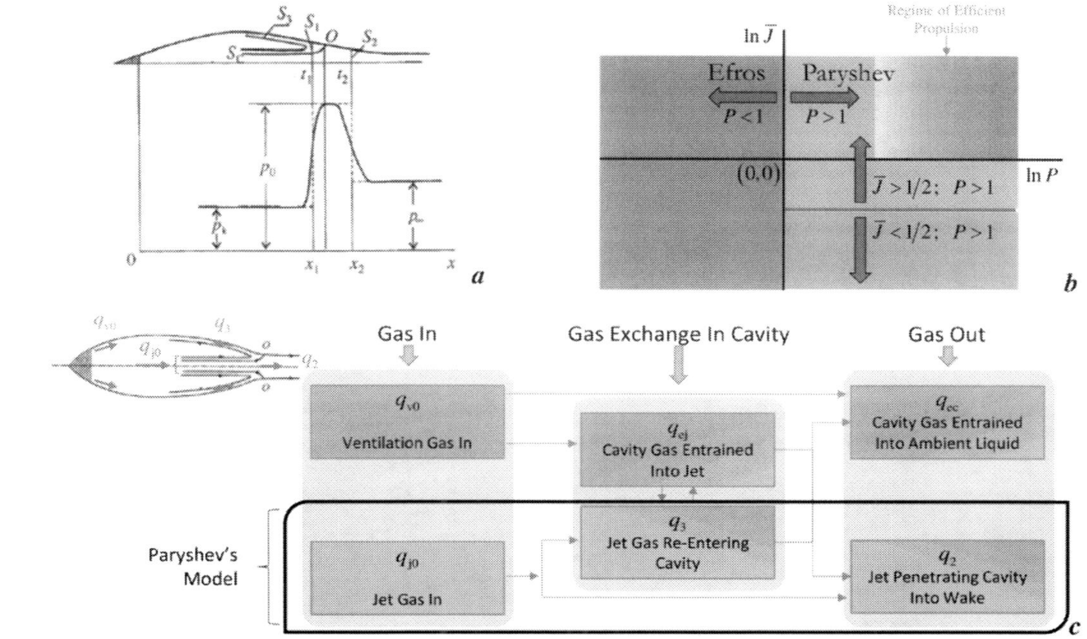


***Figure 3.*** Paryshev's closure model and the resulting regimes of flow: ***a*** – Model simplifications (from Paryshev, 2006); ***b*** – Flow regimes for cavity closure on a propulsion jet; ***c*** – Gas flow budget. The regime of efficient propulsion is discussed in a later section.

When applied in the context of a rocket free body diagram, Paryshev's model can be extended to estimate the thrust and propulsive efficiency of the rocket as a function of the same quantities. As is well known, the efficiency of a rocket propulsor is maximized when the jet exits the nozzle downstream at vehicle speed (in the vehicle reference frame), such that the gas is simply left behind the vehicle, and thrust is derived solely from the relative momentum flux of the ejecta (Sutton & Biblarz, 2001). A gas-based underwater rocket must overcome drag associated with the much heavier ambient liquid. Typically, such a system must eject mass much faster than the optimal rate, and efficiency is reduced. The following expressions for dimensionless thrust and propulsion efficiency have been derived and cast in the context of parameters characterizing supercavity closure on a jet:

$$\bar{T} \doteq T/D_n = \left(1+\frac{\Delta\sigma}{2}P\right)\bar{J} \text{ and } \eta_p = \left\{2\sqrt{\bar{\rho}_j P}\left[1+\frac{\Delta\sigma}{2}P\right]\right\} \Big/ \left\{\left[1+\frac{\Delta\sigma}{2}P\right]^2 + \bar{\rho}_j P\right\}, \tag{1}$$

where $T$ is the thrust, $\bar{\rho}_j \doteq \rho_j/\rho$ is the dimensionless jet density, and $\Delta\sigma \doteq \sigma - \sigma_j$ is the difference between the cavitation number, $\sigma \doteq (p_\infty - p_c)/(1/2\rho V^2)$, and the jet over-expansion index, $\sigma_j \doteq (p_\infty - p_j)/(1/2\rho V^2)$. Here $p_\infty$ is the depth pressure well upstream of the cavitator, $p_c$ is the ventilated cavity pressure, and $p_j$ is the pressure in the jet. Assuming the jet exits a round nozzle, its diameter can be estimated for specified operating conditions and dimensionless thrust from the first of equations (1) and the definition of $\bar{J}$. Assuming jet closure downstream of the nozzle, it must also fit within the cavity at the exit plane. Thus the jet diameter must be less than the maximum cavity diameter (or even smaller, if the nozzle is located closer to closure). This geometric constraint may be estimated using well-known approximations of cavity geometry. Also, the kinetic energy of the jet must be positive semi-definite, and it can be shown that this requires that $P \le 1/\sigma$. Thus there are two bounds on the operation of a rocket nozzle enveloped in a supercavity that limit the maximum value of stagnation pressure ratio that can be achieved.

**Results**

Example results for the required jet diameter are plotted as a function of the stagnation pressure ratio in figure 5 up to the upper bounds introduced above. Thrust and propulsion efficiency are plotted in figure 6 as functions of dimensionless jet momentum flux and stagnation pressure ratio. In both figures, the conditions are $\sigma = O(0.05)$, $\bar{\rho} = O(0.001)$, and $\sigma_J/\sigma = 1.5$.

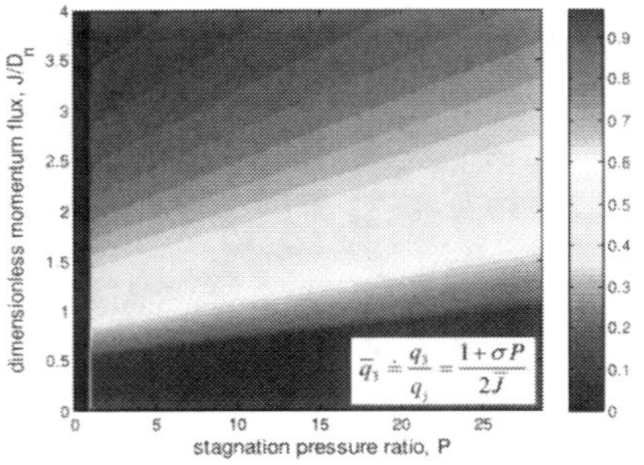


***Figure 4.*** Dimensionless ventilating volumetric flow rate from jet for $\sigma = O(0.05)$

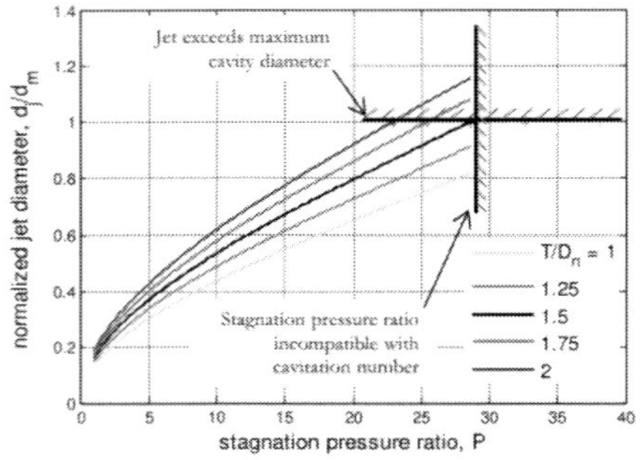


***Figure 5.*** Limitations on rocket propulsion operation for conditions stated in text.

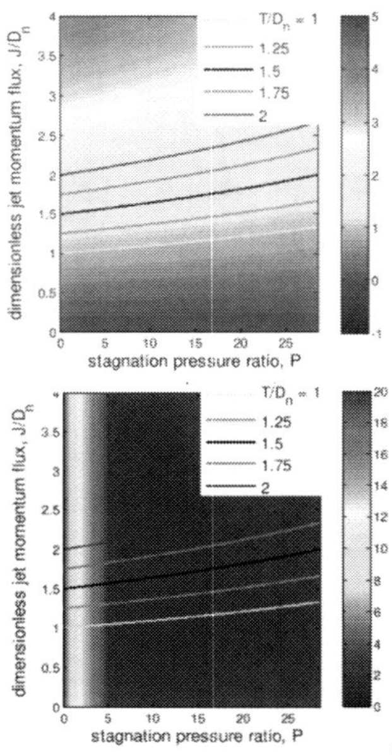


*a*

*b*

***Figure 6.*** Dimensionless thrust and propulsive efficiency for conditions stated in text: ***a*** – Dimensionless thrust, $T/D_n$ ; ***b*** – Propulsive efficiency, $\eta_P$ , in percent.

## Discussion

For the conditions of figure 6, propulsive efficiency asymptotes to a maximum value for stagnation pressure ratios greater than approximately five, and becomes nearly independent of the stagnation pressure ratio. As can be seen from the second of equations (1), efficiency is also independent of the jet momentum flux and the dimensionless thrust. The dimensionless thrust is a function of both the stagnation pressure ratio and the dimensionless jet momentum flux. A well-designed supercavitating high-speed body with minimal parasitic drag due to appendages and other effects will operate with a dimensionless thrust slightly exceeding unity.

## Summary and Conclusions

Efros-Paryshev theory of supercavity closure on a propulsion jet has been extended to explore propulsion performance of rocket-propelled supercavitating bodies. Propulsion efficiency is limited by the low density ratio of the jet relative to the ambient liquid. Efficiency can be maximized by operating in the Paryshev "soft strong" jet regime. Although it is desirable to maximize the stagnation pressure ratio, upper bounds on this quantity have been identified.

### Acknowledgments

Funding provided through ONR under contracts #N00014-12-1-0489 and #N00014-13-1-0692, Dr. Ronald Joslin, Program manager. The effort undertaken by the lead author is sponsored under a Research Subaward Agreement with The Pennsylvania State University under Grant #N00014-12-1-0489, Subaward SA12-11, Michael J. Moeny, Co-Principal Investigator with the Department of the Navy, Office of Naval Research. Any opinions, findings, and conclusions or recommendations expressed in this material are those of the authors and do not necessarily reflect the views of the Office of Naval Research or (in the case of the lead author) The Pennsylvania State University.

### References

Efros DA (1946) Hydrodynamic theory of plane cavitating flow, *DAN SSSR*, new series, **51**, 4 (in Russian).
Logvinovich GV (1969) *Hydrodynamics of Free-Boundary Flows*, Naukova Dumka, Kiev, Ukraine; translated from Russian by the Israel Program for Scientific Translation (1972), Jerusalem, Israel.
Logvinovich GV, Serebryakov VV (1975) On methods of calculation of slender axisymmetric cavities shape, *J. Hydromechanics*, **32**, pp 47-54 (in Russian).
Moeny MJ, Krane MH, Kirschner IN, Kinzel MP (2015) Jet-Supercavity Interaction: Insights from Experiments, *Proc. 9th Intl. Symposium on Cavitation (CAV2015)*, Lausanne, Switzerland.
Paryshev EV (2006) Approximate mathematical models in high-speed hydrodynamics, *J. Eng. Math.*, **55**, pp 41-64.
Sutton GP, Biblarz, O (2001) *Rocket Propulsion Elements*, Seventh Edition, John Wiley & Sons, New York, NY.

# A numerical study of unsteady cavitation on a hydrofoil by LES and URANS method

**Zi-ru Li[1] Guang-ming Zhang[1] Wei He[1] and Tom van Terwisga[2,3]**

1 Wuhan University of Technology, 430063, Wuhan, Hubei, China
2 Delft University of Technology, Delft 2628 CD, The Netherlands
3 Maritime Research Institute Netherlands, Wageningen 6700 AA, The Netherlands

Email: liziru@whut.edu.cn

**Abstract**. In this paper, the unsteady cavitation phenomena on a NACA0015 hydrofoil is numerically simulated by unsteady Reynolds-Averaged Navier-Stokes (URANS) method and Large Eddy Simulation (LES) in single-fluid approaches to multiphase modelling, respectively. It is observed that the large-scale structures and characteristic periodic shedding predicted by the URANS with the modified SST k-ω turbulence model show a good qualitative match with the experimental observations but with quantitative discrepancies, such as a different cavity length and volume, and a different location of shedding. Compared to the URANS results, the LES results reproduce more details of unsteady dynamics with an improved quantitative agreement.

## 1. Introduction

Cavitation is a complex physical phenomenon of phase change from liquid to vapor at almost constant temperature in regions where the pressure is lower than a certain critical pressure. It commonly occurs in marine propulsion systems and other hydraulic machinery. Due to the possible limitations imposed on the attainable propulsor thrust and propulsion efficiency by cavitation, it is essential to predict cavitation phenomena and make an assessment of cavitation nuisance in an early design stage.

As one of the remarkable catastrophic consequences, cavitation erosion is a great challenge to be assessed since it involves multi-scale hydrodynamic processes in combination with the response of solid material exposed to various cavitation regimes. Although much is known about the behavior of individual bubble cavitation and the material reaction, the establishment of a reliable general-purpose model for cavitation erosion assessment is still a big challenge and remains a major concern for the industry[1-4]. From a number of experimental studies, it is confirmed that a crucial phase in the process leading to cavitation erosion is the break-up of the macroscopic sheet cavity into cloudy cavities [5-7]. As a result, the capture of characteristic unsteady dynamics of sheet/cloud cavitation becomes an essential phenomenon in the assessment of the cavitation erosion risk based on computational fluid dynamic (CFD) tools and experimental observations.

Additional difficulties arise in the numerical modelling of unsteady cavitation phenomena due to the complex mechanism involving turbulent fluctuations over a wide range of length and time scales. It is noted that suitable turbulence modelling is of significant importance due to the presence of high Reynolds number and strong unsteady dynamics, such as the formation of the re-entrant jet, and the periodic shedding and collapse of cloudy cavitation [8]. This paper firstly carries out numerical simulation of the cavitating flow around the NACA0015 hydrofoil through the URANS method using a modified SST k-ω turbulence model. Subsequently, the capability of an alternative methodology for turbulent flow predictions, the LES approach, to reproduce unsteady cavitation phenomena is explored. A quasi-2D run with the 3D LES code is adopted here. It is supposed to overcome the shortcoming of the RANS method in a poor resolution of dynamic flow structures of multiple length and time scales and an over-prediction of eddy viscosity production in regions of higher vapor volume fraction. The LES results are then compared with URANS results for verification of refined resolution of transient unsteady dynamics and also improved quantitative agreement with experimental observations.

## 2. Numerical Modelling

The governing equations for the liquid/vapor two-phase flow are based on a single-fluid approach. The flow field is solved by the mixture continuity and momentum equations, which are obtained by averaging or filtering of the original Navier-Stokes equations, corresponding to URANS method and LES approach, respectively. The cavitation model developed by Schnerr and Sauer[9] has been used for the mass transfer between the vapor and liquid phases in cavitation.

For the numerical simulation conducted by URANS method, the SST k-ω turbulence model developed by Menter [10] with a modification of the turbulent viscosity is applied following the idea of Reboud et al. [11] for better simulation of the unsteady dynamics. For the LES approach, subgrid-scale (SGS) terms resulting from the filtering of the Navier-Stokes equations are modeled by the Smagorinsky-Lilly model[12].

## 3. Case description

The test geometry is a NACA0015 hydrofoil at 8 degree angle of attack with a chord length C=60mm. The computational domain is obtained by extending 3.0 chord lengths ahead of the leading edge and 5.5 chord lengths behind the trailing edge of the hydrofoil. A coarse grid with 264 edges are set on the hydrofoil for the URANS method, and a finer grid with 528 edges on the hydrofoil has been used for the LES computation, ensuring the near-wall mesh resolution along the hydrofoil surface is $y^+<1$.

A velocity inlet condition is applied at the upstream flow. A pressure outlet condition is used at the outlet boundary, where the specified pressure at the outlet can be derived from the cavitation number under consideration, $\sigma=2.2$, which is characterized by typical unsteady cavitation. The top and bottom boundaries are taken as no-slip walls. The experiments to be compared with the numerical simulations are performed by MARIN in cooperation with Lloyd's Register [7,13].

## 4. Numerical results

### *4.1. Unsteady shedding and collapse*

As shown in figure 1-(b), the observed essential features can be characterized as follows: A sheet cavity is initiated and growing from the leading edge after the detachment of last cloudy cavities; The sheet cavity grows to a certain extent and becomes unstable, simultaneously the cloudy cavities travels with the main flow towards downstream and finally collapses near the trailing edge of the hydrofoil; Re-entrant flow forms and moves upstream towards to the leading edge until it breaks the main sheet cavity into a bubbly cloud, which is shed periodically at a frequency around 210 Hz.

The URANS results basically capture the features of the unsteady cavitation, such as the break-up of the sheet cavity and the detachment of the cloudy cavities from the sheet cavity and its collapse. It shows a good qualitative match with the experimental observations but with a much smaller cavity length and volume, and different locations of the prominent shedding and collapses of the bubbly cloud, as shown in figure 1-(a). A shedding frequency of 280 Hz has been predicted.

In figure 1-(c), the LES results show a better agreement in the prediction of the cavitation behavior than the URANS results. It reproduces the unsteady shedding and collapses with larger cavity extent and closer locations of shed bubbly clouds at a lower frequency of roughly 170 Hz. This would emphasize the possible improvement by a LES approach in the assessment of cavitation erosion.

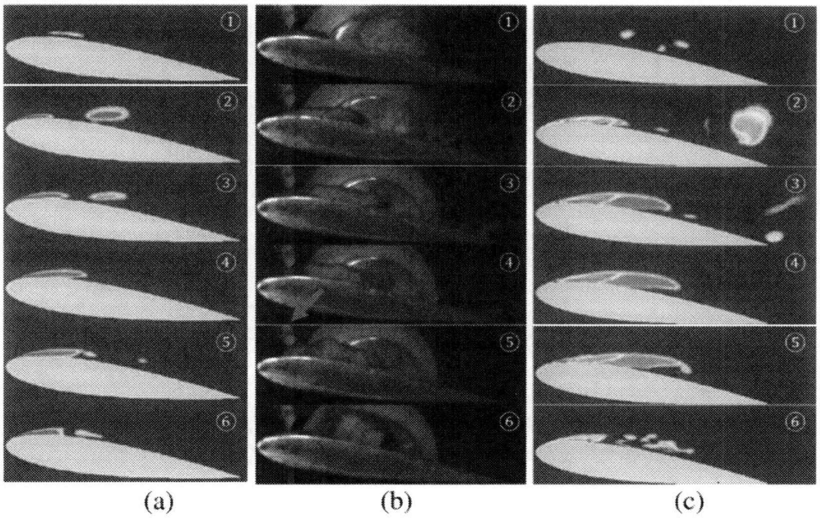


**Figure 1.** Time sequences of contours of vapor volume fraction obtained by (a) URANS method; (b) Experimental observations and (c) LES approach.

However, both approaches failed to capture the exact location where the re-entrant jet clearly cuts off the main sheet cavity just alike the experimental observations, which is almost at the leading edge, as indicated by red arrow in figure 1-(a)- ⑤. It is suggested that the numerical simulations conducted here in the 2D domain would miss the three-dimensionality effect of the flow, thus decreasing the thickness of the re-entrant jet in a weaker jet. This hypothesis has been confirmed by comparing 2D URANS results with the results from a 3D URANS computation, where the cut-off appears close to the leading edge and also clear horse-shoe shaped cloudy cavities can be reproduced[13,14].

By investigation of the characteristic shedding behavior predicted by LES approach, it is observed that the detachment of cloudy cavities is dominant by both reversed flow travels towards upstream, which will cut off cloudy cavities, and rotational motion of vortex structures, which will forms cavitating vortices at its center when the pressure falls below the vapor pressure. It confirms the statement from Hoekstra that the motion of the widely accepted re-entrant jet mechanism is a visual illusion because the motion of the cavity-liquid interface is not necessarily the same as the motion of the fluid particles[15]. In figure 2, the red arrows represent the main direction of the reversed liquid flow, and the black arrows indicate the vortex structures.

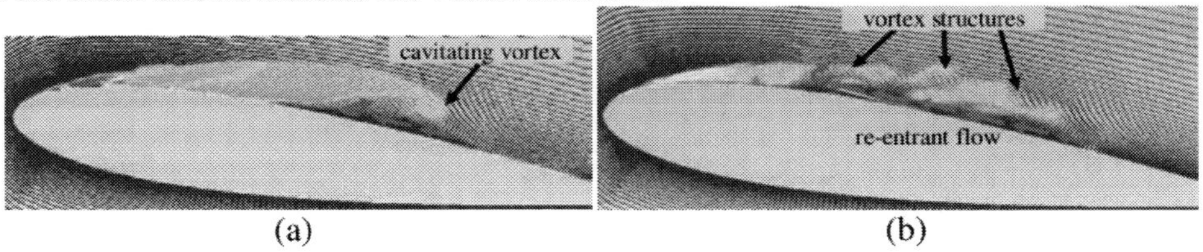


**Figure 2.** The velocity vectors colored by the vapor fraction corresponding to typical instants predicted by LES approach: (a) instant ⑤; (b) instant ⑥.

### 4.2. *Lift and drag*

The lift and drag acting on the hydrofoil is greatly affected by the unsteady behavior of the cavitating flow, which will result in fluctuating forces with sudden peaks or drops that indicates high impact may occur. For time-histories of lift and drag coefficients obtained by URANS and LES computations, periodic lift and drag signals are predicted to reveal the regularly repeated unsteady dynamics, as shown in figure 3. It is noted that more unstable fluctuating force signals have been predicted by the LES computation in combination with remarkable peaks and troughs, interpreted for a higher averaged lift and drag. The largest peak during one cycle can be correlated to the collapse of the cloudy cavities near the trailing edge, as indicated a moment between the instant ③ and ④ in figure 1-(c).

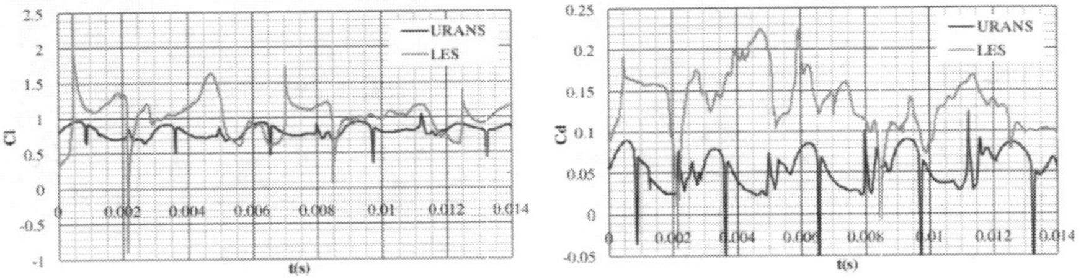


**Figure 3.** Time histories of (a) lift and (b) drag coefficients by URANS and LES computations.

## 5. Conclusions

The following conclusions can be drawn from this study:

- The LES results show better quantitative agreement with the experimental observations than the URANS computations, displaying a larger cavity extent and better prediction of the locations of the prominent shedding and collapses of the bubbly cloud.
- The inability to predict the exact location where the re-entrant jet clearly cuts off the main sheet for both approaches can be attributed to loss of three-dimensionality, which weakens the strength of the re-entrant jet.
- The break-off of the main sheet cavity and the shedding of cloudy cavities are associated with two typical mechanisms: the reversed flow travels upstream and cuts off cloudy cavities, and the inertia driven rotational motion of the flow at the closure of the sheet cavity causes shedding of vortex cavities.

**Acknowledgement**

The authors are grateful for the report of the experiments provided by the Cooperative Research Ships (CRS), administered by MARIN. This work is also financially supported by the National Natural Science Foundation of China (Project No.51309184).

## References

[1] Challlier G, Fortes-Patella R and Reboud J L 2000 *Proc. ASME Fluids Engineering Summer Conf. (Boston)*
[2] Bark G, Berchiche N and Grekula M 2004 *Application of Principles for Observation and Analysis of Eroding Cavitation-The EROCAV observation Handbook* (Sweden)
[3] Dular M, Sirok B and Sroffe B 2006 *Proc. 6th Int. Sym. Cavitation (Wageningen)*
[4] Franc J P, Riondet M, Karimi A and Chahine G L 2011 *ASME J. Fluids Eng.* **133** 121301
[5] Reisman G E, Wang Y C and Brennen C E 1998 *Journal of Fluid Mechanics* **355** pp 255-283
[6] Foeth J P, Van Terwisga T J C and Van Doorne C 2008 *ASME J. Fluids Eng.* **130** (7)
[7] Van Rijsbergen M, Foeth E J, Fitzsimmons P and Boorsma A 2012 *Proc. $8^{th}$ Int. Sym. Cavitation (Singapore, 13-16 Aug 2012)*
[8] Coutier-Delgosha O, Fortes-Patella R and Reboud J L 2003 *ASME J. Fluids Eng.* **125** pp 38-45
[9] Schnerr G H and Sauer J 2001 *Proc. 4th Int. Conf. Multiphase Flow (New Orleans)*
[10] Menter F R 1994 *AIAA Journal* **32** (8) pp 1598-1605
[11] Reboud J L , Stutz B and Coutier O 1998 *Proc. of the 3rd Int. Sym. Cavitation (Grenoble)*
[12] Smagorinsky J 1963 *Month. Wea. Rev.* **91** pp 99-164
[13] Li Zi-ru 2012 *Assessment of cavitation erosion by a multiphase RANS method* (TUDelft:VSSD)
[14] Li Zi-ru, Pourquie M and Terwisga T J C 2014 *ASME J. Fluids Eng.* **136** 041101
[15] Hoekstra M and Vaz G 2009 *Proc. 7th Int. Sym. Cavitation (Michigan, USA, 16-20 Aug 2009)*

# Finite Volume Computation of the Mitigation of Cavity Pulsation

**J Lindau, G Skidmore, T Brungart, M Moeny, and M Kinzel**
Penn State Applied Research Lab, PO Box 30, State College, PA, USA, 16804-0030
Contact E-mail: jwl10@arl.psu.edu

**Abstract.** Finite volume based modeling of ventilated supercavity pulsation and its mitigation via a priori modulation of ventilation flow was investigated. Simulated pulsation was numerically achieved, as was mitigation of pulsation via sinusoidal modulation of the ventilation flow. In addition to confirmation that the numerical approach is sufficient to capture mitigation, it was found that modulated ventilation, without altering the mean ventilation mass flow rate, results in altered cavity size, pressure, and closure condition.

## 1. Introduction

Due to their potential and proven hydrodynamic applications to high speed drag reduction and facilitation of high speed marine lifting surfaces, ventilated gas supercavities are of particular interest. The pulsation of ventilated supercavities, generated by axisymmetric cavitators, was reported by Song [1], Michel [2], and has been studied extensively by Paryshev [3], among others. Due to his contribution, and his widely cited modeling, pulsating modes of the supercavity are frequently referred to as the *Paryshev Instability*.

Control of pulsation via harmonic forcing can be attributed to Semenenko [4]. Recently, Skidmore et al. [5] conducted an investigation into potential strategies for the mitigation of cavity pulsation. This work focuses on a priori modulation of the ventilation flow. Tests were completed in a closed-loop water tunnel. Modulation was achieved using a butterfly valve. Results indicate that this strategy may, under proper application, induce the cavity to transition away from pulsation into a twin vortex condition. Although not comprehensive, the current investigation is an attempt to computationally complement those physical experiments.

## 2. Computational Physics

The computational fluid dynamics software, StarCCM+ [6], is employed as a physical modeling tool. Lindau et al. [7] have previously applied StarCCM+ to capture cavity pulsation with a finite-volume, Navier-Stokes approach. It was found that a time dependent finite volume discretization with incompressible liquid flow and isothermally compressible gas flow, without rigorous turbulence modeling, was sufficient to capture the salient physics of cavity pulsation. A standard volume of fluid approach supported by the HRIC scheme was employed (e.g. [8]). Care was taken to ensure that local convective Courant numbers were within the bounds prescribed for best use of the scheme. In [7], asymptotic numerical mesh convergence to a pulsation condition was demonstrated, supporting the current approach. A complete model (computational domain, time step size, and all numerics and physical models) found to be convergent in [7] is applied here, albeit with modulated ventilation flow.

## 3. Computational Results

Here, scaling parameters typically applied to ventilated supercavitation are used. They are the cavitation number [$\sigma \equiv 2(p_\infty - p_{cav})/(\rho U_\infty^2)$], the Froude number [$Fr \equiv U_\infty/(gD_{cav})^{1/2}$], and the ventilation flow coefficient [$C_Q \equiv Q/(U_\infty D_{cav}^2)$]. For completeness, the Reynolds number based on cavitator diameter and free stream quantities was held constant [$Re_D \equiv \rho U_\infty D_{cav}/\mu = 573{,}000$]. For the envisioned conditions, the Thoma or (vaporous) cavitation number at the modeled condition was also a constant [$\sigma_v \equiv 2(p_\infty - p_{vapor})/(\rho U_\infty^2) = 0.97$]. It should be clear that $p_{vapor}$ is much less than $p_{cav}$ for all presented results. The quantity $p_{cav}$ is the pressure inside the ventilated cavity, $D_{cav}$ is the diameter of the disk cavitator, and $Q$ is the ventilation gas volume flow rate. In addition, time scales are nondimensionalized by the quantity $D_{cav}/U_\infty$. A Detached Eddy Simulation approach based on the Spalart-Allmaras turbulence model was applied [6].

In figure 1, the computational mesh is shown. The computational domain is a cylindrical tunnel with diameter of 52.5 and length 78.7 $D_{cav}$. Tunnel walls are modeled as inviscid. The wetted face of the cavitator is concentric with the tunnel, 26.2 $D_{cav}$ downstream from the inlet. The mesh is illustrated with a plane bisecting the computational domain and on the inlet and outlet boundaries. The mesh on the cavitator surface, is also shown. Computational ventilation gas is issued via mass flux boundary conditions around the cavitator aft side and perimeter. (Isothermally compressible standard air at a reference temperature of 300K is used. The mean absolute pressure at the tunnel centerline under static conditions is 110.8kPa.) The liquid is treated as incompressible water. Water density $\rho_L$=1000kg/m$^3$.

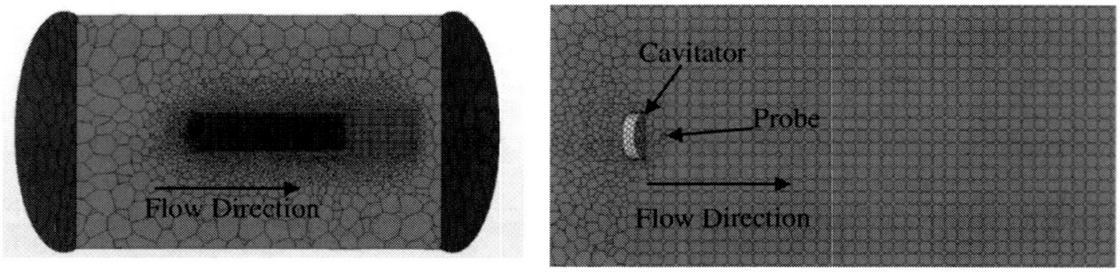


Figure 1. Computational grid.

Prior to obtaining the presented results, an initial pulsation condition was obtained based on unsteady integration with steady ventilation; the ventilation mass flow was maintained ($m_0 = 2.985 \times 10^{-3} \rho_L U_\infty D_{cav}^2$) such that the mean ventilation rate was $C_Q$=0.28. This same mass flow, $m_0$, is maintained as the mean value for all results presented here. Results from four model integration cases are presented. All cases were integrated, beginning with the same initial state, at $\Delta t = 0.0787 D_{cav}/U_\infty$ for a total integration time of $2{,}362 D_{cav}/U_\infty$. This was adequate to obtain a final stationary condition. For all cases, $Fr$=24.5. An interesting observation is that despite maintaining the same mean ventilation mass flow when modulation is introduced, the frequency content, the cavitation number, and (sometimes) the closure condition change. In other words, when modulation is begun, a transient occurs and eventually a stationary, constant mean but time varying, condition is found with a new mean cavity pressure and cavity size, often with a new closure condition. In the first of the four cases, ventilation is not modulated and the steadily pulsing condition is maintained. In the second, third, and fourth cases, different modulations are applied to the ventilation mass flow. In these modulated cases, the amplitude of modulation, $A$, is determined relative to the mean total ventilation mass flow. For instance, the ventilation mass flow modulated at $A$=0.2 varies sinusoidally between 80 and 120% of the mean ventilation mass flow.

The location of the numerical probe for determination of instantaneous cavity pressure is shown in figure 1, on the tunnel centerline and 0.66 $D_{cav}$ downstream of the cavitator face.

In figure 2, snapshots of cavities resolved with CFD, without modulation and with three different modulation conditions, from two views, port and below, are shown. In figure 3, pressure time histories and spectra from the numerical probe pictured in figure 1 are presented. The time range chosen for representation in figure 3 is sufficient to allow appreciation of the character of the pressure signal, mean value and frequency content, for each case. The snapshots in figure 2 illustrate both the general character of the pulsing cavity as well as the effect of modulated ventilation on cavity size and shape. A review of the snapshots together with the pressure histories and spectra give a fairly complete picture of the final stationary conditions. In all cases, the mean ventilation gas mass flow rate is the same and (in figure 2) $2{,}126 D_{cav}/U_\infty$ of simulation time has elapsed since numerical time integration was initiated from the common initial condition.

In the case with no modulation, the cavity continues to pulse at the nominal resonance frequency, $f_0=0.07493 U_\infty/D_{cav}$, and periodically toroidally ejected gas pockets, convected downstream, are clearly visible. The reentrant jet is clear in the translucent view from below. The mean cavity size and pulsation frequency and magnitude change little for the duration of the time integration. The pressure history, black lines in figure 3, clearly shows the nature and magnitude of the pulsing condition. It is nearly regular. The pressure spectral decomposition of the unmodulated case is also shown with solid black lines in figure 3. The peak frequency, $f_0$, is indicated with the dashed black line.

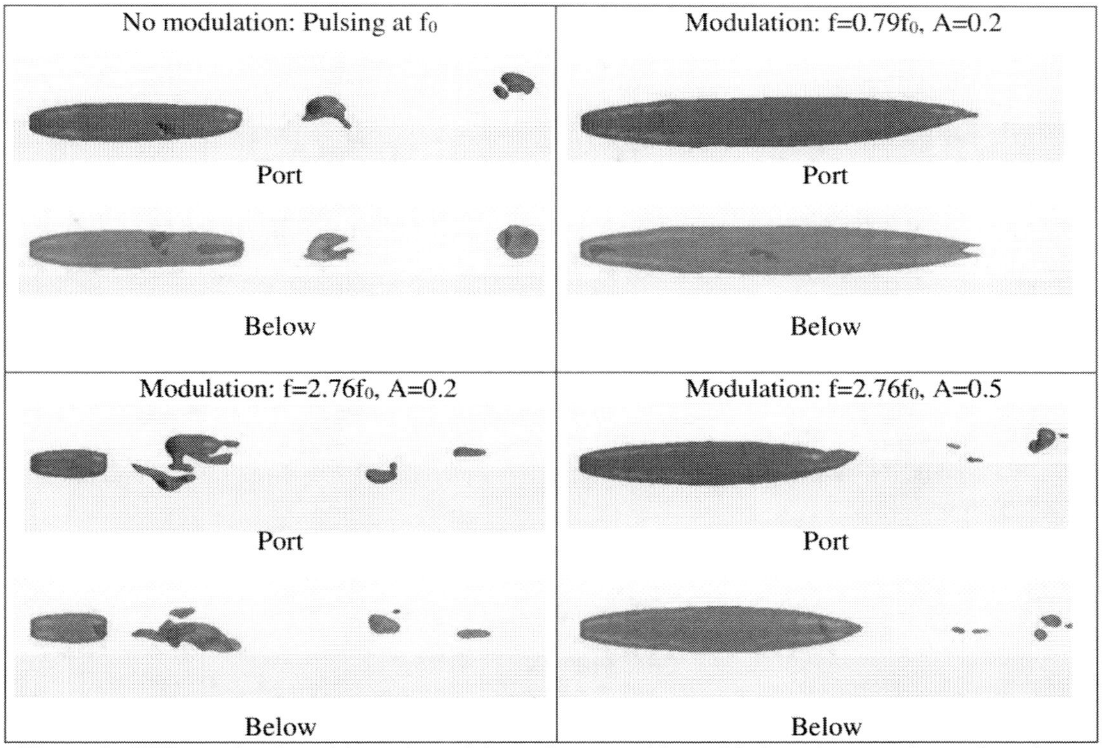


Figure 2. CFD resolved cavities, viewed from the port side and below (opaque isosurface from the side and translucent from below, both at liquid volume fraction, $\alpha_L=0.5$) with steady and modulated ventilation. All snapshots are at $2{,}126 D_{cav}/U_\infty$ after initial modulation.

With ventilation flow sinusoidally modulated, the final mean state is strongly dependent on the character of the modulation. With chosen modulation amplitude $A=0.2$, the final state can be larger and more steady, as with modulation frequency, $f=0.79f_0$ (red dashed lines, figure 3). Here, the closure appears to be twin vortex. However, with $A=0.2$ and $f=2.76f_0$ (blue dash-dot, figure 3), the cavity is reduced to a smaller, reentrant state. Re-running with $f=2.76f_0$ and $A=0.5$ (green dotted, figure 3), the cavity evolves to another pulsing state and appears to be approaching a twin vortex closure-with a smaller amplitude than the unmodulated cavity. It is noteworthy that although with $f=0.79f_0$ and $A=0.2$, the dominant pressure spectrum coincides with the modulation frequency, this is not the case at either of the higher frequency modulation conditions. The modulation frequency is discernible in the pressure spectra, but it is not dominant. In all cases, even those with apparent twin vortex closure, there is some pulsation evident in the pressure history. Also in all cases, the regular nature of pressure pulsation history is sporadically broken by disruptive events. Typically this is due to water splashing in the supercavity as gas is shed. The strongest disruptor is associated with the nearly regular reentrant jet (modulation set to $f=2.76f_0$ and $A=0.2$).

## 4. Conclusions

A computational investigation of the physical concept of pulsation mitigation via a priori modulation of ventilation flow has been presented. The nature of cavity pulsation can be strongly altered by a priori modulated ventilation flow. This was determined with water tunnel experiments by Skidmore et al. [5], and the results here confirm and expand somewhat on those findings. It was computationally demonstrated here, that a simple sinusoidal modulation of the flow can drive the pulsating cavity to either a larger twin-vortex condition or a smaller reentrant condition. It also appears that both the frequency and amplitude of the modulation have a strong effect on the final state. To further extend this kind of modeling to a specific engineering application, an in depth validation study must be done.

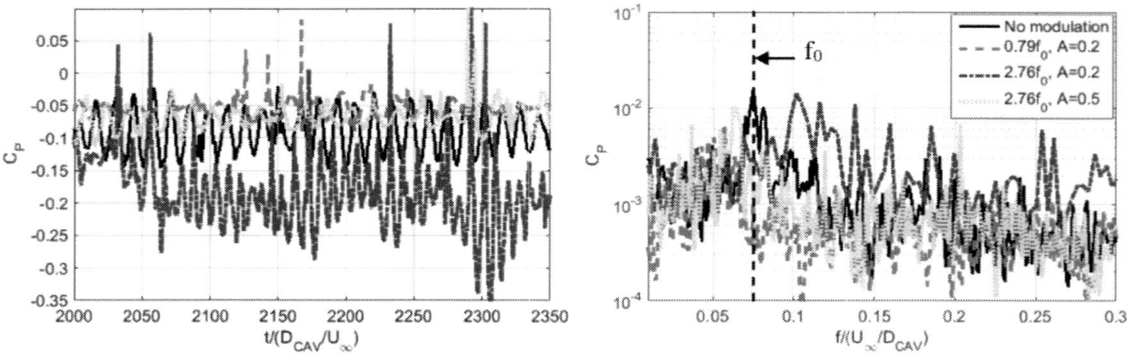


Figure 3. Cavity pressure history and pressure spectra.

## 5. Acknowledgments

This work is sponsored by the Office of Naval Research under Grant N00014-11-10322 and monitored by Dr. Ronald D. Joslin.

## References

[1] Song C. 1962 *Journal of Ship Research* **5** 1-20
[2] Michel J 1984 *Journal of Fluids Engineering*, **206** 319-326
[3] Paryshev E 2006 *Journal of Engineering Mathematics* **55** 41-64
[4] Semenenko V 1996 *International Journal of Fluid Mechanics Research* **23** 134-143
[5] Skidmore G, Brungart T, Lindau J and Moeny M 2015 A method of mitigating pulsation in ventilated supercavities *8th International Cavitation Symposium* (Lausanne, Switzerland)
[6] CD-Adapco 2014 *STAR-CCM+ 9.02.005 User Guide* (Melville, New York)
[7] Lindau J, Skidmore G, Brungart T, Moeny M and Kinzel M 2014 Finite volume, CFD based investigation of supercavity pulsations *Proceedings of 30th ONR Symposium on Naval Hydrodynamics* (Hobart, Tasmania, Australia)
[8] Demirdzic I, Muzaferija S, Peric M, Schreck E and Seid V 1999 Computation of flows with free surfaces *Scientific Computing in Chemical Engineering II* ed Kiel, Mackens, VoB and Wether (Berlin: Springer) pp 360-367

# Numerical evaluation of cavitation void ratio significance on hydrofoil dynamic response

**Xin Liu[1], Zhengwei Wang[1], Xavier Escaler[2] and Lingjiu Zhou[3]**

[1] Department of Thermal Engineering, Tsinghua University, Beijing 100084, China
[2] Center for Industrial Diagnostics, Universitat Politècnica de Catalunya, 08028 Barcelona, Spain
[3] College of Water Resources & Civil Engineering, China Agricultural University, Beijing 100083, China

Email: x-liu06@mails.tsinghua.edu.cn , wzw@mail.tsinghua.edu.cn, escaler@mf.upc.edu, zlj@cau.edu.cn

**Abstract**. The added mass effects on a NACA0009 hydrofoil under cavitation conditions determined in a cavitation tunnel have been numerically simulated using finite element method (FEM). Based on the validated model, the effects of averaged properties of the cavity considered as a two-phase mixture have been evaluated. The results indicate that the void ratio of the cavity plays an increasing role on the frequency reduction ratio and on the mode shape as the mode number increases. Moreover, the sound speed shows a more important role than the average cavity density.

## 1. Introduction

At present, FEM-based acoustic fluid approach to simulate the added mass effect of surrounding water has been widely used in hydraulic turbine runners with sufficient accuracy as for example by Liang, et al. [1]. Nevertheless, this method has not yet been used to estimate the added mass effects on hydrofoils under cavitation conditions. In fact, the experimental results indicate that the actual macroscopic hydrodynamic cavities are composed of a mixture of water and vapour phases. As a result, averaged properties must be considered in the model that are difficult to quantify and that vary from the expected effects of pure liquid water flow. Moreover, new boundary conditions at cavitation regions appear that are not fully understood.

Consequently, the research purpose of this work is to use a numerical Fluid Structure Interaction (FSI) approach for evaluation of the significance of cavitation void ratio on the dynamic response of an hydrofoil. For that, the numerical results for a NACA0009 hydrofoil with partial cavitation have been compared with the experimental ones obtained by De La Torre, et al. [2] assuming that the cavity is composed of a single fluid phase of pure vapour. With the validated model and based on the possibilities of the numerical approach, a study of the effects that the two-phase flow properties have on the hydrofoil modes of vibration (natural frequencies and mode shapes) of the coupled system have been explored.

## 2. Formulations of fluid structure interaction (FSI)

In FSI problems, the structural dynamics equation must be considered along with the Navier-Stokes (NS) and the continuity equations. The discretized structural dynamics equation can be formulated based on the structural elements. The NS and continuity equations are simplified to get the acoustic wave equation assuming that the fluid is compressible, which means that its density changes due to pressure variations (i.e. acoustic medium), and that there is no fluid flow [3].

The governing finite element matrix equations of FSI are expressed in Equation (1):

$$\begin{bmatrix} M_s & 0 \\ \rho_0 R^T & M_F \end{bmatrix}\begin{Bmatrix} \ddot{U} \\ \ddot{P} \end{Bmatrix}+\begin{bmatrix} C_s & 0 \\ 0 & C_F \end{bmatrix}\begin{Bmatrix} \dot{U} \\ \dot{P} \end{Bmatrix}+\begin{bmatrix} K_s & -R \\ 0 & K_F \end{bmatrix}\begin{Bmatrix} U \\ P \end{Bmatrix}=\begin{Bmatrix} F_s \\ F_F \end{Bmatrix} \quad (1)$$

where $M_F$, $C_F$ and $K_F$ are the fluid mass, damping and stiffness matrices, respectively. R is a coupling matrix that represents the effective surface area associated with each node on the fluid-structure interface. $F_F$ is the applied fluid pressure vector at the interface obtained by integrating the pressure over the area of the surface. P and U are the fluid pressure and the structure displacement vectors, respectively. $M_S$, $C_S$ and $K_S$ are the structure mass, damping and stiffness matrices, respectively, and $F_S$ is the structural load vector.

## 3. FEM model

The simulated case corresponds to a NACA0009 hydrofoil, as shown in **Figure 1**, under various cavitation conditions experimentally tested at the EPFL High-Speed Cavitation Tunnel. On the left of **Figure 3**, a photograph of the typical attached partial cavitation as described in De La Torre, et al. [2] is shown. The foil is made of aluminum alloy with a Young's modulus of 0.54 GPa, a Poisson's ratio of 0.33 and a density of 2770 kg/m$^3$.

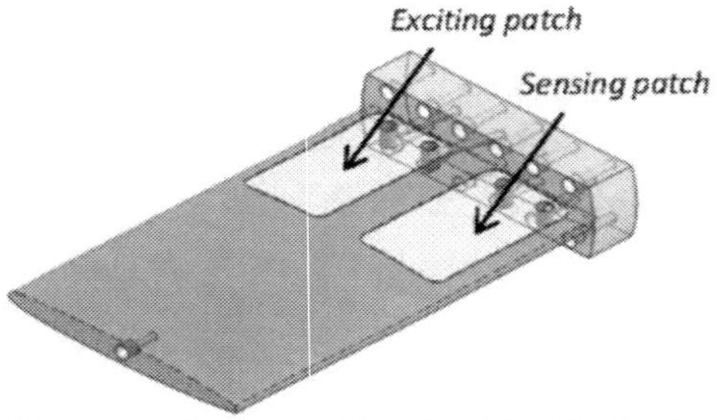


**Figure 1.** Truncated NACA0009 hydrofoil

The FSI domain, which comprises the whole test section of the cavitation tunnel, was meshed with hexahedral elements as shown in **Figure 2**. The hydrofoil was fixed with an incidence angle of 0°.

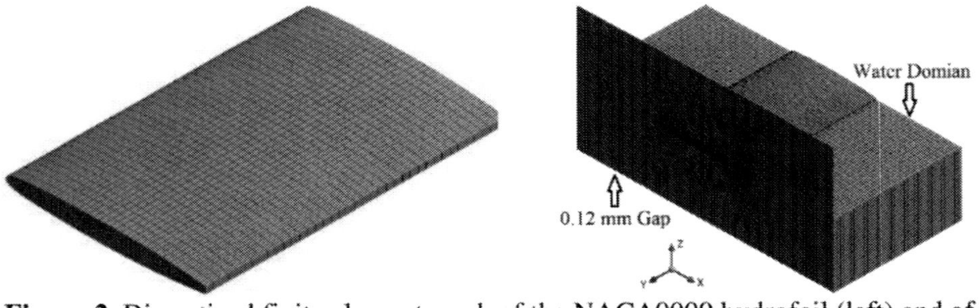


**Figure 2.** Discretized finite element mesh of the NACA0009 hydrofoil (left) and of the FSI domain with the surrounding water (right).

One lateral section of the hydrofoil was constrained in the model. The rest of hydrofoil surfaces were in contact with the fluid. A FSI condition was set at these interfaces so that the displacement of the structure is identical to that of the fluid in normal direction. Considering the coupling characteristics of vibration in water, the fluid boundaries were modeled as fully reflective. The water properties were set as 1000 kg/m$^3$ for density and 1450 m/s for speed of sound.

Regarding the modelling of the attached cavity, it was considered to be of a constant length along the whole span as shown on the right of **Figure 3**. Acoustic elements were used like for the water domain. At the interface of the vapor cavity with the surrounding water, the acoustic pressure and the normal component of the water/vapor velocity are continuous. The vapor properties were set as 0.174 kg/m$^3$ for density and 340 m/s for speed of sound.

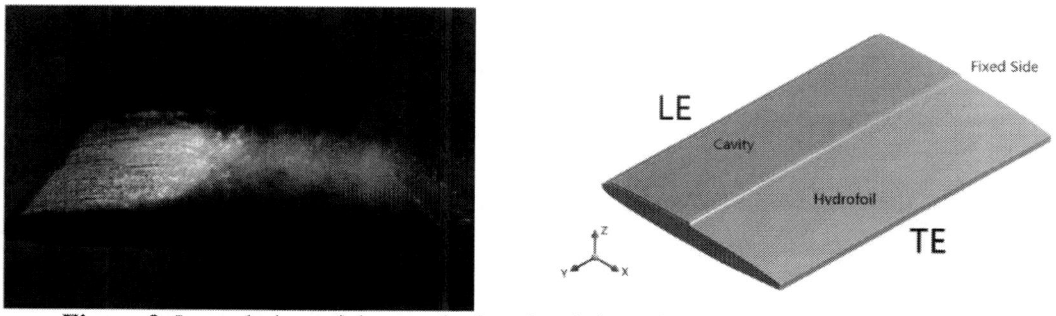


**Figure 3.** Lateral view of the attached cavity (left) and corresponding numerical model (right).

## 4. Results

The modal behaviour of the hydrofoil resembles that of a cantilever beam. The three first modes of vibration correspond to the first bending ($f_1$), the first torsion ($f_2$) and the second bending ($f_3$). To evaluate the changes of natural frequency with void ratio induced by the added mass effects, the Frequency Reduction Ratio in percent for mode *i*, *FRRi*, has been calculated as expressed in Equation (2):

$$\mathrm{FRR}_i = 100\% \cdot \frac{f_{i,\alpha=1} - f_{i,\alpha}}{f_{i,\alpha=1}} \tag{2}$$

where $f_{i,\alpha=1}$ and $f_{i,\alpha}$ are the mode $i$ natural frequencies for void ratio 1 and for any void ratio α, respectively.

As it can be observed on the left of **Figure 4**, Brennen [4] states that the cavity density and the speed of sound in a mixture of water and vapor is a function of the void ratio, $\alpha$, which is in turn the ratio between the vapor volume and the total cavity volume. Based on such data, a series of cases have been simulated considering the cavity domain void ratio in the range from 0.6 to 1, as shown on the right of **Figure 4**, while keeping the same cavity length and thickness and the same properties of the water domain around the cavity. As it can be observed from the results, both the *FRR* and the mode shape are affected by the changes in cavity properties. The significant effect of cavity void ratio on natural frequencies is parallel to a significant effect on mode shapes especially for mode $f_3$ and followed by mode $f_2$. Meanwhile, for mode f1 the effects on the shape are almost negligible.

The predicted mode evolutions are in fact induced by the particular sonic speed and averaged cavity density values considered in each case. These couple of variables are directly linked so it is difficult to determine their individual level of significance on the obtained results. To clarify this point, a series of simulations have been carried out by keeping one factor constant and changing the other although they are not physically accurate for large deviations from Brennen's data. To begin, the density has been reduced from 1000 kg/m$^3$ to 0.174 kg/m$^3$ while keeping the sonic velocity constant at 340 m/s. These results are plotted on the left of **Figure 5** and they indicate that the density of the cavity has a low influence on the natural frequency change of each mode. Only when the density is lower than 100 kg/m$^3$, the increase of the natural frequencies is larger, but still not significant. Contrarily, the role of the sonic velocity is dominant as it can be observed on the right of **Figure 5** when it is reduced from 1480 to 24 m/s while keeping the density constant at 0.174 kg/m$^3$. The results for all the modes of vibration show that the natural frequency changes significantly in a particular range of values and that outside such range it is almost constant. In particular, the frequency changes linearly with the sonic velocity following a common diagonal line for all the modes that we call "Mode Transition Line".

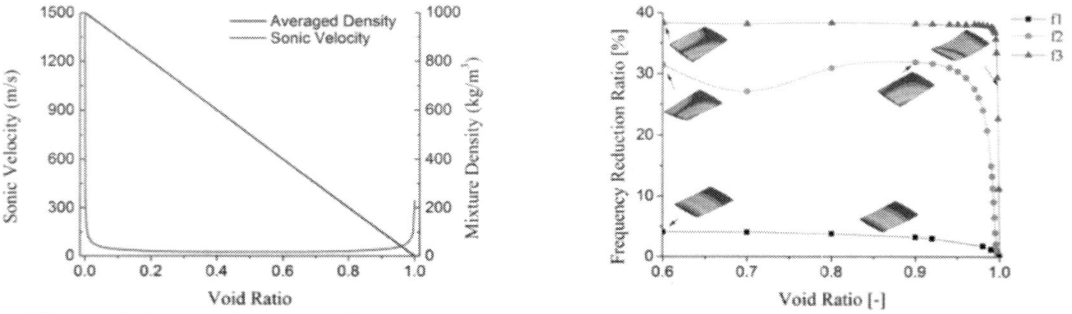


**Figure 4.** Speed of sound and cavity density as a function of void ratio (left); *FRR*'s and mode shape evolution as a function of cavity void ratio for $f_1$, $f_2$ and $f_3$ (right).

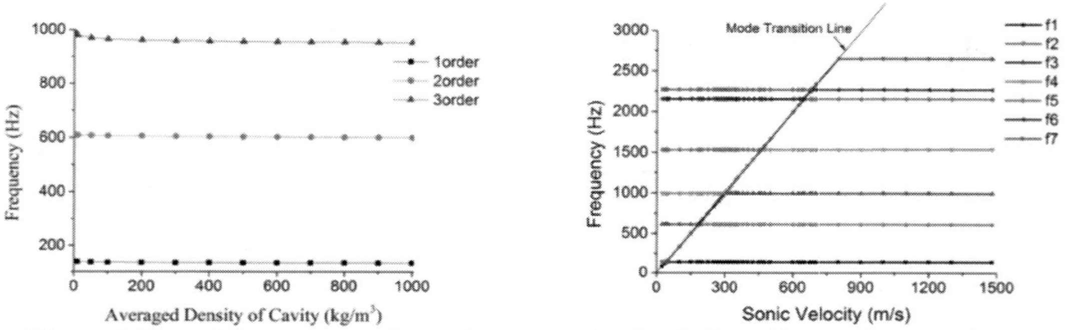


**Figure 5.** Natural frequency vs change in average density (left) and in speed of sound (right).

## 5. Conclusion

The added mass effects on a NACA0009 hydrofoil under cavitation conditions have been numerically calculated using acoustic FEM. The results show that the reduction of void ratio from 1 to 0.6 increases the added mass effects and modifies the mode shapes more significantly for higher order modes. In particular, the sonic speed is more relevant than the average density on the modal behavior.

## Acknowledgements

This work was supported by National Natural Science Foundation of China (No. 51279083).

## References

[1] Liang Q W, Rodríguez C G, Egusquiza E, Escaler X, Farhat M, and Avellan F 2007 Numerical simulation of fluid added mass effect on a francis turbine runner *Computers & Fluids* **36** 1106-18

[2] De La Torre O, Escaler X, Egusquiza E, and Farhat M 2013 Experimental investigation of added mass effects on a hydrofoil under cavitation conditions *Journal of Fluids and Structures* **39** 173-87

[3] ANSYS Inc. 2013 *Ansys theory reference* (Ansys)

[4] Brennen C E, "Homogeneous bubbly flows," in *Cavitation and bubble dynamics*, ed Oxford: Oxford University Press, 1995.

# Modelling Cavitating Flows using an Eulerian-Lagrangian Approach and a Nucleation Model

**Jingsen Ma[1], Chao-Tsung Hsiao, and Georges L. Chahine**
DYNAFLOW, INC., 10621-J Iron Bridge Road, Jessup, MD 20794, USA
E-mail: *jingsen@dynaflow-inc.com*

**Abstract.** An Eulerian/Lagrangian multi-scale two-phase flow model is developed to simulate the various types of cavitation including bubble, sheet, and tip vortex cavitation. Sheet cavitation inception, unsteady breakup, and cloud shedding on a hydrofoil are used as an example here. No assumptions are needed on mass transfer between phases; instead, the method tracks bubble nuclei, which are in the bulk of the liquid and those generated by nucleation from solid boundaries and this is sufficient to accurately capture the sheet dynamics. The multi-scale model includes a micro-scale model for tracking the bubbles, a macro-scale model for describing large cavity dynamics and a transition scheme to bridge the micro and macro scales. Nuclei are treated as flow singularities until they grow into large bubbles, which eventually merge to form a large scale discretised sheet cavity. The sheet performs large scale oscillations with a periodic reentrant jet forming under the sheet cavity, traveling upstream, and breaking the cavity. This results in bubble cloud formation and in high pressure peaks as the broken pockets shrink and collapse while travelling downstream. The results for a NACA0015 foil are in good agreement with the experimental data.

## 1. Introduction

Sheet cavitation on a lifting surface starts near the leading edge, effectuates large amplitude oscillations, and breaks up periodically releasing clouds of bubbles, which collapse violently generating high local pressures on the blade. Modelling of this process is challenging as it involves a wide range of characteristic lengths and phenomena including nucleation, large deforming free surfaces, and breakup into very fine bubble cloud, etc. Observations have indicated weak dependence of the sheet cavity dynamics on the nuclei distribution in the test facility, which has led some modellers to believe that sheet cavitation is not connected to bubble dynamics. Actually, over the years, several researchers (e.g. [1]) have observed that the presence of free field nuclei affected significantly the repeatability of the experimental results. When nuclei count was low, the value of the incipient cavitation number showed large scatter, whereas in water rich in nuclei the results were more repeatable and were not affected by further addition of nuclei. One potential explanation to these observations is that nucleation from the solid surface of the blades could play a more important role than the free field nuclei, and this has overlooked in previous studies of sheet cavitation even though bubble nucleation has been extensively investigated [2,3]. Motivated by the need for a physics-based model of sheet initiation, the present model aims at simulating cavitation starting from the dynamics of nuclei without using empirical

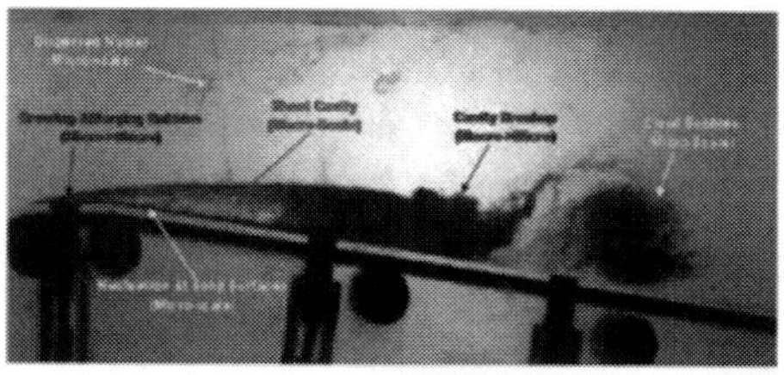

**Figure 1.** Illustration of various scales involved in sheet inception and dynamics.

[1] To whom any correspondence should be addressed.

fits for mass exchange as in conventional approaches. The model illustrated in Figure 1 addresses a macro-scale associated with the foil length and the sheet cavity, a micro-scale for the bubble nuclei, an intermediary scale for the bubble clouds, and transition and coupling procedures between these scales.

## 2. Multiscale Two-phase Model

### *2.1. Eulerian Continuum Model for Viscous Mixture Flow*

The two-phase continuum medium satisfies the following continuity and momentum equations:

$$\frac{\partial \rho_m}{\partial t} + \nabla \cdot (\rho_m \mathbf{u}) = 0, \qquad \rho_m \frac{D\mathbf{u}}{Dt} = -\nabla p + \mu_m \nabla^2 \mathbf{u} + \rho_m \mathbf{g}, \tag{1}$$

where $\mathbf{u}$ is the mixture velocity, $p$ the pressure and $\mathbf{g}$ the acceleration due to gravity. The mixture density $\rho_m$ and the mixture viscosity, $\mu_m$, are related to the void fraction, $\alpha$, and the liquid and gas properties through $\rho_m = (1-\alpha)\rho_l + \alpha\rho_g$, $\mu_m = (1-\alpha)\mu_l + \alpha\mu_g$. $\alpha$ is deduced from the location and size information of the bubbles using a Gaussian distribution scheme which smoothly "spreads" each bubble's volume over its surroundings while conserving the total bubble volumes [4].

### *2.2. Lagrangian Discrete Bubble Model for Microbubbles*

The bubble model uses a Surface Average Pressure approach [5] to average fluid quantities along the bubble surface and model its dynamics using a Rayleigh-Plesset-Keller-Herring equation:

$$\left(1-\frac{\dot{R}}{c_m}\right)R\ddot{R} + \frac{3}{2}\left(1-\frac{\dot{R}}{3c_m}\right)\dot{R} = \frac{\mathbf{u}_s^2}{4} + \frac{1}{\rho_m}\left(1+\frac{\dot{R}}{c_m}+\frac{R}{c_m}\frac{d}{dt}\right)\left[p_v + p_g - p_{enc} - \frac{2\gamma}{R} - 4\mu_m\frac{\dot{R}}{R}\right], \tag{2}$$

where $R$ is the bubble radius, $p_v$ is the liquid vapor pressure, $p_g$ is the bubble gas pressure, $\gamma$ is the surface tension, $\mathbf{u}_s = \mathbf{u}_{enc} - \mathbf{u}_b$, $\mathbf{u}_b$ is the bubble travel velocity, while $\mathbf{u}_{enc}$ and $p_{enc}$ are respectively the liquid velocity and the ambient pressure "seen" by the bubble during its travel. $c_m$ is the local sound velocity in the two phase medium. The bubble trajectory is obtained from the following bubble motion equation:

$$\frac{d\mathbf{u}_b}{dt} = \left(\frac{\rho_l}{\rho_b}\right)\left[\frac{3}{8R}C_D\,|\mathbf{u}_s|\,\mathbf{u}_s + \frac{1}{2}\left(\frac{d\mathbf{u}_{enc}}{dt} - \frac{d\mathbf{u}_b}{dt}\right) + \frac{3\dot{R}}{2R}\mathbf{u}_s - \frac{\nabla p}{\rho_l} + \frac{(\rho_b - \rho_l)}{\rho_l}g + \frac{3C_L}{4\pi}\frac{\sqrt{\nu}}{R}\frac{\mathbf{u}_s \times \mathbf{\Omega}}{\sqrt{|\mathbf{\Omega}|}}\right], \tag{3}$$

where $\rho_b$ is the bubble density, $C_D$, $C_L$ are the drag and lift coefficients and $\Omega$ is the deformation tensor.

### *2.3. Level-Set Method for Macro Cavities*

In order to simulate large free surface deformations including folding and breakup as in a sheet cavity, a Level-Set method is used. A smooth function $\varphi(x,y,z,t)$, whose zero level coincides with the liquid/gas interface when the level set is introduced, is defined in the whole physical domain (i.e. in both liquid and gas phases) as the signed distance $d(x,y,z)$ from the interface $\varphi(x,y,z,0) = d(x,y,z)$. This function is enforced to be a material surface evolving with velocity field $\mathbf{u}$ at each time step:

$$\frac{d\varphi}{dt} = \frac{\partial \varphi}{\partial t} + \mathbf{u}\nabla\varphi = 0. \tag{4}$$

### *2.4. Transition Model between Micro- and Macro-Scale*

The transformation between micro- to macro-scale and vice-versa includes two scenarios. In the first scenario, since the bubbles are tracked using the Lagrangian scheme, the bubble sizes and locations are known at every time step. A criterion based on bubble size is set to "activate" the bubbles for computation of a local distance function for neighbouring fluid cells. For each cell $i$, the distance function is defined by $\varphi_i = \min(\varphi_{LS0}, \varphi_{b,j})$; $j = 1, N_i$, where

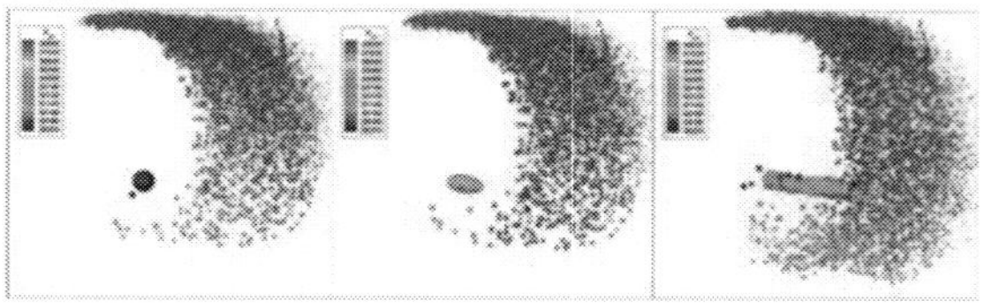

**Figure 2.** Time sequence of bubble capture in a line vortex cavitation core. The bubbles are colored by their pressure. The aqua blue iso-surface is the gas/vapor-liquid interface resulting from discrete large free surface formation.

$\varphi_{LS0}$ is the original distance cell $i$ from level set 0, $\varphi_{b,j}$ is the local distance between the center of cell $i$ and the surface of bubble $j$, and $N_i$ is the number of relevant bubbles around cell $i$. This scheme allows bubbles to merge together into a large cavity, as well as single bubbles to be absorbed by a large cavity as illustrated in Figure 2 for a sequence of bubble capture in a cavitating line vortex. Once a bubble grows explosively, its interface becomes part of the level set 0. Bubbles can coalesce and join to form a sheet or a tubular vortex cavity. The second scenario is applied when the volume of a large cavity is decreasing in time. As the large cavity collapses, the lost volume is replaced with a distribution of micro-bubbles of the same volume if the mixture normal speed is larger than the interface normal speed [6,7].

### 2.5. *Nucleation Model*

A very important aspect of our approach is to account for nucleation from the rigid boundaries [4] in addition to bubble nuclei dispersed in the liquid. The wall nucleation model involves the following parameters, which are functions of the wall material properties and the local flow conditions: a) $P_{thr}$: a nucleation pressure threshold, b) $N_s$: a number density of nucleation sites per unit area, c) $f_n$ : a nucleation rate, and d) $R_0$: the initial nuclei size(s). At each time step the pressure at a cell is compared with $P_{thr}$ to determine if nucleation should occur. Once the pressure drops below the threshold, the cell releases a total of $N$ nuclei, $N = N_s f_n \Delta t \Delta A$, where $\Delta A$ is the surface area of the grid cell and $\Delta t$ the integration time step. All above parameters are strong functions of surface roughness and temperature. In the results below, $P_{thr}$ is selected to be the vapour pressure. The initial size of the nuclei emitted is selected to be 10 μm with the base nucleation sites per unit area value selected to be $N_s = 224/\text{cm}^2$ and the base nucleation frequency selected to be $f_n = 22kHz$. These parameters are selected based on the study in [8], where the results became independent of $N_s$ and $f_n$ when larger than these values.

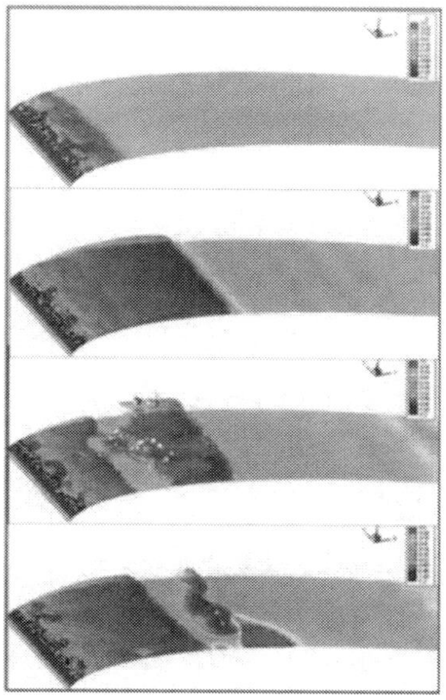

**Figure 3.** Time sequence of sheet-to-cloud cavitation modeling on

## 3. Simulation of Sheet Cavitation on Naca0015 Hydrofoil

As illustrated in Figure 3, the present model successfully captures sheet cavity initiation, dynamics, and breakup on a well-tested foil. Bubbles nucleated from the blade surface and free nuclei in the liquid were the basis of the simulations and no empirical mass exchange models were needed. Cavitation initiates on the hydrofoil leading edge when the local pressure drops below the vapour pressure. The nuclei grow, aggregate, and coalesce into a large cavity well-captured with the Level Set method. As the sheet cavity develops, its trailing edge curves into a reentrant jet, which moves upstream over the foil and into the cavity, then breaks it into a smaller cavity and a vortical large bubble cloud, which collapses generating high pressures. This periodic shedding phenomenon can be further quantified by analysing the time histories of the instantaneous pressures and stream-wise velocities in the flow. For example Figure 5 shows the pressures and velocities monitored at the location $x=0.4\ L$ downstream of the leading edge on the suction side of the foil. This is close to the maximal extent of the trailing edge of the fully developed sheet cavity. This location alternatively experiences negative (reentrant jet development) and positive (no reentrant jet) stream-wise velocities with magnitudes as high as the incoming flow $U_\infty$. The time interval between consecutive negative velocity maxima is repeatable and has a value close to 1.3 $L/U_\infty$. The corresponding oscillation frequency is about 0.78 $U_\infty/L$, which matches well with the experimentally measured shedding frequency for the same cavitation number [9]. Comparison of the numerical results with experimental observations for a large range of $\sigma$ is very satisfactory, as seen in Figure 4. Both the shedding frequency and the time-averaged cavity length compare well with the experimental measurements.

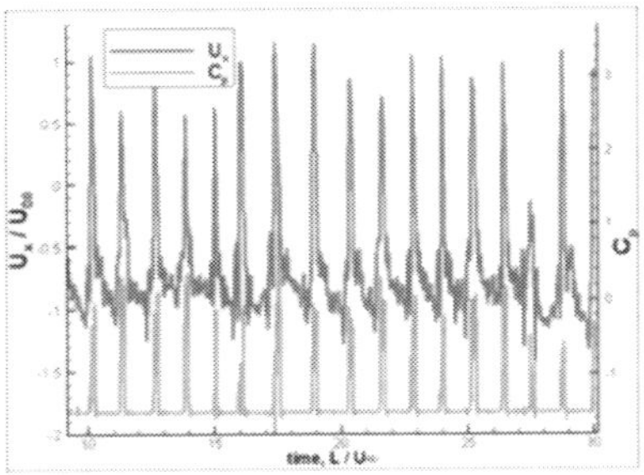

**Figure 5.** Pressures and streamwise velocity at $x = 0.4L$ on the solid wall near the cavity trailing edge. $\sigma = 1.5$

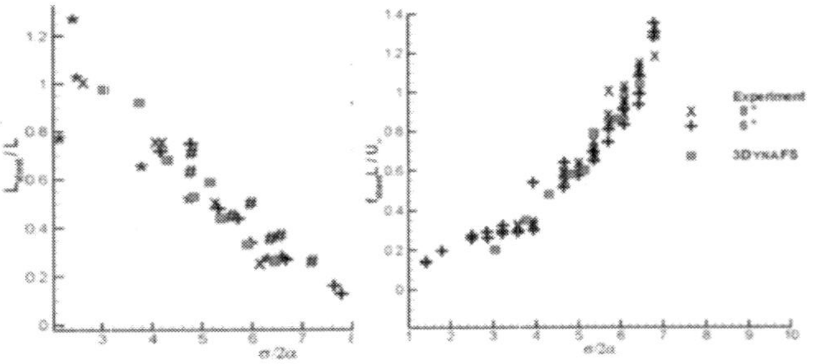

**Figure 4.** Comparison of predicted sheet cavity averaged length and shedding frequencies with data in [14].

## 4. Summary
A multi-scale framework was developed for the simulation of various forms of cavitation. The example of sheet cavitation on a hydrofoil shows how the model bridges smoothly a Level Set method for large size cavities and a Discrete Singularity Model for small bubbles. Starting with the micro-scale physics of nucleation from the solid surface and of dispersed nuclei in the bulk liquid, sheet cavitation along a NACA0015 hydrofoil is well captured by the model. The predicted sheet length and oscillation frequency match experimental data very well for a large range of cavitation numbers.

## 5. Acknowledgments
This work was supported by the Office of Naval Research under contract N00014-12-C-0382 monitored by Dr. Ki-Han Kim. This support is highly appreciated.

## 6. References
[1] Kuiper G 1985 Reflection on cavitation inception *ASME Cavitation and Multiphase Flow Forum* (Albuquerque, NM)
[2] Yount D E 1979 Skin of varying permeability: A stabilization mechanism for gas cavitation nuclei *J. Acoust. Soc. Am.* **65** 1429–39
[3] Mørch K A 2009 Cavitation Nuclei: Experiments and Theory. Journal of Hydrodynamics *J. Hydrodyn. Ser. B* **21** 176–89
[4] Ma J, Chahine G L and Hsiao C-T 2015 Spherical bubble dynamics in a bubbly medium using an Euler–Lagrange model *Chemical Engineering Science* vol 128 pp 64–81
[5] Hsiao C-T, Chahine G L and Liu H-L 2003 Scaling effect on prediction of cavitation inception in a line vortex flow *J. Fluids Eng.* **125** 53–60
[6] Ma J, Oberai A A, Drew D A, Lahey R T and Hyman M C 2011 A comprehensive sub-grid air entrainment model for RaNS modeling of free-surface bubbly flows *J. Comput. Multiph. Flows* **3** 41–56
[7] Hsiao C-T, Wu X, Ma J and Chahine G L 2013 Numerical and Experimental Study of Bubble Entrainment Due to a Horizontal Plunging Jet *Int. Shipbuild. Prog.* **60** 435–69
[8] Hsiao C T, Ma J and Chahine G L 2014 Multi-scale two-phase flow modeling of sheet and cloud cavitation *30th Symposium on Naval Hydrodynamics* (Hobart, Tasmania, Australia)
[9] Berntsen G S, Kjeldsen M and Arndt R E 2001 Numerical modeling of sheet and tip vortex cavitation with FLUENT 5 *Fourth International Symposium on Cavitation.* (Pasadena, CA)

# Effect of Non-Condensable Gas Injection on Cavitation Dynamics of Partial Cavities

**Simo A. Mäkiharju, Harish Ganesh, and Steven L. Ceccio**
University of Michigan, Ann Arbor, MI 48109, USA

E-mail: smakihar@umich.edu

**Abstract**. Partial cavities can undergo auto-oscillation causing large pressure pulsations, unsteady loading of machinery and generate significant noise. In the current experiments fully shedding cavities forming in the separated flow region downstream of a wedge were investigated. The Reynolds number based on hydraulic diameter was of the order of one million. The cavity dynamics were studied with and without injection of non-condensable gas into the cavity. Gas was injected directly into the cavitation region downstream of the wedge's apex, or into the recirculating region at mid cavity so that for the same amount of injected gas less ended up in the shear layer. It was found that relatively miniscule amounts of gas introduced into the shear layer at the cavity interface can reduce vapour production and dampen the auto oscillations, and the same amount of gas injected into the mid cavity would not have the same effect. The authors also examined whether the injected gas can switch the shedding mechanism from one dominated by condensation shock to one dominantly by re-entrant jet.

## 1. Introduction

Partial cavitation occurs in numerous industrial and naval applications. Cavitation on lifting surfaces or in cryogenic rocket motors can damage equipment, and in general be detrimental to the system performance. Much work has gone into understanding the basic physical processes involved [2, 3, 4, 8]. If a partially cavitating condition is likely to occur, it would be of use to be able to control it. Kawanami *et al.* [6] used obstacles to prevent a re-entrant from moving upstream and they were successful in preventing the generation of cloud cavitation. However, recently Ganesh *et al.* [5] discovered that under certain conditions a condensation shock can be the dominant mechanism instead of a re-entrant jet, and under these conditions an obstacle placed on the surface is not sufficient to prevent cloud shedding from partial cavities.

To investigate the nature of cloud shedding in conditions where a condensation shock occurs, we investigated the effect of non-condensable gas injection both from the apex and into the mid cavity.

## 2. Experimental setup

Experiments were carried out at the University of Michigan 9-Inch Water Tunnel. We utilized the X-ray densitometry system described by Mäkiharju *et al.* [7] to measure the time resolved 2D void fraction distribution without flow perturbations. To reduce the baseline attenuation, the water tunnel's test section was further contracted to have a $(76\text{mm})^2$ cross-section.

The wedge geometry and gas injection locations are shown in figure 1. In addition to the 1 kHz void fraction measurements, the unsteady pressure was measured using surface mounted transducers sampled at 500 kHz. The transducers were located at 32 and 50 mm downstream of the apex.

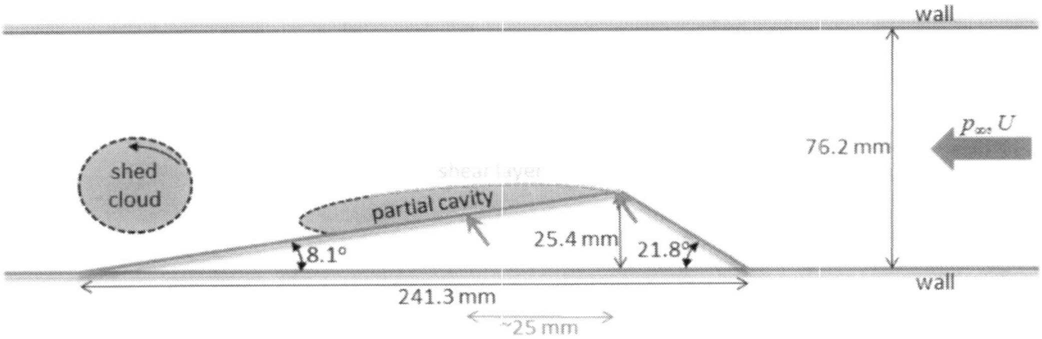


**Figure 1.** Schematic of the experimental setup. The red arrows indicate the apex and mid cavity gas injection locations, which were ~25mm apart in the streamwise direction.

## 3. Shedding partial cavitation

The cavitation regime changes as a function of the cavitation number defined as

$$\sigma_\infty = \frac{p_\infty - p_V}{\frac{1}{2}\rho U^2}$$

For the present study the free stream speed, $U$, was fixed at 8 ± 0.1 m/s and the inlet pressure kept constant at $p_\infty$ = 63.5 ± 1 kPa. Hence, the cavitation number was fixed at $\sigma_\infty$ = 1.9 ± 0.1. For the geometry and flow conditions chosen, propagation of bubbly shock waves leads to the partial cavity shedding, as shown in figure 2. In the cavity the average void fraction was $\bar{\alpha}_{MAX} \sim 40\%$.

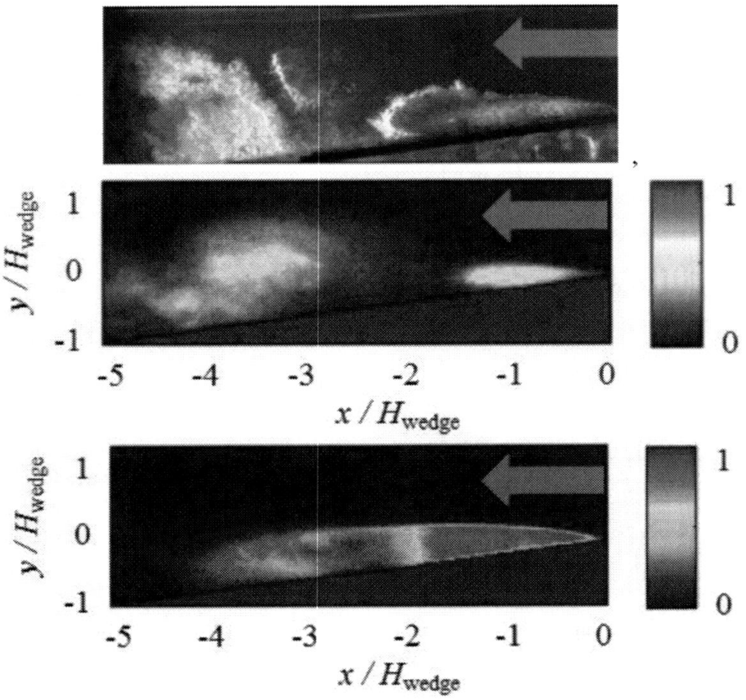


**Figure 2.** Side view of the shedding cavity as viewed in (a) visible light. In (b) and (c) we see a 2D X-ray projection of the void fraction distribution. In (c) a condensation shock is seen moving upstream.

Journal of Physics: Conference Series **656** (2015) 012161 doi:10.1088/1742-6596/656/1/012161

## 4. Effect of gas injection

For the present flow we expect most of the vapour to be created in the shear layer. Our hypothesis is that *i*) introduction of non-condensable gas into the shear layer would predominantly reduce the amount of vapor created per cycle, as the presence of the gas will increase pressure in the shear layer's vortex cores and possibly increase suction peak pressure. And, *ii*) introduction of non-condensable gas into the mid cavity would predominantly change the condensation shock speed, as a condensation shock in two phase mixture could be expected to propagate at speed $u$ given by

$$u_1^2 = \frac{p_2 - p_1}{\rho} \frac{(1-\alpha_2)}{(1-\alpha_1)(\alpha_1 - \alpha_2)}$$

This is further discussed in Brennen [1], and the above equation is the result of ignoring the bubble dynamics. Here the sub-indices 1 and 2 denote conditions up- and downstream of the shock, respectively. From this we can see if both $\alpha_1$ and $\alpha_2$ increase equally, the shock speed would increase, and if $p_2 - p_1$ decreases, the shock speed decreases.

To determine what constitutes a significant amount of gas injection, we first estimated the vapour production rate by cavitation based on the void fraction measurements obtained using X-ray densitometry. The normalized vapour production rate was defined as $C_q = Q/UL_{cav}b$, where $Q$ is volume flux of vapour at cavity pressure, $L_{cav}$ is cavity length and $b$ is the span. The value of $C_q$ for the present setup and $\sigma_\infty = 1.9$ was found to be $O(10^{-2})$, similar to that of Stutz and Rebound [8] who found $C_q \sim 0.006$. The normalized amount of injected gas was defined similarly as $C_{qI}$.

Figure 3 shows the unsteady pressure signal and their FFTs without and with gas injection. Without gas injection we observe strong shedding at St ~0.27, where the Strouhal number is defined as $St=fL_{cav}/U$. However, with injection of a relatively small amount of gas from the apex ($C_{qI}/C_q \sim 0.3$) the shedding cycle became more weakly periodic, average void fraction was reduced, as seen in figure 4, and cavity begins to be reminiscent of a transitional cavity, as discussed in [5]. Figure 4 also shows that injecting the same amount of gas into the mid cavity, has almost no observable effect. This would be consistent with suppression of vapour generation in the shear layer (*i.e.* the first hypothesis).

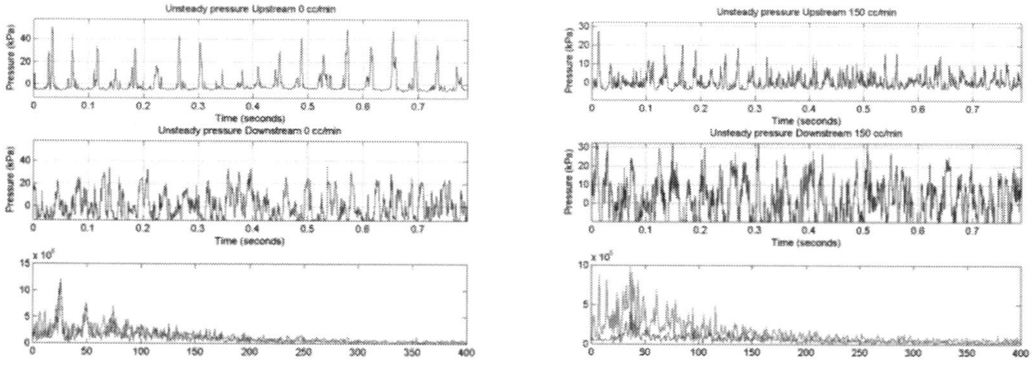


**Figure 3.** Pressure signals and their FFTs without (left) and with (right) gas injection from the apex at a rate of $C_{qI}/C_q \sim 0.3$. The top and middle figures show the pressure signals 32 and 50 mm downstream of the apex, respectively. The bottom figures show the FFTs of these signals.

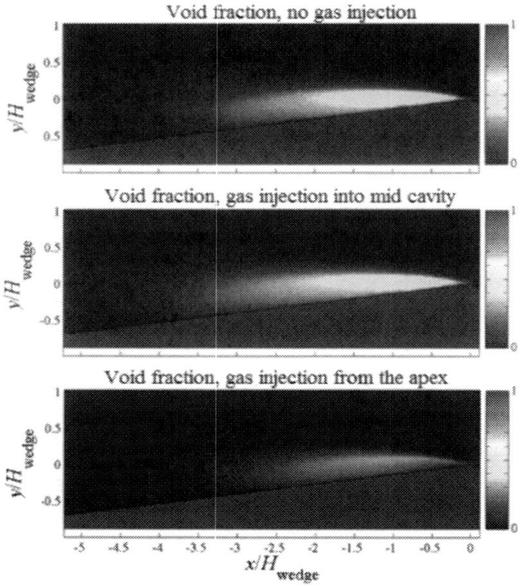


**Figure 4.** Average void fraction distributions with no gas injection (top), injection into mid cavity (middle) and injection from the apex (bottom) at $C_{ql}/C_q \sim 0.3$.

## 5. Discussion and conclusions

Injection of even a relatively small amounts ($C_{ql}/C_q \sim 0.3$) of non-condensable gas into the shear layer (*i.e.* from the apex) can dampen the shedding cycle by significantly reducing the vaporization rate of the liquid. Injecting the same amount of non-condensable gas into the mid cavity had a much weaker overall effect than injection from the apex. The authors are currently conducting additional experiments to further study effect of increased gas injection (up to $C_{ql}/C_q \sim 4$), as well as to better quantify the change in shock speed and void fraction.

**Acknowledgements**

This work was supported by the Office of Naval Research Grant N00014-14-1-0292, under program manager Dr. Ki-Han Kim.

**References**

[1] Brennen, C. E. Fundamentals of Multiphase Flows, Cambridge University Press, 2005

[2] Callenaere, M., Franc, J-P., Michel, J-M. & Riondet, M., "The cavitation instability induced by the development of a re-entrant jet", *J. Fluid Mech.* **444**, 223-256, 2001

[3] Coutier-Delgosha, O., Devillers, J.-F., and Pichon, T., "Internal structure and dynamics of sheet cavitation", *Phys. Fluids*, **18**, No. 017103, 1-12, 2006

[4] Coutier-Delgosha, O., Stutz, B., Vabre, V. & Legoupil, S., "Analysis of cavitating flow structure by experimental and numerical investigations", *J. Fluid Mech.* **578**, 171-222, 2007

[5] Ganesh, H., Mäkiharju, S.A., and Ceccio, S.L. "Bubbly shock propagation as a mechanism for sheet-to-cloud transition of partial cavities." Submitted to the Journal of Fluid Mechanics.

[6] Kawanami, Y., Kato, H., Yamaguchi, H., Tanimura, M. & Tagaya, Y., "Mechanism and control of cloud cavitation", *Trans. ASME: J. Fluids Engng.* **119**, 788-795, 1997

[7] Mäkiharju, S.A., Chang, N., Gabillet, C., Paik, B.-G., Perlin, M. & Ceccio, S.L., "Time Resolved Two-dimensional X-Ray Densitometry of a Two Phase Flow Downstream of a Ventilated Cavity", *Experiments in Fluids*, **54**, 2013

[8] Stutz, B., and Reboud, J. L. "Experiments on unsteady cavitation." *Exp. in Fluids,* 22(3), 1997

# Jet-Supercavity Interaction: Insights from Experiments

**MJ Moeny**[1], **MH Krane**[1], **IN Kirschner**[2], **MP Kinzel**[1]
[1] Applied Research Laboratory, The Pennsylvania State University, University Park, PA, USA
[2] Applied Physical Sciences, Groton, CT, USA

E-mail: mjm369@arl.psu.edu

**Abstract.** An experimental study was performed to evaluate some of the claims of Paryshev (2006) regarding changes to ventilated cavity behavior caused by the interaction of a jet with the cavity closure region. The experiments, conducted in the 1.22m dia. Garfield Thomas Water Tunnel, were performed for EDD to tunnel diameter of 0.022, $Fr$ = 14.5 and 26.2. The model consisted of a converging-section nozzle mounted to the base of a 27.9mm 37° cone cavitator placed on the tunnel centerline at the end of a 138.4mm long streamlined strut. A ventilated cavity was formed over the model, then an air jet, issuing from a converging nozzle, was initiated. Changes to cavity behavior were quantified in terms of cavitation number, thrust-to-drag ratio, and stagnation pressure ratio at the jet nozzle. The results show that, while the overall trends predicted by Paryshev were observed, the data did not fully collapse, suggesting that many of the effects neglected by Paryshev's model have measureable effect.

## 1. Introduction

This paper presents an experimental effort to evaluate model predictions of Paryshev (2006), concerning the effect of jet/closure interaction on the behavior of a ventilated cavity. That model was based on several simplifying assumptions, three of which are: (1) the jet is narrow, so the region over which jet/closure interaction occurred is short in the axial direction, (2) the jet does not significantly entrain gas from the ventilated cavity, so that the jet stagnation pressure does not vary over the length of the jet, and (3) the pressure in the ventilated cavity is uniform over the length of the cavity, upstream of the closure region, in which the pressure rises rapidly to stagnation pressure, then falls to ambient pressure at the downstream end of the cavity. Paryshev proposed that the jet/closure interaction, and its effect on the cavity as a whole, are governed by two parameters: $P = \rho_\infty U_\infty^2 / \rho_j U_j^2$, the ratio of the stagnation pressure of the fluid flow outside the cavity to that of the jet, and $\bar{J} = \rho_j U_j^2 S_j / D_n$, the ratio of jet momentum flux to the drag on the cavitator.

The combination of these two parameters determines the effect of the jet on the closure region, as shown in Figure 1. The figure shows $\bar{J}$ vs. $P$ space, with regions colored according to the behavior Paryshev predicted. In the region $P < 1$ (blue in the figure), where the jet stagnation pressure exceeds that of the pressure outside the closure region (roughly, the total pressure of the exterior flow), the jet strongly penetrates the cavity closure region, strongly modifying the closure region dynamics, increasing the rate at which gas leaks from the cavity, resulting in a smaller cavity (higher cavitation number). Paryshev notes the similarity to the *Efros-type closure*. When $P > 1$, jet stagnation pressure is always less than the exterior pressure, and there are two possibilities, depending on the value of the thrust to drag ratio $\bar{J}$. For low jet thrust, $\bar{J} < ½$, the jet effect on the closure region is minimal, and the jet acts to supplement

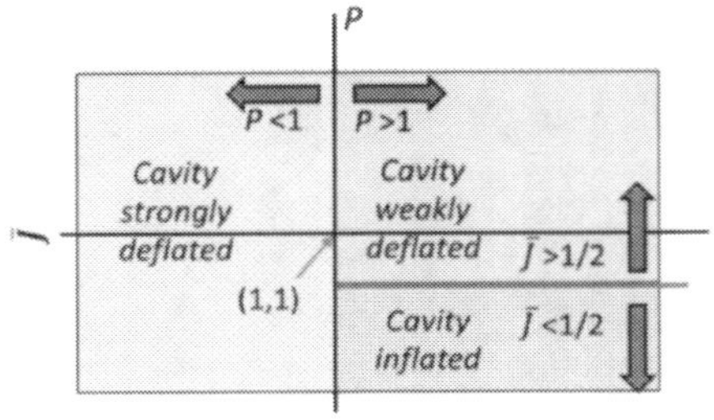


***Figure 1.*** Parameter space describing jet/cavity closure interaction regimes, after Paryshev (2006). $P < 1$ (*blue*): jet strongly deflates cavity. $P > 1$, $\bar{J} > ½$ (*orange*): jet weakly deflates cavity. $P > 1$, $\bar{J} <½$ (*green*): jet inflates cavity.

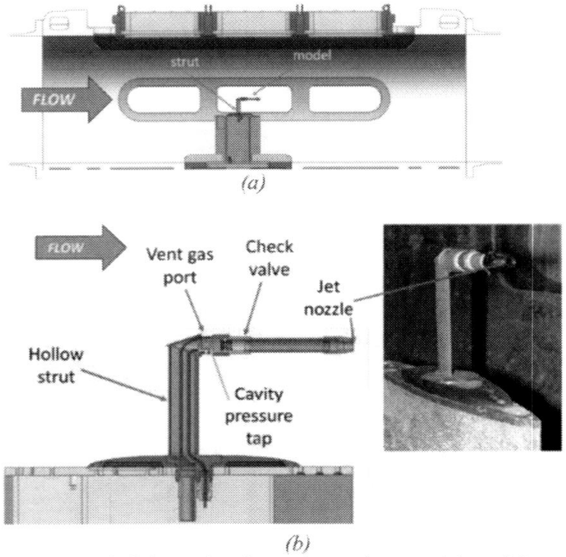


*Figure 2.* Schematic of experimental setup. (a) model installed in water tunnel test section. (b) Detail of model. Cavitator and strut jet air and cavity ventilation air, as well as the pressure line passed through the strut. A check valve prevented water from leaking back into the jet gas system. Model body consists of a short section of aluminum pipe and a converging nozzle.

cavity ventilation, enlarging the cavity (reducing cavitation number). As thrust increases to $\bar{J} > ½$, the jet more strongly modifies the closure region, though not as severely as when $P < 1$, increasing the gas leakage rate, and reducing the cavity size (increasing cavitation number).

As noted, Paryshev's model contained strong simplifications, and its predictions were compared to only a single data set, about which not much detail was given. The experiments reported here, and the parallel low-order modeling (Kirschner, *et al.*, 2015, these proceedings) and computer simulation (Kinzel, *et al.*, 2015, these proceedings) efforts were therefore undertaken to test the extent to which the behavior predicted by Paryshev's model can be observed.

## 2. Methods

Experiments were conducted in the (1.22m) diameter Garfield Thomas Water Tunnel (see, *e.g.*, Lehman, 1959; Marboe, et al., 1993) at the Penn State University Applied Research Laboratory. The closed-loop tunnel features a 4.27m long test section in which tunnel speed can be varied continuously from approximately 1 to 16m/s with independent control of test section static pressure from 14kPa to 414kPa absolute over the full velocity range.

### *2.1 Physical Model*

The experimental setup is shown in Figure 2. Figure 2a shows the model installed in the GTWT, while Figure 2b shows a detail of the model. The model consists of a strut based on a NACA 16-0021 foil of 3.51cm chord x 0.66 cm thickness and in one piece with conical cavitator, base diameter $d_n$ = 2.79cm, with a 37° included angle. Comparing the calculated $C_D$ of this cone to that of an equivalent flat disk and the water tunnel gives a ratio of equivalent disk diameter (EDD) to tunnel diameter of 0.022. Moving downstream from the base of the cone, a check valve, then a hollow aluminum cylindrical body (1.52cm OD, 1.27cm ID), and finally an aluminum converging nozzle were attached in series, using standard pipe thread. The cavitator/strut piece was fabricated as one piece from laser-sintered Inconel, permitting the strut to be low-profile and minimizing interference of its attachment to the cavitator. The hollow strut provides passage of (*i*) ventilation air through a separate line, (*ii*) a tube connecting the cavity pressure port on the side of the cavitator to a differential pressure transducer located outside the tunnel, and (*iii*) jet gas air, which flows through the remaining space in the strut. Two separate bottled high-pressure dry air reservoirs supplied air, one each to the cavity ventilation and to the jet.

### *2.2 Measurements*

A differential pressure sensor connected between the cavity pressure tap on the cavitator and to the tunnel wall at the same axial location as the model, measured $\Delta p_{cav}$, the difference between cavity and tunnel static pressure. Also measured were ventilation gas mass flow rate and jet gas mass flow rate using two thermal mass flow meters, jet gas reservoir and tunnel water temperatures using two resistance temperature detectors, and tunnel total pressure and tunnel static pressure at discrete axial locations through the test section using multiple gage-pressure sensors. Finally, a high-speed digital video camera was used to acquire video images of the cavity at a rate of 1500 frames/s.

Measured differential cavity pressure and calculated tunnel flow velocity and density were used to estimate the cavitation number, $\sigma = 2\Delta p_{cav}/\rho_\infty U_\infty^2$. The jet gas line measurements, cavity pressure and tunnel pressure were used to iteratively estimate jet gas reservoir pressure just upstream of the nozzle, along with jet speed and Mach number, as well as momentum flux at the nozzle. These were used to estimate the cavitator drag coefficient $C_d = 0.21 + 0.546\sigma$, using data from May (1975). The above quantities then returned $\bar{J}$ and $P$ for each case measured.

## 3. Results

### *3.1 Cases studied*

Each measurement consisted of establishing an over-ventilated cavity, and then introducing the jet at a target flow rate. Data was collected at two tunnel speeds, $U_\infty$ = 7.62m/s and 13.7m/s, corresponding to Froude numbers $Fr = U_\infty/(g\ d_n)^{1/2}$ equal to 14.5 and 26.2, respectively, and for two nozzle diameters, 7mm and 11mm. Target nondimensional ventilation flow rates $C_Q = Q/(U_\infty\ d_n^2)$ were $C_Q = 7$ for $Fr = 14.5$, and $C_Q = 15$ for $Fr = 26.2$. Jet mass flow rates varied from 0 to 0.11kg/s. These correspond, for the two nozzles, to a jet momentum flux range of 0 to 12N.

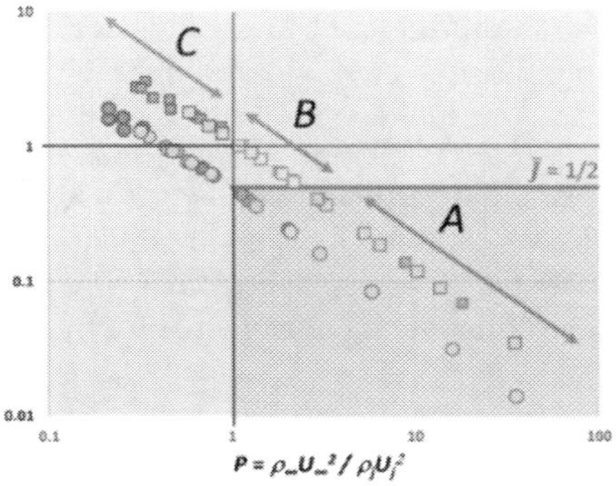


*Figure 3.* Jet/closure interaction regime map overlaid with values corresponding to measured data; jet/closure interaction regimes colored as in Figure 1. Nozzle diameter: ○ = 7mm, □ = 11mm. $Fr$ = 14.5 (dark/blue fill), $Fr$ = 26.2 (light/yellow fill). Data group by nozzle dia. Jet acts to *A*: inflate cavity, *B*: weakly deflate cavity, *C*: strongly deflate cavity.

The ranges of momentum flux, measured cavitation number, and tunnel flow speed combine to define the ranges of the two control parameters, $\bar{J}$ and $P$, identified by Paryshev, as shown in Figure 3. According to the predictions, the regime map shows that as jet momentum is increased the 7mm nozzle jet first induces cavity inflation at low jet thrust, then strong deflation beginning near $\bar{J} \approx ½$, $P \approx 1$. Behavior for the 11mm nozzle jet is more complex. As thrust increases, the cavity should transition from inflation to weak deflation at $\bar{J} \approx ½$, $P \approx 2.2$, then from weak inflation to strong deflation at $\bar{J} \approx 1$, $P \approx 1$. Experimental data overlaid on the regime map shows all three regimes are represented.

### *3.2 Observed changes to cavity behavior due to jet*

Figure 4 shows still images from the high-speed video for cases where a stable cavity was formed. In the figure, $P$ and $\bar{J}$ vary in the vertical direction, while $P$ is approximately constant horizontally. The left column contains 7mm jet nozzle data while the right column contains 11mm jet nozzle data. As discussed above, cavity size can be seen to vary with increasing $\bar{J}$. For the 7mm nozzle, data showed a maximum diameter near

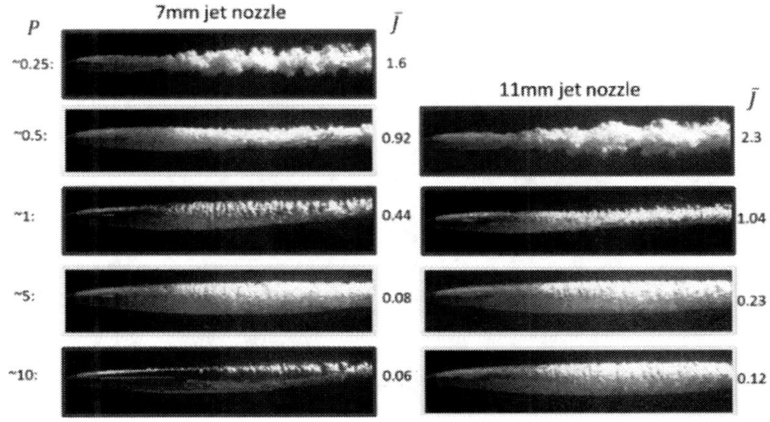


*Figure 4.* Cavity size trends, still images from high-speed video. Images at same level occurred at approximately same value of $P$, indicated to the left. Corresponding values of $\bar{J}$ are given to the right of each image. Image border color indicates tunnel flow speed – *dark/blue:* $Fr$ = 14.5, *light/yellow*: $Fr$ = 26.2.

$\bar{J} \approx ½$, $P \approx 1$, as illustrated by the image for $\bar{J} = 0.44$. The maximum cavity size for the 11mm nozzle appears to occur in the range $1 < P < 5$, $0.23 < \bar{J} < 1$. For both nozzles, cavity size decreases strongly with $\bar{J}$ for $P < 1$.

Figure 5 shows cavitation number, $\sigma$ (essentially inversely proportional to cavity size), plotted against both $P$ and $\bar{J}$. Figure 4a shows that, for decreasing $P$, $\sigma$ does not vary strongly until $P \approx 2$ for the 11mm nozzle, and until $P \approx 1$ for the 7mm nozzle. After these points, $\sigma$ rises rapidly with $P$. Figure 4b shows that, for increasing $\bar{J}$, $\sigma$ does not vary strongly until $\bar{J} \approx 0.6$, at which point it rises rapidly. Furthermore, data show a larger $\sigma$ variation for $\bar{J} > 0.6$. It should be noted that in both Fig. 4a and 4b, the highest $\sigma$ cases correspond to situations where the cavity completely collapsed due to jet action. Finally, behavior for $\bar{J} < 1/2$ appears at some variance with Paryshev's model. Cavitation number appears to reach a weak minimum at low $\bar{J}$, but, for all but one case, it occurs for $\bar{J} << 1/2$, and does not coincide to the rapid rise in cavitation number with $\bar{J}$ observed near $\bar{J} \approx ½$.

## 4. Discussion

Principle predicted trends were observed: regime change was observed for $P \approx 1, \bar{J} \approx 1$, and also at $P > 1$, $\bar{J} \approx ½$, though regime boundaries are not precise and appear to be dependent on jet nozzle diameter and $Fr$. It is speculated that physical phenomena neglected in the approximate model, especially jet entrainment, are responsible for these variations, though it should be admitted that the current data cover only a small portion of the weak deflation ($P > 1$, $\bar{J} > ½$) regime.

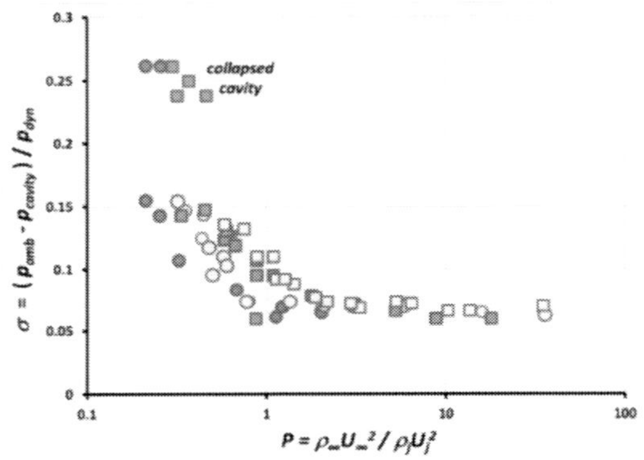


(a)

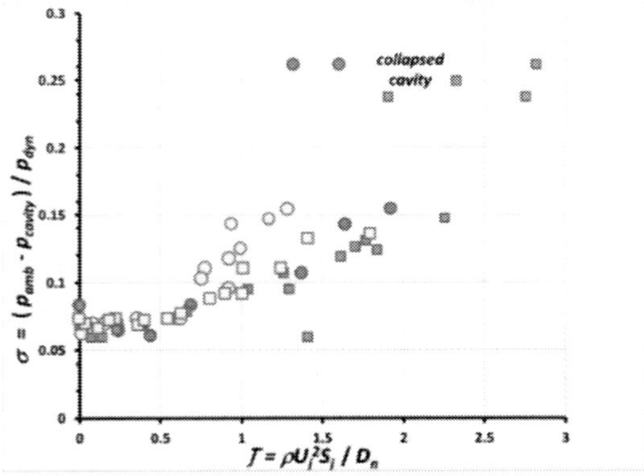


(b)

*Figure 5.* Cavitation number as function of (a) stagnation pressure ratio $P$, (b) thrust to drag ratio $\bar{J}$ and ○ - 7mm nozzle diameter, □ - 11mm nozzle diameter, dark/blue fill – $Fr = 14.5$, light/yellow fill - $Fr = 26.2$.

## 5. Summary and Conclusions

An experimental investigation was performed to test some of the predictions of the jet/closure interaction model described by Paryshev (2006). The principle predicted trends were observed, but differences motivate further study to account for the influence of effects neglected in the analysis.

## Acknowledgments

The authors gratefully acknowledge funding through ONR Grant Numbers N00014-12-1-0489 and N00014-13-1-0692, Dr. Ronald Joslin, Program Manager.

## References

Lehman AF 1959 The Garfield Thomas Water Tunnel *ARL Penn State Technical Report* No. NOrd 16597-56.

Marboe R C Weyer RM and Jonson ML 1993 Hydroacoustic Research capabilities in the Large Water Tunnel at ARL-Penn State *NCA-VOL 15/FED* **168** pp. 125-135.

May A 1975 Water Entry and Cavity-Running Behavior of Missiles *Naval Sea Systems Command Tech. Rep.* SEAHAC/TR 75-2.

Paryshev EV 2006 Approximate mathematical models in high-speed hydrodynamics *J. Eng. Math.* **55** pp 41-64.

# Cavity Dynamics behind a 2-D Wedge Analyzed by Incompressible and Compressible Flow Solvers

**Sunho Park[1], Heebum Lee[2], Hak-Kyu Choi[2], Shin Hyung Rhee[2], Yohan Choe[3], Hyeongjun Kim[3], Chongam Kim[3], Ji-Hye Kim[4], Byoung-Kwon Ahn[4], Hyoung-Tae Kim[4]**

[1] Dept. Ocean Engineering, Korea Maritime and Ocean Univ., Busan, Korea
[2] Dept. Naval Architecture and Ocean Engineering, Seoul National Univ., Seoul, Korea
[3] Dept. Aerospace Engineering, Seoul National Univ., Seoul, Korea
[4] Dept. Naval Architecture and Ocean Engineering, Chungnam National Univ., Daejeon, Korea

E-mail: shr@snu.ac.kr

**Abstract.** To study the cavity dynamics behind a two-dimensional wedge, a pressure-based numerical solver for incompressible cavitating flow and a density-based numerical solver for compressible cavitating flow solvers were developed, respectively, using a cell-cantered finite volume method. Cavity interface was captured based on an approximation of homogeneous mixture flow. Cavity dynamics analysed by the two developed solvers were compared and validated against experimental data. Cavity shape and length, re-entrant jet, and vortical cavity shedding were compared and discussed.

## 1. Introduction

Cavitation is often observed in high-speed hydrodynamic mechanical devices. Cavitation is usually treated an undesirable phenomenon, however, it is regarded as desirable phenomenon for military purposes, e.g., torpedoes. Controlled super-cavitation was used for torpedoes to reduce the skin-friction drag. To make use of super-cavitation, understanding its physics and predicting its behaviour are essential.

Many studies concerning super-cavitation have been done for decades. Ahn et al. [1] experimentally studied cavity dynamics behind the 2D wedge. Park et al. [2] developed incompressible cavitating flow solver and simulated cavitation around a 2D wedge and the wedge with a body, respectively.

The objectives of the present study were (1) to develop a pressure-based incompressible cavitating flow solver and a density-based compressible cavitating flow solver, (2) to understand the compressibility and thermodynamic effects on cavity dynamics behind a 2D wedge using the two developed solvers.

## 2. Problem Description

The 2-D wedge geometry for the present study is shown in Figure 1. The wedge shape was defined by angle ($\alpha$), chord ($c$), and depth ($d$). The wedge angle of 15 degrees, chord length of 75.96 *mm*, and depth of 20 *mm* were selected for the present study. The Reynolds number (*Re*) was based on the free-stream velocity ($U_\infty$) and the wedge depth ($d$). The cavitation number ($\sigma$) was defined as ($P_o$-

$P_v)/0.5\rho U_\infty^2$, where $P_v$ was the vapour pressure, $P_o$ was the reference pressure and $\rho$ was the fluid density. The cavitation number was in the range of 0.08 to 2.34. Cavitation begins to form in the low-pressure region behind the wedge as in the backward facing step flow and propagates downstream as shown in Figure 1.

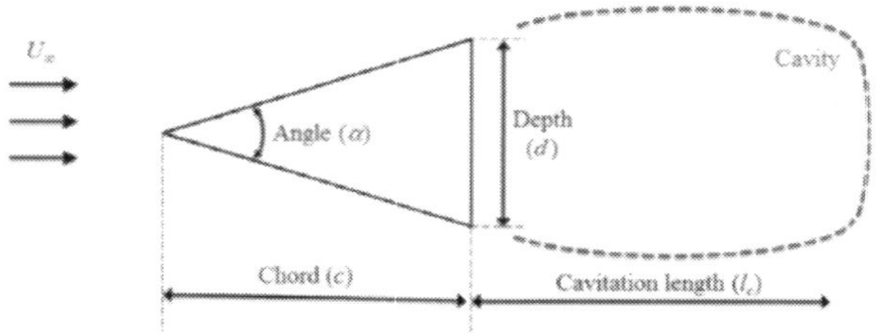


**Figure 1.** Problem description of wedge.

## 3. Computational Methods

### *3.1. Incompressible cavitating flow solver*
The governing equations consist of mass- and momentum conservation equations for the mixture fluid with a volume fraction transport equation formulated by pressure-based methods. A preconditioning is applied in the pressure correction step to cover the whole Mach-number region [2].

### *3.2. Compressible cavitating flow solver*
Assuming fully compressible flows including thermodynamic effects, the governing equations consist of mixture mass-, momentum-, and energy-conservation equations, together with a one-phase mass-conservation equation formulated by density-based methods. A preconditioning is introduced to cover the low-Mach-number region.

## 4. Model Tests
The experimental observations were carried out in the cavitation tunnel at Chungnam National University. The tunnel's test section has a cross section of 100 *mm* wide, and 100 *mm* high. The maximum speed of the tunnel free stream was 20 *m/s* and the pressure was controlled from 10 to 300 *kPa*. The cavity length and dynamics were observed for various wedge shapes. Figure 2 is a snapshot of the cavity behind the wedge for cavitation numbers of 0.35 and 0.40 with the Reynolds number of $1.7 \times 10^5$. The observed cavity lengths ($l_c/d$) were 5.45 and 3.84 for cavitation numbers of 0.35 and 0.40, respectively.

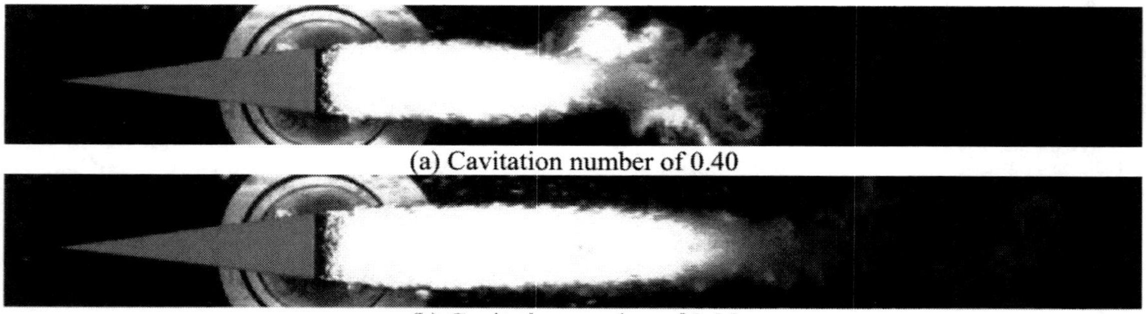

(a) Cavitation number of 0.40

(b) Cavitation number of 0.35

**Figure 2.** Snap shot of cavity behind the wedge in experiment

## 5. Results and Discussion

Figure 3 shows the volume fraction contours and streamlines around the 2D wedge. The cavitating flow shows recirculation behind the wedge, which was separated from the outer flow. Re-entrant jet was not observed in the incompressible flow, while it was clearly shown in the compressible flow. The streamlines were similar in both cases.

Figure 4 shows the contours of non-dimensionalized streamwise velocity component around the wedge. The circulating flow produced by the incompressible flow solver was stronger than that by the compressible flow solver. It was evident by the stronger negative streamwise velocity component, i.e., toward the upstream, at $y$=0 in the incompressible flow solution.

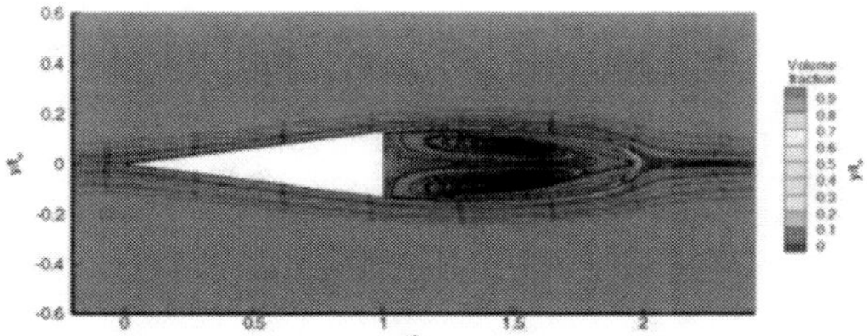
(a) Incompressible solver at σ=0.4

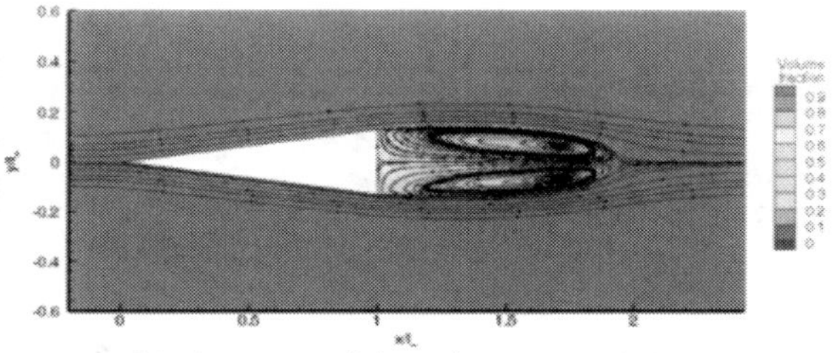
(b) Compressible solver at σ=0.4

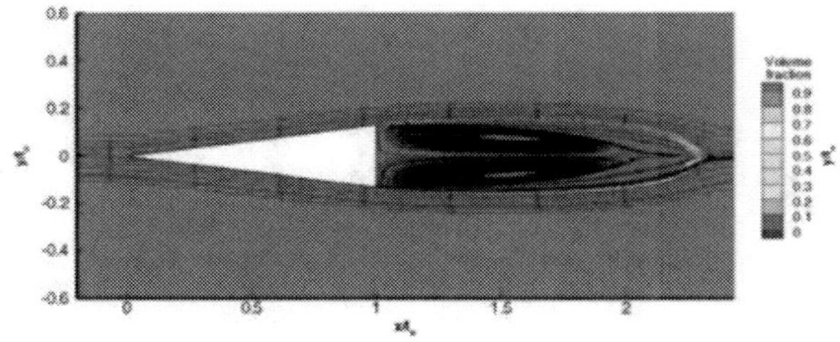
(c) Incompressible solver at σ=0.35

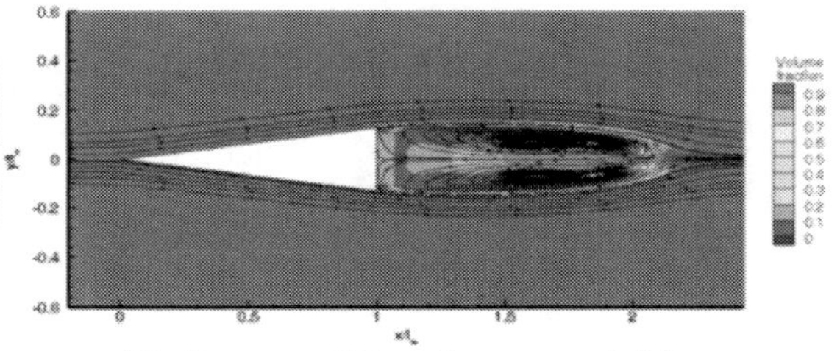
(d) Compressible solver at σ=0.35

**Figure 3.** Volume fraction contours and streamlines around wedge

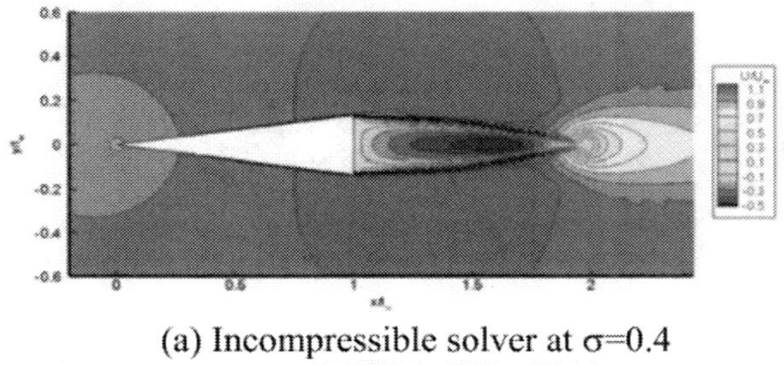
(a) Incompressible solver at σ=0.4

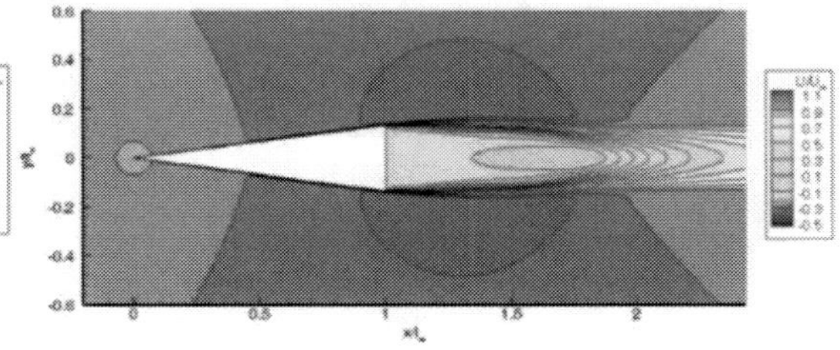
(b) Compressible solver at σ=0.4

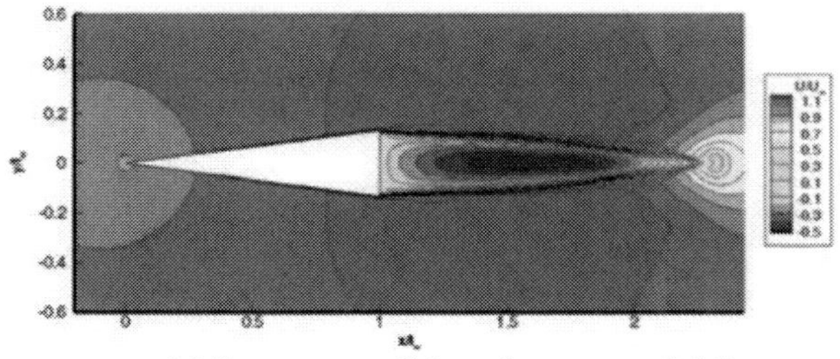
(c) Incompressible solver at σ=0.35

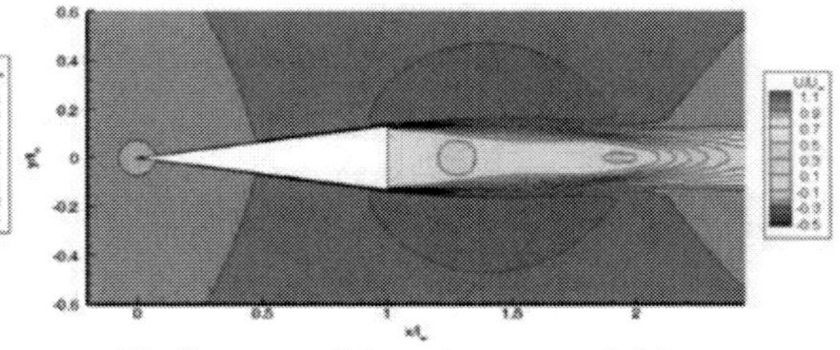
(d) Compressible solver at σ=0.35

**Figure 4.** Nondimensionalized streamwise velocity component contours around wedge

The computed cavity lengths are compared with experimental results and analytic solution [1, 2], listed in Table 1. Cavity lengths were calculated using a vapour volume fraction value of 0.5. The cavity length grew longer as the cavitation number decreased. The cavity lengths predicted by the two solvers were somewhere between the experimental data and analytic solution. The cavity lengths computed by the incompressible flow solver were longer than that by the compressible flow solver. This tendency was prominent at low cavitation number. Re-entrant jet observed in the compressible flow solver was believed to be the cause of shorter cavity length.

**Table 1.** Cavity length.

| $\pi\sigma/2d$ | Incompressible flow | Compressible flow | Experiment | Analytic solution |
|---|---|---|---|---|
| 2.1 | 5.07 | 4.73 | 5.45 | 4.42 |
| 2.4 | 3.83 | 3.65 | 3.84 | 3.56 |

Cavity shedding was observed with cavitation number of 0.83 and the Reynolds number of $1.7 \times 10^5$. On issues such as the cavity shedding, the shedding is related to the Reynolds number and the cavity is related to cavitation number. Figure 5 shows the volume fraction contours and streamlines at an instant. The cavity shedding was different for both solvers. The compressibility ad thermodynamic effects influenced cavity shedding dynamics.

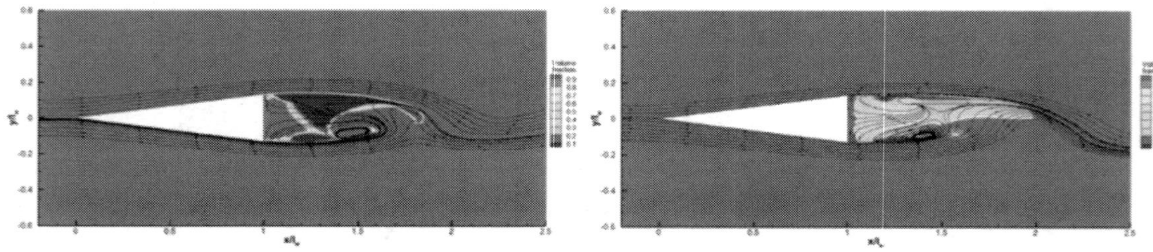

(a) Incompressible solver at $\sigma$=0.83 (b) Compressible solver at $\sigma$=0.83

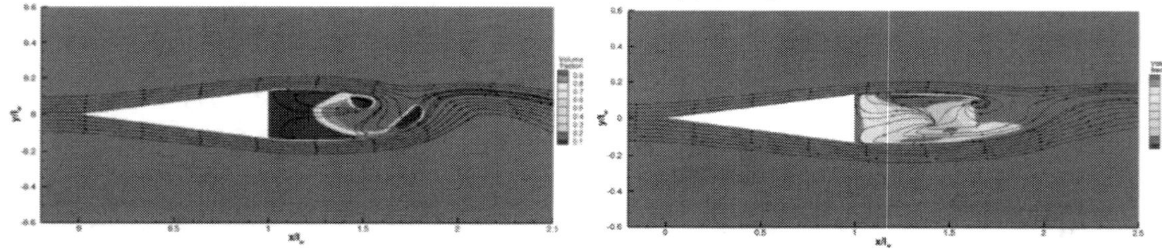

(c) Incompressible solver at $\sigma$=0.83 (d) Compressible solver at $\sigma$=0.83

**Figure 5.** Cavity shedding dynamics and streamlines around wedge at an instant

## 6. Concluding Remarks

The cavity lengths, velocity and pressure contours around a 2D wedge were studied by a pressure-based incompressible cavitating flow solver and a density-based compressible cavitating flow solver, respectively. The circulation strength and re-entrant jet were different for both solvers. The cavity shedding dynamics was weakened by the compressibility and thermodynamic effects in the compressible flow solver.

## 7. References

[1] Ahn B K, Lee C S and Kim H T 2010 Experimental and numerical studies on supercavitating flow of axisymmetric cavitators *Int J Naval Archit Ocean Eng* **2** 39–44

[2] Park S and Rhee S H 2012 Computational analysis of turbulent super-cavitating flow around a two-dimensional wedge-shaped cavitator geometry *Computers & Fluids* **70** 73-85

Journal of Physics: Conference Series **656** (2015) 012164 doi:10.1088/1742-6596/656/1/012164

# Ventilated cavity flow over a backward-facing step

**B W Pearce**[1], **P A Brandner**[1], **S J Foster**[2]

[1] Australian Maritime College, University of Tasmania, Launceston, Tasmania 7250, Australia
[2] Incat Crowther, Sydney, NSW 2084, Australia

E-mail: Bryce.Pearce@utas.edu.au

**Abstract.** Ventilated cavities detaching from a backward facing step (BFS) are investigated for a range of upstream boundary layer thicknesses in a cavitation tunnel. The upstream turbulent boundary layer thickness is varied by artificial thickening of the test section natural boundary layer using an array of transversely injected jets. Momentum thickness Reynolds numbers from 6.6 to $44 \times 10^3$ were tested giving boundary layer thickness to step height ratios from 1.25 to 3.8. A range of cavity lengths were obtained by variation of the ventilation flow rate for several freestream Reynolds numbers. Cavity length to step height ratios from 20 to 80 were achieved. Cavity length was found to be linearly dependent on ventilation rate and to decrease with increasing boundary layer thickness and/or Reynolds number. This result may have implications in the practical optimization of these flows which occur in applications such as drag reduction on marine hull forms.

## 1. Introduction

A significant portion of the propulsive power requirements for a marine vessel, up to around 60%, is required to overcome the frictional drag due to the wetted surface area. For displacement hull forms there has been recent interest in the modification of the near-hull flow field, e.g. by the addition of air bubbles/air layer [1, 2] or polymer solutions [3, 4], to reduce viscous hull resistance. A subset within the broader air bubble/air layer drag reduction field is the formation of air filled cavities from backward-facing steps (BFS) in the hull. Developmental trials and basic studies have used air cavities covering 30 to 50% of the wetted hull surface [5, 6]. An aspect of these BFS cavity flows which has not been considered in previous studies is the effect of the upstream turbulent boundary layer on the cavity physics. If significant, it may influence the efficiency of the drag reduction system and inform the optimal longitudinal location of the BFS, which determines the boundary layer thickness of the flow at the point of cavity detachment. The present work reports on aspects of an experimental investigation on the effect that an upstream boundary layer has on the geometry of a ventilated cavity forming from a BFS. A sketch of the basic flow investigated is shown in Figure 1.

## 2. Experimental Overview

Experiments were carried out in the Cavitation Research Laboratory (CRL) water tunnel at the University of Tasmania. The tunnel test section is 0.6 m square by 2.6 m long and has operating velocity and pressure ranges of 2 to 12 m/s and 4 to 400 kPa absolute respectively. The CRL water tunnel was developed with a capability to artificially thicken (or thin) the test section ceiling boundary layer to range from nominally 0.01 to 0.1 m thick within the test section

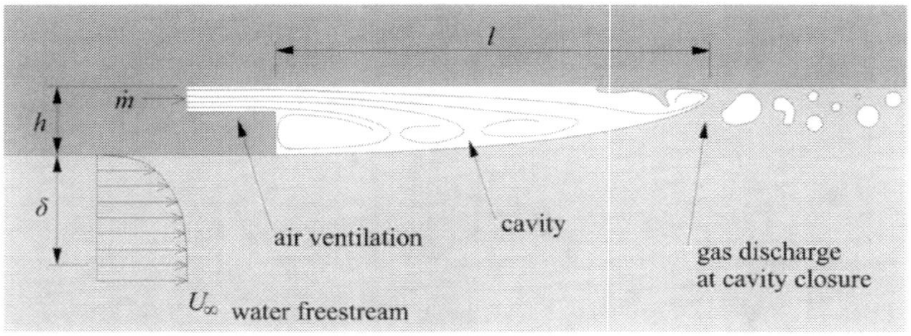


Figure 1: Sketch of a ventilated cavity detaching from a BFS with relevant flow features and parameters noted.

length. These thickened boundary layer profiles being close approximations to flat plate, zero pressure gradient, high Reynolds number turbulent boundary layers [7]. The tunnel also has ancillary systems for rapid degassing and for continuous injection and removal of nuclei and large volumes of incondensable gas. A detailed description of the facility is given in [8, 9] and detailed specifications for the associated instrumentation can be found in [10].

A 2-D plate (750 mm × 10 mm) with a 'quarter' ellipse (5:1) leading edge was attached to the test section ceiling. Compressed air is supplied through the tunnel wall into an internal manifold within the rear section of the plate from which air is admitted into the separation zone downstream of the BFS via an array of nozzles. The boundary layer profiles were measured on the test section vertical center plane, 50 mm upstream from the end of the plate, using a total head tube [11] . Images of the cavity were taken from below using a Canon EOS 50D SLR camera with a Canon EF24-70 mm zoom lens.

All tests, including boundary layer measurements, were conducted at constant Reynolds numbers ($Re$), based on test section height , of $3, 4$ & $5 \times 10^6$ (corresponding to step height based Reynolds numbers of $50, 66.7$ & $83.3 \times 10^3$). For the ventilated cavity measurements, the freestream cavitation number (i.e. that based on the liquid vapor pressure), $\sigma_v$, was kept at a constant value of 1.2. $\sigma_v = (p_\infty - p_v)/0.5\rho U_\infty^2$ where, $p_v$ is the liquid vapour pressure, $p_\infty$ the freestream reference static pressure, $U_\infty$ is the freestream reference velocity and $\rho$ is the liquid density. A cavitation number based on cavity pressure (with the cavity pressure, $p_c$ exchanged for the $p_v$ in the preceding equation) is also relevant here as the cavity is a mixture of both vapour and incondensable gas, and is denoted as $\sigma_c$. The data is presented in terms of an air entrainment coefficient, $C_Q = (\dot{m}/\rho_a)/(hwU_\infty)$, where $\rho_a$ is the air density, $\dot{m}$ is the mass flow rate into the cavity and $h$ and $w$ are the height and width of the step.

## 3. Results and Discussion

As the cavity closure is inherently unsteady, a mean cavity length was determined for each flow condition from the average of ten images with the closure oscillating typically over a length of 5 to 10% [11]. Typical images of the ventilated cavity are shown in figure 2. From the initial detachment from the step, the cavity surface has a wavy appearance attributable to the turbulence in the upstream boundary layer, with the amplitude of the disturbance growing with downstream distance, eventually contributing to the break up of the cavity. The is due to the mixing of the larger scale structures across the remnant, now decaying, boundary layer. From a previous study on base-ventilated hydrofoils [10, 12], for some conditions the cavity surface was initially undisturbed (glassy appearance) attributable to a laminar boundary layer upstream of the step. Figure 2 (a) and (c) show the increase in cavity length with a reduction in boundary layer thickness for constant $Re$ and similar $C_Q$. This is attributable the decrease in the mean kinetic energy in the overlying inter-facial layer. Similarly (b) and (d) show the decrease in cavity length with increase in $Re$ due to the greater turbulent kinetic energy in this case. With an increase in $Re$ it is observed that greater breakup occurs along the cavity surface indicating

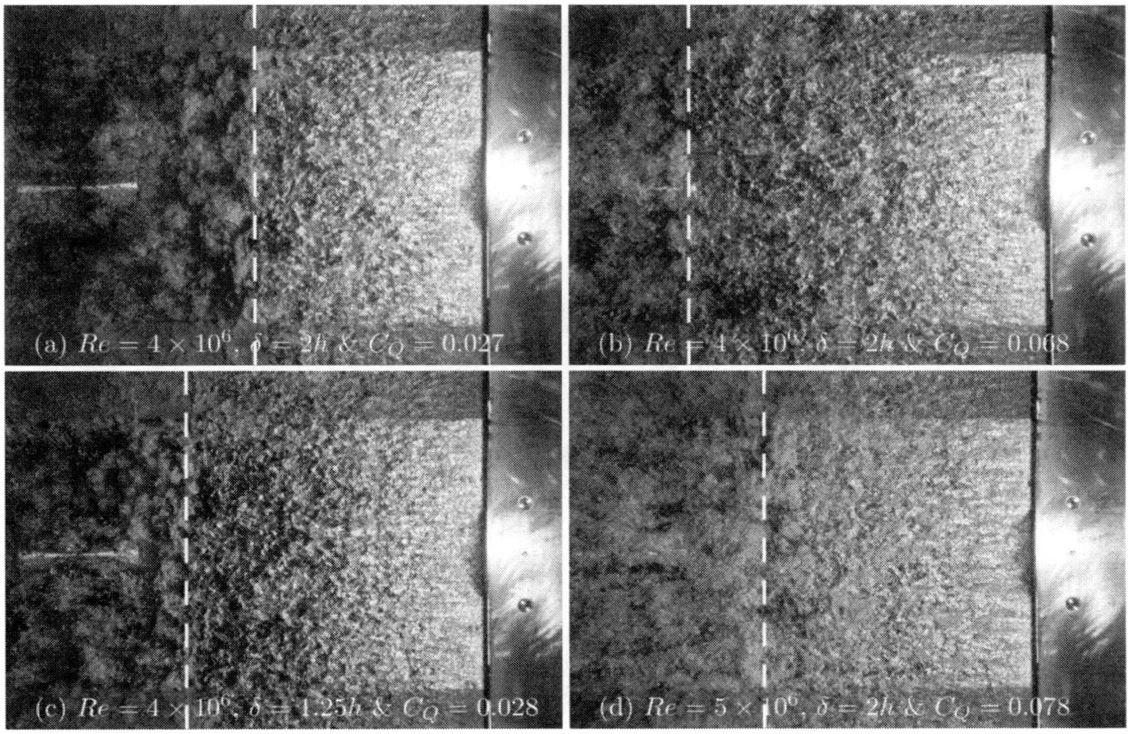


Figure 2: Typical images of the ventilated cavity detaching from a BFS for $Re = 4$ & $5 \times 10^6$ and $\delta/h = 1.25$ & $2.0$. The 2-D nature of the cavity is apparent indicating that the ventilation flow is sufficiently distributed across the span of the step. The cavity closure is identifiable (indicated by the white dashed line) with air discharging from the closure region in shed cloud cavities and individual bubbles. Flow is from right to left.

that additional air entrainment is occurring along the interface (also observed by Arndt et.al. [13]), which would contribute to a reduced cavity length.

The effect of ventilation rate, boundary layer thickness and Reynolds number on cavity length is shown in figure 3. Cavity length increases linearly with ventilation rate (see also figure 2 (a) and (b)) and reduces with increase in boundary layer thickness and/or Reynolds number. The latter effect is due to the reduced 'local' freestream velocity at the step outer edge due to the presence of the boundary layer. This consequently reduces the denominator in $\sigma_c$ resulting in an effectively lower cavitation number which implies a shorter cavity length. As discussed above, the reduced cavity length with increase in $Re$ may be attributable to the increase in air loss through the cavity surface. There is a difference in slope between the thinned and/or natural boundary layers ($\delta = 1.25h$ & $2h$) and the two thickened boundary layers for the lower Reynolds number data which may be attributable to a difference in turbulence profiles as the mean velocity profiles, except for the thickest boundary layer, are virtually identical [7, 11]. Turbulence profile measurements are planned, but have not been made for the present study.

As described in [12] the cavitation number was experimentally measured and found to be very small, $\sigma_c \ll 0.1$ which is expected for this flow geometry. The results presented here have not included $\sigma_c$ as there was a large scatter in the measurements due to the technique used discussed in more detail in the earlier publication [12].

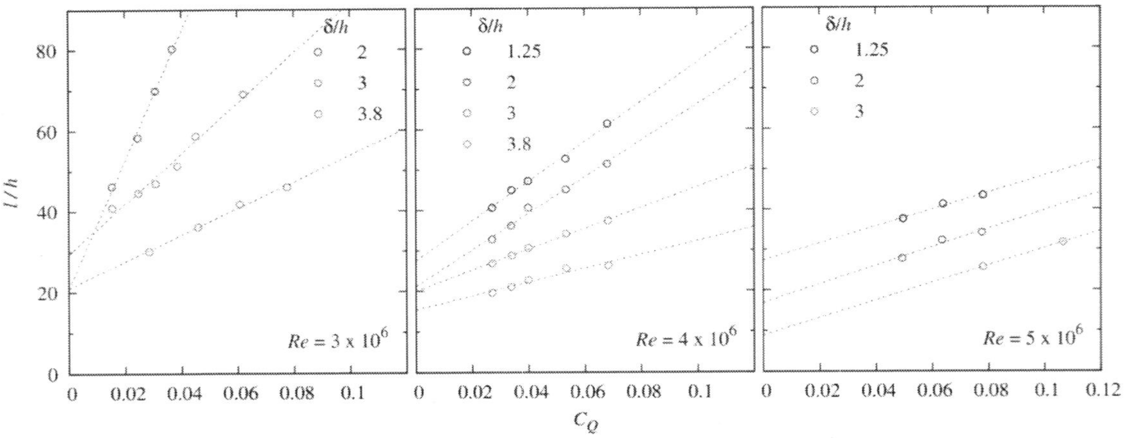


Figure 3: Cavity length versus ventilation rate for the four boundary layer thicknesses and three Reynolds numbers tested. The straight lines are linear least square fits through the experimental data indicating that the cavity length is linearly dependent on the ventilation rate.

## 4. Conclusions

The ventilated cavity flow over a BFS has been experimentally investigated as to the dependence of the cavity geometry on the Reynolds number, ventilation rate and upstream boundary layer thickness. The cavity length was found to increase linearly with ventilation rate and decreases with both increased boundary layer thickness and/or Reynolds number. This dependence of the cavity geometry on the upstream boundary layer thickness has implications for the longitudinal location of the BFS on a vessel hull employing such a device for viscous drag reduction.

## Acknowledgments

The authors wish to acknowledge the assistance of Mr Robert Wrigley in carrying out the experiments and the support of the Australian Maritime College.

## References

[1] Ceccio S L 2010 *Ann. Rev. Fluid Mech.* **42** 183–203
[2] Elbing B R, Mkiharju S, Wiggins A, Perlin M, Dowling D R and Ceccio S L 2013 *J. Fluid Mech.* **717** 484–513
[3] Procaccia I, Lvov V S and Benzi R 2008 *Rev. Mod. Physics* **80** 225–247
[4] Elbing B R, Dowling D R, Perlin M and Ceccio S L 2010 *Phys. Fluids* **22** 045102
[5] Lay K A, Yakushiji R, Mkiharju S, Perlin M and Ceccio S L 2010 *J. Ship Res.* **54** 109–119
[6] Shiri A, Leer-Andersen M, Bensow R E and Norrby J 2012 *29th Symposium on Naval Hydrodynamics* (Gothenburg, Sweden: ONR)
[7] Brandner P A, Belle A, Pearce B W and Holmes M J 2012 *18th Australasian Fluid Mechanics Conference* (Launceston, Australia: AFMS) paper 228
[8] Brandner P A, Lecoffre Y and Walker G J 2007 *16th Australasian Fluid Mechanics Conference* (Gold Coast, Australia: School of Engineering, University of Queensland) pp 630–637
[9] Brandner P A, Lecoffre Y and Walker G J 2006 *Sixth International Symposium on Cavitation - CAV2006* (Wageningen, The Netherlands: MARIN)
[10] Pearce B W and Brandner P A 2012 *Eighth International Symposium on Cavitation - Cav2012* (Singapore: Research Publishing) paper 218
[11] Pearce B W, Brandner P A, Foster S and Zarruk G A 2014 *19th Australasian Fluid Mechanics Conference* (Melbourne, Australia: RMIT University) paper 129
[12] Pearce B and Brandner P 2012 *18th Australasian Fluid Mechanics Conference* (Launceston, Australia: AFMS) paper 391
[13] Arndt R E A, Hambleton W T, Kawakami E and Amromin E L 2009 *J. Fluids Eng.* **131** 111301

# U-shaped Vortex Structures in Large Scale Cloud Cavitation

**Yantao Cao, Xiaoxing Peng[1], Lianghao Xu, Fangwen Hong**
National Key Laboratory on Ship Vibration & Noise, China Ship Scientific Research Center, Wuxi, 214082, China

E-mail: caoyantao@126.com

**Abstract.** The control of cloud cavitation, especially large scale cloud cavitation(LSCC), is always a hot issue in the field of cavitation research. However, there has been little knowledge on the evolution of cloud cavitation since it is associated with turbulence and vortex flow. In this article, the structure of cloud cavitation shed by sheet cavitation around different hydrofoils and a wedge were observed in detail with high speed camera (HSC). It was found that the U-shaped vortex structures always existed in the development process of LSCC. The results indicated that LSCC evolution was related to this kind of vortex structures, and it may be a universal character for LSCC. Then vortex strength of U-shaped vortex structures in a cycle was analyzed with numerical results.

## 1. Introduction

Cloud cavitation, named from its appearance that looks like a cloud, can be seen in many cavitating flows. Usually there are two main type of cloud cavitation, one generates from sheet cavitation shedding and exits downstream of sheet cavity, another results from vortex cavitation and occur downstream of vortex cavity. Here we refer cloud cavitation to the former one. Conventionally, cloud cavitation is not appreciated in hydraulic machines, propellers and water conservancy facilities for the undesirable effect coming with it, such as vibration, noise and even cavitation erosion. Consequently, much attention has been paid on it, as seen in the former researches. Among them, a U-shaped structure in the development of cloud cavitation has been reported by different researchers[1-3]. However, few analysis have been focused on whether it is a unique or a universal phenomenon. More detailed study may do a great help in the understanding of cloud cavitation and offering available suggestions on evaluation of cavitation erosion or its control.

In this article, cloud cavatation on different hydrofoils and a wedge were visualized by HSC. To make the special structure more clearly to see, the lens of HSC was put in back side (for the hydrofoils ) or front side ( for the wedge ) of the tested models. Then numerical simulation on one of the hydrofoils was done to help to analyse this phenomenon.

## 2. Experimental results

The experiment was conducted in cavitation mechanism tunnel located in China Ship Scientific Research Center (CSSRC). Firstly NACA16012 3D twisted hydrofoil(attack angle varies spanwise but has constant chord length) was tested in different install angles and different incoming velocities, as well as different cavitation numbers. With steady and homogeneous incoming flow, LSCC shows

[1] Correspondence author: Dr. Xiaoxing Peng, henrypxx@163.com.

periodic characteristics: Sheet cavitation forms on the leading edge of foil, then begin to grow until reach its maximum length; simultaneously re-entrant and side-entrant flow at the bottom of sheet cavitation move forward to the leading edge direction. When the reverse flow impinges on the surface of vapour sheet, the rear part of sheet structure starts to shed from the main structure and convect downstream. In this stage, the shedding sheet evolves into a cloud and become a distinct U-shaped structure eventually. Figure 1 is one example of the results, which clearly indicate that the cloud cavitation shed from the sheet cavitation is not simply a water and vapour mixed cloud, it is actually an distinct and inerratic U-shaped vortex structure ( red arrowhead denotes flow direction).

To understand U-shaped vortex structure is an occasional case on a special hydrofoil or a general phenomenon applicable to other cases, other test models are also observed. Figure 2 shows LSCC structures on a 3D NACA0012 straight hydrofoil (constant attack angle and constant chord length in spanwise). Figure 3 shows LSCC in a nonvertical (not vertical to the leading edge) incoming flow around a hydrofoil with chord length varied but attack angle fixed in spanwise, which is similar to the flow condition of a propeller. It seems that cavitation induced by this model has no difference with that on straight hydrofoil. Figure 4 is the appearance of LSCC on a wedge which may occure in many water conservancy facilities. U-shaped structures can also be seen in this flow condition, moreover, multi-U-shaped structures interact with each other and develop into complicated vortex groups or clusters.

**Figure 1.** U-shaped vortex structure around twisted hydrofoils. ( $\alpha = 0^\circ$ , $V = 14m/s$ , $\sigma = 1.0$ )

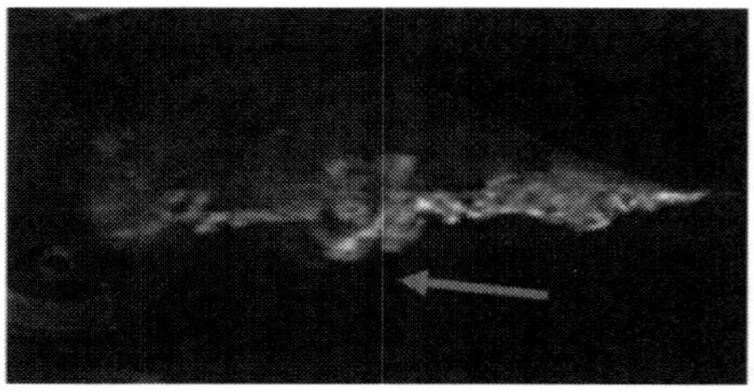

**Figure 2.** U-shaped vortex structure on straight hydrofoils. ( $\alpha = 8^\circ$ , $V = 7m/s$ , $\sigma = 1.8$ )

**Figure 3.** U-shaped vortex structure around a hydrofoil in nonvertical incoming flow.( $\alpha = 6^\circ$ , $V = 7m/s$ , $\sigma = 0.85$ )

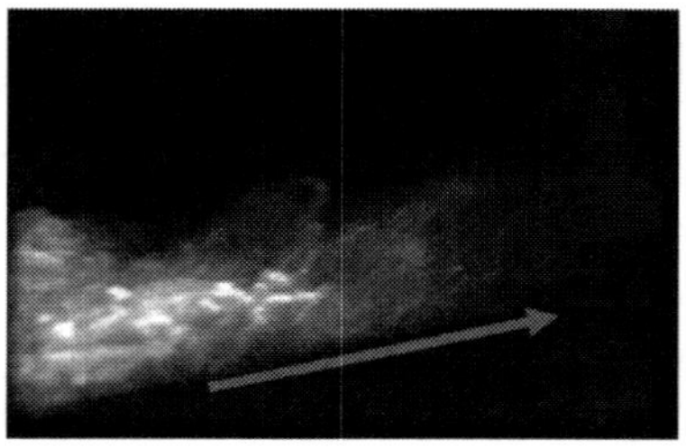

**Figure 4.** Cloud cavitation around a wedge. ( $\alpha = 0^\circ$ , $V = 7m/s$ , $\sigma = 2.0$ )

According to the results above, it can be convinced that U-shaped vortex structure is prevalent in large scale cloud cavitating flows which may exhibit a little difference on different models and in different conditions. Thereby it might be possible to assess the cavitation erosion intensity by the vortex strength of U-shaped structures with some other parameters, such as shedding frequency, nuclei density and its distribution, in the future work. Then investigation of vortex strength in the

Journal of Physics: Conference Series **656** (2015) 012165 doi:10.1088/1742-6596/656/1/012165

development of cloud cavitation is of great significance for the moment. Hence numerical simulation of cloud cavitaion around NACA16012 3D twisted hydrofoil was done to illustrate the vortex strength during the evolution process.

## 3. Numerical results and discussions

Among the tested models, a 3D twisted hydrofoil (NACA16012) was selected to be calculated for the relatively simple cloud cavitation structure on it as mentioned above. In the calculation, mixture model was employed to simulate multiphase flow, considering the mixture is composed of water and vapour with nonslip velocity. Large eddy simulation was introduced to model the turbulent flow.

Figure 5 gives numerical results of cloud cavitation evolution around the 3D twisted hydrofoil at $\sigma = 1.0$ with $V = 14m/s$ and chord length $c = 0.1m$ . Left one shows the isosurface of volume fraction 0.1 and right one shows the corresponding isosurface of Q criterion, Q=2.5$\times 10^5$). It can be seen that the calculation result is quite similar to the experimental appearance (seen in Figure1), showing that numerical methods employed in this work can capture the main features of cloud cavitation. Forthermore, the vapor structures coincident well with vortex structures distinguished by Q criterion, illuminating that cloud cavitation evolution is related to vortex structures, and the development of this cavitation cloud might be dominated by the U-shaped vortex.

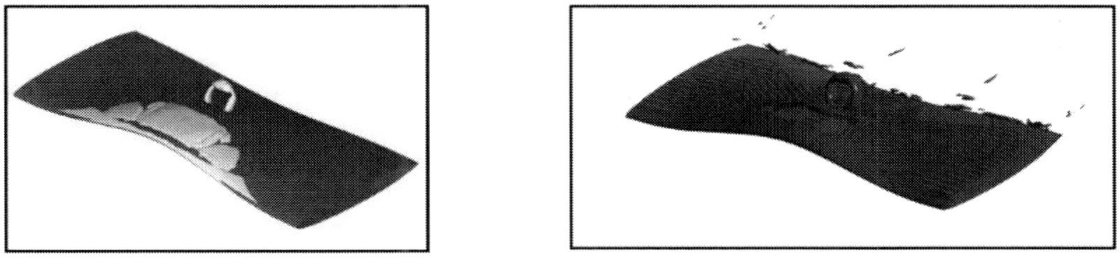

Figure 5. Numerical result of cloud cavitation evolution around NACA16012 3D twist hydrofoil. (left: isosurface of volume fraction 0.1; right: corresponding isosurface of Q criterion)

To estimate the strength of vortex of the U-shaped vortex structure we simply calculated it from velocity field in the middle span. Then velocity circulation $\Gamma$ around the U-shaped vortex section is obtained which denotes the strength of this U-shaped vortex, shown in Figure 6. It can be seen that the strength of U-shaped vortex damps when convected downstream which matches well with the experimental observation that the U-shaped structure disappears downstream. Though numerical dissipation may play a role in this process, it is a reflection of inherent physical principle that the vortex dissipate due to viscosity during transportation.

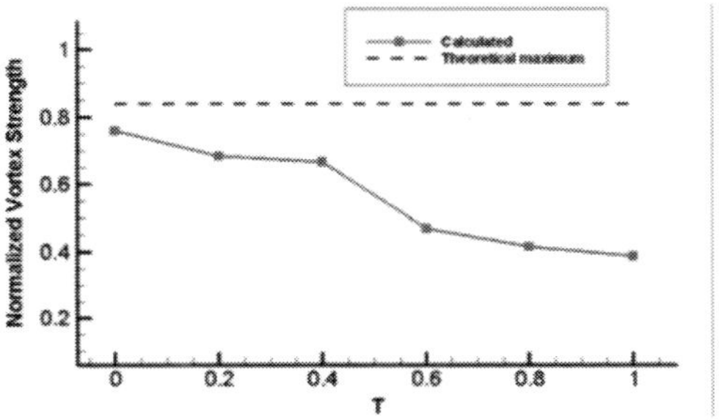


Figure 6. Normalized vortex strength (velocity circulation) in a cycle

To establish a relationship between the flow parameters and the vortex strength, an attempt was taken considering the free-stream velocity $U$ and maximum length of sheet cavitaion $L_{max}$ . Then we can acquire the maximum vortex strength $\Gamma_{theoritical}$ evaluated by the hypothesis that cloud cavitation is a result of sheet cavitation shedding in which re-entrant flow beneath the cavity sheet plays a critical role. Base on Pham et al.'s (1999) [4]research, velocity of re-entrant jet beneath the cavity is the same order of magnitude as the free-stream. With the suppose that velocity on the surface of sheet cavity is equal to the incoming flow, the maximum vortex strength of U-shaped structure can be written as,

$$\Gamma_{theoretical} = 2VL_{max}$$

Compare the velocity circulation calculated numerically at the instance of maximum sheet length with this theoretical value, it is found that they are in the same magnitude order which numerically confirms that maximum vortex strength could be assessed by free-stream velocity and maximum length of sheet cavity qualitatively. The data in Figure 6 is normalized with free-stream velocity $V$ and chord length c. According to this result, it might be possible to evaluate the maximum vortex strength with incoming flow velocity and maximum cavity length. This provides probability to assess the erosion intensity in a vortex approach with some other parameters.

## 4. Conclusions

Cloud cavitation is a complicated cavitaion type that exhibits strong instability and unsteadiness, which can bring a series of erosive effects. However its generation and evolution mechanism is still not clear at present. In this article our emphasis was put on the developing process of LSCC with experimental and numerical approaches, attempting to achieve some universal mechanisms. Firstly LSCC on diverse models was observed by HSC in experiment, then numerical analysis was conducted on a NACA16012 3D twisted hydrofoil. Some suggestions can be obtained from our present investigation:

(1) LSCC evolution is related to vortex structures which generally appear in U-shaped forms. It is convinced in the experiment that U-shaped structures existed in large scale cloud cavitating flow may be a general phenomenon despite the models and flow conditions. The evolution of cavitation cloud is governed by this kind of U-shaped vortex structures.

(2)Maximum strength of U-shaped vortex might be evaluated from the flow parameters. Vortex strength in a cycle was obtained from numerical calculation showing that its maximum value corresponds to that evaluated by theoretical analysis which can provide an access to assess the maximum vortex strength.

## Acknowledgments

The works were supported by the National Natural Science Foundation of China (Project No. 11332009 & 11072223).

## References

[1] Knapp, R.T., 1955. Recent investigations of the mechanics of cavitation and cavitation damage. Trans. ASME 77, 1045–1054.

[2] Kubota, A., Kato, H., Yamaguchi, H., Maeda, M., 1989. Unsteady structure measurement of cloud cavitation on a foil section using conditional sampling technique. J. Fluids Eng. 111, 204–210.

[3] Kawanami, Y., Kato, H., Yamaguchi, H., Tanimura, M., Tagaya, Y., 1997. Mechanism and control of cloud cavitation. J. Fluids Eng. 119, 788.

[4] Pham, T., Larrarte, F., Fruman, D., 1999. Investigation of unsteady sheet cavitation and cloud cavitation mechanisms. J. Fluids Eng. 121, 289.

Journal of Physics: Conference Series **656** (2015) 012166 doi:10.1088/1742-6596/656/1/012166

# Cavitation on a scaled-down model of a Francis turbine guide vane: high-speed imaging and PIV measurements

**K S Pervunin**[1,2], **M V Timoshevskiy**[1,2], **S A Churkin**[1,2], **A Yu Kravtsova**[1,2], **D M Markovich**[1,2,3] **and K Hanjalić**[2,4]

[1] Kutateladze Institute of Thermophysics, SB RAS, 1, Lavrentyev Ave., Novosibirsk, 630090, Russia
[2] Department of Physics, Novosibirsk State University, 2, Pirogov Str., Novosibirsk, 630090, Russia
[3] Institute of Power Engineering, Tomsk Polytechnic University, 30, Lenin Ave., Tomsk, 634050, Russia
[4] Department of Chemical Engineering, Delft University of Technology, 136, Julianalaan, Delft, 2628 BL, the Netherlands

pervunin@itp.nsc.ru

**Abstract**. Cavitation on two symmetric foils, a NACA0015 hydrofoil and a scaled-down model of a Francis turbine guide vane (GV), was investigated by high-speed visualization and PIV. At small attack angles the differences between the profiles of the mean and fluctuating velocities for both hydrofoils were shown to be insignificant. However, at the higher angle of incidence, flow separation from the GV surface was discovered for quasi-steady regimes including cavitation-free and cavitation inception cases. The flow separation leads to the appearance of a second maximum in velocity fluctuations distributions downstream far from the GV surface. When the transition to unsteady regimes occurred, the velocity distributions became quite similar for both foils. Additionally, for the GV an unsteady regime characterized by asymmetric spanwise variations of the sheet cavity length along with alternating periodic detachments of clouds between the sidewalls of the test channel was for the first time visualized. This asymmetric behaviour is very likely to be governed by the cross instability that was recently described by Decaix and Goncalvès [8]. Moreover, it was concluded that the existence of the cross instability is independent on the test body shape and its aspect ratio.

## 1. Introduction

Despite a relatively large set of common attributes, cavitating flows around bodies of quite similar shape and the same size can have different features. Investigations of those are particularly necessary in cases when the test body is a foil shaped to reproduce a 2D vane or 3D blade (even for scaled-down models) used in real full-scale hydrotechnical systems. In various experimental and numerical works, spatial structure and dynamics of cavities on generic symmetric bodies are described (e.g., [1-3]) and distributions of the mean and turbulence characteristics around them are presented (e.g., [4-6]). However, results for guide vanes and/or rotor blades are not available in literature yet.

The current work aims at a detailed analysis of high-speed imaging data on spatial structure and dynamics of gas-vapor cavities and results of PIV measurements around a cavitating scaled-down model of symmetric guide vanes used in high-pressure hydroturbines and a comparison with those for a well-studied NACA0015 series hydrofoil with the same chord length.

## 2. Experimental conditions and measurement techniques

The experiments were carried out at the Cavitation tunnel in Kutateladze Institute of Thermophysics SB RAS with the test channel cross-section ($H \times W$) of 250×80 mm. Its description can be found in [7] as well as details on experimental conditions and measurement techniques applied. In tests, the attack angle $\alpha = 0$, 3 and 9° and the cavitation number $\sigma$ were changed to register various cavitation regimes on the test bodies. The Reynolds number was between $0.9\times10^6$ and $1.7\times10^6$. The test bodies (figure 1) – NACA0015 hydrofoil and scaled-down model of a guide vane (GV) – are made of brass with the same surface roughness of 1.5 mm. Their chord $C$ and span lengths are 100 and 80 mm, respectively. The maximum width of GV is $0.2148C$ at the distance of $0.43C$ from the leading edge. The rounding radius of the leading part is 2.48 mm for the NACA0015 foil and 1.97 mm for the guide vane model.

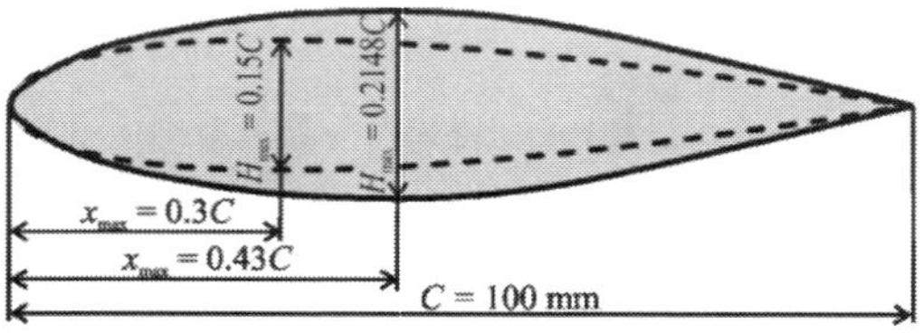


**Figure 1.** A schematic diagram of the test bodies: a NACA0015 series hydrofoil (dashed line) and scaled-down model of a guide vane (solid line).

## 3. Results

### 3.1. Visualization

At $\alpha = 9°$, cavitation on the GV section initiates at roughly $\sigma = 2.87$ as a streak array (their length is quite small ($\approx 0.02C$), whereas on the NACA0015 hydrofoil it occurs at $\sigma = 3.03$ in form of isolated and elongated bubbles (figure 2-b.1), their length is about $0.08C$. Unlike those, streaks on the GV are located very close to each other forming a cavity over the whole foil span. When $\sigma = 2.49$, the cavity length on the GV section increases up to $0.16C$ (figure 2-a.1). On the NACA foil, a decrease of $\sigma$ down to 2.48 (figure 2-b.2) leads to a change of cavitation pattern from elongated bubbles to streaks, which emerge into small clouds downstream, so that an entire cavity ($L_C/C = 0.24$) is formed.

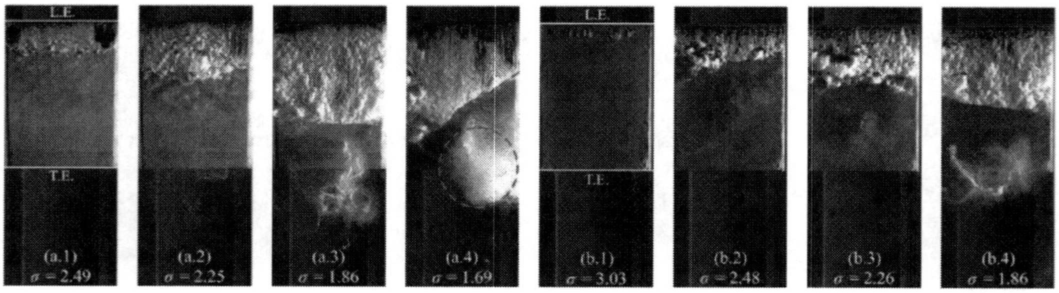


**Figure 2.** Instantaneous images of partial cavities (top view) on the suction side of (a) the GV when (a.1) $\sigma = 2.49$, $L_C/C = 0.16$ (attached cavity), (a.2) $\sigma = 2.25$, $L_C/C = 0.37$ (transitional cavitation), (a.3) $\sigma = 1.86$, $L_C/C = 0.65$, St = 0.29 (cloud cavitation), (a.4) $\sigma = 1.69$, $L_C/C = 0.8$, St = 0.27 (asymmetric cloud shedding) and (b) the NACA0015 hydrofoil when (b.1) $\sigma = 3.03$, $L_C/C = 0.075$ (isolated bubbles), (b.2) $\sigma = 2.48$, $L_C/C = 0.24$ (attached cavity), (b.3) $\sigma = 2.26$, $L_C/C = 0.33$ (transitional cavitation), (b.4) $\sigma = 1.86$, $L_C/C = 0.54$, St = 0.47 (typical cloud cavitation). $\alpha = 9°$. The flow is from above.

Transition to unsteady regimes on both foils occurs when $\sigma$ decreases down to 2.05. All unsteady regimes registered on the hydrofoils at $\alpha = 9°$ were of cloud-type. This means that intrinsic instabilities totally determine the attached cavity behaviour on both hydrofoils at large incidence angles (for details see [7]). Typical Strouhal number is 0.47 for the NACA foil when $\sigma = 1.86$ (figure 2-b.4), whereas it is 0.29 for the GV section at the same cavitation number (figure 2-a.3). Moreover, when the cavitation number was decreased down to 1.69, a variation of the attached cavity length in the spanwise direction in any phase of the cavity growth-reduction cycle was discovered for the GV section (see figure 2-a.4). The evolution of this extraordinary regime is presented in figure 3 within a full period of cavity length oscillations.

This transient, essentially 3D regime is characterized by alternating periodic asymmetric shedding of clouds from one side of the GV in spanwise direction and then from the other. Moreover, it is transient and seems to suffer a bifurcation behavior. As a result, a transition to cloud cavitation finally occurs. According to numerical calculations by Decaix and Goncalvès [8], this asymmetric behavior is very likely to be governed by the cross instability. Additionally, comparing the test object and its AR in our experiment with those in the work [8], the existence of the cross instability is highly probably independent on the test body shape and its aspect ratio.

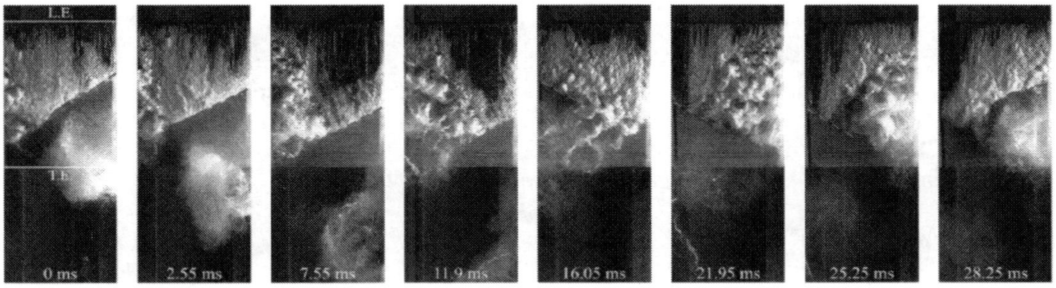


**Figure 3.** Time-series of growth-reduction cycle of unsteady cavity (top view) on the suction side of the GV at $\alpha = 9°$ for $\sigma = 1.69$ ($L_C/C = 0.8$, St = 0.27, $T = 28.35$ ms). The flow is from above.

### *3.2. Velocity distributions*

In case of the single-phase flow around the NACA foil at $\alpha = 9°$, the transverse dimension of the boundary layer is roughly $0.08C$ for $x/C = 1$ that is very close to the case at $\alpha = 3°$. However, its dimension over the GV section at $\alpha = 9°$ is about 2 times wider than that at $\alpha = 3°$ in the same cross-section (figure 4-a). Such a significant modification of the boundary layer over the GV model is caused by flow separation from its surface at $x/C \approx 0.71$, which also leads to the appearance of the second maximum in the velocity fluctuations above the trailing edge of the vane (at $y/C = 0.1$). The amplitude of turbulent velocity fluctuations rises up to $\tilde{u}/U_0 = 0.3$ in the wake ($x/C > 1$) of the GV model (figure 4-a).

When the cavitation develops (figure 4-b), the velocity distributions alter substantially and the position of the separation point on the GV changes continually. Therefore, the separation was not registered in the average velocity fields. Moreover, the cavity length on the GV model becomes equivalent to (figure 4-b) or even larger by approximately 25% than that on the NACA foil. At $x/C = 0.7$ the maxima of the velocity fluctuations are quite close for both foils ($\tilde{u}/U_0 = 0.23$) in the near-wall region, but further away from the foil surface ($y/C > 0.02$) $\tilde{u}/U_0$ is significantly higher over the NACA foil because here the boundary layer is about three times thicker ($0.12C$) than on the GV model ($0.04C$). At the trailing edge (figure 4-c), velocity fluctuations increase up to $\tilde{u}/U_0 = 0.3$ on the GV model, which is approximately 50% higher than in the case of the NACA foil. Transition to the cloud cavitation leads to the appearance of intensive quasi-periodic pulsations of the gas-vapor cavity length and probably because of this the velocity distributions become quite similar (not presented). The amplitude of velocity fluctuations grows up to $\tilde{u}/U_0 = 0.3$ over both foils ($0.4 < x/C < 0.7$) but past the foils ($x/C > 1$) the profiles almost coincide.

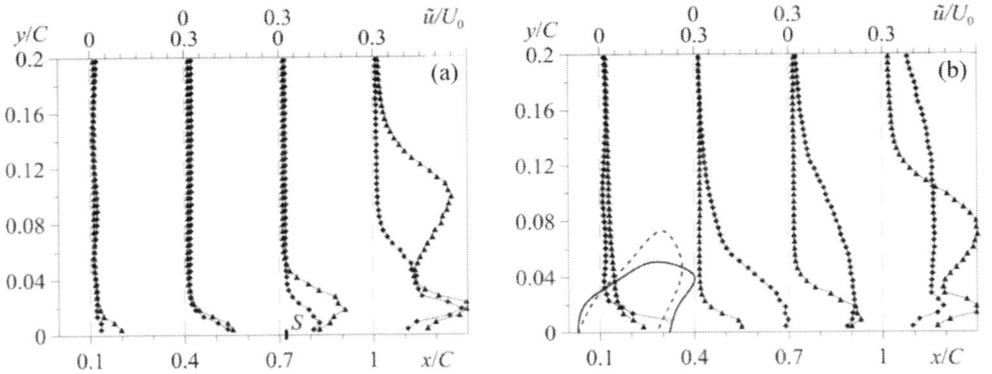


**Figure 4.** Downstream evolution of the streamwise turbulence intensity (rms-values) for (a) cavitation-free flow at $\sigma = 3.41$ (GV) and 3.35 (NACA0015) and (b) transitional cavitation at $\sigma = 2.25$ (GV) and 2.26 (NACA0015). $\alpha = 9°$. ▲ – GV and ◆ – NACA0015. Solid and dashed lines represent time-averaged interfaces of the attached cavities on the GV and NACA0015 hydrofoil, respectively. *S* denotes the separation point on the GV section. The flow direction is from the left.

## 4. Summary

At higher angles of incidence the GV produces a more intensive turbulent wake than the NACA foil due to the flow separation from the GV surface and consequent appearance of a second maximum in the fluctuating velocity profiles at quasi-stable regimes. Besides, our experimental results exhibit the axisymmetric cloud shedding process and prove the existence of the cross instability which existence seems to be independent on the test body shape and its aspect ratio.

**Acknowledgments**

The research was funded by a grant from the Russian Scientific Fund (Project No. 14-29-00203) through Novosibirsk State University.

**References**

[1] Franc J P and Michel J M 1985 Attached cavitation and the boundary layer: experimental investigation and numerical treatment *J. Fluid Mech.* **154** 63–90

[2] Foeth E-J, van Terwisga T and van Doorne C 2008 On the collapse structure of an attached cavity on a three-dimensional hydrofoil *ASME J. Fluids Eng.* **130(7)** (071303)–9

[3] Ji B, Luo X W, Arndt R E A, Peng X and Wu Y 2015 Large eddy simulation and theoretical investigations of the transient cavitating vortical flow structure around a NACA66 hydrofoil *Int. J. Multiphase Flow* **68** 121–34

[4] Kubota A, Kato H, Yamaguchi H and Maeda M 1989 Unsteady structure measurement of cloud cavitation on a foil section using conditional sampling technique *ASME J. Fluids Eng.* **111(2)** 204–10

[5] Astolfi J-A, Dorange P, Billard J-Y and Cid Tomas I 2000 An experimental investigation of cavitation inception and development on a two-dimensional Eppler hydrofoil *ASME J. Fluids Eng.* **122(1)** 164–73

[6] Huang B, Young Y L, Wang G and Shyy W 2013 Combined experimental and computational investigation of unsteady structure of sheet/cloud cavitation *ASME J. Fluids Eng.* **135(7)** (071301)–16

[7] Kravtsova A Yu, Markovich D M, Pervunin K S, Timoshevskiy M V and Hanjalić K 2014 High-speed visualization and PIV measurements of cavitating flows around a semi-circular leading-edge flat plate and NACA0015 hydrofoil *Int. J. Multiphase Flow* **60** 119–34

[8] Decaix J and Goncalvès E 2013 Investigation of three-dimensional effects on a cavitating Venturi flow *Int. J. Heat Fluid Flow* **44** 576–595

# On the calculation of forces in case of planar and axially symmetric flow around the cavitating body

**A G Petrov**[1,2]

[1] Theoretical Mechanics Department, Moscow Institute of Physics and Technology, 9 Institutskiy Per., Dolgoprudny 141700, Russia
[2] Mechanics of Systems Laboratory, Institute for Problems in Mechanics of the Russian Academy of Sciences, 101-1 Pr. Vernadskogo, Moscow 119526, Russia

E-mail: petrovipmech@gmail.com

**Abstract**. The planar and axially symmetric problems of cavitational flow around the body are considered by the Riaboushinsky scheme. The running-on flow is considered to be established for vortex-free ideal and incompressible fluid. In order to find the flow around the body the boundary elements numerical method is used which incorporates the quadrature formulas without saturation. To find the free boundary the gradient descent method based on the Riabushischinsky method is proposed. The resistance force acting on the cavitator is expressed in terms of the Riabushischinsky function, enabling us to calculate the force with rather high precision for small cavitation numbers. The dependence of the resistance coefficient for cavitators of different shape is studied: for wedge and cone, the circular arc and the spherical segment.

## 1. Introduction

The cavitation problems and the related numerical methods can be called classic. A considerable number of works is devoted to them, and their reviews may be found in monographs [1,2]. With the new technology development the more effective and precise methods for the cavitational flow problems are needed. In the present paper the planar and axially symmetric problems of the cavitational flow around the body are considered according to the Riabushischinsky scheme. The flow is considered to be established for vortex-free, ideal and incompressible fluid. Symmetric cavitators are considered and the flow around them is also supposed to be symmetric. The boundary is considered to consist of a cavitator of a given form, its mirror image and the free boundary, on which the fluid velocity is constant.

For the flow around problem the boundary elements numerical method is applied to the integral equation of force on the flow boundary. The numerical scheme of the boundary elements method is applied to the integral equation of the velocity problem.

The distinguishing features of the solution:

- when solving the integral equation the quadrature formulas without saturation for the integral with logarithmic singularity are used [3];
- The Riabushischinsky variation principle is used to find the free surface and the gradient method is used to find the function extremum;

Journal of Physics: Conference Series **656** (2015) 012167 doi:10.1088/1742-6596/656/1/012167

- The formula, that expresses the force on the cavitator through the Riabushischinsky function alows to calculate the force even at small cavitation number.

## 2. The cavitational flow on the wedge

To test the numerical scheme the exact solution of the cavitational flow around problem for the wedge according to the Riabushischinsky scheme [4] can be used. The solution is built with the help of the conformal mapping on the motion hodograph plane in the same way as the classical Riabushischinsky solution for the plate in [5] (pp. 304-306). A quarter of the flow plane CBAM in the complex plane z is mapped to the motion hodograph plane t as shown in figure 1. On the free surface MA the fluid velocity is constant $v_0 = v_\infty \cdot \sqrt{1+\sigma}$ where v∞ is the velocity in a point at infinity C, $\sigma$ is the cavitation number, figure 1. The complex potential of the flow W is expressed through the variables z and t as follows:

$$\frac{dW}{dz} = v_0 t^\alpha, \quad W = G v_0 w, \quad w = \frac{1-t^2}{\sqrt{(1/t_0^2 - t^2)(t_0^2 - t^2)}\left(1-t_0^2\right)^2 t_0}, \quad t_0 = (1+\sigma)^{-1/(2\alpha)}$$

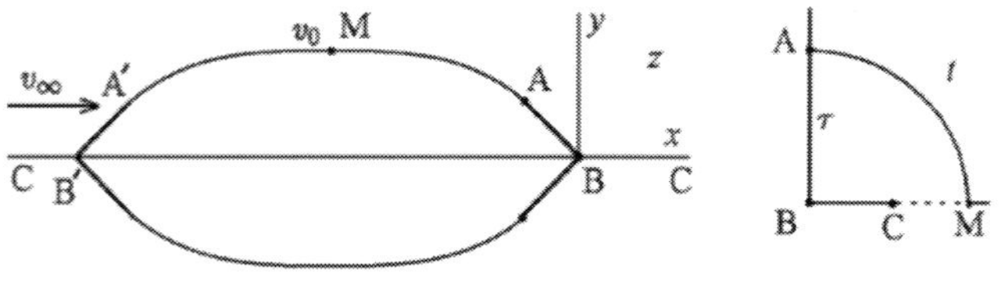


**Figure 1.** The conformal mapping scheme.

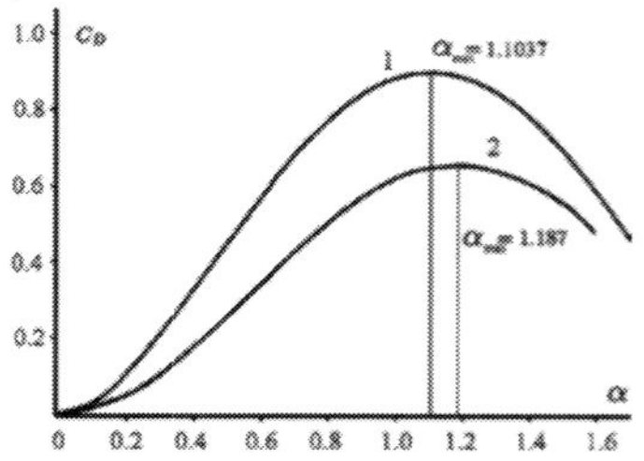


**Figure 2.** The resistance coefficient for the wedge (1) and for the cone (2).

The wedge segment AB (symmetric to the initial wedge segment A'B' ) corresponds to the segment on the imaginary axis $t=i\tau$, $0<\tau<1$. The segment AB is defined as the dependence of the complex vector $z(\tau)=i^{2-\alpha}L_\tau$ on the parameter $\tau$, the unit vector $i^{2-\alpha}$ is directed along the AB, and $L_\tau$ is the distance from the edge of the wedge to the observation point. The dependences of $L_\tau$ and the velocity $v$ on the parameter $\tau$ are:

$$L_\tau = \frac{L}{G_0}\int_0^\tau \frac{\left(1-\tau^2\right)\tau^{1-\alpha}d\tau}{\left(\tau^2+t_0^{\,2}\right)^{3/2}\left(\tau^2 t_0^{\,2}+1\right)^{3/2}}, \quad \frac{v^2}{v_0^2} = \tau^{2\alpha}, \quad G_0 = \int_0^1 \frac{\left(1-\tau^2\right)\tau^{1-\alpha}d\tau}{\left(\tau^2+t_0^{\,2}\right)^{3/2}\left(\tau^2 t_0^{\,2}+1\right)^{3/2}}$$

Here L stays for the segment AB length. The force acting on the wedge may be defined as:

$$F = v_0^2 \int_0^L \rho\left(1-\frac{v^2}{v_0^2}\right) dL_\tau \sin(\pi\alpha/2) = \rho v_\infty^2 L C_D$$

For the wedge resistance coefficient $C_D$ the exact expansion can be obtained

$$C_D = (1+\sigma)\frac{\sin(\pi\alpha/2)}{G_0}\int_0^1 \frac{(1-\tau^{2\alpha})\tau^{1-\alpha}\left(1-\tau^2\right)d\tau}{\left(t_0^2+\tau^2\right)^{3/2}\left(t_0^2\tau^2+1\right)^{3/2}} = C_{D0}(1+\sigma)(1+k(\sigma^2-\sigma^3)+O(\sigma^4)) \qquad (1)$$

By (1) the resistance coefficient can be calculated with the precision up to 5 significant digits for cavitation numbers less than 0.3. For the plate $\alpha=1$, $\sigma=0$ we obtain the known formula (see [5], p. 303):

$$C_D = 2\pi/(\pi+4) \approx 0.87980$$

For the wedge with $\alpha=1/2$ the exact value of the resistance coefficient can be found

$$C_D = 2\pi/[12+\sqrt{2}\pi+\sqrt{2}\log(3-2\sqrt{2})] \approx 0.45041$$

As the cavitation number is zero the dependences of $\alpha$ on the resistance coefficient of the wedge and cone are presented in figure 2. The wedge resistance coefficient achieves its maximum 0.89478 as $\alpha$=1.1038, which corresponds to the wedge apex angle 180°·$\alpha$=198.7°.

## 3. The flow around problem calculation scheme

Following [6] the numerical solution scheme of the flow around problem is described in the case of a two-dimensional planar or axially symmetric flow of the ideal fluid. The equation of the flow around boundary is presented in the parametric form $x(s)$, $y(s)$, where $s$ is the coordinate along the solid boundary, calculated from the critical point. According to the boundary elements method for the velocity distribution $v(s)$ on the flow around boundary the integral equation $Av(s)=2(1+n)\pi v_\infty(y(s))^{1+n}$ is solved, where $n$=0 for the planar problem, $n$=1 for the axially symmetric one, $A$ is the integral operator $Av(s)=-\int_0^l G(s,s')v(s')ds'$, $l$ – is the full length of the boundary in the planar case and the length of the generating line in the axially symmetric one. Green function for the planar problem is expressed through the distance $r=\sqrt{(x(s)-x(s'))^2+(y(s)-y(s'))^2}$ between the points of the boundary ($x(s)$, $y(s)$) and ($x(s')$, $y(s')$). For the planar problem $G(s,s')=\ln(r)$, and for the axially symmetric problem it is the flow function of the vortex ring and is expressed through elliptical integrals. To perform the numerical calculations the sampling of the boundary with a finite number of points, such that the value $\zeta=\zeta_i=i/N$, $i=1..N$ corresponds to the point $M_i$ is introduced. The parameter $\zeta$ and the coordinate $s$ are coupled by the following relation $ds=J\,d\zeta$, where $J$ characterizes the point density on the boundary. With the help of quadrature formulas [3, 6] the integral equations are reduced to a system of linear equations.

## 4. The iteration process to determine the form of the cavity

According to the Riabushischinsky principle, the solution should maximize the function $U=(p_\infty-p_0)V-T$, where $V$ is the surface or the volume, limited by the surface $\partial V$; T is the kinetic energy of the fluid when the solid in it is moving translationally with the velocity $v_\infty$, $p_\infty$ is the pressure at infinity. The free boundary is found using the gradient descent method that maximizes the function $U$. The variations of the points coordinates of the free boundary and the cavitator are set as follows $\delta x=\varepsilon x$, $\delta y=k(v_0^2-v^2)+\varepsilon x\cdot dy/dx$, $\delta l_x=\varepsilon l_x$. Then the variation of the function $U$ in the planar problem is strictly positive $\delta U=\frac{\rho}{2}\int_{\partial V} k(v_0^2-v^2)^2dx$. The parameters $k$,$\varepsilon$ can be found using that the velocity at the end-point of the cavitator is $v_0$ and the scheme stability. The formulas have the second order of accuracy by the subinterval length. The obtained forms of the cavities for spherical cavitators are shown in figure 3.

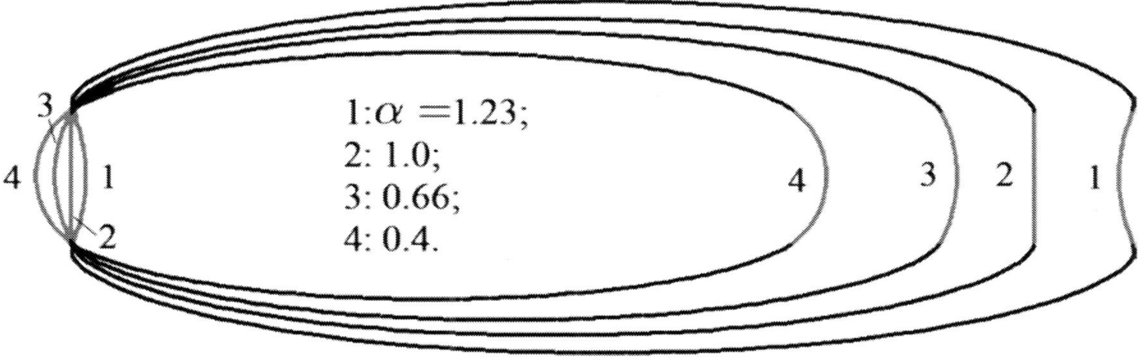


**Figure 3.** The forms of the cavities for spherical cavitators for different values of the angle parameter α. (α – is the apex angle of the descending flow divided by 180°)

## 5. The force formula

A solid flowed around by fluid with velocity $v_\infty$ is considered by the Riabushischinsky scheme. The resistance force $F=-\int_\Sigma (p-p_0)n_x dS$ can be expressed through the Riabushischinsky function. For the planar problem it is

$$F=U/l_x+\Delta F,\quad \Delta F=(1/l_x)\int_\Sigma (p-p_0)(ydx-(x+l_x)dy) \tag{2}$$

For the wedge $\Delta F$=0, and for the parabolic cavitator $x=-l_x+ky^2$ we have that $\Delta F=-(\rho/l_x)\int_0^{y_0}(v_0^2-v^2)ky^2dy$. The term $\Delta F$ is strictly negative. It constantly diminishes the resistance force for convex cavitators $k$>0 and increases it for concave ones (the cavitators in the shape of umbrellas). It is equally simple to calculate $\Delta F$ for the cavitators in the shape of circular arcs and for other shapes. In the axially symmetric case

$$F=3U/(2l_z)+\Delta F,\quad \Delta F=(1/l_z)\int_\Sigma (v_0^2-v^2)\pi r(rdz-(z+l_z)dr) \tag{3}$$

The direct calculation of the resistance coefficients by integrating the pressure on the cavitator boundary leads to substantial measurement errors for small cavitation number, as the relative number of computational points on the cavitator for the planar problem decreases proportionally to the square of the cavitation number. Formulas (2) and (3) are free of this limitation, because when they are used, the integral on the whole boundary of the whole computational domain is calculated. The extreme value of the Riabushischinsky function $U^*$, obtained in the solution of the cavitation problem allows finding the resistance coefficients with high precision.

The author gratefully acknowledges the support of this work from the Russian Foundation for Basic Research, projects 14-01-00892 and 14-01-00818.

## References

[1] Birkhoff G and Zarantonello E HY 1957 *Jets, wakes and cavities* (New York: Academic Press)
[2] Terentiev A, Kirschner I and Uhlman J 2011 *The Hydrodynamics of Cavitating Flows* (Backbone Publishing)
[3] Petrov A G 2008 Quadrature Formulas for Periodic Functions and Their Application to the Boundary Element Method *Comput. Math. Math. Phys.* **48** 1266–1283
[4] Riabushischinsky D 1927 Sur une probleme de variation *C.R. Acad. Sci. Paris* **185** 840 – 841
[5] Milne-Thomson L M 1950 *Theoretical Hydrodynamics* (New York: Macmillan)
[6] Petrov A G, Potapov I I 2015 On the Calculation of Forces that Act on Solids for Planar and Axially Symmetric Cavitation Flow Around Problems *Comput. Math. Math. Phys.* **55**

# Spatio-temporal description of the cavitating flow behavior around NACA 2412 hydrofoil

**P Rudolf [1], D Štefan [1], M Sedlář[2], J Kozák[1], V Habán[1], R Huzlík[3]**

[1]Brno University of Technology, Faculty of Mechanical Engineering, V. Kaplan Department of Fluid Engineering, Technická 2896/2, Brno, Czech Republic
[2]Centre of Hydraulic Research, J. Sigmunda 79, Lutín, 783 50, Czech Republic
[3]Brno University of Technology, Faculty of Electrical Engineering and Communication, Department of Power Electrical and Electronic Engineering, Technická 3082/12, Brno, Czech Republic

rudolf@fme.vutbr.cz

**Abstract**. Spatio-temporal description of the cavitating flow around hydrofoil with 8 degrees incidence using proper orthogonal decomposition (POD) is presented. POD is a suitable tool, which provides information not only about the flow dynamics, but also about relevance of different flow structures. POD also enables to track energy transport within the domain and energy transfer among the eigenmodes of the flow field. Analysis documents change of the flow structure for decreasing cavitation number, which can be most likely attributed to sheet/cloud cavitation transition.

## 1. Introduction

Two regimes of cavitating flow around hydrofoils are usually distinguished: sheet and cloud cavitation. Sheet grows from the hydrofoil (blade) leading edge and at some distance transition to cloud occurs. It is known that collapse of the cavitating clouds is responsible for the most severe cavitation erosion [1, 3, 12]. Strong cavitating swirling structures in form of U-shaped (horseshoe) vortices within the cloud are believed to cause the highest pressure pulses during their collapse [1]. Large pressure pulses generated by cloud cavitation are associated with shock waves, which are propagating inwards [8, 12].

Development of the cloud is connected with existence of the re-entrant jet, which is cutting off the sheet and is responsible for cloud separation. Thorough investigation of this process, its Reynolds number dependence and definition of the critical Reynolds number for transition from the sheet to cloud cavitation was carried out by Keil et al [5]. It was also pointed out by Pelz et al [7] and by Avellan et al [1] that development of the cavitation around the hydrofoil is connected with strong energy transfer from mean flow to the shedding vortices in the region of the cavity closure. Energy is stored in the sheet until the cloud is separated. It makes the cloud cavitation the most harmful cavitation regime [7]. Above described, rather complicated, cavitating flow around hydrofoil can be roughly summarized into following sentence:

Development of the unstable cavitating flow around the blade is governed by energy transfer from the mean flow into strong swirling structures. Therefore it is inevitable to understand the process of energy transport within the flow, its temporal and spatial behavior.

## 2. Experiment

An experimental visualization of the cavitating flow around NACA2412 hydrofoil was done in cavitation tunnel of Center of Hydraulic Research [11]. Hydrofoil has chord length 120 mm, test section has 150x150 mm, incidence angle is 8 degrees. Flow rate was measured with magnetic-inductive flowmeter, pressures upstream and downstream of the test section were recorded with tensometric pressure transducers. Other details of the experimental set up are in [11]. High speed camera with 3000 fps and two halogen lamps were used to visualize the cavitating flow from the top view. 21 different regimes corresponding to three different Reynolds numbers (864000; 1173600;1557600) and a range of cavitation numbers from 5.2 to 1.4 were investigated. Both Reynolds number and cavitation number are based on inlet velocity, chord length and upstream pressure.

## 3. Proper orthogonal decomposition

Since cavitating flow is highly unstable and shedding of the clouds and vortices is observed, different methods were previously applied to study the temporal evolution of the flow. Usually normal FFT analysis in individual points of the flow was applied. Kjeldsen et al [6] employed joint time-frequency analysis to investigate sheet/cloud transition. Present authors successfully applied in their previous research POD for identification of the flow structures, which carry most of the energy [9, 10] for turbine draft tubes with highly swirling flows. POD modes generally characterize the dominant unsteady flow structures, i.e. coherent structures of the flow. It can be stated that the POD modes concentrate maximum of the fluctuation power in the lowest modes, which are most energetic. Description of the method can be found in [2], its recent application on study of cavitation clouds on the side view of the cavitating hydrofoil was done by Danlos et al [4]. POD offers complete spatio-temporal description of the flow providing detection of the most energetic flow structures in whole domain and information about their spatial shape and associated frequencies. Mathematically speaking, an approximation of a function $\boldsymbol{u}(\boldsymbol{x};\ \boldsymbol{t})$ (e.g. velocity, pressure of vapor volume fraction) is searched by decomposition into set of time-dependent eigen functions $\boldsymbol{a}_i$ (temporal POD modes) and time-independent eigen functions $\boldsymbol{\phi}_i$ (spatial POD modes). Eigen functions are identified by searching for eigenvectors and eigenvalues of the temporal correlation matrix, which is built from snapshots of the flow field acquired with given sampling frequency.

$$C_{ij} = \frac{1}{\mathrm{vol}\Omega} \int_{\Omega} u(\boldsymbol{x};\ t_i)\, u(\boldsymbol{x};\ t_j) \mathrm{d}\mathbf{x} \tag{1}$$

Eigenvalues of matrix $\boldsymbol{C}_{ij}$ are then directly connected with spatial eigenmodes $\boldsymbol{\phi}_i(\boldsymbol{x})$ of the flow and magnitude of the eigenvalues corresponds to energy content of the eigenmodes. Temporal coefficients $\boldsymbol{a}_i$ are then obtained by projecting the data functions $\boldsymbol{u}(\boldsymbol{x};\ \boldsymbol{t})$ on the eigen functions $\boldsymbol{\phi}_i(\boldsymbol{x})$.

$$a_i(t) = \sum_{i=0}^{N} u(\boldsymbol{x};\ t)\phi_i(\boldsymbol{x}) \ \text{ for } i = 0, \dots, N \tag{2}$$

## 4. Results

POD analysis is based on snapshots of cavitating flow around the hydrofoil taken from the top view. Uneven lighting did not allow to process whole suction side of the hydrofoil. Rather narrow stripe along the hydrofoil axis of symmetry had to be used (evaluated domain was shrinked from 1900x1000 pixels to 500x100 pixels. Shrinked domain contains hydrofoil leading edge and extends further downstream, but does not capture whole region of cloud cavitation. Values from 0 to 255 were assigned to every pixel according to its grey level (pure vapour = white color).

Output of the POD analysis is set of figures depicting the spatial form of the eigenmodes and also frequencies associated with these eigenmodes. Example for one operating point is illustrated in figure 2. This figure depicts the most dominant modes, which can be associated with erosive process.

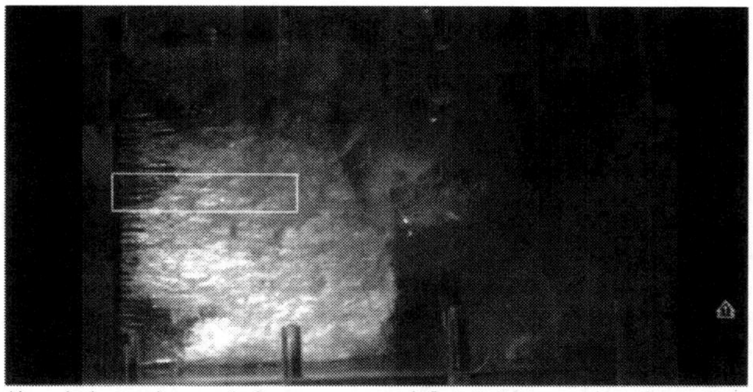

**Figure 1.** Whole visualized domain and the part used for POD evaluation (Re = 1173600, $\sigma$ = 2.75)

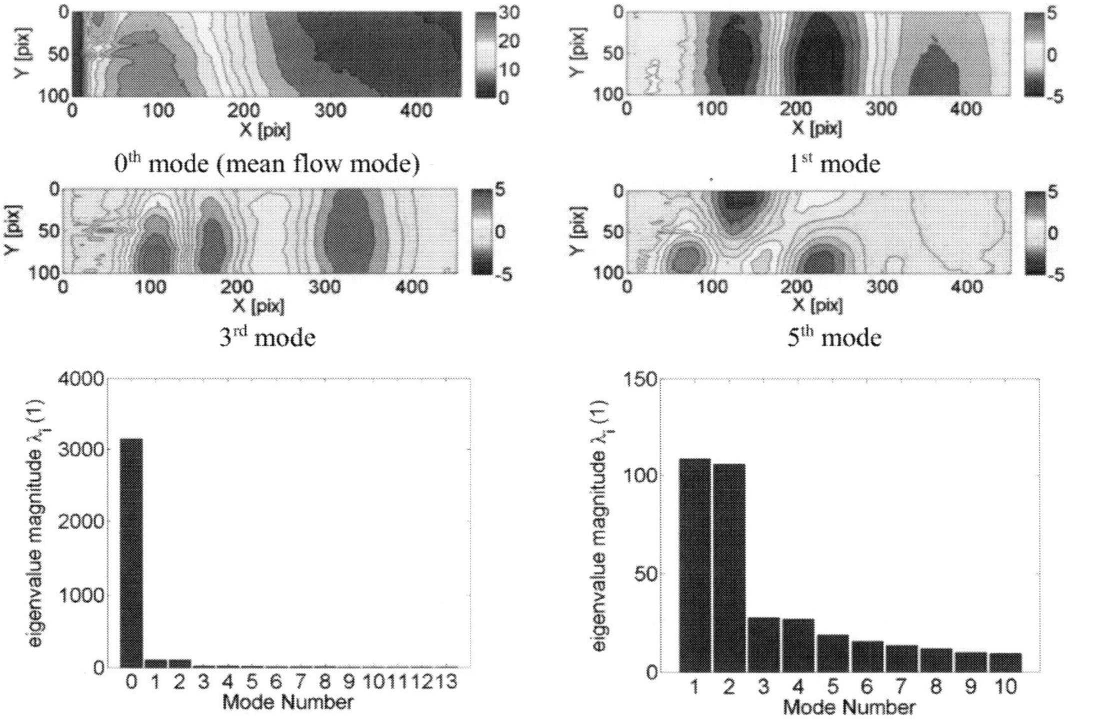


0th mode (mean flow mode)

1st mode

3rd mode

5th mode

Contribution of the modes (all modes)

Contribution of the modes (only dynamic modes)

**Figure 2.** POD analysis of operating point Re = 1173600, $\sigma$ = 3,82

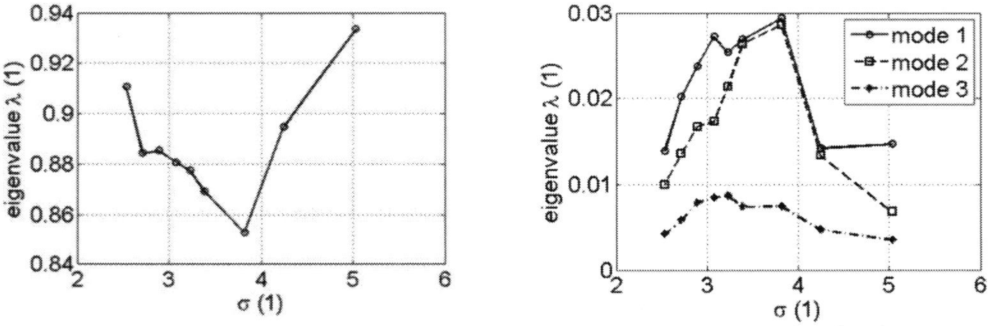


Relative power of the mean flow mode

Relative power of 1st, 2nd, 3rd eigenmodes

**Figure 3.** Relative power of eigenmodes for Re = 1173600

While figure 2 presents absolute magnitude of the eigenvalues, it is more interesting to look at relative magnitudes within each operating point characterized by unique cavitation number, see figure 3. Sudden drop of the eigenvalue of the mean flow for $\sigma$ close to 4 is observed. Simultaneously redistribution of the energy towards higher modes occurs. Enrichment of the dynamic modes (tracked up to 20$^{th}$ mode) is quite likely connected with sheet/cloud transition. Similar process was documented in [6] for pressure rms values.

## 5. Conclusion

Proper orthogonal decomposition proves to be efficient tool for study of the cavitating flow dynamics and energy transfers within the flow field. Decomposing into the very basic corner stones of the flow enables to identify the relevant modes, their frequencies and behaviour with decreasing cavitation mumber. Further investigation will focus on confirmation of the sheet/cloud transition characterized by energy transfer from mean flow to dynamic modes by rules presented in [5].

## References

[1] Avellan F, Dupont P, Ryhming I 1988 Generation Mechanism and Dynamics of Cavitation Vortices Downstream of a Fixed Leading Edge Cavity *Proc. 17$^{th}$ ONR Symp. On Naval Hydrodynamics* Hague

[2] Berkooz G, Holmes, P, and Lumley J 1993 The proper orthogonal decomposition in the analysis of turbulent flows *Annual Rev. of Fluid Mechanics 25(1):539–575*

[3] Brennen CE 2007 Bubbly Cloud Dynamics Cavitation *Invited Lecture at Acoustical Society of America Meeting* Salt Lake City

[4] Danlos A, Ravelet F, Coutier-Delgosha O, Bakir F 2014 Cavitation regime detection through Proper Orthogonal Decomposition: dynamics analysis of the sheet cavity on a grooved convergent-divergent nozzle *International Journal of Heat and Fluid Flow 47* Elsevier

[5] Keil T, Pelz PF, Buttenbender J 2012 On the Transition from Sheet to Cloud Cavitation *Proc. of the 8$^{th}$ Int. Symp. on Cavitation* Singapore

[6] Kjeldsen M, Arndt REA, Effertz M 2000 Spectral Characteristics of Sheet/Cloud Cavitation *Jour. of Fluids Eng. 122(3)*

[7] Pelz PF, Keil T, Ludwig G 2014 On the Kinematics of Sheet and Cloud Cavitation and Related Erosion *Advanced Exp. and Numer. Techniques fo Cavitation Erosion Prediction* Springer

[8] Reisman GE, Brennen CE 1996 Pressure Pulses Generated by Cloud Cavitation *ASME Fluids Eng Division Conf. Proc. Vol 1*

[9] Rudolf P, Štefan D 2012 Decomposition of the Swirling Flow Field Downstream of Francis Turbine Runner *IOP Conference Series: Earth and Environmental Science 15 (6), 062008*

[10] Rudolf P, Štefan D, Klas R 2015 Spatio-Temporal Description of Swirling Flow in Hydarulic Turbine Draft Tube, *WasserWirtschaft Extra*

[11] Sedlář M, Komárek M, Rudolf P, Kozák J, Huzlík R 2015 Numerical and Experimental Research on Unsteady Cavitating Flow around NACA 2412 Hydrofoil *IOP Conf. Series: Materials Science and Engineering 72 (022014)*

[12] Wang YC, Brennen CE 1995 The Noise Generated by the Collapse of a Cloud of Cavitation Bubbles *ASME Symp. on Cavitation and Gas-Liquid Flows in Fluid Machinery and Devices* Hilton Head

## Acknowledgement

The research was supported by project of the Czech Science Foundation No P101/13-23550S "Experimental research and mathematical modelling of unsteady phenomena induced by hydrodynamic cavitation". Research was also supported by project of Technology Agency TE02000232 (Rotary machines) and FSI-S-14-2480 (Innovative fluid machines) of Brno University of Technology, Faculty of Mechanical Engineering.

# Supercavitation: Theory, experiment and scale effects

**V V Serebryakov[1], R E A Arndt[2], and J E Dzielski[3]**

[1]Institute of Hydromechanics of NAS, 8/4 Zhelyabov Str., 03680 Kiev, Ukraine
[2]University of Minnesota St. Anthony Falls Laboratory, 2 Third Avenue SE, Minneapolis, MN 55414, USA
[3]Stevens Institute of Technology, Davidson Laboratory, 711 Hudson St., Hoboken, NJ 07030, USA

E-mail: serebryakov_vladimir@yahoo.com

**Abstract**. The article presents a number of dependences for reliable calculation of parameters required when performing supercavitation experiments. They include the equations and dependences for calculation of the shape and basic dimensions of cavities formed behind axisymmetric cavitators, influence of the free surface, cavity buoyancy, and cavitator lift, as well as compensation for certain scale effects.

## 1. Introduction

The intrinsic characteristics of supercavitating vehicles are very high flow velocities and large dimensions of the bodies. When performing experiments in cavitation tunnels or ducts, the test conditions are necessarily limited to smaller motion velocities and dimensions of the test bodies. The aim of this work is to present dependences for the experimental conditions that can be used effectively to calculate test conditions for experiments. To derive these dependences, asymptotic solutions are applied jointly with the nonlinear numerical calculation results based on the ideal incompressible liquid model, including calculation of flow around disks and cones within the framework of the symmetric Ryabushinsky scheme [1], as well as flows past thin cones with cavity closure on a disk.

## 2. Asymptotic solutions and practical dependencies

Small cavitation numbers $\sigma$ imply highly elongated cavities and comparatively small cavitators, as well as a weak dependence of the cavitator shape in the case of equal drag on the most part of the cavity form. In this case, the asymptotic second-order dependence of the cavity's maximum radius $R_k$, half-length $L_k$, and aspect ratio $\lambda = L_k / R_k$ have the known following form but with values of k and μ which are defined by dependencies on λ [2] :

$$\lambda^2 = \frac{2\mu}{\sigma},\ R_k^2 = R_n^2 \frac{c_d}{k\sigma},\ L_k = \frac{R_n}{\sigma}\left(\frac{2c_d\,\mu}{k}\right)^{1/2},\ \mu = \ln\frac{\lambda}{\sqrt{e}},\ k = \left(1+\frac{2\ln 2/\sqrt{e}}{\ln\lambda^2}\right)^{-1}. \quad (1)$$

In equation (1), $\sigma = 2\Delta P / \rho U_\infty^2$ is the cavitation number, where $\Delta P$ is difference between ambient pressure $P_\infty$ and pressure inside the cavity $P_c$, $U_\infty$ is the incident flow velocity, $\rho$ is the mass density of the liquid, and $c_d$ is the cavitator drag coefficient within the ideal fluid model framework. The parameter μ characterizes the inertial properties of cavity sections while k accounts for the small transfer of energy between cavity sections. The approximations for these parameters, which preserve the first term of the asymptotic row and extend the applicable range of these formulas to $\sigma < 0.2$, are:

$$\gamma > 1^o \div 2^o,\ \sigma < 0.2 : \mu = \ln\left(\frac{\lambda + 0.9}{\sqrt{e}}\right)\ or\ \mu = \ln\left(\frac{0.9\lambda + 1.3}{\sqrt{e}}\right);\ \mu = \ln\sqrt{\frac{1.3}{\sigma}}\ \ or\ \mu = \ln\left[\left(\frac{1.8}{\sigma}\right)^{1/2} - 0.65\right] \quad (2)$$

$$\gamma > 10^o,\ \sigma < 0.2:\ k = \left[1 + \frac{2\ln 2/\sqrt{e}}{\ln(0.6\lambda + 3)^2}\right]^{-1};\ \ k = 1 - \frac{2\ln 2/\sqrt{e}}{\ln(4/\sigma + 18)} \quad (3)$$

These formulas are applicable to conical cavitators over a wide range of it's semi-angles $\gamma$ included disc. For slender cavitators $\lambda$ corresponds to the cavity rear part after midsection. Accuracy of single-parameter approximations are $\sim 3 \div 5\%$, with two- parameters ones being $\sim 2 \div 3\%$. In view of the known principle of cavity section expansion independence [3], these dependencies can be used for approximate prediction of the steady and unsteady cavities and cavitators of different shapes.

## 3. Hydrodynamic forces at the cavitator

For cones with semi-angles $0 < \gamma \le 90^o$, $\beta(\gamma) = \gamma^o\left(90^o\right)^{-1}$, the drag coefficient $c_d = c_{do} + \kappa_\sigma \sigma$ is:

$$c_{do} = \frac{(\pi\beta)^2}{2}\ln\left[\frac{1.74\left(1 - 0.45\sqrt{\beta} + 2\beta\right)}{\pi\beta}\right] - 0.87\beta^{1.7},\ \kappa_\sigma = \left(0.46 + 0.39\beta + \frac{0.015}{\beta^{1.36} + 0.028}\right) \quad (4)$$

## 4. Axisymmetric cavity shape and dimensions

The simplest boundary-value problem for extended axisymmetric cavities [2] has the following form:

$$\mu_u \frac{d^2R^2}{dx^2} + \sigma(x) = 0,\ \left.\frac{dR^2}{dx}\right|_{x=0} = R_n\left(\frac{2(c_d - k\sigma)}{k\mu_u}\right)^{1/2},\ \left.R^2\right|_{x=0} = R_n^2,\ \mu_u = \left(\frac{2\sqrt{c_{d*}}}{\sqrt{c_{d*} - k\sigma} + \sqrt{c_{d*}}}\right)^2 \mu \quad (5)$$

Here $\mu_u \sim \mu$ for $c_{d*} \approx 0.82$ is incorporated for a slight adjustment. Equation (5) is applicable for disk-type and slender cavitators. Assume a cavity $R = R(x)$ with the maximum radius $R_k$, midsection co-ordinate $L_m$, half-length $L_k$ (length of a cavity section after the midsection), and the cavity total length $L_c$. Then at $\sigma = \text{const}$ the solution for equation (6) is derived from the following equations:

$$R^2 = R_n^2 + R_n\left[\frac{2(c_d - k\sigma)}{k\mu_u}\right]^{1/2} x - \frac{\sigma}{2\mu_u}x^2 : R_k = R_n\left(\frac{c_d}{k\sigma}\right)^{1/2},\ L_m = \frac{R_n}{\sigma}\left[\frac{2\mu_u(c_d - k\sigma)}{k}\right]^{1/2},$$
$$L_k = \frac{R_n}{\sigma}\left(\frac{2\mu_u c_d}{k}\right)^{1/2},\ L_c = \frac{R_n}{\sigma}\left(\frac{2\mu_u}{k}\right)^{1/2}\left[\sqrt{c_d - k\sigma} + \sqrt{c_d}\right] \quad (6)$$

## 5. Free-surface effect

The free-surface influence on a cavity form is modeled by known model of the influence of mirror-image flow around a cavity, which is simulated by a source and a drain within the small cavitator model framework. Here new more simple approach of the model realization is proposed. The problem solution is reduced to solving the four relationships in equation (7) for the parameters with subscript *h*, where parameters with subscript *o* correspond to a cavity without accounting for the free-surface effect. The free-surface effect is accounted for by deriving a corrected value of cavitation number $\sigma_h$ for $L_k$, the cavity half-length, $R_k$, the maximum radius, $\mu$, and $H$, the depth. In view of small variations of values $\mu_h$ and $H_h$, even the first approximation for $\sigma_h$ is quite accurate, starting from the first two equations (8) for $\bar{L}_{kh}$ and $\bar{R}_{kh}$ at values $\mu_o$ and $H_o$:

$$\bar{L}_{kh} = 1 + \frac{1-\bar{H}_h}{2\left[\bar{H}_h\left(1-2\ln 2\bar{H}_h - 4\mu_h\right)-1\right]},\quad \bar{R}_{kh}^2 = \frac{8\mu_h \bar{H}_h \bar{L}_{kh}\left(1-\bar{L}_{kh}\right)}{\bar{L}_{kh}-\bar{H}_h},$$

$$\mu_h = \ln\left[\frac{1}{\sqrt{e}}\left(0.9\sqrt{\frac{\mu_h}{\mu_o}}\frac{\bar{L}_{kh}}{\bar{R}_{kh}}\lambda_o + 1.3\right)\right], \qquad (7)$$

$$\bar{H}_h = \bar{H}\sqrt{\frac{\mu_o}{\mu_h}} :\to \sigma_h = \frac{2\mu_h}{\lambda_h^2};\ \bar{L}_{kh} = \sqrt{\frac{\mu_o}{\mu_h}}\frac{L_{kh}}{L_{ko}},\ R_{kh} = \frac{R_k}{R_{ko}},\ \bar{H} = \frac{H}{L_{ko}}$$

## 6. Cavity axis deformations

Cavity axis distortions due to gravity ($h_g$) and to cavitator lift force ($h_\alpha$) are described by the equations (8), where $c_{dy}$ is cavitator lift force [3]:

$$\frac{\mathrm{d}h_g}{\mathrm{d}x} = \frac{1}{R^2(x)}\frac{g}{U_\infty^2}\int_0^x R^2(x)\mathrm{d}x,\quad \frac{\mathrm{d}h_\alpha}{\mathrm{d}x} = -\frac{c_{dy}}{2}\frac{R_n^2}{R^2(x)},\quad \frac{U_\infty^2}{gR_n\sqrt{c_d}} > 11 \qquad (8)$$

Ranges of small deviations from the circular shape of cavity sections are estimated by inequalities (8), which are valid for a much wider range for disk-type and thin cavitators with some limitations for a disk with a nonzero angle of attack.

## 7. Modelling problems – scale effects

Both calculated and experimental results, including those depicted in Fig 1, imply that application of cavitators with a small angle of attack, in view of inequalities (8), allows one to provide a quite accurate modelling by the Froude number $Fr = U_\infty / gR_n$, which produces a nearly axisymmetric cavity with the maximum radius and length corresponding to a weightless liquid.

Figure 1 shows a photo of an experimental cavity [4] behind a cone with a semi-angle of $\gamma = 20^o$, which profile is made parallel to the free border by means of the cavitator angle of attack $\alpha = 5^o$.

a)

b)

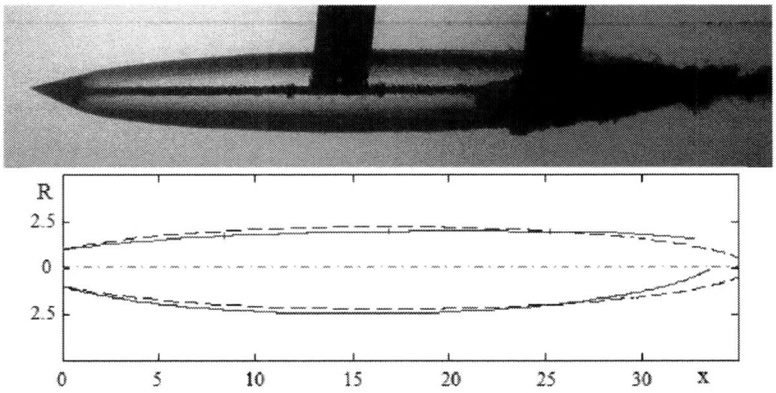


**Figure 1.** **a)** Experimental cavity behind a cone ($\gamma = 20^o$ $d = 5cm$ $\alpha = 5^o$) in the water duct at depth $H = 10cm$ moving with velocity $U_\infty = 9.75m/s$ for $\sigma = 0.039$ [4].
**b)** Calculated form of experimental cavity with normalization by $R_n$:
——— experimental cavity shape calculated via equations (5, 7, 8): $\alpha = 5^o$
– – – axisymmetric part of the cavity via equations (5 ,7) : $Fr = \infty$, $\alpha = 0^o$

The results of processing of this photo and calculations of the cavity shape and dimensions via equations (6) and then (7) for $Fr = 0$ with account of the free surface influence only yield the corrected cavitation number $\sigma_h \approx 0.0486$ (vs. experimental **exp:** 0.039), as well as cavity dimensions with account of the free surface effect: $R_k = 0.057m$ (**exp:** 0.057), $L_c = 0.90m$ (**exp:** 0.88).

The way of compensation of scale effects by the Froude number with help of lift on a slender cavitator seems to be the only option for conducting experiments in water ducts. Typical experimental curves are characterized by the minimal gas loss values at the intermediate stage between the gas loss processes via the chaotic flow in cavity wake and via vertical hollow tubes cords. Lift forces acting on a cavity, in particular, lift force at cavitator with a nonzero angle of attack, drastically affect the flow vortex system, which controls gas loss processes. The experimental studies confirm that, in case the lift force at cavitator exceeds the hydrostatic forces acting on a cavity, an additional pair of hollow vortex tubes appears in the cavity rear part. In this case, the gas loss vs. cavitator angle of attack curve exhibits two minimum and one maximum peaks [5]. If the above forces are equal, the transverse momentum of the cavity wake is zero, while the maximum gas loss, which is possible within the small cavitator model framework, is controlled by equation (9), $\bar{a} = 1 - \sqrt{1 - k\sigma / c_d}$ :

$$c_{dy}^{*} = 2\left[\frac{4}{3} - \bar{a}^2\left(1 - \frac{\bar{a}}{3}\right)\right]\left(\frac{c_d}{k}\right)^{1.5}\frac{\sqrt{2\mu}}{\sigma^2}\frac{gR_n}{U_\infty^2} \approx 2\left[\frac{4}{3} - \left(\frac{k\sigma}{2c_d}\right)^2\right]\left(\frac{c_d}{k}\right)^{1.5}\frac{\sqrt{2\mu}}{\sigma^2}\frac{gR_n}{U_\infty^2} \approx \frac{8}{3}\left(\frac{c_d}{k}\right)^{1.5}\frac{\sqrt{2\mu}}{\sigma^2}\frac{gR_n}{U_\infty^2} \quad (9)$$

$$\bar{a} = 1,\ c_d = k\sigma:\ \alpha^{*} = \frac{2\sqrt{2\mu}}{3\sqrt{\sigma}}\frac{gR_k}{U_\infty^2};\ \Gamma_c \approx gS^c / U_\infty : \Gamma_k = \frac{3}{4}\pi\alpha R_k U_\infty;\ \Gamma_c = 2\alpha R_n U_\infty \quad (10)$$

In the particular case of cavitator having a half-cavity shape ($\bar{a} = 1, c_d = k\sigma$), the value of $\alpha^{*} \approx 5^o$ coincides with the estimation and experimental data [5]. Modeling by the Froude number should provide the respective values of the transverse momentum or circulation in the cavity wake. Circulation by the cavity meridian section of area $S(x)$ is controlled by dependence (10) [3], which yields $\Gamma_k$, in case of a slender body with a half-cavity shape. Circulation $\Gamma_c$ of a slender cone with a small angle of attack and by other surfaces is also estimated. The account of scale effects in simulation of gas loss processes by the Reynolds number is more complicated. Here the known idea of J. Spurk on gas entrainment by a boundary layer can be applied.

## 8. Conclusions

The presented dependences are recommended for reliable estimation of the basic dimensions and shapes of cavities for a prompt conduction of cavitation experiments.

**Acknowledgements**

The authors would like to thank Dr. Ronald Joslin of ONR and Dr. John Zimmerman of ONRG for their support and encouragement.

## References

[1] Guzevsky L G 1983 Approximated dependence for axisymmetric cavities behind cones Set of papers: *Hydrodynamic Flows and Wave Processes* (Novosibirsk: Institute of Thermophysics RAN) pp 82-91

[2] Serebryakov V V 2002 The models of the supercavitation prediction for high speed motion in water *Proc. Int. Conf. on High Speed Hydrodynamics: HSH2002* (Cheboksary: National Academy of Sciences of Chuvash Republic) pp 71-92

[3] Logvinovich G V 1972 *Hydrodynamics of Free-Boundary Flows* (Jerusalem: Israel Program for Scientific Translations)

[4] Kochin V A, Moroz V V, Nechitailo N and Serebryakov V V 2014 Hydrodynamics of supercavitating bodies at an angle of attacks under conditions of considerable effect of liquid weightiness and closeness of free border *Proc. MAST EurAsia 2014* (Istanbul: MAST Communications Ltd) p 20

[5] Kapankin Y N and Gusev A V 1984 *J. Proc. TsAGI* **2244** pp 19-28

Journal of Physics: Conference Series **656** (2015) 012170 doi:10.1088/1742-6596/656/1/012170

# A Method of Mitigating Pulsation in Ventilated Supercavities

**Grant M. Skidmore, Timothy A. Brungart, Jules W. Lindau, Michael J. Moeny**

Contact Email: gms240@psu.edu

**Abstract.** A method for mitigating ventilated supercavity pulsation is presented. The method, which has its roots in parametric oscillators, shifts the supercavity resonance frequency by modulating its gas ventilation rate. When appropriately modulated, the supercavity is driven off resonance by the waves on the gas/water interface (that remain unchanged) and pulsation is, therefore, suppressed. Initial experimental results indicate that the gas ventilation rate modulation frequency must be sufficiently different from the supercavity resonance frequency to mitigate pulsation. If the modulation frequency is not sufficiently different from the supercavity resonance frequency, pulsation is simply shifted in frequency with a corresponding small reduction in the supercavity interior pressure spectrum level and radiated noise.

## 1. Introduction

It is possible to reduce the skin friction drag of a body, up to an order of magnitude, using supercavitation to reduce the wetted surface area. In order to develop a supercavity for many applications, gas must be ventilated in the low pressure region behind the cavitator. Thus, for high speed supercavitating vehicles, ideally only the control surfaces and the cavitator should remain in contact with the water after a supercavity is developed via ventilation. While this idea works well in theory, in practice the process of generating a supercavity is often plagued by a resonance phenomenon known as pulsation. When a supercavity pulsates, the walls of the supercavity begin to periodically expand and contract; the periodic increase and decrease of cavity size can lead to the cavity walls clipping the body, which can be an issue for the stability of the body as well as a strong source of radiated noise [1]. Oftentimes, it is possible to transition from the pulsating closure regime to either twin vortex or re-entrant closure by simply adjusting the ventilation gas flow rate, pressure/depth, or velocity. However, for cases when these parameters are fixed and pulsation occurs, an alternative means for control is desired. Thus, in this paper, a method to mitigate the pulsation phenomenon in ventilated supercavities is outlined and preliminary experimental results are presented. Please note that here, as in the remainder of this document, the terms cavity and supercavity are used interchangeably.

## 2. Background

Supercavities can be generated with natural cavitation and with gas ventilation. Ventilation provides the option for low speed supercavities, where gas is injected behind the cavitator of the body and a supercavity forms. Ventilated supercavities, when formed in a gravity field roughly orthogonal to the direction of body motion, the condition of most interest, tend to obey three main closures: re-entrant jet (also known as toroidal shedding cavities), twin vortex, and pulsation [2]. It is a fundamental requirement for all regimes that the rate of mass ingress into the cavity must match the rate of mass egress out of the cavity in order for the cavity to remain stable; thus, the difference between closure regimes is the primary mechanism employed to allow gas to exit the cavity into the water.

Supercavity pulsation was first reported by Silberman and Song [3] who found that, as the cavity gas injection rate increased, the cavitation number decreased linearly up to a certain critical cavitation number; from this point, increased gas injection lead to the cavity walls vibrating in the manner described and little correlation between ventilation rate and cavitation number was observed. Silberman and Song identified this new closure regime by a wavy cavity surface where the cavity exhibits periodic changes in both length and internal pressure. These waves travel along the cavity interface until minima, on opposing sides of the cavity, approach one another sufficiently close for "pinch off" to occur [4]. That is, pinch off occurs where the surface wave influenced local cavity diameter becomes less than the cavitator diameter. The pinch off location effectively denotes the termination of the cavity. The gas ejected from the cavity is advected with the mean flow. Hence, large portions of the gas of the supercavity are periodically ejected by this pinch off process and carried downstream by the flow. The pulsation phenomenon is depicted schematically in Figure 1a.

The first notion of control of pulsation via harmonic forcing can be attributed to Semenenko in 1996. Semenenko discussed the possibility of driving a pulsating cavity with an external forcing frequency. When the forcing frequency is close to that of the natural frequency, the pulsation is said to be "reinforced" by the forcing frequency, and when forced at a frequency away from the natural frequency the periodic mode was said to be "chaotized" [5]. A related approach, that has its roots in parametric oscillators, is to modulate the ventilation gas flow rate at frequencies other than the pulsation/resonance frequency. This, theoretically, changes the supercavity resonance frequency from its constant ventilation rate value and, as a result, can inhibit resonance excitation from the interface waves. This approach was explored numerically by solving the appropriate form of Hill's equation [6] and found it to be an effective means of shifting the supercavity resonance frequency, and thus, potentially mitigating pulsation.

## 3. Experimental Setup

The cavitator configuration used in this experiment is shown in Figure 1b; the system is similar to the one detailed by Skidmore [1]. It was installed in the test section of the ARL Penn State 0.305 m water tunnel. The cavitator itself was a truncated 15° cone with a maximum diameter of 3.175 cm and a thickness of 0.48 cm. The cavitator had a flush mounted Measurement Specialties XPM5-A1 static/dynamic pressure sensor on the downstream face for measuring the cavity interior pressure. The cavitator was oriented downstream of the sting configuration, a 14.13 cm long body with an elliptic nose that allowed for minimal flow disturbances upstream of the cavitator. The sting was, in turn, mounted to a strut with a hydrofoil cross section. The strut was bolted into the walls of the tunnel and kept the centerline of the cavitator concentric to that of the 0.305 m circular test section. A Kelly Pneumatics KPS-10-A0-4 high speed butterfly valve was used to modulate the cavity ventilation rate. A Celesco LC-10 hydrophone mounted onto the tunnel test section was used to measure the near-field noise radiated by the cavities. A Casio F1 camera was used to record images of the cavities at 600 frames per second.

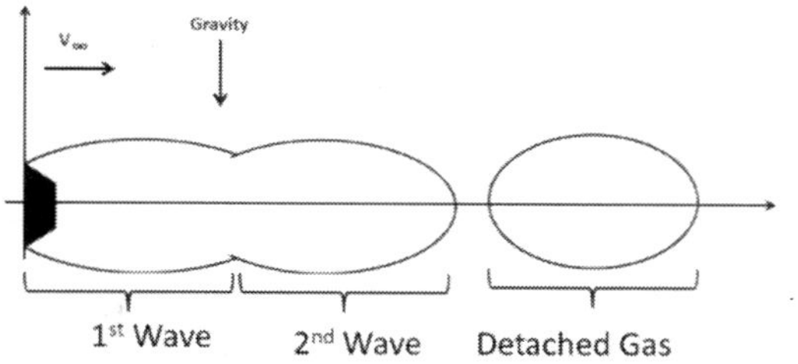


(a) Schematic of a pulsating supercavity. Note the waves on the cavity surface and the large volume of gas that is expelled from the main cavity.

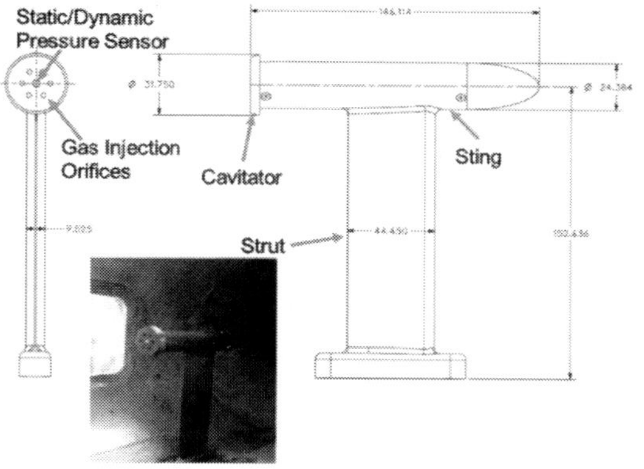


(b) Schematic of cavitator configuration used in experiment. Note that the dimensions are given in (mm).

Figure 1: Background and setup figures.

### *3.1. Experimental Procedure*

The tunnel velocity was set to 2.05 m/s (for the cavities discussed in this paper) and a pulsating supercavity was established with the high speed butterfly valve fully open. This baseline condition was established for successive tests such that a nominally fixed pulsation frequency was observed. The pulsating supercavity was given approximately 20 seconds to stabilize before the high speed butterfly valve was employed to modulate the ventilation rate. After initiating modulation of the gas ventilation rate, the cavities were observed for 30 seconds or more in order to quantify what changes, if any, occurred to the cavity.

## 4. Results

Two pulsating cavities were generated at identical test conditions and were pulsating at 38.8 Hz ($\pm$0.06 Hz) prior to modulating the gas ventilation rate. The gas ventilation rate to the first cavity was then modulated at 32.0 Hz, and within 0.3 seconds of initiating modulation, the pulsation frequency increased to 41 Hz and the amplitude of both the cavity interior pressure and the radiated sound pressure decreased 4 dB as shown in Figure 2a. The cavity was initially a $2^{nd}$ order pulsating cavity (i.e. two waves on its surface prior to pinch-off) and the spectral changes occurring 0.3 seconds after initiating modulation were accompanied by the cavity transitioning to $3^{rd}$ order as shown in Figure 2b. These results indicate that modulating the gas ventilation rate was effective at shifting the supercavity resonance frequency but not enough to mitigate pulsation.

The gas ventilation rate to the second cavity was modulated at 28.0 Hz and within 0.1 seconds of initiating modulation, the cavity ceased pulsating. When modulation was initiated, the 38.8 Hz surface traveling waves disappeared and 49.0 Hz surface traveling waves appeared. The 49.0 Hz surface traveling wave amplitude, however, decayed to a non-observable level over 1.5 seconds, at which time the cavity transitioned to the twin vortex closure regime. The internal cavity pressure and near-field noise spectra shown in Figure 3a were recorded over an approximate 23 second window prior to initiating gas ventilation modulation and 33 second after. The spectra show that the effect of modulation was to reduce the cavity pressure and radiated noise spectrum level, at the pulsation frequency 35 dB to the continuum (a near two order-of-magnitude reduction in both interior pressure and radiated sound pressure). The cavity before and after modulating the gas ventilation rate is shown in Figure 3b.

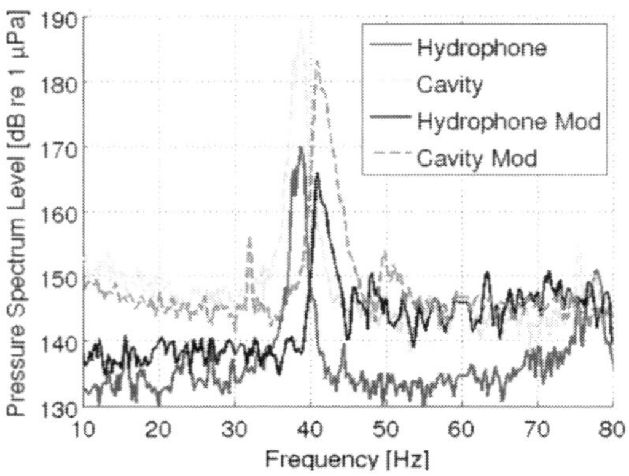


(a)

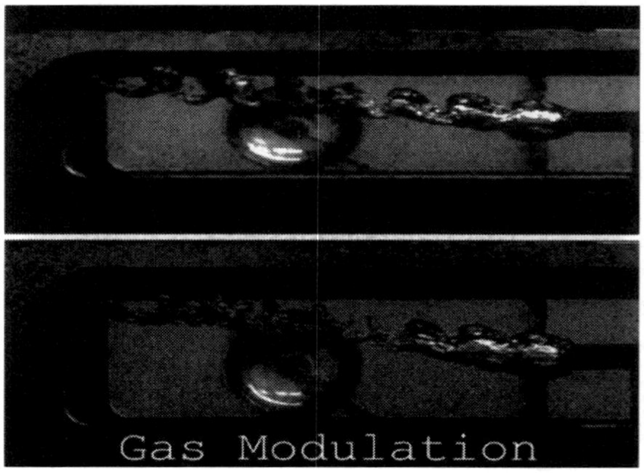


(b)

Figure 2: Subfigure 2a shows pressure spectra measured by the interior cavity pressure sensor and near-field hydrophone before and after gas modulation. Subfigure 2b shows images of the cavity before and after gas modulation

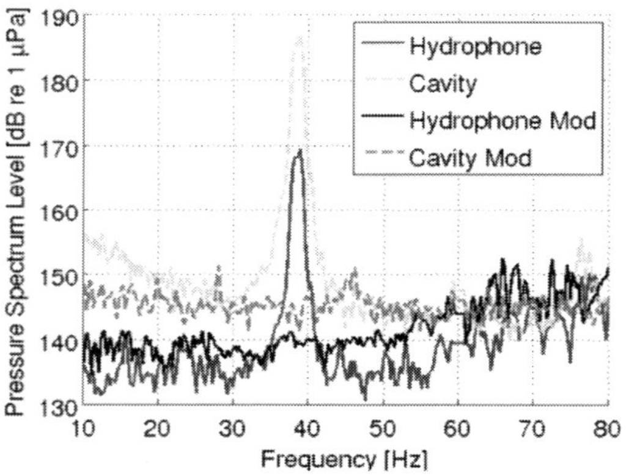


(a)

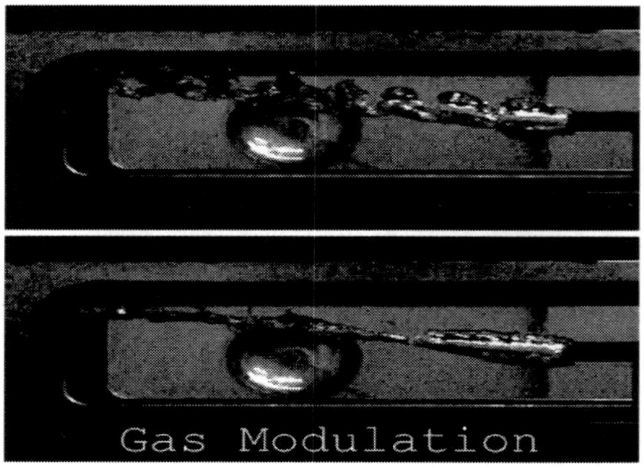


(b)

Figure 3: Subfigure 3a shows pressure spectra measured by the interior cavity pressure sensor and near-field hydrophone before and after gas modulation. Subfigure 3b shows images of the cavity before and after gas modulation

## Acknowledgement

This work is sponsored by the Office of Naval Research under Grant N-00014-1110322 and monitored by Dr. Ronald D. Joslin.

## References

[1] Skidmore G, Brungart T, Lindau J and Moeny M 2015 *Noise Control Engineering Journal* **63 (1)** 94–101
[2] Semenenko V 2001 Artificial supercavitation. physics and calculation Tech. rep. DTIC
[3] Silberman E and Song C 1959 *St. Anthony Falls Hydraulic Laboratory*
[4] Paryshev E 2003 *Fifth International Symposium on Cavitation, CAV2003* pp 1 – 18
[5] Semenenko V 1996 *International Journal of Fluid Mechanics Research* **23**
[6] Teschl G 2012 *Ordinary differential equations and dynamical systems* vol 140 (American Mathematical Soc.)

# POD study of aerated cavitation in a venturi nozzle

**P. Tomov[1], A. Danlos[2], S. Khelladi[1], F. Ravelet[1], C. Sarraf[1], F. Bakir[1]**

[1]DynFluid Laboratory, Arts et Métiers ParisTech
151, Boulevard de l'Hôpital, 75013 Paris, FRANCE
[2]National Conservatory of Arts and Crafts,
292, Saint-Martin Str., 75003 Paris, FRANCE

E-mail: Petar.TOMOV@ensam.eu

**Abstract.** The fact of injecting bubbles into a cavitating flow influences typical cavitating behavior. Cavitation and aerated cavitation experiments has been carried out on a symmetrical venturi nozzle with convergent/divergent angles of 18° and 8°, respectively. A snapshot Proper Orthogonal Decomposition (POD) technique is used to identify different modes in terms of discharge flow velocity, pressure and injected quantity of air. The energy spectrum per given mode is also presented. The first four modes are outlined in the present paper for an aerated and non-aerated cavitating flows.

## 1. Introduction

Aerated cavitating flow contains numerous coexisting flow regimes. By nature, the cavitation has an aggressive behaviour and it is capable of doing severe damage to adjacent structures. Controlling cavitation behaviour can help produce a stable regime as opposed to an unstable one. For this purpose, passive control methods based on surface roughness have been studied [1]. Another recently used technique influencing the cavitation inception is the aeration of the flow [2]. In order to reveal and explore those aerated cavitating regimes, a snapshot Proper Orthogonal Decomposition (POD) analyses have been conducted [3,4]. Different modes have been discovered, as a result. Consequently, it is possible to reproduce the factors required to rebuild the dynamic structures as a time-dependent mixture of modes. The paper is organized in four paragraphs. The experimental set-up is described in §2, followed by the results in §3 and conclusion in §4.

## 2. Experimental Set-up

### *2.1. DynFluid Laboratory small water test loop*

The experiments are conducted in a closed loop test rig (see figure 1). Two 150 - liter storage tanks supply water to the rig. The centrifugal pump impels the flow from constantly full tank 1 to tank 2. Its maximum flow rate capacity is 8 $m^3/h$. The cylindrical pipe has an inner diameter equal to $D = 40$ mm. Throughput is monitored by a turbine flow meter, followed by the air injection ring. The transparent symmetrical venturi nozzle is placed in the test section (see figure 1), between the two tanks. The converging/diverging angles are 18° and 8°, respectively. The inlet venturi height is $H_{inlet} = 30$ mm and its throat height is $H_{throat} = 10$ mm,

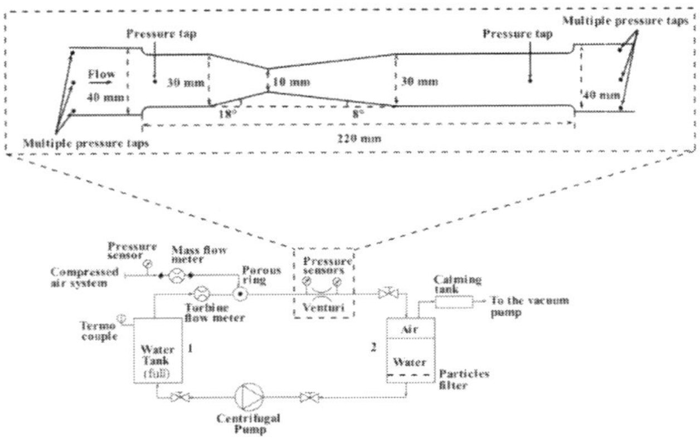


**Figure 1.** Test rig (not to scale)

which gives an aspect ratio of 3. The total length of the venturi test nozzle is 220 mm. A vacuum pump is connected to tank 2, for the purpose of mitigating the pressure at the free surface. In this way, it is possible to impact the cavitation number, defined as $\sigma = (P_{ref} - P_{vap}) / \left(1/2\rho V_{ref}^2\right)$. All positions $X^*, Y^*$ in the test section are adimensionalized by $H_{throat}$. The reference pressure $P_{ref}$ is measured at the inlet of the venturi section at $X^* = 6$ from the throat. The discharge velocity $V_{ref}$ is calculated as a function of the flow rate. The vapor pressure in operating conditions is considered 2200 Pa at 19°C.

### 2.2. *Measuring instruments and image capturing technique*

Pressure measurements are taken using two absolute pressure sensors, mounted on the inlet and outlet sections at $X^* = 6$ upstream the throat, and $X^* = 10$ downstream the throat. An absolute pressure sensor is mounted on tank 2. The range of working values is from 0 to 2 bar. The uncertainty of the measurements on $\sigma$ is $\frac{\Delta\sigma}{\sigma} = 0.22$.

Temperature is monitored by a thermocouple mounted on water tank 1. The working range is set to 0°C to 100°C. During all the experimental work, fluid temperature is monitored and maintained to 19°C ± 1°C. In the article, a temperature of 19° will be assumed, giving an uncertainty value of $\frac{\Delta P_{vap}}{P_{vap}} = \pm 0.06$.

Discharge flow rate is monitored by a turbine flow meter, with 10D straight pipe upstream and 5D downstream. The latter is calibrated for a working range of viscosities in the interval 0.6 to 10 cSt. The volume flow rate fluctuation is $\frac{\Delta Q_{liq}}{Q_{liq}} = \pm 0.08$.

Air injection is produced using a compressed air system. Two air filters have been mounted before and after the air mass flow rate meter. The filtration is equal to 5 $\mu$m. Volume flow rate fluctuations is $\frac{\Delta Q_{gaz}}{Q_{gaz}} = \pm 0.05$. It has been verified that neither the turbine flow meter nor the non-intrusive injection approach generate any flow perturbations in front of the venturi nozzle. The upstream fluid flow does not content any bubbles as a result of an early cavitation or out-gassing phenomenon.

A CamRecord 600 camera with a 100 mm Zeiss Makroplanar objective lens is mounted on a tripod to insulate it from vibrations that might cause field capturing perturbations. The frequency sample rate is 1 kHz with an exposure time set to 1/40000 s. A total number of 6554 images were taken. The flow was illuminated by a LED backlight.

## 3. Results and Discussions

Table 1 summarizes the experimental data conditions. Two cases are presented: pure cavitation and aerated cavitation. The variable $\beta$, defined as $\beta = Q_{gas}/(Q_{gas} + Q_{liq})$, describes the ratio of injected gas to total volume. The Reynolds number at the venturi throat $Re = V_{ref}.H_{throat}/\nu$ is based on the local flow velocity and the throat height. The Strouhal number $Str_L$ is based on the cavity length on the bottom part of the nozzle in aerated case. All of the images are normalized using a reference image taken at non-cavitating non-aerated flow conditions. The POD analyses in this work follows the one described in [4]. In order to minimize post-processing time, a convergence study is conducted on the minimum number of required images [4,5]. As a

**Table 1.** Aerated cavitation experimental data table

| $V_{ref}$ (m/s) | $P_{ref}$ (bar) | $P_{Tank2}$ (bar) | $\beta$ | $Re$ | $\sigma$ | $Str_L$ |
|---|---|---|---|---|---|---|
| 11.07 | 1.10 | 0.9 | 0 | $1.1 \times 10^5$ | 1.78 | 0.431 |
| 11.07 | 1.04 | 0.8 | 0.004 | $1.1 \times 10^5$ | 1.69 | 0.352 |

result, $N = 700$ is taken for the reconstruction of the flow dynamics, as it can be seen in figure 2 (the same result has been found for modes 1 to 3). The information gathered is decomposed into empirical eigenfunctions $\varphi_i$. As a result, only a small number of modes are needed to represent correct dynamical evolution of fluid flow. The energy contribution per mode is defined as $E_{mode} = \lambda_i / \sum_{i=1}^{N} \lambda_i$, where $\lambda_i$ is the eigenvalue after a singular value decomposition for mode $i$ and $N$ is the number of images. The total number of studied modes is equal to 10. The POD modes and their contributions for the reconstruction of the flow dynamics in terms of $E_{mode}$ are shown in figure 3. It can be seen that the first mode is the most important one for both of the case studies. It can also be seen that the vapour zones are symmetrical for the pure cavitating flow, which is not the case for the aerated flow. The vertical lines show the closure region of the cavitation zone. For $\beta = 0$, the closure region is at $X^* = 3.1$, where for $\beta = 0.004$, the line is in the vicinity of $X^* = 3.7$. The zone beyond the cavity closure in figures 3(a') corresponds to the bubble - vapour mixture. The bubbles at the throat do not cut the cavitating zones. Rather, they break up due to the pressure gradient at the divergent zone and spread into extremely small bubbles, which interact mostly with the upper advected cavitating vapour. This is clearly visible in figure 3 (c') — the darkened zones beyond $X^* = 3.7$. On the contrary, the bottom cavitating zone stays intact, and thus the top and bottom vapour zones are not dynamically symmetrical. Based on the plots of the phase portraits of the temporal amplitudes, the observed phenomenon in figures (b) and (c) takes the form of an ellipse. The latter can be represented as two sine waves having the same frequency, 90° out of phase. In contrast, figures (b) and (d) display a phenomena having two frequencies with similar phases. In terms of energy contribution, one can see that modes (b) and (c) are quite similar, which is also the case for (b') and (c'). It can be seen in figure 3 (b'), the cavitation zones are able to resume to their forms, once the detached air - vapour mixture is advected.

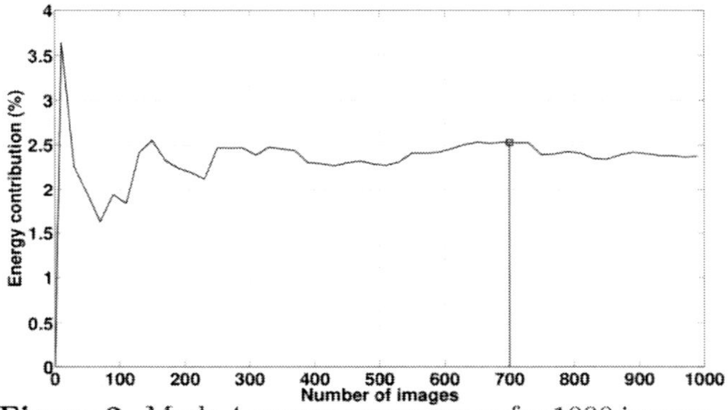

**Figure 2.** Mode 4 energy convergence for 1000 images, $\beta = 0.004$

## 4. Conclusions and Perspectives

Side view images of aerated and non-aerated cloud cavity dynamics in a horizontal convergent divergent nozzle have been studied with a snapshot POD method. The modes account for more than 86% of the total energy for the aerated flow, and nearly 90% for the pure cavitation. Based on a robust mathematical instrument, the phase portraits are efficient for the cavitation regime exploration. The aeration inevitably influences the dynamics of the flow. The bottom vapour zone manages to maintain its correct dynamics, which is not the case for the other. The upper zone undergoes much more frequent cloud detachment than the one at the bottom. Lastly, the

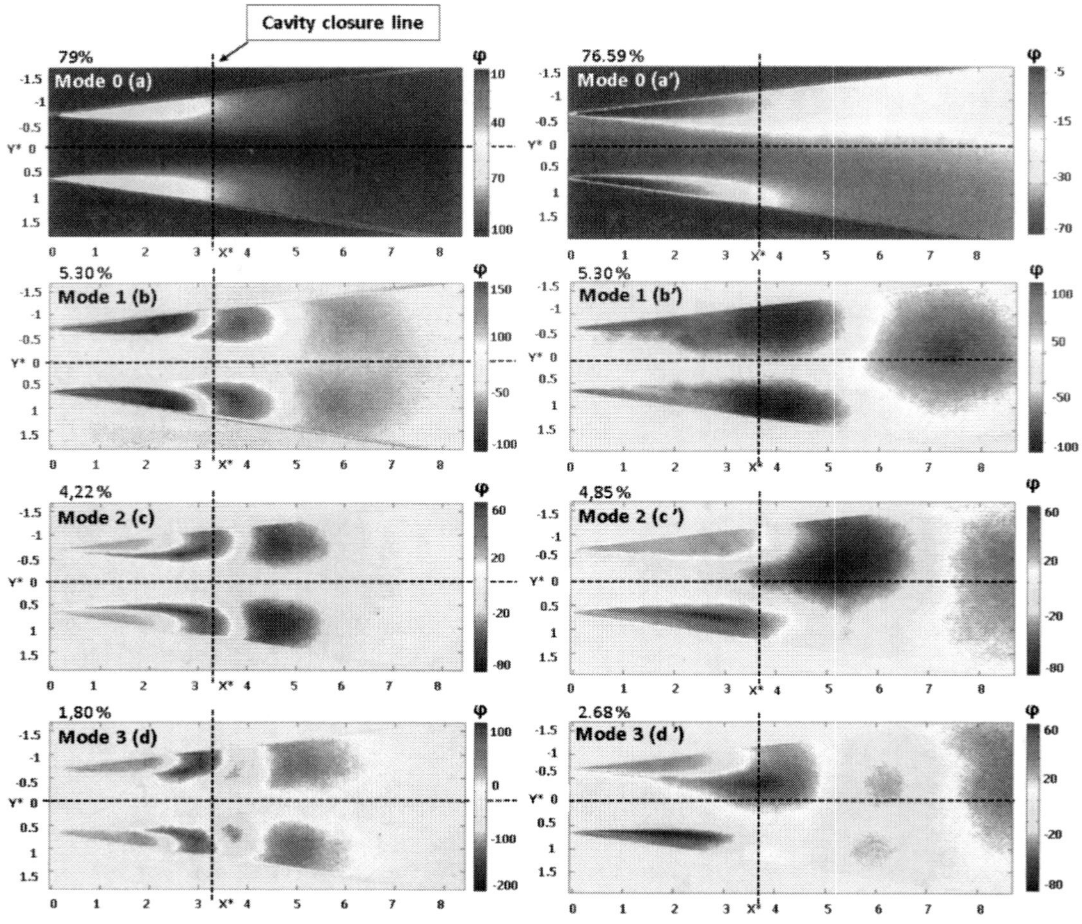


**Figure 3.** POD modes: $\beta = 0$ (left); $\beta = 0.004$ (right)

aggressiveness of the cloud cavitation, can be reduced by injecting controlled quantity of gas. In the foreseeable future, the authors plan to correlate pressure signals from the venturi throat section with the POD analysis from images.

**Acknowledgments**
The authors would like to acknowledge the financial support granted by SNECMA, part of SAFRAN group.

## References

[1] Danlos, A., Méhal, J.E., Ravelet, F., Coutier-Delgosha, O., Bakir, F., Study of the cavitating instability on a grooved Venturi profile, *Journal of Fluids Engineering*, 2014, **136**(10), pp.1—10.
[2] Dunn, P.F., Thomas, F.O., Davis, M. P., Dorofeeva, I.E., Experimental characterization of aviation-fuel cavitation, *Physics of Fluids*, **22**(2010), 117102.
[3] Berkooz, G., Holmes, P., Lumley, J., The proper orthogonal decomposition in the analysis of turbulent flows, *Annual Review of Fluid Mechanics*, **25**(1993), pp.539—575.
[4] Danlos, A., Ravelet, F., Coutier-Delgosha, O., Bakir, F., Cavitation regime detection through Proper Orthogonal Decomposition: Dynamics analysis of the sheet cavity on a grooved convergent-divergent nozzle, *International Journal of Heat and Fluid Flow*, **47**(2014), pp.9—20.
[5] Dular, M., Bachert, B., Stoffel, B., Sirok, B., Relationship between cavitation structures and cavitation damage, *Wear*, **257**(2004), pp.1176—1184.

# Bubble-induced sheet cavitation inception on an isolated roughness element

**Martijn van Rijsbergen and Jesse Slot**

Maritime Research Institute Netherlands (MARIN), P O Box 28, 6700 AA Wageningen, The Netherlands

E-mail: m.v.rijsbergen@marin.nl

**Abstract**. The nucleation process on an isolated roughness element, located at the point of minimum pressure of a NACA 0015 hydrofoil was studied experimentally and computationally. The objective of this study was to investigate the working mechanism of bubble-induced sheet cavitation inception. High-speed micro-scale observations show the generation of a streak of cavitation—attached to the roughness element—in the wake of the bubble. Below its critical diameter, the bubble can detach from the streak cavity and travel on while the streak cavity remains. The solutions of a Rayleigh-Plesset equation along a streamline extracted from a RANS calculation show strong similarities with the experimental observations, but a factor 5 to 10 higher frame rate is needed to validate the calculations.

## 1. Introduction

Sheet cavitation inception does not necessarily occur on foils and propellers when the local pressure decreases below the vapour pressure. The condition of inception is also dependent on the characteristics of the fluid, the solid surface and the local flow. Although seeding the flow with micro bubbles and applying leading edge roughness with a Reynolds-dependent grain size is adequate in most cases [1], the detailed working mechanism of the sheet cavitation inception process is not fully understood. A hypothesis formulated in [2] states that a roughness element causes a very local but significant additional pressure decrease in which a micro bubble can expand and induce a streak of cavitation. This local pressure decrease is dependent on the thickness of the boundary layer compared to the height of the roughness element and therefore also dependent on the Reynolds number [2].

High-speed micro-scale observations were made on a hydrofoil to investigate the nuclei-induced sheet cavitation inception process in more detail. The first analysis was focussed on the diameter, form and trajectories of the nuclei [3]. In the present study, a further analysis is made of the nucleation process on an isolated roughness element caused by free stream bubbles. To obtain the pressure and velocity distribution around the roughness element, the foil with the roughness element were numerically modelled and the steady wetted flow was solved using a RANS code. To study the bubble dynamics, a variant of the Rayleigh-Plesset equation has been solved along calculated streamlines passing close to the roughness element.

## 2. Experiments

The experiments were carried out in MARIN's high speed cavitation tunnel with a rectangular test section of 80 mm height and 40 mm width. A 2D NACA 0015 section with a chord length of 60 mm was placed in the test section at an angle of attack of 6.0 degrees. High-speed cameras were used to observe the leading edge of the foil from the side and the top at a frame rate of 15 kHz. The shutter times of the top and side cameras were respectively 9 and 1 μs. Further details on the test set-up can be found in [3] (corrigendum: the height of the roughness element is 60 μm).

Figures 1 and 2 show a series of combined "photo finish" images of the nucleus from the top (left) and side (right). The numbers indicate the corresponding instances of the side and top camera images. The contour of the roughness element is highlighted with a white line (right). The initial bubble diameter ($D_0$), indicated by number 1, is determined from the side camera image with an estimated uncertainty of 5 μm. The estimated uncertainty in the diameters determined from the top camera images is 60 μm. The cavitation number ($\sigma$) and Reynolds number (Re) of the two cases can considered to be equal within the estimated experimental uncertainty of 5%, only $D_0$ is different with more than a factor 2. In both cases, the bubble expands and deforms when approaching the roughness element, generates a streak of cavitation on the roughness element, expands further and passes on. In case 1, the bubble passes on the side of the roughness and generates a longer streak than in the bubble in case 2. It also stays attached to the streak for a longer time than the bubble in case 2. The measured diameters are shown in Figures 6 and 7.

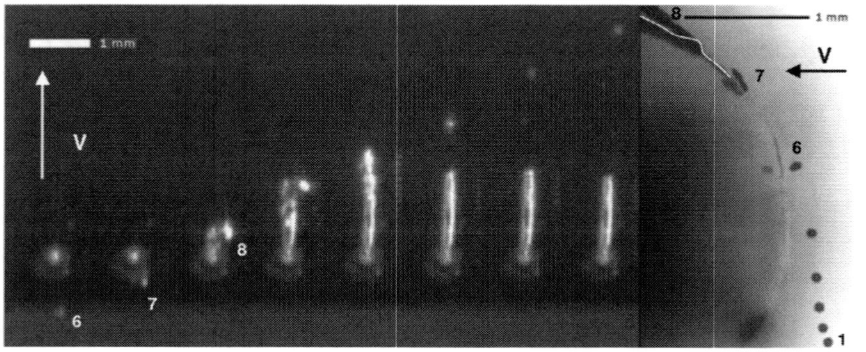


**Figure 1.** Case 1, $D_0$ = 86 μm, $\sigma$ = 2.15, Re = 5.0x10$^5$, V = 7.9 m/s

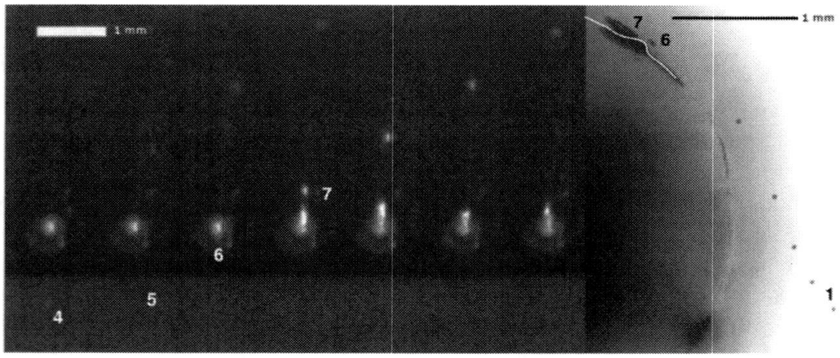


**Figure 2.** Case 2, Do = 37 μm, $\sigma$ = 2.15, Re = 5.3x10$^5$, V = 8.2 m/s

## 3. Calculations

A CAD model has been made of the test section with the hydrofoil and the roughness element as described in section 2. The 3D geometry of the roughness element was modelled as closely as possible using the available images. On the hydrofoil, as well as on the upper and lower walls of the tunnel, a no-slip condition is applied. A symmetry condition is imposed on the side walls of the tunnel as well as on the symmetry plane intersecting the roughness element. At the inlet the velocity is set. On the outflow plane a constant pressure is imposed. Figure 3 shows the grid topology around the roughness element on a very coarse grid, used only for visualization. The viscous-flow CFD code ReFRESCO has been used to solve the incompressible flow using steady RANS equations, complemented by the k-ω SST 2003 turbulence model [4]. No wall functions were used.

The numerical uncertainty of the solution for Re = $3.6 \times 10^5$ is determined using the procedure by Eça et al. [5] using 7 grids ranging in size from 6.9M to 37.2M cells. For the third finest grid with 23.4M cells, the numerical uncertainty of $C_{Pmin}$ was 0.3%. Since this is sufficiently small, this grid was chosen for the Reynolds study. Next to the numerical uncertainty there is a modelling error which might be significant on the leeward side of the roughness element. Studies on the flow around 3D hills have shown that RANS is capable of prediction the location and the magnitude of $C_{Pmin}$. But a little further downstream the RANS solution—in terms of the starting point of the flow separation, the size of the separation zone and downstream pressure recovery—deviates considerably from measurements [6], [7], [8].

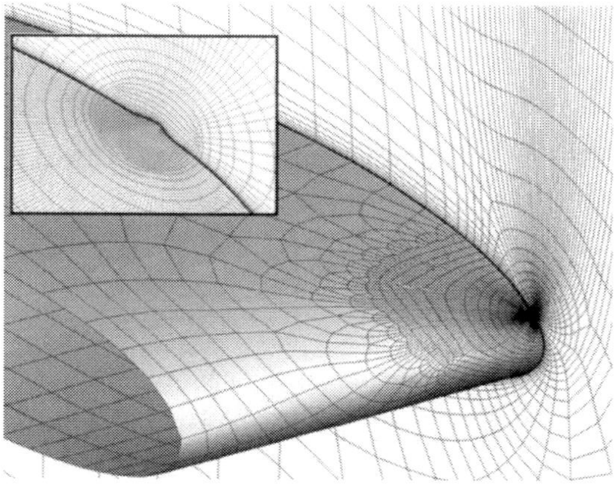

**Figure 3.** Grid distribution on hydrofoil and symmetry plane. Inset shows close-up of grid on roughness element. For illustration purposes a very coarse is shown.

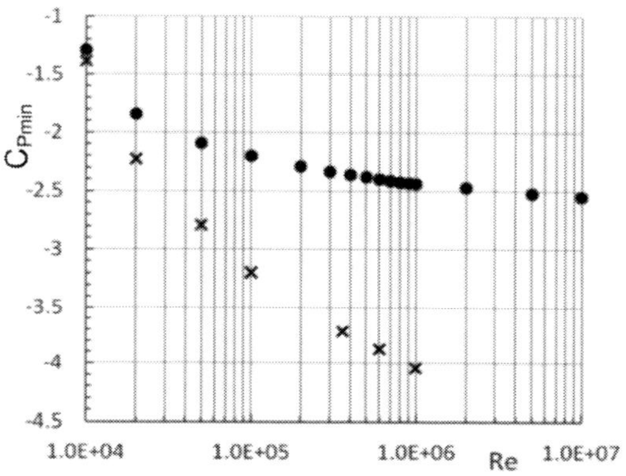


**Figure 4.** Minimum Cp as function of Reynolds number for a NACA 0015 foil in a cavitation tunnel without (dots) and with (x) roughness element.

Figure 4 shows that a roughness element can decrease $C_{Pmin}$ up to 60% at Re = $1.0 \times 10^6$. The effect reduces strongly at lower Reynolds numbers.

To analyse the behaviour of the bubbles described in section 2, the local pressure along a streamline passing at a distance of 30 μm above the roughness element has been extracted from the solution for Re = $3.6 \times 10^5$, see Figure 5. The value of 30 μm has been estimated on the basis of the side view images. The ambient pressure has been scaled to fit the experimental conditions.

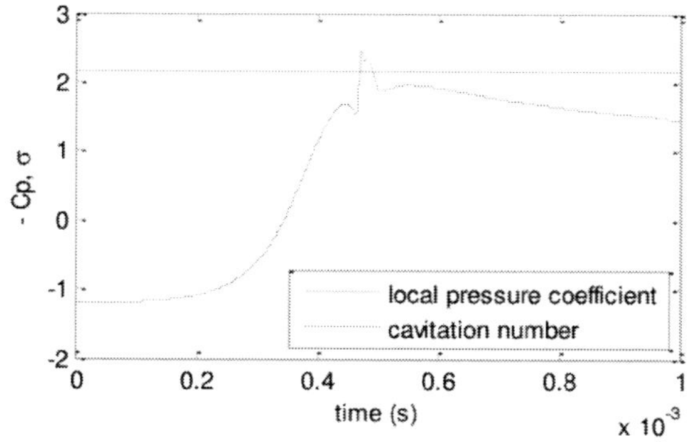


**Figure 5.** Local pressure coefficient along streamline passing 30 μm above roughness element. t = 0 corresponds with the moment when the bubble is at location 1 (see right hand sides of Figures 1 and 2). The tracer path passes over the roughness element at t = 0.47 ms.

A variant of the Rayleigh-Plesset equation—described in [9]—has been used to calculate the response of the bubble radius to the pressure history in time. The results are shown in Figure 6 and 7.

Although the local pressure is far below the critical pressures of the bubbles, they do not reach their critical diameters (respectively 835 and 250 μm for the 86 and 37 μm bubble) due to the fast pressure decrease and recovery around the roughness element. The agreement between the calculations and experiments is acceptable. The bubble resonance can clearly be seen in the calculations, but the experimental data lacks temporal resolution to resolve this phenomenon. A factor 5 to 10 higher frame rate would solve this.

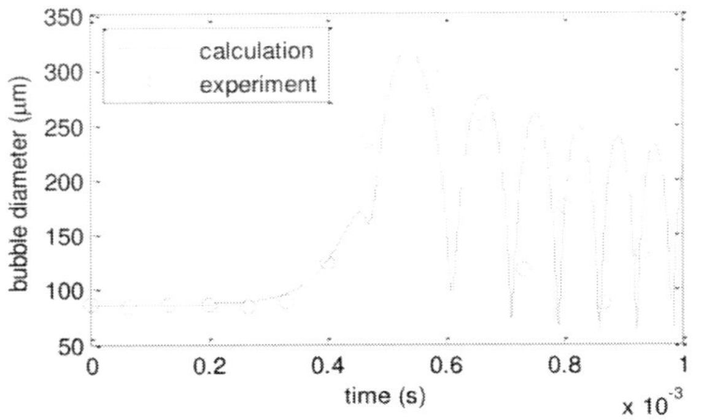


**Figure 6.** Bubble diameter as a function of time with $D_0$ = 86 μm, calculation and experiment

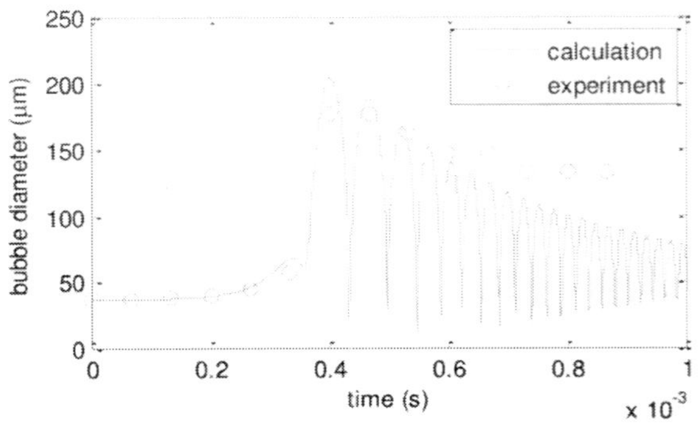


**Figure 7.** Bubble diameter as a function of time with $D_0$ = 37 μm, calculation and experiment

## 4. Concluding remarks

- The experiments show that an attached streak cavity can be formed on an isolated roughness in the wake of a free-stream bubble which travels on after the moment of inception. The largest bubble generates the largest streak.
- The calculations show that the bubble starts to resonate but does not reach its critical diameter.
- The numerical modelling can be improved by imposing a no-slip condition on the side tunnel walls and solving an equation of motion to determine the path of the bubble in the viscous flow more accurately.
- Future experiments with frame rates of 100 to 150 kHz and a factor two higher spatial resolution are expected to reveal more details on the generation of an attached streak of cavitation by a free-stream bubble and the dynamic behaviour of the bubble.

## References

[1] Ligtelijn J T, Kooij J van der, Kuiper G and Gent, W van 1992 Research on propeller-hull interaction in the depressurized towing tank *Hydrodynamics: Computations, Model Tests and Reality* Elsevier 539-55
[2] Rijsbergen M X van and Terwisga T J C van 2010 Water quality effects on sheet cavitation inception on a ship propeller model *7th Int. Conf. on Multiphase Flow* Tampa Florida
[3] Rijsbergen M X van and Terwisga T J C van 2011 High-speed micro-scale observations of nuclei-induced sheet cavitation *3rd Int. Cavitation Forum* Warwick UK
[4] Klaij C M and Vuik C 2013 *Int. J. for Num. Meth. in Fluids* **71(7)** 830-49
[5] Eça L and Hoekstra M 2009 *Computers & Fluids* **38** 1580-91
[6] Byun G 2005 Structure of three-dimensional separated flow on symmetric bumps PhD Thesis Virginia Polytechnic Institute and State University
[7] Wang C, Jang Y J and Leschziner M A 2004 *Int. J. of Heat and Fluid Flow* **25** 499–512
[8] Visbal M R, Rizzetta D and Mathew J 2007 Large-eddy simulation of flow past a 3-D bump *Proc. 45th AIAA Aerospace Sciences Meeting and Exhibit* AIAA 2007-917 Reno Nevada
[9] Hilgenfeldt S, Brenner M P, Grossmann S, Lohse D 1998 *J. Fluid Mech.* **365** 171-204

# Cavity geometry in real conditions

**Y.F Zhuravlev**[1], **A.N. Varyukhin**[2]

[1] Leading Research Fellow, TsAGI, Moscow, Russia
[2] Head of department, TsAGI, Moscow, Russia

E-mail: a.varyukhin@yandex.ru

**Abstract**. The paper presents the formulation of the problems and calculation method of the cavity geometry in real conditions. The calculating procedure used here is based on the method of small perturbations of the thin body. The calculation results and their comparison with the experimental data for the following cases are presented: cavity axis deformation under the effect of gravity forces, cavity behind elliptical cavitator, cavity behind cavitator at nonzero attack angle, the pressure pulsations in the cavity, the motion near boundaries and others.

## 1. Introduction and method description

In real condition the cavity is not axisymmetric because it is effected by different disturbing factors. These factors are design features of the moving object, effect of gravity forces, presence of solid boundary, motion parameters, and others. And they can effect simultaneously. Thus, there is the problem of developing the calculating methods that allows to define different deformations of liquid boundaries of the cavity.

The method described in the works [1] and [2] meet these requirements. The method is based on the theory of small perturbation of the known thin axisymmetric cavity ($\frac{R_0}{L_k} \sim \varepsilon \ll 1$) in an infinite ideal liquid (see figure 1).

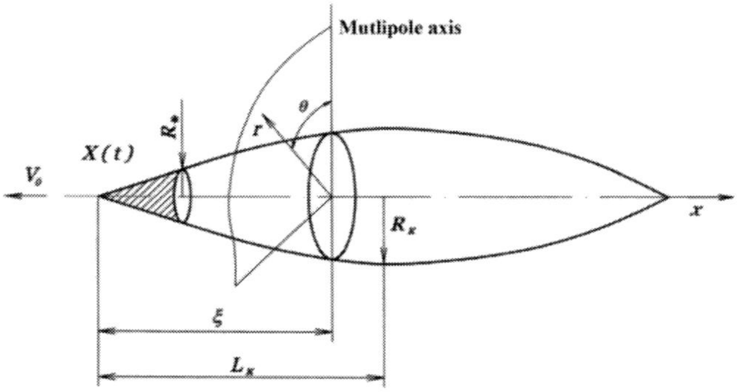

Figure 1 Thin cavity and coordinate system

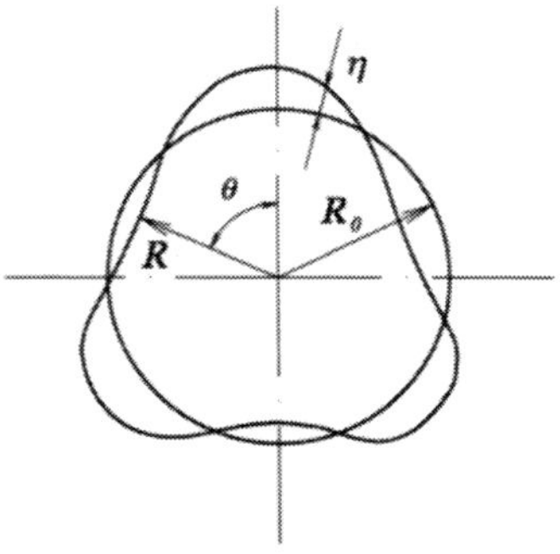

Figure 2 Cavity small deformations

We consider that the cavity shape deformations $\eta$ are small compared to its radius $R_0$ (see Figure 2).

$$R = R_0 + \eta\,,\ \frac{\eta}{R_0} \sim \delta \ll 1\,,\ \delta \sim \varepsilon\ ,\delta \ll 1\,.$$

These assumptions allow to consider the evolution of the every cavity cross-section independent from the other cross-sections. In this formulation, the calculation of cavity deformations $\eta$ can be made only in the linear approximation. The attempts to increase the accuracy of deformation calculation as proposed in [4] are ineligible, as this will lead to the need of considering the effect of longitudinal flow in the adjacent cross-sections of the cavity.

The water flow is considered as potential. In the set of equations of flow continuity and motion the terms of order of greater than $\varepsilon^2\delta$ are neglected. Small cavity shape deformations $\eta$ are decomposed in the Fourier series.

$$\eta = \sum_0^\infty \eta_j(t)\cos j\theta$$

As the result we get a set of 2nd order linear differential equations for the Fourier docomposition.

$$\begin{cases} 2\dfrac{d^2R_0\eta_0}{dt^2} = \dot{R}_0\dfrac{\partial\psi_0}{\partial r} + R_0\dfrac{\partial^2\psi_0}{\partial t\partial r} - \dfrac{\partial\psi_0}{\partial t} - U_0 - \dfrac{\Delta\sigma_0}{2} \\ \\ \ddot{\eta}_j R_0 + 2\dot{\eta}_j\dot{R}_0 - (j-1)\eta\ddot{R}_0 = (j+1)\dot{R}_0\dfrac{\partial\psi_j}{\partial r} + R_0\dfrac{\partial^2\psi_j}{\partial t\partial r} + j\dfrac{\partial\psi_j}{\partial t} - jU_j - j\dfrac{\Delta\sigma_j}{2},\ j>0 \end{cases} \quad (1)$$

Dots indicate the time derivatives. Disturbing factors that lead to deformations of the cavity cross-section are:

- the initial conditions of the fluid flow from the deformed cavitator edges:

  $$\eta(0) = \eta_* = \sum_0^\infty \eta_{*j}\cos j\theta$$

  $$\dot{\eta}(0) = \dot{\eta}_* = \sum_0^\infty \dot{\eta}_{*j}\cos j\theta$$

- cavitation number changes: $\Delta\sigma = \sum_0^\infty \Delta\sigma_j\cos j\theta$
- external fluid flow with the velocity potential: $\psi = \sum_0^\infty \psi_j\cos j\theta$
- potential function of gravity forces $U = \sum U_j\cos j\theta$

In the view of the linear formulation we can superpose small deformations generated by a variety of factors. The examples of cavity shape deformations under effect of different factors are presented below

## 2. Examples

### 2.1. *Effect of gravity forces*

The gravity forces, depending on its direction relative to the vector of cavitator movement, differently effect the cavity shape. Assume that the cavitator moves at a constant speed in a straight line inclined at $\vartheta$ angle to the horizontal. Then the potential of fluid gravity forces is:

$$U_j = \begin{cases} \mathrm{Fr}^{-2}(\mathrm{X}-\mathrm{x})\sin\vartheta, & j=0 \\ \mathrm{Fr}^{-2}\,\mathrm{R}_0\cos\vartheta, & j=1 \\ 0 & j>1 \end{cases} \quad (2)$$

Here $Fr^2 = \dfrac{V_0^2}{g L_к}$ is the Froude number. The solution of (1) with (2) allows finding a thin cavity axis curvature $\eta_1$. The results of cavity axis curvature calculations and the comparison with the experimental data are presented in Figure 3.

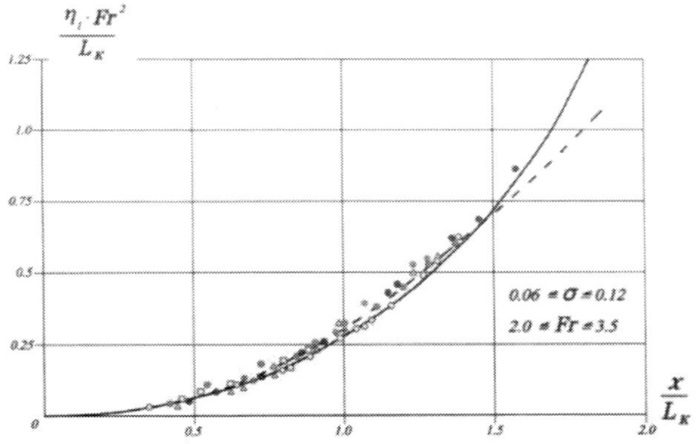


Figure 3 Cavity axis curvature under influence of gravity forces at $\vartheta = 0$

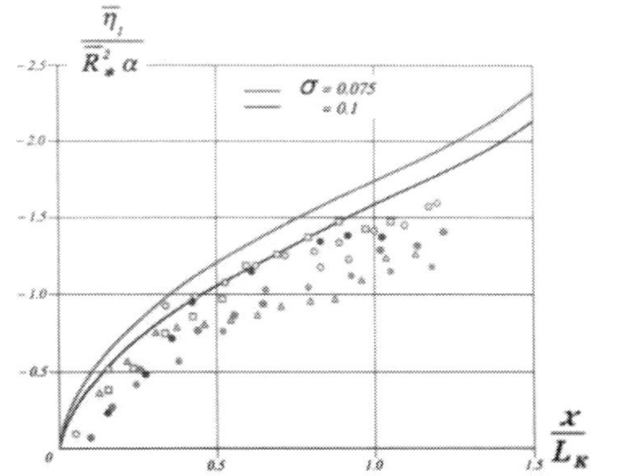


Figure 4 The cavity axis curvature when cavitator moving at nonzero attack angle

### 2.2. *The influence of the cavitator angle of attack*

The first time the problem of the cavity axis deformation when cavitator has nonzero attack angle was solved G.V. Logvinovich [3] by considering the transverse motion of the cavity circular cross-section. The system of equations (1) also allows solving this problem. Assume that the potential of the external flow in (1) is given by the expression:

$$\psi_j = \begin{cases} \alpha(x)\cdot r & j=1 \\ 0 & j\neq 1 \end{cases} \tag{3}$$

This means that in each cross section transverse speed $\dfrac{\partial \psi_1}{\partial r} = \alpha(x)$ disturbing the main flow exists.

By substituting it into the (1) we can obtain small cavity deformations:

$$\eta_1 = \alpha(x)t - \alpha(x)R_*^2 \int_0^t R_0^{-2} dt$$

Figure 4 shows the theoretical and experimental dependences of the relative cavity shape deformation when cavitator moving at attack angle $\alpha$.

### 2.3. *Cavity shape behind non-circular cavitator*

System (1) allows defining cavity shape behind non-circular cavitator. The factors disturbing the main flow are cavitator deformation in the plane of flow separation. From a mathematical point of view, these are initial conditions $\eta_j^*$ and $\dot{\eta}_j^*$ for (1).

### 2.4. *The cavity shape at motion near a plane boundary*

For solving the problems of the flow in the presence of planar boundaries it is required to find a potential harmonic function $\Phi$, which in addition to the usual conditions on the cavity surface, satisfies the conditions at a plane boundary: $\dfrac{\partial \Phi}{\partial n} = 0$ for solid boundary and $\Phi = 0$ for liquid free surface. It is known that the mirroring method, in which the features are located on the other side of the border and the flow is considered as unlimited, is widely used to solve such a problem. The potential function is $\Psi = \pm\Phi_0(x, r_1, \theta_1)$. Where sign (+) is for solid boundary, sign (-) is for free surface.

### 2.5. *Impact of gravity waves on the cavity shape*

The disturbing factor is the potential of wave flow

$$\psi = ae^{-k(H - r\cos\theta)} \sin(kx - \omega t)$$
$$\eta = \eta_0 + \eta_1 \quad (4)$$

Substituting of (4) into (1) allows to define the cavity shape.

### 2.6. *Effect of static pressure fluctuations inside the cavity*

The solution of (1) allows defining cavity shape when the pressure inside it pulses harmonically $\Delta\sigma = \frac{\Delta P}{q} = A \sin\omega t$ (see Figure 5). The Figure 6 shows a comparison of the calculated and experimental data.

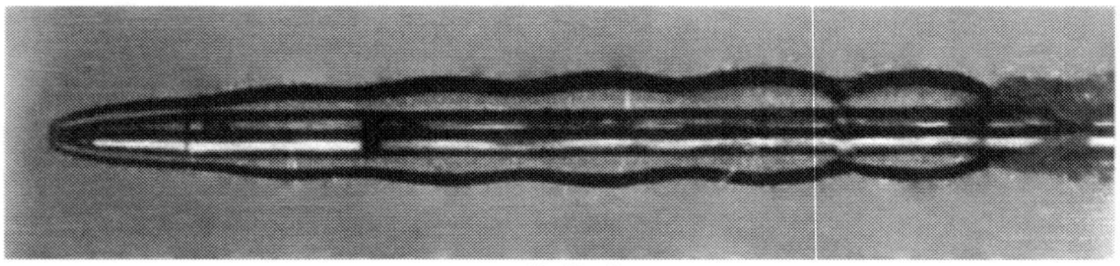

Figure 5 Photo of pulsating cavity

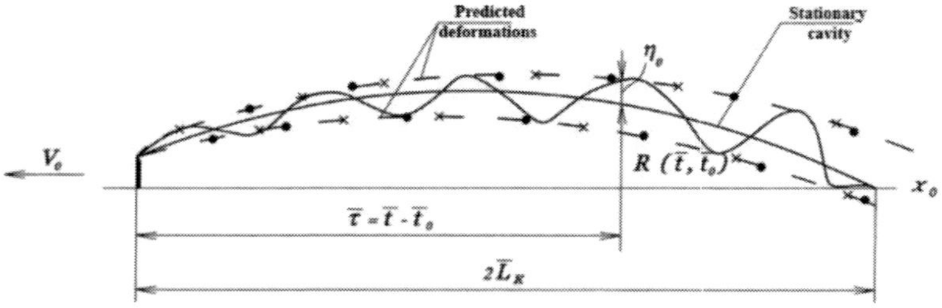


Figure 6 Comparison of the calculated profile of the pulsating cavity with experiment

## 3. Conclusions

In this paper we developed a theoretical model that allows to determine the three-dimensional form of a thin cavity under the influence of various factors such as the effects of gravity, the attack angle of cavitator etc. The model contains a relatively simple system of differential equations that does not require a lot of computational resources. Experimental studies have confirmed the sufficient accuracy of this approach.

## References

[1] Zhuravlev Y.F. 1973 Methods of perturbation theory in the flows with free surfaces. *J. of* TsAGI, 1532, (in Russian)

[2] Grumondz V.T., Zhuravlev Y.F., Paryshev E.V., Sokolyansky V.P., Shorygin O.P. 2013 Hydrodynamics and dynamics of bodies high-speed motion in liquid (Nauka, Moscow) (in Russian)

[3] Logvinovich G.V. 1969 Hydrodynamics of flow with free surfaces (Kiev, Naukova dumka) (in Russian)

[4] Buyvol G.V. 1980 Thin cavities in flows with perturbations (Kiev, Naukova dumka) (in Russian)

[5] Paryshev E.V. 2006 Approximate mathematical models in high speed hydrodynamics J. of Engineering Mathematics, 55, 41–64

Journal of Physics: Conference Series **656** (2015) 012174 doi:10.1088/1742-6596/656/1/012174

# On the flow structure of cloud cavitating flow around an axisymmetric body near the free surface

**Yiwei Wang*, Xiaocui Wu, Chenguang Huang, XianXian Yu**

Key Laboratory for Mechanics in Fluid Solid Coupling Systems, Institute of Mechanics, Chinese Academy of Sciences

No.15 Beisihuanxi Road, Beijing, China, 100190

***EMAIL:** wangyw@imech.ac.cn

**Abstract**. The influence of the free surface on the cavitating flow is an important issue involved in the design of high speed surface vehicles. In the present paper, unsteady cavitating turbulent flow around an axisymmetric body near the free surface was investigated by both launching experiment and LES simulation. The vortex motion induced by cavity shedding under the effect of the free surface is emphatically analyzed by comparing with the submerged condition. The vortex shedding process around the projectile is not synchronized, while the asymmetric characteristic in collapse process is more remarkable, with the generation of multiple vortex ring structures.

## 1. Introduction

Characteristics of cavitating flow, especially unstable structures of cloud cavitating flow, always have close relationship with the vortex motion [1]. Therefore, the turbulence resolution method plays a fatal role on the establishment of numerical approach on unsteady cavitating flow. On the simulation for engineering applications, solving Reynolds-averaged Navier–Stokes (RANS) equations and turbulence models with physical modifications is the main way. Approaches such as modified RNG k-ε turbulence model [2-6] are widely used. In the recent years, large eddy simulation (LES) method is adopted for the research on cavitating flows, which can capture considerable details of large-scale eddies in the flow field with high accuracy [7-14]. For example, an implicit large eddy simulation method was adopted to simulate dynamic cavitation behavior of a propeller by Bensow et al [7]. The cavitation shedding horse-shoe structures were captured numerically by Ji et al [8, 14].

The influence of the free surface on the cavitating flow is an important issue involved in the design of high speed surface vehicles, but relative investigations are very limited. Fatingsen et al [15] established a numerical approach based on theoretical models on the supercavitating flow near the free surface. Bal et al presented boundary element methods (BEM) [16-18] for cavitating hydrofoils near a free surface. Other similar works are mainly performed concerning stable supercavitating flow structures [19, 20]. The influence of the free surface on the unsteady cloud cavitating flow is very lacking in literature.

The present paper focuses mainly on the influence of the free surface on the flow structure of the cloud cavitating flow. Typical evolutions of nonaxisymmetrical cavity shapes around a slender body near the free surface are obtained both by LES method and a launching experiment. Numerical methods are validated by comparing results with underwater launching experiments. The vortex motion induced by cavity shedding under the effect of the free surface is emphatically analyzed by comparing with the submerged condition.

## 2. Experimental setup

The launching system is established on the basis of the SHPB technology [21]. It can transiently accelerate the projectile with slight disturbance on surrounding water. The projectile used in this study is a slender cylinder with a conical head. Its material is polished stainless steel. The total length is 246 mm, and the diameter is 37 mm. The conical angle is 90°. Typical photograph of cavitation can be obtained using a high-speed camera with 25000 fps (as shown in Fig.1). The cavitation number is 0.62, while the projectile speed is 17.8 m/s.

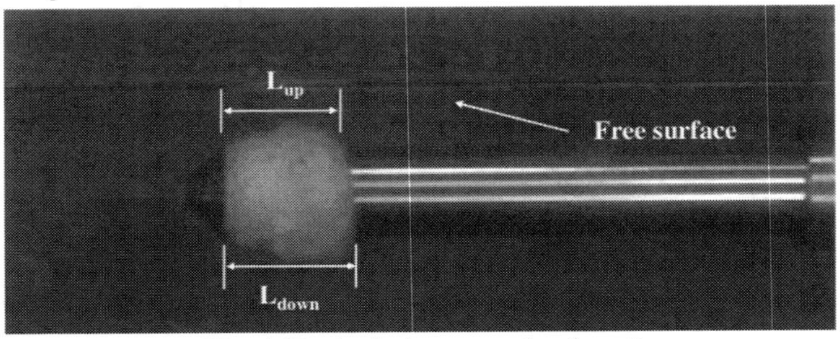


Fig. 1 Typical photograph of cavity

## 3. Numerical methods

Mixture/multiphase flow equations are adopted to simulate the motions of liquid water and vapor and air including the phase change. The mass transfer rate of evaporation and condensation are derived from the bubble dynamic equations of generalized Rayleigh-Plesset equation by Zwart et al.

The governing equations are solved by a LES approach based on Smagorinsky-Lilly model. Unsteady numerical simulations are performed on the basis of finite volume method with coupled scheme by using the commercial CFD software ANSYS-FLUENT. The volume of fluid (VOF) method and the modified high resolution interface capturing (HRIC) scheme are used to capture the interfaces of the free surface and cavity profile. The mesh used in the present research has approximately 4 million cells, which is similar with the reference [13]. The independence analysis and validation of similar grid have been performed on simulating typical evolution of cloud cavitating flow.

## 4. Results and discussion

### *4.1. Cavity evolution*

An additional case without the free surface is also performed. Typical evolution photograghs in experimental and numerical results are as shown in Fig.2. Quasi-periodical variations of cavity shapse are obtained**.** Typical evolutions of cavity includes growth, re-entry jet development, cavity shedding and collaspe, which are similar for both the process considering and neglecting the effect of the free surface. The cavity shapes affected by the free surface are non-axisymmetrical. The re-entry jet in the upper part of the cavity is generated earlier and has a slightly faster velocity (as shown in Fig.2(a)). When the re-entry jets arrive at the shoulder of the model, the cavity on the lower side is a little longer than the upper side (as shown in Fig.2(b)). The cavity sheds by the interaction between main flow and re-entry jet, and the cavity on the upper side sheds earlier (as shown in Fig.2(c)). The shedding cavity on the lower side is farer away from the main cavity, which collapses more strongly (as shown in Fig.2(d)). Cavity shapes on the lower side are similar with the case without the free surface, but cavity in the latter case is longer. Because the cavity shape is integral and like a ring, the development on the lower side is hindered by the upper side.

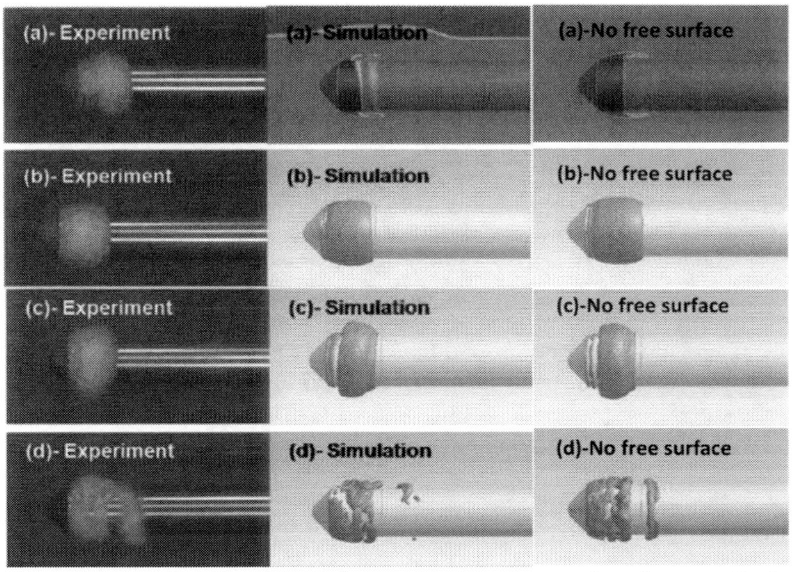


Fig. 2 Comparison of cavity shapes between experimental and numerical results

### 4.2. *Influence on shedding vortex structures*

The motion of shedding cavitation clouds has strong correlation with the vortex motion. The flow structures can be visualized based on the Q-criterion. In the present study, the iso-surface of Q=50000 is shown to represent the vortex core, and the colour represents the magnitude of the velocity component in the axial direction.

Vortex structures in the shedding process are as shown in Fig.3. For the case without the free surface in the right views, vortexes are rolled outwards gradually, and the upstream parts begin to break apart at first under the influence of the main flow. For the case with the free surface in the left views, the shedding process around the projectile is not synchronized, and the shedding vortex is inclined. The broken vortex is thinner on the upper side under the influence of the free surface.

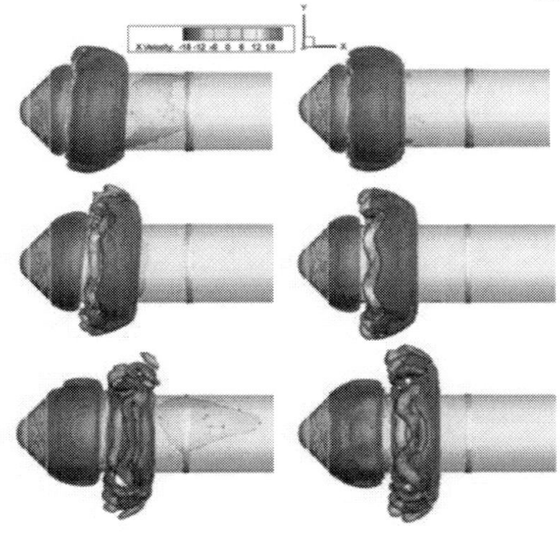

Fig. 3 Vortex structures in the cavity shedding process (Q=50000, left-with free surface, right-without free surface)

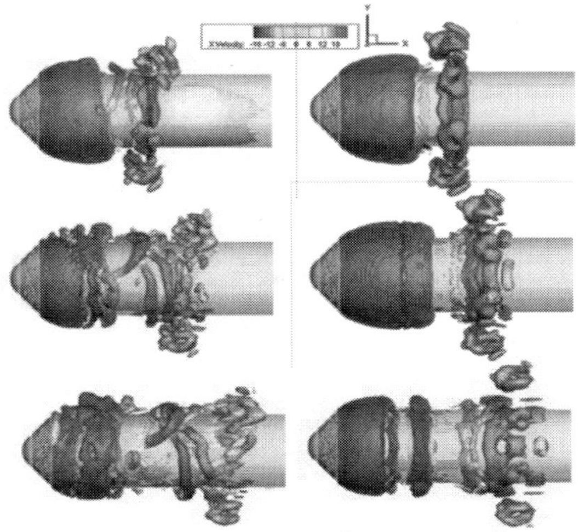

Fig. 4 Vortex structures in the cavity collapse process (Q=50000, left-with free surface, right-without free surface)

The shedding vortex breaks up strongly induced by the collapse of the shedding cavity (as shown in Fig.4), which exhibits more remarkable asymmetry during this period under the influence of the free surface. A multiple vortex ring structure is formed after cavity collapse, inducing significant non-uniform distributions of velocity and pressure, which may cause large lateral force and affect the stability of cruising.

## 5. Conclusions

Unsteady cavitating turbulent flow around an axisymmetric body near the free surface was investigated by both the launching experiment and the LES simulation. The evolution of flow structures is mainly analyzed, and several conclusions can be drawn as follows:

(1) On the upper side near the free surface, the cavity shape and cycle time are both shorter, while cavity shapes on the lower side are similar with the case without the free surface.

(2) The vortex shedding process around the projectile is not synchronized, and the broken vortex is thinner on the upper side under the influence of the free surface.

(3) The vortex collapse process is strongly asymmetric, and a multiple vortex ring structure is formed after cavity collapse. The remarkable asymmetry may cause large lateral force and affect the stability of cruising.

## Acknowledgements

The authors are grateful to the national natural science foundation of china through grant numbers 11202215&11332011 and the Youth Innovation Promotion Association of CAS (2015015).

## References

[1] Arndt R E A 2002 *Annu. Rev. Fluid Mech.* **34** 143-175
[2] Coutier-Delgosha O, Stutz B, Vabre A, Legoupil S 2007 *J. Fluid Mech.* **578** 171-122
[3] Coutier-Delgosha O, Fortes-Patella R, Reboud JL 2003 *ASME J. Fluids Eng.* **125**(1) 38-45
[4] Coutier-Delgosha O, Deniset Fo, Astolfi JA, Leroux J-B 2007 *ASME J. Fluids Eng.* **129**(3) 279-292
[5] Zhou L, Wang Z 2008 *ASME J. Fluids Eng.* **130**(1) 011302
[6] Coutier-Delgosha O, Reboud J, Delannoy Y 2003. *Int. J. Numer. Meth. Fl.* **42**(5):527-548
[7] Bensow RE, Bark G 2010 *ASME J. Fluids Eng.* **132**(4) 041302.
[8] Ji B, Luo XW, Arndt REA, Peng X, Wu Y 2015 *Int. J. Multiphas. Flow* **68** 121-134
[9] Wang G, Ostoja-Starzewski M 2007 *Appl. Math. Model.* **31**(3) 417-447
[10] Huang B, Zhao Y, Wang G 2014 *Comput. Fluids* **92** 113-124
[11] Dittakavi N, Chunekar A, Frankel S 2010 *ASME J. Fluids Eng.* **132**(12) 121301
[12] Roohi E, Zahiri AP, Passandideh-Fard M 2013. *Appl. Math. Model.* **37**(9) 6469-6488
[13] Yu X, Huang C, Du T, Liao L, Wu X, et al 2014 *ASME J. Fluids Eng.* **136**(5) 051303
[14] Ji B, Luo XW , Peng X X, Wu Y L 2013 *J. Hydrodyn.* **25**(4) 510-519
[15] Faltinsen OM, Semenov YA 2008 *J. Fluid Mech.* **597** 371-394
[16] Bal S, Kinnas SA 2002 *Comput. Mech.* **28**(304) 260-274
[17] Bal S 2011 *J. Mar. Sci. Tech. Japan* **16**(2) 129-142
[18] Bal S 2007 *Ocean Eng.* **34** 1935-1946
[19] Amromin E 2007. *Ocean Eng.* **34** 1602-1606
[20] Chen X, Lu C J, Li J, Chen Y 2011. *J. Hydrodyn.* 23(6) 730-736
[21] Wei Y P, Wang Y W, Fang X, Huang C G, Duan Z P 2011 *Chin. Phys. Lett.* **28** 024601

# An experimental study on characteristics of cavitation and ballistic of axisymmetric slender body underwater movement

**Cheng-Gong Zhao, Cong Wang, Ying-Jie Wei, Xiao-Shi Zhang**

School of Astronautics, Harbin Institute of Technology, Harbin 150001, China

E-mail: cgzhhit@gmail.com

**Abstract**. An experimental study of the axisymmetric slender body underwater movement was conducted using high-speed photography technology. From the results of the experiment, the characteristics of cavitation and ballistic of the axisymmetric, including the formation, development, evolution and collapse of the cavity, are presented in the paper. The experimental results show that the axisymmetric slender body moves in a supercavity, and the slender body rotate in the supercavity on its head at the same time due to the perturbation of launching. The supercavity wall is transparent and smooth except the tail itself. The impact between the tail of slender body and supercavity wall resulted from the slender body's rotation is termed as tail-slap which is one way to keep the stabilization of the movement. Series of different flow mechanisms and the relationship between ballistic characteristics and cavity characteristics with defferent initial velocities are discussed. The slender bodies have different accelerations and ballistics with different initial velocity which means they have different drag forces.

## 1. Introduction

Cavitation occurs in a variety of high-speed hydrodynamic systems where the absolute local pressure controlled by the flow hydrodynamics drops below the liquid saturation pressure. Cavitation is usually harmful for hydraulic machinery because it may cause material damage and force surge, etc. However, the recent research results show that the supercavitation is beneficial to reduce the skin-friction drag of the slender body travelling underwater which in turn enlarge the travelling range of the slender body [1]. The supercavitating hydrodynamics have great influence on the ballistic characteristics and structure dynamics of slender body [2].

The early research of the supercavitating problems focus on the derived analytical form of supercavitation. P. R. Garabedian studied the calculation of axisymmetric cavities [3]. A study by May focused on the variation of cavities during water entry of slender body [4-6]. Studies by Ruzzene and Edward[7,8] focused on the static and dynamic analysis of projectiles underwater movement. While Hrubes captured the high-speed supercavitating images using high-speed imaging technology [9]. In addition, Fabien Petitpas carried out the experiment of spherical projectiles underwater movement [10].

This paper describes design, setup and testing of a laboratory experiment for studying the cavity and ballistic characteristics of a high-speed free-flying slender body underwater. Moreover the paper compares the differences of cavity and ballistic characteristics with the different initial cavitation number $\sigma$ ($\sigma = \dfrac{2(p-p_v)}{\rho v^2}$, where $p$ is the reference hydrostatic pressure, $p_v$ is the vapor pressure, and $\rho$ is the mass density of the fluid ).

## 2. Experimental setup and model of slender body

The schematic of the experimental facility of the slender body underwater movement is shown in Figure 1. The experimental facility consists of four parts: (1) a horizontal launch system; (2) water tank; (3) protection and recycle system; (4) measurement system. The water tank consists of opening top and stalinite on both sides and steel plates in rest and the scale of 1.2m× 1.2m× 10m. The launch system is composed of a vessel within compressed nitrogen, light gas gun and trigger. The measurement system is a *FASTCAM SA6* high speed camera illuminated by 6× 1300W halogen lamps.

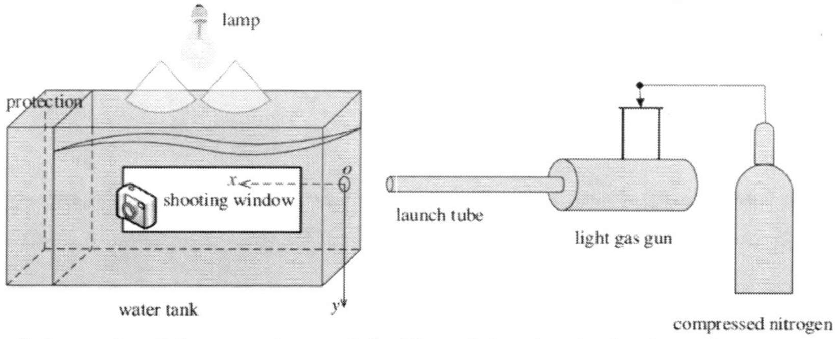


Figure 1. Schematic of the experimental facility of the slender body underwater movement

Figure 2 shows the experimental facility and model of slender body with its adapter. The model of slender body is combination of truncated cone and cylinder afterbody. The diameters of the flat head and afterbody cylinder are *d* =4mm,respectively, and the length of the slender body is 170mm. The semi-cone angle is 4 degrees. The material of the slender body is steel. The dimension of the slender body is as shown in Figure 3.

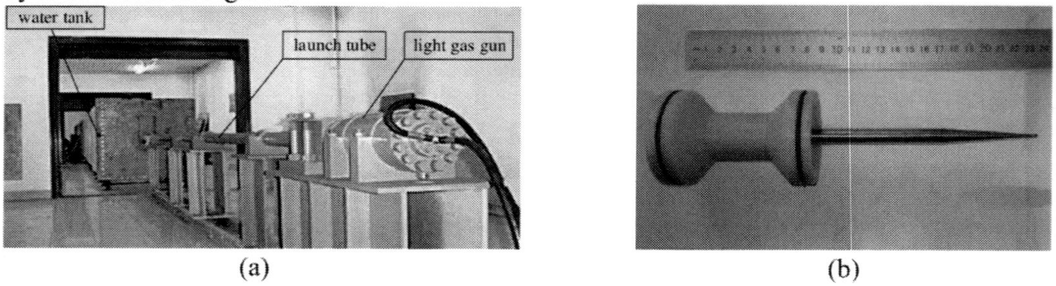


Figure 2. (a) Experimental facility (b) Model of slender body and adapter

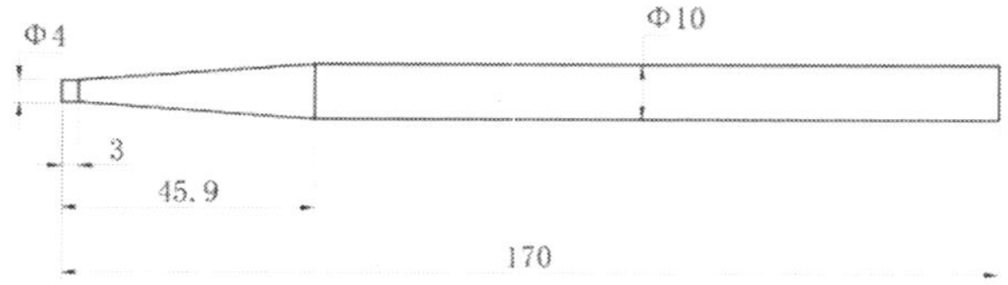


Figure 3. Dimension of the slender body

The slender body loaded in the adapter was fired horizontally into the water tank. Concurrently, the high speed camera was recording the supercavitation of slender body and its trajectory. We obtained different initial velocity and cavitation number by adjusting the pressure in the chamber of the light gas gun.

## 3. Experiment results and discussions

The coordinate system *x-O-y* used in the paper is shown in Figure 1. The *x-axis* is in direction of horizontal and *y-axis* is in direction of vertical.

### *3.1. Cavity characteristics*

Figure 4(a) shows the supercavitation of the slender body at the initial cavitation number $\sigma$=0.0044 ($v$ =212.1m/s). The supercavitation develops from the edge of the flat head of the slender body and expands along the trajectory of the flat head's centre. The wall of the supercavitation is smooth and transparent except for the contact area which can be attributable to the rotation of the slender body in the supercavitation. Figure 4(b) shows the details of the wall of supercavitation.

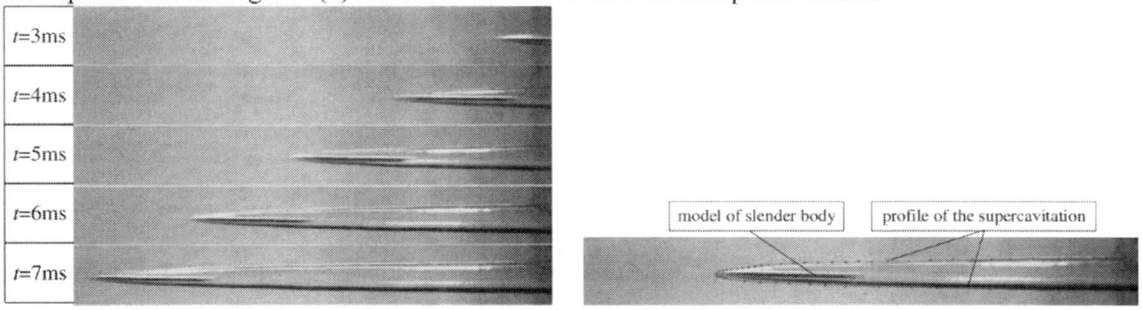


(a) (b)

Figure 4. (a) Characteristics of cavity at time $t$ =3ms-7ms (b) Profile of the supercavitation

Figure 5 shows the closure and shedding of the cavity at the initial cavitation number $\sigma$=0.0044 ($v$ =212.1m/s) at different times. The water vapor in the afterbody of the supercavitation is condensed at some extent by the high pressure caused by the severe re-entrant jet. Observe in Figure 5 that, there is mixture of water vapor and water in the tail of the supercavitation. The shedding of the vapor-water mixture produces the vortices which disappear in the flow field downstream. The number of shedding vortices decreases over time from $t$ =18ms-22ms in Figure 5 and the vibration decreases too. The decrease of the number of shedding vortices is due to the decrease of evaporation rate caused by the reduction of the slender body's velocity.

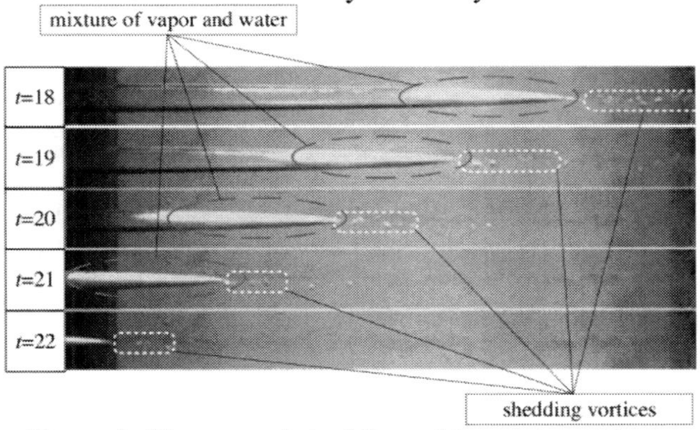


Figure 5. Closure and shedding of the cavity at time $t$ =18ms-22ms

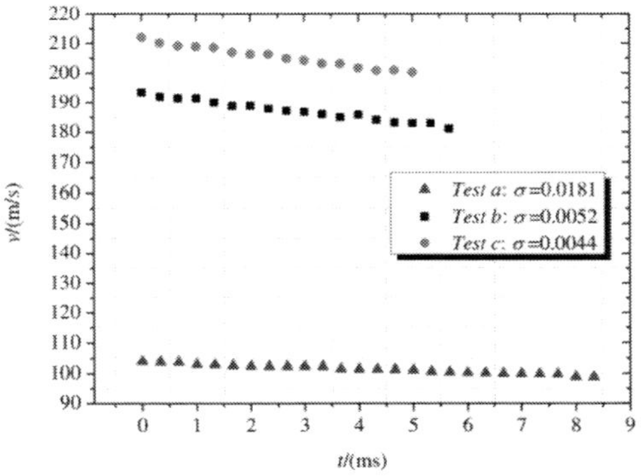


Figure 6. The variations of slender bodies with different initial cavitation numbers

### *3.2. Ballistic characteristics*

Three tests are conducted in the paper which have different initial cavitation number. Marked as follows:

*Test a*: the initial velocity of the slender body is 104.1m/s, initial cavitation number is 0.0181;

*Test b*: the initial velocity of the slender body is 193.7m/s, initial cavitation number is 0.0052;

*Test c*: the initial velocity of the slender body is 212.1m/s, initial cavitation number is 0.0044.

The variations of the velocity of slender body versus time with different initial cavitation number are shown in Figure 6. As can be seen in Figure 6, the decay rates of slender bodies with different initial cavitation number are different. The decay rates of *test b* and *test c* are high than that of *test a* because of the differences in initial cavitation numbers.

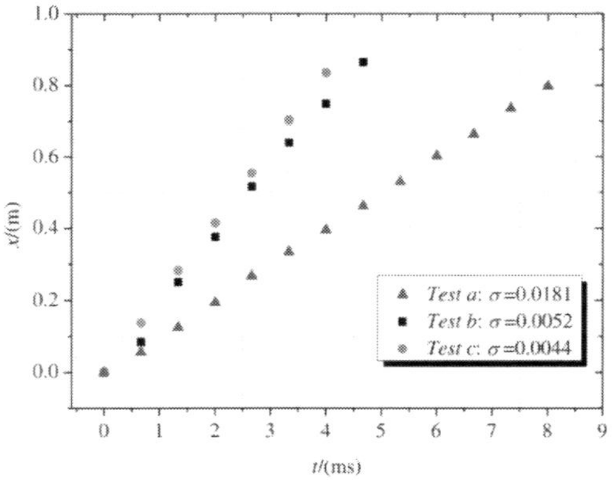


Figure 7. Trajectories in $x$ direction versus time with different initial velocities

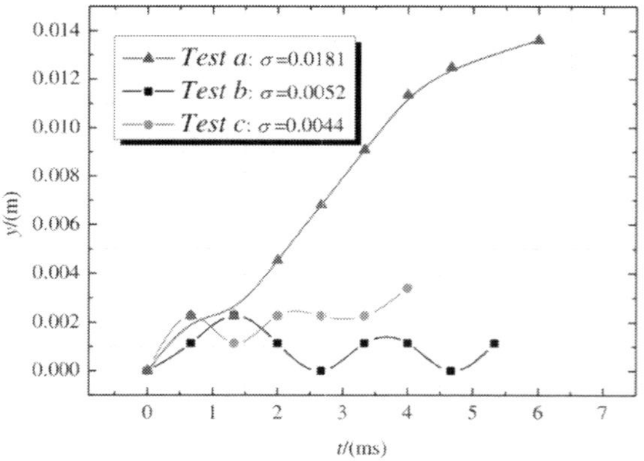


Figure 8. Trajectories in $y$ direction versus time with different initial velocities

Figure 7 and 8 show the displacements of the slender bodies in *x-direction* and *y-direction*, respectively. The effect of gravity is negligible because the initial force is greater than gravity when the slender body travel in a high-speed.

## 4. Conclusions

The experiment of supercavitating slender body travelling underwater have been presented in this paper. Section 2 focused on the setup of the experiment and model of the slender body. Section 3 presented the results of the experiment, including closure and shedding of cavity, and ballistic variation of slender body. A discussion of results shows that there is a high-pressure area in the tail of the supercavity due to the severe re-entrant jet of the supercavity. In addition, the shedding of the vapour-water mixture produces the vortices which disappear in the flow field downstream.

**Acknowledgements**

The authors would like to acknowledge the financial support of the Fundamental Research Funds for the Central Universities of Ministry of Education of China (Grant No. HIT.NSRIF.201159)

## References

[1] Savchenko Y N 2001 *VKI Special Course on Supercavitating Flows* (Brussels: St. Joseph Ottawa/Hull) p313
[2] May A 1975 AD-A020 429
[3] Garabedian P R 1955 *Pac. J. Math* **6** 611
[4] May A, Woodhull J C 1948 *J. Appl. Phys.* **19** 1109
[5] May A 1951 *J. Appl. Phys.* **22** 1219
[6] May A 1979 *J. Hydrodynamics* **4** 140
[7] Ruzzene M 2004 *Comput. Struct.* **82** 257
[8] Alyanak E, Venkayya V, Grandhi R, Penmetsa R 2005 *Finite Elem. Anal. Des.* **41** 563
[9] Hrubes JD 2001 *Experiments in Fluids* **30** 57
[10] Fabien P, Jacques M, Richard S, Emmanuel L, Laurent M 2009 *International Journal of Multiphase Flow* **35** 747

# A model for water-entry projectile with an open cavity

**W Zou and H Liu**
MOE Key Laboratory of Hydrodynamics, Department of Engineering Mechanics, Shanghai Jiao Tong University, Shanghai, China

E-mail: hopingzou@sjtu.edu.cn, hliu@sjtu.edu.cn

**Abstract.** Study on dynamics of a projectile running with a cavity in the process of water entry closely relates to the development of new generation of air-to-water vehicle. It becomes a key concern to develop a flexible model of the high-speed water-entry projectile. The projectile is modelled using the unsteady cavity model from Serebryakov in the frame of the rigid-body dynamics theory. Based on the model, without loss of generality, oblique water-entry projectile running with an open cavity is simulated, and numerical results are in good agreement with measured data available. The model proves effective and has a great potential to investigate ballistics of an air-dropped vehicle at the early stage of water entry.

## 1. Introduction

Water entry has come into researchers' sight for over 100 years of history as a classical problem, and has made some great achievements [1]. Because supercavitation drag reduction technique has drawn more and more attention, water-entry projectile at high speed always becomes a research hotspot for improving and developing air-to-water vehicle.

Some very valuable images of supercavitating projectiles [2] were obtained at the speed of sound for analysis of projectile characteristics such as flight behaviour, cavity shape, and stability mechanism. The excellent findings confirmed that projectile can lean against cavity wall to produce planning force as a stable motion mode similarly to underwater supercavitating vehicle [3]. Compared with horizontal water entry, more extensive attention focuses on water entry of air-dropped object in high-speed motion. Vertical water entry was modelled by Aristoff [4], and cavity shape and pinch-off time are formulated for cavity evolution dominated by aerodynamic pressure. Cavity model based on Logvinovich's principle [5] is also a powerful approach to study vertical water entry [6], and besides, the algorithm [7] was developed to solve Navier-Stokes equations for subsonic and supersonic projectile. It is easy to understand that oblique water entry is a more general problem. Truscott et al. [8] conducted some beneficial work, providing design foundation for oblique water-entry projectile to avoid ricochet and improving Logvinovich's model.

Although great progress has been made, trajectory characteristic, motion stability and control of high-speed projectile remain to study further, and restrict its applications, i.e., developing intelligent air-to-water vehicle. It is urgent to establish a model of high-speed water-entry projectile applied to projectile's dynamics and control effectively. Hence, the projectile is modelled using Serebryakov's unsteady cavity model [9] and the rigid-body dynamics theory. Based on the model, oblique water-entry projectile is simulated at high speed before cavity closures, and the model is validated by the measured data available [10].

## 2. Water-entry projectile model

For a projectile entering water at high speed, a cavity appears around the projectile due to violent deformation of the free surface. The mathematical formulation of projectile consists of the cavity model and the dynamics model of a projectile.

### 2.1. *Cavity model*

Considered from the Lagrangian coordinate system, cavity section is reasonably assumed to expand radially in the plane perpendicular to trajectory in water entry. In such a case, cavity is merely open to atmosphere, and has a pressure drop owing to airflow following projectile. The unsteady cavity model [9] derived by Serebryakov, validated by lots of experiments, is applied to describe cavity evolution. The model is known as a modern presentation of Logvinovich's principle with very clear physical meaning as follows:

$$\mu(s)\frac{\partial^2 R_c^2(s,t)}{\partial t^2}+\frac{2\Delta p(s,t)}{\rho_w}=0 \tag{1}$$

$$R_c^2(s,t)\Big|_{t=t_n(s)}=R_n^2(s),\qquad \frac{\partial R_c^2(s,t)}{\partial t}\Big|_{t=t_n(s)}=R_n(s)V(s)\sqrt{\frac{2(C_d(s)-k(s)\sigma_c(s))}{k(s)\mu(s)}} \tag{2}$$

where $R_c(s,t)$ and $R_n(s)$ are the radii of cavity and cavitator on the trajectory $s$; $V(s)$ is the velocity of cavitator; $\rho_w$ is the water density; $\Delta p(s,t)$ is the pressure difference inside and outside cavity, $\Delta p(s,t)=p_\infty(s,t)-p_c(s,t)$, $p_\infty(s,t)$ is the ambient pressure, $p_c(s,t)$ is the internal pressure under the influence of aerodynamic pressure [4], $p_c(s)=p_a-\mathbf{min}(C_a\rho_a V_0^2, p_a)$, $C_a$ is the coefficient of aerodynamic pressure, taken as $C_a=2.5$, $p_a$ is the atmospheric pressure, $V_0$ is the initial velocity of water entry, $\rho_a$ is the air density; $t_n(s)$ is the time when cavitation occurs; $C_d(s)$ is the drag coefficient; $\sigma_c(s)$ is the cavitation number; $k(s)$ and $\mu(s)$ are the parameters for assessments of cavity dimensions.

### 2.2. *Dynamics model of projectile*

Water-entry projectile is subject to forces on cavitator, gravity and planing force [2] for motion stability. Paryshev's planing model [11] is applied because it better agrees with experimental data. Importantly, ricochet may occur if geometry, speed and entry angle of projectile does not satisfy some criterions [8]. The case without ricochet, which is put down to reasonable designs, is discussed here.

Because projectile is generally an axisymmetric body, water-entry problem can be simplified into two-dimensional one with respect to the longitudinal plane of projectile. Considering that cavitator movement represents projectile's trajectory, it is a good strategy to build the body-fixed frame **B** at cavitator. All forces must be transformed into ones in this system by coordinate transformations. According to Newton's second law of motion and theorem of momentum moment in the inertial frame **E**, dynamics equations are written:

$$m(\dot{u}+wq-q^2x_g)=F_x \tag{3}$$

$$m(\dot{w}-uq-\dot{q}x_g)=F_z \tag{4}$$

$$I\dot{q}+m(-x_g(\dot{w}-uq)+x_g^2\dot{q})=M_y \tag{5}$$

where $u$ and $w$ are the velocity components of projectile in the longitudinal axis and its vertical axis downward; $q$ is the angular velocity in the longitudinal plane; $x_g$ is the coordinate of center of mass (**CM**) in the longitudinal axis; $I$ is the moment of inertia about the axis perpendicular to the plane; $F_x$, $F_z$ and $M_y$ are the resultant forces and moment in their respective directions, and are composed of forces on cavitator, gravity and tail wetted region, i.e., $F_x = F_{xc} + F_{xg}$, $F_z = F_{zc} + F_{zg} + F_{zp}$, $M_y = M_{yg} + M_{yp}$.

$$F_{xc} = -D_c(\mathbf{C}\delta_c\mathbf{C}\alpha_c + \mathbf{S}\delta_c\mathbf{S}\alpha_c) + L_c(\mathbf{C}\delta_c\mathbf{S}\alpha_c - \mathbf{S}\delta_c\mathbf{C}\alpha_c) \tag{6}$$

$$F_{zc} = D_c(\mathbf{S}\delta_c\mathbf{C}\alpha_c - \mathbf{C}\delta_c\mathbf{S}\alpha_c) - L_c(\mathbf{S}\delta_c\mathbf{S}\alpha_c + \mathbf{C}\delta_c\mathbf{C}\alpha_c) \tag{7}$$

$$F_{xg} = -mg\mathbf{S}\theta; \quad F_{zg} = mg\mathbf{C}\theta; \quad M_{yg} = -mgx_g\mathbf{C}\theta \tag{8}$$

$$F_{zp} = \pi\rho_w R_b^2 V^2 \mathbf{S}\alpha_p \mathbf{C}\alpha_p (1 - \frac{(R_c - R_b)^2}{(h_0 + R_c - R_b)^2}); \quad M_{yp} = -F_{zp}x_p \tag{9}$$

where $F_{xc}$ and $F_{zc}$ are the force components of cavitator in the longitudinal axis and its vertical axis; $F_{xg}$, $F_{zg}$ and $M_{yg}$ are the force components and moment of gravity; $F_{zp}$ and $M_{yp}$ are the planing force and moment; $D_c$ and $L_c$ are the drag and lift in the velocity frame of cavitator, $D_c = 0.5\rho_w V^2 S_n C_d \mathbf{C}^2\alpha_c$, $L_c = 0.5\rho_w V^2 S_n C_d \mathbf{C}\alpha_c \mathbf{S}\alpha_c$; $\delta_c$ is the deflection angle of cavitator; $m$ is the mass of projectile; $R_b$ is the radius of projectile afterbody; $h_0$ is the immersion depth in the tail; $\alpha_p$ is the angle between projectile axis and tangent of cavity axis in the tail; $x_p$ is the coordinate of acting point of planing force in the longitudinal axis.

Then according to coordinate transformation between the frame **B** and inertial frame **E**, the projectile trajectory is in the inertial frame:

$$\dot{x}_e = u\mathbf{C}\theta + w\mathbf{S}\theta; \quad \dot{z}_e = -u\mathbf{S}\theta + w\mathbf{C}\theta \tag{10}$$

where $x_e$ and $z_e$ are the trajectory coordinates; $\theta$ is the pitch angle, $\dot{\theta} = q$.

## 3. Validation of water-entry projectile model

Based on the model, oblique water-entry projectile running with an open cavity is simulated as a general case to compare with results from the experiment [10] in figure 1 to validate the model using our self-developed algorithm [3]. The cylindrical projectile model has the truncated cone-shaped head and entry angle $\theta_c = 18^0$ to prevent ricochet [8].

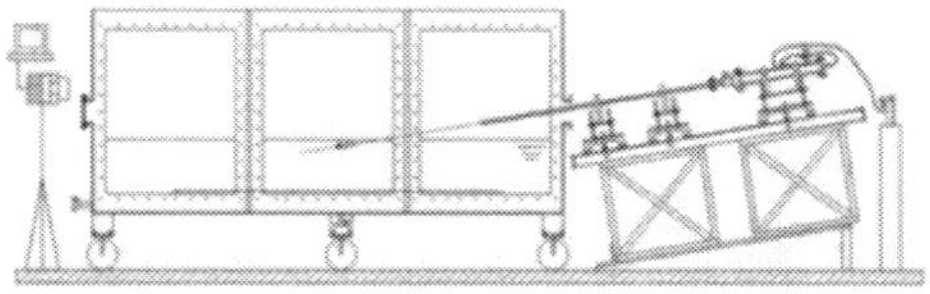

**Figure 1.** Schematic of experimental equipment [10]

Numerical results coincide well with experiment data, as is shown in figure 2, where $x_e$ and $z_e$ denote the horizontal and vertical directions. Cavity gradually elongates before closure, and its trajectory is rather stable and almost a straight line because of high speed and reasonable entry angle. Also, figure 3 shows the path lengths of head and **CM** of projectile in water, where $\bar{T}$ and $\bar{S}$ are the dimensionless time and path length, $\bar{T} = t/T_0$, $\bar{S} = s/(V_0T_0)$, $T_0$ is the total time during water entry, $T_0 = 0.01\text{s}$. The results are supported by experimental ones in the early stage of water entry, but as time goes on, some deviations display. One important reason accounts for it that viscous friction, which indeed accelerates speed attenuation and decreases path length, is not considered in the planing region. On the other hand, it is also a factor to determine a reasonable coefficient of aerodynamic pressure. On the whole, these results from the model agree with experimental data.

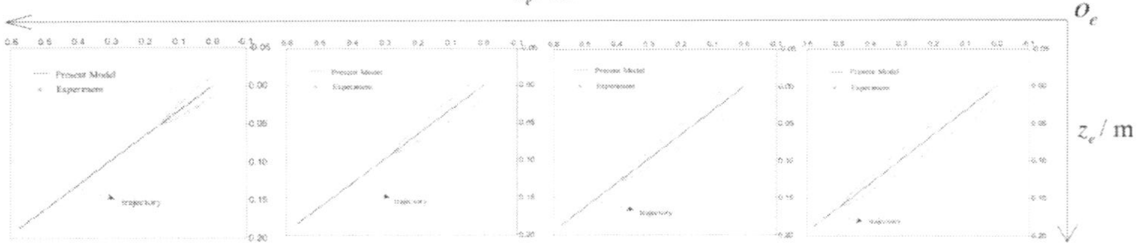


a – $t$ = 0.002s b – $t$ = 0.004s c – $t$ = 0.006s d – $t$ = 0.008s

**Figure 2.** Evolution of the cavity induced by a water-entry projectile ( $V_0 = 82.76\text{m}/\text{s}$ )

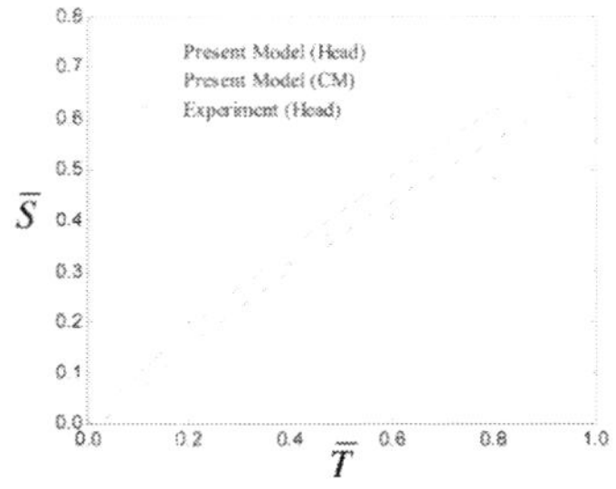


**Figure 3.** Path lengths of projectile in water

## 4. Conclusion

The model of water-entry projectile with an open cavity is proposed, and is validated by the experiment available. In our further work, viscous force in the planing region will be focused to make the proposed model more precise, and then will combine our existing supercavity control algorithm [12] to develop intelligent air-to-water vehicle at high speed.

**Acknowledgement**

This work is supported by National Natural Science Foundation of China under Grant No. 10972138 and State Key Laboratory of Ocean Engineering (SJTU, Grant No. GKZD010063).

## References

[1] Truscott T T, Epps B P and Belden J 2014 *Annu. Rev. Fluid Mech.* 46 24
[2] Hrubes J D 2001 *Exp. Fluids* 30 8
[3] Zou W, Yu K P and Arndt R *App. Math. Modell.* DOI: 10.1016/j.apm.2015.01.040, in press
[4] Aristoff J M 2009 *Proc. 7th Int. Symp. on Cavitation* 125
[5] Logvinovich G V 1969 *Hydrodynamics of Flows with Free Boundaries* (in Russian) (Kiev: Naukova Dumka)
[6] White E R and Miller T F 2010 *J. Fluids Eng.* 132 7
[7] Neaves M D and Edwards J R 2005 *J. Fluids Eng.* 128 13
[8] Truscott T T, Beal D N and Techet A H 2009 *Proc. 7th Int. Symp. on Cavitation* 100
[9] Serebryakov V V 2009 *Proc. 7th Int. Symp. on Cavitation* 169
[10] Hu Q Q 2014 *Experimental Observation and Numerical Calculation of Supercavity Flow of Blunt Body under Different Inclination Angle into the Water* (in Chinese) (Zhejiang Sci-Tech University: MS thesis)
[11] Paryshev E 2003 *Proc. 7th Int. Symp. on Cavitation* Cav03-OS-7-014
[12] Zou H, Liu H 2015 *Ocean Eng.* 101 9

# A cavity-mediated collective quantum effect in sonoluminescing bubbles

**Almut Beige and Oleg Kim**
The School of Physics and Astronomy, University of Leeds, Leeds LS2 9JT, United Kingdom
E-mail: a.beige@leeds.ac.uk

**Abstract.** This paper discusses a collective quantum effect which might play an important role in sonoluminescence experiments. We suggest that it occurs during the final stages of the collapse phase and enhances the heating of the particles inside the bubble.

## 1. Introduction

Sonoluminescence is the intriguing phenomenon of strong light flashes from tiny bubbles in a liquid [1]. The bubbles are driven by an ultrasonic wave and need to be filled with atomic species. Fig. 1 shows a typical single-bubble sonoluminescence cycle [1, 2]. For most of the cycle, the bubble radius increases isothermal and is in good agreement with the laws of classical physics [3]. However, each expansion phase is followed by a very rapid collapse phase during which the bubble becomes thermally isolated from the liquid. Close to its minimum radius of about 0.5 $\mu$m, a rapid heating of the particles inside the bubble occurs which is accompanied by the sudden emission of light. Afterwards a re-expansion phase begins in which the bubble oscillates around its equilibrium radius until it regains its stability.

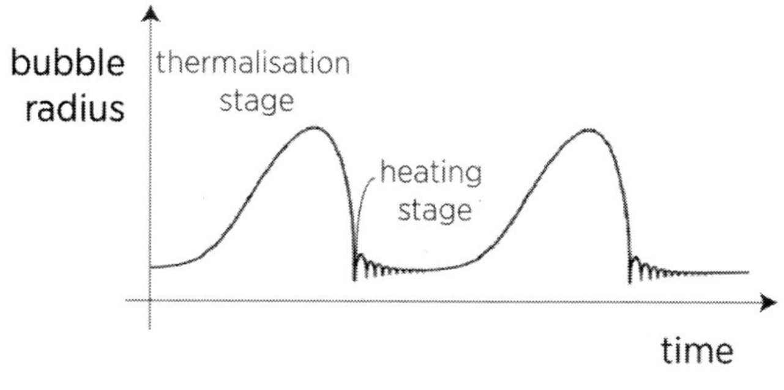


**Figure 1.** Schematic view of the time dependence of the bubble radius in a typical single-bubble sonoluminescence experiment. Most of the time, the bubble radius increases isothermal (thermalisation stage). During the collapse phase (heating stage), the temperature within the bubble increases rapidly and a strong light flash occurs.

Measuring the spectrum of the picosecond light flash and associating it with blackbody or *Bremsstrahlung* radiation indicates temperatures of at least $10^3$–$10^4$ K inside the bubble [4, 5, 6, 7, 8, 9]. It is even possible to observe light emission in the ultraviolet regime which hints at temperatures of about $10^6$ K [10]. Noteworthy is the discovery of sharp emission lines in the optical regime [1, 8, 11]. These indicate the population of highly excited energy eigenstates of noble gas and metal atoms which cannot be populated thermally but hint at the presence of a dense plasma [8, 11, 12, 13, 14, 15, 16].

Although sonoluminescence has been studied extensively, the origin of the sudden energy concentration during the final stage of the bubble collapse phase remains a mystery [8, 17]. A valid theoretical model needs to be based on the formation of a plasma and needs to include a mechanism which can increase the temperature of this plasma by at least one order of magnitude. It needs to be able to operate in a solid state-like environment and on very short time scales. Moreover, the proposed heating mechanism needs to depend strongly on the atomic species inside the cavitating bubble. For example, single bubble sonoluminescence experiments with ionic liquids reach significantly higher temperatures than experiments based on noble gas atoms. The purpose of this paper is to emphasise that such a heating mechanism could be based on a recently-identified cavity-mediated collective quantum effect [18]. More details of its relevance in sonoluminescence experiments can be found in Refs. [19, 20].

More concretely, we assume in the following that the electromagnetic field inside the bubble becomes quantised when the bubble approaches its minimum radius. During the collapse phase, the atomic particles therefore behave as if placed inside a so-called optical resonator. The walls of the cavity are formed by the walls of the collapsing bubble. As in Ref. [21], we moreover quantise the motion of the atomic species inside the bubble at this stage. The atoms are now so strongly-confined that they cannot be compressed any further. Simultaneously, the collapse of the bubble results into high enough temperatures to lead to the formation of a plasma. Once the plasma is formed, collisions create correlations between the electronic and the vibrational states of the atomic particles which can fuel a cavity-mediated collective quantum optical heating process [19, 20]. As in ion trap experiments [22] and in cavity-mediated laser cooling [18, 23], the rapid temperature changes of the confined particles are due a complex interplay between electronic and vibrational degrees of freedom combined with the spontaneous emission of photons. If repeated over many cycles, such a mechanism can have a significant effect.

## 2. A collective effect in a many-body quantum system

Before describing its relevance to sonoluminescence experiments, this section introduces a highly-simplified quantum optical toy-model. Our aim is to illustrate a collective effect in the dynamics of a many-body quantum system which changes some properties of the interacting particles very rapidly. The model is the following: Suppose $N$ particles have two different degrees of freedom which can be described by a two sets of bosonic operators $a_i$ and $b_i$ with $i = 1, ..., N$. Moreover, the particles couple to a common mode which can be modelled by another bosonic operator $c$. Then

$$\left[a_i, a_i^\dagger\right] = \left[b_i, b_i^\dagger\right] = \left[c, c^\dagger\right] \;=\; 1 \,. \tag{1}$$

In addition, suppose there is a direct coupling between the internal degrees $a$ and $b$ of the particles and the common mode $c$ which can be modelled by an interaction Hamiltonian of the form

$$H_\mathrm{I} \;=\; \sum_{i=1}^{N} \hbar\eta g \left(b_i + b_i^\dagger\right)\left(a_i c^\dagger + a_i^\dagger c\right) \,. \tag{2}$$

Here $\eta g$ is a three-body interaction constant which depends on the physical properties of the modelled quantum system. In the following, we use this Hamiltonian to analyse the dynamics of the mean average excitation number $m$ of the $b$-mode given by

$$m \equiv \frac{1}{N} \sum_{i=1}^{N} \left\langle b_i^\dagger b_i \right\rangle . \tag{3}$$

To get a closed set of rate equations, we moreover need to consider the expectation values

$$\begin{aligned} k_1 &\equiv \frac{1}{N} \sum_{i=1}^{N} \left\langle \mathrm{i} \left( b_i - b_i^\dagger \right) \left( a_i c^\dagger + a_i^\dagger c \right) \right\rangle , \\ k_2 &\equiv \frac{1}{N(N-1)} \sum_{i=1}^{N} \sum_{j \neq i} \left\langle a_i b_i^\dagger a_j^\dagger b_j - a_i b_i a_j^\dagger b_j^\dagger \right\rangle . \end{aligned} \tag{4}$$

All three expectation values are real and normalised such that they remain essentially the same in the large $N$ limit, if the number of particles in the trap changes.

Now we can use Schrödingers equation, which implies $\langle \dot{A}_{\rm I} \rangle = -(\mathrm{i}/\hbar) \langle [A_{\rm I}, H_{\rm I}] \rangle$ for the time evolution of the expectation value of any physical observable $A_{\rm I}$ in the interaction picture, to analyse the dynamics of the above introduced variables. Doing so, we find that

$$\dot{m} = \eta g \, k_1 \quad \text{and} \quad \dot{k}_1 = -2N \eta g \, k_2 \tag{5}$$

to a very good approximation. In the time derivative of $k_1$, only terms which scale as the number $N$ of particles involved have been taken into account. If $k_2$ differs from zero and $N \gg 1$, the above differential equations imply that the mean excitation number $m$ of the $b$-mode changes very rapidly in time, namely at a rate which scales as $N$. In other words, we identified a collective effect in the dynamics of a many-body quantum system with three-body interactions.

## 3. Application of the above collective quantum effect to sonoluminescing bubbles

As shown in Refs. [19, 20], when modelling the collapse phase of a sonoluminescing bubble, we use operators analogous to $a$ and $b$ to describe the electronic and the vibrational degrees of freedom of the confined atomic particles, while $c$ refers to the quantised electromagnetic field inside the bubble. When the atoms accumulate in the centre of the bubble during the final stages of the collapse phase, their dynamics is governed by an interaction Hamiltonian which is essentially the same as $H_{\rm I}$ in Eq. (2). Consequently, the mean number of excitations in the $b$-mode can change on a time scale which scales as the number $N$ of atomic particles inside the bubble. The result is a collective change of the mean number of phonons in the system, ie. very rapid heating or cooling. This process is fuelled by a non-zero expectation value $k_2$. Here $k_2$ is expected to arise during the thermalisation phase due to atomic collisions which result in the exchange of electronic and vibrational energy between the particles [19, 20]. If this process is repeated over many cycles, the described mechanism can have a significant effect on the temperature of the bubble.

Combined with the ability of optical cavities to spontaneously emit photons and the similarities of the sonoluminescing bubble with the experimental setups used for laser cooling [21, 22] and cavity-mediated laser cooling [18, 23], the above described interaction suggests that the atomic particles can change their temperature very rapidly, once the cavity approaches its minimum radius. To obtain a more intuitive picture and to show that the result is indeed heating and not cooling, notice that the frequency of the cavity field is relatively small, as long as the bubble radius is relatively large. Even when the bubble reaches its minimum radius, its frequency is always (first) below the transition frequency of the confined atomic particles. Energy conservation therefore implies that the conversion of atomic excitation into cavity photons is in general accompanied by the creation of a phonon. When the cavity photon leaks out of the resonator, thereby contributing to the strong light flash seen at the end of the collapse phase, the simultaneously created phonon remains inside the bubble. Subsequently, more phonons are created after the atoms have been re-excited during another thermalisation stage.

Finally, we remark that the strength of the above described heating mechanism depends on the physical properties of the sonoluminescing bubble, like the size of the relevant atomic transition frequency with respect to the minimum bubble radius. For noble gas atoms, the atomic transition frequency is in general very high so that the above mentioned conversion of atomic excitation into cavity photons and phonons is in general highly detuned and relatively inefficient. However, in ionic liquids, such conversions become much more likely due to being almost in resonance.

## 4. Conclusions

This paper discusses a collective quantum effect which might play an important role during the final stages of the collapse phase of sonoluminescence experiments. Although the calculations in Section 2 only provide a very rough picture of the dynamics of a sonoluminescing bubble, it captures the origin of a cavity-mediated collective quantum optical heating mechanism which we identify in Ref. [19, 20].

## Acknowledgments

AB and OK thank the UK Engineering and Physical Sciences Re- search Council EPSRC for financial support (Grant Ref. EP/H048901/1). Moreover, we acknowledge stimulating discussions with Antonio Capolupo, Andreas Kurcz, and Daniel Rumgay.

## References

[1] Brenner M P, Hilgenfeldt S, and Lohse D 2002 *Rev. Mod. Phys.* **74**, 425
[2] Lohse D 2002 *Nature* **418** 381
[3] Moss W C 1997 *J. Acoust. Soc. Am.* **101** 1187
[4] Barber B P and Putterman S J 1991 *Nature* **352** 318
[5] Hiller R, Putterman S J and Barber B P 1992 *Phys. Rev. Lett.* **69** 1182
[6] Vazquez G, Camara C G, Putterman S J and Weninger K R 2001 *Opt. Lett.* **26** 575
[7] Vazquez G, Camara C G, Puttermann S J and Weninger K R 2002 *Phys. Rev. Lett.* **88** 197402
[8] Suslick K S and Flannigan D J 2008 *Annu. Rev. Phys. Chem.* **59** 659
[9] Didenko Y T, McNamara W B, Suslick K S 2000 *Phys. Rev. Lett.* **84** 777
[10] Camara C G, Putterman S J and Kirilov E 2004 *Phys. Rev. Lett.* **92** 124301
[11] Flannigan D J and Suslick K S 2007 *Phys. Rev. Lett.* **99** 134301
[12] Flannigan D J and Suslick K S 2005 *Phys. Rev. Lett.* **95** 044301
[13] Flannigan D J and Suslick K S 2005 *Nature* **434** 52
[14] Eddingsaas N C and Suslick K S 2007 *J. Am. Chem. Soc.* **129** 3838
[15] Flannigan D J and Suslick K S 2006 *J. Phys. Chem. A* **110** 9315
[16] Didenko Y T and Gordeychuk T V 2000 *Phys. Rev. Lett.* **84** 5640
[17] Putterman S J, Evans P G, Vazquez G and Wenninger K 2001 *Nature* **409** 782
[18] Kim O and Beige A 2015 *Cavity-mediated collective laser-cooling of an atomic gas inside an asymmetric trap* Phys. Rev. Lett. (submitted); arXiv:1506.02910
[19] Kim O 2015 *Theoretical modelling of cavity cooling* Doctoral Thesis, University of Leeds (United Kingdom)
[20] Kim O and Beige A *A cavity-mediated collective heating mechanism in sonoluminescing bubbles* (to be submitted)
[21] Kurcz A, Capolupo A, and Beige A 2009 *New J. Phys.* **11** 053001
[22] Leibfried D, Blatt R, Monroe C, and Wineland D J 2003 *Rev. Mod. Phys.* **75** 281
[23] Kim O and Beige A 2013 *Phys. Rev. A* **88** 053417

# High speed synchrotron X-ray imaging of ultrasonic bubble cloud in liquid metal

**C.Wang[1], D.Tang[1], W.Zhang[1], W. Du[1], T.Connolley[2], J. Mi[1]**

[1]School of Engineering, University of Hull, Cottingham Road, Yorkshire, HU6 7RX, UK;
[2]Diamond Light Source, Didcot, Oxfordshire, OX11 0DE, UK

Chuangnan.wang@hull.ac.uk

**Abstract**. This paper presents the real-time and in-situ synchrotron X-ray high speed imaging studies of ultrasound bubbles and bubble cloud in a liquid Sn-30%Cu alloy. The collective behaviour of the ultrasound bubbles generated by ultrasound powders of 60 W and 100W were successfully captured. The number density of the individual bubbles and the density of the continuous bubble cloud were calculated from the information extracted from the images sequences and presented for the first time for the liquid Sn-30%Cu alloy.

## 1. Introduction

Ultrasound has found wide applications in science and industry, for example, ultrasound based imaging and non-destructive testing of materials, ultrasound based materials processing, cleaning or welding and ultrasound enhanced chemical reactions (sonochemistry). Among the many applications of ultrasound in a liquid medium, ultrasound induced bubbles and cavitation [1, 2] play very important roles in enhancing the transfer or exchange of mass and energy within the processed medium. However, the behaviour of ultrasonic bubbles and cavitation and their interactions with different materials is not fully understood. This is especially the case for high viscosity, high temperature, opaque liquid media, such as liquid metal alloys. Ultrasonic processing has been proven as an effective method for refining the solidification microstructures for metal alloys [3], but the underlying mechanisms are still under debate. Hence, there is ongoing, intensive worldwide research into how ultrasound bubbles and cavitation interact with liquid metals and their solidifying phases in different alloy systems [3]. Mi. *et al.* have recently used synchrotron X-rays to investigate, in real time, the interactions of ultrasound waves, bubbles with liquid metal and particles [4-6]. In these papers, the dynamics of a single bubble and its interaction with a solidifying particle at the solid-liquid interface were clearly presented. However, many ultrasound bubbles are often created simultaneously when the ultrasound power is over a certain criteria. Therefore, the collective behaviour of those bubbles has more practical significance for any potential industrial applications, and such studies in liquid metals have not been reported.

This paper presents the *in-situ* high speed synchrotron X-ray imaging studies of ultrasound bubble clouds in a liquid Sn-30%Cu alloy. The distribution of the ultrasonic bubbles and the bubble flux in liquid Sn-30%Cu alloy are presented for the first time.

## 2. Experiment

The experiment was carried out at the I12 beamline at Diamond Light Source, UK. Hull University's apparatus for *in-situ* experiments, consisting of a custom-made sample cell, furnace and ultrasound system, was placed on the beamline. Real-time, high speed X-ray images of ultrasound processing of a liquid alloy were then obtained using the beamline's imaging system.

### *2.1. Alloy, Sample and Ultrasound*

Sn-30wt%Cu alloy was chosen as the experimental alloy because Sn has a melting point of just 231.93 °C, and it is an ideal candidate as the liquid metal matrix for applying ultrasound over a wide range of temperatures, from 250 to 750 °C. Furthermore, the Sn-Cu binary system [7] is the key alloy system for lead-free soldering materials, and intermetallic compounds such as $Cu_6Sn_5$ are promising candidates for enhancing the storage capacity of Lithium ion based batteries [3].

A special quartz tube with a thin window in the middle was used to contain the liquid metal. The sonotrode was positioned at the top of the thin window of the quartz tube, and small pieces of Sn-30wt%Cu alloy were inserted inside the gap between the sonotrode and the quartz tube inner wall. The alloy was then melted in a specially designed furnace, with the temperature of the melt closely monitored by three K-type thermocouples placed at the top, middle and bottom of the thin window. A heat sink made of stainless steel rod was placed at the bottom of the quartz tube to create a thermal gradient. The details of the experiment are very similar to those described in [5, 6]. After reaching the target temperature (675 °C in this case), a trigger unit was used to send a signal to the high speed X-ray imaging camera to start image acquisition and then, after 10 ms delay, to apply ultrasound to the melt. In this way, the whole process of ultrasound bubble nucleation, growth and propagation can be captured. The ultrasound was generated using an UP50H ultrasonic processor with a MS2 sonotrode (Hielscher ultrasound technology Ltd.). Ultrasound powers of 60 W and 100W were used in the experiment to create different bubble cloud with different characteristics.

### *2.2. High speed synchrotron X-ray imaging*

The I12 Beamline has a wiggler source that provides an intense, high photon energy X-ray flux in the energy range 50-150 keV. Monochromatic and white beam modes are available. Filtered white beam was used to achieve high speed image acquisition up to 2000 frame per second (fps). (The intense X-ray beam was filtered with 4 mm of copper, to reduce heat load on the sample). The beamline's imaging system and Miro 310M high speed camera gave an image size of 800×1280 pixels with a spatial resolution of 4 μm × 4 μm per pixel. Due to the finite camera on-board memory, 4 s video was recorded in each acquisition. To correct for background noise and imperfections in the image path, flat field images (images acquired without the sample in the X-ray path), and background images (images acquired with the sample and sonotrode in the X-ray path but no ultrasound) were also obtained using the same sample setup, imaging frame rate and exposure time.

## 3. Results, analyses and discussion

### *3.1. Ultrasonic bubbles and bubble cloud*

In total, 49 video sequences were acquired. Two typical video sequences were chosen to present here. Flat field correction, noise reduction and contrast/brightness adjustments for each video sequence were performed using the ImageJ software. Figure 1 shows three typical snapshots of the collective bubbles (bubble cloud) extracted from a video sequence captured using an ultrasonic power of 60 W. It can be seen that, in the area immediately below the sonotrode tip, there is a bright region which contains many interconnected bubbles. However below this region (1.5 – 3.2 mm below the sonotrode tip), there are many individual bubbles spread out in the view field.

Figure 2 shows a cloud of bubbles captured at 100 W ultrasound power. Individual bubbles cannot be clearly seen; instead, the bubbles seem to form a stream of interconnected bubble strings, indicating that, at 2000 fps, the individual bubbles moved too fast to be captured. However, at the location away from the sonotrode such as the right bottom corner of Figure 2 (c) where the acoustic pressure is much lower than the area immediately blow the sonotrode tip, big, individual bubbles can still been seen.

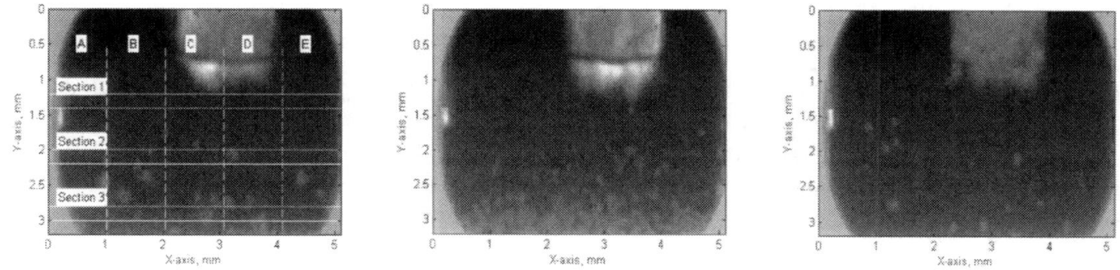


(a) (b) (c)

Figure 1 Images acquired from the liquid Sn-30%Cu alloy at 675 °C using an ultrasound power of 60 W, at (a) t0 = 5 ms (b) t1 = 25 ms (c) t2 = 45 ms, showing the ultrasonic bubble cloud immediately below the sonotrode tip and the individual bubbles at the location 1.5 – 3.2 mm below the sonotrode tip.

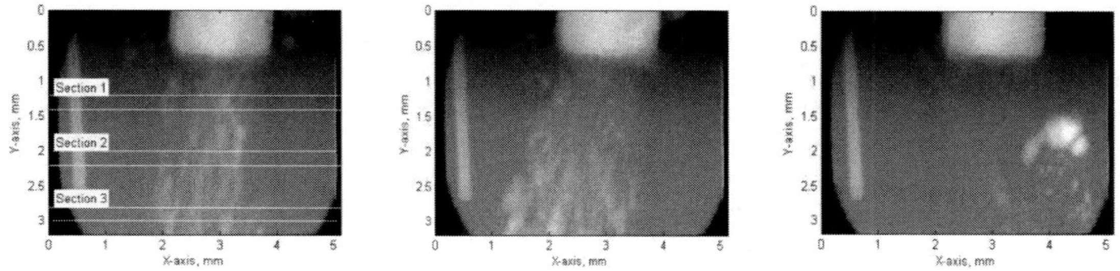


(a) (b) (c)

Figure 2 Images acquired from the liquid Sn-30%Cu alloy at 675 °C using an ultrasound power of 100 W, at (a) t0 = 25 ms (b) t1 = 35 ms (c) t2 = 45 ms, showing the ultrasonic bubble cloud.

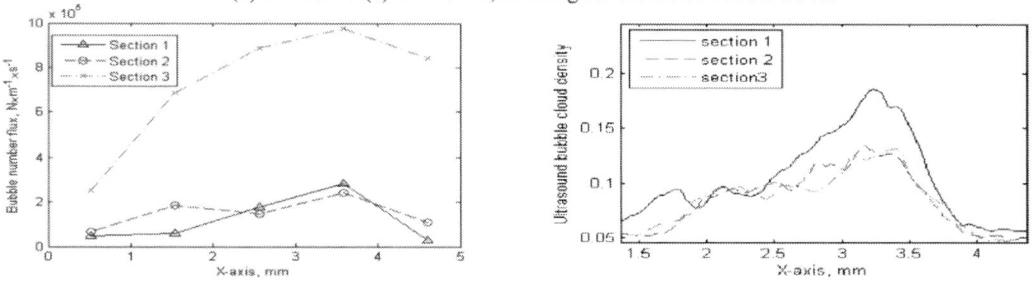


(a) (b)

Figure 3 (a) The bubble number density for the section 1, 2 and 3 showed Figure 1 (60 W), (b) the bubble cloud density for the section 1, 2 and 3 showed in Figure 2 (100 W).

### *3.2. The characteristics of the bubbles and bubble cloud*

Comparing the images shown in Figure 1 and 2, it is seen that, at 60 W, especially in the region away from the sonotrode tip, the bubbles are clearly separated from each other, and have a near spherical morphology. Bubbles generated at 100 W were elongated and interconnected together to form a continuous bubble stream. To quantify the collective behaviour of the bubbles, i.e. the spatial distribution of the bubbles and bubble density, the numbers of bubbles passing though per unit length per unit time at a defined location were measured from each image in a sequence. Figure 1 (a) shows the chosen sections, labelled 1, 2 and 3, and each section was divided into 5 areas, marked A, B, C, D, and E, with each sectional width of 1.25 mm. A total of 100 images were measured in this way to obtain sufficient statistical data which is presented in Figure 3 (a). It clearly shows that there is a larger population of bubbles passing through Section 3. It also shows the maximum bubble flux occurred in the areas C and D in Figure 2 (a).

At 100 W, the interconnection of the bubbles makes it inaccurate to measure them individually. However, the interconnected bubbles formed a nearly continuous bubble cloud and the distribution of the bubble cloud (or more accurately the density of the bubbles) passing through a sectional length can be measured by the grey level along that section. Therefore, the bubble density along the same sections 1, 2 and 3 were calculated from 20 images for the 100 W case and presented in Figure 3(b).

Before the calculation of grey level, the noise and background of the raw images were already removed by subtracting the background image. Therefore, the remaining brightness was actually due to the ultrasound bubble cloud. By subtracting the background image (obtained from imaging the liquid metal only) from the flat field (without liquid metal), the difference in grey level was obtained and used as the baseline grey level. When bubble clouds were found in the sample, the grey level will be different to the grey level of the baseline. The ratio between the bubble cloud grey level and that of the baseline actually reflects the bubble density inside the sample. These ratios for the section 1, 2, and 3 for the case of 100 W were plotted in Figure 3 (b).

Although the data showed in Figure 3(a) is the bubble number density, while in Figure 3(b) they are the bubble densities, the trends are very similar, indicating that using the grey level to represent the bubble density can capture the essential characteristics of the bubble cloud distribution and flux behaviour.

### *3.3. Discussion*

In most previous studies, the characteristics of ultrasonic bubbles and cavitation behaviours have been studied and analysed by methods such as high speed photography, foil testing, or acoustic emission spectrum measurement [8]. Those methods are generally not suitable for high temperature environments and opaque liquids. Some of them are invasive methods that may interrupt the ultrasound field, resulting in inaccurate measurement. In this paper, we demonstrated that by analysing the grey level of the captured images containing ultrasound bubbles or bubble clouds, the bubble cloud characteristics can be quantified.

## 4. Conclusion

In this paper, in-situ synchrotron X-ray imaging studies of ultrasound bubbles and bubble clouds in liquid Sn-30%Cu alloy were presented. By extracting statistical information from the acquired video sequences, the characteristics of ultrasonic bubbles and bubble cloud was quantified and discussed.

## References

[1] Lauterborn W and Kurz T 2010 Physics of bubble oscillations *Reports on Progress in Physics* **73** 106501

[2] Zhang Y-n and Li S-c 2010 Direct numerical simulation of collective bubble behavior *Journal of Hydrodynamics, Ser. B* **22** 827-31

[3] Eskin G I 1997 Principles of ultrasonic treatment: Application for light alloys melts *Advanced Performance Materials* **4** 223-32

[4] Tan D, Lee T, Khong J, Connolley T, Fezzaa K and Mi J 2015 High-Speed Synchrotron X-ray Imaging Studies of the Ultrasound Shockwave and Enhanced Flow during Metal Solidification Processes *Metall and Mat Trans A* 1-11

[5] Lee T L, Khong J C, Fezzaa K and Mi J 2013 Ultrafast X-ray Imaging and Modelling of Ultrasonic Cavitations in Liquid Metal *Light Metals Technology 2013* **765** 190-4

[6] Mi J, Tan D and Lee T 2014 In Situ Synchrotron X-ray Study of Ultrasound Cavitation and Its Effect on Solidification Microstructures *Metall and Materi Trans B* 1-5

[7] Furtauer S, Li D, Cupid D and Flandorfer H 2013 The Cu-Sn phase diagram, Part I: New experimental results *Intermetallics* **34** 142-7

[8] Moholkar V S, Sable S P and Pandit A B 2000 Mapping the cavitation intensity in an ultrasonic bath using the acoustic emission *Aiche J.* **46** 684-94

# Experimental study of the thermodynamic effect in a cavitating flow on a simple Venturi geometry

**M Petkovšek, M Dular**

Laboratory for Water and Turbine Machines, Faculty of Mechanical Engineering, University of Ljubljana, Aškerčeva 6, 1000 Ljubljana, Slovenia

E-mail: martin.petkovsek@fs.uni-lj.si

**Abstract**. The thermodynamic effects in cavitating flow are observed on a simple Venturi profile. A thorough experimental investigation of the temperature field on cavitating flow has been performed in water of 100°C at different operating conditions. Temperature measurements were performed with Infra-Red (IR) high-speed camera, while visualisation was made with conventional high-speed camera. Both, average temperature fields and temperature dynamics are presented at different operating conditions and compared with collected data in visual spectrum. In the vicinity of the throat a temperature depression up to 0.5 K was recorded.

## 1. Introduction

Cavitation is a physical phenomena simply described as a growth and collapse of multiple small vapor bubbles within a liquid by approximately constant temperature. While an exact process of cavitation development is still not fully understood and scientifically described, it is assumed that cavitation is mainly a process of vaporization by cavity growth and condensation by cavity collapse, which is a consequence of heat transfer and temperature divergence between the bulk liquid and vapor + gas inside the cavitation bubbles. Dealing with liquids like cold water, the local temperature variations caused by cavitation can be neglected, while this must not be the case in liquids, which operate close to its critical point (cryogenics), where the temperature variations are big enough to affect the cavitation development – these phenomena are described as thermodynamic effects.

Experimental studies of these phenomena are rare and most of them only observed the integral consequences (cavitation delayed inception and reduced extent) of the presence of the thermodynamic effects. Hord et al. [1-4] have made an extensive experimental observation on different geometries in liquid cryogens, while observing cavitation inception and extent at different operating conditions. Experimental studies on temperature depression measurements in individual points within cavitating flow were performed by Fruman et al. [5] and Franc et al. [6]. Rimbert et al. [7] presented experimental data of temperature depression in micro-channel cavitation with two-colors laser in several individual points, while Petkovšek & Dular [8] presented the first 2D temperature field of the cavitating flow on a small Venturi geometry with a non-invasive IR method. Dular & Coutier-Delgosha [9] used IR method to investigate thermodynamic effects on a single cavitation bubble.

Due to lack of the temperature experimental data within a cavitating flow, the improvement of the cavitation numerical models, with thermodynamic effects consideration, can not progress. To fill up the void in cavitation temperature database, the present study is performed.

## 2. Experimental set-up

Cavitation tests were performed in a cavitation tunnel at the Laboratory for Water and Turbine machines, University of Ljubljana.

The cavitation tunnel (Fig. 1 - left) has a closed circuit which enables to vary the system pressure and the temperature of the used liquid, water in our case. Circulation of the water is obtained with 4.5 kW pump (1), that enables the variation of the rotation frequency in order to set the flow rate. At the pump delivery, a tank (2) partially filled with the circulation water is used for water heating, 10 kW electric heater is installed, and for damping the periodical flow rate and pressure fluctuations due to the passage of the pump blades. Cavitation and thermodynamic effects are observed in a transparent polycarbonate test section (3). The tank downstream of the test section (4) is used for cooling of the circulation water, cooling water flows inside the tank in a secondary loop, which is connected to cold tap water. The temperature of the water is monitored with a Pt100 sensor (5) installed in the downstream tank and with a thermocouple J type directly installed in the test section. The pressure inside the cavitation tunnel can be varied with a vacuum pump connected to the downstream tank or with a compressor connected to the upstream tank in order to provide a wide range of hydrodynamic conditions. The flow rate is monitored with electromagnetic flowmeter (6).

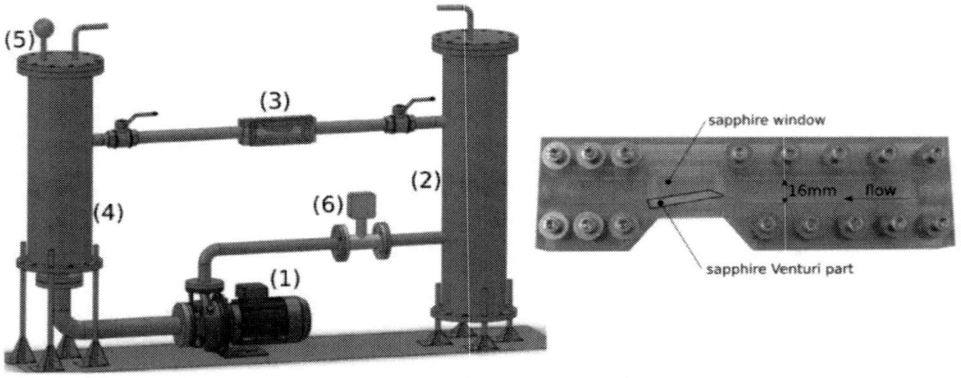


**Figure 1:** Cavitation tunnel (left), test section (right).

Specially designed test section (Fig. 1 - right), constructed out of transparent polycarbonate and sapphire glass, enables to withstand temperatures up to 120°C. Side window and part of Venturi are manufactured from sapphire glass, which enables visualization with conventional high-speed camera and thermography with IR high-speed camera. Side window allows to observe cavitation conditions from the side view, while part of Venturi constriction (part made out of sapphire, which serves as observation window) allows the view from below. The basic geometry was a 10 mm wide Venturi section with a converging angle of 18° and diverging angle of 8°. The throat cross-section dimensions were 8x10 $mm^2$.

To observe cavitation phenomena, two different camera positions were set. First, where the cameras (conventional high-speed camera and IR high-speed camera) were pointed into the side observation window and second, where cameras were pointed into bottom part of Venturi. In both cases the IR camera was perpendicular to observation surface, while the conventional camera was slightly at an angle to the IR camera.

## 3. Results and discussion

As mentioned, two types of measurements were conducted, where the thermodynamic effects were investigated from both the side and the bottom view of Venturi profile. In both observation positions, average temperature fields and temperature dynamics were observed at different cavitating conditions. All measurements were performed with tap water at approximately 100°C ± 2°C. For presented results the conventional high-speed camera operated at 10,000 fps, while IR camera operated at 830 and 920 fps for side and bottom view, respectively.

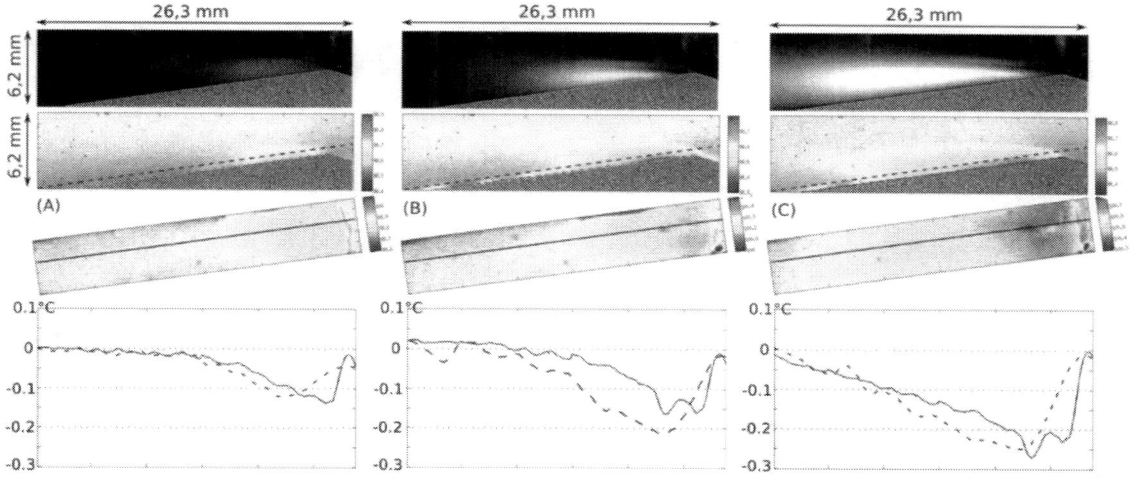


**Figure 2:** Average temperature fields for three different cavitation lengths (upper images), from side (middle images) and bottom view (bottom images). In diagrams are temperature collected from lines shown on upper temperature fields.

Figure 2 shows averaged visualization and temperature fields for the time of 1 second. The flow in presented cases goes from right to left, where the highest velocities, at the Venturi throat were between 9 m/s and 12 m/s. Images from visualization and thermography are scaled to show the same area of interest, but one must be aware, that temperature fields, which are shown on figure 2, the side view and the bottom view are not taken simultaneously. The cases were chosen from a set of measurements at different operating conditions (to chose different cavitation extent – cases A, B and C on fig. 2), where approximately the same conditions were chosen for the side and bottom view.

One can see, that in all three cases the temperature of the fluid flow starts to decrease just after the Venturi throat, due to vaporization and gas expansion, where the cavitation structures start to form. The temperature decrease for presented cases is between 0.12 K and 0.27 K, which corresponds with the results from the experiments of the Petkovšek & Dular [8] at the much smaller Venturi test section. One must be aware, that this are averaged temperature depressions and that the local temperature variations can reach up to 0.5 K in our observations. After the temperature decrease the temperature of the fluid flow returns to the initial free stream temperature. No obvious temperature increase above the free stream temperature is noticed. The magnitude and extent of temperature depression is increased with cavitation extent, as it is shown on figure 2. This means, that bigger cavitation extent cause stronger temperature depression of the flow.

Observing individual temperature fields on figure 3 gives us more thorough insight into the temperature dynamics. Figure 3 presents a short section of bottom view thermography from the case C, on fig. 2. The time step between the images is 1.1 ms. By moving along the images (fig. 3), one can see that the cold region left of the Venturi throat line is growing from image 1 to its maximum extent on image 6. After reaching its maximum extent, the cold region starts to shrinking back to to Venturi throat line. This corresponds to the cloud shedding. The smaller colder regions inside the cold region are most probably parts of cavitation cloud, which are moving closer to the Venturi channel bottom wall (observation window). Due to strong reverse flow, which appears in the Venturi test section and causes a thicker boundary layer between the observation window and observed cavitation structures, we believe, that actual temperature depression could be even higher.

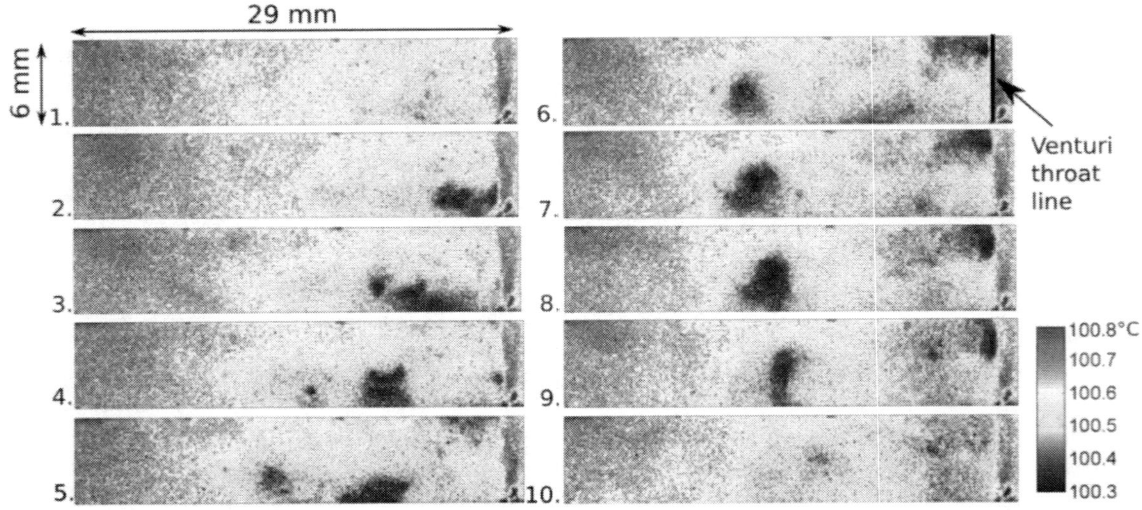


**Figure 3:** Bottom view temperature field for case of one cavitation cloud shedding.

## 4. Conclusions

The study presents a continuation of the previous work, Petkovšek & Dular [8], where the thermodynamic effects were observed in much smaller Venturi geometry. Scientific contribution is not just in scaling up the test section, but also that for the first time the thermodynamic effects were observed from the bottom view of the Venturi geometry. It was shown that the temperature depression, due to cavitation, can be measurable already in water of 100°C, by which conditions our experiments were performed.

## Acknowledgments

The study was performed under a grant by the European Space Agency (ESA).

## References

[1] Hord J, Anderson M, Hall W J 1972 NASA CR-2054.
[2] Hord J 1972 NASA CR-2156.
[3] Hord J 1973 NASA CR-2242.
[4] Hord J 1974 NASA CR-2448.
[5] Fruman D H, Reboud J L 1999 *Int. J. of Heat and Mass Transfer* 42, 3195-3204.
[6] Franc J P, Boitel G, Riondet M, Janson E, Ramina P, Rebattet C 2010 *J. of Fluid Engineering* 132, 1-9.
[7] Rimbert N, Castanet G, Funfschiling D 2012 *Proc. of the 8$^{th}$ Int. Sym. On Cav. CAV2012, Singapure.*
[8] Petkovšek M and Dular M 2013 *Int. J. of Heat and Fluid Flow* 44, 756-763.
[9] Dular M and Coutier-Delgosha O 2013 *J. Fluid Mech.* 736, 44-66.

# Prediction of cryogenic cavitation around hydrofoil by an extensional Schnerr-Sauer cavitation model

**T Z Sun, Y J Wei and C Wang**

School of Astronautics, Harbin Institute of Technology, Harbin 150001, China

sun.tiezhi.hit@gmail.com

**Abstract**. Developing a robust computational strategy to address the rich physics characteristic involved in the thermodynamic effects on the cryogenic cavitation remains a challenging problem. The objective of this present study is to model the numerical methodology to simulate the cryogenic cavitation by implanting the thermodynamic effects to the Schnerr-Sauer cavitation model, and coupling the energy equation considered the latent heat. For this purpose, cavitating flows are investigated over a three dimensional hydrofoil in liquid hydrogen and nitrogen. Experimental measurements of pressure and temperature are utilized to validate the extensional Schnerr-Sauer cavitation model. Specifically, the further analysis of the cavitation solution with respect to the thermodynamic term is conducted. The results show that the extensional Schnerr-Sauer cavitation model predicts better accuracy to the quasi-steady cavitation over hydrofoil in the two cryogenic fluids.

## 1. Introduction

Cryogenic fluids such as liquid hydrogen and liquid oxygen are being promoted for use as propellants for rocket and other aerospace equipment. It is well known that the high speed rotating turbo pump often causes a lower pressure around the inducer, and then cavitation occurs when the fluid pressure drops below the vapor pressure in the local thermodynamic state [1]. Cryogenic fluids are thermo-sensitive, and therefore, thermodynamic effects and significant variations in fluid properties can alter the cavitation properties. Hence, it is necessary to explore an effective computational strategy to capture the cavitation characteristics in cryogenic fluids.

The selection of cavitation models plays a major role on the accurate prediction of cavitating flows. In recent years, some worthy efforts have been made in the development of cavitation models in cryogenic fluids. Most of these models basically determine the liquid phase fraction by that local pressure depression. Physically, the cavitation process is governed by thermodynamics and kinetics of the phase change process. So it is important to estimate the physical characteristic for cryogenic fluids by modeling with thermodynamic effects [2-5].

In the present study, therefore, we extend the Schnerr-Sauer cavitation model, which taking into account the thermodynamic effects during the evaporation and condensation process. The extensional model is validated by predicting the quasi-steady cavitating flows around a three dimensional hydrofoil in liquid hydrogen and nitrogen.

## 2. Computational methodologies and cavitation model

The numerical simulations are carried out by using computational fluid dynamics (CFD) code CFX. The homogeneous mixture approach is used to model the two phase cavitating flows.

### 2.1. Governing equations and turbulence closure approach

The set of governing equations for cavitating flows under the homogeneous fluids modeling consists of the conservative form of continuity, momentum, energy equations, and they are given below:

$$\frac{\partial \rho_m}{\partial t}+\frac{\partial}{\partial x_j}\left(\rho_m u_j\right)=0 \tag{1}$$

$$\frac{\partial}{\partial t}\left(\rho_m u_i\right)+\frac{\partial}{\partial x_j}\left(\rho_m u_i u_j\right)=-\frac{\partial p}{\partial x_i}+\frac{\partial}{\partial x_j}\left[\left(\mu+\mu_t\right)\left(\frac{\partial u_i}{\partial x_j}+\frac{\partial u_j}{\partial x_i}\right)-\frac{2}{3}\frac{\partial u_k}{x_k}\delta_{ij}\right] \tag{2}$$

$$\frac{\partial(\rho_m T)}{\partial t}+\frac{\partial}{\partial x_j}(\rho_m u_j C_{pl} T)=\nabla\cdot(k_{eff}\nabla T)-\frac{\partial(\rho_m f_v L)}{\partial t}-\frac{\partial(\rho_m u_j f_v L)}{\partial x_j} \tag{3}$$

where $\rho_m=\rho_v\alpha_v+\rho_l(1-\alpha_v)$, the subscripts $v$ and $l$ donate vapour and liquid; $x$ and $i$, $j$ and $k$ donate coordinate axes; $u$ is the velocity; $\mu$ is the dynamic viscosity, $\mu_t$ is the turbulent viscosity; $C_{pl}$ is specific heat; $k_{eff}$ is effective thermal conductivity; $L$ is the latent heat of evaporation, $f_v$ is vapour mass fraction.

A two-equation, *k-ε* model is employed as the turbulence closure approach, as it offers a well-established regime of predictive capability.

### 2.2. Extensional Schnerr-Sauer Cavitation model

For cavitating modelling, the vapour volume fraction is solved with the following transport equation:

$$\frac{\partial}{\partial t}(\alpha_v\rho_v)+\nabla\cdot(\alpha_v\rho_v\vec{u}_m)=R_e-R_c \tag{4}$$

where the net mass transfer is expressed as $R=R_e-R_c$, and the sink term $R_e$ and source term $R_c$ represent the evaporation and condensation mass transfer rates.

Schnerr-Sauer cavitation model was derived from the bubble dynamics equation of the generalized Rayleigh-Plesset equation, and the following expressions are used to relate the mass transfer between the vapour and liquid phases in cavitation[6].

$$R_e=\frac{\rho_v\rho_l}{\rho_m}\alpha_v(1-\alpha_v)\frac{3}{\Re_B}\left(\frac{2}{3}\frac{(P_v-P)}{\rho_l}\right)^{1/2}\quad P\le P_v \tag{5}$$

$$R_c=\frac{\rho_v\rho_l}{\rho_m}\alpha_v(1-\alpha_v)\frac{3}{\Re_B}\left(\frac{2}{3}\frac{(P-P_v)}{\rho_l}\right)^{1/2}\quad P>P_v \tag{6}$$

where $\Re_B$ is the bubble radius, $P_v$ represents vapour pressure. The rate of change referring to the bubble radius caused by pressure depression in equation (5) is shown as:

$$\frac{d\Re_B}{dt}=\left(\frac{2}{3}\frac{(P_v-P)}{\rho_l}\right)^{1/2} \tag{7}$$

During bubble growth of the thermo-sensitive fluid, the heat necessary for vaporization is supposed to be supplied to the interface through the liquid. Hence, a thermal boundary layer develops on the bubble wall. The liquid temperature drops from $T_\infty$ to $T_v$ through this boundary layer. It is defined as $\sqrt{\alpha_l t}$, where $\alpha_l$ is the thermal diffusivity of the liquid. The typical temperature gradient within the boundary layer is $\Delta T/\sqrt{\alpha_l t}$ where $\Delta T=T_\infty-T_v$. According to Fourier's law, the conductive heat flux towards the interface is the order of $\lambda_l\Delta T/\sqrt{\alpha_l t}$, where $\lambda_l=\alpha_l\rho_l C_{pl}$ is the conductivity. The energy balance expresses that the heat supplied by conduction to the interface of area $4\pi\Re_B{}^2$ is used for vaporization and causes the increase of the mass of gas $4\pi\Re_B^3\rho_v/3$ inside the bubble. Hence, the energy balance is:

$$\lambda_l \frac{\Delta T}{\sqrt{\alpha_l t}} 4\pi\Re_B^2 = \frac{d}{dt}\left(\frac{4}{3}\pi\Re_B^3\rho_v\right)L \tag{8}$$

From equation (7), we can obtain:

$$d\Re_B / dt = \rho_l C_{pl}\sqrt{a_l}\Delta T / \rho_v L\sqrt{t} \tag{9}$$

As shown in equations (7) and (9), taking into account the thermal effect for cryogenic cavitation, the E-Schnerr-Sauer (Extensional Schnerr-Sauer) cavitation can be defined as follows:

$$R_e = \frac{\rho_v\rho_l}{\rho_m}\alpha_v(1-\alpha_v)\frac{3}{\Re_B}\left[\left(\frac{2}{3}\frac{(P_v(T)-P)}{\rho_l}\right)^{1/2} - \frac{\rho_l C_p\sqrt{a}(T_\infty - T_v)}{\rho_v L\sqrt{t}}\right] \quad P \le P_v(T) \tag{10}$$

$$R_c = \frac{\rho_v\rho_l}{\rho_m}\alpha_v(1-\alpha_v)\frac{3}{\Re_B}\left[\left(\frac{2}{3}\frac{(P-P_v(T))}{\rho_l}\right)^{1/2} - \frac{\rho_l C_p\sqrt{a}(T_\infty - T_v)}{\rho_v L\sqrt{t}}\right] \quad P > P_v(T) \tag{11}$$

where $P_v(T)$ represents the saturation pressure at the local temperature; The variable $t$ in these equations is replaced by a constant value which presents the bubble growth time for better prediction.

### *2.3. Numerical setup and description*

In present study, we model flow over a 3D quarter caliber hydrofoil, which is experimentally investigated by Hord [7]. The tunnel and hydrofoil geometry shown in figure 1 are used to clarify the detailed geometry structure. Both of the width and height of the tunnel are 25.4mm, and the leading edge diameter of the hydrofoil is D =7.92mm. The boundary conditions of the computational domain are also shown in figure 1. The whole flow field was divided into over 1020,000 cells and mesh refinements are performed at the leading edge and the cavitation region (see figure 2).

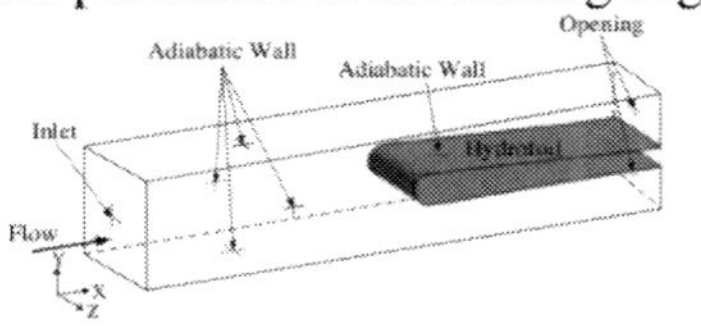


**Figure 1.** The computational domain.

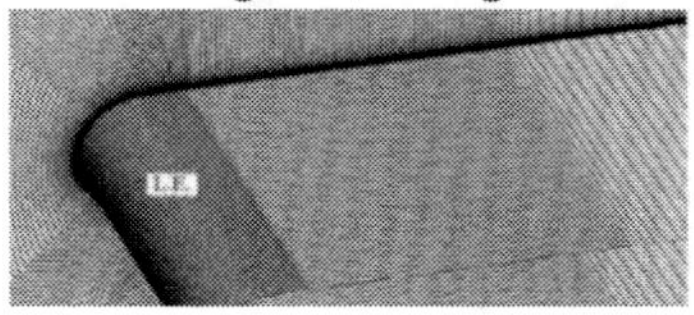


**Figure 2.** Mesh generation for hydrofoil.

## 3. Results and discussions

In present study, four cases of cavitating flows in liquid hydrogen and nitrogen with different free stream temperatures $T_\infty$ and inlet cavitation numbers $\sigma_{\mathrm{Int}}$ are selected randomly. Table 1 summarizes the run conditions of different cases.

**Table 1.** Run conditions for hydrofoil cases.

| Case | Substance | $T_\infty$(K) | $U_{\mathrm{Int}}$ (ms$^{-1}$) | $\sigma_{\mathrm{Int}}$ | *Re* |
|---|---|---|---|---|---|
| 254C | Liquid $H_2$ | 20.53 | 51.0 | 1.44 | $2.2\times10^6$ |
| 260D | Liquid $H_2$ | 20.81 | 50.2 | 1.57 | $2.1\times10^6$ |
| 290C | Liquid $N_2$ | 83.06 | 23.9 | 1.70 | $9.1\times10^6$ |
| 296B | Liquid $N_2$ | 88.54 | 23.7 | 1.61 | $1.1\times10^7$ |

Figure 3 compares the computational temperature and pressure depression with Hord's measurements data along the hydrofoil by Schnerr-Sauer and E-Schnerr-Sauer cavitation model. It is clear that temperature depression occurs at the leading edge of hydrofoil, and then it gradually recovers to the freestream temperature due to the condensation near the closure region of the cavity. As well as the behavior gradient of the pressure profiles along the hydrofoil are not constant due to the thermodynamic effects. The proposed model leads to a cavity length slightly longer than that predicted by the original Schnerr-Sauer model. This is because the revised thermal term contributed to less heat release, resulting in weaker tendency to restrain the intensity of cavitation.The performance of the E-

Schnerr-Sauer model has shown better than that of the original Schnerr-Sauer cavitation model, especially noteworthy is the fact that the prediction of thermal behavior.

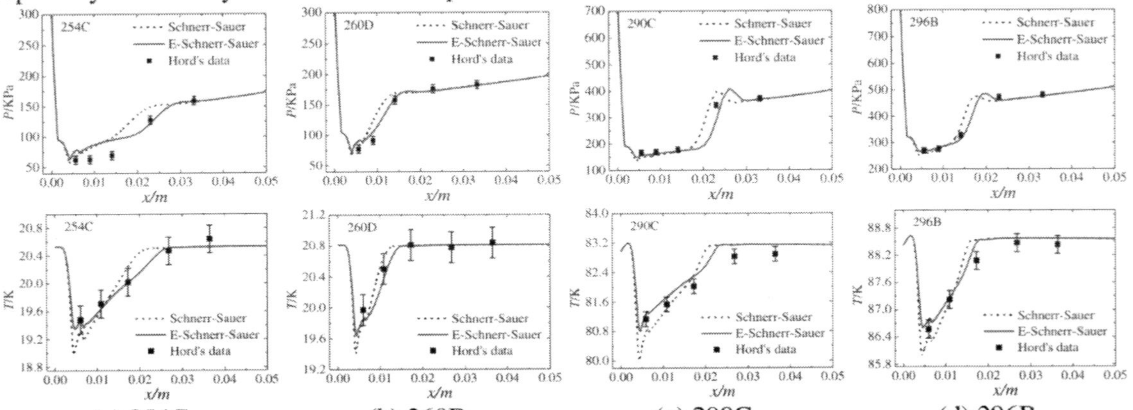


(a) 254C (b) 260D (c) 290C (d) 296B

**Figure 3.** Comparison of pressure and temperature depression to experimental data.

Figure 4 shows the contours of vapour volume fraction $\alpha_v$ , the evaporation rate $\bar{R}_e$ and condensation rate $\bar{R}_c$ (where the dimensionless mass transfer rate is defined as $\bar{R}=RD/U_{\infty}\rho_l$). Note that the evaporation and condensation contours demonstrate a mutually exclusive behavior. The thermodynamic effects term of the equation (9) causes lower evaporation rate near the leading edge and weaker condensation intensity in the closure region, hence the temperature distribution is changed inside cavity.

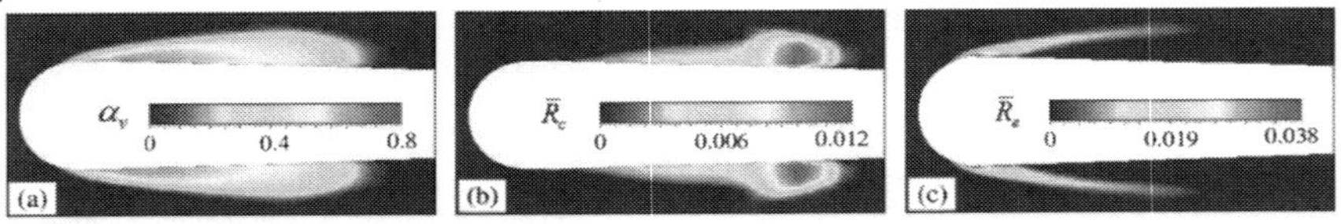


**Figure 4.** Predicted results of (a) vapor phase fraction, (b) condensation rate and (c) evaporation rate distribution in the mid-plane of the hydrofoil for case 290C.

## 4. Summary

In present study, the original Schnerr-Sauer cavitation model was extended to treat the quasi-steady cavitating flows around hydrofoil in liquid hydrogen and nitrogen. The numerical results demonstrated that the Extensional-Schnerr-Sauer model which considers the thermodynamic effects can provide an better strategy to predict the cavitation characteristics in cryogenic liquids.

**Acknowledgements**
The authors would like to acknowledge the financial support of the Natural Science Foundation of Heilongjiang Province of China (Grant NO.A201409)

**References**

[1] Ohira K, Nakayama T, and Nagai T 2012 *Cryogenics* **52** 35-44
[2] Utturkar Y, Thakur S, Shyy W 2005 *43rd Aerospace Sciences Meeting and Exhibit, Reno, Nevada*
[3] Utturkar Y, Wu J, Wang G, Shyy W 2005 *Prog Aerosp Sci* **41** 558-608
[4] Tani N, Tsuda S, Yamanishi N, Yoshida Y *2009 Proceedings of the 7th International Symposium on Cavitation, Ann Arbor, Michigan*
[5] Zhang X B, Wu Z, Xiang S J, Qin L M 2013 *Chinese Sci Bull* **58** 567-74
[6] Hord J 1973 *NASA* **CR** 2156
[7] Schnerr G H, Sauer J 2001, *4th International Conference on Multiphase Flow, New Orleans, Louisiana*

# Effects of physical properties on thermo-fluids cavitating flows

**T R Chen, G Y Wang**[1]**, B Huang, D Q Li, X J Ma and X L Li**
School of Mechanical and Vehicular Engineering, Beijing Institute of Technology, Beijing 100081, RP China
E-mail: wangguoyu@bit.edu.cn

**Abstract:** The aims of this paper are to study the thermo-fluid cavitating flows and to evaluate the effects of physical properties on cavitation behaviours. The Favre-averaged Navier-Stokes equations with the energy equation are applied to numerically investigate the liquid nitrogen cavitating flows around a NASA hydrofoil. Meanwhile, the thermodynamic parameter $\sum$ is used to assess the thermodynamic effects on cavitating flows. The results indicate that the thermodynamic effects on the thermo-fluid cavitating flows significantly affect the cavitation behaviours, including pressure and temperature distribution, the variation of physical properties, and cavity structures. The thermodynamic effects can be evaluated by physical properties under the same free-stream conditions. The global sensitivity analysis of liquid nitrogen suggests that $\rho_v$, $C_l$ and $L$ significantly influence temperature drop and cavity structure in the existing numerical framework, while $\rho_v$ plays the dominant role when these properties vary with temperature. The liquid viscosity $\mu_l$ slightly affects the flow structure via changing the Reynolds number $R_e$ equivalently, however, it hardly affects the temperature distribution.

## 1. Introduction

Cavitation is a phase-change phenomenon that occurs in liquids when the local pressure drops below the saturated vapor pressure, this phenomenon is usually assumed to be an isothermal process [1, 2]. However, the isothermal hypothesis is no longer valid in thermo-fluids cavitation for their thermo-sensitive properties. When cryogenic cavitation occurs, the temperature drop $\Delta T$ inside the cavity changes the temperature-dependent physical properties, and then cavitation dynamics is affected [3].

The cryogenic cavitation behaviours have been investigated as early as 1956 [4]. Based on a simple heat balance between two phases, Stahl et al. [4] introduced a B-factor to estimate temperature drop during phase-change process. Hord [5, 6] conducted comprehensive experiments on cavitation with liquid nitrogen and hydrogen under different conditions. Temperature and pressure data were measured at five locations on hydrofoil and ogive models, which have been commonly employed for numerical validation for cryogenic cavitation [3, 7-9].

Recently, Rodio et al. [7] modified the Rayleigh-Plesset equation by the addition of a term for convective heat transfer at the interface between the liquid and the bubble to assess the thermal effects. Experimental visualizations [5, 6] have clearly indicated a mushy nature of the cryogenic cavity. This salient feature of cryogenic cavitation was adapted into the derivation of the interfacial dynamic-based model (IDM) to yield a Mushy IDM [3]. Huang et al. [8] validate a thermodynamic cavitation model based on bubble dynamic equation and calibrate the parameters of the cavitation model for liquid hydrogen. Zhu et al. [9] developed an effective computational strategy to simulate cryogenic cavitation by implementing the "Schnerr-Sauer cavitation model", coupled with the energy equation.

[1] Corresponding author. Tel.: +86 10 68912395. E-mail address: wangguoyu@bit.edu.cn

Although the cryogenic liquid cavitating flows have received much attention in the past years, the effects of fluid physical properties on cavitation behaviours are still not well understood. The present works are to study the thermo-fluid cavitating flows and to evaluate the effects of physical properties.

## 2. Governing equations and numerical approaches

### *2.1. Governing equations*

The governing equations for thermo-fluid cavitation under the homogeneous multiphase flows consists of the continuity, momentum, enthalpy, and cavitation model equations are given in Ref. [3, 8, 9].

The $\dot{m}^-$ and the $\dot{m}^+$ in the governing equations represent for the evaporation and condensation rates during mass transfer process, the cavitation model [3, 10] are shown as following:

$$\dot{m}^- = \frac{C_{dest}\alpha_l}{t_\infty}\frac{\rho_l}{\rho_v}\frac{MIN(p-p_v(T),0)}{(0.5\rho_l U_\infty^2)}, \qquad \dot{m}^+ = \frac{C_{prod}(1-\alpha_l)}{t_\infty}\frac{MAX(p-p_v(T),0)}{(0.5\rho_l U_\infty^2)} \tag{1}$$

$C_{dest}$ and $C_{prod}$ are two empirical coefficients, defined as $C_{dest}$=3.8, $C_{prod}$ =20 respectively, $t_\infty$ is the reference time scale ($t_\infty = L_c / U_\infty$). $\alpha_l$ is liquid volume fraction, $p_v$ is the saturated vapor pressure.

The computations are performed by using the CFD code CFX to solve the Navier-Stokes equations. The $k$-$\omega$ SST (Shear Stress Transport) turbulence model is used in simulating the cavitating flows [11].

### *2.2. Numerical setup and description*

The numerical predictions are compared with the experimental measurements around a NASA hydrofoil in the liquid nitrogen [5]. The computational domain and boundary conditions are shown in figure 1. A no-slip wall is imposed on the hydrofoil surface, and no-slip symmetry is imposed on the side boundary, the lower side of the hydrofoil is not included in the domain because of its symmetry, the inlet velocity, temperature and the outlet pressure are set based on the experimental data. The thermo-sensible properties are updated from the NIST data base [12] throughout the course of computations. The mesh is composed of 19,000 elements with 220 structured elements across the foil boundary layer.

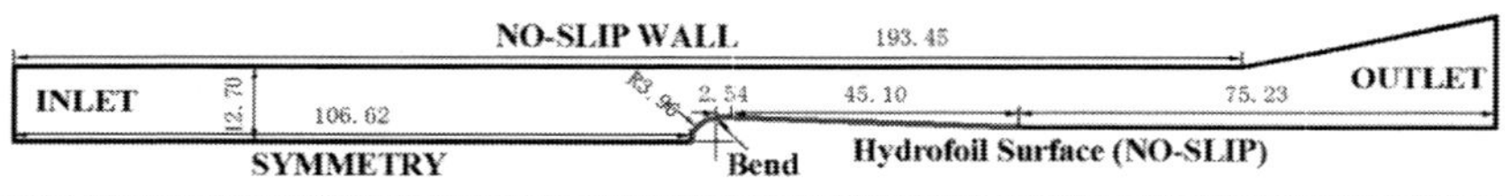


**Figure 1.** 2D computation domain and boundary conditions

## 3. Results and discussion

### *3.1. The characteristics of Thermo-fluid cavitating flows*

The thermodynamic parameter $\sum$ defined by:

$$\Sigma = \frac{\rho_v^2 L^2}{\rho_l^2 C_v T_\infty \sqrt{\varepsilon_l}}, \quad \varepsilon_l = \frac{K_l}{\rho_l C_l} \tag{2}$$

is introduced to evaluate temperature change with varying bubble diameter and is crucially important in determining the bubble dynamic behaviour [1]. Here, $\rho$ is the density, $L$ is the latent heat of vaporization, $C$ is the specific heat at constant pressure, $\varepsilon$ is the thermal diffusivity (works in unsteady heat transfer process) and $K$ is the thermal conductivity. The suffixes $v$ and $l$ indicate vapor and liquid phase. The liquid/vapor density ratio affects cavity size directly, meanwhile, larger latent heat $L$ and smaller specific heat $C_v$ lead to larger temperature drop $\Delta T$ during the phase change process.

The variations of the parameter $\sum$ for some thermo-fluids are illustrated in figure 2 based on the NIST data base [12], the values increase with increasing temperature. The values are relatively large for hydrogen and helium through the whole temperature range, which indicates strong thermodynamic effects. The values for nitrogen, oxygen and methane are similar to a certain extent, except that the value for the oxygen is extremely small when the temperature is close to its thermodynamics triple point. As to the water, the value is relatively small when the temperature is low, that is why the temperature change (thermodynamic effects) could be ignored in room temperature water. However, the value increases quickly, the magnitude of the value for the water is almost the same as nitrogen

Journal of Physics: Conference Series **656** (2015) 012181 doi:10.1088/1742-6596/656/1/012181

when temperature is close to 350K. It indicates that the thermodynamic effects on cavitating flow can be evaluated by physical properties under the same free-stream conditions.

Using the numerical method mentioned above, the liquid nitrogen cavitating flow (Case 290C in Ref. [6]) which shows quintessential characteristics of thermo-fluid cavitation behaviours [3, 7, 9] is calculated, it is selected to aid the thermo-fluid characteristic and numerical validation. Significant differences between the thermal and isothermal solutions are illustrated in figure 3, the thermal energy solution can consistently capture the main features of both temperature and pressure profiles, which shows better agreement with the experimental data. The temperature inside the cavity in figure 3(a) depresses due to evaporative cooling, and then the local saturated vapor pressure depresses, that is why the value of $p$-$p_v(T_\infty)$ in figure 3(b) is below zero in the cavity region. It is observed that a slight temperature rise above the reference temperature at the rear of cavity, which is attributed to the release of latent heat during the condensation process. Overall, thermodynamic effects significantly affect the thermo-fluid cavitation behaviours. The temperature change during the phase-change process causes the variations of thermo-sensible properties (especially the vapor pressure and density), and then the reference free-stream conditions are changed equivalently, resulting in the change of cavity structure.

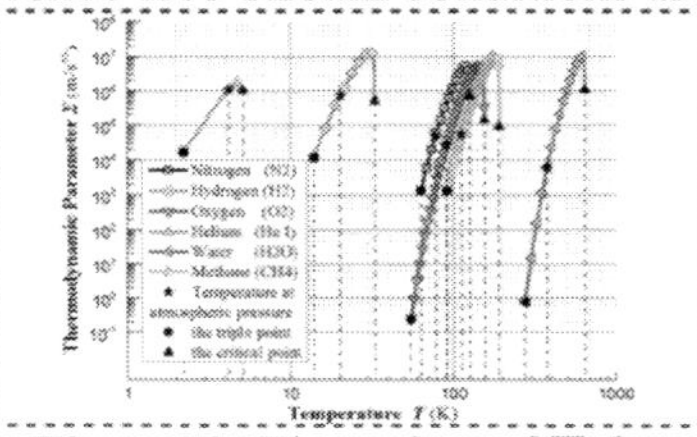


**Figure 2.** The value of $\sum$ for some thermo-fluids

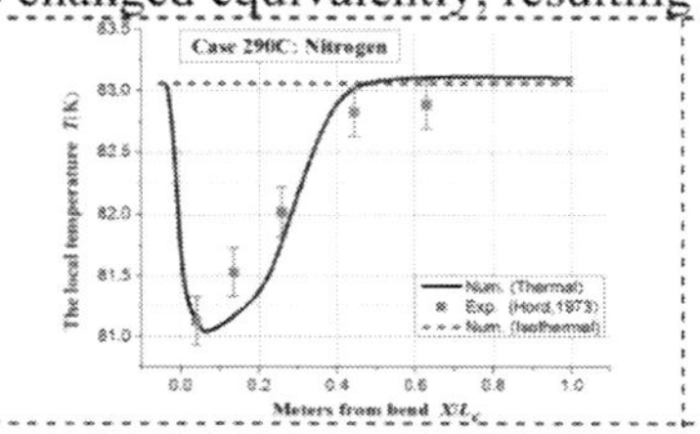


**Figure 3(a).** Temperature along the surface

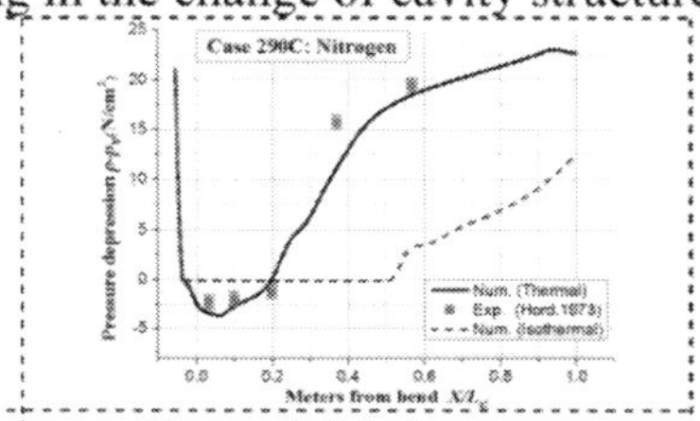


**Figure 3(b).** Pressure depression along the surface

*3.2. Assessment of the physical properties*

As previously discussed, thermo-sensible physical properties and temperature drop significantly influence the cavitation behaviours. In this section, the properties of liquid nitrogen are characterized by surrogate-based global sensitivity analysis and then the weights of each variable are evaluated.

Physical properties like liquid density $\rho_l$ and saturated vapor pressure $p_v$ could directly impact the $\sigma_\infty$, and hence they are kept constant. Thermal diffusivity $\alpha$ is not taken into consideration due to the steady computation. Since temperature change is small in cavitating flow, thermal conductivity $K$ is ignored either. The vapor density $\rho_v$ affects the mass transfer process, liquid and vapor heat capacity $C_l$ and $C_v$ influence the temperature change, and latent heat $L$ determines the energy of release during the phase change. Therefore, these four physical properties $\rho_v$, $C_l$, $C_v$, $L$ are chosen as design variables primarily, keeping $R_e$ and $\sigma_\infty$ constant for the Case 290C. Secondly, properties like liquid and gas viscosities which influence the flow structure are taken into consideration. To keep the computational expense reasonable, these properties are perturbed within ±10% as listed range 1 in table 1, it is equal for them to influence the objective. The performance of the cavitation behaviours is characterized by the predicted cavity area and the maximum temperature drop. Based on Case 290C, the responses are evaluated by using CFD simulations at 80 data points selected via Face-centered Cubic Composite Design (FCCD) and Latin Hypercube Sampling (LHS) experimental designs and the weighted average surrogate model (WAS [13]) are constructed for approximation of response.

**Table 1.** The range of the design variables based on Case 290C

| Variables | $\rho_v$ | $L$ | $C_l$ | $C_v$ | $\mu_l$ | $\mu_v$ |
|---|---|---|---|---|---|---|
| **Range 1** | ±10% | ±10% | ±10% | ±10% | / | / |
| **Range 2** | ±13.40% | ±1.13% | ±0.50% | ±1.28% | ±5.71% | ±2.09% |

Figure 4 evaluates the weights of each variable via global sensitivity analysis as pie-charts, as illustrated in figure 4(a), $C_l$ and $L$ are exceedingly significant for the cavity area, while $\rho_v$ has less contribution, and $C_v$ is the least influential parameter within the four variations. Figure 4(b) illustrates the percentage contribution for maximum temperature drop $\Delta T_{max}$, it is observed that the weights of each variable are almost the same as that in figure 4(a). These observations indicate that thermodynamic effects play the dominant role in the cavitation behaviours when $\sigma_\infty$ and $R_e$ are kept constant, it could significantly affect flow structure and temperature distribution.

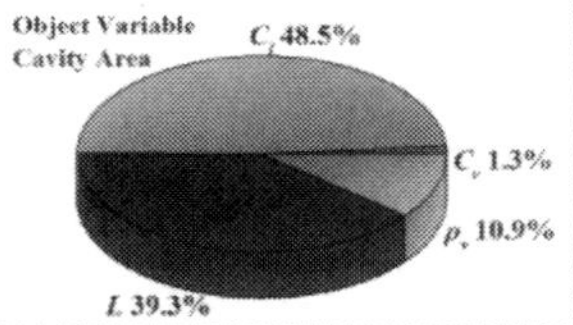


**Figure 4(a).** The sensitivity for cavity area with **range 1**.

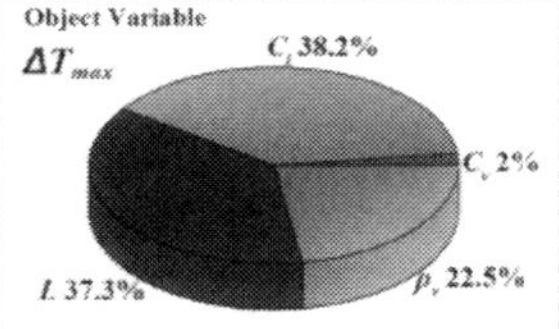


**Figure 4(b).** The sensitivity for $\Delta T_{max}$ with **range 1**.

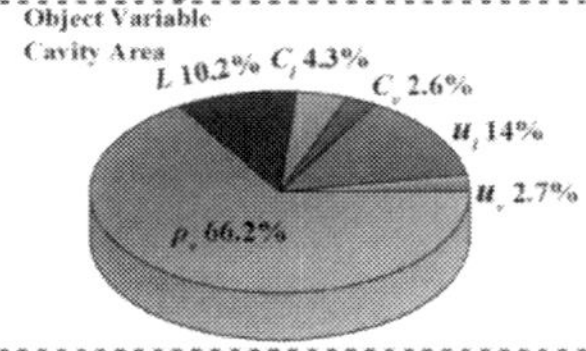


**Figure 5(a).** The sensitivity for cavity area with **range 2**.

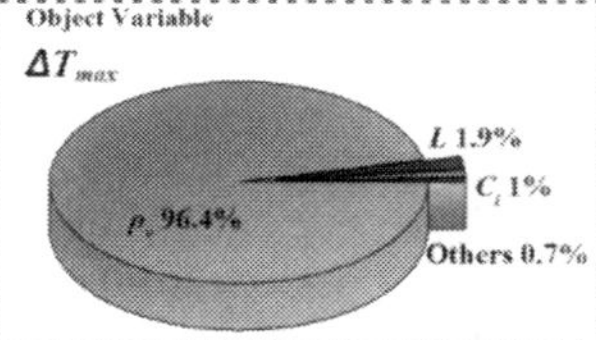


**Figure 5(b).** The sensitivity for $\Delta T_{max}$ with **range 2**.

As previously described, properties $C_l$, $L$ and $\rho_v$ have a significant influence on cavity structure, however, these properties vary differently with changing temperature. $C_l$, $C_v$ and $L$ change slowly while $\rho_v$ has higher slope of saturation curve with increasing temperature. According to the relative change ratio between each variable when liquid nitrogen temperature is 83.06K, a new range of variables named Range 2 is designed and the liquid and gas viscosities are also considered, as listed in table 1, $\rho_v$ changes a lot while other variables change a little.

The weights of each variable with the new range are shown in figure 5, which are completely different from the range 1. For the cavity area, $\rho_v$ accounts for 66.2%, $L$ and $\mu_l$ affect the cavity area slightly while $C_l$ which significantly influence the cavity area in the range 1 as well as $C_v$ and $\mu_v$ could be ignored. It is observed in figure 5(b) that $\rho_v$ plays the leading role on temperature drop, while other variables do not have noticeable contribution. It indicates that the variation of these properties could be ignored when temperature is changing. The method and results can be beneficial to select properties which should vary with temperature in thermo-fluid cavitating flows.

## 4. Conclusions

In summary, the thermodynamic effects significantly affect the thermo-fluid cavitating flow, including pressure and temperature distribution, the variation of physical properties, and cavity structures. Thermodynamic parameter $\sum$ suggests that the thermodynamic effects can be evaluated by physical properties under the same free-stream condition. $\rho_v$, $C_l$ and $L$ could significantly affect temperature drop and cavity structure in the existing numerical framework. The vapor density $\rho_v$ plays the dominant role when properties vary with temperature. The liquid viscosity $\mu_l$ slightly affects the flow structure via changing the $R_e$, however, it hardly affects the temperature distribution. The results are beneficial to select properties which should vary with temperature in thermo-fluid cavitating flows.

## Acknowledgment

This work was supportd by the National Natural Science Foundation of China (NSFC, Grant Nos. 51479002 and 51306020).

## References

[1] Brennen C E 1995 Cavitation and bubble dynamics. *Oxford University Press*
[2] Wang G, Senocak I, Shyy W, et al 2001 *Prog. Aerosp. Sci.* 37(6) 551-581
[3] Utturkar Y, Wu J, Wang G, et al 2005 *Prog. Aerosp. Sci.* 41(7) 558-608
[4] Stahl H A, Stepanoff A J 1956 *ASME J. Fluids. Eng.* 78 1691-1693
[5] Hord J 1973a *NASA* CR-2156
[6] Hord J 1973b *NASA* CR-2242
[7] Rodio M G, De Giorgi M G, Ficarella A 2012 *Int. J. Heat. Mass. Tran.* 55(23) 6538-6554
[8] Huang B, Wu Q, Wang G 2014 *Int. J. Hydrogen. Energ.* 39(4) 1698-1709
[9] Zhu J, Chen Y, Zhao D, et al 2015 *Eur. J. Mech. B-Fluid.* 52 1-10
[10] Merkle C L, Feng J, Buelow P E O 1998 Grenoble France April 7-9 307-311
[11] Menter F R 1992 *NASA* Technical Memorandu 34 103975
[12] Lemmon E W, McLinden M O, Huber M L 2002 NIST Standard Database 23 version 7.0
[13] Goel T, Haftka R T, Shyy W, Queipo N V 2007 *Struct. Multidiscip. O.* 33(3) 199-216

# Evaluation of the radial pressure distribution of vortex models and comparison with experimental data

**T Hommes[1], J Bosschers[2], H W M Hoeijmakers[1]**
[1] Engineering Fluid Dynamics Department, Twente University, Drienerlolaan 5, 7522 NB, Enschede, the Netherlands
[2] MARIN, Haagsteeg 2, 6700 AA, Wageningen, the Netherlands

E-mail: j.bosschers@marin.nl

**Abstract.** An important input parameter for a recently developed semi-empirical prediction method for the noise generated by cavitating vortices is the diameter of the cavitating vortex. This diameter may be obtained from models of non-cavitating vortices that describe the radial distribution of the azimuthal velocity, for given vortex strength and size of the viscous core. The present paper discusses several vortex models that suit this purpose and compares the distribution of the azimuthal velocity as well as that of the pressure with detailed experimental data for the tip vortex of a rectangular wing obtained from literature.

## 1. Introduction

A semi-empirical method is in development for the prediction of broadband hull pressure fluctuations and radiated noise generated by cavitating vortices on marine propellers [1]. The method requires an estimate of the cavity diameter, which can be obtained from a vortex model and a known value for the tip vortex circulation and viscous core size. The circulation will be obtained from a boundary element method and the viscous core size will be estimated empirically. The vortex model consists of the radial distribution of azimuthal velocity from which the radial distribution of the pressure can be computed. Previous research has suggested that a vortex model describing non-cavitating flow can be used to estimate the cavity diameter given the cavitation number [2]. The present paper discusses several vortex models, found in literature, that describe the distribution of the azimuthal velocity of a vortex in non-cavitating flow and that can be used in the semi-empirical method. All models require as input the tip vortex circulation and the viscous core size, while some of the models have tuning coefficients as well. The distribution of the azimuthal velocity and that of the pressure obtained from the models are compared to experimental data for a tip vortex of a half-model of a rectangular wing at ten degrees incidence as measured by Chow *et al.* [3] in a low speed wind tunnel at a Reynolds number based on chord length of $4.6\text{x}10^6$. Because the pressure has been measured for this dataset we first investigate if this pressure distribution can be reconstructed from the measured distribution of the velocity and Reynolds stresses. Data was measured at several cross-flow planes but here only results are shown for the plane at $x/c = 0.678$ with $c$ the wing chord length and with $x = 0$ located at the wing trailing edge and $x$ increasing in downstream direction.

## 2. Reconstruction of the measured pressure distribution from the measured velocity field

The measured velocity components were given in Cartesian coordinates and, after interpolation to an O-type grid, they are converted to a cylindrical coordinate system preserving the original orientation of the x-axis. The origin of this coordinate system was slightly adjusted until all azimuthal velocities are positive. A colour coded contour plot of the azimuthal velocity is given in figure 1. At $x/c = 0.678$ most of the trailing vorticity is in the tip vortex, apparent from the distribution being reasonably axisymmetric. However, further upstream where a substantial part of the vorticity is still in the wake shear layer, the distribution is far from axisymmetric.

All velocity components have been made non-dimensional by a reference velocity $V_{ref} = \Gamma_\infty/2\pi r_v$ with $\Gamma_\infty$ the vortex circulation in the far field and $r_v$ the radius of the viscous core, defined as the radius at which the azimuthal velocity has its maximum. All coordinates have been made non-dimensional with $r_v$. An overline indicates a non-dimensional parameter with $(\overline{V}_x, \overline{V}_r, \overline{V}_\theta)^T$ the non-dimensional time-averaged velocity component in axial, radial and azimuthal direction, respectively and $(\overline{v}_x, \overline{v}_r, \overline{v}_\theta)^T$ the non-dimensional turbulent velocity component in axial, radial and azimuthal direction, respectively. The pressure coefficient $C_p$, defined as $C_p \equiv (p - p_\infty)/\frac{1}{2}\rho V_{ref}^2$, can be computed from the radial component of the Reynolds-averaged Navier-Stokes equation expressed in cylindrical coordinates, assuming axisymmetric flow and constant molecular viscosity $\mu$:

$$\frac{1}{2}\frac{\partial C_p}{\partial \overline{r}} = -\overline{V}_r \frac{\partial \overline{V}_r}{\partial \overline{r}} - \overline{V}_x \frac{\partial \overline{V}_r}{\partial \overline{x}} + \frac{\overline{V}_\theta^2}{\overline{r}} + \frac{1}{\mathrm{Re}}\left\{\frac{\partial}{\partial \overline{r}}\left[\frac{1}{\overline{r}}\frac{\partial(\overline{r}\overline{V}_r)}{\partial \overline{r}}\right] + \frac{\partial^2 \overline{V}_r}{\partial \overline{x}^2}\right\} - \frac{\overline{\overline{v}_r^2}}{\overline{r}} - \frac{\partial \overline{\overline{v}_r^2}}{\partial \overline{r}} - \frac{\partial\left(\overline{\overline{v}_r \overline{v}_x}\right)}{\partial \overline{x}} + \frac{\overline{\overline{v}_\theta^2}}{\overline{r}} \tag{1}$$

in which the Reynolds number Re is defined as $\mathrm{Re} = \rho V_{ref} r_v/\mu$ and the second overline indicates the time-averaged value of a product of turbulent velocity components giving the Reynolds stresses. Equation (1) has been used to investigate how various terms contribute to the pressure distribution in the vortex. For this a distinction has been made between (i) the azimuthal velocity component $\left(\overline{V}_\theta^2/\overline{r}\right)$, (ii) all terms with a mean velocity component only and (iii) all terms including Reynolds stresses. The measured velocity components were first averaged in azimuthal direction and then numerically integrated in radial direction using a Hermite spline to interpolate between the measurement locations.

Figure 2 shows that the pressure distribution is reasonably well predicted from the azimuthal velocity component alone. The contribution of the other terms with mean velocity components in equation (1) is negligible and the result is not shown. Including the contribution of the Reynolds stresses gives some improvement near the vortex centre. The difference in the value of the minimum pressure is 16% using the azimuthal velocity component alone and 7% using all terms including Reynolds stresses. At station x/c= 0.005 the influence of the Reynolds stresses was much smaller and the difference in the value of the minimum pressure was 4% using the azimuthal velocity alone and 5% using all terms including Reynolds stresses.

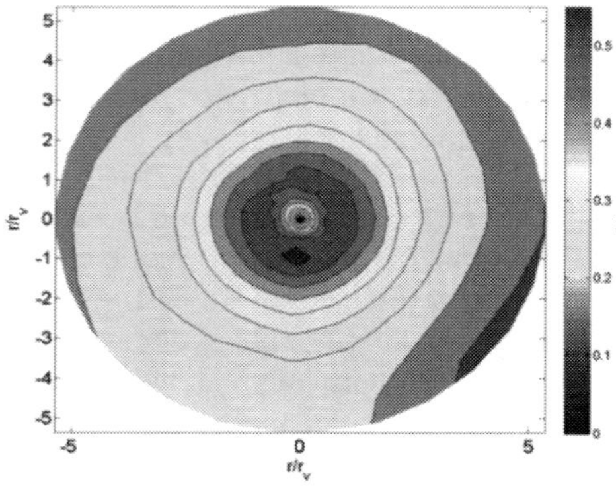


**Figure 1.** Measured $V_\theta$ interpolated to an O-type grid.

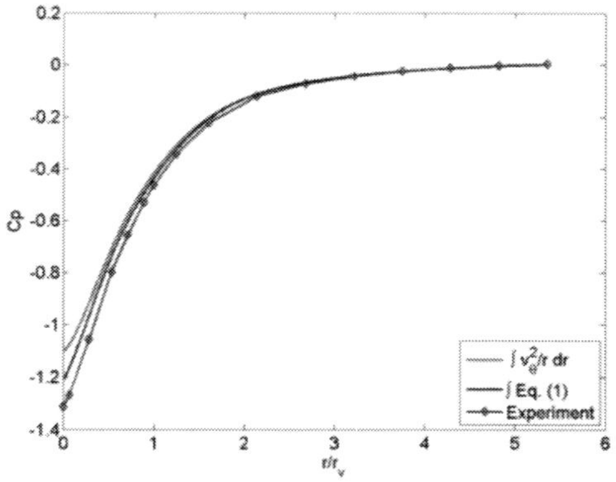


**Figure 2.** Reconstruction of the pressure distribution from various measured velocity terms in equation (1).

## 3. Comparison of vortex models

For comparison of the results of different vortex models only the azimuthal velocity is considered, presented in non-dimensional form with $\bar{V}_\theta = V_\theta / (\Gamma_\infty / 2\pi r_v)$ and $\bar{r} = r/r_v$ as

Rankine
$$\bar{V}_\theta = \begin{cases} \bar{r} & \bar{r} \le 1 \\ 1/\bar{r} & \bar{r} > 1 \end{cases} \tag{2}$$

Lamb-Oseen [4]
$$\bar{V}_\theta = \frac{1}{\bar{r}}\left(1 - e^{-1.2526\bar{r}^2}\right) \tag{3}$$

Burnham-Hallock[5]
$$\bar{V}_\theta = \frac{\bar{r}}{\bar{r}^2 + 1} \tag{4}$$

Proctor[6]
$$\bar{V}_\theta = \begin{cases} \frac{1.4}{\bar{r}}\left(1 - e^{-\beta(1/\bar{B})^{0.75}}\right) \times \left(1 - e^{-1.2526\bar{r}^2}\right) & \bar{r} \le 1 \\ \frac{1}{\bar{r}}\left(1 - e^{-\beta(\bar{r}/\bar{B})^{0.75}}\right) & \bar{r} > 1 \end{cases} \tag{5}$$

Proctor-Winckelmans[7]
$$\bar{V}_\theta = \frac{1}{\bar{r}}\left\{1 - \exp\left[-\beta_i \left(\bar{r}/\bar{B}\right)^2 \Big/ \left\{1 + \left[\left(\beta_i/\beta_o\right)\left(\bar{r}/\bar{B}\right)^{1.2}\right]^p\right\}^{1/p}\right]\right\} \tag{6}$$

Jacquin[8]
$$\bar{V}_\theta = \begin{cases} \bar{r}/\bar{r}_0^{0.5} & \bar{r} \le 1 \\ 1/\left(\bar{r}\,\bar{r}_0\right)^{0.5} & 1 < \bar{r} \le \bar{r}_0 \\ 1/\bar{r} & \bar{r} > \bar{r}_0 \end{cases} \tag{7}$$

Vatistas[9]
$$\bar{V}_\theta = \frac{\bar{r}}{\left(\bar{r}^{2n} + 1\right)^{1/n}} \tag{8}$$

In the vortex core model of Proctor and Proctor-Winckelmans the parameter $\bar{B}$ corresponds to the non-dimensional span of the wing. The parameter $\beta$ in the Proctor model is a tuning parameter with suggested value $\beta = 10$. The suggested ranges of tuning parameters in the Proctor-Winckelmans model are $p = 2-4, \beta_0 \approx 10, \beta_i = 400-800$. The parameter $\bar{r}_0$ in the Jacquin model corresponds to the radial location at which the roll-up region ends and the potential flow part of the vortex starts with $\bar{r}_0 \approx 0.1\bar{B}$. The parameter $n$ in the Vatistas model is also a tuning parameter for which $n = 2$ gave a good fit with experimental data obtained for a vortex chamber.

A comparison with the experimental data is given in figure 3. The pressure distribution inside the vortices was obtained by numerical integration (or analytical integration if possible) of the azimuthal velocity distribution using $C_p = -\int_{\bar{r}}^{\infty} 2\left(\bar{V}_\theta^2/\bar{r}\right) d\bar{r}$. For the non-dimensionalisation we have to determine $\Gamma_\infty$ and $r_v$. The parameter $r_v$ could directly be obtained from the maximum in the azimuthal velocity but for the O-type grid shown in figure 1a the circulation was not yet constant at the outer radii. The measurement grid was, however, larger for one segment of the O-type grid and at these outer radii the circulation no longer changed. The value for $\Gamma_\infty$ at these radii was about 5% larger than that obtained at the outer radius of the O-type grid.

The tuning parameters were adjusted by fitting the azimuthal velocity to the experimental data. The most accurate fit is obtained with the vortex core model of Proctor-Winckelmans, which also has the largest number of tuning parameters, closely followed by the Vatistas vortex core model, which has only one tuning parameter. Both the Rankine and Lamb-Oseen model give a very significant overprediction of the azimuthal velocity and therefore much too low values of the pressure coefficient.

The fit of the vortex core models was applied to five cross-flow planes located between $x/c = 0.005$ and $x/c = 0.678$ and the tuning parameters were adjusted for each plane. Results for the different vortex models were similar. The tuning parameter of the Vatistas model showed an almost linear variation with the reference velocity that varied between the planes while the parameters of the Proctor and Proctor-Winckelmans model showed quite some scatter. Application of the Vatistas and Proctor model to experimental data of wings of elliptical planform, published in [10], confirmed the presence of a distinct trend of the parameter of the Vatistas model with reference velocity and also confirmed the scatter of the parameter of the Proctor model.

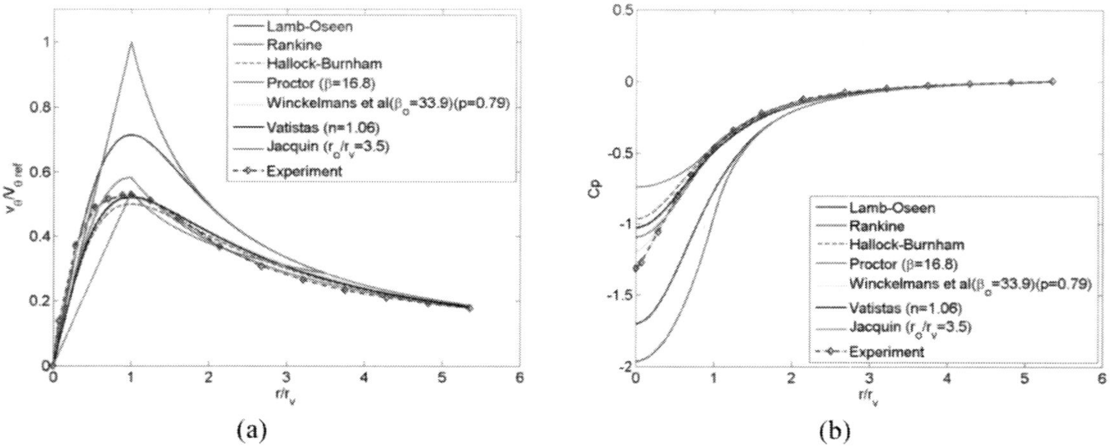


**Figure 3.** Comparison between different vortex core models and experimental data [3] at $x/c = 0.678$ for the azimuthal velocity distribution (left) and the pressure distribution (right).

## References

[1] Lafeber F H, Bosschers J and van Wijngaarden E 2015 Computational and Experimental prediction of propeller cavitation noise, *Conference Oceans'15* MTS/IEEE Genova, Italy

[2] Bosschers J 2010 On the influence of viscous effects on 2-D cavitating vortices, *9th International Conference on Hydrodynamics*, Shanghai, China

[3] Chow J S, Zilliac G G and Bradshaw P 1997 Mean and turbulence measurements in the near wake of wingtip vortex, *AIAA Journal* Vol. 35, No. 10, pp. 1561-1567

[4] Lamb H 1932 *Hydrodynamics*, Cambridge University Press

[5] Burnham D C and Hallock J N 1982 *Chicago monostatic acoustic vortex system system*, Report No. DOT-TSC-FAA-79-104.IV

[6] Proctor F H 1998 The NASA-Langley wake vortex modeling effort in support of an operational aircraft spacing system, AIAA 98-0589, *36th Aerospace Sciences Meeting & Exhibit*, Reno, Nevada

[7] Winckelmans G S, Thirifay F and Ploumhans P 2000 Effect of non-uniform wind shear onto vortex wakes: parametric models for operational systems and comparison with CFD studies, *Proc. 4th Wakenet Workshop on Wake Vortex Encounter*, NLR, Amsterdam, The Netherlands

[8] Jacquin L, Fabre D, Geffroy P and Coustols E 2001 The properties of a transport aircraft extended near field: an experimental study, *AIAA Conference Proceedings* 2001-1038

[9] Vatistas G H, Kozel V and Mih W C 1991 A simpler model for concentrated vortices, *Experiments in Fluids* 11, pp 73-76

[10] Fruman D, Dugue C, Pauchet A, Cerruti P and Briançon-Marjolet L 1992 Tip vortex roll-up and cavitation, *19th Symposium on Naval Hydrodynamics*, Seoul, Korea

# RANS computations of tip vortex cavitation

**Jean Decaix**[1], **Guillaume Balarac**[2], **Matthieu Dreyer**[3], **Mohamed Farhat**[3], **Cécile Münch**[1]

[1], University of Applied Sciences and Arts, Western Switzerland, 47 Route du Rawyl, 1950 Sion, Switzerland

[2], Univ. Grenoble Alpes, LEGI, F-38000 Grenoble, France, CNRS, LEGI, F-38000 Grenoble, France

[3], Ecole Polytechnique Fédérale de Lausanne, Laboratory for Hydraulic Machines, 33 Bis Av. De Cour, 1007 Lausanne, Switzerland

E-mail: `jean.decaix@hevs.ch`

**Abstract.** The present study is related to the development of the tip vortex cavitation in Kaplan turbines. The investigation is carried out on a simplified test case consisting of a NACA0009 blade with a gap between the blade tip and the side wall. Computations with and without cavitation are performed using a RANS modelling and a transport equation for the liquid volume fraction. Compared with experimental data, the RANS computations turn out to be able to capture accurately the development of the tip vortex. The simulations have also highlighted the influence of cavitation on the tip vortex trajectory.

## 1. Introduction

Confined tip vortices develop in the gap between a blade tip of a finite span and a side wall due to the pressure difference between the blade pressure and suction sides. Close to the leading edge, this pressure difference generates a leakage jet that rolls up to form the so-called tip-leakage vortex. The leakage jet and the vortex are influenced by the gap width as shown experimentally [1, 2] and numerically [3, 4]. Furthermore, the tip-leakage vortex promotes the cavitation inception [5] due to the low pressure reached in the vortex core. Regarding Kaplan turbines, the damages depend on the strength and the trajectory of the vortex. The prediction of these two characteristics are a challenging task for the computations.

The present paper deals with a numerical investigation of the confined tip-leakage vortex. The test case is a NACA0009 blade mounted in a rectangular channel with a gap between the blade tip and the side wall. Computations are carried out for one particular gap width, with and without cavitation. The results are analysed and compared with experimental data [2].

## 2. Experimental test case

The hydrofoil used in both the numerical and experimental cases is a blunt trailing edge NACA0009 [2] with a chord length $c = 100$ mm and a maximum thickness of 9.9 mm. It is mounted in the EPFL high-speed cavitation tunnel in a rectangular channel section of 150 mm $\times$ 150 mm $\times$ 750 mm . The foil incidence is $i = 10°$. The gap $\tau$ between the blade tip and the side wall is set to $\tau/c = 0.1$. The inlet velocity is $u_{inlet} = 10.2\ m.s^{-1}$, which leads to a Reynolds number based on the chord length $Re_c = 10^6$. The cavitation number $\sigma_u$ is computed using the

Journal of Physics: Conference Series **656** (2015) 012183 doi:10.1088/1742-6596/656/1/012183

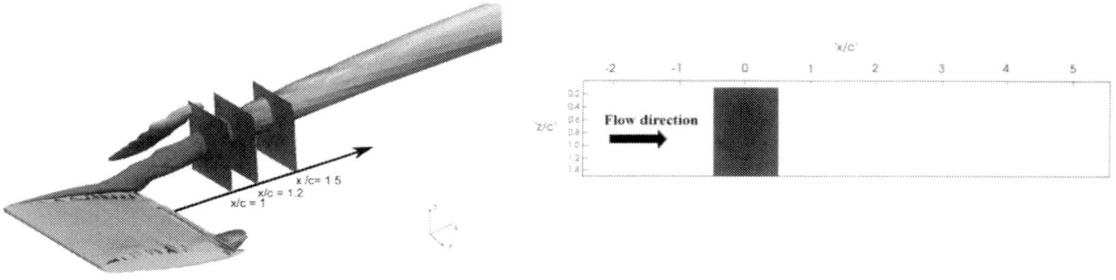


**Figure 1.** Left: position of the experimental measurement sections with the tip vortex highlighted by an iso-surface of the Q-criterion. Right: Computational domain (top view).

pressure reference located upstream of the leading edge:

$$\sigma_u = \frac{p_{ref} - p_{sat}}{\frac{1}{2}\rho u_{inlet}^2} = 2.01 \tag{1}$$

With :

$$p_{ref} = 107'000\,Pa \quad p_{sat} = 2'300\,Pa \quad \rho = 1'000\,kgm^{-3} \tag{2}$$

For the non-cavitating case, the available experimental data are the velocity fields at three cross sections downstream of the trailing edge (see figure 1 left). For the cavitating case, only flow visualisations are performed from the side of the test section.

## 3. Numerical set up

The computations are performed on a domain similar to the experimental test section except for the channel length, which extends two chords upstream of the leading edge and five chords downstream of the trailing edge (see figure 1 right). The OpenFOAM 2.1.0 software is used to solve the RANS equations. Turbulence is modelled with the $k - \omega$ SST model. Cavitation is taken into account by solving an additional transport equation for the liquid volume fraction defined by:

$$\frac{\partial \alpha_L}{\partial t} + (\vec{u} \cdot \nabla)\, \alpha_L = m_c + m_v \tag{3}$$

Where :

$$m_c = \frac{\rho}{\rho_L \rho_V} C_c \frac{\rho_V}{t_\infty} \alpha_{Llim}^2 \frac{\max(p - p_{sat}\,;\,0)}{\max(p - p_{sat}\,;\,0.01\,p_{sat})} \tag{4}$$

$$m_v = \frac{\rho}{\rho_L \rho_V} C_v \frac{\rho_V}{\frac{1}{2}\rho_L U_\infty^2 t_\infty} \min(p - p_{sat}\,;\,p_0) \tag{5}$$

The parameters are set to: $C_v = 800$, $C_c = 500$, $U_\infty = u_{inlet} = 10.2$ ms$^{-1}$ and $t_\infty = 0.005$ s.
A structured mesh containing 2 million of nodes is generated with 30 nodes along the spanwise direction in the gap. The inlet velocity profile is set by using the velocity profile provided by a RANS computation on a computational domain twice longer. The pressure level is set at the experimental probe section upstream of the blade, whereas a zero gradient pressure condition is imposed at the outlet.

**Table 1.** Dimensionless position of the vortex core.

| $x/c$ | Experiment | Non-cavitating RANS computation | Absolute error (%) |
|---|---|---|---|
| 1.0 | y/c = 0.141 | y/c = 0.132 | 6.4 |
| | z/c = 0.120 | z/c = 0.107 | 11.0 |
| 1.2 | y/c = 0.183 | y/c = 0.179 | 2.2 |
| | z/c = 0.132 | z/c = 0.122 | 7.6 |
| 1.5 | y/c = 0.303 | y/c = 0.217 | 28.4 |
| | z/c = 0.160 | z/c = 0.149 | 6.9 |

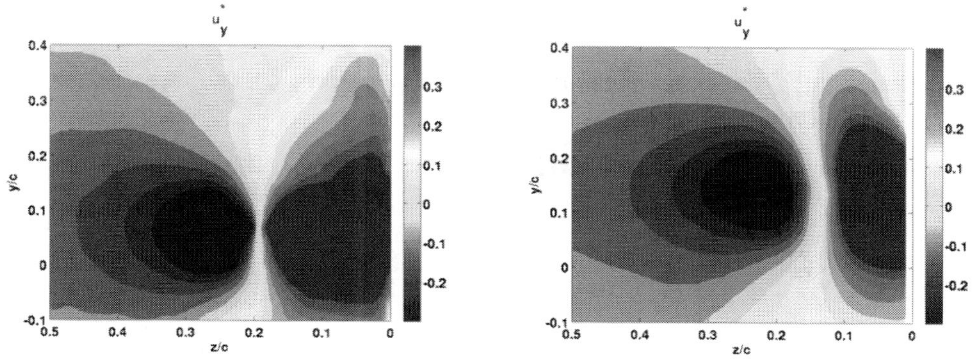


**Figure 2.** Dimensionless pitchwise velocity component at the first cross section located at $x/c = 1$. Experiment (left) and non-cavitating RANS computation (right).

## 4. Results

### *4.1. Non-cavitating case*

Figure 2 compares the experimental and computed contours of the dimensionless pitchwise component of the velocity field ($u_y^* = u_y/u_{inlet}$) one chord downstream the trailing edge. The computation agrees with the experimental both regarding the position of the contours and the magnitude of the velocity. The main difference occurs at the vortex centre where the spanwise gradient is smoother for the computation. Table 1 gathers the dimensionless pitchwise and spanwise position of the vortex core determined by the location of the maximum of axial vorticity. Except for the pitchwise position at the last section, the position of the vortex is captured with less than approximately 10% of error by the computation. Such an error represents less than 2 millimetres, which is approximately the mesh cell size. Therefore, the position of the vortex core downstream of the blade is considered to be fairly captured by the computation.

### *4.2. Cavitating case*

To compute the cavitating case, the cavitation number has been decreased to $\sigma_u = 1.3$ instead of the experimental value $\sigma_u = 2.1$. Indeed, as the RANS computation is not able to fully resolve the vortex core, the pressure drop at the vortex centre is underestimated compared to the experiment. The pressure level in the computation is thus decreased to obtain an amount of cavitation in the vortex core similar to the experimental one. Figure 3 shows the iso-surface of the liquid volume fraction provided by the computation and an instantaneous experimental picture of the cavitating tip vortex. The position and the size of the tip vortex cavitation are well captured by the computation. Downstream of the blade, due to the mesh resolution, cavitation cannot be maintained. The iso-surface of the dimensionless Q-criterion is displayed on figure 4

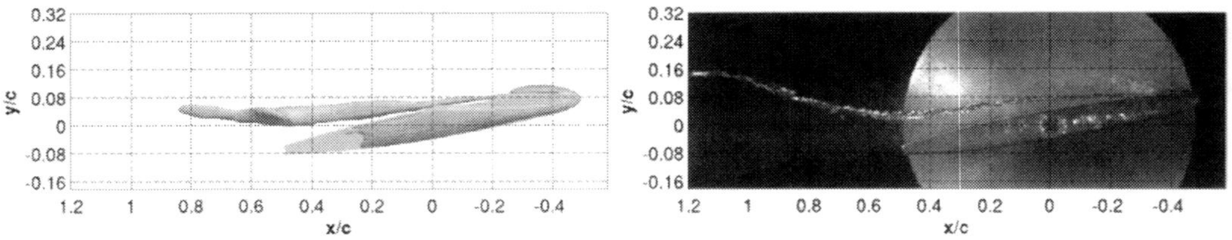


**Figure 3.** Iso-surface of the mean liquid volume fraction $\alpha_L = 0.9$ provided by the cavitating RANS computation (left) and instantaneous experimental picture (right).

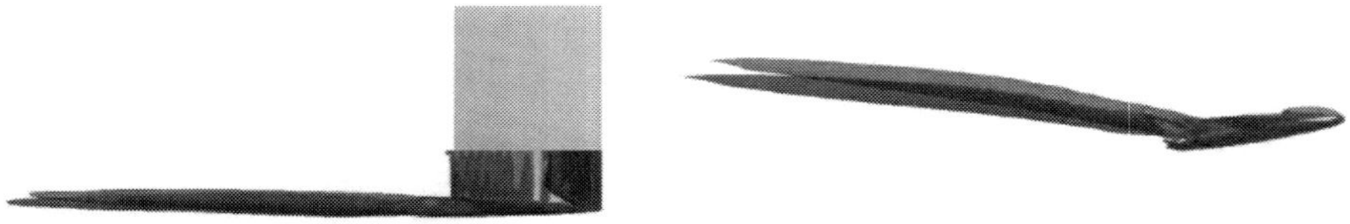

**Figure 4.** Iso-surface of the dimensionless Q-criterion. Top view (left) and side view (right). Red: non cavitating case. Blue: cavitating case.

both for the non cavitating and the cavitating cases. It is observed that the tip vortex trajectory is influenced by the cavitation. In absence of cavitation, the tip vortex moves upward and to the centre of the channel at it travels downstream. In presence of cavitation, the tip vortex moves upward more slowly and remains close to the side wall. This difference can be explained by the detachment of the boundary layer at the leading edge of the blade (not shown here) due to a thin layer of cavitation. The detachment is followed by a recirculation zone on the suction side that leads the tip vortex to move towards the blade. Furthermore, the detachment provokes a large wake downstream of the blade that confines the tip vortex close to the side wall.

## 5. Conclusion

The one-phase RANS computation is able to capture the tip vortex trajectory and the velocity field with reasonable accuracy compared to the experiment. By decreasing the pressure level in the channel, cavitation occurs in the vortex core in agreement with the experimental observations. It is shown that the tip vortex moves closer to the suction side of the blade and the side wall, which have to be taken into account regarding the cavitation erosion risk.

## Acknowledgements

The authors are very grateful to the Competence Center in Energy and Mobility (CCEM), Swisselectric Research and the foundation The Ark through the program The Ark Energy for their financial support.

## References

[1] Muthanna C and Devenport W J 2004 *AIAA Journal* **42** 2320–2331
[2] Dreyer M, Decaix J, Münch C and Farhat M 2014 *Experiment in Fluids* **55**
[3] You D, Wang M, Moin P and Mittal R 2006 *Physics of Fluids* **18**
[4] Decaix J, Balarac G, Dreyer M, Farhat M and Münch C 2015 *Journal of Turbulence* **16** 309–341
[5] Arndt R E A 2002 *Annual Review of Fluid Mechanics* **34** 143–175

# Numerical Analysis of Tip Cavitation on Marine Propeller with Wake Alignment Using a Simple Surface Panel Method "SQCM"

**T Kanemaru[1] and J Ando[1]**

[1] Faculty of Engineering, Kyushu University, 744 Motooka Nishi-ku Fukuoka, Japan

E-mail: kanemaru@nams.kyushu-u.ac.jp

**Abstract**. This paper presents the calculation method of tip cavitation with wake alignment. Tip cavitation consists of tip vortex cavitation and tip super cavitation which means the undeveloped and local super cavitation around blade tip. The feature of this study is that the method applies the wake alignment model in order to express the realistic phenomena of tip cavitation and predict the pressure fluctuation more accurately. In the present method, the wake sheet is deformed according to the induced velocity vector on the vortex lines. The singularity of the potential vortex can be removed by using the Rankine Vortex model. This paper shows the calculated results regarding cavitation pattern, pressure fluctuation etc. comparing with published experimental data and calculated results without wake alignment.

## 1. Introduction

The unsteady cavitation on marine propeller produces the considerable pressure fluctuation on ship stern, which causes the hull vibration and structural damage. Therefore, it is very important to predict the pressure fluctuation in the propeller design stage and theoretical prediction methods are indispensable for the practicality. We presented the calculation method of propeller sheet cavitation using a simple surface panel method "SQCM" which was developed by Kyushu University (Ando et al. 1998) and we obtained reasonable results regarding not only cavitation pattern but also cavity volume (Kanemaru et al. 2009). The method was extended to the calculation method of pressure fluctuation on ship stern (Kanemaru et al. 2011). Furthermore, we developed the calculation method of tip vortex cavitation and tip super cavitation which are important for the higher order frequency components prediction of the pressure fluctuation (Kanemaru et al. 2015). We named these two types cavitations 'tip cavitation' in the above mentioned paper.

Our previous methods apply the geometrical pitch model for the wake sheet arrangement. However, wake alignment is important to express the complicated phenomena of tip cavitation and obtain more accurate prediction of pressure fluctuation. In this study, we apply the wake alignment model using the induced velocity vector on the wake sheet.

## 2. Calculation Method

### 2.1. *Outline of SQCM*

SQCM (Source and QCM) uses source distributions (Hess and Smith 1964) on the propeller blade surface and discrete vortex distributions arranged on the mean camber surface according to QCM

(Quasi-Continuous vortex lattice Method) (Lan 1974), which is well known as one of lifting surface methods. The formulation of SQCM is described in the papers (Ando et al. 1998, Kanemaru et al. 2009). We applied SQCM to the calculation methods of unsteady propeller sheet cavitation which is based on the free streamline theory (Kanemaru et al. 2009). Also the method was expanded to the calculation method of pressure fluctuation induced by the calculated sheet cavitation (Kanemaru et al. 2011).

### 2.2. *Tip Cavitation*
Although the excitation by the sheet cavitation is predominant in the fluctuating pressure, tip cavitation should be considered for the accurate prediction including higher order frequency components in case that these components causes serious problem. Regarding tip vortex cavitation, several theoretical model were presented by Lee et al. (2004) and Szantyr (2007). We referred these studies in our tip vortex cavitation model. In calculation models, tip vortex cavitation means the cavitation around tip vortex line which is a free vortex arranged at the most tip side on the wake sheet. On the other hand, local and unstable tip super cavitation which exists between the trailing edge and the developed tip vortex cavitation should be considered for more accurate prediction method of pressure fluctuation. We treated these cavitation as 'tip super cavitation' and presented the practical calculation model. These details are shown in the paper (Kanemaru et al. 2015).

### 2.3. *Wake alignment*
In the calculation of panel method or lifting surface method, the wake sheet has to be arranged in advance. Generally, liner wake model based on the theoretical or empirical model. Our previous works 2.1 and 2.2 apply the geometrical pitch model using the blade geometry and the model gives reasonable results regarding thrust and torque at the design point. However, in the actual phenomena, the wake sheet geometry is formed by the velocity vector in the slipstream. In the case of tip cavitation calculation, wake alignment should be considered because the calculated cavity is affected by the wake sheet and the pressure fluctuation is very sensitive to the cavity shape and volume.

In this study, we try to deform the wake sheet according to the induced velocities at the node of the ring vortex defined by SQCM as follows:

$$\vec{W}_{\mu\ell} = \vec{W}_{\mu\ell-1} + \vec{V}_{I\mu\ell-1} \cdot \Delta t + \vec{V}_{S\mu\ell-1} \cdot \Delta t \tag{1}$$

Here, $\vec{W}_{\mu\ell}(x_{\mu\ell}, y_{\mu\ell}, z_{\mu\ell})$ is the position vector at the control point (**Figure 1**), $\vec{V}_{I\mu\ell}(\vec{V}_{Ix}, \vec{V}_{Iy}, \vec{V}_{Iz})$ is the inflow velocity vector and $\vec{V}_{S\mu\ell}(\vec{V}_{Sx\mu\ell}, \vec{V}_{Sy\mu\ell}, \vec{V}_{Sz\mu\ell})$ is the induced velocity vector by the singularities. And $\Delta t$ is the time step size. This method is based on our paper (Kanemaru et al. 2007). The paper calculates the induced velocity vector at the center of ring vortex in order to avoid the singularity by the Biot-Savart law. But in this study, the velocity vector is calculated at the node of ring vortex by applying the Rankine Vortex model which treats the flow near the vortex segment as the solid body (Kanemaru et al. 2011). It is not important to calculate the wake sheet deformation in far area from trailing edge (T.E). The present method is performed from T.E. to 45 degree position from T.E. (**Figure 2**) considering both the accuracy and the practicality.

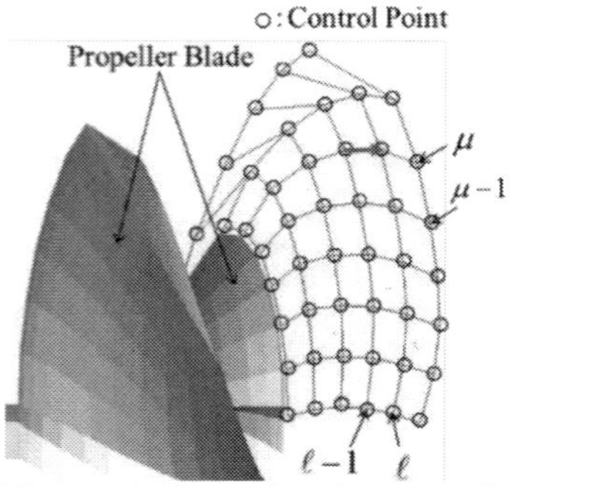

**Figure 1** Control point on wake sheet

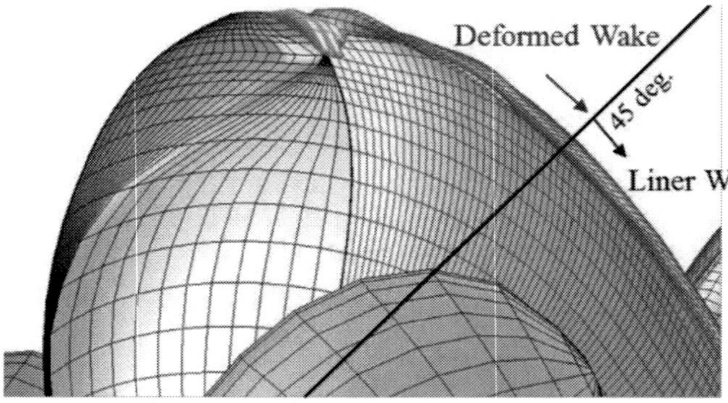

**Figure 2** Deformed wake sheet

## 3. Calculated Results

### *3.1. Calculation of Cavitation*

Seiun-Maru-I conventional propeller (CP) and highly skewed propeller (HSP) are adopted in order to compare with the experimental data at SRI (Kudo et al. 1989). The validation of sheet cavitation model is presented in our paper (Kanemaru et al. 2009). The detail of the calculation conditions and definitions are same to the paper.

**Figure 3** shows the calculated cavitation patterns at each angular position. It can be seen that the tip cavitation occurs when the blade passes the high loading position. Owing to the wake alignment, the twisting of tip vortex cavitation by the tip vortex roll up can be seen.

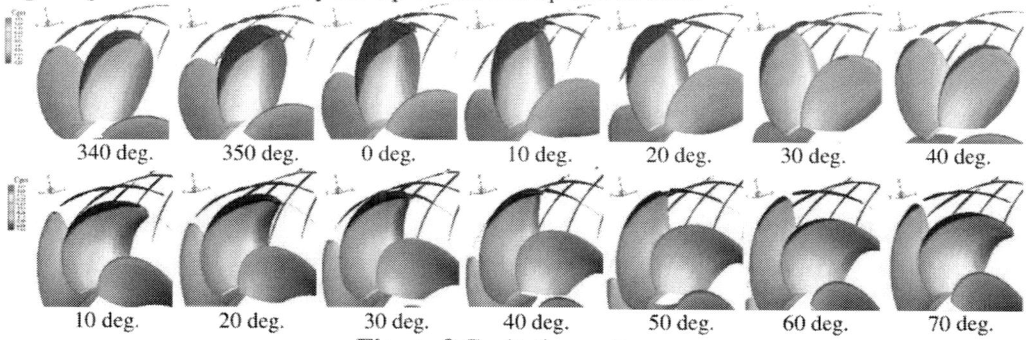


**Figure 3** Cavitation patterns
( Upper: CP ( $K_T$ =0.207, $\sigma_n$=3.06) , Lower: HSP ( $K_T$ =0.201, $\sigma_n$=2.99) )

### *3.2. Calculation of Pressure Fluctuation*

**Figure 4** shows the fluctuating pressure on the point where is right over the propeller (Kanemaru et al. 2009). The fluctuating pressure $K_p$ consists of three components as:

$$K_p = K_{p\,ShC} + K_{p\,TVC} + K_{p\,TSC} \qquad (2)$$

Here $K_{p\,ShC}$ is the result by the sheet cavitation model only which includes the components of lift and blade thickness. $K_{p\,TVC}$ and $K_{p\,TSC}$ are the components of tip vortex cavitation and tip super cavitation. $K_{p\,TVC}$ is very small because the tip vortex cavitation model is applied from the confluence point of free vortices which form the tip vortex. On the other hand, $K_{p\,TSC}$ has large fluctuation in results of both CP and HSP. The difference between w/ and w/o wake alignment can be seen in $K_{p\,TSC}$ of both CP and HSP.

**Figure 5** shows the amplitude of pressure fluctuation at the same point of **Figure 4**. In these figures, the components from 1st order blade frequency to 4th one are shown and compared with the experimental data (Kurobe et al. 1983) and the calculated results w/o wake alignment. Overall, the difference of the amplitudes by the present method and the method w/o wake alignment is not so large. However the higher order frequency components by the present method are larger than those by the method w/o wake alignment and these results are closer to the experimental data. This denotes that the treatment of wake alignment is important for accurate prediction method of pressure fluctuation.

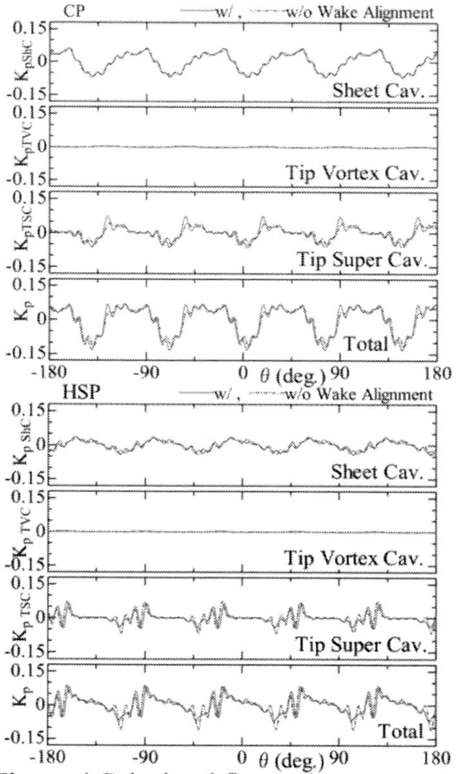


**Figure 4** Calculated fluctuating pressure
(Upper: CP, Lower: HSP)

## 4. Conclusion

The amplitudes of pressure fluctuation by the present method are closer to the experimental data comparing with the results by the method w/o wake alignment regarding the higher order frequency components. The present study shows that the exact treatment of cavitation around tip might contribute to the accuracy of prediction method. To develop more realistic model is one of our future works.

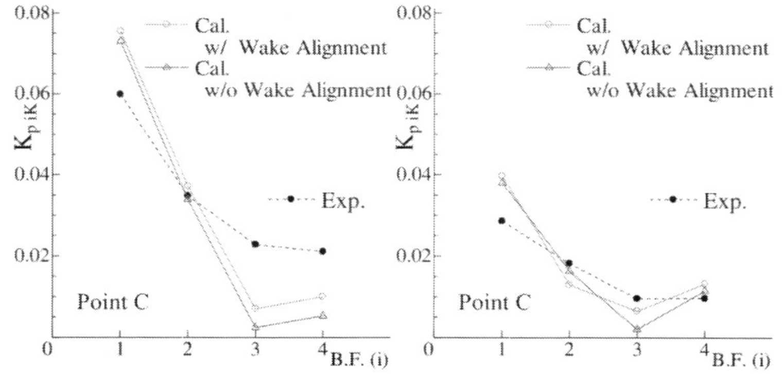


**Figure 5** Amplitude of pressure fluctuation (Left: CP, Right: HSP)

## References

[1] Ando J, Maita S and Nakatake K 1998 A New Surface Panel Method to Predict Steady and Unsteady Characteristics of Marine Propeller *Proceedings of 22nd Synposium on Naval Hydrodynamics* Washington D.C. pp 142-154

[2] Hess J L and Smith A M O 1964 Calculation of Nonlifting Potential Flow about Arbitrary Three Dimensional Bodies *Journal of Ship Research* vol 8 no 2 pp 22-44.

[3] Kanemaru T and Ando J 2007 Prediction of Unsteady Propeller Performance Including Wake Alignment *Journal of the Japan Society of Naval Archtects and Ocean Engineers* vol 18 pp 267-279

[4] Kanemaru T and Ando J 2009 A Numerical Analysis of Steady and Unsteady Sheet Cavitation on a Marine Propeller Using a Simple Surface Panel Method "SQCM" *Proceedings of 1st International Synposium on Marine Propulsor* Trondheim pp 372-379

[5] Kanemaru T. and Ando J 2011 Numerical Analysis of Pressure Fluctuation on Ship Stern Induced by Cavitating Propeller Using a Simple Surface Panel Method "SQCM" *Proceedings of 2nd International Synposium on Marine Propulsor* Hamburg pp 116-123

[6] Kanemaru T and Ando J 2011 Numerical Analysis of a Marine Propeller Performance with Unsteady Motion *Conference Proceedings of The Japan Society of Naval Architects and Ocean Engineers* vol 13 pp 29-30

[7] Kanemaru T. and Ando J 2015 Numerical Analysis of Tip Vortex Cavitation on Marine Propeller Using a Simple Surface Panel Method "SQCM" *Proceedings of 4th International Synposium on Marine Propulsor* Austin vol 3 pp 505-512

[8] Kudo T, Ukon Y, Kurobe U and Tanibayashi H 1989 Measurement of Shape of Cavity on a Model Propeller Blade *Journal of the Society of Naval Architects of Japan* vol 166 pp 93-103

[9] Kurobe Y, Ukon Y, Koyama K and Makino M 1983 Measurement of Cavity Volume and Pressure Fluctuation on a Model of the Training Ship "SEIUN-MARU" with Reference to Full Scale Measurement *Report of the SRI* vol 20 no 6

[10] Lan C E 1974 A Quasi-Vortex-Lattice Method in Thin Wing Theory *Journal of Aircraft* vol 11 no 9 pp 518-527

[11] Lee H and Kinnas S A 2004 Application of BEM in the Prediction of Unsteady Blade Sheet and Developed Tip Vortex Cavitation on Marine Propellers *Journal of Ship Research* vol 48 no 1 pp 15-30

[12] Szantyr J A 2007 Dynamic Interaction of Cavitating Propeller Tip Vortex with the Rudder *Polish maritime Research* vol 14 pp 10-14

# Vortex singing in tip vortex cavitation under the effect of water quality

**Xiaoxing Peng**[1], **Lianghao Xu, Mingtai Song, Yantao Cao**
National Key Laboratory on Ship Vibration & Noise, China Ship Scientific Research Center, Wuxi, 214082, China

E-mail: henrypxx@163.com

**Abstract.** The phenomenon of vortex singing in tip vortex cavitation is studied experimentally in this paper. It is revealed that the vortex singing can be observed during the vortex cavity transfer from one phase to another in the same incoming flow condition. The state of vortex singing can only exist in a short duration from several seconds to several minutes. The sound of the vortex singing for different water qualities, including air content and nuclei populations of water are measured by a hydrophone. It is found that the tones of vortex singings are very different under different flow conditions and water qualities.

## 1. Introduction

The tip vortex cavitation is often found in marine propellers and draws consistent attention in ship design because of its potential to produce erosion and emit noise. This topic has been studied extensively by many researches both in model scale experiments and full scale trials. Most of the existing works were focus on the formation of tip vortex cavitation and the scale effect of tip vortex cavitation inception, aiming to predict and suppress tip vortex cavitation. However, some works were made on developing vortex cavitation to explore the dynamics of tip vortex cavitation and its instability. Among these, the vortex singing in vortex cavitation is a very special phenomenon, which was first found in the 1990's.

The noise emitted by blade tip vortex cavitation was investigated on model scale by Maines and Arndt[1]. They observed the tone sound with special frequency, namely, "vortex singing". In their description, the vortex singing can occur when the vortex was attached to the blade tip and only at a small range of cavitation numbers. In their experiments, the vortex singing seems related with a special flow pattern where a part of the boundary layer can still be laminar. But the physical mechanism of this phenomenon is still not clear [2]. The recent attempt to analyze the problem was done by Bosschers[3] , who derived a theoretical formulation to predict the occurrence of the vortex singing. However, the problem is that there are few laboratories which can repeat the special phenomena up until now.

In our paper, some recent experiments for the vortex singing are reported. The occurrence process of the vortex singing and the effects of water qualities, including air content in water and nuclei populations are emphasized in the study.

---

1 Correspondence author: Dr. Xiaoxing Peng, henrypxx@163.com.

## 2. Experimental Setup

Experiments were conducted in the cavitation mechanism tunnel of China Ship Scientific Research Center (CSSRC). The tunnel has two exchanged test sections of circular and square shapes respectively as shown in Fig.1. There is a large degassing tank installed in the downstream of the test section to control the air content of water. The tunnel is also equipped with a nuclei seeding system. In this study, the test section with a square cross section of 225×225 mm is used, and the maximum incoming velocity of this section can be up to 25m/s.

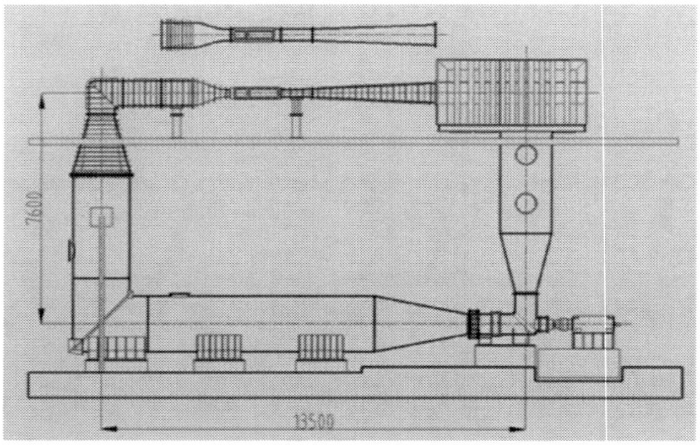


**Figure 1.** Schematic diagram of cavitation tunnel.

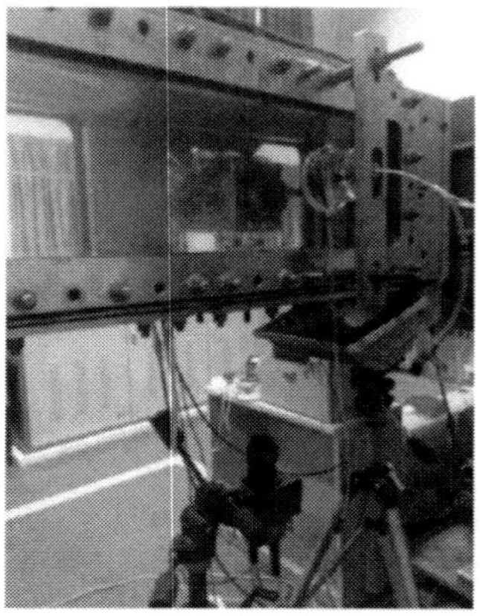

**Figure 2.** Set-up of the experiments.

The elliptic hydrofoil with section NACA $66_2$-415 was chosen as the test model. The model is 94.2mm and 112.5mm in maximum chord and half span respectively. The model was installed in the horizontal center of the test section with the tip in the centreline of the test section and the attack angle is 7° . A high-speed video camera with type of Photron APX was used to visualize the development of the tip vortex cavitation and a LED lamp was used as the light source. Both devices were set in the bottom of the test section. A hydrophone of B&K 8103 was installed in the side window with a cup filled with water. The arrangement of the experiment was shown in Fig.2.

## 3. Vortex Singing Observations

During the experiments, strong tip vortex cavitations were set first with given incoming flows, and then the pressures were decreased until the vortex singing occurred. The vortex singing happens at a specific cavitation number range, which is narrow. On the one hand, in cases when cavitation numbers are lower than that range, the vortex cavitation is quite thick, with rotation stationary and stable even in the scenario which affected by the shedding of attached cavitation in the upstream. On the other hand, in situations where cavitation numbers are higher than the vortex singing range, the vortex cavitation is quite thin and departs from the tip.

We can observe the vortex singing when we hit a specific narrow cavitation number range. During the vortex singing, we keep the cavitation number unchanged while we observe three different cavity phases as shown in Fig.3. The form of the vortex cavitation is unstable and keeps changing. In the beginning, the vortex cavitation is very similar to the pattern which usually happens at low cavitation numbers and the attached cavity around the tip is stable. In the second phase, the vortex cavitation starts to oscillate with synchronization of attached cavity and the distinct tone can be heard. This process can continue for the duration from several seconds to a few minutes with different test cases. When the vortex singing disappears, the vortex cavitation becomes very thin, which is similar to the cases which often happen at high cavitation numbers.

Noise measurements results are shown in Fig.4, where the noise spectra clearly show the characteristics of vortex singing. When vortex singing happens, resonance frequency and its second mode can be seen with very strong sound amplitude, which is over 30dB higher than non vortex singing. The interesting thing is that the two noise spectra of non vortex singing situations are almost the same while they are also consistent with the noise spectra of the vortex singing except for the resonance frequency and its second mode.

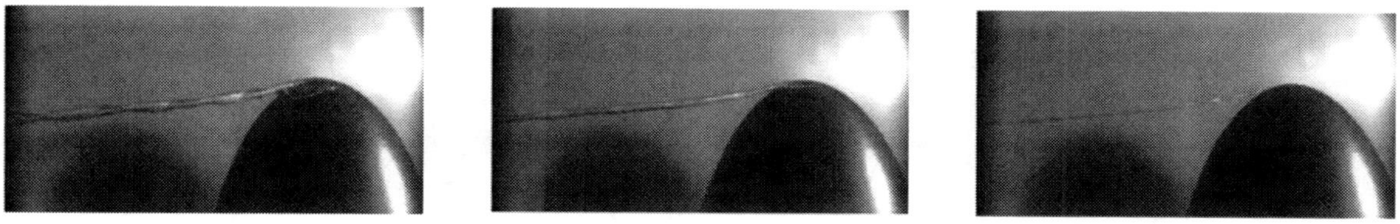

**Before singing** **In singing** **After singing**

**Figure 3.** Form of tip vortex cavitation before, in and after vortex singing.
(v=12.8m/s, $\sigma$ =1.40, $\alpha/\alpha_s$=0.77, no nuclei seeding)

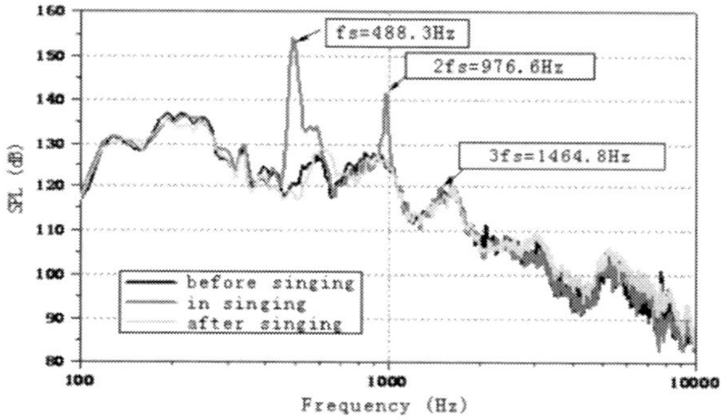


**Figure 4.** Noise spectra of vortex cavitation before, in and after vortex singing.
(v=12.8m/s, $\sigma$ =1.40, $\alpha/\alpha_s$=0.77, no nuclei seeding)

## 4. Effects of Water Quality on Vortex Singing

### *4.1. Effects of air content*

Three air content of water, 50%, 63% and 77% with different incoming velocity were tested to study the effect of air content on vortex singing. Fig.5 shows the results for the three different air content with the same incoming velocity of 12.8m/s. The cavitation numbers in the vortex singing change with different air content. With the increase of air content from 50%, 63% to 77%, the cavitation number of the vortex singing increases from 1.2, 1.33 to 1.4, while the resonance frequencies decrease from 525.3, 509.5 to 488.3Hz.

### *4.2. Effects of gas nuclei population*

Non nuclei seeding and three nuclei seeding conditions, namely "few", "medium" and "rich" were tested to study the effects of nuclei on vortex singing. Fig.6 shows the results for different nuclei populations with the same incoming velocity of 9m/s. The cavitation numbers in the vortex singing change remarkably with different nuclei populations. With the increase of nuclei populations, the cavitation number of vortex singing increase from 1.35, 1.4, 1.8 to 1.85, while the resonance frequencies do not change much.

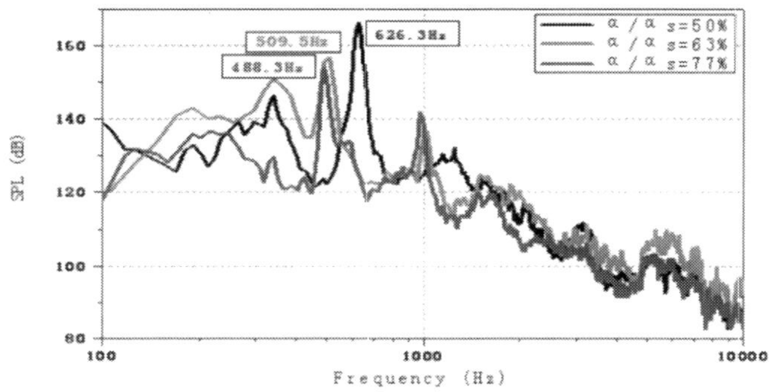


**Figure 5.** Noise spectra of vortex singing for air content 50%, 63%, 77% respectively. (v=12.8m/s)

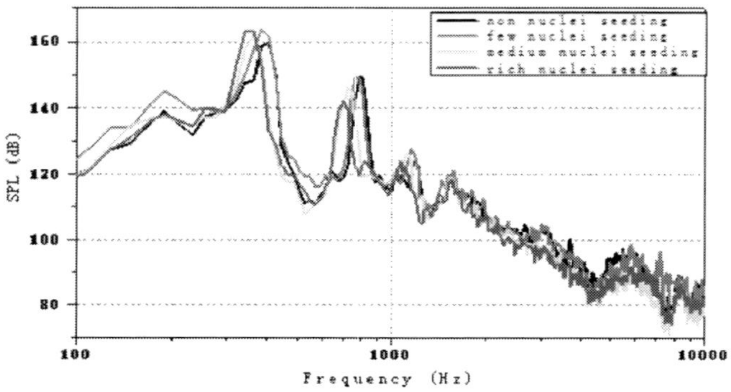


**Figure 6.** Noise spectra of vortex singing for no, few, medium and rich nuclei seeding. (v=9m/s, α /α s=63%)

## 5. Conclusions

Extensive experimental research of vortex singing in tip vortex cavitation for the elliptic hydrofoil with section NACA $66_2$-415 were carried out in this paper. Emphasis is made in the effects of air content and nuclei populations in water. Some conclusions can be drawn from the study.

1) The vortex singing can only exist in the transferring process between strong tip vortex cavitation and weak tip vortex cavitation.
2) With the increase of air content, the cavitation number of the vortex singing increases while the resonance frequency decreases.
3) With the increase of nuclei populations, the cavitation number of vortex singing increases while the resonance frequencies do not change much.

**Acknowledgments**
Authors wish to acknowledge the support from the State Key Program of National Natural Science Foundation of China (Project No. 11332009). Authors would also like to thank the anonymous reviewers for their helpful reviews and Mr. Haoruo Peng for the improvement of the writing.

## References

[1] Maines B and Arndt R E A 1997 *ASME J. Fluids. Eng.* **119** 271
[2] Arndt R E A 2001 *Ann Rev. Fluid Mech.* **87** 965
[3] Bosschers B 2009 *Proc. NAG/DAGA Int. Conf. on Acoustics (Rotterdam)* 298

# Sound signature of propeller tip vortex cavitation

**Pepijn Pennings[1,2], Jerry Westerweel[2] and Tom van Terwisga[1,3]**

1 Laboratory for Ship Hydrodynamics, Delft University of Technology, Mekelweg 2, 2628 CD, Delft, The Netherlands

2 Laboratory for Aero & Hydrodynamics, Delft University of Technology, Mekelweg 2, 2628 CD, Delft, The Netherlands

3 Maritime Research Institute The Netherlands, Haagsteeg 2, 6708 PM, Wageningen

E-mail: p.c.pennings@tudelft.nl

**Abstract.**
The design of an efficient propeller is limited by the harmful effects of cavitation. The insufficient understanding of the role of vortex cavitation in noise and vibration reduces the maximum efficiency by a necessary safety margin. The aim in the present study is to directly relate propeller cavitation sound to tip vortex cavity dynamics. This is achieved by a dedicated experiment in a cavitation tunnel on a specially designed two-bladed propeller using a high-speed video camera and a hydrophone. The sound signature of a tip vortex cavity is not evidently present in the sound spectrum above the tunnel background. The addition of a simulated wake inflow results in a high amplitude broadband sound. With a decrease in the free-stream pressure the centre frequency of this sound decreases as a result of a larger vortex cavity diameter. In the near future each blade passage in the high-speed video will be analyzed in detail. The frequency content of the cavity dynamics can then be directly related to the measured sound. An analytic model for vortex cavity dynamics resulting in a cavity eigenfrequency using a vortex velocity model can finally be evaluated as a design instrument for estimation of broadband sound from propeller cavitation.

## 1. Introduction and experimental setup

The design of an efficient propeller is often limited by various forms of cavitation. Of these forms sheet and cloud cavitation are closely related to the blade passage frequency and are relative well understood. The measures taken to prevent this type of cavitation nuisance however, appear to stimulate the occurrence of vortex cavitation. Vortex cavitation is expected to be responsible for a broadband contribution to the pressure fluctuation spectrum, typically between 40 to 70 $Hz$ for a full-scale propeller [1].

A previous study on the dynamics of steady tip vortex cavitation trailing a stationary lifting surface shows that a vortex cavity has a distinct eigenfrequency [2]. When excited by a matching frequency this has the potential to generate high amplitude sound [3]. This principally tonal component is expected to be spread over a range of frequencies mostly by signal variability between blade passages [4] and changes in vortex cavity properties.

The main aim of the experiment under consideration is to show that understanding of vortex cavity dynamics can be applied to estimate the sound production of propeller tip vortex cavitation using a few basic properties such as cavity size and vortex strength. A simultaneous registration of high speed video and hydrophone measurements is used to obtain the properties of the dynamics of the tip vortex cavity and the resulting sound. As the detailed analysis of

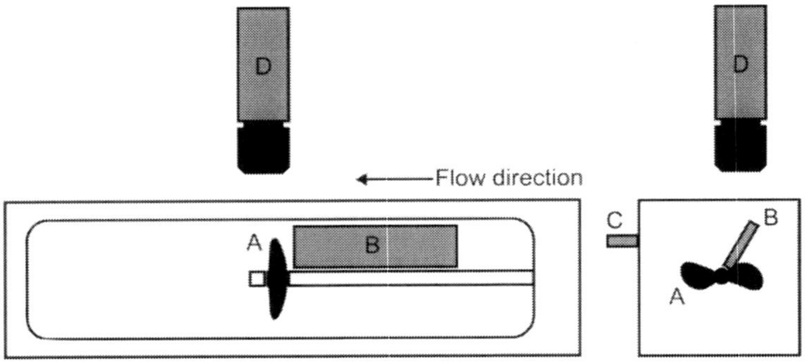


**Figure 1.** Sketch of the experimental setup with; $A$ the propeller, $B$ the wake generator, $C$ the hydrophone and $D$ the high speed camera. Left is a side view of the test section with on the right a cross section looking from upstream on the suction side of the propeller.

the high-speed video is not yet completed at present, some general trends based on the cavity images and sound spectrum are presented.

A steady tip vortex cavity trailing a propeller without excitation is not expected to produce sound with significant amplitude above the tunnel background. Therefore a simulated wake field with a sharp dip in the axial velocity component is generated upstream using a series of plates with narrow spacing. The flow aligned plates retard the flow locally resulting in a local higher loading of the propeller blade passing its wake. The consequence is a locally stronger tip vortex with a larger cavity diameter potentially followed by collapse and rebound after the blade passage. This excitation of the vortex cavity is expected to be sufficiently strong to result in audible pressure fluctuations. As the required excitation frequency is unknown the narrow wake is chosen such that it contains a broad range of frequencies. The experimental results in this study show the effect of a varying cavity size on the resulting sound by looking at the difference in sound power density spectrum between a uniform inflow and simulated wake field.

The experiments are performed in the cavitation tunnel at Delft University of Technology. The details of the tunnel can be found in the thesis by Foeth [5] while the recent modifications are described in the thesis by Zverkhovskyi [6]. The tip vortex cavity is generated by a specially designed right handed two-bladed skewed propeller with reference number 7359. The diameter is 0.15 $m$ with a constant pitch-to-diameter ratio of $P/D = 1.0$ over the span with the detailed geometry freely available upon request. The projections of the propeller geometry can be seen in the sketch of the experimental setup in figure 1.

The wake field is generated by 5 plates 2 $mm$ thick, 100 $mm$ high and 300 $mm$ long with 3 $mm$ spacing, that is mounted at 30 degrees from vertical and 10 $mm$ upstream of the propeller hub. The field of view of the high-speed video camera is chosen to include the propeller shaft and tip. The hydrophone is mounted in a water filled cup to the side window of the test section at 50 $mm$ upward from the shaft axis at the stream-wise location of the propeller tip.

## 2. Results

Figure 2 shows three experimental conditions for varying the free-stream pressure while keeping other parameters constant. At a constant blade angle the images on the left are for a uniform inflow and on the right for the simulated wake inflow. The power density spectrum of the sound is shown below the cavity images.

From the sound spectrum it is clear that a steady tip vortex cavity in uniform inflow does not contribute significantly to the sound above the level of the background noise of the cavitation tunnel. This is evident in the equal power density in the first blade passage frequency at 76 $Hz$

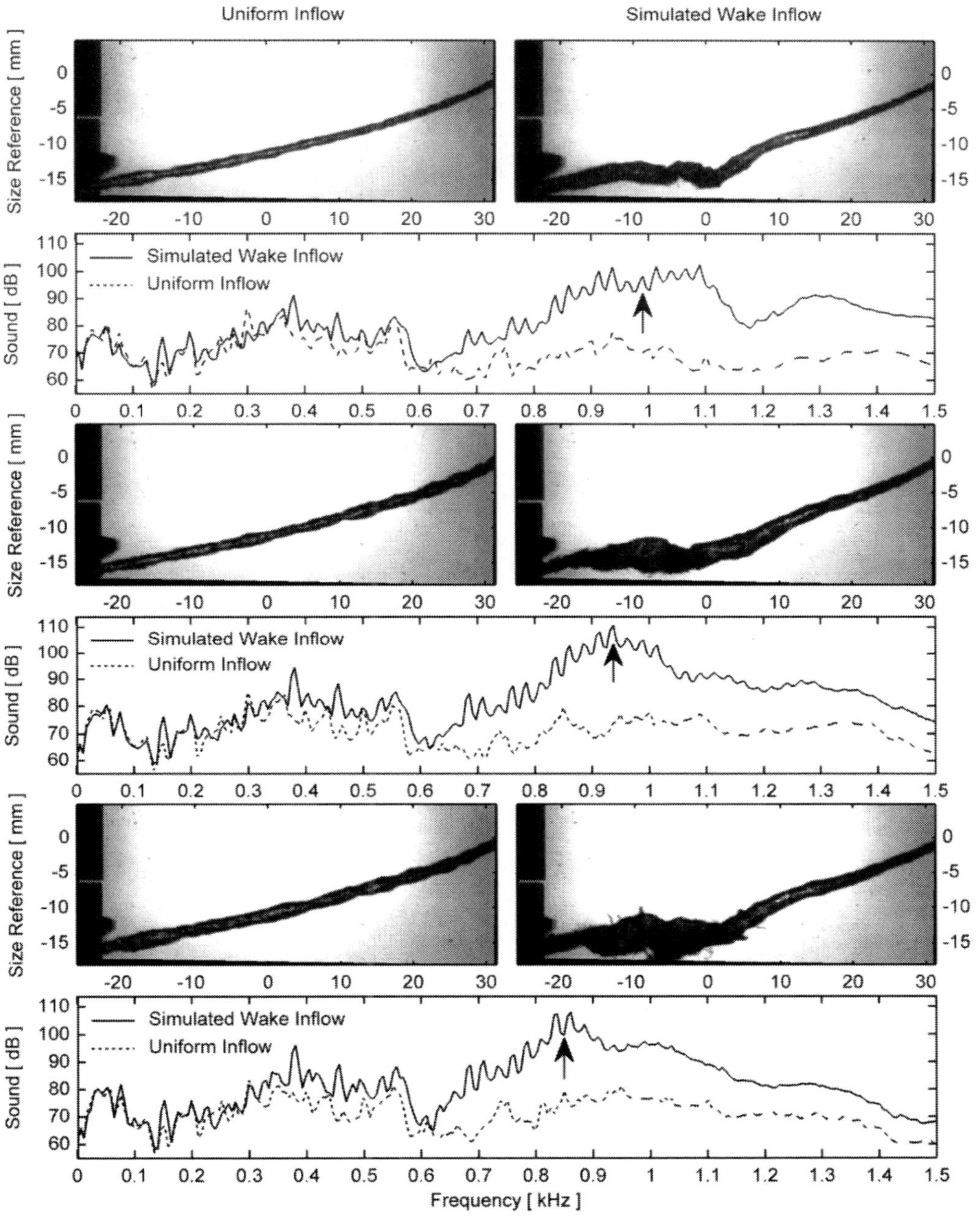


**Figure 2.** Tip vortex cavity images of uniform inflow (left) and behind a simulated wake (right), that are accompanied by a power density spectrum with reference value $10^{-12}$ $Pa^2/Hz$. Flow from bottom to top. Left in image; hubcap, propeller hub and a small portion of the propeller tip trailing edge. The gap in the shaft is 42 $mm$ downstream of the wake. Common conditions; free-stream velocity 3.2 $m/s$, rotation rate 38 $Hz$, advance ratio $J$ = 0.56, thrust coefficient $K_T = 0.18$, torque coefficient $10K_Q = 0.27$, Reynolds number at 70% radius 6.5 $\times 10^5$. Cavitation number at 70% radius varied from top to bottom to 1.16, 1.07 and 0.98. Arrows indicate high amplitude contributions that decrease in frequency consequently.

and the range between 0.6 and 1.5 $kHz$. The cavitation tunnel conditions do not allow a reference measurement without cavitation on the propeller. When the sound spectrum of the tip vortex behind the plates is considered it is observed that the cavity grows and apparently excited the rest of the tip vortex cavity. The reduced axial flow also alters the path of the vortex cavity centreline. The result in sound is a high amplitude broadband contribution as much as 30 $dB$ in power density above the uniform inflow signal. The centre frequency of this feature is related to the cavity diameter. For lower pressures thus larger cavities, a lower frequency is found. Only the 5th and 6th blade passage frequencies increase in amplitude in the presence of the wake but significantly less than the broadband part.

The broadband character of the high amplitude contribution varies with decreasing pressure. The middle case shows highest amplitude and is thus expected to contribute to the sound over a larger portion of the blade rotation resulting in a more narrow-band signature.

The simultaneous registration of the high-speed video images recorded with a frame rate slightly above one data point per degree of rotation enables a more detailed time-frequency analysis. This can be used to directly link the dynamics of the vortex cavity to sound emission.

## 3. Conclusions and future work

A tip vortex cavity in the wake of a propeller in a uniform inflow does not produce sound with significant amplitude to be detectable above the tunnel background. The addition of an obstruction in the inflow of the propeller results in a significant broadband contribution to the measured sound. The range of this frequency band strongly correlates with the size of the tip vortex cavity; for larger cavities the range shifts to lower frequencies. Finally the high-speed video is of sufficient resolution in time and space to study in detail the cavity growth, collapse and rebound of the vortex cavity behind and besides the wake plates.

The analysis described here is only a preliminary first step preceding a detailed analysis of the high speed video of the individual blade passages and the corresponding contribution to the sound by the tip vortex cavity. The main goal is to find tip vortex cavity properties which correspond to the broadband sound centre frequency. Potential candidates are the mean cavity size in the undisturbed section of the flow and the cavity size downstream of the wake plates. The frequency content of the cavity will then be compared to the time-frequency sound signal to break down the broadband sound in individual elements at several blade angle ranges. Finally the sound generating condition of an analytic cavity wave dynamics model [2] combined with a viscous vortex flow field will be used to form an engineering method for estimation of the broadband sound from tip vortex cavitation.

## Acknowledgments

This work is funded by the Lloyd's Register Foundation as part of a cooperation in the International Institute for Cavitation Research

## References

[1] Wijngaarden E, Bosschers J and Kuiper G 2005 *Proceedings of the ASME Fluids Engineering Division Summer Meeting and Exhibition* (Houston, Texas, USA)
[2] Pennings P, Bosschers J, Westerweel J and Van Terwisga T 2015 *Journal of Fluid Mechanics* DOI: 10.1017/jfm.2015.379
[3] Maines B and Arndt R 1997 *Journal of Fluids Engineering* **119** 271–276
[4] Bosschers J 2009 *Proceedings of the First Symposium on Marine Propulsors* (Trondheim, Norway)
[5] Foeth E 2008 *The Structure of Three-Dimensional Sheet Cavitation* Ph.D. thesis Delft University of Technology Delft, The Netherlands
[6] Zverkhovskyi O 2014 *Ship Drag Reduction by Air Cavities* Ph.D. thesis Delft University of Technology Delft, The Netherlands

9th International Symposium on Cavitation (CAV2015) IOP Publishing
Journal of Physics: Conference Series **656** (2015) 012187 doi:10.1088/1742-6596/656/1/012187

# Simulations of tip vortex cavitation flows with nonlinear $k$-$\epsilon$ model

**Benlong WANG, Zhihui LIU**

Engineering Mechanics, Shanghai Jiaotong University, Shanghai, 200240, China

E-mail: benlongwang@sjtu.edu.cn

**Xiaoxing PENG, Dengcheng LIU**

China Ship Scientific Research Center, Wuxi, 214081, China

**Abstract.** Effects of turbulence models on the simulations of tip vortex wetted flows and cavitating flows are studied. Two types RANS turbulence models, with and beyond the Boussinesq turbulent-viscosity hypothesis, are investigated by examining experimental results regarding strain rate and Reynolds shear stress distributions in the vortex region. The numerical results imply that the spatial phase shift between the mean strain rate and Reynolds stresses can be accurately modeled by the nonlinear $k$-$\epsilon$ turbulence model; the tip vortex cavitation region can only be predicted using the nonlinear turbulence model. The mechanism of the over-dissipation due to the turbulence model is analyzed in terms of the turbulence production, which is one of the dominant source terms in the transport equations of energy. The numerical results imply that the nonlinear $k$-$\epsilon$ model is a promising candidate for predicting tip vortex cavitation flows in practical applications at a reasonable computational cost.

## 1. Introduction

Tip vortex cavitation results in the vibration and cavitation erosion of propellers and rudders, particular for high-speed ships. Experimentation has been the primary tool for studying the formation and flow characteristics of tip vortex cavitation [1, 2, 3]. One of the most challenging task is to properly simulate the tip vortex cavitation. Using advanced computational fluid dynamics to accurately predict tip vortex flows is not simple but is possible. Great achievements have been accomplished using detached eddy simulations and large eddy simulation on cavitating flows. The Reynolds averaged Navier-Stokes equations remain the primary CFD solver used in practical applications due to their low computational cost. However, the widely used two-equation turbulence models based on the Boussinesq turbulent-viscosity hypothesis perform poorly for highly rotational flows. The intrinsic assumption of Boussinesq hypothesis that shear stress is locally determined by the mean strain rate, has no general validity [4], and fails when the flow stream is curved or strongly rotated. Chow et al. [5] and Giuni [6] measured the distribution of the strain rate and Reynolds shear stresses in the tip vortex region. It was found that the four-leaf clover pattern of these two contours was not aligned spatially, indicating that the flow is not isotropic. We denote this type of rotation as a *spatial phase shift* in the present work. These detailed experimental data imply that the turbulence models based on Boussinesq turbulent-viscosity hypothesis are not capable of describing the vortex evolution in the near wake of hydrofoils and propellers. The nonlinear relationship between the Reynolds shear stress and strain rate can fulfill the non-isotropic constitution relation by summation of strain rate tensor and rotation rate tensor. Although the theoretical foundation of the nonlinear $k$-$\epsilon$ model is significantly different from that of the standard $k$-$\epsilon$ model, the numerical implementation is straightforward in this paper by extension of the standard $k$-$\epsilon$ model, as the one proposed by Zhu & Shin [7].

In this paper we analyze the characteristics of turbulent tip vortex flow field, and verify the influence of turbulent dissipation on the formation of tip vortex cavitation flows.

## 2. Numerical method for flow solver

Turbulent viscous flows are solved using the Reynolds averaged Navier-Stokes equations. To simulate the cavitating flow around airfoils or propellers, the assumption of a homogeneous mixed fluid of water and vapor is introduced; the liquid and vapor phases are assumed to be fully mixed and share the same velocity and pressure in the flow field. A cavitation model is needed to simulate the phase transition between the liquid and vapor phases. The transport equation of Schnerr & Sauer is used in the present work [8].

The numerical models are solved using a finite volume method based on the OpenFOAM package. The PISO scheme is used to solve for velocity/pressure coupling between the momentum and continuity equations. A second-order upwind scheme is used for spatial discretization in the momentum and transport equations for $k$ and $\epsilon$, which are used to calculate the turbulent eddy viscosity according to different turbulence models. The QUICK scheme is used in the advection of the volume fraction function. The viscous terms are solved using a second-order central difference scheme. Details of the governing equations, numerical methods and mesh convergence tests will be reported elsewhere.

## 3. Numerical results and discussions for tip vortex flows

Considering wetted flow around a rounded wingtip NACA0012 hydrofoils following the setup as that in experiments [5], numerical simulation is conducted to study the spatial distribution of strain and Reynolds stresses. The number of total mesh cells is approximately 4.5M.

There is a *spatial phase shift* of the four-lobe pattern between the strain rate and the Reynolds stress [5, 6]: a phase shift of $\pi/4$ can be observed between the strain rate $S_{xy}$, $S_{yz}$ and the corresponding components of Reynolds stress; and a phase shift of $\pi/2$ can also be observed between the strain rate $S_{xz}$ and the Reynolds stress $-\overline{u'w'}$ .

Numerical results of the distribution of stain rate and Reynolds stress imply the advantages of the nonlinear $k$-$\epsilon$ model in the numerical simulations of tip vortex flows in Fig.1. The advantages of this model can be explained from the perspective of an energy budget. Defining the kinetic energy of the mean flow to be $E = u_i u_j/2$ and the turbulent kinetic energy to be $k = u'_i u'_j/2$, the transport equations of these two types of energy are:

$$\frac{D\rho_m E}{Dt} + \nabla \cdot \vec{T} = -P - \rho_m \bar{\epsilon}, \qquad \frac{D\rho_m\, k}{Dt} + \nabla \cdot \vec{T'} = P - \rho_m \epsilon \tag{1}$$

where $D/Dt = \partial/\partial t + u_i \partial/\partial x_i$ is total derivative, $\rho_m \bar{\epsilon} = \mu_m S_{ij} S_{ij}$ is the viscous dissipation rate of mean flow strain rate $S_{ij} = (u_{i,j} + u_{j,i})/2$ due to molecule viscosity, $\rho_m \epsilon = \mu_m s_{ij} s_{ij}$ is turbulent kinetic dissipation rate, relating the strain rate of fluctuate velocity field $s_{ij} = (u'_{i,j} + u'_{j,i})/2$. $\vec{T}$ and $\vec{T'}$ are convection of Reynolds stress, pressure stress and molecule viscosity by mean and fluctuated velocities.

Production is defined as $P = \tau^D_{ij} S_{ij}$ in which $\tau^D_{ij}$ is the deviatoric part of the Reynolds stress. We have $P \ll \bar{\epsilon}$ due to the fact that the turbulent eddy viscosity is much larger than the molecular viscosity $\mu_t \gg \mu_m$. Therefore, the primary source in the right hand side of Eq.1(a) is

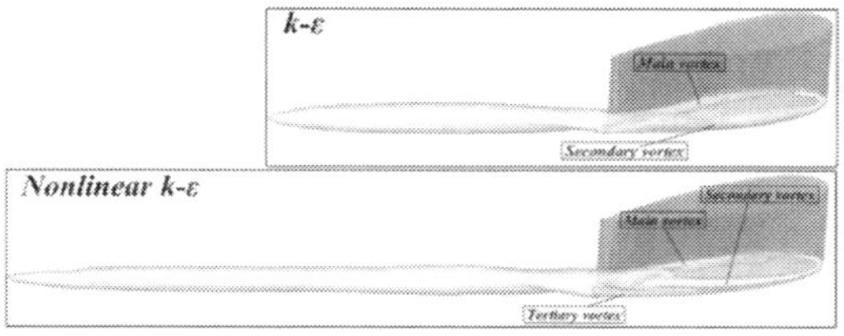


(a) isosurface of $Q = 10^3$ with the same grid

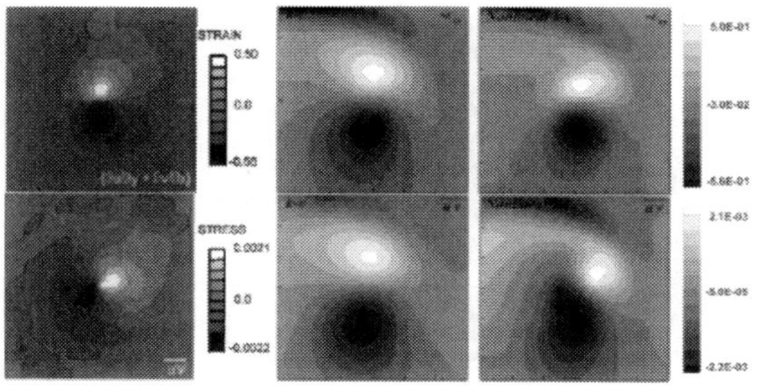

(b) $-S_{12} = -(u_y + v_x)$ and $\overline{u'v'}$

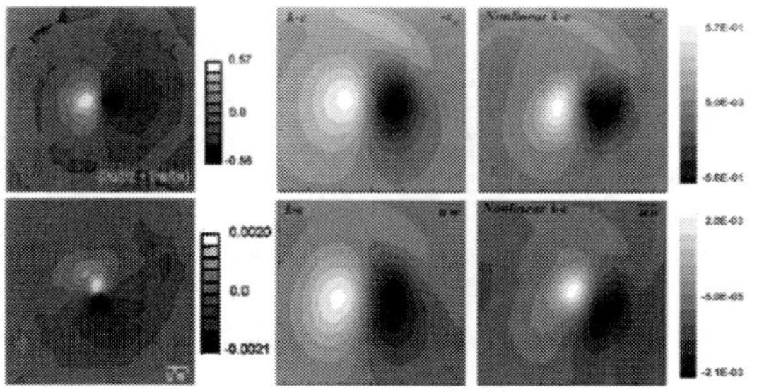

(c) $-S_{13} = -(u_z + w_x)$ and $\overline{u'w'}$

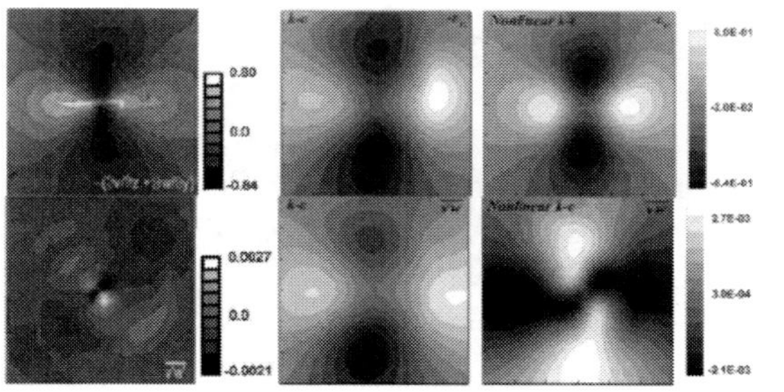

(d) $-S_{23} = -(v_z + w_y)$ and $\overline{v'w'}$

**Figure 1.** Numerical results of wetted flows around NACA0012 airfoil, attach angle 10°, Reynolds number $Re = 4.6 \times 10^6$. (a) isosurface of $Q = 10^3$. (b)-(d) Comparison of spatial distribution of negative strain rate $-S_{ij}$ and deviatoric Reynolds stress $\overline{u_i'u_j'}$ among experimental results [5], numerical results with standard $k$-$\epsilon$ and nonlinear $k$-$\epsilon$ models.

contributed by the production term $P$. The production $P$ extracts the total kinetic energy from the mean flow and transfers it to the turbulent kinetic energy. The production $P$ thus serves as a sink term in Eq.1(a) and a source term in Eq.1(b). Correct prediction of the magnitude of production $P$ influences the energy transfer between the mean and fluctuating flow. Finally, the amount of transfer energy is dissipated at the turbulent kinetic dissipation rate $\rho_m\epsilon$. For homogeneous and steady turbulent flows, we have $P \approx \epsilon$. Therefore, the turbulence dissipation may be estimated by the magnitude of production $P$. If $P$ is over estimated, the strength of the vortex will be under estimated.

To illustrate the effects of phase shift on the production term, a typical four-lobe distribution of strain rate and Reynolds stress components is plotted at a prescribed circle around the vortex center in Fig.2. For turbulence models with the Boussinesq hypothesis, the strain rate and Reynolds stress takes the same sign along the circle and thus obtains a maximum production $P = \tau_{ij}^D S_{ij}$. In the nonlinear $k$-$\epsilon$ model, there is a phase shift between the strain rate and Reynolds stress, as observed experimentally. Thus the integration of $P$ over the cross-section will be partially canceled. Therefore, the nonlinear model must predict less production and turbulence dissipation than the standard $k$-$\epsilon$ model. Numerical results shows that the integration $\iint P dydz = 4.08$ and 2.83 at section of $x = 1.45$ times chord by these two turbulence models.

## 4. Applications for cavitating flows around E779A propeller

The numerical simulations of tip vortex flows around INSEAN E779A four-blade propeller with a diameter $D = 227$ mm, nominal pitch ratio 1.1, hub diameter 45.3 mm are examined. The incoming velocity is $U_\infty$=2.808 m/s , and the advance ratio $J = U/nD = 0.71$. The cavitation number is $\sigma_n = (p - p_v)/2(nD)^2 = 1.763$. Fully structural mesh are used in these simulation. After convergence tests, the total number of computational cells number is approximately 4.5M

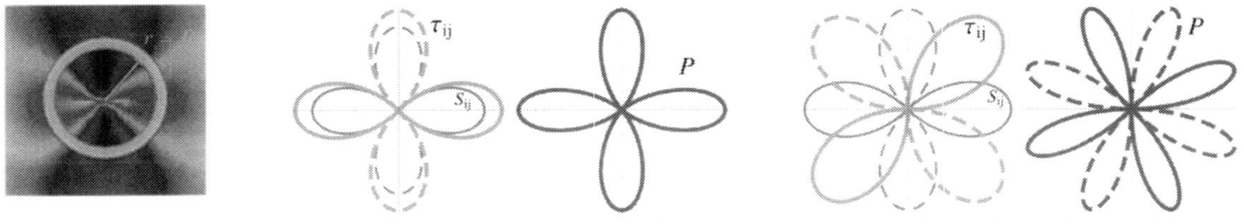


(a) interested ring region (b) standard $k$-$\epsilon$ model (c) nonlinear $k$-$\epsilon$ model

**Figure 2.** Distribution of Reynolds shear stress $\tau_{ij}^D$, strain rate $S_{ij}$ and production $P = \tau_{ij} S_{ij}$ along a narrow circular ring. ——represent positive value, – – –represent nagtive value.

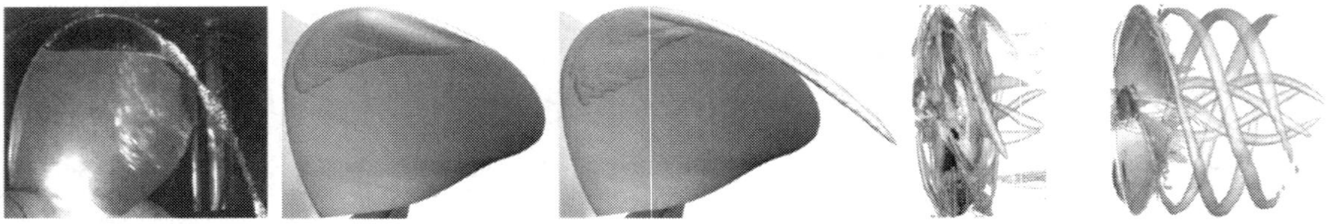

(a) isosurface of vapor valume fraction $\alpha_v = 0.1$ (b) isosurface of $Q = 10^4$

**Figure 3.** (a) Comparison of the cavitation region among: experimental results [9], numerical results with standard and nonlinear $k$-$\epsilon$ models. (b) Vortex structure predicted by standard and nonlinear $k$-$\epsilon$ models. All the comparisons are carried out at the same mesh resolution.

after Oct-tree refinement in the wake region of the propeller tip.

At the same mesh resolution, the areas of sheet cavitation predicted by the numerical simulations are qualitatively similar at the surface of blade, and both agree with the experimental results. However, the tip cavitation can only be predicted using nonlinear $k-\epsilon$ turbulence model as shown in Fig.3. The detail numerical results also confirm the turbulent energy budget analysis.

The numerical results shown in this paper imply that the nonlinear $k-\epsilon$ model is a promising candidate for predicting tip vortex cavitation flows in practical applications at a much lower computational cost compared to other advanced turbulence models, including detached eddy simulations and large eddy simulations. Finally, we need to mention that the mesh resolution was also recognized to affect the vortex wake. Using a fine mesh can resolve tip vortices further downstream. Therefore, adaptive mesh refinement in the vortex wake region is important to capture the detail flow field away from the propeller.

## Acknowledgments
Project supported by the Key Doctoral Program Foundation of Shanghai Municipality (No. B206), and the National Natural Science Foundation of China (No. 11332009).

## References
[1] Arndt REA, Arakeri VH and Higuchi H 1991 *J. Fluid Mech.* **229** 169
[2] Fruman DH, Dugue C and Cerrutti P 1992 *19th Symp. on Naval Hydrodynamic* (Washington DC)
[3] Maines BH, Arndt REA 1997 *J. Fluid Eng.* **119** 413
[4] Pope SB 2000 *Turbulent Flows* (New York: Cambridge University Press) P 126, 365
[5] Chow JS, Zilliac GG and Bradshaw P 1997 *AIAA Journal* **35**(10) 1561
[6] Giuni M 2013 *Formation and early development of wingtip vortices* (PhD thesis, University of Glasgow)
[7] Zhu J and Shin TH 1994 *Int. J. Numerical Methods in Fluids* **19**(11) 939
[8] Schnerr GH and Sauer I 2001 *Proc. 4th Int. Conf. on Multiphase Flow* (New Orleans)
[9] Salvatore F, Testa C and Greco L 2003 *Proc. 5th Int. Symp. on Cavitation* (Osaka)